2022 24th European Conference on Power Electronics and Applications (EPE'22 ECCE Europe)

Hanover, Germany
5-9 September 2022

Pages 673-1349

IEEE Catalog Number: CFP22850-POD
ISBN: 978-1-6654-8700-9

Copyright © 2022, The European Power Electronics and Drives Association
All Rights Reserved

*** *This is a print representation of what appears in the IEEE Digital Library. Some format issues inherent in the e-media version may also appear in this print version.*

IEEE Catalog Number: CFP22850-POD
ISBN (Print-On-Demand): 978-1-6654-8700-9
ISBN (Online):   978-9-0758-1539-9

Additional Copies of This Publication Are Available From:

Curran Associates, Inc
57 Morehouse Lane
Red Hook, NY 12571 USA
Phone: (845) 758-0400
Fax: (845) 758-2633
E-mail: curran@proceedings.com
Web: www.proceedings.com

2022 24th European Conference on Power Electronics and Applications (EPE'22 ECCE Europe)

Hanover, Germany
5-9 September 2022

Pages 673-1349

IEEE Catalog Number: CFP22850-POD
ISBN: 978-1-6654-8700-9

TABLE OF CONTENTS

Dynamic Power Analysis of Inverter-Fed Drives Based on the Switching Period of the Power Electronics 1

Alexander Stock

Stability Analysis in an Inverter-Dominant Microgrid Facing In-Rush Current of an Induction Machine 11

Nastaran Fazli, David Hammes, Sidney Gierschner, Hans-Gunter Eckel

Self-Oscillating Capacitive Power Transfer with Multiple Receiver Capability and Coupling Path Adaption 22

Norbert Seliger

An Electrically Driven Gas Compressor for Hydrogen Refueling Stations with Active Power Smoothing 30

Alfred Rufer

Unsymmetrical Fault Behavior of PLL Based Grid-Connected Converters 39

Philipp Hackl, Ziqian Zhang, Robert Schuerhuber

Stability Assessment and Optimization of MMC Energy Balancing for Drive Applications at Standstill using an Averaging Approach 49

Qiuye Gui, Hendrik Fehr, Albrecht Gensior

Turn-On Losses Optimization for Medium Power SiC MOSFET Half-Bridge Module 59

Pham Ha Trieu To, Felix Kayser, Hans-Günter Eckel

Oscillation Damping in a 500kW Hybrid Si/SiC Three-Level ANPC Inverter with Decoupling Capacitor 70

Pham Ha Trieu To, Hans-Günter Eckel

Multi Busbar Sub-Module Modular Multilevel STATCOM with Partially Rated Energy Storage Configured in Sub-Stacks 80

Chuantong Hao, Wenhao Ma, Michael Merlin, Paul Judge, Stephen Finney

Three-Phase ZVS Inverter with Variable and Fixed Frequency Operation Based on GaN Semiconductors 88

Benedikt Kohlhepp, Michael Lutsch, Thomas Dürbaum

Influences of Conductor Positions and Fast Rising Impulse Voltages on the Line-End Coil Based on a Three-Phase High-Frequency Model 97

Ting Helmholdt-Zhu, Volker Grabs

Simulation Tool for Optimization of Digital Active Gate Drive Sequence using Genetic Algorithm 108

Hajime Takayama, Shuhei Fukunaga, Takashi Hikihara

Analysis of Balancing Algorithms for Quasi- Two/Three-Level Single Phase Operation of a Flying Capacitor Converter 115

Stefan Mersche, Markus Bayer, Kai Rickert, Marc Hiller

Instability in Active Balancing Control of Dc Bus Voltages in VSC Converters Interconnected via Multi-Winding Transformers 125

Duro Basic, Sami Siala

Online Learning-Based Islanding Detection Scheme for Grid-Connected Systems 135
Mohammed Ali Khan, V S Bharath Kurukuru, Rupam Singh

Difference in the Design Process of LCL Filters for Grid Connected VSI When using SiC/GaN
Instead of Si Semiconductors 145
Dennis Kampen, Lukas Fräger, Niklas Badenhop, Arthur Mambetow

Analysis and Design of a Resonant DC/DC Transformer in Modular Operation 152
Abraham López, Manuel Arias, Pablo F. Miaja, Arturo Fernández

Predictive Braking Algorithm for Soft Starter Driven Induction Motors 160
Hauke Nannen, Heiko Zatocil, Gerd Griepentrog

Ambient Electromagnetic Energy Harvesting Circuit using Rectennas Manufactured with
Stereolithography Resin 169
*Xuan Viet Linh Nguyen, Tony Gerges, Jacques Verdier, Philippe Lombard, Michel Cabrera,
Bruno Allard, Jean-Marc Duchamp, Philippe Benech*

Boost/Buck-Boost Based Grid Connected Solar PV Micro-Inverter with Reduced Number of
Switches and Having Power Decoupling Capability 178
Arup Ratan Paul, Arghyadip Bhattacharya, Kishore Chatterjee

Operation and Selection of Multilevel Power Converters for Doubly Fed Induction Generator-
Based Wind Turbines 187
Kapil Jha, Joseph Banda, Hridya I, Arvind Tiwari

A Detailed View on the Trapezoidal Operation for MMC Type Braking Chopper in Medium
Voltage Application 195
Patrick Hofstetter, Viktor Hofmann, Dennis Karwatzki

Influence of Operating Frequency on High-Power Medium-Voltage Medium-Frequency
Transformers 203
Thomas B. Gradinger, Ralph M. Burkart, Marko Mogorovic

Output Power Characteristics of Isolated Secondary-Resonant SAB DC-DC Converter for Output
Voltage Variation 213
Shota Yamashita, Kohei Budo, Takaharu Takeshita

Hardware and Control Design of a High Precision Modular Power Converter Based on GaN
Technology for Particle Accelerator Magnets 223
*Thomas Margreiter, Ivan De Cesaris, Maurizio Incurvati, Sebastien Pelletier, Martin
Schiestl, Ronald Stärz*

Battery Cycler to Generate Open Li-Ion Cell Aging Data and Models 232
Matthias Luh, Thomas Blank

Function Blocks of a Highly-Integrated All-In-GaN Power IC for DC-DC Conversion 242
Michael Basler, Richard Reiner, Stefan Moench, Patrick Waltereit, Rüdiger Quay

Comparison of Redundancy Requirements for Modular Multilevel Converter Considering
Manufacturer Reliability Inputs and Mission Profile 251
Diego Velazco, Guy Clerc, Emmanuel Boutleux, Francois Wallart

Impact of Insulation and Cooling on Performance Due to Reliability-Oriented Design of Electrical
Machines 261
Lucas Vincent Hanisch, Jonas Franzki, Markus Henke

Long Switching Horizon Model Predictive Controller for High-Speed Integrated Modular Motor Drives 268

Martin Schiestl, Maurizio Incurvati, Ronald Starz, Markus Schmid

Standalone Power Management System for Flexible Piezo Electric Nano Generators (PENG) Based on the Co-Polymer P(VDF:TrFE) 279

Alexander Wölk, Mahmoud Shousha, Shashank Shekhawat Singh, Martin Haug, Lorandt Fölkel, Michael Brooks, Asier Alvarez, Andreas Petritz, Philipp Schäffner, Jonas Groten, Andreas Tschepp, Barbara Stadlober

Analysis and Estimation of Neutral-Point Voltage Balancing Ability of an Optimized Balancing Algorithm for Grid Connected Active-NPC Converter 289

Joseph Banda, Kapil Jha, Hridya Ittamveettil, Arvind Kumar Tiwari, Fernando Ramirez

A Direct Model Predictive Control Strategy of Back-To-Back Modular Multilevel Converters using Arm Energy Estimation 297

Akseli Hakkila, Antonios Antonopoulos, Petros Karamanakos

Study on Commutation Loop Inductance and Current Distribution to DC-Link Capacitors in a GaN Half-Bridge 307

Benedikt Kohlhepp, Samuel Faber, Jeremias Kaiser, Thomas Dürbaum

Cooperative Control of Online Impedance Spectroscopy Monitoring Method and Maximum Power Point Tracking Method for Photovoltaic Panels 315

Xin Wang, Zhixue Zheng, Michel Aillerie, Alexandre De Bernardinis, Jean–paul Sawicki, Marie-Cécile Péra, Daniel Hissel

Benefits of Switching from Si to SiC Modules with Further Converter Optimization 325

Antxon Arrizabalaga, Mikel Mazuela, Iosu Aizpuru, June Urkizu, Jon Aztiria

On the Reduction of Output Capacitance in Two-Level Three Phase PFC Boost Rectifier for Pulsating Loads 335

Tania C. Cano, Douglas Pedroso, Alberto Rodríguez, Ignacio Castro, Diego G. Lamar

Cognitive Insights into Metaheuristic Digital Twin Based Health Monitoring of DC-DC Converters 344

Abdul Basit Mirza, Kushan Choksi, Sama Salehi Vala, Krishna Moorthy Radha, Madhu Sudhan Chinthavali, Fang Luo

A Three-Phase Isolated Secondary-Resonant Single-Active-Bridge DC-DC Converter with a Delta-Star Connected Transformer 351

Atsushi Nishio, Kohei Budo, Mai Van Tuan, Takaharu Takeshita

A Novel Concept to Optimize Core Loss in Planar Magnetic Based on an Unbalanced-Flux-Approach 361

Sobhi Barg, Kent Bertilsson, Grover Torrico

Model Reduction using Singular Perturbation Methods for a Microgrid Application 370

Lasse Gnärig, Albrecht Gensior, Saioa Burutxaga Laza, Miguel Carrasco, Carsten Reincke-Collon

Drive Level Parameter Identification of an Induction Motor 380

Andreas Bünte, Alex Hald, Andreas Kirsch

Impedance Stability of Single-Phase LCL Grid-Connected Voltage Source Inverters with Wideband Gap Devices Under Different Control Approaches 390

Ramy Ali, Terence O'Donnell

Design and Modulation Optimization of an MMC Based Braking Chopper 400
    Viktor Hofmann, Patrick Hofstetter

Modeling the Arrangement of Drill Holes for Orthogonal Biasing in Controllable Inductors for
Power Electronic Converters 411
    Jonas Pfeiffer, Christoph Drexler, Pierre Küster, Peter Zacharias, Michael Schmidhuber

A Sectorized FCS-MPC Transformerless SST for Power Transmission Application 421
    Gabriel Gaburro Bacheti, Renner Sartório Camargo, Emilio José Bueno, Marco Liserre,
    Lucas Frizera Encarnação

Inductance Estimation for Square-Shaped Multilayer Planar Windings 432
    Theofilos Papadopoulos, Antonios Antonopoulos

Cost and Efficiency Considerations in On-Board Chargers 442
    Marija Jankovic, Christian Felgemacher, Kevin Lenz, Aly Mashaly, Abdelmouneim
    Charkaoui

A Novel Combined Control of Ground Current and DC-Pole-To-Ground Voltage in Symmetrical
Monopole Modular Multilevel Converters for HVDC Applications 451
    Pablo Briff, Amit Kumar

A PFC Boost Converter with Reduced Switching Losses Operating at a Fixed Switching Frequency 459
    Burkhard Ulrich

Predictive Control of Power Electronics Autotransformer for Mitigating Three-Phase Grid Current
Unbalance in Railway Supply Systems 468
    Tabish Nazir Mir, Faysal Hardan, Masood Hajian, Tamer Kamel, Pietro Tricoli

Parameter Sensitivity of a MRAS-Based Sensorless Control for AFPMSM Considering Speed
Accuracy and Dynamic Response at Multiple Parameter Variations 474
    Michael Brüns, Christian Rudolph, Tankred Müller

Synchronization Stability of a Grid Forming Converter Under the Effect of Current Limit in
Voltage Dips with VI Based Current Limiting Method: Analysis and Solution 484
    Siam Hasan Khan, Markel Zubiaga Lazkano, Pedro Izurza, Alain Sanchez-Ruiz, Javier Cañas
    Aceña, Joseba Arza

Analytic Calculation of Touch and Leakage Currents of Non-Isolated EV Chargers using a Fast
Common Mode Calculation Method and Non-Ideal Passive Component Models 493
    Christian Stutz, Sebastian Nielebock, Martin März

Triple-Phase-Shift Controlled Dual Active Bridge Converter with Variable Input Voltage in
Auxiliary Railway Supply 504
    Martin Scohier, Olivier Deblecker, Carlos Valderrama

Loss Characterization Methodology for Soft Magnetic Nano-Crystalline Tape Materials in Coupled
Inductors 514
    David Bohne, Valentin Wagner, Patrick Deck, Christian P. Dick

Substitution of Nanocrystalline Toroid by Laminated Ferrite Toroid in the Application of a
Common-Mode Choke 525
    Lukas Reißenweber, Fritz Wohlrath, Alexander Stadler

Direct Active Stabilization of the DC-Link in Voltage-Source Converters 534
    Matthieu Bertin, Mohamad Koteich

Hardware-In-The-Loop Control of a Modular Induction Motor Drive in Power Electronics Education 544
Jens Peter Kaerst

Design and Efficiency Analysis of an LCL Capacitive Power Transfer System with Load-Independent ZPA 554
Francesco Musolino, Ahmed Abdullah, Mario Pavone, Fabio Ferreyra, Paolo Crovetti

A Pulse Generator Based on Transmission Line Transformer for Insulation Aging Test 562
Xiao Yu, Khanh-Hung Nguyen, Peter Zacharias

Design of a Single-Phase Common Mode and Differential Mode Inductor for Interleaved Converters 572
Jonathan Robinson, Gopal Mondal, Stefan Hänsel, Matthias Neumeister

Steady-State Analysis and Comparison of SSFB, SDFB and DSFB MMC-Based STATCOM 582
Mohamed Moez Belhaouane, Pierre Vermeersch, François Gruson, Pierre Rault, Sébastien Dennetiere, Xavier Guillaud

Current Distribution Control in Parallel Connected Power Converters with Continuous Output Voltage 593
Sabrina Ulmer, Andreas Brunner, Philipp Czerwenka, Gernot Schullerus, Ertugrul Sönmez

Optimized Pulse Pattern with Half-Wave Symmetry for 5-Level Converter 604
Jonas Weires, Pedro Leal Dos Santos, Steven Liu

Characterization of Si-IGBT Crosstalk with a Concentration on Power Circuit Parasitic Elements and the Device Operation Point 614
Amir Azam Rajabian, Sadegh Mohsenzade, Javad Naghibi, Kamyar Mehran

Impact of Higher Current Harmonics on Component Current Stress and Conduction Losses of Half-Bridge-Series-Resonant-Converters in Discontinuous Conduction Mode for High-Power Applications 624
Daniel Haake, Anton Grodnichev, Fabian Schnabel, Marco Jung

Control of a Zero-Voltage Switching Isolated Series-Resonant Power Circuit for Direct 3-Phase AC to DC Conversion 634
Yusuf Kosesoy, Remco Bonten, Henk Huisman, Jan Schellekens

Design of a Robust Voltage Control for Inverters with LC Filter Based on the Internal Model Control 641
Frederik Stallmann, Axel Mertens, Lukas Fräger

Influence of Power Semiconductor Device Variations on Pulse Shape of Nanosecond Pulses in a Solid-State Linear Transformer Driver 651
Raffael Risch, Anliang Hu, Jürgen Biela

Optimal Design of Integrated Motor Drives - Comparison of Topologies (2L/3L/Modular), PWM Variants, and Switch Technologies (Si/SiC/GaN) 662
Thilo Bringezu, Jürgen Biela

Distribution Transformer Voltage Control using a Single-Phase Matrix Converter 673
Rui Wang, Henk Huisman, Korneel Wijnands

Influence of Carrier-Based PWM Techniques on the Common-Mode Voltage and Common-Mode Current of Six-Phase Full-Bridge Inverters 681
Juris Arrozy, Esin Ilhan Caarls, Henk Huisman, Jorge L. Duarte, Lorenzo Ceccarelli

Mitigation of Dead-Time Effects on Transient DC Bias Elimination in Dual Active Bridge Link Current.. 689
> MK Kharabela Mohanta, Dipankar De, Silpashree Sahu, Alberto Castellazzi

Generalized Automated Tool for Analysis and Design of Multiphase Coupled Inductor Buck Converters 698
> Rana Asad Ali, Mahmoud Shousha, Martin Haug

Experimental Study of a Directly Oil-Cooled Electrical Machine for a Full-Electric Vehicle by using Low Viscosity Oil.. 709
> Huihui Xu, Georg Tobias Götz, Shimin Zhang, Rik W. De Doncker

Development of a Family of High Voltage Gain Step-Up Multi-Port DC-DC Converters for Fuel Cell-Based Hybrid Vehicular Power Systems 719
> Pouya Zolfi, Sina Vahid, Ayman El-Refaie

Bidirectional DC Circuit Breaker with Improved Performance During Commissioning and Reclosing... 730
> Aditya Pogulaguntla, Venkata Raghavendra I, Satish Naik Banavath, Andrii Chub, T Sreekanth, Harish Sarma Krishnamoorthy

Modeling Method for Conducted Noise Flowing in Power Lines of DC/DC Converter 739
> Takato Hattori, Wataru Kitagawa, Takaharu Takeshita

High-Bandwidth Power Hardware-In-The-Loop for Motor and Battery Emulation at High Voltage Levels 749
> Manuel Fischer, Philipp Kemper, Johannes Herbold, Daniel Epping, Frank Puschmann

Analysis and Discussion of Different Three-Phase dv/dt Filter Topologies and the Influences of Their Filter Parameters on Losses and EMC 758
> Eric Fritze, Michael Meissner, Klaus F. Hoffmann, Kai-Uwe Rathjen, Stefan Dickmann, Oliver Woywode

State of Charge Prediction of Lithium-Ion Batteries Based on Artificial Neural Networks and Reduced Data 767
> Sebastian Pohlmann, Ali Mashayekh, Dominic Karnehm, Manuel Kuder, Antje Gieraths, Thomas Weyh

Investigation for Condensation Test Condition of HVIGBT Modules.. 777
> Kenji Hatori, Keiichi Nakamura, Wakana Noboru, Nils Soltau, Eugen Wiesner

Three Phase PV Inverter LCOE Optimization Considering Technological Choice 787
> Morteza Tadbiri Nooshabadi, Jean-Luc Schanen, Shahrokh Farhangi, Hossein Iman-Eini

Square Wave Operation to Reduce Pulsating Power in Isolated MMC-Based Ultrafast Chargers 798
> Ygor Pereira Marca, Maurice G. L. Roes, Jorge L. Duarte, Korneel Wijnands

Surge Current Protection for Railway Traction Applications... 805
> Michael Gleissner, Mark-M. Bakran

Impedance-Based Analysis of HVDC Converter Control for Robust Stability in AC Power Systems............ 814
> André Schön, Andreas Lorenz, Rodrigo Alonso Alvarez Valenzuela

Class-E Push-Pull Resonance Converter with Load Variation Robustness for Industrial Induction Heating 825
> Janus Dybdahl Meinert, Benjamin Futtrup Kjærsgaard, Thore Stig Aunsborg, Asger Bjorn Jorgensen, Stig Munk-Nielsen, Sune Bro Duun

Review of Power Converter Topologies for Electrochemical Impedance Spectroscopy of Lithium-Ion Batteries 833

Hamzeh Beiranvand, Julius M. Placzek, Marco Liserre, Giorgia Zampardi, Doriano Constantino Brogioli, Fabio La Mantia

Design and Experimental Validation of a Voltage Sensing-Current Cancellation Common Mode Linear Active Filter 843

B. Mohamed Nassurdine, PE Lévy, D. Labrousse, JL Schanen, X. Maynard, S. Carcouet

Partial Discharges of Insulated Wires Under Impulses from Wide Bandgap Power Electronics 854

Ting Helmholdt-Zhu, Vivien Grau, Urs Obernolte

Analysis of a Droop-Based Power Controller for Three-Phase Microgrids 865

Andrea Lauri, Hossein Abedini, Davide Biadene, Tommaso Caldognetto, Paolo Mattavelli

Efficiently Paralleling GaN-Transistors for High Current and High Frequency Applications using a Butterfly Layout 873

Martin Wattenberg, Oscar Lorenz, Juan Sanchez

Data-Driven Decentralized Volt/Var Control for Smart PV Inverters in Distribution Systems 883

Yizhou Lu, Qianwen Xu, Lars Nordström

Study of Current Ripple Generators for Accelerated Ageing of Capacitors 891

Robert Keilmann, Hendrik Schefer, Regine Mallwitz

Intra-Arm Balancing Control of Cascaded Multi-Port Converter for Whole Power Unbalance Conditions 902

Takumi Yasuda, Jun-Ichi Itoh

Investigation of Creepage Distances on Printed Circuit Boards for Avionic Applications 912

Hendrik Schefer, Zhongqing Xu, Tobias Kopp, Regine Mallwitz, Michael Kurrat

A 20 kW, 3-Level Flying Capacitor 1500 V Inverter with Characterized GaN Devices for Grid-Tie Applications 922

Van Sang Nguyen, Anthony Bier, Hajar Es-Seghier, Ulrich Soupremanien, Gérard Delette, Stephane Catellani

New Analytical Model for Calculating HF-Losses in Litz Wire Regions Located Outside the E/U-Core Window of Transformers 933

Qingchao Meng, Jürgen Biela

Fast and Accurate Soft-Switching and Hard-Switching Losses Estimation for Power Converter, Application to the Dual Active Bridge (DAB) Converter 944

Francois Boige, Nicolas Videau, Adel Ziani, Bruno Guerrero, Julien Laclaverie

Influence of an Electrical Machine on the Dimension and Packaging of Multi-Machine Systems 952

Thomas Stöckl, Hans-Georg Herzog

Design of a Serial Impingement Cooling Heatsink for a 30 kW PV String Inverter 960

Paul Bruyere, Guillaume Piquet Boisson, Gaëtan Perez

Online Junction Temperature Measurement of SiC-MOSFETs via Gate Impedance using the Gate-Signal Injection Method 971

David Hirning, Luca Bauer, Johannes Ruthardt, Jörg Haarer, Philipp Ziegler, Jörg Roth-Stielow

Powercycling Test Bench with Realistic Loss Distribution and Temperature Ripples 980
Till-Mathis Plötz, Jan Fuhrmann, Hans-Günter Eckel

Design, Implementation and Characterization of an Integrated Current Sensing in GaN HEMT
Device by using the Current-Mirroring Technique 990
*Van-Sang Nguyen, René Escoffier, Stéphane Catellani, Murielle Fayolle-Lecocq, Jérémy
Martin*

GaN-Based Modular Multilevel Converter for Low-Voltage Grid Enables High Efficiency 999
Philip Kiehnle, Patrick Himmelmann, Marc Hiller

Energy Management of Smart Homes with Electric Vehicles using Deep Reinforcement Learning............. 1006
Xavier Weiss, Qianwen Xu, Lars Nordström

Simple and Low-Computational Losses Modeling for Efficiency Enhancement of Differential
Inverters with High Accuracy at Different Modulation Schemes.. 1015
Ahmed Shawky, Mokhtar Aly, Emad M. Ahmed, Samir Kouro, José Rodriguez

Estimation of Battery Parameters in Cascaded Half-Bridge Converters with Reduced Voltage
Sensors 1025
Nima Tashakor, Bita Arabsalmanabadi, Elham Hosseini, Kamal Al-Haddad, Stefan Goetz

Method to Analyze the Influence of Switching Behavior in Hard Switching Half Bridge Topologies
for Traction Application 1036
Dominik Nehmer, Michael Gleissner, Lukas Bergmann, Mark-M. Bakran

Impact of Aluminum Casing on High-Frequency Transformer Leakage Inductance and AC
Resistance.. 1046
*Reda Bakri, Xavier Margueron, Wendell Da Cunha Alves, Xavier Cimetiere, Frédéric Gillon,
Antoine Bruyere, Lucian Vatamanu*

Neural Networks-Generalized Predictive Control for MIMO Grid-Connected Z-Source Inverter
Model 1056
Navid Salehi, Herminio Martinez-Garcia, Guillermo Velasco-Quesada

Voltage Estimation for Diode-Clamped MMCs Based on a Simplified Neural Network 1064
Nima Tashakor, Davood Keshavarzi, Shady Banana, Stefan Goetz

A Non-Cooperative Game-Theoretic Distributed Control Approach for Power Quality
Compensators 1074
*Claudio Burgos-Mellado, Victor Bucarey, Helmo K. Morales-Paredes, Diego Muñoz-
Carpintero*

A Comparative Analysis of Power Converter Topologies for Integration of Modular Batteries in
Electric Vehicles.. 1083
*Alberto Cárcamo, Aitor Vázquez, Alberto Rodriguez, Diego G. Lamar, Marta M. Hernando,
Daniel Remón*

Design of a High-Dynamic Test Bench for Accelerated Dielectric Lifetime Testing with Adjustable
Voltage Slopes and Temperatures 1094
Hendrik Schefer, Lucas Hanisch, Tim-Hendrik Dietrich, Regine Mallwitz, Markus Henke

Novel Modulation Method for Common-Mode Noise Reduction in Solid-State Transformer Based
on ISOP Configuration 1104
Naoto Kikuchi, Hiroki Watanabe, Keisuke Kusaka, Jun-Ichi Itoh

Modular STATCOM for Compensation of Reactive Power and Voltage Asymmetry in Medium-Voltage Distribution Power Grids 1114

Josef Štengl, Tomáš Kormska, Jakub Talla, Zdenek Peroutka

Novel Method for Active Short Circuit (ASC) Tests of Power Module in Automotive Traction Application 1121

Tobias Appel, Arne Bieler

Short Circuit Performance and Current Limiting Mode of a Monolithically Integrated SiC Circuit Breaker for DC Applications Up to 800 V 1128

Norman Boettcher, Taro Takamori, Keiji Wada, Wataru Saito, Shin-Ichi Nishizawa, Tobias Erlbacher

Application of a HV Bipolar Square-Wave Voltage Generator for Qualification and Assessment of Energy Equipment 1137

Rico Fischer-Baeumer, Kai Gohrmann, Konrad Domes, Benjamin Sahan, Christian Staubach

A Decentralized and Communication-Free Control Algorithm of DC Microgrids for the Electrification of Rural Africa 1147

Lucas Richard, David Frey, Marie-Cecile Alvarez-Herault, Bertrand Raison

Universal Real-Time Model for Active Rectifiers in Versatile Totem-Pole PFC Configurations 1157

Axel Kiffe, Thorben Hoffstadt

Investigation of Core-Loss Mechanisms in Large-Scale Ferrite Cores for High-Frequency Applications 1167

Michael Baumann, Christoph Drexler, Jonas Pfeiffer, Jens Schueltzke, Erwin Lorenz, Michael Schmidhuber

Generation of Methodology for Making Benchmark Microgrids and Application in ESUSCON Microgrid 1177

Oscar Dorner, Patricio Mendoza-Araya

An Overview of Grid-Connection Requirements for Converters and Their Impact on Grid-Forming Control 1187

Paul Imgart, Mebtu Beza, Massimo Bongiorno, Jan R. Svensson

Modular Battery-Integrated Power Electronics-Modelling, Advantages, and Challenges 1197

Nima Tashakor, Jan Kacetl, Tomas Kacetl, Stefan Goetz

Design of Triple-Active Bridge Converter with Inherently Decoupled Power Flows 1207

Dong-Uk Kim, Byengjoo Byen, Byunghwang Jeong, Sungmin Kim

Application of a Multi-Winding Magnetic Component Characterization Method to Optimize Cross-Regulation Performances in DCM Flyback Converters 1216

Denis Motte-Michellon, Brahim Ramdane, Yves Lembeye, Bruno Cogitore

Application of an Electrostatic Machine in a Low-Voltage Microgrid 1226

Gabriel Ramos Huerta, Patricio Mendoza-Araya

Influences of Parasitic Capacitances in Wide Bandwidth Rogowski Coils for Commutation Current Measurement 1237

Philipp Ziegler, Tobias Festerling, Jorg Haarer, Philipp Marx, David Hirning, Jorg Roth-Stielow

Systematic Analysis of Oscillations in DC-Links of Fast Switching Power Electronics 1247

Tobias Fricke, Regine Mallwitz

EMI Mitigation Induced by an IGBT Driver Based on a Controlled Gate Current Profile 1256
Daniel S. Martinez-Padron, Nicolas Patin, Eric Monmasson

An Accurate and Fast Model of Three-Level Three-Phase Dual-Active Bridge Converters in Real-Time Simulation 1266
Ming Jia, Philipp Joebges, Rik W. De Doncker

A Calorimetric and Electrical Method for Measuring Loss Energies of Half-Bridges 1277
Jörg Haarer, Mattea Eckstein, Philipp Ziegler, Philipp Marx, David Hirning, Jörg Roth-Stielow

Condition Monitoring Approach of a SiC Power Semiconductor using Turn-Off Delay with an Integration in a SiC Driver 1286
Victor Golev, Ulf Schümann, Rando Raßmann, Jan Bockholt

Measurement Results of Multilevel Hysteresis Control for Paralleled Two-Level Converters 1294
Magdalena Gierschner, Yves Hein, Hans-Günter Eckel, Christian Heien

Design and Development of a Short-Circuit Test Bench for Low-Voltage Direct Current Protection Devices 1300
Simon Ravyts, Thomas Vandenbussche, Koen Stul, Jan Cappelle

A Novel Modified-TOGI Based PLL for the Three-Phase Unbalanced and Distorted Grid Conditions 1309
Khanh-Hung Nguyen, Ahmad Ali Nazeri, Xiao Yu, Peter Zacharias

Comparison of Two and Three-Level AC-DC Rectifier Semiconductor Losses with SiC MOSFETs Considering Reverse Conduction 1319
Guangyao Yu, Thiago Batista Soeiro, Jianning Dong, Pavol Bauer

Measurement Method for Simple Determination of Sinusoidal Large Signal Losses in Inductive Components 1328
Peter Zacharias, Alejandro Aganza-Torres

A Novel Technique for the Suppression of the Displacement Current Through Power Module Base-Plate Capacitance 1336
Mahmoud Saeidi, Ahmad Ali Nazeri, Rufad Zilic, Peter Zacharias

Analysis and Implementation of Effective Placement of EMC Capacitors for WBG Modules 1343
Mahmoud Saeidi, Ahmad Ali Nazeri, Firas Jenhani, Peter Zacharias

Power Hardware-In-The-Loop Verification of a Cold Load Pickup Scenario for a Bottom-Up Black Start of an Inverter-Dominated Microgrid 1350
Mina Mirzadeh, Robin Strunk, Tobias Erckrath, Axel Mertens

Detection of Incipient Inter-Turn Short-Circuit Faults by Artificial Intelligence Classifiers 1361
Osman Örgüt, Ilker Sahin, Ece Olcay Günes

Modeling the Impact of Grid-Forming E-STATCOMs on Inter-Area System Oscillations 1371
A. Bolzoni, N. Johansson, J. P. Hasler

Combining Schwarz-Christoffel Mappings and Biot-Savart Law to Calculate the High-Frequency Current Distribution Inside a Single Slot 1381
Torben Fricke, Phil Leon Pickert, Babette Schwarz, Bernd Ponick

Standardised Switching Cell Building Block for Converter Design Optimisation with Detailed Electro-Thermal Model 1391

Georgios Papadopoulos, Jürgen Biela

Design Procedure for Transformer-Based Solid-State Pulse Modulators with Damping Network 1402

Spyridon Stathis, Juergen Biela

DC Bias Impact on Magnetic Core Losses at High Frequency 1413

Bima Nugraha Sanusi, Ziwei Ouyang

Investigation of the Short-Circuit Type II Safe Operating Area of IGBTs 1424

Madhu Lakshman Mysore, Mohamed Alaluss, Abhishek Maitra, Thomas Basler, Roman Baburske, Franz-Josef Niedernostheide, Hans-Joachim Schulze

Single Transformer, MMC Based MV Power Electronic Traction Transformer 1434

Simon Fuchs, Simon Beck, Jürgen Biela

A New Power MOSFET Technology Achieves a Further Milestone in Efficiency 1445

Ralf Siemieniec, Michael Hutzler, Cesar Braz, Tomasz Naeve, Elias Pree, Heimo Hofer, Ingmar Neumann, David Laforet

Experimental Evaluation of Battery Impedance and Submodule Loss Distribution for Battery Integrated Modular Multilevel Converters 1456

Arvind Balachandran, Tomas Jonsson, Lars Eriksson, Anders Larsson

Constant DC Power Infeed Grid Forming with Improved Ability to Ride-Through Unbalanced Low-Voltage Faults 1466

Tayssir Hassan, Malte Eggers, Huoming Yang, Peter Teske, Sibylle Dieckerhoff

Constrained Long-Horizon Direct Model Predictive Control for Grid-Connected Converters with LCL Filters 1476

Mattia Rossi, Petros Karamanakos, Francesco Castelli-Dezza

Performance Evaluation of SiC-Based Isolated Bidirectional DC/DC Converters for Electric Vehicle Charging 1486

Kaushik Naresh Kumar, Rafal Miskiewicz, Przemyslaw Trochimiuk, Jacek Rabkowski, Dimosthenis Peftitsis

Impact of Threshold Voltage Shifting on Junction Temperature Sensing in GaN HEMTs 1497

Burhan Etoz, Jose Ortiz Gonzalez, Arkadeep Deb, Saeed Jahdi, Olayiwola Alatise

Comparison of Power Cycling Results of Discrete GaN Cascodes for Automotive Power Electronics with High Temperature Swings 1506

Florian Lippold, Philipp Hauenschild, Regine Mallwitz

Current Distortion Study for Hybrid Multi-Level Grid Inverter with Active Neutral-Point-Clamped 4-Leg Topology 1515

Jonas Steffen, Matthias Klee, Fabian Schnabel, Axel Seibel, Marco Jung

Dynamic Maximum Power Point Tracking Method Including Detection of Varying Partial Shading Conditions for Photovoltaic Systems 1525

Rosalie Rouphael, Nezha Maamri, Jean-Paul Gaubert

Novel Operation Mode of the Modular Multilevel Matrix Converter Based on a Dimensioning Algorithm 1533

Rebecca Dierks, Axel Mertens

On the Cosmic Ray Influence on the Electronics Design of a High Altitude Electric Aircraft 1543
Philippe Morey, Mauro Carpita

DC-Bus Control Considerations of Asymmetrical Multilevel Inverters with Embedded Buck-Boost
Converter 1551
Theodoros P. Mouselinos, Emmanuel C. Tatakis

A Seamless Modulation Strategy for Step-Up/Down Partial Power Processing Converter (SUD-
P3C) 1561
*Chao Liu, Zhe Zhang, Ziwei Ouyang, Jiasheng Huang, Michael A. E. Andersen, Tiberiu
Gabriel Zsurzsan*

Performances Analysis of Non-Model-Based Speed Estimation Algorithms for Motor Drives 1569
*Gaetano Turrisi, Luigi Danilo Tornello, Giacomo Scelba, Giulio De Donato, Giuseppe
Scarcella*

A Method to Design Power Control System of Wayside Energy Storage System for Energy Saving
in DC-Electrified Railway 1580
Kota Sato, Keiichiro Kondo, Hiroyasu Kobayashi, Makoto Chida

A Reconfigurable Single-Stage Three-Phase Electric Vehicle DC Fast Charger Compatible with
Both 400V and 800V Automotive Battery Packs 1590
Mojtaba Forouzesh, Yan-Fei Liu, Paresh C. Sen

Efficiency Improvement of Single-Stage AC-DC LLC Converter using a Line Cycle Synchronous
Rectifier (SR) Driving Strategy 1601
Mojtaba Forouzesh, Yan-Fei Liu, Paresh C. Sen

Influence of DC Supply Voltage Unbalances on the Performance of ARCP Inverters 1611
Gholamreza Tabrizi, Sebastian Sprunck, Marco Jung

Grid-Forming Control for Enhanced Microgrid Interconnection 1620
Tobias Erckrath, Christian Bendfeld, Peter Unruh, Axel Seibel, Marco Jung

Low Phase Shift Filter for Current Sensing Based on the Difference Between AC Machine Models
with and Without Iron Losses 1631
Niklas Himker, Marcel Krümpelmann, Axel Mertens

Design and Analysis of a Voltage Clamping Active Delay Control Method for Series Connected
SiC MOSFETs 1641
Rui Wang, Asger Bjørn Jørgensen, Hongbo Zhao, Stig Munk-Nielsen

Practical Implementation of a Concept for In-Situ Detection of Humidity-Related Degradation of
IGBT Modules 1649
Benedikt Kostka, Axel Mertens

Design for Enhanced Noise Immunity of PCB Coils Used for Sensing Current Through Power
Devices 1658
Aamir Rafiq, Sumit Pramanick

Measurement Principle for Measuring High Frequency Bearing Currents in Electric Machines and
Drive Systems 1665
Benjamin Knebusch, Lennart Junemann, Pauline Holtje, Axel Mertens, Bernd Ponick

Climatically Induced Insulation Degradation in Power Semiconductor Modules of Wind Turbines 1674
Timo Lichtenstein, Sören Fröhling, Bernd Tegtmeier, Katharina Fischer

Comparison of Magnetic Noise Compensation Techniques for Dual Three-Phase Electrically Excited Synchronous Machines.. 1684

Jonas Henkenjohann, Jan Andresen, Axel Mertens

PCB Technology Comparison Enabling a 900V SiC MOSFET Half Bridge Design for Automotive Traction Inverters 1692

Matthias Spieler, Che-Wei Chang, Ayman El-Refaie, Muhammad H Alvi, Dong Dong, Rolando Burgos

Desaturated Turn-Off of Low-Saturation IGBTs with Clamping Method to Reduce Turn-Off Energy Losses.. 1703

Vishwas Acharya Nayampalli, Hans-Günter Eckel

Impact of Bond Wire Configuration on the Power Cycling Capability of Discrete SiC-MOSFET Devices 1713

Patrick Heimler, Nick Thönelt, Josef Lutz, Thomas Basler

A Low-Leakage, Low-Loss Magnetic Transformer Structure for High-Frequency Applications................. 1722

Allen Nguyen, Ajinkya Phanse, Michael Solomentsev, Alex J. Hanson

Temperature Distribution of an IGBT Chip During Repetitive Switching Events Under Consideration of Front-Side Ageing... 1733

Christian Bäumler, Bo Zhang, Maximilian Goller, Xing Liu, Thomas Basler

Boosting Pilot-Diode Reverse-Conducting IGBTs Turn-ON and Reverse-Recovery Losses with a Simple Gate-Control Technique.. 1744

Daniel Lexow, Hans-Günter Eckel

Modeling of an Interleaved DC-DC Boost Converter for a Direct Model Predictive Control Strategy... 1754

Thomas Effenberger, Hannes Böorngen, Eyke Liegmann, Michael Hoerner, Petros Karamanakos, Ralph Kennel

Static Analysis and Control Strategies of the Single Active Bridge Converter........................... 1765

Alexis A. Gómez, Alberto Rodríguez, Marta M. Hernando, Diego G. Lamar, Javier Sebastián, Ibán Ayarzaguena, Jose Manuel Bermejo, Igor Larrazabal, David Ortega, Francisco Vázquez

Multi-Port Inductive Power Transfer System Considering Charging Auxiliary Battery in EVs................... 1776

Zhuoqi Zhang, Ryosuke Ota, Ryohei Okada, Nobukazu Hoshi

Influence of IGBT and Diode Parameters on the Current Sharing and Switching-Waveform Characteristics of Parallel-Connected Power Modules... 1785

Y. Ando, J. Sakai, K. Hatori, N. Soltau, E. Wiesner

Innovative Driving Scheme for Electrical Generators in More Electric Aircrafts Employing Series Active Filtering... 1796

Nena Apostolidou, Nick Papanikolaou

Field-Measurement Based Hygrothermal Modelling of the Converter-Cabinet Climate in Wind Turbines... 1804

Katharina Fischer, Katherina Gohler

A Multi-Mode Control Based Asymmetrical Dual-Active-Bridge Series-Resonant DC-DC Converter (DABSRC) 1815

M. Yaqoob, Grover Torrico, Wang Shuqin

Extended Balancing and Dimensioning of Capacitors in MMC Double Submodules 1824
Ali Sharaf Addin, Christopher Dahmen, Thomas Brückner

Saliency Extraction and Torque Sharing Estimation of Dual Motor Drive using Special Current Sensor Configuration 1834
E. Rodriguez Montero, M. Vogelsberger, T. Wolbank

Soft-Switching Converter for Inductive Power Transfer System with Double-Sided LCC Resonant Network 1844
Ryohei Okada, Ryosuke Ota, Nobukazu Hoshi

Ultra Low Loss - MMC Submodules Favorable for SiC-FET Enabling High Functional Safety 1855
Christopher Dahmen, Rainer Marquardt

Control of an Active Gate Driver for an Electric Vehicle Traction Inverter using Artificial Neural Networks 1865
Julius Wiesemann, Jacob Dumtzlaff, Axel Mertens

Cascaded H-Bridge Converter Designs for Future Short-Range All-Electric Aircraft Propulsion 1875
Maximilian Hagedorn, Malte Lorenz, Axel Mertens

Overview and Evaluation of Energy Balancing Techniques for MMCs with Various Input and Output Frequencies 1885
Gyanendra Kumar Sah, Michael Schütt, Hans-Günter Eckel

Comparative Lifetime Estimations for IGBT Modules in Wind Turbine Converters 1895
Christian Neumann, Hans-Gunter Eckel

Single-Phase, Five-Level Inverter with SPWM-Based Neutral Point Voltage Balancing Scheme 1906
Dmytro Kondratenko, Arkadiusz Lewicki, Charles Odeh

Magnetic Core Evaluation Kit for the Comparison of Core Losses 1914
Wilmar Martinez, Xiaobing Shen, Siqi Lin, Jens Friebe

Multi-Objective Optimization of Modular Multilevel Converter Systems 1923
Nikolaus Patzelt, Christian Schlegel, Michail Vasiladiotis

Sizing of Hybrid Energy Storage System for Residential PV Applications 1933
Xiangqiang Wu, Zhongting Tang, Tamas Kerekes

DC Bias Currents in Full-Bridge DC-DC Converters in Context of WBG Semiconductors and High Switching Frequencies 1939
Niklas Badenhop, Lukas Fräger, Dennis Kampen, Sascha Langfermann, Michael Owzareck

Parameter Tuning Method for Class Φ_2 Converters for High-Frequency Wireless Power Transfer Applications 1947
Yining Liu, Prasad Jayathurathnage, Jorma Kyyrä

Inductor Design Optimization using FEA Supervised Machine Learning 1955
D. Cajander, I. Viarouge, P. Viarouge, D. Aguglia

Enabling Large-Scaled MMC EMT-RMS Co-Simulation by Data Exchange in the Loop (DXiL) 1966
Xiong Xiao, Soham Choudhury, Martin Coumont, Jutta Hanson

Advanced Low-Voltage System-In-Package Half-Bridge MOSFET with Added Protection Features 1975
S. Musumeci, V. Barba, F. Scrimizzi, C. Mistretta

Evaluation of Common-Mode Leakage Current of Aalborg-Type Transformerless PV Inverters 1985
 Georgios I. Orfanoudakis, Eftychios Koutroulis, Georgios Foteinopoulos, Weimin Wu

Multi-Frequency Traction-To-Auxiliary Integrated EV Drivetrain: Eliminating the Need for an
Auxiliary Power Module 1995
 Caniggia Viana, Mehanathan Pathmanathan, Peter W. Lehn

Potentials to Improve the Post-Fault Performance of a Fault-Tolerant Inverter System in Electrified
Aircraft Propulsion System 2003
 Yongtao Cao, Leon Fauth, Jens Friebe, Axel Mertens

Model Predictive Control-Enabled Fault Ride Through Operation Strategy for High Power Wind
Turbine 2011
 Pedro Catalán, Yanbo Wang, Zhe Chen, Joseba Arza

A Theoretical Comparison of Different Virtual Synchronous Generator Implementations on
Inverters 2021
 Patrick Körner, Andrea Reindl, Hans Meier, Michael Niemetz

Linear Flux-Switching Machine Design - A Multiobjective Optimization 2030
 Hendrik Marks, Henning Schillingmann, Sridhar Balasubramanian, Markus Henke

Single-Arm MMC-Based Converter for Transformerless Rail Interties 2038
 Simon Beck, Simon Fuchs, Jürgen Biela

Medium Voltage Diode Rectifier Design for High Step-Up DC-DC Converter 2049
 Pierre Le Métayer, Cyril Buttay, Drazen Dujic, Piotr Dworakowski

Fast Switching Planar Inductance Current Source ZETA Converter with Integrated Common Mode
Filter 2058
 Benjamin H. Zacher, Christian Schumann

System Level Simulation of Moisture Propagation and Effects in Wind Power Converters 2066
 Johannes C. Wenzel, Axel Mertens

PWM-Based Optimization-Free Active Voltage-Balancing Control of 7-Level Active Neutral-
Point-Clamped Flying-Capacitor Multicell Inverters 2073
 Vahid Dargahi

Model Predictive Power Sharing Algorithm for Fuel Cell Integration in a Dual Inverter Electric
Vehicle Drivetrain 2084
 Mehanathan Pathmanathan, Caniggia Viana, Sukhjit Singh, Peter W. Lehn

Comparative Evaluation of the 5-Phase Vienna and the 5-Phase PWM Rectifiers Under DC
Voltage Control 2092
 A. Dieng

Modelling and Control of a 50kW SiC-Based Isolated DAB Converter for Off-Board Chargers of
Electric Vehicles 2101
 Haaris Rasool, Manh Tuan Tran, Sajib Chakraborty, Joeri Van Mierlo, Thomas Geury,
 Mohamed El Baghdadi, Omar Hegazy

Impact of Cyber Attacks on Cost Oriented Power Routing Schemes in Microgrids 2110
 Kirti Gupta, Subham Sahoo, Bijaya Ketan Panigrahi, Frede Blaabjerg

Response of IGBT Chip Characteristics Due to Critical Stress 2119
 Kohei Yamauchi, Rik W. De Doncker

Mega-Hertz High-Power WPT System with Parallel-Connected Inverters using Current Balance Circuit 2127

Masamichi Yamaguchi, Keisuke Kusaka, Jun-Ichi Itoh

Investigation and Mitigation of Common-Mode Voltage in Four-Level NPC Converters Modulated by Redundant Level Modulation 2136

Jun Wang, Wei Xu, Xibo Yuan, Lihong Xie

Ferrite Optimization for a Three-Phase Wireless Power Transfer System for Electric Vehicles 2145

Shuang Nie, Mehanathan Pathmanathan, Peter W. Lehn

Frequency and Modulation Index Related Effects in Continuous and Discontinuous Modulated Y-Inverter for Motor-Drive Applications 2156

Hamzeh J. Jaber, Alberto Castellazzi

Performance Evaluation of Sinusoidal-Flux Reluctance Machine for Improving Power Density with Reduced Torque and Input-Current Ripples 2164

Kiwa Nagayasu, Masaki Iida, Kazuhiro Umetani, Mastaka Ishihara, Eiji Hiraki

Power Hardware-In-The-Loop Test of Low-Voltage Battery for a Plug-In Hybrid Electric Vehicle 2175

Ronan German, Florian Tournez, Alain Bouscayrol, Aurelien Lievre, Betty Lemaire-Semail

Stability Analysis of DFIG System Connected with High-Frequency Capacitive Grid Based on Closed-Loop Current Control and Direct Power Control 2182

Bin Hu, Heng Nian, Subham Sahoo, Frede Blaabjerg, Yaqian Zhang, Zixiao Xu

Full-Bridge Modular Multilevel Converter for the Four-Quadrant Supply of High Power Magnets in Particle Accelerators 2189

Manuel Colmenero, Ricardo Vidal-Albalate, Francisco R. Blanquez, Ramon Blasco-Gimenez

Deep Neural Network for Magnetic Core Loss Estimation using the MagNet Experimental Database 2197

Xiaobing Shen, Hans Wouters, Wilmar Martinez

Hybrid Circuit Board Structure for Power Electronics 2205

Gerrit Braun, Deniz-Heinz Moldenhauer

Active Control of Gear Mesh Vibration using a Permanent-Magnet Synchronous Motor and Simultaneous Equation Method 2211

Dominik Reitmeier

Research Laboratory for Testing Grid Connected Devices Under Grid Voltage / Grid Impedance Variations and Microgrid Conditions 2219

Swen Bosch, Jochen Staiger, Heinrich Steinhart

Reducing the Impact of Skin Effect Induced Measurement Errors in M-Shunts by Deliberate Field Coupling 2230

Hauke Lutzen, Jonas Müller, Vladimir Polezhaev, Till Huesgen, Nando Kaminski

Grid Forming Control for HVDC Systems: Opportunities and Challenges 2241

Adil Abdalrahman, Ying-Jiang Häfner, Malaya Kumar Sahu, Khirod Kumar Nayak, Ashkan Nami

A Highly Integrated and Modular High Speed Electric Drive for Lightweight Electric Mountain Bikes 2251

Matthias Hofer, Mario Nikowitz, Manfred Schrödl

Performance Enhancement of Power Conditioning Systems in More Electric Aircrafts 2257
Nick Rigogiannis, Nick Papanikolaou, Yongheng Yang

Steady State Simulations of a Hybrid HVAC/HVDC Network using OS Based ARM Devices 2266
Ioan Catalin Damian, Mircea Eremia

Experimental Comparison of FPGA-Implemented Model Predictive Voltage Control to Cascaded
Proportional Resonant Control for a Three-Phase Four-Wire Three-Level Grid-Forming Inverter of
250 kVA 2276
Jarren Lange, Dominik Schmies, Karl Stephan Stille, Joachim Böcker, Oliver Wallscheid

Experimental Study of Interleaved Y-Inverter Performance 2285
Yusuke Endo, Masataka Minami, Hamzeh J. Jaber, Alberto Castellazzi

Design of a GaN-Based Reconfigurable Resonant Converter for High Frequency On-Board
Charger of Battery Electric Vehicles 2293
*Manh Tuan Tran, Haaris Rasool, Dai Duong Tran, Mohamed El Baghdadi, Philippe Lataire,
Omar Hegazy*

Transient Liquid Phase Bond Reliability Evaluation of Die-Attach for Power Module Packaging 2301
Laxma R. Billa, Yangang Wang, Thomas Grant, Xiang Li, Harley Neal, Muhammad Morshed

Experimental Evaluation on Observer-Based Delay-Compensating Active Damping for LC-Filters 2308
Michael Schütt, Hans-Günter Eckel

Influence of Static Rotor Imbalance on the Roller Bearing Damage Due to Inverter-Induced
Bearing Currents 2316
Martin Weicker, Omid Safdarzadeh, Andreas Binder

Novel Current Balancing Method for HF Interleaved Converters with Reduced Control Effort 2327
Christian Beckemeier, Jens Friebe

dV/dt-Based Filter Design for Motor Inverters with Continuous Output Voltage 2334
Sabrina Ulmer, Stevan Bugarski, Gernot Schullerus, Ertugrul Sönmez

Evaluation of Core Losses in Transformers for Three-Phase Multi-Level DAB Converters 2344
Babak Khanzadeh, Yuriy Serdyuk, Torbjörn Thiringer

A Quasi-Offline Condition Monitoring Method of DC-Link Capacitor Banks in Accelerator Power
Converters 2355
*Timm Felix Baumann, Konstantinos Papastergiou, Raul Murillo Garcia, Dimosthenis
Peftitsis*

Minimizing Voltage Stress in Auxiliary Resonant Commutated Pole Inverters using Saturable
Inductors 2366
Markus Zocher, Norbert Grass, Ralph Kennel

Adaptive Dead-Time Control in a Resonant Wireless Power Transfer System 2375
Tim Krigar, Martin Pfost

Multilevel Battery Converter with Cascaded H-Bridges on Cell Level-Battery Management System
Or a Renewed Attempt for Power Electronic Building Blocks? 2383
*Max Rothenburger, Markus Horn, Xiao Yu, Gerold Schulze, Koenraad Muyllaert, Peter
Zacharias, Ludwig Brabetz, Hartmut Hillmer*

Design and Potential of EMI cm Chokes with Integrated DM Inductance 2392
Mohammad Ali, Rehnuma Bushra, Jens Friebe, Axel Mertens

Implementation Options of a Fully SiC Buck-CSI for Advanced Motor Drive Application......................... 2402
    Yonghwa Lee, Alberto Castellazzi

Optimized Control Scheme to Achieve ZVS for the Complete Pre-Charging Phase of
Supercapacitors with a 500 kHz SiC- And GaN-Based Dual Active Bridge 2413
    Patrick Lenzen, Martin Pfost

Fault Blocking Capability in the DC-MMC with Reduced Number of Sub-Modules................................... 2422
    J. D. Páez, F. Morel, S. Bacha, P. Dworakowski

An Open-Source FEM Magnetic Toolbox for Calculating Electric and Thermal Behavior of Power
Electronic Magnetic Components 2432
    Nikolas Förster, Jonas Hölscher, Till Piepenbrock, Philipp Rehlaender, Oliver Wallscheid,
    Frank Schafmeister, Joachim Böcker

Comparison of Dual-Active-Bridge-Based Topologies for Single-Phase Single-Stage EV On-Board
Chargers 2441
    Daniel Gaona, Denis Pauls, Eduardo Facanha De Oliveira

Design Concepts for Medium Voltage DC Networks Supplying the Future Circular Collider (FCC)........... 2451
    Manuel Colmenero, Francisco R. Blanquez, Ramon Blasco-Gimenez

A Novel Dual CC-CV Output Wireless EV Charger with Minimal Dependency on Both Coil
Coupling and Load Variation 2462
    Subhranil Barman, Kishore Chatterjee

A High-Performance EMI Filter Based on Laminated Ferrite Ring Cores 2470
    Marcin Kacki, Marek S. Rylko, John G. Hayes, Charles R. Sullivan

Investigation of the Static Performance and Avalanche Reliability of High Voltage 4H-SiC
Merged-PiN-Schottky Diodes 2477
    Chengjun Shen, Saeed Jahdi, Phil Mellor, Juefei Yang, Erfan Bashar, Jose Ortiz-Gonzalez,
    Olayiwola Alatise

On Chain-Link Based Multi-Port Converters Able to Connect HVDC and MVDC to AC
Transmission Network.. 2486
    Daniele Falchi, Oriol Gomis-Bellmunt, Eduardo Prieto-Araujo, Olivier Despouys

Voltage Control Scheme for Multilevel Interfacing PV Application: Real-Time MRAC-Based
Approach 2496
    Mohammad Sadegh Orfi Yeganeh, Mehdi Rahmani, Nenad Mijatovic, Tomislav Dragicevic,
    Frede Blaabjerg, Pooya Davari

Control Principles for Island Operation and Black Start by Offshore Wind Farms Integrating Grid-
Forming Converters... 2504
    Daniela Pagnani, Lukasz Kocewiak, Jesper Hjerrild, Frede Blaabjerg, Claus Leth Bak

Experimental Study of the Reduction and Removal of Turn-On Snubber for IGCT Based MMC
Submodule using Fast Silicon Diodes 2515
    Arthur Boutry, Cyril Buttay, Besar Asllani, Bruno Lefebvre, Eric Vagnon, Dong Dong

Characterisation of a Ferrite-Polymer Based Magnetic Material 2526
    Johan Le Leslé, Guillaume Lefevre, Julien Morand, Rémi Perrin, Pierre-Yves Pichon,
    Guillaume Regnat

Model Predictive-Based Control Technique for Fault Ride-Through Capability of VSG-Based Grid-Forming Converter.. 2537
    Mobina Pouresmaeil, Amir Sepehr, Basit Ali Khan, Jafar Adabi, Edris Pouresmaeil

Grounding Points in HV/MV Hybrid Transformer Auxiliary Converters...................................... 2544
    Adrian Wiemer, Jürgen Biela

Non-Parasitic Induced Transient Overvoltage in ANPC Topology Due to Critical Switching Sequences 2554
    Michael Geiss, Robert Kragl, Jürgen Thoma, Benjamin Volzer

Open-Delta SBC: A New Converter Topology with Low Number of Sub-Modules for MV Applications... 2564
    D. Lanzarotto, P. B Steckler, K. Vershinin, F. Morel

Characterising the Effect of an Inverter on the Regulation of the AC Voltage using a Frequency Response Identification Technique 2574
    Mohamed Aldarmon, Joan Marc Rodriguez, Adria Junyent-Ferre

Artificial-Intelligence Based DC-DC Converter Efficiency Modelling and Parameters Optimization 2581
    Fanghao Tian, Diego Bernal Cobaleda, Wilmar Martinez

Analysis of the Loss Distribution of a 6 kW Two Stage Power Supply for 600 V DC Applications............ 2588
    Lukas Fräger, Sascha Langfermann, Michael Owzareck, Dennis Kampen, Jens Friebe

Study on the Gate Loop Design and Its Impact on Switching Characteristics of GaN Transistors................ 2596
    Xiaomeng Geng, Carsten Kuring, Oliver Hilt, Mihaela Wolf, Joachim Würfl, Sibylle Dieckerhoff

Analysis of Current Sharing in the Parallel Connection of GaN Transistors 2607
    Frederik Stalleicken, Sibylle Dieckerhoff, Karsten Handt, Sebastian Nielebock

Verification of GaN-HEMT Spice Models using an S-Parameters Approach 2618
    Alonso Gutierrez, Nasri Said, Emmanuel Marcault, Mathieu Gavelle

Power Loss Modelling of GaN HEMT-Based 3L-ANPC Three-Phase Inverter for Different PWM Techniques.. 2628
    Salvatore Mita, Arjun Sujeeth, Giuseppe Aiello, Dario Patti, Francesco Gennaro, Giacomo Scelba, Mario Cacciato

Generalized Core and Winding Area Ratio - Trends for Inductors and Transformers in Power Electronics with High Switching Frequencies... 2638
    Siqi Lin, Leon Fauth, Wilmar Martnez, Jens Friebe

Active Substrate Termination of Discrete and Monolithic Bidirectional GaN HEMTs in a T-Type Inverter 2644
    Carsten Kuring, Yannic Lange, Xiaomeng Geng, Oliver Hilt, Mihaela Wolf, Joachim Würfl, Sibylle Dieckerhoff

Transformer Design Optimization and Comparison for a DC-DC Converter Used in PV Micro-Inverters.. 2655
    Tobias Manthey, Meriem Khader, Jens Friebe

Automated Gate Impedance Network Design for SiC MOSFETs using SPICE Solver Interfaced with MATLAB Environment 2661
    Pawel Piotr Kubulus, Szymon Michal Beczkowski, Stig Munk-Nielsen, Asger Bjørn Jørgensen

An Improved Multi-Loop Resonant and Plug-In Repetitive Control Schemes for Three-Phase Stand-Alone PWM Inverter Supplying Non-Linear Loads 2670
Ahmad Ali Nazeri, Peter Zacharias

High Switching Frequency Operation of a Single-Phase Five-Level Hybrid Active Neutral Point Clamped Inverter with a Model Predictive Control Approach 2682
Mohammad Najjar, Mahdi Shahparasti, Rasool Heydari, Morten Nymand

Design of Planar Coupled Inductor Applied to Zero-Current Switching Clamped Current Converter 2689
Vinicius Freire Bezerra, Tobias Manthey, Montiê Alves Vitorino, Jens Friebe

Characterization of Online Junction Temperature of the SiC Power MOSFET by Combination of Four TSEPs using Neural Network 2698
Kanuj Sharma, Simon Kamm, Kevin Muñoz Barón, Ingmar Kallfass

Novel Extended Robust Disturbance Observer for Improved Cogging Force Compensation in Permanent Magnet Linear Motors 2706
Franz Luckert, Axel Mertens

Improvement of a Self-Powered Gate Driver Power Supply 2715
Mariana Raya, Oriol Aviñó, Sergio Busquets-Monge, Xavier Perpiñá, Miquel Vellvehi, Xavier Jordà

Optimization and Scaling of a Compact High-Power IGCT Capacitor Charger Based on Simulation and Measurements with a 300 kW/3.3 kV Demonstrator 2726
Felix Haag, Fabian Albrecht, Volker Brommer, Oliver Liebfried, Klaus F. Hoffmann

Multilayer Busbars for Medium Voltage ANPC Converter Dedicated to Battery Energy Storage Systems 2736
Mamadou Lamine Beye, Luc Bimmel, Anthony Bier, Jérémy Martin

A Simulation Model for SiC MOSFET Switching Transients Controlled by an Adaptive Gate Driver with the Capability of Reducing Switching Losses and EMI Across the Full Operating Range 2744
Zheming Li, Robert W. Maier, Mark-M. Bakran, Franz-J. Niedernostheide, Daniel Domes

Phase-Shift Modulation for Flying-Capacitor DC-DC Converters 2754
Philipp Rehlaender, Frank Schafmeister, Joachim Böcker

An EV Integrated Isolated DC Charger using a Six-Phase Synchronous Machine 2763
Sukhjit S Ghumman, Mehanathan Pathmanathan, Peter W Lehn

Configurable ISOP-IPOP DC-DC Converter for Universal Solid-State Transformer 2773
Pramod Apte, Jens Friebe, Lukas Fräger

Using System-On-Chip Boards for the Deployment of Controller for Verification and Prototyping 2780
Adeel Jamal, Gerd Griepentrog

Utilizing the Reactive Current Control Capability of an MMC-Fed AC/DC Converter for Volt-Second Balancing in Medium Frequency Transformers 2788
Kaveh Pouresmaeil, Maurice Roes, Jorge Duarte, Korneel Wijnands, Nico Baars, George Papafotiou

Cost Comparison for Different PV-Battery System Architectures Including Power Converter Reliability 2795
Martijn Deckers, Leander Van Cappellen, Glenn Emmers, Fereshteh Poormohammadi, Johan Driesen

Insulation Design and Analysis of a Medium Voltage Planar PCB-Based Power Bus Considering Interconnects and Ancillary Circuit Integration 2806
Joshua Stewart, Rolando Burgos, Dushan Boroyevich

Modular Multilevel Converter Control with using a General Space Vector PWM Method in Medium Voltage Hydro Power Application.. 2813
Chengjun Tang, Torbjörn Thiringer

A Technical Overview of Single-Stage Three-Port DC-DC-AC Converters 2824
Sebastian Neira, Zoe Blatsi, Michael M. C. Merlin, Javier Pereda

Common-Mode EMI Noise Modeling of Three-Level T-Type Inverter for Adjustable Speed Drive Systems.. 2835
Vefa Karakasli, Abdelmoumin Allioua, Gerd Griepentrog

A Condition Monitoring Scheme for Semiconductor Devices in Modular Multilevel Converters with Cascaded H-Bridge Submodules... 2843
Mohsen Asoodar, Mehrdad Nahalparvari, Christer Danielsson, Hans-Peter Nee

Particular Requirements on Drive Inverters for Safe and Robust Operation on an Open Industrial DC Grid... 2852
Simon Puls, Jan-Niklas Koch, Martin Ehlich, Holger Borcherding

Investigation About Operation and Performance of Gate Drivers for Power Electronics Converters for Cryogenic Temperatures... 2860
Mustafeez-Ul-Hassan, Yuxuan Wu, Vyacheslav Solovyov, Fang Luo

Synchronization Angle Determination in DVCSFO of DFIM Naval Propulsion... 2869
Youssef Drimizi, Maria Pietrzak-David, Pascal Maussion

Power Control of LCR-DAB Converter with Phase Shift in Fixed Switching Frequency 2877
Seung-Hyuk Baek, Jaehong Lee, Seung-Hwan Lee, Sungmin Kim

A Simplified Braking Method for Direct Matrix Converter-Fed PMSM Drives with Consideration of Avoiding Regenerative Energy 2885
Jun Xie, Dustin Henneberg, Martin Suberski, Thomas Ellinger, Uwe Radel, Jürgen Petzoldt

Inverter-Machine Parametric Co-Design for Energy Efficient Electric Drives.. 2893
Jaedon Kwak, Alberto Castellazzi

Bidirectional Cuk Converter in Partial-Power Architecture with Current Mode Control for Battery Energy Storage System in Electric Vehicles 2903
J. S. Artal-Sevil, J. Anzola, V. Ballestín-Bernad, I. Aizpuru

Design Space Exploration for a Capacitive 36V, 4A, 4:1 DCDC Converter with GaN Switches using a Performance-Cost-Matrix Including Uncommon Topologies.. 2912
Adrian Gehl, Malte Kempchen, Simon Disselkamp, Markus Olbrich, Bernhard Wicht

A Fast Control for a Three-Switch Multi-Input DC-DC Converter... 2919
Simone Cosso, Andrea Formentini, Mario Marchesoni, Massimiliano Passalacqua, Luis Vaccaro

Impact on the Torque and on the Copper Losses Under Fault-Tolerant Control of 5-Phase PMSG 2930
A. Dieng

Weighting Factor Design for FS-MPC in VSCs: A Brain Emotional Learning-Based Approach 2939
Mohammad Sadegh Orfi Yeganeh, Arman Oshnoei, Saeed Peyghami, Nenad Mijatovic, Tomislav Dragicevic, Frede Blaabjerg

A Strategy for Smooth Microgrid Transitions Without Phase Misalignment and Voltage Mismatch 2948
Gabriel Silva Rocha, Amiron Wolff Dos Santos Serra, Cesar Augusto Santana Castelo Branco, Hercules Araujo Oliveira, Jose Gomes De Matos, Luiz Antonio De Souza Ribeiro

Subtle Design and Performance Comparison of WF-FSM and DC-VRM for Large-Scale Direct-Drive Wind Power Generation 2958
Udochukwu B. Akuru, Maarten J. Kamper, Zi-Qiang Zhu

Analysis and Implementation of Different Non-Isolated Partial-Power Processing Architectures Based on the Cuk Converter... 2967
J. S. Artal-Sevil, J. Anzola, V. Ballestín-Bernad, J. L. Bernal-Agustín

GaN HEMT and SiC Diode Commutation Cell Based Dual-Buck Single-Phase Inverter with Premagnetized Inductors and Negative Gate Driver Turn-Off Voltage 2977
Tobias Brinker, Hendrik Gräber, Jens Friebe

Determination of Optimal Associated Discrete Circuit Switch Model Parameters for Real-Time Simulation of Dual-Active Bridge Converters 2985
Marija Stevic, Ravinder Venugopal

Integrated Motor Drive: A Multidisciplinary Approach.. 2996
Betty Lemaire-Semail, Nadir Idir, Eric Semail, Souad Harmand

Hardware in the Loop Test of an Electric Aircraft Powertrain... 3005
Sebastian Mönninghoff, Moritz Scholjegerdes, Kay Hameyer

A Multi-Port Smart Transformer for Green Airport Electrification 3014
Giampaolo Buticchi, Giovanni De Carne, Thiago Pereira, Kangan Wang, Xiang Gao, Jiajun Yang, Youngjong Ko, Zhixiang Zou, Marco Liserre

Improvement of EMI Filter Attenuation using Shielding.. 3022
Mohammad Ali, Rehnuma Bushra, Jens Friebe, Axel Mertens

Implementation of Onsite Junction Temperature Estimation for a SiC MOSFET Module for Condition Monitoring.. 3031
Farzad Hosseinabadi, Shahid Jaman, Sachin Kumar Bhoi, Md. Mahamudul Hasan, Sajib Chakraborty, Mohamed El Baghdadi, Omar Hegazy

Energy Storage Systems for Airborne Wind Generators.. 3037
Bakr Bagaber, Axel Mertens

Design Interactions of AC- And DC-Side Filters for Traction Drives with SiC Inverters 3048
Hedieh Movagharnejad, Benjamin Knebusch, Axel Mertens, Bernd Ponick

Investigation of an Interleaved Current-Fed Single Active Bridge DC-DC Converter for PV Applications.. 3059
Lucas Vinícius De Araújo Gomes, Tobias Manthey, Montiê Alves Vitorino, Jens Friebe

Real-Time Thermal Characterization of Power Semiconductors using a PSO-Based Digital Twin Approach 3067
Johannes Kuprat, Yoann Pascal, Marco Liserre

Self-Sensing Design and Control for an Induction Machine with an Additional Short-Circuited Rotor Coil 3075
    Stefan Luecke, Axel Mertens

Calculating the Tractive Power and Power Conversion Efficiency of Battery Electric Vehicles using a Global Navigation Satellite System and a Road Elevation Database 3084
    Shinichi Domae, Alberto Castellazzi, Hamzeh J. Jaber, Tenghui Dong, Taketsune Nakamura

PCB Layer Optimization of Planar Medium Frequency Transformer for On-Board EV Chargers 3092
    Fabian Groon, Hamzeh Beiranvand, Thiago Pereira, Görkem Can, Marco Liserre

Fault Current Capability Assessment of Low-Voltage Side Inverters in Smart-Transformers 3101
    Thiago Pereira, Luis Camurca, Francisco Santos, Marco Liserre

Adaptive Resonant-Valley Switching for a GaN HEMT Direct AC-AC Auxiliary Resonant Commutated Pole Converter 3112
    Kyle Steyn, Johan Beukes

The Variation of Core Loss in High-Frequency Transformers Under Different Load Conditions 3120
    Navid Rasekh, Jun Wang, Xibo Yuan

A Complete PFC Inductor Design for Lighting Equipment Applications 3130
    Wai Keung Mo, Kasper M. Paasch, Thomas Ebel

Automatic Generation Control-Based Charging/Discharging Strategy for EV Fleets to Enhance the Stability of a Vehicle-To-Weak Grid System 3140
    Majid Mehrasa, Mehrdad Gholami, Reza Razi, Khaled Hajar, Antoine Labonne, Ahmad Hably, Seddik Bacha

Model-Based Converter Control for the Emulation of a Wind Turbine Drive Train 3149
    Alexander Ernst, Wilfried Holzke, Dawid Koczy, Nando Kaminski, Bernd Orlik

A Novel Grid-Demanded Power Point Tracking (GPPT) Control Method for Wind Turbines to Preserve Grid Stability with High Wind Energy Penetration 3159
    David Matthies, Alexander Ernst, Henning Sauerland, René Reimann, Wilfried Holzke, Bernd Orlik

Extension and Implementation of a Model-Based Lifetime Monitoring System with Parallel Calculation of Multiple Power Semiconductors 3169
    Steffen Menzel, Wilfried Holzke, Michael Hanf, Holger Groke, Bernd Orlik, Nando Kaminski

Smart Charging Strategy for Electric Vehicles using an Optimized Fuzzy Logic System 3179
    M. Gholami, M. Mehrasa, R. Razi, K. Hajar, A. Hably, S. Bacha, A. Labonne

Analysis and Discussion of a Concept for an Adjustable Inductance Based on an Impact of an Orthogonal Magnetic Field 3188
    Guido Schierle, Michael Meissner, Klaus F. Hoffmann

A Field Programmable and Dynamic Configurable Power Electronic Converter Concept 3198
    Bjarte Hoff

DAB Converter Discrete ADRC Control into Real-Time CHIL Simulation of a MVDC/LVDC Power Grid 3206
    Alessio Clerici, Riccardo Chiumeo, Diego Raggini, Alessandro Veroni

SNNFT: Sequential Neural Network-Fuzzy Thermal Early Warning System for Lithium-Ion Batteries 3215
    Marui Li, Chaoyu Dong, Yunfei Mu, Qian Xiao, Jingming Cao, Hongjie Jia

Fine-Grained Dynamics Representation and Stability Analysis for MMC-Based Hybrid AC/DC Power Systems 3225
    Jingming Cao, Chaoyu Dong, Qian Xiao, Marui Li, Xiaodan Yu, Hongjie Jia

Adaptive Pontryagin's Minimum Principle-Inspired Supervised-Learning-Based Energy Management for Hybrid Trains Powered by Fuel Cells and Batteries 3235
    Hujun Peng, Feifei Li, Zhu Chen, Kai Deng, Sebina Jeschke, Kay Hameyer

A Case Study of Pole-Phase Changing Induction Machine Performance 3246
    Konstantina Bitsi, Sjoerd G. Bosga

New Topology of Superconducting Fault Current Limiter with Bypass Resistor 3254
    D. Baimel, Eli Barbi, S. Bronstein, N. Baimel, A. Kuperman

A Pre- And Discharge Unit for Capacitive DC-Links Based on a Dual-Switch Bidirectional Flyback Converter 3262
    Madlen Hoffmann, Martin März

Control and Integration of a Multiphase Brushless Wounded Synchronous Motor Drive 3272
    Remi Perrin, Guilherme Bueno-Mariani

A Way Forward to Achieve Interoperability in Multi-Vendor HVDC Systems 3282
    Adil Abdalrahman, Ying-Jiang Häfner, Philippe Maibach, Christoph Haederli

Model Predicitve Position Control of Electrical Drives on an Industrial PC 3292
    Fabian Karau, Michael Leuer

Bidirectional Active EMC Filter for Industrial Power Converters 3301
    Bernhard Wunsch, Stanislav Skibin, Ville Forsstrom

A General Method to Measure Parasitic Capacitance of Transformer using Guarding Technique 3309
    Shaokang Luan, Stig Munk-Nielsen, Bruce Wakelin, Magnus Hortans, Jan Schupp, Hongbo Zhao

Inductance Analysis of Electric Machines by Classical and Numerical Methods 3318
    J. J. Germishuizen, T. J. E. Miller

Dynamic Wireless Power Transfer DWPT Time Domain Model: Xyz Position and Speed Coupling Effect 3327
    Iosu Aizpuru, Eneko Agirrezabala, Mikel Mazuela, Unai Iraola, Estanis Oyarbide, Carlos Bernal

Dynamic Average Small Signal Model of the SAB Converter 3336
    Alexis A. Gómez, Alberto Rodríguez, Marta M. Hernando, Diego G. Lamar, Javier Sebastián, Ibán Ayarzaguena, Jose Manuel Bermejo, Igor Larrazabal, David Ortega, Francisco Vázquez

Algorithm for Optimal Selection of Drive Motor Transmission Combination 3344
    Santiago Ramos Garces, Dries Jacques, Stijn Derammelaere, Simon Houwen, Nick Van Oosterwyck, Bart Vanwalleghem

Evaluation of Drain-Source Voltage in Switch Transient Time Intervals as Gate Oxide Degradation Precursor of SiC Power MOSFETs 3353
    Javad Naghibi, Sadegh Mohsenzade, Kamyar Mehran, Martin P. Foster

Active Output LLC Converter Topology 3362
   Hannes Börngen, Eyke Liegmann, Sriram Jagannath, Ralph Kennel

Short Circuit Type II and III Behavior of 1.2 kV Power SiC-MOSFETs.. 3373
   Xing Liu, Xupeng Li, Thomas Basler

Analog MPPT Comparison for Interplanetary Small Satellites Missions 3382
   C. Torres, A. Garrigós, J. M. Blanes, P. Casado, D. Marroquí, C. Orts

Feasibility Assessment of Variable-Speed Generator Set Concepts with Focus on Rating of Power
Electronic Equipment 3391
   Hendrik Fehr, Albrecht Gensior, Andreas Möckel, Frank Atzler, Tilo Roß, Carsten Reincke-
   Collon

Bus Voltage Regulation using Sequentially Switched ZVZCS Converters for Spacecraft Power
Systems.. 3401
   A. Garrigós, C. Orts, D. Marroquí, J. M. Blanes, C. Torres, P. Casado

A Standardized and Modular Power Electronics Platform for Academic Research on Advanced
Grid-Connected Converter Control and Microgrids 3411
   Frank S. R., Schulz D., Stefanski L., Schwendemann R., Hiller M.

Gate Input Capacitance Characterization for Power MOSFETs using Turn-On and Turn-Off
Switching Waveforms 3420
   Yota Nishitani, Michiko Inoue, Takashi Sato, Michihiro Shintani

AC Battery: Modular Layout with Cell-Level Degradation Control 3429
   Claudio Burgos-Mellado, Marcos Orchard, Diego Muñoz-Carpintero, Tomislav Dragicevic,
   Lorenzo Reyes-Chamorro, Jacqueline Llanos

Analysis of Test Methods for Measurement of Leakage and Magnetising Inductances in Integrated
Transformers 3440
   Sajad A. Ansari, Jonathan N. Davidson, Martin P. Foster, David A. Stone

A Topology-Morphing Series Resonant Converter for Photovoltaic Module Applications........................ 3450
   Grigorios Sergentanis, Liliana De Lillo, Lee Empringham, C. Mark Johnson

A Novel Parameter for the Evaluation of Protective Circuits for IGBT Explosion Protection in
Submodules of MMC 3460
   Christoph Junghans, Hans-Guenter Eckel

Sub-Modules Switching Algorithms for Dual Active Bridge Modular Multilevel Converters to
Optimize Capacitor Voltage Deviation Versus Power Efficiency... 3470
   Peizhou Xia, Chuantong Hao, Stephen Finney, Michael Merlin

Systematic Adaptive Robust State Feedback Control for Active Front-End Rectifiers 3480
   Aidar Zhetessov, Giri Venkataramanan

An Optimized Compensation Strategy of Direct Matrix Converter-Fed PMSM Drives with Field
Weakening Under Unbalanced Supply Conditions 3491
   Jun Xie, Dustin Henneberg, Martin Suberski, Manuel Kusebauch, Uwe Rädel, Jürgen
   Petzoldt

Double Inverter Concept for High-Speed Drives Without Motor Filters 3501
   Henning Kasten, Stephan Beineke, Matthias Bachmann

A Universal Single Stage Current-Fed Bidirectional Converter with Both AC and DC Input Power Source Compatibility.. 3511
Manish Kumar, Sumit Pramanick, Bijaya Ketan Panigrahi

Optimization of Electric Vehicle Charge Scheduling with Consideration of Battery Degradation................ 3518
Raka Jovanovic, Sertac Bayhan, Islam Safak Bayram

Onboard ESU Sizing and Dynamic IPT Charging Scenarios for a Tramway Application.............................. 3529
Endika Bilbao Muruaga, Irma Villar, Florian Legay, Pierre Prenleloup, Jean-François Reynaud

Investigations on the Active Reduction of Common Mode Noise with Opposing Noise Sources 3536
Philipp Marx, Felix Seybold, Philipp Ziegler, David Hirning, Jörg Roth-Stielow

Knowledge Based Grey Box Modeling of Inaccessible Circuits for System EMC-Simulation in Time Domain... 3545
Jan-Philipp Roche, Jens Friebe, Oliver Niggemann

Novel Quasi-Direct Rotor Position Estimator for Permanent Magnet Synchronous Machines Based on the Back-Electromotive Force using Current Oversampling... 3555
Georg Lindemann, Viktor Willich, Axel Mertens

Design Considerations for Fast On-State Voltage Measurement Circuits.. 3565
Mathias C. J. Weiser, Manuel Rueß, Ingmar Kallfass

Analytical, FEM and Experimental Study of the Influence of the Airgap Size in Different Types of Ferrite Cores 3574
Asier Arruti, Francisco Jose Perez-Cebolla, Jon Anzola, Iosu Aizpuru, Mikel Mazuela

Design Method of a High Frequency GaN-Based Half-Bridge with Bottom-Side Cooled Transistors using Multi-PCB Assembly... 3582
Loris Pace, Florian Chevalier, Thierry Duquesne, Nadir Idir

A 30 kW Dynamic Wireless Inductive Charging System for EVs 3590
Zariff Meira Gomes, José Renes Pinheiro, Gilney Damm, Karim Kadem, Hassan Moussa

Dynamic Control of the Switching Behavior of SiC MOSFETs in Converter Operation 3599
Jochen Henn, Laurids Schmitz, Rik W. De Doncker

A Series Resonant Balancing Converter for Bipolar DC Grids on Ships.. 3607
Sachin Yadav, Zian Qin, Pavol Bauer

A V2G-Enabled Seven-Level Buck PFC Rectifier for EV Charging Application 3615
Anekant Jain, Ritika Agarwal, Krishna Kumar Gupta, Sanjay K. Jain

Experimental Demonstration of a 2.2kW Active-Clamp Converter for High-Current Wide-Voltage-Transfer Ratio Applications 3625
Philipp Rehlaender, Bastian Korthauer, Frank Schafmeister, Joachim Böcker

A Simplified Model for the Battery Ageing Potential Under Highly Rippled Load 3636
Tomáš Kacetl, Jan Kacetl, Nima Tashakor, Stefan Goetz

System Modeling and Design of a Hybrid Renewable Energy System for a Cable Network Head-End Station in Rural Area... 3646
Tobias Schillinger, Thomas Schuhmann, Martin Eckart

Comparison of System-Level Availability in Industrial Grids 3655
G. Emmers, J. Driesen

Ageing Mitigation and Loss Control in Reconfigurable Batteries in Series-Level Setups 3665
Tomáš Kacetl, Jan Kacetl, Nima Tashakor, Stefan Goetz

Characterization of Conventional and Advanced Current Measurement Techniques Suitable for
WBG Semiconductor Devices 3676
Severin Klever, André Thönnessen, Rik W. De Doncker

Zero-Sequence Voltage Reduces DC-Link Capacitor Demand in Cascaded H-Bridge Converters for
Large-Scale Electrolyzers by 40% 3686
Roland Unruh, Frank Schafmeister, Joachim Böcker

Thermal Behavior Impact on the Electric Motor Shape Multi-Objective Optimization............................ 3696
Aissam Riad Meddour, Anthony Babin, Nassim Rizoug, Christopher Vagg, Richard Burke,
Laid Degaa

Modelling Approaches of Power Systems Considering Grid-Connected Converters and Renewable
Generation Dynamics 3704
Jaume Girona-Badia, Vinícius Albernaz Lacerda, Eduardo Prieto-Araujo, Oriol Gomis-
Bellmunt, Stephan Kusche, Florian Pöschke, Horst Schulte

Efficiency and Lifetime Analysis of Several Airborne Wind Energy Electrical Drive Concepts 3711
Bakr Bagaber, Daniel Heide, Bernd Ponick, Axel Mertens

Design and Performance Analysis of Single-Phase Axial Flux Permanent Magnet Motor for
Coaxial Cascade 3722
Chu Wang, Xiaowei Hu, Xiaoya Wang, Weiwei Geng, Qiang Li, Jingning Hou

Comparison of Pulse Current Capability of Different Switches for Modular Multilevel Converter-
Based Arbitrary Wave Shape Generator Used for Dielectric Testing of High Voltage Grid Assets 3729
Dhanashree Ashok Ganeshpure, Ajeeth Phrassanna Soundararajan, Thiago Batista Soeiro,
Mohamad Ghaffarian Niasar, Peter Vaessen, Pavol Bauer

Accurate Modeling of IGBT-Based Converters in PLECS 3740
Anne Von Hoegen, Philipp Tillmann, Tetsuya Kojima, Rik W. De Doncker

Novel Analytical Method for Estimating the Junction-To-Top Thermal Resistance of Power
MOSFETs.. 3750
José Miguel Sanz-Alcaine, Francisco Jose Perez-Cebolla, Carlos Bernal-Ruiz, Asier Arruti,
Iosu Aizpuru

DC-Side Impedance for Handling Interoperability of Multi-Vendor Multi-Terminal HVDC
Systems... 3757
Ashkan Nami, Adil Abdalrahman, Ying-Jiang Häfner, Malaya Kumar Sahu, Khirod Kumar
Nayak

Utilizing the Electroluminescence of SiC MOSFETs as Degradation Sensitive Optical Parameter 3766
Lukas A. Ruppert, Michael Laumen, Rik W. De Doncker

Characterization of GaN-On-AlN/SiC Transistors Towards Monolithic Integrability 3775
Nick Wieczorek, Xiaomeng Geng, Carsten Kuring, Oliver Hilt, Frank Brunner, Mihaela Wolf,
Joachim Würfl, Sibylle Dieckerhoff

Optimal Frequency for Dynamic Wireless Power Transfer 3786
Mincui Liang, Khalil El Khamlichi Drissi, Christophe Pasquier

A Wide-Input-Voltage-Range 50W Series-Capacitor Buck Converter with Ancillary Voltage Bus for Fast Transient Response in 48V PoL Applications.. 3796
  Nameer Khan, James Xu, Gerard Villar Piqué, John Pigott, Henk Jan Bergveld, Alaa El Sherif, Olivier Trescases

Four-Level Boost Inverter Based on ANPC Topology with Switched-Capacitor Branch........................... 3804
  Robert Stala, Adam Penczek, Stanislaw Piróg, Aleksander Skala, Andrzej Mondzik, Zbigniew Waradzyn, Krishna Kumar Gupta, Pallavee Bhatnagar, Sanjay K. Jain, Kasinath Jena

Comparative Evaluation of Partially-Rated Energy Storage Integration Topologies for High Voltage Modular Multilevel Converters.. 3813
  Zoe Blatsi, Sebastian Neira, Stephen Finney, Michael M. C. Merlin

Influence of Current Collapse Due to V_{ds} Bias Effect on GaN-HEMTs I_d-V_{ds} Characteristics in Saturation Region 3822
  Xuyang Lu, Arnaud Videt, Ke Li, Soroush Faramehr, Petar Igic, Nadir Idir

Deep-Learning Fault Detection and Classification on a UAV Propulsion System 3831
  Pierre-Yves Brulin, Fouad Khenfri, Nassim Rizoug

A Compact Solid State Transformer for Replacing Conventional Medium Power Transformer in Weight-Critical Applications.. 3838
  Leon Fauth, Felix Willer, Jens Friebe

Comparative Study of Single-Phase and Three-Phase DAB for EV Charging Application........................... 3846
  Nicola Blasuttigh, Hamzeh Beiranvand, Thiago Pereira, Marco Liserre

Dynamic Load Emulation for Automotive Power IC Robustness Validation 3855
  Alexander Ulbing, Daniel Kostynski, Markus Sievers

DAB Frequency Decoupling Control with Current Minimization 3862
  Simon Uicich, Jean-Yves Gauthier, Xuefang Lin-Shi, Bruno Allard, Arnaud Plat

Design and Performance Analysis of a Modified Proportional Multi-Resonant (PMR) Controller for Three-Phase Voltage-Source Inverters 3871
  Ahmad Ali Nazeri, Mahmoud Saeidi, Peter Zacharias

Proposition and Comparison of Several Solutions for High Induced Voltage Across Inactive Transmitting Coils in a Series-Series Compensation DIPT System... 3883
  Wassim Kabbara, Tanguy Phulpin, Mohamed Bensetti, Antoine Caillierez, Serge Loudot, Daniel Sadarnac

Modeling and Measuring the Bearing Capacitance of Radially Loaded Bearings 3893
  Stefan Quabeck, Daniel C. Rodriguez, Rik W. De Doncker

Comprehensive Control of Matrix Converters in On-Board Electric Drive Applications............................. 3903
  Galina Mirzaeva

Power System Simulation Tool for Quick Benchmarking of Innovative MVDC Grids in E-Mobility Applications.. 3910
  Daniel Siemaszko, Philippe Noisette

An Artificial Intelligence Pipeline for Critical Equipment Thermal Conditioning System Design 3920
  Raik Orbay, Athanasios Tzanakis, Inko Marcaide, Jonas Löfgren, Torbjörn Thiringer, Thomas Bernichon

Aspects of Stability Issues of HVAC/HVDC Coupled Grids.. 3928
  Gianni Bakhos, Kosei Shinoda, Juan-Carlos Gonzalez-Torres, Abdelkrim Benchaib, Luigi
  Vanfretti, Seddik Bacha

Measurement of Coss-V Characteristic of the 1.7kV/900A SiC Power Module and Estimation of
the Channel Current... 3938
  Jacek Rabkowski, Fernando Gonzalez-Hernando, Mariusz Zdanowski, Irma Villar, Uxue
  Larrañaga

In-Slot Cooling of Electrical Machines using Traditional Techniques and Additive Manufacturing 3947
  Ahmed Hembel, Gokhan Cakal, Bulent Sarlioglu

Comparison of High-Power 2-Level and 3-Level Converters in Terms of Power Density, Costs and
Performance.. 3957
  Ludwig Schlegel, Wilfried Hofmann

Autonomous Characterization of Lithium-Ion Battery Model Parameters Utilizing a Mathematical
Optimization Methodology 3966
  Hamzeh Beiranvand, Helge Krüger, Sandra Hansen, Marco Liserre, Christian Werlig,
  Andreas Würsig

SOC Governed Algorithm for an EV Cascaded H-Bridge Connected to a DC Charger 3975
  Giulia Tresca, Andrea Formentini, Filippo Gemma, Federico Lusardi, Riccardo Leuzzi,
  Pericle Zanchetta

Shaping the Transition from Si-Based Power Devices to SiC MOSFETs and GaN HEMTs 3984
  Gerald Deboy

Reinventing Batteries Through Nanotechnology 3986
  Yi Cui

Advancing GaN Power ICs: Efficiency, Reliability & Autonomy.. 3987
  Dan Kinzer

Electrification Strategy of Volkswagen Group... 3989
  Alexander Krick

Make it Fly — the Future of Sustainable Aviation.. 3991
  Tanja Neuland

The Instrumental but Extremely Challenging Role of Hydrogen Towards a Decarbonized Society 3992
  Stefan Linder

Short Circuit Behavior of Dual Three-Phase Permanent Magnet Synchronous Motors with
Different Mutual Inductance in Electric Propulsion Application 3993
  Yinghui Yang, Georg Möhlenkamp

Hybrid Silicon-SiC Inverter – Combining the Best of Both Worlds 4003
  Hans-Günter Eckel, Felix Kayser, Pham Ha Trieu To

Robustness of SiC Trench MOSFETs 4004
  Christian Felgemacher

3D Predictive Fatigue Modeling of Power Modules 4005
  Ben Samples, Brandon Passmore

Heterogeneous Integration of Power Conversion using Power Supply on Chip and Power Supply in Package.. 4006
Cian Ó Mathúna, Seamus O'Driscoll

Driving Innovations for Power Electronics with Integratable and Sustainable Magnetics........................... 4008
Matt Wilkowski

Impact of Package Technology on the Switching Behavior of High-Voltage GaN FETs............................ 4011
Sebastian Klötzer

Impact of Power Electronics on Battery Operation 4012
Dirk Uwe Sauer

Trends in Power Electronics and Batteries for Electrified Vehicle Infrastructure...................................... 4013
Torsten Leifert

Impact of High Frequency Current Pulses on Battery Ageing 4014
Julia Kowal

Aircraft Electrification – System-Level Potentials for Aviation Decarbonization 4015
Kathrin Ebner, Antoine Habersetzer, Arne Seitz

About Power Electronics Challenges in Aviation 4016
Marco Bohllaender

Development of Electric Motors for Aircraft Applications.. 4017
Simon Wolfstädter

Powertrain Trends in Electric Trucks.. 4018
Luciana C. Afonso

Modulation Strategy Impact of BEV Inverters on the Voltage Ripple and the High-Voltage Traction System Stability 4019
Cornelius Rettner

Zero Emission Trucks & Bodies 4020
Martin Glaser

Integrating Offshore Wind & Hydrogen - An Operator's View 4021
Florian Gremme

Status Quo and Future Prospects of Power Electronic Solutions for Electrolysis Plants 4022
Sven Schumann

Modular Power Supply System for Large Scale Water Electrolyzers 4023
Ralf Juchem, Klaus Rigbers

Properties of a Lithium-Ion Battery as a Partner of Power Electronics... 4025
Alexander Blömeke, Katharina Lilith Quade, Dominik Jöst, Weihan Li, Florian Ringbeck, Dirk Uwe Sauer

Author Index

Distribution transformer voltage control using a single-phase matrix converter

Rui Wang, dr. ir. Henk Huisman, prof. ir. Korneel Wijnands
Eindhoven University of Technology
P.O. Box 513, 5600 MB Eindhoven, The Netherlands Eindhoven, the Netherlands
Email: r.wang.1@tue.nl

Keywords

≪On-load tap changer(OLTC)≫, ≪Hybrid transformer≫,≪AC-AC converter≫, ≪Single-phase matrix converter≫.

Abstract

A new topology is proposed for fully-electronic tap changers in distribution transformers. In every single phase, the voltage is regulated by a 2×1 matrix converter, which can be extended using the concept of multilevel converters. The topology allows reaching the desired functionality with one single tap, which reduces the cost of the transformer.

Introduction

Medium-voltage (MV) to low-voltage (LV) distribution transformers are used to step down the MV grid voltage to a suitable level for residential consumers. LV side end-users shall receive a stable voltage with $\pm 10\%$ tolerance typically. However, users can receive different voltage levels due to the voltage drop across the line impedance. As a trend, residential load variations are increasing due to the growing penetration of electrical vehicles and heat pumps. Apart from these, the installation of solar panels can also cause bidirectional power flow, which roughly doubles the voltage swing at consumer side. As a result, voltage fluctuations become larger and happen more frequently.

To maintain a relatively stable LV side voltage, tap changers are used to vary the effective transformer turns ratio. In principle, tap changers can be placed at either MV or LV side of the transformer; the most common is at the MV side. Fig. 1 shows a simplified circuit of a distribution transformer with tap changers. The transformer is in Delta-Wye connection; it has $(n_{1a} + n_{1b})$ turns at the MV side and n_2 turns at the LV side. The tap winding is located at the MV side, in this example, it has n_{1b} turns in total and five tap positions. By switching to different positions, different voltage levels can be obtained at the LV side.

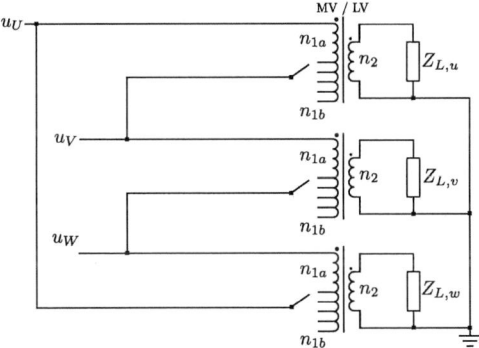

Fig. 1: Distribution transformer with mechanical tap changers at the MV side.

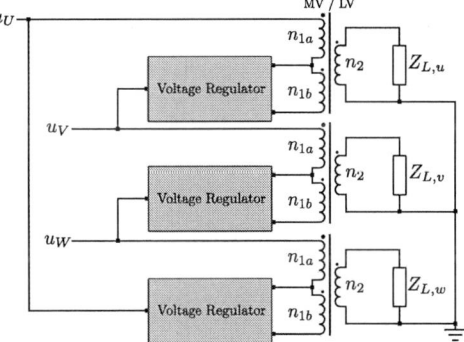

Fig. 2: Proposed voltage regulators for a distribution transformer.

Conventional tap changers use mechanical switches, which limit the speed of the tap-changing process. The resolution of the voltage regulation depends on the number of tap positions. The lifetime of mechanical switches is significantly impacted by arcs during switching. To reduce maintenance cost, electronically-assisted tap changers use both mechanical and semiconductor switches. Semiconductor switches are used in the process of switching between different tap position, which avoids arcing and thereby extends the lifetime of the mechanical switches. However, the time of the tap-changing process (between two taps) is still $100 \sim 200$ ms [1, 2, 3]; and jumping taps is not allowed as this would lead to large circulating current. Fully-electronic tap changers replace mechanical switches with semiconductor switches, which bring many advantages compared to mechanical switches [2]:

- Low maintenance cost: The arcs caused by mechanical switches disappear.
- Fast tap-changing process because of the high switching speed of semiconductor devices.
- Better controllability: The LV side voltage can be controlled continuously, which suggests that these tap changers can be used in power quality management.

Considering the switching and conduction losses, efficiencies of fully-electronic tap changers are lower than of their mechanical counterparts. Manufacturing and component costs are higher for fully-electronic tap changers; however, their maintenance costs will be significantly lower compared to designs using mechanical switches. Therefore, fully-electronic tap changers are very promising compared to conventional products.

Previous research proposed fully-electronic tap changers by inserting parallel half-bridge circuits between multiple tap positions [5, 6, 7]. In this paper, a new topology is proposed for fully-electronic tap changers as shown in Fig. 2. The tap changer requires only one tap connection. For simplicity, the following discussion will focus on the VR in one single phase of the distribution transformer as shown in Fig. 3. The transformer is assumed ideal; for brevity, the filter design will not be covered in this paper.

Voltage regulator

Fig. 4 provides an equivalent circuit of the VR. In this circuit, the VR is presented as three voltage sources u_x, u_z and u_y. According to coupling of the transformer:

$$u_y = u_z - u_{UV} + u_L \frac{n_{1a} + n_{1b}}{n_2} \tag{1}$$

u_y can be defined as a function of inputs u_x and u_z:

$$u_y = \frac{1}{2} \left((u_x + u_z) + K(u_x - u_z) \right) \tag{2}$$

where K is a control variable to realize different LV side voltages and $-1 \leqslant K \leqslant 1$. The effective MV side winding turns $n_{1,e}$ is defined as:

$$\frac{n_{1,e}}{n_2} = \frac{u_{UV}}{u_L} \tag{3}$$

By combining (1) and (2), $n_{1,e}$ can be obtained:

$$n_{1,e} = n_{1a} + \frac{1}{2} (1 - K) n_{1b} \tag{4}$$

For $K = 1$, the tap winding n_{1b} is open and only n_{1a} is used; for $K = -1$, the tap winding is fully connected and $(n_{1a} + n_{1b})$ winding turns are used at the MV side. The effective MV side turns can be controlled to any value in between n_{1a} and $(n_{1a} + n_{1b})$ by setting K.

For a conventional mechanical tap changer, the grid voltage is applied to the main transformer winding n_{1a} and part of the tap winding $\frac{1}{2}(1 - K) n_{1b}$; whereas the rest of the tap winding $\frac{1}{2}(1 + K) n_{1b}$ is open and there is no current flowing. However, for the electronic VR, a current i_z flows through the entire tap winding n_{1b}.

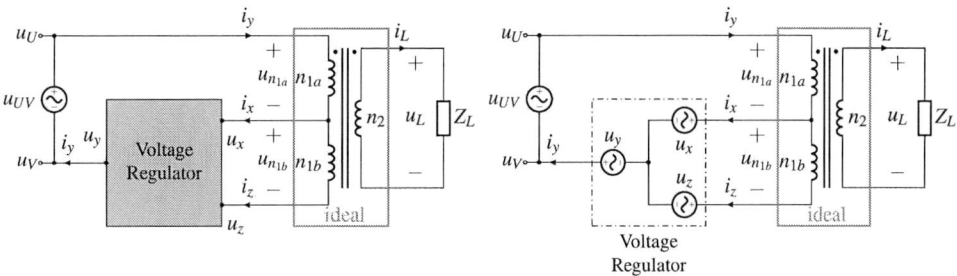

Fig. 3: Voltage regulator of one phase.   Fig. 4: Simplified single-phase circuit.

Fig. 5 and Fig. 6 show two basic topologies for the VR. For the half-bridge topology in Fig. 5, the VR is directly switching between the tap winding terminals. The average voltage of u_y can be set by the duty cycle of the pulse-width modulation. Bidirectional voltage-blocking switches are required due to the AC voltage. The advantage of this circuit is its simplicity. However, these bidirectional switches need to be controlled very accurately in the commutation process.

Fig. 6 shows a different solution using the matrix topology, it contains three switching legs X, Y, Z and a DC link capacitor C_{dc}. Phases X and Z are connected to the tap winding; they control the circulating current between the VR and the tap winding. The LV side voltage is determined by the duty cycle of phase Y. Because of the DC link, bidirectional voltage-blocking switches are unnecessary; standard IGBTs or MOSFETs with diodes can be used. As in the case of mechanical switches, the VR does not need to store energy. The DC link voltage is constant in the steady state; and as a consequence, the capacitor can be small in principle. This circuit needs two more switches compared to the AC half bridge. Thus, the matrix circuit has a higher manufacturing cost. However, the implementation of control is easier because bidirectional voltage-blocking switches are not required.

Considering MV applications, the high voltage is challenging for IGBTs/MOSFETs. For the half bridge circuit, a possible solution would be using series connected switches to reduce the voltage stress per switch. However, this requires a lot of efforts in hardware design. For the matrix converter, there are more multilevel solutions available, such as modular multilevel converters (MMC). This topic will be covered in future research.

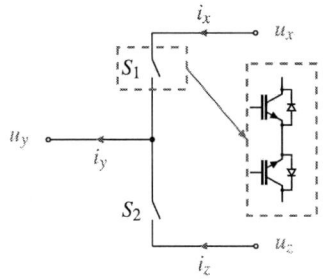

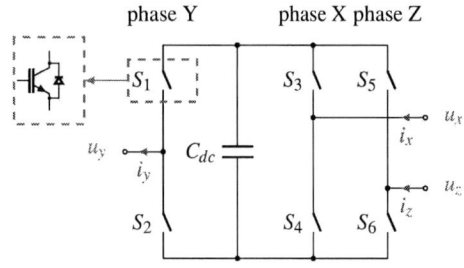

Fig. 5: Half bridge   Fig. 6: Matrix converter

Matrix converter control

A cascade control strategy is used for the matrix circuit. The inner current loop controls the circulating current to guarantee the energy balance; and the voltage loop keeps the DC link voltage constant.

The current loop is shown in Fig. 8. The circulating current is defined as:

$$i^\Delta = i_x - i_z \qquad (5)$$

The reference of circulating current is derived from the energy balance of the VR:

$$\int_{t_0}^{t_0+\frac{1}{f_0}} \left(i_x u_x + i_z u_z - i_y u_y \right) \mathrm{d}t = 0 \tag{6}$$

where f_0 is the fundamental frequency of the grid voltage. This equation can be simplified to:

$$\int_{t_0}^{t_0+\frac{1}{f_0}} \left(i^\Delta - K i_y \right) \left(u_x - u_z \right) \mathrm{d}t = 0 \tag{7}$$

According to (7), the circulating current reference i_{ref}^Δ should be

$$i_{\mathrm{ref}}^\Delta = K i_y + i^{\Delta\perp} \tag{8}$$

where the term $i^{\Delta\perp}$ only contributes to reactive power. This orthogonal term $i^{\Delta\perp}$ can have a 90° phase difference with respect to the voltage $(u_x - u_z)$. Alternatively, $i^{\Delta\perp}$ can also contain harmonics at different frequencies; this will not be discussed in this paper. A proportional-resonant (PR) controller C_{i_Δ} is used in the current loop due to the sinusoidal current reference. In practice, an additional integrator (PIR) might be needed to prevent DC offset in the current.

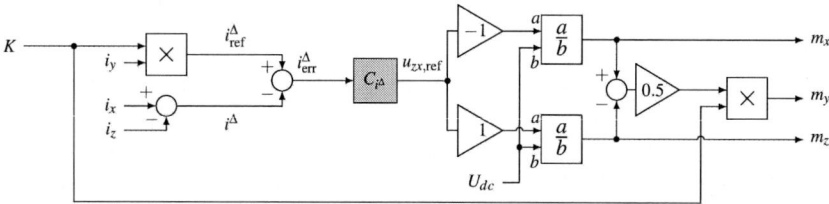

Fig. 7: Current loop control.

The control diagram including the voltage loop is shown in Fig. 8. A PI controller $C_{U_{dc}}$ is used to keep a constant DC link voltage. The output of the PI controller is multiplied by the signal $0.5(m_x - m_z)$, in order to obtain a sinusoidal reference $i_{\mathrm{ref},2}^\Delta$ to be added to the current loop.

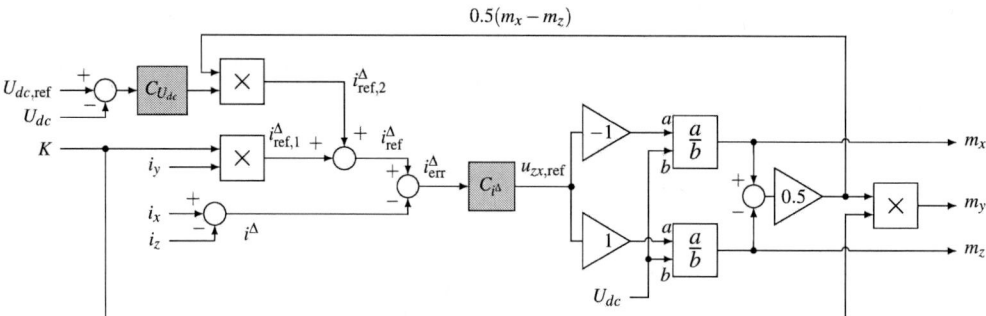

Fig. 8: Control diagram for the matrix baseline.

Scaled-down test

The purpose of this test is to verify the operating principle of the proposed matrix topology. Fig. 9 shows the lab setup; the schematic is shown in Fig. 10. The input voltage u_{in} is 240 Vrms, 50Hz. The VR circuit uses an existing 3-phase IGBT inverter with a 2.45 mF DC link capacitor. The IGBTs are switching at 20kHz with $2\mu s$ dead time. The high frequency components of the modulated voltages are eliminated by LC filters. The selected transformer (USTE 1000/2x115) has a rated power of 1 kVA and the turns ratio is $n_{1a} : n_{1b} : n_2 = 2.02 : 0.48 : 1$. The tap winding n_{1b} is arranged at the primary side. A 60Ω resistor R_L is placed at the secondary side as load.

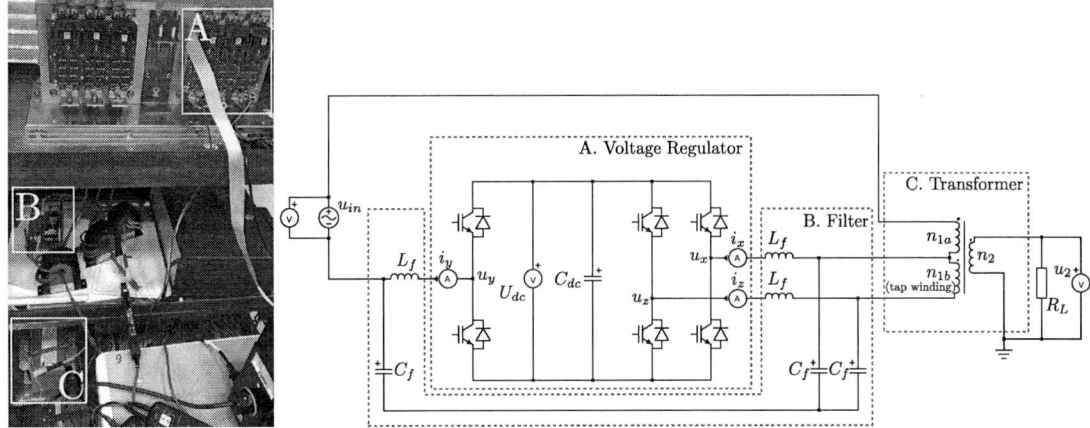

Fig. 9: Lab setup   Fig. 10: Schematic of the test setup

The control is realized using a dSPACE MicrolabBox 1202/1302. The DC link voltage U_{dc} should be higher than the tap winding peak voltage in order to prevent over modulation. Initially, U_{dc} is controlled at 80V to just satisfy this requirement. In the current loop, the orthogonal term $i^{\Delta\perp}$ is set to zero.

Results

Fig. 11 shows the results of the operation when K equals 0.5. The input voltage u_{in} and regulated transformer secondary voltage u_2 are shown in Fig. 11a; and the resulting VR currents are shown in Fig. 11b. Fig. 11c shows the DC link voltage of the VR; u_{dc} was controlled at 80V with a very small ripple. In Fig. 11d, the circulating current i^{Δ} is tracking the reference. However, the actual current was slightly larger than its ideal reference (Ki_y). This difference could be related to the considerable dead time, which was not concerned in the ideal situation. The modulation errors caused by the dead time are nonlinear functions of currents. In the VR, currents in switching legs X, Z and Y have different amplitudes, which lead to different errors of u_x, u_z and u_y. As a result, the circulating current reference needs to be corrected to maintain the energy balance. Besides, the converter loss is also a concern. Especially the switching loss can be considerable due to the hard switching.

The current flow to the input voltage source i_y and also the load current i_2 were measured using an oscilloscope as shown in Fig. 12. Spikes appeared in the unfiltered waveform of i_y due to the hard switching of IGBTs. After filtering, harmonics at the switching frequency were mostly eliminated. In terms of the load current i_2, the amplitude of the ripple was less than 3%. The ripple and harmonic distortion can be further reduced by optimizing the filter design if necessary.

The dynamic performance was investigated by using step references for K. Results are shown in Fig. 13 and Fig. 14. In both test scenarios, the current loop immediately followed new references. Due to the disturbance in the current loop, the DC link voltage changed in the beginning and was gradually controlled back to its reference. Especially, in Fig. 14, the step of K happened when the input voltage was at its peak value. Though oscillations were observed initially, controllers regulated the system to the steady state in a short time.

According to (4), the effective number of turns n_{1e} at the primary side of the transformer can be more than $(n_{1a}+n_{1b})$ if $K < -1$, or less than n_{1a} if $K > 1$. However, when K is set beyond the normal operation range of $-1 \leqslant K \leqslant 1$, the amplitude of the VR output voltage u_y will increase. Thus, a higher DC link voltage is required to prevent overmodulation. In Fig. 15, the DC link voltage was increased to 120V. In the normal operation, the waveform of u_2 was limited to the area between $u_2|_{K=-1}$ and $u_2|_{K=1}$. When K was set to +/-1.2, u_2 resulted in slightly larger/smaller amplitudes.

Fig. 16 summarizes the voltage regulation range as a function of the control variable K. The voltage base

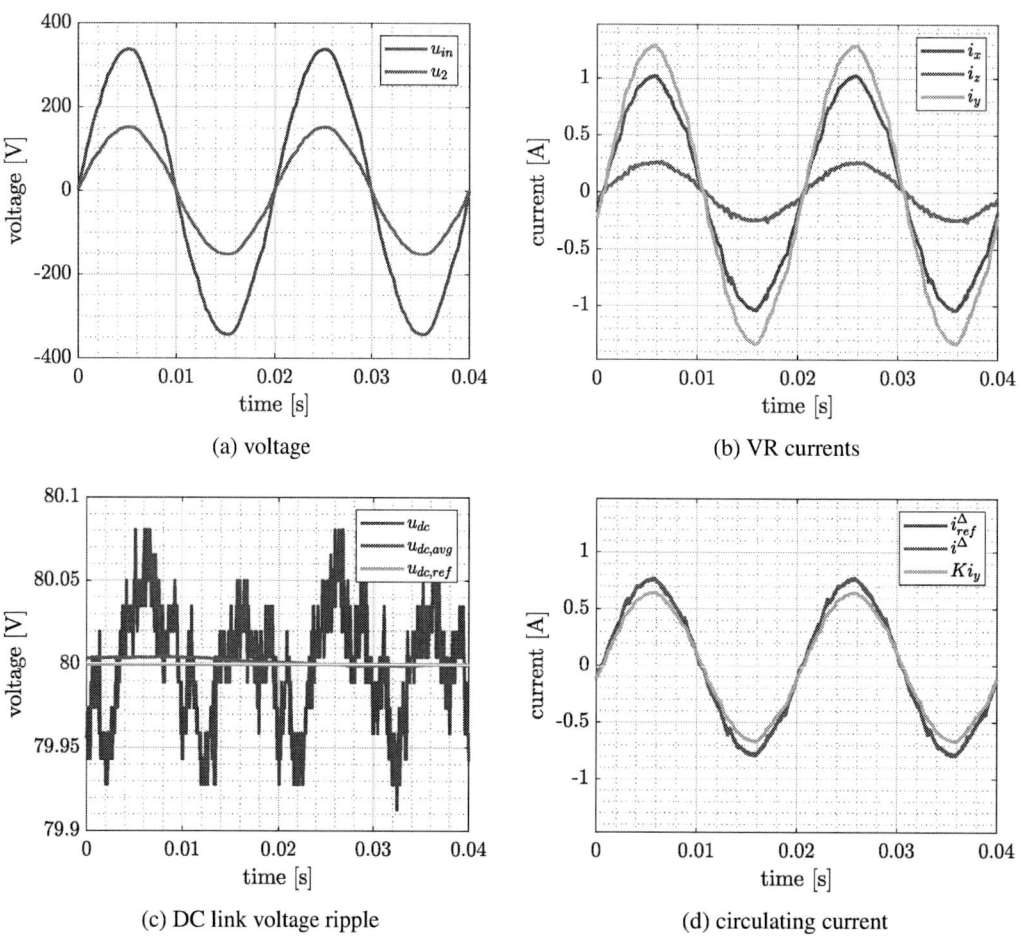

Fig. 11: Test results at $K = 0.5$.

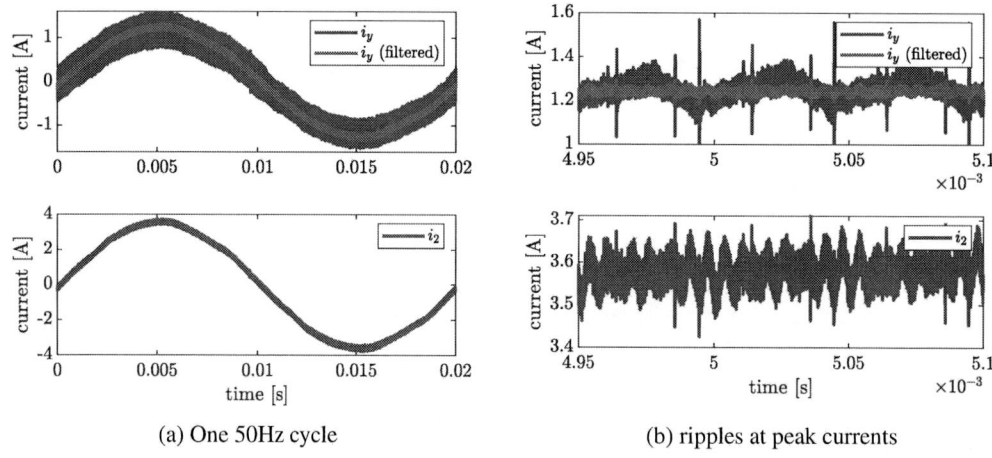

Fig. 12: Current i_y and i_2, measured by oscilloscope ($K = 0.5$).

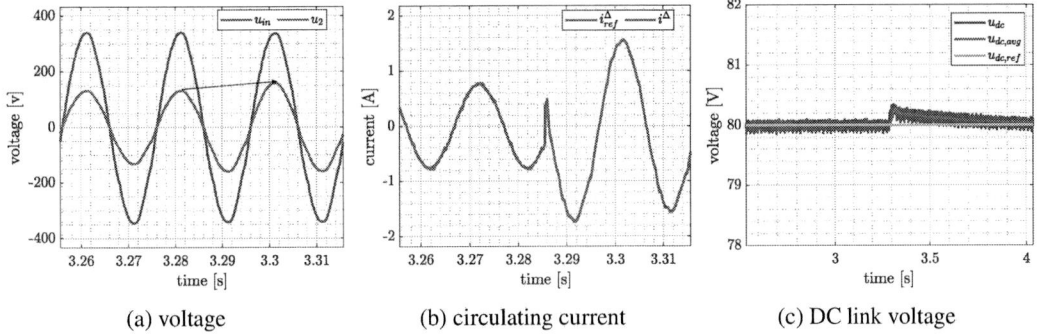

(a) voltage   (b) circulating current   (c) DC link voltage

Fig. 13: Step response: K steps up from -1 to 1.

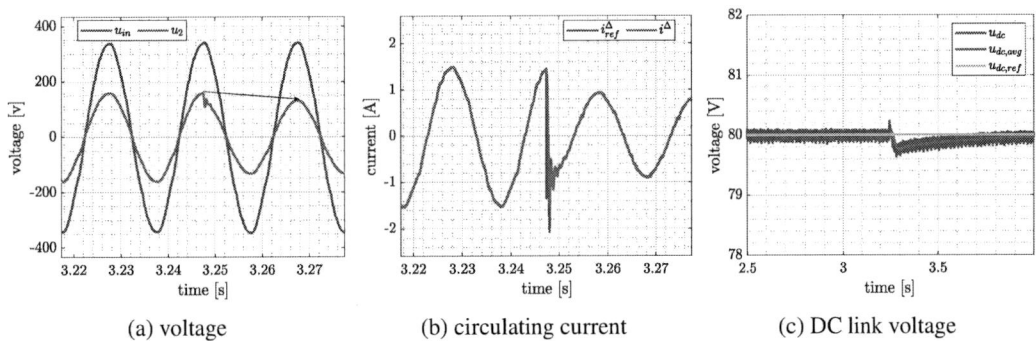

(a) voltage   (b) circulating current   (c) DC link voltage

Fig. 14: Step response: K steps down from 1 to -1.

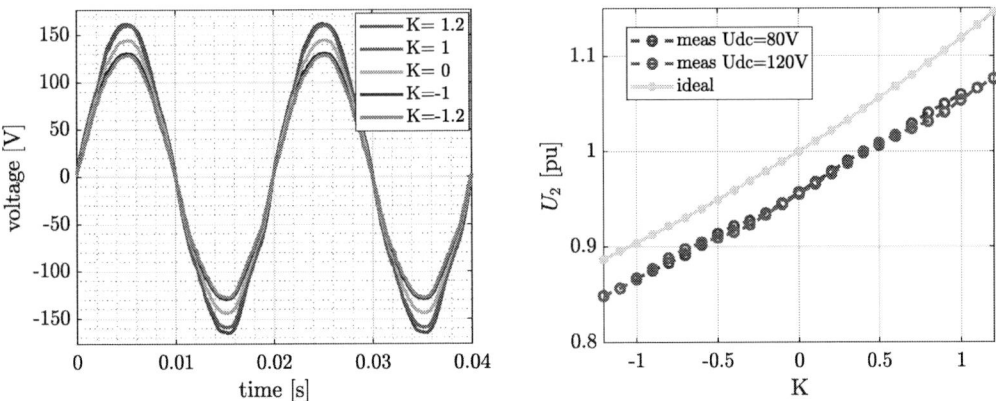

Fig. 15: u_2 wrt. K values, $U_{dc} = 120$V   Fig. 16: Voltage regulation range

is chosen as the ideal voltage of U_2 when $K = 0$, and the input grid voltage is 240Vrms.

$$U_{base} = 240 \times \frac{n_2}{n_{1a} + \frac{1}{2} n_{1b}} = 106.2\,\text{V}. \tag{9}$$

The ideal curve is estimated by:

$$U_2\,[pu] = \frac{240\,\text{V}}{U_{base}} \times \frac{n_2}{n_{1a} + \frac{1}{2}(1 - K)\,n_{1b}}. \tag{10}$$

During the test, u_2 was measured when a different DC link voltage U_{dc} was applied. In normal operation, U_2 did not show a dependency on U_{dc}. However, operation at $K < -1$ or $K > 1$ was not possible when U_{dc} was 80V. As shown in Fig. 16 , the measured secondary voltages were smaller than their ideal values. To explain this difference, aside from the dead time distortion as mentioned before, other reasons could be the impacts of the filter and also the leakage components of the transformer. This will be investigated in the future.

Conclusion and future work

In this paper, a new power-electronic tap changer (VR) is proposed for distribution transformers. The voltage regulation is realized by a single-phase matrix type converter. In previous research, the voltage regulation range was often limited by the length of the tap winding; whereas this proposed VR is capable to realize a wider range of effective turns coupled to the input voltage (with $K > 1$ and $K < -1$). The functionalities of the proposed matrix type circuit are verified in a scaled-down test. However, there were some mismatches between ideal estimations and measurements. Further study needs to include more details in modelling: for instance, the three-winding transformer model including leakage components and also considering the converter loss in the energy balance calculation.

The voltage of the MV grid is normally 10kV or 23kV in Europe. Thus, extending the proposed concept to multilevel/multi-cell topologies (e.g. modular multilevel converters) is needed. This will be addressed in future research.

References

[1] G. R. C. Mouli, P. Bauer, T. Wijekoon, A. Panosyan, and E.-M. Bärthlein, "Design of a power-electronic-assisted oltc for grid voltage regulation," *IEEE Transactions on Power Delivery*, vol. 30, no. 3, pp. 1086–1095, 2014.

[2] J. Faiz and B. Siahkolah, *Electronic tap-changer for distribution transformers*. Springer Science & Business Media, 2011, vol. 2.

[3] H. Jiang, R. Shuttleworth, B. A. Al Zahawi, X. Tian, and A. Power, "Fast response gto assisted novel tap changer," *IEEE Transactions on Power Delivery*, vol. 16, no. 1, pp. 111–115, 2001.

[4] J. V. López, J. C. Rodríguez, S. M. Fernández, S. M. García, and M. P. García, "Analysis of fast onload multitap-changing clamped-hard-switching ac stabilizers," *IEEE transactions on Power Delivery*, vol. 21, no. 2, pp. 852–861, 2006.

[5] P. Bauer, S. De Haan, and G. Paap, "Electronic tap changer for 10 kV distribution transformer," in *European Conference on Power Electronics and Applications*, vol. 3. Proceedings published by various publishers, 1997, pp. 3–1010.

[6] J. de Oliveira Quevedo, F. E. Cazakevicius, R. C. Beltrame, T. B. Marchesan, L. Michels, C. Rech, and L. Schuch, "Analysis and design of an electronic on-load tap changer distribution transformer for automatic voltage regulation," *IEEE Transactions on Industrial Electronics*, vol. 64, no. 1, pp. 883–894, 2016.

[7] S. P. Engel, "Thyristor-based high-power on-load tap changers: control under harsh load conditions," Ph.D. dissertation, RWTH Aachen University, 2017.

Influence of Carrier-Based PWM Techniques on the Common-Mode Voltage and Common-Mode Current of Six-Phase Full-Bridge Inverters

Juris Arrozy, Esin Ilhan Caarls, Henk Huisman, Jorge L. Duarte, Lorenzo Ceccarelli
Eindhoven University of Technology
5612AZ Eindhoven
Eindhoven, The Netherlands
Email: j.arrozy@tue.nl

Keywords

≪Pulse Width Modulation (PWM)≫, ≪EMC/EMI≫, ≪Leakage current≫, ≪Multiphase drive≫, ≪Open-end windings≫

Abstract

This paper compares several carrier-based PWM techniques to see their impact on the common-mode voltage (CMV) and common-mode current (CMC) of a six-phase full-bridge inverters. Fast Fourier Transform (FFT) analysis displays the dominant harmonics spectrum of each carrier-based PWM technique. The influence of dead-time on the CMV and CMC is also addressed. Simulation and experimental results are included.

Introduction

Multiphase machine drives have recently gained attention because of their potential candidacy in various propulsion applications such as electric vehicles, electric ship propulsion, and more-electric aircraft [1, 2]. This is mainly because of their advantages such as higher torque density [3], better fault-tolerant operation [4], and reduced input current ripple [5]. For this reason, many aspects of multiphase machine drives such as modelling, control, and modulation techniques has received much attention [2, 6, 7].

The open-winding multiphase machine offers a variant to the traditional wye/delta configuration. This yields better dc voltage utilization at the cost of doubled switch count [8]. This also allows more flexibility in the selection of modulation techniques such as space-vector modulation [9] and carrier-based PWM techniques [10].

Carrier-based PWM techniques are considered to be more suited to multiphase machine drive applications because of their simplicity [10]. There are already several papers that address the influence of carrier-based PWM techniques on the input current ripple [11] and output voltage and current total harmonic distortion (THD) performance [12] in the context of open-winding machine drives. However, the influence of carrier-based PWM techniques on the common-mode voltage (CMV) and common-mode current (CMC) of the motor-inverter system have not been thoroughly addressed yet. This is especially important because the CMV and CMC generated in the machine can lead to EMI-related problems such as bearing degradation and high-frequency noise.

Therefore, this paper compares the influence of several carrier-based PWM techniques on the CMV and CMC performance of the open-winding machine drive. The case study chosen here is a six-phase full-bridge inverter supplying a six-phase open-winding permanent magnet synchronous motor (PMSM). Several carrier-based PWM techniques are considered, namely unipolar modulation, phase-shifted PWM (PS-PWM), and level-shifted PWM (LS-PWM) for the cases of double three-phase and symmetrical six-phase systems. Fast Fourier Transform (FFT) is used to inspect the dominant harmonics spectrum of the

CMV and CMC. The influence of dead-time is also addressed. Simulation and experimental results are included to see which carrier-based PWM techniques generate the highest rms value of common-mode current to the six-phase full-bridge inverters.

CMV and CMC of Six-Phase Full-Bridge Inverters

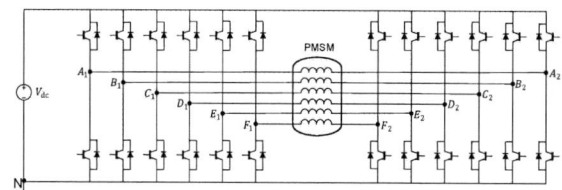

Fig. 1: Six-phase full-bridge inverter supplying a six-phase PMSM

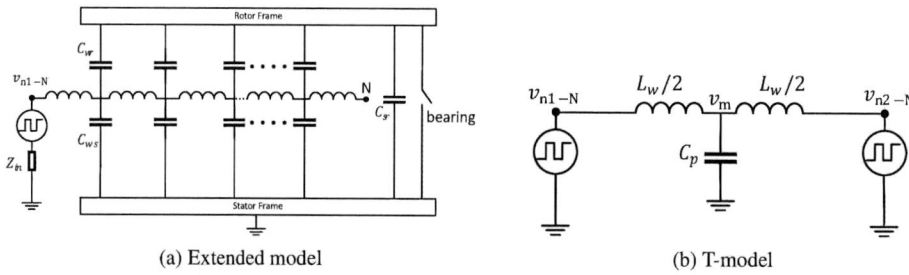

(a) Extended model   (b) T-model

Fig. 2: Single winding high-frequency-EM model for the CMC path

Fig. 1 shows a six-phase full-bridge inverters supplying a six-phase open-winding PMSM. When the inverters legs are switching, a rapid change of voltage in the voltage node occurs. This causes a leakage current via the machine's parasitic capacitance, which leads to EMI-related problems such as bearing degradation and high-frequency noise.

There are several ways to model the leakage current path. One of them is the extended model, which models the parasitic capacitance between winding-rotor (C_{wr}), winding-stator (C_{ws}), and rotor-stator (C_g), in addition to the bearing that is modeled as a switch [13]. The extended model is shown in Fig. 2a.

Since the interest of this paper is just to see the influence of carrier-based PWM techniques on the CMV and CMC of the six-phase full-bridge inverters, the extended model is not favourable for it complicates the circuit beyond necessity. Thus, the T-model as in Fig. 2b is proposed as a simplification of the extended model. The leakage current path is provided by the parasitic capacitance (C_p) from the winding to the ground (i.e. housing of the PMSM). In this model, the square-wave input voltage is applied. The leakage current depends on the change of the midpoint voltage of the winding. The steady-state midpoint voltage (v_m) of a winding is given as

$$v_m = \frac{v_{n1\text{-}N} + v_{n2\text{-}N}}{2} \qquad .$$
(1)

Taking (1) into account, the CMV of six-phase full-bridge inverter is formulated as

$$v_{cm} = \frac{1}{6} \sum_{n=A...F} \frac{v_{n1\text{-}N} + v_{n2\text{-}N}}{2}$$
(2)

Ignoring the influence of L_w, the CMC of the six-phase full-bridge inverter is approximated as

$$i_{cm} = C_p \frac{d}{dt} \left(\sum_{n=A...F} \frac{v_{n1\text{-}N} + v_{n2\text{-}N}}{2} \right)$$
(3)

It follows from (2) and (3) that to minimize the CMC, v_{cm} should remain as constant as possible with minimum level jumps. This depends on the switching sequence of each H-bridge, which in its turn depends on the carrier-based PWM technique used.

Carrier-Based PWM Techniques

Four carrier-based PWM techniques are considered in this paper: bipolar modulation, unipolar modulation, phase-shifted PWM (PS-PWM), and level-shifted PWM (LS-PWM). In addition, double three-phase and symmetrical six-phase systems are considered. These are shown in Fig. 3 and Fig. 4, respectively.

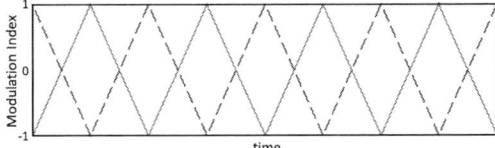 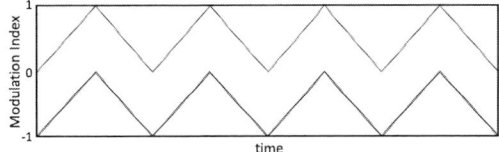

Fig. 3: Carrier Signal for: Phase-shifted PWM / PS-PWM (left) and Level-shifted PWM / LS-PWM (right)

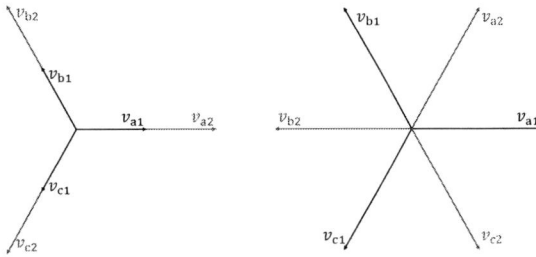

Fig. 4: Double three-phase (left) and symmetrical six-phase systems (right)

Since in bipolar modulation and PS-PWM v_m is always half of the dc link voltage (V_{dc}), the CMV is always constant and hence the CMC produced will be theoretically zero. For unipolar and level-shifted modulation, there are cases when both of the high-side switches in an H-bridge are on, making $v_m = V_{dc}$. Thus, there will be CMC induced because of the voltage change. For a double three-phase system, v_m over a carrier period for unipolar modulation and LS-PWM is depicted in Fig. 5.

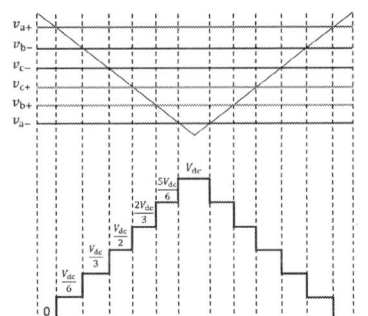 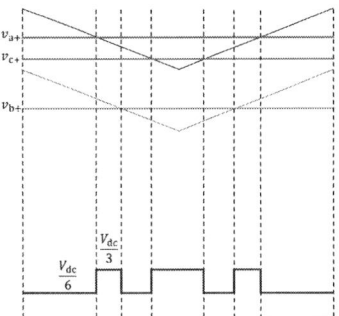

Fig. 5: Example of midpoint voltage (v_m) over a carrier period in an H-bridge for: unipolar modulation (left) and LS-PWM (right)

Simulation Results

Table I: System parameters for six-phase full-bridge inverter

Parameters	Value
V_{dc}	400V
R_w	1Ω
L_w	4mH
R_{damp}	600Ω
C_p	4.5nF
f_s	20kHz
m	0.7

Table I shows the system parameters of the six-phase full-bridge inverters and the motor. Each winding is modeled as an inductance (L_w), a resistance (R_w), and a leakage capacitance in the center of the winding to the housing (C_p). A damping resistance (R_{damp}) is included in series with the C_p to model the skin effect of the winding resistance so that the oscillation due to the series connection of L_w and C_p is damped. For the test case of f_s=20kHz and m=0.7, the results are shown in Fig. 6, Fig. 7, and Table II:

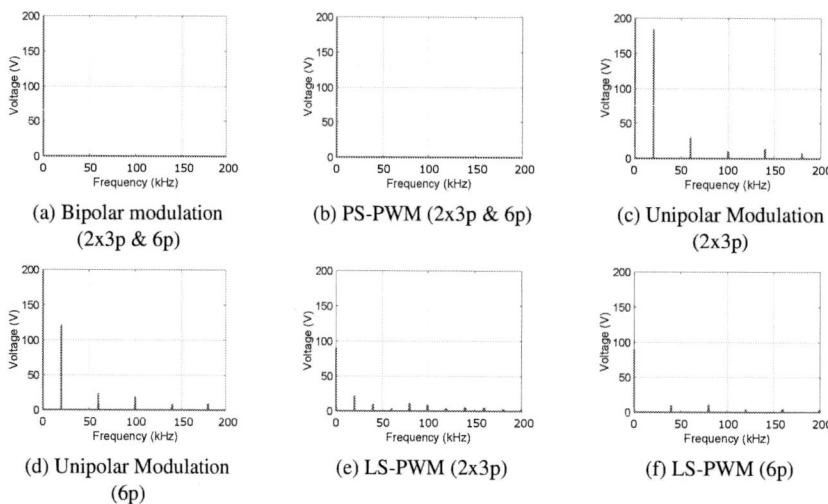

(a) Bipolar modulation (2x3p & 6p) (b) PS-PWM (2x3p & 6p) (c) Unipolar Modulation (2x3p)

(d) Unipolar Modulation (6p) (e) LS-PWM (2x3p) (f) LS-PWM (6p)

Fig. 6: CMV of six-phase full-bridge inverter for various carrier-based PWM techniques

Table II: RMS value of the CMC of six-phase full-bridge inverters for various modulations

	CMC (Arms)	
	2x3p	6p
Bipolar	0	0
PS-PWM	0	0
Unipolar	0.807	0.744
LS-PWM	1.269	1.243

As shown in Fig. 6a, the bipolar modulation and LS-PWM result in a constant CMV of 200V (half of V_{dc}). This is because in bipolar modulation and LS-PWM the midpoint voltage (v_m) is always $V_{dc}/2$ due to the complementary switching pattern of the two legs in an H-bridge. As a result, there is no CMC induced in such modulation techniques (see Fig. 7a).

It is shown in Fig. 6c & Fig. 6d that odd harmonics occur in the CMV of the inverter for unipolar modulation. This is because the resulting CMV waveform meets the requirement of odd and half-wave

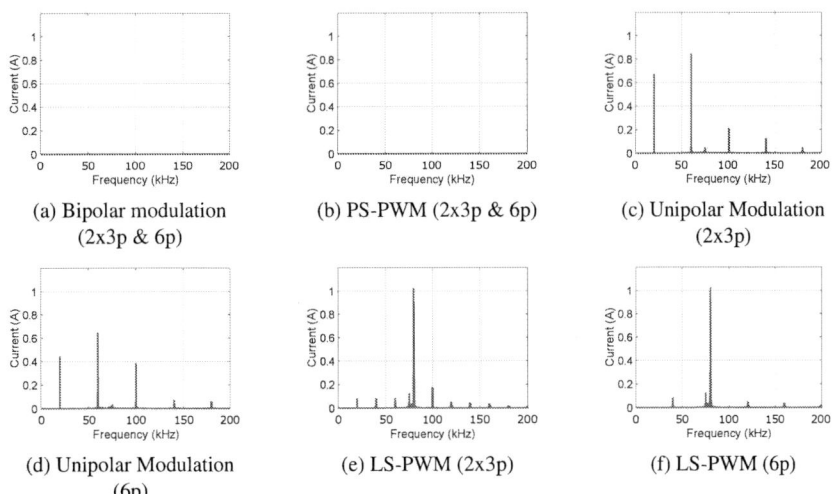

(a) Bipolar modulation (2x3p & 6p)

(b) PS-PWM (2x3p & 6p)

(c) Unipolar Modulation (2x3p)

(d) Unipolar Modulation (6p)

(e) LS-PWM (2x3p)

(f) LS-PWM (6p)

Fig. 7: CMV of six-phase full-bridge inverter for various carrier-based PWM techniques

symmetry, as indicated in Fig. 5 (left).

In the case of LS-PWM, different harmonics occur depending on the current phasors of the six-phase system. For the double three-phase system (Fig. 6e), the fundamental frequency of the CMV is the switching frequency. Since the CMV waveform does not meet the requirements of half-wave symmetry, both odd and even harmonics occur in the FFT result. For the symmetrical six-phase system (Fig. 6d), only even harmonics appear in the FFT spectrum. This is because of the combination of v_m produced by two opposing phases, which results in CMV with the first harmonic being twice the switching frequency. Therefore, the spectrum are the twofold multiplication of the original spectrum in Fig. 6e.

For the CMC spectrum in unipolar modulation and LS-PWM, it is shown that the 80kHz component shows the highest amplitude of CMC. This is because 80kHz is close to the resonance frequency between the parallel connection of two $\frac{L_w}{2}$ and C_p. It is shown in Table II that double three-phase LS-PWM yields the highest rms value of CMC out of all PWM tecnhiques. This is because in double three-phase LS-PWM all the odd and even harmonics appear, which adds up to the total rms value of CMC in the six-phase full-bridge inverters. Another reason is because in LS-PWM, 80kHz frequency appears in the harmonic spectrum of the CMV, which induces higher CMC to the circuit.

Influence of Dead-time on Bipolar Modulation and PS-PWM

The dead-time of $2\mu s$ is added to the circuit to see its influence on the bipolar modulation and PS-PWM. The results are shown in Fig. 8, Fig. 9, and Table III.

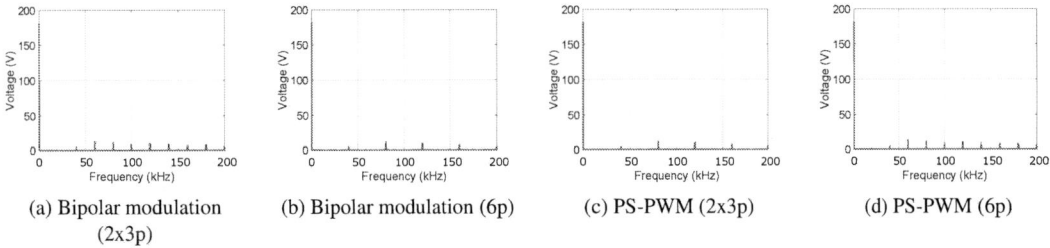

(a) Bipolar modulation (2x3p)

(b) Bipolar modulation (6p)

(c) PS-PWM (2x3p)

(d) PS-PWM (6p)

Fig. 8: CMV of six-phase full-bridge inverter for bipolar modulation and PS-PWM considering the effect of dead-time

From the results above, it is shown that introducing dead-time causes the bipolar modulation and PS-PWM to have considerable CMV and CMC. This is because the dead-time makes the v_m in an H-bridge

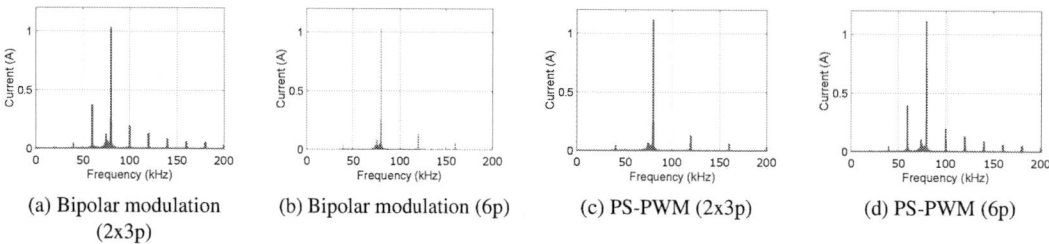

(a) Bipolar modulation (2x3p)

(b) Bipolar modulation (6p)

(c) PS-PWM (2x3p)

(d) PS-PWM (6p)

Fig. 9: CMC of six-phase full-bridge inverter for bipolar modulation and PS-PWM considering the effect of dead-time

Table III: RMS value of the CMC of six-phase full-bridge inverters for various modulations

	CMC (Arms)	
	2x3p	6p
Bipolar	0.916	0.801
PS-PWM	0.801	0.916

equal to zero when all the switches are turned off. This introduces dv/dt, which in turn induces CMC in the circuit. Additionally, it is also shown that the CMV and CMC results between bipolar modulation double three-phase and PS-PWM symmetrical six-phase (and those of bipolar modulation symmetrical six-phase and PS-PWM double three-phase) are similar. This is because the 180° phase-shifted carrier causes the double three-phase system to behave like a symmetrical six-phase system and vice versa.

Experimental Results

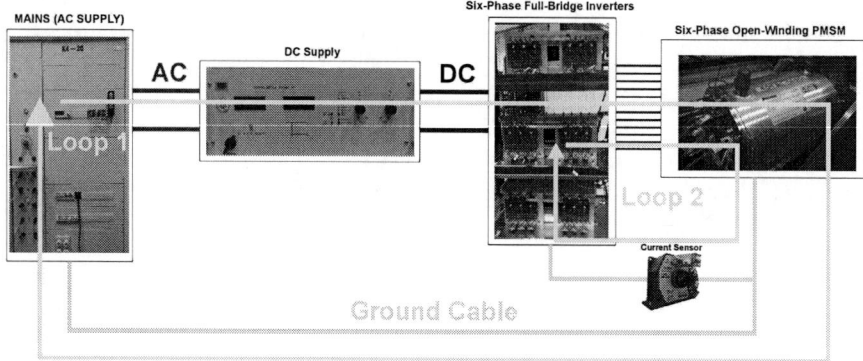

Fig. 10: Experimental Setup for the CMC measurement of Six-Phase Full-Bridge Inverters

Table IV: System parameters for six-phase full-bridge inverter

Parameters	Value
V_{dc}	100V
R_w	0.2Ω
L_w	4mH
C_p	4.5nF
f_s	20kHz
m	0.7
dead-time	$2\mu s$

Table V: RMS value of the CMC of six-phase full-bridge inverters for various modulations (experimental results)

	CMC (Arms)	
	2x3p	6p
Bipolar	0.077	0.063
PS-PWM	0.049	0.023
Unipolar	0.086	0.078
LS-PWM	0.1	0.073

Fig. 10 and Table IV show the system parameters for the experimental setup of six-phase full-bridge inverters. It is shown in Fig. 10 that the CMC measurement takes place in the ground cable connecting the PMSM housing and frame and heatsink of the inverter. The CMC loop (see loop 1 in Fig. 10) consists of the motor housing, frame and heatsink of the inverter, and go back to the machine via the cable connection. The second CMC path (loop 1 in Fig. 10) via the mains to the inverter via the DC source is ignored here because 1) the loop is bigger than loop 2; 2) the DC supply is an isolated one.

Table V shows the experimental results for the CMC measurement of six-phase full-bridge inverters for various carrier-based PWM techniques. It is shown that the double three-phase LS-PWM yields the highest rms value of CMC, as also shown in the simulation results. In addition, it is shown that bipolar modulation and PS-PWM also yield a considerable amount of CMC. This is mainly caused by the presence of dead-time in the experimental setup.

Conclusion

This paper compares several carrier-based PWM techniques to see their effect on the CMV and CMC of a six-phase full-bridge inverters. The CMV and CMC of the inverters is important because it can lead to EMI-related problems such as bearing degradation and high-frequency noise. In the given test bench, simulation and experimental results show that double three-phase LS-PWM yields the highest rms value of CMC on the six-phase full-bridge inverters. This is because in double three-phase LS-PWM all the odd and even harmonics spectrum appear in the CMC, which adds up to the total rms value. Bipolar modulation and PS-PWM does not induce CMC on the inverter circuit because the midpoint voltage is always half of the dc-link voltage. However, the inclusion of dead-time in the circuit causes both bipolar modulation and PS-PWM to generate CMC to the inverter circuit.

For future research, the CMV and CMC model will be included with the influence of switches parasitic capacitance, switching transient, and propagation delay. This will yield more accurate CMV and CMC estimation of the six-phase full-bridge inverters.

References

[1] E. Levi, "Multiphase Electric Machines for Variable-Speed Applications," in *IEEE Transactions on Industrial Electronics*, vol. 55, no. 5, pp. 1893-1909, May 2008, doi: 10.1109/TIE.2008.918488.

[2] E. Levi, "Advances in Converter Control and Innovative Exploitation of Additional Degrees of Freedom for Multiphase Machines," in *IEEE Transactions on Industrial Electronics*, vol. 63, no. 1, pp. 433-448, Jan. 2016, doi: 10.1109/TIE.2015.2434999.

[3] K. Wang, Z. Q. Zhu and G. Ombach, "Torque Improvement of Five-Phase Surface-Mounted Permanent Magnet Machine Using Third-Order Harmonic," in *IEEE Transactions on Energy Conversion*, vol. 29, no. 3, pp. 735-747, Sept. 2014, doi: 10.1109/TEC.2014.2326521.

[4] J. Arrozy, D. V. Retianza, J. L. Duarte and H. Huisman, "Fault-Tolerant Control of Series Connectable Modular Full-Bridge Inverter Mitigating Open Switch Faults," *2020 22nd European Conference on Power Electronics and Applications (EPE'20 ECCE Europe)*, 2020, pp. P.1-P.9, doi: 10.23919/EPE20ECCEEurope43536.2020.9215820.

[5] L. Jin, S. Norrga, H. Zhang and O. Wallmark, "Evaluation of a multiphase drive system in EV and HEV applications," *2015 IEEE International Electric Machines & Drives Conference (IEMDC)*, 2015, pp. 941-945, doi: 10.1109/IEMDC.2015.7409174.

[6] F. Barrero and M. J. Duran, "Recent Advances in the Design, Modeling, and Control of Multiphase Machines—Part I," in *IEEE Transactions on Industrial Electronics*, vol. 63, no. 1, pp. 449-458, Jan. 2016, doi: 10.1109/TIE.2015.2447733.

[7] M. J. Duran and F. Barrero, "Recent Advances in the Design, Modeling, and Control of Multiphase Machines—Part II," in *IEEE Transactions on Industrial Electronics*, vol. 63, no. 1, pp. 459-468, Jan. 2016, doi: 10.1109/TIE.2015.2448211.

[8] Z. Liu, Y. Li and Z. Zheng, "A review of drive techniques for multiphase machines," in *CES Transactions on Electrical Machines and Systems*, vol. 2, no. 2, pp. 243-251, June 2018, doi: 10.30941/CES-TEMS.2018.00030.

[9] E. Levi, I. N. W. Satiawan, N. Bodo and M. Jones, "A Space-Vector Modulation Scheme for Multilevel Open-End Winding Five-Phase Drives," in *IEEE Transactions on Energy Conversion*, vol. 27, no. 1, pp. 1-10, March 2012, doi: 10.1109/TEC.2011.2178074.

[10] N. Bodo, E. Levi and M. Jones, "Investigation of Carrier-Based PWM Techniques for a Five-Phase Open-End Winding Drive Topology," in *IEEE Transactions on Industrial Electronics*, vol. 60, no. 5, pp. 2054-2065, May 2013, doi: 10.1109/TIE.2012.2196013.

[11] J. Arrozy, H. Huisman and J. L. Duarte, "Input Current Ripple Analysis of Six-Phase Full-Bridge Inverters," *2021 IEEE 12th Energy Conversion Congress Exposition - Asia (ECCE-Asia)*, 2021, pp. 131-136, doi: 10.1109/ECCE-Asia49820.2021.9479174.

[12] F. Patkar, A. Jidin, E. Levi and M. Jones, "Performance comparison of symmetrical and asymmetrical six-phase open-end winding drives with carrier-based PWM," *2017 6th International Conference on Electrical Engineering and Informatics (ICEEI)*, 2017, pp. 1-6, doi: 10.1109/ICEEI.2017.8312446.

[13] Shaotang Chen, T. A. Lipo and D. Fitzgerald, "Modeling of motor bearing currents in PWM inverter drives," *IAS '95. Conference Record of the 1995 IEEE Industry Applications Conference Thirtieth IAS Annual Meeting*, 1995, pp. 388-393 vol.1, doi: 10.1109/IAS.1995.530326.

[14] A. Boglietti, A. Cavagnino and M. Lazzari, "Experimental High-Frequency Parameter Identification of AC Electrical Motors," in *IEEE Transactions on Industry Applications*, vol. 43, no. 1, pp. 23-29, Jan.-feb. 2007, doi: 10.1109/TIA.2006.887313.

[15] A. Boglietti, E. Carpaneto (2001) An Accurate Induction Motor High-Frequency Model for Electromagnetic Compatibility Analysis, *Electric Power Components and Systems*, 29:3, 191-209, DOI: 10.1080/153250001300006626

Mitigation of Dead-Time Effects on Transient DC Bias Elimination in Dual Active Bridge Link Current

MK Kharabela Mohanta, Dipankar De, Silpashree Sahu
School of Electrical Sciences
Indian Institute of Technology Bhubaneswar
Jatni, Odisha-752050, India
email: dipankar@iitbbs.ac.in

Alberto Castellazzi
Faculty of Engineering
Kyoto University
18 Gotanda-cho, Yamanouchi
Ukyo-ku, Kyoto, Japan, 615-8577
email: alberto.castellazzi@kaus.ac.jp

Keywords

≪Dual Active Bridge (DAB)≫, ≪Dead-time≫, ≪Converter control≫, ≪Compensation≫

Acknowledgement

This work is supported by Department of Science and Technology (DST), India under project grant ECR/2017/001079.

Abstract

Transient DC bias elimination in the link current of the dual active bridge DC-DC converter under power transients including dead-time effect is investigated in this paper. When a power change command is given to a dual active bridge, the modulation instants should be adjusted in an adequate manner in order to avoid over current during transients. The paper summarizes the adverse effect of the dead-time in applying DC-bias elimination techniques by dividing the entire operating region into six different zones based on operating phase angles and in terms of dead-time. An improved compensation method (based on pre-, post- transition current waveform computation) to mitigate the dead-time effect is suggested. A detailed mathematical analysis is carried out to select the suitable switching instants. The proposal is demonstrated through simulation studies and through experimental verification.

Introduction

One of the main challenges in designing modulator for Dual Active Bridge DAB converter [1, 2] is the transient DC bias current through the link inductance. The DC bias elimination technique is reported in [3–9] where extensive simulation and/or experimental results were presented to reduce the unwanted current stress on the power converter components. On the other hand, different optimization techniques are reported in the literature for DAB converter to improve its steady state performance. The optimization techniques (reported in [10, 11]) explain about the RMS link current and reactive power optimization respectively. In this work the optimization method for steady state performance is combined with the transient DC bias elimination algorithm. The second concept that this paper focuses is the effect of dead-time on the transient DC-bias elimination method. There are a good number of reported works in the literature to investigate and compensate the effect of dead-time to improve the performance of the dual active bridge converter [12–14]. In [15], a technique to eliminate DC bias current including dead-time is reported. A DC bias current elimination technique (with accurate computation of intermediate phase shifts during transients) is proposed in this work to mitigate the effect of dead-time and the corresponding mathematical/analysis is investigated. The proposed method unlike the previously reported method considers adverse effect on the dead time especially when the switching takes place closed to zero link current or link current reaches to zero during dead-time. The entire operating zone is classified in six sub-zones based on operating phase angles and in terms of dead-time (not in the form of power directly). The verification of the proposed concept is presented along with an optimization technique and the improvements achieved by this method are highlighted.

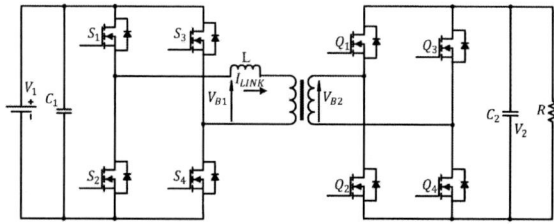

Fig. 1: Dual active bridge converter topology connected in power circulation mode

Transient DC Bias Without Dead-time Effects

Fig. 1 shows the basic circuit diagram of dual active bridge connected with source V_1 at the input side and load resistance R at the output. Let us consider a basic transient DC bias control technique to start with ignoring the effect of dead-time. The operation of dual active bridge usually has two phase shifts namely, the phase shift between the two full bridges (ϕ) and one inner phase shift (α) in the primary bridge (between the two legs of the converter). Hence, the transient operation can be divided into two zones based on $\phi < \alpha$ (zone-1) and $\alpha < \phi$ (zone-2).

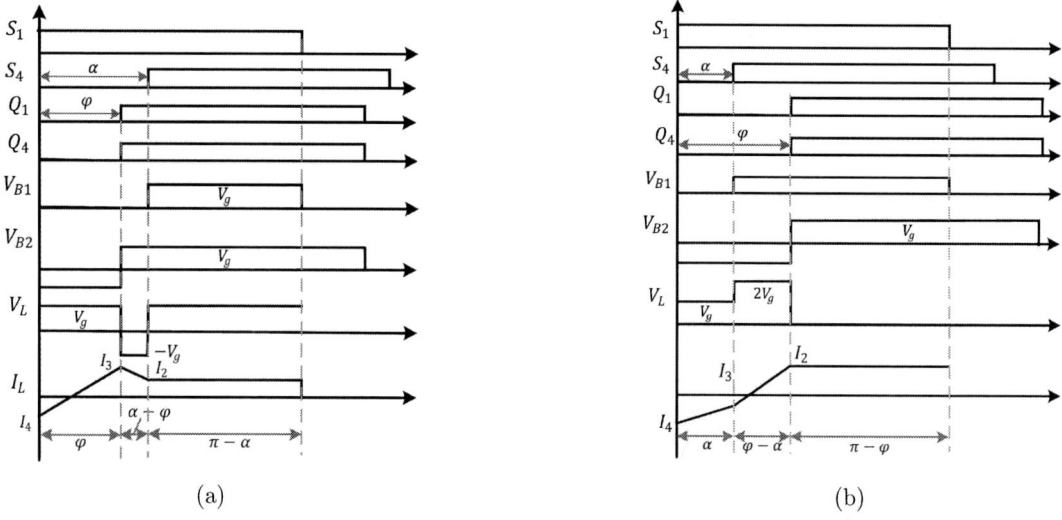

Fig. 2: Waveforms of zone-1 (a) and Waveforms of zone-2 (b)

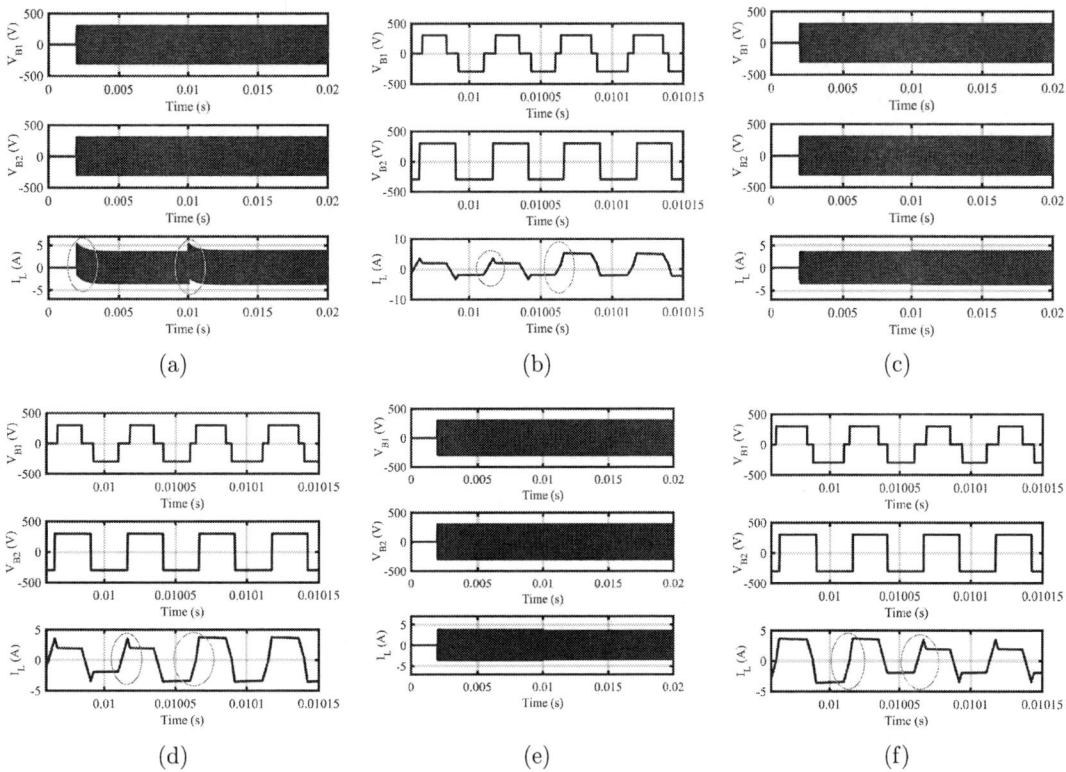

Fig. 3: Simulated Waveform for transition from Zone-1 (400W) to Zone-2 (800W): (a) complete transitions (b) zoomed at the power transition instant without DC-bias Elimination (without considering the effect of dead-time); Simulated Waveform for transition from Zone-1 (400W) to Zone-2 (800W): (c) complete transitions, (d) zoomed at the power transition instant with DC-bias Elimination techniques (without considering the effect of dead-time); Simulated Waveform for transition from Zone-2 (800W) to Zone-1 (400W): (e) complete transitions, (f) zoomed at the power transition instant with DC-bias Elimination techniques (without considering the effect of dead-time)

Fig. 2 shows the various typical waveforms such as primary voltage, secondary voltage, link voltage and link current for zone 1 (top figure) and for zone 2 (bottom figure). As the number of zones are two, there are total four possible combination of transitions possible when there is a transient change in the power reference. Fig. 3(a) shows the simulated result of transient DC bias current without any additional DC-bias elimination technique for zone 1 to zone 2 transition. The power transition instant is shown by the arrow mark and it can be observed that there is some DC bias in the link current at the power transition instant Fig. 3(b) zoomed at the transient instant.

Transient DC Bias Control Without Dead-time Effects

A transient DC-bias elimination technique is used to remove the DC Bias current during the transient process for all four cases. The main objective is that the current during its transition from one zone of operation to another zone should complete within half switching cycle (with the starting and end corner points of the link current as the steady state values of the respective zones). The mathematical description of one of the transitions (zone 1 to zone 2) is given in the following. Let, I_{41} is the negative steady state point for the pre-transition operating point for power P_1 and I_{22} is the positive steady state point for the post transition operating point for power P_2.

$$I_{41} = \frac{V_g T_s}{2L\pi}(-\phi_1 + \frac{\alpha_1}{2}); \quad I_{22} = \frac{V_g T_s}{2L\pi}(\phi_2 - \frac{\alpha_2}{2}) \tag{1}$$

The transient equations can be obtained by assuming the transitions are taking place in zone-2 and intermediate corner point has 50% value of I_{41} ($I_{32m} = x \times I_{41}$ with $x = 0.5$):

$$I_{41} + \frac{V_g T_s}{2L\pi}(\alpha_m) = I_{32m}; \quad I_{32m} + (\phi_m - \alpha_m)\frac{-2V_g T_s}{2L\pi} = I_{22} \tag{2}$$

Where, I_{31m} is the intermediate value and (α_1, ϕ_1), (α_2, ϕ_2) and (α_m, ϕ_m) are the pre transition, post transition and intermediate (for the transition period of half cycle) phase shifts respectively. Solving the above equations, we can obtain the intermediate phase shift needed for a smooth and DC bias free transition.

$$\phi_m = \frac{I_{22}}{2k} + \frac{I_{41}}{4k}; \quad \alpha_m = \frac{I_{41}}{2k} \tag{3}$$

Here, $(V_g T_s)/(2L\pi) = k$. It can be noted that the condition of having transition points in zone-2, $\alpha_m < \phi_m$ or $I_{22} > 0.5I_{41}$ should be satisfied. Otherwise, the intermediate angle values are to be determined assuming the the transitions are taking place in zone-1. Alternatively, the fraction x can be chosen such that it satisfies the zone-2 conditions. Fig. 3(c) shows the simulated waveforms for the transition from zone-1 to zone-2. The waveforms at the instant of power transients are zoomed and are shown Fig. 3(d). It can be clearly observed that during the transitions no additional DC bias present. An optimization block takes the power command P and a fraction F for the optimization (similar to the method that reported in [16–18]). The details of the optimization block will be explained in the subsequent sections. Similarly, Fig. 3(e), (f) show the simulated waveforms for the transition from zone-2 to zone-1.

Effect of Dead-time on DC Bias Current

With the incorporation of dead-time each zone can further be subdivided into 3 different sub-zones depending on link current direction and magnitude. These zones are shown in Table I. Case-A, Case-B and Case-C are the 3 sub-zones of zone-2 and Case-D, Case-E and Case-F are the 3 sub-zones of zone-1. Table. I provides the conditions or relationship between various phase angles for different cases/zones. Fig. 4 shows the effect of dead time on the link current corner points at various zones (Mode-A, Mode-B, Mode E are shown) and it can be seen that if the technique is not updated accordingly the performance of the DC bias elimination get affected and an unwanted deviation appears in the link current during the transients (as it can be seen from Fig. 5(a), (b) where case-A to case-B transition is shown without any dead-time compensation). The expression of the corner point current magnitudes at Mode-A, Mode-B, Mode E (as in Fig. 4) are summarized as follows:

- Case-A:

$$I_3 = \frac{V_g T_s}{2L\pi}(\phi - \frac{\alpha}{2}) = -I_1; \quad I_2 = \frac{V_g T_s}{2L\pi}(-\phi - 1.5\alpha) \tag{4}$$

- Case-B:

$$I_3 = \frac{V_g T_s}{2L\pi}(\phi - \frac{\alpha}{2} - \frac{D_T}{2}) = -I_1; \quad I_2 = \frac{V_g T_s}{2L\pi}(-\phi - 1.5\alpha + 1.5D_T) \tag{5}$$

- Case-E:

$$I_3 = \frac{V_g T_s}{2L\pi}(\phi - \frac{\alpha}{2} - \frac{D_T}{2}) = -I_1; \quad I_2 = \frac{V_g T_s}{2L\pi}(\frac{\alpha}{2} - \frac{D_T}{2}) \tag{6}$$

Table I: Conditions of Different Cases

Zones	Conditions	Parameters	Specifications
A	$\phi > \alpha,\ -\phi + 1.5\alpha < -2D_T$	V_1	300 V
B	$\phi > \alpha,\ -\phi + 1.5\alpha > -0.5D_T$	V_2	300 V
C	$\phi > \alpha,\ -2D_T < -\phi + 1.5\alpha < -0.5D_T$	Switching Frequency	20 kHz
D	$\alpha > \phi,\ -\phi + 0.5\alpha < -1.5D_T$	Inductor (L)	360 μH
E	$\alpha > \phi,\ -\phi + 0.5\alpha > -D_T$	Dead time	1 μs
F	$\alpha > \phi,\ -1.5D_T < -\phi + 0.5\alpha < -D_T$	Transformer	$1:1$

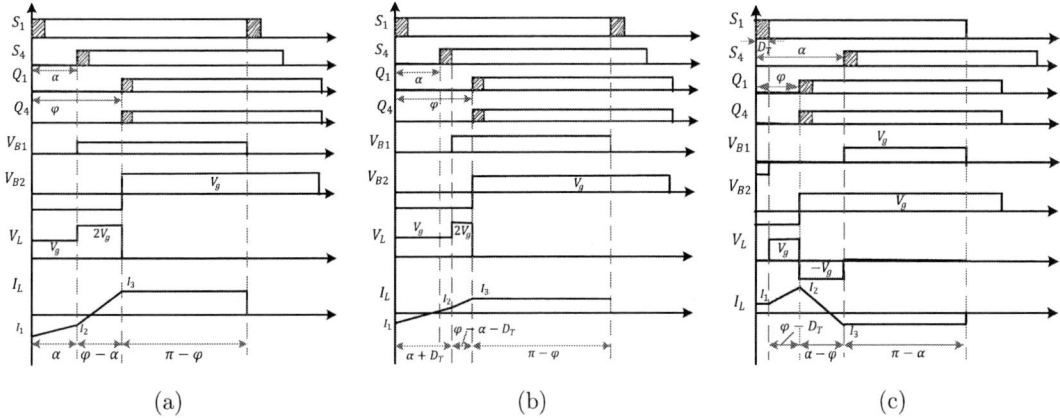

(a)   (b)   (c)

Fig. 4: Illustrative wave forms for current corner points considering the effect of dead-time (a) Case-A, (b) Case-B, (c) Case-E

Proposed Mitigation Techniques

Mathematical Expressions for Different Zones

The modified values of the phase shift can be adjusted by incorporating the effect of dead time in the computation. The mathematical expressions for two cases are shown in this section even though the implementation involves all the possible configurations. The transition from Case-A to Case-B can be described as follows. The transient equations in this case can be written as:

$$I_{1A} + \frac{V_g T_s}{2L\pi}(\alpha_{mA}) = I_{mA}; \quad I_{mA} + \frac{2V_g T_s}{2L\pi}(\phi_{mA} - \alpha_{mA}) = I_{3B} \tag{7}$$

Where, I_{1A} is the negative steady state corner point of the link current with the system is operating in Case-A condition and I_{3B} is the positive steady state corner point of the link current with the system is operating in Case-B condition.

$$I_{1A} = -\frac{V_g T_s}{2\pi L}(\phi_1 - \alpha_1); \quad I_{3B} = \frac{V_g T_s}{2\pi L}\left(\phi_2 - \frac{\alpha_2}{2} - \frac{D_T}{2}\right) \tag{8}$$

I_{mA} is the intermediate operating point. By solving above equations with assumption $I_{mA} = 0.5I_{1A}$ (ensures the transition interval within zone-A), we get,

$$\alpha_{mA} = \frac{I_{1A}}{k} - \frac{I_{mA}}{k} = 0.5\frac{I_{1A}}{k}; \quad \phi_{mA} = \frac{I_{3B}}{2k} - \frac{3I_{mA}}{2k} + \frac{I_{1A}}{2k} = \frac{I_{3B}}{2k} + \frac{I_{1A}}{4k} \tag{9}$$

Table I shows the power circuit parameters used for simulation/experiments and different power values considered in different zone for validation of the proposed concepts. Fig. 5(c), (d) show

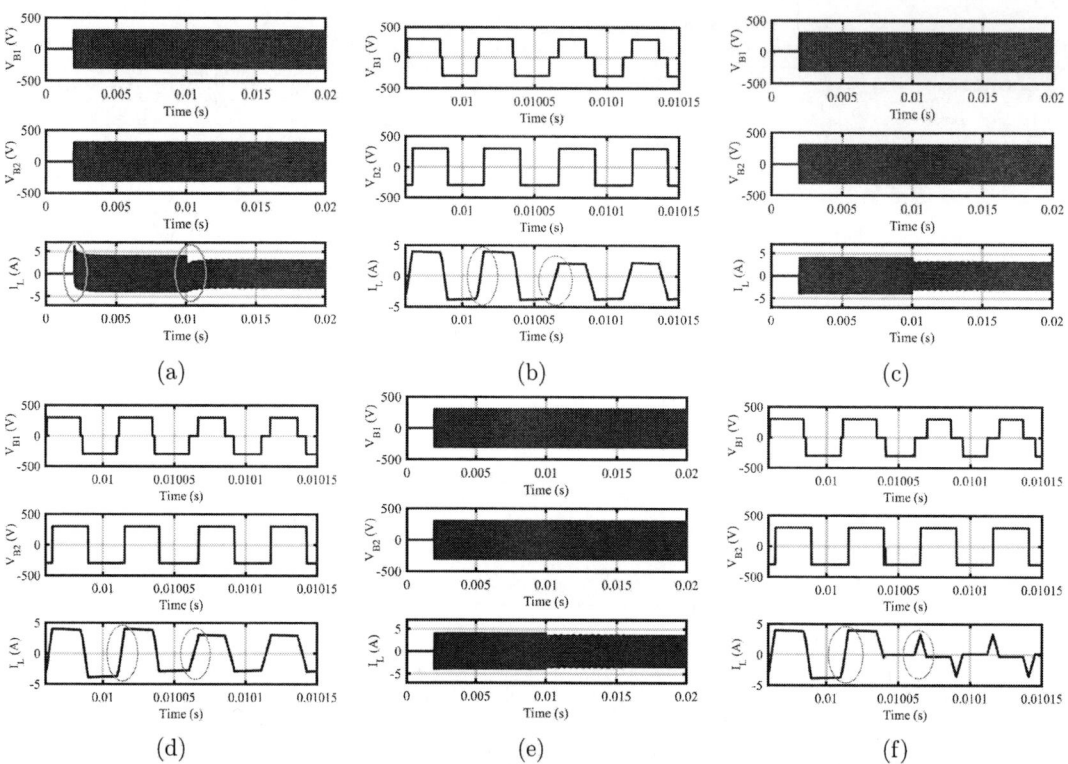

Fig. 5: Simulated Waveform for transition from Case-A to Case-B: (a) complete transitions. (b) zoomed at the power transition instant with DC-bias Elimination techniques (no compensation associated with dead-time); Simulated Waveform for transition from Case-A to Case-B: (c) complete transitions, (d) zoomed at the power transition instant with the proposed DC-bias Elimination techniques; Simulated waveform for Transition from Case-A to Case-E (e) complete transitions, (f) zoomed at the power transition instant with the proposed DC-bias Elimination techniques

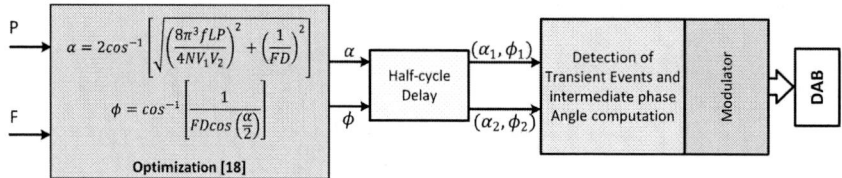

Fig. 6: Complete Control Diagram for Modified for DC Bias Current Elimination

the transient result where there is a transition from Case-A to Case-B. The power transition is made at 0.01s and it can be observed that with proposed algorithm the dead time does not affect the link current during transients. The encircled portion in the figure indicates the two steady state transients before and after the power transition.

Similarly, for the transition from A to E, the corner points in each zone can be obtained as,

$$I_{1A} = \frac{V_g T_s}{2L\pi}\left(-\phi_1 + \frac{\alpha_1}{2}\right); \quad I_{3E} = \frac{V_g T_s}{2L\pi}\left(\phi_2 - \frac{\alpha_2}{2} - \frac{D_T}{2}\right) \tag{10}$$

The transient equation and the phase shift angles by taking α_m is equal to ϕ_m can be written as

$$I_{1A} + \frac{V_g T_s}{2L\pi}(2\phi_m - \alpha_m - D_T) = I_{3E}; \quad \alpha_m = \phi_m = \frac{I_{3E}}{k} - \frac{I_{1A}}{k} + D_T \tag{11}$$

Fig. 5(e), (f) show the transient result where there is a transition from Case-A to Case-E with the dead-time compensation logic.

Complete Block Diagram with Modified DPS for DC Bias Elimination

Fig. 6 summarizes the complete block diagram of the proposed Modified DPS for DC Bias current elimination along with optimization block. This optimization ensures improved performance of the converter for given loading condition. The first step is to get the optimized α and ϕ value by taking power (P) and F as input as described in [18]. The power (P) is obtained from the output from DC voltage controller or it can be set to a particular value for a specified power transfer from input to output side. In the optimization block, by varying F different optimization in DAB performance can be achieved. For the minimization of fundamental component of link RMS current, $F = 1$ and for minimization of the fundamental component of reactive power flowing back to source, $F = 2$ [18]. D is the voltage ratio of the dual active bridge converter converter. After obtaining α and ϕ, these angles are passed through a delay block (half the switching period) to detect any power transient command. Once the α_1, α_2, ϕ_1 and ϕ_2 values are obtained, in case of any transition the zone of operation (the previous zone and the next zone) is detected first and then, α_m and ϕ_m are computed for DC bias elimination as discussed in the previous section.

Experimental Results

In order to verify the concept experimentally, the experimental lab prototype was built and the experimental set-up is shown in Fig. 7(a). The control signals are generated using digital signal controller TMS320F28335. A 1:1 HF transformer having 46 turns in both primary and secondary sides (E65/32/27 Ferrite Core) is used in the set-up. The DAB is connected in the circulating mode configuration, which circulates power in 300 V DC bus. The experimental results with proposed method with a transition from case-A 880W to case-B 730W, from case-B 730W to case-A 880W, from Case A 880W to Case E 50W are depicted respectively in Fig. 7(b), Fig. 7(c), Fig. 7(d). These figures show that the instant power change command (in channel-1) and the high frequency link current for the DAB converter (in channel-2). The instant of transition (of the power) within half the switching period are highlighted in these figures. The results show a smooth transition at power change. It can be noted that due to the computation of the pre-transient and post-transient instants (current corner points) and the required intermediate phase shift there is a one switching period delay in the execution of the transient events.

Conclusions

This work presents the effect of dead time on the transient DC-bias in the link current of DAB converter and a novel DC bias elimination technique is proposed incorporating the effect of dead time. The transitions between different zones are analysed mathematically. The expressions of various switch timings including the effect of dead time is derived based on different zones of operation in the DAB operation. The detailed simulation studies and experimental results are presented in the selected cases including the effect of unavoidable dead-time effects. The presented simulation and experimental results show the effectiveness of the (DAB link current) DC Bias elimination method. The immediate future work lies in extending the above study for any possible transition.

References

[1] Mi C., Bai H., Wang C., and Gargies S.: Operation, design and control of dual H-bridge based isolated bidirectional DC-DC converter, IET Power Electronics, Vol.1, No. 4, pp. 507-517, March

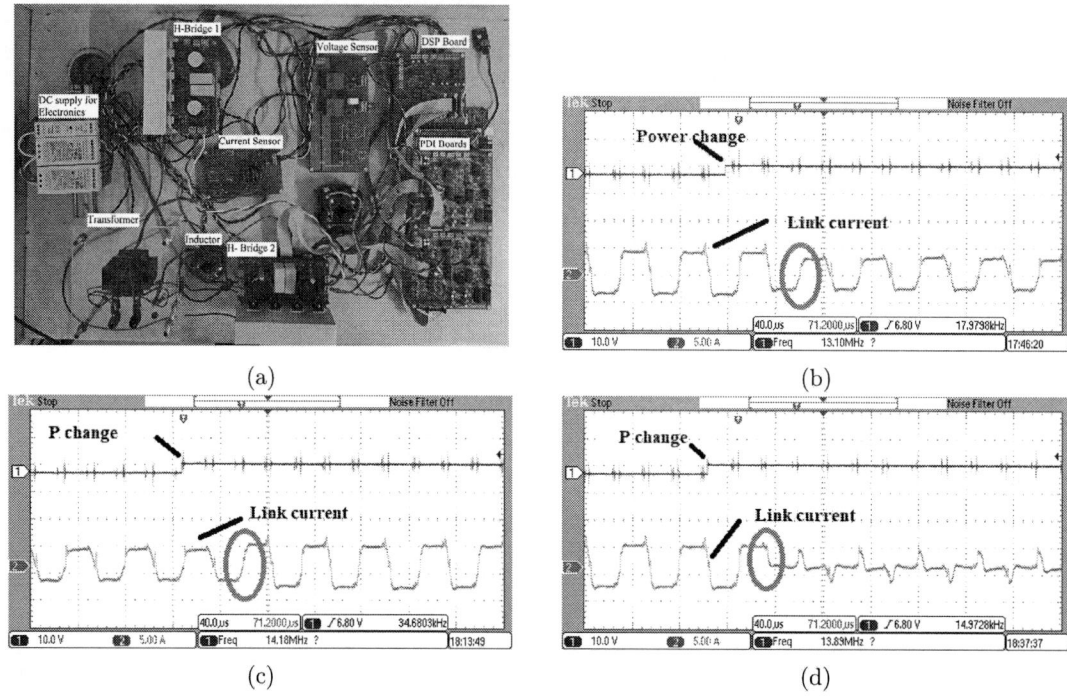

Fig. 7: (a) Photograph of the laboratory set up for DAB; (b) Experimental result of Link current (5A/div) with DPS: Transition from Case A -880W to Case B -730W, Time (20 μs/div); (c) Experimental result of Link current (5A/div) with DPS: Transition from Case B -730W to Case A -880W, Time (20 μs/div); (d) Experimental result of Link current (5A/div) with DPS: Transition from Case A-880W to Case E -50W, Time (20 μs/div)

2008.

[2] Tong A., Hang L., Li G., Jiang X. and Gao S.: Modeling and Analysis of a Dual Active-Bridge-Isolated Bidirectional DC/DC Converter to Minimize RMS Current with Whole Operating Range, IEEE Trans. Power Electronics, vol. 33, no. 6, pp. 5302-5316, June 2018.

[3] Bu Q., Wen H., Wen J., Hu Y., Du Y.: Transient DC Bias Elimination of Dual-Active-Bridge DC-DC Converter with Improved Triple-Phase-Shift Control, IEEE Trans. Industrial Electronics, vol. 67, no. 10, October 2020.

[4] Zhang Z., Sun J., Wang P., Cai Z., Kong J., Bai X., Ma D.: An Improved DC Bias Elimination Strategy with Extended Phase Shift Control for Dual-Active-Bridge DC-DC, CAC 2019, pp. 4274-4279.

[5] Bu Q., and Wen H., Control Strategies for DC-bias Current Elimination in Dual-Active-Bridge DC-DC Converter: An Overview, ICPS Asia 2020, pp. 1155-1162.

[6] Pena-Alzola R., Mathe L., Liserre M., Blaabjerg F. and Kerekes T.: DC-bias cancellation for phase shift controlled dual active bridge, IECON 2013, pp. 596-600.

[7] Zhao B., Song Q., Liu W. and Zhao Y.: Transient DC Bias and Current Impact Effects of High-Frequency-Isolated Bidirectional DC–DC Converter in Practice, IEEE Trans. Power Electronics, vol. 31, no. 4, pp. 3203-3216, April 2016.

[8] Takagi K. and Fujita H.: Dynamic Control and Performance of a Dual-Active-Bridge DC–DC Converter, IEEE Trans. Power Electron, vol. 33, no. 9, pp. 7858-7866, Sept. 2018.

[9] Su J., Luo S. and Wu F.: Improvement on Transient Performance of Cooperative Triple-Phase-Shift Control for Dual Active Bridge DC-DC Converter, ECCE 2019, pp. 1296-1301.

[10] Li Z., Wang Y., Shi L., Huang J., Lei W.: Optimized modulation strategy for three-phase dual-active-bridge DC-DC converters to minimize RMS inductor current in the whole load range, IPEMC-ECCE Asia 2016, pp. 2787-2791.

[11] Shi H., Wen H., Chen J., Hu Y., Jiang L. and Chen G.: Minimum-Reactive-Power Scheme of Dual-Active-Bridge DC-DC Converter with Three-Level Modulated Phase Shift Control, IEEE Trans. Industry Applications , vol. 53, no. 6, pp. 5573-5586, Nov.-Dec. 2017.

[12] Hu J., Yang Z. and De Doncker R. W.: A Comprehensive Dead Time Compensation Method for a Three-Phase Dual-Active Bridge Converter with Hybrid Modulation Schemes, IEEE IPEC, pp 1073 - 1079, 2018.

[13] Zhao B., Song Q., Liu W. and Sun Y.: Dead-Time Effect of the High-Frequency Isolated Bidirectional Full-Bridge DC-DC Converter: Comprehensive Theoretical Analysis and Experimental Verification, IEEE Trans. Power Electron., Vol. 29, No.4, pp.1667-1680, 2014.

[14] Takagi K. and Fujita H.: Dynamic Control and Dead-Time Compensation Method of an Isolated Dual-Active-Bridge DC-DC Converter,IEEE ECPA 2015, pp-1-10.

[15] Luo S., Wu F. and Gang W.: Effect of Dead Band and Transient Actions on CTPS Modulation for DAB DC-DC Converter and Solutions, IEEE Trans. on Transportation Electrification, vol. 7, no. 4, pp. 949-957.

[16] Mukherjee S., Dash A., De D. and Castellazzi A.: Study of Dual Active Bridge with Modified Modulation Techniques for Harmonic Reduction in AC Link Current, ICSETS 2019, pp. 144-149.

[17] Maharana S., Mukherjee S., De D. and Castellazzi A.: Dead-Time Compensated Dual Active Bridge with Online Hybrid Optimized Operation, ICPEE 2021, pp. 1-6.

[18] Mukherjee S., Dash A., De D. and Castellazzi A.: Trade-off in Minimization of Fundamental Link Current and Reactive Power using a Novel Online Calculation based Triple Phase Shift Modulator for Dual Active Bridge, EPE ECCE Europe 2019, pp. P.1-P.10

Generalized Automated Tool for Analysis and Design of Multiphase Coupled Inductor Buck Converters

Rana Asad Ali, Mahmoud Shousha, Martin Haug
MagI³C PU, Würth Elektronik eiSos Group
Garching bei München, Germany
E-Mail: mahmoud.shousha@we-online.de
URL: Würth Elektronik eiSos Group

Acknowledgements

This work has received support from the European Union's Horizon 2020 program, for the project EleGaNT under the Grant Agreement 101004274.

Keywords

«DC-DC converter», «Coupled inductor», «Modelling», «Design», «Multiphase converter», «Paralleling», «Simulation»

Abstract

This paper presents the implementation and validation of a tool for analyzing multi-phase coupled inductor buck converters. In this type of converter, the equivalent inductance seen by each phase changes significantly within one switching cycle and the number of equivalent inductances increases significantly as the number of phases increase. The manual analysis of these converters is prone to errors and using available simulation tools does not result in a symbolic closed-form solution for the equivalent inductance versus the number of phases. The proposed tool not only considers the coupled inductor design with symmetric inductances and coupling coefficients but also the asymmetric design parameters. Moreover, the possibility of specifying the winding directions of the coupled inductor makes it more suitable for practical applications. The tool is benchmarked against simulation and experimental setup by designing three winding symmetric and asymmetric coupled inductors for three phase buck converters. The tool has an error ranging from 0.0288% to 4.661%.

Introduction

With the increasing power demand of low-voltage and high-current systems such as today's microprocessors, single-phase buck converters pose a challenge to meet steady state requirements for efficiency, output voltage ripples, and thermal performance as well as transient requirements such as output voltage deviation and recovery time. In addition, if the transient response is improved by reducing the inductance value, the system's efficiency degrades due to the high current ripples which also result in high output voltage ripples, making such types of converters ill-fitting to the targeted applications. Moreover, to meet the high-current demand of such systems, single-phase buck converters require an inductor with high inductance and saturation current values, which eventually leads to a big solution size. Therefore, multi-phase buck converters (MPBC) are the solution which address the above-mentioned challenges by having multiple buck converters which operate in parallel at a phase shift of $360°/N$, where N denotes the number of phases. In addition to this, each stage carries lower amounts of current in comparison to the total current, which relaxes the component sizing at the expense of a larger number of components. As far as thermal management is concerned, the thermal hot spots are distributed among the number of phases and hence the system efficiency improves. The multiphase buck converter is different from single phase in a way that output ripple current gets reduced in comparison to phase ripple current due to ripple cancellation effect. The effective frequency at the output of MPBC is multiplied by the factor of N, which eases the design of output filter and reduces the output voltage ripples.

Instead of using discrete inductors per phase in MPBC, the coupled inductors could be used in such applications where the steady-state response and dynamic response require separate optimization by controlling the effective inductance, which is possible in integrated or coupled inductors [1]. These converters are known as multi-phase coupled inductor buck converters (MPCIBC).

Previous works [2]-[3] only analyzed two-phase coupled inductor converters without extending the concept to N-number of phases. Analysis of four-phase boost and three-phase buck converters were shown in [4]-[5]. The authors did not consider the difference in mutual inductances between phases, assuming a single value in the analysis. In addition, the direction of winding was not covered in their analysis, resulting in the same equivalent inductance value for both cases. The authors of [6] presented the analysis of two phase-coupled inductors but the analysis is only valid for loosely coupled inductors. The authors of [7]-[9] provide a solution for two equivalent inductances only for N-number of phases, however as the number of phases increases, the mode of operation increases, and the number of the equivalent inductances increases. The main goal of this paper is to provide a closed form symbolic solution for MPCIBC, covering all modes of operation for any number of phases considering symmetrical design, asymmetrical design, and winding direction of every phase. A tool is developed on MATLAB allowing users to enter their conditions and giving them the symbolic or numerical equivalent inductances and current-slopes per mode within each case.

Modeling of the Coupled Inductor

A coupled inductor is one which contains a single core with more than one coil wound around it. Due to the presence of a common core, the flux links with other windings, unlike in discrete inductors. This flux linkage changes the voltage appearing across each winding within a switching cycle, hence the analysis of multiphase converters with coupled inductors (MPCIBC) differs from a system with discrete inductors. Figure 1 describes that a coupled inductor can be modeled by $L_{1,lk}$, $L_{2,lk}$, L_m and an ideal transformer. $L_{1,lk}$ and $L_{2,lk}$ are known as leakage inductances, and magnetizing inductance is represented as L_m. [8]-[10].

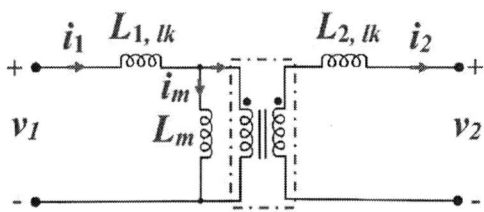

Fig. 1: T model of coupled inductor.

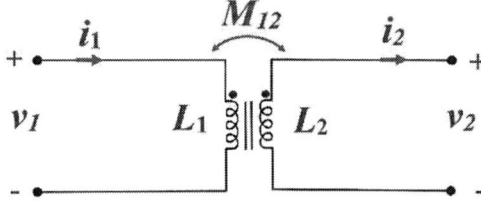

Fig. 2: Modeling of coupled inductor for analysis.

The modelling of the coupled inductor as shown in Fig. 1 is based on leakage L_{lk} and magnetizing inductances. The L_{lk} plays an important role in analysis and operation of power electronics converters such as MPCIBC [8]-[9], [11]. Usually, a short circuit test is performed for measuring the leakage inductance, however, due to high leakage and large air gaps in the magnetic structures of coupled inductors of multiphase converters, it is prone to inaccuracy [12]. Due to this reason, the standard T-model of a coupled inductor as given in Fig. 1 is replaced by the model as shown in Fig. 2. The self-inductances labelled in Fig. 2 as L_1 and L_2 are summation of L_{lk} and L_m. This inductance is measured through a standard open circuit test. The inductance which exists between two magnetically coupled coils is called mutual inductance, labelled as M_{12}. Hence, the model as illustrated in Fig. 2 based on M_{12}, L_1 and L_2 will be used for further analysis. These mutual and self-inductances are the key parameters for defining the design of coupled inductors in the designed tool and simulation.

In [12], the accurate method of series-coupling for calculating the mutual inductance is mentioned. Figure 3 explains it by showing two test circuits, in Fig. 3 (a) where two dots are aligned, and this configuration is known as directly coupled, differential or series-opposing. The flux generated by two coils connected differentially opposes each other in comparison to series-aiding, cumulative configuration, as shown in Fig. 3 (a) and (b) respectively. The series aiding and opposing inductances L_{opp} and L_{aid} are measured by the test circuits as shown in Fig. 3 are given as:

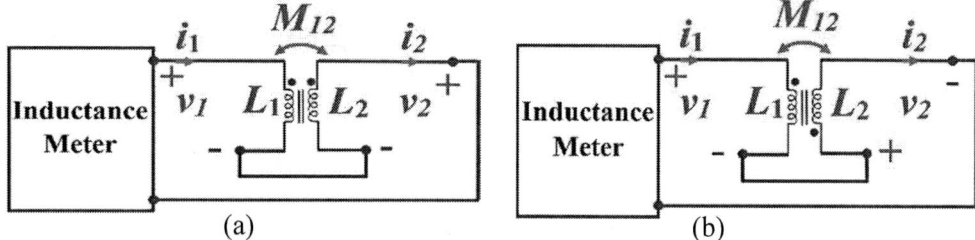

(a)   (b)

Fig. 3: Series (a) differential and (b) cumulative, coupling test.

$$L_{opp}=L_1+L_2-2M_{12} \qquad (1)$$
$$L_{aid}=L_1+L_2+2M_{12} \ , \qquad (2)$$

After measuring L_1 and L_2 using open circuit as well as L_{opp} and L_{aid} using series-coupling tests, the mutual inductance can be calculated as follows [11]:

$$M_{12}=\frac{L_{aid}-L_{opp}}{4} \ , \qquad (3)$$

The coupling coefficient that is denoted by k relates mutual inductance M_{ji} and self-inductances L_j, L_i of a coupled inductor as given in eq. (6), where k ranges from $-1 \le k \le 1$.

$$k_{ji}=\frac{M_{ji}}{\sqrt{L_j L_i}} \qquad (4)$$

Analysis of a Multiphase Buck Converter with Coupled Inductor

Figure 4 depicts the schematic of a N-phase buck converter that contains a coupled inductor with N number of windings wound on a common core each having self-inductances of L_1, L_2, L_3, ..., L_N. The mutual inductance between the inductors is mentioned as M_{ji}, where I ranges from 1 to N-1 and j ranges from 1 to N. The objective of the following analysis is to investigate the effective inductance and current-slopes of each mode of operation within a switching cycle.

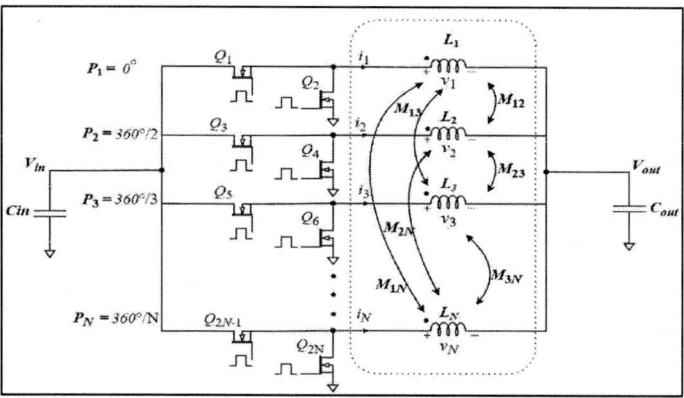

Fig. 4: N-phase buck converter with coupled inductors.

The voltages across the N windings are related to their respective current-slopes as follows:

$$v_1=L_1\frac{d_{i1}}{dt}+M_{12}\frac{d_{i2}}{dt}+M_{13}\frac{d_{i3}}{dt}+ \cdots\cdots +M_{1N}\frac{d_{iN}}{dt}$$
$$v_2=M_{12}\frac{d_{i1}}{dt}+L_2\frac{d_{i2}}{dt}+M_{23}\frac{d_{i3}}{dt}+ \cdots\cdots +M_{2N}\frac{d_{iN}}{dt} \qquad (5)$$
$$\vdots \qquad\qquad \vdots \qquad\qquad \vdots \qquad\qquad \vdots$$
$$v_N=M_{1N}\frac{d_{i1}}{dt}+M_{2N}\frac{d_{i2}}{dt}+M_{3N}\frac{d_{i3}}{dt}+ \cdots\cdots +L_N\frac{d_{iN}}{dt}$$

Writing the system of equation, as given in eq. (5), as a matrix yields:

$$
A = \begin{bmatrix} L_1 & +M_{12} & +M_{13} & \cdots & \cdots & +M_{1N} \\ +M_{12} & L_2 & +M_{23} & \cdots & \cdots & +M_{2N} \\ +M_{13} & +M_{23} & L_3 & \cdots & \cdots & +M_{3N} \\ \vdots & \vdots & \vdots & \ddots & \ddots & \vdots \\ +M_{1N} & +M_{2N} & +M_{3N} & \cdots & \cdots & L_N \end{bmatrix}, \quad \dot{X} = \begin{bmatrix} d_{i1}/dt \\ d_{i2}/dt \\ d_{i3}/dt \\ \vdots \\ d_{iN}/dt \end{bmatrix}, \quad B = \begin{bmatrix} v_1 \\ v_2 \\ v_3 \\ \vdots \\ v_N \end{bmatrix}, \tag{6}
$$

To solve the matrix $\dot{X}$ which contains the slope of the currents through each winding of the coupled inductor, the equation as given in eq. (6) can be written as:

$$
\dot{X} = A^{-1} . B, \tag{7}
$$

When all the self and mutual inductance values are the same in the inductance matrix A, the situation is referred to as having symmetrical inductance. Other, the situation is called asymmetrical inductance. The inductance matrix must be a singular matrix, which means that the coupling coefficient must be restricted within the range of $-1 < k < 1$. The mutual inductances between the windings of the coupled inductor depend upon the direction of the winding; if two corresponding windings have the same direction, as illustrated in Fig. 3, with the dots on the same side, it is indicated by a positive M_{ji} while opposing winding directions are represented by a negative M_{ji} with dots on the opposite sides. Therefore, the direction of each two corresponding windings can easily be entered in matrix A in the presented tool as shown in Fig. 5

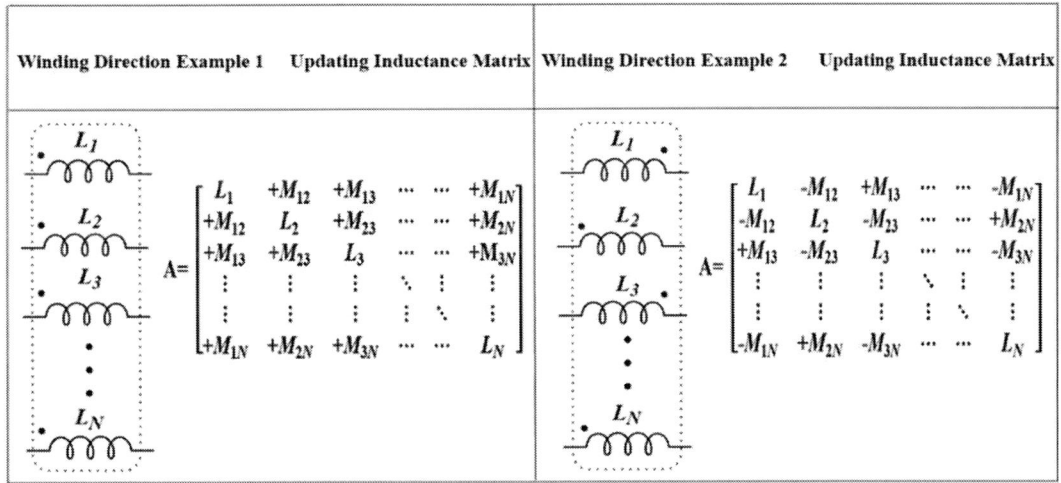

Fig. 5: Changing the winding directions and updating the mutual inductances accordingly in matrices.

The duty cycle range has been divided into different cases, which depend on N. During each case, the voltages across the windings remain the same hence, so do the effective inductances. Within each case, the switching cycle contains $N \times 2$ distinct modes or effective inductances. As it can be seen in Fig. 9 and 10, for $N = 3$, there are six modes such as: $m_1, m_2, m_3, \ldots, m_6$. Each mode corresponds to a distinct value of effective inductance with its own associated winding current. As N increases, the number of cases increases and the number of effective inductances also increase, indicated in table I. It is worth mentioning here that the analysis was started from $N = 2$, however; due to space constraints only the relation for $N=3$ has been mentioned. The effective inductances and slopes of current through the winding have been calculated using the tool, which matches with the results derived manually, as shown in Table II.

Table I: Effectives inductances for $N = 3$ across the whole duty cycle range

Modes	Derivation of Effective Inductances Manually		
	Case 1: $0 < D < \frac{1}{3}$	Case 2: $\frac{1}{3} \leq D < \frac{2}{3}$	Case 3: $\frac{2}{3} \leq D < 1$
1	$L_{1,C1,m1} = \dfrac{(L-M)(L+2M)}{L + \left(1+2\dfrac{D}{D'}\right)M}$	$L_{1,C2,m1} = \dfrac{(L-M)(L+2M)}{L + \dfrac{D}{D'}M}$	$L_{1,C3,m1} = L+2M$
2	$L_{2,C1,m2} = L+2M$	$L_{2,C2,m2} = \dfrac{(L-M)(L+2M)}{L + \left(1+2\dfrac{D}{D'}\right)M}$	$L_{2,C3,m2} = \dfrac{(L-M)(L+2M)}{L + \dfrac{D}{D'}M}$
3	$L_{3,C1,m3} = \dfrac{(L-M)(L+2M)}{L + \dfrac{D'}{D}M}$	$L_{3,C2,m3} = \dfrac{(L-M)(L+2M)}{L + \dfrac{D}{D'}M}$	$L_{3,C3,m3} = L + 2M$
4	$L_{4,C1,m4} = L+2M$	$L_{4,C2,m4} = \dfrac{(L-M)(L+2M)}{L + \dfrac{D'}{D}M}$	$L_{4,C3,m4} = \dfrac{(L-M)(L+2M)}{L + \dfrac{D}{D'}M}$
5	$L_{5,C1,m5} = \dfrac{(L-M)(L+2M)}{L + \dfrac{D'}{D}M}$	$L_{5,C2,m5} = \dfrac{(L-M)(L+2M)}{L + \left(1+2\dfrac{D'}{D}\right)M}$	$L_{5,C3,m5} = L+2M$
6	$L_{6,C1,m6} = L+2M$	$L_{6,C2,m6} = \dfrac{(L-M)(L+2M)}{L + \dfrac{D'}{D}M}$	$L_{6,C3,m6} = \dfrac{(L-M)(L+2M)}{L + \left(1+2\dfrac{D'}{D}\right)M}$

It can be observed from Table II that the derivation of these relations manually is quite time-consuming and prone to error even in the symmetrical design case. Hence, the level of difficulty rises exponentially with asymmetrical inductances and increased number of phases. Therefore, the automation of deriving and solving the above-mentioned relationships is highly advantageous, a task which is addressed by the presented tool.

Development of the Tool

Figure 6 illustrates the flowchart of the presented tool implemented on MATLAB. It starts from taking the user input of the number of phases, N, and duty cycle, D. Based on the value of D, it determines the respective case and initializes the matrix accordingly. It contains the voltages that appear across windings within each mode of switching cycle. With the help of symbolic substitution in MATLAB the respective equations of current-slopes and effective inductances are derived. The tool not only derives all the generalized equations of effective inductances and slopes of all the phase currents but also calculates their numerical values by specifying the input voltage, output voltage and design parameters of the coupled inductor, such as self and mutual inductances.

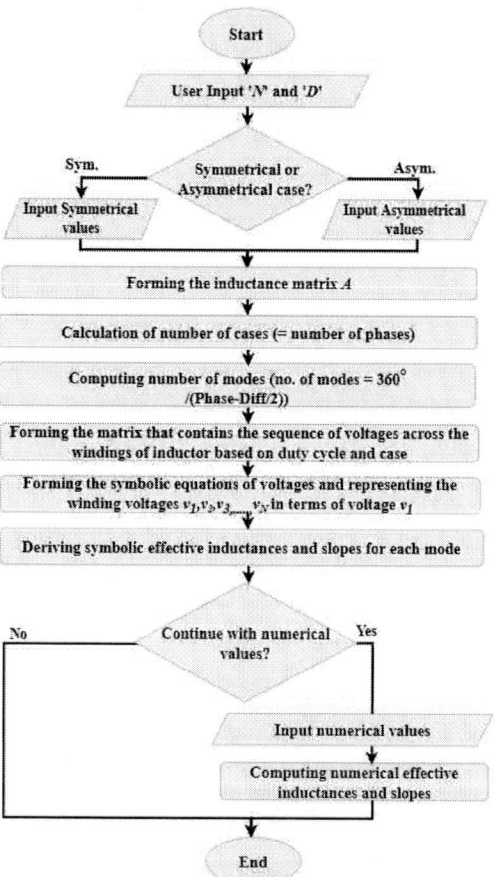

Fig. 6: Flowchart of the presented tool.

It is essential to highlight the fact that available controllers for multiphase converters, used for high performance microprocessors and FPGAs, usually handle up-to 16 phases. In this case, the significance of the presented tool is evident as it automates the analysis of a converter with 16 different cases across the entire duty cycle range. Each case has 32 (=16×2) effective inductances with 512 (=16^2×2) total inductances for the system across the entire range of the duty cycle.

Experimental Validation

For experimental verification of the results, a three-phase buck converter was implemented using a single coupled inductor containing three windings. Figures 7 shows the simplified diagrams of the chosen coupled inductor. They depict the self-inductances L_1, L_2, L_3 and the mutual inductances M_{12}, M_{13}, M_{23} present between the respective windings. Based on the values of self and mutual inductances, which control the coupling coefficients, the coupled inductors are classified into symmetrical and asymmetrical designs.

In Fig. 7 (a), the coupled inductor is designed by interleaving the windings so that homogenous flux can be ensured. Due to interleaving, the distance between windings and core is similar which makes L_{lk} and L_m of three windings uniform. The rod-shaped core makes sure the presence of leakage inductance in the design. Therefore, overall identical values of k exist between windings.

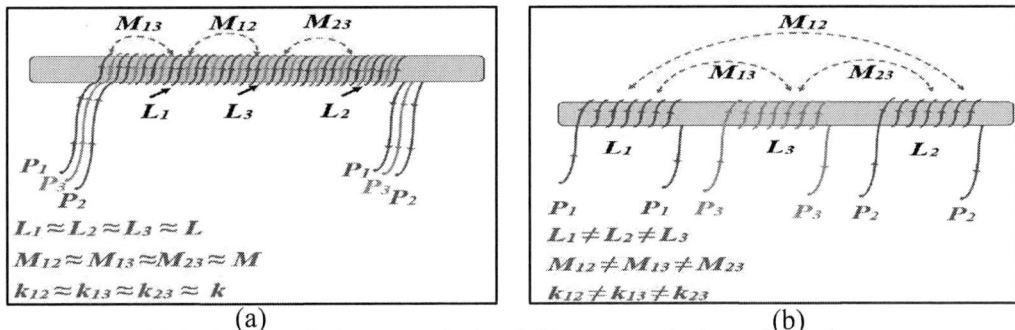

(a)   (b)

Fig. 7: Simplified diagram of (a) symmetrical and (b) asymmetrical coupled inductor.

The complexity of achieving symmetrical k increases rapidly with the increase in number of phases. It is evident from Table II by looking at design parameters of the symmetrical coupled inductor. The values of self, mutual inductances and coupling coefficients are unequal even for symmetrical design. Therefore, asymmetrical design of coupled inductor is close to practical scenarios. To benchmark the effectiveness of the tool against the design parameters of asymmetrical coupled inductor, such inductor is designed by winding the coils in the manner as explained in Fig. 7 (b). The mismatch in distances between the windings, different numbers of turns, unequal effective core area used by each winding and lack of interleaving while winding the coils all contribute to unequal L_{lk} and L_m. Therefore, the design differs from the symmetrical case due to a mismatch of mutual self-inductances and k. Table II summarizes the design values of symmetrical and asymmetrical coupled inductors measured by the methods mentioned earlier.

Table II: Design parameters of three phase symmetrical and asymmetrical coupled inductors

Design Parameters	Symbols	Symmetrical Design	Asymmetrical Design
Self-Inductances	L_1	4.2540 µH	7.2670 µH
	L_2	4.1660 µH	8.7520 µH
	L_3	4.0320 µH	5.4387 µH
Mutual Inductances	M_{12}	3.3500 µH	3.2004 µH
	M_{13}	2.9790 µH	3.7930 µH
	M_{23}	3.1295 µH	4.2510 µH
Coupling- Coefficient	k_{12}	0.7957	0.4012
	k_{13}	0.7175	0.6005
	k_{23}	0.7654	0.6191

Experimental Setup

The experimental setup consists of three phase buck converter with coupled inductor and their associated controllers. The output of each phase is regulated by a peak current mode controller to guarantee a desired level of accuracy in current sharing as can be seen in Fig. 8.

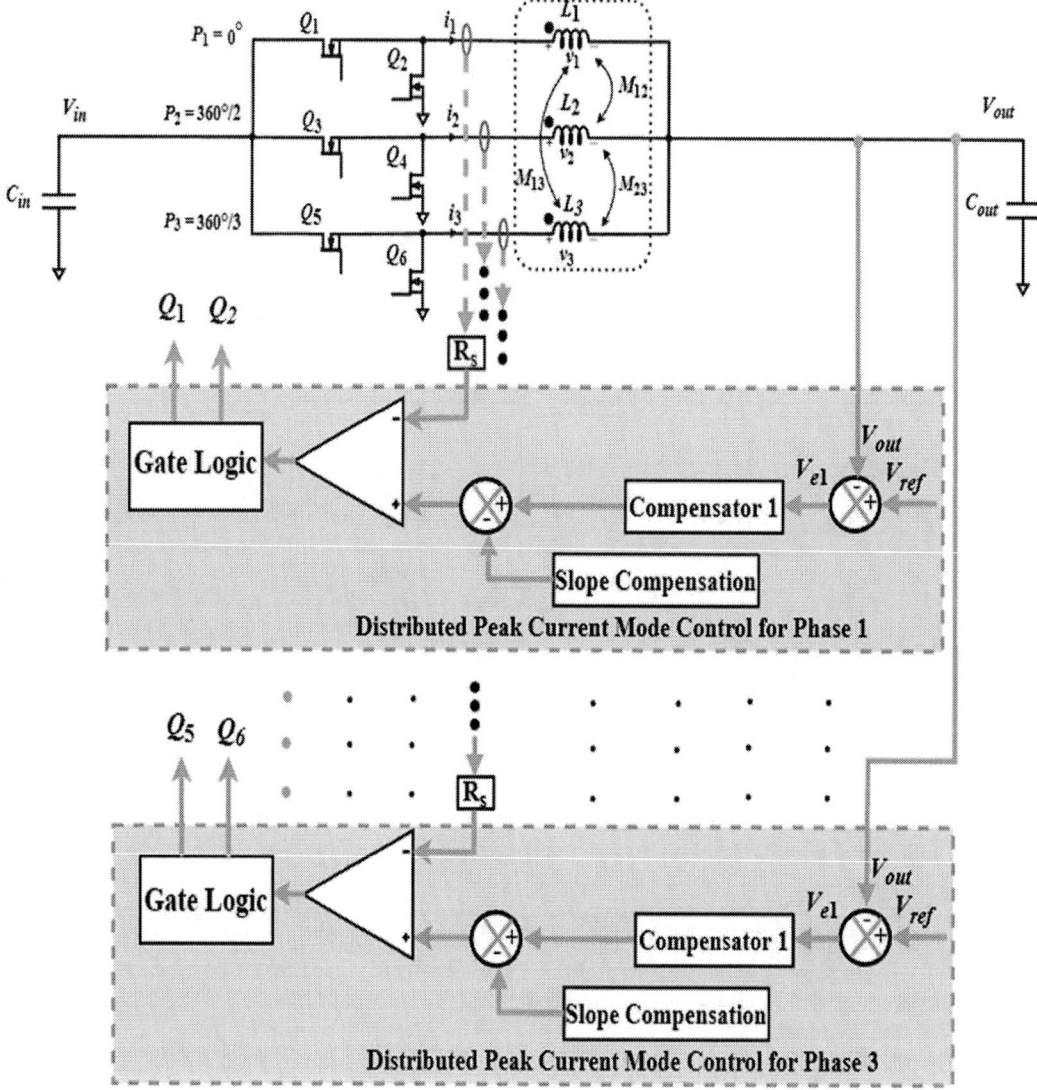

Fig. 8: Block diagram of experimental setup

In Fig. 9(a-c) the inductor currents through each phase of the converter are shown. To benchmark the effectiveness of tool, results from an LTspice simulation and an experimental setup are compared. It can be observed that within each duty-cycle-dependent case, phase currents have similar patterns and rates of change irrespective of their peak values. Due to the phase-shifted control signals, a phase shift of 120° can be seen in phase currents. As the duty cycle changes, the phase current-slopes change accordingly. The current-slopes within each case are highlighted by red dotted lines in Fig. 9 and 10, indicated as m_1, m_2, m_3, ..., m_6.

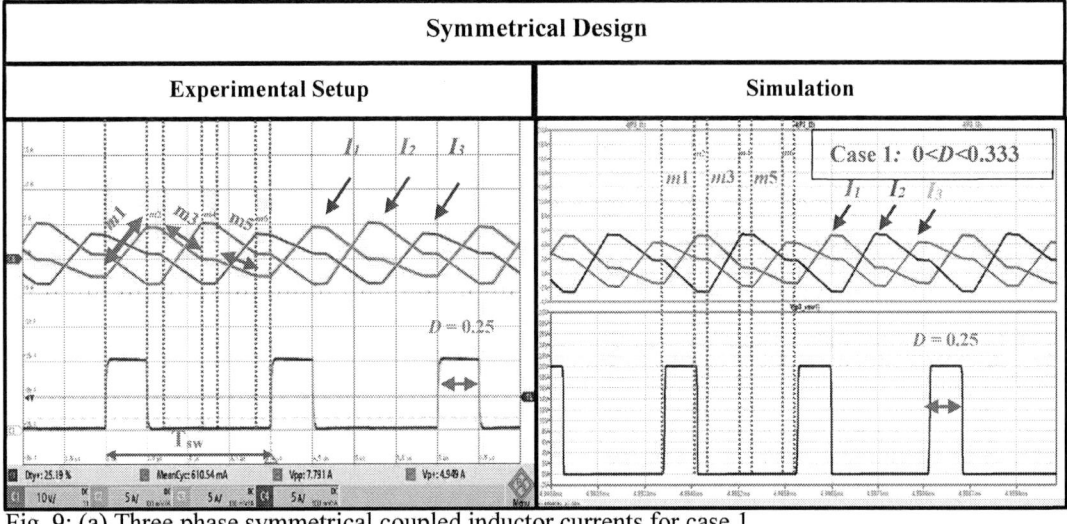

Fig. 9: (a) Three phase symmetrical coupled inductor currents for case 1.

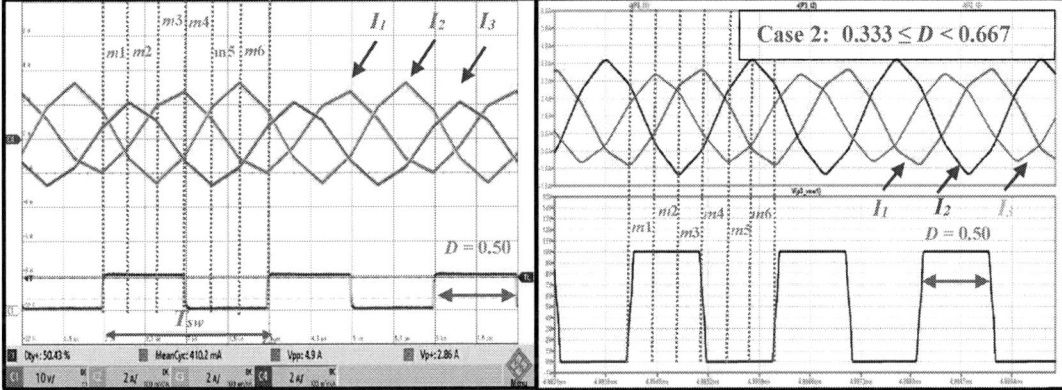

Fig. 9: (b) Three phase symmetrical coupled inductor currents for case 2.

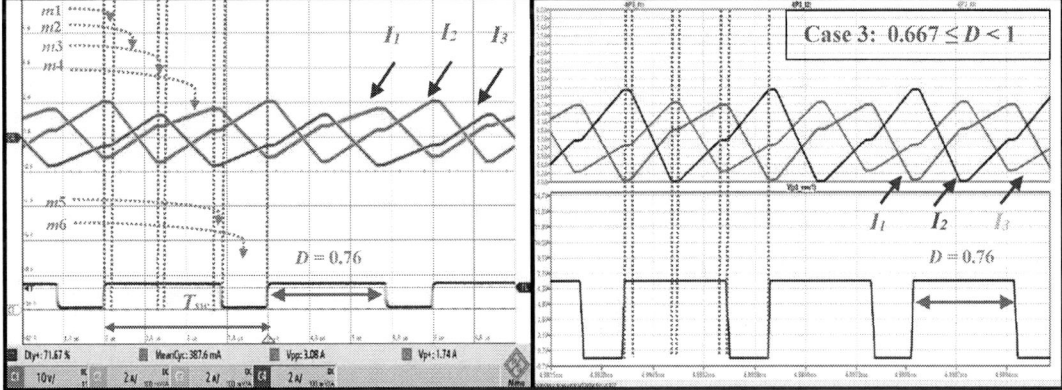

Fig. 9: (c) Three phase symmetrical coupled inductor currents for case 3.

Figure 10 (a-c) shows the operation of the converter with an asymmetrical coupled inductor. The significant difference between Fig. 9 and 10 is that due to the asymmetrical design, the current-slopes, and waveforms of the phase currents in each case have significant disparity than the symmetrical ones. In Fig. 9 (a-c) the phase currents have similar shapes; however, in Fig. 10 (a-c) each phase current waveform has different pattern. Therefore, when analyzing the current-slopes of each case, all the phases must be considered for the asymmetrical design, while in the symmetrical design only phase one is used due to symmetrical slopes of the other two phases.

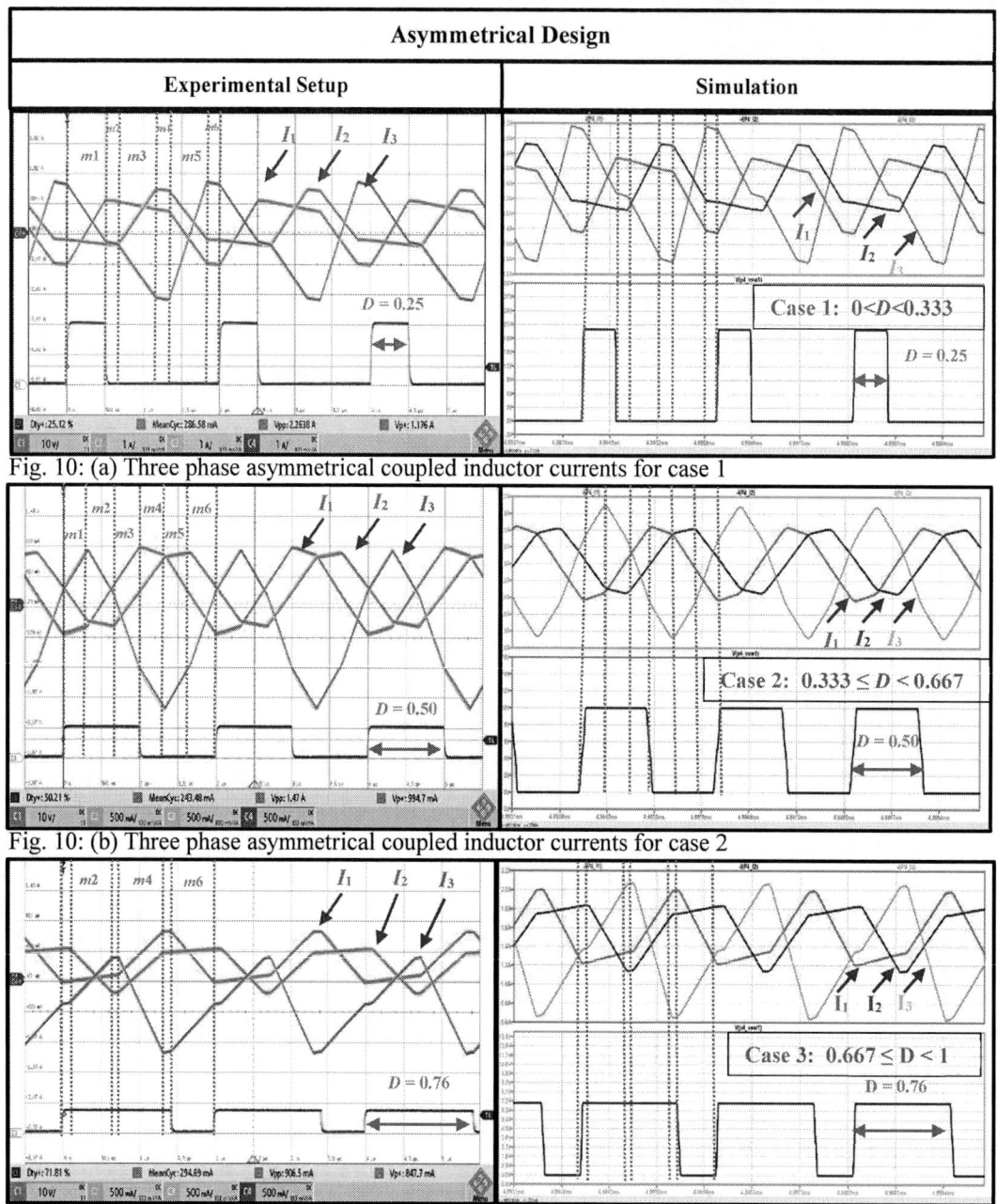

Fig. 10: (a) Three phase asymmetrical coupled inductor currents for case 1

Fig. 10: (b) Three phase asymmetrical coupled inductor currents for case 2

Fig. 10: (c) Three phase asymmetrical coupled inductor currents for case 3.

The accuracy of the presented tool against the simulation and experimental results is benchmarked by observing each slope controlled by its respective effective inductance within a switching cycle as illustrated in Fig. 9 and 10. In Fig. 11, the percentage errors between the simulation-tool, simulation-experimental and tool-experimental setup are presented for symmetrical and asymmetrical coupled inductors. All three phases are analyzed, accounting for the six duty-cycle-dependent modes for the asymmetrical inductor case. Due to the symmetric design of the coupled inductor, only a single phase is presented for comparative analysis. The minimum and maximum percentage error for all the comparison cases ranges from 0.0288 % to 4.661 %. It can be concluded that the error between the tool-experimental setup is more than that of the simulation-tool or the simulation-experimental setup. The reason being that during small periods, measurements of current-slopes can be challenging and prone to inaccuracy as compared to the period where slope is higher with sharp rise and fall. For instance, high error appears in the asymmetrical design for phase 1 during mode 5 and case 3 where the rate of change of current is not that significant, as can be seen in Fig. 10 (c) and 11. In addition, the root-cause of error also lies in the fact that MOSFETs with parasitic parameters and on-state resistances have been used in the SPICE model. Therefore, due to the voltage drops across the $R_{DS,ON}$ and the voltage across windings, the results are not the same as those found in the theoretical analysis and hardware setup.

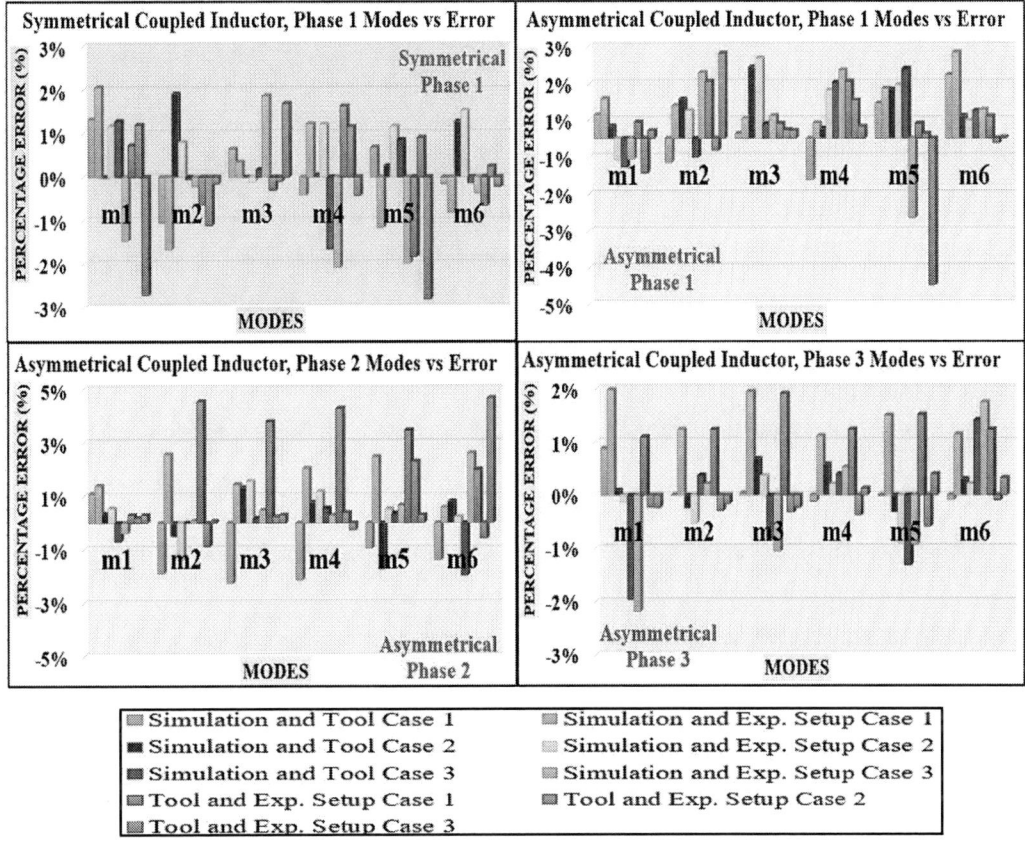

Fig. 11: Comparative analysis of the results of tool, simulation, and experimental setup for symmetrical and asymmetrical designs.

Conclusions

This paper introduces the challenges of analyzing the MPCIBC and proposes a tool to address them. MPBCs, which are widely used in microprocessors and other high-current low-voltage applications, meet the critical efficiency and voltage-regulation requirements by using multiple phases. This paper offers a tool which automates the process of deriving such inductances and current slopes taking into account different design parameter such as inductance symmetry, number of phases and windings direction. The tool produces symbolic representation and values of the effective inductance of each mode of operation in addition to phase current-slopes, which help designers in analyzing and designing the power stage of MPCIBC.

References

[1] Wong PL, Xu P, Yang P, Lee FC.: Performance improvements of interleaving VRMs with coupling inductors, IEEE Transactions on Power Electronics. 2001 Jul;16(4):499-507.

[2] Kroics K.: Design of Interleaved GaN Transistor Based Buck Converter with Directly Coupled Foil Winding Inductor, InCIPS 2020; 11th International Conference on Integrated Power Electronics Systems 2020 Mar 24 (pp. 1-6). VDE.

[3] Wibowo SA, Ting Z, Kono M, Taura T, Kobori Y, Onda KI, Kobayashi H.: Analysis of coupled inductors for low-ripple fast-response buck converter, IEICE transactions on fundamentals of electronics, communications and computer sciences. 2009 Feb 1;92(2):451-5.

[4] Kroics K, BRAZIS US.: Design of coupled inductor for interleaved boost converter, diM. 2014;3(2):2.

[5] Pan T, Wang Y, Qu Z, Tao W.: Topology optimisation and current sharing strategy of interleaved bidirectional dc/dc converter with coupling technique, IET Power Electronics. 2018 Dec;11(15):2470-80.

[6] Lee JP, Cha H, Shin D, Lee KJ, Yoo DW, Yoo JY. Analysis and design of coupled inductors for two-phase interleaved DC-DC converters. Journal of power electronics. 2013;13(3):339-48.

[7] Shi, Meng.: Design and analysis of multiphase DC-DC converters with coupled inductors, PhD diss., Texas A&M University, 2007.

[8] Liu, Jie.: Investigation of Multiphase Power Converter using Integrated Coupled Inductor Regarding Electric Vehicle Application, PhD diss., 2016.

[9] Dong, Yan.: Investigation of multiphase coupled-inductor buck converters in point-of-load applications, PhD diss., Virginia Tech, 2009.

[10] Shi, Meng. "Design and analysis of multiphase DC-DC converters with coupled inductors." PhD diss., Texas A&M University, 2007

[11] Kosai, Hiroyuki, Seana McNeal, Austin Page, Brett Jordan, Jim Scofield, and Biswajit Ray. "Characterizing the effects of inductor coupling on the performance of an interleaved boost converter." *Proc. CARTS USA 2009* (2009): 237-251.

[12] Hayes, John G., Neil o'Donovan, Michael G. Egan, and Terence O'Donnell. "Inductance characterization of high-leakage transformers." In *Eighteenth Annual IEEE Applied Power Electronics Conference and Exposition, 2003. APEC'03.*, vol. 2, pp. 1150-1156. IEEE, 2003

Experimental Study of a Directly Oil-Cooled Electrical Machine for a Full-Electric Vehicle by Using Low Viscosity Oil

Huihui Xu[1], Georg Tobias Götz[1], Shimin Zhang[2], Rik W. De Doncker[1]

[1]Institute for Power Electronics and Electrical
Drives (ISEA), RWTH Aachen University
Jaegerstrasse 17/19
Aachen, Germany
Phone: +49 (0)241 80-99562
Email: post@isea.rwth-aachen.de
URL: https://www.isea.rwth-aachen.de/

[2]TotalEnergies
Centre de Recherche de Solaize,
Chemin du Canal - BP 22
Solaize, France
Phone: +33 (0) 4 78 02 61 75
Email: shimin.zhang@totalenergies.com
URL: http://www.totalenergies.fr

Acknowledgments

The experimental work for this paper has been carried out within a research project collaborated with TotalEnergies. Special thanks are given to Shimin Zhang and Grégoire Roux of TotalEnergies for their contribution and guidance in the project. The authors also gratefully acknowledge the contribution of Claas Ehrenpreis of the Institute of Heat and Mass Transfer (WSA) RWTH Aachen involved in this work.

Keywords

≪Thermal model≫, ≪Permanent magnet motor≫, ≪Electrical drive≫, ≪Electric vehicle≫, ≪Cooling≫.

Abstract

The thermal performance of an integrated electric drive unit, that consists of a directly oil-cooled electrical machine and a transmission, is experimentally studied. The cooling performance using two types of oil with different viscosities is analyzed with the help of empirical formulas and a lumped-parameter thermal model. The resulting thermal model with a single parameter set models the thermal behavior with the two different oil types. This approach allows the separate study of the drive unit and the oil performance with respect to different oil viscosities.

Introduction

In electric vehicles (EVs), there is a development trend towards highly integrated drive units. In EVs such as the Toyota Prius, Chevrolet Bolt and Tesla Model 3 the electrical machine and the transmission are integrated in the same housing. Therefore, the transmission oil that lubricates the gears also absorbs the heat losses generated in the electrical machine. This can contribute to an improved thermal performance of the complete system [1], a compact design of the electric drive unit [2] and a weight reduction of the vehicles [3]. The most common lubrication fluid, which is in this case also the cooling medium for the electric traction motor, is automatic transmission fluid (ATF). Depending on the requirements of applications in vehicles, the combination of components can be changed to modify certain properties of the ATF, such as the kinematic viscosity [4]. A comprehensive study on the cooling effect on directly oil-cooled end windings of an electric motor is presented in [5], which investigates the influence of the flow rate and the temperature of oil, the rotor speed and the design of injection on the cooling performance. In [6], the cooling performance of ATF in comparison to the high viscous manual transmission fluid in an oil-cooled induction motor is experimentally studied, whereby no significant difference was observed.

In this work, the thermal behavior of the drive unit of the EV Chevrolet Bolt from General Motors is investigated. Rather than redesigning or optimizing the cooling system [5, 7], the focus of this study is the influence of oil properties on the cooling performance. Here, two types of oil with different viscosities are experimentally studied. With help of the measurement results, the direct oil-cooling is modeled based on a lumped-parameter thermal model and empirical formulas for convective heat transfer correlations.

Electric Drive Unit and Experimental Setup

The electric drive consists of the electrical machine and the transmission unit, as shown in Fig. 1. The speed of the machine is reduced by the transmission to adapt to the driving speed of vehicles.

Fig. 1: 3D view of the drive unit

Fig. 2: Cross section schematic of the electrical machine

The cross section view of the studied electrical machine in the drive unit is shown in Fig. 2. It is an interior permanent magnet synchronous machine (IPMSM) with eight poles and V-shaped magnets in the rotor and six-layer hairpin windings in the stator. The stack length of the stator is 125 mm and the outer diameter of the machine is 204 mm. The maximum output power is 150 kW. [8, 9]

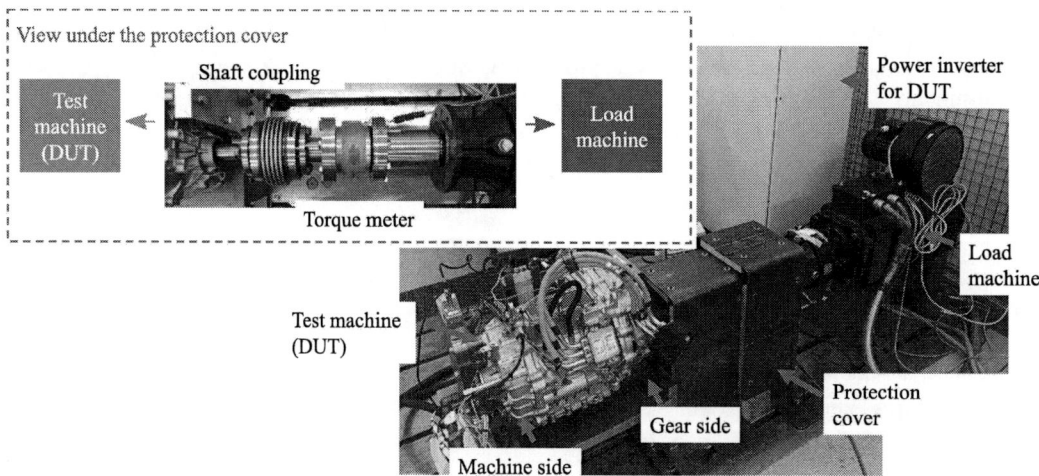

Fig. 3: Picture of the experimental test bench

For the laboratory study, measurements are conducted at a maximum output power of 58.6 kW due to the speed and torque limitations of the load machine. However, sufficient temperature developments can still be observed in the reduced operation region. The setup of the experimental test bench can be seen in Fig. 3. The test machine or device under test machine (DUT) is mechanically connected to the load machine by a shaft coupling. The torque and speed are measured by a torque meter. The power inverter for the DUT is connected to a dc-link providing a constant voltage source. The output connectors of the power inverter are connected to the three terminals of the DUT without filtering circuits.

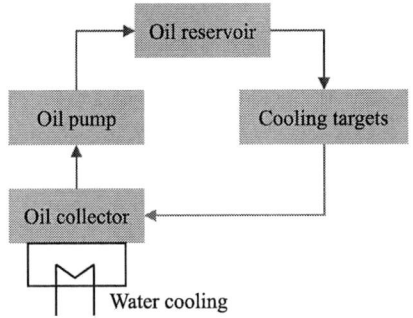

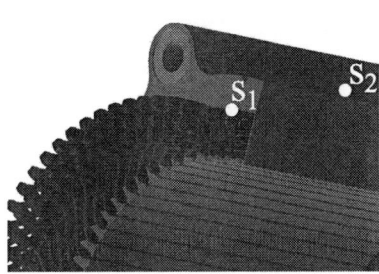

Fig. 4: Oil and water loop of the cooling system   Fig. 5: Sensor position on the end winding (s_1) and at outer surface of the stator (s_2)

The cooling system of the drive unit consists of two cooling circuits as depicted in Fig. 4. In the primary cooling circuit, oil is used as cooling medium. It is circulated by an electric gear pump and passes through the internal components of the machine and transmission, whereby the end windings in the stator especially in the top sector are directly oil-cooled. Because the oil reservoir is located on top of the machine and the oil distributes and flows into the drive unit from there. The secondary cooling is accomplished by a water cooling circuit. Here, the water indirectly cools down the oil in a heat exchanger. In comparison to real cooling conditions in vehicles, the variation of the ambient temperature and the air flow at different driving velocities are not emulated in the laboratory experiments. Instead, the drive unit is tested under a constant ambient temperature of 20 °C and it is not additionally ventilated.

For the temperature measurements, thermocouples are attached to the outer surface of the stator and to the end windings. To study the cooling by the oil explicitly, the stator segment at around 0° (see Fig. 2) and the attached sensors (see Fig. 5) are selected for the analysis in this work. The stator and the end winding in the following text refer to the two measurement points in Fig. 5. Furthermore, the oil temperature in the oil reservoir (see Fig. 4) and seven points on the housing are measured.

During the experiments, the flow rate and the inlet temperature of the water cooling circuit are kept constant. The gear pump supporting the oil circulation is pulse-width modulation (PWM) controlled. The supply voltage and the duty cycle of the PWM are constant. Two oils with different viscosities denoted as oil A and oil B are investigated. Applying these two oils, the cooling performance in the electrical machine is studied.

Experimental Results

To gain applicable measurements for the parameter identification of the thermal model, operating points at different torque and speed values are kept constant for a long time (from 15 min to 110 min). Three measured operating points are given in Table I. They demonstrate high, medium and low power operation of the drive unit. The experiments are started from room temperature at 20 °C. The DUT runs at the given operating points and the temperatures at aforementioned locations are measured and monitored. When any of the sensors reaches the predefined temperature limit of 150 °C, the torque is reduced to zero.

Table I: Selected operating points

Operating point	Torque in Nm	Speed in min^{-1}	Power in kW
1	125	4000	52.3
2	75	4000	31.4
3	75	2500	19.6

Due to the demagnetization of the permanent magnet in the rotor and the increased machine losses as the total temperature increases, the torque produced by the DUT decreases. As shown by the plots of filtered data in Fig. 6, the speed and current are kept constant. While the machine temperature increases over time, a torque reduction is observed. To compensate this thermal effect, a torque feedback compensation is added to the current controller. Hence, the torque stays constant even under temperature variations. That means, a larger current is provided to maintain a constant torque as the temperature increases [10].

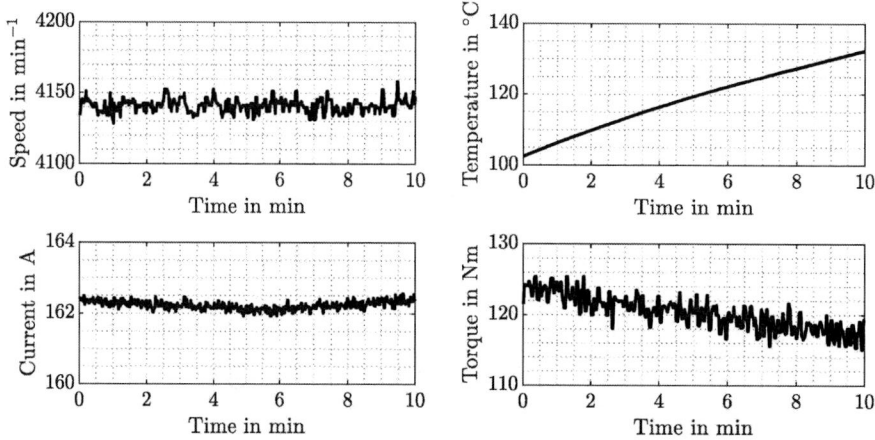

Fig. 6: Torque reduction in experiment without compensation control

a) Thermal Interaction Between Machine and Transmission

To illustrate the thermal effect of the transmission on the machine, the measured temperatures on two sides of the machine (as labeled by machine side and gear side in Fig. 3) are shown in the following. In one experiment with oil B, the DUT is operated at a torque of $125\,\mathrm{N\,m}$ and a speed of $4000\,\mathrm{min^{-1}}$ as given in Fig. 7. The current profile is the blue curve in Fig. 8.

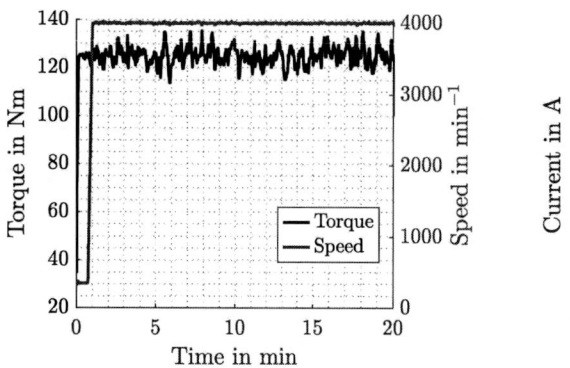

Fig. 7: Speed and torque profiles

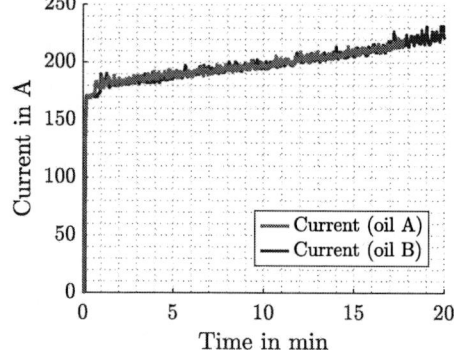

Fig. 8: Phase currents (rms value) at $125\,\mathrm{N\,m}$ and $4000\,\mathrm{min^{-1}}$

The thermocouples attached on the housing show two different temperature ranges on the machine side (measured by three sensors) and the gear side (measured by four sensors). As given in Fig. 9, the housing temperature measured on the machine side is higher than that on the gear side. This is attributable to the fact that a major part of the losses in the drive unit is generated by the machine under normal operations. The temperature ranges measured on the end windings of the machine are given in Fig. 10. The temperature band gained on the gear side is narrower compared to the machine side, which means a more homogeneous temperature distribution. This can be caused by the better oil distribution on the gear side considering the oil distributed by the gearwheels.

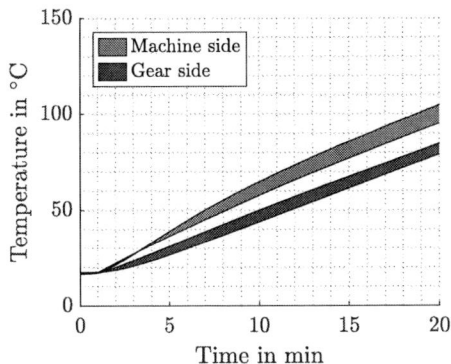
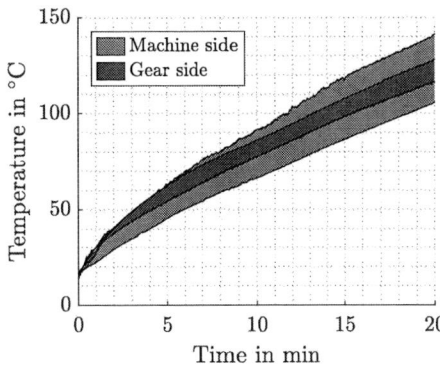

Fig. 9: Measured temperature on the housing   Fig. 10: Measured temperature on the end windings

b) Cooling Performance Using Low Viscous Oil

The comparison of the normalized kinematic viscosity at $40\,°C$ and $100\,°C$ is shown in Fig. 11, which shows the different material properties between the investigated two oils over a large temperature range. The differences in other physical properties, such as density, specific heat capacity and thermal conductivity are relatively small and therefore are not discussed in this work.

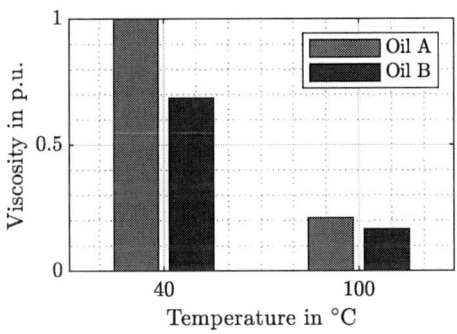
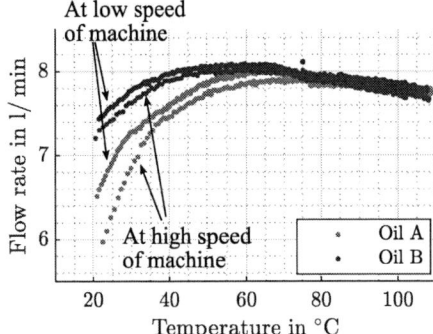

Fig. 11: Normalized kinematic viscosity   Fig. 12: Oil flow rate over temperature for oil A and B

The flow rate, in dependence of oil temperature, is shown in Fig. 12. At the same temperature, the low viscous oil B has a higher flow rate than A until a temperature of about $70\,°C$. At higher temperatures, the flow rates are similar. It is worth mentioning that the oil behavior is also influenced by the machine speed. The impact of the rotational speed of the DUT is observed by the two curves for each oil in Fig. 12. This could be caused by the change of internal pressure in the machine at different speeds.

The cooling performance using oil B with low viscosity is analyzed in comparison to that using oil A. The measured rms phase currents at $125\,\mathrm{N\,m}$ and $4000\,\mathrm{min}^{-1}$ in these two experiments are shown Fig. 8, which demonstrate that the injected currents are almost identical during the first 17.6 min. The corresponding temperature measurements are shown in Fig. 13, whereby the following features are identified. (a) The temperature profiles measured by the same sensor in two experiments show a similar development. (b) The improved cooling effect by using the low viscous oil B can be clearly observed. At $t = 17.6\,\mathrm{min}$, a temperature reduction of 9.7 K using oil B with respect to A is detected for the stator. For the end winding, a reduction of 6 K is observed.

The comparisons at $75\,\mathrm{N\,m}$ and $4000\,\mathrm{min}^{-1}$ are shown in Fig. 14. The specified temperature limit of $150\,°C$ is not reached. Instead, a quasi-steady state after nearly two hours of operation is detected.

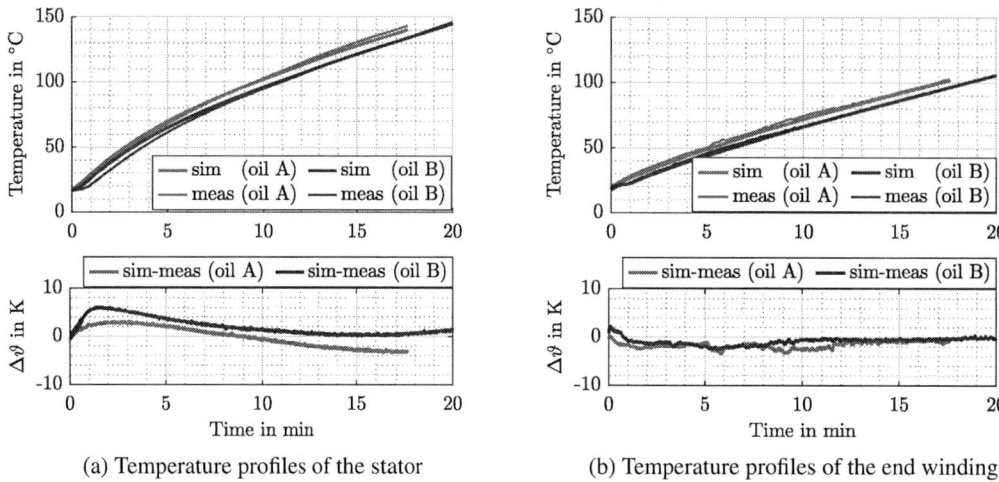

(a) Temperature profiles of the stator

(b) Temperature profiles of the end winding

Fig. 13: Temperature profiles at 125 N m and 4000 min^{-1}

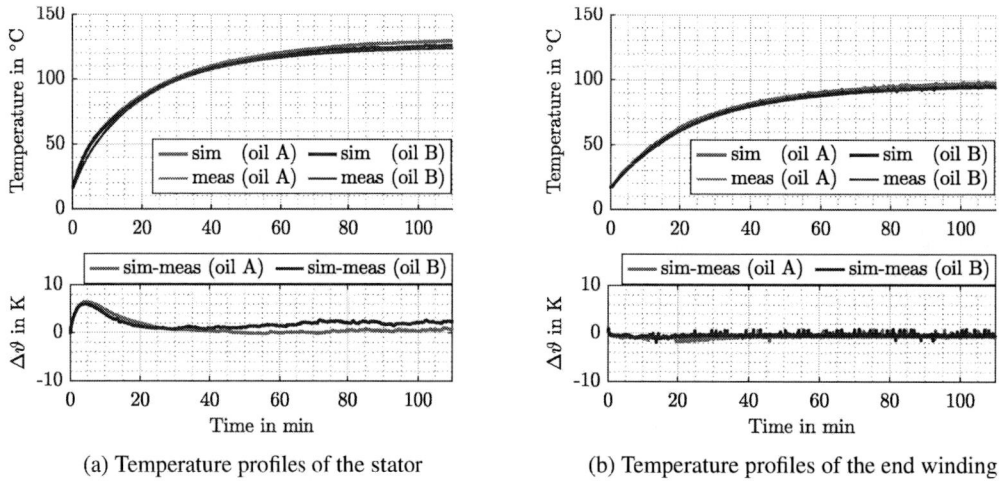

(a) Temperature profiles of the stator

(b) Temperature profiles of the end winding

Fig. 14: Temperature profiles at 75 N m and 4000 min^{-1}

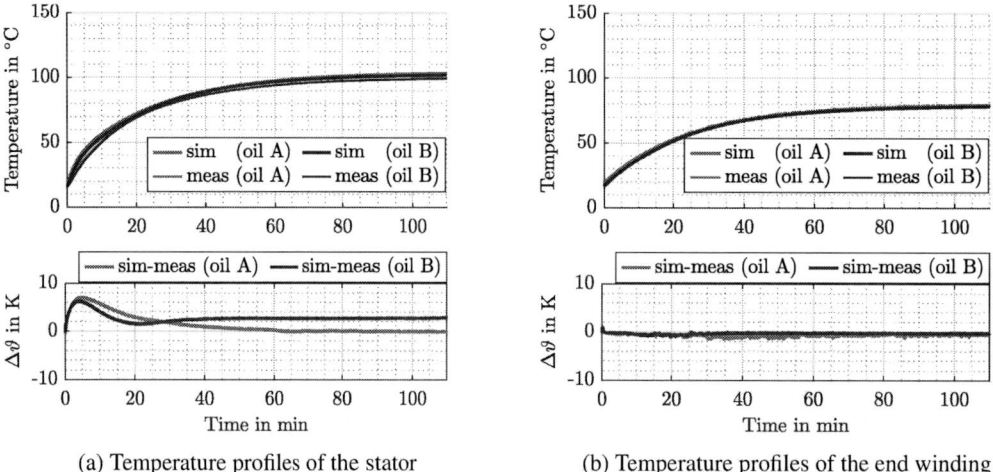

(a) Temperature profiles of the stator

(b) Temperature profiles of the end winding

Fig. 15: Temperature profiles at 75 N m and 2500 min^{-1}

At $t = 110\,\text{min}$, using the low viscous oil B, temperature reductions of $5.2\,\text{K}$ on the stator and $3\,\text{K}$ on the end winding are measured, with respect to oil A.

For the operating point at $75\,\text{N}\,\text{m}$ and $2500\,\text{min}^{-1}$ as shown in Fig. 15, the experiment duration is also approximately two hours until the quasi-steady state is achieved. As improved cooling effect using oil B, temperature reductions at $t = 110\,\text{min}$ are $4.3\,\text{K}$ for the stator and $1.1\,\text{K}$ for the end winding.

Thermal Modeling

The influence of oil properties on the cooling performance is analyzed with help of the presented measurements. For this purpose, a lumped-parameter thermal model of the electrical machine is built by using the method introduced in [11–14]. A segment of the machine as shown in Fig. 2 is simulated for simplicity with respect to the number of stator slots and winding wiring. In case a complete thermal view of the machine is required, the model should include the axial and tangential dimension as well [12, 15].

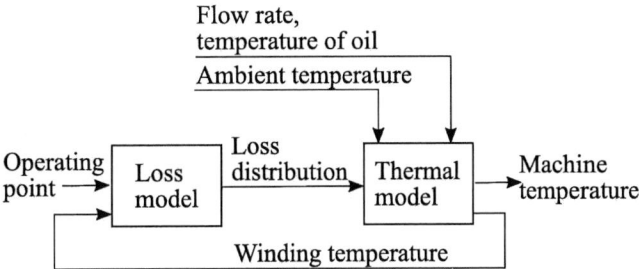

Fig. 16: Block diagram of the thermal simulation

The block diagram in Fig. 16 shows the essential steps in the simulation. The power loss distribution determined by the loss model, the oil temperature, oil flow rate and the ambient temperature are specified as inputs of the thermal model. The different oil temperatures in the machine are not explicitly modeled. Instead, the input oil temperature is applied as the only oil temperature in the model. The outputs are the temperatures of selected thermal nodes. It is worth mentioning that the oil behavior is affected not only by the heat exchange with the drive unit, but also by the rotational speed (see Fig. 12), considering the pressure change in the interior space in the drive unit. For this reason, the oil behavior is not modeled.

The power losses and their spatial distribution in the machine are studied with finite element analysis (FEA). Thereby, ideal sinusoidal currents are defined as machine inputs for simplicity. The copper losses include the dc losses and the variations considering the temperature dependent coil resistance. The sum of the losses is verified by the measured total losses. The distribution of iron losses is a challenge by the loss study. The iron losses in the stator and rotor can be derived from the FEA. However, the analytical determination of the loss ratio between the yoke and teeth is not straightforward. Therefore, this loss ratio is defined as a parameter and is optimized together with the thermal parameters in the model.

The direct oil cooling on the end winding is described by an empirical correlation of the convection heat transfer. The heat transfer coefficient h_{th} is calculated by

$$h_{\text{th}} = Nu\frac{k}{d_{\text{c}}}, \tag{1}$$

where d_{c} is the characteristic length, k is the thermal conductivity of the oil and Nu the dimensionless Nusselt number. Calculating Nu from the Prandtl number Pr and Reynolds number Re [12, 16] and determining the oil velocity v from the flow rate and the geometric dimensions of the oil nozzles, (1) can be rewritten using v and the oil temperature ϑ. To reduce the number of parameters in the model, only exponential terms are applied. The final formula becomes

$$h_{\text{th}} = K \cdot v^{\alpha} \cdot \vartheta^{\beta}, \tag{2}$$

where the correlation constant K and the exponents α and β are identified from measurements including torque settings at $75\,\mathrm{N\,m}, 125\,\mathrm{N\,m}$ and $140\,\mathrm{N\,m}$ and speed settings at $1500\,\mathrm{min}^{-1}, 2500\,\mathrm{min}^{-1}$ and $4000\,\mathrm{min}^{-1}$. As the stack length of the stator is small, the oil flow can enter the space between the stator and the housing. Therefore, the outer surface of the stator is assumed to be cooled down in a similar manner as the end winding. Its corresponding heat transfer coefficient is calculated by (2) as well. Performing the parameter identification from the measurements using oil A, the resulting parameters K, α and β are given in Table II.

Table II: Convective heat transfer correlation parameters

Sensing point	K	α	β
End winding	101	0.36	0.84
Stator outer	5.26	0.46	0.44

Validation and Analysis of Thermal Simulations

Applying the flow rate and temperature of the coolant as thermal model inputs, the simulation with a single parameter set is able to characterize the cooling performance applying differently viscous oils. As shown in Fig. 13a, at $125\,\mathrm{N\,m}$ and $4000\,\mathrm{min}^{-1}$, the simulated temperature profiles for the stator have a maximum error of $6.1\,\mathrm{K}$. The error is observed at the beginning of the measuring sequence, which can be related to the so-called sensor quality [13] and the collection of oil. In this work, the thermocouples are considered as ideal, neglecting their heat capacity. However, real thermocouples act like a low pass and show a filtered temperature profile. Besides, as the oil flows over the component surface, a small amount of oil remains there. This increases the entire thermal capacity of the sensor and the sensing target. Furthermore, the influence of those factors differs with respect to (a) the contact quality between sensors and sensing targets; (b) the roughness, shape and area of the component surface; (c) the dominance of thermal behavior of sensing targets compared to that of sensors. As shown in Fig. 13b, for the end winding, the influences of the sensor and oil remains are low, where the simulations match the measurements with nearly zero error irrespective of the measurement distortion.

The results (see Fig. 13-15) show that the improved cooling observed from experiments can be recognized from simulations as well. An overview of the simulation analysis is given in Table III. The calculated temperatures at the end winding are more accurate compared to that of the stator. At low-power operating points, the introduced modeling method is also capable of yielding the insignificant cooling improvements.

Table III: Improved cooling given by temperature reduction $\Delta\vartheta$ using oil B with respect to A

Operating point	Time point	$\Delta\vartheta$ on stator in K		$\Delta\vartheta$ on end winding in K	
		Measurement	Simulation	Measurement	Simulation
$125\,\mathrm{N\,m}, 4000\,\mathrm{min}^{-1}$	17.6 min	9.7	6.7	6	5.2
$75\,\mathrm{N\,m}, 4000\,\mathrm{min}^{-1}$	110 min	5.2	3.5	3	2.8
$75\,\mathrm{N\,m}, 2500\,\mathrm{min}^{-1}$	110 min	4.3	1.5	1.1	1

Furthermore, the simulation error is low with respect to measurements with both oils. The thermal model does not consider the exact oil properties, as their physical appearance is explicitly included. As the temperature changes, the temperature-dependent density, specific heat capacity and thermal conductivity of the oil are taken into account in the model by the term ϑ^β in (2). The oil flow rate, which is strongly affected by the viscosity, is included in the term v^α. Thus, considering the major difference between those two oils and taking the oil flow rate and oil temperature as model inputs, the value of α and β does not need to be recalculated for oil A and B. That is why a single parameter set can fit the measurements for the investigated two oils.

In addition to the aforementioned constant operating points, simulations with dynamic driving profiles are also conducted. The plots in Fig. 17 show good simulation results even when changing the operating point. With oil A as the cooling fluid, the speed remains at $4000\,\text{min}^{-1}$ while the torque follows the given profile. The simulated temperatures of the stator and end winding can represent the measurements well. High deviations of up to $6.5\,\text{K}$ are only observed at the beginning and after the torque steps.

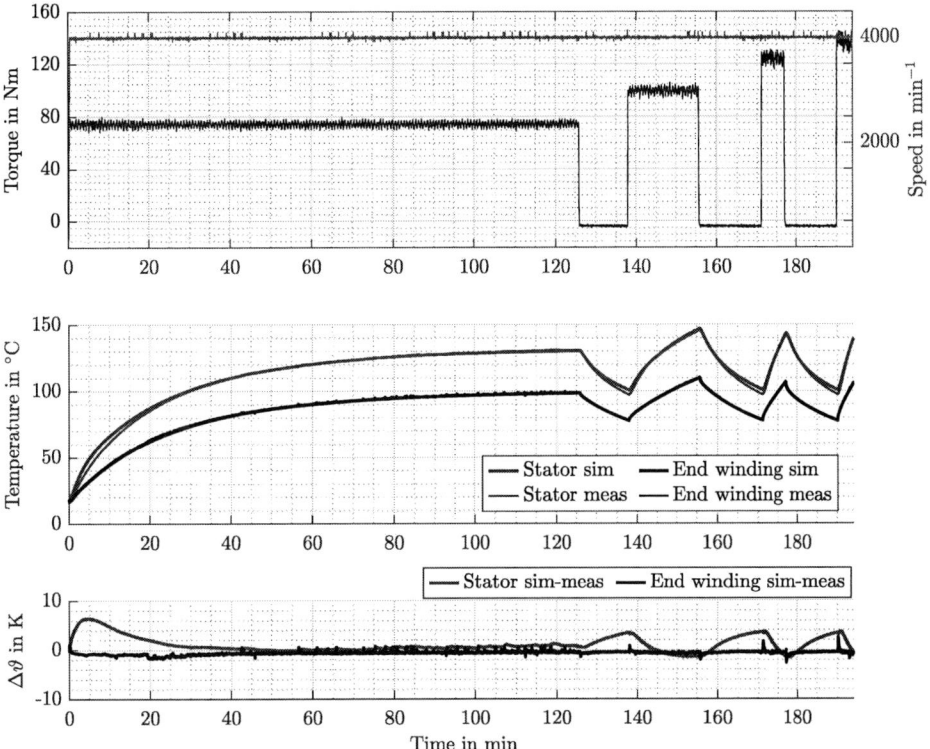

Fig. 17: Model validation at $4000\,\text{min}^{-1}$ and varying torque

It is worth mentioning that the cooling performance results from an interaction of different aspects, such as influences of the oil viscosity on the transmission losses, the power losses of the machine, the torque compensation in dependence of thermal conditions, etc., which can only be studied by multiphysics simulations.

Summary

In this work, the thermal performance of a directly oil-cooled IPMSM integrated in the transmission housing is studied. Lower temperatures and a more homogeneous temperature distribution are noticed on the transmission side. With the low viscous oil, an improved cooling effect is observed, especially on the cooling targets. A lumped-parameter thermal model is developed including the convective heat transfer for the oil cooling based on empirical formulas. Applying the oil flow rate and oil temperature as inputs, the thermal model of the electrical machine and the oil performance can be individually studied. As a result, the flow rate and temperature are recognized as the most relevant parameters of the coolant with respect to cooling performance. This outcome can help to modify the oil properties considering the cooling capability in addition to the lubrication of transmissions. In the future work, thermal modeling of cooling circuits can be conducted considering the oil properties and the drive unit operation to predict the thermal performance of the system completely from simulations.

References

[1] H. Meinert, T. Senger, N. Wiebking, and C. Diegelmann, "Die plug-in-hybridtechnologie im neuen BMW x5 eDrive," *MTZ Motortechnische Zeitschrift*, vol. 76, pp. 16–21, mar 2015.

[2] G. Mühlberg, W. Hackmann, and K. Buzziol, "Highly integrated electric powertrain," *ATZelektronik worldwide*, vol. 12, pp. 42–45, aug 2017.

[3] M. Erriquez, T. Morel, P.-Y. Moulière, and P. Schäfer, "Trends in electric-vehicle design," tech. rep., McKinsey & Co., McKinsey Center for Future Mobility, Oct. 2017.

[4] R. Sindjui, G. Zito, and S. Zhang, "Experimental study of systems and oils for direct cooling of electrical machine," *Journal of Thermal Science and Engineering Applications*, vol. 14, aug 2021.

[5] T. Davin, J. Pellé, S. Harmand, and R. Yu, "Experimental study of oil cooling systems for electric motors," *Applied thermal Engineering*, vol. 75, pp. 1–13, jan 2015.

[6] B. Assaad, K. Mikati, T. Tran, and E. Negre, "Experimental study of oil cooled induction motor for hybrid and electric vehicles," in *2018 XIII International Conference on Electrical Machines (ICEM)*, IEEE, sep 2018.

[7] C. Liu, Z. Xu, D. Gerada, J. Li, C. Gerada, Y. C. Chong, M. Popescu, J. Goss, D. Staton, and H. Zhang, "Experimental investigation on oil spray cooling with hairpin windings," *IEEE Transactions on Industrial Electronics*, vol. 67, pp. 7343–7353, sep 2020.

[8] F. Momen, K. M. Rahman, Y. Son, and P. Savagian, "Electric motor design of general motors' chevrolet bolt electric vehicle," *SAE International Journal of Alternative Powertrains*, vol. 5, pp. 286–293, apr 2016.

[9] J. Liu, M. Anwar, P. Chiang, S. Hawkins, Y. Jeong, F. Momen, S. Poulos, and S. Song, "Design of the chevrolet bolt EV propulsion system," *SAE International Journal of Alternative Powertrains*, vol. 5, pp. 79–86, apr 2016.

[10] T. Sebastian, "Temperature effects on torque production and efficiency of PM motors using NdFeB magnets," *IEEE Transactions on Industry Applications*, vol. 31, no. 2, pp. 353–357, 1995.

[11] P. Mellor, D. Roberts, and D. Turner, "Lumped parameter thermal model for electrical machines of TEFC design," *IEE Proceedings B (Electric Power Applications)*, vol. 138, no. 5, p. 205, 1991.

[12] S. Nategh, Z. Huang, A. Krings, O. Wallmark, and M. Leksell, "Thermal modeling of directly cooled electric machines using lumped parameter and limited CFD analysis," *IEEE Transactions on Energy Conversion*, vol. 28, no. 4, pp. 979–990, 2013.

[13] F. Qi, *Online model-predictive thermal management of inverter-fed electrical machines.* Dissertation, RWTH Aachen University, 2019.

[14] O. Wallscheid, "Thermal monitoring of electric motors: State-of-the-art review and future challenges," *IEEE Open Journal of Industry Applications*, vol. 2, pp. 204–223, 2021.

[15] H. Xu, K. Lin, C. Ehrenpreis, G. Roux, and R. W. D. Doncker, "Thermal modeling of electrical machines with advanced fluid cooling," in *2020 19th IEEE Intersociety Conference on Thermal and Thermomechanical Phenomena in Electronic Systems (ITherm)*, IEEE, jul 2020.

[16] C. Ehrenpreis, H. E. Bahi, H. Xu, G. Roux, R. Kneer, and W. Rohlfs, "Physically-motivated figure of merit (FOM) assessing the cooling performance of fluids suitable for the direct cooling of electrical components," in *2020 19th IEEE Intersociety Conference on Thermal and Thermomechanical Phenomena in Electronic Systems (ITherm)*, IEEE, jul 2020.

Development of A Family of High Voltage Gain Step-Up Multi-Port DC-DC Converters for Fuel Cell-based Hybrid Vehicular Power Systems

Pouya Zolfi, Sina Vahid, Ayman EL-Refaie
Werner's Sustainable Energy Lab - MARQUETTE UNIVERSITY
Engineering Hall (EH261), 1250 W Wisconsin Ave, 53233
Milwaukee, WI, USA
Tel.: +1 (414) 288-1940
E-Mail: pouya.zolfi@marquette.edu

Keywords

«DC-DC power converter», «Electric vehicle», «Energy storage», «Fuel Cell», «Power management»

Abstract

Battery assisted fuel cell based vehicular power systems are feasible solutions for transportation electrification. Step up DC-DC converters play an important role in energy conversion process of these systems. Due to uncertain nature and low voltage level of fuel cells, large voltage gain converters are desired. This paper proposes a family of non-isolated step-up DC-DC multi-port converters for hybrid FC + battery vehicular power systems with smart grid services capability. Fewer active and passive components compared to other step-up topologies in the literature, extendibility, simple power flow management, and low current and voltage ripples are among the benefits of this multi-port converter family. Design and control of these converters are discussed, and the simulations are conducted for one of the proposed topologies. In the simulation scenario, a 60-kW FC + battery vehicle system is used as the baseline for real-world application. The simulation results demonstrate the effectiveness of this step-up multi-port converter and the peak efficiency of 95.6% was recorded. Also, a lab-scale prototype of the converter is built and tested under various power flow scenarios to validate the simulation results.

Introduction

Step-up DC-DC converters play an important role in energy conversion process of renewable energy systems (RESs) and energy storage systems (ESSs). Due to low voltage levels of the mentioned systems, high voltage gain converters are required to reach system level voltages. Furthermore, low-ripple input current should be provided to increase power quality on both generation and consumption sides. High step-up DC-DC topologies are divided into two main categories: isolated and non-isolated. The main demerit of isolated high step-up converters is their transformers' high weight and volume [1]. Non-isolated high step-up converters are mostly based on coupled inductor technologies. By adjusting turn ratio of the coupled inductor and implementation of switched capacitor technique, these converters can achieve high voltage gains [2]. Effects of the leakage inductance on power switches' stress, reverse recovery problem of diodes, and high volume and cost of the converter are regarded as the main drawbacks of the coupled inductor-based high gain step-up topologies [3]. In [4], a quadratic coupled inductor-based high gain step-up DC-DC converter is proposed for RESs applications. High number of circuit components is the main demerits of this circuit. In [4], an extendable non-isolated ultra-high step-up topology is discussed for renewable energy applications. High number of components in [5] has led to a bulk and complex structure. Other non-isolated topologies are presented in [6-8], which suffer from one or more of the mentioned issues. The other type of the non-isolated high step-up DC-DC converters is based on cascading basic topologies and switched capacitor/switched inductor techniques. Large gain step-up non-isolated converters based on geometric structures are discussed in [9].

Benefiting from fewer active and passive components, multi-port converters (MPCs) are considered an effective solution for the integration of RESs and ESSs and can similarly be categorized into isolated and non-isolated types. In [10, 11], two new isolated topologies for three-port DC-DC converters (TPCs) are proposed for DC microgrid application but has high number of power switches. A new MPC based on push-pull converter is introduced in [12] for the auxiliary power unit of refrigerated vehicles with assistance of three inductors (cores) and four power switches which leads to a bulk topology in high power levels. A non-isolated coupled inductor-based TPC is presented in [13] for fuel-cell electric vehicle (FCEV) applications. Apart from the coupled inductor-initiated problems, this converter is not flexible and/or extendable and the energy management scheme is not discussed. Another non-isolated MPC is proposed by authors in [14] for hybrid streetcar application. The comprehensive energy management scheme is introduced, but the converter suffers from high current ripple in at least one of the input ports. In order to develop MPCs, this study uses systematic design approaches presented in [15, 16] which is suitable for a wide range of applications.

This paper presents a family of low switch count non-isolated continuous input current step-up DC-DC converters for single-input single-output (SISO) and multi-input multi-output (MIMO) systems. Although the major focus of this paper is on FC-based e-mobility, the proposed MPC is capable of being used in a wide range of applications. A battery-assisted hybrid vehicular FC power system is considered, with mileage extension and smart grid services. The baseline and MPC-based FC + battery systems are shown in Fig. 1(a) and (b), respectively. Proposed topologies provide high voltage gains with proper duty cycle values in addition to low ripple input current/output voltage and high efficiency values. Simulation results using PLECS® software are validated with the lab-scale converter prototype.

The rest of the paper is organized as follows: The development process of the proposed converters is discussed in the upcoming section. Integration scenarios and energy management scheme are discussed next. Simulation and experimental verifications and finally the conclusions are presented.

Development and Analysis of The Converter Topologies

Fundamental Step-up Converter Topology (Single Input Single Output - SISO)

The fundamental high gain step-up converter is a single switch topology that consists of two inductors, four capacitors and five diodes as shown in Fig.2. L_1 is the input side inductor and guarantees the continuity of input current that is regarded as an important factor for renewable energy integration. This topology is derived using boost and voltage multiplier building blocks (BBs) and presents a wide range voltage conversion ratio with proper switching duty cycle values. C_4 is considered as the output side capacitor and provides nearly-zero ripple voltage for load terminals. Output side diode D_5 also prevents

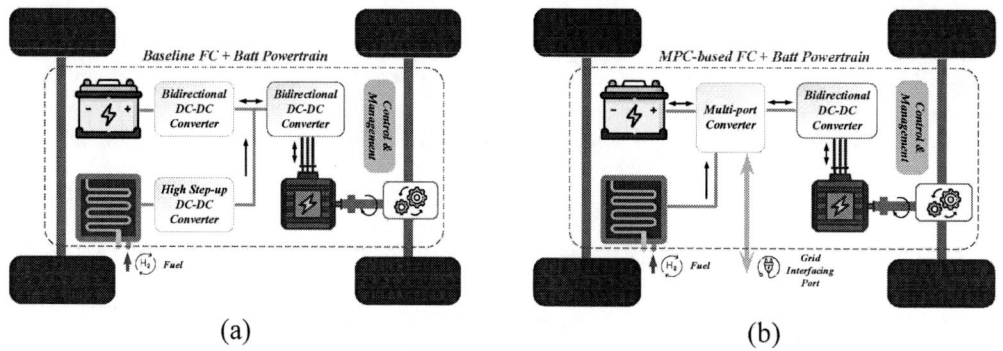

Fig. 1: A battery-assisted hybrid vehicular FC power system (a) baseline design and (b) MPC-based

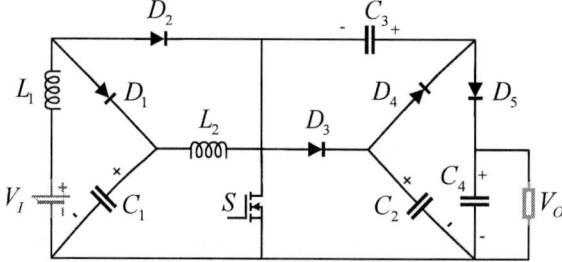

Fig. 2: Baseline fundamental SISO topology

unwanted power flow from load side to the input side in the case of active load. To simplify the analysis, following assumptions are considered:

- Capacitors C_1, C_2, C_3, and C_4 are large and their voltages are considered to be constant during one switching period, and
- ESR resistances of capacitors and inductors are neglected in steady state analysis but are considered for efficiency studies.

The operating modes of the high step-up SISO converter and the related waveforms during continuous conduction mode (CCM) are shown in Fig. 3. The converter has two operating modes in CCM which are discussed as follows:

Mode 1 [$0 \leq t < DT_S$] This mode starts by turning on the power MOSFET. L_1 is charged by V_I through D_2 and S. C_1 that has been charged from previous switching period is discharged to L_2 through S. Furthermore, C_2 is discharged through D_4 to charge C_3. D_1, D_3, and D_5 are reverse biased in this mode and output capacitor (C_4) feeds the load. This mode ends when the power MOSFET is turned off. Following equations are derived for this operation mode:

$$V_{L1} = V_I \tag{1}$$
$$V_{L2} = V_{C1} \tag{2}$$
$$V_{C2} = V_{C3} \tag{3}$$

Mode 2 [$DT_S \leq t < T_S$] In this mode, the power MOSFET is off and D_1 and D_3 are forward biased. C_1 is charged by V_I and L_1 through D_1. L_2 charges C_2 through D_3. D_5 is conducting in this mode and provides a path from C_3 to output for charging C_4 and feed the load side. This mode continues until power MOSFET is switched on. Equations related to this mode are presented as follows:

$$V_{L1} = V_I - V_{C1} \tag{4}$$
$$V_{L2} = V_{C1} - V_{C2} \tag{5}$$
$$V_{C2} = V_O - V_{C3} \tag{6}$$

By applying volt-seconds balance principle on the inductors, following equations are derived which leads to the CCM voltage gain equation of the SISO topology.

$$DV_I + (1-D)(V_I - V_{C1}) = 0 \quad \rightarrow \quad V_{C1} = \frac{V_I}{1-D} \tag{7}$$

$$DV_{C1} + (1-D)(V_{C1} - V_{C2}) = 0 \quad \rightarrow \quad V_{C2} = \frac{V_{C1}}{1-D} = \frac{V_I}{(1-D)^2} \tag{8}$$

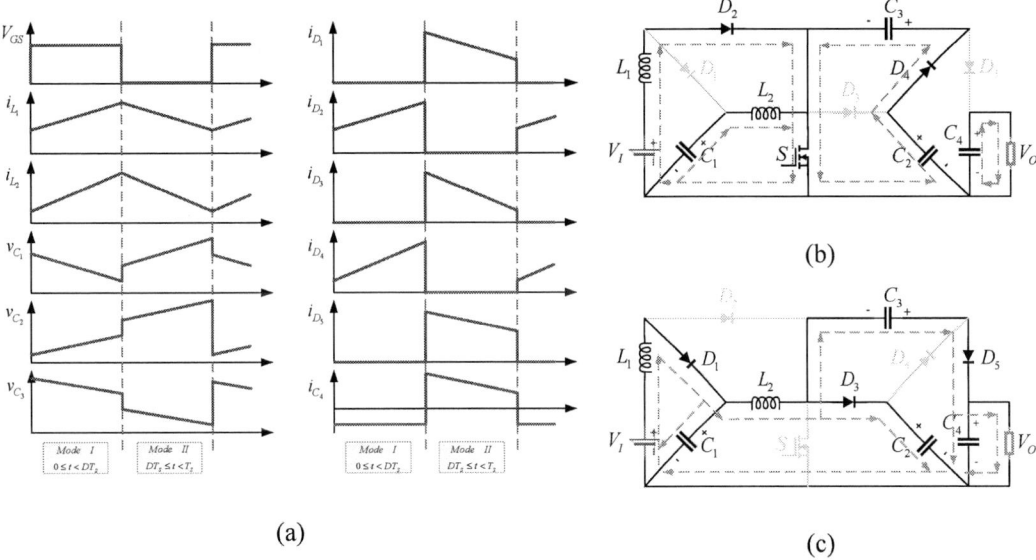

Fig. 3: (a) CCM waveforms, (b) mode I and (c) mode II for the SISO topology during CCM

$$\begin{cases} V_{C3} = V_{C2} \\ V_O = V_{C2} + V_{C3} \end{cases} \rightarrow M_{CCM} = \frac{V_O}{V_I} = \frac{2}{(1-D)^2} \tag{9}$$

Similar analysis can be provided for discontinuous conduction mode (DCM) operation of the SISO converter. The converter has one additional operating mode during DCM which is when all semiconductor switching components are off, i.e. all diodes are reverse biased in this mode and power MOSFET is also in off state. Load is fed only by output capacitor C_4, that has been charged from previous operating mode. This mode ends when the power MOSFET is turned on in the next switching period. D' is considered as the time interval that inductors' currents decrease from their maximum value to zero. By applying volt-seconds balance principle on the inductors, following equations are derived:

$$DV_I + D'(V_I - V_{C1}) = 0 \;\rightarrow\; V_{C1} = (1 + \frac{D}{D'})V_I \tag{10}$$

$$DV_{C1} + D'(V_{C1} - V_{C2}) = 0 \;\rightarrow\; V_{C2} = (1 + \frac{D}{D'})^2 V_I \tag{11}$$

$$\begin{cases} V_{C3} = V_{C2} \\ V_O = V_{C2} + V_{C3} \end{cases} \rightarrow M_{DCM} = \frac{V_O}{V_I} = 2(1 + \frac{D}{D'})^2 \tag{12}$$

Average currents of the capacitors in one switching period are zero. Also, average current of the diode D_5 is equal to the average output (load) current, hence:

$$I_O = \frac{1}{2} D' I_{D5} \tag{13}$$

$$I_O = \frac{V_O}{R_L} \tag{14}$$

$$I_{D5} = \frac{V_I}{L_T f_S} \tag{15}$$

$$L_T = L_1 + L_2 \tag{16}$$

By replacing (14) in (13) and then (13) in (15), respectively, following equations are obtained:

$$D' = M_{DCM} \frac{2 L_T f_S}{D R_L} \tag{17}$$

$$\tau = \frac{L_T f_S}{R_L} \tag{18}$$

Where τ is the normalized time constant of the inductor. By replacing (17) in (12), simplification and using (18), a new form of voltage gain relation during DCM is derived as follows:

$$M_{DCM} = \sqrt{1 + 4\frac{D^2}{\tau}} \tag{19}$$

Boundary conduction mode (BCM) is the common mode between CCM and DCM. Therefore, by equalizing (9) and (19), boundary normalized time constant of the inductor, τ_B is obtained as follows.

$$M_{CCM} = M_{DCM} \Rightarrow \tau_B = \frac{(1-D)^4 . 4D^2}{4 - (1-D)^4} \tag{20}$$

τ_B curve in terms of duty cycle is presented in Fig.4 for the conventional boost converter, converter presented in [3] (as an example of modern single-switch high step-up converters) and the baseline SISO converter to evaluate the ability of converters for operation in CCM. Suggested converter would operate in CCM – that is favorable – for $\tau > \tau_B$. This figure shows that the high step-up SISO converter operates in CCM for a broader range of duty cycle values in comparison with other topologies.

The values of inductors and capacitors are calculated based on steady-state analysis to guarantee CCM operation and ripple requirements of the SISO topology. Using (18) and (20), following equation for equivalent inductance is derived. During the design process, L_2 is obtained based on the load requirements, as presented in (22). Then L_1 can be calculated based on (16).

$$L_T \geq \frac{R_L}{f_S} \frac{4D^2(1-D)^4}{4-(1-D)^4} \tag{21}$$

$$L_2 \geq \frac{R_L}{f_S} \frac{D(1-D)^2}{4} \tag{22}$$

Output side capacitor value for CCM operation is obtained by following equations:

$$V_{C_4}(DT_S) = V_{C_4}(0) + \frac{1}{C_4} \int_0^{DT_S} i_{C_4}(t).dt \Rightarrow \Delta V_{C_4} = \frac{DV_O}{C_4 R_L f_S} \Rightarrow C_4 \geq \frac{DV_O}{\Delta V_{C_4} R_L f_S} \tag{23}$$

Similarly, by deriving current equations for the capacitors C_1, C_2, and C_3 the capacitor values can be calculated as shown in (24) and (25).

$$C_1 \geq \frac{2DV_O}{\Delta V_{C_1}(1-D)(1-2D)R_L f_S} \tag{24}$$

$$C_{2\&3} \geq \frac{V_O}{\Delta V_{C_{2\&3}}(1-2D)R_L f_S} \tag{25}$$

Discussed SISO topology provides higher voltage gain for a specific duty cycle value in comparison to similar high step-up topologies. Table I shows a thorough comparison between suggested converter and similar converters which are introduced in literature. Fig.5 provides gain curves for the mentioned topologies. This figure shows that for $D \geq 0.5$, voltage gain of the discussed topology is more outstanding in comparison with other structures.

Step-up Multi-port Converter Topologies (Multiple Input Multiple Output – MIMO)

A family of step-up MPCs can be developed based on the SISO topology. Fig. 6(a) illustrates a unidirectional MIMO topology with only one power switch. This topology is capable of utilizing two input ports and two output ports to integrate RESs. Power flow is unidirectional in the abovementioned converter which makes it more suitable for low- to medium-power renewable energy systems. By replacing one of the power diodes with the combination of diode-power MOSFET, bidirectional power flow can be achieved. Fig. 6(b) shows the bidirectional MIMO MPC, which can be used in ESS applications and is the main topology considered in this study for the battery-assisted hybrid vehicular FC power system. As discussed before, the proposed converter is developed based on the fundamental DC-DC BBs, thus, this family of step-up MPCs is capable of being extended by adding/replacing the BBs. The possible operating modes and related voltage gain relationships for the proposed bidirectional

Table I: Comparison between the SISO topology and other similar converters

Converter Parameter	Boost	SISO Converter	[3]	[17]	[18]	[19]	[20]
CCM Gain	$\frac{1}{1-D}$	$\frac{2}{(1-D)^2}$	$\frac{3+D}{2(1-D)}$	$\frac{2+n}{1-D};n=2$	$\frac{3+D}{1-D}$	$\frac{1+D}{1-D}$	$\frac{n}{D(1-D)};n=2$
# Inductor core	1	2	2	2	2	2	2
# Diode	1	5	4	3	5	4	4
# Switch	1	1	1	1	1	1	2
# Capacitor	1	4	4	4	4	1	4
Maximum Input Current Ripple %	25	20	22.5	20	200	75	20
Peak η %	93.4	95.6	93.1	97	95	83	97

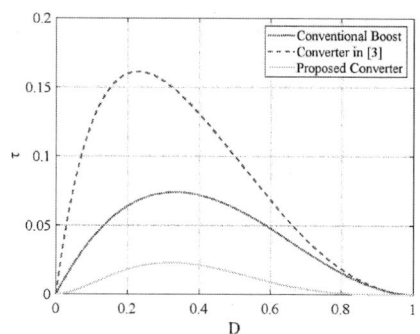

Fig. 4: Curve for the normalized time constant of inductor in terms of duty cycle values

Fig. 5: Voltage gain comparison graphs

MIMO topology are presented in Fig.7 and Table II, respectively.

Integration Scenarios and Energy Management Scheme

A variety of source-load integration modes are possible by implementing the proposed MIMO topology. Seven modes of integration are defined for the MIMO topology in a hybrid FC + battery vehicular power

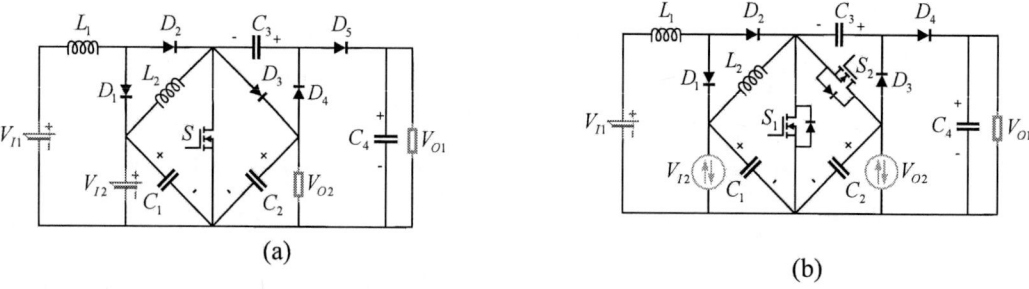

Fig. 6: MIMO topologies (a) unidirectional, and (b) bidirectional

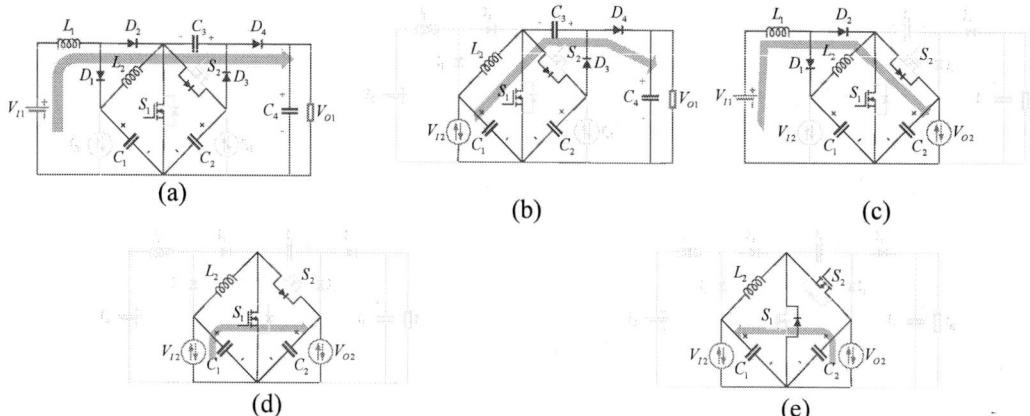

Fig. 7: Operating modes for the bidirectional MIMO topology

Table II: Voltage gain equations for the bidirectional MIMO topology

Mode	M_a	M_b	M_c	M_d	M_e
Voltage Gain	$\dfrac{V_{O_1}}{V_{I_1}} = \dfrac{2}{(1-D)^2}$	$\dfrac{V_{O_1}}{V_{I_2}} = \dfrac{2}{1-D}$	$\dfrac{V_{O_2}}{V_{I_1}} = \dfrac{1}{(1-D)^2}$	$\dfrac{V_{O_2}}{V_{I_2}} = \dfrac{1}{1-D}$	$\dfrac{V_{I_2}}{V_{O_2}} = D$

system. Fig. 8(a) presents the system configuration implementing the bidirectional MIMO converter and Fig. 8(b) summarizes the possible modes among the main ports and the reconfigurable converter types.

Scenario 1 [FC to Motor]: This is the main operating scenario and occurs when the FC tank is full. As shown in Fig. 7(a), switch S_1 is the only active power switch during this mode and the proposed MIMO acts as a high step-up converter with full voltage gain utilization.

Scenario 2 [FC to Battery]: In the no load condition, when the FC has the capability of providing excessive power and the battery is fully depleted, this scenario occurs. The battery is charged by C_1 through L_1 and D_1.

Scenario 3 [FC to Grid]: This mode has the lowest priority and occurs only in the no load condition where the battery is fully charged, and grid demands power, i.e. vehicle to grid (V2G) services. This scenario is shown in Fig. 7(c). The electronic contactor is in position A during this mode.

Scenario 4 [Battery to Motor]: When the FC is not available, the backup battery provides power to the load side. As shown in Fig. 7(b), S_1 is the active switch during this mode. It has to be mentioned that the backup battery can feed the load for a limited period of time due to the small capacity.

Scenario 5 [Motor to Battery]: Bidirectional port of the MIMO converter (V_{O2}) is utilized to charge the back-up battery pack during regenerative braking (electronic contactor in position B). This mode is shown in Fig. 7(e), where S_2 is the active switch and converter behaves as a step-down topology.

Scenario 6 [Grid to Battery]: The proposed bidirectional MIMO topology is capable of providing grid to vehicle (G2V) services. The backup battery can be charged by the grid interface connection. This mode is similar to the previous scenario, except here the electronic contactor in position A.

Scenario 7 [Battery to Grid]: This scenario also addresses the G2V service providing capability of the proposed converter. This mode is shown in Fig. 7(d), where the battery is connected to the grid interface through a boost building block via the electronic contactor in position A.

Energy management scheme for the MIMO converter-based FC + battery vehicular power system is presented in Fig. 9. At the first stage, the motor status is checked to determine its operating mode. Next stage determines the availability of the FC stacks. If available, FC is the main power source and is preferred over the other resources. In the third stage, state of charge (SOC) of the battery is obtained by

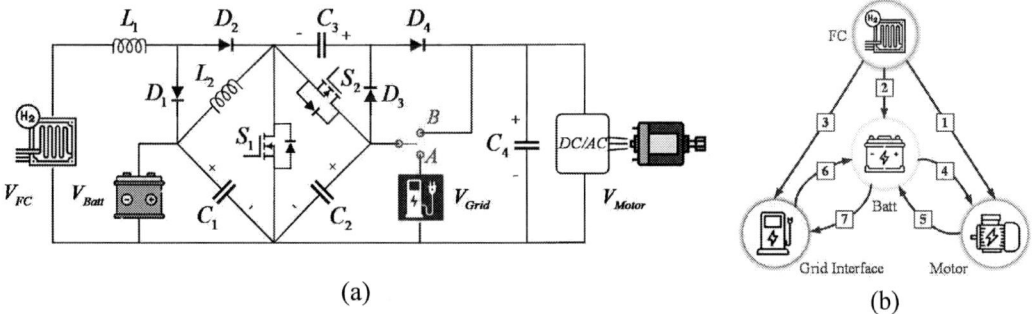

(a)   (b)

Fig. 8: (a) Proposed converter for FC-based vehicular powertrain (b) Integration scenarios

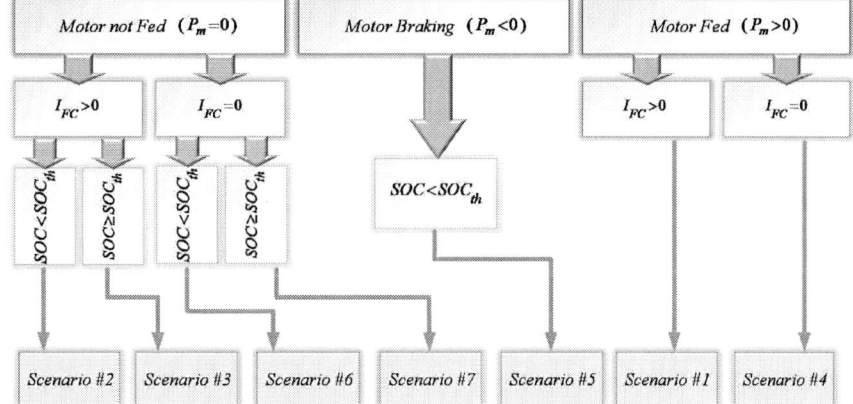

Fig. 9: Decision chart for the proposed MIMO-based power system

its current and voltage, and then the SOC is given to the control unit to make the final decision and determine the scenario of operation. The battery is the emergency power provider. Charging and discharging processes are defined and the suitable control decision can be made by a comparison between the battery's SOC and the threshold SOC (SOC_{th}).

Simulation and Experimental Validation

Simulation of the proposed converter using PLECS® software demonstrates the effectiveness of the theoretical analysis. Simulations are conducted for a 60-kW system which is typical for a medium size FCEV. The switching frequency is 50kHz in this simulation. The FC stacks and motor side DC link voltages are considered to be 100 V and 800 V, respectively. A 200 V – 60 kWh battery pack is modeled for the FCEV system. Grid interface port voltage is 400 V which is selected based on the voltage level standards in DC distribution networks [21]. Fig. 10(a) and 10(b) present the voltage and current waveforms for the FC and motor side DC link, respectively, with 50% duty cycle. Proposed converter is capable of providing nearly-zero input/output current/voltage ripples which increases FC stack lifetime [22] and simplifies the motor speed control procedure. High frequency current ripple leads to sharp increase in the high-frequency resistance (HFR) which is identified as a severe degrading condition for FC stacks. A maximum efficiency of 95.6% is recorded for the simulations (scenario 7 shown in Fig. 7(d)) considering the thermal models of semiconductors and ESR resistances of the passive components. Loss distribution among various components during this mode is shown in Fig. 11.

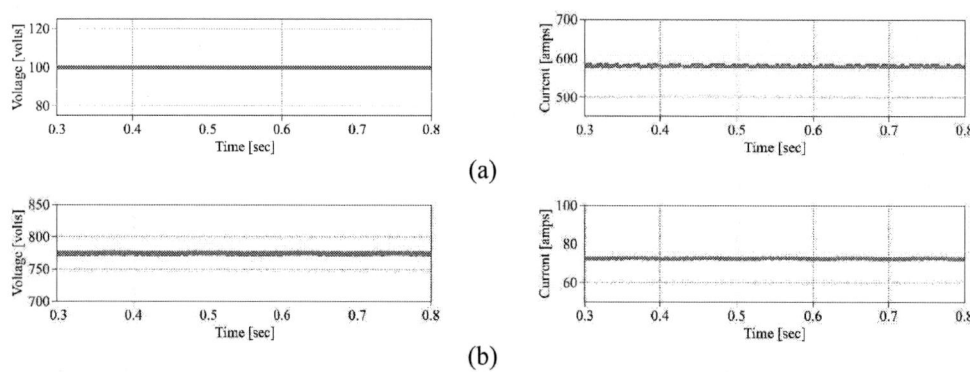

Fig. 10: Simulation results for (a) FC and (b) Motor side

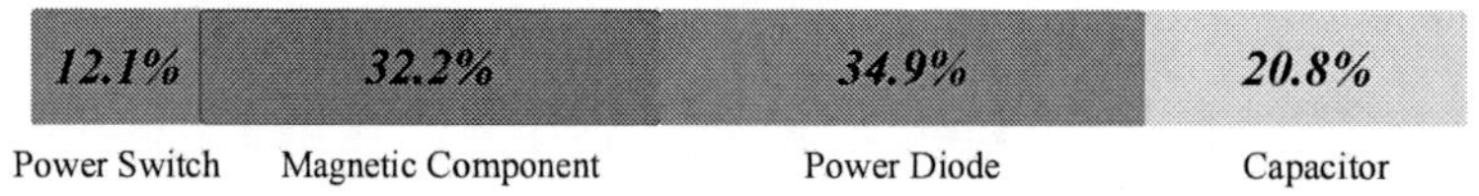

12.1%	32.2%	34.9%	20.8%
Power Switch	Magnetic Component	Power Diode	Capacitor

Fig. 11: Loss distribution diagram for the peak efficiency case (scenario 7)

Fig. 12: 1-kW prototype of the bidirectional MIMO

Table II: Prototype Specifications

Spec/ Component	Value/ Type
V_{I1}, V_{I2}	30 V, 60 V
V_{O1}, V_{O2}	240 V, 120 V
L_1, L_2	174 μH, 311 μH
C_1	200 V – 100 μF
C_2, C_3	400 V – 680 μF
C_4	600 V – 220 μF
D_1, D_2	V30202C-M3/4W
D_{S1}, D_{S2}, D_3	SBR40U300CT
D_4	SBR40U300CT-G
S_1	IXFH120N30X3
S_2	IXFQ72N30X3

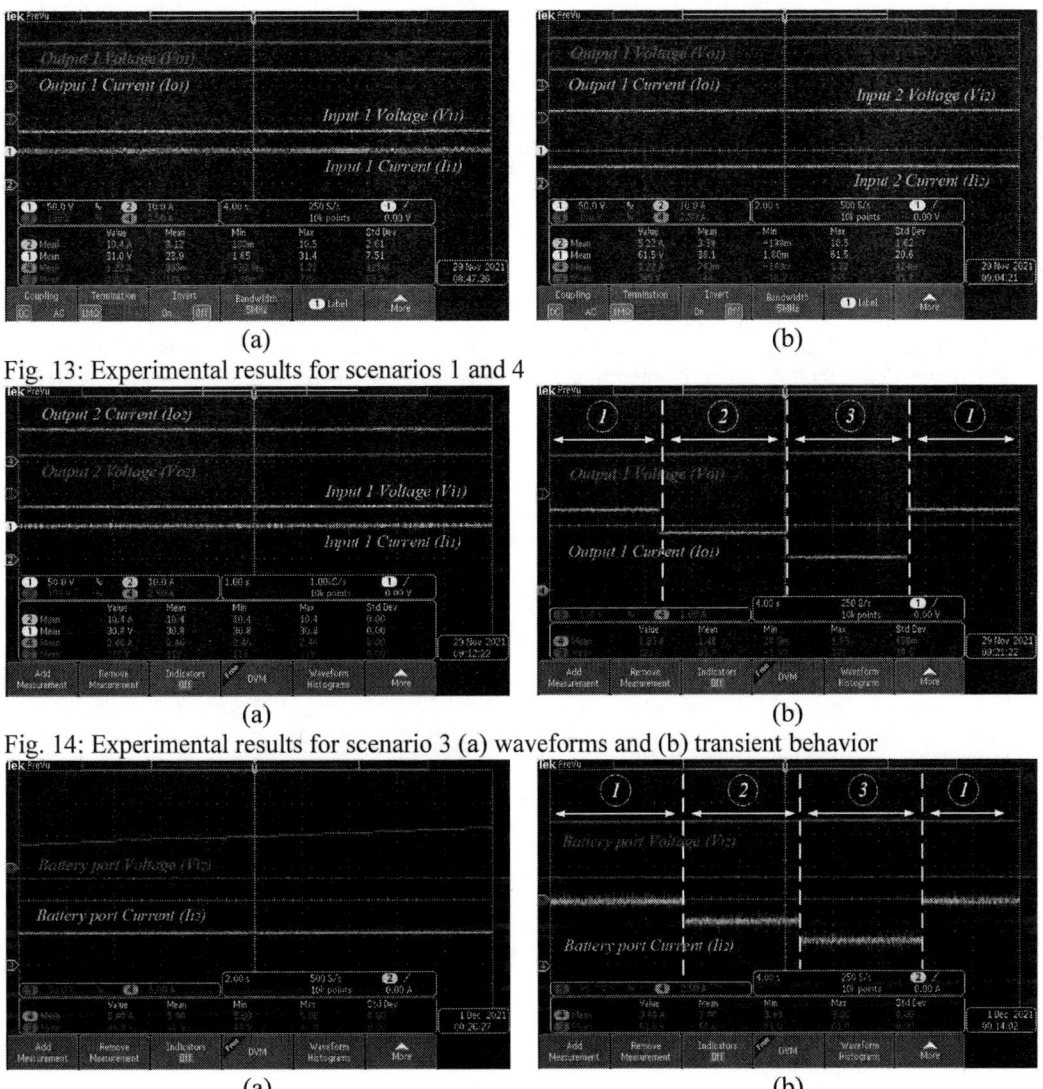

Fig. 13: Experimental results for scenarios 1 and 4

Fig. 14: Experimental results for scenario 3 (a) waveforms and (b) transient behavior

Fig. 15: Experimental results for battery charging (a) CC and (b) CV scenarios

A lab-scale prototype of the proposed converter is built and tested for various integration scenarios and is shown in Fig. 12. The implemented converter is tested for 300 W power level (Peak efficiency of 96.7%). Table III provides the detailed specification of the developed converter prototype. Fig.13(a) and 13(b) present the experimental results for scenarios no.1 (FC to motor; based on the diagram presented in Fig. 8(b)) and no.4 (battery to motor), respectively. For FC to grid-side scenario, experimental voltage and current waveforms are recorded and presented in Fig.14(a), while the transient behavior of the proposed MPC for a variable current case is presented in Fig.14(b). The converter is capable of supplying the load with a constant low-ripple voltage which improves the power quality of the grid interaction. Fig.15(a) shows the results for a constant current (CC) battery charging scenario (integration scenario no.6), where the battery voltage increases gradually as its SOC increases. On the other hand, the constant voltage (CV) case scenario is also tested and is shown in Fig.15(b).

Conclusion

In this paper, a family of high voltage gain step-up DC-DC MPCs is introduced, briefly analyzed, simulated and tested. SISO, unidirectional MIMO, and bidirectional MIMO extensions are discussed. Proposed topologies have lower number of power switches in comparison to the similar structures in the literature. Furthermore, high simulation efficiency (95.6% peak value), high voltage gain, nearly-zero voltage and current ripple in both input and output ports, extendable architecture for a wide range of applications, and a simple management scheme are regarded as the main advantages of the proposed converters. Bidirectional MIMO topology was considered as the main interface in a battery-assisted FCEV powertrain application and respective simulations performed for a 60-kW system which indicate the effectiveness of the proposed converter. A lab scale 300 W prototype is implemented to verify the operation of the proposed power converter. Peak efficiency of 96.7% is recorded for this low-power prototype. Several integration scenarios are tested, and results are compatible with the simulation and analytical analysis.

References

[1] S. S. Sayed and A. M. Massoud, "Review on State-of-the-Art Unidirectional Non-Isolated Power Factor Correction Converters for Short-/Long-Distance Electric Vehicles," in IEEE Access, vol. 10, pp. 11308-11340, 2022, doi: 10.1109/ACCESS.2022.3146410.

[2] F. Barmoudeh, P. Zolfi, S. Karami and A. Ajami, "A Novel High Power High Step up DC-DC Converter with Parallel Structure", *order*, vol. 2, pp. D3.

[3] S. Saravanan and N. R. Babu, "Design and Development of Single Switch High Step-Up DC–DC Converter," in IEEE Journal of Emerging and Selected Topics in Power Electronics, vol. 6, no. 2, pp. 855-863, June 2018, doi: 10.1109/JESTPE.2017.2739819.

[4] S. A. Modaberi, B. Allahverdinejad and M. R. Banaei, "A Quadratic High Step-up DC-DC Boost Converter Based on Coupled inductor with Single Switch and Continuous Input Current," 2021 12th Power Electronics, Drive Systems, and Technologies Conference (PEDSTC), 2021, pp. 1-6, doi: 10.1109/PEDSTC52094.2021.9405958.

[5] P. Mohseni, S. Rahimpour, M. Dezhbord, M. R. Islam and K. M. Muttaqi, "An Optimal Structure for High Step-Up Non-Isolated DC-DC Converters with Soft-Switching Capability and Zero Input Current Ripple," in IEEE Transactions on Industrial Electronics, doi: 10.1109/TIE.2021.3080202.

[6] X. Zhu, B. Zhang and K. Jin, "Hybrid Nonisolated Active Quasi-Switched DC-DC Converter for High Step-up Voltage Conversion Applications," in IEEE Access, vol. 8, pp. 222584-222598, 2020, doi: 10.1109/ACCESS.2020.3043816.

[7] S. -W. Seo, J. -H. Ryu, Y. Kim and H. H. Choi, "Non-Isolated High Step-Up DC/DC Converter With Coupled Inductor and Switched Capacitor," in IEEE Access, vol. 8, pp. 217108-217122, 2020, doi: 10.1109/ACCESS.2020.3041738.

[8] S. Seo, D. Lim and H. H. Choi, "High Step-Up Interleaved Converter Mixed With Magnetic Coupling and Voltage Lift," in IEEE Access, vol. 8, pp. 72768-72780, 2020, doi: 10.1109/ACCESS.2020.2983757.

[9] K. Li, Y. Hu and A. Ioinovici, "Generation of the Large DC Gain Step-Up Nonisolated Converters in Conjunction With Renewable Energy Sources Starting From a Proposed Geometric Structure," in IEEE Transactions on Power Electronics, vol. 32, no. 7, pp. 5323-5340, July 2017, doi: 10.1109/TPEL.2016.2609501.

[10] P. Zolfi, S. Vahid and A. EL-Refaie, "A Novel Three-Port DC-DC Converter for Integration of PV and Storage in Zonal DC Microgrids," 2021 22nd IEEE International Conference on Industrial Technology (ICIT), 2021, pp. 285-291, doi: 10.1109/ICIT46573.2021.9453479.

[11] P. Zolfi, A. Ajami and V. Behjat, "An Isolated Three Port DC-DC Converter for Energy Management in Zonal DC Microgrids", *The 33rd International Power System Conference (PSC2018)*, October 2018.

[12] S. Vahid, P. Zolfi, J. Land and A. EL-Refaie, "An Isolated Step-Down Multi-Port DC-DC Power Converter for Electric Refrigerated Vehicles Auxiliary Power Unit System," 2022 IEEE Applied Power Electronics Conference and Exposition (APEC), 2022, pp. 1133-1140, doi: 10.1109/APEC43599.2022.9773710.

[13] P. Zolfi, A. Ajami, " A Novel Three Port DC-DC Converter for Fuel Cell based Electric Vehicle (FCEV) Application" in Renewable Energies and Distributed Generation, The 6th Iranian Conference on (ICREDG2018), 2018.

[14] P. Zolfi, S. Vahid and A. EL-Refaie, "A Novel Non-Isolated Multi-Port DC-DC Converter for Hybrid Streetcar Application," IECON 2021 – 47th Annual Conference of the IEEE Industrial Electronics Society, 2021, pp. 1-7, doi: 10.1109/IECON48115.2021.9589573.

[15] S. Vahid and A. El-Refaie, "Generalized Systematic Approach Applied to Design a Novel Three-Port Power Converter," 2020 IEEE International Conference on Industrial Technology (ICIT), 2020, pp. 542-548, doi: 10.1109/ICIT45562.2020.9067254.

[16] S. Vahid and A. EL-Refaie, "General Approach to Synthesize Multi-Port Power Converters for Hybrid Energy Systems," IECON 2021 – 47th Annual Conference of the IEEE Industrial Electronics Society, 2021, pp. 1-7, doi: 10.1109/IECON48115.2021.9589531.

[17] H. Ardi, A. Ajami and M. Sabahi, "A Novel High Step-Up DC–DC Converter With Continuous Input Current Integrating Coupled Inductor for Renewable Energy Applications," in IEEE Transactions on Industrial Electronics, vol. 65, no. 2, pp. 1306-1315, Feb. 2018.

[18] C. Y. Chan, S. H. Chincholkar and W. Jiang, "Adaptive Current-Mode Control of a High Step-Up DC–DC Converter," in IEEE Transactions on Power Electronics, vol. 32, no. 9, pp. 7297-7305, Sept. 2017.

[19] M. E. S. Ahmed, M. Orabi and O. M. Abdelrahim, "Two-stage micro- grid inverter with high-voltage gain for photovoltaic applications," in IET Power Electronics, vol. 6, no. 9, pp. 1812-1821, November 2013.

[20] A. Rajaei, R. Khazan, M. Mahmoudian, M. Mardaneh and M. Gitizadeh, "A Dual Inductor High Step-Up DC/DC Converter Based on the Cockcroft–Walton Multiplier," in IEEE Transactions on Power Electronics, vol. 33, no. 11, pp. 9699-9709, Nov. 2018, doi: 10.1109/TPEL.2018.2792004.

[21] S. Anand and B. G. Fernandes, "Optimal voltage level for DC microgrids," IECON 2010 - 36th Annual Conference on IEEE Industrial Electronics Society, 2010, pp. 3034-3039, doi: 10.1109/IECON.2010.5674947.

[22] F. Parache, H. Schneider, C. Turpin, N. Richet, O. Debellemanière, É. Bru, AT. Thieu, C. Bertail, C. Marot, "Impact of Power Converter Current Ripple on the Degradation of PEM Electrolyzer Performances," *Membranes*, 2022, 12(2):109, doi: 10.3390/membranes12020109

Bidirectional DC Circuit Breaker with Improved Performance during Commissioning and Reclosing

Aditya Pogulaguntla[*1], Venkata Raghavendra I[*2], Satish Naik Banavath[*3], Andrii Chub[*4],
T Sreekanth[†5] and Harish Sarma Krishnamoorthy[®6]

[*]Department of Electrical Engineering, Indian Institute of Technology Dharwad,
Dharwad, Karnataka, India, 580011.
[*]Department of Electrical Power Engineering and Mechatronics,
Tallinn University of Technology, Tallinn, Estonia.
[†]Department of Electrical and Computer Engineering,
University of Minnesota, Twin Cities, MN, USA.
[®]Department of Electrical and Computer Engineering,
University of Houston, Houston, TX, USA.
Email: [1]pogulaguntla1999@gmail.com, [2]raghavendra.iv@gmail.com, and [3]satish@iitdh.ac.in.,
Email: [4]andrii.chub@taltech.ee, [5]tsreekan@umn.edu, and [6]hskrishn@central.uh.edu

Acknowledgments

This research is based upon work supported by the SERB, Department of Science and Technology (DST), India, under SIRE: SIR/2022/000211.

The work of Andrii Chub was supported by the Estonian Research Council (Grant PSG206).

Keywords

≪DC Microgrid≫, ≪Z-source Circuit Breaker≫, ≪Bidirectional DCCB≫, ≪Fault Interruption≫.

Abstract

DC circuit breakers play a major role in developing and expanding electric power system networks by protecting the network elements from short circuits and overloads. This paper proposes a modified bidirectional z-source dc circuit breaker (MBZSDCCB) by retaining features such as common ground between source and load side terminals, low pass filtering, minimised reflected fault current, and incorporates the features of reduced starting current, elimination of negative current flow during starting and reclosing, and bidirectional power flow capability. The sizing of the components used in the proposed dc circuit breaker is analysed in detail. The MBZSDCCB operation during a short circuit, and commissioning/reclosing instances is demonstrated in SPICE simulation. The issues with the existing z-source circuit breaker are identified by an experimental prototype with a system rating of 120V/3.5A. The performance testing of the proposed MBZSDCCB is validated using simulations in spice and also by developing an experimental prototype of a system rating 400V/10A.

Introduction

Among various dc circuit breaker topologies, the solid-state dc circuit breakers (SSCBs) use power semiconductor devices promoting features such as fast reaction (less interruption time), arc-free operation and ease to realise by simple control logic. These dc circuit breakers can be a promising technology in the protection of modern dc distribution networks, especially in low-voltage (LV) dc and medium-voltage (MV) dc applications such as standalone solar PV systems, wind farm power generation systems and also electric transportation systems [1–3]. Although SSCBs provide tremendous advantages, they have high conduction losses in semiconductor devices, require additional sensing and control circuitry, lack galvanic isolation, and have low fault current capability. To overcome these issues, significant research is in progress to develop a highly efficient and reliable SSCB.

Various power semiconductor devices like fully-controlled devices (IGBT, MOSFET), semi-controlled devices (SCR), and the development of semiconductor devices based on wide-bandgap materials such as SiC and GaN are under investigation. Among them, the fully controlled devices have high conduction resistance, and wide-bandgap devices have the advantages of low losses that increase the efficiency of SSCBs, but they are not yet commercially economical. Semi-controlled devices such as thyristors have high power efficiency, high power density, high short circuit withstand capability, and are economical. Thereby thyristors are the best suitable semiconductor devices for the design of SSCBs. The circuit breaker technology requires additional circuitry for the commutation of a thyristor. To avoid the additional sensing and control circuitry z-source concept was adopted from the z-source inverter [4]. The galvanic isolation is provided by using a low-speed switch (LS).

For fault identification, sensing and control circuitry is usually used in most circuit breakers. However, z-source dc circuit breakers (ZSCBs) do not require sensing and control circuits. ZSCB uses z-source network elements such as inductors (L) and capacitors (C) to automatically sense and trip the fault. The evolution of z-source dc circuit breakers took place with the classical z-source dc circuit breaker [5] that has the disadvantage of the non-common ground connection. Various issues of ZSCBs such as non-common ground, high reflected fault current, and load change withstand capability are addressed in [6–8]. Another class of ZSCBs, namely, series ZSCB has various features such as common ground, less reflected fault current, inbuilt low pass filter, and reduced footprint size. The initially proposed series ZSCB has only the unidirectional protection capability [7]. Later many topologies were introduced to incorporate bidirectional protection capability [9–11]. The power rating of the circuit breakers is extended to MVDC by using multilevel stages in series [12] and modular structures in parallel [13, 14]. Most of the articles have analysed and proposed their circuit breaker operation during only faulty conditions. However, in this paper, the authors have analysed the ZSCB topology during starting and reclosing and its operation during faulty conditions. In the existing very popular series ZSCB, a few issues such as high starting current and negative load current during starting/reclosing have been observed. The negative load current flow would lead to a severe threat to loads meant to handle power in only one direction. Also, the mechanical loads will experience jerks and precision errors. To overcome these issues reference [15] has already proposed a dc circuit breaker for unidirectional power flow networks. This paper presents a bidirectional ZSCB (MBZSDCCB) with a symmetrical configuration for system applications with bidirectional power flow capability, such as the battery charge-discharge circuit in an electric transport system or Photovoltaic system. The component selection is done by the simulation results, and the design equation for discharging resistance has been presented. Detailed experimental validation is done for different scenarios of the proposed MBZSDCCB for a system rating of 400V/10A.

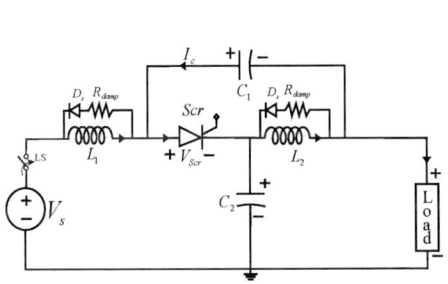

(a) Schematic diagram of ZSCB-unidirectional.   (b) Schematic diagram of ZSCB topology-bidirectional.

Fig. 1: Schematic diagrams of series z-source dc circuit breaker.

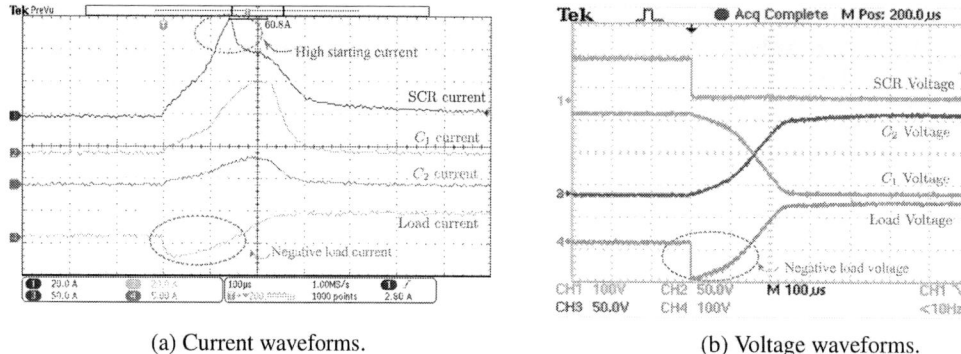

(a) Current waveforms.   (b) Voltage waveforms.

Fig. 2: Experimental current and voltage waveforms of the series ZSCB topology showing negative current issue in the load, and high inrush current through the SCR during starting and reclosing process.

Issues Present in the Conventional Series ZSCB Topology

The schematic circuit diagram of the series ZSCB topology is shown in Fig. 1a. Its application is limited to only unidirectional power flow systems [7]. The circuit schematic shown in Fig. 1b is developed for bidirectional power flow applications [16] using symmetrical circuit arrangement. For better understanding of the problem, the unidirectional topology of ZSCB given in Fig. 1a is initially considered. During the fault, the capacitor C_2 starts discharging through a path consisting of (C_2 - Scr - C_1 - $fault$), which results in a net-zero current or zero current crossing in the main thyristor (Scr) and forces it to go into an off state. As a consequence, the fault current is interrupted in the circuit. However, the resonant transients take place after the fault has been interrupted. It further charges the capacitor C_1 to the grid voltage and simultaneously discharges the capacitor C_2. After the fault is completely interrupted, the thyristor needs to be turned on to reinstate the power flow to the load. During the circuit breaker reclosing, fully charged capacitor C_1 discharges through a path ($C_1 \rightarrow Scr \rightarrow C_2 \rightarrow Load$), results in a reverse current flow in the load, and most of the DC loads do not support the reverse current. Additionally, as the capacitor C_1 discharges through the SCR, it exhibits higher inrush current and demand for a high-current rated thyristor. The above-said issues have been experimentally validated by developing an experimental prototype of the series ZSCB topology of 120V/3.5A rating in the laboratory. The experimental results showing the negative current issue through the load during starting/reclosing process are elucidated in Fig. 2a and Fig. 2b. To mitigate the negative current flow in the load, a modified bidirectional z-source DC circuit breaker is proposed in this paper.

Proposed Modified Bidirectional ZSCB Topology

The schematic diagram of the proposed bidirectional z-source dc circuit breaker is shown in Fig. 3(a). To eliminate the negative current flow in the load and reduce the current stress on the thyristor during starting, the charged capacitors C_2 or C_3 (depending on the direction of power flow) are discharged before the reclosing. The components that participate actively when the power flows from the left to right are indicated by blue colour and while the components that participate actively when the current flow from right to left are indicated by green colour as illustrated in Fig. 3(a).

Modes of operation

The modes of operation of the proposed MBZSDCCB for the power flow from the left to the right side are discussed below. Due to the symmetry of the circuit topology, the bidirectional operation is not discussed.

1. Under normal operation, the circuit components that are active are illustrated in Fig. 3(b). During steady-state, the thyristors T_1 and T_2 are triggered and are in on-state. The capacitor C_1 is charged to DC grid voltage, capacitor C_2 is discharged and the IGBT (Q_1) is turned on. The IGBTs Q_1 and Q_2 work in a complementary manner, so during this mode of operation Q_2 is in off state.

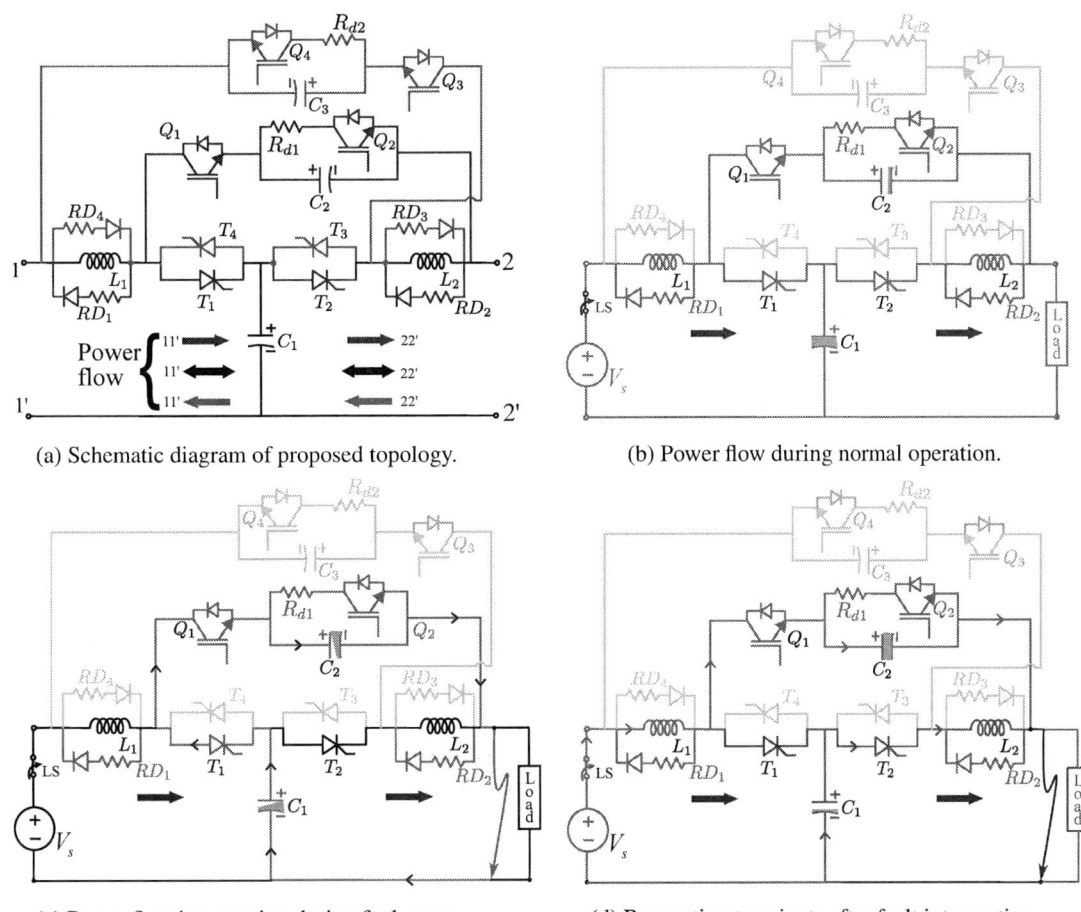

(a) Schematic diagram of proposed topology.

(b) Power flow during normal operation.

(c) Power flow interruption during fault state.

(d) Resonating transients after fault interruption.

Fig. 3: Schematic and modes of operation of proposed MBZSCB topology.

2. Whenever there is a fault on the load side, the capacitor C_1 discharges and circulates current in the opposite direction to that of the thyristor T_1 current. At the instant, when the net current flowing through the thyristor T_1 is zero, the fault current is interrupted and gets bypassed to the alternate path that contains capacitor C_2. This current flow path is illustrated in Fig. 3(c).

3. As soon as the thyristor T_1 gets interrupted, the fault current from the source gets diverted to the capacitor C_2 through IGBT (Q_1). This charges the capacitor C_2 to DC grid voltage, and finally, the current from the source gets interrupted. In the other path, the capacitor C_1 discharges through the inductor L_2 into the fault and reaches a steady state once it is completely discharged. These current flow paths and the status of capacitors charge are shown in Fig. 3(d).

4. This is a freewheeling mode. In this mode, the storage elements discharge their energy to enter a steady state and get ready for reclosing. The magnetic energy stored in the inductors L_1 and L_2 freewheels through the parallel-connected diodes and gets dissipated in the high resistances RD_1 and RD_2 in the freewheeling path. Also, during this mode, the IGBT Q_1 is turned off, and the IGBT Q_2 is turned on. This dissipates the charge stored in the capacitor C_2 through R_{d1}.

Design of discharge resistor

The commissioning of the circuit breaker is to be done once the capacitor C_2 is completely discharged. As soon as the capacitor C_2 voltage reaches the grid voltage, the IGBTs are triggered complementary during which IGBT Q_2 is on. When IGBT Q_2 is on and IGBT Q_1 is off, the capacitor forms an independent circuit with discharge resistor R_{d1} and dissipates the capacitor charge through R_{d1}. The discharge resistance R_{d1} value completely depends on the capacitor C_2 maximum current capability and its dis-

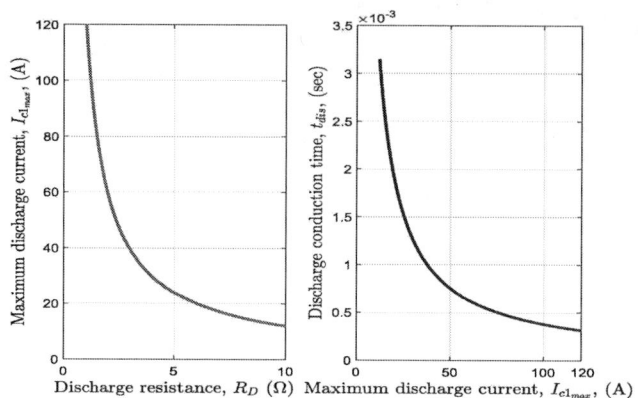

Fig. 4: Selection of discharge resistance R_D.

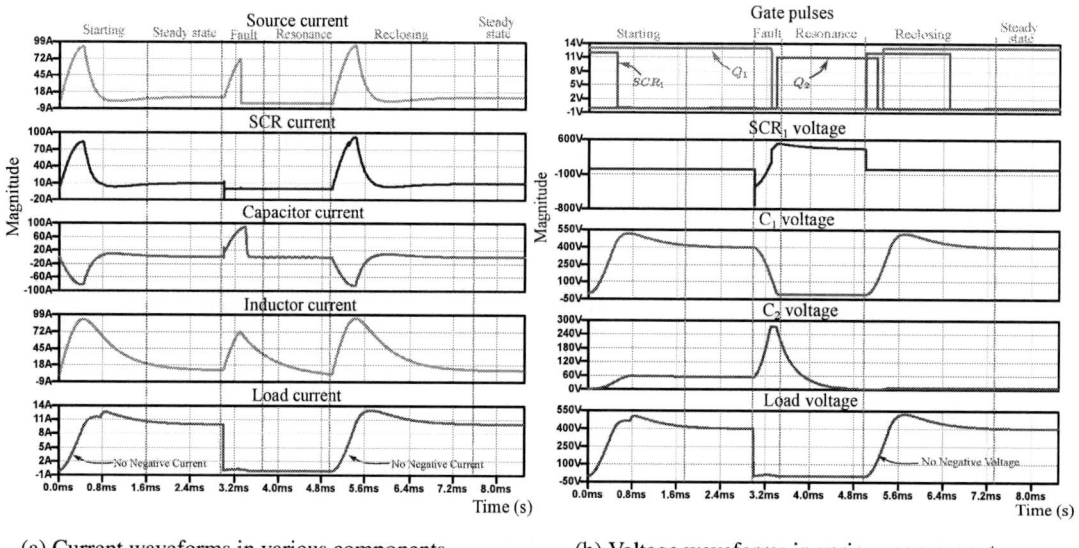

(a) Current waveforms in various components.   (b) Voltage waveforms in various components.

Fig. 5: Results showing the operation of the proposed bidirectional dc circuit breaker.

charge rate. The expressions of capacitor C_2 voltage and current during its discharge are derived as in Eq. (1), and Eq. (2) respectively.

$$v_{c1} = V_s e^{-\frac{t}{R_D C_1}} \tag{1}$$

$$i_{c1} = \frac{V_s}{R_D} e^{-\frac{t}{R_D C_1}} \tag{2}$$

Simulation results

From the system parameters, the value of discharge resistances can be chosen from Fig. 4. The design value is dependent on the trade-off between the discharge conduction time and peak discharge current.

Results and Discussion

The proposed bidirectional dc circuit breaker is validated using simulations by considering the instance of starting, fault and reclosing in a single frame. Later an experimental prototype has been developed in the laboratory and each instance is verified separately and the results are produced in this section.

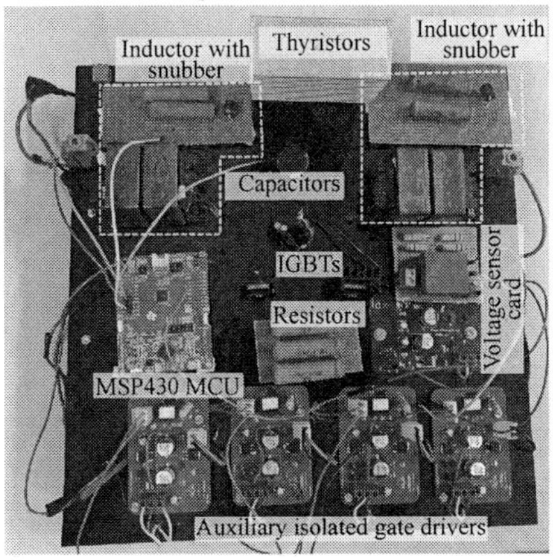

Fig. 6: Photograph of the experimental prototype.

Table I: System parameters.

S. No	Parameter	Value/Rating	Make/Part No
1	DC-link voltage (V_s)	400V	Enarka-400V/25A
2	Z-Source inductors (L_1, and L_2)	1.2mH	EE-42/21/20
3	Z-Source capacitors (C_1, and C_2)	$100\mu F$	Kemet, 400V
4	Load resistance (R_l)	40Ω	Resistive
5	IGBTs (Q_1 & Q_2)	1200V, 20A	H20R1203
6	Thyristors (T_1 & T_2)	1200V, 40A	40TPS12A
7	Discharge resistor (R_{d1} & R_{d2})	$4.5\ \Omega$	Wire wound

The proposed bidirectional dc circuit breaker is simulated using a spice tool for a system rating of 400V/10A when the power flow is from the left to right side. The corresponding current and voltage waveforms across various components are shown in Fig. 5. The circuit breaker is turned on at $t = 0ms$, and it can be observed that there is no negative current flow in the load during starting as the capacitor C_2 is initially not charged. Later a fault is created at $t = 3.2ms$, which turns off the thyristor (SCR_1) with a reverse recovery followed by resonance in the capacitors. This leaves the capacitor C_1 uncharged and capacitor C_2 charged to the DC grid voltage. After the resonance reaches its steady state, the IGBT (Q_2) is turned on such that the capacitor C_2 gets discharged. After the fault is extinguished and the capacitor C_2 is completely discharged, the circuit breaker is ready for reclosing. At $t = 5ms$, the circuit breaker is reclosed by giving gate pulse to both thyristors SCR_1 and SCR_2. It can also be observed from the load current waveform during reclosing that the negative current flow does not occur. Also, the capacitor C_2 has been discharged before reclosing, hence the initial inrush current flowing through the thyristors is significantly reduced. The reverse current flow problems and high inrush currents in the ZSCBs reported in literature have been eliminated in the proposed circuit breaker topology.

Experimental results

The experimental validation of the proposed bidirectional dc circuit breaker is done for a system rating of 400V/10A. A photo of the experimental prototype is given in Fig. 6. The prototype contains auxiliary isolated gate driver boards to drive the IGBTs and thyristors. The TLP250 optocoupler is employed in the gate driver boards. The control logic is implemented using the MSP430 microcontroller. The capacitor C_2 dissipates the stored energy in the discharging resistance, while its voltage is monitored using the voltage sensor card LV25-P. The capacitors used are off-the-shelf capacitors from Kemet. Inductors

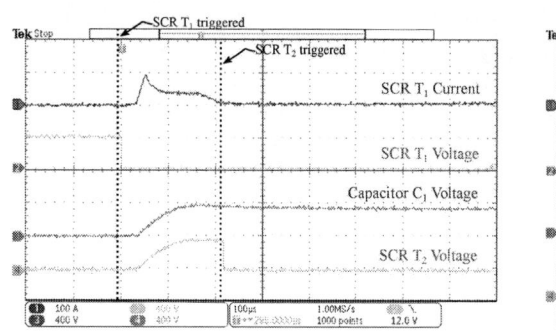 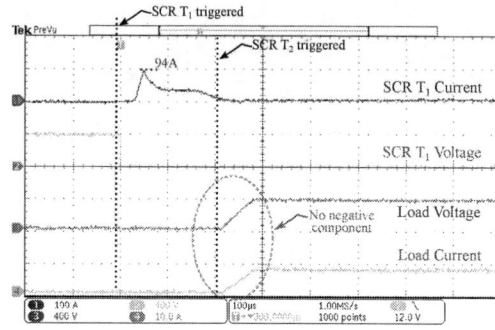

(a) Thyristors triggering during system reclsoing.   (b) Results showing "no negative component".

Fig. 7: Waveforms of components during reclosing.

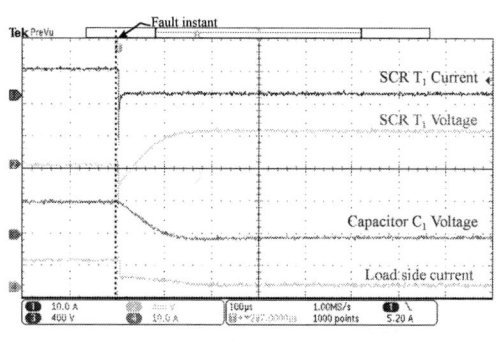 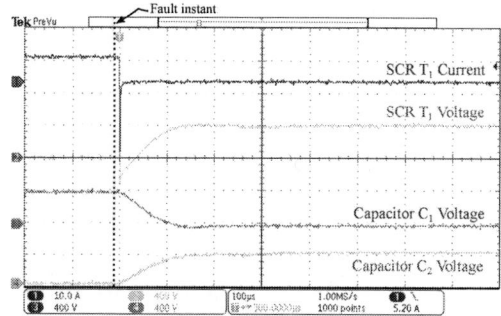

(a) Fault interruption during a fault.   (b) z-source network charge discharge - during a fault.

Fig. 8: Waveforms of various components of MBZSDCCB during the fault.

are wounded in the laboratory by stacking two EE-42/21/20 cores. The RD snubbers are used across the inductors to discharge the energy stored during the transient fault occurrence. The rating of the components used in the prototype is provided in the Table. I.

Initially, the MBZSDCCB is tested for system reclosing. The capacitor C_2 is already discharged during the transient of fault occurrence and the IGBTs (Q_1 & Q_2) are operated in a complementary fashion. At the reclosing moment, the IGBT Q_1 is in on-state to make the MBZSDCCB ready for fault interruption. The trigger pulse is given to thyristor T_1 to charge the capacitor C_1 and later thyristor T_2 is also triggered to restore the normal power flow to the load. The waveforms of the input current, thyristor voltages and capacitor C_1 charging are shown in Fig. 7(a). During this reclosing instant, it can be observed that the load voltage and load current do not contain any negative component. The corresponding waveforms are shown in Fig. 7(b). Also, it is evident from Fig. 3, that the IGBT Q_1 is off during reclosing. This reduces the starting current in the main thyristor T_1 during reclosing.

A short circuit fault is created by using a mechanical circuit breaker (MCB) in parallel to the load. The moment the MCB is turned on, the z-source network detects the fault and turns off the thyristors. The thyristor T_1 current, along with its reverse recovery, can be observed in Fig. 8(a). Also the thyristor T_1 voltage, capacitor C_1 discharging and load side current can be seen in Fig. 8(a). The z-source network detects the sudden change in the load impedance and discharges the capacitor C_1 to charge the capacitor C_2. This z-source network current flows in the opposite direction to the main thyristor T_1 current. This creates a reverse voltage across the thyristor and a zero current flow. The thyristors are turned off, and the fault current is interrupted after a transient. The capacitor C_2 charging and capacitor C_1 discharging waveforms are shown in Fig. 8(b).

Conclusions

This paper presents a modified series bidirectional z-source dc circuit breaker that eliminates the issues of unwanted negative current flow in the loads during starting and reclosing processes. The paper initially presents the existing series ZSCB topology problems through an experimental study. The proposed topology uses thyristor as the main switch for the dc fault interruption, as SCRs present the advantage of high short-circuit withstand capability and are also cost-optimal solution. The proposed dc circuit breaker also eliminates the issues of high inrush currents, which are usually observed in series ZSCBs. Proposed bidirectional ZSCB uses active discharging of the capacitor that further eliminates the negative current flow in the load. A detailed analysis and operation of the proposed MBZSCB with its validation through simulation results are presented for a system rating of $400V/10A$. An experimental prototype of the dc circuit breaker has been developed and tested for various scenarios including reclosing and the fault with a system rating of 400V/10A. The experimental result justify the proposed ZSCB for bidirectional dc distribution applications.

References

[1] R. Rodrigues, Y. Du, A. Antoniazzi, and P. Cairoli, "A review of solid-state circuit breakers," *IEEE Transactions on Power Electronics*, vol. 36, no. 1, pp. 364–377, 2020.

[2] A. Shukla and G. D. Demetriades, "A survey on hybrid circuit-breaker topologies," *IEEE Transactions on Power Delivery*, vol. 30, no. 2, pp. 627–641, 2014.

[3] H. Schefer, L. Fauth, T. H. Kopp, R. Mallwitz, J. Friebe, and M. Kurrat, "Discussion on electric power supply systems for all electric aircraft," *IEEE Access*, vol. 8, pp. 84188–84216, 2020.

[4] F. Z. Peng, "Z-source inverter," *IEEE Transactions on industry applications*, vol. 39, no. 2, pp. 504–510, 2003.

[5] K. A. Corzine and R. W. Ashton, "A new z-source dc circuit breaker," *IEEE Transactions on Power Electronics*, vol. 27, no. 6, pp. 2796–2804, 2011.

[6] K. Corzine and R. W. Ashton, "Structure and analysis of the z-source mvdc breaker," in *2011 IEEE Electric Ship Technologies Symposium*, pp. 334–338, IEEE, 2011.

[7] A. H. Chang, B. R. Sennett, A.-T. Avestruz, S. B. Leeb, and J. L. Kirtley, "Analysis and design of dc system protection using z-source circuit breaker," *IEEE Transactions on Power Electronics*, vol. 31, no. 2, pp. 1036–1049, 2015.

[8] A. Maqsood and K. Corzine, "Z-source dc circuit breakers with coupled inductors," in *2015 IEEE Energy Conversion Congress and Exposition (ECCE)*, pp. 1905–1909, IEEE, 2015.

[9] D. J. Ryan, H. D. Torresan, and B. Bahrani, "A bidirectional series z-source circuit breaker," *IEEE Transactions on Power Electronics*, vol. 33, no. 9, pp. 7609–7621, 2018.

[10] Y. Wang, W. Li, X. Wu, and X. Wu, "A novel bidirectional solid-state circuit breaker for dc microgrid," *IEEE Transactions on Industrial Electronics*, vol. 66, no. 7, pp. 5707–5714, 2019.

[11] D. Keshavarzi, T. Ghanbari, and E. Farjah, "A z-source-based bidirectional dc circuit breaker with fault current limitation and interruption capabilities," *IEEE Transactions on Power Electronics*, vol. 32, no. 9, pp. 6813–6822, 2016.

[12] T. Pang and M. D. Manjrekar, "A surgeless diode-clamped multilevel solid-state circuit breaker for medium-voltage dc distribution systems," *IEEE Transactions on Industrial Electronics*, vol. 69, no. 7, pp. 7329–7339, 2022.

[13] I. Venkata Raghavendra, S. N. Banavath, C. Ajmal Muhammed, and A. Ray, "Modular bidirectional solid-state dc circuit breaker for next-generation electric aircrafts," *IEEE Journal of Emerging and Selected Topics in Power Electronics*, pp. 1–1, 2022.

[14] S. Nandakumar, I. Venkata Raghavendra, C. N. Muhammed Ajmal, S. N. Banavath, and K. Rajashekara, "A modular bidirectional solid state dc circuit breaker for lv and mvdc grid applications," *IEEE Journal of Emerging and Selected Topics in Power Electronics*, pp. 1–1, 2022.

[15] V. Raghavendra, B. S. Naik, and T. Sreekanth, "Modified z-source dc circuit breaker with zero negative current in the load," in *2021 IEEE 12th Energy Conversion Congress & Exposition-Asia (ECCE-Asia)*, pp. 1548–1553, IEEE, 2021.

[16] S. Savaliya and B. Fernandes, "Comparative analysis and coordination study of bi-directional z-source breaker with reclosing capabilities," in *2017 19th European Conference on Power Electronics and Applications*, pp. P–1, IEEE, 2017.

Modeling method for conducted noise flowing in power lines

of DC/DC converter

Takato Hattori, Wataru Kitagawa and Takaharu Takeshita
Dept. of Electrical and Mechanical Engineering, Graduate school of Engineering,
Nagoya Institute of Technology
Gokiso, Showa
Nagoya, 466-8555 Japan
Tel.: +81 (52) 735-5441
Fax: +81 (52) 735-5432
E-Mail: 33413105@stn.nitech.ac.jp, kitagawa.wataru@nitech.ac.jp
URL: http://motion.web.nitech.ac.jp

Acknowledgements

A part of this work was supported by JSPS KAKENHI Grant Number JP22K04043

Keywords

«EMC», «Noises», «Modeling», «Simulation», «DC/DC converter»

Abstract

This paper presents the modeling method of the conducted noise flowing in the input and output side power lines of DC/DC converter. The noise evaluation of DC-DC converters for automotive is based on the leakage current flowing in the power lines. The equivalent circuit of the conducted noise is derived by measuring and frequency analysis the leakage current flowing in the power lines. The noise evaluation simulation is performed by using the equivalent circuit. The usefulness of this proposed method for noise evaluation simulation has been confirmed by comparing the noise reduction effect of EMI filters in simulation and experiment.

Introduction

Recently, due to global environmental issues, carbon neutrality is desired. One approach to this problem is the development and widespread use of HEV (hybrid electric vehicle) and EV (electric vehicle) that can reduce CO_2 emissions. In such vehicles, motor drive and battery charging are performed by power electronics devices, and the demand for power electronics devices is increasing. Nowadays, the power electronics devices are improved power density by downsizing. [1] In order to reduce the size of passive components of power electronics devices such as inductors and capacitors, the switching frequency of power semiconductors (MOSFET, IGBT, etc.) in the power electronics devices has been increased. However, since higher switching frequencies of power semiconductors lead to shorter voltage and current turn-on and turn-off times in the power semiconductors, electromagnetic interference (EMI) is increased. The conducted noise generated by switching may cause malfunctions in peripheral devices, and EMI countermeasure is an important technical issue.[2]-[4]

Current EMI countermeasures use a method of optimizing noise filters in experiments, which is time-consuming and expensive. In order to achieve efficient EMI countermeasures, the conducted noise must be evaluated and countermeasures using simulation. [5]-[11] Conventionally, by modeling the leakage currents flowing to the LISN (Line Impedance Stabilization Network) [12]-[17] or measuring the impedance of the noise path [18]-[20], the noise evaluation simulation by disturbance voltage has been performed. The noise evaluation method based on the disturbance voltage can only evaluate the noise in one direction where the LISN is inserted. The DC-DC converters for EV and HEV need to be evaluated for noise in both input and output directions to convert power between two batteries in both directions. Evaluating the noise flowing in the power lines on the input and output sides make it possible evaluate the noise in bidirectional. Therefore, the noise modeling method that enables bidirectional noise simulation is needed.

This paper presents the modeling method of conducted noise flowing in the power lines of DC-DC converter. The noise modeling is performed by deriving equivalent circuits from measurements and analysis leakage current. The noise evaluation simulation is performed using equivalent circuit. Furthermore, comparing experimental and simulation result with Y-capacitor as EMI filter show the usefulness of this modeling method.

Evaluation Method of Conducted Noise

Fig. 1 shows the noise evaluation experimental system with DC-DC converter. In this circuit, the voltage of the regulated power supply is stepped down by the DC-DC converter and flows to the load. This DC/DC converter is used to convert the voltage between batteries in HEV and EV. The voltage is converted by switching MOSFETs. This DC-DC converter has four phases MOSFETs, and each switching frequency is 200 kHz. The LISN is inserted between the stabilized power supply and DC-DC converter on the input side to suppress external noise. The LISN, surface layer and GNDlow terminal of the DC-DC converter are connected to the copper plate in the experimental system. The leakage current occurred by the MOSFETs switching flows to the copper plate (i_l). The leakage current on the copper plate flows to the input side power lines through the LISN (i_{Vhigh} and $i_{GNDhigh}$) or to the output side through the LOAD (i_{Vlow} and i_{GNDlow}). These four currents flowing on the input and output sides are measured with current probe. To evaluate the noise on the input and output side of the DC/DC converter, performing FFT analysis on the four measured leakage currents. The parameters of noise evaluation experiment are indicated in TABLE 1.

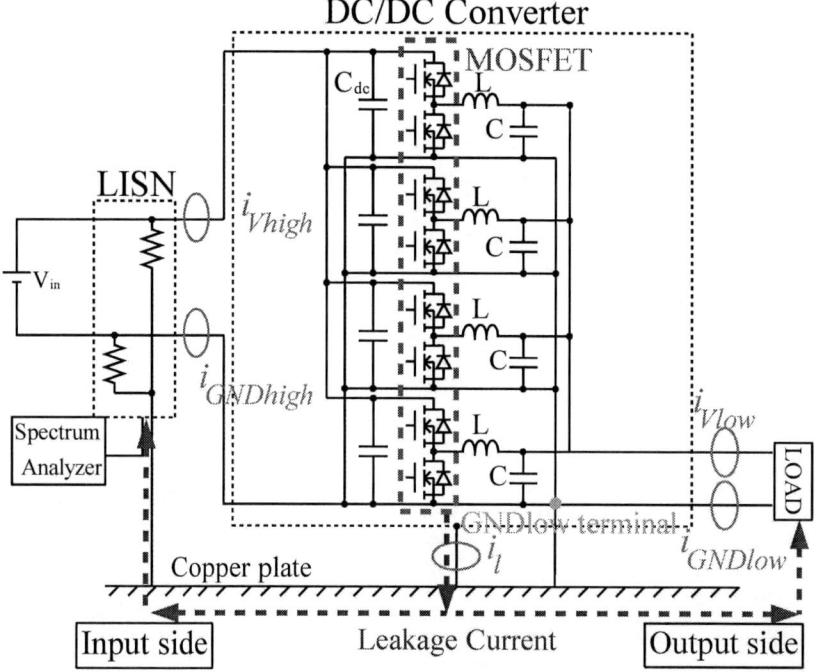

Fig. 1: Experimental system

Table I: Experimental condition

Input voltage V_{in}	35.0 V
Input current I_{in}	2.06 A
Output voltage V_{out}	12.0 V
Output current I_{out}	6.00 A
Output power P_{out}	72.0 W
LOAD R	2.00 Ω
Switching frequency f_{sw}	200 kHz

Experimental result

Fig. 2 shows the experimental FFT analysis result of the leakage current flowing to the copper plate from the surface layer of DC-DC converter (i_l). This leakage current consists of frequencies of 23.5, 28.3, 30.1 and 38.0 MHz. Fig. 3 shows the experimental result of the leakage current flowing in the input and output side power lines of DC/DC converter (i_{Vhigh}, $i_{GNDhigh}$, i_{Vlow} and i_{GNDlow}). The leakage currents in the power lines on the input and output sides are damped and vibrating. Fig. 4, Fig. 5, Fig. 6 and Fig. 7 shows the experimental FFT analysis result of the four leakage currents (i_{Vhigh}, $i_{GNDhigh}$, i_{Vlow} and i_{GNDlow}) in the input and output side power lines. The leakage currents in the power lines on the input and output sides consist of frequencies from 20 to 40 MHz and need to be reduced for EMI. Comparing Fig. 2, Fig. 4, Fig. 5, Fig. 6 and Fig. 7, the frequencies that constitute the leakage currents flowing into each of the power lines match the frequencies that constitute the leakage currents flowing from the substrate surface layer of the DC-DC converter to the copper plate. The conducted noise generated by MOSFET switching flows from the surface layer of the DC-DC converter board to the copper plate and then to the power lines on the input and output sides. In order to focus on the conducted noise leaking from the substrate surface layer of the DC-DC converter, in this paper 20 to 40 MHz bandpass filter is applied to the measured result of the leakage currents in the input and output sides power lines.

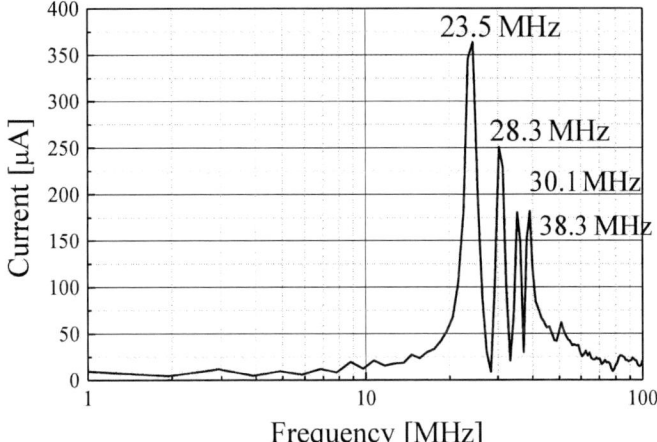

Fig. 2: Experimental FFT analysis result of the leakage current flowing to copper plate (i_l)

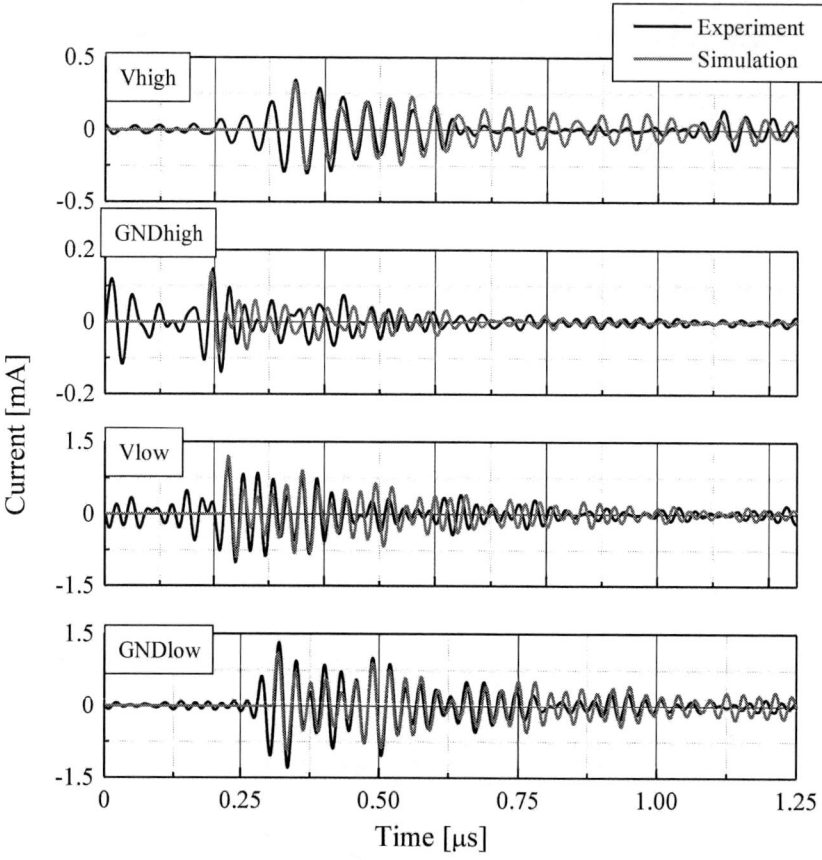

Fig.3: Experimental and simulation result of the leakage current flowing to the power lines

Simulation

The conducted noise flowing in the power lines on the input and output sides of DC-DC converters must be counteracted, and the noise modeling method is proposed to enable noise evaluation in simulation for efficient noise countermeasures. The leakage currents are modeled with a simple equivalent circuit from the measurement results.

Modeling Method

To simulate the conducted noise in the power lines, the leakage current equivalent circuit is derived by using experimental result of the four leakage currents. From Fig. 3, the waveform of i_{Vhigh}, i_{Vhigh}, i_{Vlow} and i_{GNDlow} are damped vibration. Therefore, the leakage current equivalent circuit of the four leakage currents is approximated by simple RLC series circuits. The equation of the damped vibration current after applying the step voltage E_c to the RLC series circuit is shown by the following equation.

$$i(t) \cong \frac{E_c}{Z_0} e^{-\zeta \omega_n t} \sin \omega_n t \tag{1}$$

$$Z_0 = \frac{E_c}{i_{peak}}, \qquad \zeta = \frac{R}{2Z_0}, \qquad \omega_n = 2\pi f \tag{2}$$

In the eq. (1) and (2), the parameter Z_0 is the characteristic impedance, the parameter ω_n is the eigen frequency, the parameter i_{peak} is the maximum value of the leakage current and the parameter ζ is the attenuation coefficient. If the step voltage E_c is known, the eigen frequency and characteristic impedance can be derived from the peak current as in equation (2). The eigen frequency ω_n and the characteristic impedance Z_0 are obtained from the following equations for L and C of the resonant circuit.

$$\omega_n = \frac{1}{\sqrt{LC}}, \quad Z_0 = \sqrt{\frac{L}{C}} \tag{3}$$

In the eq. (3), the parameter L and C are inductance and capacitance of RLC series circuit. The parameters L and C are decided by the following equation.

$$L = \frac{Z_0}{\omega_n}, \quad C = \frac{1}{\omega_n Z_0} \tag{4}$$

The parameter R of RLC series circuit is determined by the state of the attenuation. Therefore, the leakage current equivalent circuit is completed when the parameters of RLC are determined with accuracy.

In this paper, the parameters of the equivalent circuit of the leakage current in Fig. 3 are determined by this modeling method. The step voltage E_c is 35 V, the maximum leakage current amplitude i_{peak} is referenced from Fig. 3, the resonance frequency is extracted from the results of each leakage current FFT analysis Fig. 4 - Fig. 7. Since the leakage current is generated by the sum of leakage currents of several frequencies, the high frequency equivalent circuit of the leakage current is composed of three or four parallel RLC circuits. The parameters of the equivalent circuit are calculated by using eq. (2) and eq. (4). The parameters are shown in TABLE II.

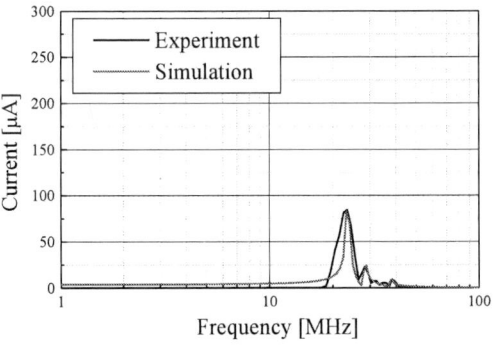

Fig. 4: Experimental and simulation FFT analysis result of i_{Vhigh}

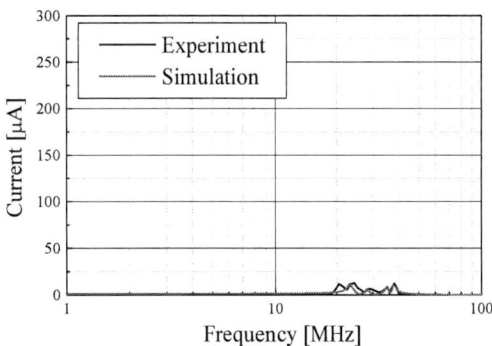

Fig. 5: Experimental and simulation FFT analysis result of $i_{GNDhigh}$

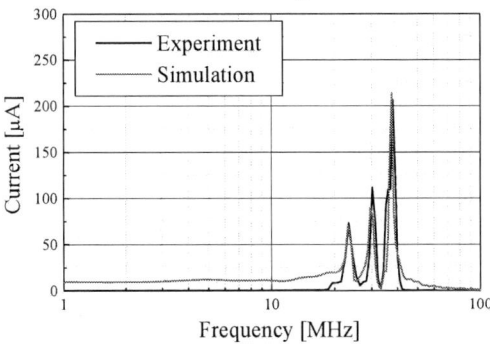

Fig. 6: Experimental and simulation FFT analysis result of i_{Vlow}

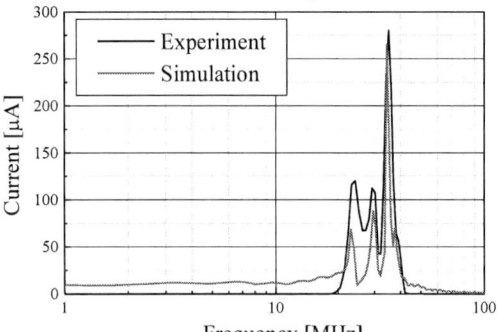

Fig. 7: Experimental and simulation FFT analysis result of i_{GNDlow}

TABLE II: The parameters of the high frequency equivalent circuits

i_{Vhigh}

f [MHz]	23.5	28.3	38.3
E_C [V]	35	35	35
i_{peak} [μA]	255	69	15.4
ω_n [Mrad/s]	148	178	239
Z_0 [kΩ]	137	507	2270
L_{a1},L_{a2},L_{a3} [μH]	927	2850	9490
C_{a1},C_{a2},C_{a3} [fF]	49.4	11.1	1.84
R_{a1},R_{a2},R_{a3} [kΩ]	3.0	9.0	2.0

$i_{GNDhigh}$

f [MHz]	23.5	28.3	35.1	38.3
E_C [V]	35	35	35	35
i_{peak} [μA]	48.6	23.6	31.8	46
ω_n [Mrad/s]	148	178	220	239
Z_0 [kΩ]	720	1490	1100	760
$L_{b1},L_{b2},L_{b3},L_{b4}$ [mH]	4.87	8.35	4.99	3.18
$C_{b1},C_{b2},C_{b3},C_{b4}$ [fF]	9.41	3.79	4.12	5.51
$R_{b1},R_{b2},R_{b3},R_{b4}$ [kΩ]	30	50	30	20

i_{Vlow}

f [MHz]	23.5	28.3	38.3
E_C [V]	35	35	35
i_{peak} [μA]	257	399	743
ω_n [Mrad/s]	148	189	239
Z_0 [kΩ]	136	87.7	47.1
L_{c1},L_{c2},L_{c3} [μH]	920	463	197
C_{c1},C_{c2},C_{c3} [fF]	49.8	60.3	88.9
R_{c1},R_{c2},R_{c3} [kΩ]	4.5	3.0	1.0

i_{GNDlow}

f [MHz]	23.5	30.1	35.1
E_C [V]	35	35	35
i_{peak} [μA]	255	399	743
ω_n [Mrad/s]	148	189	220
Z_0 [kΩ]	135	87.6	47.1
L_{d1},L_{d2},L_{d3} [μH]	927	463	214
C_{d1},C_{d2},C_{d3} [fF]	49.9	60.3	96.3
R_{d1},R_{d2},R_{d3} [kΩ]	3.0	2.8	0.6

Simulation Result

The conducted noise evaluation simulation circuit is shown in Fig. 8. The derived leakage currents equivalent circuit is inserted between the LISN and the LOAD part. The equivalent circuit configuration and parameters of the LISN are from the data sheet. [21] In addition, the equivalent circuit configuration and parameters of the LOAD and between LOAD and copper plate are decided from measurement result by impedance analyzer. Fig. 3 show simulation results of the leakage current flowing in the input and output side power lines (i_{Vhigh}, $i_{GNDhigh}$, i_{Vlow} and i_{GNDlow}). In Fig. 3, the attenuation of the leakage current flowing in the power line can be simulated. From Fig. 4 to Fig. 7 show the FFT analysis results of the simulation leakage currents (i_{Vhigh}, $i_{GNDhigh}$, i_{Vlow} and i_{GNDlow}). Compared the simulation result with the experimental result, the frequencies and amplitudes of the leakage currents in power lines are well simulated. Therefore, the conducted noise flowing in the input and output sides of the DC-DC converter is reproduced on the simulation.

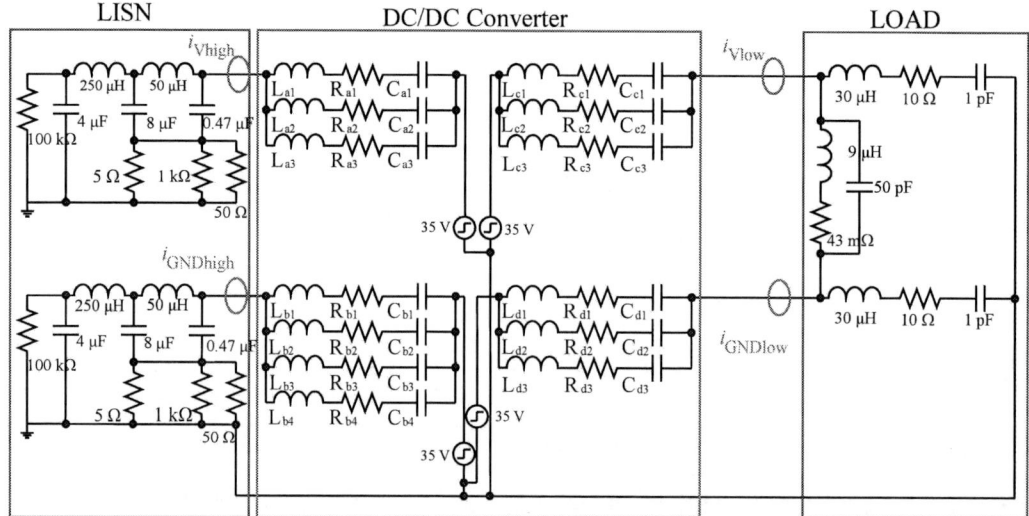

Fig. 8: Noise evaluation simulation circuit

Noise Evaluation Simulation and Experiment with EMI Filter

To confirm the usefulness of the proposed modeling method in this paper, the noise reduction effect of the EMI filter is simulated and then compared with the noise reduction effect of the EMI filter in the experiment.

Modeling EMI Filter

In this paper, Y-capacitors are used as EMI filters. Y-capacitor is inserted between the DC-DC converter and the LOAD to reduce conducted noise in the power line on the output side (i_{Vlow} and i_{GNDlow}). Inserting the Y-capacitor reduces the leakage current flowing in the output side power lines because the leakage current path is created through the capacitor instead of the load. Fig. 9 shows the impedance frequency characteristics of the Y-capacitor. The Y-capacitor with an impedance resonance point at 68 MHz is used for simulation and experiment. The impedance characteristics of Y-capacitor in the high frequency range can be modeled by RLC series circuit. The parameters of the RLC series circuit are matched from the frequency response data of the impedance of the capacitor. The parameters of Y-capacitor are as R_y = 3.00 Ω, C_y = 900 pF, L_y = 9.33 nH.

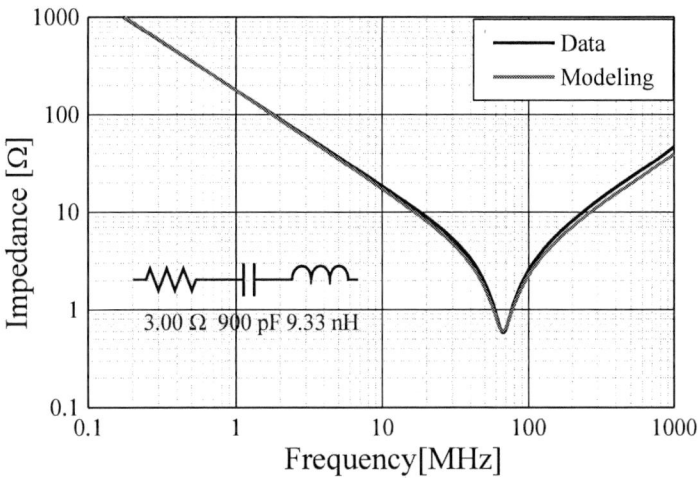

Fig. 9 Impedance Frequency Response of Y-Capacitor

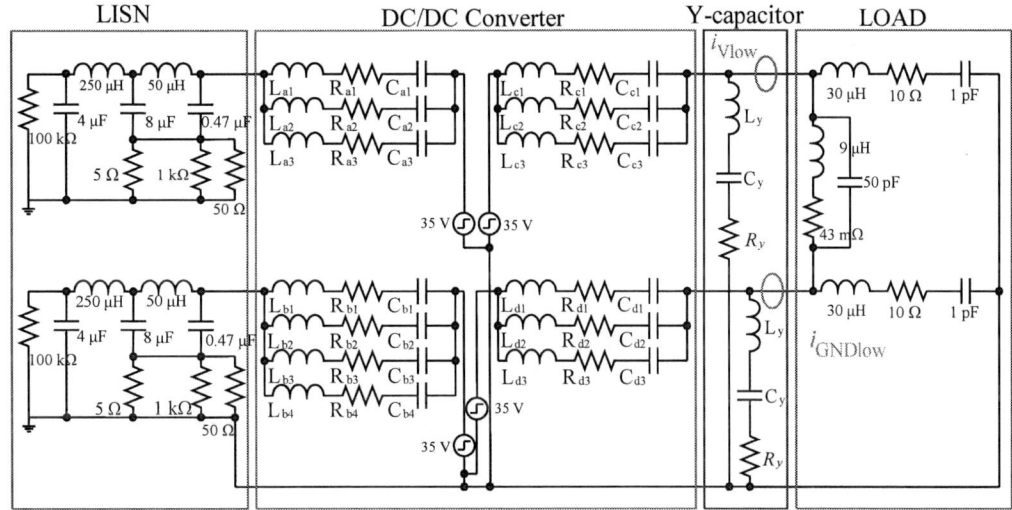

Fig. 10: Noise evaluation Simulation circuit with Y-capacitor

Noise Evaluation Simulation and Experiment Result

Fig. 10 shows the noise evaluation Simulation circuit with Y-capacitor. The Y-capacitor is inserted between DC-DC converter and LOAD part. The simulation results of the FFT analysis of the conducted noise in the output side power lines when the Y-capacitor is inserted are shown in Fig. 11 and 12 (i_{Vlow} and i_{GNDlow}). The Y-capacitor reduces the between 20 and 40 MHz conducted noise in the output side power lines in the simulation. The reduced effect of 200 µA is observed at the most reduced point. Next, the noise evaluation experiment is performed by inserting the Y-capacitor into the experimental circuit. Fig. 13 shows the experimental circuit when the Y-capacitor is inserted. As in the simulation, the Y-capacitor is inserted between the DC-DC converter and LOAD to reduce conducted noise on the output side power lines. Fig. 11 and 12 shows the experimental result of the FFT analysis of the conducted noise in the output side power lines when the Y-capacitor is inserted. From Fig. 11 and Fig.12, the Y-capacitor reduces the between 20 and 40 MHz conducted noise in the output side power lines in the experiment.

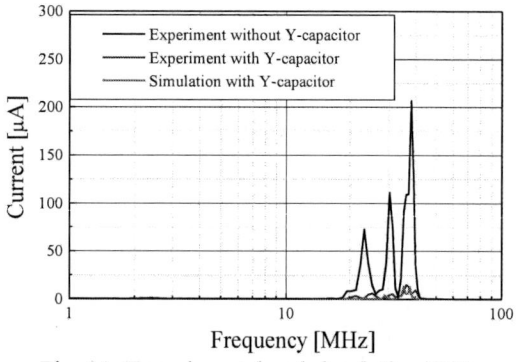

Fig. 11: Experimental and simulation FFT analysis result of i_{Vlow} with Y-capacitor

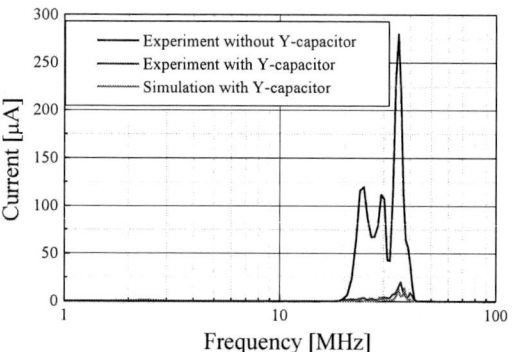

Fig. 12: Experimental and simulation FFT analysis result of i_{GNDlow} with Y-capacitor

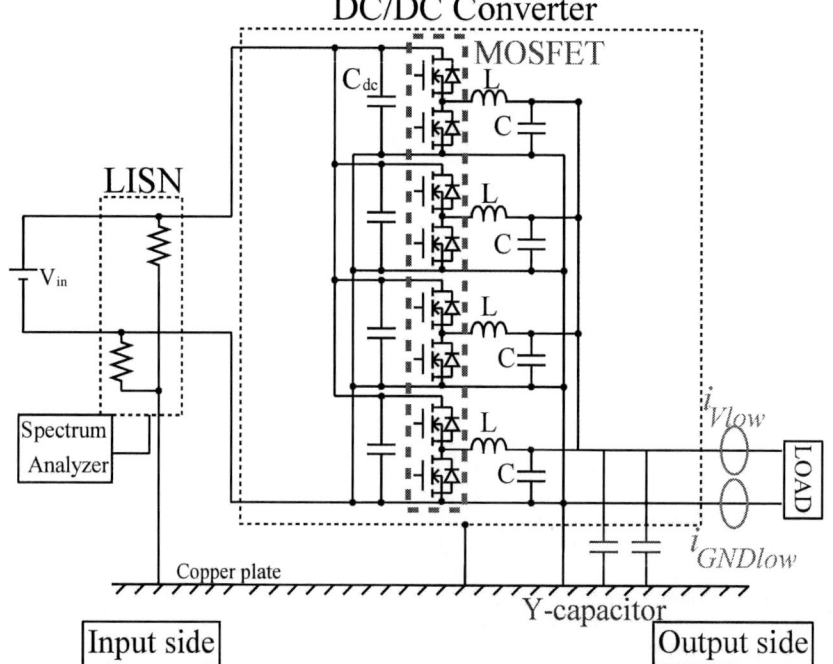

Fig. 13: Experimental circuit with Y-capacitor

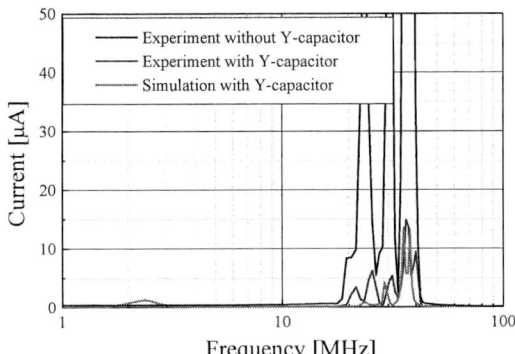

Fig. 14: Enlarged view of Experimental and simulation FFT analysis result of i_{Vlow} with Y-capacitor

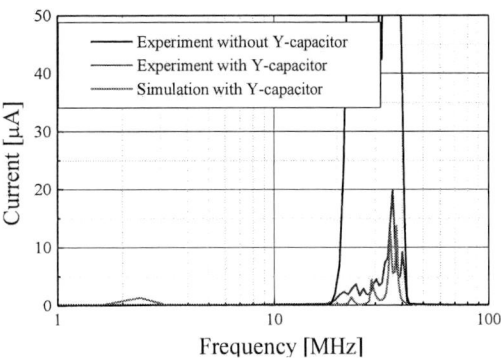

Fig. 15: Enlarged view of Experimental and simulation FFT analysis result of i_{GNDlow} with Y-capacitor

Comparing the simulation and experimental results of the leakage current FFT analysis in the output side (i_{Vlow}) in Fig. 11, the error in the reduction ratio is 0.3% at the most affected noise frequency of 35 MHz. In the same way, comparing the simulation and experimental results of the leakage current FFT analysis in the output side (i_{GNDlow}) in Fig. 12, the error in the reduction ratio is 2.6% at the most affected noise frequency of 38 MHz. Based on these results, simulations using the modeling method proposed in this paper are useful in confirming the effectiveness of noise reduction. Fig. 14 and Fig. 15 show the enlarged views of the FFT analysis results of the simulation and experiment of the leakage current in the output side power line of the DC-DC converter showed in Fig. 11 and 12. Comparing the simulation with the experiment when the Y-capacitor is inserted, there are frequencies of the leakage current that have been measured experimentally that have not been reproduced in the simulation. This is due to the lack of accuracy of the parameters of the leakage current equivalent circuit and the circuit configuration and circuit parameters of the load portion in the simulation circuit. Thus, the accuracy of the simulation can be enhanced by improving these simulation circuits.

Conclusion

This paper describes the modeling method of the conducted noise in the power lines of DC-DC converter. The noise evaluation by the leakage currents flowing in the input and output sides power lines enables bidirectional noise evaluation, which was not possible with the conventional noise evaluation of noise disturbance voltage. The noise evaluation simulation of leakage current flowing in power lines is performed by deriving the equivalent circuit from the leakage current measurement results. In the simulation results, the amplitude spectrum of the leakage current flowing in the power line (i_{Vhigh}, $i_{GNDhigh}$, i_{Vlow} and i_{GNDlow}) is well simulated. Furthermore, the simulation and experimentation were performed when Y-capacitor was inserted on the output side. The similar noise reduction effect with Y-capacitor in simulation and experiment with Y-capacitor. The usefulness of the simulation of noise evaluation using the modeling method proposed in this paper was confirmed. However, there is a difference in the peak frequency of the leakage current between simulation and experiment when a Y capacitor is inserted. In order to make the simulation using the modeling method in this paper practical in the future, the configuration and parameters of the simulation circuit should be reviewed to improve the accuracy.

References

[1] T. Kitamura, M. Yamada, S. Harada, M. Koyama, "Development of Highpower-density Interleaved DC/DC Converter with SiC Devices", IEEJ Transactions on Industry Applications, Vol. 134, No. 11, pp. 956-961, 2014
[2] T. Shimizu, "Historical Review of EMI Measures in the Field of Power Electronics", Annual Meeting Record, I.E.E Japan (CD-ROM), ROMBUNNO.S12-8, 2021

[3] C. Nan, R. Ayyanar and Y. Xi, "High frequency active-clamp buck converter for low power automotive applications," 2014 IEEE Energy Conversion Congress and Exposition (ECCE), pp. 3780-3785, 2014

[4] L. Zhai, G. Hu, M. Lv, T. Zhang and R. Hou, "Comparison of Two Design Methods of EMI Filter for High Voltage Power Supply in DC-DC Converter of Electric Vehicle," in IEEE Access, vol. 8, pp. 66564-66577, 2020

[5] M. Terasaki, Y. Oohashi, Y. Masuyama and T. Sudo, "Design and analysis for noise suppression of DC/DC converter," 2014 IEEE Electrical Design of Advanced Packaging and Systems Symposium (EDAPS), pp. 109 - 112, 2014

[6] S. Maekawa, J. Tuda, A. Kuzumaki, S. Matsumoto, H. Mochikawa, H. Kubota, "EMI Prediction Method for SiC Inverter by Developing an Accurate Model of Power Device ", IEEJ Transactions on Industry Applications, Vol. 134, No. 4, pp. 461-467, 2014

[7] E. Rondon-Pinilla, F. Morel, C. Vollaire and J. Schanen, "Modeling of a Buck Converter with a SiC JFET to Predict EMC Conducted Emissions," in IEEE Transactions on Power Electronics, vol. 29, no. 5, pp. 2246-2260, May 2014

[8] D. Drozhzhin, V. Karakasli and G. Griepentrog, "Comprehensive Analysis of Converter Output Voltage for Conducted Noise Simulation," 2019 International Symposium on Electromagnetic Compatibility – EMC EUROPE, pp. 42-47, 2019

[9] I. A. Makda and M. Nymand, "Common-mode noise analysis, modeling and filter design for a phase-shifted full-bridge forward converter," 2015 IEEE 11th International Conference on Power Electronics and Drive Systems, pp. 1100-1105, 2015

[10] Y. Ishii et al., "Accurate Conducted EMI Simulation of a Buck Converter With a Compact Model for an SiC-MOSFET," 2020 IEEE Applied Power Electronics Conference and Exposition (APEC), pp. 2800-2805, 2020

[11] S. Takahashi, K. Wada, H. Ayano, S. Ogasawara, "Review of Modeling and Suppression Techniques for Electromagnetic Interference in power Conversion Systems", 2022 IEEE Journal of Industry Applications, Vol. 11, No. 1, pp. 7-19 ,2022 The Institute of Electrical Engineer of Japan

[12] H. Tanaka, K. Suzuki, W. Kitagawa, T. Takeshita, "Conducted Noise Reduction on AC/DC Converter using SiC-MOSFET ", 2016 IEEE International Conference on Renewable Energy Research and Applications (ICRERA), pp. 341-346, 2016

[13] H. Tanaka, K. Suzuki, W. Kitagawa and T. Takeshita,"Design for conducted noise reduction on AC/DC converter using SiC-MOSFET," 2016 19th International Conference on Electrical Machines and Systems (ICEMS), pp. 1-6, 2016

[14] Y. Kawamura, H. Tanaka, K. Suzuki, W. Kitagawa and T. Takeshita, "Investigation of Modeling for Conducted Noise Reduction on Isolated AC/DC Converter using SiC Devices," 2018 20th European Conference on Power Electronics and Applications (EPE'18 ECCE Europe), pp. 1- 10, 2018

[15] Y. Kawamura, H. Tanaka, K. Suzuki, W. Kitagawa and T. Takeshita, "Consideration of conducted noise reduction on isolated AC/DC converter using SiC devices," 2017 IEEE 6th International Conference on Renewable Energy Research and Applications (ICRERA), pp. 359-364, 2017

[16] K. Kuwana, Y. Kawamura, W. Kitagawa and T. Takeshita, "Modeling for Conducted Noise Simulation Considering Switching Characteristics on AC/DC Converter," 2019 10th International Conference on Power Electronics and ECCE Asia (ICPE 2019 - ECCE Asia), pp. 927-932, 2019

[17] W. Kitagawa, T.Kutsuna, K. Kuwana, Y. Kawamura, T. Takeshita, "Conducted Noise Simulation on AC/DC Converter using SiC-MOSFET," in IEEE Transactions on Industry Applications, vol. 57, no. 2, pp. 1644- 1651, March-April 2021

[18] K. Mitani, W. Kitagawa and T. Takeshita, "Modeling for Conducted Noise on AC/DC Converter by Using Impedance Characteristic," 2019 22nd International Conference on Electrical Machines and Systems (ICEMS), pp. 1-6, 2019

[19] K. Mitani, Y. Kawamura, W. Kitagawa and T. Takeshita, "Circuit Modeling for Common Mode Noise on AC/DC Converter Using SiC Device," 2019 21st European Conference on Power Electronics and Applications (EPE '19 ECCE Europe), pp. P.1-P.10, 2019

[20] W. Kitagawa, K. Mitani, Y. Kawamura and T. Takeshita, "Modeling and Simulation for Conducted Noise on AC/DC Converter Using SiCDevice," 2019 10th International Conference on Power Electronics and ECCE Asia (ICPE 2019 - ECCE Asia), pp. 1-6, 2019

[21] Line Impedance Stabilisation Network data sheet, NSLK 8126, SCHWARZBECK MESS-ELEKTRONIK OHG, 2021. [online]. Available: http://www.schwarzbeck.de/Datenblatt/k8126.pdf

High-Bandwidth Power Hardware-in-the-Loop for Motor and Battery Emulation at High Voltage Levels

Manuel Fischer, Philipp Kemper, Johannes Herbold, Daniel Epping, Frank Puschmann
dSPACE GmbH
Real-Time Test & Development Solutions
Paderborn, Germany
Tel.: +49 5251 1638-0
E-Mail: MaFischer@dspace.de
ORCiD: 0000-0002-5499-697X
URL: http://www.dspace.com

Keywords

«Power Hardware-in-the-Loop», «Machine emulation», «Modular converter», «Interleaved converters», «Test bench».

Abstract

Power hardware-in-the-loop (PHIL) emulation of batteries and electric machines is an efficient method to accelerate the development and testing process of traction inverters and their control systems at different operating conditions. This paper presents the overall setup of a high-bandwidth, high-voltage PHIL system. Real-time machine and battery models, a control unit for the power electronics device and the power electronic device itself are highlighted as the most relevant parts of the PHIL system. Multiple approaches fulfilling the requirements of these parts are explained in detail. Finally, this paper determines the most suitable approaches to setting up a modular PHIL system that is scalable in terms of power range and number of output connectors.

The overall hardware of the PHIL system with the described components is set up. Concluding measurement results prove its performance.

Introduction

Electromobility is one of the mega trends in the automotive industry. The drivetrain of an electrically powered car generally consists of a high-voltage (HV) battery, a traction inverter and an electric machine. Throughout the development process, testing plays a decisive role. In addition to proving the entire system's functionality, reliability and safety, testing can also improve efficiency by enhancing the control. These hardware tests must proceed in conformance with common standards, such as ISO 26262, ISO 21498 and ISO 21782.

Typically, a dynamometer test bench is set up for this purpose, see Fig. 1. The investigated machine is coupled to a load machine in order to set several mechanical operating points. The device under test (DUT) inverter converts the battery's DC voltage to the required voltage courses feeding all the phases of the investigated machine. These dynamometer test benches are expensive and challenging to set up. Moreover, they cannot flexibly used for different machine settings or battery behavior. Alternatively, the DUT inverter can be tested by using power hardware-in-the-loop (PHIL) simulators. For this, the DUT inverter's input and output terminals are connected to a power electronics device which has the same electrical characteristic as the investigated battery and machine, see Fig. 1.

The PHIL test bench consists of three main components:
- A battery and a machine model, which calculate the investigated battery's and machine's electrical behavior in real-time depending on the DUT inverter's switching behavior,
- A control unit, which controls the power electronic devices in a way that the same load currents occur as in the real-world dynamometer test bench,
- The power electronics devices itself, which adjust the desired load currents.

The use of a common switched-mode operated power electronics device in PHIL application leads to additional disturbing ripple content in the output current which differs from the reference current. The following concepts have already been investigated and offer a solution that minimizes the additional ripple content. Generally, a setup using a multi-branch inverter with an interleaved modulation technique [1-5] is recommended. This topology leads to an increasing effective switching frequency at the output terminals and thereby decreases the additional ripple content. Adding a passive LC filter at the output terminals of a switched-mode operated two-level or multi-level inverter decreases the additional ripple as well, but has to work with a lower bandwidth of the output voltage [6-12]. In [13,14] a linear amplifier is suggested, albeit only for medium and low voltage levels. Specialized concepts are a four-level inverter with variable output voltages and a modular multilevel converter [15-17].

This paper presents the setup of an entire PHIL system. A modular concept is used for the power electronics devices emulating the battery and the investigated machine. These power electronics modules are based on multilevel half bridge branches connected in parallel which fulfill the requirements for high voltage, high bandwidth as well as a small additional emulator current ripple. Moreover, two different control schemes – voltage-based and current-based – are presented and implemented on the control unit. The theory of the machine and the battery model as well as measurement results of the PHIL system complete this work.

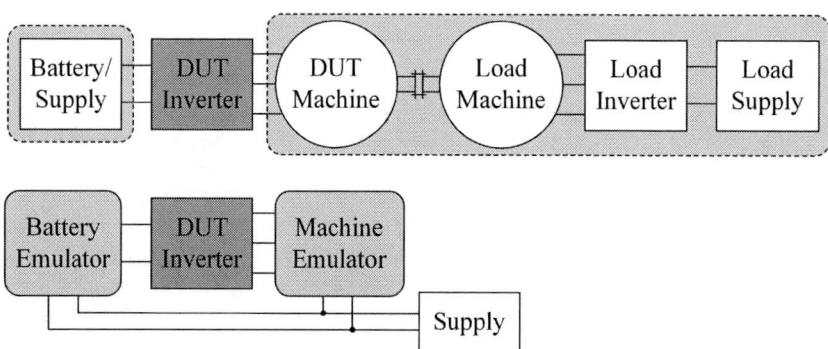

Fig. 1: Schematic diagram of a common dynamometer test bench (top) compared to a PHIL test bench (bottom)

Overview of the PHIL System

Fig. 2 depicts a schematic diagram of the entire PHIL system. The DUT inverter is connected to the common DC link $V_{DC,Sup}$ via power electronics (PE) modules. The PE modules at the DUT inverter's DC terminals emulate the behavior of the battery. By measuring the input currents i_{Bat+} and i_{Bat-} the battery model can evaluate the actual state and especially the output voltage $v_{Bat,ref}$ of the emulated battery. This reference output voltage is passed to the device control unit which determines the switching commands SC for the dedicated PE modules on the battery side. The PE modules at the three DUT inverter's AC terminals emulate the behavior of the investigated machine. By measuring the output voltages $v_{DUT,x}$ (with x = U,V,W), the machine model can calculate the actual state of the emulated machine and thereby especially the machine's input currents $i_{M,ref,x}$. In addition, the emulator outputs all relevant interfaces, for example, communication or position sensor emulation. The reference currents are passed to the device control unit which determines the switching commands SC for the PE modules on the machine side. When the virtual machine is in motor mode, the battery PE modules feed the DUT inverter and the machine PE modules feed the inverter's output power back into the common DC link (and vice versa when the virtual machine is in generative mode). Consequently, the power flow circulates and the supply unit feeds only the power losses of the system. In this case, no bidirectional grid connection or galvanically isolated DC/DC converters are required. In the following sections, the relevant parts of the PHIL system are explained in detail.

High-Bandwidth Power Hardware-in-the-Loop for Motor and Battery Emulation at
High Voltage Levels   FISCHER Manuel

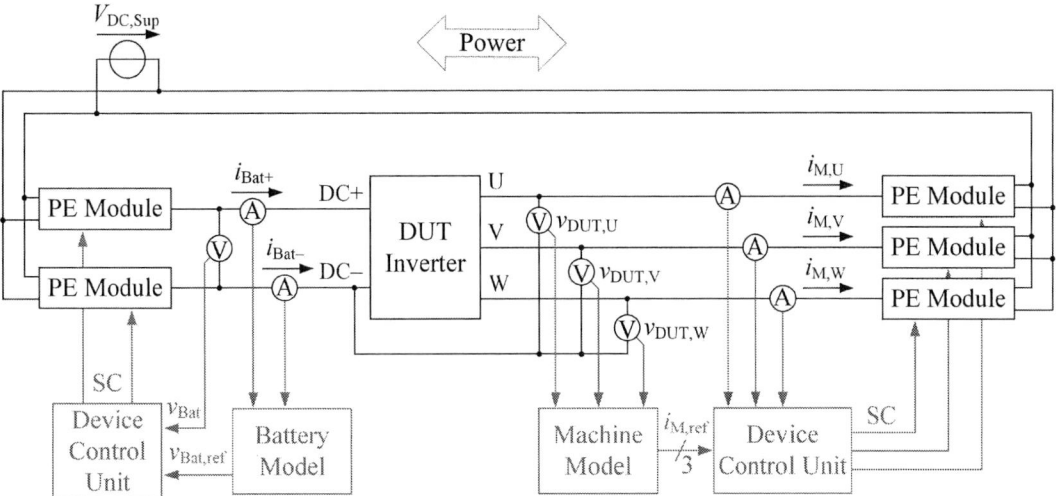

Fig. 2: General overview diagram of the entire PHIL system

Power Electronics Modules

In a PHIL application, the PE modules have to overcome multiple challenges: the HV level with DC voltages above 1,000 V, the demand for a high output current with a bandwidth of a few kHz, the capability for allowing bidirectional power flow, insertion of additional current ripple that has to be as low as possible, and modularity so it can operate multiple modules in parallel. Moreover, the electronic loads must be able to fulfill any conceivable electrical operation point. Therefore, systems with any kind of derating functions, like power hyperbola, do not meet the motor emulation demands.

A topology which fulfills these requirements is a multi-branch three-level NPC inverter, see Fig. 3. Additional output voltage levels compared to a conventional two-level inverter decrease the currents' total harmonic distortion. Moreover, in every switching state the DC link voltage $V_{DC,Sup}$ drops over at least two switches. Hence, each switch only has to feature an electric strength of half the DC link voltage. The switches are executed as fast-switching silicon-carbide MOSFETs. On each module, three NPC half bridge branches are connected in parallel. Consequently, the output current can be three times higher than the maximum current of a single branch. The three branches are each coupled in pairs by three differential-mode chokes. If the current is symmetrically distributed between the three branches, the magnetic flux inside the differential-mode chokes will be nearly zero and only the choke's magnetic leakage flux will be effective. Otherwise, in case of asymmetric distribution, the occurring magnetic flux leads to an inductance, which counteracts the asymmetry. Consequently, depending on the

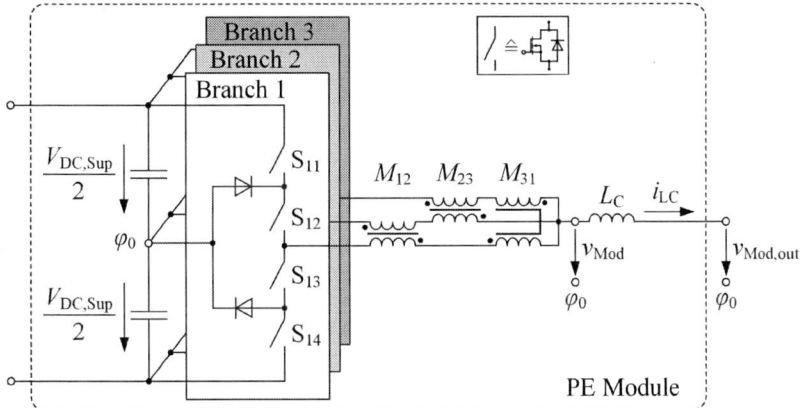

Fig. 3: Schematic of an implemented PE module

EPE'22 ECCE Europe   751

switching states of the three branches, the output voltage v_{Mod} at the common coupling point can assume seven discrete voltage states:

$$v_{\text{Mod}} = \pm \frac{k}{6} \cdot v_{\text{DC,Sup}} \quad \text{with } k = \{0, 1, 2, 3\} \tag{1}$$

After the coupling point, a further coupling inductance L_{C} is connected in series, which acts as controlled system between the DUT inverter and the PE modules.

Machine and Battery Model

The accuracy of the machine and battery models defines the quality of the emulation results. Nevertheless, one basic requirement has to be fulfilled: Both models have to run in real-time. In this paper, a three-phase, permanent magnet synchronous machine is emulated. Possible real-time models are a common or an extended fundamental component model [18,19], a magnetic equivalent circuit (MEC) model [20] or a simplified analytical model [21]. All of them use the machine's terminal voltages as input variables and calculate the three load currents. An efficient way to simulate nonlinearities, e.g., in the inductances due to saturation effects, and harmonic content is to use models in which these nonlinearities are stored in several look-up tables [22-24]. Usually, the inductance or the magnetic flux and the torque, each dependent on the machine currents and the rotor position, are stored. This way, results of an FEM model can be implemented and executed in real-time.

The setup in this paper uses the nonlinear model presented in [22-24] for emulating a three-phase, permanent magnet synchronous machine (PMSM) without limiting the capability of emulating machines with an arbitrary number of phases or other machine types. Using any explained real-time model would be also easily possible. The model's electrical part is depicted in Fig. 4. The DUT inverter's output voltages $v_{\text{DUT,x}}$ are measured and transformed into the field-oriented, rotating dq-reference frame. The model considers the voltage drops over the constant windings' resistance R_{S} on the one hand and the variable inductances L_{d} and L_{q} in direct and quadrature direction and the flux of the permanent magnet Ψ_{PM} on the other hand, all depending on the motor currents i_{d} and i_{q}. These dependencies are stored in three 3-D look-up tables. The discrete elements of the look-up tables are linearized. After transforming them back into the UVW reference frame, the three machine currents serve as reference currents for the device control unit on the machine side.

For the calculation of the inductive voltage drops and the transformation angle, the machine's mechanical behavior is also calculated. The corresponding part of the model is shown in Fig. 5. The electrical circular

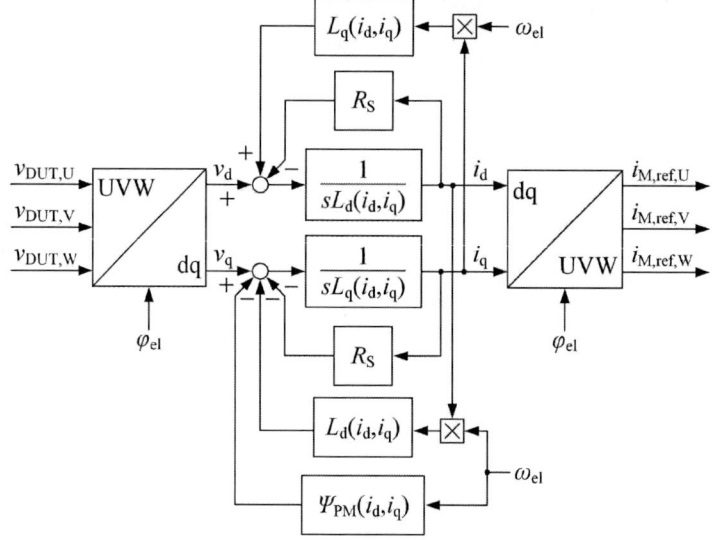

Fig. 4: Block diagram of nonlinear machine model (electrical part)

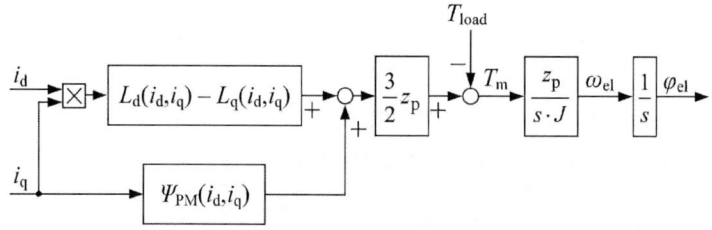

Fig. 5: Block diagram of nonlinear machine model (mechanical part)

frequency ω_{el} and the transformation angle φ_{el} are evaluated by using the machine's number of pole pairs z_{p}, the drivetrain's inertia J, and if present, an outer load torque T_{load}.

The easiest way to model the battery is as an ideal DC voltage source. The inner resistance and the dynamic behavior of the

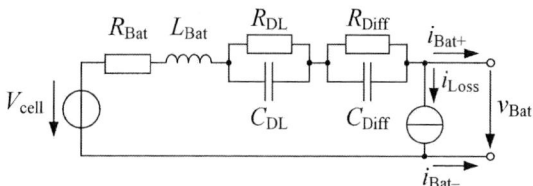

Fig. 6: Circuit diagram of the dynamic battery model

battery can be considered by simulating additional equivalent resistors and capacitors in series and in parallel to the ideal voltage source [25]. The actual battery current leads to voltage drops over the equivalent elements. Depending on the battery type, its behavior can be simulated more precisely by using more equivalent elements and a look-up table, where the values of the equivalent elements are stored as a function of the state of charge (SOC) and the state of health (SOH) [26]. The implemented equivalent circuit diagram of the used model is shown in Fig. 6. The voltage V_{cell} represents the inner cell voltage that directly depends on the SOC. Furthermore, the equivalent circuit model considers the resistance R_{Bat} and the inductance L_{Bat} of the battery, the dynamic behavior of its diffusion (R_{Diff}, C_{Diff}) and its double layer (R_{DL}, C_{DL}), and additional losses (i_{Loss}).

Device Control Units

The device control units have to calculate switching commands for the dedicated PE modules, so that the actual battery voltage v_{Bat} follows the course of the desired battery voltage $v_{\text{Bat,ref}}$, and the actual machine currents $i_{\text{M,x}}$ follow the courses of the reference currents $i_{\text{M,ref,x}}$. For this purpose, two approaches are possible: a voltage-based one and a current-based one.

In the voltage-based approach, the output voltages of the PE modules are adjusted by an open-loop controller. A multicarrier pulse width modulation (PWM) unit with six equidistant carrier signals evaluates the desired output voltages and determines the corresponding switching commands for the three half bridges per module [27]. Thereby, the average of the module's output voltage over one pulse period follows its desired value.

On the battery side, the reference battery voltage can be realized directly if the coupling inductance is $L_{\text{C}} = 0$ or the current is in the steady state. For the real setup, where dynamic processes are common, a closed-loop voltage control system must be implemented. The control path for that is the resonant circuit consisting of the modules' coupling inductances and the DC link capacitance of the DUT inverter. A state controller is recommended.

On the machine side, currents are used as the reference value. Therefore, the coherence between the control path and the current change must be analyzed in order to calculate the required countervoltages. These countervoltages are adjusted by the PE modules. The ideal control path per phase consists of the coupling inductance L_{C}. Its voltage equation is

$$v_{\text{Mod}} - v_{\text{Mod,out}} = L_{\text{C}} \cdot \frac{\mathrm{d}i_{\text{LC}}}{\mathrm{d}t} .$$
(2)

When referencing to the DUT inverter's output voltages $v_{\text{DUT,x}}$ and using the reference currents $i_{\text{M,ref,x}}$, the module must supply the following output voltage at the coupling point:

$$v_{\text{Mod,x}} = v_{\text{DUT,x}} - L_{\text{C}} \cdot \frac{\mathrm{d}i_{\text{M,ref,x}}}{\mathrm{d}t}$$
(3)

Generally, the control path differs from the ideal coupling inductor and the currents will differ from their reference currents. Hence, an additional current controller is required which can be a proportional controller due to integral behavior of the coupling inductance [8,16].

In this paper, a current-based approach is chosen. Each single module's output current i_{LC} is controlled by the model-predictive controller published in [28]. This establishes a high-bandwidth control strategy and leads to the advantage that every module is an independent current source. Moreover, this offers the option to connect and operate two or any number of modules in parallel in order to increase the overall ampacity. The current controller for each module is implemented on an FPGA, which is located decentralized on the corresponding PE module. Hereinafter, the realized entity of PE module and decentralized FPGA is called HV module.

The switching commands themselves are determined by a multi-carrier PWM and a contracyclical interrupt, called 'DUT state detection'. Both select one of the seven possible output voltage states. The DUT state detection responds to changes in the DUT output voltages immediately within 500 ns in order to enable a highly dynamic response of the PHIL system.

Four out of seven possible module output voltages ($k = 1$ or $k = 2$ in eq. 1) can be adjusted each by three different switching states. This additional degree of freedom is used to ensure a symmetric current distribution between the three branches per PE module and avoids saturation in the differential-mode chokes. Each of the three possible switching states effects different slew rates and especially directions of change in the courses of the three branch currents i_1, i_2 and i_3, see Table I. In this table, state "1" means that the branch's output potential is connected to the upper DC link potential (and vice versa for state "−1" and the lower DC link potential). State "0" means, that the branch's output potential is connected to the neutral potential φ_0.

Cyclically, with a higher frequency than the PWM frequency, the device control unit decides by means of the measured values of the branch currents which switching state is most suitable in order to counteract asymmetry. For example, if an output voltage $v_{\text{Mod}} = 2/6 \cdot V_{\text{DC,Sup}}$ is required and the branch current i_1 is higher than the currents i_2 and i_3, the device control unit selects switching state [0 1 1] in order to decrease current i_1. As soon as i_1 is smaller than i_2 and i_3 and i_2 becomes the highest of these

Table I: Possible switching states

v_{Mod}	Sw. State	di_1/dt	di_2/dt	di_3/dt
$V_{\text{DC,Sup}}/2$	[1 1 1]	0	0	0
$2/6 \cdot V_{\text{DC,Sup}}$	[1 1 0]	> 0	> 0	< 0
	[1 0 1]	> 0	< 0	> 0
	[0 1 1]	< 0	> 0	> 0
$1/6 \cdot V_{\text{DC,Sup}}$	[1 0 0]	> 0	< 0	< 0
	[0 1 0]	< 0	> 0	< 0
	[0 0 1]	< 0	< 0	> 0
0 V	[0 0 0]	0	0	0
$- 1/6 \cdot V_{\text{DC,Sup}}$	[−1 0 0]	< 0	> 0	> 0
	[0 −1 0]	> 0	< 0	> 0
	[0 0 −1]	> 0	> 0	< 0
$- 2/6 \cdot V_{\text{DC,Sup}}$	[−1 −1 0]	< 0	< 0	> 0
	[−1 0 −1]	< 0	> 0	< 0
	[0 −1 −1]	> 0	< 0	< 0
$- V_{\text{DC,Sup}}/2$	[−1 −1 −1]	0	0	0

three currents, the switching state changes at the next cyclical interrupt to [1 0 1] without affecting the output voltage.

The PHIL implementation on the machine side is quite simple. The calculated reference currents per phase, each divided through the number of modules connected in parallel if necessary, are sent to the controllers on the HV modules. They adjust very dynamically so that the actual phase currents follow the courses of the reference currents.

In case of battery voltage emulation, the HV modules have to adjust the output voltages. To this end, an output capacitance is required as a controlled system. In some cases, the DUT inverter's DC link capacitance is sufficient, alternatively an additional capacitor must be added between the output terminals of the HV modules emulating the battery. At the same time, an additional voltage control is implemented, which overlays the modules' current controllers. The reduction of control speed, which is caused by an overlain control system, is not a concern because the course of the battery voltage does not include high frequency harmonics.

Implementation in Hardware and Measurement Results

The overall PHIL system is implemented in hardware and can be operated at DC link voltages above $V_{\text{DC,Sup}} = 1,000$ V. Fig. 7 shows the hardware setup of one HV module. Each HV module is designed to carry currents up to 75 A_{RMS} and reach slew rates up to 5 A/µs. If the device under test (DUT) requires higher currents or higher slew rates of the current, any number of modules can be operated in parallel to enlarge these characteristic numbers. The HV modules are equipped with a decentralized FPGA performing the current control and the switching state selection part of the

Fig. 7: Setup of the realized HV module

Coupling inductor
Differential-mode choke
FPGA
SiC MOSFETs
DC input
Water cooling

device control unit. To avoid a high amount of switching interference, the PWM's carrier frequency is set to $f_{PWM} = 200$ kHz in this paper. However, variable switching frequencies up to 800 kHz are possible due to the DUT state detection and the switching state selection.

Both the machine and the battery model are implemented on a high-performance real-time FPGA board with very fast I/O interfaces. Only the reference currents are transmitted from this real-time FPGA to the decentralized FPGAs placed on the HV modules. A real-time processor serves as an interface between the real-time FPGA and the control computer. Thereby, model parameters can be changed during run time. The real-time FPGA has a sample time of 8 ns.

For the measurements results in this paper, the setup of Fig. 2 with a single module per battery pole and per machine phase is set up in hardware. The DUT inverter is a three-phase machine inverter. As the focus is on machine currents, the battery voltage is controlled to a constant value. The parameters of the DUT inverter and the PHIL system are summarized in Table II.

The virtual PMSM is adjusted to a constant speed of $n = 1,000$ rpm. The DUT inverter operates in current control mode. Its reference currents are set to $i_{DUT,ref,d} = 0$ A in direct direction and $i_{DUT,ref,q} = 50$ A in quadrature direction. Thus, a nearly constant virtual torque of $T_m = 195$ Nm occurs. The measurement results of DUT inverter's phase U output current $i_{M,U}$ and its reference value $i_{M,ref,U}$ from the machine model are shown in Fig. 8. The closed-loop control of the DUT inverter leads to a sinusoidal current waveform with an amplitude of 50 A as desired. The real load current follows its reference values from the machine model without any visible deviation. Hence, a detailed view of the currents is observed, see Fig. 9. Due to its high bandwidth, the PHIL system is able to emulate the reference current accurately, even the ripple behavior caused by the DUT inverter's switched-mode operation. Only a small amount of additional current ripple is observed and results the HV modules' own switching behavior. If the PHIL system's DC link voltage is higher than 1,000 V, the magnitude of the additional disturbance ripple content will

Table II: Parameters of the hardware setup

PHIL System Parameters	
DC link voltage $V_{DC,Sup}$	800 V
HV module half bridges' switching frequency	≤ 800 kHz
HV modules per battery pole	1
HV modules per motor phase	1
DUT Inverter Parameters	
Battery voltage v_{Bat} (constant)	500 V
Switching frequency	5.2 kHz
Machine Model Parameters	
Machine type	PMSM
Number of pole pairs z_p	3
Windings resistance R_S	0.01 Ω
Inductance in direct direction L_d	2.7 mH
Inductance in quadrature direction L_q	2.7 mH
Flux of the permanent magnet Ψ_{PM}	0.87 Vs

increase, but the emulation of the desired currents' slew rate is still more accurate than known from other motor emulators.

Conclusion

This paper presents the entire setup of a powerful PHIL system which is suitable for high-voltage and high-frequency application. The setup of an HV module is introduced which behaves as a modular, high-quality current source with a high bandwidth. Each HV module is current-controlled, so multiple modules can be operated in parallel to increase the overall system power. Moreover, two different kinds of control schemes are presented. For the final implementation, the more sophisticated control approach is chosen, which selects the most appropriate switching state in order to counteract an asymmetric current distribution within an HV module. The reference values of the battery and machine currents are calculated by real-time machine and battery models.

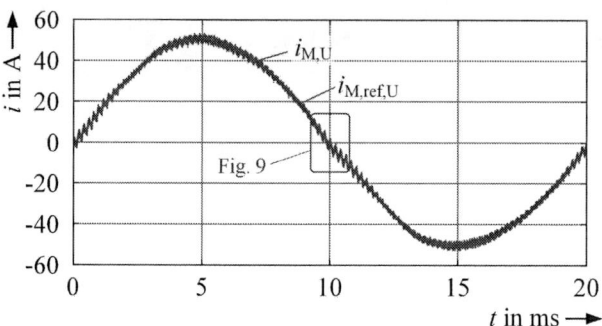

Fig. 8: Measurement of the phase current

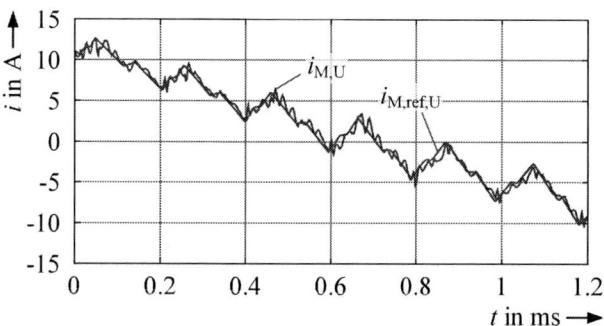

Fig. 9: Detailed view of the measured phase current

For this purpose, multiple real-time capable models are introduced. For the machine model in particular, a nonlinear approach using look-up tables is used in the final implementation.

The entire PHIL system is set up in hardware. Its functionality is proven by measurement results, which highlight the strong performance regarding real-time capability and emulation accuracy.

References

[1] C. Nemec and J. Roth-Stielow, "Ripple current minimization of an interleaved-switched multi-phase PWM inverter for three-phase machine-emulation," 14th European Conference on Power Electronics and Applications, 2011.

[2] A. Schmitt, J. Richter, M. Braun, and M. Doppelbauer, "Power Hardware-in-the-Loop Emulation of Permanent Magnet Synchronous Machines with Nonlinear Magnetics – Concept & Verification," PCIM Europe, pp. 393-400, 2016.

[3] M. Fischer, S. Petzner, J. Ruthardt, J. Schuster, S. Bintz, and J. Roth-Stielow, "Current Control for a Multiphase Interleaved-Switched Inverter Using Field Oriented Coordinates," 20th European Conference on Power Electronics and Applications, 2018.

[4] A. Schmitt, J. Richter, M. Gommeringer, T. Wersal, and M. Braun, "A Novel 100 kW Power Hardware-in-the-Loop Emulation Test Bench for Permanent Magnet Synchronous Machines with Nonlinear Magnetics," 8th IET International Conference on Power Electronics, Machines and Drives, 2016.

[5] M. Fischer, J. Ruthardt, M. Nitzsche, P. Ziegler, and J. Roth-Stielow, "Investigation on Carrier Signals to Minimize the Overall Current Ripple of an Interleaved-Switched Inverter," PCIM Europe, pp. 1644-1649, 2019.

[6] R. Sudharshan Kaarthik, and P. Pillay, "Emulation of a Permanent Magnet Synchronous Generator in Real-time Using Power Hardware-in- the-loop," International Conference on Power Electronics, Drives and Energy Systems, 2016.

[7] M. Fischer, F. Gliese, J. Ruthardt, M. Zehelein, and J. Roth-Stielow, "Investigation on a Three-Phase Inverter with LC Output Filter for Machine Emulation," 21st European Conference on Power Electronics and Applications, 2019.

[8] M. Fischer, D. Erthle, P. Ziegler, J. Ruthardt, and J. Roth-Stielow, "Comparison of Two Power Electronic Topologies for Power Hardware in the Loop Machine Emulator," IEEE Applied Power Electronics Conference and Exposition (APEC), pp. 2950-2954, 2020.

[9] S. Lentijo, S. D'Arco, and A. Monti, "Comparing the Dynamic Performances of Power Hardware-in-the-Loop Interfaces," in IEEE Transactions on Industrial Electronics, vol. 57, no. 4, pp. 1195-1207, 2010.

[10] G. Si and R. Kennel, "Switch mode converter based high performance power-hardware-in-the-loop grid emulator," 2016 IEEE 2nd Annual Southern Power Electronics Conference (SPEC), pp. 1-6, 2016.

[11] A. Monti, S. D'Arco, and A. Deshmukh, "A new architecture for low cost Power Hardware in the Loop testing of power electronics equipments," 2008 IEEE International Symposium on Industrial Electronics, Cambridge, pp. 2183-2188, 2008.

[12] S. L. Baciu, S. Trabelsi, B. Amlang, and W. Schumacher, "Linverter a low-harmonic and high-bandwidth inverter based on a parallel multilevel structure," in 2004 IEEE 35th Annual Power Electronics Specialists Conference, vol. 5, no. 5, pp. 3927–3931, 2004.

[13] M. Fischer, R. Malic, N. Tröster, J. Ruthardt, and J. Roth-Stielow, "Design of a Three-Phase 100 Ampere Linear Amplifier for Power-Hardware-in-the-Loop Machine-Emulation," IET The Journal of Engineering, no. 17, pp. 4041-4044, 2019.

[14] K. S. Amitkumar, R. Thike, and P. Pillay, "Linear Amplifier-Based Power-Hardware-in-the-Loop Emulation of a Variable Flux Machine," IEEE Transactions on Industry Applications, vol. 55, no. 5, pp. 4624–4632, 2019.

[15] M. Fischer, J. Ruthardt, V. Ketchedjian, P. Ziegler, M. Nitzsche, and J. Roth-Stielow, "Four-Level Inverter with Variable Voltage Levels for Hardware-in-the-Loop Emulation of Three-Phase Machines," 22nd European Conference on Power Electronics and Applications, Lyon, France, pp. 1-8, 2020.

[16] M. Fischer, Y. Hu, J. Ruthardt, P. Ziegler, J. Haarer and J. Roth-Stielow, "Comparison of an Interleaved Multi-Branch Inverter and a Four-Level Inverter with Variable Voltage Levels for Emulation of Three-Phase Machines," 2021 IEEE Energy Conversion Congress and Exposition (ECCE), pp. 2570-2575, 2021.

[17] M. Schnarrenberger, L. Stefanski, C. Rollbühler, D. Bräckle, and M. Braun, "A 50 kW Power Hardware-in-the-Loop Test Bench for Permanent Magnet Synchronous Machines based on a Modular Multilevel Converter," 20th European Conference on Power Electronics and Applications, 2018.

[18] P. Pillay and R. Krishnan, "Modeling of permanent magnet motor drives," in IEEE Transactions on Industrial Electronics, vol. 35, no. 4, pp. 537-541, Nov. 1988.

[19] S. Decker, J. Stoss, A. Liske, M. Brodatzki, J. Kolb, and M. Braun, "Online Parameter Identification of Permanent Magnet Synchronous Machines with Nonlinear Magnetics based on the Inverter Induced Current Slopes and the dq-System Equations," EPE '19 ECCE Europe, 2019.

[20] R. Manju Bhashini and K. Ragavan, "Magnetic Equivalent Circuit for Surface-Mounted PM Motor," 2018 IEEE International Conference on Power Electronics, Drives and Energy Systems (PEDES), pp. 1-5, 2018.

[21] A. L. Rodríguez, D. J. Gómez, I. Villar, A. López-de-Heredia, and I. Etxeberria-Otadui, "Improved analytical multiphysical modeling of a surface PMSM," 2014 International Conference on Electrical Machines (ICEM), pp. 1224-1230, 2014.

[22] M. Boesing, M. Niessen, T. Lange, and R. De Doncker, "Modeling spatial harmonics and switching frequencies in PM synchronous machines and their electromagnetic forces," 2012 XXth International Conference on Electrical Machines, pp. 3001-3007, 2012.

[23] M. Plöger and M. Deter, "Highly precise real-time simulation of E-motors," in ATZ Elektronik, 2013.

[24] A. Schmitt, J. Richter, U. Jurewitz and M. Braun, "FPGA-Based Real-Time Simulation of Nonlinear Permanent Magnet Synchronous Machines for Power Hardware-in-the-Loop Emulation Systems," in Industrial Electronics Society, IECON 2014 - 40th Annual Conference of the IEEE, Dallas, 2014.

[25] W. S. Putra, B. R. Dewangga, A. Cahyadi, and O. Wahyunggoro, "Current estimation using Thevenin battery model," Proceedings of the Joint International Conference on Electric Vehicular Technology and Industrial, Mechanical, Electrical and Chemical Engineering (ICEVT & IMECE), 2015, pp. 5-9.

[26] X. Zhang, W. Zhang, and G. Lei, "A Review of Li-ion Battery Equivalent Circuit Models," Transactions on Electrical and Electronic Materials, 17, pp. 311-316, 2016.

[27] T. Prathiba and P. Renuga, "Multi Carrier PWM based Multi Level Inverter for High Power Application," in International Journal of Computer Applications, 2010.

[28] G. Meyer, "Enhanced Power Electronics System for High-Performance Testing of Motor Control Units in a Power HIL Environment," PCIM Asia 2017; International Exhibition and Conference for Power Electronics, Intelligent Motion, Renewable Energy and Energy Management, pp. 1-8. 2017.

Analysis and Discussion of Different Three-Phase dv/dt Filter Topologies and the Influences of Their Filter Parameters on Losses and EMC

[1]Eric Fritze, [1]Michael Meissner, [1]Klaus F. Hoffmann,
[1]Kai-Uwe Rathjen, [1]Stefan Dickmann, [2]Oliver Woywode
[1]HELMUT SCHMIDT UNIVERSITY
[2] PHILIPS MEDICAL SYSTEMS DMC GMBH
[1]Holstenhofweg 85
[1]Hamburg, Germany
Phone: +49 (0) 40 6541 2713
Fax: +49 (0) 040 6541 2018
Email: eric.fritze@hsu-hh.de
URL: https://www.hsu-hh.de/lek/en/

Acknowledgements

This work has been supported by the Federal Ministry of Education and Research of Germany (BMBF grant FKZ 13GW0326E). The responsibility for the content of this paper lies with the authors.

Keywords

≪Passive filter≫, ≪EMC/EMI≫, ≪Wide bandgap devices≫, ≪Reliability≫, ≪Efficiency≫, ≪Electrical machine≫, ≪Automotive application≫, ≪Converter machine interactions≫.

Abstract

In this paper two modified three-phase dv/dt filter networks are analysed. They're investigated in a fast switching SiC-MOSFET inverter system with the primary purpose of reducing the steep voltage slopes caused by the wide bandgap semiconductor devices at the output of the inverter. This should preserve the insulation system of a connected electrical load machine from partial discharges and deterioration. The necessity of damping down the voltage slopes outside of the inverter can arise from the possibility of lower switching losses and more favourable temperatures within the inverter. An additional filter of course will cause power losses depending on it's parameters. Hence, the influence of the passive filter parameters on those losses are an integral part of the analysis. In addition to the slope-damping of the inverter output voltage, the two discussed filter topologies should also challenge the conducted EMI of the system. Therefore, both of the circuits use special modifications to reduce the interference levels introduced by the fast switching inverter. Furthermore, a diode clamping of the filter output voltage to the DC-link potentials is implemented and it's impact on the dv/dt reduction, filter losses and EMI is discussed. Generally, a higher filter inductance, meaning a lower necessary capacitance, will reduce the extra filter losses at the expense of a higher filter volume as well as additional oscillations of the output voltage. The diode clamping leads to a significantly less overshoot and ringing of the resulting voltages, but to an increase of the measured losses. Finally, examining the influences of the different filter topologies, their parameters and the diode clamping on the conducted EMC behaviour, it will be shown that both of the modified topologies have the ability to reduce EMI levels in certain areas.

Introduction

The development and commercial availability of wide bandgap devices based on e. g. silicon carbide (SiC) and gallium nitride (GaN) enhances the possibilities of designing more efficient, compact and high-performance systems. Due to higher switching speeds, lower switching losses combined with low on-resistance and higher thermal conductivity, both high-temperature applications and high switching frequencies can be addressed as well [1, 2]. The wide field of applications empowers their usage in DC-DC converters and DC-AC traction inverters, for example in the automotive industry.

However, the combination of modern fast switching inverter topologies and electric machines recently causes several technical challenges, such as cable reflection phenomena due to long stator cables and problems concerning EMI, shaft voltages and bearing currents [3, 4]. But also using short cable lengths, the permitted rise times of the inverter output voltages have to be limited to prevent partial discharges and premature insulation failure of the electric machines' stator windings [5]. One way to face those problems while maintaining high switching speeds of the used semiconductors and favoring lower power loss-based heating are $\mathrm{d}v/\mathrm{d}t$ filter networks in order to damp the voltage gradients to a value which is non-critical for the machine insulation system and to reduce EMI. Although losses are induced by the usage of $\mathrm{d}v/\mathrm{d}t$ damping networks as well, the benefit of lower semiconductor temperatures at comparable efficiencies is remarkable concerning the durability and reliability of the devices [6].

In this paper two $\mathrm{d}v/\mathrm{d}t$ filter topologies are evaluated. For this, a fast switching, three-phase 600 V SiC-MOSFET inverter is used, which generates voltage slopes of up to $50\,\mathrm{V/ns}$. The filters are designed to damp the voltage transients at the filter output to a maximum of $10\text{-}15\,\mathrm{V/ns}$. Therefore, different *LCR*-parameter-combinations are analysed. Besides their effectiveness in reducing high voltage transients, the influence of parameters like the size of the filter inductance on occuring filter losses and the efficiency of the system is examined. To prevent overvoltages a diode clamping of the filter output voltage to the DC-link potentials is implemented. Finally, the impact of various filter topology setups and extensions as well as the influence of the diode clamping on the conducted EMC behaviour is presented.

Measurement setups

In previous publications, passive filter topologies have already been discussed concerning their advantages in both differential mode $\mathrm{d}v/\mathrm{d}t$ reduction and common mode (CM) voltage suppression [7, 8, 9]. Lots of them have used conventional IGBT inverter technology with medium to low voltage transients compared to the dynamic switching abilities of modern wide bandgap devices such as SiC MOSFETs. In this paper, those filter topologies will be investigated in a wide bandgap setup using a fast switching three-phase SiC inverter.

Figure 1 shows the standard *LCR*-filter topology connected between a three-phase SiC MOSFET inverter and a machine load. The used additional diode clamping of the filter output voltage to the DC-link voltage (marked in grey) should prevent a high voltage overshoot. In figure 2 the above mentioned extensions of the standard topology are presented to challenge both common and differential mode EMI by using a simple $\mathrm{d}v/\mathrm{d}t$ filter network. The filter structure in figure 2a uses an accessible midpoint of the DC-

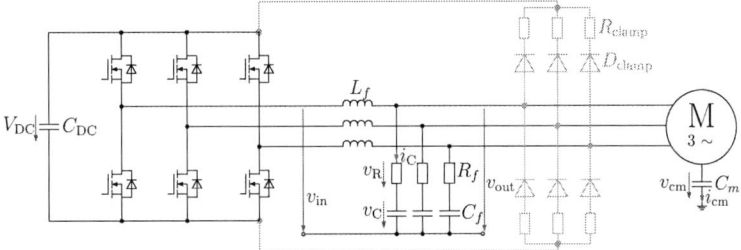

Fig. 1: Standard three-phase *LCR* $\mathrm{d}v/\mathrm{d}t$ filter topology

link (marked as O), which is connected to the neutral point of the three *RC*-legs of the filter to create a common mode path for high-frequency noise caused by the SiC inverter [8]. This is implemented to additionally reduce the amount of common mode noise leaving the drive system for example through a parasitic capacitance C_m from the machine load to ground. For the same purpose, the filter topology shown in figure 2b uses a symmetrical connection of two three-phase RC-legs to both of the DC-link potentials to reduce the dv/dt at each switching operation [9]. An advantage of this topology, despite the higher amount of filter components needed, is that the DC-link positive and negative potentials are generally more easily accessible than a DC-link midpoint of an inverter.

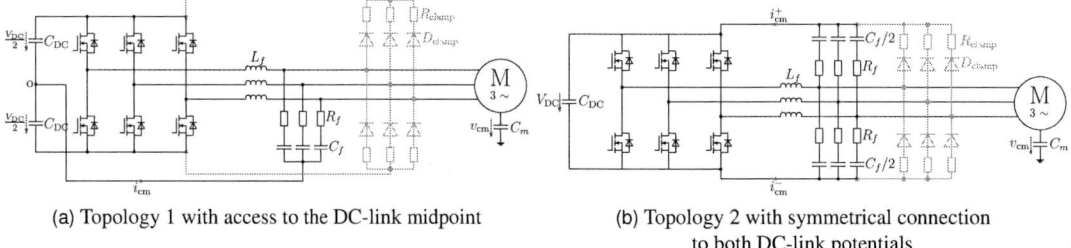

(a) Topology 1 with access to the DC-link midpoint

(b) Topology 2 with symmetrical connection to both DC-link potentials

Fig. 2: Filter topologies to challenge common and differential mode EMI

Dimensioning and filter design

The LCR-parameter-combinations were chosen in order to meet the dv/dt-damping requirements of reducing a voltage slope of $50\,\text{V/ns}$ down to a maximum of $10\,\text{-}15\,\text{V/ns}$ and to keep the filter losses manageable. According to [10], the losses dissipated by the damping resistors of one phase can be estimated by the equation

$$P_{\text{loss}} = f_{\text{load}} \cdot C_f \cdot \Delta V_C^2, \tag{1}$$

which means that they are proportional to the size of the filter capacitor. It should be considered that ΔV_C is the phase-to-neutral voltage of the inverter in this three-phase filter topology. Another analytical approach to calculate the losses of a three-phase filter is given in [4]. The correlation between L_f and C_f in order to achieve the required resonance frequency for the dv/dt low-pass filter ($\omega_0 = 1/\sqrt{L_f C_f}$) describes that by reducing C_f to minimize the filter losses, L_f has to be increased. This would lead to higher space requirements of the filter as well as a higher voltage drop across the inductor, reducing the effective exploitation of the DC-link voltage transferred to the load [4]. Hence, a compromise between capacitance and inductance has to be found. For the filter dimensioning in this paper, at first the desired inductance value L_f was selected, mainly by availability and size. Afterwards, the values of C_f and R_f were determined by an optimising algorithm following the Nelder–Mead method in order to meet the slope-damping criterion. To configure the filter parameters analytically, an alternative approach is also given e.g. in [5]. The chosen dimensioning can be checked previous to measurement by solving the differential equations of a one-phase LCR-circuit, which are excited by an equivalent ramp function. The following equations

$$\frac{\Delta V_{\text{in}}}{\Delta T} \cdot t = L_f \cdot \frac{di(t)}{dt} + R_f \cdot i(t) + \frac{1}{C_f} \int i(t) dt + K_1 \qquad \Delta V_{\text{in}} = 600\,\text{V} \qquad \Delta T = 12\,\text{ns} \tag{2}$$

and

$$V_{\text{in}} = L_f \cdot \frac{di(t)}{dt} + R_f \cdot i(t) + \frac{1}{C_f} \int i(t) dt + K_2 \qquad V_{\text{in}} = 600\,\text{V} \tag{3}$$

therefore consider a linear part and a constant part of an excitatory ramp function, respectively. The voltage and current curves resulting from this are exemplarily depicted in figure 3. The characteristics of the filter output voltage were calculated according to the case of a damped harmonic oscillator. Due to more parasitic passive elements in the real setup, increased ringing and a higher overshoot of the voltage can be expected. Furthermore, the real inverter output voltage will not occur as a perfect ramp with constantly high slope, but the presented method is still suitable for the parameter dimensioning. Another aspect to mention is that this approach doesn't include the diode clamping. The mathematical behaviour of the diodes would introduce non-linearity to the differential equations presented in equation (2) and (3). Therefore, the behaviour of the filter constellations utilizing a diode clamping was checked with an appropriate simulation model in Simulink / PLECS Blockset. Both of the discussed filter topologies were analysed using four parameter combinations with different inductances, see table I. For the clamping, SiC Schottky diodes are used with a resistance of $R_{\text{clamp}} = 7.5\,\Omega$ in series (see figure 2).

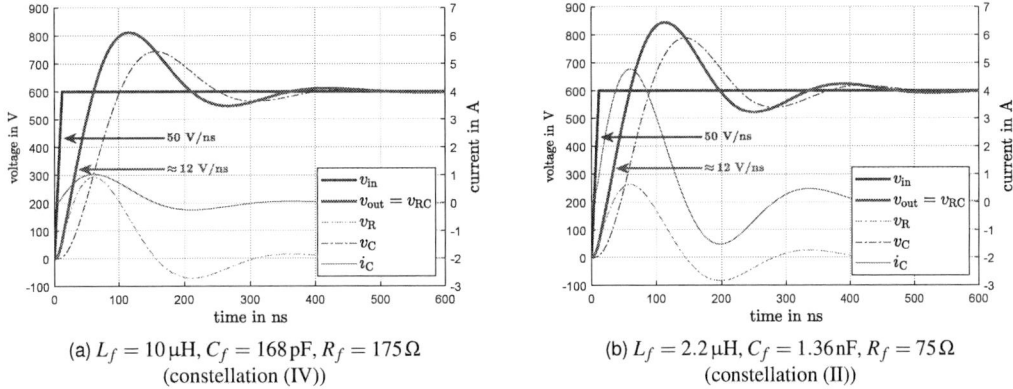

(a) $L_f = 10\,\mu\text{H}$, $C_f = 168\,\text{pF}$, $R_f = 175\,\Omega$
(constellation (IV))

(b) $L_f = 2.2\,\mu\text{H}$, $C_f = 1.36\,\text{nF}$, $R_f = 75\,\Omega$
(constellation (II))

Fig. 3: Voltage and current curves of the standard LCR-circuits based on analytical solutions of the differential equations excited by a $50\,\text{V/ns}$ ramp function (see figure 1)

The $\mathrm{d}v/\mathrm{d}t$-filter PCB layouts are given in figure 4. In order to investigate different LCR-combinations with the same circuit board, various components and packages were included on the PCBs, dependent on the desired constellation. They were designed as symmetrical as possible to achieve similar behaviours on all three phases.

(a) PCB-layout of filter topology 1

(b) PCB-layout of filter topology 2

Fig. 4: Designed $\mathrm{d}v/\mathrm{d}t$-filter PCBs for investigation of the different topologies and LCR-combinations

Table I: L, C, R parameters used for the measurements

Constellation	Filter topology 1			Filter topology 2		
	L_f	C_f	R_f	L_f	$c_f/2$	R_f
(I)	$1.2\,\mu\text{H}$	$1.33\,\text{nF}$	$15\,\Omega$	$1.2\,\mu\text{H}$	$680\,\text{pF}$	$15\,\Omega$
(II)	$2.2\,\mu\text{H}$	$780\,\text{pF}$	$33\,\Omega$	$2.2\,\mu\text{H}$	$470\,\text{pF}$	$33\,\Omega$
(III)	$4.2\,\mu\text{H}$	$470\,\text{pF}$	$75\,\Omega$	$4.2\,\mu\text{H}$	$200\,\text{pF}$	$75\,\Omega$
(IV)	$10\,\mu\text{H}$	$168\,\text{pF}$	$175\,\Omega$	$10\,\mu\text{H}$	$100\,\text{pF}$	$175\,\Omega$

Measurement results

For the performance analysis of the filter setups as well as for the EMC measurements, a DC-link voltage of 600 V and a switching frequency of 30 kHz were defined. The fundamental output frequency of the inverter was set to 1.5 kHz. The electrical machine was replaced by an equivalent RL-load to achieve a RMS load current of about 11.7 A at steady-state. To prevent overheating, the inverter baseplate was cooled down to 20 °C using a water-cooled heat sink and external fans were used to cool the damping resistors and components of the filter. In addition to the modified output dv/dt-filters, an input common mode filter was used. It consisted of a 5 mH and a 0.9 mH common mode choke in series as well as a 47 nF parallel capacitance, resulting in a resonance frequency of the filter of about 9.56 kHz.

Performance analysis and losses

Measurement results that confirm the functionality of the dv/dt-filters in terms of slope reduction are given in figure 5, for constellation (I) of filter topology 1 and constellation (III) of filter topology 2, respectively. To analyse the maximum dv/dt of all the constellations accordingly, the observed timespan ΔT was set fix to 30 timesteps of the discretely sampled signal, meaning 4.8 ns. The dv/dt values were then calculated as $\Delta V/\Delta T$. In general, the voltage slopes and curves show a good fit compared to the parameter layout and preliminary expectations. The highest dv/dt values reached by the inverter of about 52 V/ns are damped down to a maximum of about 12 V/ns at the output of the filter networks. The shown pulses are representative for the remaining constellations, which all showed a similar dv/dt reduction in the desired area. However, it has to be mentioned that the diode clamping has a big influence concerning the occuring overvoltages and ringing of the filter output. The overshoot (marked in green) compared to the measurements without clamping is remarkably lower. Omitting the diode clamping results in overvoltages of up to 1 kV. This value is expected to be problematic for the insulation system of a prospective electrical machine connected to the filter network [11]. The high amount of voltage ringing can furthermore contribute to higher amounts of EMI caused by the inverter system.

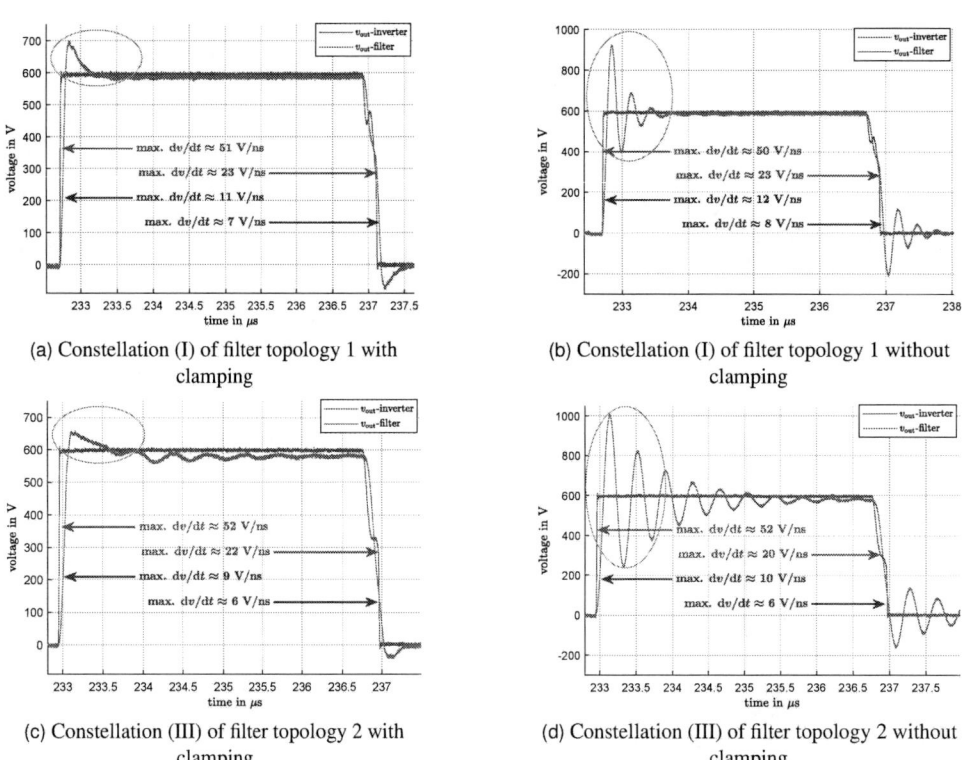

(a) Constellation (I) of filter topology 1 with clamping

(b) Constellation (I) of filter topology 1 without clamping

(c) Constellation (III) of filter topology 2 with clamping

(d) Constellation (III) of filter topology 2 without clamping

Fig. 5: Measurement results of inverter and dv/dt-filter output voltages regarding representative phase-to-phase voltage pulses

In figure 6, each oscillogram presents three representative output voltage pulses to further investigate the impact of the diode clamping as well as the filter parameters in terms of ringing. It is clearly observable that an increase of the filter inductance L_f, meaning also a decrease of capacitance C_f, leads to higher ringing of the filter output voltage in both setups with and without diode clamping. For the constellation shown in figure 6d, regarding the first falling slope, this means that the oscillation is not decayed until the next positive voltage pulse. This behaviour can be referred to a higher parasitic capacitance of the larger coil as well as the lower voltage-stabilizing capacitance and was also observed for filter topology 2. Moreover, a higher steady-state voltage difference from inverter output voltage to filter output voltage is noticeable with increasing filter inductance due to a higher voltage drop across L_f. Referring to

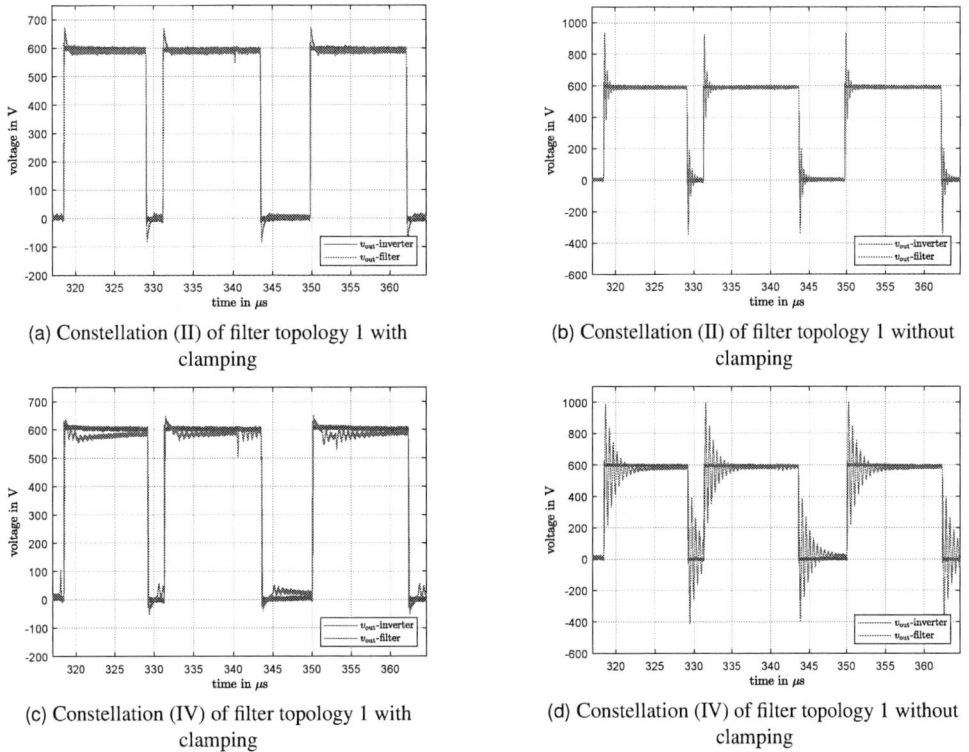

(a) Constellation (II) of filter topology 1 with clamping

(b) Constellation (II) of filter topology 1 without clamping

(c) Constellation (IV) of filter topology 1 with clamping

(d) Constellation (IV) of filter topology 1 without clamping

Fig. 6: Measurement results showing three pulses of inverter and filter output voltages

the filter constellations of table I and the previously mentioned performance settings of the inverter, also the power losses P_{loss} of the different filter circuits were analysed. They are specified in table II. For reference, the inverter PCB, including the input CM-filter, was also tested without any dv/dt-filter leading to power losses of 86 W at an efficiency of 98.75 %. Those losses are implied in table II and III. It can be seen clearly that the size of the inductance behaves inversely proportional to the capacitance value and therefore to the occuring losses that are dissipated mainly by the damping resistors. The increase of inductance from constellation (I) to (IV) of 8.8 µH comes along with a decrease of capacitance by almost 90 % and reduces the system losses significantly by 29 W for topology 1. Regarding topology 2, the filter capacitance decreases by 85 % and the losses by 23 W.

Table II: Filter power losses and total system efficiencies of the filter circuits with diode clamping

Constellation	Filter topology 1		Filter topology 2	
	P_{loss}	η	P_{loss}	η
(I)	136 W	97.96 %	131 W	98.12 %
(II)	128 W	98.07 %	121 W	98.25 %
(III)	117 W	98.22 %	111 W	98.40 %
(IV)	107 W	98.35 %	108 W	98.43 %

Concerning the impact of the diode clamping on arising losses, table III shows the measurement results of all previously mentioned constellations without clamping of the output voltage. Here, a noticeable reduction of power losses is observable regarding every setup. This can be referred to the omitted losses dissipated by the clamping diodes and resistors. However, considering the measurements in figure 6, this reduction of losses comes along with rising oscillations of the filter output voltage.

Table III: Filter power losses and total system efficiencies of the filter circuits without diode clamping

Constellation	Filter topology 1		Filter topology 2	
	P_{loss}	η	P_{loss}	η
(I)	129 W	98.07 %	129 W	98.15 %
(II)	117 W	98.23 %	108 W	98.44 %
(III)	105 W	98.40 %	87 W	98.74 %
(IV)	93 W	98.57 %	90 W	98.69 %

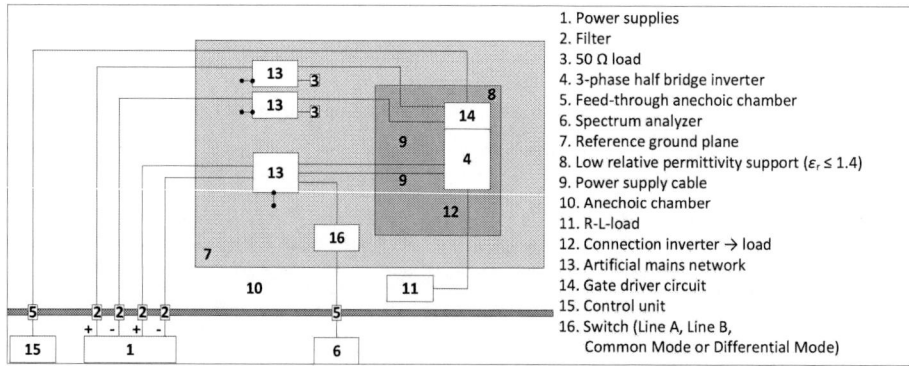

1. Power supplies
2. Filter
3. 50 Ω load
4. 3-phase half bridge inverter
5. Feed-through anechoic chamber
6. Spectrum analyzer
7. Reference ground plane
8. Low relative permittivity support ($\varepsilon_r \leq 1.4$)
9. Power supply cable
10. Anechoic chamber
11. R-L-load
12. Connection inverter → load
13. Artificial mains network
14. Gate driver circuit
15. Control unit
16. Switch (Line A, Line B, Common Mode or Differential Mode)

Fig. 7: Measurement setup for the common mode EMI measurements

EMC measurement

The measurement setup for the common mode EMI measurements is depicted in figure 7. As frame condition and in order to investigate the EMC effects under a standard-conditioned environment, the standard *CISPR25* was the used basis [12]. In the measurements the inverter, the driver circuit board as well as the input and output filter constellations were included. The results are presented in figure 8 and 9. To compare the impact of the modified dv/dt-filters on the common mode interference levels, the measurement of the inverter and input filter without any output filter is depicted in black in each figure.

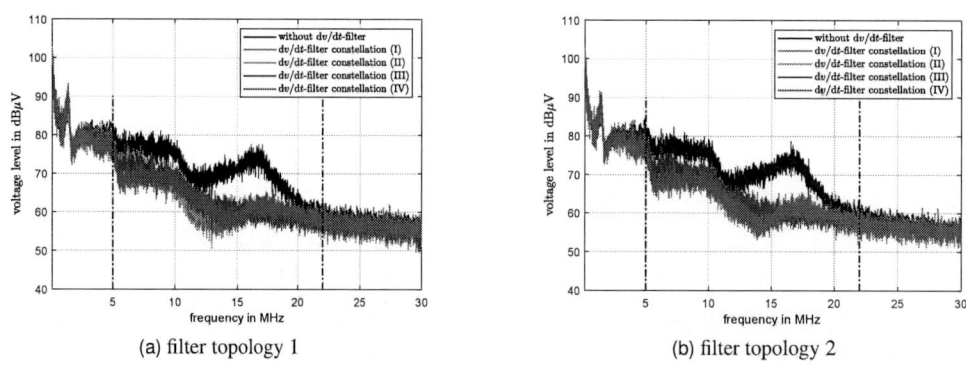

(a) filter topology 1

(b) filter topology 2

Fig. 8: Common mode EMI measurement results of the modified dv/dt-filter circuits with diode clamping of the output voltage

At lower frequencies of up to 2 MHz the voltage curves don't show noticeable differences, which is the reason why a linear display of the frequency domain was chosen instead of a logarithmic display. Despite that, in general the measurement without output filter shows remarkably higher voltage levels than the measurements with the modified dv/dt-filter networks, meaning that the modifications for a better common mode behaviour indeed show a positive impact. This applies for both presented filter topologies, especially in the wide interval of 5 MHz to 22 MHz. All four LCR-constellations are characterised by an overall similar behaviour for filter topology 1. Constellation (IV) shows a more advantageous behaviour regarding filter topology 2, in the area around 14 MHz, compared to other constellations.

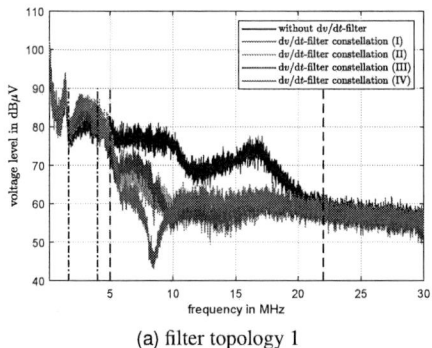
(a) filter topology 1

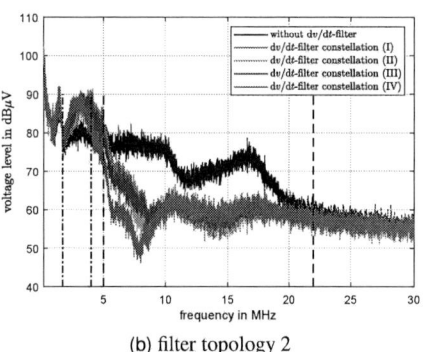
(b) filter topology 2

Fig. 9: Common mode EMI measurement results of the modified dv/dt-filter circuits
without diode clamping of the output voltage

Regarding figure 9, remarkable differences between the measurements with and without diode clamping are observable. In the interval of 1.7 to 4 MHz, the interference levels of the setups without diode clamping are even higher than in the measurement without any dv/dt-filter. This is a clear disadvantage of those constellations and can be referred to the high-frequency ringing of the filter output voltage, as seen e. g. in figure 5d. Another perceivable difference can be seen in both figures 9a and 9b concerning filter constellation (IV). With a negative peak in the interval around 8 MHz the common mode voltage level drops below 50 dBμV, which is a significant difference compared to any other measurement setup. This effect is a little more considerable for filter topology 1. However, the filter constellations with the additional diode clamping lead to more favourable conditions and a better overall dv/dt-filter performance.

Conclusion

In this paper two different three-phase dv/dt-filter topologies, which have been previously presented in other publications, are investigated and discussed in a fast switching, wide bandgap inverter environment. Besides a dv/dt damping of the steep inverter output voltage pulses, they should also challenge the common mode interference behaviour of the system. The slope reduction of the inverter output voltage in the desired area was verified successfully with both topologies and all four LCR-combinations. Furthermore, the influence of the inductor size on occuring losses was clearly observable. Increasing the inductance leads to a reduction of the required filter capacitance and therefore to lower system power losses. Although, this also introduces higher oscillations of the filter output voltage and a higher filter volume due to larger coils. Regarding the utilization of an additional diode clamping of the filter output voltage to the DC-link potentials, positive effects on the investigated objectives of this paper were detectable. While preserving the requested damping of the inverter's voltage slopes, the overshoot of the filter output voltage is reduced drastically. Moreover, it leads to a less oscillating voltage. Nevertheless, it induces some additional losses due to the power dissipation of the diodes and clamping resistors. An analysis of the conducted EMI levels shows an advantageous behaviour of the dv/dt filter topologies concerning the common mode interference, as it was desired by their modifications. In a wide frequency range, the common mode voltage levels are considerably lower than in the measurement without the output dv/dt-filters. Omitting the diode clamping increases the amount of CM noise in certain areas, which can be referred to the higher amount of oscillation of the filter output voltage in this frequency domain.

References

[1] J. Millán, P. Godignon, X. Perpiñà, A. Pérez-Tomás, and J. Rebollo, "A Survey of Wide Bandgap Power Semiconductor Devices," IEEE Transactions on Power Electronics, vol. 29, no. 5, pp. 2155– 2163, 2014. DOI: 10.1109/TPEL.2013.2268900.

[2] C. M. DiMarino, R. Burgos, and B. Dushan, "High-Temperature Silicon Carbide: Characterization of State-of-the-Art Silicon Carbide Power Transistors," IEEE Industrial Electronics Magazine, vol. 9, no. 3, pp. 19–30, 2015. DOI: 10.1109/MIE.2014.2360350.

[3] A. von Jouanne and P. Enjeti, "Design Considerations for an Inverter Output Filter to Mitigate the Effects of Long Motor Leads in ASD Applications," IEEE Transactions on Industry Applications, vol. 33, no. 5, pp. 1138–1145, 1997. DOI: 10.1109/28.633789.

[4] J. He, C. Li, A. Jassal, N. Thiagarajan, Y. Zhang, et al., "Multi-Domain Design Optimization of dv/dt Filter for SiC-Based Three-Phase Inverters in High-Frequency Motor-Drive Applications," in 2018 IEEE Energy Conversion Congress and Exposition (ECCE), 2018, pp. 5215–5222. DOI: 10.1109/ECCE. 2018.8557859.

[5] H. Kim, A. Anurag, S. Acharya, and S. Bhattacharya, "Analytical Study of SiC MOSFET Based Inverter Output dv/dt Mitigation and Loss Comparison with a Passive dv/dt Filter for High Frequency Motor Drive Applications," IEEE Access, vol. 9, pp. 15 228–15 238, 2021. DOI: 10.1109/ACCESS.2021.3053198.

[6] F. Broecker, K. Hoffmann, H. Solmecke, and M. Grimmig, "Analysis of Power Losses within a SiC-MOSFET- Inverter with Passive dv/dt-Damping Network for Reduced Voltage Slopes at Inductive Loads," in PCIM Europe digital days 2020, 2020, pp. 1–6.

[7] C. Choochuan, "A Survey of Output Filter Topologies to Minimize the Impact of PWM Inverter Waveforms on Three-Phase AC Induction Motors," in 2005 International Power Engineering Conference, 2005, pp. 1–544. DOI: 10.1109/IPEC.2005.206967.

[8] D. Rendusara and P. Enjeti, "An Improved Inverter Output Filter Configuration Reduces Common and Differential Modes dv/dt at the Motor Terminals in PWM Drive Systems," IEEE Transactions on Power Electronics, vol. 13, no. 6, pp. 1135–1143, 1998. DOI: 10.1109/63.728340.

[9] L. Palma and P. Enjeti, "An Inverter Output Filter to Mitigate dv/dt Effects in PWM Drive System," in APEC. Seventeenth Annual IEEE Applied Power Electronics Conference and Exposition (Cat. No.02CH37335), vol. 1, 2002, 550–556 vol.1. DOI: 10.1109/APEC.2002.989298.

[10] F. Broecker, P. J. Andres, K. F. Hoffmann, H. Solmecke, and M. Grimmig, "Modular Silicon Carbide Inverter for Drive Applications with High Voltage Slopes - Challenges Concerning Conducted EMC," in 2019 21st European Conference on Power Electronics and Applications (EPE '19 ECCE Europe), 2019, P.1–P.8. DOI: 10.23919/EPE.2019.8915141.

[11] F. Pauli, L. Yang, M. Schröder, K. Hameyer, Lebensdauerabschätzung von Wicklungsisolierstoffsystemen in SiC-betriebenen elektrischen Niederspannungsmaschinen. e & i Elektrotechnik und Informationstechnik. 136, 175–183 (2019). https://doi.org/10.1007/s00502-019-0711-2.

[12] CISPR25, "Vehicles, Boats and Internal Combustion Engines – Radio Disturbance Characteristics - Limits and Methods of Measurement for the Protection of On-Board Receivers", IEC, Standard Edition 4.0 2016-10, 2016.

State of Charge Prediction of Lithium-Ion Batteries Based on Artificial Neural Networks and Reduced Data

Sebastian Pohlmann[1], Ali Mashayekh[2], Dominic Karnehm[1], Manuel Kuder[2],
Antje Gieraths[1], Thomas Weyh[2]

[1]University of the Bundeswehr Munich, Institute of Distributed Intelligent Systems

[2]University of the Bundeswehr Munich, Institute of Electrical Energy Systems

Werner-Heisenberg-Weg 39

85577 Neubiberg, Germany

Phone: +49 89-6004-4378

Email: sebastian.pohlmann@unibw.de

Acknowledgments

This research is funded by dtec.bw Digitalization and Technology Research Center of the Bundeswehr which we gratefully acknowledge [project MORE].

Keywords

≪Battery≫, ≪State of charge≫, ≪Machine learning≫, ≪Deep Learning≫, ≪Battery Management Systems (BMS)≫.

Abstract

Lithium-ion batteries (LIBs) are the key technology for the electrification of the transport sector. Since LIBs have a complex, electrochemical structure, it is a challenge to accurately determine the condition, which is crucial for safety and efficiency during operation. This paper presents the forecasting of the state-of-charge (SOC) of a LIB based on machine learning (ML) algorithms. Data from battery simulation and augmented data are additionally used to train the models. To reduce the dimension of the feature matrix, a singular value decomposition is performed. A multi-layer perceptron (MLP) and a convolutional neural network (CNN) are compared to a linear regression. The impact of the augmented data on the prediction accuracies and the reliabilities of the models is analyzed. The lowest test error is achieved using the CNN with augmented data with a root mean square error (RMSE) of 1.78 %. The results show the applicability of data-driven models for the SOC prediction and the optimization potential using data augmentation techniques.

Introduction

The need to reduce greenhouse gas emissions and, consequently, the government policy requirements substantiate the relevance of the electrification of the transport sector. The move towards renewable energy is a key challenge for leading automotive manufactures [1]. In this context, lithium-ion batteries (LIBs) have taken a predominant role in the automotive industry due to their high energy density and their long lifespan [2, 3]. The dominant role in automotive applications over the next decade is predicted [4]. To reduce costs, achieve a high driving range, and increase the overall efficiency of the battery, it is crucial to determine the condition of the battery [5]. In operation, this is taken over by the Battery Management System (BMS), which ensures the safety and balanced charge and discharge cycles by monitoring and controlling the battery cells [6]. Among other functions, the BMS is responsible for the estimation of the state of charge (SOC), which is the ratio of the remaining capacity to its full capacity [7]. A direct measurement is not possible due to the complex electrochemical structure of a LIB and the varying charateristics under different working conditions [8]. The approaches to estimate the SOC can be separated in three areas. The first one are physical models like coulomb counting [9], open-circuit voltage (OCV) correction method [10, 11], or other physical models to approximate the electrochemical connections [12]. The second one are model-based estimations with mainly the usage of Kalman filters [13, 14]. The third one are data-driven models by means of machine learning (ML) or other statistical algorithms. Applied methods are, for example, support vector machines (SVMs) [15] and primarily artificial neural networks (ANNs) [16]. The models vary between deep neural networks [17] over recurrent neural networks (RNN) [18, 19] to convolutional neural networks (CNNs) [20, 21]. The data-driven models show a high accuracy without the need for a detailed electrochemical approximation. While keeping the computing costs moderate, the correlation between battery parameters and the SOC can be determined. Nevertheless, ML models are highly dependent on their input data. As the working conditions of a battery are extremely diverse, it is important to have a sufficient data basis. The costly and highly time-consuming battery tests exacerbate the problem. Possible relief, instead of real world tests, might come in the form of simulation data and data augmentation.

In this work, simulation data of a electrical equivalent model (ECM) and methods of data augmentation are used to enrich the data basis for training different ML models. Further, the dimensionality of the input data is reduced to optimize the models. A linear regression is used as the benchmark model and compared to different forms of ANNs. Based on the test results with a real-world data set, the most suitable model to predict the SOC is identified.

Methodology

Data pre-processing

According to the knowledge discovery in databases process (KDD), one important step is the data pre-processing, which includes data collection, cleaning, accounting for time-sequence, and transformation [22]. Different sources of data are combined in this work. A battery cell test system is used to obtain voltage and current values of a LIB of the type Samsung INR18650-30Q. This data is complemented by simulation data with an ECM with a voltage source U_{OCV}, the internal ohmic resistance R_i, and two RC-pairs [23]. The model is selected for high accuracy simulating the battery cell while keeping the computing time low. The ECM is shown in Fig. 1.

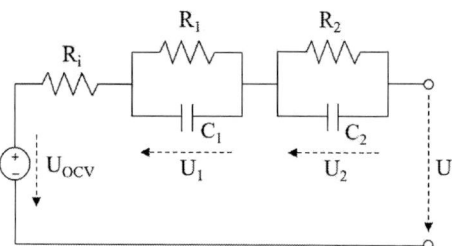

Fig. 1: Second order electrical equivalent model of a battery cell.

During the simulation, the C-rates are varied between 0.1 C and 2 C. To improve the simulation results, the simulation data is transformed by means of the end-of-charge voltage and the cut-off voltage of the real battery. Both values are used for a better approximation of the battery behavior, which can be seen in Fig. 2.

Additionaly, a data augmentation method is applied. While the current values are kept constant, the voltage is randomly varied. As starting point, a random difference using a Gaussian distribution is used. All further points are influenced by a random uniform distribution, the current real value, and the previous deviations. Using the Savitzky-Golay-Filter [24], the difference curve is smoothed and added to the initial discharge curve. Following this procedure, a confidence interval around the initial curve is created, where the voltage is randomly varied. Thereby, the differences between the cells should be reproduced.

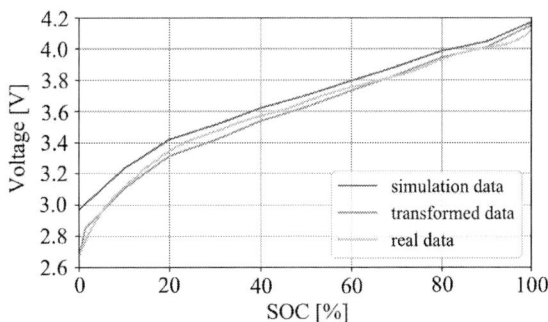

Fig. 2: Data transformation of simulation data with end-of-charge and cut-off voltage in comparison to the real data.

Before the data is further processed, the data is normalized using the standard scaler. Therefore, the standard deviation σ and the mean values $\bar{x}$ of the data distribution are used to calculate the transformed values x_i^n, see (1) [25].

$$x_i^n = \frac{x_i - \bar{x}}{\sigma} \tag{1}$$

With a higher number of time steps used for the SOC prediction, the number of features for one data point and, thereby, the dimensionality of the problem is growing. To adequately describe a high dimensional data distribution, the needed data size is increasing exponentially with the dimension. This is referred to as the "curse of dimensionality" [26]. A possible solution are dimensionality reduction techniques. The data is preprocessed using a singular value decomposition (SVD), which is factorization of a matrix M in two orthogonal matrices U and V, and a diagonal matrix Σ, as can be seen in (2) [27].

$$M = U \Sigma V^\top \tag{2}$$

In other words, a matrix $M \in \mathbb{R}^{m \times n}$ is transformed using a rotation, scaling, and another rotation in the following form, displayed in Fig. 3. μ_m are the singular values of the matrix M [28].

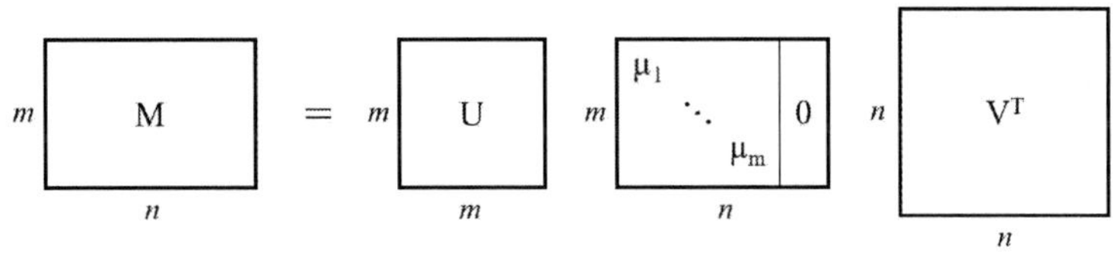

Fig. 3: Graphical representation of a singular value decomposition of a $m \times n$ matrix M in two orthogonal and one diagonal matrices.

Based on the used singular values, the number of features can be reduced significantly. The SVD is used for the reduction of the dimension of the input data for the ML models.

Machine learning methods

Linear regression

As benchmark model, a linear regression is implemented. Aim is to obtain the linear dependence between the input features based on voltage and current, and the target values SOCs. Weight parameters w_i in combination with the input features x_i are used to calculate the target value y^*, see (3) [29].

$$y^* = w_0 + \sum_{i=1}^{n} w_i x_i \tag{3}$$

The weight parameters are calculated by minimizing a loss function. The loss function L consists mainly of the squared error between the predicted and the real target values, see (4) [30].

$$L(w) = \frac{1}{2n} \sum_{i=1}^{n} (y^*(x_{ij}) - y_i)^2 \tag{4}$$

The data points are separated in two intervals for the prediction. The mean value between end-of-charge and cut-off voltage is used as the threshold. Both intervals are trained individually and then combined for the SOC prediction.

Artificial neural networks

Different types of artificial neural networks (ANNs) are created to predict the SOC. The initial structure was influenced by the information transfer and processing in a biological brain. Artificial neurons are arranged in layers and share information through connections [31]. Based on the structure and the layer formation, different types of ANNs can be separated. A simple form is the multi-layer-perceptron (MLP), which is a feed-forward ANN. The information flow is in one direction from input to output layer by means of several mathematical functions. The information in a neuron is summed up and then used in an activation function, where the output of the neuron is calculated [29]. An artificial neuron is connected to all neurons in the next layer. Two hidden layers are used with eight neurons in the first, and four neurons in the second layer. As activation function, the Rectified Linear Unit (ReLU) function is used for a time efficient training phase.

Another form of ANNs are convolutional neural networks (CNNs), which main application field is image processing, but also the usage for time series analysis is increasing [32]. It consists of a convolution, where a kernel function is used to convolve the input and mostly a pooling layer, where several data points can be pooled in a single data point [31]. The structure is shown in Fig. 4 [33].

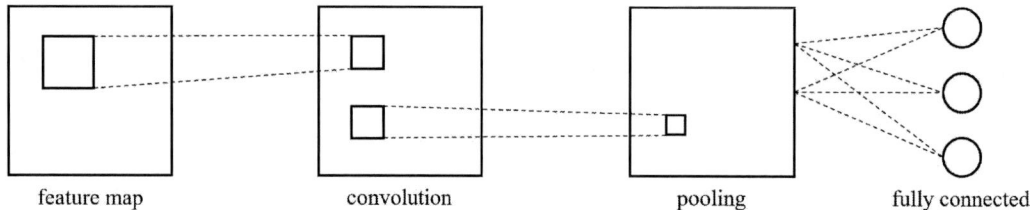

feature map   convolution   pooling   fully connected

Fig. 4: Traditional structure of a CNN with feature map, convolution layer, pooling layer, and the fully connected layer.

A 1-D CNN is used for the SOC prediction. The convolution filters are moved along the temporal dimension, while analyzing the time-series data. In comparison to the MLP, other processing steps are added to the model, even though the basic structure remains similar. For both types of ANNs, several hyperparameters have to be determined. The hyperparameter space is searched systematically by using a grid search. The parameter combination which minimizes the loss function is selected for further analysis [34].

To evaluate the prediction accuracy, the root mean square error (RMSE) is used for training and test results. Thereby, the target values y are compared to the predicted values y^*, see (5) [25].

$$RMSE = \sqrt{\frac{1}{n}\sum_{i=1}^{n}(y_i - y_i^*)^2} \tag{5}$$

Results

Different ML models are trained using simulation and augmented data. These models are tested on a real world dataset to determine the prediction accuracy. The results are evaluated by means of the RMSE with the overall aim to reduce the test error. The models are retrained five times, and the mean errors with the according standard deviation are stated. To reduce the dimension of the input data, a SVD is performed. The dimension of the initial data was 80 and is reduced to 14. This dimension was chosen based on the singular values, which are more similar to each other in higher dimensions. The deviation and, consequently, the impact on the transformation is lower with an increasing number of features. A linear regression is used as the benchmark model and is compared to different types of ANNs. The linear model is separated in two voltage intervals for an improved SOC prediction. The mean train error is 4.52 % with a standard deviation of 0.04, while the test error is 3.80 % with a deviation of 0.08. The mismatch between train and test error can be explained by the self defined current patterns used in the battery simulation and the overall number of outliers in the training set. As explained, the data driven model highly depend on the input data. Nevertheless, battery tests are expensive and time consuming. In reality, expert knowledge is necessary to assess if data points should be included or sorted out. However, it could be shown that the outliers have a higher impact in the evaluation than affecting the model training in a negative way. The results are plotted in Fig. 5. For reasons of comprehensibility, the results picture the SOC plotted over the voltage of the current time step. This is included as one feature in the initial feature map before the dimensionality reduction using the SVD, but it is not the only feature as indicated in the figure.

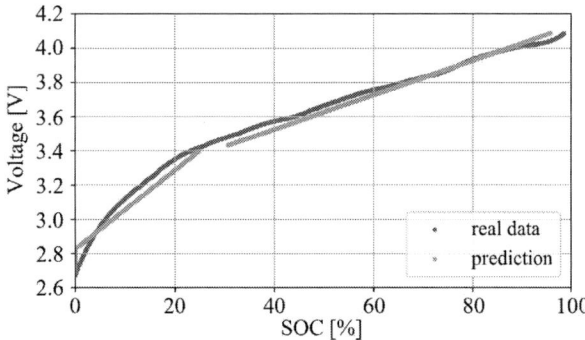

Fig. 5: Test results of the linear regression model separated in two voltage intervals.

Next to the linear regression, a MLP and a CNN are implemented to predict the SOC. The MLP is trained without augmented data, with augmented data five times the initial data, and ten times the initial data. The data is preprocessed using a standard scaler and afterwards a SVD. The training and test errors of the MLP are summarized in Table I.

Table I: Training and test results of the MLP with and without augmented data. The results are mean values of five experiments with the according standard deviation.

	Train	Test
without augmented data	4.40 (1.38) %	2.45 (0.32) %
with augmented data (5x)	4.71 (0.39) %	2.28 (0.32) %
with augmented data (10x)	2.09 (0.09) %	1.83 (0.17) %

Even the model without augmented data has test errors below 3 %. All in all, the test errors can be reduced using augmented data. Especially with ten times the initial data, the RMSE could be decreased to 1.83 % in comparison to a RMSE of 2.45 % without augmented data. Additionaly, the model is more robust, which is apparent analyzing the standard deviation of the models. Using more data, the deviation is decreasing and the models became more reliable. To further reduce the prediction error, a CNN is implemented to better approximate the time-series data. The results are shown in Table II.

Table II: Training and test results of the CNN with and without augmented data. The results are mean values of five experiments with the according standard deviation.

	Train	Test
without augmented data	2.70 (0.18) %	2.26 (0.49) %
with augmented data (5x)	2.48 (0.35) %	2.17 (0.31) %
with augmented data (10x)	2.61 (0.46) %	1.78 (0.34) %

An improvement in comparison to the MLP becomes apparent for all types of input data, even though the impact is small. A lowest test error can be identified using the CNN with ten times the initial data with a RMSE of 1.78 %. The standard deviation is slighlty higher compared to the MLP, but the trend for lower errors while using more data can be determined. The time frame used for the models is relatively small. A considerable improvement analyzing a larger area of past data is expected for the CNN. The MLP with a comparable simpler architecture fits the needs for this problem. However, the CNN shows more potential for optimazation. A comparision of the results of the ANNs is shown in Fig. 6.

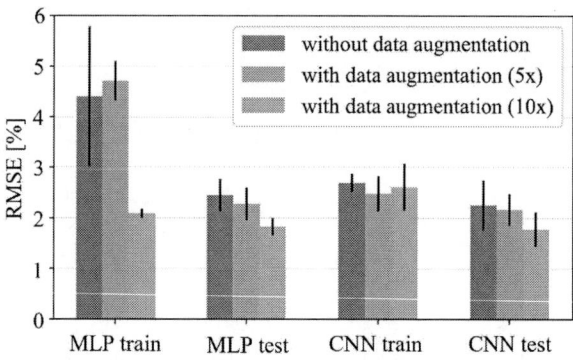

Fig. 6: Results of the ANNs for different input data compared for training and test phase with the corresponding standard deviation.

The MLP without augmented data and with five times the initial data have similar training errors to the linear regression and are not sufficient to predict the SOC. Further, the deviation between training and test error is higher for these models. The MLP with augmented data (ten times) shows a good tradeoff between prediction accuracy and structure complexity compared to the CNN. The overall better prediction ability of the CNN can be identified using the models without augmented data. Both models show the trend of improvement while using a larger data basis. The CNN is more suitable for predicting the SOC of a LIB; however, the optimized MLP shows similar results.

Another considered aspect is the convergence of the models and the impact of data augmentation. The course of the RMSE over the epochs is plotted in Fig. 7.

It is striking that the more data is used, the earlier convergence is reached. While the convergence values are similar, an earlier drop of the model error can be identified using augmented data. A higher improvement can be observed from the change of no augmented data to five times the initial data. The further improvement using ten times the data is also significant. On the one hand, the data augmentation shows potential for optimizing the ML models for SOC prediction, resulting in more robust models and a

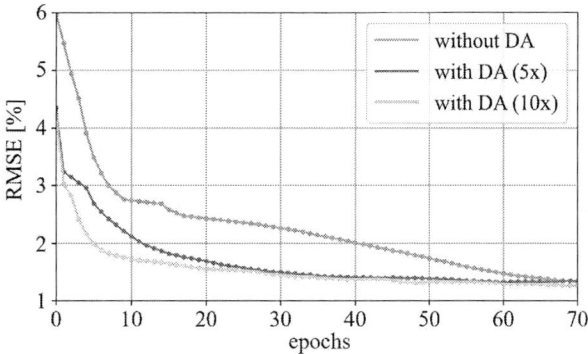

Fig. 7: Comparison of the convergence without augmented data (without DA), with five times the initial data (with DA (5x)), and ten times the initial data (with DA (10x)).

higher prediction accuracy. On the other hand, data augmentation in the field of battery analysis demands expert knowlegde and is more complicated than conventional methods. Benefits of data augmentation are the higher robustness of the models and the improved prediction accuracies. The impact is greater for the MLP with the more simple structure, but the trend for model improvement can also be identified using the CNN.

Conclusion

The move to renewable energy is crucial for environmental protection. LIBs are a key technology to electrificate the transport sector. A challenge in operation is the determination of the battery condition. For a safe and efficient usage, it is important to accurately assess the state. Among others, data-driven models show high accuracies predicting the SOC. Disadvantages of these models are the high dependence on their data basis. Because battery tests are time and cost consuming, data augmentation techniques can be used to enrich the data basis. To reduce the experiments to a minimum, simulation and augmented data are used to train different types of ANNs and a linear benchmark model. The data is preprocessed using a standard scaler, and, further, the dimension of the features is reduced using a SVD. All in all, the data augmentation method leads to more reliable models and reduced model errors. The CNN with augmented data reaches a lowest test RMSE of 1.78 % and is therefore slightly favorable in comparison to the MLP with a test error of 1.83 %.

In future work, it is planned to analyze advanced data augmentation methods for a larger timeframe. The optimization potential is expected to be higher for the CNN. Further, it is planned to test the algorithms in an experimental prototype to analyze the applicability in the real world and the transferability for other cell chemistries.

References

[1] J. Buberger, A. Kersten, M. Kuder, R. Eckerle, T. Weyh, and T. Thiringer, "Total co2-equivalent life-cycle emissions from commercially available passenger cars," *Renewable and Sustainable Energy Reviews*, vol. 159, p. 112158, 2022.

[2] Ralph J. Brodd, *Batteries for Sustainability: Selected Entries from the Encyclopedia of Sustainability Science and Technology.* New York, NY: Springer New York, 2013.

[3] S. Stock, S. Pohlmann, F. J. Guenter, L. Hille, J. Hagemeister, and G. Reinhart, "Early quality classification and prediction of battery cycle life in production using machine learning," *Journal of Energy Storage*, vol. 50, p. 104144, 2022. [Online]. Available: https://www.sciencedirect.com/science/article/pii/S2352152X22001785

[4] R. Schmuch, R. Wagner, G. Hörpel, T. Placke, and M. Winter, "Performance and cost of materials for lithium-based rechargeable automotive batteries," *Nature Energy*, vol. 3, no. 4, pp. 267–278, 2018.

[5] Y. Shi, S. Ahmad, Q. Tong, T. M. Lim, Z. Wei, D. Ji, C. M. Eze, and J. Zhao, "The optimization of state of charge and state of health estimation for lithium-ions battery using combined deep learning and kalman filter methods," *International Journal of Energy Research*, vol. 45, no. 7, pp. 11 206–11 230, 2021.

[6] Weihan Li, Monika Rentemeister, Julia Badeda, Dominik Jöst, Dominik Schulte, and Dirk Uwe Sauer, "Digital twin for battery systems: Cloud battery management system with online state-of-charge and state-of-health estimation," *Journal of Energy Storage*, vol. 30, p. 101557, 2020. [Online]. Available: https://www.sciencedirect.com/science/article/pii/S2352152X20308495

[7] Guanyong Zhang, Bizhong Xia, Jiamin Wang, Bo Ye, Yunchao Chen, Zhuojun Yu, and Yuheng Li, "Intelligent state of charge estimation of battery pack based on particle swarm optimization algorithm improved radical basis function neural network," *Journal of Energy Storage*, vol. 50, p. 104211, 2022. [Online]. Available: https://www.sciencedirect.com/science/article/pii/S2352152X22002420

[8] M. A. Hannan, M. S. H. Lipu, A. Hussain, P. J. Ker, T. M. I. Mahlia, M. Mansor, A. Ayob, M. H. Saad, and Z. Y. Dong, "Toward enhanced state of charge estimation of lithium-ion batteries using optimized machine learning techniques," *Scientific Reports*, vol. 10, no. 1, p. 4687, 2020.

[9] K. S. Ng, C.-S. Moo, Y.-P. Chen, and Y.-C. Hsieh, "Enhanced coulomb counting method for estimating state-of-charge and state-of-health of lithium-ion batteries," *Applied Energy*, vol. 86, no. 9, pp. 1506–1511, 2009. [Online]. Available: https://www.sciencedirect.com/science/article/pii/S0306261908003061

[10] Vaclav Knap and Daniel-Ioan Stroe, "Effects of open-circuit voltage tests and models on state-of-charge estimation for batteries in highly variable temperature environments: Study case nano-satellites," *Journal of Power Sources*, vol. 498, p. 229913, 2021. [Online]. Available: https://www.sciencedirect.com/science/article/pii/S0378775321004444

[11] H. Chaoui and S. Mandalapu, "Comparative study of online open circuit voltage estimation techniques for state of charge estimation of lithium-ion batteries," *Batteries*, vol. 3, no. 2, 2017. [Online]. Available: https://www.mdpi.com/2313-0105/3/2/12

[12] James Marcicki, Marcello Canova, A. Terrence Conlisk, and Giorgio Rizzoni, "Design and parametrization analysis of a reduced-order electrochemical model of graphite/lifepo4 cells for soc/soh estimation," *Journal of Power Sources*, vol. 237, pp. 310–324, 2013. [Online]. Available: https://www.sciencedirect.com/science/article/pii/S0378775313000694

[13] Y. Luo, P. Qi, Y. Kan, J. Huang, H. Huang, J. Luo, J. Wang, Y. Wei, R. Xiao, and S. Zhao, "State of charge estimation method based on the extended kalman filter algorithm with consideration of time–varying battery parameters," *International Journal of Energy Research*, vol. 44, 2020.

[14] Jaemoon Lee, Oanyong Nam, and B.H. Cho, "Li-ion battery soc estimation method based on the reduced order extended kalman filtering," *Journal of Power Sources*, vol. 174, no. 1, pp. 9–15, 2007. [Online]. Available: https://www.sciencedirect.com/science/article/pii/S0378775307007112

[15] J. C. Álvarez Antón, P. J. García Nieto, C. Blanco Viejo, and J. A. Vilán Vilán, "Support vector machines used to estimate the battery state of charge," *IEEE Transactions on Power Electronics*, vol. 28, no. 12, pp. 5919–5926, 2013.

[16] C. Vidal, P. Malysz, P. Kollmeyer, and A. Emadi, "Machine learning applied to electrified vehicle battery state of charge and state of health estimation: State-of-the-art," *IEEE Access*, vol. 8, pp. 52 796–52 814, 2020.

[17] D. N. T. How, M. A. Hannan, M. S. H. Lipu, K. S. M. Sahari, P. J. Ker, and K. M. Muttaqi, "State-of-charge estimation of li-ion battery in electric vehicles: A deep neural network approach," *IEEE Transactions on Industry Applications*, vol. 56, no. 5, pp. 5565–5574, 2020.

[18] Meng Jiao, Dongqing Wang, and Jianlong Qiu, "A gru-rnn based momentum optimized algorithm for soc estimation," *Journal of Power Sources*, vol. 459, p. 228051, 2020. [Online]. Available: https://www.sciencedirect.com/science/article/pii/S0378775320303542

[19] S. Li, C. Ju, J. Li, R. Fang, Z. Tao, B. Li, and T. Zhang, "State-of-charge estimation of lithium-ion batteries in the battery degradation process based on recurrent neural network," *Energies*, vol. 14, no. 2, 2021. [Online]. Available: https://www.mdpi.com/1996-1073/14/2/306

[20] Cheng Qian, Binghui Xu, Liang Chang, Bo Sun, Qiang Feng, Dezhen Yang, Yi Ren, and Zili Wang, "Convolutional neural network based capacity estimation using random segments of the charging curves for lithium-ion batteries," *Energy*, vol. 227, p. 120333, 2021. [Online]. Available: https://www.sciencedirect.com/science/article/pii/S036054422100582X

[21] Chong Bian, Shunkun Yang, Jie Liu, and Enrico Zio, "Robust state-of-charge estimation of li-ion batteries based on multichannel convolutional and bidirectional recurrent neural networks," *Applied Soft Computing*, vol. 116, p. 108401, 2022. [Online]. Available: https://www.sciencedirect.com/science/article/pii/S1568494621011571

[22] U. Fayyad, G. Piatetsky-Shapiro, and P. Smyth, "From data mining to knowledge discovery in databases," *AI Magazine*, vol. 17, no. 3, p. 37, 1996. [Online]. Available: https://ojs.aaai.org/index.php/aimagazine/article/view/1230

[23] Xiaofeng Ding, Donghuai Zhang, Jiawei Cheng, Binbin Wang, and Patrick Chi Kwong Luk, "An improved thevenin model of lithium-ion battery with high accuracy for electric vehicles," *Applied Energy*, vol. 254, p. 113615, 2019. [Online]. Available: https://www.sciencedirect.com/science/article/pii/S0306261919312899

[24] R. W. Schafer, "What is a savitzky-golay filter? [lecture notes]," *IEEE Signal Processing Magazine*, vol. 28, no. 4, pp. 111–117, 2011.

[25] M. G. Pecht and M. Kang, *Prognostics and Health Management of Electronics: Fundamentals, Machine Learning, and the Internet of Things*, ser. IEEE Press. Wiley, 2018. [Online]. Available: https://books.google.de/books?id=vitpDwAAQBAJ

[26] I. Narsky and F. C. Porter, *Statistical Analysis Techniques in Particle Physics: Fits, Density Estimation and Supervised Learning*, 1st ed. Weinheim: Wiley-VCH, 2013.

[27] M. Kern, *Numerical Methods for Inverse Problems*, 1st ed. New York, NY: John Wiley & Sons, 2016.

[28] H. Yanai, K. Takeuchi, and Y. Takane, *Projection Matrices, Generalized Inverse Matrices, and Singular Value Decomposition*, 1st ed., ser. Statistics for Social and Behavioral Sciences. New York, NY: Springer New York, 2011.

[29] A. V. Joshi, *Machine Learning and Artificial Intelligence*. Springer International Publishing, 2020.

[30] F. Zhang, T. L. Lai, B. Rajaratnam, N. R. Zhang, and Stanford University. Department of Statistics, *Cross-validation and Regression Analysis in High-dimensional Sparse Linear Models*. Stanford University, 2011.

[31] G. Rebala, A. Ravi, and S. Churiwala, *An Introduction to Machine Learning*, 01 2019.

[32] W. Pedrycz and S.-M. Chen, *Interpretable Artificial Intelligence: A Perspective of Granular Computing*, 1st ed., ser. Studies in Computational Intelligence. Cham: Springer International Publishing, 2021.

[33] D. Graupe, *Principles of Artificial Neural Networks*, 4th ed. WORLD SCIENTIFIC, 2019.

[34] H. Alibrahim and S. A. Ludwig, "Hyperparameter optimization: Comparing genetic algorithm against grid search and bayesian optimization," in *2021 IEEE Congress on Evolutionary Computation (CEC)*, 2021, pp. 1551–1559.

Investigation for Condensation Test Condition of HVIGBT Modules

Kenji HATORI*, Keiichi NAKAMURA*, Wakana NOBORU*
Nils SOLTAU**, Eugen WIESNER**
*Mitsubishi Electric Corporation, 1-1-1 Imajukuhigashi, Nishi-Ku, Fukuoka, Japan
**Mitsubishi Electric Europe B.V., Germany
Tel.: +81-92-805-3406
Fax: +81-92-805-3676
E-Mail: Hatori.Kenji@dx.MitsubishiElectric.co.jp
URL: https://www.mitsubishielectric.com/semiconductors/products/powermod/index.html

Keywords

«IGBT», «Humidity», and «Reliability» of the official keywords list.

Abstract

Humidity robustness is one of the main concerns of IGBT modules since the modules are not hermetically sealed. IGBT modules are generally filled with silicone gel which has a filter effect. Condensation amount at various condensation test conditions is described in this paper based on silicone gel humidity absorption behavior investigation. At first, the humidification condition impact is investigated. As a result, it is confirmed that condensation amount depends on dissolved humidity in silicone gel, not on the absolute humidity of ambient air. Secondly, cooling speed impact is investigated, and it is confirmed that humidification at lower temperature is effective to obtain stable results independently from cooling speed. Thirdly, it is confirmed that humidification at higher temperature is effective to shorten test time. Next, it is confirmed that humidification at lower temperature is effective to obtain stable results independently from gel thickness of the module. Finally, the calculation result is verified with the experiment.

Introduction

The power modules for today's modern HVIGBTs are not hermetically encapsulated, resulting in condensation occurring inside or around the power semiconductor in some cases. Therefore, HVIGBTs are required to have excellent performance and high reliability even under such conditions. Hence, the SCC (Surface Charge Control) technology [1], [2] and other unique technologies [3], [4], [5] are proposed to improve the robustness against humidity. Also, the humidity acceleration model [6] or humidity lifetime model [7] are proposed. However, previous studies are based on static tests with stable operational and environmental conditions. Whereas, in the field, conditions are usually not stable. In addition to humid condition, condensation is considered as one of the harshest environments as well. In order to confirm the robustness against condensation, the condensation tests of HVIGBT modules are proposed [8], [9]. However, it has not been clarified how frequently condensation occurs around IGBT chips during field operation.

IGBT modules are generally filled with silicone gel which has a filter effect against vapor diffusion. Therefore, it takes time to transfer the ambient environment to the local environment around semiconductor chips. This transfer function is very important to understand condensation amount during condensation test, which is related with severity of the test.

Considering previous works, reference [10], regarding the humidity absorption behavior of the whole IGBT module, has contributed to this issue. Moreover, the humidity absorption behavior of silicone gel has been analyzed in [11]. Based on the previous work about silicone gel humidity absorption behavior, this paper evaluates the condensation amount at various condensation test conditions.

Humidity absorption behavior of silicone gel

Generally, IGBT modules are filled with silicone gel. Polymer resin-like silicone gel absorbs humidity depending on the ambient environment. This humidity absorption takes time and then it also takes time for humidity to reach IGBT chips. Humidity absorption behavior of polymer resin can be explained by diffusion and dissolution of vapor [12], [13].

The correlation between vapor concentration c and diffusion coefficient D is defined with the following equation of Fick's laws of diffusion. The below equation is simplified with one dimension: location parameter is x here. The diffusion coefficient D follows the Arrhenius equation: $D0$ is the pre-exponential factor, E_D is activation energy and R is the universal gas constant (8.314 J/K·mol). The previous study confirms the activation energy $E_D = 20.8$ kJ/mol and pre-exponential factor $D_0 = 40.3$ mm^2/s [11].

$$\frac{\partial c}{\partial t} = \frac{\partial}{\partial x}\left(D\frac{\partial c}{\partial x}\right) \qquad (1)$$

$$D = D_0 \exp\left(-\frac{E_D}{RT}\right) \qquad (2)$$

The correlation between vapor concentration c and solubility coefficient S is defined with the following equation: P_v is defined as vapor pressure. The solubility coefficient S also follows the Arrhenius equation: S_0 is the pre-exponential factor, ΔH_S is enthalpy and R is the universal gas constant (8.314 J/(K·mol)). The previous study confirms the enthalpy $\Delta H_S = -26.8$ kJ/mol and pre-exponential factor $S_0 = 4.12 \times 10^{-12}$ mg/(mm^3·Pa) [11].

$$c_{(t=\infty)} = S \cdot P_v \qquad (3)$$

$$S = S_0 \exp\left(-\frac{\Delta H_S}{RT}\right) \qquad (4)$$

Condensation is considered as the most severe humid condition. In order to confirm the robustness against condensation, a condensation test is proposed [8][9]. In the condensation test, firstly, rapid cooling is applied after humidification. Subsequently, DC voltage is applied. Rapid cooling causes condensation in the silicone gel. The amount of condensate in silicone gel affects the condensation test result. Thus, it should be investigated how much condensation occurs in silicone gel depending on the test condition.

Humidification condition impact on the condensation test

Condensation test result strongly depends on the pre-humidification condition that is applied as pretreatment. Then relative humidity during humidification is one of the biggest factors. Based on equations (1), (2), (3) and (4), condensation amount during the condensation test is calculated with various humidification conditions as shown in Fig. 1 and Fig. 2.

As shown in Fig. 2, condensation amount is well related to humidification condition. Humidification condition of 75°C40%RH and 35°C80%RH brings same condensation amount of 200 g/m^3. These two humidification conditions of 75°C40%RH and 35°C80%RH shows the same dissolved humidity (SH) of 644g/m^3 as shown in Fig. 3 which is calculated by equation (4) [11]. Therefore, the humidification condition dependency on condensation amount can be described with dissolved humidity, not with absolute humidity of ambient air.

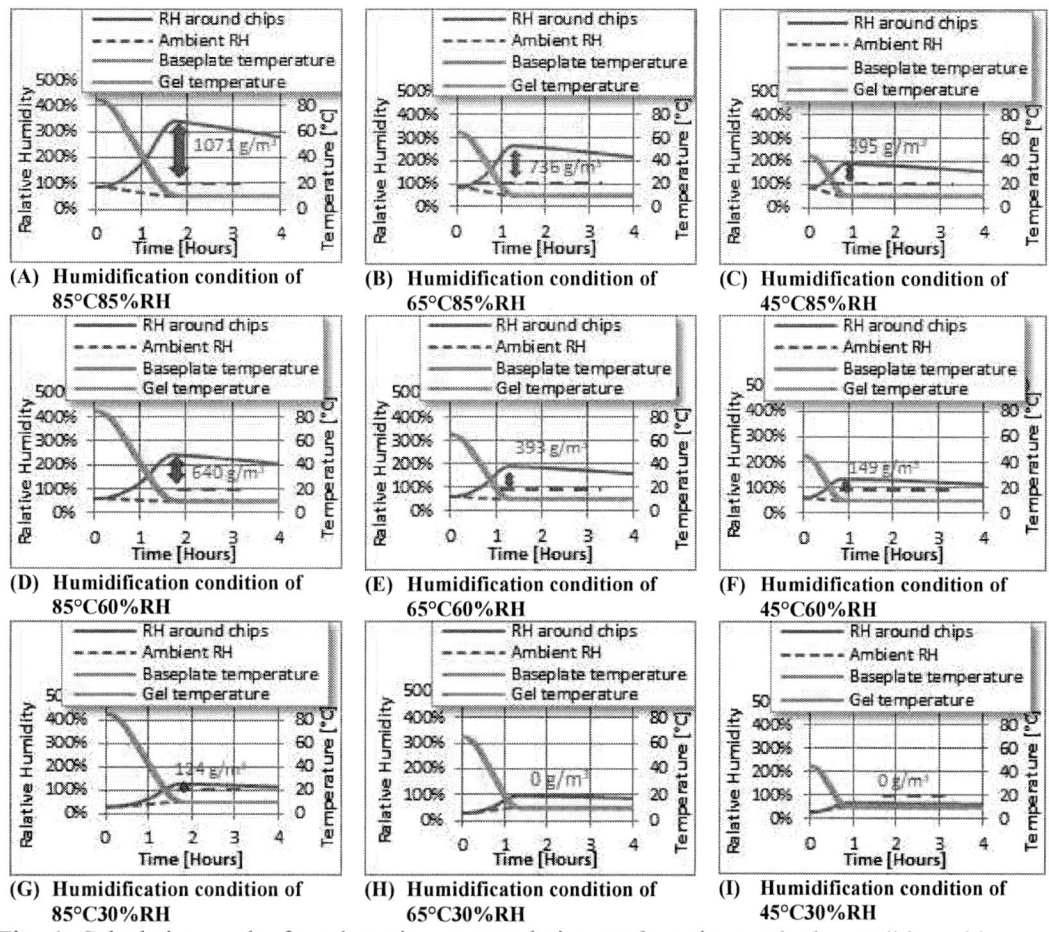

Fig. 1: Calculation result of condensation amount during condensation test in the condition with cooling to 10°C50%RH, dT/dt=-1K/min, saturated humidification duration, silicone gel thickness=20 mm, time constant of gel temperature=5min

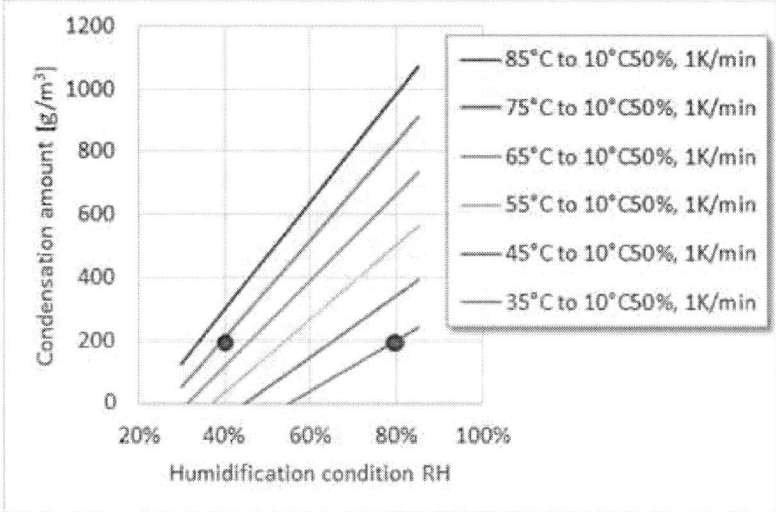

Fig. 2: Humidification condition impact

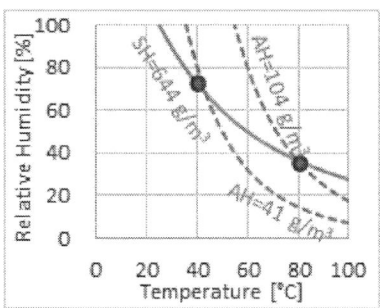

Fig. 3: Dissolved humidity (SH) & Absolute humidity (AH)

Cooling speed impact on the condensation test

Cooling speed is also important to decide condensation amount. Condensation amount at the condensation test is calculated with various cooling speeds and various temperatures at humidification as shown in Fig. 4 and Fig. 5.

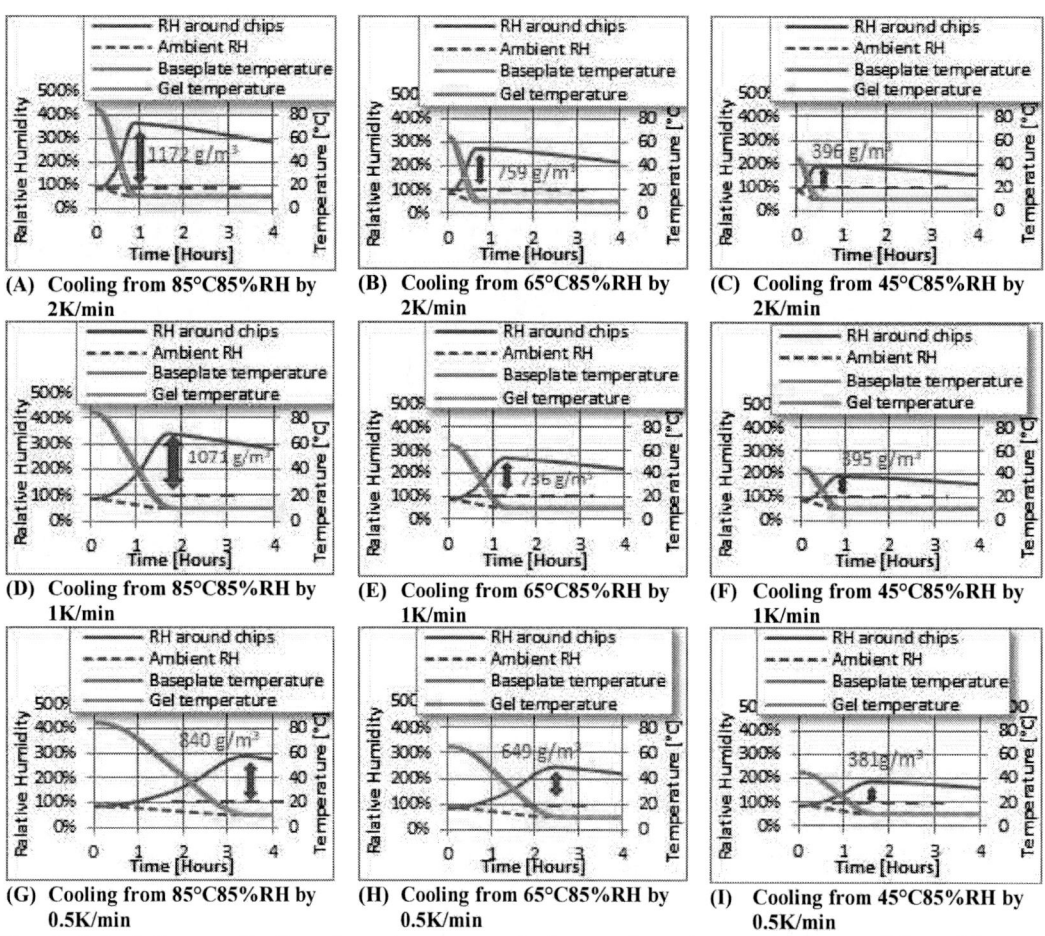

Fig. 4: Calculation result of condensation amount during condensation test in the condition with cooling to 10°C50%RH, saturated humidification duration, silicone gel thickness=20mm, time constant of 5min for gel temperature

Diffusion coefficient is larger at higher temperature according to equation (2). Hence, vapor is diffused out faster. As a result, vapor in silicone gel is dried out during ramping down phase. Thus, as shown in Fig. 5, cooling speed influences on condensation amount at higher temperature, but does not influence at lower temperature. It is shown that humidification at lower temperature is effective to obtain stable result independently from cooling speed.

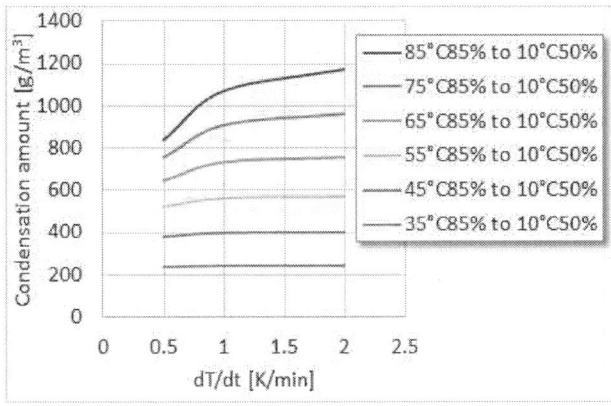

Fig. 5: Cooling speed impact

Humidification duration impact on the condensation test

Condensation amount at the condensation test is calculated with various humidification duration and various temperature at humidification as shown in Fig. 6 and Fig. 7.

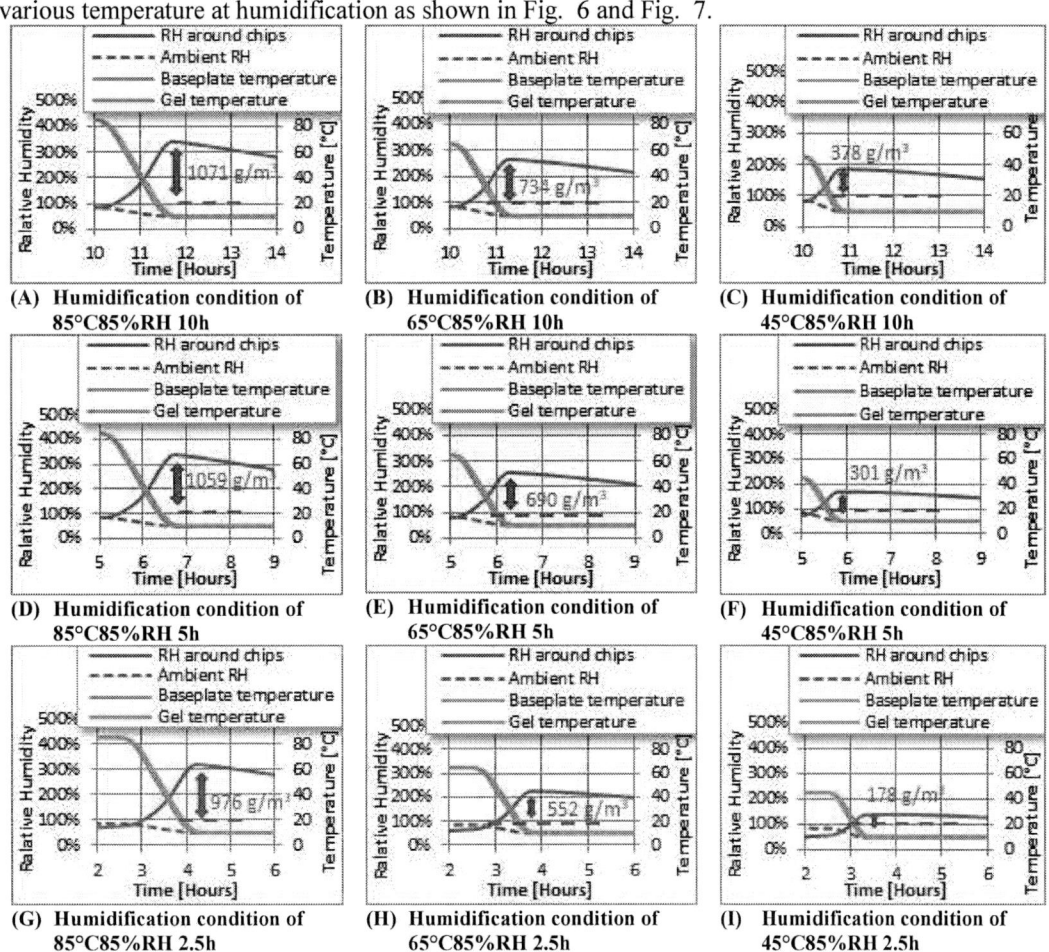

Fig. 6: Calculation result of condensation amount during condensation test in the condition with cooling to 10°C50%RH, dT/dt=-1K/min, silicone gel thickness=20mm, time constant of 5min for gel temperature

EPE'22 ECCE Europe

Humidification speed becomes higher at higher temperature because diffusion coefficient is larger according to equation (2). As a result, humidification becomes saturated faster at higher temperature but slower at lower temperature. Then, as shown in Fig. 7, humidification duration difference does not have impact on condensation amount with range of more than 5 hours humidification at higher temperature. On the other hand, humidification duration should be cared at lower temperature. It is shown that humidification at higher temperature is effective to shorten test time.

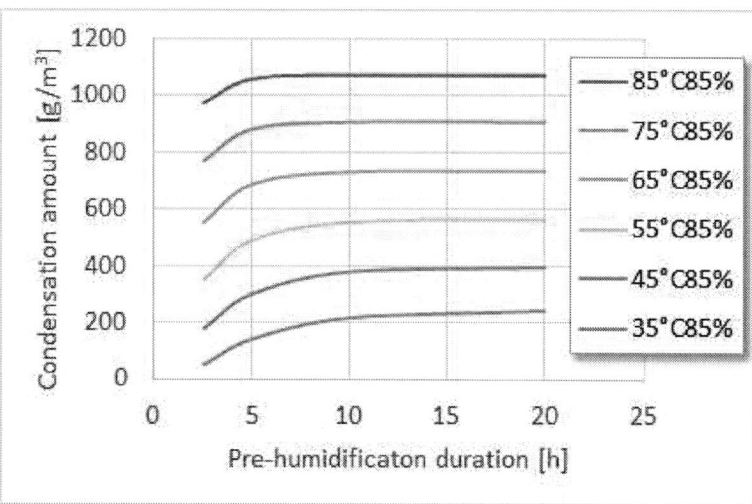

Fig. 7: Humidification duration impact

Gel thickness impact on the condensation test

Condensation amount at the condensation test is calculated with various gel thickness, various temperature, and various humidity as shown in Fig. 8 and Fig. 9. In this calculation, sufficient humidification duration is defined to obtain saturated vapor absorption.

Vapor is diffused into the air through gel during cooling phase. In case of thinner gel, the vapor is diffused into the air more quickly. As a result, thicker gel brings larger amount of condensation than thinner gel. This diffusion effect is stronger at higher temperature according to equation (2). Thus, gel thickness dependency is bigger at higher temperature. In contrast, gel thickness dependency is negligible at lower temperature. Also, relative humidity does not influence on the gel thickness dependency on condensation amount. Then, it is shown that humidification at lower temperature is effective to obtain stable results independently from gel thickness of the module.

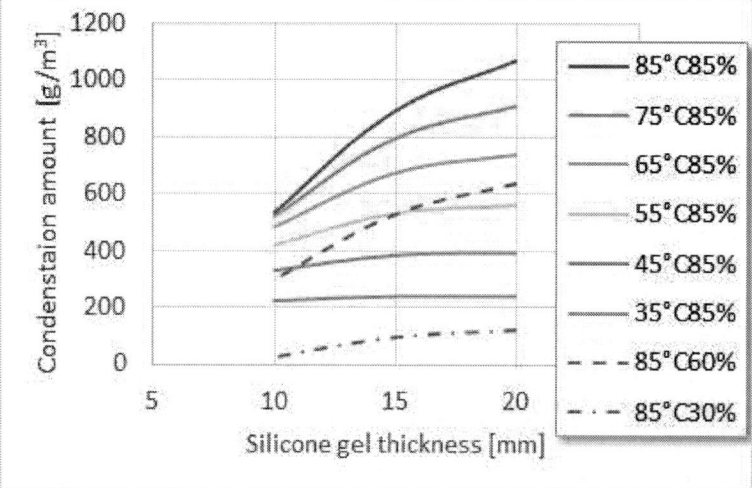

Fig. 8: Gel thickness impact

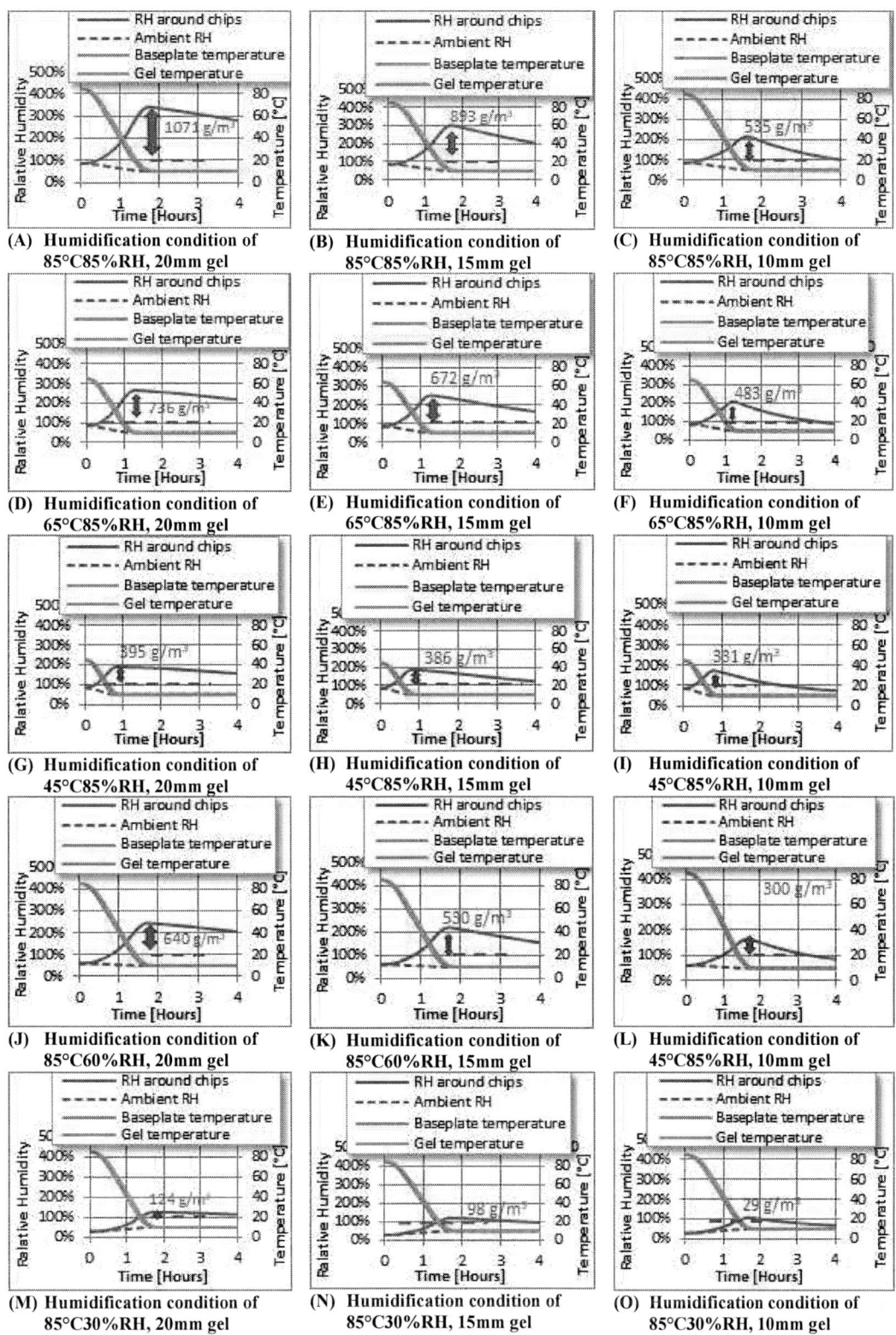

Fig. 9: Calculation result of condensation amount during condensation test in the condition with cooling to 10°C50%RH, dT/dt=-1K/min, time constant of 5min for gel temperature

Verification by experiment

The calculation result was verified by the experiment. The gel was filled in the glass cylinder with 30 mm diameter and the gel thickness is 10 mm. These gel samples were humidified with two conditions of 85°C85%RH and 45°C85%RH. The humidification duration varied from 40 min to 60 min for 85°C85%RH experiment and 155 min to 170 min for 45°C85%RH experiment. The cooling profile is shown in Fig. 10. As shown in Fig. 10, cooling speed is around 40 K/min for 85°C85%RH experiment and 20 K/min for 45°C85%RH experiment in early phase. Temperature sensor was located on the bottom of the cylinder. If condensation duration is too short, enough vapor does not reach the bottom of the cylinder. As a result, condensation appears only on the upper side of the gel in the cooling phase. The border between condensation area and the non-condensation area is marked in the experiment as shown in Fig. 11 and Fig. 12.

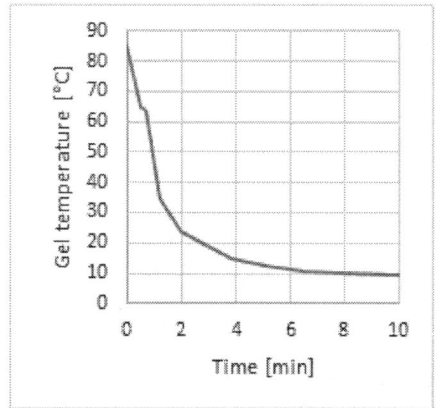

(A) Humidification condition of 85°C85%RH

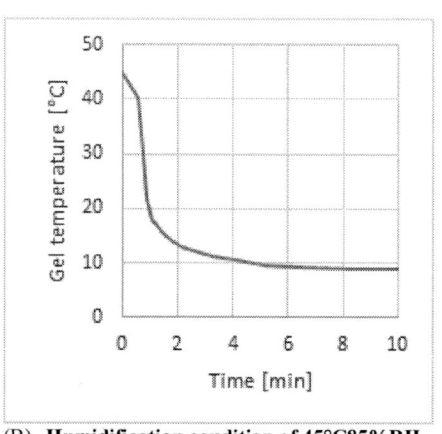

(B) Humidification condition of 45°C85%RH

Fig. 10: The cooling profile of the experiment

(A) Humidification duration of 40 min

(B) Humidification duration of 51 min

(C) Humidification duration of 55 min

(D) Humidification duration of 60 min

Fig. 11: The experiment result with 85°C85%RH humidification condition

As shown in Fig. 11, the border between condensation area and non-condensation area was observed if 85°C85%RH humidification is less than 55 min. On the other hand, condensation reached the bottom of the cylinder in the case of longer humidification than 60 min with the condition of 85°C85%RH.

In the case of 45°C85%RH humidification, the border between condensation and non-condensation was observed if humidification is less than 165 min. In the case of longer humidification than 170 min, condensation reached the bottom of the cylinder.

Then, 45°C85%RH humidification condition requires longer humidification to have condensation on the bottom of cylinder compared to the 85°C85%RH humidification condition. This result verified the calculation result which indicated higher temperature condition helped to reduce humidification duration for creating condensation on the bottom of the gel.

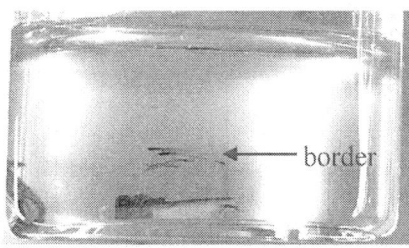

(A) Humidification duration of 155 min

(B) Humidification duration of 160 min

(C) Humidification duration of 165 min

(D) Humidification duration of 170 min

Fig. 12: The experiment result with 45°C85%RH humidification condition

Summary

As discussed above, humidification condition influences the condensation event. In order to establish the condensation test method, humidification condition is one of the most important parameters. Table 1 summarizes the findings of this work.

As described here, high temperature humidification condition is good to shorten test duration. However, it increases risk of cooling speed deviation and dependency on gel thickness. On the other hand, more stable condensation can be expected with humidification at lower temperature, since dependency on cooling speed and gel thickness is reduced. However, longer test duration is required.

Table 1: Humidification temperature impact on the condensation test

	Cooling speed dependency	Pre-humidification duration	Gel thickness dependency
High temperature humidification	Dependent	Enough with short time	Dependent
Low temperature humidification	Independent	Long time is required	Independent

Conclusion

Based on the humidity absorption model of silicone gel, condensation amount is calculated at various condensation-test conditions. At first, it is confirmed that condensation amount depends on dissolved humidity in silicone gel, not on absolute humidity of ambient air. Also, it is confirmed that humidification at lower temperature is effective to obtain stable results independently from cooling speed. Thirdly, it is confirmed that humidification at higher temperature is effective to shorten test time. Next, it is confirmed that humidification at lower temperature is effective to obtain stable results independently from gel thickness of the module.

Finally, the calculation result is verified by experiment.

It can be concluded that high temperature humidification condition is good to shorten test time, but low temperature humidification condition is good to obtain stable test results on the other hand.

Therefore, humidification temperature should be carefully chosen to meet condensation test requirement.

More research is required on the experiment in terms of quantification of condensation amount. This is desired to further reconfirm today's working results.

References

[1] S. Honda et al, "High Voltage Device Edge Termination for Wide Temperature Range plus Humidity with Surface Charge Control (SCC) Technology," *28th International Symposium on Power Semiconductor Devices and ICs (ISPSD)*, 2016, pp. 291-294.

[2] N. Tanaka et al, "Durable Design of the New HVIGBT Module," *PCIM Europe 2016; International Exhibition and Conference for Power Electronics, Intelligent Motion, Renewable Energy and Energy Management*, 2016, pp. 1-7.

[3] S. Kremp, et al, "Humidity robustness for high voltage power modules: Limiting mechanisms and improvement of lifetime," Microelectronics Reliability, vol. 88–90, Sep. 2018, pp. 447-452.

[4] C. Papadopoulos, et al, "The influence of humidity on the high voltage blocking reliability of power IGBT modules and means of protection," Microelectronics Reliability, vol. 88–90, Sep. 2018, pp. 470-475.

[5] C. Papadopoulos, et al, "Humidity Robustness of IGBT Guard Ring Termination," *PCIM Europe 2019; International Exhibition and Conference for Power Electronics, Intelligent Motion, Renewable Energy and Energy Management*, 2019, pp. 1-8.

[6] Christian Zorn, Nando Kaminski, "Acceleration of Temperature Humidity Bias (THB) Testing on IGBT Modules by High Bias Levels," *2015 IEEE 27th International Symposium on Power Semiconductor Devices & IC's (ISPSD)*, 2015, pp. 385-388.

[7] Y. Kitajima et al, "Lifetime estimation model of HVIGBT considering humidity," *PCIM Europe 2017; International Exhibition and Conference for Power Electronics, Intelligent Motion, Renewable Energy and Energy Management*, 2017, pp. 1-6.

[8] N. Tanaka et al, "Robust HVIGBT module design against high humidity," *PCIM Europe 2015; International Exhibition and Conference for Power Electronics, Intelligent Motion, Renewable Energy and Energy Management*, 2015, pp. 368-373.

[9] K. Nakamura et al, "The test method to confirm robustness against condensation", *2019 21st European Conference on Power Electronics and Applications (EPE '19 ECCE Europe)*, 2019, pp. P.1-P.8.

[10] S. Kremp et al, "Realistic climatic profiles and their effect on condensation in encapsulated test structures representing power modules," Microelectronics Reliability, vol. 76–77, Sep. 2017, pp. 409-414

[11] K. Hatori et al, "Humidity Absorption Behavior of Silicone Gel in HVIGBT Modules," *2021 23rd European Conference on Power Electronics and Applications (EPE'21 ECCE Europe)*, 2021, pp. 1-8.

[12] T. Mizutani, "Warpage Analysis in LCD Panel under Moisture Diffusion and Hygroscopic Swelling," Journal of Japan Institute of Electronics Packaging, Vol. 12, 2009, pp. 144-153

[13] K. Nagai, "Barrier Technology," 62-63, Kyoritsu Shuppan Co., Ltd. 2014 Print

Three phase PV inverter LCOE optimization considering technological choice

Morteza Tadbiri Nooshabadi[1,2], Jean-Luc Schanen[1], Shahrokh Farhangi[2], Hossein Iman-Eini[2]
[1]Univ. Grenoble Alpes, CNRS,
Grenoble INP, G2Elab,
38000 Grenoble, France
jean-luc.schanen@grenoble-inp.fr
[2]University of Tehran
School of Electrical and Computer Engineering
Tehran, Iran
m.tadbiri@ut.ac.ir ,Imaneini@ut.ac.ir

Abstract

This work uses design by optimization of power electronics converter to achieve the best Levelized Cost of Energy in a PV application. The methodology uses detailed models of power electronics active and passive components to determine the cost and performances of the solid-state energy conversion, and connect them to the system level vision. The deterministic algorithm used for converter sizing allows taking into account a large number of variables and constraints. Methodology, models and some illustrations of the results are provided in this paper.

Keywords

«Design optimization», «Photovoltaic», «Levelized cost of energy», «Reliability», «Cost analysis»

I.   Introduction

Photovoltaic (PV) technology requests a high efficiency power conversion in order to achieve acceptable price per produced kWh. Indeed the cost of power converters is not negligible in the installation. Both cost and performances of the power electronics conversion depends on the choice of components and the converter sizing: for instance, SiC devices are more expensive than Si IGBTs, but exhibit lower losses. The choice is therefore not straightforward. On the other hand, it is well known that the efficiency of a power converter depends on its nominal power [1], as well as its cost. Therefore, the design of a converter and the associated components ratings becomes a crucial issue.

In order to quantify and compare the cost for different energy technologies, Levelized Cost of Energy (LCOE) index is generally used [2]. LCOE represents the price at which the electricity is generated from a specific energy source over the whole lifetime of the generation unit. The index is expressed by:

$$LCOE = \frac{Total\ life\ cycle\ cost}{Total\ lifetime\ energy\ production} \tag{1}$$

Based on above expression, a cost-effective grid-connected PV systems can be obtained by minimizing the initial investment cost which is included the cost of the PV system components (e.g., PV modules, DC/AC inverters, etc.), maximizing the amount of energy injected into the grid, and increasing its reliability. The lifetime of the PV system components are indeed very important since any failure in operational time causes missing PV energy [3]. The injected energy into the grid is upped by maximum power point tracker (MPPT) control algorithm [4].

The rule of thumb for solar inverter overclocking is that solar panel capacity should not be more than roughly 30% greater than inverter capacity. More scientific work has already been done on optimal sizing of PV inverters, using various models and algorithms [5]. On the modeling point of view, database of existing inverters, simulations or simple analytical models have been reported [5-6]. Obviously, the mission profile (irradiation, local climate [7]) is always taken into account in these kinds of studies.

However, converter-level analytical models are only representative of a global behavior, and cannot reflect precisely the impact of technological choice and component design. Database of existing hardware is by definition limited to available technologies and cannot be used to investigate potential breakthrough or unconventional design. Precise simulation of the power electronics converter can of course be used to obtain the performances depending on the technological choices and inverter design, but it is very long and not really compatible with optimization, especially if various technological or structural options are considered.

Therefore, this paper proposes a methodology which is clearly optimization-oriented, based on component models to obtain the minimum Levelized Cost of Energy (LCOE) of the power electronics part only. Each part of the PV inverter and MPPT boost converter are considered, and the global performances of the conversion take therefore into account the components behavior and sizing. Several constraints are addressed in the optimization: device-level constraints (as the semiconductor maximum temperature), as well as system level constraints (as THD on the AC side). Section II will illustrate the interest of having a precise representation of the converters performances, based on a case study using three different manufacturers. Section III will then provide all models used in the converters optimization, as well as the optimization methodology, which is based on a deterministic algorithm. The lifetime prediction is also evaluated in this section. Section IV will present some optimization results for various cases.

II.   LCOE of industrial inverters: case study

The evaluation of the performances of a PV inverter has to be achieved with respect to the balance between the investment cost (price of the inverter if we focus on this part of the PV system only) and the amount of energy produced in the product lifetime. In order to quantify and compare the cost for different situations, the Levelized Cost of Energy (LCOE) index will be used for the converters. For this purpose, the mission profile of the PV inverter has first to be defined, and the efficiency of the inverter vs power to be considered. Fig. 2 shows three cases studies of the same power (20kVA) obtained from manufacturer datasheets [8-10]. To enlarge the study, two different locations were considered (Grenoble (France) and Tehran (Iran)), with different irradiation characteristics.

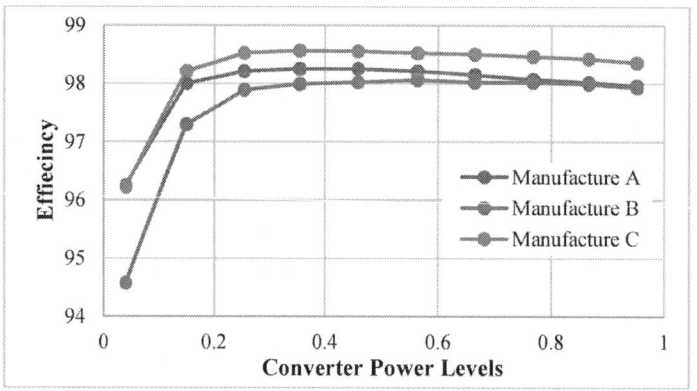

Fig. 2: The inverter efficiency curve for 3 different 20 kVA inverters

The mission profile was developed based on local measurements. Data points were taken every ten minutes, corresponding to the 10-minutes average of irradiance and ambient temperature. The mean daily profiles, averaged over the duration of the considered data in Grenoble, are shown in Fig 3. It was then split into 10 steps for operational phases and one step for dormant phase (Fig. 4), for an example of application which is a 20 kW installation, composed of 4 strings of 16 * 320W panels (Fig. 5). By using [11] method, Table.1 shows the mission profile data at Grenoble in each step.

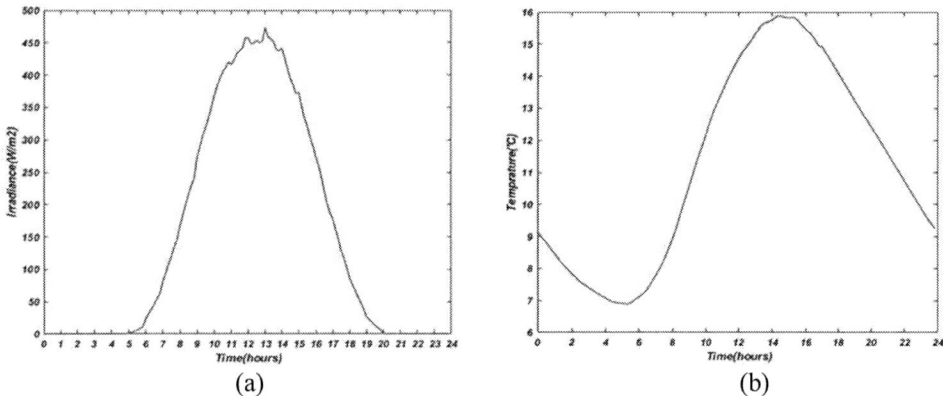

(a)　　　　　　　　　　　(b)

Fig. 3: Mean diurnal profiles in Grenoble. (a) Irradiance, (b) Temperature.

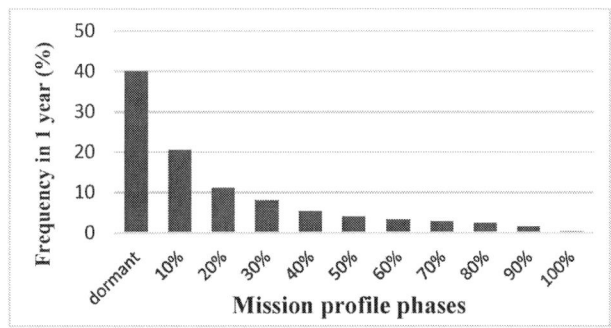

Fig. 4: Mission profile expressed in 10 different phases. Percent length of each phase, at location Grenoble

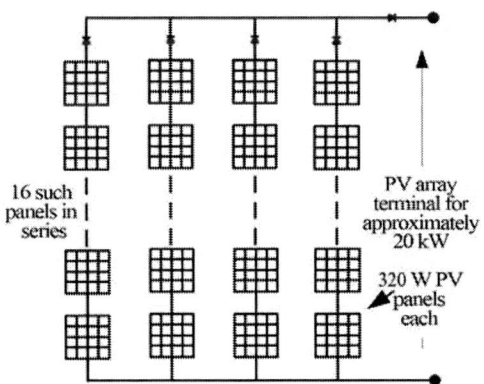

Fig. 5: PV panel arrangement to make it an array of 20 kW.

TABLE I: GRENOBLE MISSION PROFILE

	Phase	hL(hour)	TL(°C)	Vpv	Ppv
	dormant	3503.667	10.35695	0	0
	10%	1796.333	11.53196	610.10	823.28
	20%	974.6667	13.44483	639.32	3031.9
	30%	708.3333	15.28658	641.26	5088.7
	40%	468.1667	17.10864	638.29	7095.6
Pmax = 20.533 kW	50%	359.8333	18.66902	632.91	9194.6
	60%	295.1667	19.83923	628.72	11270
	70%	253.8333	20.89177	623.22	13305
	80%	222.6667	21.70667	618.04	15351
	90%	136.8333	22.02976	613.16	17328
	100%	40.5	22.07141	611.07	19017

By combining the mission profile data with the efficiency curve of inverter, the total amount of energy is obtained as eq. (2), for a duration of 25 year as follow (this duration is considered as useful lifetime of solar panel and industrial inverters are guaranteed by manufactures to work without any problem in this time duration).

$$E(MWh) = 25 \, yr * \sum_{i=1}^{10} P_{PV_i}(W).\eta_i.t_i(h).10^{-6} \qquad (2)$$

Referring to the price of each inverter [12] leads to the LCOE of each inverter, in €/MWh (Fig.6). From this figure, it is clear that the efficiency difference (Fig. 2), which is due to different technological and design choices, clearly impacts the LCOE. Regardless the inverter lifetime, which will be addressed in section III of the paper, manufacture C inverter seems to be best choice from an LCOE perspective.

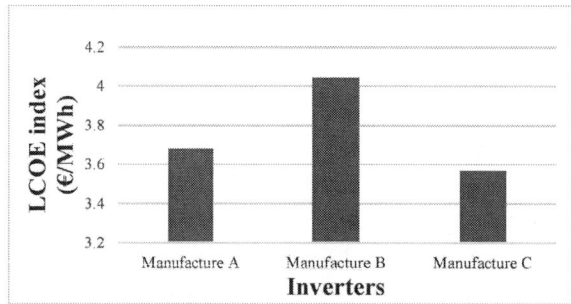

Fig. 6: The LCOE index in each inverter for location Grenoble.

Another study case consists in choosing various sizing powers from the same manufacturer (Manufacture C inverters in this case). By applying the mission profile to each of them, the efficiency curves (Fig.7) and finally the LCOEs are obtained. The usual PV panel degradation [13] has not be considered here for simplicity. Results, shown in Fig. 8 for location Tehran, show that a 17kW sizing is the best. The same approach for localization Grenoble shows that a 12kW choice would be better, according to LCOE index applied to the inverter. That is notable that there is a PV array oversizing in inverters with nominal power less than 20 kW which causes to some of PV energy produced by panels have been lost and, in this condition, inverter works in its nominal power. Even though some of PV energy have been lost, this method is used to gain more energy during the low solar irradiance [3]. So, the LCOE index shows the best choice for inverter in a fixed installed PV array condition.

This simple illustration on existing PV inverters shows that both technological choice and converter design impact the LCOE, and have to be considered when installing a PV inverter in a given location.

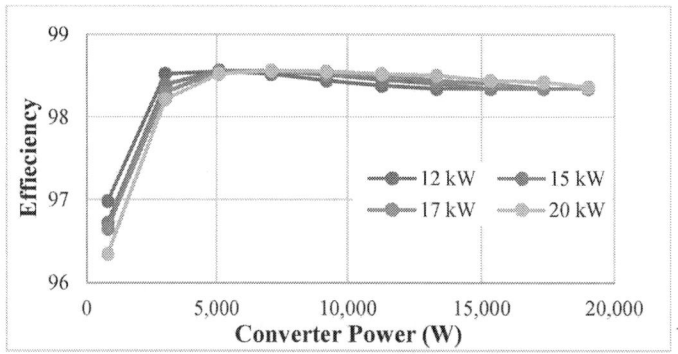

Fig. 7: Efficiency curves for four different power levels

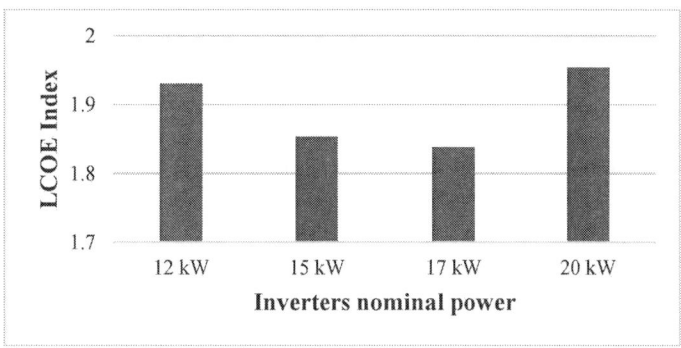

Fig. 8: LCOE index for manufacture C inverters at location Tehran

As mentioned before, lifetime is a mandatory part of LCOE calculation, and has not been included in the previous study, since no information is provided in commercial inverters catalogs. For obtaining the previous results, it was assumed that each inverter was working at least 25 years, based on manufacturer claiming. However, lifetime evaluation needs a comprehensive knowledge on inverter components. This will be obtained through a detailed analysis of the converter design, using simple but quite accurate models for pre-design. This pre-design step will also allow evaluating the inverter cost, based on the bill of materials (BoM). The next section will illustrate all models used for inverter pre-design.

III. Models for inverter pre-sizing

Fig. 9 shows the hardware components of a PV-System from the solar array to the grid. The grid-connected photovoltaic system has two main parts: the maximum power point tracker (MPPT) and the grid-connected inverter. The MPPT is responsible for maintaining the solar array at its maximum power as well as supplying the DC link voltage in the specified value. The inverter connected to the grid is responsible for supplying the sine current injected to the grid according to the existing standards, what leads to the necessity of output filters L_1-C-L_2. Each component of the two converters (active and passive) are modeled quite accurately.

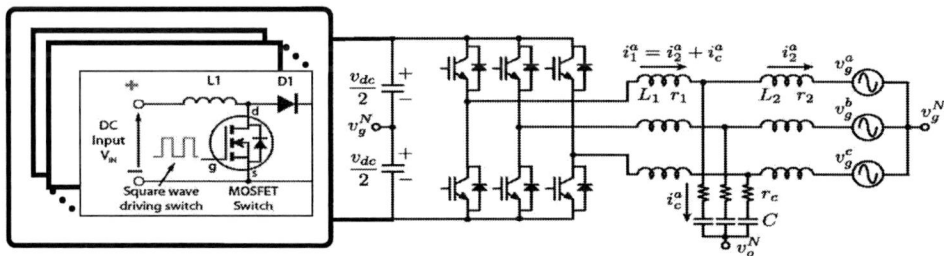

Fig. 9: Different components of the grid-connected photovoltaic system

A. AC Filter

In case of PWM inverters, the design of the AC inductor is particularly critical, as it concentrates a large part of losses of the whole system [14]. This inductor should have a significant value to decrease the ripple at switching frequency, but should also fulfill a thermal constraint. The saturation phenomenon may decrease the inductor value during the current peak and affect the effectiveness of the filter. Therefore, it is taken into account. Design procedure of inductor and LC filter with saturation consideration is explained in [15] and has been adapted to LCL filter. It considers the material choice, the core and wiring size. The capacitors are designed according to the needed capacitance, voltage and RMS current. To do this, the ripple current is assumed to be sinusoidal, at the switching frequency, and centered on the low frequency current. These assumptions are translated into eq. (3) as follows:

$$I_{L1,AC}(t) = i_1 + i_{ripple} = \sqrt{2}I_{L1,AC}\sin(\omega t + \varphi_1) + \frac{\Delta I_{L1,AC}(t)}{2}\sin(\omega_{sw}t) \qquad (3)$$

$\Delta I_{L1,AC}(t)$ is the current ripple in the inductor at the inverter side that is variable due to the saturation effect. Thus, for simplicity, a mean ripple is defined:

$$\Delta I_{L1,AC,mean} = \frac{1}{N_p} \sum_{i}^{N_p} \Delta I_{L1,AC}(t_i) \tag{4}$$

The ripple currents in L_2 and C obtain by putting L_1 ripple current in filter transfer function as follows:

$$\Delta I_{L2,AC}(t) = \left| \frac{r_C + \dfrac{1}{j\omega_{sw}C}}{R_f + \dfrac{1}{j\omega_{sw}C} + j\omega_{sw}L_2} \right| \Delta I_{L1,AC}(t) \tag{5}$$

$$\Delta I_{C_f,AC}(t) = \left| \frac{j\omega_{sw}L_2}{R_f + \dfrac{1}{j\omega_{sw}C} + j\omega_{sw}L_2} \right| \Delta I_{L1,AC}(t) \tag{6}$$

Therefore, knowing the filter parameters, the other steps of [15] can be continued. Also, the THD of injected current that should be limited in standard margin, is calculated by:

$$THD = 100 \frac{\dfrac{\Delta I_{L2,AC,mean}}{2\sqrt{2}}}{I_{L2,AC}} \tag{7}$$

B. Inverter Losses

A bipolar PWM pulse, computed by a comparison of a sine-wave and a triangular carrier, controls the commutation of the switches. The power losses in the inverter depend on the current and voltage patterns across the switches and on the switches sizing (i.e. voltage and current rating), which is also a design parameter. Indeed, higher current capability leads to reduced conduction losses but increased switching losses. In accordance with [16], the RMS and average current flowing through each branch of inverter is analytically calculated by computing switching angles. This calculation takes into account the AC current ripple, which depends on the AC output filter. Losses in inverter are calculated by [17]. The switching frequency is also a design variable. As in [17], the losses are stated based on pure sine current, in order to consider the ripple current in calculations, the ripple current in summing in fundamental current as below:

$$i(t) = \sqrt{I_{m1}^2 + \frac{\Delta I_{L1,AC,mean}^2}{4}} \sin(\omega t + \varphi_i) \tag{8}$$

In switching losses, the switch turns on in $i_1 - \frac{\Delta I_{L1,AC,mean}}{2}$ current and turns off in $i_1 + \frac{\Delta I_{L1,AC,mean}}{2}$ current. Also, the diode turns off in $i_1 - \frac{\Delta I_{L1,AC,mean}}{2}$ current. Therefore, energy losses in the MOSFET and diode are modified as below:

$$E_{on} = E_{on}\left(V_{dc}, i_1 - \frac{\Delta I_{L1,AC,mean}}{2}\right) \tag{9}$$

$$E_{off} = E_{off}\left(V_{dc}, i_1 + \frac{\Delta I_{L1,AC,mean}}{2}\right) \tag{10}$$

$$E_{d(off)} = E_{d(off)}\left(V_{dc}, i_1 - \frac{\Delta I_{L1,AC,mean}}{2}\right) \tag{11}$$

C. Boost Converter

Boost converter is composed of inductor and switches. By using the same method as in previous sections in inductor design and also the switch current calculation, the output spectrum of current in boost converter is calculated. In the purpose of boost inductor waveform modeling, like as the current in AC filter, the ripple current is assumed to be sinusoidal at the switching frequency and superimposed on the dc part current. These assumptions are translated into eq. (12).

$$I_{Lb}(t) = I_{in} + \frac{\Delta I_{Lb}}{2} sin(\omega_{sw,b} t) \tag{12}$$

The comparison between the modeled L_b current and the simulation result is shown in Fig. 10.

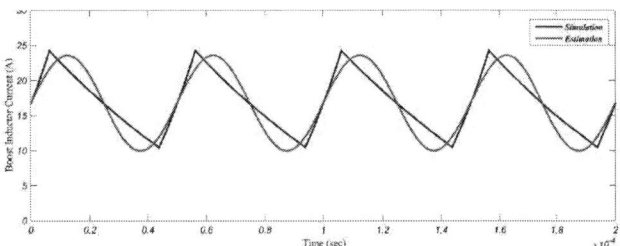

Fig. 10: Modelled boost converter current waveform and simulation validation

Beside of current waveforms, the losses in inductor and switches are calculated. The switching frequency is again a design variable, as well as the semiconductor sizing. In the same way, by considering current as a ramp, the switch turns on in $I_{in} - \frac{\Delta I_{Lb}}{2}$ current and turns off in $I_{in} + \frac{\Delta I_{Lb}}{2}$ current. Also, the diode turns off in $I_{in} - \frac{\Delta I_{Lb}}{2}$ current. Therefore, energy losses in the switch and diode are calculated as inverter losses.

D. DC-link Capacitor

DC-link capacitor is designed based on the maximum allowed ripple voltage, but also on the maximum RMS current passing through the capacitor. This current is defined in the frequency domain by $i_C = i_{boost} - i_{inv}$ of the inverter input current i_{inv} and the output current i_{boost} of the boost converter which are obtained in previous steps. The steps for calculating the RMS current of the DC-link are explained in [16] and [18].

E. Thermal Model

Beside of dynamic and impedance modelling mentioned in [19], since the studying is based on steady-state conditions, the system is modeled by the thermal resistance (Rth). Junction temperature of switches and diodes are calculated based on thermal resistance from junction to ambient, and limited by constraints. By supposing that all of semiconductors are located on a unique heat-sink, the thermal model of the case studied photovoltaic inverter is shown in Fig. 11. Each loss is evaluated from sections B and C. The thermal resistances are evaluated according to a low power MOSFET and diode as reference semiconductors which can be paralleled by N_{sw} and N_d. In this case, the equivalent thermal resistance is calculated by dividing the reference value by the number of parallels semiconductors.

$$R_{th,eq,MOSFET(diode)} = R_{th,MOSFET(diode),ref}/N_{sw(diode)} \tag{13}$$

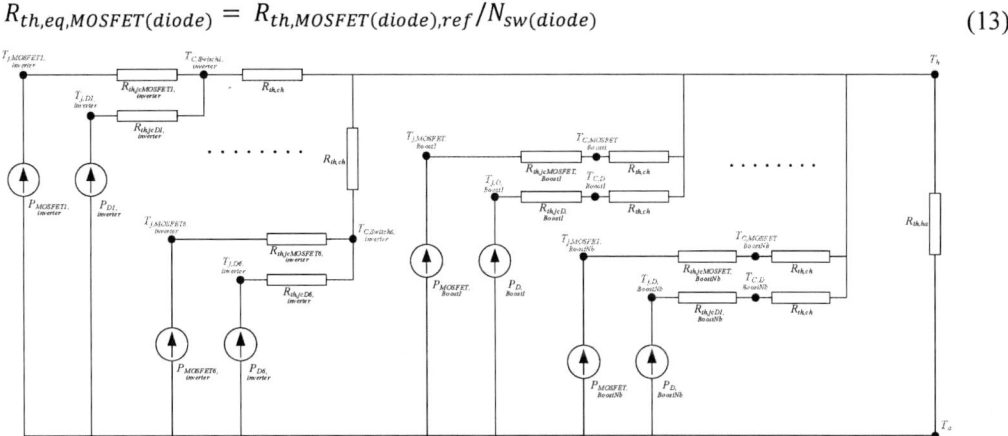

Fig. 11: Thermal model of the case studied photovoltaic inverter.

Using the thermal model of Fig. 11 allows evaluating the temperatures of each device, being the input of reliability models. Furthermore, the maximum allowed temperature is the sizing constraint for the heatsink.

F. Heatsink model

The heatsink is studied from the data of the [20] which has a large range of heatsinks in its productions. Fig. 12 shows the interpolation of a large amount of different heatsinks.

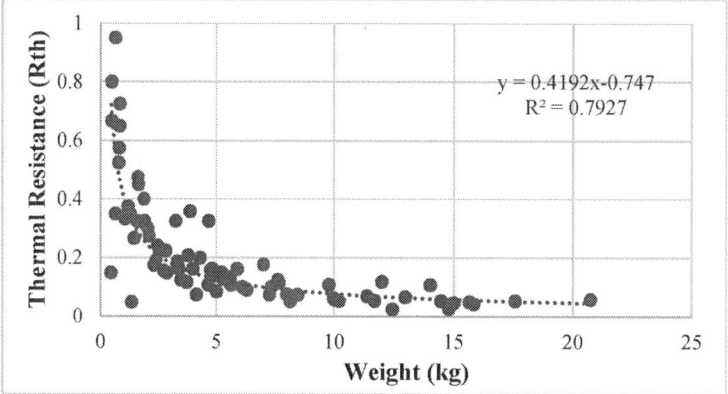

Fig. 12: Thermal resistance of the heatsink according to its weight

The thermal resistance parameter of the heatsink is expressed by its weight according to eq. 14.

$$Rth_{hs} = 0.4192W_{hs}^{-0.747} \tag{134}$$

G. Reliability Model

In order to obtain longer operational lifetime, the most fragile components must be specified, and reliability evaluation methodologies can be used for this approach. For this approach, FIDES methodology is used which is based on Physics-of-Failure (PoF) and developed by a French companies consortium. FIDES Methodology is supported by the analysis of test data, field returns and existing modelling. According to a previous evaluation, the FIDES methodology provides better results in compare with the observed ones [11]. FIDES is a standard which defines mathematical formula for evaluating the life-time of devices as a function of various parameters like as such as operating temperature, voltage and also manufacturing and mechanical stress [21]. Nevertheless, the most significant parameter in reliability and life-time evaluating is temperature of devices. Following the steps in [21] to evaluate the failure rate of each component in each phase of operation (λ_i^{phj}), and then each component failure rate (λ_i), the overall system failure rate is obtained by multiplying the failure rate of each component as follows:

$$\lambda_{total} = \prod_i \lambda_i \tag{15}$$

H. Cost Model

Despite the poor availability of cost data in academia, cost models are very difficult to establish, therefore, only the individual cost of components (semiconductors, capacitors, core and windings for inductors), based on manufacturer prices for large ordering quantities, numerical values for the cost model parameters are taken account. Of course, this is not representative of the actual cost of a PV inverter, but it is sufficient for comparison purpose. In addition to the analysis of the cost of components in the markets, it can be considered by the models that are proposed in [22].

IV. LCOE estimation for various case study

All previous models have been developed taking care to their derivability, in order to be used with a deterministic algorithm (Sequential Quadratic Programming method (SQP) [23]). This presents the advantage of being very effective in finding quickly the optimum in a large space of solution, with large amount of variable and constraints. The model indeed is composed of more than 25 parameters and 50 constraints (at component or system level). The design framework used is able to perform automatically the derivation of all equations, what leads to a significant time saving. The objective function is to reach the minimum cost for a given sizing power. In optimization process, the reliability is force to be at least more than 25 years to be insured that the inverter could work in PV panels lifetime. For each optimized inverter, the mission profile is applied. Of course, the maximum power is limited to the sizing power during operation. To be noticed that overload capability has not be considered since the models cannot handle the consequence of this kind of overloads. Consequently, the converter is exactly sized regarding nominal power, without any margin. Fig. 13 shows the design level of optimization procedure where the design variables in each iteration are chosen. More than 25 variables are considered including filter parameters (core size and turn number for inductors), switching frequencies (inverter and boost converters), number of switches and …. The set of design variable should satisfy some constraints on THD, junction temperature of switches and diodes and rise temperature in capacitors. then, the LCOE index is evaluated by combining results of sections G and H and then the best set of design variables choose. The flowchart of LCOE optimization is presented in Fig. 14.

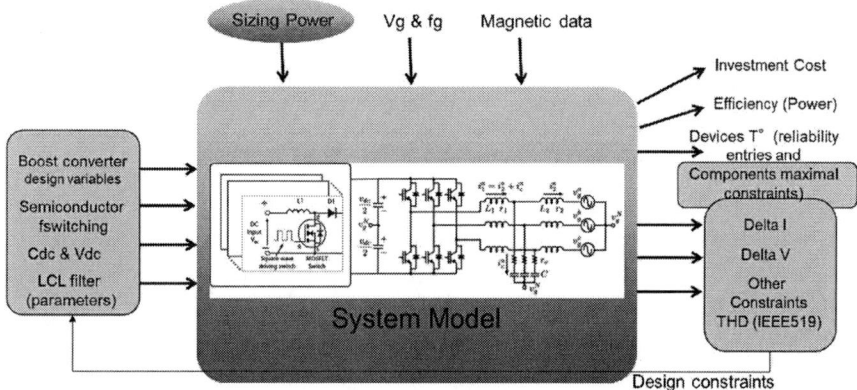

Fig. 13: the design level of optimization procedure with considering design variables as input.

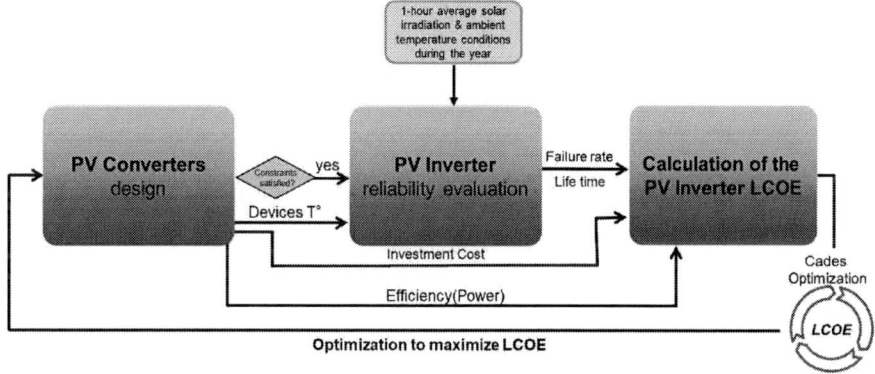

Fig. 14: The flowchart of LCOE optimization.

Results of LCOE for Grenoble and Tehran are displayed in Fig. 15. It is worth noting that the optimal sizing power is different for the two locations, according to the different solar irradiation. This confirms the results from previous work, which showed the interest of downsizing the nominal power to gain more energy yield during the low solar irradiance conditions [3]. Note that the LCOE index is applied to inverter only, and that the inverter cost is reduced to the sum of components cost, therefore absolute values of sizing power are not to be considered as a strict result, but only for comparison purpose (see conclusion).

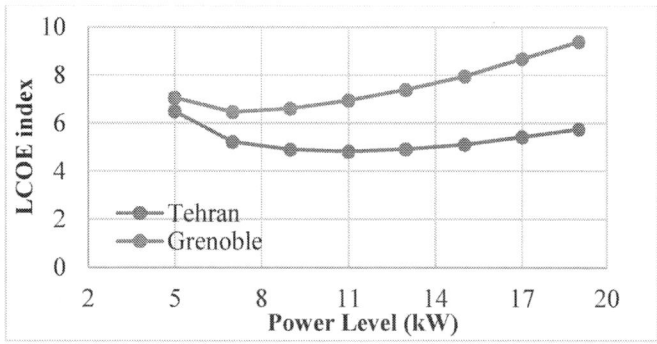

Fig. 15: LCOE index comparison for Grenoble and Tehran

Another interest of the approach is that it allows investigating the impact of some technological choices. For instance, using SiC MOSFET or Si IGBT has been illustrated in Fig 16, for location Tehran. It is worth noting that IGBT, despite is lower cost leads to higher LCOE. Indeed, its higher switching losses lead to reduced switching frequency (16 kHz roughly for each design, which was the minimum imposed for audition limit), what leads to higher cost for passive components, specifically to meet the THD constraint.

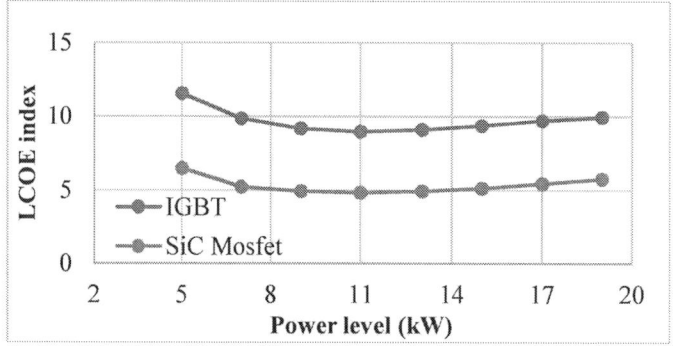

Fig. 16: LCOE index Si vs SiC (Tehran)

Another technological choice that can be considered is the comparison of the wire material between copper and aluminum. The LCOE index for both material is presented in Fig. 17 where this index is almost the same for both cases while the aluminum is cheaper than copper in the markets but the resistivity coefficient of aluminum (ρ) is higher than that of copper [24] and it causes more power losses in the system. Thus, the injected energy is lower than the other case. It should be noted that the temperature coefficient (α) is not considered in this study and since this coefficient is higher in aluminum than in copper the LCOE can be expected to be higher than the one presented.

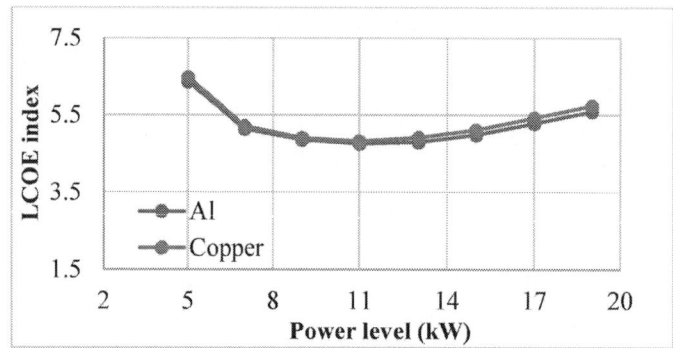

Fig. 17: LCOE index Copper vs Aluminium (Tehran)

V. Conclusions

Considering the actual sizing and technological choice for PV inverters is determining to solve the tradeoff between inverter cost and its performances during its lifetime. This paper illustrates first the impact of these parameters with existing inverters, using the LCOE index (applied to power electronics only). Then an optimization method is proposed, using detailed models of components, in order to obtain the best sizing power for the inverter, according to LCOE index and a given mission profile. The design method accounts for components and system level constraints. It is also applied to investigate the impact of semiconductor and material choice.

References

[1] L. Keller and P. Affolter, "Optimizing the panel area of a photovoltaic system in relation to the static inverterU° practical results," Solar Energy, vol. 55, no. 1, pp. 1–7, 1995.
[2] E. Koutroulis and F. Blaabjerg, "Design Optimization of Transformerless Grid-Connected PV Inverters Including Reliability," in IEEE Transactions on Power Electronics, vol. 28, no. 1, pp. 325-335, Jan. 2013, doi: 10.1109/TPEL.2012.2198670.

[3] A. Sangwongwanich, Y. Yang, D. Sera, and F. Blaabjerg, "Mission Profile-Oriented Control for Reliability and Lifetime of Photovoltaic Inverters," 2018 Int. Power Electron. Conf. IPEC-Niigata - ECCE Asia 2018, pp. 2512–2518, 2018, doi: 10.23919/IPEC.2018.8507577.

[4] R. Kadri, J.-P. Gaubert, and G. Champenois, "An improved maximum power point tracking for photovoltaic grid-connected inverter based on voltage-oriented control," IEEE Trans. Ind. Electron., vol. 58, no. 1, pp. 66–75, Jan. 2011.

[5] R. Nasiri, M. Khayamy, M. Rashidi, A. Nasiri and V. Bhavaraju, "Optimal Solar PV Sizing for Inverters Based on Specific Local Climate," 2018 IEEE Energy Conversion Congress and Exposition (ECCE), Portland, OR.

[6] G. Velasco, F. Guinjoan, R. Pique, A. Conesa and J. J. Negroni, "Inverter Power Sizing Considerations in Grid-Connected PV Systems," 2007 European Conference on Power Electronics and Applications, Aalborg,

[7] M.G.Kratzenberg, E.M.Deschamps, L.Nascimento, R.Rüther, H.H.Zürn, "Optimal Photovoltaic Inverter Sizing Considering Different Climate Conditions and Energy Prices",Energy Procedia, Volume 57, 2014, Pp 226-234

[8] S. M. A. Solar and T. Ag, "Integrated service for ease and comfort," pp. 8–11.

[9] Fronius, "Fronius Symo Datasheet," pp. 6–11, [Online]. Available: http://www.fronius.com/.

[10] Huawei, "Smart String Inverter," pp. 2–3, 2019.

[11] S. E. De Leon-Aldaco, H. Calleja, and J. Aguayo Alquicira, "Reliability and mission profiles of photovoltaic systems: A FIDES approach," IEEE Trans. Power Electron., vol. 30, no. 5, pp. 2578–2586, 2015, doi: 10.1109/TPEL.2014.2356434.

[12] "Solar inverters | Solar Inverter for PV System | Europe Solar Store." [Online]. Available: https://www.europe-solarstore.com/solar-inverters.html?inverter_power=24.

[13] A. Sangwongwanich, Y. Yang, D. Sera, F. Blaabjerg, "Lifetime Evaluation of Grid-Connected PV Inverters Considering Panel Degradation Rates and Installation Sites," IEEE Trans. Pow. Electr., vol. 33, no. 2, 2018.

[14] T. Orlowska-Kowalska and M. Kaminski, Advanced and Intelligent Control in Power Electronics and Drives, vol. 531. 2014.

[15] A. Voldoire, J. L. Schanen, J. P. Ferrieux, C. Gautier, C. Saber, "Optimal Design of an AC Filtering Inductor for a 3-Phase PWM Inverter Including Saturation Effect," PEDSTC 2019, Shiraz, Iran

[16] A. Voldoire, J. L. Schanen, J. P. Ferrieux, C. Gautier, and C. Saber, "Analytical calculation of dc-link current for N-Interleaved 3-Phase PWM inverters considering AC current ripple," 2019 21st Eur. Conf. Power Electron. Appl. EPE 2019 ECCE Eur., p. P.1-P.10, 2019, doi: 10.23919/EPE.2019.8915183.

[17] M. H. Ahmed, M. Wang, M. A. S. Hassan, and I. Ullah, "Power Loss Model and Efficiency Analysis of Three-Phase Inverter Based on SiC MOSFETs for PV Applications," IEEE Access, vol. 7, pp. 75768–75781, 2019.

[18] J. W. Kolar and S. D. Round, "Analytical calculation of the RMS current stress on the DC-link capacitor of voltage-PWM converter systems," IEE Proceedings - Electric Power Applications, vol. 153, no. 4, p. 535, 2006, doi: 10.1049/ip-epa:20050458.

[19] Y. Shen, S. Song, H. Wang, and F. Blaabjerg, "Cost-Volume-Reliability Pareto Optimization of a Photovoltaic Microinverter," in 2019 IEEE Applied Power Electronics Conference and Exposition (APEC), Mar. 2019, pp. 139–146. doi: 10.1109/APEC.2019.8722043.

[20] "Wakefield Thermal Air Cooled Thermal Extrusions." https://wakefieldthermal.com/thermal-solutions/air-cooled/thermal-extrusions/

[21] "FIDES guide 2009. Edition A. Reliability Methodology for Electronic Systems," 2010. [Online]. Available: www.fides-reliability.org.

[22] R. Burkart and J. W. Kolar, "Component cost models for multi-objective optimizations of switched-mode power converters," in 2013 IEEE Energy Conversion Congress and Exposition, Sep. 2013, pp. 2139–2146.

[23] P. T. Boggs and J. W. Tolle, "Sequential Quadratic Programming," Acta Numer., vol. 4, pp. 1–51, 1995

[24] Giancoli, Douglas C., Physics, 4th Ed, Prentice Hall, (1995).

Square wave operation to reduce pulsating power in isolated MMC-based ultrafast chargers

Ygor Pereira Marca, Maurice G. L. Roes, Jorge L. Duarte and Korneel Wijnands
Eindhoven University of Technology
Electromechanics and Power Electronics Group
P.O. Box 513, 5600MB
Eindhoven, The Netherlands
Phone: +31 (0) 61-377-4267
Email: y.pereira.marca@tue.nl

Acknowledgments

This publication is part of the project NEON (with project number 17628 of the research programme Crossover which is (partly) financed by the Dutch Research Council (NWO)).

Keywords

≪Charger≫, ≪Full-bridge sub-modules≫, ≪Medium-frequency transformer≫, ≪Medium-voltage grid≫, ≪Modular multilevel converter≫, ≪Square wave≫.

Abstract

This paper presents an application of modular multilevel converters to reduce pulsating power, and therefore sub-modules in ultrafast chargers. The converter's analysis and a control scheme were presented to realize bidirectional power transfer between a three-phase medium-voltage grid and a single-phase medium-frequency transformer with square-shaped voltage, successfully reducing power fluctuation.

Introduction

Due to its modularity, output voltage quality, and high efficiency, the modular multilevel converter (MMC) has been widely used for medium-voltage and high-power conversion since its first publication in [1]. Its configuration is scalable to satisfy different voltage and power demands. Furthermore, the voltage over semiconductors is minimized because sub-modules are connected in series [2].

Recently, ultrafast chargers implementing ac/ac MMC are being researched to reduce cost and size in medium-voltage applications. As shown in Fig. 1, a medium-frequency transformer can be integrated into the MMC-based charger to substitute the bulky line-frequency transformer that is normally used for electrical isolation [3]. Furthermore, by use of a proper transformer current control, switching losses in the ac/dc converter (Fig. 1) can be reduced [4, 5]. Yet, the single-phase output of the MMC inherently leads to power fluctuation, which increases the medium-frequency transformer losses and size of capacitor [6, 7]. Therefore, it is appropriate to evaluate and propose a strategy to decrease pulsating power in such a system.

Previous research has shown the possibility of creating a square instead of a sinusoidal voltage waveform on the MMC single-phase terminals [8,9]. Square-shaped current and voltage quantities reduce pulsating power in medium-voltage chargers [10]. Therefore, this paper presents a strategy to implement the square-shaped voltage waveform in the MMC single-phase terminals. The proposed system increases the efficiency and power density of ultrafast charging stations.

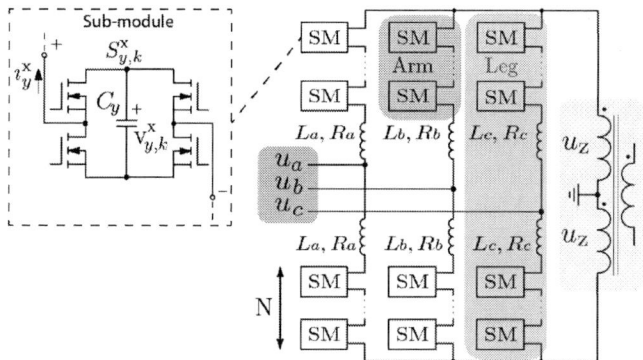

Fig. 1: MMC-based ultrafast charger with a medium-frequency transformer.

Modular multilevel converter

For each of the three phases indexed by $y \in \{a, b, c\}$, there is an upper and a lower arm indicated by $x \in \{u, \ell\}$. The arms are composed of N series-connected sub-modules, all of which contain a full-bridge converter and a capacitor as shown in Fig. 2. The MMC based on full-bridge sub-modules can generate any periodic voltage waveform over the single-phase medium-frequency transformer terminals [11]. However, the amplitude is limited by the converter's voltage rating, which can be chosen to obtain high efficiency [3].

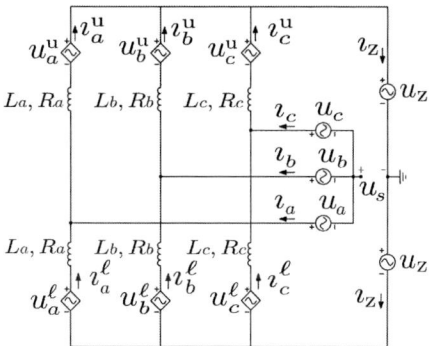

Fig. 2: Three-phase to single-phase ac/ac MMC with full-bridge sub-modules.

Fundamentals

Controllable average voltage sources can be used to represent the average behavior of all sub-modules in an arm to simplify the analysis, as shown in Fig. 3. In addition, each arm has a series inductance to allow current control [2].

Fig. 3: Three-phase to single-phase ac/ac MMC equivalent circuit.

As described in [3], it is favorable to separate the arm currents into components, namely, common-mode current ($\imath_y^{\Sigma}$) and differential-mode current ($\imath_y^{\Delta}$), as

$$\imath_y^{\Sigma} = \frac{1}{2}\left(\imath_y^{\mathrm{u}} + \imath_y^{\ell}\right), \qquad (1) \qquad\qquad \imath_y^{\Delta} = \frac{1}{2}\left(\imath_y^{\mathrm{u}} - \imath_y^{\ell}\right). \qquad (2)$$

In the same manner, the arm voltages are split into common-mode (u_y^{Σ}) and differential-mode voltage (u_y^{Δ}), as

$$u_y^{\Sigma} = \frac{1}{2}\left(u_y^{\mathrm{u}} + u_y^{\ell}\right), \qquad (3) \qquad u_y^{\Delta} = \frac{1}{2}\left(u_y^{\mathrm{u}} - u_y^{\ell}\right). \qquad (4)$$

Using these definitions, the circuit differential equations can be written with common-mode and differential-mode components. Then, Kirchhoff's voltage law applied to the upper and lower loops results in

$$L_y\frac{d}{dt}i_y^{\Sigma} + R_y i_y^{\Sigma} = u_y^{\Sigma} - u_z, \qquad (5) \qquad L_y\frac{d}{dt}i_y^{\Delta} + R_y i_y^{\Delta} = u_y^{\Delta} + u_y. \qquad (6)$$

The arm voltages u_y^{x} can be used to control the common-mode and differential-mode currents. These voltages are the result of the output voltages of N series-connected sub-modules as $u_y^{\mathrm{x}} = \sum_{k=1}^{N} S_{y,k}^{\mathrm{x}} v_{y,k}^{\mathrm{x}}$, with the kth sub-module's switching function given by $S_{y,k}^{\mathrm{x}} \in \{-1,0,1\}$ and v_y^{x} the kth sub-module's capacitor voltage. In addition, a control scheme is required to balance sub-modules' capacitor voltage [2].

Power distribution

For stable operation, each converter leg must deliver or receive a constant average power to the three-phase or single-phase terminals, regardless of their voltage and current waveforms. Furthermore, the average power difference between the upper and lower arms must be zero to keep the converter balanced.

Based on the instantaneous power $p_y^{\mathrm{x}} = -u_y^{\mathrm{x}} i_y^{\mathrm{x}}$ processed by each controllable voltage source in Fig. 3, it is convenient to define in line with (1)-(2),

$$p_y^{\Sigma} = \frac{1}{2}\left(p_y^{\mathrm{u}} + p_y^{\ell}\right), \qquad (7) \qquad p_y^{\Delta} = \frac{1}{2}\left(p_y^{\mathrm{u}} - p_y^{\ell}\right). \qquad (8)$$

The differential-mode components are related to the grid with frequency ω_1 [3]. Assuming that the grid voltages and currents are sinusoidal, we have that

$$-u_y^{\Delta} = \sqrt{2}U_y^{\Delta}\cos\left(\omega_1 t + \theta_y^{\Delta}\right), \qquad (9) \qquad i_y^{\Delta} = \sqrt{2}I_y^{\Delta}\cos\left(\omega_1 t + \varphi_y^{\Delta}\right). \qquad (10)$$

Then, after some manipulations (7) and (8) result in

$$p_y^{\Sigma} = 2U_y^{\Delta}\cos\left(\omega_1 t + \theta_y^{\Delta}\right) I_y^{\Delta}\cos\left(\omega_1 t + \varphi_y^{\Delta}\right) - u_y^{\Sigma} i_y^{\Sigma}, \qquad (11)$$

$$p_y^{\Delta} = \sqrt{2}U_y^{\Delta}\cos\left(\omega_1 t + \theta_y^{\Delta}\right) i_y^{\Sigma} - \sqrt{2}I_y^{\Delta}\cos\left(\omega_1 t + \varphi_y^{\Delta}\right) u_y^{\Sigma}. \qquad (12)$$

This common-mode and differential-mode power must be controlled, to have in steady-state operation the averages $P_y^{\Sigma} = P_y^{\Delta} = 0$.

Efficiency

According to [3], an estimation of the charger efficiency can be determined by considering the switch conduction losses. The required number of sub-modules per arm and the derated maximal voltage of the switches are given by

$$N = \left\lceil \frac{\widehat{U}_y + \widehat{U}_z}{V_{sw}} \right\rceil, \qquad (13) \qquad V_{sw} = V_{ds}D_F, \qquad (14)$$

where V_{ds} and D_F are the semiconductor device breakdown voltage and derating factor, respectively. Switching losses may be disregarded in efficiency evaluation when considering SiC MOSFETs that are operated at sufficiently low switching frequencies. Therefore, given the RMS arm currents [3], $I_y^x = \sqrt{\frac{I_y^2}{4} + \frac{I_y^2}{9}}$, the total conduction losses and power efficiency of the ac/ac MMC are approximated by

$$P_{\text{loss}} = 12 R_{ds} N (I_y^x)^2, \qquad (15) \qquad \eta = \left(1 - \frac{P_{\text{loss}}}{P}\right) \cdot 100, \qquad (16)$$

where R_{ds} is the on-state resistance of each switch in a sub-module and P is the charger's nominal processed power.

MMC control

The proposed scheme controls the bidirectional power flow between the three-phase and single-phase terminals of the ac/ac MMC as well as the sub-modules capacitor voltages. The power exchange with the grid can be regulated through a PQ controller of frequency ω_1 [12], and the power exchange with the transformer terminals can be controlled through a PQ controller of frequency ω_2 or as presented in [3].

Internal and square wave current control

Maintaining all capacitor voltages ($v_{y,k}^x$) balanced around a reference is a requirement for the MMC's stable operation [12, 13]. As analyzed in (11) and (12), two independent components must be introduced in u_y^Σ, and therefore also in i_y^Σ, with frequencies ω_1 and ω_2 for decoupled control. Frequency ω_1 is related to the three-phase grid, and frequency ω_2 corresponds to the (fundamental) frequency of the single-phase medium-frequency transformer terminal voltage. So, common-mode components are introduced as

$$u_y^\Sigma = \sqrt{2} U_{y1}^\Sigma \cos\left(\omega_1 t + \theta_{y1}^\Sigma\right) + \sqrt{2} U_{y2}^\Sigma \cos\left(\omega_2 t + \theta_{y2}^\Sigma\right), \qquad (17)$$

leading to

$$i_y^\Sigma = \sqrt{2} I_{y1}^\Sigma \cos\left(\omega_1 t + \varphi_{y1}^\Sigma\right) + \sqrt{2} I_{y2}^\Sigma \cos\left(\omega_2 t + \varphi_{y2}^\Sigma\right). \qquad (18)$$

While components with frequency ω_1 are necessary to regulate the imbalance in the capacitor voltages that occurs due to the three-phase grid currents, components with frequency ω_2 are needed to control power transfer with single-phase terminals and the imbalance that this creates in the MMC. Therefore, given that $\varphi_{y1}^\Sigma = \theta_{y1}^\Sigma - \frac{\pi}{2}$ due to the MMC arm inductance, and imposing $\theta_{y1}^\Sigma = \theta_y^\Delta$, the substitution of (17) and (18) into (11) and (12) results in the average common-mode and differential-mode power

$$P_y^\Sigma = U_y^\Delta I_y^\Delta \cos\left(\theta_y^\Delta - \varphi_y^\Delta\right) - U_{y2}^\Sigma I_{y2}^\Sigma \cos\left(\theta_{y2}^\Sigma - \varphi_{y2}^\Sigma\right), \qquad (19)$$

$$P_y^\Delta = -U_{y1}^\Sigma I_y^\Delta \cos\left(\theta_y^\Delta - \varphi_y^\Delta\right). \qquad (20)$$

The average power P_y^Σ in (19) is equal to the difference between the three-phase and single-phase terminal powers. This difference should be made equal to zero during steady-state by controlling the power transfer. In addition, P_y^Δ in (20) will be regulated by U_{y1}^Σ. In steady-state $U_{y1}^\Sigma = 0$ and $I_{y1}^\Sigma = 0$. Therefore, in view of (19) and (20), u_y^Σ and u_y^Δ can be controlled through P_y^Σ and P_y^Δ. Numerous control schemes are capable of balancing the MMC arms' average power, as presented for example in [3, 4, 12, 13]. However, it is not covered in this paper. The diagram presented in Fig. 4 is responsible for shaping the voltage and current quantities of the single-phase transformer terminals into square waveforms to mitigate pulsating power in ultrafast chargers. The influence of shaping sinusoidal into square waves on the capacitor voltage control is negligible because the average remains $P_y^\Sigma = P_y^\Delta = 0$ after modifying (17) in Fig. 4.

The references $u_{y1}^{\Sigma*}$ and $u_{y2}^{\Sigma*}$ in Fig. 4 are given by an internal balance control needed to regulate capacitors' voltage [3, 12]. Then, to obtain a medium-frequency transformer with square-shaped quantities,

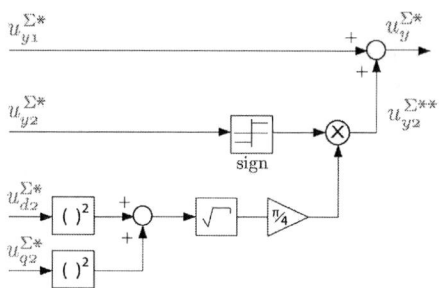

Fig. 4: Control scheme for a square-shaped medium-frequency transformer voltage.

$u_{y2}^{\Sigma*}$ is modified as presented in Fig. 4. The sign function transforms the sinusoidal waveform with angle θ_{y2}^{Σ} into a square wave. The dq voltages $u_{d2}^{\Sigma*}$ and $u_{q2}^{\Sigma*}$ from the PQ control of the single-phase terminals give the square wave amplitude. As a result, both $u_{y1}^{\Sigma*}$ and $u_{y2}^{\Sigma**}$ create the common-mode voltages and currents that modify (17) and (18), and therefore (19) and (20), successfully resulting in a square-shaped voltage and current on the single-phase side.

Simulation results

Simulations were conducted with the averaged equivalent circuit shown in Fig. 3. The capacitors in each arm are assumed to be equal, therefore the total equivalent arm capacitance C_σ is equal to $C_\sigma = \frac{C_y}{N}$, where C_y is the capacitance of each sub-module. The capacitors' voltage is stabilized around a reference by means of control. The main system parameters are described in Table I.

Table I: Ratings of the simulated three-phase to single-phase ac/ac MMC.

Description	Variable	Value	Unit
Charger nominal power	P	1	MW
Phase-to-neutral RMS voltage	U_y	$25\sqrt{1/3}$	kV
Grid frequency	f_1	50	Hz
Transformer RMS voltage	U_z	5	kV
Transformer frequency	f_2	1	kHz
Arm capacitance	C_σ	250	uF
Arm inductance	L_y	1	mH
Arm resistance	R_y	0.05	Ω
MOSFET on-resistance	R_{ds}	20	mΩ
MOSFET breakdown voltage	V_{ds}	1.2	kV
MOSFET derating factor	D_F	75	%

According to (3) and (4), the arm voltages contain components with frequencies ω_1 and ω_2 from the grid and medium-frequency transformer terminals, respectively. Fig. 5 illustrates the arm voltages produced by the common-mode and differential-mode voltages. Moreover, the square-shaped transformer current control can be made by generating common-mode voltage references $u_y^{\Sigma*}$ as presented in Fig. 4 and attested in Fig. 6. However, square wave voltage reduces the possible number of voltage levels in the MMC, leading to higher dv/dt on sub-modules and medium-frequency transformer.

Fig. 7 illustrates the stabilized capacitor voltages that act as energy buffers between both ac terminals and Fig. 8 shows purely sinusoidal three-phase grid currents, validating the steady-state operation of the ac/ac MMC. Since both frequencies are decoupled in the common-mode and differential-mode components, the medium frequency ω_2 is not reflected to the grid terminals, and therefore, filters are not required [14].

The amplitude and RMS values of a square-shaped voltage are equal, while the amplitude of a sinusoidal

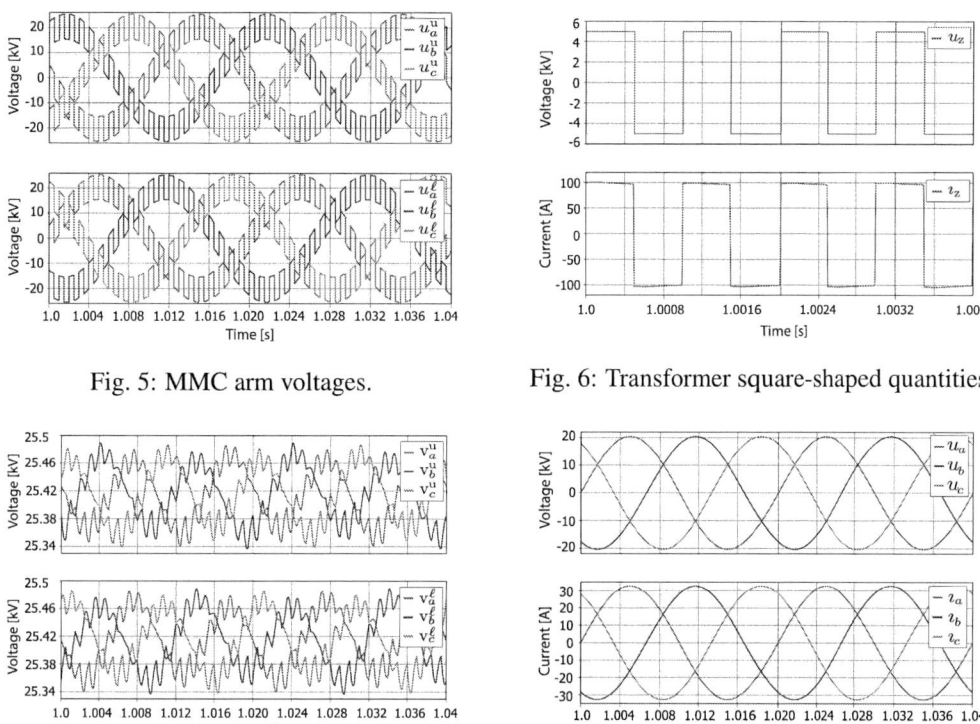

Fig. 5: MMC arm voltages.

Fig. 6: Transformer square-shaped quantities.

Fig. 7: Summed capacitor voltage of MMC arms.

Fig. 8: Three-phase grid voltages and currents.

wave is higher than its RMS value. Therefore, the instantaneous power of square-shaped quantities is equal to the average, while sinusoidal-shaped power has a higher pulsation [10]. In addition, square-shaped voltage increases the utilization of components in MMC, because as presented in (13), square wave quantities require fewer sub-modules than sinusoidal waves, meaning higher efficiency and power density. For instance, (13) and (16) are presented in Fig. 9 for both square-shaped and sinusoidal single-phase voltages with ratings shown in Table I. Note that above $U_z = 5$kV, conduction losses represent less than 1% of the processed power and sub-modules are reduced with square-shaped voltage.

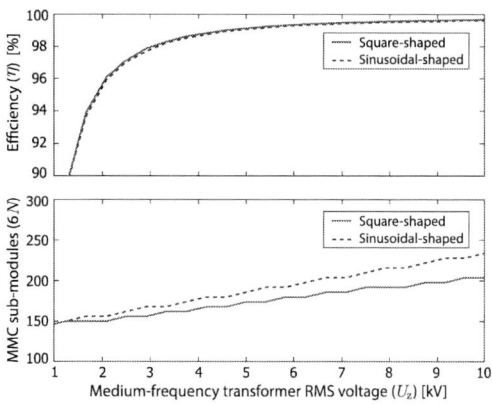

Fig. 9: MMC efficiency and number of sub-modules in relation to the transformer RMS voltage.

Conclusion

In order to decrease pulsating power in ultrafast charging stations, an ac/ac MMC connected to a medium-frequency transformer with a square wave voltage output was proposed in this paper. Using the presented strategy, a three-phase grid voltage can be converted into a medium-frequency square-shaped voltage. As a result, the application decreases cost and volume of high-power chargers by replacing line-frequency with medium-frequency transformers. However, the application has the disadvantage of increasing dv/dt on the MMC sub-modules and transformer terminals.

Simulations show the power exchange with the single-phase medium-frequency transformer using square wave voltage output, while keeping the capacitor voltages regulated. Transferring power with square-shaped voltage and current reduces pulsating power in single-phase systems. In this paper, the pulsating power reduction is quantified by the decrease in sub-modules of MMC-based chargers. Furthermore, as the model decouples frequencies from the grid and transformer terminals, filters are not necessary to comply with harmonic emission requirements.

References

[1] R. Marquardt, A. Lesnicar, J. Hildinger *et al.*, "Modulares stromrichterkonzept für netzkupplungsanwendung bei hohen spannungen," *ETG-Fachtagung, Bad Nauheim, Germany*, vol. 114, 2002.

[2] A. Antonopoulos, L. Angquist, and H.-P. Nee, "On dynamics and voltage control of the modular multilevel converter," in *2009 13th European Conference on Power Electronics and Applications*. IEEE, 2009, pp. 1–10.

[3] Y. P. Marca, M. G. L. Roes, J. L. Duarte, and C. G. E. Wijnands, "Isolated MMC-based ac/ac stage for ultrafast chargers," in *IEEE-ISIE2021-30th International Symposium on Industrial Electronics-Kyoto*. Institute of Electrical and Electronics Engineers, 2021.

[4] M. Schnarrenberger, F. Kammerer, M. Gommeringer, J. Kolb, and M. Braun, "Current control and energy balancing of a square-wave powered 1AC-3AC modular multilevel converter," in *2015 IEEE Energy Conversion Congress and Exposition (ECCE)*. IEEE, 2015, pp. 3607–3614.

[5] M. Schnarrenberger, F. Kammerer, D. Bräckle, and M. Braun, "Cell design of a square-wave powered 1AC-3AC modular multilevel converter low voltage prototype," in *2016 18th European Conference on Power Electronics and Applications (EPE'16 ECCE Europe)*. IEEE, 2016, pp. 1–11.

[6] R. Agarwal, S. Martin, Y. Shi, and H. Li, "High frequency transformer core loss analysis in isolated modular multilevel DC-DC converter for MVDC application," in *2019 IEEE Energy Conversion Congress and Exposition (ECCE)*. IEEE, 2019, pp. 6419–6423.

[7] R. Mo, H. Li, and Y. Shi, "A phase-shifted square wave modulation (PS-SWM) for modular multilevel converter (MMC) and dc transformer for medium voltage applications," *IEEE Transactions on Power Electronics*, vol. 34, no. 7, pp. 6004–6008, 2018.

[8] M. Glinka and R. Marquardt, "A new AC/AC-multilevel converter family applied to a single-phase converter," in *The Fifth International Conference on Power Electronics and Drive Systems, 2003. PEDS 2003.*, vol. 1. IEEE, 2003, pp. 16–23.

[9] M. Glinka, "Prototype of multiphase modular-multilevel-converter with 2 MW power rating and 17-level-output-voltage," in *2004 IEEE 35th Annual Power Electronics Specialists Conference (IEEE Cat. No. 04CH37551)*, vol. 4. IEEE, 2004, pp. 2572–2576.

[10] K. Pouresmaeil, J. L. Duarte, C. G. E. Wijnands, M. G. L. Roes, and N. H. Baars, "Single-phase bidirectional ZVZCS AC-DC converter for MV-connected ultra-fast chargers," in *PCIM Europe 2022*. VDE Verlag, 2022, pp. 124–130.

[11] G. Mondal and S. Nielebock, "Control of M2C direct converter for AC to AC conversion with wide frequency range," in *2016 18th European Conference on Power Electronics and Applications (EPE'16 ECCE Europe)*. IEEE, 2016, pp. 1–10.

[12] L. Bessegato, K. Ilves, L. Harnefors, S. Norrga, and S. Östlund, "Control and admittance modeling of an ac/ac modular multilevel converter for railway supplies," *IEEE transactions on power electronics*, vol. 35, no. 3, pp. 2411–2423, 2019.

[13] K. Sharifabadi, L. Harnefors, H.-P. Nee, S. Norrga, and R. Teodorescu, *Design, control, and application of modular multilevel converters for HVDC transmission systems*. John Wiley & Sons, 2016.

[14] F. Rojas, M. Diaz, M. Espinoza, and R. Cárdenas, "A solid state transformer based on a three-phase to single-phase modular multilevel converter for power distribution networks," in *2017 IEEE Southern Power Electronics Conference (SPEC)*. IEEE, 2017, pp. 1–6.

Surge current protection for railway traction applications

Michael Gleissner, Mark-M. Bakran

UNIVERSITY OF BAYREUTH
DEPARTMENT OF MECHATRONICS
CENTER OF ENERGY TECHNOLOGY
Universitaetsstrasse 30
Bayreuth, Germany
Phone: +49 (0) 921-55 7804
Email: michael.gleissner@uni-bayreuth.de
URL: http://www.mechatronik.uni-bayreuth.de

Acknowledgments

This work was supported by the Federal Ministry for Economic Affairs and Climate Action on the basis of a decision by the German Bundestag.

Keywords

≪Over-current protection≫,≪Protection device≫, ≪Fault handling strategy≫,≪Faults≫, ≪Railway traction system≫, ≪Silicon Carbide (SiC)≫

Abstract

Reducing the transformer impedance for higher efficiency results also in a higher surge short-circuit current for railway traction applications. This paper addresses the problem of an I^2t-value higher than the nominal rating of SiC MOSFET body diodes. Circuit options for protecting the intact components after a single failure are presented and examined.

Introduction

In railway traction, by increasing the switching frequency of the input converter, it is possible to reduce the impedance of the transformer and thus to reduce losses. The free-wheeling diodes of the power modules must be designed for short-term, high current loads. If a switch in the four-quadrant input converter fails as a short-on or if the DC-link circuit is short-circuited, this leads to a high, mostly recurring surge current load on the still functioning diodes from the line via the transformer (see Fig. 1). Furthermore, in the event of an active bridge short-circuit of a B6 pulse-controlled inverter, the electrical machine can feed back, which also results in a high current load on the diodes. In the data sheets, the I^2t-value is usually specified by a sinusoidal current half-wave with a length of 10 ms, which the component can withstand as a maximum [1, 2]. The measurement of 10 ms results from half the period duration in the 50 Hz network. Since the AC railway supply network in Germany and other European countries operates at 16.7 Hz, the data sheet values must be evaluated accordingly. Since the diode is integrated in SiC MOSFETs, the additional module freewheeling diodes for normal operation can be saved. The requirements of surge current robustness are transferred to the body diodes. In the case of voltage classes below 3.3 kV or with Si technology, the components can handle the previously expected maximum surge current strengths of usually less than $1\,\mathrm{MA}^2\mathrm{s}$ [3, 4]. If their surge short-circuit current strength is not sufficient, appropriate protection concepts must be applied, which will be presented and discussed in this paper. As far as the authors are aware, there are no relevant and freely accessible references on this very specific topic of surge current protection for railway traction applications due to short-on failures.

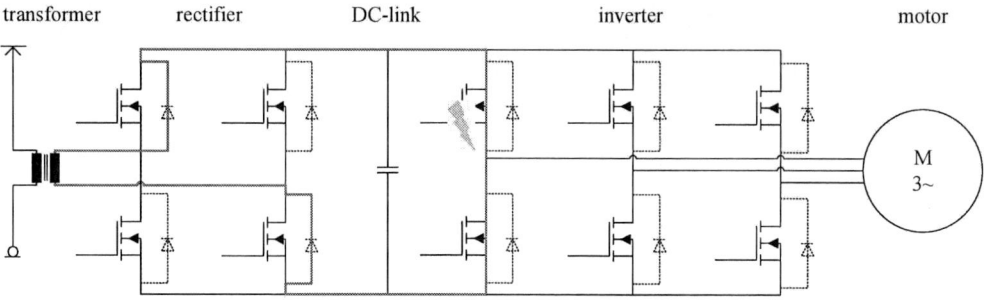

Fig. 1: Line-fed transformer, four-quadrant converter as rectifier, DC-link circuit, inverter and electrical machine. The dashed diodes indicate the body diodes of SiC-MOSFETs. Illustrated in red is a surge current due to a DC-link short-on failure. The body diodes within the current path have to be protected from secondary failure caused by a too large I^2t-value.

Basics on I^2t-value

The I^2t-value, which corresponds to the time integral of the squared current, is used to characterize the surge current robustness of components [5]:

$$I^2t = \int i^2(t)\mathrm{d}t \tag{1}$$

The equivalent circuit diagrams in Fig. 2 can be used to estimate the surge short-circuit current and I^2t-value for DC-link as well as single switch short-on failures. The transformer is represented by an ideal AC voltage source with transformer leakage inductance and transformer resistance. All sizes are related to the secondary side to which the four-quadrant converter is connected.

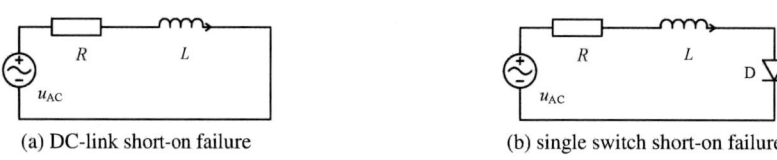

(a) DC-link short-on failure   (b) single switch short-on failure

Fig. 2: Simplified equivalent circuit diagrams for the line-side surge current

The related short-circuit voltage of the transformer depends on its leakage inductance and resistance. The lower the related short-circuit voltage, the lower the leakage inductance and resistance of the transformer. This in turn leads to a higher current in the event of a failure and thus I^2t-value. If the failure occurs in the voltage zero crossing, the worst case is obtained with the maximum amount of the current. If the fault occurs in a voltage maximum or voltage minimum, the best case results with the minimum amount of current. In the event of a single-chip short-on failure, the surge current can only flow in one direction and a mean value in the current that is dependent on the time of the failure cannot be reduced over several periods, as is the case with an DC-link circuit short-on. Therefore, in the event of a shutdown after several periods, a higher I^2t-value for the single-chip short-on failure results. The shape of the first period is identical for both failure scenarios. Fig. 3 shows the simulated voltage and current curves for DC-link and single chip short-on failure in the worst and best case as well as the associated I^2t-value. The failure occurs at time $t = 0\,\mathrm{s}$.

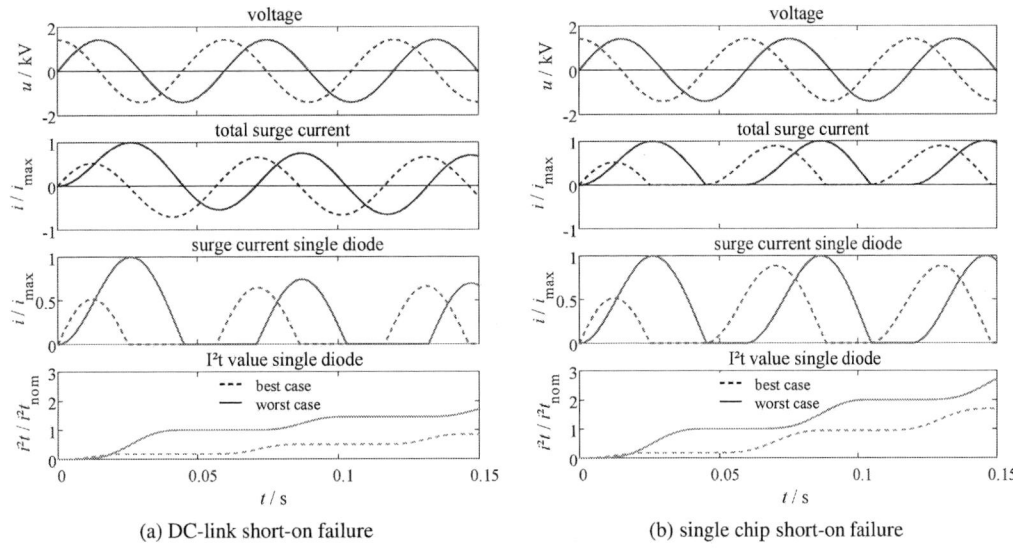

(a) DC-link short-on failure   (b) single chip short-on failure

Fig. 3: Simulated voltage and current curves in the worst and best case as well as the associated I^2t-value

Protective measures

Protective measures are necessary to minimize the stress of still healthy semiconductor devices within the surge current path, e.g. SiC MOSFET body diodes. Technical solutions can be based on switches, diodes or a combination of both. Their working principle and characteristics are given in the following subsections.

Mechanical circuit breaker

In previous applications, surge currents due to short-on failures are recognized by the central control unit and switched-off by opening the mechanical main switch on the primary side of the transformer in a time range of typically 60 to 80 ms, whereby the current can only be stopped after the subsequent zero current crossing due to an arc occurring [6] (see Fig. 4). With 16.7 Hz line frequency, this can be only after one or even two full sine half-waves depending on the failure scenario and time (best/worst case). The simulation results for the failure scenarios in the best and worst case vary from 3 to 20 MA^2 s. These I^2t-values are too large and better protection measures are required.

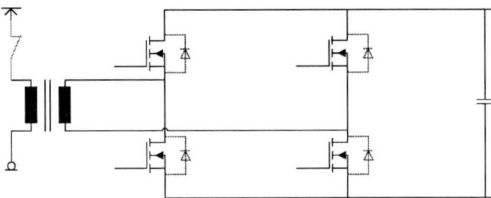

Fig. 4: Circuit breaker on AC-primary side (either mechanical or semiconductor based)

Fast circuit breaker

Instead of a mechanical disconnection switch, which requires at least a half-period to extinguish due to the arc, a faster switch can be used that can safely disconnect even during a current half-wave. Possible approaches to a solution result from electronic switches or from the use of pyro switches. Depending on the I^2t-robustness of the components, fast switches with a separation time below 10 ms are required.

Crow-bar

A crow-bar on the secondary side of the transformer can be used to relieve the rest of the circuit. A crow-bar typically consists of a thyristor, which enables after firing a bidirectional parallel path for the surge current. With a crow-bar, technically faster switching times are possible than with a disconnection element, e.g. $100\,\mu s$ with a thyristor (see Fig. 5a). This is more than sufficient, because the same tripping times as with a fast disconnector are necessary to avoid a too large I^2t-value.

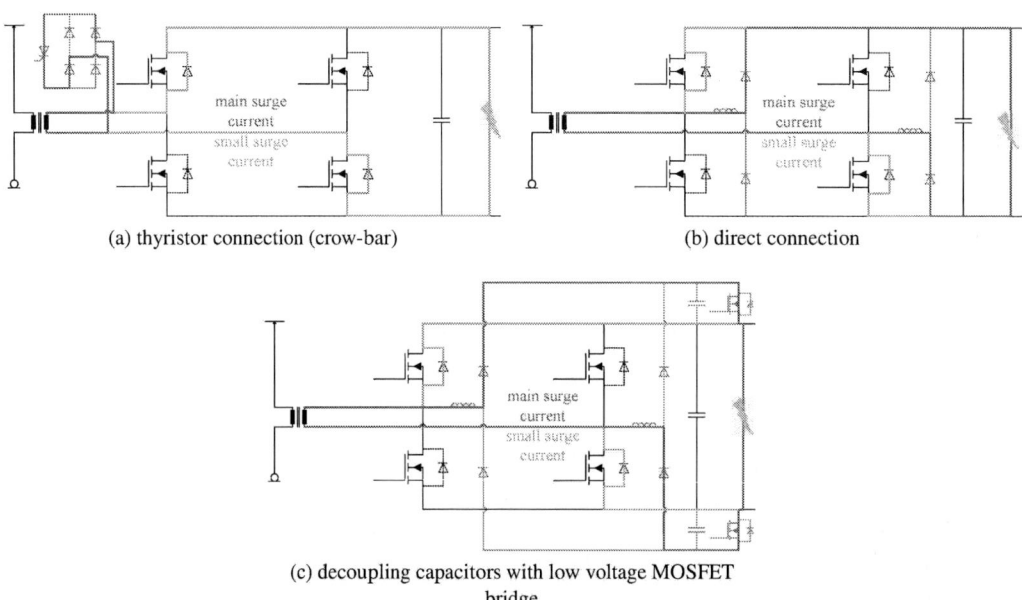

(a) thyristor connection (crow-bar)   (b) direct connection

(c) decoupling capacitors with low voltage MOSFET bridge

Fig. 5: Connection of paralleled Si rectifier diodes (marked in green) for surge short-circuit current relief after DC-link short-on failure

Paralleled Si rectifier diodes

Silicone rectifier diodes connected in parallel offer another option for reducing the surge short-circuit current stress. Their I^2t-robustness is typically in the range of 5 to $10\,MA^2 s$. They must be connected anti-parallel to the SiC MOSFETs in order to reduce the flow of current via the internal body diode of the SiC MOSFETs in the event of a failure, thereby relieving them and protecting them from destruction. Depending on the distance between the connection, a parasitic connection inductance arises, which can also be intentionally increased in order to influence the current distribution between Si rectifier diodes and SiC MOSFET body diodes. In normal operation, the Si diodes should only carry a small amount of current or no current at all in order to generate little or no switching losses. In the event of a surge current failure, however, they should take on as much current as possible. The variant of the direct connection is shown in Fig. 5b. A prerequisite for a successful implementation of this direct connection without additional switches is that the Si diodes have only very low reverse recovery loss. The variant of the diode activation by means of a thyristor is shown in Fig. 5a. The thyristor is activated in the event of a failure and thereby switches on the parallel Si rectifier diodes. These are switched-off in normal operation and cannot carry any current and thus cause no or only little loss. The thyristor must be designed for the full system voltage. It is a large and expensive component. Instead of the thyristor, the Si diodes can also be decoupled by capacitors during normal operation. In the event of a failure, these decoupling capacitors are bypassed by low-voltage MOSFETs and in this way switch-on the Si rectifier diodes (see Fig. 5c). For this circuit, it must be clarified which capacity the decoupling capacitors must have, what voltage they must be pre-charged to and how many low-voltage MOSFETs must be connected in parallel in order to be able to safely carry the entire surge short-circuit current in the event of a worst-case failure.

Analysis of paralleled Si rectifier diodes

Reverse recovery characteristic

In order to test the use of the direct connection, the reverse recovery behavior of a Si rectifier diode has been measured with a double-pulse circuit. The results in Fig. 6 indicate that even a very small load current of a few amperes leads to a very large reverse current peak. With a system voltage of 1.8 kV and a switching frequency of 3 kHz, this is associated with unacceptable losses. Therefore, the variant of the direct connection of the Si rectifier diodes is not further considered.

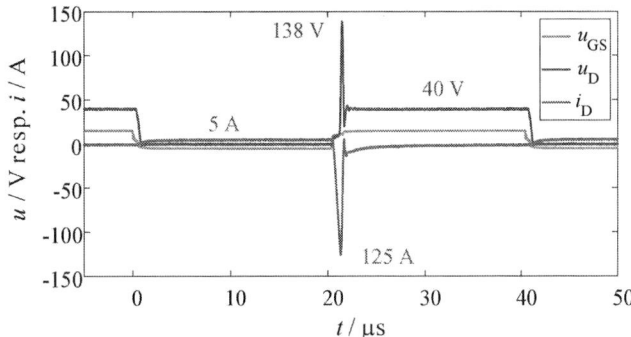

Fig. 6: Measurement of large reverse recovery peak even at low load current and low blocking voltage of diode DZ1070N22K

Loss due to du/dt-stress

If the Si rectifier diodes are connected by the circuits shown in Fig. 5a and Fig. 5c, there are no reverse recovery losses because the diodes do not conduct during normal operation, but they still see a du/dt stress from the SiC MOSFETs which operate in parallel. The gradient of the voltage when switching can be adjusted using the gate resistance. The measured voltage at the diode and the capacitive charge reversal can be seen in Fig. 7 for various voltage slopes. The peak value of the capacitive recharging current increases linearly with the voltage gradient and results in a capacitance of the Si diode of about 2 nF. This capacitive recharging current leads to a power loss of about 10 W at a switching frequency of 3 kHz and a system voltage of 1.8 kV. This power loss is acceptable.

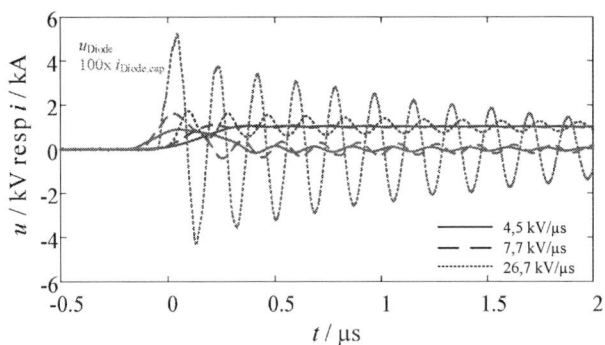

Fig. 7: Measurement of du/dt test for capacitance determination of Si diode DZ1070N22K

Peak rectifier characteristic of diode types

The diodes have to be decoupled during normal operation to avoid loss. The characteristic of Si diodes and fast diodes in a peak rectifier circuit has been tested with the circuit shown in Fig. 8a. The measured voltages after turn-on of S_1 and S_4 are shown in Fig. 8 for a fast diode and a Si diode. With the fast diode, the phase voltage u_{out} is below the capacitor voltage u_C after the switching process, which means that the diode is not conducting. The initial capacitor voltage is increased. The Si diode shows a voltage u_{out}, with is equal or above to the capacitor voltage and thus results in a conducting mode. The initial capacitor voltage is not increased and results in repetitive conduction mode of the Si diode and thus unwanted reverse recovery loss. But the fast diode has unfortunately only a small I^2t-robustness and cannot be employed for surge current release. Consequently, the Si diodes have to be decoupled by capacitors, which are actively charged to a voltage higher than the peak system voltage in order to avoid loss.

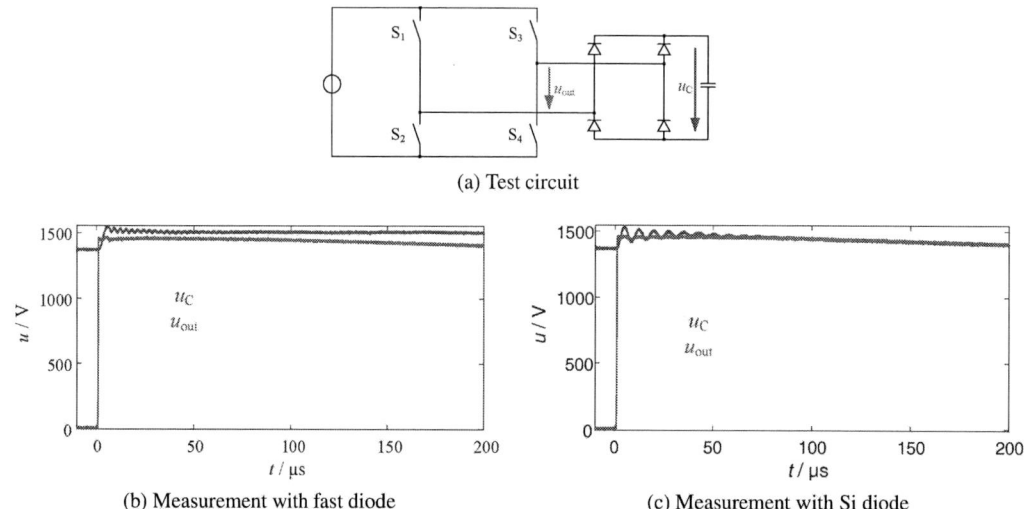

(a) Test circuit

(b) Measurement with fast diode

(c) Measurement with Si diode

Fig. 8: Peak rectifier test to check decoupling of diodes. Initially all switches are turned-off. S_1 and S_4 are turned-on at $t = 0\,\text{s}$.

Design of decoupling capacitor

For the protection circuit depicted in Fig. 5c, the starting voltage and capacitance of the decoupling capacitor as well as the connection inductance must be dimensioned appropriately. As a first test, the measurement setup shown in Fig. 9a was implemented. The starting voltage of the capacitor is particularly relevant, which must be sufficiently high so that the Si diode does not conduct and thus does not cause any reverse recovery losses. In the first test, the Si rectifier diode and SiC MOSFETs are not yet properly scaled to one another. In addition, this test has not yet been carried out at a system voltage of 1.8 kV, but only at 100 V for a first function test. The measurement results of the decoupling circuit show that the Si rectifier diode conducts and causes reverse recovery if the capacitor is not charged or is insufficiently charged (see Fig. 9b). From a starting voltage of the capacitor of 9 V, the diode no longer conducts and you can only see the capacitive recharging current (see Fig. 9c).

This capacitive charge reversal current continues to charge the capacitor for each switching process. It is therefore necessary to provide a charging / discharging circuit for the decoupling capacitors, which precharges them sufficiently when the system is started up and continuously discharges them again during operation in order to keep the voltage stable.

Low-voltage MOSFETs in the voltage class of 100 V or less can be used as bypass switches in the event of a failure.

Surge current robustness of bypass low-voltage MOSFETs

The surge current robustness of single low-voltage MOSFETs in PG-HSOF-8 package has been investigated by means of destruction tests with the surge current test circuit shown in Fig. 10. The period and amplitude of the surge current can be adjusted by L_{surge}, C_{surge} and the initial voltage. Switch T_1 decouples and protects the voltage source and switch T_2 triggers the surge current sine half-wave. The diodes block the negative half-wave and thus prevent an oscillation. In case of an open-circuit failure of the DUT several serial connected diodes enable an additional current path. It should be noted that the total

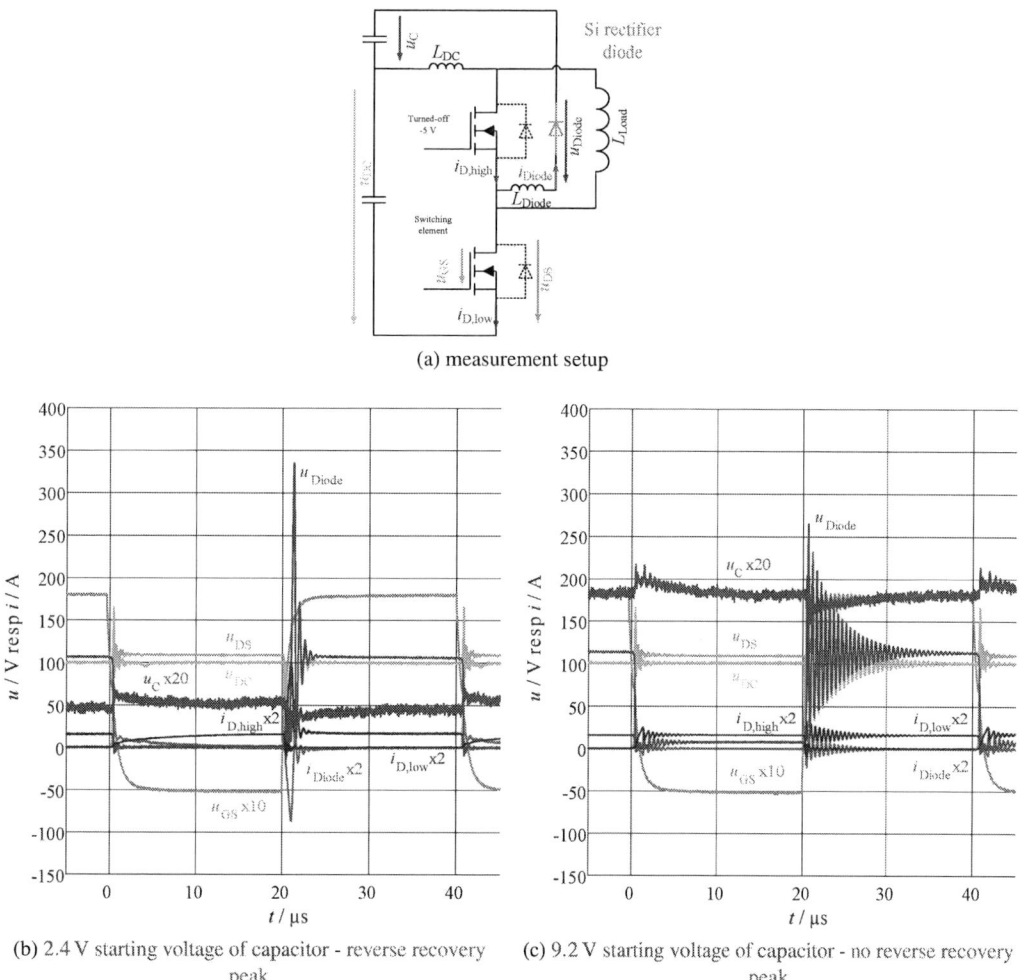

(a) measurement setup

(b) 2.4 V starting voltage of capacitor - reverse recovery peak

(c) 9.2 V starting voltage of capacitor - no reverse recovery peak

Fig. 9: Low voltage test of required pre-voltage u_C of capacitor to decouple Si rectifier diode

resistance of the circuit is less than the resistance for the damping case of the LCR series connection, so that a half-sine-wave oscillation is still produced. In particular, for longer durations of the half-sine oscillation in the range of 40 ms, the structure must have a very low resistance.

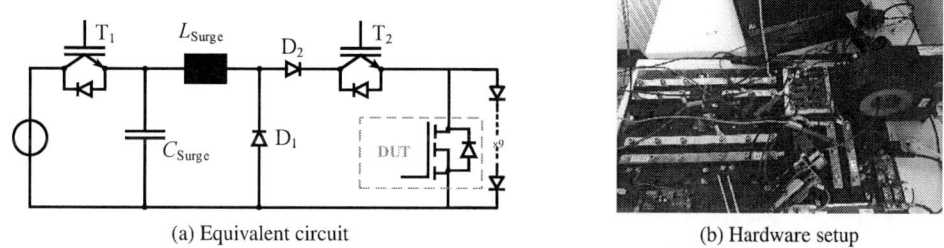

(a) Equivalent circuit

(b) Hardware setup

Fig. 10: Surge current test circuit

The last measurement before failure and the failure measurement are shown in Fig. 11. The failure limit of a single device is roughly at a peak current of just under 700 A. The tested switches have failed with low resistance in relation to the drain-source path, which is advantageous for later use as a bypass in the event of a failure.

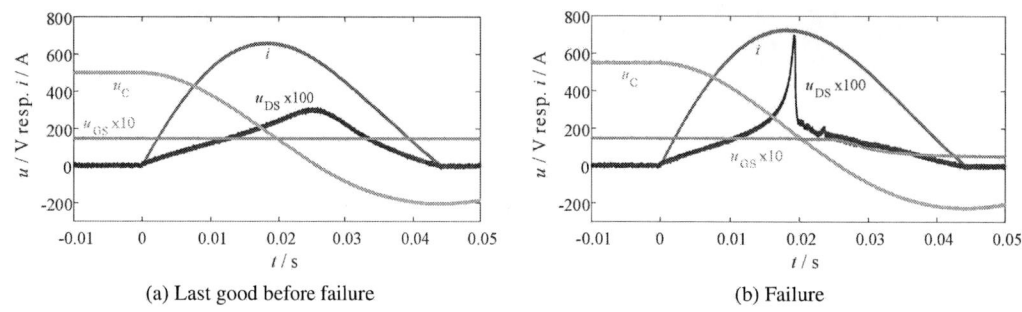

(a) Last good before failure   (b) Failure

Fig. 11: Surge current test measurement for single low-voltage MOSFET

The measured power loss was corrected by the circuit board resistance to get the power loss of the chip. This power loss has been applied to a thermal network in a simulation model, which corresponds to the data sheet value for the thermal impedance from junction to ambient. The failure temperature is in the range above 600 °C.

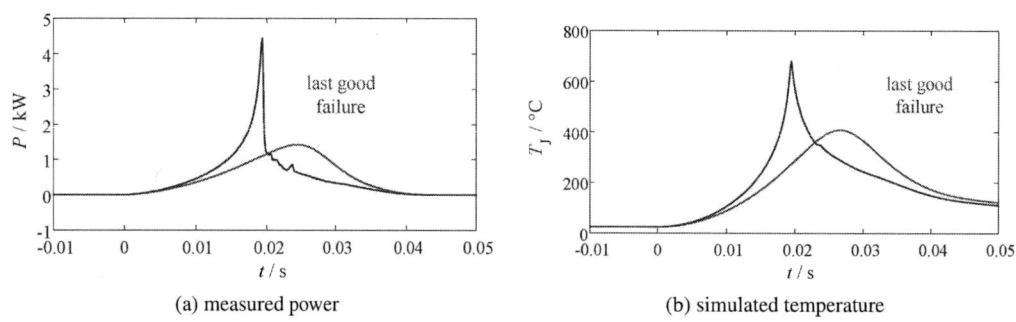

(a) measured power   (b) simulated temperature

Fig. 12: Calculated power based on measurement and simulated junction temperature based on data sheet thermal impedance for last good and failure measurement

The permissible individual load results in around less than 50 low-voltage MOSFETs to be switched in parallel in order to reach sufficient surge current capability. To evaluate the effects when several dozens of MOSFETs are paralleled, a sample board (see Fig. 13) was designed and exposed to a surge current pulse with a peak value of 14 kA. The resistance of the paralleled MOSFETs on the board including contacts is less than 200 μΩ.

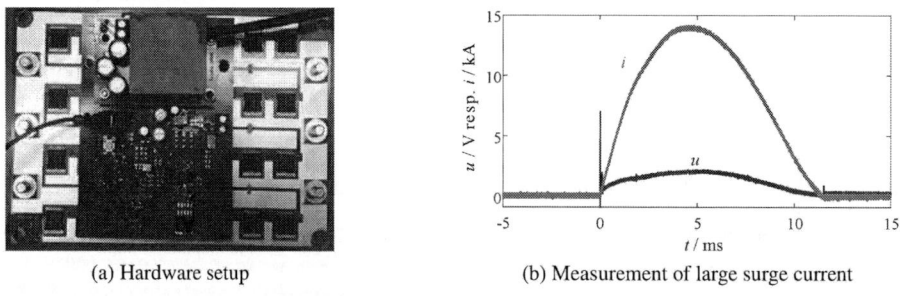

(a) Hardware setup   (b) Measurement of large surge current

Fig. 13: Low-voltage bypass switch consisting of several paralleled low-voltage MOSFETs

Conclusion

Simple simulation models were used to estimate which I^2t-values can be expected and which remedial measures appear sensible for the surge current in railway traction applications after short-on failures. Mechanical switches are not sufficient for a faster disconnection of the line side current. To limit the I^2t-value in the event of a failure, electronic or pyro switches with a separation time of below 10 ms are required. The use of a crow-bar directly at the input of the converter with thyristors also enables fast switching times and can achieve low enough I^2t-values. By connecting Si rectifier diodes with a high current carrying capacity in parallel with the SiC MOSFETs, according to the simulation, sufficient relief can be achieved in the event of a failure. A direct connection is not possible due to the reverse recovery losses of the Si diodes. The voltage slope load of the Si diodes, both with the connection by means of a thyristor and the low voltage decoupling capacitor circuit, is acceptable. The decoupling capacitors have to be charged to a voltage level higher than 9 V so that the Si diodes are decoupled during normal operation. In the event of a failure, these can be bypassed, for example, by a sufficient number of paralleled low-voltage MOSFETs. This represents an alternative to the previously known solution using a thyristor as an activation element. A prove of concept implementation of the solution was successfully exposed to a large surge current.

References

[1] P. Hofstetter and M.-M. Bakran, "Comparison of the Surge Current Ruggedness between the Body Diode of SiC MOSFETs and Si Diodes for IGBT," in *CIPS 2018; 10th International Conference on Integrated Power Electronics Systems*, 2018.

[2] ——, "Predicting Failure of SiC MOSFETs under Short Circuit and Surge Current Conditions with a Single Thermal Model," in *Proceedings of the 20th European Conference on Power Electronics and Applications (EPE ECCE Europe)*, 2018.

[3] F. Carastro, J. Mari, T. Zoels, B. Rowden, P. Losee, and L. Stevanovic, "Investigation on diode surge forward current ruggedness of Si and SiC power modules," in *Proceedings of the 18th European Conference on Power Electronics and Applications (EPE ECCE Europe)*, 2016.

[4] Z. Dong, R. Ren, F. Wang, Z. Dong, R. Ren, and F. Wang, "Evaluate I2t Capability of SiC MOSFETs in Solid State Circuit Breaker Applications," in *IEEE Energy Conversion Congress*, 2020, pp. 6043–6048.

[5] R. H. Kaufmann, "The Magic of I2t," *IEEE Transactions on Industry and General Applications*, vol. IGA-2, no. 5, pp. 384–392, 1966.

[6] M.-M. Bakran, H.-G. Eckel, M. Helsper, and A. Nagel, "Next generation of IGBT-modules applied to high-power traction," in *Proceedings of the 12th European Conference on Power Electronics and Applications (EPE ECCE Europe)*, 2007.

Impedance-based analysis of HVDC converter control for robust stability in AC power systems

André Schön, Andreas Lorenz, Rodrigo Alonso Alvarez Valenzuela
SIEMENS ENERGY GLOBAL GMBH & CO. KG
schoen.andre@siemens-energy.com

Keywords

≪Converter control≫, ≪HVDC≫, ≪MMC≫, ≪Impedance analysis≫, ≪Stability≫

Abstract

Ensuring robust control stability to varying grid conditions is a key requirement when installing a new HVDC station to the AC grid. With its very accessible criteria for stability margins, the Nyquist theorem for single input/single output systems is commonly used to determine the robustness of the control system of HVDC converter stations and tune it accordingly. However, three-phase AC system are not single input/single output systems and the validity as well as the underlying assumptions of this investigation has rarely been under scrutiny. In this paper, the commonly used impedance based approach to assess stability and robustness of HVDC converter stations is reviewed from a control theoretical point of view. Starting with a clarification on the properties of commonly used reference system transformations, the repercussions for robustness investigations in a multivariable control environment are discussed. Based on that, possible shortcuts to allow classical single input/single output investigations as well as limitations to that approach are derived and explained in detail on a generic control model.

1 Introduction

AC power systems with a high penetration of power electronic (PE), actively controlled components face new challenges assessing the overall system stability. The impedance based analysis of AC power systems offers a comprehensive approach, where each component of the grid is modelled by its terminal behavior. Each subsystem can be described as an independent black box, with no necessary information on the internal physical or electrical structure. These models can be determined by analytical calculations or even by frequency sweep measurements[1–9]. Interactions of PE devices with the grid and with each other can then be investigated from a top level perspective.

The general requirements for PE components are primarily given by specific control goals for the stationary operation and dynamic requirements, like the fault ride through behavior or specific step responses. To avoid harmful interactions, the requirements to the frequency domain input impedance often include positive damping for a very wide frequency range and even certain phase margins or adjustable damping in specific frequency areas with the goal of ensuring robust, stable operation for unknown and varying grid conditions [10–12].

In this paper, the validity of the commonly used approach, to investigate stability and control robustness based solely on the main diagonal elements of the converter input admittance (e.g. [5–7, 13–15]) is dissected. With the help of control theoretical concepts for multi variable feedback systems, boundary conditions to the calculation method and to the grid conditions for the validity of this approach are derived.

In Section 2 the general approach for an analytical description of the converter input impedance as well as its challenges are outlined based on a generic control structure. Section 3 reviews common reference systems for the description of the converter input impedance and their (dis)advantages. Section 4 reviews the stability problem from a control theoretical perspective and shows possible shortcuts based on an advantageous reference system selection and boundaries to the grid impedance. Section 5 combines the

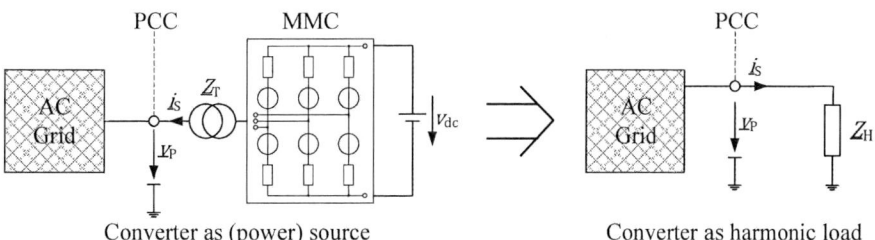

Converter as (power) source   Converter as harmonic load

Fig. 1: Impedance based analysis - the source converter is modelled as a harmonic load to the grid

results of the previous sections into an evaluation of an exemplary converter system and describes the conditions for a classical stability analysis as well as the relations between the converter input impedance and the performance requirements of the converter.

2   Calculation of the converter input impedance

From a control perspective, any perturbation in the voltage at the point of common coupling (PCC) is a disturbance to the stationary control targets of the converter control. Any voltage perturbation leads to a corresponding response in the line current due to the internal control. To evaluate the feedback effects between an AC grid and a converter station, the converter is modelled as a harmonic load at the PCC (see Fig. 1). Since HVDC converter station transformers usually decouple the zero sequence, any zero sequence interaction between converter and AC grid is limited to parasitic effects which, for the sake of clarity, are neglected here. Hence, for this paper, any 3-by-3 AC impedances matrix is reduced to a 2-by-2 matrix. The analysis will only be shown in the $\alpha\beta$ and the dq frame as well as in the sequence domain (PN). Whenever a specific reference frame is regarded, the variables will be denoted with the indices $\alpha\beta$, dq or PN. Variables without denotation regard to any reference frame.

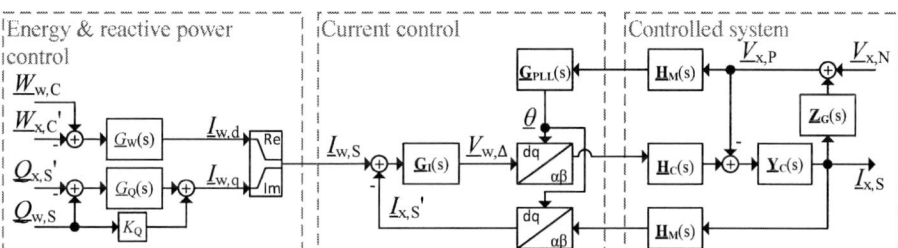

Fig. 2: Generalized MMC control diagram

In general, there are two options to model the converter input impedance. One option is to calculate the disturbance response of the converter control analytically, the other option is to measure the input impedance directly via a voltage (or current) perturbation frequency sweep directly at the PCC in a simulation model or if possible, given the physical limitations, at the real device. Analytical approaches often face difficulties with non-linearities or with selecting the right reference frame. They also often lead to models with a very high order, especially when dead times are involved, which can increase the computational time drastically. Measured models on the other hand lead to reduced accuracy and tend to neglect narrow banded effects depending on the frequency resolution of the measurement sweep. They are not generally suited for stability analysis, can however validate analytical models. Fig. 2 shows a generalized control model [9, 15–17] including an inner current controller $\underline{\mathbf{G}}_I$ and energy as well as a reactive power controller $\underline{\mathbf{G}}_{WQ}$ forming secondary control layers. The controlled electrical system is divided into a summarized grid impedance $\underline{\mathbf{Z}}_G$ and the electrical components of to converter station up to the PCC $\underline{\mathbf{Y}}_C$, containing e.g. the converter transformer and the arm impedances. The model also contains a PLL $\underline{\mathbf{G}}_{PLL}$ and a measurement system $\underline{\mathbf{H}}_M$ including dead times. The generation of the physical output voltage is combined in $\underline{\mathbf{H}}_C$. Due to the high number of voltage levels and the resulting high effective switching frequency in each converter arm, for Modular Multilevel Converters (MMC) in an HVDC application, the output voltage can be modelled as a delayed ideal voltage source. Due to the non-linearities, the linearized small signal converter model is highly operation point dependent. For the simplified control structure of Fig. 2, the input impedance calculates to

$$\underline{\mathbf{Y}}_H = -\frac{\vec{I}_{x,S}}{\vec{V}_{x,P}} = \left(\mathbf{I} + \underline{\mathbf{Y}}_C \underline{\mathbf{H}}_C \underline{\mathbf{G}}_I \left(\mathbf{I} + \tfrac{3}{2}\underline{\mathbf{G}}_{WQ}\underline{\mathbf{V}}_{P0}\right)\underline{\mathbf{H}}_M\right)^{-1} \cdot \underline{\mathbf{Y}}_C \left(\mathbf{I} + \underline{\mathbf{H}}_C \left((\underline{\mathbf{G}}_I \mathbf{I}_{Sm,0} - \underline{\mathbf{V}}_{\Delta,0})\,\underline{\mathbf{G}}_{PLL} + \tfrac{3}{2}\underline{\mathbf{G}}_I \underline{\mathbf{G}}_{WQ}\mathbf{I}_{S,0}\right)\underline{\mathbf{H}}_M\right) \quad (1)$$

with the voltage and current operation point

$$\mathbf{I}_{Sm,0} = \begin{bmatrix} I_{Sm,d0} & I_{Sm,q0} \\ I_{Sm,q0} & -I_{Sm,d0} \end{bmatrix}, \ \mathbf{I}_{S,0} = \begin{bmatrix} I_{S,d0} & I_{S,q0} \\ -I_{S,q0} & I_{S,d0} \end{bmatrix}, \ \underline{\mathbf{V}}_{\Delta,0} = \begin{bmatrix} V_{\Delta,d0} & -V_{\Delta,q0} \\ V_{\Delta,q0} & V_{\Delta,d0} \end{bmatrix} \text{ and } \underline{\mathbf{V}}_{P,0} = \begin{bmatrix} V_{P,d0} & V_{P,q0} \\ V_{P,q0} & -V_{P,d0} \end{bmatrix}.$$

The generic control structure, shown in Fig. 2 has only three main control loops, energy respectively reactive power and AC current. The AC current controller bandwidth is up to 30 Hz, the energy controller bandwidth up to 10 Hz, which is also the frequency range, where the PLL is active and the reactive power controller is designed slow with up to 1 Hz bandwidth. Hardware design was chosen to operate at a maximum power of 800 MW as an example, which is also the operation point (rectifier) shown in Fig. 5. The assumed dead times are approximated $200\,\mu$s in the current controller loop. This configuration as well as the control structure given in Fig. 2 serve just as a generic example to facilitate the investigations of the following sections. The specific implementation and parametrization are of no concern to the understanding of this paper.

3   Reference System Selection — $\alpha\beta$, dq and PN

In this section we investigate the I/O map of linear systems under commonly used signal transformations. A linear I/O map can be described through a transfer function matrix (TFM) in the Laplace domain, denoted by $\underline{\mathbf{G}}(s)$ which maps the input signal $\vec{u}(t) \circ\!\!-\!\!\bullet \underline{U}(s)$ to the output signal $\vec{y}(t) \circ\!\!-\!\!\bullet \vec{Y}(s)$. Application of a regular linear coordinate transformation $\underline{\mathbf{T}}(t) \in \mathbb{C}^{n \times n}$ to the in- and output signals then result in

$$\vec{y} = \mathbf{T}^{-1} \cdot \mathcal{L}^{-1}\left\{\mathbf{G} \cdot \mathcal{L}\{\mathbf{T} \cdot \vec{u}\}\right\} \quad (2)$$

The transformed I/O map given by Eq.(2) is in general no longer a linear map and therefore cannot be described by a TFM in the Laplace domain. However, for the special case of a constant transformation matrix $\mathbf{T}$ Eq.(2) becomes $\vec{Y} = \mathbf{T}^{-1}\mathbf{GT} \cdot \vec{U}$ and the coordinate transformation induces the transformation

$$\tilde{\mathbf{G}} := \mathbf{T}^{-1}\mathbf{GT} \ \leftrightarrow \ \mathbf{G} = \mathbf{T}\tilde{\mathbf{G}}\mathbf{T}^{-1} \quad (3)$$

for the TFM, which immediately implies that the new I/O map is again linear with the new structure $\tilde{\mathbf{G}}$. This transformation does not change the eigenvalues of the I/O map [18]. These preliminaries and notations now enable us to describe the influence of often used signal transformations to the I/O map of linear systems.

3.1   Input/Output map under *PN* transformation

The PN transformation, which associates real signals with complex signals in the time domain, is a constant, linear, complex signal transformation and therefore induces a transformation to the TFM given by Eq.(3):

$$\underline{\mathbf{G}}_{PN} = \overbrace{\frac{1}{2}\begin{bmatrix} 1 & j \\ 1 & -j \end{bmatrix}}^{\mathbf{T}_{PN}} \begin{bmatrix} \underline{G}_{11} & \underline{G}_{12} \\ \underline{G}_{21} & \underline{G}_{22} \end{bmatrix} \overbrace{\begin{bmatrix} 1 & 1 \\ -j & j \end{bmatrix}}^{\mathbf{T}_{PN}^{-1}} = \frac{1}{2}\begin{bmatrix} \underline{G}_{11}+\underline{G}_{22}-j(\underline{G}_{12}-\underline{G}_{21}) & \underline{G}_{11}-\underline{G}_{22}+j(\underline{G}_{12}+\underline{G}_{21}) \\ \underline{G}_{11}-\underline{G}_{22}-j(\underline{G}_{12}+\underline{G}_{21}) & \underline{G}_{11}+\underline{G}_{22}+j(\underline{G}_{12}-\underline{G}_{21}) \end{bmatrix} \quad (4)$$

$$\mathbf{G} = \frac{1}{2}\begin{bmatrix} 1 & 1 \\ -j & j \end{bmatrix} \begin{bmatrix} \underline{G}_{PP} & \underline{G}_{PN} \\ \underline{G}_{NP} & \underline{G}_{NN} \end{bmatrix} \begin{bmatrix} 1 & j \\ 1 & -j \end{bmatrix} = \frac{1}{2}\begin{bmatrix} \underline{G}_{NN}+\underline{G}_{NP}+\underline{G}_{PN}+\underline{G}_{PP} & j(\underline{G}_{PP}+\underline{G}_{NP}-\underline{G}_{PN}-\underline{G}_{NN}) \\ j(\underline{G}_{NN}+\underline{G}_{NP}-\underline{G}_{PN}-\underline{G}_{PP}) & \underline{G}_{NN}-\underline{G}_{NP}-\underline{G}_{PN}+\underline{G}_{PP} \end{bmatrix} \quad (5)$$

P and *N* stand for the positive phase sequence (PPS) and the negative phase sequence (NPS) which is a well known transformation [19], especially in the $\alpha\beta$ frame. Nevertheless, *PN* transformation can be used on any 2-by-2 TFM. If the scalar transfer functions (TF) $\underline{G}_{ij}$ are conjugate symmetrical ($\underline{G}_{ij}(s^*) = \underline{G}_{ij}^*(s)$ $\Leftrightarrow$ Laplace transforms of real signals), then the scalar TF in $\underline{\mathbf{G}}_{PN}$ have complex coefficients, and therefore the conjugate symmetry in the Laplace domain is lost. But then these TF fulfill $\underline{G}_{PN}(s) = \underline{G}_{NP}^*(s^*)$ and $\underline{G}_{PP}(s) = \underline{G}_{NN}^*(s^*)$, which can be easily verified by direct computation from Eq.(4).

3.2 Input/Output map under Park transformation

The Park transformation from the $\alpha\beta$ to the dq reference frame is a multiplication in the time domain with a rotation matrix $\mathbf{T_P}$ to rotate the input vector by a time varying angle θ. Therefore, it changes the I/O map of the system as described by Eq.(2). By careful application of the Laplace transformation rules to Eq.(2), like e.g. stated in [20], it can be seen that the transformed I/O map only remains linear, if the TFM $\underline{\mathbf{G}}$ has the following structure:

$$\underline{\mathbf{G}} = \begin{bmatrix} \underline{G}_M & \underline{G}_O \\ -\underline{G}_O & \underline{G}_M \end{bmatrix} \text{ with } \mathbf{T_P} = \begin{bmatrix} \cos\theta(t) & \sin\theta(t) \\ -\sin\theta(t) & \cos\theta(t) \end{bmatrix} \text{ and } \mathbf{T_P}^{-1} = \begin{bmatrix} \cos\theta(t) & -\sin\theta(t) \\ \sin\theta(t) & \cos\theta(t) \end{bmatrix} \tag{6}$$

The resulting map between in and output signals given by Eq.(3) can be described by the induced TFM $\underline{\mathbf{G}}_{dq}$ or $\underline{\mathbf{G}}_{\alpha\beta}$ of the form:

$$\alpha\beta \to dq : \underline{\mathbf{G}}_{dq} = \begin{bmatrix} \underline{G}_M^r - \underline{G}_O^i & \underline{G}_M^i + \underline{G}_O^r \\ -\underline{G}_M^i - \underline{G}_O^r & \underline{G}_M^r - \underline{G}_O^i \end{bmatrix} \text{ and } dq \to \alpha\beta : \underline{\mathbf{G}}_{\alpha\beta} = \begin{bmatrix} \underline{G}_M^r + \underline{G}_O^i & -\underline{G}_M^i + \underline{G}_O^r \\ \underline{G}_M^i - \underline{G}_O^r & \underline{G}_M^r + \underline{G}_O^i \end{bmatrix} \tag{7}$$

$$\text{with} \quad \underline{G}_x^r = \frac{1}{2}\left(\underline{G}_x(s+j\omega_0) + \underline{G}_x(s-j\omega_0)\right) \quad \text{and} \quad \underline{G}_x^i = \frac{j}{2}\left(\underline{G}_x(s+j\omega_0) - \underline{G}_x(s-j\omega_0)\right)$$

3.3 Input/Output map under combined Park and *PN* transformation

Section 3.1 shows that the *PN* transformation maintains the LTI system property and Section 3.2 shows under which conditions the Park transformation applied to real time domain signals maintains the linear I/O map. Now we investigate the change of the I/O map if the Park-transformation is applied to complex valued time domain signal in *PN* coordinates. From Eq.(2) and successive application of the signal transformation it can be obtained:

$$\alpha\beta_{PN} \to dq_{PN} : \vec{\underline{y}}_{PN}^{dq} = \underline{\mathbf{T}}_{PN}\mathbf{T_P}(t)\underline{\mathbf{T}}_{PN}^{-1} \cdot \mathcal{L}^{-1}\{\underline{\mathbf{G}}_{PN}^{\alpha\beta}(s) \cdot \mathcal{L}\{\underline{\mathbf{T}}_{PN}\mathbf{T_P}^{-1}(t)\underline{\mathbf{T}}_{PN}^{-1} \cdot \vec{\underline{u}}_{PN}^{dq}(t)\}\} \tag{8}$$

$$dq_{PN} \to \alpha\beta_{PN} : \vec{\underline{y}}_{PN}^{\alpha\beta} = \underline{\mathbf{T}}_{PN}\mathbf{T_P}^{-1}(t)\underline{\mathbf{T}}_{PN}^{-1} \cdot \mathcal{L}^{-1}\{\underline{\mathbf{G}}_{PN}^{dq}(s) \cdot \mathcal{L}\{\underline{\mathbf{T}}_{PN}\mathbf{T_P}(t)\underline{\mathbf{T}}_{PN}^{-1} \cdot \vec{\underline{u}}_{PN}^{\alpha\beta}(t)\}\} \tag{9}$$

with $\underline{\mathbf{T}}_{PN}\mathbf{T_P}^{-1}(t)\underline{\mathbf{T}}_{PN}^{-1} = \begin{bmatrix} e^{j\omega_0 t} & 0 \\ 0 & e^{-j\omega_0 t} \end{bmatrix}$ and $\underline{\mathbf{T}}_{PN}\mathbf{T_P}(t)\underline{\mathbf{T}}_{PN}^{-1} = \begin{bmatrix} e^{-j\omega_0 t} & 0 \\ 0 & e^{j\omega_0 t} \end{bmatrix}$

$$\vec{\underline{Y}}_{PN}^{dq} = \begin{bmatrix} \underline{G}_{PP}^{\alpha\beta}(s+j\omega_0) & 0 \\ 0 & \underline{G}_{NN}^{\alpha\beta}(s-j\omega_0) \end{bmatrix} \begin{bmatrix} \underline{U}_P^{dq}(s) \\ \underline{U}_N^{dq}(s) \end{bmatrix}$$
$$+ \begin{bmatrix} 0 & \underline{G}_{PN}^{\alpha\beta}(s+j\omega_0) \\ \underline{G}_{NP}^{\alpha\beta}(s-j\omega_0) & 0 \end{bmatrix} \begin{bmatrix} \underline{U}_P^{dq}(s-j2\omega_0 t) \\ \underline{U}_N^{dq}(s+j2\omega_0 t) \end{bmatrix} \tag{10}$$

$$\vec{\underline{Y}}_{PN}^{\alpha\beta} = \begin{bmatrix} \underline{G}_{PP}^{dq}(s-j\omega_0) & 0 \\ 0 & \underline{G}_{NN}^{dq}(s+j\omega_0) \end{bmatrix} \begin{bmatrix} \underline{U}_P^{\alpha\beta}(s) \\ \underline{U}_N^{\alpha\beta}(s) \end{bmatrix}$$
$$+ \begin{bmatrix} 0 & \underline{G}_{PN}^{dq}(s-j\omega_0) \\ \underline{G}_{NP}^{dq}(s+j\omega_0) & 0 \end{bmatrix} \begin{bmatrix} \underline{U}_P^{\alpha\beta}(s+j2\omega_0 t) \\ \underline{U}_N^{\alpha\beta}(s-j2\omega_0 t) \end{bmatrix} \tag{11}$$

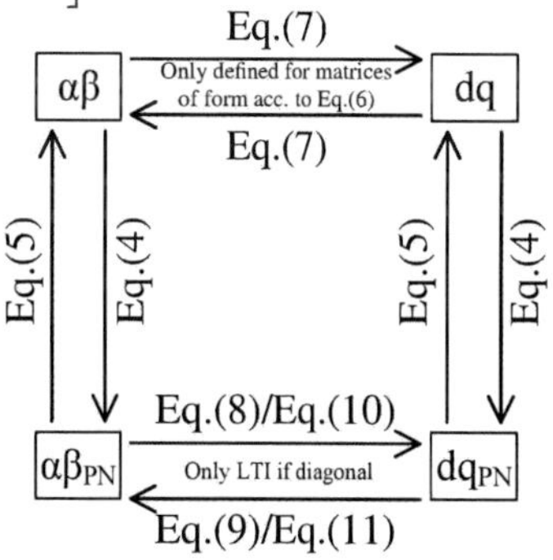

Evaluation of Eq.(10) and Eq.(11) reveals that the I/O map of the system has to be diagonal to maintain the LTI property because off-diagonal elements of $\underline{\mathbf{G}}_{PN}$ lead to system responses at shifted frequencies compared to the input signal $\vec{\underline{u}}$. Eq.(4) to Eq.(11) build a commutative diagram of successive signal transformations and ensure that the coordinates can be switched for a convenient system description (compare to e.g. [1, 5, 8, 9, 18, 21–23]).

Fig. 3: Commutative graph for transfer function matrix transformation between $\alpha\beta$, dq and *PN* frame

4 Impedance based stability analysis and requirements to the grid impedance

This chapter, relates the established impedance-based stability analysis to the well-known control theoretical concepts of passivity and the small gain theorem. These concepts are introduced by using of the Nyquist theorem, beginning with the SISO case. Also, the necessity to extend these stability analysis methods to the MIMO system case for HVDC applications will be motivated and the obstacles that arise thereby will be pointed out. Finally, a direct Nyquist array method to reduce the MIMO case to multiple SISO systems is utilized, circumventing the problems described and restoring the benefits of standard control design methods in the frequency domain for the MIMO case.

4.1 Control theoretical interpretation of the stability problem in electrical systems

The general control representation of the converter grid interconnection at a PCC is given in Fig. 4 [4, 17]. The converter station is represented by its input admittance $\underline{\mathbf{Y}}_H$ and the AC grid by its impedance $\underline{\mathbf{Z}}_G$. The feedback interconnection of both systems is given by

$$\underline{\mathbf{G}}_Z = (\mathbf{I} + \underline{\mathbf{Y}}_H \underline{\mathbf{Z}}_G)^{-1} \underline{\mathbf{Y}}_H \tag{12}$$

In general, both $\underline{\mathbf{Y}}_H$ and $\underline{\mathbf{Z}}_G$ are fully set 3-by-3 matrices in the natural RST domain, which due to the fact, that converter transformer usually decouples the zero sequence, can be reduced to 2-by-2 matrices in $\alpha\beta$ and respectively in dq domain. This is a special system class of MIMO systems, further referred to as TITO (two inputs two outputs) system, where many problems can be solved analytically and become therewith

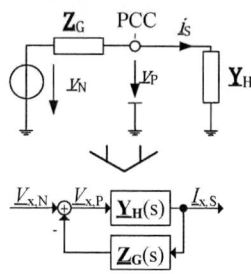

Fig. 4: Control representation of a simplified electrical circuit

directly accessible. For HVDC converters, the goal of stability analysis is often to evaluate the influence of a well known converter input admittance to unknown and varying grid impedances, hence stating robustness to control an unknown system.

4.2 Impedance based stability analysis and their relation to SISO stability theorems

The Nyquist criterion as a frequency domain stability theorem was originally derived for the SISO case and later extended to MIMO systems. As it is well known, its transparent design benefits only hold for the SISO case and much of them are lost during the extension to MIMO systems. Since the Nyquist theorem for the SISO and MIMO case is well known we only refer to e.g. [24–27] for a comprehensive summary, especially for the perquisites on the system transfer function in the MIMO case.

The Nyquist theorem in the SISO case (SNC) directly introduces two measures of robustness, the phase and the gain margin. A stable SISO system is called passive, iff the phase remains between $\pm$ 90°. From the SNC, the conclusion follows, that two passive systems under unity feedback must be stable, independent of the loop gain. The feedback system has infinite gain margin. However, delays in the feedback loop may overcome the limited phase margin and destabilize the system, nonetheless.

The small gain theorem addresses the gain margin and therefore complements the passivity approach. The small gain theorem for the SISO system states, that the feedback connection of two stable SISO systems $\underline{A}$ and $\underline{B}$ is stable, iff $|\underline{A}| \cdot |\underline{B}| < 1 \; \forall \, \omega \in (-\infty, \infty)$. The feedback system has infinite phase margin.

Both concepts establish conditions for so-called structural stability. This means stability is guaranteed as a consequence of the system class (e.g. the class of passive systems). However, model variations often change the system class, since the vast majority of transfer functions only remain passive or fulfill the small gain theorem for a certain range of their parameters and even then only within certain frequency ranges. This observation is very important for the application of passivity theorem-based stability analysis in the HVDC environment since it directly impacts the strength and applicability of the result in real-world applications. The knowledge of structural stability properties and deficiencies of the HVDC input admittance directly leads to boundaries for the AC grid impedance (and vice versa) in which stability and robustness can be guaranteed. E.g., as a consequence of the small gain theorem, if the magnitude of the grid impedance is lower than that of the HVDC within a certain frequency range, the interconnected system cannot become unstable in that range. The same goes for the passivity criterion.

The classical impedance based analysis considers the real part of the converter admittance and demands that

$$\mathbf{Re}\{\underline{Y}_H(j\omega)\} \geq 0 \; \forall \, \omega \in [-\infty, +\infty] \Leftrightarrow \arg(\underline{Y}_H(j\omega)) \in [-90°, +90°], \tag{13}$$

which is equivalent to the converter admittance being passive, given it is stable [16, 26, 28–30]. If the converter admittance is passive in a certain frequency range, where the grid impedance is known with certainty to be passive as well, the system cannot become unstable in that range. However, even a passive converter admittance does not necessarily lead to a stable system, when the AC grid has non-passive regions e.g. because of other actively controlled devices, which is a given at least in the frequency range around the nominal grid frequency. Moreover, condition (13) is very conservative and cannot be fulfilled over the whole frequency range in a real world application control system. It must be ensured, that the open loop suffices the Nyquist theorem, by adapting the range of allowable phase variations of the grid as well. Instead of purely relying on the passivity of the converter and the grid, mixed boundaries in phase and gain within the framework of the SNC can also be used to achieve less conservative stability conditions.

4.3 MIMO theorems for closed loop stability

The Nyquist theorem in the MIMO case loses two of the main benefits given in the SISO case: First, the clear dependency between open loop transfer functions and the Nyquist curve and secondly the easy definition of the stability and robustness margins. The MIMO Nyquist theorem, also known as generalized Nyquist theorem (GNC)[1, 31, 32], involves the locus of the determinant of the feedback difference matrix (FDM) $\underline{F} = I + \underline{G}_O$ and its correct encirclement of the origin [26, 28]. The definition of the phase and gain margin are no longer meaningful applicable. Due to these facts, the MIMO Nyquist criterion can be used to determine the stability of a given open loop, but not to give robustness measures in terms of phase and gain margins. This is especially disadvantageous, since the robustness for varying grid conditions is an essential design criteria for the HVDC control.

There are equivalents to SISO passivity and small gain theorem for MIMO [25, 33]. However, strict preconditions must apply. Moreover, they provide no guarantees about robustness against phase distortions and model parameter variations. Since the grid conditions are mostly only known by statistics and with the rising number of active components, like HVDC and SVC converters, these MIMO theorems are not as useful as in the SISO case.

As stated, MIMO stability criteria have certain drawbacks, when it comes to evaluating the robustness of a certain control system to unknown or varying plants. Therefore, the next section identifies circumstances, under which the MIMO system can be handled by solving adequate multiple SISO problems to restore their benefits and circumvent the design problems introduced by the MIMO criteria.

4.4 Diagonal dominance and it's application to the GNC

In this section the concept of generalized diagonal dominance (GDD) is introduced. This concept belongs to the so-called direct Nyquist array methods. The common idea of these concepts is the reduction of the MIMO problem into multiple SISO problems, which can be solved with the classical SISO open-loop shaping methods. Therefore, this methodology is called a quasi-classical approach and dates back to the 1970th with the works of MacFarlane and Rosenbrock [24].

The GNC problem is stated as the locus of the determinant of the FDM, which can also be given by the eigenvalues of this matrix:

$$\det(\underline{F}) = \det(I + \underline{G}_O) = \prod_i^n \lambda_i(\underline{F}) = \prod_i^n (1 + \lambda_i(\underline{G}_O)) \tag{14}$$

With Eq.(14) the MIMO problem is already decomposed into n SISO problems, for TITO system $n = 2$. Since the grid impedance is mostly unknown, the loci of the eigenvalues of $\underline{G}_O = \underline{Y}_C\underline{Z}_G$ as a function of $j\omega$ cannot be exactly calculated. However, the eigenvalues of a TFM can be approximated by the main diagonal elements, if that matrix is diagonal dominant (DD) or generalized diagonal dominant (GDD).

A matrix $\underline{A} \in \mathbb{C}^{n \times n}$ is DD if it is row or column dominant, meaning

$$\forall i : |\underline{a}_{ii}| > \sum_{j;j \neq i}^n |\underline{a}_{ji}| \quad \text{or} \quad \forall i : |\underline{a}_{ii}| > \sum_{j;j \neq i}^n |\underline{a}_{ij}|. \tag{15}$$

This translates to the easy condition in the TITO case, that the absolute value of each diagonal element has to be greater than the absolute value of its related row or column off-diagonal element $\forall \omega \in (-\infty, \infty)$. The Gershgorin theorem now assures, that the eigenvalues of $\underline{A}$ rest within circles given by the midpoint $\underline{a}_{ii}$ and radii given by $\sum_{j;j \neq i}^n |\underline{a}_{ji}|$ or $\sum_{j;j \neq i}^n |\underline{a}_{ij}|$ for row- or column dominance, respectively. As one can easily see, the approximations exclude the origin of the complex plane and therefore the continuous change in phase of the loci of the eigenvalues and of the diagonal elements are the same. Hence, the diagonal elements are a sufficient approximation for the GNC criterion.

A matrix $\underline{A} \in \mathbb{C}^{n \times n}$; $\underline{A}$ irreducible, is GDD if there is an invertible (and hence non-singular) diagonal transformation matrix $\underline{R}$ such that $\underline{\tilde{A}} := \underline{R} \cdot \underline{A} \cdot \underline{\tilde{R}}$ is DD. The similarity transformation with a diagonal matrix ensures, that the eigenvalues AND the diagonal elements of $\underline{A}$ and $\underline{\tilde{A}}$ are the same. Therefore, the transformation only affects the radii of the Gershgorin bounds. Hence, it follows immediately, the GDD criterion is less restrictive than the DD criterion. Nonetheless, both conditions imply that the Gershgorin bounds do not include the origin and therefore are equivalent in terms of the application to the GNC.

The matrix $\underline{\mathbf{A}}$ is GDD iff $\lambda_P\{\mathbf{C}(\underline{\mathbf{A}})\} < 1$, where λ_P denotes the Perron-Root of

$$\mathbf{C}(\underline{\mathbf{A}}) := \mathrm{diag}\left(\frac{1}{|\underline{a}_{ii}|}\right) \cdot \begin{bmatrix} 0 & |\underline{a}_{12}| & \cdots & |\underline{a}_{1n}| \\ |\underline{a}_{21}| & 0 & \cdots & |\underline{a}_{2n}| \\ \vdots & \vdots & \ddots & \vdots \\ |\underline{a}_{n1}| & |\underline{a}_{n2}| & \cdots & 0 \end{bmatrix}. \tag{16}$$

The Gershgorin theorem can also be stated in relation to the Perron-Root $\lambda_P\{\mathbf{C}(\underline{\mathbf{A}})\}$ with $|\underline{a}_{ii}| \cdot \lambda_P$ as bounds for the Gershgorin circles. This also shows directly, that the bound does not include the origin. The condition that $\underline{\mathbf{A}}$ is irreducible implies that the matrix $\mathbf{C}$ is also irreducible, then since $\mathbf{C}$ is positive real by construction and irreducible by assumption, the Perron-Frobenius theorem ensures the existence of the Perron-Root. For TITO systems the assumption that $\underline{\mathbf{A}}$ is irreducible is not necessary since if $\underline{\mathbf{A}}$ is reducible, it is of triangular form and the eigenvalues are give directly by the main diagonal elements. GDD is sufficient to justify SISO stability analysis. However, in frequency ranges where the Perron-Root is close to one, robustness criteria like phase and gain margin might fail.

4.5 Requirements to the grid impedance to ensure the validity of SISO stability analysis

In the previous section conditions to the FDM, DD and GDD, have been derived, that would allow the common SISO approach (e.g.[5–7, 13–15]) for converter $\leftrightarrow$ grid stability analysis. That approach is only valid iff the feedback difference is DD or GDD. In the next chapter it will be shown, that the converter input admittance, by itself is never DD or GDD for the whole frequency range, hence there are boundaries to the grid impedance that have to be met, to ensure DD or GDD for the FDM and to be able to use SISO criteria to investigate system stability.

During project execution, the grid impedance is often given from transmission system operators to the vendor for a set of frequencies as polygons in the complex plain where the grid impedance most like would rest during converter runtime. Hence, the exact impedance is unknown and stability and interactions have to be investigated based on the converter input impedance alone. For that, two assumptions are crucial: the grid impedance is symmetrical (hence, diagonal and identical in the $\alpha\beta$ frame) and passive. These precondition seem restrictive, are however reasonable for high voltage AC grids. In addition, from the previous section another limitation arises: The FDM must be diagonal dominant.

The converter admittance can only be calculated in a dq frame and cannot be transferred to $\alpha\beta$ without losing its LTI property, due to lack of skew symmetry. In Section 3 is also shown, that a diagonal and identical $\alpha\beta$ frame grid system matrix is fully coupled in the dq frame, with a maximum coupling in the frequency range below $50\,\mathrm{Hz}$. Since its specific values are unknown, diagonal dominance cannot be ensured. Hence, this would not be usable for a SISO investigation. However, transferred in the dq_{PN} frame, the grid impedance matrix is diagonal again. This operation is also possible on the given impedance polygons since it is a simple frequency shift from the $\alpha\beta$ frame (see Eq.(10)).

The requirement for GDD for a given converter input admittance and a grid impedance, only known by its structural properties in the dq_{PN} frame denotes to:

$$\mathbf{F}_{\mathrm{PN}}^{\mathrm{dq}} = \mathbf{I} + \underline{\mathbf{Y}}_{\mathrm{C,PN}}^{\mathrm{dq}} \underline{\mathbf{Z}}_{\mathrm{G,PN}}^{\mathrm{dq}} = \begin{bmatrix} 1 + \underline{Y}_{\mathrm{C,PP}}^{\mathrm{dq}} \underline{Z}_{\mathrm{G,PP}}^{\mathrm{dq}} & \underline{Y}_{\mathrm{C,PN}}^{\mathrm{dq}} \underline{Z}_{\mathrm{G,NN}}^{\mathrm{dq}} \\ \underline{Y}_{\mathrm{C,NP}}^{\mathrm{dq}} \underline{Z}_{\mathrm{G,PP}}^{\mathrm{dq}} & 1 + \underline{Y}_{\mathrm{C,NN}}^{\mathrm{dq}} \underline{Z}_{\mathrm{G,NN}}^{\mathrm{dq}} \end{bmatrix} \tag{17}$$

$$\lambda_P\{\mathbf{C}(\underline{\mathbf{F}}_{\mathrm{PN}}^{\mathrm{dq}})\} = \sqrt{\frac{|\underline{Y}_{\mathrm{C,PN}}^{\mathrm{dq}} \underline{Z}_{\mathrm{G,NN}}^{\mathrm{dq}}|}{|1 + \underline{Y}_{\mathrm{C,PP}}^{\mathrm{dq}} \underline{Z}_{\mathrm{G,PP}}^{\mathrm{dq}}|} \cdot \frac{|\underline{Y}_{\mathrm{C,NP}}^{\mathrm{dq}} \underline{Z}_{\mathrm{G,PP}}^{\mathrm{dq}}|}{|1 + \underline{Y}_{\mathrm{C,NN}}^{\mathrm{dq}} \underline{Z}_{\mathrm{G,NN}}^{\mathrm{dq}}|}} \overset{!}{<} 1 \tag{18}$$

With a conservative approximation, that condition (18) is fulfilled, if either fraction is less than 1, direct boundaries for $\underline{Z}_{\mathrm{G,PP}}$ and $\underline{Z}_{\mathrm{G,NN}}$ can be derived.

$$|\underline{Z}_{\mathrm{G},ii}| \overset{!}{<} \min\left(1 / \left(-|\underline{Y}_{\mathrm{H},ii}|\cos\Sigma\varphi \pm \sqrt{|\underline{Y}_{\mathrm{H,od}}|^2 - |\underline{Y}_{\mathrm{H},ii}|^2 \sin^2\Sigma\varphi}\right)\right) \quad i \in \{\mathrm{P,N}\}, \Sigma\varphi \in [-\pi,\pi], \forall\mathrm{j}\omega$$

where $\Sigma\varphi = \arg(\underline{Y}_{\mathrm{H},ii}) + \arg(\underline{Z}_{\mathrm{G},ii})$ and $\underline{Y}_{\mathrm{H,od}}$ is the off-diagonal element of line or column i $\quad$ (19)

If the condition of Eq.(19) is fulfilled for either both lines or both columns, the system is GDD and stability can be analyzed using the common SISO tools and if the condition of Eq.(19) is fulfilled with enough margin, even the SISO robustness criteria, like phase and gain margin, are valid. Otherwise, the evaluation of the MIMO GNC is necessary. Hence, the condition of Eq.(19) can be used to indicate areas of the given grid impedance polygons where SISO analysis is possible and areas where the SISO approach is not valid.

5 Properties of and requirements to the converter input impedance as well as the underlying control design

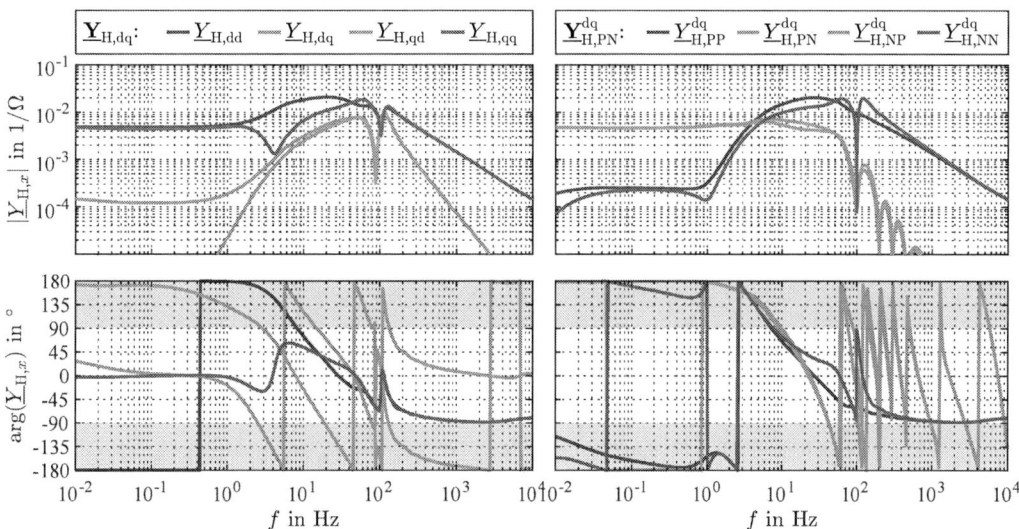

Fig. 5: Calculated input admittance in a dq (left) and a dq_{PN} (right) frame for maximum power rectifier operation

In this section, the findings of the previous sections are brought together on an exemplary converter input admittance, calculated according to Section 2. Fig. 5 shows the converter input admittance in the dq and the dq_{PN} coordinate system.

For $f < 80\,\text{Hz}$ the converter input admittance is highly operation point dependent. While the admittance itself shows strong diagonal dominance in dq below 90 Hz and above 200 Hz, the admittance in dq_{PN} shows strong diagonal dominance above 10 Hz, with an exception at 100 Hz, and no diagonal dominance below. The main diagonal admittance in both frames is passive above 10 Hz. It is noteworthy, that for the higher frequencies, this is mainly because of the real part of the converter transformer impedance. Due to unavoidable dead times, the control impedance can never be passive at higher frequencies. Hence, to avoid negative damping in this area, the controller proportional gain must be significantly smaller than the real part of the transformer impedance. This is especially challenging with transformers optimized for their losses.

The negative damping area in the low frequency range of the converter input admittance is directly given by the bandwidth of the controllers integral parts. Any conventional control system containing integrators shows this behavior. The higher the dynamic of these controllers, the higher is the crossover frequency, where the real part of the control impedance becomes positive. Hence, the requirement for a passive converter input impedance is diametral to its performance requirements (compare to [14–16]).

The coupling between the PPS and the NPS is mainly a result of the separated active (energy) and reactive power control loops as well as the PLL dynamics, since e.g. any oscillation in the energy system in combination with the Park transformation directly convolves to a PPS and NPS response. This could only be mitigated by a strong limitation of the bandwidth of these controllers as well as the PLL, which again is diametral to their performance requirements.

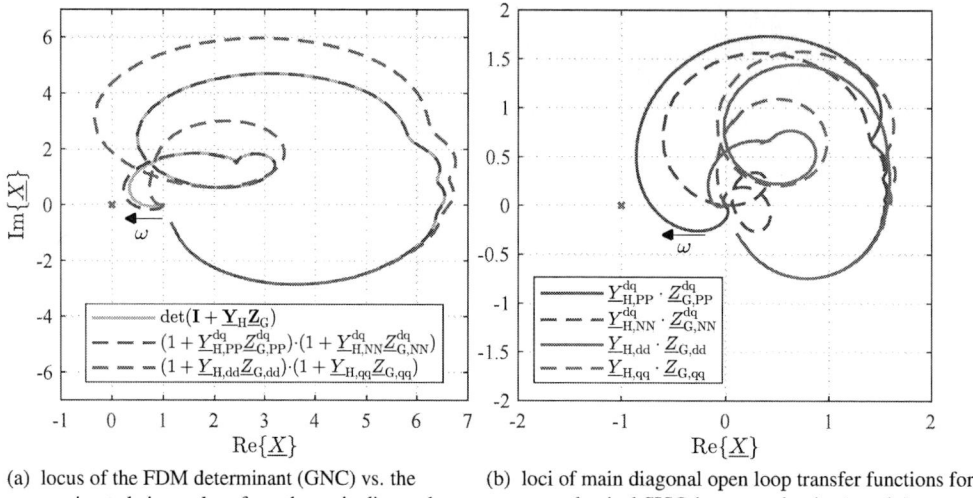

(a) locus of the FDM determinant (GNC) vs. the approximated eigenvalues from the main diagonal elements in dq and dq_{PN}

(b) loci of main diagonal open loop transfer functions for a classical SISO interpretation in dq and dq_{PN}

Fig. 6: Comparison of the evaluation of the eigenvalue locus according to the MIMO GNC with the approximated loci of the eigenvalues by the main diagonal elements of the FDM containing a thevenin source grid impedance for a short circuit level of $3\,\text{GVA}$

Fig. 6a shows the GNC locus evaluated for the full FDM in comparison to its approximation by the diagonal elements in the dq and dq_{PN} frame. The deviation between the GNC locus and the approximated loci provides a measure for the quality of the approximation. Due to the strong coupling of the grid impedance between the d and q axis in the dq frame the approximation of the FDM by its main diagonal elements is insufficient, especially for lower frequencies. Evaluation of stability margins would be way too optimistic in that frame as shown in Fig. 6b. In the dq_{PN} frame, the eigenvalue approximation is a near perfect fit to the locus of the FDM GNC. While *PN* transformation does not affect the eigenvalues of the FDM, its positive influence on the diagonal dominance becomes obvious. Hence, the main diagonal elements in the dq_{PN} frame are suitable for SISO stability and robustness investigation.

Fig. 7 shows the evaluation of Eq.(19) for the dq_{PN} converter input admittance shown in Fig. 5. In addition, for each main diagonal element, the nominal thevenin grid impedance in dq_{PN} for a short circuit level of $3\,\text{GVA}$ is plotted as a reference. Eq.(19) gives an upper bound for the magnitude of each main diagonal element of the grid impedance in the dq_{PN} frame, within which the requirement for GDD is fulfilled. Obviously, a smaller Perron-Root implies a stronger GDD property, since possible deviations between the GNC locus and its approximation are bounded narrower, which improves the validity of

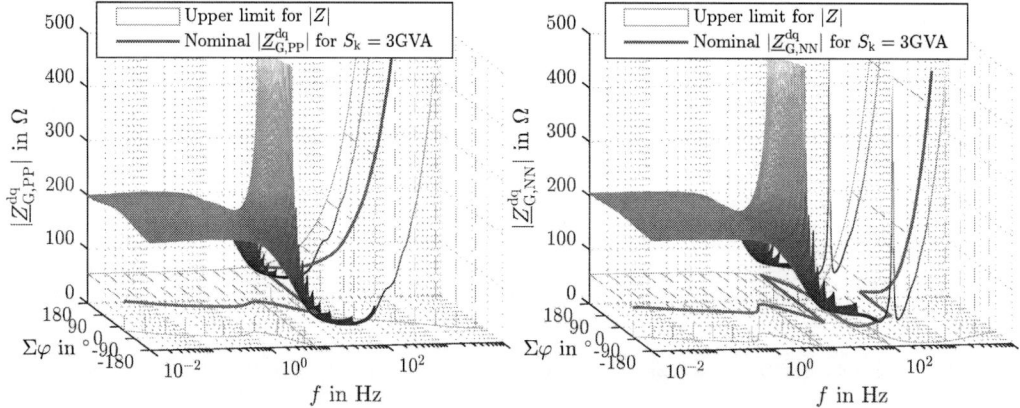

Fig. 7: Upper bounds of the grid impedance for the converter admittance of Fig. 5 and comparison to a nominal grid impedance at a short circuit level of $3\,\text{GVA}$

SISO open loop investigations. The surface given by Eq.(19) opens for frequencies above approximately $10\,\mathrm{Hz}$, meaning that there is no limit to the magnitude of the grid impedance in that area. Only where the summarized angle of $Y_{\mathrm{C,PP/NN}}^{\mathrm{dq}}$ and $Z_{\mathrm{G,PP/NN}}^{\mathrm{dq}}$ is close to $\pm 180°$, hence at the passivity limit of the feedback system, there is still an upper bound. In that frequency range the coupling between the axis of the converter input admittance is very low (in all reference frames) and the restrictions on the system class of the AC grid impedance could be weakened.

Since Fig. 7 shows only the evaluation for each fraction of Eq.(19), it seems that GDD is violated for $|Z_{\mathrm{G,PP}}^{\mathrm{dq}}|$ between 6 and 8 Hz. However, since the product of both fractions must be less than 1, only DD is violated while the requirements for GDD are still met with the nominal grid impedance.

For the given example control and configuration, in the frequency range up to about $10\,\mathrm{Hz}$ SISO stability investigation is only valid up to a grid impedance magnitude of about $200\,\Omega$. Obviously any grid resonance in that frequency range would exceed this limit. Hence, very detailed models and most likely a MIMO GNC investigation are necessary, for e.g. sub-synchronous torsion interaction and power oscillation damping investigations.

6 Conclusion

Ensuring robust stability when installing a new HVDC station to the AC grid is a fundamental design criterion. The commonly used approach for this investigation is based on the SISO Nyquist theorem, considering only the main diagonal elements of the converter input admittance to draw conclusions on the stability margins for the interconnected feedback system with the AC grid. This is clearly motivated by the obvious robustness criteria given by the SISO Nyquist theorem. In this paper, a control theoretical foundation for the validity of this SISO approach has been derived. At first by showing the limitations to the linearity of reference system transformations and the limitations of the SISO approach applied to MIMO systems, due to the cross coupling between e.g. the d- and q-axis within the converter admittance. By introducing the MIMO criteria for diagonal dominance and more important generalized diagonal dominance, a comprehensive and sufficient boundary can be drawn, separating systems classes, where a SISO investigation is justified and where a MIMO stability analysis via the GNC is advisable. This requires knowledge from both, the converter and the AC grid system. Based on that, it has been shown, that this criterion can only be implemented for a certain class of AC systems, namely passive, symmetrical systems where the α and β grid impedance components are identical. A symmetrical AC grid in the $\alpha\beta$-frame is diagonal in the dq_{PN} frame, allowing the assessment of GDD solely based on the structural properties of the converter input admittance. Passivity of the AC system is a necessary condition to enable stability evaluation via the phase margin of the converter input admittance.

With a generic converter control system with only basic control loops for converter current, energy and reactive power an exemplary converter input impedance was calculated. Based on that, it has been demonstrated, that a SISO robustness investigation based on the main diagonal elements of the converter admittance in the dq frame is most likely insufficient due to the strong coupling in the grid impedance in that frame. However, in the dq_{PN} frame, the systems eigenvalues can be matched with good accuracy by the main diagonal elements, as long as the criteria for GDD are fulfilled. Based on the calculated converter input admittance exact boundaries to the grid impedance magnitude can be derived. Within these boundaries, stability and robustness investigation based on the SISO Nyquist theorem is justified. It has been shown, that especially in the high frequency range due to strong diagonal dominance in the converter admittance the validity of SISO robustness investigation can be ensured. However, in the low frequency range careful evaluation of and especially a deeper knowledge on the grid impedance is necessary to evaluate robustness.

References

[1] L. Harnefors. "Modeling of Three-Phase Dynamic Systems Using Complex Transfer Functions and Transfer Matrices". In: *IEEE Transactions on Industrial Electronics* 54.4 (2007), pp. 2239–2248.

[2] Y. A. Familiant, J. Huang, K. A. Corzine, et al. "New Techniques for Measuring Impedance Characteristics of Three-Phase AC Power Systems". In: *IEEE Transactions on Power Electronics* 24.7 (2009), pp. 1802–1810.

[3] G. Francis, R. Burgos, D. Boroyevich, et al. "An algorithm and implementation system for measuring impedance in the D-Q domain". In: *IEEE Energy Conversion Congress and Exposition*. 2011, pp. 3221–3228.

[4] R. Turner, S. Walton, and R. Duke. "A Case Study on the Application of the Nyquist Stability Criterion as Applied to Interconnected Loads and Sources on Grids". In: *IEEE Transactions on Industrial Electronics* 60.7 (2013), pp. 2740–2749.

[5] M. Cespedes and J. Sun. "Impedance Modeling and Analysis of Grid-Connected Voltage-Source Converters". In: *IEEE Transactions on Power Electronics* 29.3 (2014), pp. 1254–1261.

[6] X. Wang, F. Blaabjerg, and W. Wu. "Modeling and Analysis of Harmonic Stability in an AC Power-Electronics-Based Power System". In: *IEEE Transactions on Power Electronics* 29.12 (2014), pp. 6421–6432.

[7] B. Wen, D. Dong, D. Boroyevich, et al. "Impedance-Based Analysis of Grid-Synchronization Stability for Three-Phase Paralleled Converters". In: *IEEE Transactions on Power Electronics* 31.1 (2016), pp. 26–38.

[8] I. Vieto and J. Sun. "Sequence Impedance Modeling and Converter-Grid Resonance Analysis Considering DC Bus Dynamics and Mirrored Harmonics". In: *2018 IEEE 19th Workshop on Control and Modeling for Power Electronics (COMPEL)*. 2018, pp. 1–8.

[9] X. Wang, L. Harnefors, and F. Blaabjerg. "Unified Impedance Model of Grid-Connected Voltage-Source Converters". In: *IEEE Transactions on Power Electronics* 33.2 (2018), pp. 1775–1787.

[10] C.M. Wildrick, F.C. Lee, B.H. Cho, et al. "A method of defining the load impedance specification for a stable distributed power system". In: *IEEE Transactions on Power Electronics* 10.3 (1995), pp. 280–285.

[11] H. Saad, Y. Fillion, S. Deschanvres, et al. "On Resonances and Harmonics in HVDC-MMC Station Connected to AC Grid". In: *IEEE Transactions on Power Delivery* 32.3 (2017), pp. 1565–1573.

[12] Cigré WG B4.67. "AC side harmonics and appropriate limits for VSC HVDC". In: *Techn. Brochure 754* (2019).

[13] H. Zhang, L. Harnefors, X. Wang, et al. "Stability Analysis of Grid-Connected Voltage-Source Converters Using SISO Modeling". In: *IEEE Transactions on Power Electronics* 34.8 (2019), pp. 8104–8117.

[14] C. Hirsching, S. Wenig, S. Beckler, et al. "Passivity-Based Sensitivity Analysis of the Inner Current Controller in Grid-Following MMC-HVdc Applications - An Overview". In: *IECON 2020 The 46th Annual Conference of the IEEE Industrial Electronics Society*. 2020, pp. 1412–1417.

[15] C. Hirsching, A. Bisseling, S. Wenig, et al. "On the impact of controller implementations on passivity and damping properties in grid-following MMC-HVdc applications". In: *accepted for the 47th Annual Conference of the IEEE Industrial Electronics Society*. 2021.

[16] L. Harnefors, M. Bongiorno, and S. Lundberg. "Input-Admittance Calculation and Shaping for Controlled Voltage-Source Converters". In: *IEEE Transactions on Industrial Electronics* 54.6 (2007), pp. 3323–3334.

[17] A. Bayo-Salas, J. Beerten, J. Rimez, et al. "Impedance-based stability assessment of parallel VSC HVDC grid connections". In: *11th IET International Conference on AC & DC Power Transmission*. 2015, pp. 1–9.

[18] A. Rygg, M. Molinas, C. Zhang, et al. "A Modified Sequence-Domain Impedance Definition and Its Equivalence to the dq-Domain Impedance Definition for the Stability Analysis of AC Power Electronic Systems". In: *IEEE Journal of Emerging and Selected Topics in Power Electronics* 4.4 (2016), pp. 1383–1396.

[19] G. Herold. *Drehstromsysteme, Leistungen, Wirtschaftlichkeit*. 3rd ed. Vol. 1. Elektrische Energieversorgung. Wilburgstetten: J. Schlembach, 2011.

[20] D.N. Zmood, D.G. Holmes, and G.H. Bode. "Frequency-domain analysis of three-phase linear current regulators". In: *IEEE Transactions on Industry Applications* 37.2 (2001), pp. 601–610.

[21] J. Sun. "Small-Signal Methods for AC Distributed Power Systems–A Review". In: *IEEE Transactions on Power Electronics* 24.11 (2009), pp. 2545–2554.

[22] A. Rygg, M. Molinas, C. Zhang, et al. "On the Equivalence and Impact on Stability of Impedance Modeling of Power Electronic Converters in Different Domains". In: *IEEE Journal of Emerging and Selected Topics in Power Electronics* 5.4 (2017), pp. 1444–1454.

[23] C. Zhang, X. Cai, A. Rygg, et al. "Sequence Domain SISO Equivalent Models of a Grid-Tied Voltage Source Converter System for Small-Signal Stability Analysis". In: *IEEE Transactions on Energy Conversion* 33.2 (2018), pp. 741–749.

[24] J. Raisch. *Mehrgrössenregelung im Frequenzbereich*. Methoden der Regelungs- und Automatisierungstechnik. Oldenbourg Wissenschaftsverlag, 1994.

[25] B. Brogliato, R. Lozano, B. Maschke, et al. *Dissipative systems analysis and control*. en. 2nd ed. Communications and Control Engineering Series. London, England: Springer, 2007.

[26] J. Lunze. *Regelungstechnik 2 Mehrgrößensysteme, Digitale Regelung*. 8th ed. Springer, 2014.

[27] J. Sun. "Impedance-Based Stability Criterion for Grid-Connected Inverters". In: *IEEE Transactions on Power Electronics* 26.11 (2011), pp. 3075–3078.

[28] H. Unbehauen. *Regelungstechnik I*. 15th ed. Wiesbaden: Vieweg + Teubner, 2008.

[29] L. Harnefors, X. Wang, A. G. Yepes, et al. "Passivity-Based Stability Assessment of Grid-Connected VSCs — An Overview". In: *IEEE Journal of Emerging and Selected Topics in Power Electronics* 4.1 (2016), pp. 116–125.

[30] A. J. Agbemuko, J. L. Domínguez-García, O. Gomis-Bellmunt, et al. "Passivity-Based Analysis and Performance Enhancement of a Vector Controlled VSC Connected to a Weak AC Grid". In: *IEEE Transactions on Power Delivery* 36.1 (2021), pp. 156–167.

[31] C. Desoer and Y.-T. Wang. "On the generalized nyquist stability criterion". In: *IEEE Transactions on Automatic Control* 25.2 (1980), pp. 187–196.

[32] J. Samanes, A. Urtasun, E. L. Barrios, et al. "Control Design and Stability Analysis of Power Converters: The MIMO Generalized Bode Criterion". In: *IEEE Journal of Emerging and Selected Topics in Power Electronics* 8.2 (2020), pp. 1880–1893.

[33] S. Skogestad and I. Postlethwaite. *Multivariable feedback control*. Chichester, England: John Wiley & Sons, 1996.

Class-E Push-Pull Resonance Converter with Load Variation Robustness for Industrial Induction Heating

Janus Dybdahl Meinert, Benjamin Futtrup Kjærsgaard, Thore Stig Aunsborg
Asger Bjørn Jørgensen, Stig Munk-Nielsen
Department of Energy, Aalborg University
Pontoppidanstræde 111
9220 Aalborg Øst, Denmark
Email: {jdm, bfk, tsu, abj, smn}@energy.aau.dk
URL: https://www.energy.aau.dk

Sune Bro Duun
Topsil GlobalWafers A/S
Siliciumvej 1
3600 Frederikssund, Denmark
Email: sdu@gw-topsil.com
URL: http://www.topsil.com

Acknowledgments

Support has been received from the MVolt and CoDE projects. The MVolt project is co-funded by the Department of Energy Technology of Aalborg University, Innovation Fund Denmark, Siemens Gamesa Renewable Energy, Vestas Wind Systems, and KK Wind Solutions. The CoDE project is funded by the Poul Due Jensen Grundfoss Foundation.

Keywords

≪Resonant converter≫, ≪Zero-voltage switching≫, ≪Wide bandgap devices≫, ≪Radio frequency (RF)≫ , ≪Current Source Inverter (CSI)≫, ≪Simulation≫

Abstract

Emerging wide bandgap devices are extending the operating frequency range and power handling capability of solid state based resonant power converter solutions. Presently, resonant power converters for industrial induction heating are using vacuum-tubes, achieving efficiencies of 50-60 %. By replacing the prevalent vacuum tube technology with a solid state based solution, the efficiency of the industrial induction heating processes is expected to be increased. A design of a Class-E Push-Pull resonance converter using silicon carbide MOSFETs is proposed. A prototype, operating at 2.5 MHz, has been built showing a proof-of-concept of the topology at 4 kVA, achieving an efficiency of 91.8 % with a representative industrial induction heating load.

Introduction

Resonant converters operating in a high frequency high power range are used in a wide range of industrial applications including dielectric and inductive heating. Dielectric heating includes various drying processes, whereas inductive heating is used for e.g. sawblade hardening and float zone processes [1–4]. The float zone processing industry currently uses vacuum-tube technology for the resonance converters with efficiencies ranging from 50-65 % [5, 6]. However, with the higher breakdown voltage, lower on-resistance, higher thermal conductivity and lower gate charge of the silicon carbide (SiC) MOSFETs compared to their silicon counterparts [7,8], high power radio frequency (RF) applications based on solid state technology is enabled [9–13]. Thus, by using wide bandgap (WBG) devices as a replacement for the vacuum-tubes, the efficiency of the float zone process is expected to be increased significantly.

In this paper, a proposed prototype resonance converter using SiC MOSFETs will be built and demonstrated for 2.5 MHz induction heating applications. The next section will present the intended converter design, its functionalities and design considerations. An experimental demonstration is given and a comparison between experiments and digital twin simulations is presented in the following two sections. Lastly the most relevant findings of this paper are summarized.

Proposed Topology

The chosen converter topology for the proposed design is a Class-E Push-Pull resonance current source converter, illustrated in Fig. 1a. Recent research utilizing the Class-E Push-Pull converter for inductive power transfer applications has shown high frequency oscillations and voltage spikes during switching instances, which is found to be caused by the stray inductance as presented in [13,14].
The proposed design intends to utilize the stray inductance by having two distinct resonance loops; (1) The drain-source resonance loop, consisting of the stray inductance L_{stray} and the respective drain-source capacitance C_{ds} visualized by the red and green colored areas in Fig. 1a. (2) The load resonance tank, illustrated by the blue colored area, consisting of the resonance capacitor C_r and the single turn resonance induction coil L_r which through inductive power transfer will dissipate power in the form of iron losses heating the load object represented by R_{iron}. A sketch of the used single turn coil is shown in Fig. 1b.

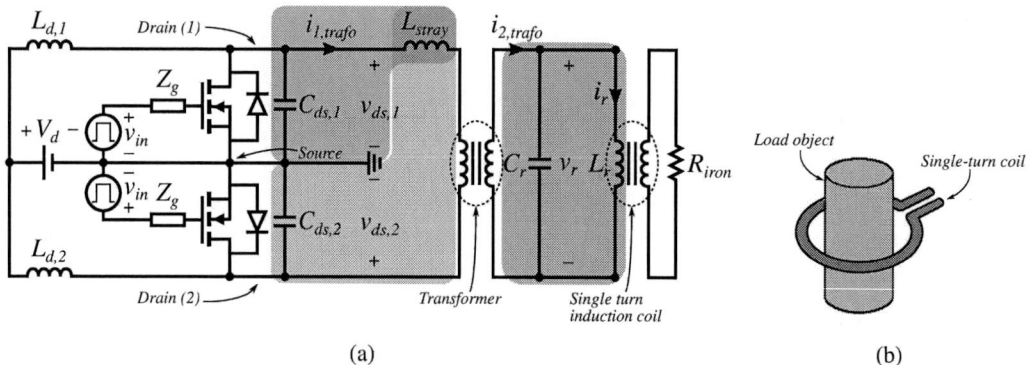

(a)   (b)

Fig. 1: (a) Circuit schematic of proposed topology with colored areas illustrating different resonance loops. (b) Sketch of a single turn induction coil.

The concept of the design is to ensure zero-voltage switching (ZVS) over a wide range of load resonance frequencies f_r by controlling the drain-source resonance frequency f_{ds}. This is obtained with the following two design considerations; (1) Utilizing the leakage inductance of the transformer $L_{lk,trafo}$ the trace/wire inductance from the converter to the resonance load $L_{trace,\sigma}$ is dominated. (2) Inserting sufficiently large drain-source capacitors $C_{ds,ext}$ in parallel with the MOSFETs, reduces the influence of the voltage dependent intrinsic output capacitance $C_{ds,\sigma}$ of the MOSFETs.
The two design considerations are summarized in (1) and (2).

$$L_{stray} = L_{lk,trafo} + L_{trace,\sigma} \simeq L_{lk,trafo}, \qquad\qquad L_{lk,trafo} \gg L_{trace,\sigma} \qquad (1)$$
$$C_{ds} = C_{ds,ext} + C_{ds,\sigma} \simeq C_{ds,ext}, \qquad\qquad C_{ds,ext} \gg C_{ds,\sigma} \qquad (2)$$

By the designers choice the drain-source resonance frequency f_{ds} can be varied by controlling the size of the transformer leakage inductance and the inserted drain-source capacitance.

$$f_{ds} = \frac{1}{2\pi \cdot \sqrt{L_{stray} \cdot C_{ds}}} \simeq \frac{1}{2\pi \cdot \sqrt{L_{lk,trafo} \cdot C_{ds,ext}}} \tag{3}$$

By choosing the drain-source resonance frequency f_{ds} higher than the switching frequency f_{sw}, ZVS of the MOSFETs are ensured for a wide range of load variations if the constraint in (4) is satisfied.

$$f_{sw} = f_r < f_{ds} \tag{4}$$

The mode of operation of the proposed topology is similar to a single-ended Class-E where the turn-OFF of a MOSFET triggers the drain-source resonance circuit and a half sine-wave voltage is generated across the drain-source terminals of the MOSFET [15]. Since the drain-source resonance frequency is higher than the switching frequency, the drain-source capacitor will discharge to 0 V before the MOSFET turns ON, leading to a time period where the body diode is conducting. In Fig. 2a the expected waveforms for the topology are seen, where the drain-drain voltage is defined as $v_{dd} = v_{ds,1} - v_{ds,2}$.

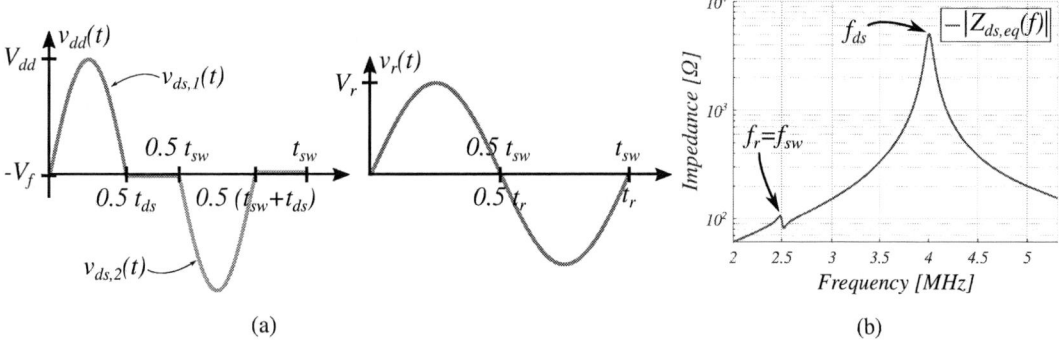

(a)   (b)

Fig. 2: (a) Class-E Push-Pull expected voltage waveforms and (b) The absolute value of the equivalent drain-source impedance illustrating the two resonance frequencies of the converter, f_r and f_{ds}.

As seen from Fig. 2a two time periods are defined; t_{ds} and t_{sw} which represent the drain-source resonance period and switching period respectively, with the switching period being equal to the load resonance period. The two different resonance frequencies of the converter are illustrated as an impedance curve in Fig. 2b. The analytical expressions for the voltage waveforms in Fig. 2a are shown in (5) and (6).

$$v_{dd}(t) = \begin{cases} V_{ds} \cdot \sin(2\pi f_{ds} \cdot t), & 0 < t \le 0.5\, t_{ds} \\ -V_f, & 0.5\, t_{ds} < t \le 0.5\, t_{sw} \\ -V_{ds} \cdot \sin(2\pi f_{ds} \cdot t), & 0.5\, t_{sw} < t \le 0.5\,(t_{sw} + t_{ds}) \\ V_f, & 0.5\,(t_{sw} + t_{ds}) < t \le t_{sw} \end{cases} \tag{5}$$

$$v_r(t) = V_r \cdot \sin(2\pi f_r \cdot t + \theta) \tag{6}$$

Using the analytical expressions in (5) and (6), the proposed topology shown in Fig. 1a can be reduced to the equivalent circuit diagram in Fig. 3, where $L_{eq,dd}$ is the total equivalent inductance between the drain-terminals and resonance load.

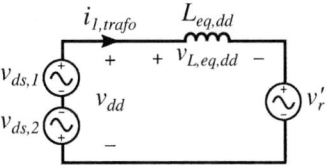

Fig. 3: Equivalent Class-E Push-Pull circuit diagram.

As depicted in Fig. 3, the current injected to the load resonance circuit is dependent on the voltage drop across the equivalent inductance. This can be utilized for controlling the output power of the converter by controlling the phase of this current, similar to a current controlled voltage source converter.

Experimental Verification

The functionality of the proposed topology is confirmed through experimental verification. A prototype converter is built as shown in Fig. 4.

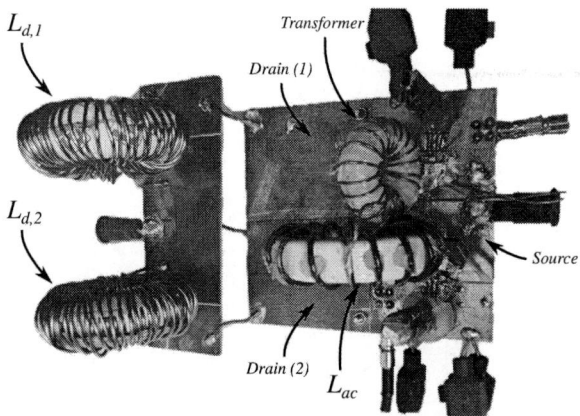

Fig. 4: Built Class-E Push-Pull prototype converter with a transformer between the converter and load.

As observed an AC-inductance L_{ac} is introduced in series with the transformer leakage inductance, thus the stray inductance will be given as $L_{stray} = L_{ac} + L_{lk,trafo} + L_{trace}$. The AC-inductance is needed to obtain the desired drain-source resonance frequency, without exceeding the 12 A current limit of the chosen Wolfspeed C3M0160120J SiC MOSFETs [16]. The commutation loops of the two MOSFETs are designed with a focus on symmetry to achieve similar impact from parasitic inductance and capacitance on the drain-source voltage waveforms $v_{ds,1}$ and $v_{ds,2}$. All magnetic components are designed using core materials with approximately constant permeability in the 2.5 MHz range. The DC-inductors are designed to achieve near constant DC-supply current. Due to the inductors being self-wound a slight deviation in DC inductance is observed. Using the IXDN614YI [17] gate driver, a low inductive hard switched gate driver design has been achieved through symmetrical layout, allowing for fast switching performance without gate-source voltage oscillations while having no external gate resistance. Circuit parameters are given in Table I.

Table I: The circuit parameters used in the built Class-E Push-Pull resonance converter.

			Converter side			
Components	V_d	$L_{d,1}$	$ESR_{Ld,1}$	$L_{d,2}$	$ESR_{Ld,2}$	L_{ac}	$C_{ds,ext}$
Value	30 V	145 µH	1 Ω	144 µH	1.4 Ω	1.6 µH	440 pF

			Load side			
Components	$L_{lk,trafo}$	$L_{m,trafo}$	$R_{c,trafo}$	n_{trafo}	L_r	C_r	Q_r
Value	442 nH	4.2 µH	30 kΩ	0.353	93 nH	46 nF	61

The performance of the converter is tested by connecting the secondary side of the transformer to the external resonance tank with a representative single turn induction coil for induction heating purposes designed in [13]. The driver signals for the two MOSFETs are supplied complimentary with a 50% duty cycle. The switching frequency is adjusted until the resonance frequency of the load is met at 2.45 MHz. The resulting experimental results are shown in Fig. 5.

As observed ZVS is achieved with a drain-source resonance frequency of 3.9 MHz ($T_{ds} = 256\,\text{ns}$). This leaves a tolerable margin of load resonance frequency variations greater than 1 MHz for which ZVS can be ensured. The drain-source voltage amplitude of 163 V yields a significantly higher voltage gain than the usual Class-E voltage gain of approx. 3.6, which is due to the volt-second balance of the DC-inductor yielding a voltage gain dependent on the ratio of the switching and drain-source resonance frequencies as shown in (7) [15, 18].

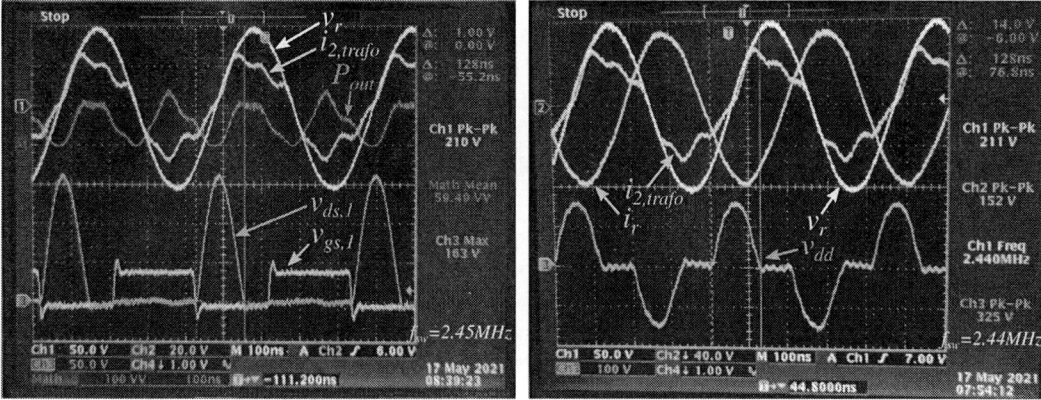

Fig. 5: Experimental results demonstrating the performance of the built prototype converter, with the drain-source resonance frequency being higher than the switching frequency, thus achieving ZVS. The MOSFETs are operated at the resonance frequency of the load, seen from v_r and $i_{2,trafo}$ being in phase. The drain-drain voltage v_{dd} is following the expected trend derived in (5).

$$V_{ds} = (V_d - V_{ESR,Ld}) \cdot 3.562 \cdot \frac{T_{sw}}{T_{ds}} = \big(30\,\text{V} - (1.08\,\text{A} \cdot 1\,\Omega)\big) \cdot 3.562 \cdot \frac{408\,\text{ns}}{256\,\text{ns}} = 164\,\text{V} \tag{7}$$

Where the term $V_{ESR,Ld}$ takes into account the DC-voltage drop across the ESR of the DC-inductor, with an observed DC-current of 1.08 A read from the DC-supply.

The current i_r in the single turn induction coil and the voltage v_r of the resonance capacitor are observed to be approx. 90° phase-shifted due to the reactive nature of the resonance tank. An apparent power of 4 kVA is observed in the resonance tank, which with a supply input power of only 64.8 W implies a reactive power of approx. 4 kVAr.

From Fig. 5 it is observed that the secondary side transformer current $i_{2,trafo}$ is in phase with the resonance voltage v_r, due to switching at the load resonance frequency. By multiplying these two signals the instantaneous output power is obtained, for which the mean active power transferred to the load object through the single turn induction coil is calculated to 59.49 W with an input DC-voltage of 30 V. Thus, the efficiency of the built prototype converter is 91.8 %.

Analysis of Design Robustness

A SPICE model of the proposed design is developed to analyze whether the design has obtained the desired robustness and drain-source resonance frequency predictability in regards to the influence of parasitic inductance and capacitance on the drain-source resonance circuit. Three different SPICE model levels are created with increasing complexity; (*Model 1*) an ideal circuit model, (*Model 2*) a model including the parasitic inductance between the circuit elements and (*Model 3*) a model including the parasitic inductance and capacitance between the circuit elements. A 3D model of the circuit board is shown in Fig. 6 and imported into ANSYS Q3D Extractor to extract the mentioned parasitics. This method has in previous studies proven valuable in determining the parasitic influence on the performance for a given circuit design [19].

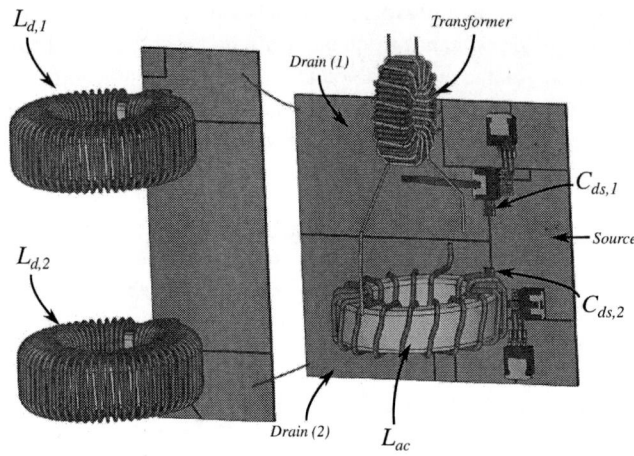

Fig. 6: 3D model of the built Class-E Push-Pull prototype converter.

The three SPICE models are simulated at an input voltage of 30 V and compared to experimental measurements as shown in Fig. 7.

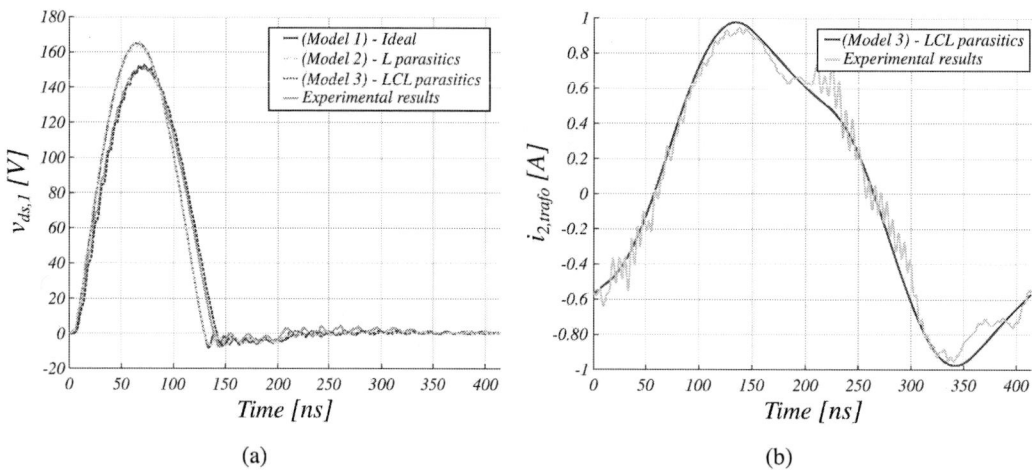

Fig. 7: Experimental and simulated waveforms with an input voltage of 30 V and a switching frequency of 2.415 MHz. (a) shows the drain-source voltage and (b) shows the secondary side transformer current.

It is observed that the drain-source voltage is similar for both SPICE model 1 and 2. This implies that the drain-source resonance loop in the present design is not influenced by the parasitic inductance between the circuit elements, implying that the first design consideration seen in (1) is satisfied. This outcome is expected as the extracted parasitic inductances between the circuit elements are between 10-30 nH which is 2 orders of magnitude smaller than the inserted stray inductance.

However, the two drain-source voltages from SPICE model 1 and 2 do not match the experimental drain-source voltage. A good similarity is obtained between SPICE model 3 and the experimental measurements, implying that the drain-source capacitance of the present design is influenced by the physical circuit board parasitic capacitance. The parasitic drain-source capacitance is mainly contributed from the capacitance to the bottom side of the PCB, as this is connected to the source plane. The size of the extracted equivalent parasitic drain-source capacitance is read to approximately 80 pF, which is in the same order of magnitude as the external drain-source capacitance of 440 pF. Thus, the constraint in (2) is not satisfied leading to a decrease in design robustness in regards to drain-source capacitance, which is also observed from the deviation between the drain-source voltage waveforms in Fig. 7.

Investigating the secondary side transformer current shown in Fig. 7 it is seen that a great resemblance between SPICE model 3 and the experimental measurements is obtained. This indicates that the impedance of both the transformer and the single turn induction coil is modelled adequately which together with the matching drain-source voltage waveforms enables the use of SPICE model 3 for performance predictability of future designs.

Concerning the design considerations in (1) and (2), the present design has only achieved complete robustness in regards to the parasitic inductance. Ideally the inserted inductance and capacitance should both have been order of magnitudes higher than the parasitic impedances for a fully robust design. Due to the size difference between the stray inductance and the parasitic inductance, it is a possibility for future designs to reduce the inserted stray inductance while increasing the external drain-source capacitance in order to achieve the desired drain-source capacitance robustness. Simultaneously the design constraint from (4) has to be considered in order to maintain a drain-source resonance frequency higher than the switching frequency to ensure ZVS. For the present design the current limitation of the MOSFET's has been a constraint which lead to a minimum allowable stray inductance. Due to this minimum allowable inductance it has not been possible to insert a sufficiently large drain-source capacitor while still satisfying the design constraint in (4) to ensure ZVS. Ultimately, for future designs a trade-off between robustness in terms of capacitance and inductance has to be made, while ensuring ZVS by satisfying the design constrain in (4). Additionally the current limitation of the MOSFET's has to be addressed, which will also aid in the scalability of the proposed design.

Conclusion

In this paper a proof-of-concept of a 4 kVA, 2.5 MHz prototype Class-E Push-Pull resonance converter for industrial induction heating has been demonstrated with an efficiency of 91.8 %. It is shown how the addition of a transformer and an external drain-source capacitor can provide load variation robustness and ZVS capability over a wide range of load resonance frequencies. A digital twin simulation based on parasitic extractions is showing a good agreement between experiment and simulation, which for future designs allows for predictability of the converter performance in new design domains.

References

[1] Topsil GlobalWafer A/S: Preferred Float Zone (PFZ) Silicon for Power Electrioncs, url: http://www.topsil.com/media/56273/pfz_application_notelong_version_september_2014.pdf, Application Note 2010

[2] Aunsborg T. S., Duun S. B., Uhrenfeldt C., Munk-Nielsen S.: Challenges and Opportunities in the Utilization of WBG Devices for Efficient MHz Power Generation, IECON 2019 - 45th Annual Conference of the IEEE Industrial Electronics Society, 14-17 October, Lisbon, Portugal, pp. 5107-5113

[3] Simon C., Eizaguirre S., Denk F., Heidinger M., Kling R., Heering W.: SiC 2.5 MHz Switching Mode Resonant Halfbridge Inverter, PCIM Europe 2018; International Exhibition and Conference for Power Electronics, Intelligent Motion, Renewable Energy and Energy Management, 5-7 June, Nuremberg, Germany, pp. 1644-1651

[4] Tomljenovic J.: 200 kW High Frequency Press for Dielectric Heating, url: https://www.plustherm.com/uploads/5/5/2/0/55207703/_____hes07tomljenovic.pdf

[5] Nair U. R., Munk-Nielsen S., Jørgensen A. B.: Performance Analysis of Commercial MOSFET Packages in Class E Converter Operating at 2.56 MHz, EPE 2017 ECCE Europe - 19th European Conference on Power Electronics and Applications, 11-14 September, Warsaw, Poland, pp. 1-9

[6] Gupta A., Arondekar Y., Ravindranath S. V. G., Krishnaswamy H., Jagatap B. N.: A 13.56 MHz High Power and High Efficiency RF Source, 2013 IEEE MTT-S International Microwave Symposium Digest, 2-7 June, Seattle, WA, USA, pp. 1-4

[7] Lucia O., Sarnago H., Burdío J: Design of Power Converters for Induction Heating Applications Taking Advantage of Wide-Bandgap Semiconductors, COMPEL - International Journal of Computations and Mathematics in Electrical and Electronic Engineering, pp. 483-488

[8] Baliga B. J.: Power Semiconductor Device Figure of Merit for High-Frequency Applications, IEEE Electron Device Letters, vol. 10, no. 10, pp. 455-457

[9] Denk F., Haehre K., Simon C., Eizaguirre S., Heidinger M., Kling R., Heering W.: 25 kW High Power Resonant Inverter Operating at 2.5 MHz based on SiC SMD Phase-leg Modules, PCIM Europe 2018; International Exhibition and Conference for Power Electronics, Intelligent Motion, Renewable Energy and Energy Management, 5-7 June, Nuremberg, Germany, pp. 1-7

[10] Guo S., Liu P., Yu R., Zhang L., Huang A. Q.: Analysis and Loss Comparison of Megahertz High Voltage Isolated DC/DC Converters Utilizing Integrated SiC MOSFET Module, 2016 IEEE WiPDA - 4th Workshop on Wide Bandgap Power Devices and Applications, 7-9 November, Fayetteville, AR, USA, pp. 291-296

[11] Ghodke D. V., Khachane P., Senecha V. K., Kulkarni V., Joshi S. C.: Simulation & Development of High Power Class-D, 2 MHz, 4 kW RF Source for RF based H-ion Source, 2016 ICDCS - 3rd International Conference on Devices, Circuits and Systems, 3-5 March, Coimbatore, India, pp. 10-13

[12] Choi J., Tsukiyama D., Rivas J.: Comparison of SiC and eGaN Devices in a 6.78 MHz 2.2 kW Resonant Inverter for Wireless Power Transfer, 2016 IEEE ECCE - Energy Conversion Congress and Exposition, 18-22 September, Milwaukee, WI, USA, pp. 1-6

[13] Aunsborg T. S., Duun S. B., Munk-Nielsen S., Uhrenfeldt C.: Development of a Current Source Resonant Inverter for High Current MHz Induction Heating, IET Power Electronics, vol. 15, no. 1, pp. 1-10

[14] Alonso J. M., Garcia J., Calleja A. J., Ribas J., Cardesin J.: Analysis, Design, and Experimentation of a High-Voltage Power Supply for Ozone Generation based on Current-fed Parallel-Resonant Push-Pull Inverter, IEEE Transactions on Industry Applications, vol. 41, no. 5, pp. 1364-1372

[15] Kazimierczuk M. K., Czarkowski D.: Resonant Power Converters, Wiley 2011

[16] Cree Inc.: C3M0160120J Datasheet, url: https://www.mouser.dk/datasheet/2/90/Cree_Inc_C3M0160120J-1846636.pdf, url-date: 2021-12-02

[17] IXYS Integrated Circuits Division: IXDN614YI Datasheet, url: https://www.ixysic.com/home/pdfs.nsf/www/IXD_614.pdf/$file/IXD_614.pdf, url-date: 2021-12-02

[18] Thrimawithana D. J., Madawala U. K.: Analysis of Split-Capacitor Push–Pull Parallel-Resonant Converter in Boost Mode, IEEE Transactions on Power Electronics, vol. 23, no. 1, pp. 359-368

[19] Jørgensen, A. B., Nair, U. R., Stig, M.N., Uhrenfeldt, C.: A SiC MOSFET Power Module With Integrated Gate Drive for 2.5 MHz Class E Resonant Converters, 2018 CIPS - 10th International Conference on Integrated Power Electronics Systems, 20-22 March, Stuttgart, Germany, pp. 128-133

Review of Power Converter Topologies for Electrochemical Impedance Spectroscopy of Lithium-Ion Batteries

Hamzeh Beiranvand[1,2], Julius M. Placzek[1], Marco Liserre[1,2]

1. Chair of Power Electronics, KIEL UNIVERSITY
2. Kiel Nano, Surface and Interface Science KiNSIS, KIEL UNIVERSITY
Kaiserstraße 2, 24143 Kiel
Kiel, Germany
Phone: +49 431 880-6100
Fax: +49 431 880-6103
Email: {hab,jmpl,frha,ml}@tf.uni-kiel.de
URL: http://www.pe.tf.uni-kiel.de

Giorgia Zampardi[3], Doriano Constantino Brogioli[3], Fabio La Mantia[3,4]

3. BREMEN UNIVERSITY
Bibliotheksstrasse 1, 28359 Bremen
4. Fraunhofer Institute for Manufacturing Technology and Advanced Materials – IFAM
Wiener Strasse 12, 28359 Bremen
Bremen, Germany
Phone: +49 421 2246-7331
Fax: +49 421 2246-300
Email: {zampardi,brogioli,lamantia}@uni-bremen.de
URL: https://www.esecs.uni-bremen.de/

Acknowledgment

Funded by the European Union - European Regional Development Fund (EFRE), the German Federal Government and the State of Schleswig-Holstein (LPW-E/1.1.2/1486).

Keywords

≪Battery≫, ≪Condition Monitoring≫, ≪Impedance Measurement≫, ≪DC-DC Converters≫, ≪DC-AC Converters≫.

Abstract

Frequency domain impedance of Li-ion batteries contains valuable information about the state of charge (SOC) and state of health (SOH). Normally, electrochemical impedance spectroscopy (EIS) is performed during the relaxation of battery cells. However, performing EIS during the batteries operation has been achieved through switching power converters. This paper reviews the power converter topologies for both online and offline Electrochemical Impedance Spectroscopy (EIS) characterization of batteries. The information that can be extracted from EIS Nyquist plots are discussed. Comparative analysis between converter topologies is presented. Finally, challenges are identified and new converter topologies are proposed for further consideration in online/offline EIS characterization.

I. Introduction

The demand for battery storage systems is increasing in many applications and precise parameter estimation is essential for optimal use of the remaining battery capacity. Even though machine learning

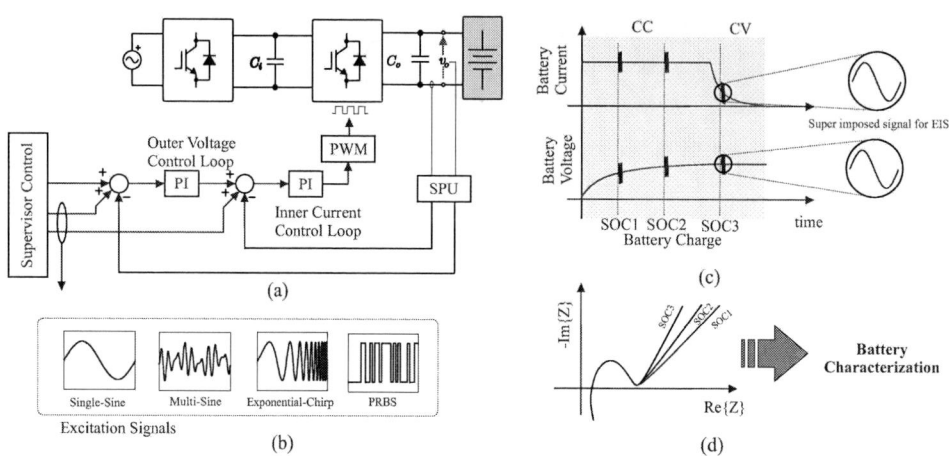

Fig. 1: Concept of EIS through power electronic converters: (a) circuit and control topology, (b) excitation signals, (c) superimposed signals, and (d) extracted EIS results

(ML) techniques significantly improve the parameter estimation of batteries, nevertheless, the capacity fade is quite nonlinear and regular adjustments of the estimator algorithms are required. Electrochemical Impedance Spectroscopy (EIS) is proven to be an effective and comprehensive measurement technique to supply the required data for adjusting the estimator's parameters [1, 2].

Normally, EIS is achieved using linear amplifiers [3] when the battery is relaxed. Therefore, a high-precision EIS can be performed in laboratory environment [4]. Conversely, battery storage systems are interfaced with a load/grid by power electronic converters in the respective applications of electric vehicles (EVs) and stationary storage systems. Power electronic converters have sufficiently flourished in the last decade both in terms of topology and semiconductor technology. On one hand, modular and non-modular converter topologies have emerged in many applications, on the other hand, wide bandgap semiconductor technologies such as SiC and GaN can reach high switching frequencies without sacrificing efficiency. These advancements make power converters a promising tool for performing EIS. However, power converters are not adequately developed for EIS purposes [5]. A comprehensive review is required to unveil the obstacles and draw a road-map for developing EIS as an embedded function in power converters. Although reviews on the state-of-the-art of EIS methodologies are present (see for example [6]), a comprehensive perspective and road-map is still missing.

This paper reviews the current status and addresses the challenges and future developments of power electronic converters for EIS. Frequency-dependent electrochemical behavior of batteries, power converter topologies, control strategies, and signal processing for EIS are compared. Challenges originated from application types are identified at battery cell, module and pack levels. New possibilities to obtain online EIS by power electronic converters are described such as partial power processing, isolated multiport and modular topologies.

II. Principles of EIS and Applications

Conventionally, EIS is done by applying a periodic signal current with known frequency contents and measure its voltage response where the Li-ion battery impedance is defined by $Z(\omega) = \frac{V(\omega)}{I(\omega)}$. Excitation signals are applied to the battery through a linear amplifier [3] while the battery is relaxed (i.e. disconnected from source/load) to achieve EIS. Switching power semiconductors can also be used for EIS as shown in Fig. 1 [5]. Subfigures (a) to (d) show a conventional control strategy, excitation signals, voltage and current profile during charging, and EIS graph, respectively. The signal processing unit (SPU) measures current and voltage, filters and converts signals, and extracts $Z(\omega)$ for battery characterization in the controller.

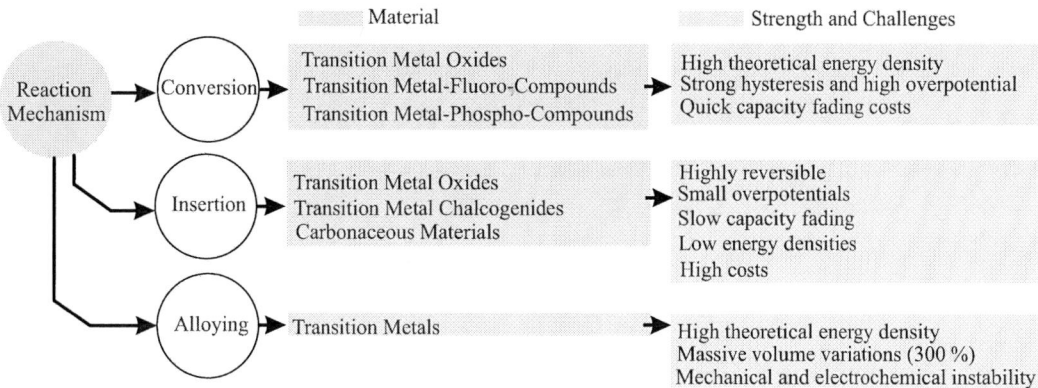

Fig. 2: Different reaction mechanisms, possible battery material and characteristics.

II.1 Electrochemistry

The electro-chemical response of batteries to an excitation depends on their individual chemistry which can be characterized by $Z(\omega)$ obtained from EIS. Li-ion batteries work following the three main reaction mechanisms insertion, conversion and alloying. Insertion reactions allow the accommodation of Li-ions within a host material, whose crystalline structure does not change significantly with the lithium content. In general, insertion reactions are highly reversible, and show very little overpotential and slow capacity fading [7]. Conversion reactions, are displacement reactions where the lithium ions take the place of a species of a binary compound [8]. This kind of reaction mechanism leads to the formation of new phases and it is often characterized by a strong hysteresis in the charge/discharge profile and by a quick capacity fading [9]. Alloying reactions involve a significant change of the original microstructure of the electrode, thus leading to the formation of new phase(s) in dependence of the lithium content (and therefore the potential applied to the electrode) [8]. Metals such as Si, Sn and Sb working through alloying mechanism are typically used as negative electrodes (i.e. anodes) in a Li-ion battery, and upon cycling they go through massive volume variations, higher than 300 % [8, 10]. Fig. 2 summarizes the reaction mechanisms along with their involved materials and advantages/disadvantages.

II.2 EIS Measurement

It is shown in [11] that EIS and other measurement techniques are identical and EIS provides more comprehensive results. Fig. 3 (a) shows the DC pulse test where different components of the resistance inducing from electronic and ionic resistances, charge transfer resistance, and polarization resistance which are shown by R_O, R_{CT}, and R_P respectively. These variables can be also directly obtained from the EIS impedance map as depicted in Fig. 3 (b). Moreover, conventional AC tests which are mainly carried out 1 kHz can be considered as a subset of EIS. Therefore, EIS is the most comprehensive impedance characterization technique for batteries.

A correct measurement of EIS requires special care about battery cell connections as well as the signal to noise ratio (SNR) and the nonlinear behavior of the battery cell. To minimize the error and avoid extra impedance of the connections, the EIS device shall be as close as possible to the battery [18]. Usually, a 4 terminal connection or Kelvin connection results in minimum external impedance [19] at cell level. An optimum current excitation should results in 10 mV voltage amplitude [20]. This selection is a trade off between SNR and non-linearity of the electrochemical reaction as illustrated in Fig. 3 (c). To achieve this voltage amplitude, excitation current should be applied considering the internal impedance of the battery. Fig. 3 (d) shows the required excitation current for NiMnCo (NMC) chemistry versus different battery capacities. It can be observed that the required excitation current amplitude increases linearly proportional to the battery cell capacity.

II.3 EIS Applications

EIS can be used for many purposes in the BMS and as well as battery analysis and modeling [21]. Normally EIS is performed when the cell is relaxed and the Gibbs energy approaches its minimum [22].

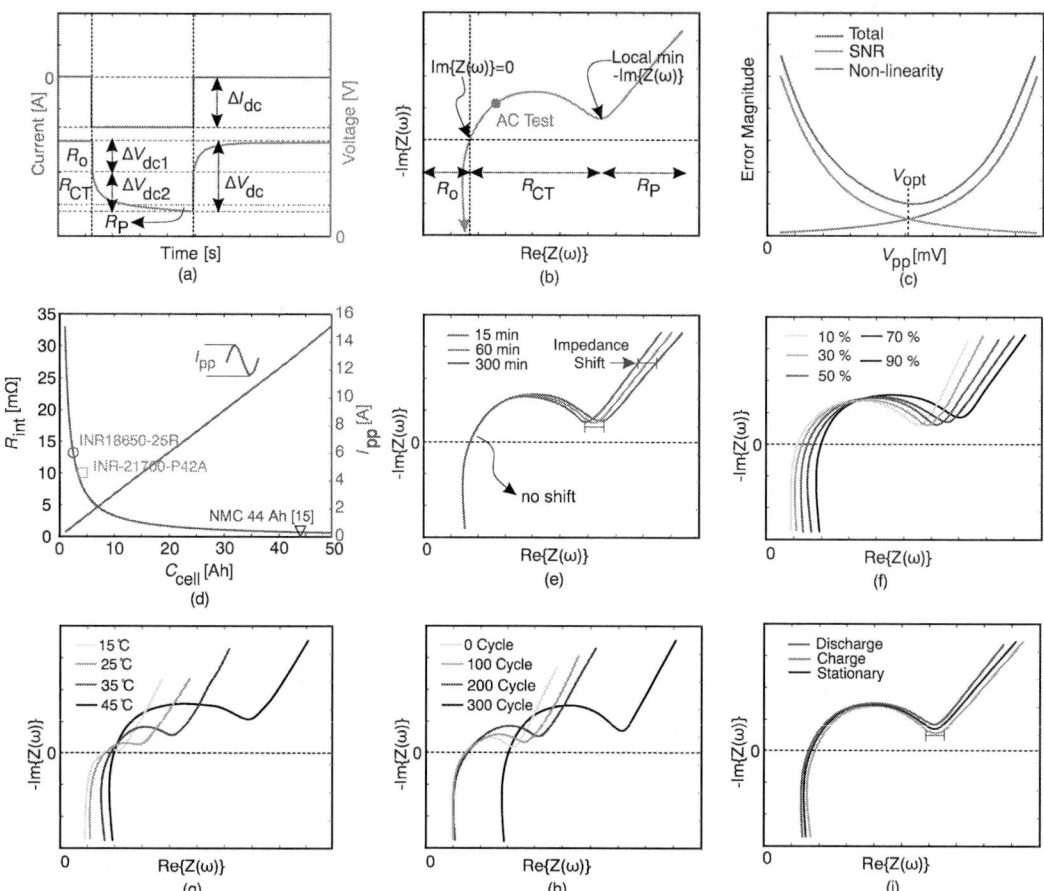

Fig. 3: EIS measurement: (a) DC pulse current for measuring the internal impedance [11], (b) EIS graph in complex coordinate system and an exemplary AC test point [11], (c) optimum voltage magnitude as a compromise between SNR and nonlinearity, (d) required excitation current for Li-ion NMC chemistry, (e) effect of the rest time on the EIS [12, 13], (f) SOC influence on the NMC cell impedance spectra at the same temperature [14], (g) deformation of the EIS Nyquist as a function of the temperature [14], (h) aging of the NMC cells [15, 16], and (i) DC offset during online EIS can displace the EIS Nyquist plot depending on the cell being charged or discharged [17].

Duration of the rest time causes an impedance shift to the right side at lower frequencies around 1 Hz [12, 13]. However, at high frequencies near 1 kHz, the impedance behaves independent to the rest time. The effect of the rest time on the EIS Nyquist plot is shown in Fig. 3 (e).

A preliminary requirement of any BMS is the SOC estimation [23]. Conventionally, EIS data are fitted to an equivalent circuit model (ECM). If EIS is performed online, then the equivalent circuit can be used for online SOC and SOH estimation employing Kalman filter techniques [1, 24–27]. A review of SOC estimation techniques can be found in [23, 28]. Impacts of the SOC on the EIS Nyquist plot are shown in Fig. 3 (f) [14]. There is a shift toward right side and up as the SOC increases. The impedance spectra become more irregular at SOC equal to 0 and 100 %. Battery impedance highly depends on the battery temperature and therefore it should be considered in SOC estimation. EIS Nyquist plots easily reflect this dependency and specially at low temperature the shape of the graph deforms such that the charge transfer region disappears gradually (see Fig. 3 (g)) [2].

Similar to SOC, variation of the EIC spectra during the aging of an NMC cell is shown in Fig. 3 (h) [15, 16]. Aging not only shifts the spectra toward the up right side of the complex plain but also deforms the shape of the Nyquist plot at particularly low frequencies. And finally, online EIS or dynamic EIS spectra differ from the stationary EIS due to the DC offset and as well as zero rest time [17]. Depending

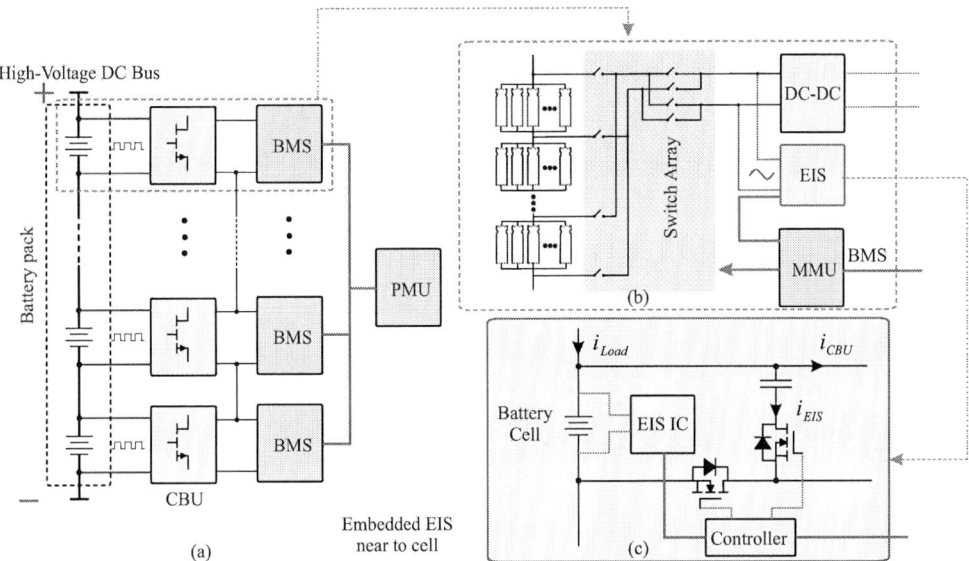

Fig. 4: Online EIS Implementation at cell level [18]: (a) battery pack structure of an EV, (b) battery module and its BMS including EIC subsystem, (c) Circuit topology enabling online EIS

on the state of the cell operation (charging/discharging) the EIS spectra can be slightly higher/lower than that of stationary EIS as demonstrated in Fig. 3 (i). It can be concluded that EIS is a solution for monitoring and state estimation of lithium ion batteries.

III. Power Converter Topologies

EIS can be done by BMS or power converter. The general structure of a battery pack comprising of multiple battery modules each equipped with a BMS is depicted in Fig. 4 (a). Fig. 4 (b) demonstrates how online EIS can be implemented as close as possible to the battery cell [18] to minimize the unwanted additional impedance. An integrated circuit (IC) for online EIS is proposed in [18] and is shown in Fig. 4 (c). Since EIS might be carried out at very short time intervals, the required high current can be provided by a supercapacitor. In [29], an online EIS embedded in the BMS is proposed using minimal hardware where the importance of the correct measurement and signal processing is highlighted. Nonetheless, the proposed method is implemented into the passive cell balancing and its utilization in the active cell balancing still remains questionable.

Normally, a battery pack can be connected to grid or the drive train of EVs using a power electronic converter which might comprise of one or multiple conversion stages. DC-AC and DC-DC converters are frequently used as excitation sources for EIS as well as conversion stages. Also, power electronic converters can be separately used to excite batteries for EIS as shown in Fig. 5 (a). Galvanic isolation is required in many battery storage applications and isolated DC-DC converters can be used for EIS as in Fig. 5 (b) [30]. The AC-DC converter in Fig. 5 (c) achieves EIS by superimposing the excitation signals on the controller reference [31]. Similarly, different non-isolated DC-DC converters perform EIS as shown in Fig. 5 (d) to (g) [5, 32–35]. Most of these researches are focusing on performing EIS on one cell, which limits their usage in a grid or EV battery application. Moreover, the possibility to scale up the proposed EIS to module or pack level is not clarified. Table I provides a brief overview of power electronic converters used for EIS.

In [32], the EIS excitation signal is added to the modulation directly in a feed-forward mode. This strategy might result in instability for high excitation currents comparable to converter rated current. Injecting excitation signal through converters reference signals is the most common way in the literature.

VI. Challenges and Opportunities

This section describes the associated challenges and opportunities with EIS.

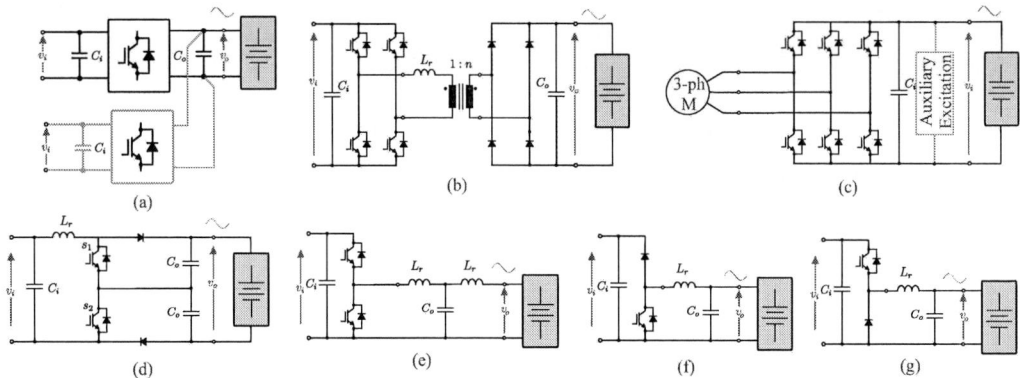

Fig. 5: Converter topologies: (a) separate excitation concept, (b) full-bridge, (c) three-phase drive, (d) three-level DC-DC, (e) synchronous, (f) boost, and (g) buck converters.

VI.1 Challenges in EIS Measurements

EIS has a challenging nature as a consequence of its sensitivity to many parameters as discussed. From measurement point of view, offline EIS is operated after relaxing the battery cell. EIS at low frequency signals in range of mHz is time consuming. Therefore, EIS can only be done in fewer occasions during the battery operational life. Conversely, for online EIS, the accuracy might be impacted and the power converter design and control become complex [36, 37]. Moreover, impedance measurement behind the internal voltage is challenging in comparison to a passive component, because the impedance is very small and needs a large injected current to raise the voltage across the impedance to an acceptable signal-to-noise ratio (SNR) [4].

From the hardware point of view, EIS of battery packs and modules where many cells are connected in series and parallel is challenging. Even if the EIS is isolated from the rest of the cells, the paralleled cells result in smaller sensible voltage rise and consequently a high amount of AC current is required [37]. The extra impedance of terminals, connections and cell balancing system impact on the measured impedance. Therefore, the EIS unit must be as close as possible to cells.

Another main challenge is the sensitivity of the battery cell impedance to operating conditions. The EIS graph changes versus temperature, SOC and SOH and suitable estimation methodologies are required to extract the correct information from the acquired EIS [26].

VI.2 Challenges of EIS Implementation in Power Converters

The mostly used power converters for EIS are non-isolated DC-DC converters owing to their simplicity and minimal requirements. Nevertheless, applying a variable voltage and current to the battery might impact the consumer. Therefore, extra circuits are required for compensating the impact of online EIS [38].

Table I: Comparison of converter topologies for EIS.

Ref.	Converter type	EIS level	Signal Type	Scalability	N. Semiconductors
[5]	DC-DC Synchronous	Cell	Control Ref.	No	2
[18]	DC-DC Cuk	Cell	Control Ref.	Yes	2
[29]	DC-DC Chopper	Cell	Control Ref.	Yes	1
[31]	AC-DC 3-phase	Pack	Multisine/Noise	Yes	6
[32]	DC-DC 3-level Boost	Module	Feed-forward	No	2
[33]	DC-DC Synchronous	Cell	Control Ref.	No	2
[34]	DC-DC Boost	Cell	Control Ref.	No	1
[35]	DC-DC Buck	Cell	Control Ref.	No	1
[36]	Ladder Converter	Module	Control Ref.	Yes	$n+1$ per n cells

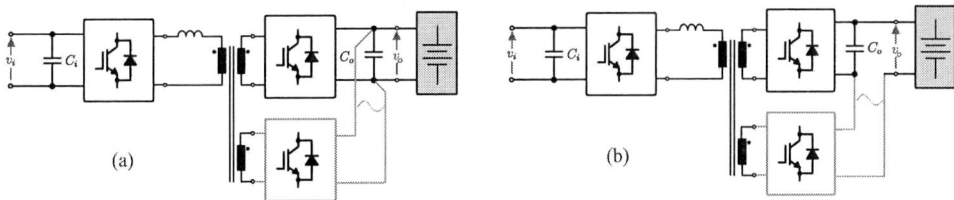

Fig. 6: EIS using isolated multiport DC-DC converters: (a) parallel connection and (b) serial connection.

EIS usage has been limited to laboratory as well as low voltage applications [4]. While in EVs, the pack voltage can be higher than 800V [39] and high voltage converters are required for excitation of EIS. Imposing an AC signal on the base DC current, increases the peak current level in the switches and batteries and therefore the used semiconductors might be overrated.

For a cell connected inside the battery pack, online EIS is influenced by the connected circuits and thereby the measured results need to be compensated [36] or extra circuitry must be utilized to isolate each cell during EIS as in [18]. Since the degradation of the cells could be different, the most promising way is to apply EIS to individual cells or parallel connected cells in modules which results in higher final cost of the product.

Emerging battery technologies, such as solid-state silicon batteries in [40], easily achieve 20 C-rate without significantly losing capacity after 20,000 cycles. Therefore, such a battery storage can be charged and discharged 10 times per hour which leads to maximum 14,400 cycles per year. In such a system, reliability of the power electronic converter becomes a concern rather than the battery itself due the large thermal cycling stress imposed on the semiconductors.

VI.3 Opportunities

Emerging WBG devices allow for high frequency switching without sacrificing the efficiency. Therefore, EIS can be achieved to reasonably high frequencies in the range of hundreds of kHz for a short period of time. Combined with signal processing techniques, up to MHz can be obtained. The AC excitation unit for EIS can be separated from the charging unit. Such a topology can be realized by partial processing converters with maximum efficiency and power density [41, 42].

Galvanic isolation is mandatory in many battery storage applications. In isolated topologies, multiport isolated DC-DC converters provide numerous advantages for realizing online/offline EIS with minimum required components and highest possible safety and fault tolerance [43]. Fig. 6 shows possible configuration of multiport isolated converters for conducting EIS.

Lithium batteries are sensitive to temperature variation. In such a case, multiport isolated topologies can be utilized to achieve electrochemical-thermal impedance spectroscopy as the AC component can implement internal heating and therefore an external heater can be omitted [44].

Modular and multilevel converter topologies [45], such as modular multilevel converter (MMC) and cascaded H-bridge (CHB), can contribute to the field by implementing independent EIS function in the sub-modules with minimum interaction with nearby sub-modules. For instance, a CHB converter can be used as an EV drive train which has a charging/discharging current containing both low and high order harmonics as shown in Fig. 7. Since the same current flows through all the cells, EIS can be achieved utilizing suitable voltage measurement and signal processing unit for each cell [46].

V. Conclusion

Electrochemical impedance spectroscopy (EIS) techniques extract the frequency domain features of the device under test. The frequency response of Li-ion batteries contains SOC and SOH information which can be strongly used in condition monitoring and improved state-estimation of batteries. This paper reviews the literature on power electronic converters used for EIS, especially for Li-ion batteries. Moreover, a brief explanation of the EIS Nyquist plots which can be used in battery parameter estimation are presented for NMC chemistry. Comparative studies among different converter topologies are carried

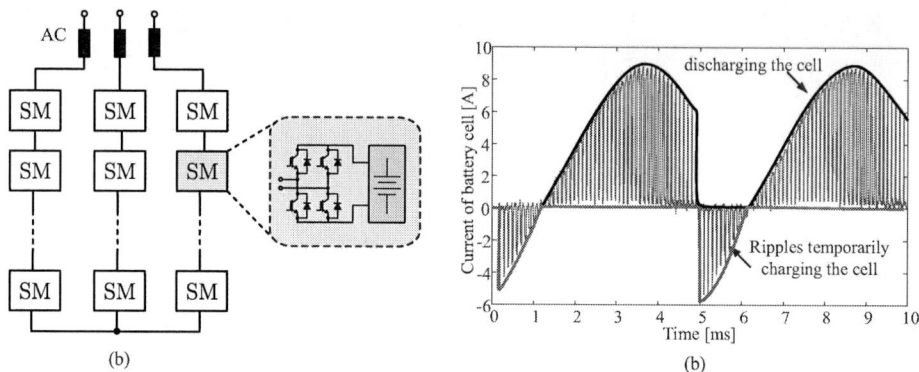

(b)   (b)

Fig. 7: CHB converter operation for EV application: (a) topology, (b) current waveform containing hamonics which can online EIS [46].

out. Challenges and opportunities for utilizing power converters as EIS perturbation source are outlined. Multiple issues of the already utilized converters are addressed. New multifunctional converter topologies are introduced which can improve the performance of EIS by implementing internal heating and EIS of Li-ion batteries in the same time.

References

[1] A. Guha and A. Patra, "Online estimation of the electrochemical impedance spectrum and remaining useful life of lithium-ion batteries," *IEEE Transactions on Instrumentation and Measurement*, vol. 67, no. 8, pp. 1836–1849, 2018.

[2] D. Li, L. Wang, C. Duan, Q. Li, and K. Wang, "Temperature prediction of lithium-ion batteries based on electrochemical impedance spectrum: A review," *International Journal of Energy Research*, 2022.

[3] U. Troeltzsch and O. Kanoun, "Miniaturized impedance measurement system for battery diagnosis," *Proceedings SENSOR 2009, Volume I*, pp. 251–256, 2009.

[4] H. Homayouni, J. DeVaal, F. Golnaraghi, and J. Wang, "Voltage reduction technique for use with electrochemical impedance spectroscopy in high-voltage fuel cell and battery systems," *IEEE transactions on transportation electrification*, vol. 4, no. 2, pp. 418–431, 2018.

[5] R. Koch, R. Kuhn, I. Zilberman, and A. Jossen, "Electrochemical impedance spectroscopy for online battery monitoring-power electronics control," in *2014 16th European Conference on Power Electronics and Applications*. IEEE, 2014, pp. 1–10.

[6] M. A. Varnosfaderani and D. Strickland, "A comparison of online electrochemical spectroscopy impedance estimation of batteries," *IEEE Access*, vol. 6, pp. 23 668–23 677, 2018.

[7] M. Winter, J. O. Besenhard, M. E. Spahr, and P. Novak, "Insertion electrode materials for rechargeable lithium batteries," *Advanced materials*, vol. 10, no. 10, pp. 725–763, 1998.

[8] R. Huggins, *Advanced batteries: materials science aspects*. Springer Science & Business Media, 2008.

[9] R. Malini, U. Uma, T. Sheela, M. Ganesan, and N. Renganathan, "Conversion reactions: a new pathway to realise energy in lithium-ion battery," *Ionics*, vol. 15, no. 3, pp. 301–307, 2009.

[10] V. Kuznetsov, A.-H. Zinn, G. Zampardi, S. Borhani-Haghighi, F. La Mantia, A. Ludwig, W. Schuhmann, and E. Ventosa, "Wet nanoindentation of the solid electrolyte interphase on thin film si electrodes," *ACS applied materials & interfaces*, vol. 7, no. 42, pp. 23 554–23 563, 2015.

[11] A. Barai, K. Uddin, W. Widanage, A. McGordon, and P. Jennings, "A study of the influence of measurement timescale on internal resistance characterisation methodologies for lithium-ion cells," *Scientific reports*, vol. 8, no. 1, pp. 1–13, 2018.

[12] M. S. Hosen, R. Gopalakrishnan, T. Kalogiannis, J. Jaguemont, J. Van Mierlo, and M. Berecibar, "Impact of relaxation time on electrochemical impedance spectroscopy characterization of the most common lithium battery technologies—experimental study and chemistry-neutral modeling," *World Electric Vehicle Journal*, vol. 12, no. 2, p. 77, 2021.

[13] M. Messing, T. Shoa, and S. Habibi, "Electrochemical impedance spectroscopy with practical rest-times for battery management applications," *IEEE Access*, vol. 9, pp. 66 989–66 998, 2021.

[14] T. Stanciu, D.-I. Stroe, R. Teodorescu, and M. Swierczynski, "Extensive eis characterization of commercially available lithium polymer battery cell for performance modelling," in *2015 17th European Conference on Power Electronics and Applications (EPE'15 ECCE-Europe)*. IEEE, 2015, pp. 1–10.

[15] A. Maheshwari, M. Heck, and M. Santarelli, "Cycle aging studies of lithium nickel manganese cobalt oxide-based batteries using electrochemical impedance spectroscopy," *Electrochimica Acta*, vol. 273, pp. 335–348, 2018.

[16] Y. Zhang, Q. Tang, Y. Zhang, J. Wang, U. Stimming, and A. A. Lee, "Identifying degradation patterns of lithium ion batteries from impedance spectroscopy using machine learning," *Nature communications*, vol. 11, no. 1, pp. 1–6, 2020.

[17] J. Huang, Z. Li, and J. Zhang, "Dynamic electrochemical impedance spectroscopy reconstructed from continuous impedance measurement of single frequency during charging/discharging," *Journal of Power Sources*, vol. 273, pp. 1098–1102, 2015.

[18] Z. Gong, Z. Liu, Y. Wang, K. Gupta, C. Da Silva, T. Liu, Z. Zheng, W. Zhang, J. M. van Lammeren, H. Bergveld *et al.*, "Ic for online eis in automotive batteries and hybrid architecture for high-current perturbation in low-impedance cells," in *2018 IEEE Applied Power Electronics Conference and Exposition (APEC)*. IEEE, 2018, pp. 1922–1929.

[19] N. Meddings, M. Heinrich, F. Overney, J.-S. Lee, V. Ruiz, E. Napolitano, S. Seitz, G. Hinds, R. Raccichini, M. Gaberšček *et al.*, "Application of electrochemical impedance spectroscopy to commercial li-ion cells: A review," *Journal of Power Sources*, vol. 480, p. 228742, 2020.

[20] N. Lohmann, P. Weßkamp, P. Haußmann, J. Melbert, and T. Musch, "Electrochemical impedance spectroscopy for lithium-ion cells: Test equipment and procedures for aging and fast characterization in time and frequency domain," *Journal of Power Sources*, vol. 273, pp. 613–623, 2015.

[21] U. Westerhoff, K. Kurbach, F. Lienesch, and M. Kurrat, "Analysis of lithium-ion battery models based on electrochemical impedance spectroscopy," *Energy Technology*, vol. 4, no. 12, pp. 1620–1630, 2016.

[22] D. del Olmo, M. Pavelka, and J. Kosek, "Open-circuit voltage comes from non-equilibrium thermodynamics," *Journal of Non-Equilibrium Thermodynamics*, vol. 46, no. 1, pp. 91–108, 2021.

[23] D. N. How, M. Hannan, M. H. Lipu, and P. J. Ker, "State of charge estimation for lithium-ion batteries using model-based and data-driven methods: A review," *Ieee Access*, vol. 7, pp. 136 116–136 136, 2019.

[24] L. Ran, W. Junfeng, W. Haiying, and L. Gechen, "Prediction of state of charge of lithium-ion rechargeable battery with electrochemical impedance spectroscopy theory," in *2010 5th IEEE Conference on Industrial Electronics and Applications*. IEEE, 2010, pp. 684–688.

[25] J. Zhang, P. Wang, Y. Liu, Z. Cheng *et al.*, "Variable-order equivalent circuit modeling and state of charge estimation of lithium-ion battery based on electrochemical impedance spectroscopy," *Energies*, vol. 14, no. 3, p. 769, 2021.

[26] I. Babaeiyazdi, A. Rezaei-Zare, and S. Shokrzadeh, "State of charge prediction of ev li-ion batteries using eis: A machine learning approach," *Energy*, vol. 223, p. 120116, 2021.

[27] A. La Rue, P. J. Weddle, M. Ma, C. Hendricks, R. J. Kee, and T. L. Vincent, "State-of-charge estimation of lifepo4–li4ti5o12 batteries using history-dependent complex-impedance," *Journal of The Electrochemical Society*, vol. 166, no. 16, p. A4041, 2019.

[28] R. Xiong, J. Cao, Q. Yu, H. He, and F. Sun, "Critical review on the battery state of charge estimation methods for electric vehicles," *Ieee Access*, vol. 6, pp. 1832–1843, 2017.

[29] M. Koseoglou, E. Tsioumas, D. Papagiannis, N. Jabbour, and C. Mademlis, "A novel on-board electrochemical impedance spectroscopy system for real-time battery impedance estimation," *IEEE Transactions on Power Electronics*, vol. 36, no. 9, pp. 10 776–10 787, 2021.

[30] A. Narjiss, D. Depernet, D. Candusso, F. Gustin, and D. Hissel, "On-line diagnosis of a pem fuel cell through the pwm converter," *Proceedings of FDFC 2008*, 2008.

[31] D. A. Howey, P. D. Mitcheson, V. Yufit, G. J. Offer, and N. P. Brandon, "Online measurement of battery impedance using motor controller excitation," *IEEE transactions on vehicular technology*, vol. 63, no. 6, pp. 2557–2566, 2013.

[32] O. M. Faloye and P. Barendse, "A three level dc-dc converter for battery impedance spectroscopy," in *2019 IEEE Energy Conversion Congress and Exposition (ECCE)*. IEEE, 2019, pp. 2682–2689.

[33] T.-T. Nguyen, V.-L. Tran, and W. Choi, "Development of the intelligent charger with battery state-of-health estimation using online impedance spectroscopy," in *2014 IEEE 23rd International Symposium on Industrial Electronics (ISIE)*. IEEE, 2014, pp. 454–458.

[34] M. A. Varnosfaderani and D. Strickland, "Online impedance spectroscopy estimation of a battery," in *2016 18th European Conference on Power Electronics and Applications (EPE'16 ECCE Europe)*. IEEE, 2016, pp. 1–10.

[35] E. Sadeghi, M. H. Zand, M. Hamzeh, M. Saif, and S. M. M. Alavi, "Controllable electrochemical impedance spectroscopy: From circuit design to control and data analysis," *IEEE Transactions on Power Electronics*, vol. 35, no. 9, pp. 9933–9942, 2020.

[36] E. Din, C. Schaef, K. Moffat, and J. T. Stauth, "A scalable active battery management system with embedded real-time electrochemical impedance spectroscopy," *IEEE Transactions on Power Electronics*, vol. 32, no. 7, pp. 5688–5698, 2016.

[37] Z. Gong, B. A. C. van de Ven, K. M. Gupta, C. da Silva, C. H. Amon, H. J. Bergveld, M. C. F. T. Donkers, and O. Trescases, "Distributed control of active cell balancing and low-voltage bus regulation in electric vehicles using hierarchical model-predictive control," *IEEE Transactions on Industrial Electronics*, vol. 67, no. 12, pp. 10 464–10 473, 2020.

[38] N. Katayama and S. Kogoshi, "Real-time electrochemical impedance diagnosis for fuel cells using a dc–dc converter," *IEEE Transactions on Energy Conversion*, vol. 30, no. 2, pp. 707–713, 2015.

[39] A. Poorfakhraei, M. Narimani, and A. Emadi, "A review of multilevel inverter topologies in electric vehicles: Current status and future trends," *IEEE Open Journal of Power Electronics*, vol. 2, pp. 155–170, 2021.

[40] L. Ye and X. Li, "A dynamic stability design strategy for lithium metal solid state batteries," *Nature*, vol. 2, pp. 155–170, 2021.

[41] H. Beiranvand, F. Hoffmann, F. Hahn, and M. Liserre, "Impact of partial power processing dual-active bridge converter on li-ion battery storage systems," in *2021 IEEE Energy Conversion Congress and Exposition (ECCE)*, 2021, pp. 538–545.

[42] F. Hoffmann, J. Person, M. Andresen, M. Liserre, F. D. Freijedo, and T. Wijekoon, "A multiport partial power processing converter with energy storage integration for ev stationary charging," *IEEE Journal of Emerging and Selected Topics in Power Electronics*, pp. 1–1, 2021.

[43] T. Pereira, F. Hoffmann, R. Zhu, and M. Liserre, "A comprehensive assessment of multiwinding transformer-based dc–dc converters," *IEEE Transactions on Power Electronics*, vol. 36, no. 9, pp. 10 020–10 036, 2021.

[44] X. Hua, Y. Zhenga, D. A. Howeyb, H. Perezc, A. Foleyd, and M. Pechte, "Battery warm-up methodologies at subzero temperatures for automotive applications: Recent advances and perspectives," *Progress in Energy and Combustion Science*, vol. 77, p. 100806, 2020.

[45] A. Kersten, M. Kuder, W. Han, T. Thiringer, A. Lesnicar, T. Weyh, and R. Eckerle, "Online and on-board battery impedance estimation of battery cells, modules or packs in a reconfigurable battery system or multilevel inverter," in *IECON 2020 The 46th Annual Conference of the IEEE Industrial Electronics Society*, 2020, pp. 1884–1891.

[46] F. Chang, F. Roemer, and M. Lienkamp, "Influence of current ripples in cascaded multilevel topologies on the aging of lithium batteries," *IEEE Transactions on Power Electronics*, vol. 35, no. 11, pp. 11 879–11 890, 2020.

Design and experimental validation of a Voltage Sensing-Current Cancellation Common Mode Linear Active Filter

B.Mohamed Nassurdine[1,3,4], PE.Lévy[1], D.Labrousse[2], JL.Schanen[3], X.Maynard[4], S.Carcouet[5]

[1] Univ. Paris-Saclay, ENS-Paris Scalay, SATIE, F-91190 Gif-sur-Yvette, France
[2] Le Cnam, SATIE, UMR 8029, F-75003 Paris, France, HESAM Université
[3] Univ. Grenoble Alpes, CNRS, Grenoble INP, G2Elab, F-38000 Grenoble, France
[4] CEA-Liten, F-38000 Grenoble, France
[5] CEA-Leti, F-38000 Grenoble, France
E-Mail : bacar.mohamed_nassurdine@ens-paris-saclay.fr, pierre-etienne.levy@ens-paris-saclay.fr, denis.labrousse@satie.ens-cachan.fr, jean-luc.schanen@grenoble-inp.fr, xavier.maynard@cea.fr, sebastien.carcouet@cea.fr

Keywords

«DC-DC converter», «Electromagnetic Interference (EMI) », «Active filter», «Boost», «Switched-mode power supply»

Abstract

In recent years, the study and design of linear active EMI filters (AEF) have been the subject of many research papers. Different modeling approaches to study the performances of the AEF have been proposed. The aim of this paper is to introduce a methodology to design a Common Mode (CM) AEF able to cancel the CM electromagnetic interference noise in a given frequency range. The active filter is analyzed based on the CM noise model of a boost converter. The first part of the study is devoted to the modeling of the electrical structure of the filter and the definition of component values. In this part, attenuation, design rules and stability of AEF are addressed in detail. The second part of the study is dedicated to the simulations and experiments to validate the proposed methodology. This step allows realizing a functional prototype that complies with the DO160 standard (aeronautics) up to 1 MHz

I. Introduction

Nowadays, our daily life can no more be dissociated from the use of Switched-Mode Power Supply (consumer electronics, transportation …). These converters are sources of electromagnetic interferences (conducted and radiated) due to the high gradients of voltages and currents [1], [2]. With the coexistence of various electronic apparatus within the same system, Electromagnetic Compatibility (EMC) has become a major issue for manufacturers and users.

Usually, EMC conformity including aeronautic conformity (DO160 standard) in the 'RF' (Radio Frequency) range (150 kHz – 30 MHz) can be met using passive filters, for differential-mode (DM) and common-mode (CM) noise. However, passive filters are quite expensive and bulky, and may represent up to 30% of the cost, weight, and volume of a power converter . Therefore, hybrid filtering (active plus passive) appears to be a promising alternative to reduce the size, weight, and cost of converters. A volume reduction greater than 40% has been reported in [3],[4]. In hybrid filtering, low frequency interferences are filtered by AEF and high frequency interferences by passive filter. The principle of AEF consists of detecting an output quantity (voltage or current) at the converter's input and comparing it to a zero reference (ground). The error is processed in an analog- (Op amp) or digital- (DSP or FPGA) way to obtain the necessary gain to reject noises. In the literature, two main structures of AEF have been reported: the feedback and the feedforward topologies. A comparative analysis of these two topologies was made in [5] and it was concluded that the feedback topologies is more easily controllable than the feedforward. Indeed, for efficient filtering, the feedforward gain must be equal to one [3], [5]: this condition is hard to achieve due to the non-linearity and parasitic elements of the passive components used for the detection and injection in the AEF. On the other hand, four feedback topologies are available depending on the sensing and the cancelling methods.

Noise sensing topologies involve noise current and noise voltage sensing. Current sensing uses a very high bandwidth current transformer, which can become bulky and can also saturate depending on noise level, contrary to voltage sensing, which requires only RC circuit. The current sensing however has the interest of galvanic isolation. Noise injection techniques can use either current or voltage injection. In current cancellation, the active circuit drives a voltage across an injection capacitor resulting in cancellation current injected in the circuit. In voltage cancellation, the active circuit drives a voltage injection transformer that injects the cancellation voltage in the main circuit. Some of the commonly used noise sensing and noise injecting topologies are given in [6] and shown in Fig.1. In this figure, the CM noise generation of the converter is represented using a Norton equivalent circuit. An experimental study in [6] shown that current injection topologies provide better attenuation than voltage injection for CM noise filtering. On the other hand, voltage injection attenuates more than current injection topologies for DM noise filtering. An analytical study validates these results.

As current injection topologies provide better attenuation than voltage injection for common mode filtering and taking into account its simplicity and lightness, a voltage sensing-current cancellation common mode topology (VSCC) has been chosen and studied in this paper for a boost power converter (14V/42V - 115 kHz – 27W).

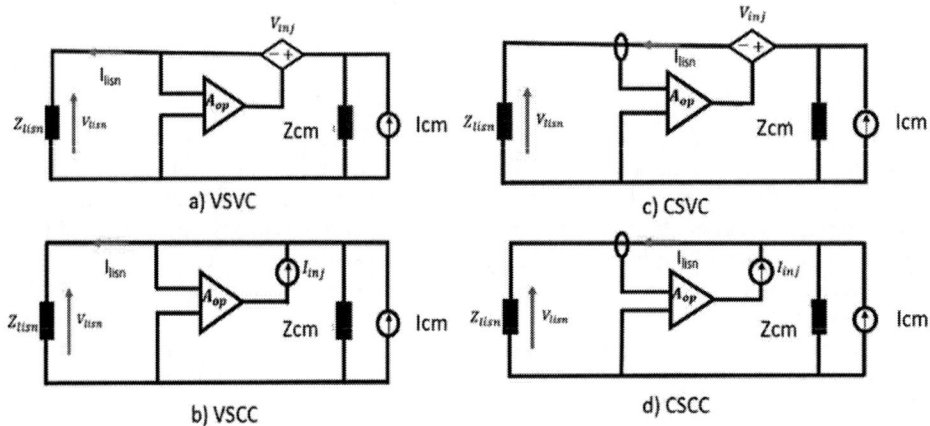

Fig. 1: Four active filter topologies. (a) Voltage sensing-voltage cancellation (VSVC). (b) Voltage sensing-current cancellation (VSCC). (c) Current sensing-voltage cancellation (CSVC). (d) Current sensing-current cancellation (CSCC)

II. VSCC AEF Analysis and modeling

The equivalent circuit model of the feedback VSCC common mode AEF is depicted in Fig. 1b. The basic principle of the AEF can be illustrated by (1), which is simply derived from Kirchhoff laws. Considering that in the frequency range of interest (150 kHz – 30 MHz), the common mode impedance, Z_{cm}, is greater than the impedance of the noise receiver, Z_{lisn}, the Line Impedance Stabilization Network (LISN) current I_{lisn} is given by equation (1) where A_{op} is the absolute closed loop transfer function of the Op Amp.

$$I_{lisn} = \frac{1}{1 + Z_{lisn}A_{op}} I_{cm} \tag{1}$$

From equation (1), it is obvious that, ideally, an infinite gain of the transfer function A_{op} allows completely rejecting the noise. However, this equation does not reflect the real behavior of AEF because it does not take into account the sensing and injection transfer functions. Fig. 2a. Shows the proposed common mode AEF connected between the LISN and the converter's input. Common mode noises are sensing by a high pass RC circuit (resistor R_{sen} and two capacitors $C_{sen}/2$) and injecting through R_{inj} in series with two capacitors $C_{inj}/2$.

The operation of the AEF at the two power lines is symmetric, so the two power lines are in parallel with regard to the earth ground. An equivalent circuit model with an explicit reference is shown in Fig.2b. Based on Fig.2b, above equations can be briefly written.

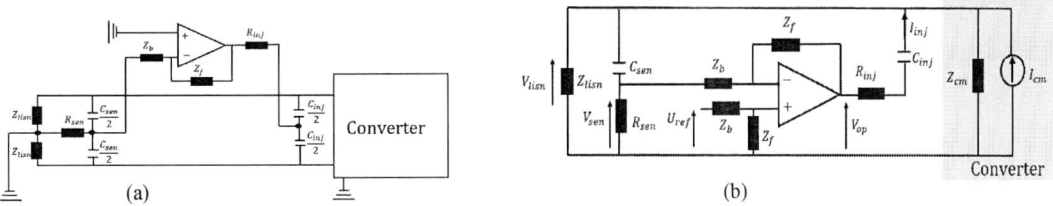

Fig.2: (a) Schematic of the proposed AEF, (b) Equivalent circuit of the proposed AEF with explicit reference

$$A_{sen} = \frac{R_{sen}}{R_{sen} + Z_{Csen}} \tag{2}$$

$$Z_{inj} = Z_{Cinj} + R_{inj} \quad A_{op} = \frac{Z_f}{Z_b} \tag{3}$$

$$A_{op} = \frac{Z_f}{Z_b} \tag{4}$$

$$V_{sen} = A_{sen} V_{lisn} \quad \text{with} \quad R_{sen} \ll Z_b \tag{5}$$

$$I_{inj} = \frac{V_{op} - V_{lisn}}{Z_{inj}} = \frac{A_{op}(U_{ref} - A_{sen} V_{lisn}) - V_{lisn}}{Z_{inj}} \tag{6}$$

$$V_{lisn} = \frac{Z_{lisn}}{Z_{lisn} + Z_{cm}} (V_{cm} + Z_{cm} I_{inj}) \tag{7}$$

Sensing noises are compared to zero reference, $U_{ref} = 0$, (U_{ref} grounded), so expressions (5), (6) and (7) allow writing the closed loop transfer function of the system (8).

$$\frac{V_{lisn}}{I_{cm}} = \frac{\dfrac{Z_{cm} \cdot (Z_{lisn}//Z_{inj})}{Z_{lisn}//Z_{inj} + Z_{cm}}}{1 + \dfrac{Z_{lisn}//Z_{cm}}{Z_{lisn}//Z_{cm} + Z_{inj}} A_{op} A_{sen}} \tag{8}$$

A_{op} is the absolute closed loop transfer function gain of the inverting operational amplifier, and A_{sen} is the sensing transfer function. From equation (8), an infinite gain of A_{op} allows completely rejecting EMI noise ($V_{lisn} \approx 0$).

A. Feedback Loop Gain and atteunation of VSCC AEF

The AEF is basically a feedback system with an analog-input and an analog-output, and its stability should be carefully designed and guaranteed. If the system is unstable, the system can oscillate even when the converter is not supply and common mode noise can be amplified. The stability study is done with the open loop gain transfer function, T_o, which can be obtained from the closed loop transfer function. Assuming that $Z_{cm} \gg Z_{lisn}$ in the interest frequency range, the open loop gain, T_o, is given by equation (9)

$$T_o = \frac{Z_{lisn}//Z_{cm}}{Z_{lisn}//Z_{cm} + Z_{inj}} A_{op} A_{sen} = \frac{Z_{lisn}}{Z_{lisn} + Z_{inj}} A_{op} A_{sen} \tag{9}$$

From equation (9), we can see that if the condition $Z_{cm} \gg Z_{lisn}$ is satisfied, stability and design of AEF is independent of common mode noise source contrary to the passive filter that the knowledge of the common mode noise is essential for design. Moreover, the performance of an EMI filter is quantified by its attenuation, which is its ability to mitigate interferences. The attenuation is defined as the ratio of the voltage across Z_{lisn} without the active filter to that with the active filter. The voltage across Z_{lisn} with the active filter is obtained from equation (8) and that without filter is given by equation (10).

$$V_{lisn,wo\ filter} = \frac{Z_{lisn} Z_{cm}}{Z_{lisn} + Z_{lisn}} I_{cm} \tag{10}$$

Then, the attenuation of VSCC AEF is rewritten as

$$A_{tt} = \frac{V_{lisn,wo\ filter}}{I_{cm}} \frac{I_{cm}}{V_{lisn,with\ filter}} = (1 + T_o)\left(1 + \frac{Z_{lisn}}{Z_{inj}}\right) \tag{11}$$

From equation (11), we can conclude that the attenuation is proportional to the feedback loop gain T_o. Therefore in the design of the VSCC active filter, a compromise between stability and high attenuation is necessary because both depend on T_o. Substituting equations (9) in (11), the attenuation can be rewrite as equation (12)

$$A_{tt} = (1 + T_o)\left(1 + \frac{Z_{lisn}}{Z_{inj}}\right) = 1 + \frac{Z_{lisn}}{Z_{inj}}\left(1 + A_{sen} A_{op}\right) \tag{12}$$

From equation (12), a typical model of the LISN is necessary to calculate the impedance Z_{lisn}. The LISN used is the DO160 ComPower LI-325. An external 10uF capacitor is necessary in power port side to satisfy the requirement specified in the DO160 standard. The impedance between the LISN converter port and ground is measured with the external 10uf capacitor connected.

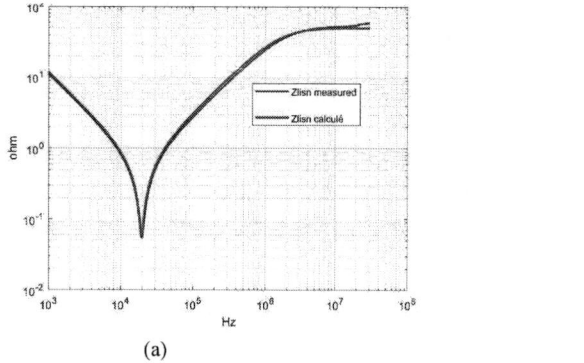
(a)

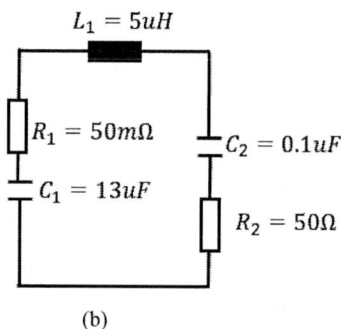
(b)

Fig 3: (a) Comparison of the impedance Li-325 DO160 LISN calculated and measured, (b) Circuit model of LISN

Fig.3a shows the impedance of LISN measured and modeled. We can see that the model matches well with the measurement. The circuit model is depicted in Fig.3b where the external 10uF capacitor is taking into account in the capacitor C_1. However, in common mode model, there are two LISN in parallel, which can be modeled as a single equivalent impedance. Also within the concerned 'RF' frequency range, the impedances of the two capacitors 13uF and 0.1uF can be ignored, so the equivalent CM impedance of LISN can be approximately modeled as an inductance of 2.5uH in parallel with a resistance of 25Ω as depicted in Fig.3c.

$$Z_{lisn} = \frac{sL_{lisn} R_{lisn}}{sL_{lisn} + R_{lisn}} \quad \text{with } L_{lisn} = 2.5\mu H \text{ and } R_{lisn} = 25\Omega \tag{13}$$

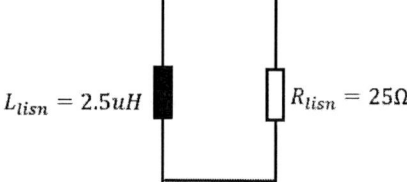

Fig 3c: Equivalent common mode impedance of LISN ignoring capacitors

Substituting equations (2), (3) and (13) in (12) allow to rewrite

$$A_{tt} = 1 + \frac{sL_{lisn}R_{lisn}}{sL_{lisn} + R_{lisn}} \frac{sC_{inj}}{1 + sR_{inj}C_{inj}} \left(1 + \frac{sR_{sen}C_{sen}}{1 + sR_{sen}C_{sen}} A_{op}\right) \tag{14}$$

B. Design guidelines of the proposed VSCC common mode AEF

In this section, the practical design guidelines for the proposed VSCC common mode AEF are developed with consideration of performance and stability.

a) operating frequency range and maximum attenuation

The first step of the design an EMI filter is to determine the required attenuation and the operation frequency range of filter. This approach remains valid in the case of AEF. Keep in mind that a high closed loop gain of the inverting amplifier is necessary to reject EMI noise. From equation (14), the closed loop transfer function A_{op} can provide the necessary closed loop gain and compensates the pole of the detection circuit ($R_{sen}C_{sen}$). Assuming the amplifier gain is designed to be equation (15) with ω_k is the cutoff frequency at which the **closed loop gain** of opamp start to roll off which can be found in the datasheet of the operational amplifier. The attenuation becomes independent of the sensing circuit.

$$A_{op} = \left| -\frac{1 + sR_fC_f}{sR_bC_f\left(1 + \frac{s}{\omega_k}\right)} \right| = \frac{1 + sR_fC_f}{sR_bC_f\left(1 + \frac{s}{\omega_k}\right)} \quad \text{with } R_fC_f = R_{sen}C_{sen} \tag{15}$$

Then, the attenuation becomes

$$A_{tt} = 1 + \frac{sL_{lisn}R_{lisn}}{sL_{lisn} + R_{lisn}} \frac{sC_{inj}}{1 + sR_{inj}C_{inj}} \left(1 + \frac{R_f}{R_b\left(1 + \frac{s}{\omega_k}\right)}\right) \tag{16}$$

For convenience, the minimum operation frequency, f_{min}, is proposed as the lowest frequency boundary that the attenuation is positive and its can be found from the low frequency model approximation of equation (16). In low frequency ω_k can be ignored and equation (16) becomes equation (17).

$$A_{tt} = \frac{s^2 + 2m\omega_n.s + \omega_n^2}{\omega_n^2} \frac{1}{\left(1 + s\frac{L_{lisn}}{R_{lisn}}\right)(1 + sR_{inj}C_{inj})} \tag{17}$$

With

$$\omega_n = \sqrt{\frac{R_{lisn}}{L_{lisn}C_{inj}\left[R_{inj} + R_{lisn}\left(1 + \frac{R_f}{R_b}\right)\right]}} \quad \text{and} \quad m = \frac{L_{lisn} + R_{lisn}R_{inj}C_{inj}}{2\omega_n}$$

The minimum operation frequency is obtained as

$$f_{min} = \frac{\omega_n}{2\pi} = \frac{1}{2\pi}\sqrt{\frac{R_{lisn}}{L_{lisn}C_{inj}\left[R_{inj} + R_{lisn}\left(1 + \frac{R_f}{R_b}\right)\right]}} \tag{18}$$

To obtain the full operating frequency range of the AEF, the high frequencies boundary and final value must be determined. They are obtained from the high frequency approximation of equation (16). In high frequency, the feedback capacitor of inverting amplifier can be ignored (the zero of inverting opamp transfer function is in low frequency). On other hands, the impedances of the LISN and injection circuits are respectively equal to R_{lisn} and R_{inj}. Therefore, the maximum operating frequency and the high frequency final value can be determined based on the high frequency approximation of the attenuation given by equation (19)

$$A_{tt} = \left[1 + \frac{R_{lisn}}{R_{inj}}\left(1 + \frac{R_f}{R_b}\right)\right] \frac{1 + \frac{s}{\omega'_k}}{1 + \frac{s}{\omega_k}} \quad ; \text{with} \quad \omega'_k = \frac{1 + \frac{R_{lisn}}{R_{inj}}}{\omega_k\left[1 + \frac{R_{lisn}}{R_{inj}}\left(1 + \frac{R_f}{R_b}\right)\right]} \tag{19}$$

The maximum operation frequency and final value are obtained as

$$f_{max} = \frac{\omega_k}{2\pi} = \quad \text{and} \quad A_{tt,HF} = 1 + \frac{R_{lisn}}{R_{inj}} \tag{20}$$

Now, an interesting asymptotic behavior of attenuation can be extracted and illustrated in Fig.4. The necessary frequencies boundaries and gain are determined.

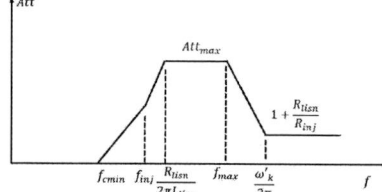

Fig 4: Ideal asymptotic behavior of attenuation

b) Sensing circuit

The sensing circuit is a high pass RC circuit composed of a sensing capacitor C_{sen} and resistor R_{sen}. From equation (16), after pole compensation, attenuation is independent from the sensing parameters but the open loop gain of the AEF depends. Therefore, the detection circuit can be used to sensing noise and avoid stability issue in low frequency without affecting attenuation. By fixing a sensing frequency close to the switching frequency and choosing R_{sen} according to equation (5), the sensing capacitor is given by equation (21).

$$C_{sen} = \frac{1}{2\pi f_{sen} R_{sen}} \tag{21}$$

c) Amplifier circuit

The minimum amplification gain, R_f/R_b, required to achieve the necessary attenuation is obtained from equation (16) and it is given by (22)

$$\left.\frac{R_f}{R_b}\right|_{dB} = A_{tt,dB} + \left|\frac{Z_{inj}\left(1 + \frac{s}{\omega_k}\right)}{Z_{lisn}}\right|_{dB} \tag{22}$$

Moreover, the operational amplifier is the central element of the VSCC common mode AEF. The choice of this active device is important since its limitations will greatly affect the performances of the AEF. It must be high speed and operate in linear mode and its current capacity should be sufficient to avoid saturation. A push-pull current amplification can be used if the Op-amp has a low output current. In addition, it should have a high frequency bandwidth (GBP $\geq$ 30MHz).

d) Injection circuit

The minimum operating frequency f_{min} depends on the cut-off frequency of the injection circuit. By fixing a lowest boundary frequency f_{min} of the operation frequency range of the filter, and choosing an injection resistor R_{inj}, the injection capacitor is determined based on equation (23) . Keep in mind that a high injection resistor decreases the maximum attenuation of filter and a lowest value cannot be chosen due to the stability and capability of opamp to drive a capacitive load (injection capacitor).

$$C_{inj} = \frac{R_{lisn}}{L_{lisn}\left[R_{inj} + R_{lisn}\left(1 + \frac{R_f}{R_b}\right)\right](2\pi f_{min})^2} \tag{23}$$

II. Implementation and validation

To verify the validity of the proposed methodology, the test setup is established on 14V/42V – 27W Boost operating at 115 kHz switching frequency. The transfer functions previously calculated are validate in both simulations and measurements. The performance of the proposed AEF is then by spectrum analyzer measurements.

a) Common mode current measurement without filter

The common mode current spectrum are measured and compared to the DO160 standard to determine the operating frequency range and the required maximum attenuation. Fig. 5a and 5b depict the experimental temporal and spectrum common mode noise without filter.

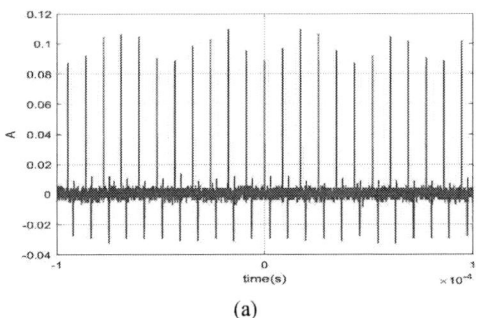
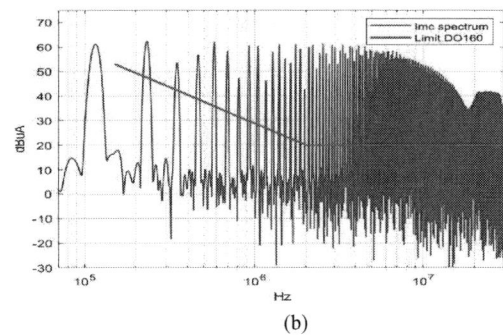

(a)   (b)

Fig 5 (a) Temporal common mode noise current without filter, (b) Spectrum of common mode noise without filter

Based on Fig. 5b, a common mode EMI Filter is necessary over the entire 'RF' frequency range. To satify the standard in low frequency, an active filtering is proposed in the frequency range below 1MHz. Above 1MHz, a passive filter can be used to achieve the required attenuation and satisfy the DO160 standard. To determine the required minimum amplification, the maximum attenuation in the active filter operating frequency range is determined by subtract the 8th harmonic to the DO160 limit.

$$A_{ttmax} = I_{mc} - Limit(DO160) = 60dB\mu A - 30dB\mu A = 30$$

As we want to improve the effectiveness of AEF in low frequency, the lowest and highest operating frequency of AEF are respectively fixed at $f_{cmin} = 30$ kHz and $f_{cmax} = 1$ MHz. Even if the switching frequency is at 115kHz, the minimum frequency boundary is chosen too low in case it needs to use this AEF for application where the switching frequency would be lower. Moreover, as the injection amplifier supply the injection current, for an ideal filter $I_{inj} = I_{cm}$. Therefore, the output current of opamp must be at least equal to I_{mc}. Based on Fig. 5a, the minimum output current capability of the opamp is 120 mA.

b) Sensing , injection and amplifier circuits and stability

The injection resistor should not be high to reduce the maximum attenuation but also a reasonable value must be chosen to assure stability of opamp driving capacitive load. So a value of 6.8Ω is chosen. Then the injection capacitor is determined based on (23) and equal to **0.2uF**. The minimum required amplification to achieve the required attenuation is calculated from (22) and shown in Fig.6a.

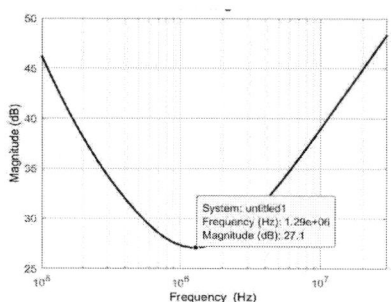

Fig 6a: Minimum required amplification

Based on Fig.6a, to achieve an attenuation of 30dB at 1MHz, a minimum amplification of 27dB is required. Moreover, Opamps are generally of two types: voltage feedback opamp (VFA) and current feedback opamp (CFA). For a VFA, its gain-bandwidth product is constant, so the gain is inversely proportional to the bandwidth. On other hands, CFA has a bandwidth independent of its gain, so it can achieve a wider bandwidth than VFA. So a CFA is chosen in this application. Due to the high common mode current, a Texas Instrument CFA, THS3121, has been selected. It exhibits a 475mA output current capability as well as a large bandwidth of xxxxMHz.

The attenuation is independent of the sensing parameters but the open loop gain of the AEF depends. From (16) the pole of detection must be compensated by the zero of the CFA. So, Sensing frequency can be chosen close to switching frequency 115kHz, and it is fixing at 70kHz and sensing resistor is equal to 110Ω, the component values are given in Table 1.

Table 1: Component Value

Component	Value
R_f	33kΩ
C_f	68pF
R_b	1 kΩ
R_{sen}	110
C_{sen}	2 x 10nF
R_{inj}	6.8 Ω
C_{inj}	2x0.1uF
CFA	THS3121

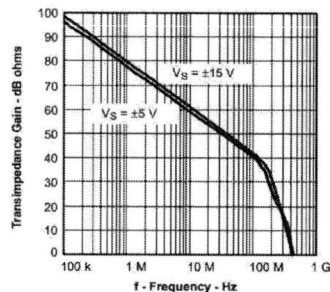

Fig 6b: Transimpedance of THS3121

The open loop transimpedance gain Z of THS3121 is given by Fig.6b. In frequency below 30MHz it can be expressed as

$$Z = \frac{Z_{dc}}{1 + \frac{s}{\omega_p}} \quad \text{with } \omega_p = 2\pi(100\text{kHz})$$ (24)

Z_{dc} is the DC transimpedance. As the transimpedance at the cutoff frequency $f_p = 100kHz$ is equal to 100dB then it can conclude that $Z_{dc} = 103dB$. For a CFA, the absolute closed loop gain of an inverting opamp is expressed as

$$A_{op} = \frac{\frac{Z_f}{Z_b}}{1 + \frac{Z_f}{Z}} \approx \frac{Z_{dc}}{R_b} \frac{1 + sR_fC_f}{\left(1 + \frac{s}{\omega_1}\right)\left(1 + \frac{s}{\omega_k}\right)}$$ (25)

Where ω_1 is a very low frequency pole, which approximatively located at $(R_f + Z_{dc})C_f$ and ω_k is the high frequency closed loop gain pole, which define the closed loop bandwidth of the opamp. As the DC transimpedance Z_{dc} is very high, we fund again the equation (15) from equation (25) of the absolute closed loop in our frequency range by taking the behavior beyond ω_1. Based on Fig.7b, the calculated expression of A_{op} matches well with measurement. As mentioned before, the stability should be carefully designed and guaranteed to avoid oscillation or damage the filter.

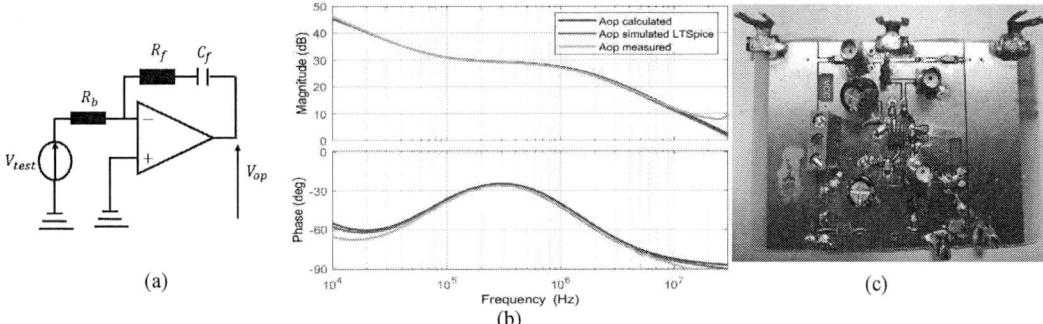

(a)   (b)   (c)

Fig 7. (a) Absolute Transfer function Aop =Zf/Zb measurement, (b) Comparison between Aop calculated, simulated and measured, (c) Propsed common mode AEF

Experimentally, to measure the loop gain from the circuit model, the feedback loop is disconnected at the opamp input and a test voltage is applied in the input of the amplifier from the disconnected node, while the converter is not supplied based on Fig.8a

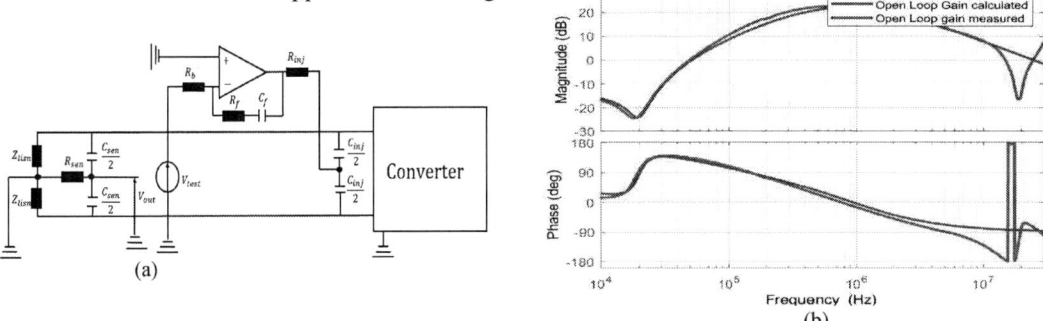

(a)   (b)

Fig.8 (a) Open loop gain measurement To= -Vout/Vtest, (b) Comparison between calculated and measured open loop gain transfer function

Based on Fig.8b, the loop transfer function, T_o, measured and calculated matche. There is a slight difference in high frequency above 10MHz, where a resonance due to common mode impedance occurs. The loop transfer function crosses unity gain both when the loop gain is rising and falling. For such a system, the closed loop system is stable if the open loop transfer function has an amplitude less than unity at all frequencies corresponding to the phase equal to $-180 - nx360$, where $n = 0,1,...\infty$. Based on Fig.8b, the stability criteria mentioned above is satisfied in low frequency contrary to high frequency around 15MHz.

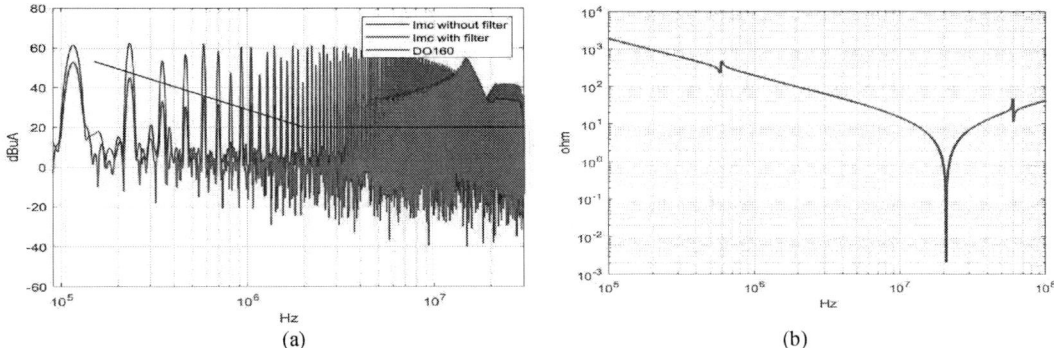

(a)   (b)

Fig 9: (a) Comparison between common mode noise spectrum with and without AEF, (b) Common mode impedance

Based on Fig.9a, in the operating frequency range of the AEF, the standard is met. An attenuation of 10dB is reached at 115kHz. Above the high frequency boundary of the AEF, 1MHz, a passive filter can be used to reach the standard in whole the 'RF' frequency range. On the other hand, the high frequency instability results an amplification of common mode noise around 15MHz. This frequency of 15MHz is higher than the operating frequency range of the AEF, a small passive filter can be used to attenuate the noise amplification. To understand well the origin of the resonance at 15MHz, the common mode impedance is measured and shown in Fig.9b. It. is not capacitive in the whole 'RF' frequency range and the assumption $Z_{lisn}//Z_{cm}$ equal Z_{lisn} is not valid.

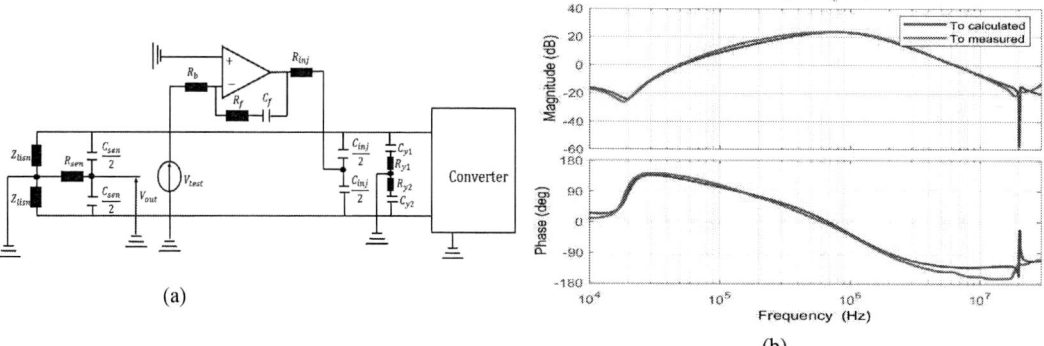

Fig.10 (a) Open loop gain measurement with Cy, (b) Comparison between calculated and measured open loop gain transfer function with Cy

To limit the effect of common mode noise, two capacitors Cy of 4.7nF in series with 1Ω have been used based on fig.10(a). So, the expression of open loop gain (9) becomes (26) and no resonance appears in high frequency as shown in Fig.10b. The stability criteria is satisfied and noises are not amplified as shown in Fig.11b.

$$T_o = \frac{Z_{lisn}//(Z_{cm}//Z_{CY})}{Z_{lisn}//(Z_{cm}//Z_{CY}) + Z_{inj}} A_{op} A_{sen} \tag{26}$$

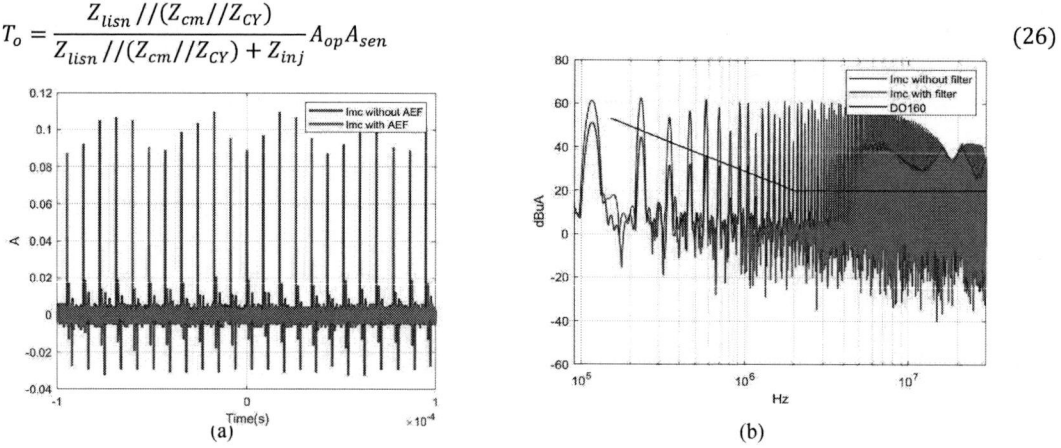

Fig 11a: Comparison between Imc with filter and without filter, (b) Comparison between common mode noise spectrum with and without filter

III. Conclusion

In this paper, the study and design of a VSCC AEF were discussed. A design guideline has been established and validated by simulation and experimental measurements. Stability has been checked and the effect of parasitic inductance of common mode impedance has shown. Based on the design guideline components value are calculated and we can see that the DO160 standard was met up to 1MHz. The filter still attenuated above 1 MHz but not sufficiently to comply with the standard. Therefore a passive filter could be added to comply the standard over frequency range [115kHz – 30MHz]. This study also shows us the role of active filter in hybrid filtering. In fact, active filter essentially aims to attenuate low frequencies and allow the size of the passive filter to be reduced, which would complete the filtering for very high frequencies. An attenuation of 10dB is obtained at 115kHz

IV. REFERENCE

[1] X. Pei, J. Xiong, and J. Chen, "Analysis and Suppression of Conducted EMI Emission in PWM inverter," p. 6.

[2] T. Ninomiya, M. Shoyama, C.-F. Jin, and G. Li "EM1 ISSUES IN SWITCHING POWER CONVERTERS," p. 6.

[3] M. Ali, E. Laboure, and F. Costa, "Integrated Active Filter for Differential-Mode Noise Suppression," *IEEE Trans. Power Electron.*, vol. 29, no. 3, pp. 1053–1057, Mar. 2014

[4] J. Biela, A. Wirthmueller, R. Waespe, M. L. Heldwein, K. Raggl, and J. W. Kolar, "Passive And Active Hybrid Integrated EMI Filters," *IEEE Trans. Power Electron.*, vol. 24, no. 5, pp. 1340–1349, May 2009

[5] M. Pasko and M. Szymczak, "Analysis and simulation of the basic structures of active EMI filters," p. 14.

[6] W. Chen, W. Zhang, X. Yang, Z. Sheng, Z. Wang, "An Experimental Study of Common-and Differential-Mode Active EMI Filter Compensation Characteristics," *IEEE Trans on Electromagnetic Compatibility*, vol. 51, no. 3, p. 9, 2009

Partial Discharges of Insulated Wires under Impulses from Wide Bandgap Power Electronics

Ting Helmholdt-Zhu[1], Vivien Grau[2], Urs Obernolte[3]

[1] Leibniz University Hannover	[2] RWTH Aachen University	[3] Lenze SE
Welfengarten 1	Templergraben 55	Breslauer Str. 3
30167, Hannover	52062, Aachen	32699, Extertal
Germany	Germany	Germany

Acknowledgments

The work presented in this publication was supported by the research project Verse, funded by the German Federal Ministry of Education and Research (BMBF, support code 16EMO0278). The responsibility for the content of this publication lies with the authors.

Keywords

≪Partial discharge≫, ≪Insulation≫, ≪Filtering≫, ≪Pulse Width Modulation (PWM)≫, ≪Wide bandgap devices≫.

Abstract

Due to the high voltage slopes from the wide bandgap (WBG) power electronics, which generate high frequency (HF) electromagnetic noises, the identification of partial discharges (PD) becomes very cumbersome. In this paper, a validated PD detection system is utilized to decouple PD signals from those HF noises. In addition, the influences of different insulation systems, grade of insulated wires, steepness of the voltage slope as well as voltage overshoot because of various cable length on the PD events are also presented.

1. Introduction

PD detection is essential to the evaluation and qualification of insulation systems in electrical machines, especially for those which are restrained by the norm DIN EN 60034-18-41 (Partial discharge free electrical insulation systems) [1]. However, due to the steep-edged voltage impulses from the new generation of WBG power electronics, the identification of PD is more cumbersome [2], as the impulse contains HF components and generates electromagnetic noises, which share a similar frequency spectrum like the PD signals [3, 4]. In this paper, a PD detection system, which is validated through two different measurement setups, is illustrated to decouple the PD signals from HF noises, effectively. In addition, influences of the rise-time, the DC-link voltage and the peak value of the voltage impulses on PD events are also studied and presented by different samples of insulated wires.

2. PD Detection System Validation

A schematic description of the developed PD detection system is presented in Fig. 1. The function of a PD Ring-Sensor is to receive all electromagnetic signals, for example, from the transmission cables (noises from switching operations and PD events) as well as the background noises (for example, from radios, mobile phones). In order to ensure, that measurement results are reproducible, the transmission cables are located in defined position through the Ring-Sensor.

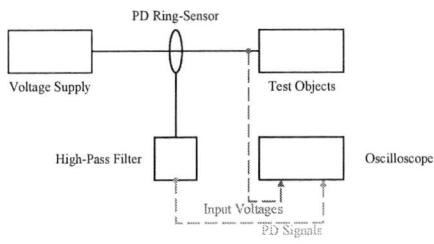

Fig. 1: Schematic description of the PD detection system

Afterwards, all the captured signals are passed through a High-Pass (HP) filter, which has a high corner frequency at about 730 MHz, as illustrated in Fig. 2. This high corner frequency enable a selective filtering between PD signals and switching noises.

The validation of this PD detection system is conducted through the following measurements.

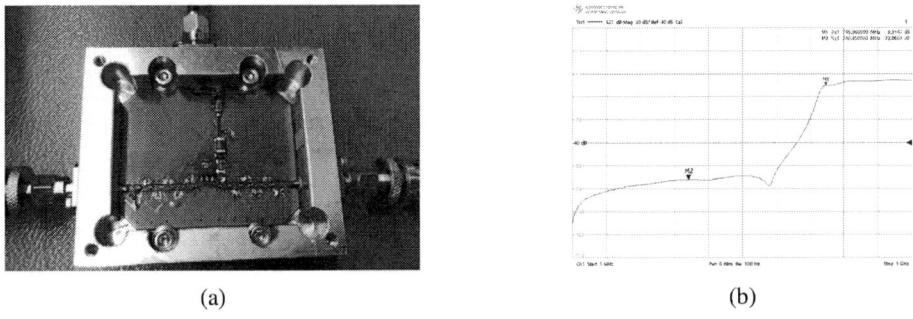

(a)   (b)

Fig. 2: HP Filter: (a) Circuit board of HP filter, (b) Frequency spectrum

2.1 Validation with Twisted Pair

The first validation measurement is conducted with a silicon carbide (SiC)-based inverter (CREE C2M004-5170D transistors) as a voltage supplier, a twisted pair with 0.75 mm copper diameter, grade 2 as test objects, a camera as a real-time monitoring system and a PD detection system, which was designed for insulated gate bi-polar transistors (IGBT)-based inverters [5], as a reference object (see Fig. 3).

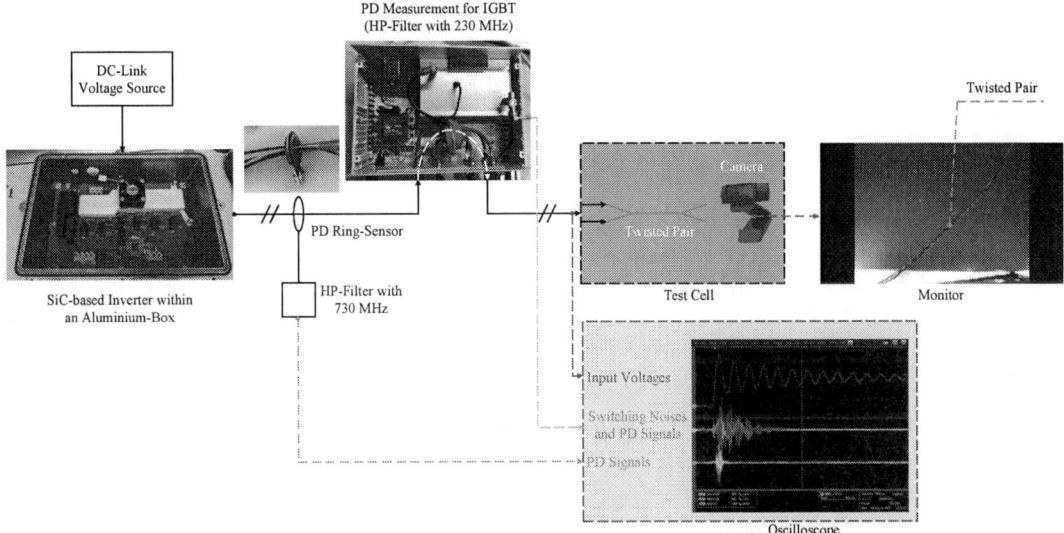

Fig. 3: Schematic circuit diagram for the validation measurement with twisted pair

The purpose and intention of this validation measurement are:

1. With twisted pair as test objects, the detection of PD events is much easier, as it can be directly observed in real-time through the camera.
2. The ability of this new PD detection system, especially the corner frequency, in terms of decoupling PD signals from switching noises of the SiC-based inverter can be examined.

The results of both with and without PD events are illustrated in Fig. 4. The evidence, that there are PD events in Fig. 4 (b) is, that, firstly, the PD lights are observed from the display monitor; secondly, the oscillation of the input voltage converge faster than the 'healthy' voltage curve (Fig. 4 (a)). In other words, the voltage vibration under PD condition cannot reach the double DC-link voltage, which means that there are extra energy losses during this period of time and it is caused by the PD events. It can be noticed, that a corner frequency of 230 MHz is no longer sufficient for the PD detection with SiC-based inverter, as the switching noises contain higher spectrum of frequency noises and cannot be filtered out. Therefore, the corner frequency needs to be increased and the experiments with twisted pair demonstrate, that the chosen of 730 MHz is efficient to filter the noises from the SiC-inverter.

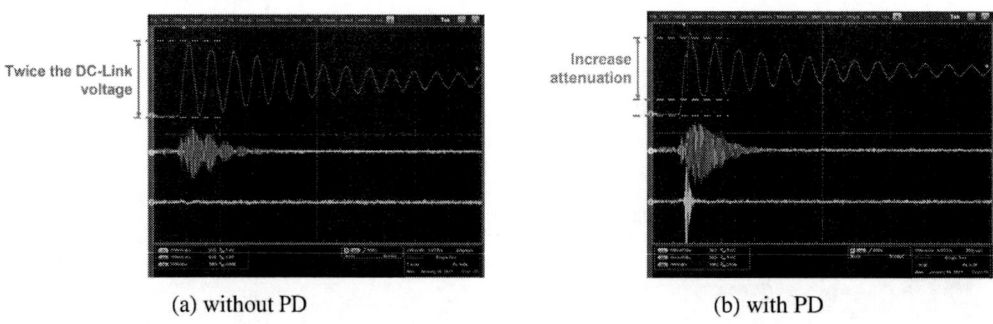

(a) without PD   (b) with PD

Fig. 4: Signal records from the oscilloscope: input voltage of the twisted pair (green line); output signal from the PD detection system for IGBT (yellow line); output signal from the new PD detection system (blue line)

2.2 Validation through a Commercial Surge Voltage Tester

Based on the validation measurement with twisted pair, it is proved, that this PD detection system is able to detect PD signals in a SiC-based application environment. The next step is to verify the PD-sensitivity of this system with a qualified commercial surge voltage tester (see Fig. 5).

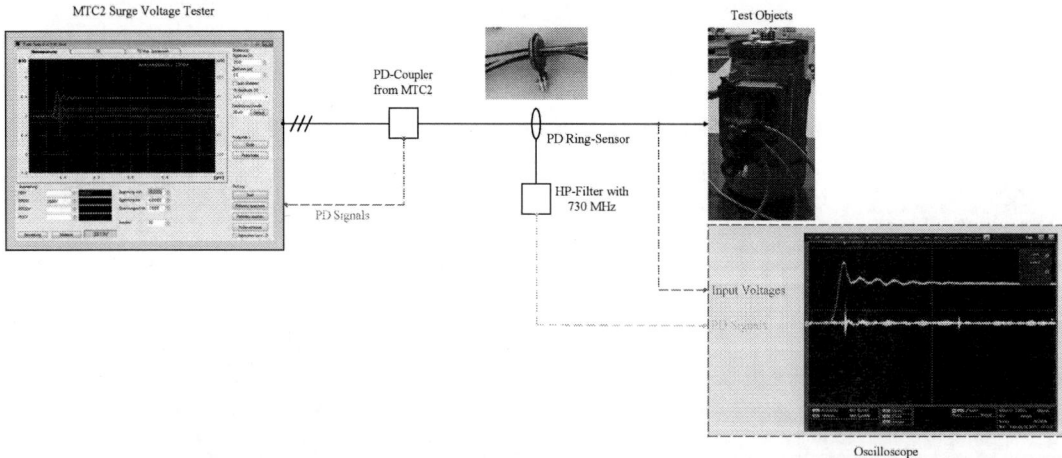

Fig. 5: Schematic circuit diagram for the validation measurement with the surge voltage tester

The surge tester produces surge voltages and using its own PD-coupler to detect the PD signals. Meanwhile, the developed PD detection system with PD Ring-Sensor and HP-Filter is also integrated into the measurement system to detect the same PD signals. The test object is a three-phase asynchronous machine. The surge voltage and recorded PD signals are illustrated on the monitor of the surge tester and the oscilloscope (see Fig. 6).

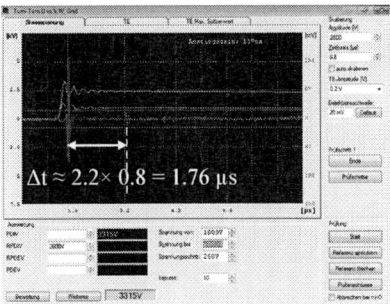
(a) display interface of the surge tester

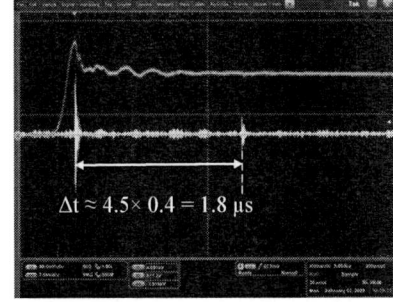
(b) display interface of the oscilloscope

Fig. 6: Recorded surge voltages and PD signals from (a) the surge tester and (b) the oscilloscope: (a) PD signals (red line), surge voltage (yellow line), the reference surge curve (green line); (b) PD signals (yellow line), surge voltage (green line)

The surge tester is programmed to stop, after the PD inception voltage (PDIV) is detected and at the mean time, the oscilloscope shows only the triggered event. In this way, both systems detect the same signal (the first PD event). It can be seen, that both systems show the same number of recorded PD signals and the time intervals between adjacent PD illustrate also a high agreement. This process is repeated several times with similar results.

3. PD Measurements

The ability of decoupling WBG-related HF noises and PD-sensitivity of the developed PD detection system are validated through SiC-based inverter and surge voltage tester, respectively. The following PD measurements utilize this verified detection system to investigate the influences of the DC-link voltage, cable length, rise times, impregnation on the repetitive PDIV (RPDIV) and PD pattern of enameled wires.

3.1 Specimens

The test objects are three different types of enameled wires with the same copper diameter of 0.85 mm and are of the same manufacturer. Two of them are of insulation grade 2 (A and B) where one of them features a PD-resistant additive (B). The third type (C) is of insulation grade 3. All specimens are twisted according to the standard DIN IEC 60851-5 [6] with eight twists. The standardized twisted pair specimens, as well as the twisted pair specimens with impregnated insulation resin are both considered, as proposed in [2]. The respective designations of the specimens are listed in Table I. Hence, overall, there are six variants and each of them has five samples: without impregnation (A1, B1 and C1); with impregnation (A2, B2 and C2). In this way, the effects of insulation film thickness, corona-resistant insulation film and the impregnation process can be investigated.

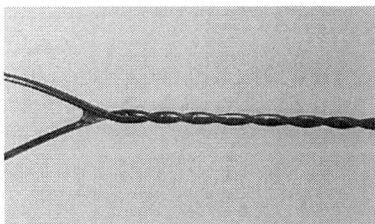

Fig. 7: In insulation resin impregnated twisted pair of enameled wire

For impregnation, the test specimens are first immersed in a coating resin. To extract air voids from the resin coating, the specimens are placed in a vacuum chamber for 20 min. Subsequently, the resin is precured for 10 min at 50°C before being fully cured for 60 min under UV irradiation. Figure 7 shows a section of an impregnated twisted pair specimen.

Table I: Parameters of enameled wires

Specimen	Type of Wire	Resin Impregnation
A1	grade 2	no
A2	grade 2	yes
B1	grade 2 PD resistant	no
B2	grade 2 PD resistant	yes
C1	grade 3	no
C2	grade 3	yes

3.2 Test Circuits

The PD measurements are conducted with the surge tester and SiC-based inverter, respectively.

3.2.1 Surge Tester

The applied test voltage of the surge tester with Step-by-Step (SBS) methods [7] starts from 400 V and stops automatically, when the peak-to-peak RPDIV is detected. The voltage rises 100 V on each step and each step contains 10 impulses. The threshold of PD detection is defined at 60 mV according to the system and background noises. Figure 8 shows the peak-to-peak RPDIV of each specimen.

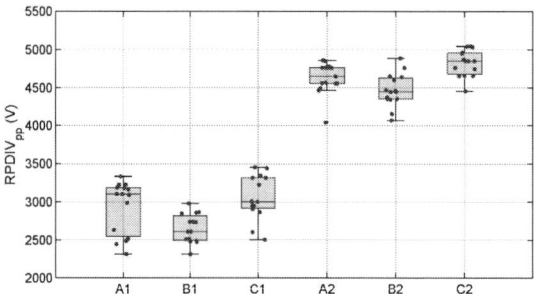

Fig. 8: RPDIV of each specimen under surge tests

It is shown, that peak-to-peak RPDIV of impregnated specimens are almost twice as high as the specimens without impregnation. As a results of that, the maximum allowed DC-link voltage of the SiC-based inverter (950 V_{DC}) cannot reach the PDIV of impregnated twisted pair. Hence, for the test with SiC-based inverter only A1, B1 and C1 are used as test objects.

3.2.2 SiC-Based Inverter

Similarly, the test voltage of SiC-based inverter is, as well, produced in terms of SBS method. It starts also at 400 V and rises step by step (100 V each step) until RPDIV is detected. The measurement setup is demonstrated in Fig. 9.

In this paper, due to the limitation of memory space and the storage dead-time of the oscilloscope (Tektronix MSO5104B) the meaning of RPDIV [8] needs to be redefined. The oscilloscope is triggered through an external equipment (Keysight 33500B), which produces a train of square impulse with a frequency of 0.5 Hz, 50 % duty cycle and 15 cycles. In other words, during each voltage step, the oscilloscope will save the input voltage and PD signals 15 times, which means that each voltage level lasts 30 s. The time period of each saved file is 1 ms. At the meantime, the inverter produce Phase-to-Phase voltage impulses at a frequency of 16 kHz. Hence, each storage contains 32 bipolar-impulses. The definition of $RPDIV_{DC}$ is the minimum DC-link voltage, at which the sum number of recorded PD events is greater than one half of the sum number of the recorded voltage impulses within one voltage step:

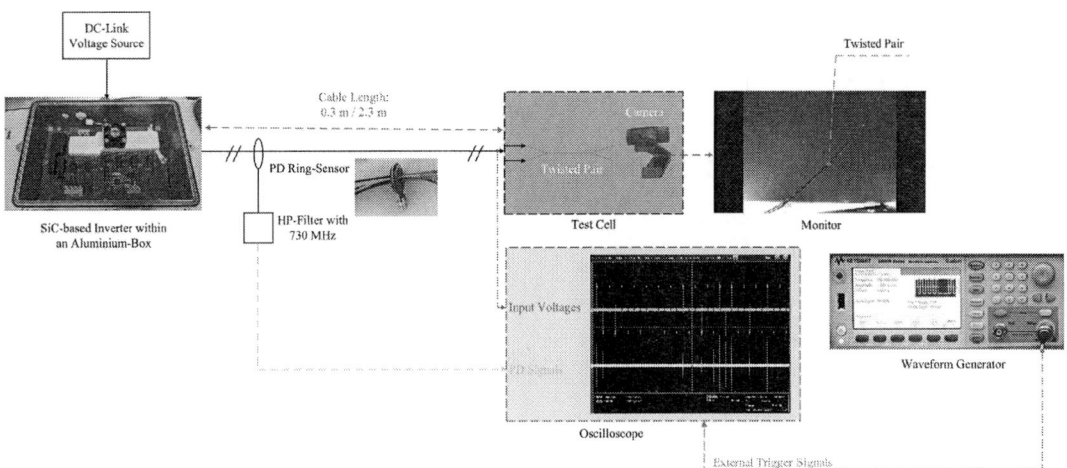

Fig. 9: Measurement setup with SiC-based inverter

$$\text{RPDIV}_{DC}(\text{PD}_{sum}, \text{Imp}_{sum}) = U_{DC,\,min} \tag{1}$$

subject to $\text{PD}_{sum} > 0.5 \cdot \text{Imp}_{sum}$ and $\text{Imp}_{sum} = 32 \cdot 15 = 480$, where 15 is the number of trigger as well as storage times within one voltage step.

In order to study the effects of peak voltage and the slew rate (dU/dt) the cable between the specimen and the SiC-inverter is varied between 0.3 and 2.3 m (see Fig. 9) and two different gate resistances ($20\,\Omega$, $0.5\,\Omega$) are utilized. The rise time t_r is defined as the time for the voltage rise from $0.1U_p$ to $0.9U_p$, where U_p is the maximum voltage overshoot [9]

$$dU/dt = \frac{0.9U_p - 0.1U_p}{t_r} \tag{2}$$

After the measurements, the slew rates under different gate resistance and cable length can be summarized as: $20\,\Omega$ with $55\,\text{V/ns}$; $0.5\,\Omega$ with $120\,\text{V/ns}$.

3.3 Measurement Results with SiC-Based Inverter

The measurement results are analyzed from the aspects of RPDIV_{DC}, the time lag between adjacent PD signals and 2D-density plots presenting the pattern of PD events during transient oscillation phases. Each type of enameled wires has five samples and each sample is tested three times.

3.3.1 RPDIV$_{DC}$

The RPDIV_{DC} is measured with different cable lengths and gate resistances (see Fig. 10). As mentioned, the maximum DC-link voltage is $950\,\text{V}_{DC}$. However, some tests unsatisfied the condition of RPDIV_{DC} (defined in 3.2.2), which means that for these tests their RPDIV_{DC} are higher than $950\,\text{V}_{DC}$. In order to present this undetected RPDIV_{DC}, the voltage is set to be $1000\,\text{V}_{DC}$ (for example, RPDIV_{DC} at $20\,\Omega$ with 0.3 m cable length).

It can be seen, that, in general, the RPDIV_{DC} with shorter cable length is higher than those with 2.3 m cable length. Similarly, this tendency can also be observed under larger gate resistance. However, a long cable length weaken the influence of gate resistance, as the peak value of the transient oscillation plays a more dominant role for RPDIV_{DC} than the voltage slew rate. In addition, the RPDIV_{DC} of C1 in Fig. 10 (a) is not illustrated, as all fave samples of C1 present a higher dielectric strength, which can resist the steep voltage impulse under 0.3 m connection cable. On the other hand, with long cable length

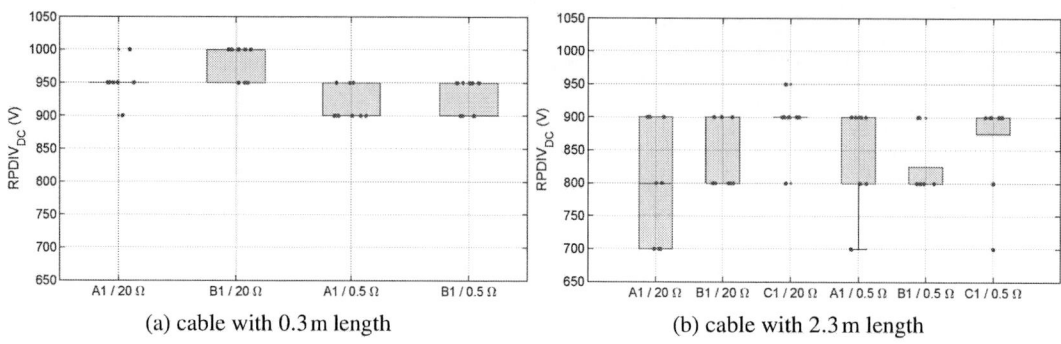

(a) cable with 0.3 m length   (b) cable with 2.3 m length

Fig. 10: $RPDIV_{DC}$ (a) with 0.3 m connection cable and (b) with 2.3 m connection cable

it is capable to reach the $RPDIV_{DC}$ of C1 due to the much higher peak voltage amplitude (see Fig. 10 (b)).

3.3.2 Time Lag between Adjacent PD Signals

The influence of voltage slew rate and cable length on the time interval Δt between adjacent PD events are demonstrated in Fig. 11. The maximum DC-link voltage with 0.3 m cable length is capable to reach $950\,V_{DC}$ (see Fig. 11 (a)(b)), on the other side, with 2.3 m cable length the maximum value is set to be $800\,V_{DC}$ (see Fig. 11 (c)(d)) due to the much higher voltage overshoot.

As is shown, the time interval Δt is impacted markedly by the voltage slew rate, DC-link voltage and cable length. The values are mostly concentrated at $3\,\mu s$, $28\,\mu s$, $31\,\mu s$ and multiples of $31\,\mu s$, which correspond to the time intervals between the four edges of the applied test voltage (see Fig. 12):

1. $31\,\mu s$: PD appears in the early stage only at positive and negative rise edges. This phenomenon is explained in [9] in terms of the charge accumulation.
2. $3\,\mu s$ and $28\,\mu s$: PD appears no longer only at positive and negative rise edges, but all four edges due to higher electrical field strength caused by higher voltage slew rate or longer cable length.
3. Multiples of $31\,\mu s$: After the first PD event, the successive discharge appears not immediately at the next rise edge, but after some periods.
4. $\approx 0\,\mu s$: In contrast, this pattern shows, after the first PD event, the successive discharge appears immediately afterwards.

Overall, in terms of the PD event distribution, the influences of voltage slew rate and overshoot on all three types of wires are similar. At $20\,\Omega$, the occurrence of PD events is mostly triggered at rise edges (except Fig. 11 (c) at $800\,V_{DC}$, as the electrical field strength is high enough to trigger the PD under each voltage edge) and after the first PD often follows another PD event, immediately. With the increase of DC-link voltage, the shortly followed successive discharge happens more frequently. On the contrary, with smaller gate resistance $(0.5\,\Omega)$, the PD distribution is changed and PD events appear now at all edges, as illustrated in Fig. 11 (b)(d). Besides, with the increase of cable length, the most successive PD occur rather after some periods than shortly afterwards.

3.3.3 2D-Density Plot

From the analysis of section 2.3.2, it is clearly demonstrated, that the PD often occurs at the impulse voltage edges. Hence, in order to study the PD pattern during these transient oscillation periods, a 2D-density plot presentation is introduced (see Fig. 13 and Fig. 14). As the pattern with longer cable (2.3 m) is similar to that with 0.3 m and it is almost independent of wire types, the recorded data of wire A1, number 5, is used to represent the discharge distribution as a function of the voltage slew rate and DC-link voltage. The orange dot representing the recorded discharges during the repeated measurements, indicates the normalized magnitude, the point of time of each PD event and are overlapped in the plots with different test voltages (the blue line). All voltages are normalized by the maximum magnitude of the voltage at $0.5\,\Omega$, $950\,V_{DC}$.

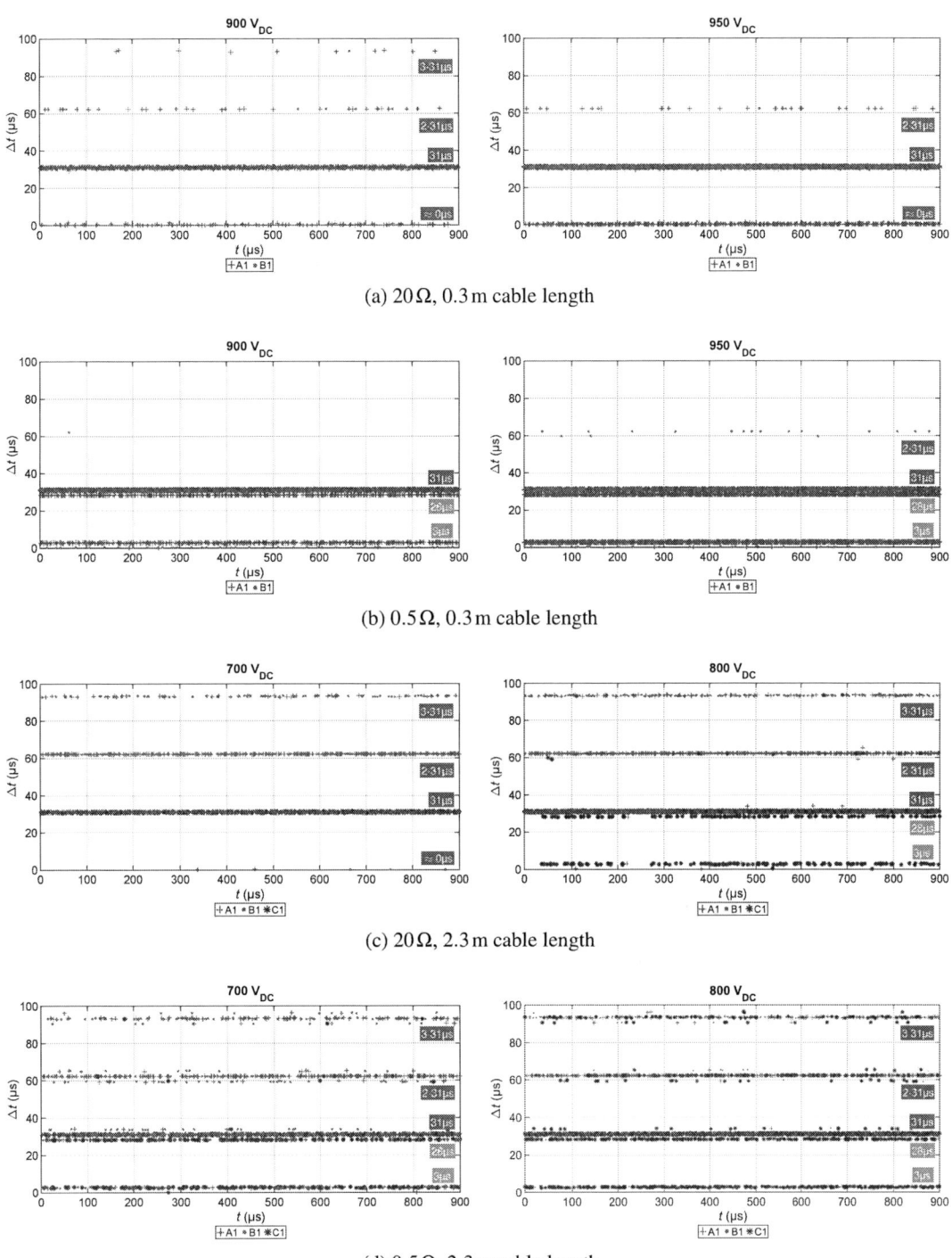

(a) 20 Ω, 0.3 m cable length

(b) 0.5 Ω, 0.3 m cable length

(c) 20 Ω, 2.3 m cable length

(d) 0.5 Ω, 2.3 m cable length

Fig. 11: Time interval Δt between two adjacent PD events under different slew rate (20 Ω: 55 V/ns, 0.5 Ω: 120 V/ns) and peak voltage amplitude in terms of cable length (0.3 m, 2.3 m)

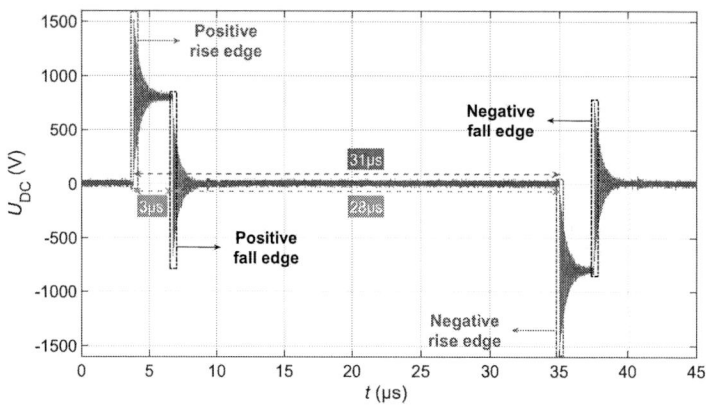

Fig. 12: Four edges: positive and negative rise as well as fall edges

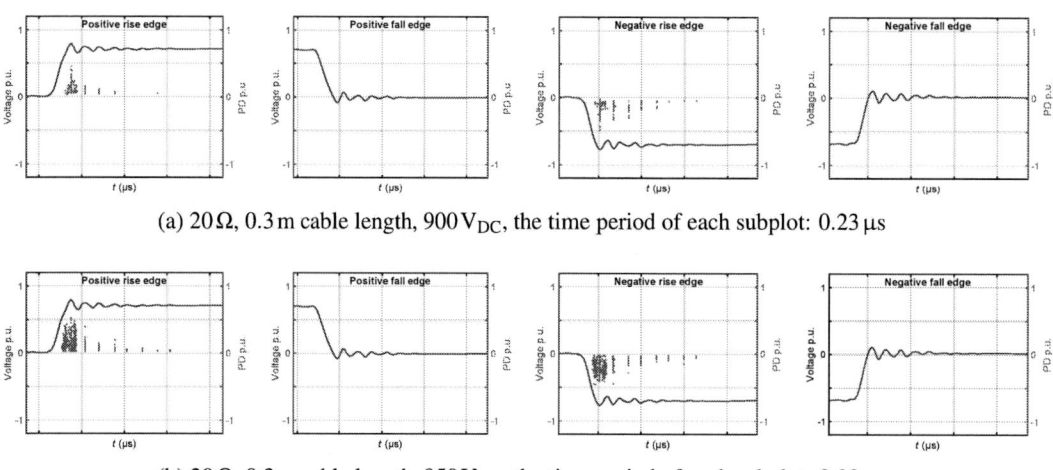

(a) 20 Ω, 0.3 m cable length, 900 V_DC, the time period of each subplot: 0.23 μs

(b) 20 Ω, 0.3 m cable length, 950 V_DC, the time period of each subplot: 0.23 μs

Fig. 13: 2D-density plot: illustrating PD patterns during transient voltage oscillations by 20 Ω

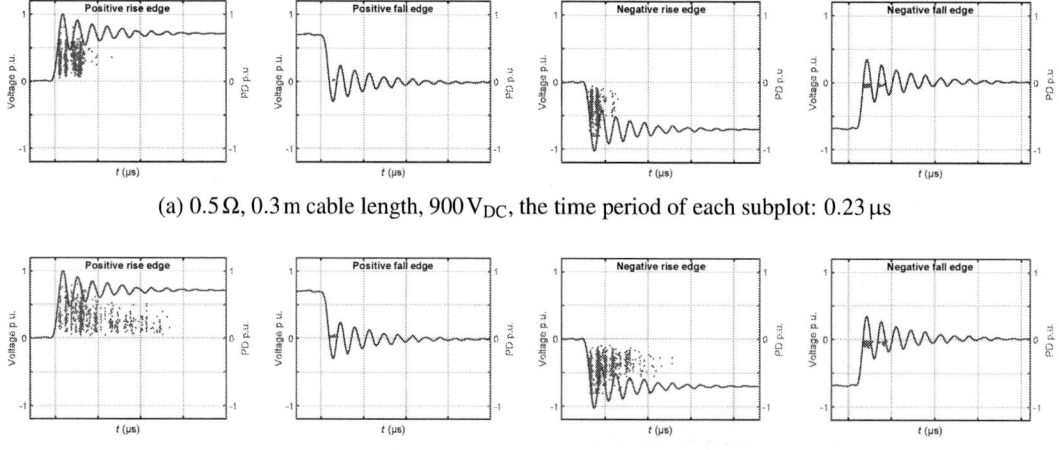

(a) 0.5 Ω, 0.3 m cable length, 900 V_DC, the time period of each subplot: 0.23 μs

(b) 0.5 Ω, 0.3 m cable length, 950 V_DC, the time period of each subplot: 0.23 μs

Fig. 14: 2D-density plot: illustrating PD patterns during transient voltage oscillations by 0.5 Ω

Combining with the previous results from the time lag between adjacent PD signals, the same conclusions could be drawn from the 2D-density plot. At $20\,\Omega$, PD occur only at both rise edges. With the increase of voltage slew rate (at $0.5\,\Omega$), there are a few PD appeared at the fall edges, however, with much smaller discharge magnitude and PD counts compared to the PD at rise edges. Along with the rising of DC-link voltage, the PD counts enhance markedly. Moreover, it is obvious, that:

1. At $20\,\Omega$, the PD concentrate under every maximums with declined magnitude and intensity along the oscillation period.
2. On the other hand, at $0.5\,\Omega$, the PD pattern is changed. Instead of concentrating under the maximums, the discharges locate at the left and right sides of the peaks, separately. And the occurrence of PD is more scattering distributed along the whole vibration zone than at $20\,\Omega$.
3. The magnitude of PD at $20\,\Omega$ is generally lower than those at $0.5\,\Omega$.

4. Conclusion

In this paper, a cable-connected PD-detection system is introduced and validated through both PD tests with SiC-based inverter as well as the commercial surge tester considering the ability of HF Electro-Magnetic Interference rejection and PD-sensitivity. In addition, this system is utilized to investigate the influences of steep-edged voltage impulses on different insulated wires with or without impregnation. The recorded PD events are analyzed through the aspects of $RPDIV_{DC}$, the time lag between adjacent PD signals and the 2D-density plot.

The results show, in general, with impregnation the $RPDIV_{DC}$ of wires is much higher than that without impregnation. Similarly, the wire samples with thicker isolation film (type C) also have considerable improved dielectric strength. Secondly, the overshoot of the test voltage plays a more dominant role than the voltage slew rate in terms of $RPDIV_{DC}$.

With regard to the time lag, a significant effects of voltage slew rate and voltage overshoot on the time interval Δt between adjacent PD events can be observed. Normally, the PD events occur only at rise edges, which corresponds to the value of $31\,\mu s$. With the enhancement of the electrical field strength, via gate resistance, DC-link voltage or cable length, more discharges are triggered at the fall edges ($3\,\mu s$ and $28\,\mu s$), which demonstrate an increasement of the PD counts and an acceleration of the altering process.

Additionally, these patterns are also validated in the 2D-density plot. Furthermore, from the density plot, it can also be noticed, that the PD concentrate firstly under each maximum during the vibration phase. With the rising voltage overshoot, the location of discharges is divided into right and left parts around the maximum and no longer directly under the peaks, which is similar to the PD pattern under AC voltage [10]. This PD pattern and the influence of voltage slew rate as well as the voltage overshoot on the $RPDIV_{DC}$ should be further studied with actual stators.

References

[1] Rotating Elect. Machines - Part 18-41: Partial discharge free electrical insulation systems (Type I) used in rotating electrical machines fed from voltage converters - Qualification and quality control tests, 2014
[2] Grau, Vivien: Development of a Test Bench to Investigate the Impact of Steep Voltage Slopes on the Lifetime of Insulation Systems for Coil Windings, 2021
[3] M. Fuerst and M. Bakran: An Advanced Filtering Method for Partial Discharge Measurement in the Presence of High dV/dt, PCIM Europe 2019; International Exhibition and Conference for Power Electronics, Intelligent Motion, Renewable Energy and Energy Management, 2019, pp. 1-7
[4] T. Billard, T. Lebey and F. Fresnet: Partial discharge in electric motor fed by a PWM inverter: off-line and on-line detection, in IEEE Transactions on Dielectrics and Electrical Insulation, vol. 21, no. 3, pp. 1235-1242, June 2014, doi: 10.1109/TDEI.2014.6832270
[5] Kai Mueller: Entwicklung und Anwendung eines Messsystems zur Erfassung von Teilentladungen bei an Frequenzumrichtern betriebenen elecktrischen Maschinen, 2003
[6] Winding Wires - Test methods - Part 5: Electrical Properties (IEC 55/1072/CD:2008), 2008
[7] Rotating Elect. Machines - Part 27-5: Off-line partial discharge tests on winding insulation of rotating electrical machines during repetitive impulse voltage excitation, 2018
[8] Rotating Elect. Machines - Part 18-41: Partial discharge free electrical insulation systems (Type I) used in rotating electrical machines fed from voltage converters - Qualification and quality control tests, 2014

[9] N. Driendl, F. Pauli and K. Hameyer: Influence of Ambient Conditions on the Qualification Tests of the Interturn Insulation in Low-Voltage Electrical Machines, in IEEE Transactions on Industrial Electronics, vol. 69, no. 8, pp. 7807-7816, Aug. 2022, doi: 10.1109/TIE.2021.3108721

[10] Rotating Elect. Machines - Part 27-1: Off-line partial discharge measurements on the winding insulation, 2017

Analysis of a Droop-Based Power Controller for Three-Phase Microgrids

Andrea Lauri, Hossein Abedini, Davide Biadene, Tommaso Caldognetto, Paolo Mattavelli
University of Padova
Stradella San Nicola, 3
Vicenza, Italy
Email: name.surname@unipd.it

Acknowledgements

This work was supported in part by the project ADPE funded by the Department of Management and Engineering (DTG), University of Padova, Vicenza, Italy, and in part by the research project "Interdisciplinary Strategy for the Development of Advanced Mechatronics Technologies (SISTEMA)", DTG, University of Padova - Project code CUP-C36C18000400001.

Keywords

≪Grid-connected inverter≫, ≪grid-forming converter≫, ≪parallel operation≫,≪seamless transfer≫.

Abstract

A droop-based controller for three-phase converters is described herein. It allows independent control of the converter power at each phase and smooth transitions to the islanded operation. Both three-phase four-wire and three-phase three-wire connections are considered, with a particular focus on the latter configuration. The islanded condition is detected automatically, without the need of communication with other units, and the transition toward the islanded operation mode is performed without interruptions of supply for the energized loads and resources. The considered application scenario is the one of smart microgrids, where power control flexibility and islanded capabilities are features of paramount importance to provide demand-response and uninterrupted operation. A systematic analysis of the approach is presented and validated herein.

Introduction

Low-voltage microgrids provide by means of distributed electronic power converters (EPC) advanced services to the users and upstream grids and continuity of service during emergencies [1]. Examples of crucial features are *i*) uninterrupted supply via islanded operation in response to adverse localized events affecting mainstream electricity supply, *ii*) participation to transactive energy markets [2] by exploiting flexible power flow control, and *iii*) improved power quality in terms of power factor and balanced power absorption [3].

The many requirements are commonly accommodated by means of control hierarchies in which the *P-f* droop control constitutes the primary layer [4], as displayed in Fig. 1. The advantages of droop control include grid voltage support and the capability of adapting the voltage references of EPCs to automatically share the power needs in islanded grids [5]. On the other hand, provisions are necessary to achieve the capability of tracking set-points of output power, which is required for power flow control and for providing services like demand-response.

Then, two opposite needs are present: *i*) support the grid voltage by adapting the inverter output power according to the droop laws, useful especially during islanded operation, *ii*) make the output power fixed and independent from grid voltages and loading conditions to allow power tracking. Remarkably, *P-f*

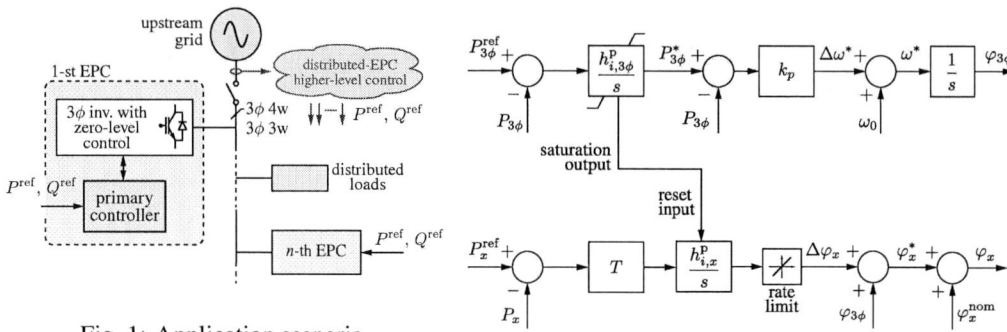

Fig. 1: Application scenario.

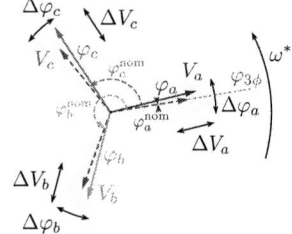

Fig. 2: Quantities involved in Fig. 3-Fig. 4.

Fig. 3: Proposed primary controller for P-control.

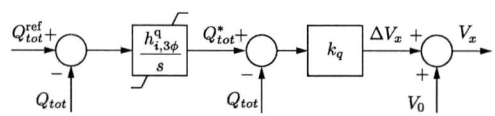

Fig. 4: Proposed primary controller for Q-control.

Table I: Comparison with other approaches

Control kind	islanded operation	grid-tied / islanded transitions	three-phase power-tracking	per-phase power tracking
Grid-feeding	-	-	+	+
Traditional droop	+	+	-	-
Droop with 3φ $P\&Q$ track	+	+	+	-
Approach analyzed herein	+	+	+	+

droop control does not allow *independent* output power control of *each* of the phases of a three-phase converter because it would lead to different frequencies among the phases, which is not acceptable. However, independent power control of each phase is necessary in several circumstances (see, for example, [6, 7, 8, 3, 9, 10, 11]).

A control scheme addressing the three needs discussed above, namely, *i*) control of total output power, *ii*) independent power tracking at converter phases, *iii*) seamless transition toward islanded operation, is proposed and analyzed herein for both three-phase with neutral connection (3φ-4w) and *three-phase without neutral* (3φ-3w) connection, *iv*) parallel islanded operation with other converters. The challenge of jointly providing all the above features was tacked in [12] considering three-phase four-wire connections to the grid, while the case of absence of the neutral wire, which imposes additional constraints and considerations for the implementation of the control, is discussed herein.

The proposed control scheme is useful in microgrid applications of generic network structures, that is, with or without the availability of the neutral connection. The involved quantities and control scheme are shown in Fig. 2-Fig. 4 and briefly outlined in the following. A comparison with other representative approaches is given in Table I, to highlight the features of this solution as compared to other methods currently available in the literature. Grid-feeding converters, for example, require the presence of the main grid or the presence of other voltage-forming units in order to operate in an islanded microgrid [13]. In traditional droop control, where droop laws are defined as

$$
\begin{cases}
\omega = \omega_0 - k_p P \\
V = V_0 - k_q Q
\end{cases}
\tag{1}
$$

island operation is possible, however total active and reactive power regulation during grid-tied operation is achievable by modifying the droop laws as in [4] and [14]. In this case of droop with 3ϕ active and reactive power tracking, however, per-phase power regulation is left unexplored. In fact, it is not possible to independently impose droop laws on each phase: this would lead to different frequencies for each phase of the three-phase system.

Of course, a number of other relevant contributions are present in the literature falling within the control kinds referred to in Table I, of which only a few representative papers are mentioned in the brief discussion above.

Basics of Per-Phase Power Control for 3ϕ Networks with Neutral Connection

Per-phase control considering the case of 3ϕ-4w connection is first reviewed next. Operation without the neutral connection (i.e., 3ϕ-3w) is then analyzed in the subsequent section.

The controller is composed of a synchronization loop, displayed in Fig. 3-top, that aligns the instantaneous phase $\varphi_{3\phi}$ of the three-phase inverter voltage to the one of the grid voltage, producing a suitable phase-shift based on the total active power reference signal. Then, the active and reactive power of each phase is regulated by *independent* controllers, in Fig. 3-bottom and Fig. 4. Active power control regulation at each converter phase is achieved by adjusting the phase displacement $\Delta\varphi_x$ of the considered x-th phase with respect to the three-phase $\varphi_{3\phi}$. In order for the regulator in Fig. 3-bottom to process exclusively the differential power (e.g., for phase-a, $P_a - \sum P_x/3$), matrix T is employed, defined as

$$T \triangleq \begin{bmatrix} 2/3 & -1/3 & -1/3 \\ -1/3 & 2/3 & -1/3 \\ -1/3 & -1/3 & 2/3 \end{bmatrix} \tag{2}$$

and consequently the common-mode power (i.e., $\sum P_x/3$) is controlled only by the regulator in Fig. 3-top. This allows to independently set the common mode and differential mode power control bandwidth. Reactive power control is achieved by regulating the amplitude of the generated phase voltages, which can be done phase-by-phase in the case of 3ϕ-4w. To ensure the balance between generated and absorbed power, power control is no longer possible when the system is operating in island conditions: converters must supply the power absorbed by loads connected to the grid. This situation gradually leads to the saturation of the active and reactive control loops, smoothly changing the control structure to grid-forming traditional *P-f* droop structure.

For the sake of clearness, it is worth noticing that by this control scheme active and reactive power control are achieved via phase and amplitude regulation, respectively, which relies on mainly inductive interconnection impedances among the converter and the outer sources. This is a common condition that can be conveniently imposed by a proper zero-level inverter control design.

Per-Phase Power Control for 3ϕ Networks *without* Neutral Connection

Let us consider the power exchange of a single-phase inverter connected to the grid. Be $V_i \angle \varphi_i$ the inverter voltage phasor, and V_g the grid voltage phasor. Considering Thevenin's model of the inverter, namely a voltage generator with a series impedance, connected to the grid, the equations for active and reactive power exchange can be derived. Assuming $\omega = 2\pi f$ the grid frequency, mainly-inductive interconnection impedances, and small voltage and phase differences, the following linearized equations yield [15, 16]:

$$P \simeq \gamma_p \varphi_i, \quad Q \simeq \gamma_q (V_i - V_g), \quad \text{with} \quad \gamma_p \triangleq \frac{V_g^2}{\omega L}, \ \gamma_q \triangleq \frac{V_g}{\omega L} \tag{3}$$

Such relations hold independently for each of the phases of a 3ϕ-with-N connection. In this case, there is no interdependence between the phases, that is, the power exchange at one phase does not affect the power exchange of the other phases. This is not the case in 3ϕ-without-N connection, which is originally analyzed herein for the controller in Fig. 3 and Fig. 4.

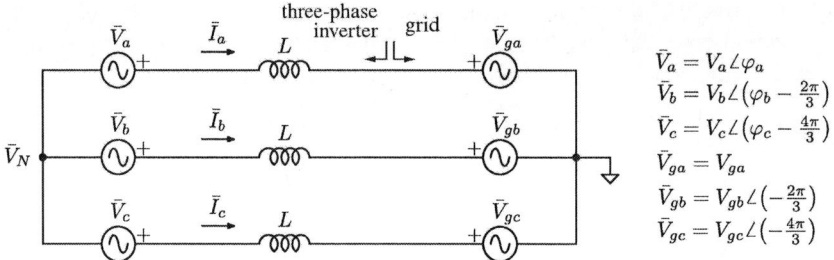

Fig. 5: Thevenin model of a grid-connected three-wires three-phase voltage inverter.

Consider the 3φ-without-N connection of Fig. 5 with the indicated nomenclature for voltage phasors. In this case, full independent power control is not physically possible. That is, the power exchanged by one generator is affected by the phase and voltage differences imposed by the others, because the neutral point voltage results $\bar{V}_N = -\left(\bar{V}_a + \bar{V}_b + \bar{V}_c\right)/3$—under the hypothesis of a symmetric grid voltage.

By exploiting the linearized power exchange equations from the Thevenin's model (3), the following matrix representation can be achieved:

$$\begin{bmatrix} P_a \\ P_b \\ P_c \\ Q_a \\ Q_b \\ Q_c \end{bmatrix} = \frac{1}{6} \underbrace{\begin{bmatrix} 4\gamma_p & \gamma_p & \gamma_p & 0 & \sqrt{3}\gamma_q & -\sqrt{3}\gamma_q \\ \gamma_p & 4\gamma_p & \gamma_p & -\sqrt{3}\gamma_q & 0 & \sqrt{3}\gamma_q \\ \gamma_p & \gamma_p & 4\gamma_p & \sqrt{3}\gamma_q & -\sqrt{3}\gamma_q & 0 \\ 0 & -\sqrt{3}\gamma_p & \sqrt{3}\gamma_p & 4\gamma_q & \gamma_q & \gamma_q \\ \sqrt{3}\gamma_p & 0 & -\sqrt{3}\gamma_p & \gamma_q & 4\gamma_q & \gamma_q \\ -\sqrt{3}\gamma_p & \sqrt{3}\gamma_p & 0 & \gamma_q & \gamma_q & 4\gamma_q \end{bmatrix}}_{M} \cdot \begin{bmatrix} \varphi_a \\ \varphi_b \\ \varphi_c \\ \Delta V_a \\ \Delta V_b \\ \Delta V_c \end{bmatrix} \qquad (4)$$

where $\Delta V_x = V_x - V_{gx}$. It can be shown that the rank of $M \in \mathbb{R}^{6\times6}$ is 4, meaning that only four variables can be arbitrarily regulated, while the total six active and reactive powers for phases a, b, and c can not. A possibile choice may be the control of P_a, P_b, Q_a, Q_b, or the control of P_a, P_b, P_c, $Q_{3\phi} = \sum Q_x$. Let us consider the latter option. Then, control of P_a, P_b, P_c, $Q_{3\phi}$ can be achieved with a single voltage amplitude reference signal, the same for all the three-phases of the inverter:

$$\begin{bmatrix} P_a \\ P_b \\ P_c \\ Q_{3\varphi} \end{bmatrix} = \frac{1}{6} \begin{bmatrix} 4\gamma_p & \gamma_p & \gamma_p & 0 \\ \gamma_p & 4\gamma_p & \gamma_p & 0 \\ \gamma_p & \gamma_p & 4\gamma_p & 0 \\ 0 & 0 & 0 & 18\gamma_q \end{bmatrix} \cdot \begin{bmatrix} \varphi_a \\ \varphi_b \\ \varphi_c \\ \Delta V \end{bmatrix} \qquad (5)$$

Analysis and Design of the Controller for the 3φ-without-N Connection

The proposed controller can be analyzed by deriving its state-space representation in the form:

$$\begin{cases} \dot{x} = Ax + Bu \\ y = Cx + Du \end{cases} \qquad (6)$$

which allows design and additional analyses (e.g., stability analysis, state-in-mode participation factors analysis, etc.). Let state vector $x = \left[\varphi_{3\varphi}, \Delta\varphi_a, \Delta\varphi_b, \Delta\varphi_c, P^*_{3\varphi}, Q^*_{3\varphi}\right]^T$, output vector $y = [P_a, P_b, P_c, Q_a, Q_b, Q_c]^T$ and input vector $u = \left[P^{ref}_a, P^{ref}_b, P^{ref}_c, Q^{ref}_{3\varphi}\right]^T$.

Parameter		Value	
P-f droop coefficient	k_p	0.2094	mHz/W
3-phase P saturation limit	$\pm P_{3\varphi}^{*\text{sat}}$	± 6	kW
Q – V droop coefficient	k_q	0.9167	mV/VAr
3-phase Q saturation limit	$\pm Q_{3\varphi}^{*\text{sat}}$	± 6	kVAr
3-phase P contr. integ. gain	$h_{i,3\varphi}^{\text{p}}$	4.3566	1/s
per-phase P contr. integ. gain	$h_{i,x}^{\text{p}}$	0.6283	mrad/Ws
3-phase Q contr. integ. gain	$h_{i,3\varphi}^{\text{q}}$	16.923	1/s
nominal voltage amplitude	V_g	$110\sqrt{2}$	V
nominal frequency	ω	$2\pi 50$	rad/s
inductive impedance	L	2.8	mH

Table II: System parameters. 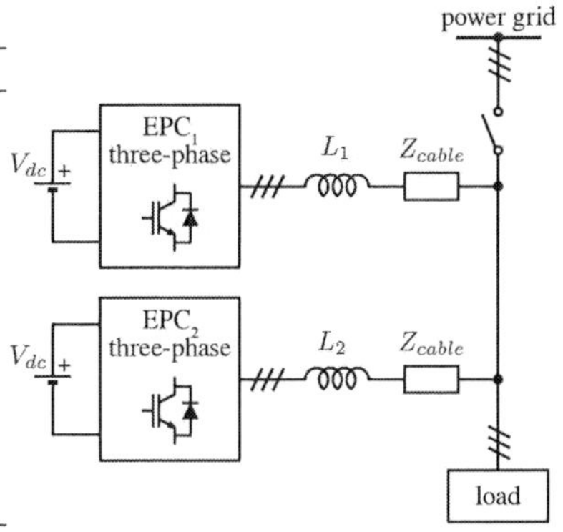
Fig. 6: Structure of the prototype.

The state-space matrices result:

$$A = \begin{bmatrix} -3\gamma_p k_p & 0 & 0 & 0 & k_p & 0 \\ 0 & -\frac{1}{2}h_{i,a}^{\text{p}}\gamma_p & 0 & 0 & 0 & 0 \\ 0 & 0 & -\frac{1}{2}h_{i,b}^{\text{p}}\gamma_p & 0 & 0 & 0 \\ 0 & 0 & 0 & -\frac{1}{2}h_{i,c}^{\text{p}}\gamma_p & 0 & 0 \\ -3h_{i,3\varphi}^{\text{p}}\gamma_p & 0 & 0 & 0 & 0 & 0 \\ 0 & 0 & 0 & 0 & 0 & -h_{i,3\varphi}^{\text{q}}\frac{3\gamma_q k_q}{1+3\gamma_q k_q} \end{bmatrix}$$

$$B = \begin{bmatrix} 0 & 0 & 0 & 0 \\ \frac{2}{3}h_{i,a}^{\text{p}} & -\frac{1}{3}h_{i,b}^{\text{p}} & -\frac{1}{3}h_{i,c}^{\text{p}} & 0 \\ -\frac{1}{3}h_{i,a}^{\text{p}} & \frac{2}{3}h_{i,b}^{\text{p}} & -\frac{1}{3}h_{i,c}^{\text{p}} & 0 \\ -\frac{1}{3}h_{i,a}^{\text{p}} & -\frac{1}{3}h_{i,b}^{\text{p}} & \frac{2}{3}h_{i,c}^{\text{p}} & 0 \\ h_{i,3\varphi}^{\text{p}} & h_{i,3\varphi}^{\text{p}} & h_{i,3\varphi}^{\text{p}} & 0 \\ 0 & 0 & 0 & h_{i,3\varphi}^{\text{q}} \end{bmatrix} \quad C = \begin{bmatrix} \gamma_p & \frac{1}{2}\gamma_p & 0 & 0 & 0 & 0 \\ \gamma_p & 0 & \frac{1}{2}\gamma_p & 0 & 0 & 0 \\ \gamma_p & 0 & 0 & \frac{1}{2}\gamma_p & 0 & 0 \\ 0 & 0 & -\frac{\sqrt{3}}{6}\gamma_p & \frac{\sqrt{3}}{6}\gamma_p & 0 & \frac{\gamma_q k_q}{1+3\gamma_q k_q} \\ 0 & \frac{\sqrt{3}}{6}\gamma_p & 0 & -\frac{\sqrt{3}}{6}\gamma_p & 0 & \frac{\gamma_q k_q}{1+3\gamma_q k_q} \\ 0 & -\frac{\sqrt{3}}{6}\gamma_p & \frac{\sqrt{3}}{6}\gamma_p & 0 & 0 & \frac{\gamma_q k_q}{1+3\gamma_q k_q} \end{bmatrix}$$

(7)

while $D = [0]_{6\times4}$.

State-space representation allows to study system stability through its eigenvalues, and eventually to design the regulators through pole allocation. To this purpose, it is possible to compute in closed-form eigenvalues of matrix A as the roots of $\det[sI - A]$. The characteristic polynomial yields:

$$\psi(s) = \left(s + h_{i,3\varphi}^{q}\frac{3\gamma_q k_q}{1+3\gamma_q k_q}\right)\left(s + \frac{1}{2}\gamma_p h_{i,x}^{\text{p}}\right)^3\left(s^2 + 3\gamma_p k_p s + 3\gamma_p k_p h_{i,3\varphi}^{\text{p}}\right) \tag{8}$$

By (8), controller design can be performed by allocating the eigenvalues of matrix A, on the basis of, for example, time response specifications. The last factor in (8) relates to a complex-conjugate pair of eigenvalues, whose damping factor can be adjusted by control. For example, the critically damped solution (i.e., two coincident real poles) is obtained for $h_{i,3\varphi}^{p} = \frac{3}{4}\gamma_p k_p$, while the damping factor of $\xi = \frac{1}{\sqrt{2}}$ is obtained for $h_{i,3\varphi}^{p} = \frac{3}{2}\gamma_p k_p$. In the following, system parameters in Table II are considered. In this case, $\xi = \frac{1}{\sqrt{2}}$, while the time constants are chosen to be all the same.

Simulations and experimental results

Simulations and experimental tests were performed to validate the control scheme proposed. For experimental tests, the laboratory-scale prototype displayed in Fig. 7 has been employed, implementing the circuit shown in Fig. 6. The controller parameters used are listed in Table II.

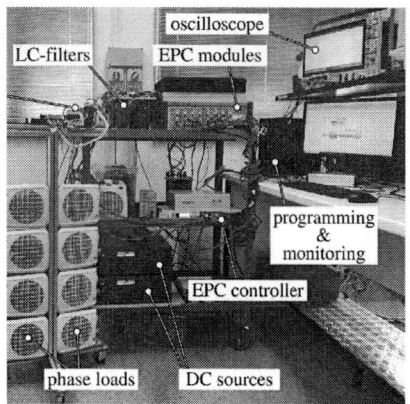

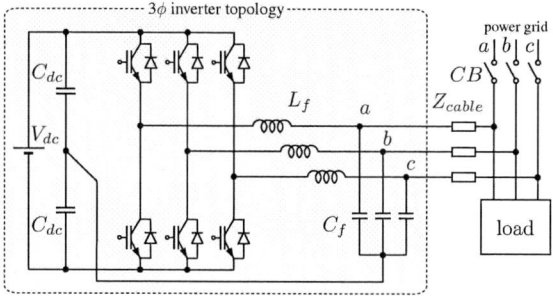

Fig. 7: Laboratory-scale prototype employed in experimental tests.

Fig. 8: Three-phase inverter implemented for validation.

Simulation results

Fig. 9 shows the system response of a single EPC to step variations of power references. At $t = 1$ s the first step variation is performed, changing $P_a, P_b, P_c : 0 \rightarrow 500$ W. Then at $t = 3$ s, $Q_{3\phi} : 0 \rightarrow 1.5$ kW. At $t = 5$ s the system is led to unbalanced operating condition by changing $P_c : 500 \rightarrow 750$ W. It is possible to see that power reference signals are tracked with zero steady-state error, as expected. Fig. 10 shows the behavior of two different EPCs connected to the grid, both designed according to Table II. A three-phase 10 Ω resistive load is also connected to the grid. Initially, EPC$_1$ is supplying 1.5 kW active power, while EPC$_2$ outputs zero active power. Both EPCs supply zero reactive power. At $t = 1$ s the main grid is disconnected: output power regulation is no more possible, thus regulation loops gradually saturate. Between 2 s and 3 s saturation limits are reached, and the controllers smoothly change their operation mode from output power regulation to voltage-forming mode in a droop-like operation.

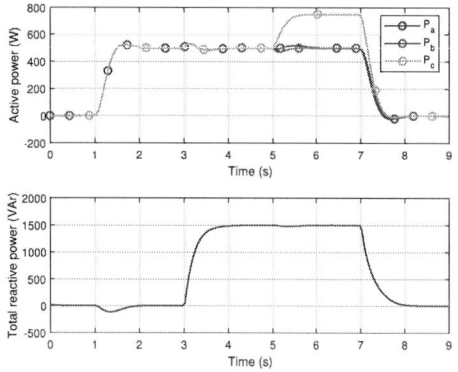

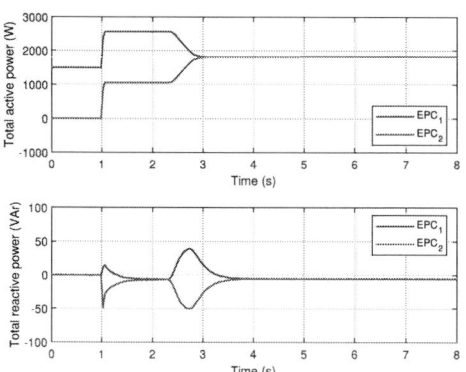

Fig. 9: Step responses of EPC$_1$ while operating grid-tied.

Fig. 10: Grid-connected to islanded transition of EPC$_1$ and EPC$_2$ of Fig. 6.

Experimental results

Fig. 11 shows system response to an unbalanced step change of reference signals. In particular, the figure on the left shows output power waveforms P_a, P_b, P_c, and $Q_{3\phi}$, while instantaneous phase voltages and currents are shown on the right. At $t = 0$ the step-change $P_c : 0 \rightarrow 1$ kW is performed. The system correctly tracks the reference signal with zero steady-state error. There is an overshoot visible on P_a and P_b, due to the fact that also total active power is changed. Thus, both synchronization and per-phase control loops are involved in the regulation. The undershoot in $Q_{3\phi}$ is due to the increase of output active current that causes a voltage drop on the resistive part of the line impedance and leads to an initial reactive power absorption, which is eventually compensated by the controller.

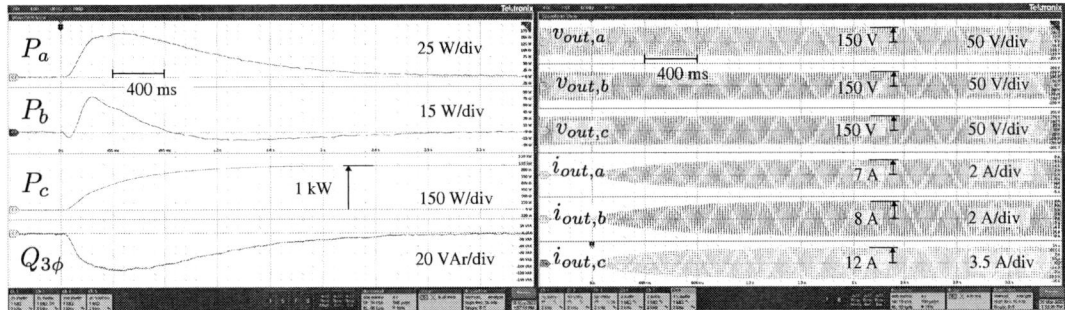

Fig. 11: Unbalanced power reference step-change $P_c : 0 \rightarrow 1\,\text{kW}$ is performed. Output power waveforms are on the left, while instantaneous output voltages and currents are displayed on the right.

Conclusions

A controller achieving per-phase output power regulation in three-phase three-wires systems is proposed herein. In three-phase three-wires systems output active and reactive power can not be independently controlled at each phase: a suitable control scheme is required to perform per-phase output power control with the limited degrees of freedom imposed by the absence of the neutral connection. Together with per-phase output power control, the possibility of island operation is fundamental for uninterrupted power supply in microgrids in case of disconnection from the main grid. The provided experimental validation shown that the proposed controller succeeds in harmoniously integrating phase-by-phase output power tracking and the capability of smoothly transitioning into the islanded operation.

References

[1] S. Chandak and P. K. Rout, "Microgrids During the Outbreak of COVID-19," *IEEE Smart Grid Newslett.*, July 2020.

[2] M. Shahidehpour, M. Yan, P. Shikhar, S. Bahramirad, and A. Paaso, "Blockchain for peer-to-peer transactive energy trading in networked microgrids: Providing an effective and decentralized strategy," *IEEE Electrif. Mag.*, vol. 8, no. 4, pp. 80–90, 2020.

[3] A. S. Vijay, S. Doolla, and M. C. Chandorkar, "Unbalance mitigation strategies in microgrids," *IET Power Electron.*, vol. 13, no. 9, pp. 1687–1710, 2020.

[4] J. M. Guerrero, J. C. Vasquez, J. Matas, L. G. de Vicuna, and M. Castilla, "Hierarchical Control of Droop-Controlled AC and DC Microgrids—A General Approach Toward Standardization," *IEEE Trans. Ind. Electron.*, vol. 58, no. 1, pp. 158–172, 2011.

[5] J. Rocabert, A. Luna, F. Blaabjerg, and P. Rodríguez, "Control of Power Converters in AC Microgrids," *IEEE Trans. Power Electron.*, vol. 27, no. 11, pp. 4734–4749, 2012.

[6] H. Abedini, T. Caldognetto, P. Mattavelli, and P. Tenti, "Real-Time Validation of Power Flow Control Method for Enhanced Operation of Microgrids," *Energies*, vol. 13, no. 22, 2020. [Online]. Available: https://www.mdpi.com/1996-1073/13/22/5959

[7] J. Wang, N. Zhou, Y. Ran, and Q. Wang, "Optimal Operation of Active Distribution Network Involving the Unbalance and Harmonic Compensation of Converter," *IEEE Trans. Smart Grid*, vol. 10, no. 5, pp. 5360–5373, 2019.

[8] F. H. M. Rafi, M. Hossain, M. S. Rahman, and S. Taghizadeh, "An overview of unbalance compensation techniques using power electronic converters for active distribution systems with renewable generation," *Renewable and Sustainable Energy Reviews*, vol. 125, p. 109812, 2020.

[9] P. Tenti and T. Caldognetto, "On Microgrid Evolution to Local Area Energy Network (E-LAN)," *IEEE Trans. Smart Grid*, vol. 10, no. 2, pp. 1567–1576, 2019.

[10] D. I. Brandao, L. S. Araujo, A. M. S. Alonso, G. L. dos Reis, E. V. Liberado, and F. P. Marafão, "Coordinated Control of Distributed Three- and Single-Phase Inverters Connected to Three-Phase Three-Wire Microgrids," *IEEE Trans. Emerg. Sel. Topics Power Electron.*, vol. 8, no. 4, pp. 3861–3877, 2020.

[11] C. Burgos-Mellado, R. Cárdenas, D. Sáez, A. Costabeber, and M. Sumner, "A Control Algorithm Based on the Conservative Power Theory for Cooperative Sharing of Imbalances in Four-Wire Systems," *IEEE Trans. Power Electron.*, vol. 34, no. 6, pp. 5325–5339, 2019.

[12] T. Caldognetto, H. Abedini, and P. Mattavelli, "A Per-Phase Power Controller for Smooth Transitions to Islanded Operation," *IEEE Open Journal of Power Electronics*, vol. 2, pp. 636–646, 2021.

[13] A. Timbus, M. Liserre, R. Teodorescu, P. Rodriguez, and F. Blaabjerg, "Evaluation of Current Controllers for Distributed Power Generation Systems," *IEEE Trans. Power Electron.*, vol. 24, no. 3, pp. 654–664, 2009.

[14] S. Lissandron and P. Mattavelli, "A controller for the smooth transition from grid-connected to autonomous operation mode," in *2014 IEEE Energy Convers. Congr. & Expo.*, 2014, pp. 4298–4305.

[15] K. De Brabandere, B. Bolsens, J. Van den Keybus, A. Woyte, J. Driesen, and R. Belmans, "A Voltage and Frequency Droop Control Method for Parallel Inverters," *IEEE Trans. Power Electron.*, vol. 22, no. 4, pp. 1107–1115, 2007.

[16] W. Yao, M. Chen, J. Matas, J. M. Guerrero, and Z. Qian, "Design and Analysis of the Droop Control Method for Parallel Inverters Considering the Impact of the Complex Impedance on the Power Sharing," *IEEE Trans. Ind. Electron.*, vol. 58, no. 2, pp. 576–588, 2011.

Efficiently Paralleling GaN-Transistors for High Current and High Frequency Applications Using a Butterfly Layout

Martin Wattenberg, Oscar Lorenz, Juan Sanchez
Infineon Technologies AG
Siemensstraße 2
Villach, Austria
E-Mail: martin.wattenberg@infineon.com
URL: http://www.infineon.com

Keywords

«Gallium Nitride (GaN)», «Device application», «Discrete power device», «High frequency power converter», «High power density systems»

Abstract

This paper presents a scalable design for up to 8 GaN transistors per switch in a three-phase motor drive. To minimize the layout related issues, a fully symmetrical "butterfly" layout was considered. An experimental three-phase prototype with a total of 8 chips per switch was built to verify the concept. Due to the highly symmetrical nature, this work focuses on four half-bridges in parallel, i.e. 4 chips per switch. For evaluation, the design is operated as a 100 kHz buck converter with 50% duty cycle. At 24 and 48 V output currents up to 100 and 60 A respectively are achieved reliably. Measured voltage transition times of 6 ns for all devices and minimal variations between device's temperature (ΔT_j) of 12 and 5 K for 24 V/100 A and 48 V/ 60 A respectively confirm a correct operation.

Introduction

In order to achieve high-current converters the use of power switches in parallel configurations is the most common approach both for Si as well as Gallium Nitride High Electron Mobility Transistors (GaN HEMTs). Due to the typically high-frequency operation in GaN-based applications, getting optimal designs with devices in parallel is a challenging task. Many GaN HEMT-based studies have evaluated the circuit parasitic influence on the switching behavior of GaN transistors in parallel. Others investigated the effects of non-identical electrical properties of those devices [1,2]. The study in [3] showed that board level parasitics had a stronger impact on power loss than device's parameter variation, which was within 10% for V_{th} and 25% for $R_{DS(on)}$ for investigated commercial devices.

GaN HEMTs are majority carrier devices like silicon power MOSFETs (Si MOSFETs), thus similar design considerations for paralleling apply. A comprehensive overview of the fundamental considerations for parallel operation of Si MOSFETs is given in [4]. Some causes for current unbalance like gate driver and layout mismatch can be avoided or minimized by using a single gate driver and a symmetrical layout respectively. Other effects like threshold voltage variation need to be accounted for. However, it should be noted that some imbalance may in fact be acceptable, depending on the relationship between switching loss P_{sw} and the conduction loss P_c as well as the devices safe operating area. Passive balancing method based on cross-coupled transformer feedback presented either in [5] for the phase current or in [6] in the gate path are an interesting solution if the application profile exhibits a high relationship of P_{sw}/ P_c. For motor drive applications this is often not the case.

The work presented in [7] uses four 7 mΩ GaN HEMTs in parallel to achieve 35 A output current. A similar layout concept is used in [8] and the output current capability is increased to 70 A_{RMS}. However, the implementation is more complex and requires a larger footprint on the PCB. A separate isolated supply provides bipolar gate voltage to suppress parasitic Miller turn-on caused by the limited hold-down capability of the gate driver used. For the same reason, the turn-on speed needed to be reduced by an additional 6 Ω resistor in the turn-on gate path.

The presented "butterfly" layout concept extends paralleling to 8 devices. Current loops are separated by design. The layout benefits for a more suitable compact gate driver with 4 A peak current capability and integrated charge pump for generation of negative rails. The concept is discussed in detail and experimentally verified.

Impact of Parameter Variation for Paralleled Power Devices

Similar to Si MOSFETs, GaN HEMTs exhibit a positive temperature coefficient for $R_{DS(on)}$ as shown in Fig. 1 (a). The $R_{DS(on)}$ is shown versus junction temperature T_j for a gate source voltage V_{GS} of 5 and 10 V for the HEMT and MOSFET respectively. Comparing the $R_{DS(on)}$ at 25 and 150°C, an increase of 120 versus 91% are observed for the HEMT over the MOSFET. For paralleling, the increased resistance is beneficial as it provides a negative feedback mechanism where hotter devices will conduct less current due to higher resistance compared to cold devices. Similarly, the transfer characteristic of HEMTs is also more strongly influenced by increased temperature than of a MOSFET as can been seen in Fig. 1 (b). Compared to Si MOSFETs, transconductance ($\Delta I_D/\Delta V_{GS}$) for GaN HEMT falls more strongly when the device heats up. This is especially beneficial during switching transients where hotter devices will conduct less current than cold devices for the same V_{GS}. Based on this data GaN HEMTs already show similar negative feedback mechanisms for paralleling but with strong temperature dependency.

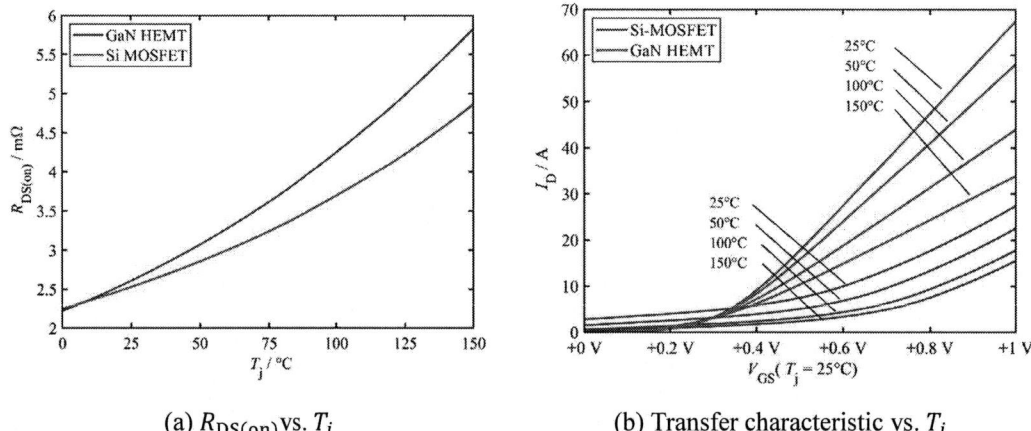

(a) $R_{DS(on)}$ vs. T_j (b) Transfer characteristic vs. T_j

Fig. 1: Comparison of $R_{DS(on)}$ in (a) and transfer characteristic in (b). For comparison axis of (b) has been normalized, see Fig. 2

In addition to $R_{DS(on)}$ and the transfer characteristic another important aspect is temperature dependency of V_{th}. For Si MOSFETs this strong temperature dependency can often be a reason for thermal runaway where a hotter device will have a lower V_{th}, conduct more current, have higher losses and heat up further. Even though a similar temperature dependency exists for GaN, the impact is much less pronounced as Fig. 2 shows. For the GaN HEMT V_{th} only drops by about 1 mV/K whereas for the considered MOSFET close to 4 mV/K.

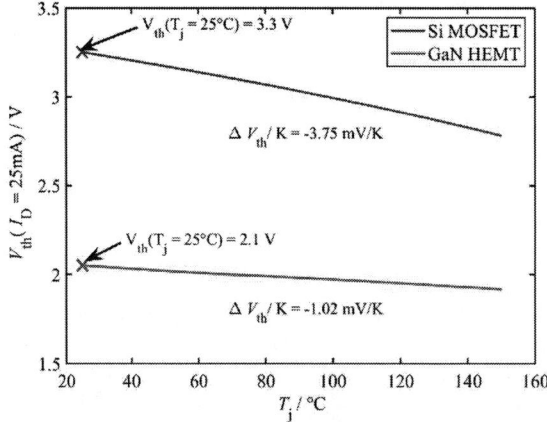

Fig. 2: Threshold voltage V_{th} versus junction temperature for a 100 V 3.1 mΩ GaN HEMT and 100 V 3 mΩ state-of-the-art Si MOSFET.

Design Consideration for a GaN-based Motor Drive

DC-link Capacitance

The DC-link plays a major part in the size of the final inverter. Following the considerations of [9] the current ripple depends on the modulation index M as well as on the $\cos\phi$ of the motor. It is independent of the switching frequency f_{sw}. For a target design with phase current I_0 of 100 A$_{rms}$ the RMS DC-link current I_{DC} can be approximated as follows:

$$I_{DC} = 0.65 \cdot I_0. \tag{1}$$

This leads to a worst-case RMS current of 65A which can be easily met with multi-layer ceramics capacitors (MLCCs). Providing a similar amount of bulk capacitance compared to electrolytic capacitors can be challenging. However, the voltage ripple $\Delta v_{pp,max}$ of the DC-link is inversely proportional to f_{sw}. Due to the lower switching losses of GaN HEMTs it is possible to meet the requirements of $\Delta v_{pp,max}$ at higher f_{sw} and lower bulk capacitance without sacrificing efficiency. In this case $\Delta v_{pp,max} = 1.5$ V the input capacitance C_{DC} is calculated as described in [10] as

$$C_{DC} \geq \frac{I_{0,pk}}{4 f_{sw} \Delta v_{pp,max}}. \tag{2}$$

For a switching frequency of 100 kHz this results in 233 µF of capacitance that needs to be distributed across the board. Care must be taken to consider the de-rating when DC-bias is applied. Therefore, some over-provisioning is recommended. Additionally, distributed capacitance produces a lower inductance connection to the DC-link thanks to using many capacitors in parallel.

Low impedance connection to the DC-link

Even though GaN HEMTs exhibit a positive temperature coefficient for $R_{DS(on)}$ and are self-balancing in steady state, this condition is not immediately reached after the transition occurs. Instead, depending on how uneven the current distribution between devices during the switching transition was, the current will follow an exponential approximation towards steady state with a time constant τ_{DC} defined as

$$\tau_{DC} = \frac{L_{DC}}{R_{DS(on)} + R_{DC} + R_{CU}}. \tag{3}$$

As (3) shows, τ_{DC} is proportional to the parasitic inductance of L_{DC} and reciprocal to the effective $R_{DS(on)}$, ESR of the DC-link capacitor R_{DC} and parasitic resistance of the copper traces R_{CU}. This additional time constant further exacerbates the imbalance in device loss that can occur during the switching transition. Increasing the R_{DC} or R_{CU} in favor of a smaller settling time constant is not practical. Instead, the design should minimize L_{DC} as much as possible by alternating layers between V_{in} and GND in the layer stack up.

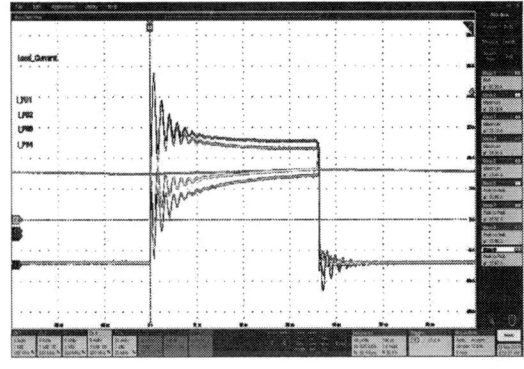

(a) Experimental results based on earlier design with four Si MOSFETs in parallel.

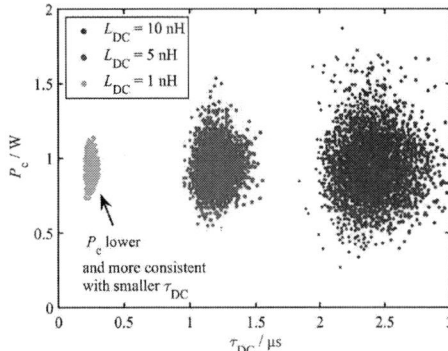

(b) Conduction energy distribution for a typical set of 4 paralleled devices.

Fig. 3: Settling of current waveform in (a) and difference in conduction loss P_c caused by settling time constant τ_{DC} in (b).

Fig.3 (a) shows captured waveforms from previous work done on a non-optimized design using Si MOSFETs. The relatively large value for τ_{DC} limits the maximal attainable switching frequency because the impact of asymmetry will grow with shorter periods. For comparison Fig.3 (b) shows a simulated distribution of P_c for four paralleled devices similar to Fig.3 (a). For the $R_{DS(on)}$ and initial current after switching a normal distribution has been assumed and the system has been simulated several times for different values of L_{DC}. A smaller τ_{DC} due to lower L_{DC} ensures lower and more homogenous distribution of P_c. A multi-layered PCB design with interleaved V_{in} and GND planes combined with distributed capacitance is therefore preferred.

Switching Loss Considerations

As stated in the introduction for a motor drive the conduction loss P_C typically dominates. Based on previous work shown in [11] a loss breakdown has been derived by running a non-paralleled motor drive under the same conditions but at different switching frequencies using an inductive-resistive load. Fig.4 (a) shows the measured losses of the inverter at frequencies ranging from 20 to 100 kHz in 20 kHz increments. Thanks to large inductances, current ripple was negligible in all cases. Based on this data a loss model has been derived to distinguish individual loss components. The quiescent power describes the power draw of all auxiliary components as well as gate driver losses when switching at 100 kHz with no V_{DS} applied. Conduction and deadtime loss were derived from measured device parameters. The remaining difference in loss was attributed to switching loss.

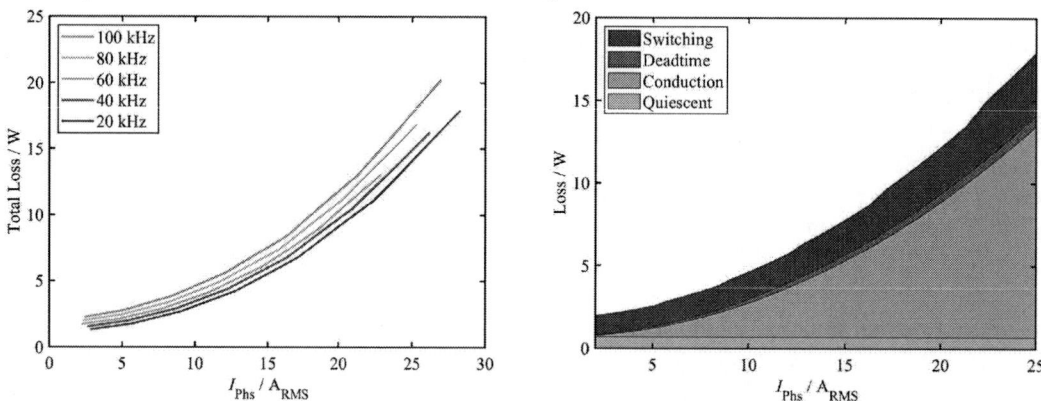

(a) Measured inverter losses from 20 to 100 kHz. (b) Approximated loss breakdown at 100 kHz.
Fig. 4: Sweep of inverter loss and approximated breakdown of inverter losses at 100 kHz.

Fig. 4 shows that even for a highly optimized GaN inverter switching loss is not negligible. However, at full-load the dominant loss mechanism is conduction loss with a contribution close to 75% of the total losses. Thus, some imbalance during switching can still be accepted as long as the device safe operating area is not violated.

Overview of the Proposed Layout with Symmetrical PCB Parasitics

It is of critical importance to minimize PCB parasitics to realize the full potential of fast switching GaN HEMTs. Designing the high-frequency (HF) power loop so that the current can flow vertically to under the devices can significantly reduce the loop inductance over a more traditional horizontal power loop design ([12]). However, layer stack-up and prepreg thickness becomes critical. Fig. 5 shows the cross section of the proposed stack-up for one half-bridge. Using the first inner layer as a ground return creates a tight coupling in the HF power loop. A total loop inductance estimate of about 400 pH is obtained applying the approximated expressions given in [12] for a conductor width of 5, a length of 12 and a height of 0.1 mm. This very low inductance results in a minimal amount of overshoot and satisfies the requirements for τ_{DC}.

Equally important is the tight differential coupling of the high-side (HS) and low-side (LS) gate loop. In particular for the low-side the layout needs to be done carefully as it is referenced to the same ground potential as the power loop. When both loops intersect and share common traces the resulting common source inductance can introduce problematic voltage spikes leading to increased switching loss or even device failure. The high-side devices can exhibit similar issues with respect to the switch node. Thus, a tight coupling of the gate loops with low impedance connection between devices and separated HF power loops are critical aspects of the design.

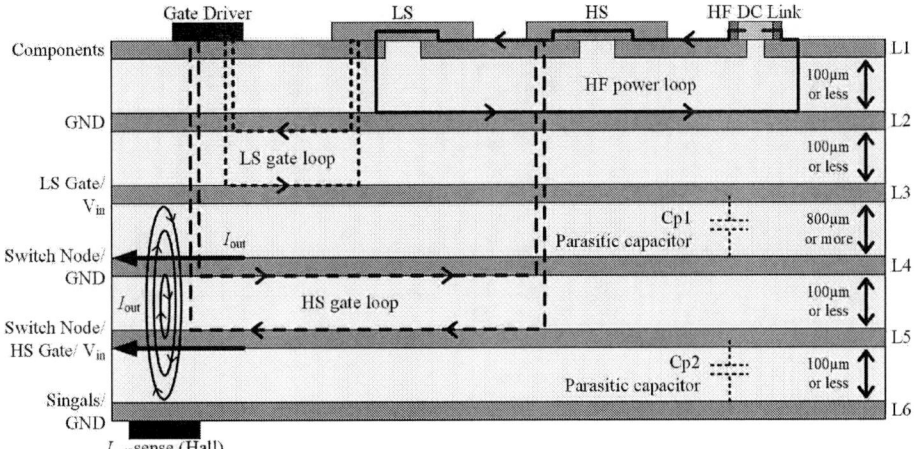

Fig. 5: Layer stack-up for the proposed layout.

Using an identical layout for the switch node on layers 4 and 5 avoids any capacitive coupling between the two layers. However, some parasitic capacitance exists to toward layer 3 and 6. The capacitance can be approximated as a plate capacitor and calculated with

$$C = \varepsilon_0 \varepsilon_r \frac{A}{d}. \tag{4}$$

For a total area of (30mm)² the total capacitance is equivalent to $C_{p1} = 40$ pF and $C_{p2} = 320$ pF. The total effective output capacitance at the switch node is close to 10 nF, thus the additional capacitance introduced by the layout is less than 4% and acceptable for the design.

Instead of one large half-bridge with a single power loop the proposed layout mitigates the challenges of asymmetrical inductances by paralleling half-bridges with separate HF loops and one common gate driver in the center. Fig. 6 shows a simplified overview of the proposed layout. The power and gate loops are indicated.

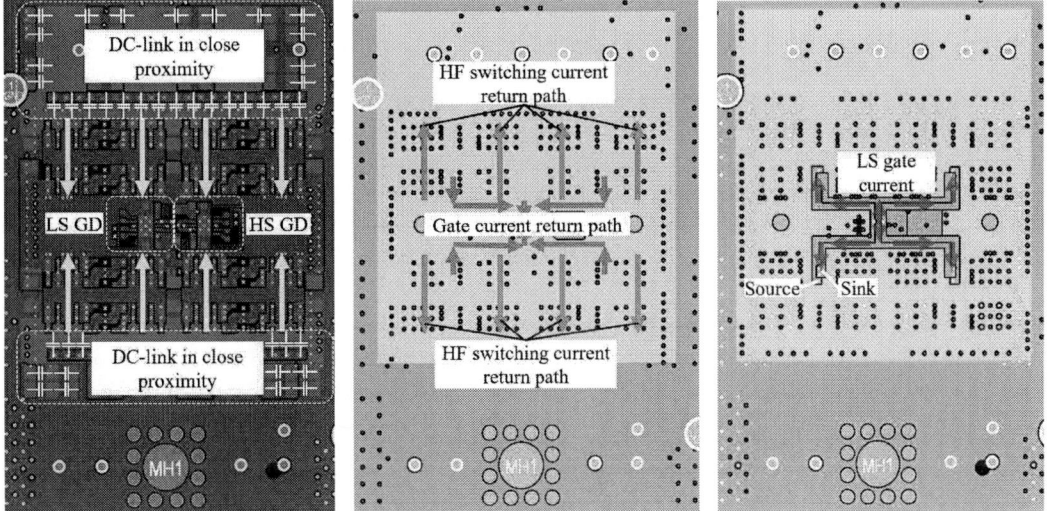

Layer 1: Components and V_{in}.   Layer 2: Ground plane for power loop and gate return.   Layer 3: V_{in} and gate signals.

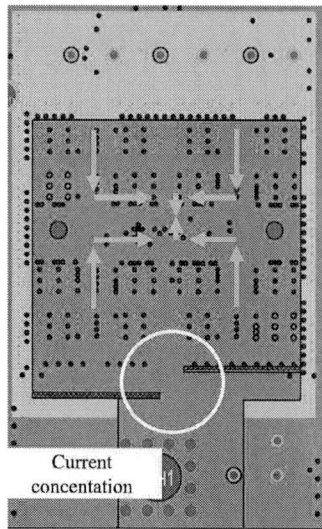 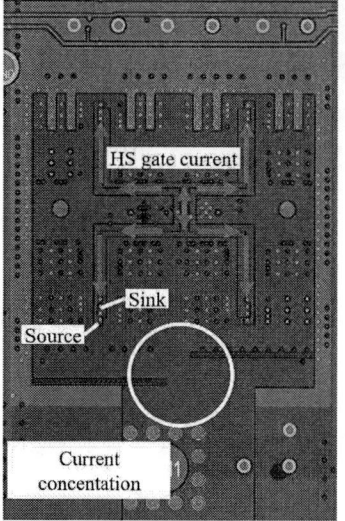 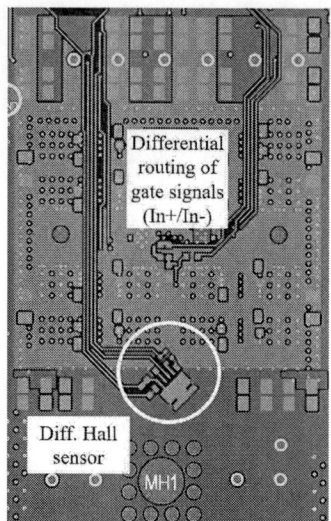

| Layer 4: Switch node, HS gate return and V_{in}. | Layer 5: Switch node, HS gate and ground. | Layer 6: Signals, aux. components and ground. |

Fig. 6: Overview of each layer for one phase of the motor drive.

A Hall sensor, shown on layer 6 of Fig. 6, measures the phase current for current control in a DC/DC or motor drive application. In order to maximize the magnetic field intensity, the current is locally concentrated as can be seen in layer 4 and 5 of Fig. 6. Overall this can provide lower insertion resistance and better scalability when addressing large currents with additional layers. However, thermal imaging does show a slight impact on device temperature for HEMTs located close to the current concentration point. For practical applications this is irrelevant but when looking at device temperature as an indication of current sharing performance this effect must be considered.

Top-side cooling allows to decouple the thermal from the electrical path. This way additional PCB layers can be stacked alternating between V_{in} and GND and reducing the parasitic inductance to the DC-link. In addition, thermal coupling between devices can be improved not only via the PCB but also via the heatsink.

Implementation on Prototype

Following the approach presented in Fig. 5 and 6, a prototype has been built and is shown in Fig.6. The devices used are 100 V Schottky-gate GaN HEMTs with a $R_{DS(on),max}$ of 3.1 mΩ in a 3 x 5 mm package. One 1EDN7116G gate driver is used to drive the gates of the high and low-side switches respectively. The driver features an integrated charge pump capable of generating a negative V_{GS} of up to -3 V and is capable of providing up to 4 A source current with a 5 A miler clamp to avoid parasitic turn-on. In addition, its differential input stage combined with a symmetrical layout of the logic signals significantly reduce the risk of unintended switching events that could otherwise be caused by coupled noise. To sense the phase current, the TLE4972 is used. It features a differential Hall sensor element as well as a differential output. Both significantly mitigate the associated issues of sensing sensitive signals and transmitting analog signals over a comparatively large PCB. The design is copied three times on the circuit board to build a three-phase motor drive.

The use of MLCCs significantly reduces the inverter volume and results in a very low-profile design compared to a more conventional approach with fewer, large electrolytic capacitors. MLCCs are especially attractive for lower voltages where a large capacitance values are available. Optionally, for higher voltages where MLCC often compare unfavorably in energy density the design includes the possibility to use leaded electrolytic capacitors.

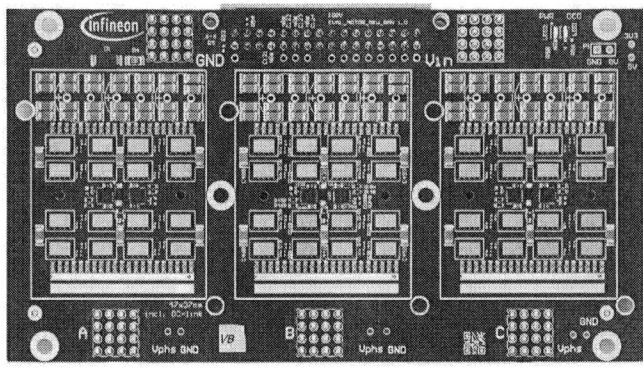 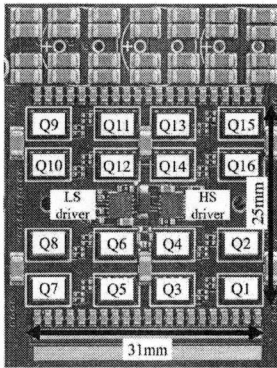

(a) Overview of the three-phase design.   (b) Close-up of single phase.

Fig. 7: Image of the implemented prototype using the proposed layout structure for 8 chips in parallel.

The design in Fig. 7 uses large, low-impedance planes to connect all devices to a common reference (ground or switch node), thus reducing voltage differences as much as possible. However, this approach makes quantifying the symmetry of current sharing impractical. For this reason, a second prototype (not shown) that allows current measurement on each of the eight half-bridges was built. That approach presents its own challenges, some of the low-impedance planes have to be split up and is in general not recommended. The experimental verification was carried out on half of the populated devices. Given the symmetry of the design this does not have any impact on the validity of the results.

Experimental Results

The experimental verification of the design focuses on operating a single phase as a buck converter. If not stated otherwise, a 50% duty cycle at 100 kHz is used.

Symmetry in Transient Waveforms

To judge the symmetry in gate-source voltage waveforms the low-side devices Q6, Q8, Q10 and Q12 as well as the switch node voltage V_{sw} have been measured with 1 GHz passive probes each. For further verification one high-side device (Q7) has been captured with an optically isolated probe (Tek IsoVu) and the total phase current I_{out} was measured with a current probe. Fig. 8 (a) and (b) show the captured waveforms for rising and falling switch-node voltage for 48 V and 60 A output current. The switch-node voltage V_{sw} shows a very clean behavior without overshoot. As initially discussed a stronger gate driver can benefit the design by further bringing down switching losses.

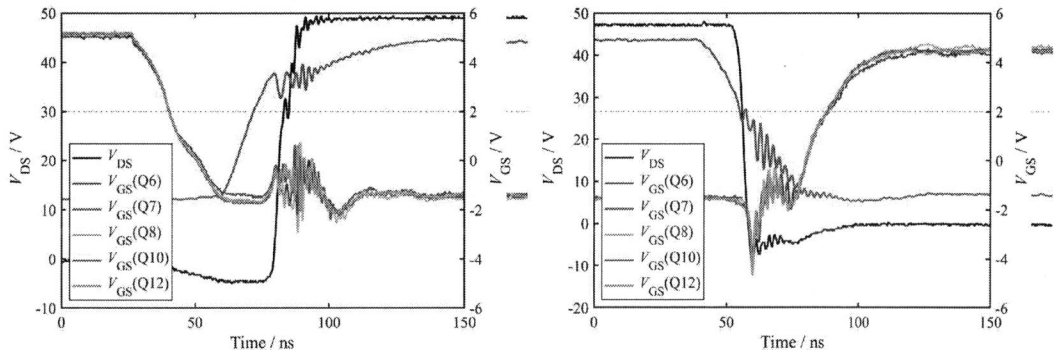

(a) Low side V_{GS} during turn-off.   (b) Low side V_{GS} during turn-on.

Fig. 8: Comparison of switching waveform for gate-source and drain-source.

Fig 8 (a) gives a closer look at low-side turn-off and high-side turn-on transitions. On close inspection minor differences in V_{GS} can be seen for the low-side devices. The captured V_{GS} for Q7 shows minor oscillations during the voltage transition. Prior studies indicate that some of these oscillations are artifacts introduced by the twisted pair cables used to connect probe and PCB. This is further supported by Fig. 8 (b) where oscillations are observed at the threshold level but the switch node smoothly commutates to zero (while the low-side devices are OFF).

Peak Low-Side Gate Voltage During Turn-on and Off

If the voltage on the gate is not well-controlled it can lead to false turn-on or even to failure should the device ratings be exceeded. However, at the same time the undershoot during falling V_{DS} must respect device rating. Therefore Fig. 9 evaluates the peak negative voltage during soft transition on the synchronous low-side devices. The peak negative voltage is highlighted in Fig. 9 (a) for 24, 36 and 48 V at 60 A. This spike correlates with the falling voltage of the switching node. The trend of the peak negative voltage is shown in Fig. 9 (b). One clearly sees that it gets larger with both increasing output current and higher voltages.

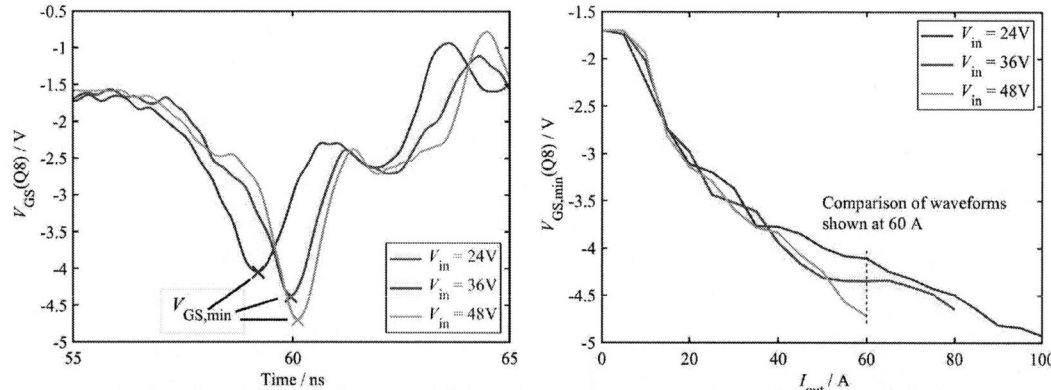

(a) Low-side V_{GS} waveform after turn-off of complementary high-side device.

(b) Peak negative voltage during falling V_{DS} as a function of output current

Fig. 9: Comparison of low-side V_{GS} during soft commutation for 24, 36 and 48 V at 60 A in (a). The trend of peak negative voltage is shown in (b).

Even though a clear trend towards higher negative voltage spike can be observed, it must be noted that the observed levels remain uncritical. The devices rating allows for up to -6.5 V transient gate voltage, thus significant margin still exist in the design.

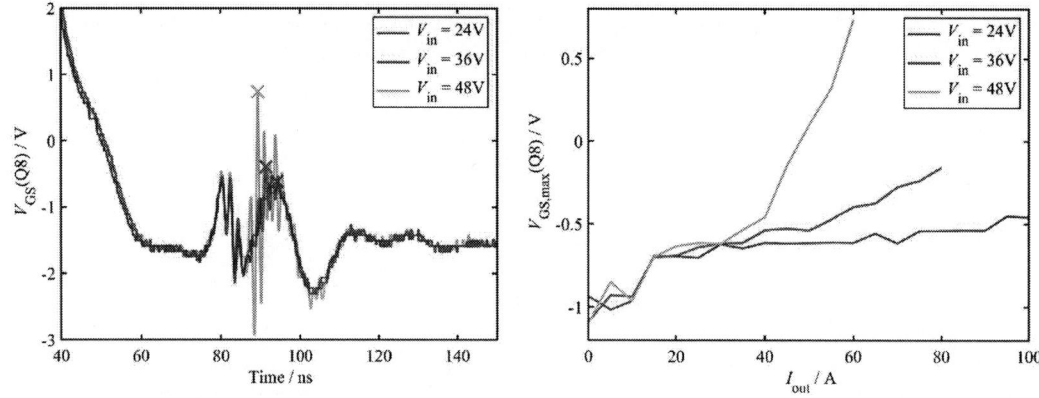

(a) Low-side V_{GS} waveform during turn-on of complementary high-side device.

(b) Peak positive voltage during turn-on of complementary high-side device.

Fig. 10: Low-side V_{GS} during high-side turn-on / rising V_{DS}.

Similar to the negative voltage spike in Fig. 9 a positive spike on V_{GS} maybe be introduced during turn-on of the complementary switch as shown in Fig. 10.

Symmetry in Current Sharing

To evaluate the symmetry of current sharing individual phase currents are measured on the separate design variant. The average output current of each of the four phases are shown as a function of total output current in Fig 11. The current difference is less than 8% in one of the phases and indistinguishable for the others.

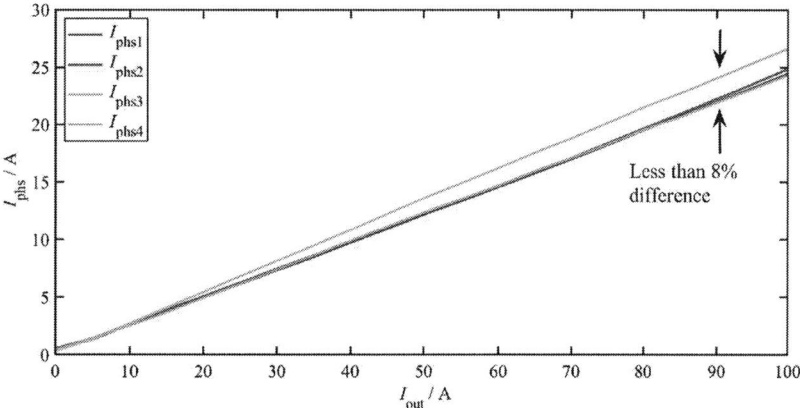

Fig. 11: Current of the four investigated phases.

In addition, the temperature distribution of the devices was captured as a secondary way of evaluating current distribution in paralleled designs. The exposed top-side of the die was covered in black paint and temperatures were observed with a thermal camera. Under normal operation conditions a heatsink can be placed directly on the HEMTs to lower thermal resistance. Fig. 12 (a), (b) and (c) show the temperature for all considered devices for 48, 36 and 24 V respectively. When operating at $V_{in} = 36$ or 48 V the output power is 1.44 kW. For the operation at 24 V the output power is limited to 1.2 kW due to a thermal limit of 130°C. It should be noted that some thermal difference is created due to the current concentration in the layout nearby the Hall sensor. Slightly elevated temperatures in the vicinity around Q5 are therefore expected.

When operating at 48 V with $I_{out} = 60$ A the temperature distribution across all the devices is very even. The peak temperature is observed for the high-side at Q5 with 61°C and the lowest for the low-side at Q8 at 55°C. Lowering V_{in} to 36 V and increasing the output current to 80 A produces a very similar temperature distribution. The peak temperature remains at 82°C on Q5.

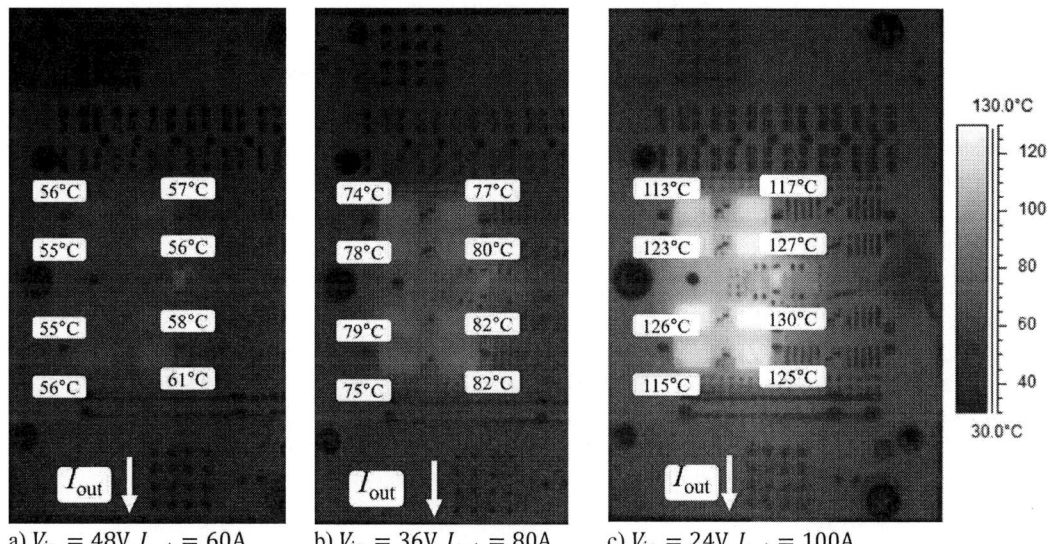

a) $V_{in} = 48V$, $I_{out} = 60A$ b) $V_{in} = 36V$, $I_{out} = 80A$ c) $V_{in} = 24V$, $I_{out} = 100A$

Fig. 12: Temperature distribution for different input voltage levels at maximum investigated current.

Conclusion

A scalable layout for paralleling a large number of power transistors with a focus on Gallium-Nitride (GaN) transistors is proposed. Compared to Si, GaN devices exhibit a stronger dependency of both $R_{DS(on)}$ and transconductance with temperature. This stronger dependency leads to an enhanced self-balancing effect. Although the threshold voltage V_{th} of GaN is significantly lower than that of silicon devices it has a weaker temperature dependency (1.02 mV/K vs. 3.75 mV/K), therefore experiencing a reduced positive feedback effect for current unbalance.

Although not negligible, even at increased carrier frequencies of 100 kHz the switching losses in motor drives applications are shown to play a secondary role with a contribution close to 25% to the total power semiconductor losses. Therefore, some imbalance during the switching transition can be acceptable and the above stated technology parameters considerations hold to be the most relevant.

Still, the fast switching speed of GaN mandates a symmetrical layout with low parasitics in order to ensure a fair spread of current through the PCB board and paralleled devices. In addition, the design benefits from a negative V_{GS} generated directly by the gate driver used. The proposed layout effectively establishes high frequency switching loops of high-side and low-side pairs rather than a single common shared loop for all the paralleled devices. This approach is demonstrated to result in a very even current sharing with less than 8% difference between phases. It achieves as well a uniform temperature distribution with measurements showing less than 5 K of difference between the hottest and the coldest devices at a nominal 48 V input voltage.

References

[1] Wickramasinghe T.: An investigation of current distribution over four GaN HEMTs in parallel configurations, WiPDA, Oct. 2019

[2] Wu Y.-F.: Paralleling High-speed GaN Power HEMTs for Quadrupled Power Output, APEC, Mar. 2013

[3] Wickramasinghe T.: Electrical property variability of GaN transistors in parallel and their impact on fast switching operations, EPE'20 ECCE Europe, Sept. 2020

[4] Forsythe J.: Paralleling of Power MOSFETs For Higher Power Output, International Rectifier, 1996

[5] Lu S.: A Passive Transient Current Balancing Method for Multiple Paralleled SiC-MOSFET Half-Bridge Modules, APEC, Mar. 2019

[6] Hackel J.: A Novel Gate Driving Approach to Balance the Transient Current of Parallel-Connected GaN-HEMTs, CIPS Mar. 2018

[7] Reusch D.: Effectively Paralleling Gallium Nitride Transistors for High Current and High Frequency Applications, APEC, Mar. 2015

[8] Burkhard J.: Paralleling GaN switches for low voltage high current half-bridges, ECCE, Sept. 2019

[9] Kolar J.W.: Paper on DC link, IEEE Transaction on Industry Applications, Mar. 2021

[10] Vujacic M.: Analysis of dc-Link Voltage Switching Ripple in Three-Phase PWM Inverters, Energies, Nov. 2018

[11] Wattenberg M.: A Low-Profile GaN-Based Integrated Motor Drive for 48V FOC Applications, PCIM, May 2021

[12] Sun B.: Research of Power Loop Layout and Parasitic Inductance in GaN Transistor Implementation, IEEE Transaction on Industry Applications, Mar. 2022

Data-driven decentralized volt/var control for smart PV inverters in distribution systems

Yizhou Lu, Qianwen Xu, Lars Nordström
KTH Royal Institute of Technology
Email: yizhoul@kth.se

Acknowledgments

This work was supported by Digital Futures on project "Autonomous coordination and control of smart converters for sustainable power systems".

Keywords

≪Voltage regulation≫,≪Decentralized control structure≫, ≪Neural network≫, ≪Reactive power≫, ≪Renewable energy systems≫

Abstract

The growing penetration of renewable energy sources (RES) in modern grids may result in severe voltage violation problems due to high stochastic features. Conventional centralized approaches could provide optimal solutions for voltage regulation while with great communication burdens. Control methods based on local information usually have non-optimal results and cannot always guarantee voltage security. This paper proposes a neural network-based decentralized strategy for volt/var control using inverter reactive power capacity. Learning from optimal power flow (OPF) results of historical data, the developed controller can provide optimal results approximate to centralized solutions and outperform local control methods in minimizing the power loss. The proposed method is tested on the IEEE 33-bus system and simulation results illustrate the effectiveness in voltage regulation and loss minimization.

Introduction

Renewable energy resources (e.g., solar PV and wind) have been deeply participated in modern distribution power system development, playing essential roles in achieving the sustainable goal. However, due to their high stochastic and uncertain nature, the increased penetration of renewables brings voltage stability issues and has attracted global attention in recent years [1].

Voltage stability is one of the necessary prerequisites and basic guarantees for secure power grid operations. Voltage/Var control (VVC) can mitigate the voltage violation and reduce the power loss. Some devices are developed to provide reactive power support, such as static Var compensators (SVCs) and static Var generations (SVGs) [2]. As the interface of renewables, inverters are proved to be a cost-effective and flexible device for reactive power support and voltage regulation [3] [4], which are controlled based on power injection demand.

Centralized control methods are widely used for voltage control with PV inverters. However, the determined optimal setpoint may change once the system power varies. Then the optimal power flow (OPF) needs to be recalculated, which brings heavy communication and computational burdens to power grids [5].

Some distributed control strategies are developed to alleviate these burdens. Under designed communication topology, agents can achieve desired goals based on neighbor information [6]. The alternative path in the topology can also avoid single-point failure occuring in conventional centralized methods. But it still needs frequent power flow calculations and considerable communication resources.

Decentralized control approaches are developed to further reduce the demand in communication, which only uses local information to operate without mutual interactions. The droop control is proposed through the approximately linear relationship of reactive power and voltage amplitude [7]. Under scenarios with fluctuating power flow, it can react quickly to local bus voltages without considering changes of other buses. However, the general linear droop curve cannot give the optimal response of reactive power support under local voltage change. The results are non-optimal, and voltage security may not be guaranteed.

In recent years, data-driven methods have attracted much attention, like data-driven OPF [8], deep reinforcement learning [9]. But most of these works are centralized methods with a high communication burden. Ref [10] proposes a data-driven local control method where the local voltage/reactive power relationship is assumed as a piecewise linear function to be identified with historical data. However, a piece-wise linear function cannot perfectly describe the underlying voltage/reactive power relationship. Thus the developed local control method cannot achieve optimal results.

Based on the above analysis, this paper proposes a decentralized method to regulate voltages, which can provide approximate optimal solutions under fluctuating loads and PV generations. The OPF calculation is performed in a centralized manner using historical PV and load data. Then multiple optimal power settings and corresponding voltages are obtained for different PV systems located at various buses. The neural network is trained by these optimal scatters (Q, V) to find underlying relationships between local bus voltage and reactive power support. Next, the trained network functions as a local controller, which gives out optimal local voltage control curves. At last, using developed local controllers, simulations on IEEE 33-bus system are presented to show the effectiveness of proposed method, which can achieve secure voltage regulation and perform comparably to centralized OPF.

This paper is organized as follows. Section II gives the problem formulation and presents conventional voltage control methods. Section III introduces the proposed method and the operation process. Section IV performs the simulation based on the IEEE 33 bus case with actual fluctuating PV and load data, showing the effectiveness of the approach in voltage regulation. At last, the conclusion is given in Section V.

Problem Formulation and Conventional Method

This section presents the Voltage/Var control principle with the implementation of PV inverters. Then, the conventional centralized approach is introduced. It provides the optimal power setting for reference by calculating OPF on historical data.

Voltage/Var Control Using PV Inverter

The impedance in transmission lines will cause voltage drops during power flow. To avoid the risk of severe voltage deviation under fluctuated power flow, volt/var control is used to mitigate the violation. The principle is as follows [1]. As the branch shown in Fig. 1, the voltage drop across the series impedance can be expressed as

$$\Delta V = \left(\frac{P+jQ}{V_j}\right)^* (R+jX) = \frac{PR+QX}{V_j} + j\frac{PX-QR}{V_j} \tag{1}$$

with i and j of the send and to buses. This paper analyzes the High Voltage (HV) level system with a low R/X ratio, so the effect of resistance can be ignored, then it obtains

$$\Delta V = \frac{QX}{V_j} + j\frac{PX}{V_j} \tag{2}$$

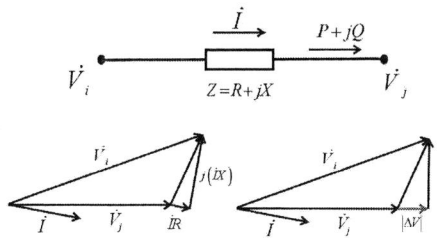

Fig. 1: Branch power flow and voltage triangle.

As the approximation of voltage triangle shown in figure 1, effects of the imaginary component can be omitted, and it gets

$$|\Delta V| = \left(\frac{PR + QX}{V_j}\right) \tag{3}$$

From equation (3), the voltage deviation is affected by both active and reactive power. But due to the low R/X ratio, the reactive power contributes more to the voltage drop. So reactive power can be used to regulate voltage.

Different devices can be applied to provide reactive power support. Compared with other var support devices, inverters have already been added to grids as the interface for PV generators, so the inverter-based volt/var control is more cost-effective and flexible[3] [4]. Since the PV usually works on Max Power Point Tracking (MPPT) mode for more generation, assume the active power output of the inverter is P_{pv} with power capacity S_{inv}, then the reactive power capacity can be determined by

$$Q_{pv} = \sqrt{S_{inv}^2 - P_{pv}^2} \tag{4}$$

where the Q_{inv} is the Qg_{max} of PV generator outputs.

The consideration of the capacity constraint on inverter reactive power is reasonable. The fluctuated PV active power output affects the reactive power that the inverter can supply. Therefore, inverter-based volt/var control has dynamic reactive power capacity compared to devices with fixed one.

The volt/var control principle and the reactive power support from inverters have been presented above. Next, it will introduce conventional centralized OPF to obtain the reference optimal power settings based on historical data.

Conventional Centralized Approach

Conventional centralized control can achieve optimal results with global communication. To implement OPF on historical data, the selected power flow model and its calculation process are presented below.

Assume there are N buses in the grid, and their voltage amplitudes are denoted as V_i ($i = 1, 2, ..., N$). The *DistFlow* model is uesd to describe the AC power flow in general network as follows [11].

$$p_i = \sum_{j \in B_i^D} P_{ij} - \sum_{k \in B_i^U} (P_{ki} - R_{ik} l_{ki}) \tag{5}$$

$$q_i = \sum_{j \in B_i^D} Q_{ij} - \sum_{k \in B_i^U} (Q_{ki} - X_{ik} l_{ki}) \tag{6}$$

$$v_i = v_j + 2(R_{ij} P_{ij} + X_{ij} Q_{ij}) - (R_{ij}^2 + X_{ij}^2) l_{ij} \tag{7}$$

$$v_i l_{ij} = P_{ij}^2 + Q_{ij}^2 \tag{8}$$

with $k \in B_i^U$, $j \in B_i^D$. B is the feeder bus set, B_i^D and B_i^U are respectively the downstream and upstream bus sets of bus i. p_i and q_i represent the power injections of bus i, and are equal to $Pg_i - Pd_i$, Qg_i respectively.

Power injection from generators are denoted as Pg_i, Qg_i. Active power load demand is noted as Pd_i. The branch current, active and reactive power flow on the branch ij are noted as I_{ij}, P_{ij}, Q_{ij}. And $v_i = V_i^2$, $l_{ij} = |I_{ij}|^2$. Line impedances are described as R_{ij}, X_{ij} which induce line losses. The target of centralized OPF is to minimize the line loss.

$$\min_{Pg,Qg,Pd} \sum_{i=1}^{N} \sum_{j \in B_i^D} R_{ij} l_{ij} \tag{9}$$

There are boundary limits of variables

$$V_{\min} \leq V_i \leq V_{\max}, \quad l_{ij} \leq l_{ij\max} \tag{10}$$

$$P_{ij}^2 + Q_{ij}^2 \leq S_{ij\max}^2 \tag{11}$$

$$Pg_{i\min} \leq Pg_i \leq Pg_{i\max}, \quad -Qg_{i\max} \leq Qg_i \leq Qg_{i\max} \tag{12}$$

The upper and lower limits of voltage amplitude are denoted as V_{max} and V_{min}. The square of current magnitude should be less than the maximum value $l_{ij\max}$. Also, the line power flow should not exceed the upper limit $S_{ij\max}$. From equation (4), the reactive power support from the PV has a capacity limit Qg_{max}, which varies with PV active power output.

Based on these constraints and with desired optimization target, conventional centralized OPF can be conducted to generate optimal power settings for PV volt/var control in grids. But this approach requires extensive communication resources and cannot provide the optimal set point for real-time local control when global communication is not triggered. As analyzed above, there are certain relationships between bus power injection and bus voltage deviation. Meanwhile, certain rules of electricity consumption exist, and so does the sunshine. So OPF results on historical data similar to desired scenarios may provide information to find local volt/var control curve approximate to the centralized performance. And the neural network has a solid ability to dig and fit the underlying relationship. Therefore, a neural network-based decentralized approach will be developed for effective local volt/var control in the following section.

The Proposed Data-driven Decentralized Approach

Motivated by recent advancements in machine learning techniques and based on the centralized OPF described above, we will develop a data-driven decentralized approach to achieve near-optimal results without communication. The principle is to use neural networks to identify the underlying relationship between local voltage and reactive power support from optimal operations. Then the trained networks function as local controllers for PV inverters.

Principle of Neural Net Fitting

The basic structure of a neural network is as follows. There are input, hidden, and output layers. When inputs enter the hidden layer, the data is weighed, and biases are added. These sums are the input of the activation function. The output of the nonlinear activation function is the transferred signal intensity of input data, which is distributed over a limited range, such as [0,1]. The sigmoid function is used here.

$$f(x) = \frac{1}{1 + e^{-x}} \tag{13}$$

with the $f(x) \in (-1, 1)$ for $x \in (-\infty, +\infty)$. It restricts the hidden neurons' output to a limited extent. The activation function itself needs to be differentiable for reverse optimization of neural parameters using strategies such as gradient descent.

There can be multiple hidden layers. But theoretically, a neural network with a single hidden layer is enough to approach arbitrary functions. This fitting capability is enabled by the nonlinear activation

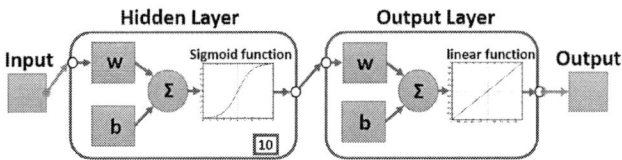

Fig. 2: Two-layer feed-forward neural network structure.

function, through which the nonlinear relationships between system input and output can be mapped successfully. So the neural network can work effectively in function approximation. As shown in Figure 2, the neural network used has a single hidden layer with 10 neurons. The training process can be conducted conveniently by powerful modern software, such as MATLAB.

Based on the above analysis, though there are complex relationships between bus voltage and optimal reactive power setting, the neural net fitting provides a competent approach to fit them.

The Proposed Approach

In this section, the proposed method will be introduced, along with the functioning process.

First, it uses the conventional centralized method to calculate OPF for historical data. Then the neural network is applied to fit the underlying relationships between local voltage and optimal reactive power setpoint of PV inverters. So each PV inverter will obtain a local neural network to provide the approximate optimal reactive power setpoint under the variation of bus voltage due to the fluctuations of load and PV. At last, the local controller based on the trained neural network can achieve near-optimal performance in a decentralized way. The whole process is presented in Fig. 3.

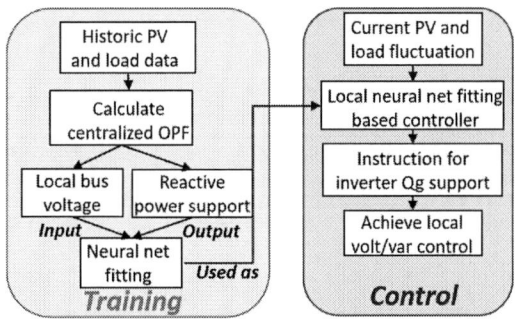

Fig. 3: Process of proposed data-driven decentralized control.

Some details are as follows. The neural network for each PV inverter is trained based on centralized OPF results performed on historical PV and load data. The local voltage of the inverter is as input and the corresponding optimal power setting is as the output. The trained neural network determines the inverter Q_g support only upon local voltage amplitude in the real-time operation.

Therefore, the proposed method has both advantages of centralized and local methods, with solutions near centralized optimal results but without communication requirements.

Case Study

To demonstrate the effectiveness of proposed data-driven decentralized control method, the IEEE 33 bus case (case33) from matpower is studied. Comparisons are made for the proposed method with the conventional centralized and local control methods.

The PV and load data of one month are from [12]. The load data is from the 'Office 2' in the 'Consumption' module, and the PV data is from 'PV GECAD LASIE' in the 'PV Generation' module. 15

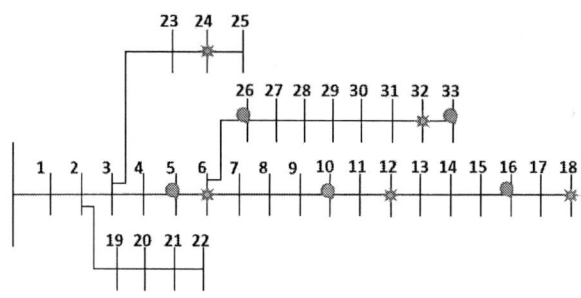

Fig. 4: The 33 bus case topology

weekdays are selected as historical data to perform centralized OPF and generate the training set. Remained 5 weekdays are applied as current data to test the performance of the proposed method.

Based on the IEEE 33-bus system, changing loads are added to several buses (5, 10, 16, 26, 33) and vary from 0.02 MVA to 0.12MVA in every bus. PVs are installed in another buses (6, 12, 18, 24, 32). They are connected to the grid by inverters, enabling 1.6 MVA power capacity for reactive power support of each connected bus. The grid topology is shown in Fig. 4, where blue denotes varying loads and the sun for fluctuated PVs. The load and PV test data are demonstrated in Fig. 5, fluctuating during the 120 hours (5 weekdays).

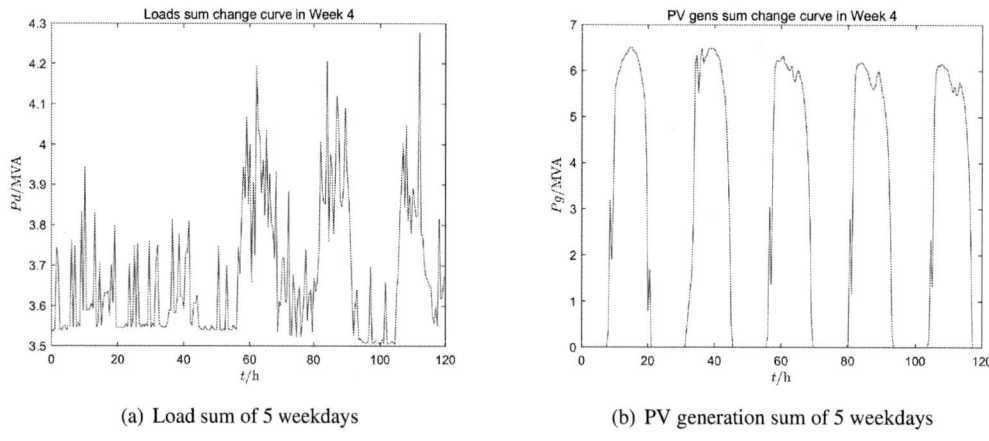

(a) Load sum of 5 weekdays

(b) PV generation sum of 5 weekdays

Fig. 5: Power fluctuation in the grid.

Fig. 6 shows the voltage control results of OPF without reactive power support from PV inverter (denoted as w/o Q), centralized approach (denoted as OPF), and proposed approach (denoted as NN), and the detail in Fig. 8. It can be observed that, without the reactive power support from PV inverters (i.e., w/o Q curve), the voltage security boundary will be violated. Also, by comparing the results of OPF and NN, it shows that the proposed method can effectively control voltage under its boundary without violation, and the performance is close to the centralized method. So the proposed approach can guarantee voltage security, and the outcome approximate centralized optimization results.

Fig. 7 shows the results of the reactive power output of PV inverters under the centralized approach (OPF) and the proposed method (NN). Based on the voltage/var control principle, the reactive power output effectively supports the voltage regulation. It can be observed that the reactive power output of each inverter under the proposed method is approximate to that under the centralized method. This illustrates that the proposed decentralized approach achieves close results as the centralized control method due to similar reactive power generation for support.

Fig. 9 demonstrates results of line loss under the centralized approach (OPF), droop control method

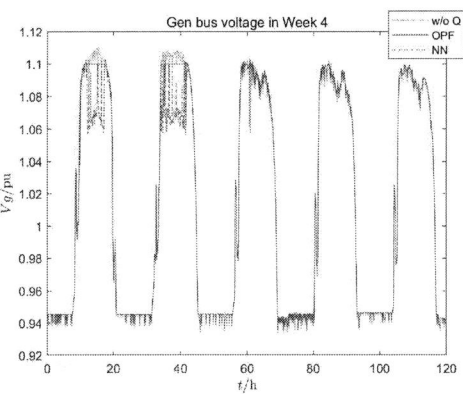

Fig. 6: Overall voltage change curves

Fig. 7: Q_g change curves in one day of Week 4.

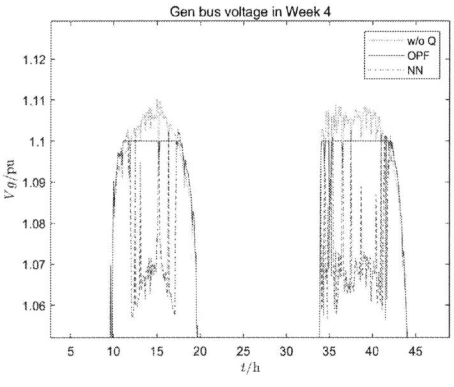

(a) Detail in upper voltage boundary (zoomed in)

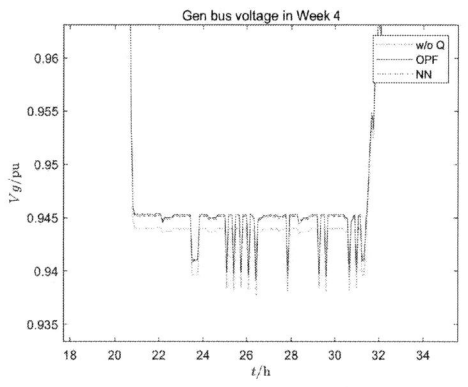

(b) Detail in lower voltage boundary (zoomed in)

Fig. 8: Gen bus voltage in bus 18 of Week 4 under 3 control approaches

(Droop) and the proposed method (NN). The droop control is based on $V_m = V_m^* - m\left(Q_g - Q_g^*\right)$, with V_m^* and Q_g^* as the setting point [13]. Since the inverter reactive power capacity ΔQ varies with time, adaptive control is used here for comparison, which has a changed slope $m = \frac{\Delta Q}{\Delta V}$. It shows that the proposed method achieves similar line loss as the centralized method, while droop control, the conventional decentralized method, has much higher line loss. Therefore, the proposed method can achieve optimal results in a decentralized manner, while droop control cannot.

The above simulation results comparisons show that the proposed approach has approximate performances of centralized OPF but works in a decentralized way. It can mitigate the voltage violation and guarantee voltage security while minimizing the line loss. Therefore, those advantages enable this decentralized method to provide near-optimal solutions effectively, under constantly changing loads and PVs, and without communication.

Conclusion

This paper proposes a neural network-based decentralized approach for volt/var control by inverter reactive power capacity. Neural net fitting is used to train local controllers from centralized OPF results on historical data. As neural networks can describe underlying relationships between voltage and optimal reactive power support, approximate optimal solutions will be provided by local controllers with fluctuating loads and PVs. In this way, inverter reactive power setpoints are optimized without communication. Simulations are conducted based on the IEEE 33 bus case, with fluctuated PVs and loads. The obtained results illustrate the effectiveness of the proposed method in both voltage regulation and loss minimiza-

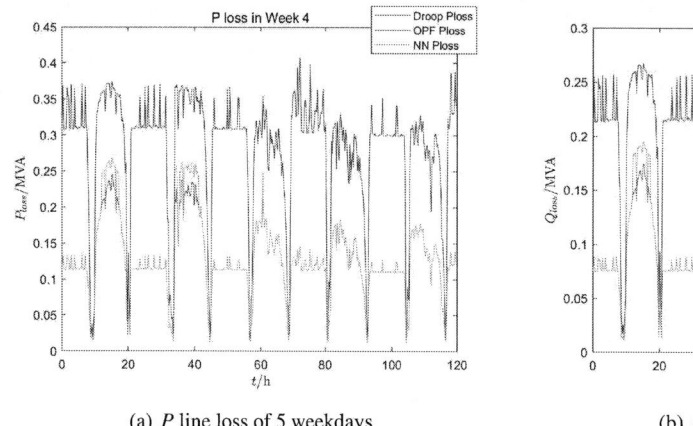
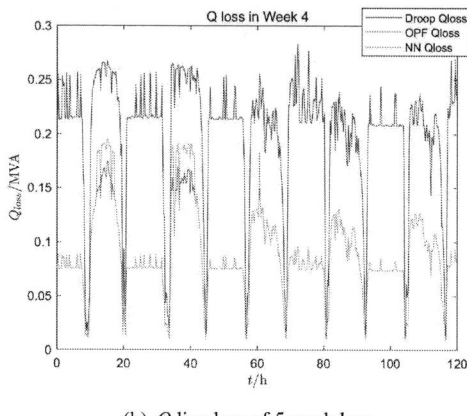

(a) *P* line loss of 5 weekdays   (b) *Q* line loss of 5 weekdays

Fig. 9: Line losses comparison of 3 control approaches

tion. Overall, this decentralized approach provides near-optimal solutions for voltage regulation using inverters, which are cost-effective and easy to implement.

References

[1] H. Sun et al., "Review of Challenges and Research Opportunities for Voltage Control in Smart Grids," in IEEE Transactions on Power Systems, vol. 34, no. 4, pp. 2790-2801, July 2019.

[2] H. Zhao, Q. Wu, J. Wang, Z. Liu, M. Shahidehpour and Y. Xue, "Combined Active and Reactive Power Control of Wind Farms Based on Model Predictive Control," in IEEE Transactions on Energy Conversion, vol. 32, no. 3, pp. 1177-1187, Sept. 2017.

[3] A. Singhal, V. Ajjarapu, J. Fuller and J. Hansen, "Real-Time Local Volt/Var Control Under External Disturbances With High PV Penetration," in IEEE Transactions on Smart Grid, vol. 10, no. 4, pp. 3849-3859, July 2019.

[4] S. Li, Y. Sun, M. Ramezani and Y. Xiao, "Artificial Neural Networks for Volt/VAR Control of DER Inverters at the Grid Edge," in IEEE Transactions on Smart Grid, vol. 10, no. 5, pp. 5564-5573, Sept. 2019.

[5] N. A. Awadhi and M. S. E. Moursi, "A Novel Centralized PV Power Plant Controller for Reducing the Voltage Unbalance Factor at Transmission Level Interconnection," in IEEE Transactions on Energy Conversion, vol. 32, no. 1, pp. 233-243, March 2017.

[6] Z. Fan, B. Fan and W. Liu, "Distributed Control of DC Microgrids for Optimal Coordination of Conventional and Renewable Generators," in IEEE Transactions on Smart Grid, vol. 12, no. 6, pp. 4607-4615, Nov. 2021.

[7] X. Meng, J. Liu and Z. Liu, "A Generalized Droop Control for Grid-Supporting Inverter Based on Comparison Between Traditional Droop Control and Virtual Synchronous Generator Control," in IEEE Transactions on Power Electronics, vol. 34, no. 6, pp. 5416-5438, June 2019.

[8] Lei, Xingyu, et al. "Data-driven optimal power flow: A physics-informed machine learning approach," in IEEE Transactions on Power Systems 36.1 : 346-354, 2020.

[9] Wang S, Duan J, Shi D, et al. "A data-driven multi-agent autonomous voltage control framework using deep reinforcement learning," in IEEE Transactions on Power Systems, 35(6): 4644-4654, 2020.

[10] S. Karagiannopoulos, P. Aristidou and G. Hug, "Data-Driven Local Control Design for Active Distribution Grids Using Off-Line Optimal Power Flow and Machine Learning Techniques," in IEEE Transactions on Smart Grid, vol. 10, no. 6, pp. 6461-6471, Nov. 2019.

[11] Q. Li and V. Vittal, "Convex Hull of the Quadratic Branch AC Power Flow Equations and Its Application in Radial Distribution Networks," in IEEE Transactions on Power Systems, vol. 33, no. 1, pp. 839-850, Jan. 2018.

[12] IEEE Open Data Sets, https://site.ieee.org/pes-iss/data-sets/.

[13] J. M. Guerrero, J. C. Vasquez, J. Matas, L. G. de Vicuna and M. Castilla, "Hierarchical Control of Droop-Controlled AC and DC Microgrids—A General Approach Toward Standardization," in IEEE Transactions on Industrial Electronics, vol. 58, no. 1, pp. 158-172, Jan. 2011.

[14] Vanderkeyn Ralf W.: Example of fast switching component, EPE Journal Vol 20 no 5, pp. 48- 56

[15] Deboe B. D.: A novel type of grid converter, EPE 2013-ECCE Europe, paper 0321

Study of Current Ripple Generators for Accelerated Ageing of Capacitors

Robert Keilmann[1], Hendrik Schefer[1,2], Regine Mallwitz[1,2]

INSTITUTE FOR ELECTRICAL MACHINES, TRACTION AND DRIVES (IMAB)[1]

CLUSTER OF EXCELLENCE SE²A – SUSTAINABLE AND ENERGY-EFFICIENT AVIATION[2]

TU Braunschweig

Braunschweig, Germany

E-Mail: h.schefer@tu-braunschweig.de

URL: https://www.tu-braunschweig.de/imab

URL: https://www.tu-braunschweig.de/se2a

Acknowledgements

This work was supported by the Deutsche Forschungsgemeinschaft (German Research Foundation, DFG) through Germany's Excellence Strategy-EXC 2163/1-Sustainable and Energy Efficient Aviation under Grant 390881007.

Keywords

≪Test bench≫, ≪Reliability≫, ≪Passive component≫, ≪Resonant converter≫, ≪Wide bandgap devices≫

Abstract

Upcoming developments in aerospace power applications demand the use of WBG semiconductors. They offer higher switching frequencies resulting in high-frequency ripple currents in the DC-link. These currents may lead to excessive stress on the DC-link capacitors. This paper proposes a modular ripple current generator for the accelerated ageing of capacitors using high-frequency ripple currents.

I Introduction

Research shows that many applications benefit from the use of fast-switching, wide bandgap (WBG) semiconductors. There are countless studies on the use of novel semiconductors in grid feed applications. This especially includes renewable energy systems like photovoltaics and wind turbines respectively wind parks. But there are also mobile applications that benefit from the use of WBG semiconductors. These are, for example, electrified cars and trains. The scope of current research also includes applications in the aviation sector. Researchers found that using WBG semiconductors might be advantageous for all-electric aircraft [1].

Capacitors are critical components that are frequently affected by failures [2]. Therefore, the studies of [3], [4] or standards such as EIA IS-749 provide guidance to develop power electronics with lifetime based approaches as well as to induce accelerated ageing. A novel way to approach a variety of test frequencies using a modular concept will be shown to induce accelerated ageing under challenging climatic conditions for different capacitor types.

This paper is structured as follows. Section II discusses the ageing mechanisms known from the literature for the types of capacitors typically used in power electronics. Section III illustrates the overall problem using the example of a three-phase inverter and describes two approaches for the targeted accelerated ageing of capacitors. In Section IV, the advantages and disadvantages of three circuit variants for generating ripple currents are worked out in a topology study. Section V specifies the method and devised approach of the modular and flexible series resonant converter. Section VI deals with the parameters influencing the overall system as well as the design rules of the ripple current generator. The working prototype is presented in section VII. The fact that first studies have been successfully carried out with the novel approach is then illustrated in conclusion in section VIII.

II Ageing of Capacitors

Aluminium-Electrolytic-Capacitor (Al-Caps), Metallized-Polypropylene-Film Capacitors (MPPF-Caps) and Multi-Layer-Ceramic-Capacitors (MLC-Caps) have different voltage ratings and capacity ranges; therefore, various possible use-cases. The paper [4] compares these capacitor types against each under three relative performance indices.

Aircraft applications require different types of capacitors and their properties. There is no capacitor type for all kinds of converters and inverters. Furthermore, co-existences in DC-links are often required. The substitution of conventional propulsion systems in aircraft needs high voltage ranges, high energy densities and high frequencies under high-reliability requirements [1].

Table I summarizes the external stress factors of the chosen capacitor types. Al-Caps have reliability-related issues to electrolyte evaporation driven by thermal conditions, self-heating mechanism (ripple current and ESR) and normal ageing [4]. The applied DC voltage exerts dielectric stress [4]. Mechanical stress can lead to electrical disconnections [5]. MPPF-Caps have an additional lifetime-related parameter. There is reversible and irreversible humidity absorption [4]. Over a humidity-exposure time, degradation of metallised layers occurs [6]. In the dielectric of MLC-Caps, there can occur some ceramic micro-cracking due to mechanical oscillations [7].

Table I: Capacitor Stress Focus Point Matrix based on [4]

	Al-Caps	MPPF-Caps	MLC-Caps
Critical Stressors	Voltage, Temperature, Ripple Current		
	Vibration	$\frac{dU}{dt}$, Humidity	Vibration/Shock
Most Critical Stressors	Voltage and Temperature		
	Ripple Current	Humidity	Vibration/Shock

The replication of the stress factors requires a current ripple generator with temperature, voltage, and humidity adjustment opportunities. Mechanical stress is not taken into account. The number of stress factors necessitates the design of experiments; therefore, it requires many test slots for different test conditions and statistical significance.

III Approaches for test benches and definition of requirements

Depending on the target topology, different current characteristics occur in capacitors. The current shapes differ, the amplitudes are largely determined by the required power and the difference between the input voltage and output voltage. A 100 kW inverter with a 50 kHz switching frequency serves as an illustration. The ripple current depends on many parameters, such as modulation degree M, power factor $\cos\varphi$, line frequency f_s, the number of parallel-switched capacitor strings, etc. The publication [8] shows analytical equations to calculate the root-mean-square DC-link ripple current $I_{c,rms}$ for a three-phase inverter system. Eight capacitor strings lead to a per-string current $I_{c,rms} = 8.4\,\mathrm{A}$ at the doubled order of the switching frequency $f_\mu = 100\,\mathrm{kHz}$.

To obtain information about the expected service life of the capacitors used, they can be subjected to accelerated ageing in the laboratory. In the literature, two possibilities are usually shown for how to generate ripple currents for accelerated ageing. On the one hand, it is possible to actively induce a current into the capacitor [9] which represents a topology-independent approach. A differing approach is the targeted-topology approach where the capacitor is operated in the target topology directly [10].

A Targeted-Topology Approach

This method [10] offers the advantage that the expected current waveforms and amplitudes occur and the capacitor undergoes the same ageing processes in the laboratory as it does in the field. But this method has also several disadvantages.

A practical disadvantage arises in target applications with high power ratings. Here, an electronic load is needed where electric energy is being fed back into the grid or the DC-link. Especially, multi-phase AC inverters require multi-phase loads with massive filter efforts. In [11] this problem is mitigated by the use of a scaled-down version of the inverter. There, the DC-link capacitor is exposed to a high DC-link voltage and high ripple currents in a very similar manner to the full-scale inverter. Due to the scaling, the miniature inverter in [11] has a power rating of 640 VA instead of the full-scale version with a power rating of 6400 VA.

Secondly, one is now restricted to a single topology. This means that systematic testing of different capacitors for a wide range of applications and topologies is not possible. Additionally, accelerated testing could be limited to power electronic housing. A crucial parameter is humidity for MPPF-caps, as mentioned above in Table I. Other electronic components have to withstand the conditions in a case of an open power electronic housing. To extend the DC-link outside the electronic housing is not a good solution because an additional impedance leads to unrealistic waveforms. Fig. 1 illustrates a possible approach. Here, the power electronics are positioned in a climate chamber. It can be addressed, controlled and monitored during operation via appropriate interfaces. The power electronics located in the climate chamber must be connected to a high-voltage source and the load via a suitable cable connection. If water cooling is required, supply lines to the chiller provided for this purpose must be implemented.

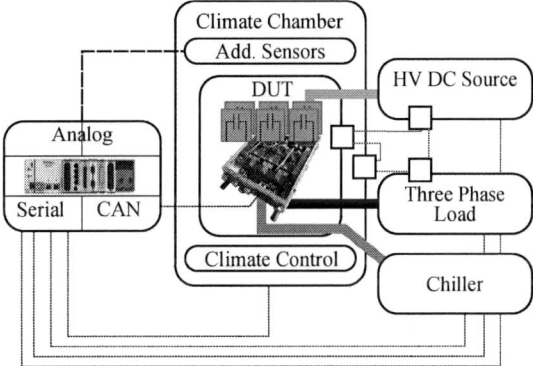

Fig. 1: Topology-dependent approach for accelerated ageing of components

B Topology-independent Approach

In own studies, the possibility to induce the real ripple currents (example: trapezoidal, triangle, etc.) of the DC-link capacitor according to the real waveforms and amplitudes was investigated. For this purpose, investigations and efforts were made to develop a push-pull output stage (Class G Amplifier) capable of reproducing the expected DC-link capacitor currents in shape and amplitude. However, limitations quickly arise here that prevent use in the intended form.

IV Topology Study

In advance, three circuit types were subjected to a topology study. Namely, these are the Push-Pull Output Stage (Class G Amplifier), Class D Amplifier and the Series Resonant Converter (see Fig. 2 a)-c)). Table II illustrates their main characteristics.

The Class G Amplifier's ability to amplify arbitrary waveforms is advantageous. This amplifier has a higher efficiency than the Class A, Class B or Class AB amplifier [12]. As mentioned above, in a practical design study a Class G Amplifier was designed with high dynamics and power ratings to recreate the

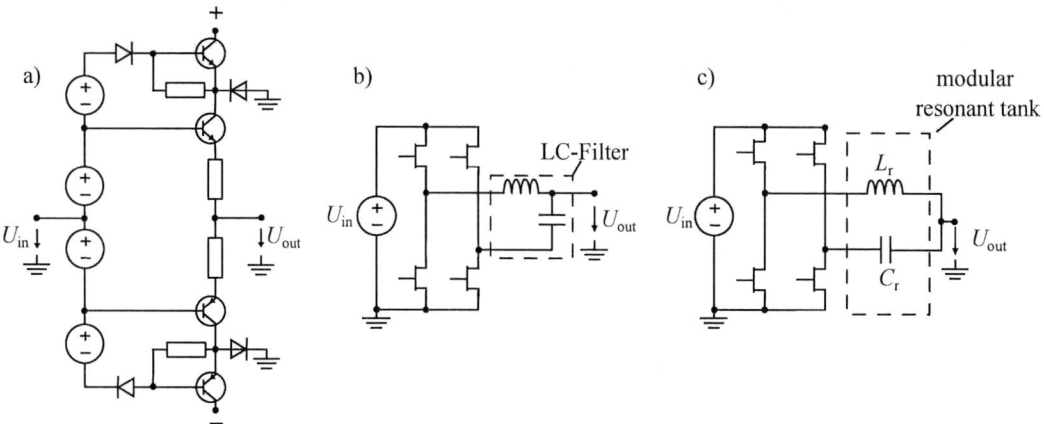

Fig. 2: From left to right: a) Push-Pull Output Stage, b) Class D Amplifier, c) Series Resonant Converter

real waveforms that occur at the DC-link. It was found that the averaged power dissipation in the output stage transistors is in the range of 385 W. Thus, a water cooling would be necessary and in the context of a statistical study, where multiple ripple current generators have to be operational for a long period of time, high energy costs would be inevitable due to the high power losses. It was also found that the cable and transformer impedances have the consequence that originally sharp edges in the waveforms are being smoothed out. With these findings in mind, it must be concluded that the Class G Amplifier is not suitable as a ripple current generator.

As an alternative, the Class D Amplifier shifts into focus. Its switching semiconductors promise higher efficiency but also make an output filter necessary. At first glance, this topology seems very advantageous. The frequencies could simply be specified via software. To reduce the EMC difficulties that come with this topology at least a filter second order is recommended [12]. There are major challenges in the dimensioning of this output filter, since it depends on the load impedance [12]. The proposed Class D Amplifier in [13] delivers an output power of 300 W at an output signal frequency of 13.56 MHz. For this, the output filter is a resonant circuit with fixed and known values for the resonant inductance L, resonant capacitance C as well as a load resistance R_L. Also [13] approaches a high-frequency Class D Amplifier delivering 1 W at 1 MHz output signal frequency using a resonance circuit. A high power Class D Amplifier is proposed in [14] which is also based on a resonant circuit with a rated output power of 1600 W at a resonant frequency of 34.6 kHz.

In summary, it can be stated that the load impedance is critical for the design of the Class D Amplifier. In the publications presented above, a series resonant circuit with well-known properties is used to obtain sinusoidal signals. For the use of the Class D Amplifier in a ripple current generator, this means that the passive elements involved would have to be adapted to the connected load which are the capacitors to be aged.

The discussion of the Class D Amplifier shows that technical useful solutions rely on a series resonant network. So does the proposed solution in [9] where a Series Resonant Converter is commissioned. Analyzing the waveforms given in the publication, it delivers an effective current of at least 5.3 A at 10 kHz. The frequency is determined by the components that form the resonant circuit. That means, that a change in the test frequency makes it necessary to make changes in the used components. This is not a topology-specific drawback because these changes would be necessary for a Class D Amplifier, too.

So the goal is to develop a compact generator that is as easy to modify as possible. With the help of a holistic system approach, rules for dimensioning the resonance components are to be derived to realize test frequencies of interest for the systematic ageing of capacitors. The discussion above requires a wide frequency range and opportunities to adjust the temperature, humidity, DC voltage and a huge number of DUTs for the sake of modularity.

Table II: Compact Representation of the Topology Study

	Class G Amplifier	**Class D Amplifier**	**Series Resonant Converter**
greatest advantage	specifiability of the real current forms	high flexibility by specifying the switching pattern	widely pure sinusoidal currents
greatest drawback	limited dynamics, very high power dissipation	harmonics and EMI due to switching pattern [12], therefore high filter effort	frequency determined by resonant tank, impedances of the transformer, cables, DUTs, ...
current form and maximum test frequency	any (up to 50 kHz)	sinusoidal (MHz range) depending on switching pattern and filter design [13], [14], [15]	sinusoidal (MHz range)
approach	topology-independent induce expected waveforms	topology-independent test frequency components individually	topology-independent test frequency components individually

V Proposed Approach

Fig. 3 illustrates the proposed approach to consider the ripple current. The current waveforms occurring at the DC-link capacitor in any target topology can be easily decomposed into their spectral components in a circuit simulation. The Fourier Transform then provides information about the occurring current amplitudes over the frequency range of interest.

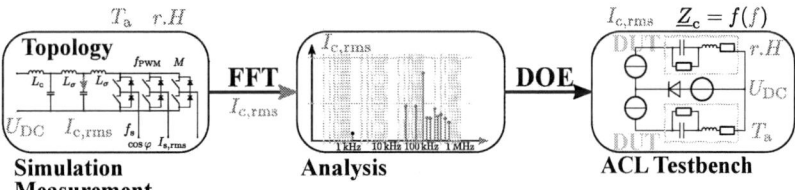

Fig. 3: Methodology of the proposed approach

If sinusoidal currents are generated by the current ripple generator to be developed, the interfering frequencies can be tested individually and their influence on the ageing behaviour of the capacitors can be analyzed. Unlike the push-pull output stage, there is no disturbing distortion of the output signal since here only one frequency is impressed and no superposition of many. In this approach, the generator can stay outside the climatic chamber due to the current sine wave. However, this setup has to cover a wide frequency range for applications with WBG semiconductors, as shown in Fig. 3.

Fig. 4 shows the modular concept. In a ventilated housing, multiple ripple current generators can be operated in the laboratory environment. The devices under test (DUTs) can be placed into a climate chamber where environmental conditions like temperature and humidity can be controlled. The electrical stressors are then provided by the modular ripple current generators.

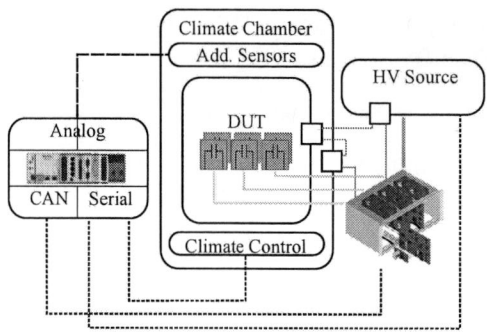

Fig. 4: Proposed modular, topology-independent approach

VI Design of a modular and flexible Series Resonant Converter

A modular resonant tank is proposed which can realize frequencies in the range of 10 kHz to 500 kHz. Gallium nitride (GaN) semiconductors are used here. With a view to future applications with higher frequencies and due to the low input voltage, these are ideally suited for the design of the ripple current generator. Further requirements are given in Table III.

Table III: Requirements List

transferable power	100 W
maximum effective current	10 A
DUT capacitance range	2 μF ... 200 μF
frequency range	10 kHz ... 500 kHz

A transformer is needed to connect to the high voltage side where the DUTs are exposed to a voltage of up to 1 kV.

For current regulation, a two-stage approach is used. Using a buck converter, the voltage of the input side can be adjusted in reference to the secondary side current through the DUTs. For this purpose, the current in the resonant circuit is measured and compared with the set reference value. A control algorithm is used to adjust the input voltage of the resonant converter so that the sinusoidal current maintains its amplitude over the entire course of the test. With this approach, a status monitoring of the corresponding circuit can be realized and errors can be reacted to.

A Subsystem Analysis

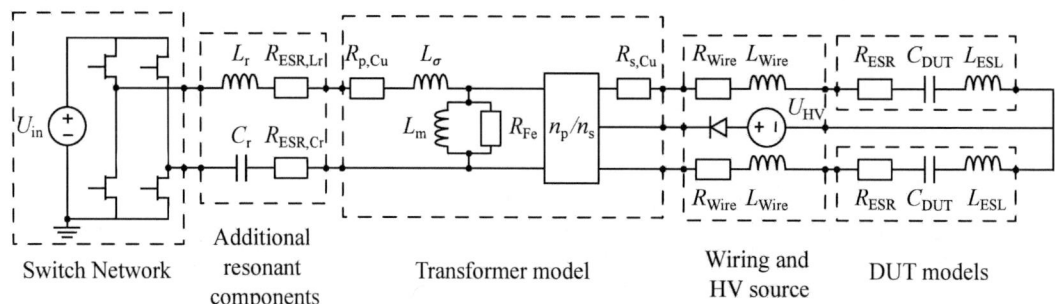

Fig. 5: Circuit model of the ripple current generator

The resonant system consists of five subsystems which are visible in Fig. 5. The input quantity is impressed by the Switch Network with the switching frequency f_{sw} and the duty cycle $D = 0.5$. Using the

First Harmonics Approximation, the input voltage of the resonant system can be calculated with

$$u(t) \approx \frac{4 \cdot U_{\text{in}}}{\pi} \cdot \sin(\omega t) \qquad (1)$$

where $\omega = 2\pi f_{\text{sw}}$. The system's resonant frequency is dependent on the subsystems consisting of the additional resonant components, the transformer, the wiring and the DUTs. A practical approach for the design of the ripple current generator is to start on the right-hand side of Fig. 5.

This example considers two capacitors connected in series. They can be modelled with their parasitic components so that each DUT is formed by a series connection of the capacitor's equivalent series resistance R_{ESR}, the capacitance C_{DUT} as well as the equivalent series inductance L_{ESL}.

The second subsystem that needs to be considered is the wiring. The primary line constants of the cables used cannot be ignored. In this work, cables are going to be approximated by an ohmic-inductive load. This model is a result of measurements using an impedance analyzer. Two coaxial cables are used for the test fixture. These are specified for the high ambient temperatures and operating voltages. While the inner conductor carries the ripple current in each case, the outer conductors ensure the determination of the potential specified by the voltage source U_{HV}.

The transformer also affects the resonant system. This subsystem can be modelled using a simplified equivalent circuit of the transformer that is also given in Fig. 5. Were $R_{\text{p,Cu}}$ is the ohmic resistance of the primary winding at a given frequency. The leakage inductance L_σ contributes dominantly to the properties of the resonant circuit. With decreasing transformer size also the magnetizing inductance L_{m} decreases. When L_σ and L_{m} are in the same order of magnitude, the magnetizing current cannot be neglected. Thus, especially for higher frequency ranges, the magnetizing inductance L_{m} needs to be considered. On the transformer's secondary side there $R_{\text{s,Cu}}$ is the ohmic resistance of the secondary winding.

Lastly, there are the additional resonant components that are modelled in Fig. 5. Because the other components that contribute to the overall resonant frequency are fixed by the properties of the used subsystems (DUTs, wiring and transformer) the additional resonant components are being used to tune the resonant frequency to the desired value.

B   Reducing the complexity of the model and formulation of design rules

The complex system shown in Fig. 5 can be transformed into a more compact one if related elements and impedances are combined at suitable points. This summary of the components is shown in Fig. 6 for the subsystems to the right of the Switch Network. Here, secondary-side impedances Z_{s} are taken into

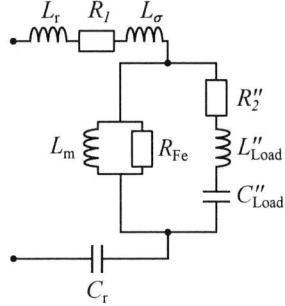

Fig. 6: Equivalent circuit of the resonant circuit

account by the transformation

$$Z_{\text{s}}'' = \left(\frac{n_{\text{p}}}{n_{\text{s}}}\right)^2 \cdot Z_{\text{s}} \qquad (2)$$

on the primary side using the number of primary turns n_{p} as well as the number of secondary turns n_{s}. With all relevant resonance frequency-determining elements identified, the network model given in Fig. 6

can be derived. Analyzing the overall impedance of the network model and taking into account that at the resonance frequency the imaginary part of the impedance needs to be zero the resonance frequency can be calculated using

$$f_{\text{res}} = \frac{1}{2\pi\sqrt{\left(L_{\text{r}} + L_{\sigma} + \frac{L_{\text{Load}}'' \cdot L_{\text{m}}}{L_{\text{Load}}'' + L_{\text{m}}}\right) \cdot \frac{C_{\text{r}} \cdot C_{\text{Load}}''}{C_{\text{r}} + C_{\text{Load}}''}}}. \tag{3}$$

At resonance frequency, only the purely real, ohmic resistance

$$R_{\text{sum}} = R_1 + R_2'' \tag{4}$$

is still effective when the losses in the core materials are neglected by stating $R_{\text{Fe}} \to \infty$. With the target frequency f_{target} and using the definition of the quality factor

$$Q_{\text{target}} = \sqrt{\frac{L_{\text{target}}}{C_{\text{target}}}} \cdot \frac{1}{R_{\text{sum}}} \tag{5}$$

the target values of the overall equivalent inductance L_{target} and capacitance C_{target} can be calculated using

$$L_{\text{target}} = \frac{1}{2\pi f_{\text{target}}} \cdot (Q \cdot R_{\text{sum}})^2 \tag{6}$$

respectively

$$C_{\text{target}} = \frac{1}{2\pi f_{\text{target}}} \cdot \frac{1}{(Q \cdot R_{\text{sum}})^2}. \tag{7}$$

The values of the additional resonant components L_{r} and C_{r} that need to be placed into the resonant path are determined by

$$L_{\text{r}} = L_{\text{target}} - L_{\sigma} - \frac{L_{\text{Load}}'' \cdot L_{\text{m}}}{L_{\text{Load}}'' + L_{\text{m}}} \tag{8}$$

as well as

$$C_{\text{r}} = \frac{C_{\text{target}} \cdot C_{\text{Load}}''}{C_{\text{Load}}'' - C_{\text{target}}} \tag{9}$$

where $L_{\text{r}}, C_{\text{r}} \geq 0$ must be fulfilled.

VII Experimental setup

Fig. 7: Modular ripple current generator with adapter boards

Fig. 7 shows the developed ripple current generator with adapter boards to set the resonant frequency using L_{r} and C_{r}. These adapter boards also carry the transformer. The proof of function is to be shown

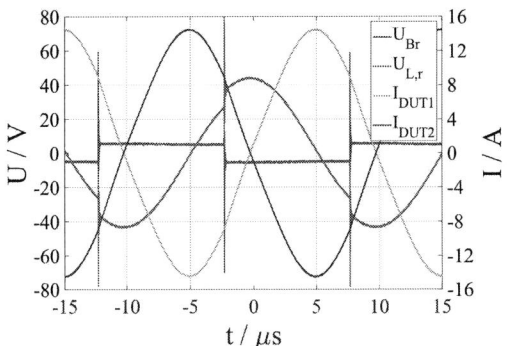 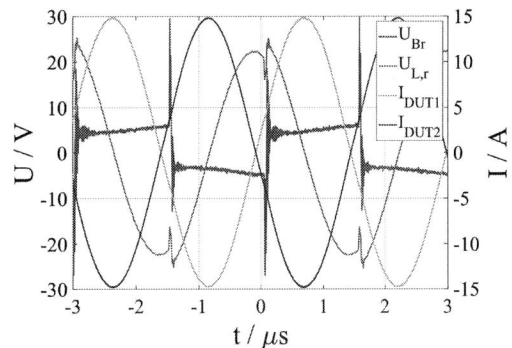

a) Waveforms of the 50 kHz resonance circuit with planar transformer

b) Waveforms of the 327 kHz resonance circuit with conventional transformer

Fig. 8: Waveforms captured during the commissioning of the proposed ripple current generator

for two frequencies. The first frequency is $f_1 = 50\,\text{kHz}$ which is the fundamental frequency in the example application of the three-phase inverter. The second frequency $f_2 = 327\,\text{kHz}$ is the self-resonance frequency of the DUTs which were previously identified using an impedance analyzer.

Three transformers are designed and used in this work from which one is a planar transformer. With a small number of transformers with well known properties, a high degree of modularity can be achieved. The researcher can choose the right component from a well documented pool of transformers. In this context, the planar transformer can significantly contribute to a further increase in modularity. Challenges arise concerning the development time of such a device. The benefits are the repeatability and multiple ways of the winding realization.

An oscillogram with measured quantities on the 50 kHz circuit is given in Fig. 8a. The measured quantity U_{Br} is the voltage at the input terminals of the resonant circuit shown in Fig. 6. The voltage $U_{\text{L,r}}$ is the voltage across the resonance inductor, while I_{DUT1} and I_{DUT2} are a representation of the current through the capacitors to be aged. While I_{DUT1} flows out of the ripple current generator, I_{DUT2} is the current flowing back into the device. Because of the planar transformer's direction of winding the positive half-wave of I_{DUT1} appears during the period where U_{Br} is negative.

Fig. 8a shows sinusoidal currents through the capacitors to be aged. Analyzing the current's amplitude, the goal to imprint an effective current of 10 A is achieved.

The voltages shown in Fig. 8a exhibit high voltage peaks at the switching moment, accompanied by oscillations. These are significantly favoured by two circumstances.

First, it can be seen that the resonant frequency of the system does not match the switching frequency completely. Component tolerances, which also drift over the temperature, are primarily responsible for this.

The second reason is associated with the semiconductor switches used. The GaN switches exhibit a very steep switching behaviour and challenging reverse conduction characteristics when reverse biased [16]. This favours the observed behaviour. As a result, additional losses occur in the inductor, but these can be well controlled with the measures taken.

In conclusion, it can be stated that the voltage stress on the components, which is determined by Q, is in the expected range. Thermal tests show that the ripple current generator and the designed resonance circuits are thermally safe for continuous operation.

The commissioning of the 327 kHz resonance circuit shows similar behaviour. Fig. 8b shows sinusoidal currents with an effective current of 10 A. Because the resonance frequency of the resonant circuit is closer to the switching frequency, the voltage peaks at the switching moment are lower in their amplitude. This leads to lowered losses in the inductor. Thus, this resonance circuit is thermally safe for continuous operation, too.

VIII Experimental results

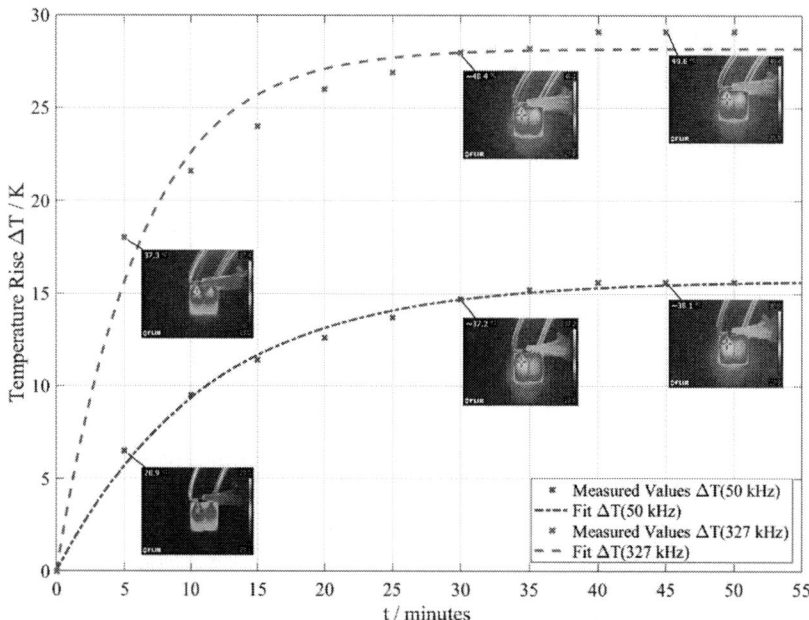

Fig. 9: Temperature rise of the tested capacitor at two different test frequencies

During commissioning, it is observed that the capacitors heat up to different degrees depending on the frequency of the ripple current, as shown in Fig. 9. At a frequency of 50 kHz, they heat up by 15.7 K compared to room temperature, while at 327 kHz a heating of 28.2 K is detectable. The temperature is measured in each case with the aid of a thermal imaging camera. The capacitor's thermal time constant is $\tau \approx 11.1$ min. For future investigations at further frequencies, a more suitable measurement method must be selected. For example, thermocouples can be used here, which may be attached to the capacitors.

IX Conclusion

Reliable electrical drive and supply systems for aviation are getting more and more important. New developments make use of fast-switching WBG semiconductors. Thus, new approaches must be found to determine the reliability and service life of capacitors which are exposed to a harsh environment and high frequent ripple currents of rising frequencies. For this purpose, they have to be systematically subjected to a ripple current under fixed (climatic and electrical) conditions. This arises the need for modular and compact systems for accelerated ageing of capacitors in higher frequency ranges. Current publications and standards do not provide a satisfactory solution for this issue. This paper addresses some approaches to accelerated ageing of capacitors known from the literature. For the generation of high-frequency sinusoidal ripple currents, dimensioning rules of a modular ripple current generator based on a resonant converter are presented. The ripple current generator dimensioned with these rules is presented and its commissioning is documented.

References

[1] H. Schefer, L. Fauth, T. H. Kopp, R. Mallwitz, J. Friebe and M. Kurrat, "Discussion on Electric Power Supply Systems for All Electric Aircraft," in IEEE Access, vol. 8, pp. 84188-84216, 2020, https://doi.org//10.1109/ACCESS.2020.2991804

[2] S. Yang, A. Bryant, P. Mawby, D. Xiang, L. Ran and P. Tavner, "An industry-based survey of reliability in power electronic converters," 2009 IEEE Energy Conversion Congress and Exposition, 2009, pp. 3151-3157, https://doi.org/10.1109/ECCE.2009.5316356

[3] H. Wang, H.Wang, Z. Shen, "Reliability of Capacitors and Magnetic Components in Power Electronic Applications," in CIPS 2020; 11th International Conference on Integrated Power Electronics Systems, 2020, pp. 1-6, https://ieeexplore.ieee.org/document/9097729, accessed: 17.11.2021

[4] H. Wang and F. Blaabjerg, "Reliability of Capacitors for DC-Link Applications in Power Electronic Converters—An Overview," in IEEE Transactions on Industry Applications, vol. 50, no. 5, pp. 3569-3578, Sept.-Oct. 2014. https://doi.org/10.1109/TIA.2014.2308357

[5] Nippon Chemi-Con Corporation, "Judicious use of aluminium electrolytic capacitor," Technical Note, https://www.chemi-con.co.jp/e/catalog/pdf/al-e/al-sepa-e/001-guide/al-technote-e-2020.pdf, accessed: 12.11.2021

[6] M. Makdessi, A. Sari, P. Venet, "Metallized polymer film capacitors ageing law based on capacitance degradation", Microelectronics Reliability, Volume 54, Issues 9–10, 2014, Pages 1823-1827, ISSN 0026-2714, https://doi.org/10.1016/j.microrel.2014.07.103.

[7] W. Minford, "Accelerated Life Testing and Reliability of High K Multilayer Ceramic Capacitors," in IEEE Transactions on Components, Hybrids, and Manufacturing Technology, vol. 5, no. 3, pp. 297-300, September 1982, https://doi.org/10.1109/TCHMT.1982.1135974

[8] J.W. Kolar, S.D.Round, "Analytical calculation of the RMS current stress on the DC-link capacitor of voltage-PWM converter systems", IEEE Proceedings - Electric Power Applications, Volume 153, Issue 4, Pages 535-543, ISSN 1359-7043, https://doi.org/10.1049/ip-epa:20050458

[9] M. Makdessi, A. Sari, P. Venet, P. Bevilacqua, C. Joubert, "Accelerated Ageing of Metallized Film Capacitors Under High Ripple Currents Combined With a DC Voltage," in IEEE Transactions on Power Electronics, vol. 30, no. 5, pp. 2435-2444, May 2015, https://doi.org/10.1109/TPEL.2014.2351274

[10] C. Kulkarni, G. Biswas, X. Koutsoukos, J. Celaya, and K. Goebel, "Experimental Studies of Ageing in Electrolytic Capacitors," in Annual Conference of the Prognostics and Health Management Society, Portland, OR, October 2010, https://citeseerx.ist.psu.edu/viewdoc/download?doi=10.1.1.300.1607&rep=rep1&type=pdf, accessed: 15.11.2021

[11] K. Hasegawa, I. Omura and S. -i. Nishizawa, "A new evaluation circuit with a low-voltage inverter intended for capacitors used in a high-power three-phase inverter," 2016 IEEE Applied Power Electronics Conference and Exposition (APEC), 2016, pp. 3032-3037, https://doi.org/10.1109/APEC.2016.7468295

[12] Douglas Self. Audio power amplifier design handbook. 3. ed., reprinted. Amsterdam: Newnes, 2003. ISBN: 0750656360.

[13] S.-A. El-Hamamsy, "Design of high-efficiency RF Class-D power amplifier," in IEEE Transactions on Power Electronics, vol. 9, no. 3, pp. 297-308, May 1994, https://doi.org/10.1109/63.311263

[14] H. Koizumi, T. Suetsugu, M. Fujii, K. Shinoda, S. Mori and K. Iked, "Class DE high-efficiency tuned power amplifier," in IEEE Transactions on Circuits and Systems I: Fundamental Theory and Applications, vol. 43, no. 1, pp. 51-60, Jan. 1996, https://doi.org/10.1109/81.481461

[15] N. -J. Park, D. -Y. Lee and D. -S. Hyun, "A Power-Control Scheme With Constant Switching Frequency in Class-D Inverter for Induction-Heating Jar Application," in IEEE Transactions on Industrial Electronics, vol. 54, no. 3, pp. 1252-1260, June 2007, https://doi.org/10.1109/TIE.2007.892741

[16] E. A. Jones, F. F. Wang and D. Costinett, "Review of Commercial GaN Power Devices and GaN-Based Converter Design Challenges," in IEEE Journal of Emerging and Selected Topics in Power Electronics, vol. 4, no. 3, pp. 707-719, Sept. 2016, https://doi.org/10.1109/JESTPE.2016.2582685

Intra-arm Balancing Control of Cascaded Multi-Port Converter for Whole Power Unbalance Conditions

Takumi Yasuda and Jun-ichi Itoh
NAGAOKA UNIVERSITY OF TECHNOLOGY
1603-1, Kamitomioka-machi, Nagaoka, Niigata, Japan
Tel.: +81 / (258) – 47.9533.
E-Mail: t_yasuda@stn.nagaokaut.ac.jp, itoh@vos.nagaokaut.ac.jp
URL: http://itohserver01.nagaokaut.ac.jp/itohlab/en/index.html

Keywords

«Modular Multilevel Converters (MMC)», «Load imbalance», «Capacitor voltage balancing», «AC-DC converter», «Charging infrastructure for EV´s»

Abstract

This paper proposes a capacitor voltage balancing control for load imbalance among cells in a cascaded multi-port converter. The input/output power of the cell is controlled by adjusting the ac-side voltage of the cell to balance the capacitor voltage even when the load imbalance occurs in the arm. The current of the multi-port converter with the conventional controller can distort due to overmodulation of the cell under the large load imbalance. The proposed controller prevents overmodulation with the improved generation method of the ac-side voltage, which extends the operation region with respect to the power imbalance. Furthermore, the proposed controller injects minimal additional circulating current to achieve the operation for whole power conditions. The experimental results reveal that the proposed controller improves the total harmonic distortion of the grid current by 2.9p.t. with minimal circulating current while achieving the capacitor voltage balancing.

Introduction

Recently, cascaded multi-port converters have been actively researched as a decentralized battery energy storage system [1]-[6], an integrated PV power conditioner [7]-[10], and so on [11], [12]. The cascaded configuration of cells realizes a grid connection without a bulky line frequency transformer [13], [14]. In addition, a filter on the ac-side is downsized due to the multilevel operation compared to conventional 2-level converters [2], [15].

Fig. 1 shows a cascaded multi-port converter. The multi-port converter in this paper is composed of the cascaded chopper cells and has no high-voltage dc-link port. The loads, such as the battery and EV, are assumed to be connected in parallel to the cell capacitor. Although the rated voltage and power of the cells are identical, a power imbalance among the cells can appear due to the different operation points of the loads caused by factors such as a slight difference in parasitic parameters and operating environments. The balancing controller of the multi-port converter is required to balance the capacitor voltage of the cells in spite of the power imbalance. In other words, the controller for the multi-port converter distributes the different

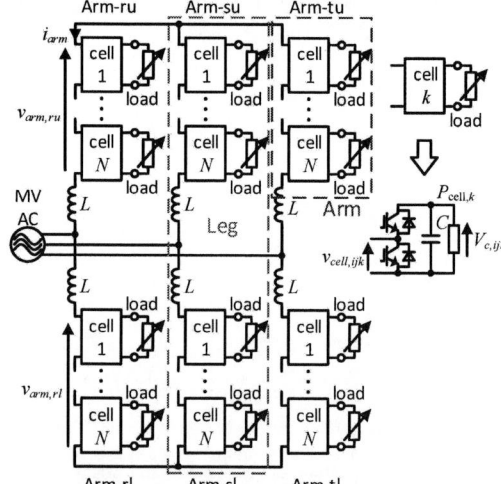

Fig. 1. Multi-port converter using modular multilevel cascaded converter with double-star chopper cells.

amounts of power to each cell according to the states of the load.

A circulating current is utilized to interact the power between the arms when a power imbalance occurs among the arms [4], [12]. Besides, a different ac-side voltage of each cell is applied to distribute the desired power when loads of cells within the arm are not identical. However, there is an operational limit with respect to the power imbalance because the ac-side voltage of the cells has to avoid overmodulation to suppress current distortion [5]-[8]. The additional circulating current (intra-arm balancing current) prevents the overmodulation because the larger arm current reduces the required amplitude of the ac-side voltage of the cell to gain the desired power. However, the past literature seems that the intra-arm balancing current controllers have not been proposed with a discussion of a hardware design of the multi-port converter, including the intra-arm balancing current effect.

This paper proposes an intra-arm balancing controller for the multi-port converter with a bidirectional operation, which operates in all possible power conditions with the compensation voltage injection to the ac-side voltage of the cell and the intra-arm balancing current control. The proposed controller minimizes the intra-arm balancing current while balancing all the capacitor voltage in the converter. In addition, the cell of the converter is designed to achieve the operation in whole power conditions taking into account the effect of the intra-arm balancing current. This paper is organized as follows; first, the control strategy for the multi-port converter is proposed. Next, the multi-port converter is designed. Finally, the proposed controller is evaluated by the simulation and the experiment with a miniature model. The experimental results show that the proposed controller prevents the overmodulation of the cells by the minimal intra-arm balancing current and improves the total harmonic distortion (THD) compared with the conventional controller. Moreover, the intra-arm balancing current is minimized by the proposed controller.

Controllers for multi-port converter

The objectives of the controller for the multi-port converter are to shape the grid current and to balance all the cell capacitor voltage in spite of the load imbalance among the cells. In order to achieve that, the balancing controllers are installed to distribute a different amount of power among the cells.

In this paper, a voltage redundancy ρ is defined as a ratio of the peak-to-peak value of the grid phase voltage to the maximum available voltage of the arm in the steady-state as

$$\rho := \frac{NV_c}{2\sqrt{2}V_g} \tag{1}$$

where N is the number of cells in one arm, V_c is the capacitor voltage of the cell, and V_g is the RMS value of the grid phase voltage.

Current controllers and inter-arm balancing controllers

Fig. 2 depicts the proposed control block diagram for the multi-port converter. The grid-side controller controls the average capacitor voltage in the converter by controlling the input and output power of the converter and shapes the grid current. The power is distributed evenly to each cell by the grid-side controller. The balancing controllers redistribute the power to the cells according to the state

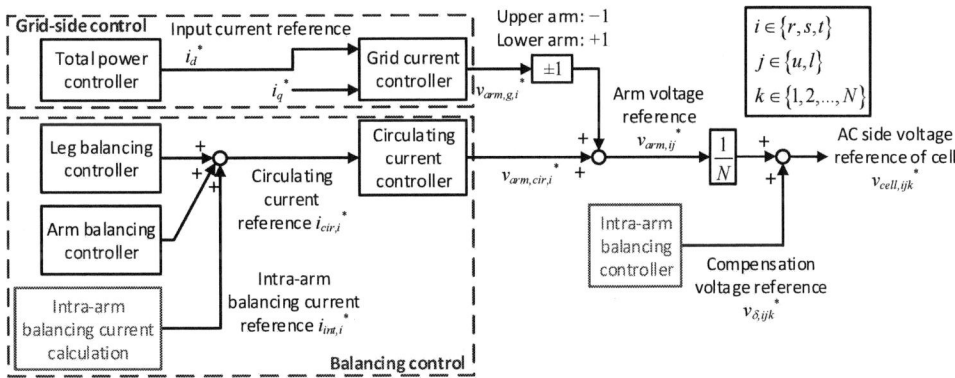

Fig. 2. Proposed control strategy of multi-port converter

of the loads in order to control all the capacitor voltages of the cells to the command value. The leg balancing controller regulates the averaged capacitor voltage of the phase by the power distribution utilizing a dc circulating current. The arm balancing controller utilizes the circulating current of the fundamental frequency to interchange the power between the upper and lower arm. The proposed intra-arm balancing current calculation block generates the intra-arm balancing current command, which is explained

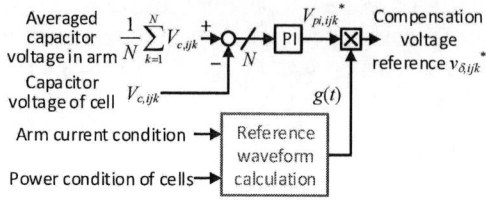

Fig. 3. Block diagram of proposed intra-arm balancing controller with optimized reference waveform.

in detail in the following subsection. A sum of the output of the grid current controller and the circulating current controller is the arm voltage reference, which is assumed as

$$v_{arm,ij}(t) = \frac{NV_c}{2} - K_j \sqrt{2} V_g \cos\left(\omega_g t - \phi_i\right) \quad ...\left(i \in \{r,s,t\}, j \in \{u,l\}\right) \tag{2}$$

where $K_u = 1$, $K_l = -1$, ω_g is the angular grid frequency, ϕ_i is the phase angle ($\phi_r = 0$, $\phi_s = 2\pi/3$, $\phi_t = 4\pi/3$). Note that (2) neglects the voltage drop on the arm inductor L. This paper defines the arm current as

$$i_{arm,ij}(t) := I_{0,i} + \sqrt{2} I_{1,ij} \cos\left(\omega_g t - \phi_{1,ij} - \phi_i\right) + i_{int,i}(t) \tag{3}$$

where $I_{0,i}$ is the dc circulating current of phase i, $I_{1,ij}$ and $\phi_{1,ij}$ is the RMS value and the phase angle of the fundamental frequency component of the arm current, respectively, which is the sum of the grid current and the circulating current for the upper/lower arm balancing, and $i_{int,i}$ is the intra-arm balancing current. Since there is no high-voltage dc-link port in the multi-port converter, as shown in Fig. 1, the circulating current circulates through three phase legs. Thus, the circulating current of the three phases can be different when a power imbalance occurs among the cells.

The sum of the arm voltage reference and the compensation voltage from the proposed intra-arm balancing controller is the ac-side voltage reference of the cells. This paper adopts the phase-shifted PWM (PS-PWM), which is suitable for decentralized control [16], [17].

Intra-arm balancing controller

The intra-arm balancing controller regulates the capacitor voltage error among the cells within the arm by injecting the compensation voltage to the ac-side voltage reference of the cells.

Fig. 3 shows the proposed intra-arm balancing controller. The controller feedbacks the capacitor voltage and regulates the capacitor voltage deviation within the arm to zero. The reference waveform is obtained by multiplying a reference waveform $g(t)$ to an output by the PI controller. Since the command of the intra-arm balancing controller is the averaged capacitor voltage of the cell within the arm, the sum of the outputs of the PI controllers over the arm is always zero, that is,

$$\sum_{k=1}^{N} v_{\delta,ijk}^*(t) = g(t) \sum_{k=1}^{N} V_{pi,ijk}^* = 0 . \tag{4}$$

Eq. (4) implies that the arm voltage, which is the sum of the ac-side voltage of the cells in the arm, is not affected by the compensation voltage regardless of any selection of the reference waveform $g(t)$. Therefore, $g(t)$ is designed to guarantee the operation without the overmodulation over wide power conditions. The boundary of the overmodulation for the compensation voltage is expressed as

$$-\frac{v_{arm,ij}(t)}{N} \le v_{\delta,ijk}(t) \le V_{c,ijk} - \frac{v_{arm,ij}(t)}{N} . \tag{5}$$

The maximum amplitude of the compensation voltage is calculated from the arm voltage reference and the capacitor voltage. Besides, the compensation power $P_{\delta,ijk}$, which is the power obtained by the compensation voltage, is calculated as

$$P_{\delta,ijk} = \frac{1}{T_g} \int_0^{T_g} v_{\delta,ijk}(t) i_{arm,ij}(t) \, dt \tag{6}$$

where T_g is a fundamental period. Here, a maximum and minimum compensation power $P_{\delta ij,max}$ and $P_{\delta ij,min}$ are defined as a maximum and a minimum power in arm-ij which is generated by the intra-arm balancing controller without the overmodulation with an arbitrary reference waveform $g(t)$, respectively.

In order to prevent diverging the capacitor voltage, the intra-arm balancing controller has to satisfy

$$P_{\delta,ij,min} \le \min_k \left[P_{cell,ijk} - \frac{1}{N} \sum_{k=1}^{N} P_{cell,ijk} \right], \quad P_{\delta,ij,max} \ge \max_k \left[P_{cell,ijk} - \frac{1}{N} \sum_{k=1}^{N} P_{cell,ijk} \right]. \tag{7}$$

For example, the reference waveform $g(t)$ should be adjusted so that the $P_{\delta,ij,max}$ becomes large when there is a significantly heavier load on the cell than the typical load in the arm. On the other hand, $g(t)$ should be adjusted to decrease $P_{d,ij,min}$ for the significantly lighter load than the typical load. Therefore, $g(t)$ is decided by the loaded conditions of the cells in the arm as,

$$P_{\delta,ij,max} : P_{\delta,ij,min} = \max_k \left[P_{cell,ijk} - \frac{1}{N} \sum_{k=1}^{N} P_{cell,ijk} \right] : \min_k \left[P_{cell,ijk} - \frac{1}{N} \sum_{k=1}^{N} P_{cell,ijk} \right]. \tag{8}$$

The proposed intra-arm balancing controller calculates $g(t)$, fulfilling (5) and (8) online.

Fig. 4 shows an example of the ac-side voltage of the cell generated by the proposed intra-arm balancing controller. The voltage varies discontinuously according to the polarity of the arm current. The amplitude of the compensation voltage is decided from the instantaneous arm voltage reference.

The operation region with respect to the power imbalance is extended by the proposed intra-arm balancing controller with the optimized reference waveform. However, the capacitor voltage balance is not necessarily achieved in whole power conditions even though the proposed reference waveform is applied. For example, it is obvious that the compensation voltage does not contribute to the power distribution, and the capacitor voltage would diverge in the case of $i_{arm,ij}(t)=0$ as shown in (6). It means that the intra-arm balancing current is required under some conditions.

Intra-arm balancing current calculation

The larger compensation power $P_{\delta,ij,max}$ and the smaller $P_{\delta,ij,min}$ reduce the possibility of overmodulation. The larger intra-arm balancing current increase $P_{\delta,ij,max}$ and decrease $P_{\delta,ij,min}$, which result in the extension of the operation region. However, the large intra-arm balancing current injection causes the large switching and conduction losses. Therefore, the proposed intra-arm balancing current calculation part in Fig. 2 minimizes the intra-arm balancing controller online to suppress the loss.

Fig. 5 shows the flowchart of the proposed intra-arm balancing current calculation. The controller minimizes the intra-arm balancing current in the steady-state. First, the proposed controller calculates the maximum and the minimum compensation power $P_{\delta,ij,max}$ and $P_{\delta,ij,min}$ from the arm current with the discrete integral. The arm current for the calculation is the command value generated by the other balancing controllers in this paper. Then, the controller minimizes the intra-arm balancing current with

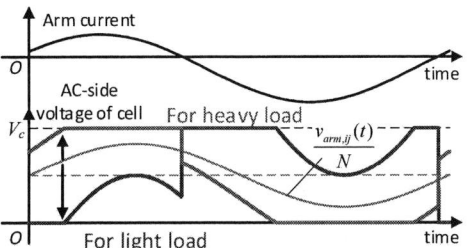

Fig.4. Example of the ac-side voltage reference of the cell with the compensation voltage reference. The loaded condition in the arm is that the magnitude of the maximum load imbalance in the arm is half of that of the minimum load imbalance in the arm.

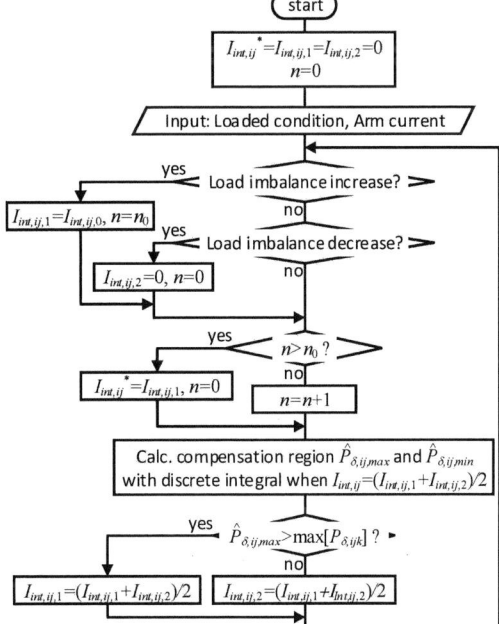

Fig. 5 Minimization strategy of intra-arm balancing current using bisection and discrete integral. The maximum value of derived $I_{int,ij}{}^*$ in six arms is applied to the converter as the intra-arm balancing current reference.

the bisection by comparing the calculated $P_{\delta,ij,max}$ and $P_{\delta,ij,min}$ to the loaded condition. The minimization is performed in each arm. Finally, the maximum value of the derived $I_{int,ij}{}^*$ for six arms is applied as the intra-arm balancing current $I_{int}{}^*$. The intra-arm balancing current $I_{int,ij}{}^*$ is only updated when the derived $I_{int,ij}$ satisfies (7) and is smaller than $I_{int,ij}{}^*$, which is the minimum intra-arm balancing current derived in the last calculation. An update frequency of the intra-arm balancing current reference is set to be sufficiently low not to interfere with the intra-arm balancing controller in Fig. 3. In addition, a low pass filter with a cut-off frequency of 6.7 Hz is installed on the output of the intra-arm balancing current calculation to prevent a step change of the current reference. In this paper, the intra-arm balancing current reference input to the circulating current controller is 1.1 times larger than the derived value considering the calculation error because the proposed algorithm works as a feed-forward controller.

$I_{int,ij}{}^*$ is increased temporarily to a sufficiently large value $I_{int,0}$ to satisfy (7) when the power imbalance increases as a result of the load variation. In this paper, $I_{int,0}$ is decided from the rated current of the system and the given arm current condition as

$$I_{int,ij,0} = \sqrt{I_{arm,n}{}^2 - \left(I_{0,i}{}^2 + I_{1,ij}{}^2\right)} \qquad (9)$$

where $I_{arm,n}$ is a rated current of the arm. The multi-port converter operates without the overmodulation with $I_{int,ij,0}$ in any loaded conditions if the rated current is designed properly, as in the following section. The initial interval of the bisection is set to $I_{int,ij,0}$ and $I_{int,ij}{}^*$, which are (9) and the minimum intra-arm balancing current derived by the last calculation, respectively. The amplitude of the intra-arm balancing current can be decreased. Thus, the bisection interval is changed to $I_{int,ij}{}^*$ and zero when the power imbalance decreases due to the load variation. Note that this paper assumes that the controller can detect the loaded condition. This is because the applications of the multi-port converter are the integrated PV power conditioner and the battery energy storage system, which usually measures the load voltage and the current to perform the maximum power point tracking or control a state-of-charge.

The frequency of the intra-arm balancing current I_{int} has to be chosen not to interfere with other controllers. Too low frequency of I_{int} increases the capacitor voltage ripple [18], [19]. On the other hand, too high frequency of I_{int} increases the voltage drop on the arm inductor, which may result in the saturation of the arm voltage. Therefore, this paper adopts the double-frequency component as I_{int} as

$$i_{int,i}(t) := -\sqrt{2}I_{int}\sin\left(2\omega_g t - 2\phi_i\right). \qquad (10)$$

Note that the phase angle (10) is based on (2).

Fig. 6 compares the intra-arm balancing current I_{int} required to operate in the power imbalance. The conventional controller in Fig. 6 utilizes the polarity of the arm current as the reference waveform $g(t)$ [3], [4]. Required I_{int} becomes larger as the load imbalance increases. The proposed controller reduces the intra-arm balancing current compared to the conventional controller. It reveals that the proposed intra-arm balancing controller helps to reduce the arm current of the converter.

Design of switching devices and cell capacitors

In this section, the switching devices and the cell capacitors are designed considering the intra-arm balancing. As described above, the intra-arm balancing current is derived by the nonlinear calculation with the discrete integral and the bisection. Therefore, this section presents the maximum arm current of the converter with the proposed controller calculated numerically. 64 million loaded conditions were analyzed.

Fig. 7 shows the maximum intra-arm balancing current and the maximum arm current of the system, which is derived as a result of the numerical calculation. The current is normalized by the rated grid current. The maximum intra-arm balancing current decrease as the voltage redundancy ρ, which is defined in (1), increases. Since the large ρ allows the cell to inject a larger

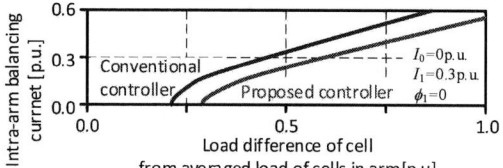

Fig. 6. Comparison of required intra-arm balancing current for operation in a power imbalance. The magnitude of the minimum load difference is set to double that of the maximum load difference. The intra-arm balancing current and the power imbalance are normalized by the rated grid current and the averaged power of cells in the arm, respectively.

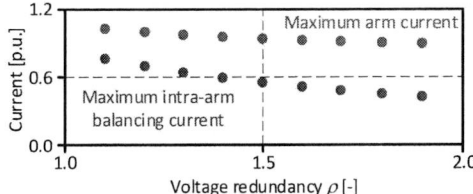

Fig. 7. Maximum intra-arm balancing current
and arm current.

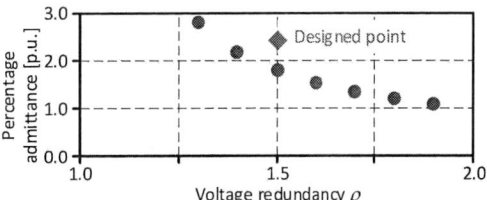

Fig. 8. Minimum capacitance for operation
without overmodulation.

amplitude of the compensation voltage, the required amplitude of the intra-arm balancing current decreases. As a result, the maximum arm current is reduced. Fig. 7 reveals that almost double arm current in the full loaded operation is required for the system when $\rho=1.5$, which is the condition in the following simulation and experiment. $I_{arm,n}$ in (9) is designed as 0.94 times as rated grid current.

Fig. 8 shows the minimum capacitance to avoid the overmodulation in whole power conditions. The required capacitance decreases as ρ increases. Since the maximum arm current is reduced by the large ρ, the capacitor voltage ripple is reduced as ρ increases. In addition, the cell can operate without the overmodulation even though the capacitor voltage contains larger fluctuation when ρ is large. Practically, the cell capacitance is required to be designed with redundancy. The indication written as "Designed point" in Fig. 8 indicates a cell capacitance for the simulation and the experiment in the following section.

Simulation results

Table I shows the simulation conditions. In this simulation, both the cell with the power of 1p.u. and the cell with the power of -1p.u. are in the same arm, which is the largest power imbalance for the converter. One calculation of the discrete integral in Fig. 5 is performed every 1 ms, and the update frequency of the intra-arm balancing current command is set to 6.7 Hz, which is also applied to the practical controller.

Fig. 9 shows the simulation results of the steady-state operation with the load imbalance. Fig. 9(a) shows modulation waveforms of the cells and the arm current in the corresponding arm. The proposed controller injects different compensation voltages. A large second harmonic component of 20.8 A is injected as the intra-arm balancing current i_{int} in order to operate in the severe power imbalance, which is 13.4% larger than the theoretical minimum value. The calculation error is 3.4% because a 10% larger i_{int} of the originally derived value by the controller is applied to the system, as mentioned above. The error is caused by the error between the current used for the calculation in the proposed controller and the true value. Fig. 9(b) shows the grid current. The converter operates with a THD of

Table I. Simulation conditions

Parameter	Symbol	Value
Rated power of cell	P_{cell}	2.2 kW
Grid line-line voltage	$\sqrt{3}V_g$	6.6 kV (RMS)
Grid frequency	f_g	50 Hz
Number of cells per arm	N	40
Rated capacitor voltage	V_c	400 V
Arm inductance	L	158 mH (0.10p.u.)
Cell capacitance	C	1.87 mF (68 mJ/VA)
Carrier frequency	f_{car}	6.0 kHz

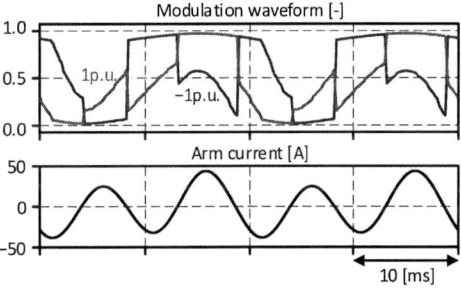

(a) Modulation waveforms and arm current of
upper arm in r-phase.

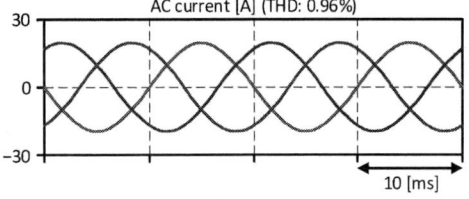

(b) Grid current

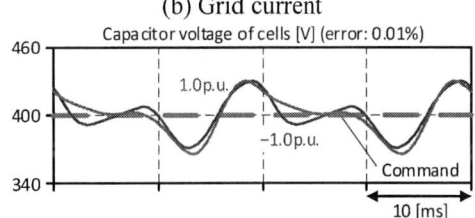

(c) Capacitor voltage of cells.

Fig. 9. Simulation results of steady-state
operation with load imbalance.

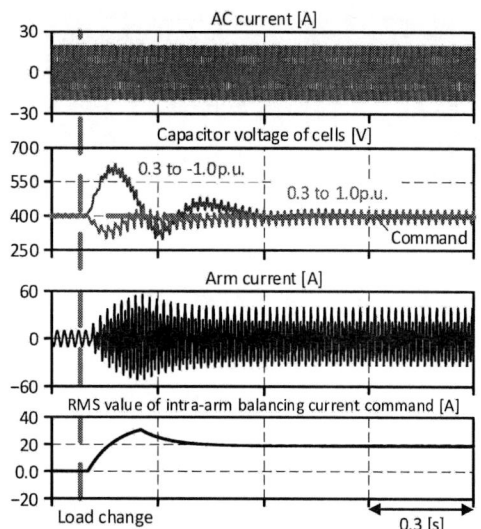

Fig. 10. Simulation result of transient response under step increase of load imbalance.

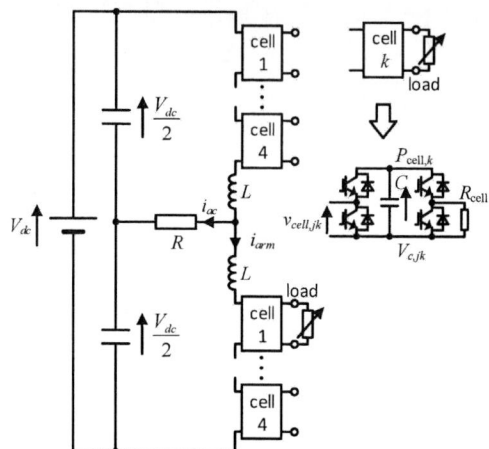

Fig. 11. Scaled model for experimental verification of proposed intra-arm balancing controller.

0.96% despite the largest power imbalance. Fig. 9(c) shows the capacitor voltages of the cells in the same arm. The capacitor voltages are balanced thanks to the proposed controller.

Fig. 10 shows the transient response under the load variation. Before the load change, there is no power imbalance in the arm and no intra-arm balancing current i_{int} flows. After the load change, one cell is fully loaded, and a load of another cell becomes -1p.u. The command of i_{int} increases rapidly after the load change. Then, i_{int} is decreased thanks to the proposed controller. The capacitor voltage of the cells and the grid current is balanced.

Table II. Experimental conditions.

Parameter	Symbol	Value
Rated power of cell	P_{cell}	169 W (R_{cell}=100 Ω)
AC line-line voltage	$\sqrt{3}V_g$	125 V (RMS)
DC-link voltage	V_{dc}	520 V
Grid frequency	f_g	50 Hz
Number of cells per arm	N	4
Rated capacitor voltage	V_c	130 V
AC resistance	R	50 Ω
Arm inductance	L	5.0 mH (0.07 p.u.)
Cell capacitance	C	1.36 mF (68 mJ/VA)
Carrier frequency	f_{car}	6.0 kHz

Experimental results

Fig. 11 and Table II show the experimental circuit and the conditions. The experimental multi-port converter is composed of single-phase cascaded choppers, and the dc voltage source is utilized instead of the other two phases. The variable load is installed in only one cell to emulate the condition that there is a fully loaded cell and an unloaded cell in the same arm.

Fig. 12 shows the block diagram of the controller for the experiment. The output of the total power controller is dc circulating current command, unlike Fig. 2. The grid current is controlled at a constant value of I_d=2.5 A and I_q=0 A to keep the voltage redundancy ρ constant. The proposed controller was tested under the condition that the dc and the fundamental frequency component of the arm current flow.

Fig. 13 shows experimental waveforms in the steady-state operation with the power imbalance when the proposed controller is applied. The different modulation waveforms and the relatively large intra-arm balancing current i_{int} of 1.93 A are applied, as shown in Fig. 13(a). The error of i_{int} from the theoretical minimum value is 2.38%. Fig. 13(b) shows the grid current with a THD of 3.98% even under the power imbalance. The capacitor voltage is balanced with a maximum error of 1.3% as in Fig. 13(c).

Fig. 14 shows experimental waveforms in the steady-state operation with the power imbalance when the conventional controller [3], [4] is adopted. The conventional controller does not inject the i_{int}. The overmodulation occurs in the cells on account of the large amplitude of the compensation voltage by the conventional intra-arm balancing controller. As a result, the arm current and the grid current distort. THD of the grid current deteriorates by 2.9p.t. compared with the proposed controller.

Fig. 15 shows the transient behavior of the converter with the proposed controller when the load in

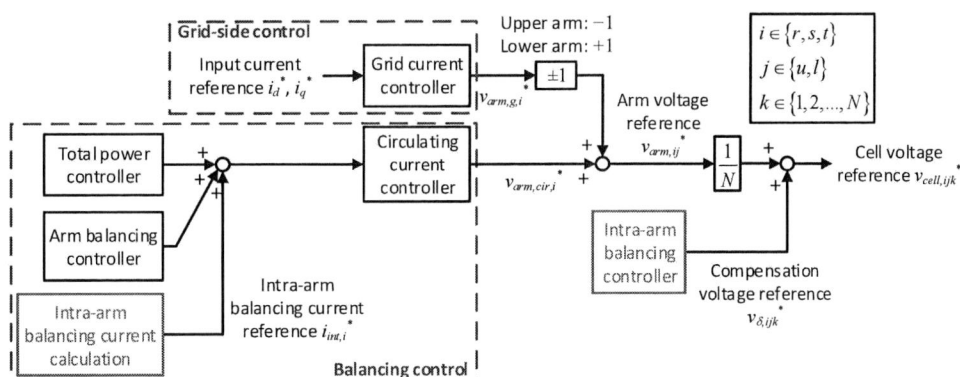

Fig. 12. Control strategy for experimental circuit.

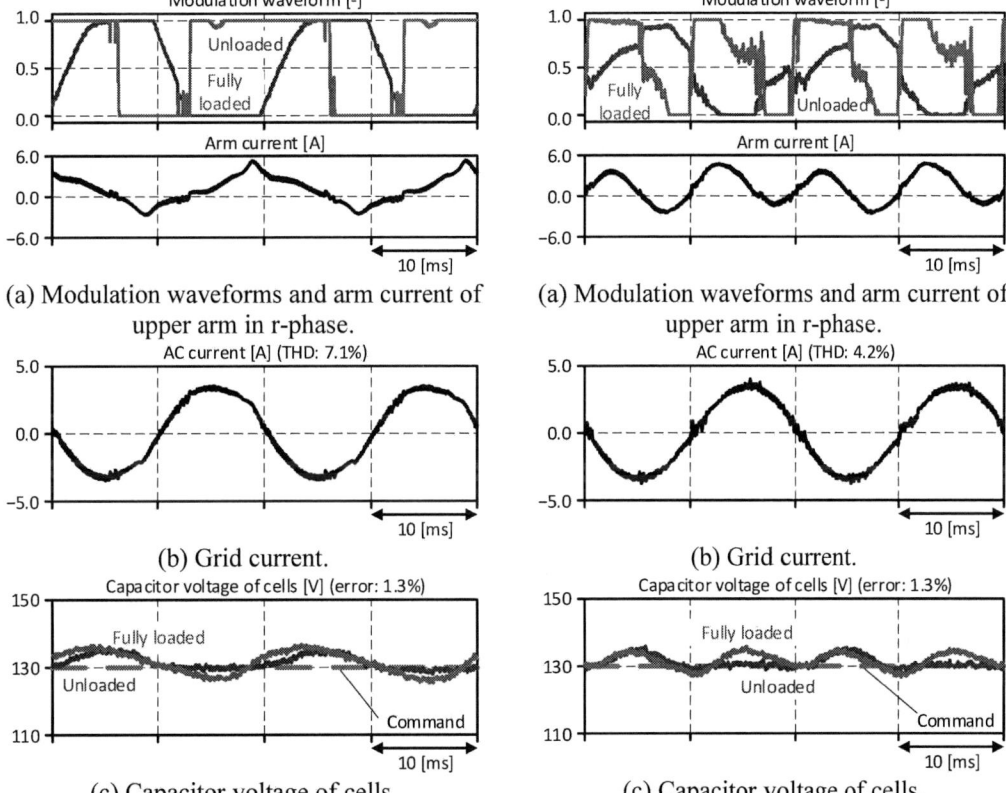

(a) Modulation waveforms and arm current of upper arm in r-phase.

(a) Modulation waveforms and arm current of upper arm in r-phase.

(b) Grid current.

(b) Grid current.

(c) Capacitor voltage of cells.

(c) Capacitor voltage of cells.

Fig. 14. Experimental results of steady-state operation with conventional controller in load imbalance.

Fig. 13. Experimental results of steady-state operation with proposed controller in load imbalance.

one cell increases from 0p.u. to 1p.u. The proposed controller decreases i_{int} and converges to 2.21 A, which is 11% larger than the theoretical minimum value, although i_{int} increases after the load variation. The capacitor voltage drops by approximately 30% of its nominal value after the load change. After that, the capacitor voltage converges to the command value of 130 V within 1 s. The update of i_{int} causes a small voltage drop of 12% on the cell capacitor at 0.15 s after the load change. However, the voltage fluctuation on the cell capacitor does not affect the grid current.

Fig. 16 shows the transient response when the load in one cell increases from 1p.u. to 0.5p.u. i_{int} decreases to zero after the load changes, and the power imbalance in the arm decreases. The capacitor voltage increases by 23% at maximum after the load change. The capacitor voltage converges to the command value within 0.5 s, even though the change of i_{int} causes the overshoot of the voltage. The grid

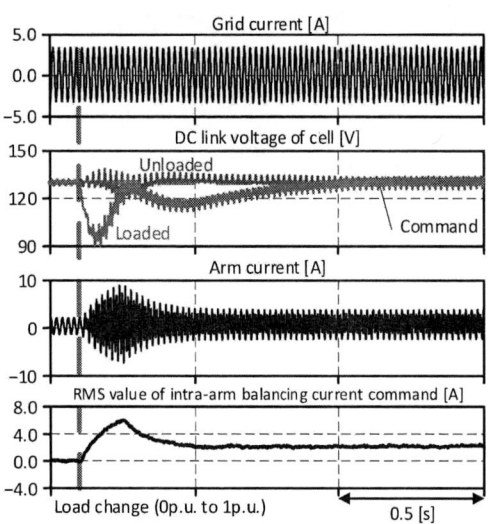

Fig. 15. Experimental results of transient behavior under step increase of load imbalance.

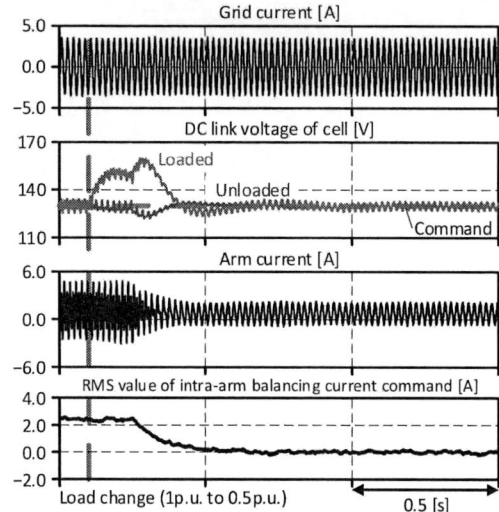

Fig. 16. Experimental results of transient behavior under step decrease of load imbalance.

current is little affected by the load variation. These results verify that the proposed controller achieves operation in whole power conditions with the minimum intra-arm balancing current.

Conclusion

This paper proposed the intra-arm balancing controller to achieve the capacitor voltage balancing under the power imbalance with the minimum current injection. In addition, the maximum arm current and the cell capacitors were designed to ensure the operation in all power conditions considering the effect of the intra-arm balancing current. The experimental result with the designed multi-port converter achieved both the balanced grid current and the balanced capacitor voltages of all cells. The proposed controller improved the grid current THD by 2.9pt compared with that of the conventional controller. In addition, the intra-arm balancing current was minimized with an error of 2.38% compared with the theoretical minimum current.

References

[1] N. Kawakami, S. Ota, H. Kon, S. Konno, H. Akagi, H. Kobayashi, and N. Okada, "Development of a 500-kW Modular Multilevel Cascade Converter for Battery Energy Storage Systems," in IEEE Transactions on Industry Applications, vol. 50, no. 6, pp. 3902-3910, Nov.-Dec. 2014.

[2] G. Wang et al., "A Review of Power Electronics for Grid Connection of Utility-Scale Battery Energy Storage Systems," in IEEE Transactions on Sustainable Energy, vol. 7, no. 4, pp. 1778-1790, Oct. 2016.

[3] Z. Wang, H. Lin, Y. Ma and T. Wang, "A prototype of modular multilevel converter with integrated battery energy storage," 2017 IEEE Applied Power Electronics Conference and Exposition (APEC), Tampa, FL, 2017, pp. 434-439.

[4] W. Zeng, R. Li and X. Cai, "A New Hybrid Modular Multilevel Converter with Integrated Energy Storage," in IEEE Access, vol. 7, pp. 172981-172993, 2019.

[5] T. Soong and P. W. Lehn, "Assessment of Fault Tolerance in Modular Multilevel Converters With Integrated Energy Storage," in IEEE Transactions on Power Electronics, vol. 31, no. 6, pp. 4085-4095, June 2016.

[6] G. Liang et al., "Analytical Derivation of Intersubmodule Active Power Disparity Limits in Modular Multilevel Converter-Based Battery Energy Storage Systems," in IEEE Transactions on Power Electronics, vol. 36, no. 3, pp. 2864-2874, March 2021.

[7] S. Yang et al., "Quantitative Comparison and Analysis of Different Power Routing Methods for Single-Phase Cascaded H-Bridge Photovoltaic Grid-Connected Inverter," in IEEE Transactions on Power Electronics, vol. 36, no. 4, pp. 4134-4152, April 2021.

[8] P. Sochor and H. Akagi, "Theoretical and Experimental Comparison Between Phase-Shifted PWM and Level-Shifted PWM in a Modular Multilevel SDBC Inverter for Utility-Scale Photovoltaic Applications," in

IEEE Transactions on Industry Applications, vol. 53, no. 5, pp. 4695-4707, Sept.-Oct. 2017.

[9] L. Liu, H. Li, Y. Xue and W. Liu, "Reactive Power Compensation and Optimization Strategy for Grid-Interactive Cascaded Photovoltaic Systems," in IEEE Transactions on Power Electronics, vol. 30, no. 1, pp. 188-202, Jan. 2015.

[10] V. Sridhar and S. Umashankar, "A Comprehensive Review on CHB MLI based PV Inverter and Feasibility Study of CHB MLI based PV-STATCOM," in Renewable and Sustainable Energy Reviews, vol. 78, pp. 138-156, Oct. 2017.

[11] Y. Su, P. Wu and P. Cheng, "Design and Evaluation of a Control Scheme for the Hybrid Cascaded Converter in Grid Applications," in IEEE Transactions on Power Electronics, vol. 35, no. 3, pp. 3139-3147, March 2020.

[12] G. Guidi, S. D'Arco, K. Nishikawa and J. A. Suul, "Load Balancing of a Modular Multilevel Grid-Interface Converter for Transformer-Less Large-Scale Wireless Electric Vehicle Charging Infrastructure," in IEEE Journal of Emerging and Selected Topics in Power Electronics, vol. 9, no. 4, pp. 4587-4605, Aug. 2021.

[13] J. Sastry, P. Bakas, H. Kim, L. Wang, and A. Marinopoulos, "Evaluation of Cascaded H-Bridge Inverter for Utility-Scale Photovoltaic Systems," in Renewable Energy, vol. 69, pp. 208-218, Sept. 2014.

[14] S. Alotaibi and Ahmed Darwith, "Modular Multilevel Converter for Large-Scale Grid-Connected Photovoltaic Systems: A Review," in Energies, vol. 14, p. 6213, Sept. 2021.

[15] Y. Zhong, N. Roscoe, D. Holliday, T. C. Lim and S. J. Finney, "High-Efficiency mosfet-Based MMC Design for LVDC Distribution Systems," in IEEE Transactions on Industry Applications, vol. 54, no. 1, pp. 321-334, Jan.-Feb. 2018.

[16] S. Yang, Y. Tang and P. Wang, "Distributed Control for a Modular Multilevel Converter," in IEEE Transactions on Power Electronics, vol. 33, no. 7, pp. 5578-5591, July 2018.

[17] B. Çiftçi, S. Schiessl, J. Gross, L. Harnefors, S. Norrga and H. -P. Nee, "Wireless Control of Modular Multilevel Converter Submodules," in IEEE Transactions on Power Electronics, vol. 36, no. 7, pp. 8439-8453, July 2021.

[18] L. Harnefors, A. Antonopoulos, S. Norrga, L. Angquist and H. Nee, "Dynamic Analysis of Modular Multilevel Converters," in IEEE Transactions on Industrial Electronics, vol. 60, no. 7, pp. 2526-2537, July 2013

[19] H. Fujita, M. Hagiwara, and H. Akagi, "Power Flow Analysis and DC-Capacitor Voltage Regulation for the MMCC-DSCC," in Electrical Engineering in Japan, vol. 193, no. 1, pp. 1-9, 2015.

Investigation of Creepage Distances on Printed Circuit Boards for Avionic Applications

Hendrik Schefer[1,3], Zhongqing Xu[1], Tobias Kopp[2,3], Regine Mallwitz[1,3], Michael Kurrat[2,3]
INSTITUTE FOR ELECTRICAL MACHINES, TRACTION AND DRIVES (IMAB)[1]
INSTITUTE FOR HIGH VOLTAGE TECHNOLOGY AND POWER SYSTEMS (elenia)[2]
CLUSTER OF EXCELLENCE SE²A – SUSTAINABLE AND ENERGY-EFFICIENT AVIATION[3]
TU Braunschweig
Braunschweig, Germany
Email: h.schefer@tu-braunschweig.de
URL: https://www.tu-braunschweig.de/imab
URL: https://www.tu-braunschweig.de/elenia
URL: https://www.tu-braunschweig.de/se2a

Acknowledgments

This work was supported by the Deutsche Forschungsgemeinschaft (German Research Foundation, DFG) through the Germany's Excellence Strategy-EXC 2163/1-Sustainable and Energy Efficient Aviation under Grant 390881007.

Keywords

≪Insulation≫, ≪Reliability≫, ≪HVDC≫, ≪High voltage power converters≫, ≪Aerospace≫

Abstract

Avionic power electronic applications need enormous power densities ($10\,\frac{kW}{kg} <$) under high reliability to enable a high level of electrification. An increase in the DC bus voltage and high switching frequencies due to WBGs is unavoidable. An improved level of integration of these WBGs is needed to meet the aviation specific requirements regarding weight. Thus, this paper focuses on determining the breakdown voltage induced by DC voltages on printed circuit boards to show new design possibilities for future .

I Introduction

The electrification of aerospace is one opportunity to fulfil societal goals to reduce CO_2 emissions, for example, European Flightpath 2050. Power electronics in aircraft are subject to challenging requirements[1]. The conventional drive train must be replaced by electrical components and withstand environmental conditions in flight heights up to 13.000 m. For a sufficient overall design which considers in particular functional safety and to exhaust all opportunities of an electrical drive train, the power electronic should not be dependent on air-conditioning [1].

In the last decade, wide-bandgap semiconductors have offered new technology approaches to increase the bus voltage (up to 7.5 kV) at high switching frequencies (20 kHz <) and high efficiencies in mobile power applications. Due to the high switching frequencies at higher powers, passive components gain an enormous weight reduction compared to conventional semiconductors, especially magnetics profit. Magnetics are needed as filter compounds in drives or DC/DC converters connecting batteries or fuel cells. Furthermore, higher junction temperatures enable further advantages in near future [1].

A compact design requires deep knowledge about insulation coordination in power dense avionic power electronics under challenging environmental conditions. Therefore, the application of the well-known

Paschen's law is questionable for creepage distances, due to the gas-solid interface and its influence on the breakdown voltage. This paper focuses on determination of the breakdown voltage induced by DC voltages on printed circuit boards (PCBs). It research the dielectric basis and is thus crucial for further studies, for example to understand the ageing of creepage distances due to high dynamic rectangular voltage shapes ($\frac{du}{dt}|_{max} > \frac{60\,kV}{\mu s}$).

There are also standards, for example for equipment connected to low voltage systems, IEC 60 664 "Insulation coordination for equipment within low-voltage supply systems", which includes considerations of altitudes up to 2000 m, DC voltages up to 1500 V, AC rated voltage of 1000 V, sine wave frequencies up to 30 kHz, the Comparative Tracking Index (CTI) and different pollution degrees. Creepage distances from the IEC 60 664 are often used for electric vehicles [2]. Other standards like IEC 62368-1 [3] (Subsequent standard from IEC 60950 since 2020) have a similar restriction considering the voltage height of the voltage supply. High voltage standards like the IEC 60815 are specially designed for ceramic and glass insulators and are far away from the application of PCBs. In aviation, the MIL-HDBK-5400 is often used, but do not consider stresses of WBGs.

Thus, the higher the switching frequency, high voltages and shorter rise times, the more likely occur partial discharge. In publications [4], comparable investigations show interesting results and test bench ideas. This paper distinguishes the use of a Design of Experiments (DoE) to get a statistical significance and the preparation for rectangular stress in further studies. Other publications [5] and [6] pursue the same research objectives.

The paper is structured into an investigation methodology in section II, the description of the DoE in section III, test bench design in IV & VI, the data analysis in section V & VII and the conclusion in VIII.

II Methodology of Investigations

The International Standard Atmosphere [7] describes the environmental conditions over altitude. The humidity, pressure and temperature decrease with the altitude. Otherwise, the temperature inside the housing can reach higher values because of the losses and the high-required power density. Additional, outer conditions at sea level have to be taken into account. For example, landing, taxing and take-off are other interesting operation points. Airports, depending on their geographical location, have extremely various outer conditions. A housing can protect the power electronics for external conditions, like humidity, temperature and pressure. Power electronic integration in the fuselage of the propulsor, like the drive inverter, requires deep understanding. Tab. I and II compare the environmental conditions and test bench boundaries. The test benches cover the environmental requirements.

Table I: Environmental conditions

		Temperature °C	rel. Humidity %	Pressure $\frac{N}{m^2}$	Voltage V
Environmental	min	-56.50	1	61943	
	max	+150	90	101325	

Table II: Test bench boundaries [* climate control, ** temperature control, TB: test bench]

		Temperature °C	rel. Humidity %	Pressure $\frac{N}{m^2}$	Voltage V
Climate TB	min	-70** / +5*	10*		0
	max	+180** / +95*	95*		5000
Pressure TB	min			100	0
	max			101325	12000

By the literature review, a lot of factors could be identified which have an impact on the breakdown voltage induced by DC voltages. The study indicates the temperature, humidity, pressure, electrode distance, the existence of the solder resist and pollution. Therefore, the analysis requires massive tests, which the DoE can reduce.

III   Design of Experiments (DoE)

DoE is a scientific method that reduces experimental errors and R&D costs by optimizing the DoE protocols, reducing experimental workload, and scientifically analyzing experimental results [8].

For the climate test, a design of experiments is used to reduce the number of trials. Otherwise, the pressure test is conducted with the single factor method (single varying of one parameter set, pressure, distance).

A   Design of the Devices under Test (DUTs)

Tests can be done at small probes and in suitable test benches. PCB probes with two contact surfaces and different electrode distances/geometries offer good opportunities. The distances between the electrodes are available from 0.2 mm to 0.5 mm with and without solder resist. These PCBs are designed to have a breakdown in the limits of the voltage source and in future studies for accelerated altering due to rectangular voltage shapes. Fig. 1 shows examples of the designed Devices under Tests (DUTs). This study focuses on the breakdown voltage and serves primary data for the accelerated testing.

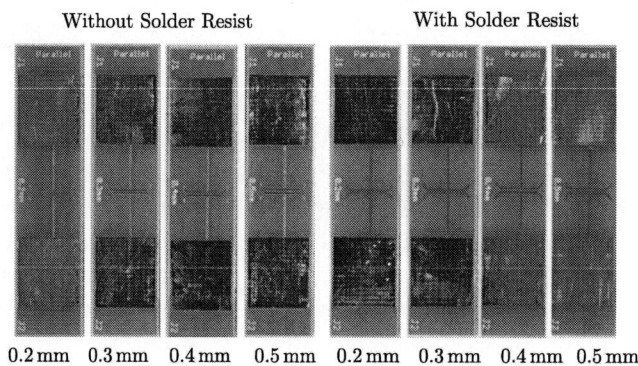

Fig. 1: DUTs with/without solder resist (0.2 mm to 0.5 mm) with the same geometry

B   Factors of Climate Tests

The test bench has physical boundaries, as mentioned above in Tab. II in section II. For the presented studies, the number of factors is reduced. In the Tab. III below, the considered factors are given.

Table III: Factors of climate tests

Temperature T	rel. Humidity $r.H$	Distances d	Solder Resist
30 °C	30 %	0.2 mm	
50 °C	55 %	0.3 mm	
70 °C	80 %	0.4 mm	without
		0.5 mm	

C   D-optimal Design of Experiments for Climate Tests

Compared with full factorial design, D-optimal design uses as few experiments as possible to obtain as much information as possible. In the experimental response model Eq. 1, the covariance of the x matrix is proportional to $(X'X)^{-1}$, so it needs to be minimized $(X'X)^{-1}$. The D-optimal can equate this problem to solving the determinant problem, i.e., maximizing $|X'X|$ [9]. The coefficients (β_0 to β_{11}) are output parameters of the DoE.

$$U_{\text{Break}} = \beta_0 + \beta_1 \cdot T + \beta_2 \cdot d + \beta_3 \cdot r.H + \beta_{11} \cdot T^2 + \varepsilon \tag{1}$$

The test matrix shown in Tab. IV is calculated and optimised using a data analysis software, called "Cornerstone". A randomised experimental design avoids systematic errors. The software is also helpful in evaluating the experiments.

Table IV: Output of the DoE (test matrix)

Nr.	T	$r.H$	d	Nr.	T	$r.H$	d
	°C	%	mm		°C	%	mm
1.	30	55	0.2	8.	70	80	0.4
2.	50	55	0.4	9.	70	55	0.3
3.	70	30	0.4	10.	30	30	0.3
4.	70	30	0.2	11.	30	80	0.4
5.	50	80	0.2	12.	30	55	0.5
6.	50	30	0.5	13.	50	80	0.3
7.	70	80	0.5				

IV Climate Test Bench for Determining the Breakdown Voltage induced by DC voltages

As shown in Fig. 2, the breakdown experiments of the PCB specimen were carried out in the automated climate test bench. The computerised climate chamber test bench consists of a climate chamber, a HV source, real-time embedded industrial controller (Ni cRIO) and a test bench computer. A serial communication interface enables the controlling of the high voltage source and temperature and humidity of the climate chamber. For safety reasons, a contact switch shuts the voltage source down in case of an open door.

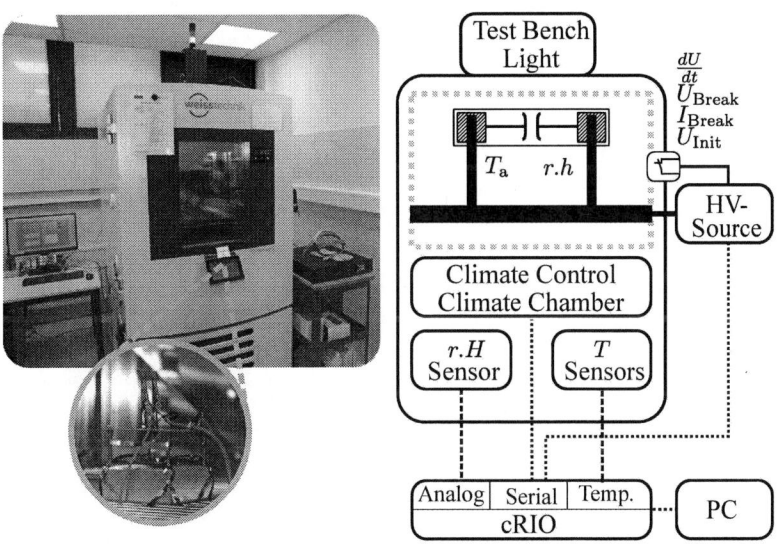

Fig. 2: Test Bench for realising climate factors

External sensors inside the climate chamber check the climate behaviour and enable capturing metadata. Crucial experimental test bench parameters are the voltage ramp, the threshold current (linked to carbonisation) and the initialisation voltage; therefore, the user can adjust these essential parameters in the graphical user interface.

After equipping the self-build HV interface, the climate test bench controls the environmental conditions (humidity and temperature) to the set values. The cRIO captures humidity, temperature, and time during the controlling time because the DUT faces the conditions from the beginning of the tests. After the predefined environmental conditions are reached, the voltage ramp starts. Getting the threshold condition (threshold current), the test bench stops, and the HV voltage supply switches off. Before exchanging the DUT, the user must determine the absence of voltage. The procedure begins from the starting point. The test bench is also designed for metadata analysis.

A pressure chamber can replace the climate chamber, which guarantees avionic pressure ranges (down to 1 mbar).

V Data Analysis of Climate Tests

The data originates from the test with the climate test bench. There are five and ten replicate experiments (Tab. V), and preliminary studies using data analysis with Matlab (Lilliefors) state a normal distribution. Due to the help of the standard deviation, it is noticeable that increasing the replicate experiments does not lead to better results at some measurement points. Especially the high humidity of $r.H = 80\%$ produces straying. However, a decrease in the standard deviation can be recognised at most of the measurement points.

Table V: Climate measurement data

Nr.	Five Replicate Experiments		Ten Replicate Experiments	
	$\overline{U}_{\mathrm{Break}}/\mathrm{V}$	$\sigma_U/\%$	$\overline{U}_{\mathrm{Break}}/\mathrm{V}$	$\sigma_U/\%$
1	1409.8	19.35	1440.7	15.58
2	2241.2	17.02	2209.9	11.99
3	2326	5.59	2302	6.23
4	1428.4	13.37	1497.6	12.29
5	935	6.2	992.9	11.08
6	2828.6	5.88	2920.5	5.52
7	1678.2	14.68	1674.7	15.53
8	1381.2	16.41	1339.1	11.81
9	1644.2	12.8	1724	11.94
10	1974	5.33	2000.7	6.86
11	1736	9.69	1701.6	11.39
12	2642	14.54	2625.7	11.14
13	1554.8	7.89	1450	14.4

The Tab. VI & VII illustrate the model parameter of $n = 5$ and $n = 10$. In comparing both models, an additional quadratic dependency of the temperature is added to the model ($n = 10$). Therefore, the temperature coefficient β_1 and the offset change in both models β_0.

Table VI: Model parameter (n=5)

β_0	β_1	β_2			β_3		β_{11}
V	$\mathrm{V}\cdot\mathrm{K}^{-1}$	mm	$\mathrm{V}\cdot\mathrm{mm}^{-1}$	%	$\mathrm{V}\cdot\mathrm{r.H}^{-1}$		$\mathrm{V}\cdot\mathrm{K}^{-2}$
2179	-6.50	0.2	-595.4	30	318.1		
		0.3	-129.2	55	91.5		
		0.4	195.7	80	-409.6		
		0.5	529.4				

Table VII: Model parameter (n=10)

β_0	β_1	β_2				β_3		β_{11}
V	$V \cdot K^{-1}$	mm	$V \cdot mm^{-1}$	%		$V \cdot r.H^{-1}$		$V \cdot K^{-2}$
993	46.13	0.2	-562.6	30		351.6		-0.516
		0.3	-148.1	55		117.1		
		0.4	176.7	80		-468.7		
		0.5	534					

The Tab. VIII expresses the fitting of the model to the captured data. As shown in Tab. VIII, for the PCB breakdown voltage predictive model ($n = 5$), the R^2 (goodness of fit in linear regression) is 0.9473, and the adjusted R^2 (corrected goodness of fit) is 0.8946. The RMSE (root mean square error) is 176.67; for the PCB breakdown voltage model ($n = 10$), R^2 is 0.9838, and the adjusted R^2 is 0.9611. Compared with the predictive model ($n = 5$), the PCB predictive breakdown voltage model ($n = 10$) has a better fit. Both R^2 of the deviation model of model ($n = 5$) and model ($n = 10$) are less than 0.6, and the deviation model fit is poor. However, the fit of the deviation model of model ($n = 10$) is slightly better than that of the predictive model ($n = 5$).

Table VIII: Goodness of fit [R^2: goodness of fit in linear regression, adjusted R^2: corrected goodness-of-fit, RMSE: root mean square error]

Model	Predictive PCB Breakdown Model			Deviation Model		
	R^2	adjusted R^2	RMSE	R^2	adjusted R^2	RMSE
$n = 5$	0.9473	0.8946	176.67	0.4916	0.3899	3.84
$n = 10$	0.9838	0.9611	108.32	0.5553	0.4663	2.36

Fig. 3 compares the predictive model ($n = 5$) with its confidence interval (98 %) with five fractional experiments to verify the quality of the model. Between different measurement values, a second order regression fits the data. The predictive model differs slightly from the fractional experiments; it can be recognised in the temperature dependency.

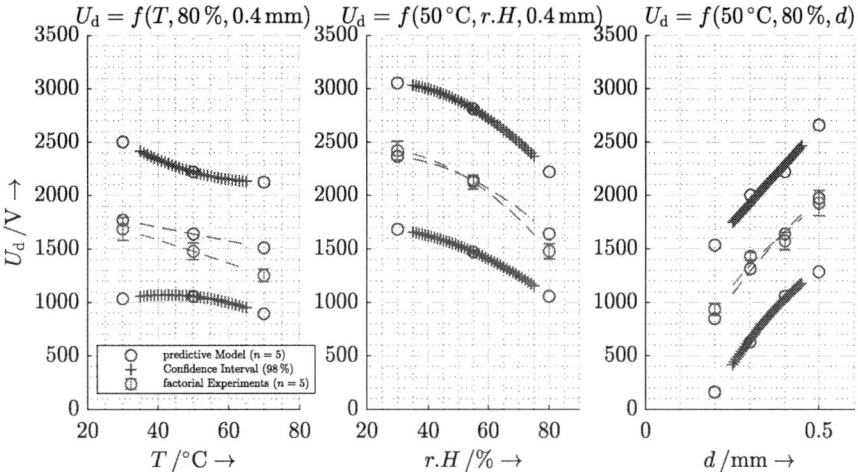

Fig. 3: Comparison fractional experiments vs. DoE predictive model ($n = 5$)

Fig. 4 evaluate the predictive model ($n = 10$) with its confidence interval (98 %) with the same five fractionals as in the study above. The five fractionals do not fit so well in the model ($n = 10$) as in the model ($n = 5$). The reason is the derivation of the fractional experiments. However, the derivation between both models is improved, as analysed in Tab. VIII, and a higher number of fractional tests will improve the derivation of the validating tests.

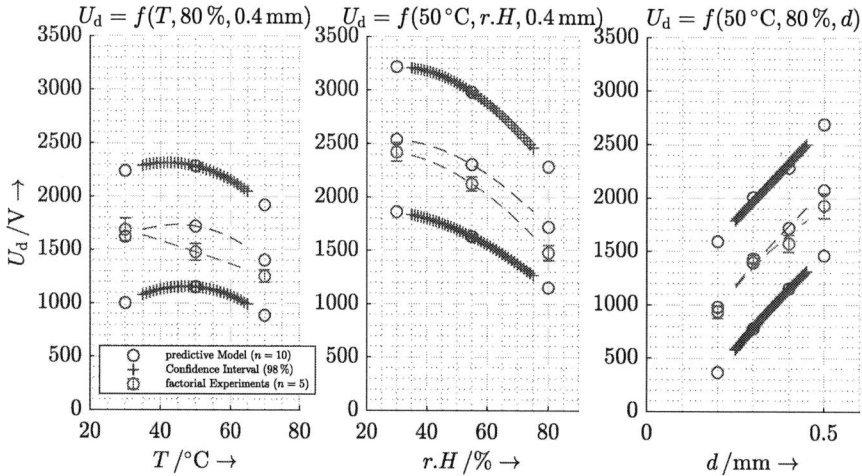

Fig. 4: Comparison fractional experiments vs. DoE predictive model ($n = 10$)

A Pareto analysis explains that distances have the most significant impact, second the humidity and third the temperature. A study with ten trials per test point offers a decrease in the derivation.

VI Pressure Test Bench for Determining the breakdown Voltage induced by DC voltages

Fig. 5: Pressure chamber

A pressure chamber (Fig. 5) replaces the climate chest, which guarantees aeronautical pressure ranges (down to 1 mbar). The low-pressure chamber offers many features for insulation studies, such as high voltage bushings up to 12 kV (AC, DC) and high current bushings up to 400 A continuous and 5 kA peak. In addition, the low-pressure chamber has a viewing window made of borosilicate. This window makes the chamber suitable for analysing components using various optical methods such as spectrography or high-speed cinematography. Furthermore, the chamber is equipped with a large number of low voltage electrical bushings and several BNC bushings for required measurement equipment. Also, fluid feed-throughs are available.

VII Data Analysis of Pressure Test

During a meeting of the ETG department Q2 (German expert group for Materials, Insulation Systems, Diagnostics) in 2022 a demand for an improved knowledge of high voltage performance for creepage distances in the pressure range of aviation was visible. Moreover standards like the IEC 62368-1 [3] only defines altitude dependencies for air gaps, a separate definition for creepage distances can be deduced from these values. However, these deduced values may not consider special breakdown behaviour of creepage distances for lower pressures.

As a contribution to the ongoing discussion, measurements of the creepage distance under a varied pressure will be presented within the pressure chamber, Fig 6. Currently, a combination with temperature and humidity will not be part of the results. The combination of these environmental conditions is challenging and still under development. It will be part of future investigations and is an important investigation target, as regarded in V.

Measurements of the breakdown voltage U_d were performed with the presented circuit board design (Fig. 6) using a DC Generator (Type: GLP1-g HV-DC 4 kV 10 mA, Company: Schleich GmbH).

For each data-point the breakdown voltage of 8 circuit boards were investigated. The here shown data represents the extended measurement uncertainty with a confidence interval of 98% of the first breakdown voltage of each PCB, Fig. 6.

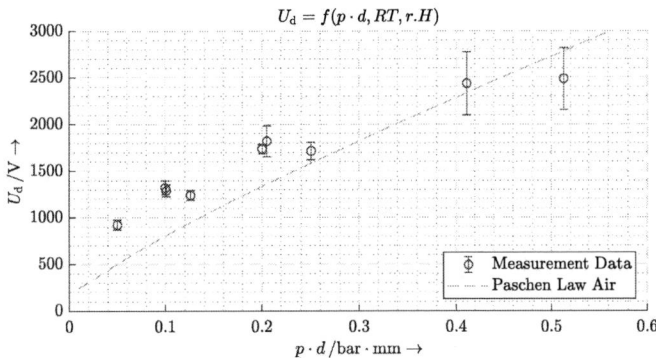

Fig. 6: Fractional pressure dependence breakdown-voltage for PCBs ($T =$room temperature, $r.H = 80\%$)

For this data analysis a single-factor research is done due to the relative small parameter area. The pressure is adjusted in the range of 1 bar, 0.5 bar and 0.25 bar to investigate an important area for the aviation. However, pressures down to 0.044 bar have to be tested according to [10]. Thus, this area should be integrated in future research. The gap distance on the PCB is chosen to 0.5 mm, 0.4 mm and 0.2 mm to match the measurements above, chapter 2.

In accordance with the well-known Paschen's law (dashed lines, Fig. 6) the x-Axis is the product of distance and pressure. The curve of Paschen's Law is according to [11] presented in this Figure with the parameters for:

- $A = 1130\,\mathrm{mm}^{-1}\cdot\mathrm{bar}^{-1}$

- $B = 27.4\,\mathrm{kV}\cdot\mathrm{mm}^{-1}\cdot\mathrm{bar}^{-1}$

- $\gamma = 0.025$

In this first overview differences to the Paschen's law are obvious. In the higher areas of $p\cdot d > 0.5\,\mathrm{bar}\cdot\mathrm{mm}$ the breakdown voltage of the PCBs seems to be slightly under the curve. A lower dielectric strength could be assumed here. In contrast to this behaviour the area of $p\cdot d < 0.25\,\mathrm{bar}\cdot\mathrm{mm}$ of the breakdown voltage of the PCB seems to be more and more underrated by the Paschen's law. This result could lead to an assumption that a better dielectric strength is available. Would this be confirmed in further research, PCB designs could be adapted slightly within this design area.

Additionally, during repetitive measurements with these circuit board a dependence of the breakdown voltage regarding previous breakdowns on this board are observed. A permanent influence on the PCB material can be deduced, thus a breakdown leads to a damaged PCB with undefined properties. Resulting in the need of a change of PCBs with a single breakdown occurrence with no exceptions.

VIII   Conclusion

The paper describes an investigation of creepage distances on printed circuit boards induced by DC voltages. This study in the article is one of the upcoming insulation coordination studies for power electronic devices in aircraft applications. For upcoming life time testing loaded by rectangular stress, this investigation delivers a first understanding of the environmental conditions' influence on creepage distances. This study offers experiment-driven investigations, the d-optimal design is presented, the test bench is explained, and the results are discussed in detail.

It presents a mathematical model of the creepage distance dependent on climate parameters and analyses the goodness of fit. A confidence interval of 98 % offers trustable results. One measurement point for climate test in the experimental matrix takes 40 minutes; therefore, a further increase in the number of experiments isn't practical. Life time testing loaded by rectangular stress is more critical.

Additionally, the dependency of pressure tests are investigated and the correlation with Paschen's law are discussed. The breakdown voltage is affected by the distance, the temperature, the humidity and the pressure. The most likely effect next to the distance of the electrodes is the pressure. However, as seen in the presented data, each environmental condition has an influence and thus the interaction of these influences should be researched carefully. Especially because they occur simultaneously during the application and could be system critical.

The environmental condition is crucial for isolation coordination for a power electronic device for aviation applications. This paper shows mathematical models based on breakdown tests on PCBs due to an applied DC voltage. Therefore, proven power electronic housing concepts with a particular focus on the temperature are required because the temperature has the lowest impact of all investigated factors. Integrated solutions without totally encapsulated housing (for example, propulsor, electrical machine, power electronic inverter) must consider the challenging environmental conditions at their mounting location. Nevertheless, the time and the fast-changing voltage shapes are other exciting parameters. After this experience has been gained, design recommendations for power-dense power electronic components can be developed.

References

[1] H. Schefer, L. Fauth, T. H. Kopp, R. Mallwitz, J. Friebe and M. Kurrat, "Discussion on Electric Power Supply Systems for All Electric Aircraft," in IEEE Access, vol. 8, pp. 84188-84216, 2020. https://doi.org//10.1109/ACCESS.2020.2991804

[2] R. Foley, R. Nagappala, G. Ressler, P. Andres, B. Martel, "Application of Insulation Standards to High Voltage Automotive Applications", ISSN: 0148-7191, e-ISSN: 2688-3627. https://doi.org/10.4271/2013-01-1528 Published April 08, 2013 by SAE International in United States

[3] Audio/video, information and communication technology equipment - Part 1: Safety requirements, IEC 62368-1:2018, Oct. 2018.

[4] Q. Zhou, M. Wen, T. Xiong ,T. Jiang, M. Zhou,X. Ouyang and L. Xing,"Study on Insulation Breakdown Characteristics of Printed Circuit Board under Continuous Square Impulse Voltage", Energies 2018, 11(11), 2018. https://doi.org/10.3390/en11112908

[5] W. Li, I. Cotton and R. Lowndes, "Development of a Test Method for Validation of Creepage Distances in High Voltage Aerospace Power Systems," 2020 IEEE Electrical Insulation Conference (EIC), 2020, pp. 517-520, https://doi.org/10.1109/EIC47619.2020.9158720.

[6] Y. Gao, Y. K. Men and B. X. Du, "Effect of relative humidity on surface dielectric breakdown of epoxy based nanocomposites under repeated pulses," Proceedings of 2014 International Symposium on Electrical Insulating Materials, 2014, pp. 69-72, https://doi.org/10.1109/ISEIM.2014.6870722.

[7] International Standard Atmosphere, DWD, Access: 05.12.21. http://www.iup.uni-bremen.de/emerge/home/pressure_altitude.html

[8] C. Sun and C. Lu, "Design and Development of an Experimental Design and Evaluation System", 2014 Fourth International Conference on Instrumentation and Measurement, Computer, Communication and Control, 2014, pp. 175-179. https://doi.org//doi:10.1109/IMCCC.2014.44

[9] P.F. de Aguiar, B. Bourguignon, M.S. Khots, D.L. Massart, R. Phan-Than-Luum, "D-optimal designs Chemometrics and Intelligent Laboratory Systems", Volume 30, Issue 2, 1995, Pages 199-210. https://doi.org/10.1016/0169-7439(94)00076-X

[10] Environmental Conditions and Test Procedures for Airborne Equipment, EUROCAE ED-14F, 2008. https://do160.org/rtca-do-160g/

[11] A. Kuechler, "High Voltage Engineering", Schweinfurt, Germany: Springer Vieweg, 2018.

A 20 kW, 3-level flying capacitor 1500 V inverter with characterized GaN devices for grid-tie applications

Van Sang NGUYEN, Anthony BIER, Hajar ES-SEGHIER, Ulrich SOUPREMANIEN,
Gérard DELETTE, Stephane CATELLANI

CEA - French Alternative Energies and Atomic Energy Commission
50 avenue du Lac Léman
Le Bourget du Lac - 73375, France
Tel.: +33 / (0) 4 79 79 27 50
E-Mail: van-sang.nguyen@cea.fr
URL: www.cea.fr

Acknowledgements

The authors acknowledge CEA/LITEN - CARNOT project "FlyGaN" for funding the activities presented in this paper.

Keywords

«DC-AC converter», «Flying Capacitor Boost Converter», «Gallium Nitride (GaN)», «Photovoltaic» «Renewable energy systems»

Abstract

This work presents the static and dynamic characterizations of high voltage GaN power devices (GaN FET 900 V and GaN HEMT 1200 V) in order to implement a 3-level flying capacitor 1500 V_{DC} inverter for high power density grid-tie applications with renewable energy sources such as solar and hydrogen energy. In the first part, the static characterizations are shown for two selected GaN power devices. Then these GaN devices were placed in a double-pulse-test-bench dedicated to the dynamic characterizations intended to observe the switching behaviors of the devices under the nominal voltage and current. Finally, in order to demonstrate the compactness of the converter, these GaN devices were implemented in a 20 kW, 3-level flying capacitor 1500 V_{DC} inverter with the full-custom suitable passive elements of the output filters connected to the 800 V_{AC} 3-phase grid.

Introduction

Wide-band-gap power devices take an important place in the power electronics systems design. Today, SiC devices become mature and emerge in many high power electronics products [1]. On the other hand, lower voltages GaN devices (≤ 650 V) are widely used for low power consumer applications, thanks to their efficient and compact designs [2]. Due to their thermal and blocking voltage constraints, GaN devices are still in optimizing stage for achieving high power applications, in particular, for 800 V 3-phase grid-tie applications with renewable energy power systems. Using 650 V GaN devices for applications with DC bus of 1500 V_{DC} requires the consideration of a high multi-level topology, where challenges are: the reliability, the control and the auxiliary circuitry of the systems [3]. Following up a recent work of the flying-capacitor topology on a SiC based design [4] and a GaN based design with a DC bus of 800 V_{DC} [5]; this article presents a design of a 3-level flying capacitor inverter working at 20 kW tied to the 800 V 3-phase grid where the nominal input voltage is 1500 V_{DC}. This GaN based design's approach gives a potential compact solution to several renewable energy systems like solar, hydrogen or even battery power systems. Today's trend in the large PV plant is the increasing in the power of the string by using smart panels [6]. Another work in parallel is ongoing where the GaN based power electronics parts are integrated inside the junction box of the PV panel, which allows installing much more number of photovoltaic (PV) panels in a single string. In a conventional 1500 V string, the output power is typically at 10 kW, the power of the 1500 V smart string could excess 20 kW thanks to the higher number of the PV panels [6].

Working with 1500 V_{DC} bus voltage, GaN devices have to perform consequently with a nominal drain-source blocked voltage of 750 V in steady-state operations. This work analyses and challenges the GaN FET 900 V devices and the GaN HEMT 1200 V in such operating conditions. The comparison of these two GaN devices firstly takes place on a Keysight B1506A power device analyzer for the static characterizations, and then dynamic test-bench is used to qualify these selected GaN devices for estimating the switching losses under the nominal configurations. This article includes the design of the grid-filters inductors with thermal consideration and the simulations of this 1500 V inverter. Finally, this work shows the experimental results of this GaN based 3-level flying capacitor inverter until 20 kW characterizations.

High voltage GaN devices: static and dynamic characterizations

Static characterizations

By using a standard industrial test-bench (Keysight B1506A), the I/V curves, the threshold voltage and the parasitic capacitances of the selected GaN devices are extracted under the similar configurations. Fig. 1 shows large differences on the capacitance's characterization of DUT1 and DUT2 up to a drain-source voltage of 900 V. At this high voltage level, DUT1 has 958 pF of Ciss meanwhile this value of Ciss for DUT2 is 251 pF. A high frequency design up to a switching frequency of several MHz need to consider these capacitances; it is event more critical for the thermal issue in a high power design.

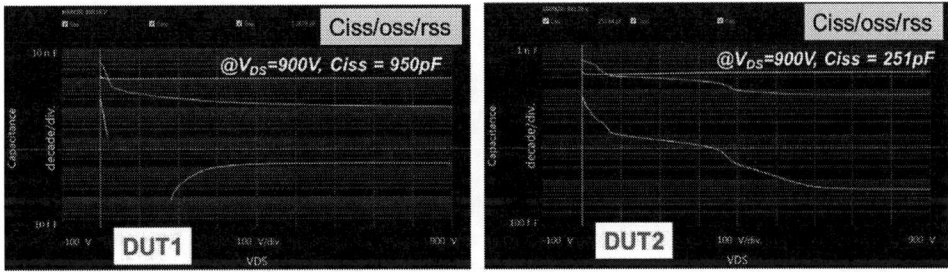

Fig. 1: Static characterizations of capacitances for GaN FET 900V (DUT1) and GaN HEMT 1200V (DUT2)

Under the similar configurations of the drain-source current, the threshold voltage of DUT1 is approximately 3.4 V and for DUT2 about 1.5 V. As well known, this value is extremely critical for GaN based high power applications where the transient voltage dv/dt is usually very high due to the high voltage applied on the devices. This dv/dt might be the main reason at the origin of short-circuit within an inverter-leg where there are complementary transistors located at high side and low side. Design of the grid-tie flying-capacitor inverter need to take into account carefully this issue due to the DC bus voltage of 1500 V.

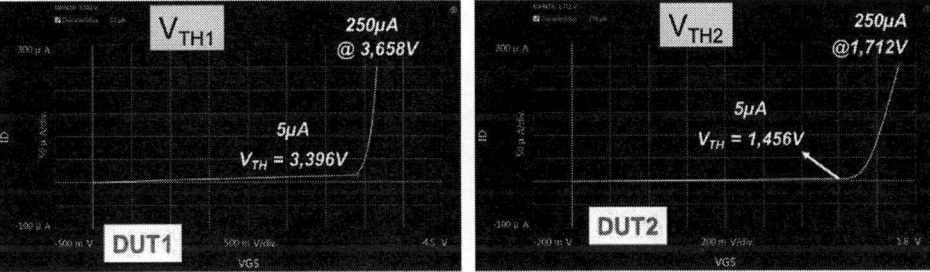

Fig. 2: Threshold voltages of GaN FET 900 V (DUT1) and GaN HEMT 1200 V (DUT2)

Fig. 3 and Fig. 4 show the static resistances and the current/voltage curves of these two GaN devices, the DUT1 has a higher capacitance; however, its resistance value is much smaller than the DUT2. These static results suggest the correct driving voltage of each transistors in order to obtain the best performance of the devices.

Under a drain current (at 25°C) of 30 A with different driving voltages, the on-state resistance of DUT1 varies between 36 mΩ and 48 mΩ where the typical value on the datasheet is 50 mΩ. This value of second device (DUT2) under the similar conditions is comprised between 76 mΩ and 80 mΩ where the maximum static on-state resistance value in its preliminary datasheet is 75 mΩ.

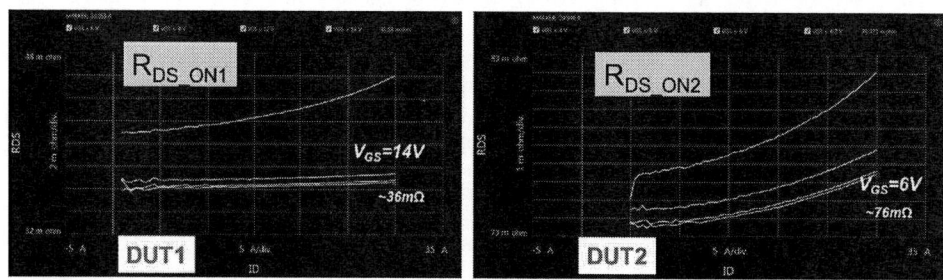

Fig. 3: Static resistances of GaN FET 900 V (DUT1) and GaN HEMT 1200 V (DUT2)

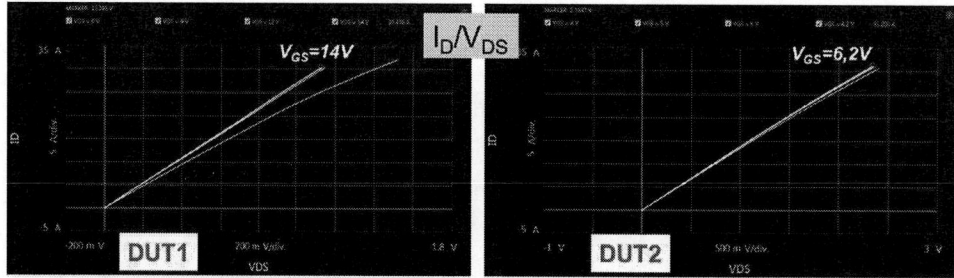

Fig. 4: Current/voltage characterizations of GaN FET 900 V (DUT1) and GaN HEMT 1200 V (DUT2)

Dynamic characterizations

After these static characterizations, the dynamic characterizations use the high bandwidth voltage and current probes [7], which are suitable for high-speed GaN devices. As in Fig. 5, an IZM current sensor with low insertion inductance of 0.3 nH and a bandwidth of 500 MHz [8] from Fraunhofer-IZM, [9] allows to record precisely the rising and falling edges of the drain-source current of the device. An IsoVU probe [8] with a bandwidth of 800 MHz is used to measure the drain-source voltage, a second low voltage probe with a bandwidth of 1 GHz measures the gate-source voltage.

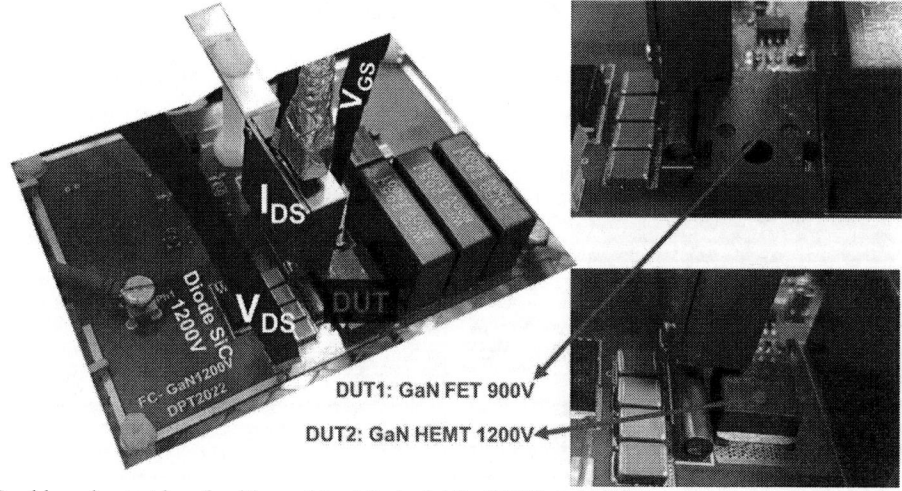

Fig. 5: Double-pulse-test bench with combined footprint for DUT1 and DUT2

From these measurements, the turn-on and turn-off losses can be calculated. During the experiments, difficulties were noticed to perform the dynamic characterization of DUT2 at the nominal voltage, we found out the problem on the gate driver. The set-up of the dynamic characterizations for DUT2 is ongoing. In this article, only experimental results of the dynamic characterizations on the DUT1 are presented.

Fig. 6 shows the experimental results of DUT1, which are the principal waveforms of the DPT set-up: the control signal on the gate V_{GS} of 0 V/9 V, the rising - falling edges from IZM sensor, the drain-source voltage V_{DS} of 0 V/760 V and the current on the inductance.

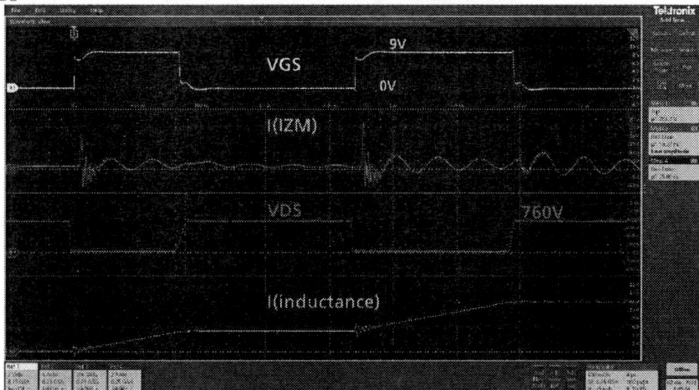

Fig. 6: Experimental waveforms of the DPT test

In Fig. 7, GaN devices (FET 900 V) turns on from a blocking voltage of 760 V, a peak slew rate dv/dt of 100 V/ns is observed.

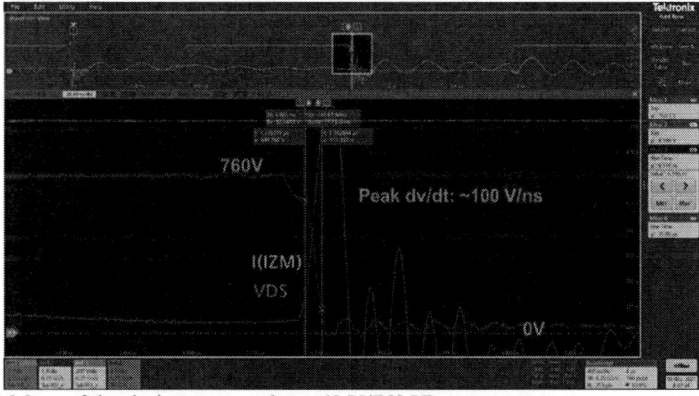

Fig. 7: Fall time of 6.8 ns of the drain source voltage (0 V/760 V) at turn-on

In Fig. 8, GaN FET 900 V devices turns off to block a voltage of 760 V, the peak slew rate dv/dt of 50 V/ns is observed in this rising edge.

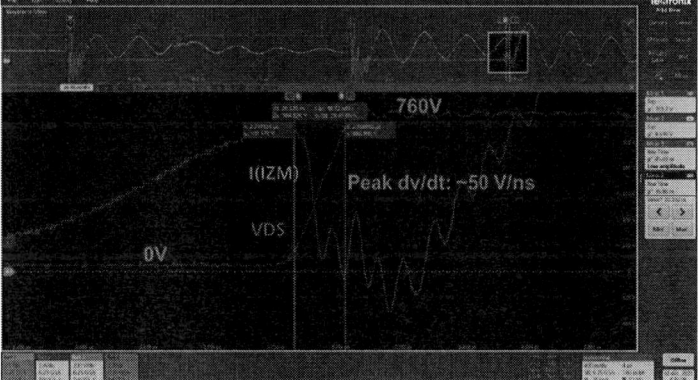

Fig. 8. Rise time of 20 ns of V_{DS} at turn-off

Simulations of the flying-capacitor 3-ph inverter

The electrical behavior of the 3-phase flying capacitor inverter is simulated using PLECS software. Fig. 9 shows a one-phase schematic representation model of the converter power stage. A 1500 V_{DC} voltage source is placed at the input and the inverter is connected to the 800 V_{AC} grid via a LCL-type filter. The current flowing through inductor L_1 is controlled within a PI-based control. The current control loop and the grid voltage synchronization is implemented considering the rotating reference frame (DQ domain).

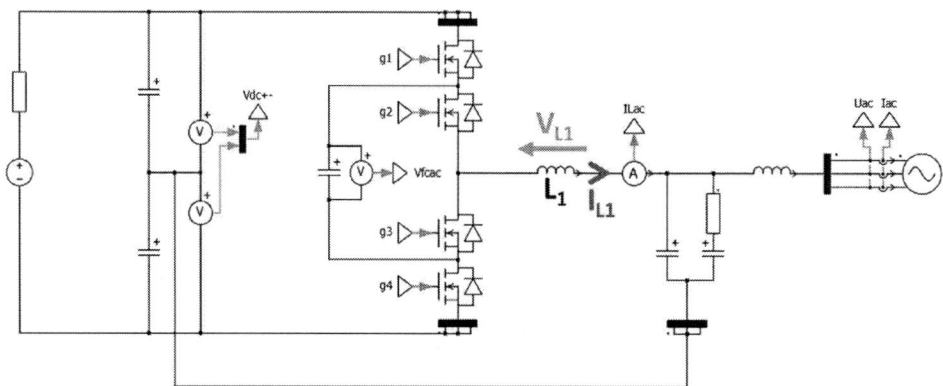

Fig. 9: PLECS simulation of the 3-phase flying-capacitor inverter

Simulation results are focused in Fig. 10 on the voltage and the current waveform of the inductor L_1. For a 20 kW 3-phase inverter, the peak value of the 50 Hz fundamental frequency sinusoidal AC current is 20 A and the RMS value of the signal 14.4 A. A 200 kHz triangular current ripple can be observed, which is twice the working frequency of the GaN devices. For the inductor design, its voltage must also be taken into account. The challenge in the design of the output filter, for a high power application, is the compliance to the high current, high voltage and high frequency at the same time.

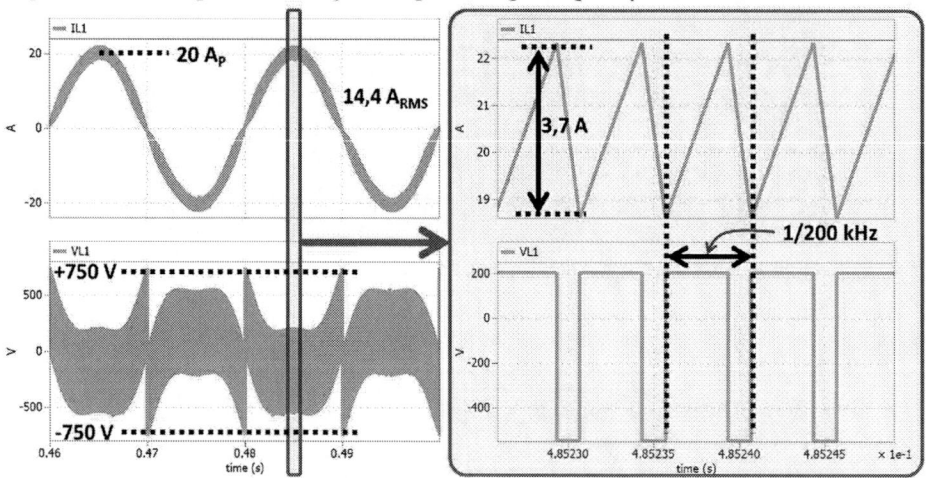

Fig. 10: Simulation results: waveforms on the output filters with twice higher working frequency

Output filters design

In this work, the GaN based 3-level flying capacitor 1500 V inverter operates with an output LCL passive filter configuration. As we can see on the simulation, in this flying-capacitor topology, the output filters operate at a double switching frequency compared to GaN power devices. That is why the magnetic materials of these filters should be able to operate at high frequency and at the same time, the associated windings should have a low AC loss in high frequency

The targeted inductance value for the larger inductive filter element is 200 μH and should be held constant during operation within a wide range of current values. This performance are obtained providing that the magnetic core works in a linear regime, i.e. the partial saturation of the core has to be avoided. The inductance design has been performed using the standard rules with large margin regarding the maximum average induction allowed in the core. A special attention has been paid to the thermal heating control of both the core and winding. The temperatures have been calculated at the nodal position over a 3D meshing by considering the core and winding losses as heat sources and the cooling by natural convection regime at boundary condition. The thermal resistance network and its numerical implementation were described in a previous work [10]. We considered two kinds of ferromagnetic material:

(i)   a MnZn ferrite from TDK (N87)

(ii)   a powder core from Magnetics (Koolμ60)

Commercial core geometries from the supplier catalogs have been screened, and for each material, a tradeoff has been found between the compactness and the heating. The characteristics of the designed inductors are listed in Table I.

Table I: Characteristics of the designed inductors

	Ferrite core (N87)	Powder Core (Koolμ60)
Core + winding volume	421 cm³	315 cm³
Turn number with 729 x 0.071 mm Litz wire	6	22
Gap	0.5 mm	1 mm
Iron loss	4.5 W	3 W
Winding loss	10 W	14 W

The calculated temperature maps are plotted on Fig11. It can be seen that the configuration with the Koolμ60 core is more prone to the winding heating due to a larger turn number compared to the N87 inductor. This is a consequence of the lower permeability of the powder core (μr=60) vs the ferrite one (μr = 1700).

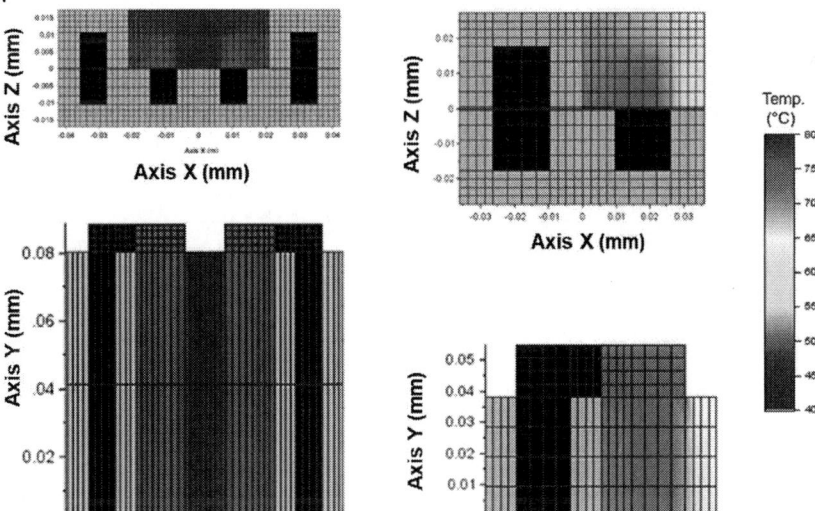

Fig. 11: Temperature maps for Ferrite (left) and Koolμ (right) solutions

In the same time, cooling of the ferrite solution was improved thanks to a larger external surface. Finally, with a maximal operating temperature of 80°C, the Koolμ60 core volume is 25 % lower than the volume of its ferrite counterpart.

Fig. 12 and Fig. 13 shows that the inductance value measured on inductors prototypes remains steady up to 40 A for the two selected materials.

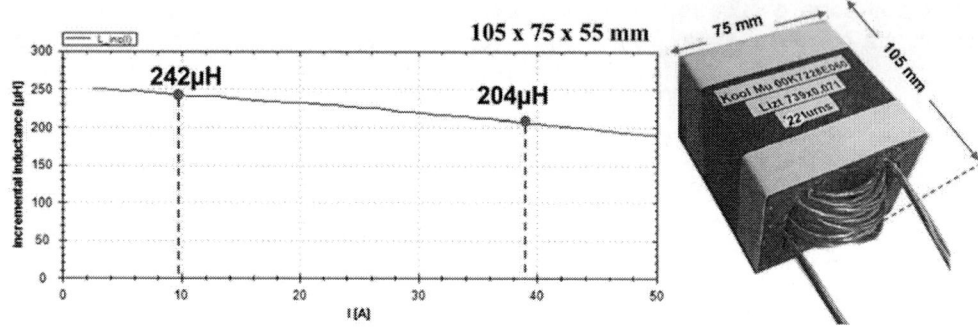

Fig. 12: Kool-Mµ based output filter with 22 turns of Litz wires: 105 x 75 x 55 mm

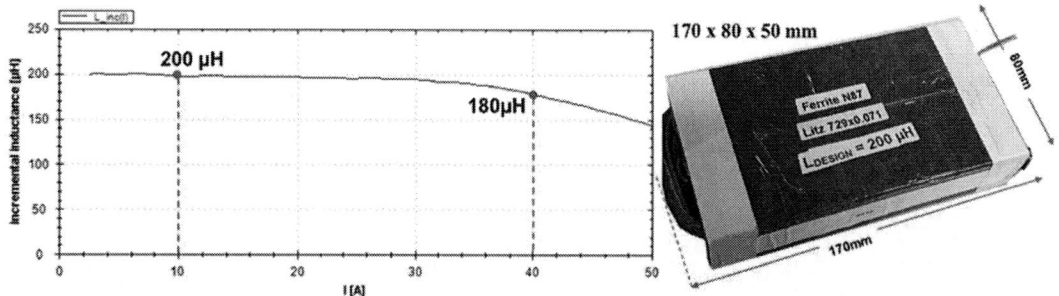

Fig. 13: Ferrite N87 based output filter with 6 turns of Litz wires: 150 x 80 x 50 mm

Experimental results on a flying capacitor inverter with GaN devices

In this part, the experimental results with GaN devices are given within the 3-level 1500 V inverter. Fig. 14 shows the schematic of the 3-phase flying-capacitor inverter on the left side where the input voltage is 1500 V_{DC} and the output is 3-phase 800 V_{AC}.

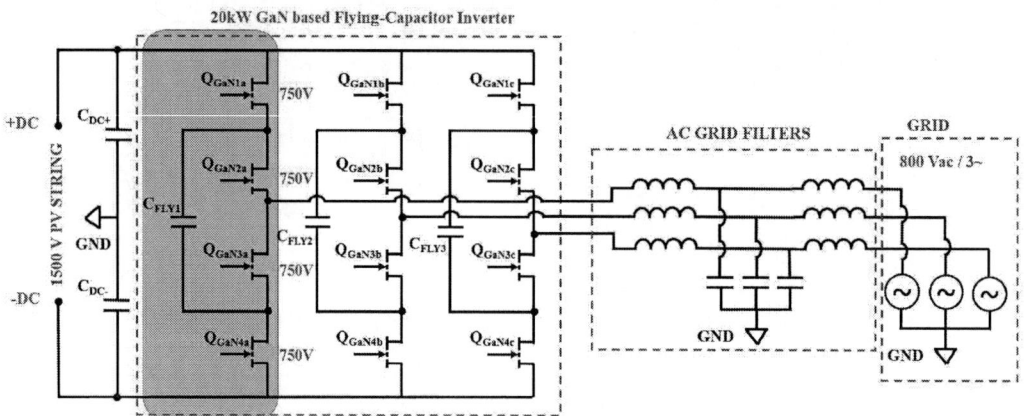

Fig. 14: Schematic of the 3-phase flying-capacitor inverter

On the right side of Fig. 13, GaN based active parts for one phase are demonstrated with GaN HEMT 1200V devices, so-called DUT2 in the previous sections, integrating gate driver Hey-1011 dedicated to GaN devices, GaN transistors are driven by V_{GS} of 0/6 V. In order to mitigate the influences of the EMI perturbations, the optical communications (optical controls and optical measurements) with an Imperix controller were implemented, and the batteries supply all the auxiliary circuits. The size of this part is 150 x 150 x 80 mm including the underneath heatsink. This is the first implementation of the GaN based Flying-capacitor inverter within this article; we call it version 1 (V1). In this implementation, we could not increase the DC bus voltage up to the nominal value; we are working on the modification of this version for a further experiment, in particular, driving GaN HEMTs by negative base voltage to avoid the spike on the gate when the high *dv/dt* occurs.

In the middle of Fig. 15, the second version of the GaN based Flying-capacitor inverter (V2) with GaN FET 900 V; the gate drivers Si8271 with the isolated power supplies generate V_{GS} of 0/9 V. Due to high *dv/dt*, once again, this set-up could not perform at the nominal DC bus voltage. On the left side of Fig. 3, the third version of the GaN based Flying-capacitor inverter (V3) with GaN FET 900 V; the gate drivers Si8271 with the isolated power supplies generate V_{GS} of -5/14 V. To simplify the implementation, the Imperix's voltage and current sensors replace the optical onboard measures. By using this set-up, the GaN devices could perform at nominal input voltage, and generate the suitable sinusoidal output voltage for 3-phase grid of 800 V. In the designs with GaN FET 900 V, the GaN power in packaging TO-247 are between the board of the auxiliary circuits and the heatsink.

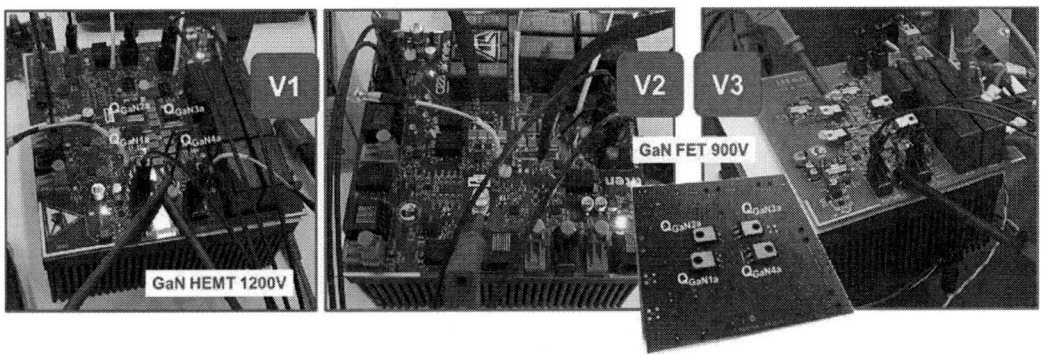

Fig. 15: Three different versions of GaN based active parts for the implementation of one-phase

For one-phase implementation in Fig. 16, bulky input capacitors (900 µF/1500 V) have been used to mitigate the 100 Hz fluctuant voltage on the input DC bus voltage, these bulky capacitors do not exist in the final 3-phase inverter. Using these high voltage GaN devices where GaN HEMT are still prototype devices for these implementations, the authors preferred to implement one-phase experiment before going further for the full implementation of the 3-phase inverter.

Fig. 16: Experiment of one-phase GaN FET 900V based inverter leg

Fig.17 shows the 100 kHz control signals of four GaN devices in a leg of the flying-capacitor topology, also the 3-level voltage at the middle point of the leg connected to the AC grid filter. With all the three versions of the implementations, these control signals and the 3-level voltage perform correctly at the low DC bus voltage. The differences appear at a higher DC bus voltage when the *dv/dt* becomes critical to the low threshold voltage of GaN devices.

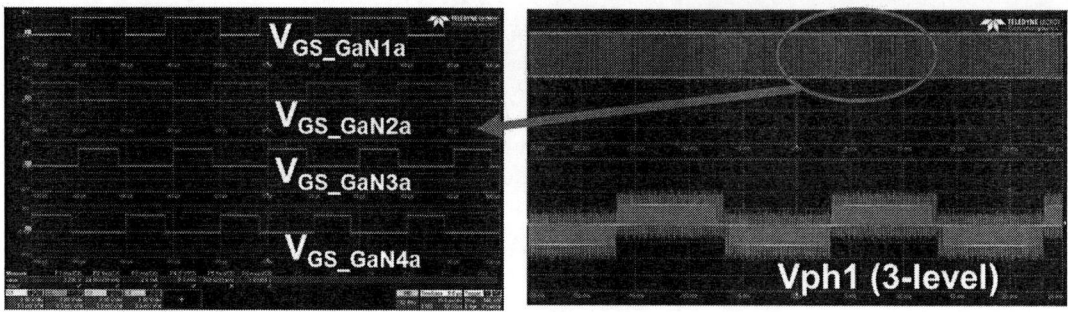

Fig. 17: Control signals of four GaN devices in a leg and the 3-level voltage of the leg

Finally, the experimental results in the nominal DC-Bus voltage are given in the Fig. 18 where the input DC-bus voltage is 1420 V (on channel 4 – green color), the flying-capacitor voltage is approximately 710 V (channel 2 – red color). At the output on a resistive load for this experiment, the output sinusoidal RMS voltage (on channel 3 – cyan color) is 462 V_{AC}, this voltage is equal to 800 V/√3 as line to neutral value. The output RMS current is 25.6 A (on channel 1 – yellow color), where the maximum current goes through the output filter is up to 40 A. The output power in this single-phase or so-called one-phase experiment is 11.8 kW with a fan underneath the heatsink. Based on these experimental results, in the full configuration, the power of the 3-phase GaN based flying-capacitor could reach almost 36 kW.

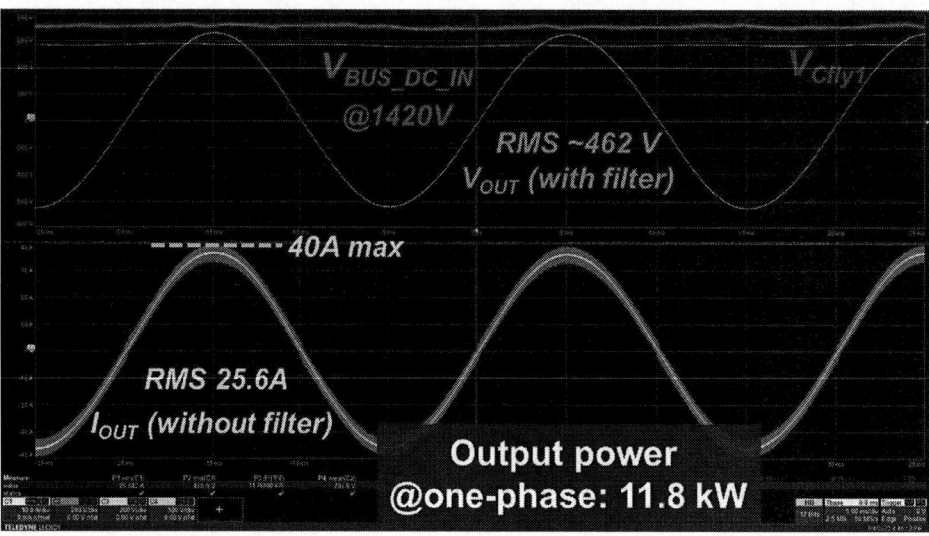

Fig. 18: Waveforms of a one-phase inverter under DC bus voltage of 1420V

Fig. 19 shows the thermal result of this one-phase inverter leg under 11.8 kW active output power, for a short duration of the experiment, the temperature on the packaging of GaN devices reach 40°C and keep increasing.

We are working to demonstrate the complete 3-phase inverter with both GaN HEMT 1200 V and GaN FET 900 V with modifications on the gate driver circuitry for GaN HEMT 1200 V, on the thickness of PCB and betterments on the thermal perspectives. The grid-tie protocols implement as our control loops for a grid-tie three-phase current source inverter [11-12]. Eventually, in this demonstration of the complete 3-phase inverter, we are going to measure the stabilized temperature on the GaN power devices and the efficiency of the inverter.

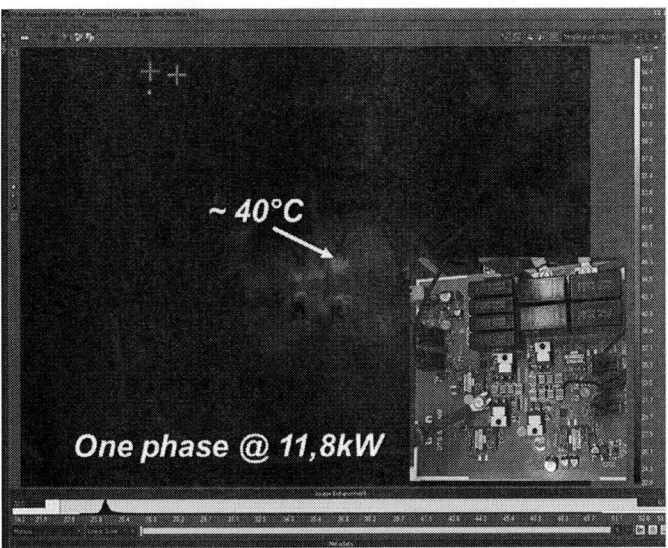

Fig. 19: Thermal image of one-phase under the output power of 11.8 kW

Conclusion

This work presents the design of a 3-level flying capacitor 1500 V_{DC} inverter with characterized GaN devices for 3-phase grid-tie applications where GaN devices perform under 750 V in steady-state operation. The static and dynamic characterizations take place to ensure the operation of the selected high voltage GaN devices in a nominal configuration. The paper include the PLECS simulations for this flying capacitor topology to identify the different working frequency of the elements in the inverter. The current and the voltage across the elements are presented in order to design the passive devices of the inverter; in particular, the output inductor filter is 3D-designed, optimized, characterized in this work. In addition, the design, the detailed characterizations of this compact GaN based flying-capacitor inverter with its dedicated output filters are given. We show several difficulties due to the GaN devices themselves and issues on the design of gate driver and their power supplies. Finally, an experiment was performed at the nominal voltage of the PV string, 1500 V. The nominal power announced is 20 kW as the power of a smart PV string [6], however the experimental characterizations show that the output power of this flying capacitor 1500 V_{DC} inverter can reach almost 36 kW. The realization of a complete 3-phase inverter is ongoing with improvements on the mechanical and thermal point-of-views.

References

[1] Yole report on Power SiC 2020: Materials, Devices and Applications, Yole 2020

[2] Yole report GaN power 2021: epitaxy, devices, applications and technology trends, Yole 2021

[3] M. Farhangi, Y-P. Siwakoti, R. Barzegarkhoo, S. Ul Hasan, D. Lu, D. Rogers "A Compact Design Using GaN Semiconductor Devices for a Flying Capacitor Five-Level Inverter" 2021 IEEE Energy Conversion Congress and Exposition (ECCE), nov 2021

[4] L-G. Alves Rodrigues, G. Perez "A 200 kW Three-level Flying Capacitor Inverter using Si/SiC based Devices for Photovoltaic Applications" PCIM Europe digital days 2021, May 2021

[5] E. Bunin (VisIC) "Performance and cost benefits of D3gaN in 3-level 800V EV inverters" PCIM Europe digital days 2021, May 2021

[6] S. Catellani, V-S. Nguyen, A. Bier, T. Delaplagne, P. Merhej, F. Bizzarri "GaN based panel-integrated, high-efficiency DC/DC optimizer for maximizing the yield of the large photovoltaic power plant" PVSEC 2022

[7] V. S. Nguyen, A. Bier, R. Escoffier, S. Catellani, J. Martin, C. Gillot "A high precision dynamic characterization bench with a current collapse measurement circuit for GaN HEMT operating at 175°C" PCIM Europe digital days 2021, May 2021

[8] K. Klein, D. E. Hoene, and D. K.-D. Lang, "Comprehensive AC Performance Analysis of Ceramic Capacitors for DC link usage," p. 7, 2017.

[9] K. Klein, D. E. Hoene, and D. K.-D. Lang, "Power module design for utilizing of WBG switching performance," p. 8, 2019

[10] G. Delette, U. Soupremanien and S. Loudot, "Thermal Management Design of Transformers for Dual Active Bridge Power Converters," in IEEE Transactions on Power Electronics, vol. 37, no. 7, pp. 8301-8309, July 2022

[11] A. Bier: Three-phase grid-tied current-source inverter sizing and control for photovoltaic application, SPEEDAM, June 2016

[12] A.Bier, V-S. Nguyen, S. Catellani, J. Martin " Control of a single-phase grid-tied GaN based solar micro-inverter" EPE 2020 (22nd European Conference on Power Electronics and Applications, EPE'20 ECCE Europe), Sept 2020

New Analytical Model for Calculating HF-Losses in Litz Wire Regions Located Outside the E/U-Core Window of Transformers

Qingchao Meng and Jürgen Biela
Laboratory for High Power Electronic Systems (HPE), ETH Zurich
meng@hpe.ee.ethz.ch   http://www.hpe.ee.ethz.ch

Keywords

≪Modelling≫, ≪Litz wire≫, ≪Transformer≫, ≪Magnetic device≫

Abstract

Litz wire is essential for reducing high-frequency losses in medium frequency transformers, which are built often with 'E' or 'U' shaped cores. Many analytical models focus only on the winding section inside the core window, where the winding is completely enclosed by the core. The losses of the winding section outside the core window are not determined separately. This paper compares the losses of both winding sections and provides an accurate and fast loss model for the winding section outside the core window. The error of the model is less than 5%, which is much smaller than the 1D field model (up to 30%). Moreover, to compute the total winding losses the model requires about 90 µs, which is comparable to the time consumed by the 1D field model (72 µs on a standard laptop with a 4.8 GHz, 4-core CPU and a 16 GB RAM).

1 Introduction

Isolated DC-DC converters are widely used in power supplies of modern data centers, electric vehicle charging, renewable energy industry, etc. due to their high efficiency and power density. As a key component in these isolated DC-DC converters, the medium frequency transformer (MFT) enables a simple voltage level adaption by the turn number and the galvanic isolation between different power stages. As the winding losses increase rapidly with increasing operating frequencies due to the skin and the proximity effect, litz wire (LW) is often used to reduce the winding losses in MFTs. LW is made of numerous thin strands, which are twisted to avoid the bundle level high frequency (HF) effect [1]. Because of the complex geometry of LWs, an accurate loss calculation is challenging.

3D FEM is an accurate method for calculating winding losses of LW, since it takes all geometrical details into account. However, the 3D structure of LW in transformers is typically very complex, so that 3D FEM simulations are too time consuming for converter optimization routines, since the losses need to be calculated very often during the optimization routine. A method, which combines 3D FEM and analytical models is presented in [1], [2]. There, the LW is replaced either by a winding region with a

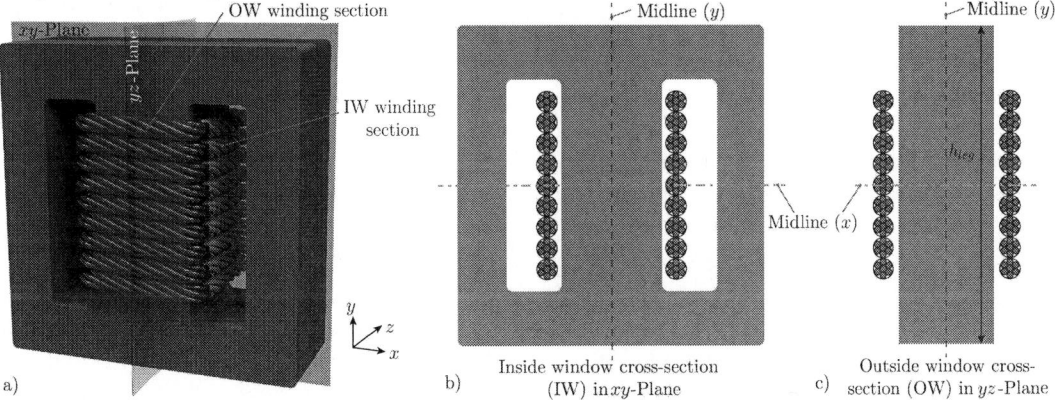

Fig. 1: Transformer with an 'EE' core and a single layer LW winding as an example for describing the double-2D concept. The cross-section of the transformer in the xy- and the yz-plane is shown in b) and c)

constant current density or by an equivalent round wire and then the magnetic field is calculated by a 3D FEM for this "replacement setup". With this magnetic field, the winding losses are calculated with analytical models using the magnetic field calculated by the 3D FEM. Although this method considers the 3D geometry of the transformer in the loss calculation, the computational time is still too long for converter optimization routines.

As an alternative to 3D FEM, often 2D FEM are used. With the assumption of a perfect twisting of the LW, 2D FEM for a single cross-section of transformers are a feasible method to calculate the winding losses. However, with 2D FEM for a single cross section the winding losses of LW in transformers that use 'E' or 'U' cores can not be calculated accurately, since such transformers are neither rotationally symmetrical nor can such transformers be obtained by extending a single 2D cross-section in one direction. To solve this issue, the double-2D method is proposed in [3], where 2D FEM simulations are performed for two different planes, which include the winding part in the inside window cross-section (IW winding section) and in the outside window cross-section (OW winding section) as shown in Fig. 1. By multiplying the resulting losses per unit length of the IW and the OW winding section by the corresponding length, the total winding losses can be calculated. Although this method is more accurate than the 2D FEM only based on a single cross-section and faster than the 3D FEM, it is still too time consuming to be used in optimization routines, where fast analytical models are usually required. However, no analytical model for such a double-2D approach is given in the literature.

Many analytical loss models for LW [4]-[7] focus on the IW winding section, where the winding is completely enclosed by the core and a 1D magnetic field with a constant amplitude is assumed. However, the magnetic fields caused by the OW winding section are strongly 2D, and the 1D field models can result in relatively larger errors for the winding losses caused by the OW winding section. Furthermore, a large share of the winding losses is caused by the OW winding section in transformers based on 'E' or 'U' cores. For accurately calculating the winding losses in the OW winding section, a fast and accurate model with closed-form is presented in this paper.

This paper is arranged as follows: The losses of the IW and the OW winding section are compared in section 2 based on the results of 2D FEM. In order to accurately calculate the losses of the OW winding section, the calculation of the external magnetic field for the OW winding section is described in section 3. Based on the magnetic field model in section 3, the winding loss calculation is presented in section 4. To evaluate the accuracy of this proposed model, a numerical validation is performed in section 5.

2 Comparison of losses of IW and OW winding section

MFTs based on 'E' or 'U' shaped cores typically have a simple design and a small volume. Furthermore, a wide range of available core sizes and wire types offers a large number of design possibilities and reduces the development effort. However, the winding losses of such transformer can not be fully modelled with a single 2D plane. In Fig. 1a), a transformer with an 'EE' core and a single layer winding is given. As can be seen, about half of the winding is located inside the window and the other half is located outside the window. In order to take both parts into account, a double-2D concept is proposed in [3]. Based on this double-2D concept, the winding losses of the transformers can be calculated based on two cross-sections, which are in the xy- and the yz-plane as shown in Fig. 1b) and 1c).

In order to evaluate and compare the losses of the IW and the OW winding section, the losses of 13 different transformer designs are determined. The detailed parameters of the 13 designs can be found in [8] and the winding losses per unit length of these designs are calculated by 2D FEM. Since the cross-sections in the xy- and the yz-plane of the transformers are all symmetric to the midline (y) as shown in Fig. 1b) and 1c), only half of the cross-sections are considered in the 2D FEM simulations. The geometries for calculating the losses of the IW and the OW winding section are presented in Fig. 2 as small sketches. The dashed lines show the core without center core leg of the IW winding section. In total, 2×13 FEM simulation setups are investigated in frequency range $0.1 \leq \Delta = \frac{d_s}{\delta} \leq 1.2$, where d_s is the diameter of a single strand and δ is the skin depth. In Fig. 2, the difference between the IW and the OW winding section losses is calculated by $\frac{P_{IW} - P_{OW}}{P_{OW}} \times 100\%$, where P_{IW} represents the losses of the IW winding section and P_{OW} represents the losses of the OW winding section. In addition, the error of 2D FEM simulations based only on the inside window cross-section with respect to the double-2D method

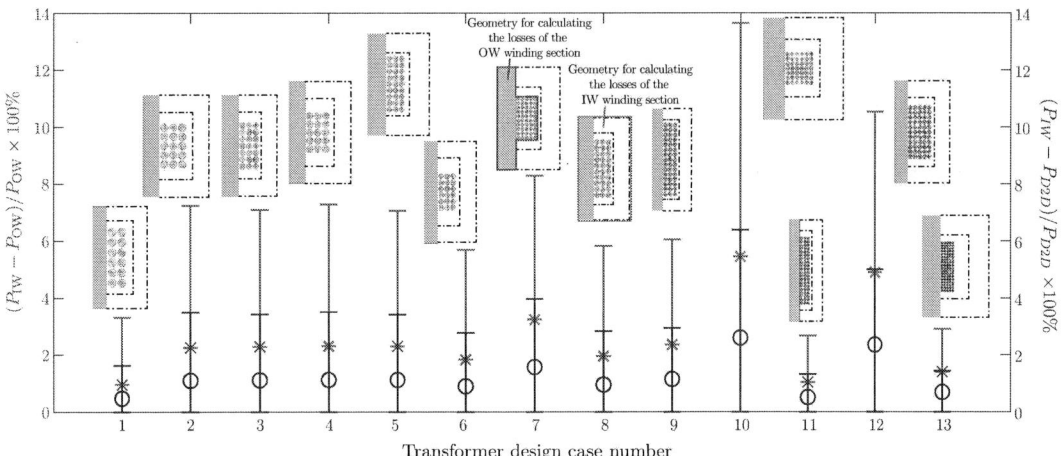

Fig. 2: Loss differences between the IW and the OW winding section and between the IW winding section and the double 2D method are shown. The geometry with and without dashed lines represents the IW and the OW winding section respectively. The detailed geometry parameters can be found in [8].

is calculated by $\frac{P_{IW}-P_{D2D}}{P_{D2D}} \times 100\%$. In the loss calculation $P_{D2D} = 0.5(P_{IW} + P_{OW})$, the loss contribution of the IW and the OW winding section are both considered to be approximately 50%, as all the transformers are based on 'EE' cores.

As shown in Fig. 2, the loss difference between the IW and the OW winding section is between 3% and 14%. There, the loss difference is always positive, since the losses caused by the IW winding section are always higher than the losses caused by the OW winding section due to the high magnetic field. The main difference between the IW and the OW winding section is the external magnetic field distribution, which affects primarily the winding losses caused by the external proximity effect. Since the core is assumed to be ideal ($\mu_r \to \infty$), the magnetic field inside the core is zero. The IW winding section is completely enclosed by the core and the magnetic field is only non-zero in the core window as shown in Fig. 3a). The

OW winding section is located next to the center core leg and the magnetic field profile is different from the one of the IW winding section. The magnetic field path can either be through the core leg or be completely closed in air as shown in Fig. 3b). Therefore, the effective field paths are longer in the OW winding section than them in the IW winding section and the amplitude of the magnetic field of the OW winding section is according to Ampere's law smaller than the one of the IW winding section. The smaller external magnetic field results in smaller winding losses caused by the external proximity effect, which explains why the losses of the IW winding section are higher.

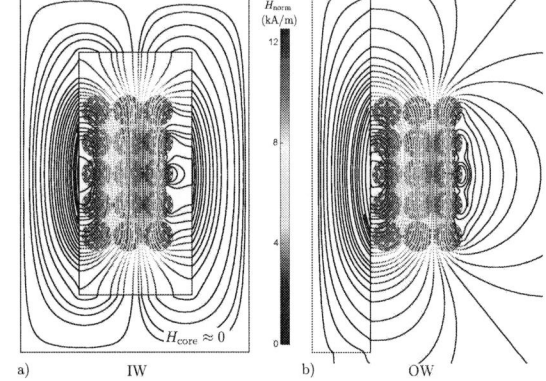

Fig. 3: H-field distribution caused by the IW and the OW winding section of transformer case 2.

3 Magnetic field calculation

In this section, the 2D external H-field, which could be split into H_x and H_y is calculated. As presented in [4], the HF-losses in LW consist of 3 types of losses, including losses caused by the skin (P_{Skin}), the internal ($P_{p,int}$), and the external proximity effect ($P_{p,ext}$), which can be calculated by using (1) for a single turn. In (1), the parameters n_s, $F(f)$, r_{Li} and $G(f)$ are the number of strands, the resistance factor for the skin effect, the radius of LW, and the resistance factor for the proximity effect.

$$P = \underbrace{\frac{F(f)I^2}{n_s}}_{P_{Skin}} + \underbrace{n_s G(f)\frac{I^2}{8\pi^2 r_{Li}^2}}_{P_{p,int}} + \underbrace{n_s G(f)H^2}_{P_{p,ext}} \tag{1}$$

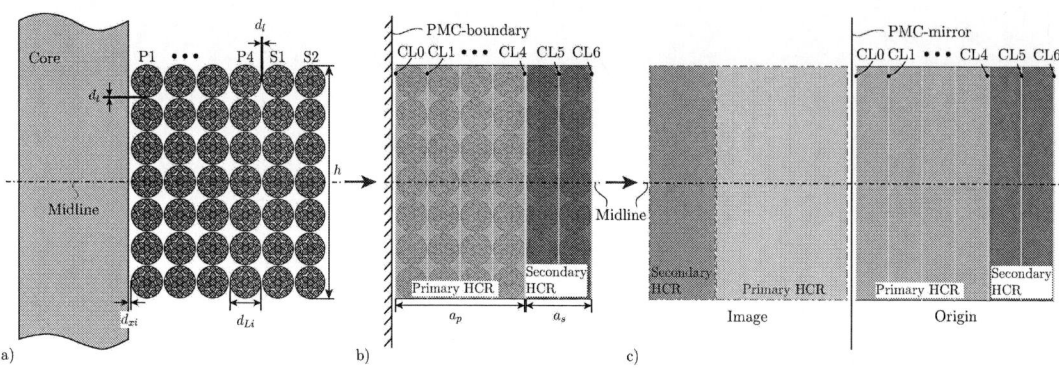

Fig. 4: a). Transformer design case 10 with a 4 layer primary and a 2 layer secondary winding is chosen as an example. b) The core and the windings in a) are replaced by a PMC-boundary (perfect magnetic conductor boundary) and homogeneously distributed current density regions (HCRs). c) The PMC-boundary is replaced by the images of the primary and the secondary winding, which are mirrored at the PMC-boundary.

As LW reduces HF losses only in the low range of penetration ratio $\Delta = d_s/\delta$, i.e. typically in a range $\Delta < 1$, this paper focuses on the penetration ratio range $0.1 \leq \Delta \leq 1.2$. In this penetration ratio range, the losses caused by the skin effect excluding DC losses ($P_{Skin} - \frac{1}{2}I^2 R_{DC}$) are often relatively small and the losses $P_{p,int}$ are zero at DC and increase with frequency much slower than the losses $P_{p,ext}$. Moreover, the losses caused by the skin and the internal proximity effect can be relatively accurately calculated by (1) for a perfectly twisted LW. As can be seen in (1), P_{Skin} and $P_{p,int}$ depend only on the parameters of the LW itself, such as radius of LW r_{Li}, number of strands n_s, etc., rather than on the assembly parameters of a transformer, e.g. the winding arrangement, the presence of the ferrite core, etc. In contrast, $P_{p,ext}$ depends not only on the parameters of LW itself, but also on the assembly parameters, which affect the external magnetic field distribution. Furthermore, $P_{p,ext}$ increases rapidly with frequency and is the dominant share of the total losses with increasing frequency. Many analytical models, such as [4]-[7], are based on the assumption of 1D external magnetic fields, and are not suitable for calculating $P_{p,ext}$ in the OW winding section, as the magnetic fields are strongly 2D. Since only $P_{p,ext}$ is different in the OW winding section, this paper focuses on the calculation of $P_{p,ext}$. The key to an accurate estimation of $P_{p,ext}$ is the 2D H-field calculation that is presented in the following section.

3.1 Method of mirror image

As the OW winding section is located next to the center core leg of the core and is not enclosed by the core, the magnetic field is strongly 2D. To calculate such a 2D magnetic field, the method of mirror image (MMI) is used, because the core is only on one side of the windings, so that the primary and the secondary winding only need to be mirrored once, which results in relatively low computational effort. To apply MMI for the LW winding, 3 assumptions are made.

- **A1:** The LW is considered to be perfectly twisted, so that each strand conducts the same current and the eddy current circulates only within the same strand rather than in loops of different strands. Therefore, the bundle level skin and proximity effect are considered to be perfectly cancelled.

- **A2:** In the penetration ratio range $0.1 \leq \Delta \leq 1.2$, the current density distribution in each strand is relatively uniform. Moreover, the distance between two adjacent turns d_t and layers d_l are typically much smaller than the diameter d_{Li} of the LW and the height h of each layer in the primary and the secondary winding is assumed to be identical, as shown in Fig. 4a). Therefore, the primary and the secondary winding can be replaced by two rectangular regions with homogeneous current density distribution (HCRs) as shown in Fig.4b).

- **A3:** The core is considered to be ideal ($\mu_{core} = \infty$), and is replaced by an infinitely long perfect magnetic conductor (PMC) boundary as shown in Fig. 4b). Based on this and assumption A2, the PMC-boundary can be further replaced by the images of the primary and the secondary HCR, which are mirrored at the PMC-boundary as shown in Fig. 4c).

Based on A2, the basic element for calculating the magnetic field distribution is a rectangular conductor. The total magnetic field distribution of the OW winding section is equal to the superposition of the magnetic fields caused by all the 4 HCRs as shown in Fig. 4c). Also in [9] and [10], MMI is used to calculate the magnetic field as well as the winding losses. However, the assumed basic element for calculating the magnetic field is a single turn, which is a round conductor rather than a rectangular HCR. Consequently, more turns result in a longer computational time. In contrast, the computation time of the proposed method is independent of the number of turns, as the windings are replaced by a rectangular HCRs. However, for transformers with complex winding arrangements, where assumption A2 can not be applied, more HCRs are needed for calculating the magnetic field. Since MMI is only applied to the HCRs rather than to each strand or turn, the proposed method is named by MMI for HCRs (MMIH).

3.2 Magnetic field calculation for a rectangular conductor

The magnetic field of a rectangular conductor has been derived in [11]. The definition of the angles for calculating the magnetic field $\theta_1 \dots \theta_4$ is given in Fig. 5a). Based on this definition, angles $(\theta_1 \dots \theta_4)$ are defined for points with coordinates $x > a, y > b$ as shown in Fig. 5a). However, for points for example located inside the conductor, the angles need to be redefined as shown in Fig. 5b).

The basic idea for calculating the magnetic field in [11] is to replace the rectangular conductor by an infinite number of line current elements. In the following, the magnetic field calculation is treated as a magnetostatic problem. To calculate the x and the y component of the magnetic field, the vector potential **A** ($\mathbf{B} = \nabla \times \mathbf{A}$) is used. By using Stokes' theorem as given in (2), it can be concluded, that the integral of **A** over a closed loop is equal to the flux Φ through that loop as illustrated in Fig.5c).

$$\iint \nabla \times \mathbf{A} \cdot d\mathbf{S} = \iint \mathbf{B} \cdot d\mathbf{S} \rightarrow \oint \mathbf{A} \cdot d\ell = \iint \mathbf{B} \cdot d\mathbf{S} = \Phi \tag{2}$$

For a line current, the magnetic field $H(r)$ at the lower red point as shown in Fig. 5d) can be derived by using Ampere's law as given by (3).

$$\oint \mathbf{H} \cdot dl = \iint \mathbf{J} \cdot d\mathbf{S} = I \rightarrow H = \frac{I_z}{2\pi r} \tag{3}$$

From (2) and (3), the vector potential **A** can be derived as given in (4). If the current I_z is equal to 0, the vector potential A_z should be equal to 0 everywhere in the space, therefore the coefficient C is equal to 0.

$$A_z(r+\Delta r)\Delta l - A_z(r)\Delta l = B_\phi \Delta r \Delta l \rightarrow \Delta A_z(r) = B_\phi \Delta r \rightarrow A_z(r) = \int B_\phi dr = \frac{\mu_0 \mu I_z}{2\pi} \ln r + (C = 0) \tag{4}$$

As the rectangular conductor consists of an infinite number of such line currents, the vector potential caused by the rectangular conductor at an arbitrary point (x,y) is equal to the integral of the vector potential of the line current over the rectangle as given by (5), where $r = \sqrt{(x-x')^2 + (y-y')^2}$. As the current in the rectangular conductor is considered to be uniform, the amplitude of the line current is equal to $I_z = \frac{I_R}{4ab}$, where I_R is the amplitude of the current carried by the rectangular conductor.

$$A_z(x,y) = \frac{\mu \mu_0}{4\pi} \frac{I_R}{4ab} \int_{-b}^{b} \int_{-a}^{a} \ln\left((x-x')^2 + (y-y')^2\right) dx' dy' \tag{5}$$

The magnetic fields H_x and H_y of the rectangular conductor at an arbitrary point (x,y) are given by (6).

$$H_x = \frac{1}{\mu_0 \mu} \frac{\partial A_z}{\partial y}, \quad H_y = \frac{1}{\mu_0 \mu} \frac{\partial A_z}{\partial x} \tag{6}$$

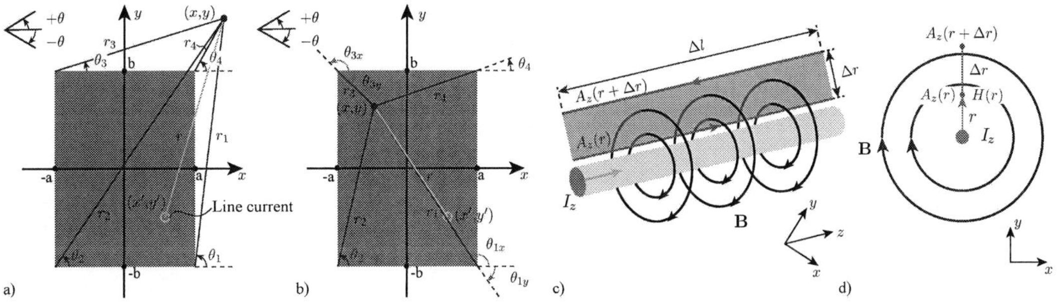

Fig. 5: a)/ b) Coordinate systems for calculating the magnetic field at an arbitrary point (x,y) outside and inside the conductor. If both θ_x and θ_y are positive, they are represented by a single angle θ. c) Flux density **B** and vector potential **A** distribution for a short segment of an infinitely long line current. d) 2D illustration of c).

The expressions for H_x and H_y are given by (7) and (8). Note that the equations for the angles $\theta_1 \ldots \theta_4$ given in [4] are unfortunately wrong. The correct expressions are given in (9) based on the coordinate system in Fig. 5 a) and b). The angles for calculating H_x and H_y are separately defined as $\theta_{1x} \ldots \theta_{4x}$ and $\theta_{1y} \ldots \theta_{4y}$ as given in (9), which are different from the definitions in [11] and [4]. The ranges of angles for calculating H_x and H_y are in the intervals $[0, \pi]$ and $[-\frac{\pi}{2}, \frac{\pi}{2}]$.

$$H_x(x,y) = \frac{I_R}{8\pi ab}\left[(y+b)(\theta_{1x}-\theta_{2x}) - (y-b)(\theta_{4x}-\theta_{3x}) + (x+a)\ln(\frac{r_2}{r_3}) - (x-a)\ln\frac{r_1}{r_4}\right] \quad (7)$$

$$H_y(x,y) = \frac{I_R}{8\pi ab}\left[(x+a)(\theta_{2y}-\theta_{3y}) - (x-a)(\theta_{1y}-\theta_{4y}) + (y+b)\ln(\frac{r_2}{r_1}) - (y-b)\ln\frac{r_3}{r_4}\right] \quad (8)$$

$$r_1 = \sqrt{(x-a)^2+(y+b)^2} \quad r_2 = \sqrt{(x+a)^2+(y+b)^2} \quad r_3 = \sqrt{(x+a)^2+(y-b)^2} \quad r_4 = \sqrt{(x-a)^2+(y-b)^2}$$

$$\theta_{1x} = \arctan\tfrac{y+b}{x-a}\,(\mathrm{mod}\,\pi) \quad \theta_{2x} = \arctan\tfrac{y+b}{x+a}\,(\mathrm{mod}\,\pi) \quad \theta_{3x} = \arctan\tfrac{y-b}{x+a}\,(\mathrm{mod}\,\pi) \quad \theta_{4x} = \arctan\tfrac{y-b}{x-a}\,(\mathrm{mod}\,\pi) \quad (9)$$

$$\theta_{1y} = \arctan\tfrac{y+b}{x-a} \qquad \theta_{2y} = \arctan\tfrac{y+b}{x+a} \qquad \theta_{3y} = \arctan\tfrac{y-b}{x+a} \qquad \theta_{4y} = \arctan\tfrac{y-b}{x-a}$$

The validation of the proposed method with 2D FEM is presented in the following section.

3.3 Numerical evaluation for magnetic field calculation

To evaluate the proposed method, the transformer design case 10 illustrated in Fig. 2 is selected as an example. The primary and the secondary winding are considered as 2 HCRs as shown in Fig. 6a). The magnetic field distribution is equal to the superposition of the magnetic field of the 2 original HCRs and the 2 mirrored HCRs as given by (10).

$$H_x = H_{xp,ori} + H_{xs,ori} + H_{xp,ima} + H_{xs,ima} \quad H_y = H_{yp,ori} + H_{ys,ori} + H_{yp,ima} + H_{ys,ima} \quad (10)$$

In Fig. 6b) the magnetic field calculated with 2D FEM for LW windings is shown. In general, the magnetic field calculated by MMIH (Fig. 6a) matches the one calculated by 2D FEM for LW (Fig. 6b). However, the magnetic field in Fig. 6a) is "smoother" than the one in Fig. 6b) in the winding region. This is especially true in layers P1 and S2, which are the transition region between the winding and air/core, as the magnetic field close to the strands/LW is bent, because of the curvature of the strands/LW.

To compare the magnetic field calculated by MMIH and by FEM quantitatively, the magnetic field distributions on the cut lines CL0 ... CL6 as shown in Fig. 4b) are calculated. As shown in Fig. 7a) and b), H_x along cut lines CL0 ... CL6 is symmetric to the origin point (0,0), as $H_x(x,-y) = -H_x(x,y)$. Because of the curvature of the LW, the curves for H_x calculated by FEM "oscillate/fluctuate" periodically around the curves calculated by MMIH, which is relatively smooth, as the current density is homogeneously distributed in the HCRs.

In Fig.7c) and d), the curves for H_y calculated by FEM "oscillate/fluctuate" periodically around the curves calculated by MMIH on cut lines CL0, CL4, and CL6 which is the transition region between the winding and air/core and the transition region between the primary and the secondary winding. However, along CL1 ... CL3 and CL5, the curves of H_y calculated by FEM and MMIH are all very smooth and

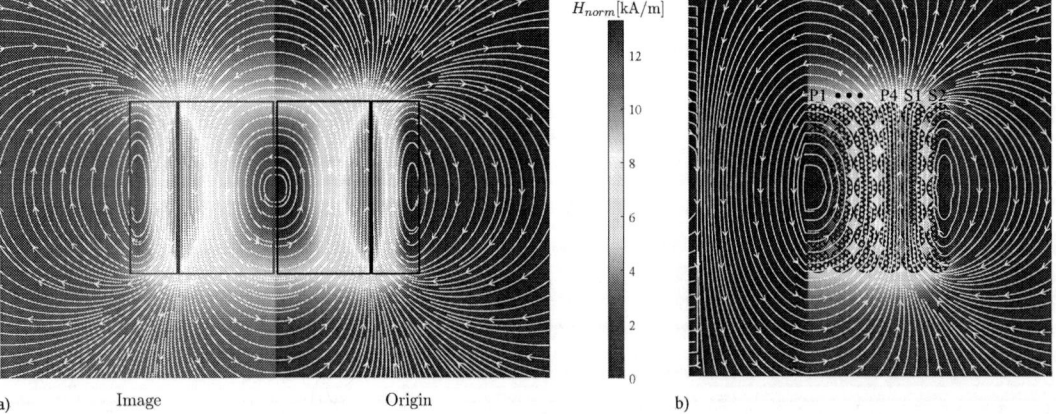

Fig. 6: Comparison of magnetic fields calculated by MMIH a) and by FEM for the LW b). Both colormap of $H_{norm} = \sqrt{H_x^2 + H_y^2}$ and the magnetic field are shown.

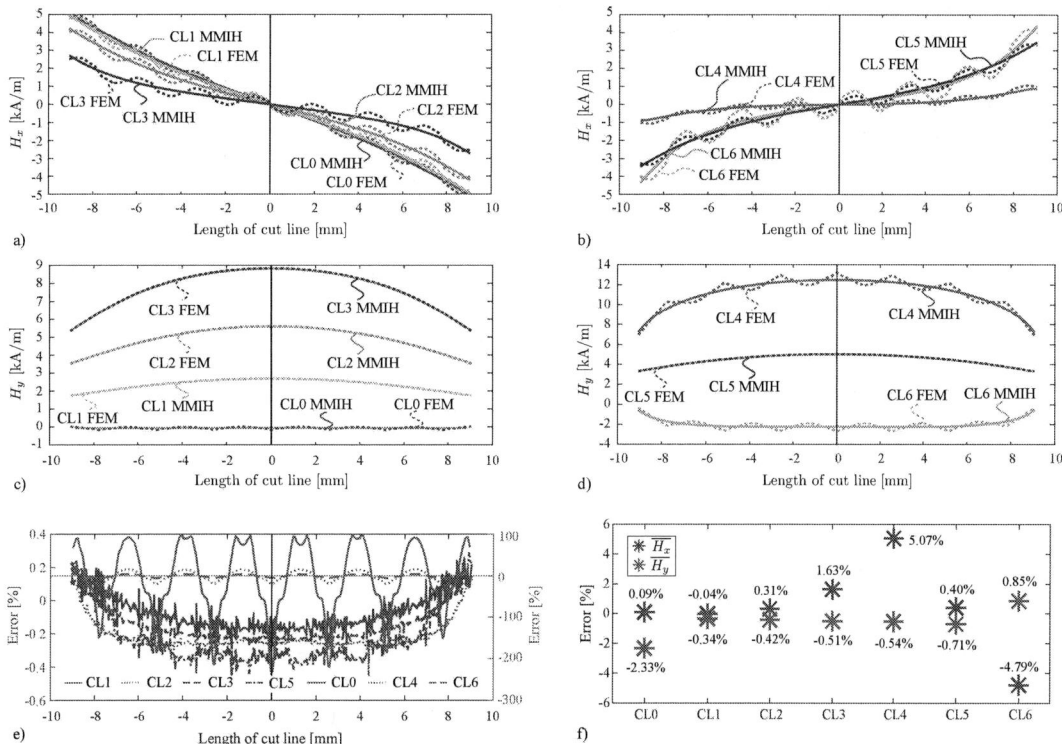

Fig. 7: a)~d) show the magnetic field distribution, which is calculated by MMIH and by FEM for LW for H_x and H_y, on cut lines CL0 ... CL6 as shown in Fig. 5b). e) Differences of the magnetic field calculated by FEM and MMIH for H_y on cut lines CL0 ... CL6 is shown. f) Error of the average value of $|H_x|$ and H_y calculated by FEM and MMIH on cut lines CL0 ... CL6 is shown, where the average value for both methods are calculated by using (11) and the error is equal to $Error = \frac{\overline{H_{x/y,MMIH}} - \overline{H_{x/y,FEM}}}{\overline{H_{x/y,FEM}}} \times 100\%$.

match very well. The differences between the magnetic fields calculated by these two methods on cut lines CL0 ... CL6 are shown in Fig. 7e). Note that the error for cut lines CL1 ... CL3 and CL5 is calculated by $Error = \frac{H_{y,MI} - H_{y,FEM}}{H_{y,FEM}} \times 100\%$. It can be directly seen, that the magnetic field H_y calculated by MMIH on cut lines CL1 ... CL3 and CL5 is very accurate, as the error is smaller than 0.5%. Furthermore, the error curves are relatively noisy in Fig. 7e), which is caused by the meshing and linear interpolation between mesh nodes. As the magnetic field calculated by FEM on cut line CL0 fluctuates around $H_y = 0$ as shown in Fig. 7c), the error for cut line CL0 is calculated by $Error = \frac{H_{y,FEM} - H_{y,MMIH}}{H_{y,MMIH}} \times 100\%$, to avoid a division by zero. The same error equation is used for cut lines CL4 and CL6, which fluctuate around the lines for H-field calculated by the MMIH method.

As in the loss calculation the average value or the RMS value of the magnetic field is required, it is important to evaluate how accurate can those values be calculated with the proposed method. In this paper, only the average values on cut lines CL0 ... CL6 are compared, as they are used in the loss calculation. The average value $\overline{H_x}$ and $\overline{H_y}$ on cut lines CL0 ... CL6 can be calculated by (11) and the error, which is

$$\overline{H_x} = \frac{1}{h} \int_{-\frac{h}{2}}^{\frac{h}{2}} |H_x| \mathrm{d}y = \frac{2}{h} \int_{0}^{\frac{h}{2}} H_x \mathrm{d}y \quad \overline{H_y} = \frac{1}{h} \int_{-\frac{h}{2}}^{\frac{h}{2}} H_y \mathrm{d}y = \frac{2}{h} \int_{0}^{\frac{h}{2}} H_y \mathrm{d}y \tag{11}$$

shown in Fig. 7f), is equal to $Error = \frac{\overline{H_{x/y,MI}} - \overline{H_{x/y,FEM}}}{\overline{H_{x/y,FEM}}} \times 100\%$. Note that for H_x the integral is applied for $|H_x|$, since H_x is an odd function, the integral of H_x over the cut line is 0. Furthermore, the average value of the magnetic fields on the cut lines can be calculated by only integrating the magnetic fields over half of the cut lines, if the windings are symmetric to the midline as shown in Fig. 4. The integral of the magnetic field calculated by FEM on each cut line is calculated by using the trapezoidal method for discrete data. The integral of the magnetic field calculated by MMIH on each cut line is calculated by (17), (18) and (19) given in section 4.

As can be seen in Fig. 7 f), the error of $\overline{H_x}$ and $\overline{H_y}$ calculated by MMIH is smaller than 1% for most cut lines. The biggest error for $\overline{H_x}$ occurs on cut line CL4 with 5.07%, however, the amplitude of the magnetic field on CL4 is the smallest as can be seen in Fig. 7b) and consequently has the smallest impact on the loss calculation. The biggest error for $\overline{H_y}$ occurs on cut line CL0, where the amplitude of magnetic field H_y is also the smallest as shown in Fig. 7c).

4 Loss calculation

Since as explained the losses caused by the skin and the internal proximity effect are independent of the external magnetic field and can be relatively accurately calculated, this section focuses on the losses caused by the external proximity effect $P_{p,ext}$. The losses $P_{p,ext}$ can be calculated by using the equation given in (12) for a single strand. Note that both x and y components need to be considered in the loss calculation, since H_x and H_y are both perpendicular to the strands and generate losses. The factor $G(f)$ is either based on Bessel functions or hyperbolic functions. The detailed expression can be found in [12].

$$\frac{P_{p,ext}}{S} = \frac{G(f)H^2(x,y)}{S} = \frac{1}{S}G(f)\left(H_x^2(x,y)+H_y^2(x,y)-(2H_xH_y\cos(\theta_{xy})=0)\right) \quad P_{p,ext}=\iint \frac{P_{p,ext}}{S}\mathrm{d}S \quad (12)$$

In Fig. 8, four different methods for calculating $P_{p,ext}$ of a strand in the n-th winding layer are shown. If the magnetic field in a strand $H_s(x,y)$ is known, the losses $P_{p,ext}$ of this strand can be calculated by (13). However, since the winding regions are considered as HCRs in this paper, the RMS value of the magnetic field over the complete HCR is used for calculating $P_{p,ext}$, as given by (14). Since a closed-form of the required antiderivative of $H_x^2(x,y)$ and $H_y^2(x,y)$ is complicated, the integral of $H_x^2(x,y)$ and $H_y^2(x,y)$ can only be calculated numerically, which takes several minutes.

For simplifying the calculation, $P_{p,ext}$ can approximately be calculated by using the average value of the magnetic field. There, it is assumed that the losses caused by a non-uniform distributed magnetic field are approximately equal to the losses caused by a uniform distributed magnetic field, of which the amplitude is equal to the average amplitude over the HCR area. In Fig. 8c), the losses $P_{p,ext}$ are calculated by using the average H-field value over the area of HCR as given by (15). However, the double integrals of $H_x(x,y)$ and $H_y(x,y)$ are still relatively complicated, and it takes several seconds or minutes to compute the integral. To further reduce the computational effort, the average value of the magnetic field over the HCR area is replaced by the average value on the two cut lines CL$(n-1)$ and CL(n) for the n-th layer as given by (16). The cut lines are the boundary lines of the winding layer, which are as high as the winding layers as shown in Fig. 8d). Consequently, the double integral is now reduced to a one-dimensional integral. A closed-form of the antiderivative of H_y with respect to y is given in (17), which is valid in the whole 2D space. Note that (17) is calculated at $x = a + c$ for simplicity and is based on the coordinate system shown in Fig. 9.

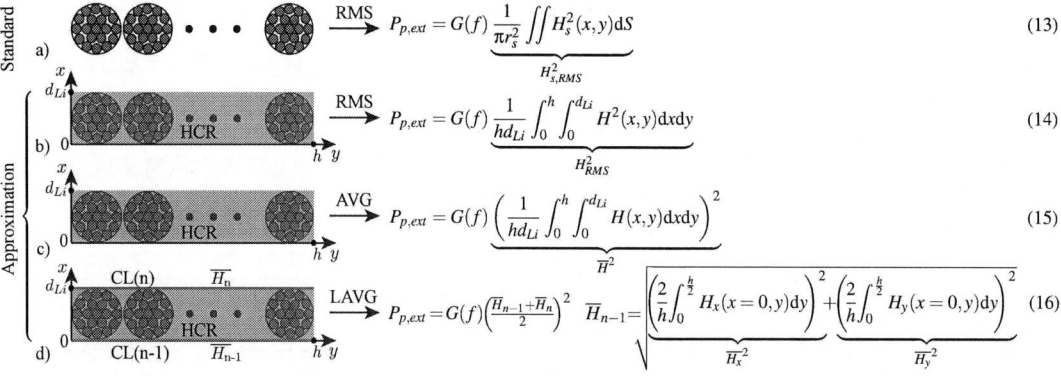

Fig. 8: Calculation of losses caused by the external proximity effect $P_{p,ext}$ for the n-th layer by using different average and RMS values of magnetic fields H where $H = \sqrt{H_x^2 + H_y^2}$. a) Using the RMS value of H over each strands to calculate $P_{p,ext}$. b) Using the RMS value of H over the assumed HCR of the winding layer to calculate $P_{p,ext}$. c) Using the average (AVG) value of H over the assumed HCR to calculate $P_{p,ext}$. d) Using the average value of the magnetic field $\overline{H} = \sqrt{\overline{H_x}^2 + \overline{H_y}^2}$ over cut line CL$(n-1)$ and CL(n) (LAVG) to calculate $P_{p,ext}$.

$$\int_{-b}^{b} H_y(x=a+c,y)\mathrm{d}y = \frac{I_R}{4\pi ab}\left[\frac{\left(4b^2-(2a+c)^2\right)\ln\left(4b^2+(2a+c)^2\right)}{4} + \frac{\left(c^2-4b^2\right)\ln\left(4b^2+c^2\right)}{4}\right.$$
$$\left. -\frac{c^2\ln\left(c^2\right)}{4} + 2\left(a+\frac{c}{2}\right)^2\ln(2a+c) + 4b\left(a+\frac{c}{2}\right)\arctan\left(\frac{2b}{2a+c}\right) - 2cb\arctan\left(\frac{2b}{c}\right)\right] \tag{17}$$

The antiderivative of H_x with respect to y needs to be derived individually for different regions as shown in Fig. 9, because the antiderivatives of H_x differ from region to region. Since the magnetic fields on the cut lines are equal to the superposition of the magnetic fields of the 4 HCRs as shown in Fig. 4c), the magnetic field of each HCR on the cut lines need to be calculated. Cut lines CL0 … CL4 and cut lines CL5 & CL6 are located inside the original primary and secondary HCR, but outside the other HCRs. Therefore, the region of interest for calculating the losses $P_{p,ext}$ is $-b < y < b$, which is the region between the two dashed lines in the coordinate system given in Fig. 9. Furthermore, the region $-b < y < b$ can be further divided into 3 subregions R1, R2 and R3, based on the values of the angles $\theta_{1x}\ldots\theta_{4x}$ in each region. The values of angles $\theta_{1x}\ldots\theta_{4x}$ and the closed-form antiderivative of H_x is given in Table I. It can be concluded that in the regions (R1 & R3) outside the conductor, the antiderivative is equal to (18) and in the region (R2) inside the conductor, the antiderivative is equal to (19).

$$2\int_0^b H_x(x=a+c,y)\mathrm{d}y = \frac{I_R}{4\pi ab}\left[(2a+c)b\ln\left(\frac{(2a+c)^2+4b^2}{(2a+c)^2+b^2}\right) - cb\ln\left(\frac{4b^2+c^2}{b^2+c^2}\right) + \left(c^2-b^2\right)\arctan\left(\frac{b}{c}\right)\right.$$
$$\left. + \left(2\left(a+\frac{c}{2}\right)^2-2b^2\right)\arctan\left(\frac{2b}{2a+c}\right) + \left(b^2-4\left(a+\frac{c}{2}\right)^2\right)\arctan\left(\frac{b}{2a+c}\right) + \left(2b^2-\frac{c^2}{2}\right)\arctan\left(\frac{2b}{c}\right)\right] \tag{18}$$

$$2\int_0^b H_x(x=a+c,y)\mathrm{d}y = \frac{I_R}{4\pi ab}\left[(2a+c)b\ln\left(\frac{(2a+c)^2+4b^2}{(2a+c)^2+b^2}\right) - cb\ln\left(\frac{4b^2+c^2}{b^2+c^2}\right) + \left(c^2-b^2\right)\arctan\left(\frac{b}{c}\right)\right.$$
$$\left. + \left(2\left(a+\frac{c}{2}\right)^2-2b^2\right)\arctan\left(\frac{2b}{2a+c}\right) + \left(b^2-4\left(a+\frac{c}{2}\right)^2\right)\arctan\left(\frac{b}{2a+c}\right) + \left(2b^2-\frac{c^2}{2}\right)\arctan\left(\frac{2b}{c}\right) + b^2\pi\right] \tag{19}$$

Table I: The angle values and the antiderivative of H_x in regions R1 … R3.

	θ_{1x}	θ_{2x}	θ_{3x}	θ_{4x}	$\int H_x \mathrm{d}y$
R1 $(x < -a)$	$\theta_{1y}+\pi$	$\theta_{2y}+\pi$	θ_{3y}	θ_{4y}	(18)
R2 $(-a \le x \le a)$	$\theta_{1y}+\pi$	θ_{2y}	$\theta_{3y}+\pi$	θ_{4y}	(19)
R3 $(x > a)$	θ_{1y}	θ_{2y}	$\theta_{3y}+\pi$	$\theta_{4y}+\pi$	(18)

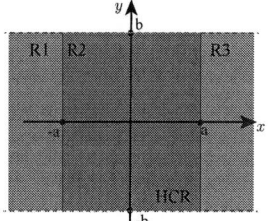

Fig. 9: Regions of interest

5 Numerical validation

In order to evaluate the accuracy of the proposed model, the losses of the OW winding section calculated by 2D FEM simulations for the 13 transformer design cases shown in Fig. 2 are used as a reference. The losses are also calculated with the proposed model based on the hyperbolic and the Bessel functions presented in [4] & [7] and are compared to the values computed by FEM. The error is defined by $\frac{P_{model}-P_{FEM}}{P_{FEM}} \times 100\%$. In addition, the error for the 1D field model given in [4] is also calculated for comparison to the standard model. As can be seen in Fig. 11, except for transformer cases 9 and 11, the absolute error of the proposed model based on the hyperbolic functions (MMIH+Hyperbolic) is lower than 2.2%, which is much smaller than the 1D field model, which results in errors up to 30% (e.g. transformer case 6). The error for transformer cases 9 and 11 is relatively large because the center core leg to winding height ratio is lower than the one of the other transformer cases as shown in Fig. 2. As an infinite long PMC-boundary is assumed in assumption A3, all the magnetic fields either enter the PMC-boundary or are closed on the right-hand side of the PMC-boundary as shown in Fig. 4b). However, in reality, the center core leg is

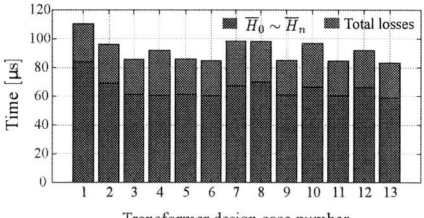

Fig. 10: Calculation time of proposed model

not infinite long so that a certain amount of magnetic field lines enter the upper and the outer edge of the core leg as highlighted in red in Fig. 3b). This affects the magnetic field distribution inside the winding as well as the losses $P_{p,ext}$. Nevertheless, the center core leg is always higher than the winding, so that the impact of assumption A3 on the model accuracy is limited.

In addition to being accurate, the proposed model is also computationally very efficient. The calculation time for each transformer case is obtained by using the function "timeit" in Matlab on a laptop with 16GB RAM and a CPU type that has 4 cores and 4.8 GHz turbo frequency. As the average magnetic field on each cut line needs to be calculated before the loss calculation, the whole procedure of the loss calculation can be basically divided into two steps, calculation of the average value of H-field ($\overline{H}_0 \dots \overline{H}_n$) for the n-layer winding, and the total HF-loss calculation. In Fig. 10, the calculation time of these 2 steps for the 13 transformer design cases are shown. Note that the time for the total loss calculation is performed for the method based on the hyperbolic function (MMIH+Hyperbolic), as this method is the most accurate one. The calculation of the average magnetic field value $\overline{H}_0 \dots \overline{H}_n$ consumes the most time which is around 65 μs on average, whereas the calculation time of the total losses is around 25 μs as shown in Fig. 10. Therefore, the total time for the loss calculation of the proposed model is around 90 μs, which is comparable to the 1D field model, which has a calculation time of approximately 72 μs as given in [12].

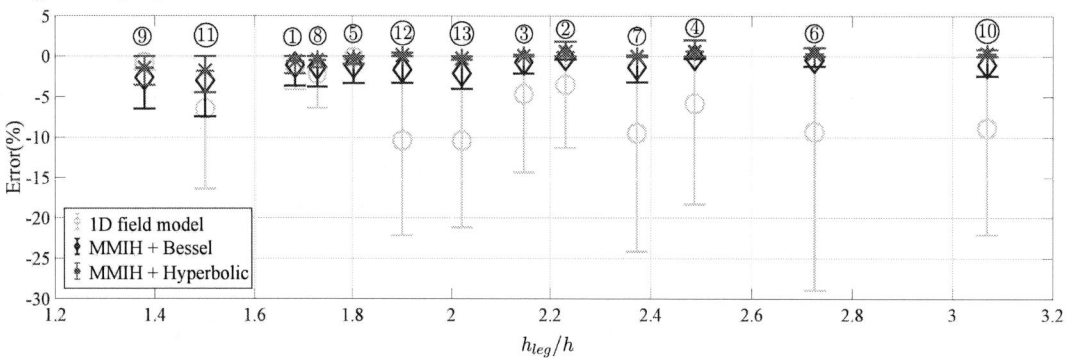

Fig. 11: Error of the proposed and 1D field model compared to 2D FEM results for the OW winding section based on 13 transformer design cases given in Fig. 2. The error is calculated by $\frac{P_{model} - P_{FEM}}{P_{FEM}} \times 100\%$.

Conclusion

This paper compares the HF-losses of the IW and the OW winding section of transformers with LW windings. The losses determined with 2D FEM simulations for 13 different transformer design cases indicate that the loss difference between the IW and the OW winding section is between 3% and 14%. A large loss difference occurs for transformers, where the windings are so wide, that a significate amount of magnetic field lines are not through the center core leg. In order to calculate the losses of the OW winding section accurately, an analytical model with closed-form equations is presented. This model can accurately estimate the magnetic field distribution as well as the winding losses. The error of the proposed model is smaller than 5%, which is much smaller than the error calculated with a standard 1D field model, which results in an error up to 30%. Moreover, the proposed model is computationally very efficient. The average calculation time on a laptop, including magnetic field and loss calculation is around 90 μs, which is comparable to the calculation time of the 1D field model, which takes 72 μs. The proposed model is not only suitable for the OW winding section in transformers but also suitable for the OW winding section in inductors without air gaps.

References

[1] C. R. Sullivan, "Computationally efficient winding loss calculation with multiple windings, arbitrary waveforms, and two-dimensional or three-dimensional field geometry," *IEEE Transactions on Power Electronics*, vol. 16, no. 1, pp. 142–150, Jan. 2001.

[2] A. Roßkopf, C. Joffe, and E. Bär, "Calculation of ohmic losses in litz wires by coupling analytical and numerical methods," in *Proc. Int. Electric Drives Production Conf. (EDPC)*, Sep. 2014, pp. 1–6.

[3] R. Prieto, J. Cobos, O. Garcia, P. Alou, and J. Uceda, "Study of 3-d magnetic components by means of "double 2-d" methodology," *IEEE Transactions on Industrial Electronics*, vol. 50, no. 1, pp. 183–192, 2003.

[4] J. A. Ferreira, *"Electromagnetic Modelling of Power Electronic Converters"*. Boston, MA: Springer US, 1989.

[5] M. Bartoli, N. Noferi, A. Reatti, and M. Kazimierczuk, "Modeling litz-wire winding losses in high-frequency power inductors," *IEEE Power Electron. Spec. Conf. (PESC)*, 2020.

[6] F. Tourkhani and P. Viarouge, "Accurate analytical model of winding losses in round litz wire windings," *IEEE Trans. on Magnetics*, vol. 37, no. 1, p. 538–543, 2001.

[7] R. Wojda and M. Kazimierczuk, "Winding resistance of litz-wire and multi-strand inductors," *IET Power Electron.*, vol. 5, no. 2, p. 257, 2012.

[8] Q. Meng and J. Biela, "Survey and comparison of 1d/2d analytical models of HF losses in litz wire," in *Euro. Conf. on Power Electron. and Appl. (EPE ECCE Europe)*. IEEE, Sep. 2020.

[9] J. Muhlethaler, J. W. Kolar, and A. Ecklebe, "Loss modeling of inductive components employed in power electronic systems," *Int. Conf. on Power Electron. - ECCE Asia*, 2011.

[10] M. Jaritz and J. Biela, "Optimal design of a modular series parallel resonant converter for a solid state 2.88 MW/115 kV long pulse modulator," *IEEE Trans. on Plasma Science*, vol. 42, no. 10, pp. 3014–3022, Oct. 2014.

[11] K. Binns and L. P.J., *Analysis and computation of electric and magnetic field problems*, ser. International series of monographs on physics. Pergamon Press, 1973.

[12] Q. Meng and J. Biela, "New method for calculating hf-losses in litz wire caused by 2d magnetic fields," in *23rd European Conference on Power Electronics and Applications (EPE'21 ECCE Europe)*, 2021, pp. P.1–P.11.

Fast And Accurate Soft-Switching And Hard-Switching Losses Estimation For Power Converter, Application To The Dual Active Bridge (DAB) Converter

Francois Boige, Nicolas Videau, Adel Ziani, Bruno Guerrero, Julien Laclaverie
Gamma Technologies
601 Oakmont Ln, Westmont, IL 60559, US
f.boige@gtisoft.com
www.gtisoft.com

Keywords

«Converter Design», «Resonant Converter», «Soft Switching», «Zero-voltage Switching», «Dual Active Bridge (DAB)»

Abstract

This paper presents an integrated method and workflow to estimate Dual Active Bridge (DAB) DC/DC converter semiconductor loss in losses both in soft-switching including Zero-Voltage Switching (ZVS), Zero-Current Switching (ZCS) and hard-switching. The method uses soft-switching and hard-switching energy look-up table with a frequential solver to estimates DAB loss. In case, the switching energies are not available, a virtual double pulse simulation is performed based on the datasheet characteristic. The comparison with the literature display a very fast computation time with a reasonable trade-off in accuracy.

Introduction

Electric vehicle, large-scale storage of electrical energy in batteries, photovoltaic energy, aircraft onboard power grid, etc. require the use of galvanically insulated electrical energy converters to guarantee the safety of goods and people. To accompany the growth of these applications the converters must be as much as possible cost-effective with a high efficiency. Faced with these objectives and regulatory constraints, isolated converters such as DAB converters (Fig. 1a) appears as good technical solutions [1] especially with the bidirectional power transfers and the possible operating in ZVS/ZCS (Zero-Voltage Switching/Zero-Current Switching) depending on the operating point and the command used. However, the important operating switching frequencies of these converters impose a strict knowledge of the switching losses to avoid overheating. Consequently, it is vital to possess an accurate and quick tool to estimate the switching losses for any converter configuration and operating point.

The methods used today for semiconductor loss estimation are always a trade-off between accuracy of results versus computation time and the model building time. In this paper, a novel and generalist method based on mixed simulation [2] is presented to estimate the switching losses applied to DAB converters. It is based on perfect switches to simulate the waveforms and the allocate the switching energy loss stored in a lookup table. This approach is used in GT-PowerForge, an innovative software to design power converters [3].

The proposed methodology is an enhancement of the procedure presented in [2] and is illustrated in Fig. 1b. The main issue with this methodology is the inability to handle ZVS and hard-switching losses estimation within a given period. Indeed, in ZVS the switches are only controlled for turn-off while the turn-on is caused by the diode "natural" turn-on in parallel with the transistor. In hard switching, the switches are controlled for both turn-on and turn-off. However, in switching energy map based approach such as presented in [2] or used in PSIM or PLECS software, the turn-off energy is tabulated and used the same way for both hard and soft switching. Losses energies are typically provided in the semiconductor datasheet or obtained by double pulse test (measured or virtually obtained). This test is known to overestimate the turn-off value by measuring the stored energy into the output capacitance in addition to the dissipated energy [4]–[8]. Other measurements need to be implemented to obtain the soft-switching energies as presented in [5], [6]. Two different datasets of hard and soft switching energies are needed to estimate the losses as the two kinds of switching can coexist in a given operating sequence.

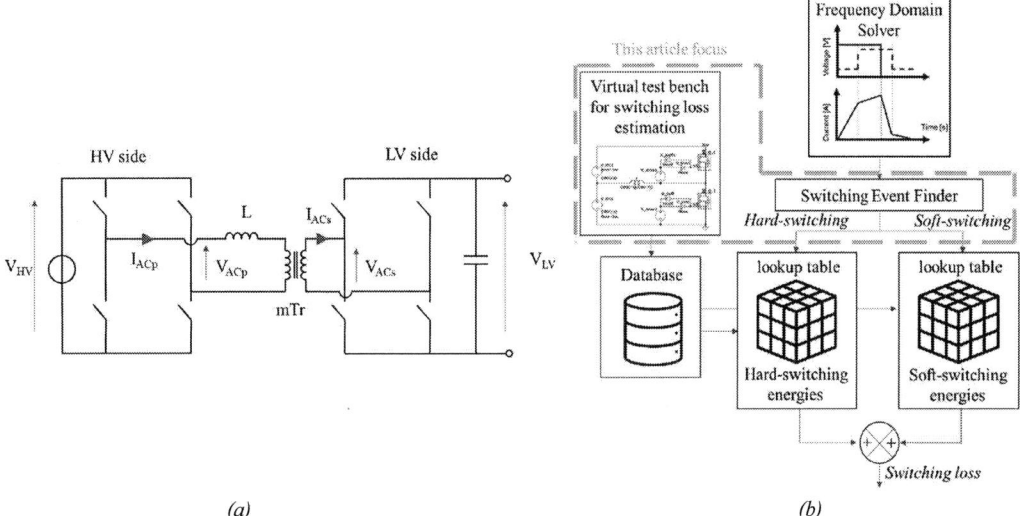

Figure 1: (a) Dual Active Bridge converter with H-bridge on both sides (b) switching losses estimation procedure to handle both hard and soft switching data.

In a first part, the motivation to propose this method is discussed. In a second part, loss estimation algorithm is presented. In a third part, the switching energies estimation is explained. In a fourth part, the results are compared to the literature and the limitations discussed.

Switching losses estimation procedure

The procedure to handle the two switching types (hard and soft switching) is as follow: at first, the waveforms are generated with a fast frequential MNA solver presented in [9] with perfect switch model. Then, the switching instants, the type of switching, and the switched voltage and current are estimated based on the waveforms. The switching losses at each switching instants are estimated based on this information by using lookup tables of the switching energies stored into a database, one for hard-switching and the other for soft-switching. The soft switching energies estimation will be discussed in the next section. Finally, all the losses are summed for each switching instant. This algorithm is also included into an electro-thermal convergence loop as discussed in [2].

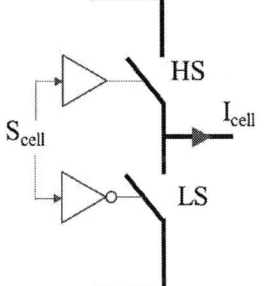

Table I: Controlled switching events

	$S_{cell}\uparrow$	$S_{cell}\downarrow$
$I_{cell} > 0$	HS turn-on	HS turn-off
$I_{cell} < 0$	LS turn-off	LS turn-on

Figure 2: Switching cell conventions

To find the type of switching (hard or soft) a switching event algorithm has been developed and is based on the switching cell concept introduced back in the 70s [10]. The Low Side switch (LS) and the High

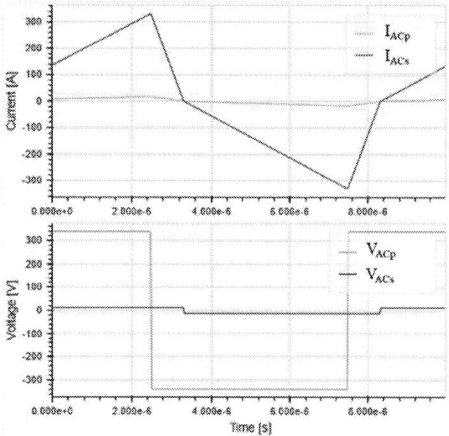

Figure 3: Simulated waveforms at $V_{LV}=12V$, $V_{HV}=340V$, $P=2kW$, $L=26.7\mu H$, $mTr=19$.

Side switch (HS) are commanded with the signal S_{cell} as presented in Fig. 2. If $I_{cell} > 0$ and S_{cell} present a rising edge ($S_{cell}\uparrow$), HS will have a controlled switch-on. This approach is generalized for all the cases in Table 1. The "natural" turn-on of a diode is not accounted as it is supposed lossless. The natural turn-off loss caused by diode recovery is accounted at the same time as the controlled turn-on loss. In a case of the usual rotation of controlled turn-on/turn-off, the hard-switching energies lookup table is used to estimate the loss. Indeed, as explained above, the classical double pulse test energies (Eon, Eoff) can only the used in pair because of the Coss stored energy. However, in a case of more than one consecutive controlled turn-off without controlled turn-on, the output capacitance energy is given back to the system, but it is still accounted as a loss in the double pulse measured turn-off energy (Eoff). In this case, the converter is in a ZVS condition, and the soft-switching data is chosen. The main advantage of this method lays in the flexibility in switching energy data handling and fast simulation capabilities. However, getting soft-switching data can be a bottleneck to this approach as it is not a data supplied by the manufacturer. Furthermore, the near zero current switching can imply incomplete soft-switching which is not handled at this level of the procedure but can be handled at the soft-switching energies data level. In the next section, the loss estimation algorithm will be validated by using measured soft-switching losses extracted form [11]. To address these issues, a simulation approach to estimate soft-switching energies for MOSFET will be presented in a further section.

The switching energies are stored in lookup tables, $E_{off|on|rr}(I, V, T_j)$ for hard-switching energies and $E_{soft}(I,V,T_j)$ for soft-switching energies. The data is interpolated or extrapolated linearly to estimate the energies at every switching instant.

The proposed procedure is compared to the literature [11] in two steps: first, the semiconductor loss estimated are compared with [11] using the switching energy curves presented in [11]. This comparison is done in the next section. In a second step, the losses are compared with energy curves estimated from simulation.

Semiconductor losses estimation for Dual Active Bridge compared to literature with the same loss curves

Table II: Switching losses comparison with same energies lookup tables

Results owners	LV Cond. loss	LV Sw. loss	LV total loss	HV Cond. loss	HV Sw. loss	HV Total loss	Total Loss HV+LV
Ref [11] (W)	25	6	31	15	9	24	55
With same energy data (W)	21.1	7.5	28.6	13.2	11.4	24.5	53.2
Difference (W)	3.9	-1.5	2.4	1.8	-2.4	-0.5	1.8
Difference (%)	15%	-25%	8%	12%	-26%	-2%	3%

A DAB with the topologies displayed in Fig.1a is compared with $V_{LV}=12V$, $V_{HV}=340V$, $P=2kW$, $L=26.7\mu H$, transformer ratio=19 with the semiconductor package imposed at 25°C. The LV side includes 8 IRF2804 [12] in parallel and for the HV side the SPW47N60CFD [13]. The switching frequency is 100 kHz. The simulation is done with GT-PowerForge solver. The waveforms are displayed in Fig.3. The results are presented in Table II. The estimated total semiconductor loss is 3% lower than

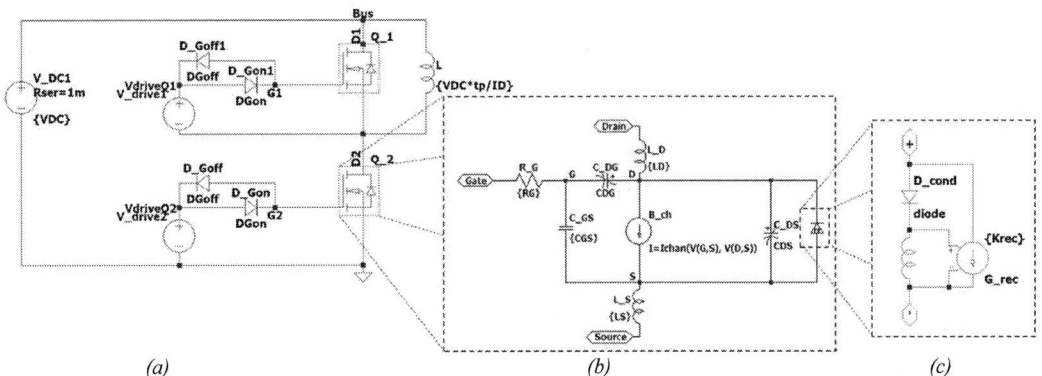

Figure 4: Proposed virtual double pulse test bench. (a) double pulse circuit. (b) MOSFET Model. (c) Body diode model.

the reference, the error being small, the proposed method is considered to be validated. In detail, the conduction losses and switching losses compensate themselves. The switching losses are overestimated which is coherent as the author in [7] proposed a method lowering the estimated switched current energies as the one used in the proposed simulation. The conduction losses are underestimated compared to the literature. The proposed results seem closer to reality as it is based on a more complex model as explained and compared in [2]. It can be noted that the converter losses are compared to the simulation done by the author in [11] and not to measurement. However, the author in [11] estimates the all-converter losses to be 6% higher than the measurement.

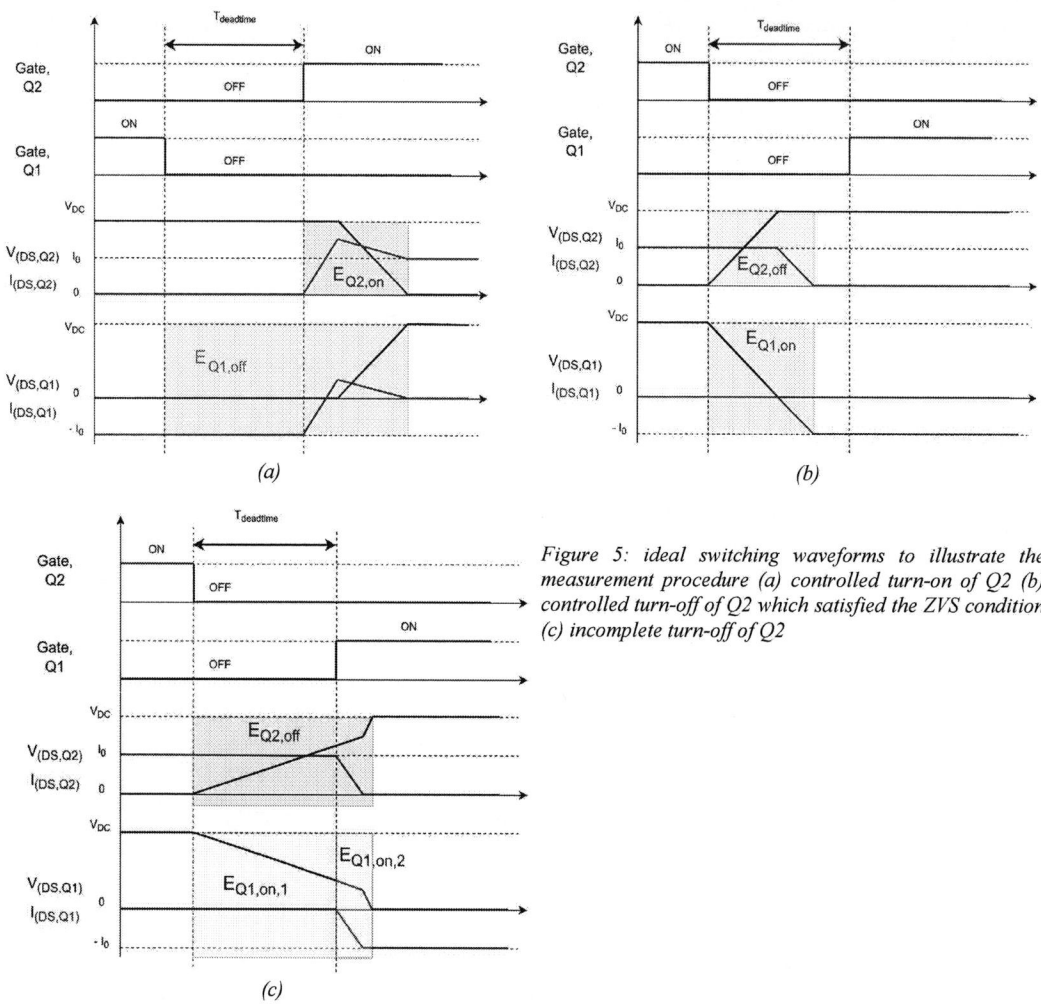

Figure 5: ideal switching waveforms to illustrate the measurement procedure (a) controlled turn-on of Q2 (b) controlled turn-off of Q2 which satisfied the ZVS condition (c) incomplete turn-off of Q2

Switching energies estimation based on double pulse simulation

The data can be extracted from manufacturer datasheets or from measurement results. However, when the data is not available, double pulse simulations are performed for MOSFET devices based on a LTSPICE MOSFET model as explained in [2] and displayed in Fig. 4b and Fig. 4c. The LTSPICE model parameters are fitted on the datasheet, especially for transconductance and diode parameters. The parasitic capacitances values are lookup table extracted from the same datasheet. The parasitic common source inductance which is known to be a key parameter for switching losses estimation [14], [15] is estimated based on the package used [16]–[18]. It can be noted that this virtual test bench can be used with LTSPICE model given by semiconductor device manufacturers, but convergence issues can happen.

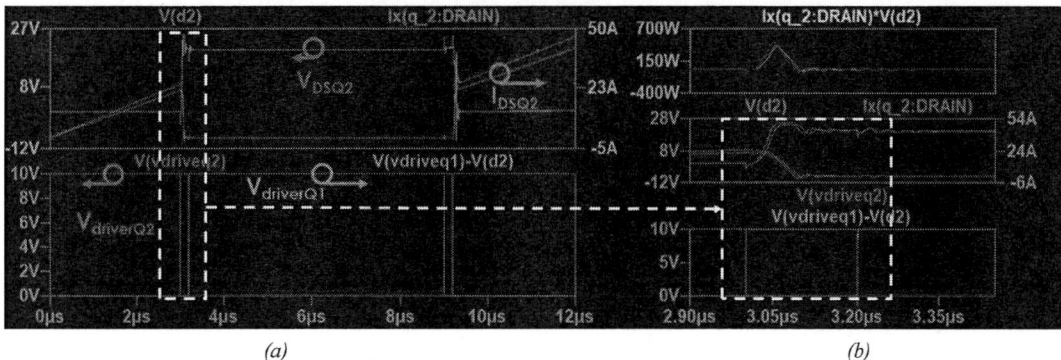

(a)   *(b)*

Figure 6: (a) simulated waveforms for multiple target current, drain bias and temperature. (b) zoom in the switching instant with the deadtime illustration as well as the measured turn-off power.

For soft-switching energy measurement, a modified double pulse measurement is used as explained in [11], the circuit is displayed in Fig. 4a. At the start, Q2 is ON and Q1 OFF, the current will rise to the target switched current at the turn-off time for Q2. After a deadtime, the device Q1 is switching on. During device switch-off, the energy is measured for both Q1 and Q2 devices as presented in Fig.5b if the turn-of ends before Q1 is turned on after the deadtime (ZVS condition) as a result, Esoft is defined as Esoft = EQ2,off + EQ1,on. The simulated waveforms obtained with LTSPICE in Fig. 6 for 2 different current I_{DS}. However, if the ZVS condition is not fulfilled, an incomplete turn-off [6] append as presented in Fig. 5c and Esoft is defined as Esoft = EQ2,off + EQ1,on,1 + EQ1,on,2. The effect of deadtime on the blocking energy (Esoft) for a given switched current is shown in Fig.7a. With a sufficient deadtime, the energy is not impacted by it.

During the double pulse simulation, the turn-on and recovery losses energies are also estimated as

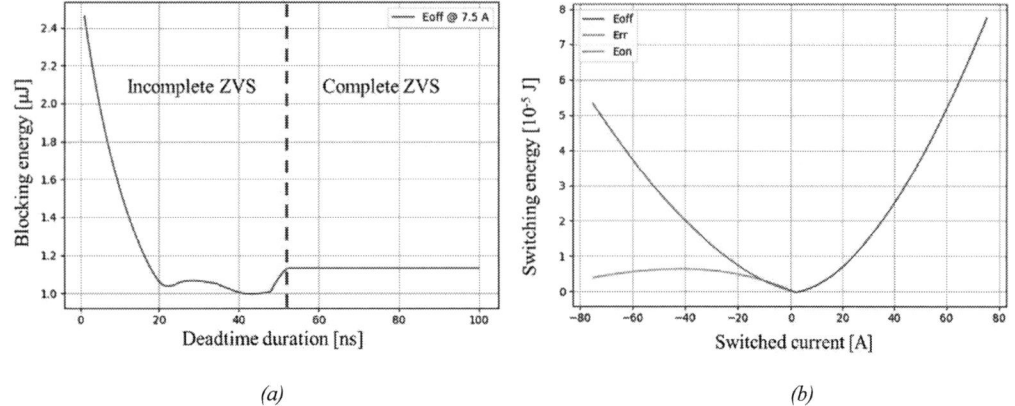

(a)   *(b)*

Figure 7: (a) Effect of deadtime over the blocking energy for a given switched current (7.5A) (b) switching energies simulated for the device IRF2804 with a deadtime of 200 ns.

presented in Fig. 7b and are defined as Eon = EQ2,on and Err = EQ1,off, it can be noted that Err depends on the deadtime as conduction loss occurs during the deadtime.

Semiconductor losses estimation for Dual Active Bridge compared to literature with loss curves extracted from virtual double pulse test

Based on the procedure presented in Fig.1b, the waveform generated by GT-PowerForge frequential solver in Fig.3 and the switching energies data estimated based on the double pulse simulation presented in the previous section, the DAB loss are estimated and compared again against [11], the results are summarized in Table III. The total semiconductor losses are 16% lower with the results presented in this article. The conduction losses are identical as the results presented in Table II as the data used are the same. However, the error for switching losses are more pronounced (79% and 56%). Especially for LV switching loss the difference can be explained as the switching energies are measured for 8 devices directly in parallel in [11] while the simulation performed in the previous section is done with 1 device in parallel as this kind of simulation is very much dependent on the parasitic elements (especially the common source inductance) put 8 devices in parallel in a PCB will certainly impact a lot the results. A way to reduce this difference would be to recalibrate the LTPSICE model with the appropriate parasitic element or improve the used SPICE model. This model recalibration will be addressed in future work. However, proposed model has the great advantage to only rely on public available data to build the model, allowing a fair benchmark of the available semiconductor device in the market and identify the right trade-off for power converter design.

Table III: Switching losses comparison with energies lookup tables from LTSPICE virtual test bench

Results owners	LV Cond. loss	LV Sw. loss	LV total loss	HV Con. loss	HV Sw. loss	HV Total loss	Total Loss HV+LV
Ref [11]**(W)**	25	6	31	15	9	24	25
With energies from simulation (W)	21.0	1.3	22.3	13.1	14.0	27.1	21.0
Difference (W)	4.0	4.7	8.7	1.9	-5.0	-3.1	4.0
Difference (%)	16%	79%	28%	13%	-56%	-13%	16%

Conclusion

In conclusion, to obtain results more faithful to reality for converters operating in both hard and soft switching, a novel and integrated approach to estimate soft-switching losses is presented and compared to the literature for a DAB converter. The results are aligned with what is expected and the simulation is performed in less than 4s. The goal of obtaining a fast and more accurate method to estimate power converter switching losses in hard and soft switching is fulfilled. The proposed model and virtual double pulse setup allow the fair comparison of the semiconductor device removing the tedious and costly process of hardware measurement in a pre-design phase at the cost of a loss in precision in the results. This method is generic for all kinds of devices if the soft-switching data is available. However, only the MOSFET models is currently supported by the virtual test bench and it needs to be extended to other devices such as GaN devices and IGBTs as their behavior is strongly different from the MOSFET in soft-switching [19].

References

[1] F. Krismer, J. Biela, and J. W. Kolar, "A comparative evaluation of isolated bi-directional DC/DC converters with wide input and output voltage range," in *Fourtieth IAS Annual Meeting. Conference Record of the 2005 Industry Applications Conference, 2005.*, Hong Kong, China, 2005, vol. 1, pp. 599–606. doi: 10.1109/IAS.2005.1518368.

[2] G. Fontes, F. Boige, A. Morentin, G. Delamare, T. Meynard, and N. Videau, "Semiconductor loss estimation in an innovative global power converter designer," in *PCIM Europe digital days 2020; International Exhibition and Conference for Power Electronics, Intelligent Motion, Renewable Energy and Energy Management*, Jul. 2020, pp. 1–7.

[3] "PowerForge - Software for Power Converters," *PowerForge*, Dec. 20, 2019. http://www.powerdesign.tech/ (accessed Dec. 20, 2019).

[4] S. Lefebvre, F. Costa, and F. Miserey, "Influence of the gate internal impedance on losses in a power MOS transistor switching at a high frequency in the ZVS mode," *IEEE Trans. Power Electron.*, vol. 17, no. 1, Art. no. 1, Jan. 2002, doi: 10.1109/63.988667.

[5] D. Rothmund, D. Bortis, and J. W. Kolar, "Accurate Transient Calorimetric Measurement of Soft-Switching Losses of 10-kV SiC mosfets and Diodes," *IEEE Trans. Power Electron.*, vol. 33, no. 6, Art. no. 6, Jun. 2018, doi: 10.1109/TPEL.2017.2729892.

[6] M. Kasper, R. Burkat, F. Deboy, and J. Kolar, "ZVS of Power MOSFETs Revisited," *IEEE Trans. Power Electron.*, pp. 1–1, 2016, doi: 10.1109/TPEL.2016.2574998.

[7] F. Krismer and J. W. Kolar, "Accurate Power Loss Model Derivation of a High-Current Dual Active Bridge Converter for an Automotive Application," *IEEE Trans. Ind. Electron.*, vol. 57, no. 3, Art. no. 3, Mar. 2010, doi: 10.1109/TIE.2009.2025284.

[8] J. C. Brandelero, "Conception et réalisation d'un convertisseur multicellulaire DC/DC isolé pour application aéronautique," phd, 2015. Accessed: May 18, 2022. [Online]. Available: http://ethesis.inp-toulouse.fr/archive/00003097/

[9] G. Fontes *et al.*, "Fast Solver to Get Steady-State Waveforms for Power Converter Design," in *PCIM Europe 2018; International Exhibition and Conference for Power Electronics, Intelligent Motion, Renewable Energy and Energy Management*, Jun. 2018, pp. 1–7.

[10] S. Cuk and R. Middlebrook, "A general unified approach to modelling switching DC-tO-DC converters in discontinuous conduction mode," in *1977 IEEE Power Electronics Specialists Conference*, Palo Alto, CA, USA, Jun. 1977, pp. 36–57. doi: 10.1109/PESC.1977.7070802.

[11] F. Krismer, "Modeling and optimization of bidirectional dual active bridge DC-DC converter topologies," Doctoral Thesis, ETH Zurich, 2010. doi: 10.3929/ethz-a-006395373.

[12] I. T. AG, "IRF2804 - Infineon Technologies." https://www.infineon.com/cms/en/product/power/mosfet/n-channel/irf2804/ (accessed Dec. 06, 2021).

[13] I. T. AG, "SPW47N60CFD - Infineon Technologies." https://www.infineon.com/cms/en/product/power/mosfet/n-channel/500v-950v/spw47n60cfd/ (accessed Dec. 06, 2021).

[14] D. Reusch, "Impact of Parasitics on Performance," p. 5, 2020.

[15] Bo Yang and J. Zhang, "Effect and utilization of common source inductance in synchronous rectification," in *Twentieth Annual IEEE Applied Power Electronics Conference and Exposition, 2005. APEC 2005.*, Austin, TX, USA, 2005, vol. 3, pp. 1407–1411. doi: 10.1109/APEC.2005.1453213.

[16] "Features and Benefits of 650 V CoolMOS™ C6 / E6." Infineon application note.

[17] "650 V CoolMOS™ C7 Gold in TOLL package." Infineon Product Brief.

[18] J. Zhang, "Choosing The Right Power MOSFET Package." International Rectifier application note.

[19] G. Ortiz, H. Uemura, D. Bortis, J. W. Kolar, and O. Apeldoorn, "Modeling of Soft-Switching Losses of IGBTs in High-Power High-Efficiency Dual-Active-Bridge DC/DC Converters," *IEEE Trans. Electron Devices*, vol. 60, no. 2, Art. no. 2, Feb. 2013, doi: 10.1109/TED.2012.2223215.

Influence of an Electrical Machine on the Dimension and Packaging of Multi-Machine Systems

Thomas Stöckl, Hans-Georg Herzog
Renk Group, Technical University of Munich
Gögginger Strasse 73
Augsburg, Germany
Phone: +49 (0) 821-5700-1272
Email: thomas.stoeckl@renk.biz
URL: https://www.renk-group.com

Keywords

≪Electrical machine≫, ≪Multi-machine system≫, ≪Packaging≫, ≪Standardization≫, ≪Marine≫, ≪Ship≫

Abstract

The design of multi-machine systems requires the estimated installation space of the e-machines. Developed models calculate the space for different speeds and power concepts. Hereby, the advantages of combining several drives is shown. The paper illustrates the importance of machine parameters and number of units for a multi-machine system. The outlook gives an impression on how the efficiency of the e-machine can be improved by the operation management.

Introduction

High power density is an important design goal for power train applications [3]. Typically, a combination of an e-machine and a gearbox reduces space requirements. The gear ratio is the main design parameter in single-machine applications and it is used to optimize the systems installation space. RENK's AED concept is an example of electromechanical drive trains for ships, consisting of an asynchronous machine and a transmission gearbox [4]. In the industrial and automotive sector, there are concepts that combine several electric machines onto one gearbox [5], [6]. This leads to more compact and lighter e-machines. Higher reliability [7], standardization and design cost savings are further advantages compared to other concepts [8], [9].

However, it remains the question how multi-machine drives with maximum power density look like. Of specific interest is the empirical model of the machine's installation space. As a contribution to answer this question, this paper focuses on the development of growth models. Based on the design equations and empirical modelling, the installation space of an e-machine for a special speed and shaft power is calculated. The corresponding power curve shows that the nominal shaft power decreases for higher speeds due to electromagnetic limits. This is one reason, why several machines must be combined to reach a system power of some megawatts. These models are combined with a gear model and they are used for multi-motor system concept comparisons. The models illustrate how the number, speed and power of the machines influence the dimensions and packing density.

Structure of multi-machine drives

The typical structure of a multi-machine drive with a single-stage gearbox is shown in figure 1. Several e-machines drive the central wheel via a pinion. The torque at the output shaft is the sum of all pinions times the gear ratio. The required volume corresponds to a cylinder with the overall diameter D (red) and length l according to:

$$V = \frac{\pi}{4} \cdot D^2 \cdot l \tag{1}$$

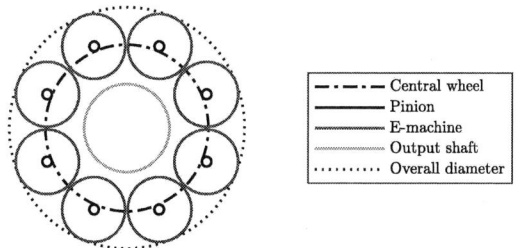

Fig. 1: Front view of a multi-motor system with eight e-machines

Due to the square in equation 1, the diameter dominates over the length. This is the central parameter for the design space. As figure 1 shows, the machines are typically arranged equally spaced in a circuit. For the arrangement of k machines with a flange diameter d in a closed circle, the overall diameter D must be at least [10], [11]:

$$D = d + \frac{d}{sin(\pi/k)} \tag{2}$$

Connecting equations 1 and 2 shows that the space requirement of multi-machine system mainly depends on the flange diameter d of the machine and number of machines k.

Growth model of an electrical machine

The model is based on having the same conditions for the type, voltage and cooling of the e-machines. This paper uses permanent magnet synchronous machines with low voltage up to 690 Volts and a water jacket cooling. The installation space for a specific power and speed is determined via design equation 3. In the literature [12], [13] the shaft power of an e-machine P_m is calculated from the mechanical Esson's utilization factor C_m, the stator bore diameter D_i, the length l_i and the speed n:

$$P_m = C_m \cdot D_i^2 \cdot l_i \cdot n \tag{3}$$

Since the rotor diameter is much larger than the air gap, the bore diameter D_i roughly equals the rotor outer diameter D_r, i.e. $D_i = D_r$. For the design of the system, the maximum possible diameter of a machine is the greatest challenge. Consequently, the maximum circumferential speed is used to determine the largest rotor and stator diameter:

$$D_i = \frac{v_{max}}{\pi \cdot n} \tag{4}$$

The diameter D_i is thus indirectly proportional to the machine speed:

$$D_i \sim \frac{1}{n} \tag{5}$$

Equation 4 is the basis of all following diameters – housing, flange and shaft size. Figure 2 illustrates the curve shape for the rotor and the estimated housing. Theses model was validated with reference e-machines. Basically, the e-machine becomes slimmer by increasing speed.

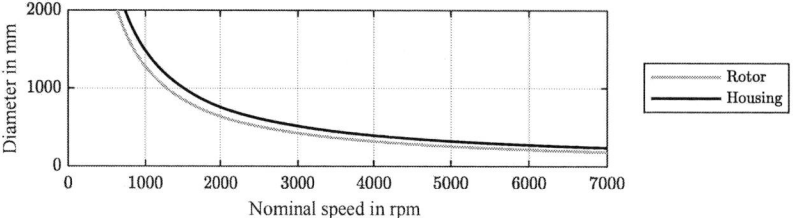

Fig. 2: Speed-dependent diameter curve of the e-machine

The total length of the e-machine is composed of the active iron length, bearing shields and winding heads. Basically thermal and mechanical limits determine the rotor length and the whole machine. Müller and Ponick shows that the relation between length and diameter can be reduced and assumed as [12], [13]:

$$l \sim D_i \tag{6}$$

The resulting housing length is therefore speed-dependent. Manufacturers usually group machines with the same diameter but different lengths into series. The range in length is given in the diagram below.

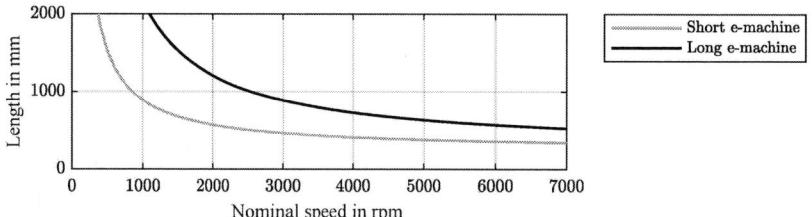

Fig. 3: Speed-dependent max e-machine length

As already shown above, the diameter and length of an e-machine decreases with increasing speed. A high-speed machine is therefore more compact; the correlation is expected to be $V \sim n^{-3}$. If the same electromagnetic conditions are set and the volume of the active part decreases, the maximum possible power declines to the same extent [12], [13]:

$$P \sim D_i^3 \sim n^{-3} \tag{7}$$

The figure 4 gives an overview of the speed-dependant power of a standard e-machine series. The curve underlines the equation 7.

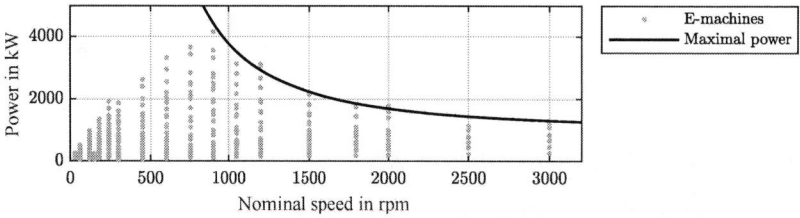

Fig. 4: Course of the maximum power of an e-machine over the speed

System Behavior

A higher speed of the e-machines leads to lower shaft power. For this reason the number of machines must increase, see Fig. 5. The figure underlines the importance of multi-machine concepts. Several machines must be combined to reach a system power of a few megawatts.

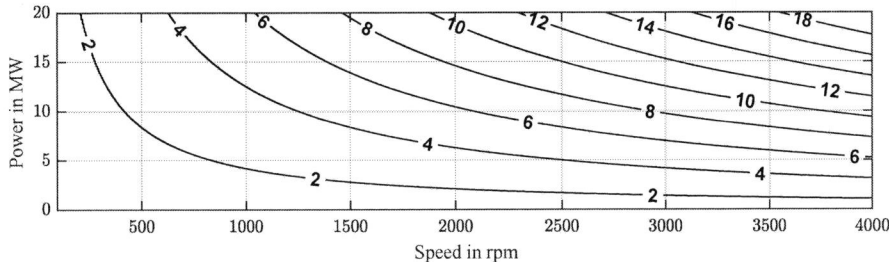

Fig. 5: Required number of standard machines as a function of speed and power

During the design of a multi-motor system there are a few degrees of freedom:

- gear ratio i
- number of pinions z_p
- number of e-machines z_e

These design parameters define the power and speed of the e-machine and its installation space. The gear ratio i of the summation gear determines the e-machine's speed by multiplying the gear ratio i and the speed of the output shaft. The second degree of freedom is the number of pinions z_p, which are driving to one central wheel. Additionally, one or two e-machines can be mounted on one pinion. Therefore, the number of installed e-machines z_e is the third degree of freedom.

The figure 6 shows diameters curves for different system concepts for an output shaft power of 5 megawatts at speed of 200 rpm. There is one installed machine per pinion. Due to the interaction of e-machine and gearbox model, two trends can be seen. Firstly, the curve drops steeply for small gear ratios, since the machine's diameter dominates. Secondly, the curve rises for increasing gear ratios, because the central wheel gets bigger and therefore the gearbox dominates. According to the equation 2, the solutions are differentiated between placing the machines in a common pitch circle, solid lines, or not, dashed lines. The curves suggest a minimum with 8 pinions, which is set as reference for the axes. To sum it up, the curves suggest a minimum of the diameter for all numbers of drives.

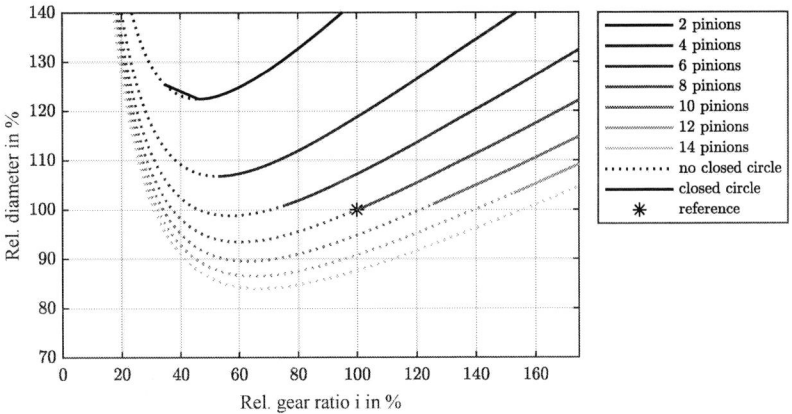

Fig. 6: Comparison of different concepts with one e-machine per pinion

A higher number of e-machines leads to a more compact machine but a higher gear ratio is needed to mount the machines in a common pitch circle. All in all a reduced diameter results when the amount of machines increases. That is why the number of 8 pinions seems to be a optimum solution concerning the diameter.

The length of the overall system with one e-machine per pinion is the sum of e-machine length and the gear unit width. Figure 7 shows the technically maximum lengths. Due to the relation $V \sim D^2 \cdot l$, the length plays a subordinate role in the installation space consideration. In additional, multi-motor applications usually use higher gear ratios and the relative difference in length between different speeds gets very small.

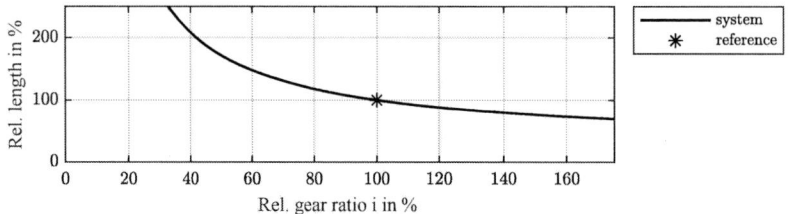

Fig. 7: Length of the total system

Another possibility to reduce the system diameter is to mount two e-machines per pinion. In this way, the torque requirement of one e-machine is put to half. According to equation 4, the torque or P/n-ratio essentially determines the diameter. Depending on the physical limits, the diameter D_i can be reduced and power P is delivered by two smaller e-machines. The speed n is not changed. For equal electromagnetic and thermal boundary conditions the Essons's utilization factor C_m is identical for one or two e-machines per pinion. The following equation shows, that it results a smaller diameter of round about 70 percent.

$$D_i = \sqrt{\frac{P_m}{n \cdot C_m \cdot 2 \cdot l_i}} \simeq \sqrt{\frac{1}{2}} \tag{8}$$

A smaller diameter of the machines makes the system more compact, see equation 1. Figure 8 illustrate the relative diameter advantage compared to the reference system with 8 pinions and 8 machines. For the same machine speed, the diameter can be reduced by nearly 10 percent. In addition, the figure shows a new minimum (circle) for a concept with 8 sprockets and 16 e-machines.

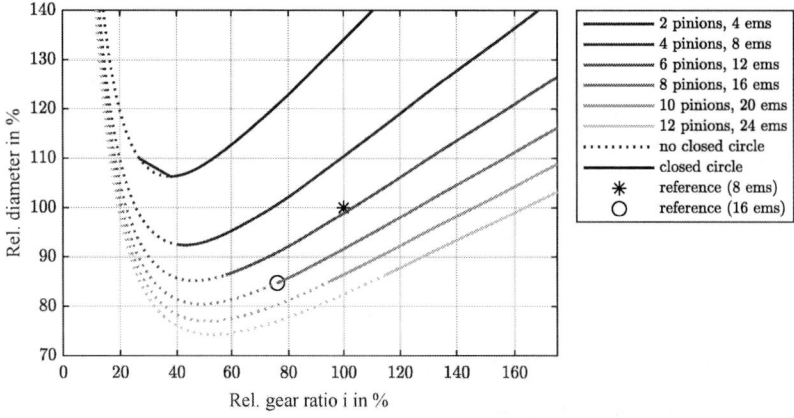

Fig. 8: Systems lengths for concepts with two e-machines per pinion

The diameter for the new minimum can be reduced to about 85 percent of the reference concept. This new minimum confirms that 8 pinions driven by two e-machine per pinion are useful. Further, this concept uses less gear ratio, round about 75 percent of the reference.

The use of two e-machines per pinion makes the system about twice as long, see figure 9. This is essentially because the length of the e-machine is much greater than the width of the gearbox.

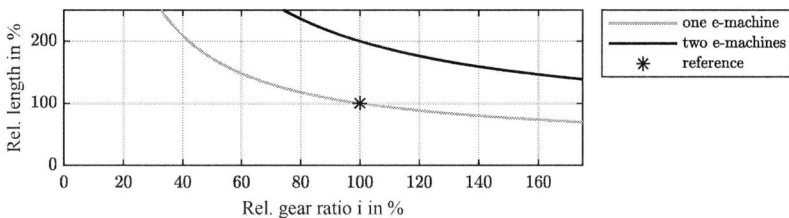

Fig. 9: Length of the total system

In ships with a narrow hull, the width of the drive system is limited and therefore concepts with two e-machines per pinion are particularly interesting. In these cases, the length is less critical. The system model shows a diameter reduction of approximately 10 percent but a volume increase of approximately 62 percent must be accepted.

Outlook

In addition to the installation space and weight advantages, the operating strategies can improve the efficiency of e-machines in multi-machine drivetrains. Figure shows three different operating strategies for speed-dependent power curve $(P \sim v^3)$ of ships. One possibility is, that all e-machines run actively with identical load. Another solution is to operate the plant with the minimum number of e-machines. The system can be use the minimum required number of e-machines. Thirdly, an operating strategy is shown where efficiency has been improved.

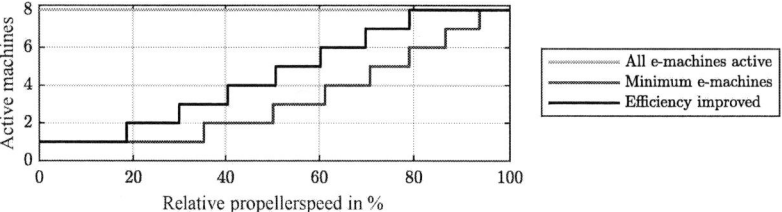

Fig. 10: Different operating strategies of the e-machines

The comparison of the three operating strategies shows the potential for improvement in efficiency of the e-machines, figure 11. Basically, the efficiencies differ considerably in the lower speed range. The efficiency when operating all e-machines is lower in relation to the other operating strategies. Switching off the e-machines brings some advantages, see orange and blue curves. By adjusting the switching points, the efficiency can additionally be tuned a little bit.

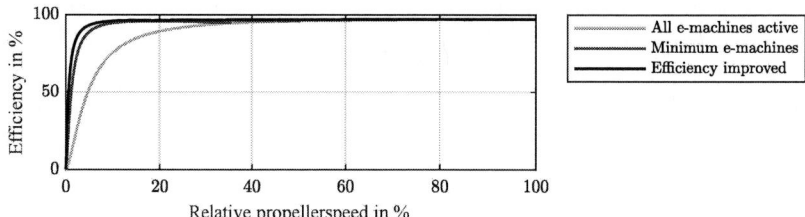

Fig. 11: Comparison of the e-machines' efficiency with different management strategies

Conclusion

The models demonstrate that multi-machine systems provide advantages in the megawatt range. Moreover, using multi-machine drives improve packaging and power density depending on the number of e-machines and the gear ratio. As the outlook shows, the models can also be used to investigate and improve the system efficiency.

References

[1] Vanderkeyn Ralf W.: Example of fast switching component, EPE Journal Vol 20 no 5, pp. 48- 56

[2] Deboe B. D.: A novel type of grid converter, EPE 2013-ECCE Europe, paper 0321

[3] R. K. Srinivasa, P. J. Chauhan, S. K. Panda, G. Wilson, X. Liu and A. K. Gupta, "An exercise to qualify LVAC and LVDC power system architectures for a platform supply vessel", Proc. IEEE Conf. Expo Transp. Electrific. Asia–Pacific (ITEC Asia-Pacific), pp. 332-337, Jun. 2016.

[4] Renk Group (2017). RENK AED – Advanced Electric Drive The ultimate propulsion for quiet ships

[5] S. Cui, S. Han and C. C. Chan, "Overview of multi-machine drive systems for electric and hybrid electric vehicles", IEEE Conference and Expo Transportation Electrification Asia-Pacific (ITEC Asia-Pacific), 2014, pp. 1-6

[6] E. Thöndel, "Modelling and simulation of a 6 of motion platform with permanent magnet linear actuators for testing in wind tunnel", Modelling and Simulation 2014 - European Simulation and Modelling Conference ESM 2014, pp. 403-408, 2014.

[7] I. Bolvashenkov, H-G. Herzog, F. Ismagilov, V. Vavilov, L. Khvatskin, I. Frenkel, A. Lisnianski (2020). Fault-Tolerant Traction Electric Drives: Reliability, Topologies and Components Design (1. edition). Springer Nature Singapore

[8] T. Glasberger, Z. Kehl, T. Kosan and J. Molnar, "Design of modular low-profile frequency converter for multi-motor manipulators", 22nd European Conference on Power Electronics and Applications (EPE'20 ECCE Europe), 2020, pp. 1-6

[9] Kumar D., Zare F. and Ghosh A., "DC Microgrid Technology: System Architectures, AC Grid Interfaces, Grounding Schemes, Power Quality, Communication Networks, Applications, and Standardizations Aspects", IEEE Access, vol. 5, pp. 12230-12256

[10] S. Kravitz, Packing cylinders into cylindrical containers, Math. Mag. 40 (1967), 65–71.

[11] R.L. Graham, Sets of points with given minimum separation (Solution to Problem El921), Amer. Math. Monthly 75 (1968), 192–193.

[12] G. Müller, B. Ponick and K. Vogt (2012). Berechnung elektrischer Maschinen (6. edition). WILEY-VCH Verlag GmbH

[13] G. Müller and B. Ponick (2014). Grundlagen elektrischer Maschinen (10. edition). WILEY-VCH Verlag GmbH

Design of a serial impingement cooling heatsink for a 30 kW PV string inverter

Paul BRUYERE[1], Guillaume PIQUET BOISSON[1], Gaëtan PEREZ[2]
[1]Univ. Grenoble Alpes, CEA, LITEN, Campus INES, 73375 Le Bourget du Lac, France.
[2]Univ. Grenoble Alpes, CEA, LITEN, 38000 Grenoble, France.
Tel.: +33 (0)4 79 79 21 64.
E-Mail: Guillaume.PiquetBoisson@cea.fr

Keywords

«Thermal design», «Discrete power device», «Silicon Carbide (SiC)», «DC-AC converter», «Grid-connected inverter».

Abstract

The design of a cooling solution for a 30 kW string inverter is detailed in this paper. As opposed to traditional solutions, a novel approach has been proposed, consisting of three pinfin copper heatsinks tightly arranged with mutual airflow. This solution has been characterized on a dedicated test bench, partly based on a $T_j(I_{DS}, V_{DS})$ TSEP itself precisely calibrated, both being described in this paper. Comparisons between traditional solutions, expected solution, characterized solution, and perspective solutions have been conducted based on the CSPI indicator, reaching up to 34 $W.K^{-1}.L^{-1}$.

Introduction

Historically, Photovoltaic (PV) string inverters have been dedicated to low power installations, central inverters being of use for large-scale PV plants. However, in the recent years, the market of PV string and multi-string inverters has been increasing and these are now installed in a wide range of applications. Initial projects such as domestic production and low/medium power installations (canopies...) have been joined by large scale PV plants in the applications integrating string inverters. Huawei foresees an increase of this trend for the next years [1].

In addition to the increase of efficiency, the reduction of volume is critical for string inverters. Indeed, as these inverters are usually installed on the string frames, a smaller inverter allows saving time and manpower during the installation process, subsequently leading to a cost reduction for the installation. The heatsink being responsible, along with the passive components, of a huge part of the volume of converters, the thermal design of a string inverter has been addressed in this paper. Forced convection with off-the-shelf heatsinks have been considered, and all solutions are compared by using the Cooling System Performance Index (CSPI) [2]. This study is done for a 30 kW string inverter used to connect a 1217 V_{OC} PV array to a three phase 800 V_{LL} grid; the design of the full converter is not detailed in this paper, but for the sake of demonstration is oriented toward highest performances rather than cost management. A two-stage inverter topology has been selected, whose schematic diagram is shown in Fig. 1. Three PV strings of 10.4 kW_C (26 modules of 400 W_C per string) are connected to a capacitive DC bus fixed to 1400 V through a DC/DC stage; a single DC/DC with three paralleled PV strings or a multi-string DC/DC (one independent DC/DC per string) are possible. One 30 kW 3-level Flying Capacitor inverter is used to inject the power from the DC bus to the grid. This paper is focusing on the thermal design of the Flying Capacitor inverter. Regarding the semiconductors, as commercial power modules dedicated to the Flying Capacitor topology of this power range are not yet available, discrete 1200 V / 32 mΩ silicon carbide (SiC) MOSFETs (C3M0032120K from Wolfspeed) have been selected. Analytical calculations of the semiconductor losses to be dissipated are presented in the first section of this paper. The second section will address a comparison of the possible thermal designs. The experimental characterization of the 12 MOSFETs electrical parameters is developed in the third section. Finally, experimental validation of the selected design and perspectives are discussed in the last section.

EPE'22 ECCE Europe

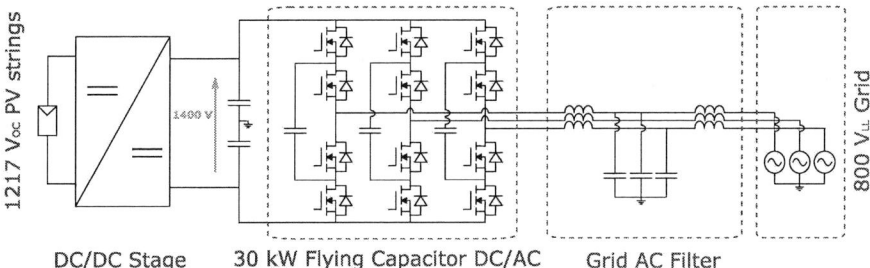

Fig. 1: Diagram of the considered PV inverter, this thermal design regarding its DC/AC stage

I.   Analytical calculation of semiconductor losses

In many power electronics applications, semiconductor devices are the major contributors to overall power losses. For that reason, the cooling system must be carefully designed to ensure the safe thermal operation of these devices. However, semiconductor losses estimation is not a trivial problem and needs to be carried out with precision for the future thermal management. Switching and conduction losses estimations are described in two subsections, where both corresponding methods are exposed.

A.   Switching losses estimation

The purpose of this subsection is to estimate the semiconductors switching losses. Predicting these losses is no trivial task, since they depend on numerous parameters like the Printed Circuit Board (PCB) and semiconductor packaging parasitics, gate driver circuit, etc. However, some methods derived from the manufacturer's datasheets have been proposed to efficiently estimate these semiconductor losses [3]. Another solution, widely used in literature, is to experimentally test the semiconductor devices by using the Double Pulse Test (DPT) [4]. This method is reliable but time consuming, on top of requiring dedicated and expensive measurement setup, especially for wide bandgap devices. Analytical calculation of losses being faster and easier to implement, this method have been retained in this work.

The flying capacitor voltage is fixed to 700 V (half of the DC bus voltage) to have the same voltage constraints on all MOSFETs; each one will switch a constant voltage of 700 V. Since the switching losses from the datasheet are given solely for 600 V and 800 V, a methodology, detailed in [3], has been used to interpolate the losses at 700 V. This method is quite precise since it considers various commutation mechanisms such as capacitive losses (voltage-dependent) and crossover losses (voltage and current -dependent). As the example in this reference is dedicated to a current source inverter, the equations have been adapted to the Flying Capacitor inverter, where a constant voltage and a sinus-modulated current are switched by the semiconductors. The datasheet's curve representing the switching energy as a function of the switched current is fitted by a second order polynomial equation. The switching energy E_{sw}, calculated over a grid period and for the two reference switched voltages from the datasheet, is calculated by using equation (1); $I_{sw}(t)$ being the sinusoidal grid current given by equation (2). A grid RMS current of 21.65 A per phase is considered with nominal grid voltage.

$$E_{sw}(V_{REFi}, I_{sw}) = \int_0^{T_{grid}} [a.I_{sw}(t)^2 + b.I_{sw}(t) + c]\, dt \tag{1}$$

$$I_{sw}(t) = I_{grid_{RMS}}\sqrt{2}\sin(\omega t) \tag{2}$$

Once the switching energy is calculated for the two voltage references, the switching energy at 700 V is interpolated as proposed in [3]. A switching energy of 137.5 µJ is finally calculated, which, given the 40 kHz switching frequency, yields an estimated 5.5 W of switching losses for each SiC MOSFET.

B.   Conduction losses estimation

Regarding the conduction losses, electrothermal coupling has to be considered due to the strong non-linearity of SiC MOSFET R_{DSon} against T_j. Such a temperature consideration was not needed for switching losses calculation of unipolar devices since they are not substantially impacted by the junction temperature. This electrothermal consideration is applied by using equation (3). P_{sw} are the switching losses calculated in the previous subsection, P_{cond} are the conduction losses, $I_{sw,RMS}$ is the RMS current in one MOSFET, $R_{th,jc}$ is the junction to case thermal resistance of one MOSFET defined in its datasheet (0.45 K.W^{-1}) and T_c is the case temperature. To avoid a too high junction temperature of the MOSFETs and in turn to limit their conduction losses and maximize converter efficiency and reliability, the maximum case temperature is defined to be 70°C. The RMS current being the same for all MOSFETs in the Flying Capacitor topology and being equal to $I_{grid,RMS}/\sqrt{2}$ [5], the calculated conduction losses are 8.1 W per MOSFET, regardless of its position. This total of 13.6 W per MOSFET gives a junction temperature of 76°C under these conditions. The total semiconductor losses to be dissipated for the three phases, $P_{SC,3\varphi}$ is therefore 163.2 W. These results are summarized in Table 1.

$$T_j = (P_{sw} + P_{cond}).R_{th,jc} + T_c = \left(P_{sw} + (\alpha.T_j^2 + \beta.T_j + \gamma).I_{sw,RMS}^2\right).R_{th,jc} + T_c \qquad (3)$$

Table I: Results of calculations for SiC MOSFET C3M0032120K from Wolfspeed

Conditions	Per MOSFET	Full inverter (12 MOSFETs)
F_{sw} = 40 kHz	P_{sw} = 5.5 W	$P_{sw,3\varphi}$ = 66.1 W
T_j = 76 °C ; T_c = 70 °C	P_{cond} = 8.1 W	$P_{cond,3\varphi}$ = 97.2 W
V_{DC} = 1400 V ; $I_{grid,RMS}$ = 21.65 A	$P_{SC,1MOSFET}$ = 13.6 W	$P_{SC,3\varphi}$ = 163.2 W

II. Design and comparison of relevant thermal solutions

A. Choices and scope for the heatsink design

As semiconductors packaged in TO-247 have been chosen, the design of the converter's heatsink is quite open, as opposed to the design of a power module-based converter. As a consequence, several types of heatsinks have been considered, with the objective of minimizing the overall volume of the cooling solution, within defined thermal limits. So as to compare these solutions, the CSPI factor of merit, which is inversely proportional to heatsink plus fans volume and to baseplate-to-ambient thermal resistance, will be used. The higher the CSPI is, the higher the heatsink plus fans thermal performances are. Throughout this approach, it has been deemed necessary to retain several criteria that ensure the applicative validity of the designed cooling system.

Regarding the electrical insulation of each device, EN62109-2 standard applies given that the inverter under development is meant for use in PV systems. With the middle point of DC bus linked to heatsinks, the maximum DC voltage to be considered is 700 V (case of TO-247 baseplate being connected to the drain of the MOSFET), leading to a required creepage distance of 3.55 mm. Given the exact package of C3M0032120K, and given available off-the-shelf insulators, a Fischer Elektronik (AOS 218 247) Al$_2$O$_3$ ceramic insulator has been chosen, giving 4.6 mm of creepage (3 mm vertically through the insulator's hole, and 1.6 mm radially toward the discrete's baseplate). A thin layer of thermal paste (50 μm of considered thickness) with a thermal conductivity of 10 W.m^{-1}.K^{-1} will be applied both on heatsink-side and on TO-247-side of the ceramic insulator.

It has been previously defined that the MOSFETs case temperature T_c will have to be kept below 70°C. The applicative in-cabinet maximum ambient temperature T_a is considered to be 50°C. These temperature limits and all thermal elements but the heatsinks being henceforth fixed, equation (4) gives $R_{th,sa,3\varphi}$, the allowed equivalent sink to ambient thermal resistance per 12 MOSFETs (forming the three phases). It has been chosen, from experience, to consider an equivalent spreading angle of 60° through the insulator in order to calculate equivalent surfaces receiving the thermal flux. The ensuing three-phases case to sink thermal resistance $R_{th,cs,3\varphi}$ is 32.6 mK.W^{-1}. Which, given the total semiconductor losses calculated in section I, $P_{SC,3\varphi}$ = 163.2 W, finally yields a maximum $R_{th,sa,3\varphi}$ of 90 mK.W^{-1}. The corresponding $R_{th,sa,1\varphi}$, for a single phase, is of 270 mK.W^{-1}.

$$R_{th,sa,3\varphi} = \frac{T_c - T_a}{P_{SC,3\varphi}} - R_{th,cs,3\varphi} \qquad (4)$$

In order to restrict the design space, off-the-shelf heatsink references have been solely considered. Their masses are deemed not relevant, as a modern 30 kW PV inverter will generally fall far below the single person lifting limit, therefore allowing both aluminum and copper.

B. Comparison of linear fin and pin fin heatsink designs

Traditionally, for the thermal management of medium power converters, designers use linear fin heatsinks, associated with fans in case of forced convection. Benefits of this linear fin approach include the possible optimization of fins (height, width and spacing), in order to obtain better thermal performances. This type of heatsink relies on air-cooling the fins, which are heated by the baseplate, itself receiving and spreading energy from the components on its opposite side. As the same airflow orientation is applied to fins and to the baseplate itself, and due to fins/baseplate areas ratio, little power is drained directly from the baseplate. Fin thermal resistance is therefore a major contributor of a linear fins heatsink total thermal resistance and can hinder its performances.

An alternative to the usual linear fins heatsinks is known as pinfin, and quite rarely seen in power electronics. This second type of heatsink consists of a baseplate associated with column fins having cylindrical or polygonal base forms. Despite the fans disposition being quite open with this type of heatsink, optimal performances are generally achieved when using an airflow normal to the baseplate. The ensuing impingement effect allows to directly cool the baseplate, on top of cooling the fins, hence benefiting total thermal resistance.

Different thermal designs will now be compared in order to select the optimal design for our application.

The first design, quite classic, represented in Fig. 2.a, consists of reference LAV8-100-12, a 100x188x74 mm cooling aggregate with axial fans (3x 60x60x25 mm) and air buffer from Fischer Elektronik, hosting all three phases. It yields a thermal resistance of 67 mK.W^{-1} according to datasheet values, with a CSPI calculated at 7.5. The total footprint of components (as 12 Al$_2$O$_3$ ceramic insulator surfaces) representing 34% of heatsink's baseplate, with 6300 mm², this solution is quite sub-optimal in terms of baseplate use.

Another approach with a linear fin heatsink per phase may help respect the criterion of compactness. One design represented in Fig. 2.b consists of W80-45W, an aluminum 80x80x45 mm linear heatsink from Alpha Novatech associated with two 40x40x15 mm fans, results in a thermal resistance of 260 mK.W^{-1} at approximately 16 cfm. This solution matches with the one phase limit of 270 mK.W^{-1} and yields a CSPI of 11.4. In this design, the surface occupation stays sub-optimal, at roughly 33%.

A third solution consists in considering a single pinfin heatsink ref. 3-575725RFA from Cool Innovations, a black anodized aluminum 145x145x63 mm heatsink, which, associated with a slim 120x120x25 mm fan placed normal to the baseplate (in impingement), displays 84 mK.W^{-1} at 75 cfm, and gives a CSPI of 7.0. This solution is represented on Fig. 2.c and has a 30% surface occupation.

A fourth design, depicted on Fig. 2.d, uses a 4-626207U from the same manufacturer, a copper 156x156x18 mm heatsink, with a high performance 120x120x38 mm fan, yields somewhat less than 90 mK.W^{-1} at 200 cfm, and gives a CSPI of 11.3. Its surface occupation is of 26%. In this last design, the fan represents 56% of the cooler volume, as opposed to respectively 13%, 14%, and 21% in designs a, b, and c.

Fig. 2: Four possible designs

The drawback of pinfin heatsinks regarding compactness is therefore mainly their form factor, where the fan roughly matches the greatest surface of the heatsink, leading to lower CSPI and to generally higher fan consumption. To circumvent this drawback, it is proposed in this paper to use pinfin heatsinks in series, where the airflow from the fans benefits more than one heatsink, and where available baseplate surface can therefore exceed fans' footprint by a certain factor. A corresponding structure is proposed in Fig. 4, where each phase of the inverter (4 MOSFETs each) is cooled by one of the three heatsinks. The airflow is full in the central heatsink while it is distributed between the two external ones. Since the airflow gets divided between the two external heatsinks, these are the ones to be considered for the design of the cooling system. In this solution, the airflow is produced by three 38x38x28 mm high performance fans, which at 19 cfm apiece bring the 4-451508U (38x114x20 mm) Cool Innovations copper heatsink to 150 mK.W^{-1}, whereas the external heatsinks deliver 280 mK.W^{-1} with the split airflow. The calculated total equivalent CSPI is 35.7, with the fans occupying 31% of the cooler volume. Another drawback however lies in the serial structure of this solution, where on top of being split, the airflow is already preheated when it reaches the external heatsinks. The extent of this issue can be asserted as limited, since a 3x19 cfm airflow absorbing a phase worth of losses (54 W) will undergo less than 2°C of temperature rise.

All of these designs are compared on Fig. 3. Focusing on volume, thermal resistance and therefore CSPI, the last design, in red, seems to be the best option. This innovative solution has various pros and cons, that need to be quantified in order to validate or not the presented theoretical design. This study is presented in section IV of this paper.

Fig. 3: Radar plot comparison of the five proposed three phases thermal designs

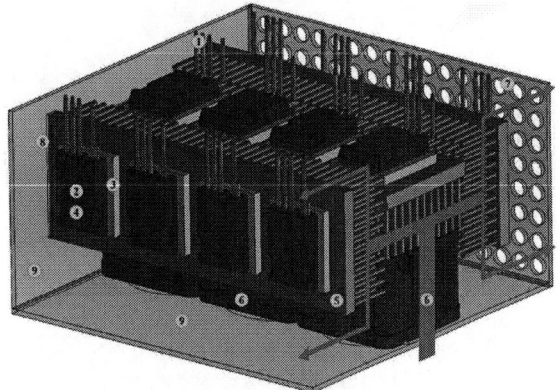

① : All 12 MOSFETs are to be soldered on a single PCB

② : 4 TO-247 MOSFETs are cooled by each heatsink

③ : Each discrete device is insulated from the heatsink by means of an Al$_2$O$_3$ plate and screws insulators

④ : The discretes are to be screwed to the heatsinks

⑤ : 3 copper pinfin heatsinks are used (38x114x20 mm, with 363 cylindrical pins of 1.7 mm each)

⑥ : 3 high performance 38 mm fans, sucking air from the bottom of the inverter are used; airflow's distribution is represented in blue

⑦ : Airflow outputs on 2 sides of the inverter

⑧ : Lateral plates, which can be PCBs, support the heatsinks and channel the airflow

⑨ : Components may be placed in hollow areas, flying or soldered on PCBs, to gain space and maximize airflow use

Fig. 4: Overall view and description of the proposed serial impingement cooling heatsink

III.   Static characterization of discrete SiC MOSFETs against temperature

A. Junction temperature deduction for discrete power devices

As mentioned before, using discrete power devices gives some flexibility as of thermal management. One difficulty with discrete packages is however die temperature (T_j) estimation, whereas power modules may allow infrared (IR) measurements, or the placement of sensors on dies (or directly include a sensor on substrate, albeit with imprecision). As a result, an alternative method for estimating this parameter must be used. Several workarounds are identified in literature for estimating T_j in discrete devices, [6] summarizes some of these methods. Among those, a commonly cited and accessible approach, with IR measurements, can be ruled off for it is too imprecise with discrete components (measure on top of the package). Another approach, using temperature probes, is likewise ruled off. Temperature Sensitive Electrical Parameters (TSEP) therefore constitute an interesting alternative.

Three distinct TSEP categories can be found in literature, off-line TSEP which implies shutting down the converter, dynamic TSEP where a switching event of semiconductor is necessary (among which the temperature-induced variation of threshold voltage $V_{GS(th)}$), and finally static TSEP under normal operation. This last category, often used in power electronics field, consists in calibrating an electrical parameter of the component which is directly linked with T_j. The most classical static TSEP is the ON-state resistance R_{DSon}. New TSEPs emerge regularly in literature, where these continue to be compared against physical measurements [7]. Some authors emphasize the fact that static TSEPs require very precise calibration, which entails the development of dedicated test benches and of the ensuing characterization campaigns.

Whereas T_a can be directly measured at the inlet, it has therefore been chosen to deduce T_j from a TSEP, among which R_{DSon} is the best candidate for it is easily measurable and is highly sensitive against junction temperature [8]. Despite this TSEP being often described as subject to strong deviations depending on the measurement setup [9], it has been evaluated that in this static characterization, it would be possible to account for lead resistances and to set in place very precise measurements of V_{DS} and I_{DS}, in turn yielding precise R_{DSon} values. A characterization of this TSEP has therefore been conducted for each of the 12 MOSFETs of our setup. The corresponding test bench and experimental results are presented in the next sub-section.

B. Development of a test bench and presentation of experimental results

In order to assess the R_{DSon} change against T_j and I_{DS}, an accurate test bench has been developed and applied to the components that will be used during thermal characterization and final converter operation. Its purpose has been to calibrate this TSEP and secondly to quantify intrinsic differences between components. Using MOSFETs in a low $d(R_{DSon}/T_j)$ portion of their characteristic, underlines the necessity of very precise calibration, since the smallest increment in resistance value induces a substantial difference in junction temperature. Characterization test bench consists in measuring $R_{DSon}(I_{DS})$ with a high accuracy Keysight B1506A power device analyzer. During the measures, the 12 MOSFETs are placed in a custom enclosure both highly thermally capacitive (high inertia) and highly isolated (thermally) to regulate very precisely and homogeneously their temperature at several values of interest. Fig. 5 presents a synoptic diagram of the developed test bench, where one copper bar hosts the 12 MOSFETs on its upper face, and is heated on its bottom face by power resistors. Type K thermocouples and PT100 Resistance Temperature Detectors (RTD) are distributed on several locations in the enclosure (on copper bar, on packaging of MOSFETs, on power resistances, etc.), and are used to monitor temperature and regulate it thanks to power resistors.

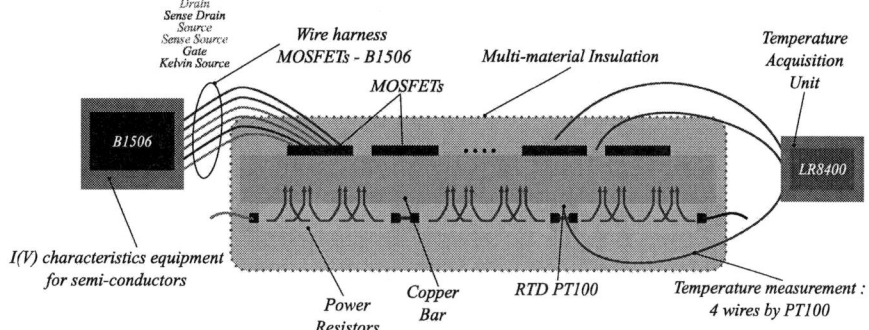

Fig. 5: MOSFETs static characterization against temperature test bench and description

This custom setup allowed the static characterization of the 12 MOSFETs at eight different temperature points, each regulated with a deviation lower than 1°C between probes. That, combined with the low duty cycle of current pulses (100µs pulses separated by 100ms pauses) which leads to negligible die self-heating, allows us to conclude that copper bar and die temperatures are nearly identical. This calibration has then been put into equations thanks to an interpolation between T_j, R_{DSon} and I_{DS}. The implementation of equation (5) in an optimization algorithm minimizing the 12 variables in orange provides the results of Fig. 6. With an average error on all 12 MOSFETs of 0.83°C (and a max of 3.35°C) between experimental measurement and optimization results, interpolation of $T_j(V_{DS}, I_{DS})$ is deemed correct. Another technique, not detailed in this paper, based on 4th order bi-polynomial interpolation gave approximately the same results regarding differences on measured points, but its highly dynamic behavior between points led us not to retain it.

$$T_j = a + b(R')^d + (e + f.R')I_{DS}^{(g+h.R')} + (i + j.R')I_{DS}^2 + k(R'.I_{DS})^l \; ; \; \text{with } R' = R_{DSon} - c \qquad (5)$$

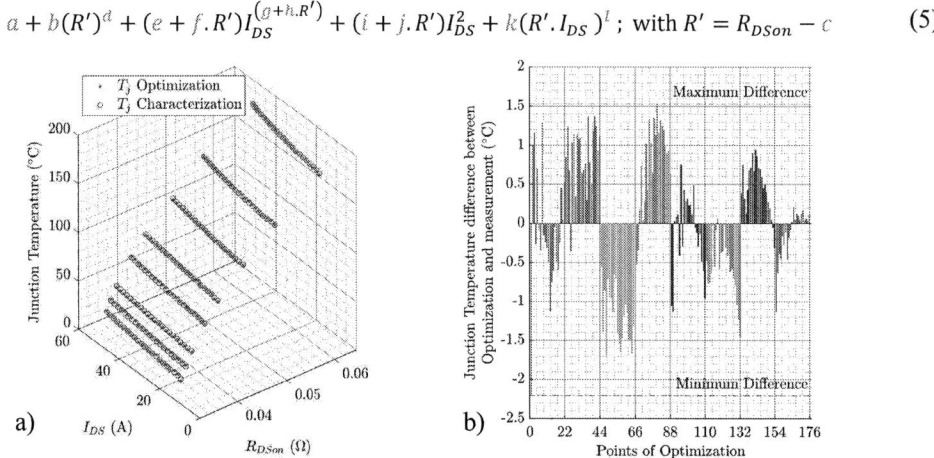

Fig. 6 : a) 3D plot results of static characterization and optimization algorithm for MOSFET no. 1 and b) residuals between characterization and optimization results for this MOSFET

IV. Thermal characterization of the serial impingement cooling solution

A. Presentation of the thermal characterization test bench

The thermal characterization of the presented cooling solution is based on the precise evaluation of the effective junction to ambient thermal resistance of the assembly, under nominal power and under augmented power. The required thermal flux is generated through the injection of a chosen DC current through the 12 MOSFETs temporarily placed in series. This current is chosen, and later adjusted, to be representative, once multiplied by each measured V_{DS}, of the per-MOSFET losses calculated in section I. Electrical properties of each MOSFET are measured using precise equipment, 34465A Keysight digital multimeters for voltages, with an added Burster precise shunt for I_{DS}. Several power supplies are used for generating the constant current (thermal flux), the PWM and power supply of fans, and the insulated gate supplies. The mechanical and electrical design of the thermal assembly, not detailed in this paper, is to be kept for the final converter. It is based on PCBs on all 6 sides : the upper PCB provides the connection of the 12 MOSFETs (in series here), the bottom PCB hosts the fans, and lateral PCBs either insure the connections between upper and lower PCBs or act as an IP2x air outlet. This test bench is described in Fig. 7 below.

① : IR camera
② : 20V-38A power supply
③ : MOSFETs control and V_{DS} measure selection daughter board
④ : Pressure measurements daughter board
⑤ : Fans PWM control and speed measurement daughter board
⑥ : Acquisition units for temperature probes
⑦ : Precision multimeters
⑧ : Hot air exhaust ducts
⑨ : Precision shunt for I_{DS} measurement
⑩ : Arbitrary waveform generator for PWM generation
⑪ : Thermal solution under characterization, based on 6 PCBs
of 1.6mm-thick FR4 :
- Top PCB, 4-layer : 70μm external and 35μm internal;
- Lower PCB, 4-layer : 70μm external and 35μm internal;
- Front/back PCBs, 2-layer : 35μm;
- Lateral PCBs, not plated.

Fig. 7: Thermal characterization test bench for the proposed cooling solution and its description

In order to perform the chosen thermal characterization, and to avoid thermal runaway, several steps have been sequentially followed. Fans are powered and adjusted, MOSFETs are turned ON with V_{GS}=15V, DC current is injected and precisely adjusted with respect to equation (6) at thermal equilibrium, where the values of generated thermal flux are respected with a tolerance of 1‰ with respect to Table II. As the design has been constrained by restrictive ambient temperature and efficiency objectives (described in first sections), it becomes possible to carry out additional qualifications at augmented power (x times nominal injected power is denoted xPIn), which in turns should bring more precision to the calculations through the amplification of the thermal gradients. The 4 thermal fluxes are combined with 3 fan speeds (20, 50, 100%) to produce 12 possible campaigns (within safe limits).

Table II: Thermal flux level for thermal characterization

Naming	Values (in W)
0.5PIn	81.6
1PIn	163.2
2.1PIn	326.4
4PIn	571.8 to 649.2 (limits apply)

$$B_1 < \sum_{i=1}^{12} V_{DS_i} * I_{DS} < B_2 \; ; \; B = \begin{bmatrix} xPIn * (1 - 0.001) \\ xPIn * (1 + 0.001) \end{bmatrix} \quad (6)$$

B. Experimental results of the thermal characterization

As described in the previous sub-section, the main objective of this characterization is to obtain the thermal resistances of the cooling solution. Deducing these is however quite complex due to the disposition of the 12 MOSFETs on 3 partly-serial heatsinks. Our method consists in calculating 12 thermal resistances, each regarding a slice (from die to inlet air) of the cooling solution. Each slice can be different, the measures being carried out under nominal and therefore naturally (un)balanced thermal flux: i.e. a thermal resistance is calculated for each slice through the die-inlet ΔT and the die-injected power (V_{DS} times I_{DS}) whatever the thermal behaviour below. Each phase (a heatsink) is therefore composed of 4 slices (albeit different), and phase thermal resistance can easily be deduced. In this approach, T_j is estimated with the TSEP previously defined and inlet temperature is calculated according to equations (7) and (8) where Q represents the total airflow.

$$T_{air,in,heatsink_{center}} = T_{air,in,fans} + \frac{P_{fans}}{Q * C_{v,air}} \quad (7)$$

$$T_{air,in,heatsink_{lateral}} = T_{air,in,heatsink_{center}} + \frac{P_{center\,phase}}{Q * C_{v,air}} \quad (8)$$

All thermal boundaries and electrical values of interest being obtained, and using datasheet's value for $R_{th,jc}$ and previously calculated value for $R_{th,cs}$, it is now possible to calculate the equivalent thermal resistance per phase (per heatsink) based on equation (9).

$$R_{th,sa,1\varphi} = \left[\sum_{i=1}^{4} \left(\frac{1}{R_{th,sa_i}} \right) \right]^{-1} = \left[\sum_{i=1}^{4} \left(\frac{V_{DS_i} * I_{DS}}{T_{s_i} - T_{air,in,heatsink}} \right) \right]^{-1} \quad (9)$$
$$\text{where } T_{s_i} = T_{j_i} - \left(V_{DS_i} * I_{DS} \right) * \left(R_{th,jc} + R_{th,cs} \right)$$

Fig. 8 presents experimental results for impingement-oriented airflow, for one example MOSFET (no. 1), with a plurality of thermal fluxes (0.5PIn to 4PIn along graph's X-axis) and fan speeds (100%, 50%, 20% in resp. blue, green, yellow) as described in the previous sub-section. On top of the trivial observation of the reduced thermal resistance with an increasing airflow, a remarkable thermal behaviour can be witnessed. As depicted by the fitted non-linear curves, there is a decrease of $R_{th,sa,MOS1}$ when the injected power increases (at constant airflow). For instance, $R_{th,sa,MOS1}$ decreases by 20% with a doubled fan speed (50% to 100%) at 1PIn, but it also increases again by 52% with half the injected power (1PIn to 0.5PIn) at 100% speed.

This behaviour has yet to be thoroughly analyzed, as no clear explanation emerges. Indeed, the whole assembly is constituted of solid elements, measures have low tolerances, and usual parasitic phenomena (additional convection, radiative exchange…), even with exaggerated scopes, may not produce this much of an effect. Several leads are explored, such as an adverse effect of impingement, where strong airflow turbulences appear between fins, and might have a non-linear behaviour against temperature.

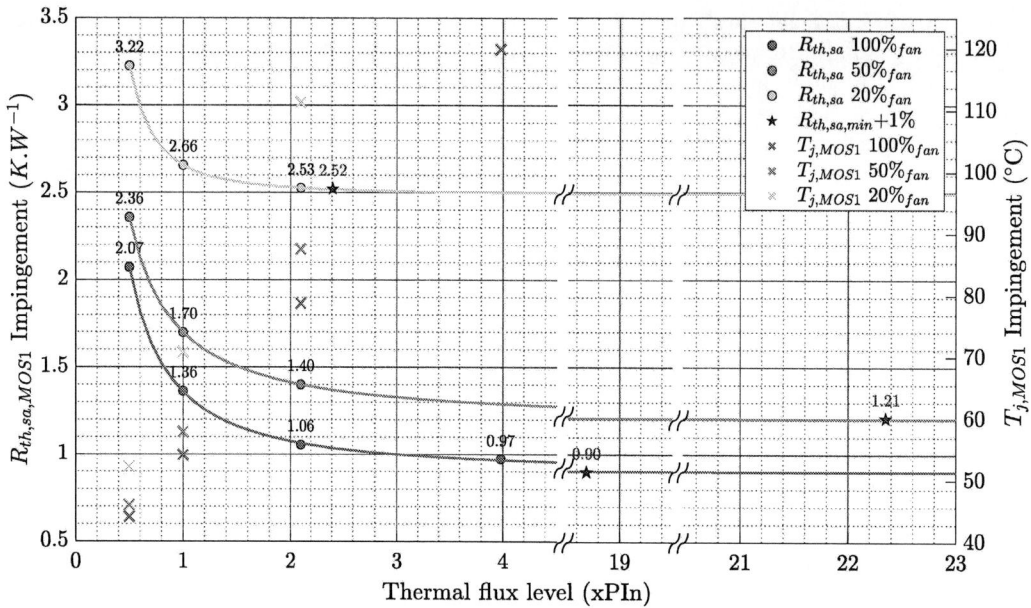

Fig. 8: Experimental results for MOSFET no. 1: $R_{th,sa}$ and T_j against injected power for three airflows

In the meanwhile, it has been chosen to study this system with regard to its asymptotic performances (injected-power-wise). The corresponding thermal resistances are depicted with red stars, at 1% above the asymptote of fitted functions of $R_{th,sa,MOS1}$ against injected power for each fan speed. These extrapolated values are reached for 2.5PIn to 22PIn (caused by the choice of only 1% above minimum).

Albeit solely shown for MOSFET no. 1, the above results are similar for the others. Fig. 9 illustrates the equivalent thermal resistances for each heatsink of the structure, obtained with equation (9). The color code is kept for the fan speeds, and the spreads of thermal resistances between the three heatsinks are illustrated with different markers (the difference between the best and the worst heatsink being shown). The asymptotic performances are also depicted on the right, whatever the actual value of PIn_{max}.

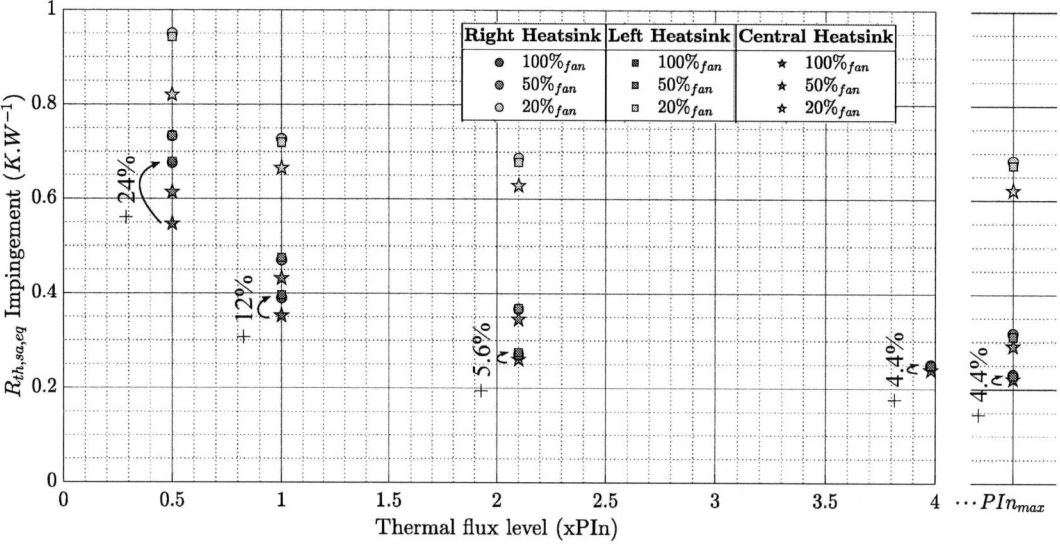

Fig. 9: Experimental per-phase results: Rth,sa,1φ against injected power for three airflows

In order to compare these experimental results to the theoretical design, we can refer to Table III which shows the corresponding per-phase thermal resistances. One can see that the central heatsink does performs better than the outer ones, albeit with a lower difference than expected. Indeed, even with half the airflow, the added $R_{th,sa,1\varphi}$ on lateral heatsinks is of +12% at PIn and of +4% asymptotically, against the +87% that was expected in theory. This difference might be explained by the simplified approach used in subsection II.B. Indeed a halved airflow has been considered to be applied on the whole surface of each outer heatsink, but in reality each sees a concentrated airflow on part of its surface (an average of 7.3 m.s^{-1} on 42% of baseplate surface, see assembly on Fig. 4). This surface hosting the MOSFETs (in the middle of heatsink's height), and pinfins being aligned between heatsinks, it is probable that this lower-than-expected difference emerges from a better-than-expected use of the halved airflow.

Table III: Thermal resistances and CSPI : theoretical, experimental and asymptotical results

In mK.W^{-1} at 100% airflow	$R_{th,sa,1\varphi,theory}$	$R_{th,sa,1\varphi,1PIn}$	$R_{th,sa,1\varphi,asymptotic}$
Central Heatsink	150	352	221
Right Heatsink	280	390	231
Left Heatsink	280	396	228
Total CSPI	35.7	20.8	34.2

The experimental thermal resistances themselves stand higher than expected, with an excess of 39% to 135% at 1PIn compared to theory. The corresponding CSPI, adapted to multiple-heatsinks systems, is computed at 20.8 as indicated on Table III, standing 42% lower than its theoretical counterpart. It can however be highlighted that the asymptotical values are closer to our expectations, nay exceed them with a nearly perfect balance between all three heatsinks, in turns yielding a CSPI of 34.2, on par with theory. Reaching sufficiently low thermal resistances (and high CSPI) could involve obvious solutions such as increasing the airflow, or could focus on the asymptotic approach, an increase in thermal flux level having been shown to benefit thermal performances. However it also leads to higher junction temperatures and subsequent low efficiencies, which highlights the existence of an optimum between thermal system performances and inverter performances. Our scope for this design might well be too efficiency-oriented (with ensuing low losses) to fully benefit from this cooling system design.

Whereas the higher-than-expected performances of lateral heatsinks have already been detailed, the underperformances of the central heatsink conjure several leads, related to airflow inhomogeneity and turbulency or to power sources repartition. Both airflow problematics would be caused by the high-performances fans (with large hubs and high swirl) and might be hindering heatsink's performances by providing airflow solely to certain sections of it, which airflow might also not be normal to the baseplate thence creating more turbulence along pinfins. This could be solved by inserting an air buffer between fans and fins, in the form of a box with internal features, which could both allow airflow to spread below the hubs and straighten airflow's direction (the CSPI would be impacted by the augmented volume and the potentially reduced thermal resistance). The discrete repartition of power on baseplate's surface brings to the table the spreading contribution to the thermal resistances, which has been neglected in section II for lack of information. Potential countermeasures would include convection spreaders (the copper baseplate being already an excellent conductor), such as caloducs (placed along heatsink's length) or vapor chambers (covering the heatsink); or could include those consisting in using the free surfaces of the heatsink by bringing thermal fluxes from other components (capacitors or inductors).

Conclusion

The design of a cooling system for a Flying Capacitor inverter dedicated to PV applications has been described in this paper. After an estimation of switching and conduction losses of the topology, different thermal designs have been presented and compared, mainly based on CSPI figure of merit. A novel structure, based on a serial arrangement of pinfin heatsinks, has been proposed and characterized. In order to deduce thermal resistances through a precise estimation of die temperatures, a static characterization of a MOSFET TSEP has been conducted on a custom test bench. The cooling solution has been characterized on a dedicated test bench representing the real converter layout, at real and augmented thermal fluxes from the MOSFETs. Results have been discussed based on the satisfaction of each criterion and perspectives have been proposed for current drawbacks. In particular, it has been highlighted that although quite insufficient at nominal power, the asymptotic performances of this cooling system are up to our initial expectations, and could be well suited to another design scope, with higher losses to dissipate.

References

[1] Huawei, "How String Inverters Are Changing Solar Management on the Grid."

[2] U. Drofenik, G. Laimer and J. W. Kolar: Theoretical Converter Power Density Limits for Forced Convection Cooling, PCIM 2005, official proceedings book, pp. 608 – 619

[3] G. Lefevre, A. Bier and S. Catellani: A cost-controlled, highly efficient SiC-based Current Source Inverter dedicated to Photovoltaic applications, EPE'18 ECCE Europe

[4] L. G. Alves Rodrigues and G. Perez: A 200 kW Three-level Flying Capacitor Inverter using Si/SiC based Devices for Photovoltaic Applications, PCIM Europe digital days 2021

[5] J. Azura Anderson, L. Schrittwieser, M. Leibl and J. W. Kolar: Multi-Level Topology Evaluation for Ultra-Efficient Three-Phase Inverters, 2017 IEEE International Telecommunications Energy Conference (INTELEC)

[6] N. Baker, M. Liserre, L. Dupont and Y. Avenas: Improved Reliability of Power Modules: A Review of Online Junction Temperature Measurement Methods, IEEE Industrial Electronics Magazine, vol. 8, n° 3, p. 17-27, sept. 2014

[7] L. Dupont and Y. Avenas: Preliminary Evaluation of Thermo-Sensitive Electrical Parameters Based on the Forward Voltage for Online Chip Temperature Measurements of IGBT Devices, IEEE Transactions on Industry Applications, vol. 51, n° 6, p. 4688-4698, nov. 2015

[8] L. Zhang, P. Liu, S. Guo, A. Q. Huang: Comparative Study of Temperature Sensitive Electrical Parameters (TSEP) of Si, SiC and GaN Power Devices, IEEE WiPDA 2016

[9] N. Baker, M. Liserre, L. Dupont, Y. Avenas: Junction temperature measurements via thermo-sensitive electrical parameters and their application to condition monitoring and active thermal control of power converters, IEEE IECON 2013

Online Junction Temperature Measurement of SiC-MOSFETs via Gate Impedance Using the Gate-Signal Injection Method

David Hirning, Luca Bauer, Johannes Ruthardt, Jörg Haarer, Philipp Ziegler,
Jörg Roth-Stielow
INSTITUTE FOR POWER ELECTRONICS AND ELECTRICAL DRIVES
University of Stuttgart
Pfaffenwaldring 47
Stuttgart, Germany
Tel.: +49 / (711) – 685 67371
E-Mail: david.hirning@ilea.uni-stuttgart.de
URL: http://www.ilea.uni-stuttgart.de

Keywords

«TSEP», «Silicon Carbide (SiC)», «Reliability», «Junction Temperature Measurement»

Abstract

This paper presents a method for junction temperature monitoring of SiC-MOSFETs based on a high-frequency gate-signal injection. The signal is injected during steady state (e.g. off-state) resulting in a current response, which depends on the temperature dependent gate impedance. The external gate resistor is used as a current shunt to capture the current response. The resulting signal contains the junction temperature information due to the temperature dependency of the gate impedance. This paper focuses on a sinusoidal approach to overcome the challenges due to the temperature dependent parasitic capacitance of the gate circuit. Measurements show the proof of concept, however, there are still challenges to face.

Introduction

In recent years, there has been a trend in power electronics towards more compact and faster switching designs. This development was made possible by constantly improved and partly new types of power transistors. Due to this trend power transistors are also facing new challenges. As a result of the higher switching frequencies and the higher integration level, the thermal stress of the power transistors is increasing.

In order to counteract this and increase the reliability, a high dynamic junction temperature measurement creates the possibility of detecting over temperature and an online health monitoring [1] or even temperature control [2, 3, 4].

The gate-signal injection method has been successfully implemented and investigated for Si-IGBTs [1, 5–9]. Applying this method to SiC-MOSFETs is more complex due to the temperature dependent parasitic capacitance of the gate circuit, by which a sign change of the sensitivity of the temperature dependent impedance of the gate circuit can appear, as shown in [10, 11]. Through a sign change in the sensitivity an ambiguous relationship between the temperature and the gate impedance can occur, which makes it impossible to infer the temperature from the impedance [10]. Because the sensitivity as well as the impedance is strongly dependent on the frequency, a simplified rectangular signal injection can have a negative influence on the temperature measurement due to its harmonics at higher frequencies, as described in [10]. This is why in this paper a more complex sinusoidal approach is selected, similar to the approach described in [7] for Si-IGBTs. It is investigated how the sinusoidal approach can improve the accuracy of the measurement.

This paper presents a gate-signal injection method, which overcomes the challenges due to the temperature dependent parasitic capacitance of the gate circuit, which will be validated by measurements.

Approach

Fig. 1 (left) shows a simplified circuit of the gate-driver circuit of the SiC-MOSFET. The voltage v_{dr} is the voltage applied by the gate driver, R_G is the external gate resistor, L_P the parasitic inductance of the MOSFET as well as of the gate-driver circuit, R_{Gi} the temperature dependent internal gate resistance and C_P the parasitic temperature dependent input capacitance of the MOSFET [10, 12, 12]. Due to the temperature dependent elements within the gate-driver circuit the impedance $\underline{Z}_G$ of the gate-driver circuit is temperature dependent as well, see (1).

$$\underline{Z}_G(\omega_i, T_j) = R_G + R_{Gi}(T_j) + j\omega_i L_P + \frac{1}{j\omega_i C_P(T_j)} \tag{1}$$

Since the temperature dependency cannot be measured directly, a high frequency sinusoidal signal is injected into the gate circuit, while the MOSFET is in off-state, see Fig. 1 (right) [8]. Due to the injection the operation of the MOSFET is limited, because the time in off-state of the MOSFET t_{off} needs to be at least as long as the injection takes place. The injected signal has the frequency ω_i and the amplitude v_i. This signal causes a gate current which again leads to a voltage drop across the external gate resistor. By means of a voltage divider, as seen in equation (2), the voltage drop across the external gate resistor is described as a function of the frequency of the injected signal as well as of the junction temperature [1, 7, 8].

$$v_{RG} = v_i \cdot \frac{R_G}{\left|Z_G(\omega_i, T_j)\right|} \tag{2}$$

If the relationship between the temperature and the voltage drop across the external gate resistor is biunique, the junction temperature can be inferred by the voltage drop across the external gate resistor [10], hence it is mandatory to find a frequency of the injected signal, at which this relationship is biunique.

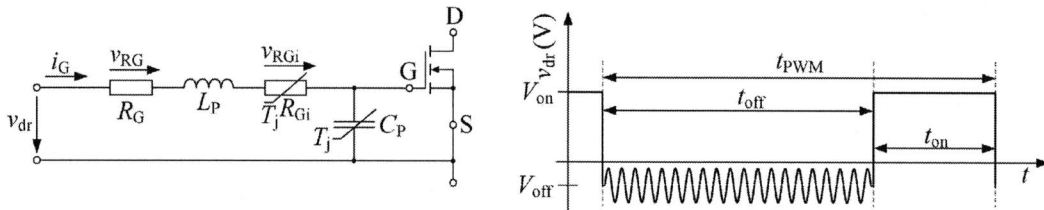

Fig. 1: Equivalent circuit diagram of the gate-driver circuit (left). Gate-driver voltage v_{dr} (right)

Furthermore the sensitivity, which is the derivative of the voltage drop across the external gate resistor with respect to the temperature, see (3), should be as high as possible for a precise and robust measurement [1, 7, 8].

$$s_v = \frac{dv_{RG}}{dT_j} = \frac{d}{dT_j} v_i \cdot \frac{R_G}{\left|Z_G(\omega_i, T_j)\right|} \tag{3}$$

Hence a frequency not only with a biunique relationship, but also with the highest possible sensitivity should be selected [10]. Therefor a profound investigation of the gate circuit of the SiC-MOSFET is mandatory.

Investigations of the gate circuit

To find a frequency, matching the requirements, a half-bridge SiC-module (BSM120D12P2C005 by Rohm) is investigated with a network analyzer (Bode100, Omicron). For the investigations, only the low side of the module is considered, since the test setup is a buck converter, which only uses the low side MOSFET. The MOSFET is investigated within a gate-driver circuit as in Fig. 1 (left).

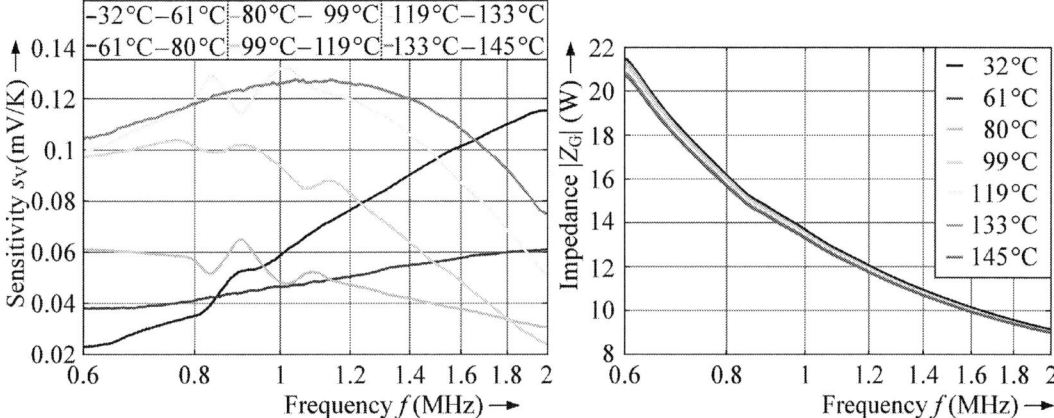

Fig. 2: Sensitivity of the voltage drop across the external gate resistor depending on the frequency (left). Temperature dependency of the impedance of the gate driver circuit depending on the frequency (right).

As already mentioned, a biunique relationship between the voltage drop and the temperature is mandatory, thus the sensitivity s_v must not change its sign over the target temperature range at the preferred frequency. Fig. 2 shows the sensitivity s_v as well as the absolute value of the impedance $\underline{Z}_G$ depending on the temperature and the frequency. It can be seen, that the impedance decreases with increasing temperature within the depicted frequency range. Furthermore, the temperature dependency of the impedance is biunique for the shown frequencies, which can also be seen within the sensitivity, because the sensitivity does not change its sign in this frequency range. The sensitivity s_v of the voltage drop across the external gate resistor is calculated by differentiation with respect to the junction temperature and the parameters from Tab. I, as seen in (3). The voltage V_{DS} is set to $V_{DS} = 30$ V, since the change of the input capacitance of the MOSFET is neglectable above a drain-source voltage of $V_{DS} = 10$ V [13]. Because the measurement is not continuous, the difference quotient of the measured data is used, see (4).

$$ s_v = \frac{dv_{RG}}{dT_j} \approx \frac{\Delta v_{RG}}{\Delta T_j} = \frac{\Delta}{\Delta T_j} v_i \cdot \frac{R_G}{\left| Z_G(\omega_i, T_j) \right|} \tag{4} $$

Table I: Parameters of the impedance measurement and for the calculation of the sensitivity

Parameter	Value
V_{GS}	-5 V
V_{DS}	30 V
T_j	32°C – 145°C
f_i	0.6 MHz – 1 MHz
R_G	5.6 Ω

The sensitivity varies strongly with the temperature and the frequency. For higher temperatures the sensitivity has its maximum between 1 MHz and 1.2 MHz. For temperatures in the middle range the sensitivity decreases with an increasing frequency. In contrast to this the sensitivity increases with increasing frequency for lower temperatures. Depending on the field of application a frequency can be selected with the optimal sensitivity. To find a frequency with the best results over the whole temperature range the voltage difference Δv_{RG}, see (5), between the lowest and highest temperature is considered, see Fig. 3. The voltage difference has its maximum between 1 MHz and 1.2 MHz, so the linearized sensitivity over the whole temperature range has its optimum within these frequencies.

$$\Delta v_{RG} = v_{RG}(145°C) - v_{RG}(32°C) \tag{5}$$

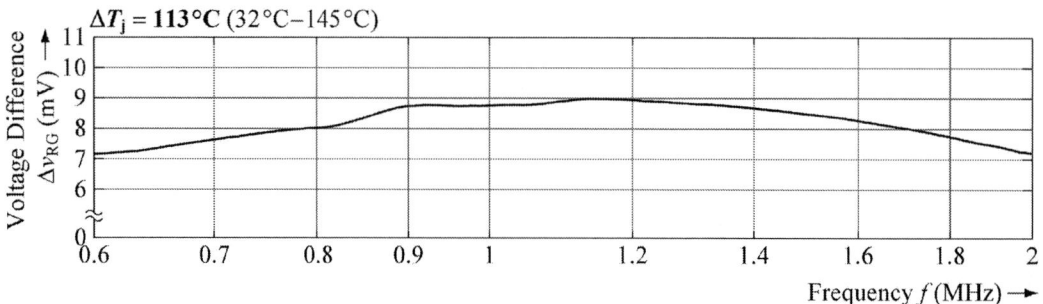

Fig. 3: Change in voltage depending on the frequency over the complete temperature range from 32°C to 145 °C.

Realization

The superimposed sinusoidal signal v_i is provided by a signal generator and is superimposed by means of a transformer, as seen in Fig. 4. The capacitance C_d decouples the injection path from the DC-voltage. To turn off the MOSFET a current can flow through the diodes D_1 and D_2, but the added forward voltage of the diodes is high enough to block the injected signal. Without the added forward voltage the injected signal would be shorted. The voltage $v_{off,i}$ is the sum of the turn off voltage v_{off} and the injected voltage v_i as seen in Fig. 1 (right).

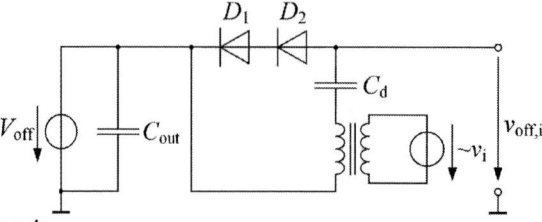

Fig. 4: Signal injection circuit

The voltage $v_{off,i}$ is connected to the driver IC, which again applies this voltage into the gate-driver circuit, see Fig. 5 (left). For the measurement of the voltage drop across the external gate resistor a differential amplifier is used. Fig. 5 (right) shows the measurement circuit as well as the signal processing of the measured voltage. By means of an IC the RMS value of the voltage drop across the external gate resistor is formed. This signal again is filtered, amplified and adjusted to the input voltage range of an analog-digital converter, which is read out by a microcontroller, where the calibration curve is lodged, to infer to the junction temperature [7]. The value of the voltage v_{ADC} is converted every PWM cycle.

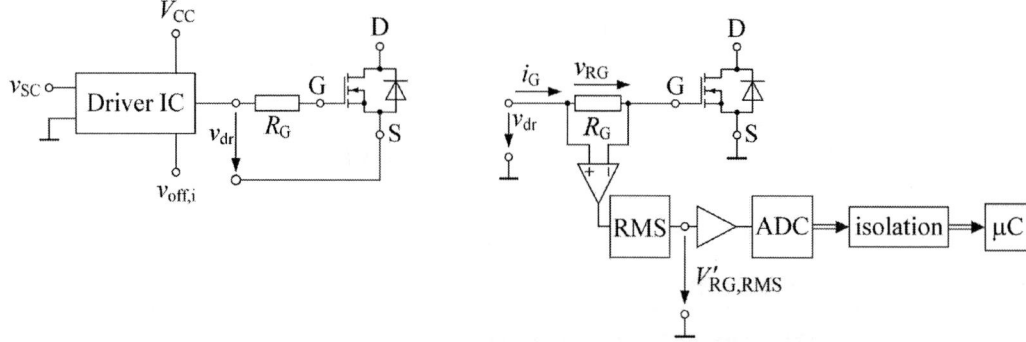

Fig. 5: Gate-driver circuit with driver IC (left). Signal processing of the measured voltage drop across the external gate resistor (right).

Fig. 6 shows the gate-driver voltage v_{dr}, the RMS-value $V'_{RG,RMS}$ of the voltage across the external gate resistor as well as the input voltage v_{ADC} of the analog-digital converter, during operation of the MOSFET with no drain-source voltage applied. Looking at the gate-driver voltage, the superimposed sinusoidal signal can be seen while the MOSFET is in off-state. The injected signal has a frequency f_i of $f_i = 1$ MHz. The RMS-value of the voltage drop is filtered by a second-order low pass filter to minimize noise. Due to the low sensitivity and the small change in voltage over the entire temperature range, as seen in Fig. 2 and 3, the RMS-value has to be highly amplified. The RMS-value can be seen in Fig. 6, it seems to be in steady state at about 20 μs. But in the amplified and adjusted voltage v_{ADC}, it is clear to see, that the voltage signal is still in its transient response. Even at the end of the signal injection the voltage signal is not in steady state. Since the voltage v_{ADC} is converted each PWM-cycle at the same time this will be neglected for the measurements.

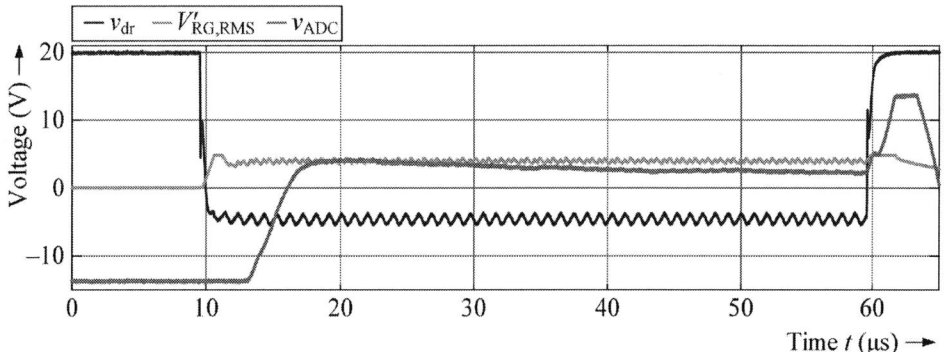

Fig. 6: Gate-driver voltage, measured RMS-value of the voltage drop and the filtered and amplified RMS-value before the analog-digital converter.

Measurements

In a first experiment the MOSFET was heated twice with a heat gun. A significant change in the voltage drop across the external gate resistor occurred, see Fig. 7. The heating and cooling phase of the MOSFET can be clearly seen. The MOSFET is mounted to a heat sink. Furthermore, the presented measurement principle is investigated in an operating test setup as seen in Fig. 8.

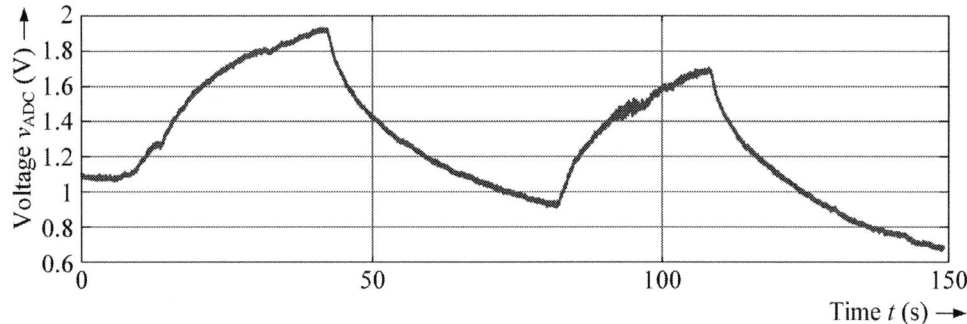

Fig. 7: Measured voltage across the external gate resistor while heating the MOSFET with a heat gun twice.

The MOSFET is used in a buck converter with an ohmic-inductive load. The temperature is monitored with an infrared camera as a reference to the gate-signal injection method. Therefore, the SiC-module is opened and blackened so the infrared camera captures the junction temperature as precise as possible [8]. The parameters of the test setup can be found in Tab. II.

To infer to the junction temperature a calibration must be executed. Due to the high frequency signal injection the range of adjustment of the duty cycle is limited. Since the injection period is set to $t_i = 50$ μs the maximum settable duty cycle is 50 %. For the calibration this duty cycle is set and the MOSFET is heated. While heating the MOSFET the voltage across the external gate resistor as well as the junction temperature captured by the infrared camera is recorded, see Fig. 9. The calibration curve, Fig. 9, shows a biunique relationship between the junction temperature and the converted voltage across the external gate resistor from 40°C to about 105°C. Within this temperature range it is possible to infer to the junction temperature by means of the calibration curve. For temperatures above 105°C the voltage v_{ADC} no longer increases with increasing temperature, so the relationship is no longer biunique. The investigations of the gate-driver circuit (Fig. 2) showed a biunique relationship for temperatures until 145°C, however, this behavior cannot be verified within the calibration. An assumption for this is a temperature drift within the signal processing of the measured voltage drop across the external gate resistor, which could be neglected by a differential measurement approach. This has to be further investigated.

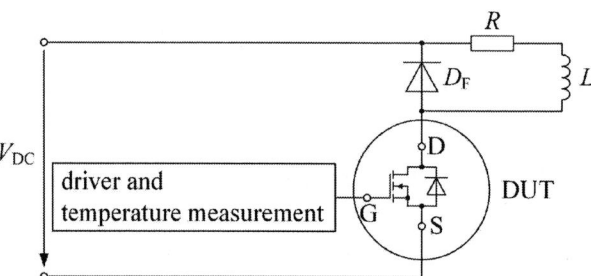

Fig. 8: Test setup of the junction temperature measurement.

Table II: Parameters of the test setup.

Parameter	Value
R	1.5 Ω
L	610 µH
V_{DC}	215 V
f_{PWM}	10 kHz
f_i	1 MHz
R_G	5 Ω

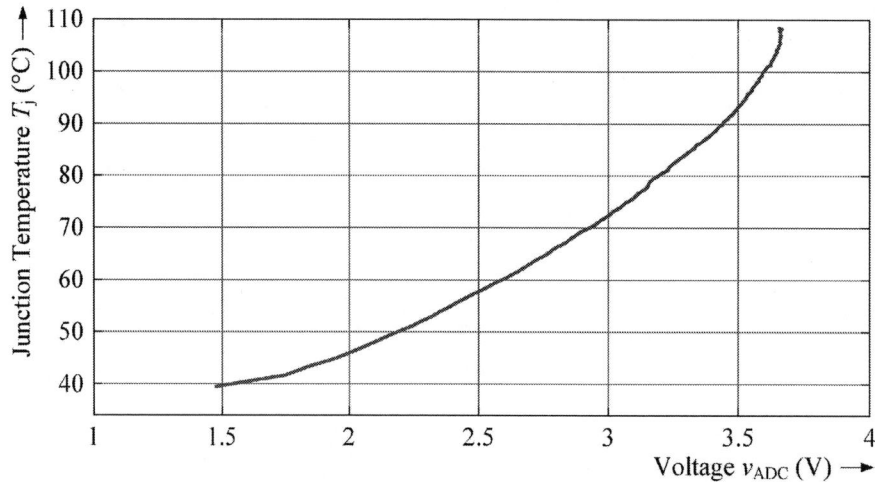

Fig. 9: Calibration curve.

In a first measurement the test setup is operated in the same operation point as for the calibration curve. The duty cycle is set to 50%, the result is depicted in Fig. 10. In the first plot in blue the by means of the gate-signal injection method inferred junction temperature and in red the reference temperature from the infrared camera is shown. In the second plot the deviation between the reference and the TSEP-method can be seen. The deviation is throughout the whole measurement range less than 2 K, so in this operation point the presented signal injection method provides good results.

$$\Delta T_j = T_{j,\text{reference}} - T_{j,\text{TSEP-method}} \qquad (6)$$

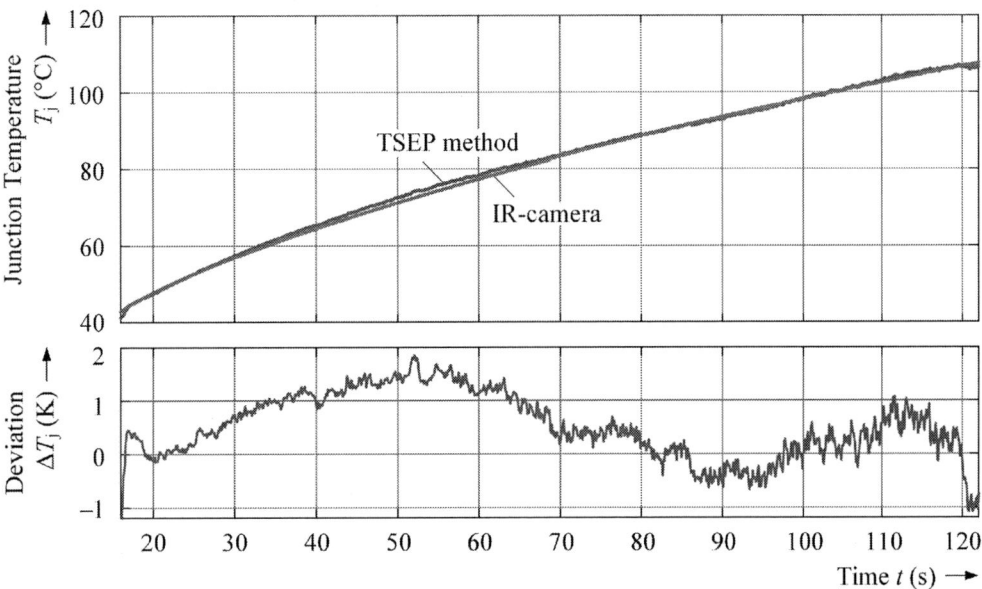

Fig. 10: Inferred and reference junction temperature for a duty cycle of 50% (top) and deviation from the reverence temperature (bottom).

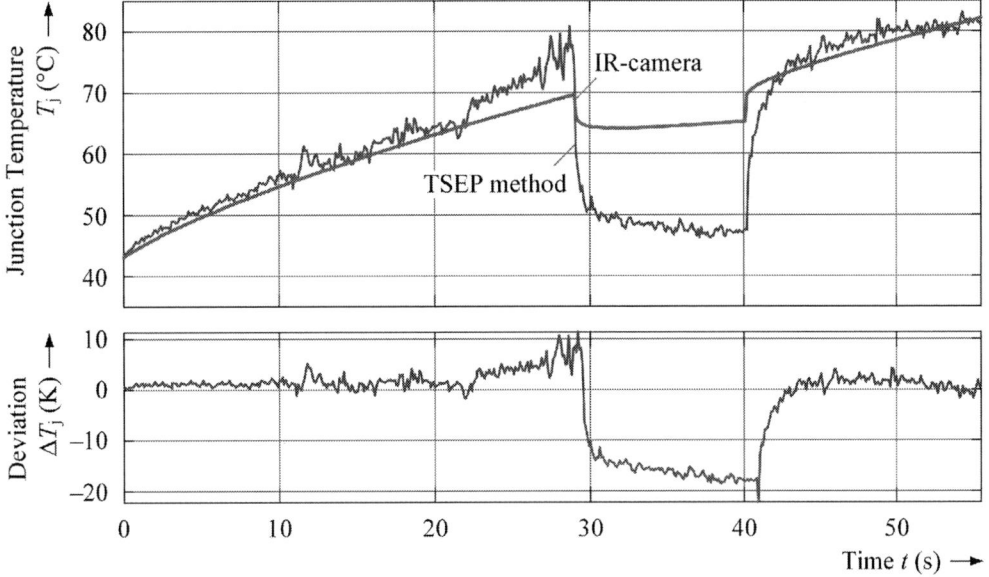

Fig. 11: Inferred and reference junction temperature for different duty cycles (50%, 35%, 50%) (top) and deviation from the reverence temperature (bottom).

In a second measurement, Fig. 11, the duty cycle is varied. In the first interval the duty cycle is set to 50% as in the calibration until about $t = 29$ s. In the second interval the duty cycle is set to 35% for about 11s and in the third and last interval the duty cycle is set to 50% until the end of the measurement. For both intervals with a duty cycle of 50% the deviation is less than 10 K, most of the time even less than 5 K. As seen in the first measurement the method provides good results within the operation point of the calibration. For the second interval the deviation is up to 20 K, it can be seen, that the method, implemented as presented, only provides usable results within the operation point of the calibration. By varying the duty cycle, respectively the load current, the inferred temperature deviates highly from the actual junction temperature. However, these measurements provide a general proof of concept, but for an actual usage of the presented method further investigations have to be made.

Conclusion

This paper presents a junction temperature measurement setup for the gate-signal injection method using a high frequent sinusoidal signal to overcome the challenge due to the temperature dependent parasitic capacitance of the gate circuit. An investigation on the gate-driver circuit of the MOSFET shows, that a junction temperature measurement via the temperature dependent gate impedance is possible. Therefore, a signal injection circuit as well as a measurement circuit is shown. This setup is tested in an initial experiment with a heat gun and shows promising results. For further validation a test setup is introduced and within this setup measurements are performed. The high frequency gate-signal injection method provides good results within the electric operation point of the calibration. However, for other operation points there is a high deviation between the actual junction temperature and the junction temperature inferred by means of the presented TSEP-method. For an actual usage of the presented method further investigations have to be made.

References

[1] M. Denk, "In-Situ-Zustandsüberwachung von IGBT-Leistungshalbleitern mittels Echtzeit-Sperrschichttemperaturmessung," 2016. [Online]. Available: https://eref.uni-bayreuth.de/32289/

[2] Ruthardt Johannes, Schulte Hendrik, Ziegler Philipp, Fischer Manuel, Nitzsche Maximilian, and Roth-Stielow Jörg, "Junction Temperature Control Strategy for Lifetime Extension of Power Semiconductor Devices," in *2020 22nd European Conference on Power Electronics and Applications (EPE'20 ECCE Europe)*, 2020, pp. 1–9.

[3] J. Ruthardt *et al.,* "Closed Loop Junction Temperature Control of Power Transistors for Lifetime Extension," in *2020 IEEE Applied Power Electronics Conference and Exposition (APEC)*, 2020, p. 2955.

[4] van der Broeck Christoph H., Polom Timothy A., Lorenz Robert D., and De Doncker Rik W., "Real-Time Monitoring of Thermal Response and Life-Time Varying Parameters in Power Modules," *IEEE Transactions on Industry Applications*, vol. 56, no. 5, pp. 5279–5291, 2020, doi: 10.1109/TIA.2020.3001524.

[5] Denk Marco and Bakran Mark-M., "Junction Temperature Measurement during Inverter Operation using a TJ-IGBT-Driver," in *Proceedings of PCIM Europe 2015; International Exhibition and Conference for Power Electronics, Intelligent Motion, Renewable Energy and Energy Management*, 2015, pp. 1–8.

[6] Denk Marco and Bakran Mark-M., "Comparison of UCE- and RGi-based junction temperature measurement of multichip IGBT power modules," in *2015 17th European Conference on Power Electronics and Applications (EPE'15 ECCE-Europe)*, 2015, pp. 1–11.

[7] Denk Marco and Bakran Mark-M., "An IGBT Driver Concept with Integrated Real-Time Junction Temperature Measurement," in *PCIM Europe 2014; International Exhibition and Conference for Power Electronics, Intelligent Motion, Renewable Energy and Energy Management*, 2014, pp. 1–8.

[8] Ruthardt Johannes *et al.,* "Online Junction Temperature Measurement via Internal Gate Resistance Using the High Frequency Gate Signal Injection Method," in *PCIM Europe 2019; International Exhibition and Conference for Power Electronics, Intelligent Motion, Renewable Energy and Energy Management*, 2019, pp. 1–7.

[9] Denk Marco and Bakran Mark-M., "IGBT Gate Driver with Accurate Measurement of Junction Temperature and Inverter Output Current," in *PCIM Europe 2017; International Exhibition and Conference for Power Electronics, Intelligent Motion, Renewable Energy and Energy Management*, 2017, pp. 1–8.

[10] J. Ruthardt *et al.,* "Investigations on Online Junction Temperature Measurement for SiC-MOSFETs Using the Gate-Signal Injection Method," in *2021 IEEE Energy Conversion Congress and Exposition (ECCE)*, 2021, pp. 5354–5359.

[11] T. Kestler and M.-M. Bakran, "Expansion of the Junction Temperature Measurement via the Internal Gate Resistance to a wide range of Power Semiconductors," in *2019 21st European Conference on Power Electronics and Applications (EPE '19 ECCE Europe)*, 2019, pp. 1–9.

[12] S. M. Sze, *Physics of semiconductor devices,* 3rd ed. Hoboken, N.J: Wiley-Interscience, 2007.

[13] Rohm Co., Ltd., *SiC Power Module BSM120D12P2C005 Datasheet.* [Online]. Available: https://fscdn.rohm.com/en/products/databook/datasheet/discrete/sic/power_module/bsm120d12p2c005-e.pdf

Powercycling Test Bench with Realistic Loss Distribution and Temperature Ripples

Till-Mathis Plötz, Jan Fuhrmann, Hans-Günter Eckel
University of Rostock
Albert-Einstein-Str. 2
Rostock, Germany
Phone: +49 (0) 381-498-7135
till-mathis.ploetz@uni-rostock.de

Keywords

≪Power cycling≫, ≪Test bench≫, ≪Switching losses≫, ≪Reliability≫, ≪Leakage current≫

Abstract

An innovative test bench is presented, which allows the powercycling for semiconductors under a superposition of temperature ripples with different frequencies. In addition to the implementation of switching losses, this creates loadpatterns which are similar to the application. Further, the leakage current of the semiconductors as a thermo-sensitive parameter is introduced. First results validate the concept.

Introduction

Accurate predictions on the lifetime of power semiconductors are crucial for economical converter designs. An early failure due to power cycling will lead to costly standstill periods. This is especially true for wind turbines, where precious yield is lost. To prevent this, the converter can easily be designed with a large safety margin in lifetime. The overdesign in lifetime is achieved by additional semiconductors, which consequently leads to additional investment costs.

In the following, a powercycling test bench is presented, which enables the testing of power semiconductors under an application-based superposition of different temperature ripples and loss sharing.

In the classical semiconductor power cycling test, the semiconductors are stressed with forward conduction losses. This is achieved by connecting a low voltage current source with the devices and changing the conducting device periodically. External switches switch the load current. To finish the test in a feasible time frame highly accelerated life test (HALT) are used. However, in the application a significant

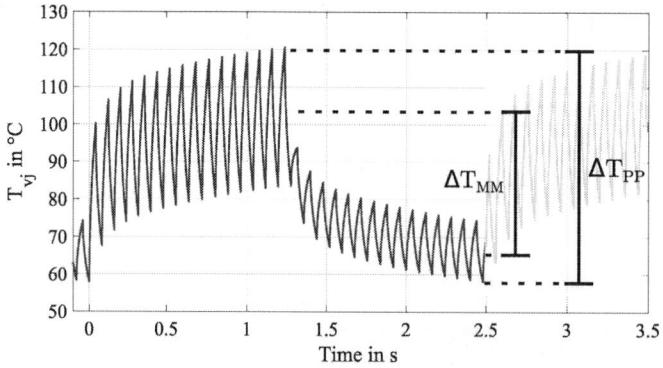

Fig. 1: Possible temperature pattern of innovative test bench

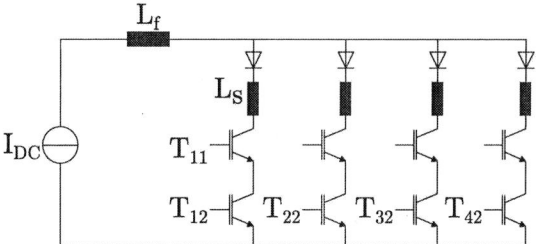

Fig. 2: Simplified electrical circuit diagram of the powercycling test bench

amount of losses are switching losses. A much higher forward current compared to the application has to be used to generate comparable temperature ripples with conduction losses only. The high forward current tends to stress the bondwire more due to self-heating. This problem intensifies with the need for highly accelerated tests.

Another mismatch from the application is how the semiconductor is stressed with different temperature swings. Recent publications on the linear cumulative damage theory show that a general addition of different temperature swings and their respective lifetime consumption is valid [1]-[3], as long as the same failure mechanism is stimulated. This concept, known as the Miners Rule [4], is limited to time serially events. However, in a complex mission profile, many different temperature ripples occur simultaneously. Different ripples with different period times superimpose upon each other. One single test for this has been done in [5], here the big and slower cycle was dominant for the failure. In lifetime calculations the superpositon is generally answered by a simulation with different time resolutions [6]-[7]. Often, a low resolution is used for load changes. Here, neither the frequency of the loadcurrent nor the thermal impedance can be rendered.

In [8], it is shown that the used resolution has an influence on the calculated lifetime. The large temperature ripple due to fluctuations in the wind can have a significant share of the overall calculated lifetime consumption [9]. However, this superposition of different temperature swings is not tested in standard power cycling test. A potential simulated temperature pattern of the test bench is shown in Figure 1. The pattern consists of two superimposed temperature ripples. A classical rainflow-algorithm will count the big ripple as the peak to peak value T_{pp}. Although, a calculation with a low resolution for slow temperature swings would calculate more like the middle to middle value T_{MM}. The question arises what the effective lifetime consuming temperature-ripple is and how the prevailing frequencies influence this.

Test Bench Design

A simplified electrical circuit diagram of the test bench is shown in Figure 2. The test bench consists of a low voltage current source (I_{DC}), a filter inductance (L_f), four switching inductances (exemplary L_s) and four IGBT half-bridge modules, with the lowside (LS) switches T_{12}-T_{42}. The concept of adding switching losses to power cycling tests with inductances was published and discussed by Herold and Lutz in [10] and [11]. These additional inductances cause voltage peaks at the turn-off of the IGBTs. The voltage peaks, in addition to the slopes in the current, lead to switching losses inside the devices. The current and voltage for one turn-off event during power cycling is depicted in Figure 3.

Three frequencies are used to control the test bench and create temperature patterns that consist of slow and fast ripples at the same time:

- Frequency of switching between any LS-Switch f_{sw}
- Frequency of load alternating between T1 and T2, or T3 and T4 with f_{fast}
- Frequency of varying the modulation index of each device f_{slow}

The usages of these frequencies are further explained in Figure 4. The shown pattern starts with an alternating switching between T_1 and T_2. This switching is done with f_{sw}. With this switching, the switching losses are introduced to the modules. T_1 is on 90% of the time, while T_2 is 10% of the time

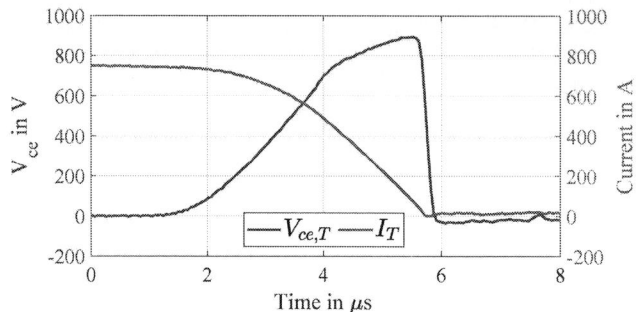

Fig. 3: Measurement: Voltage and Current during switching-event, $L_s \approx 1.3\,\mu H$

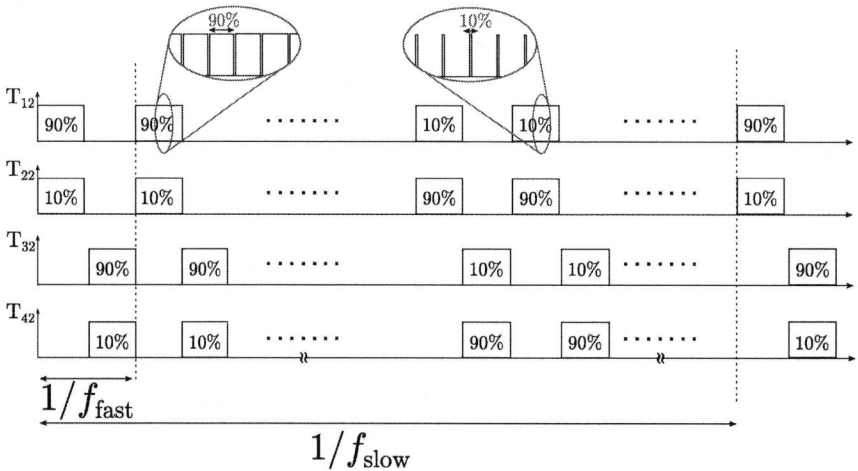

Fig. 4: Exemplary control Signals of LS-Switches

(the modulation indices of 90% and 10% are just an example to illustrate the functionality). After half of the fundamental period $1/f_{\text{fast}}$ the current is switched between T_3 and T_4. This continues until half of the period time of f_{slow} is reached. Here the modulation index is changed. So for example, T_1 is now on only 10% of the time. The change in the modulation index forms the large temperature ripple with f_{slow}. The alternation between switching the current and not switching forms the second temperature ripple with $1/f_{\text{fast}}$.

With these frequencies, a complex loadpattern can be reproduced. Hence, a rotor-side converter semiconductor of a windturbine can be reproduced, for example. With a change of f_{sw} a realistic loss distribution can be achieved. The second frequency f_{fast} can be set to the frequency of the nominal current. The third frequency f_{slow} can be adjusted to reproduce turbulent wind behavior. Additionally to the frequencies the value of the DC-Current (I_{DC}) and the modulation index (a) are degrees of freedom in the design of the testpattern. The test bench is not limited to four devices, the pattern can also be designed for six or more devices.

Additional Leakage Current Measurement

The leakage currents of power semiconductors are highly temperature-dependent. The positive feedback loop between power losses and temperature increase is known as thermal runaway, which affects high-voltage devices in the blocking state if the cooling is insufficient [12].

The physical principles for the temperature dependence will be briefly summarized in the following. The leakage current density j_r originates from charge carriers swept across the depletion region. It can be split into the diffusion current density, caused by the concentration difference of the PN Junction j_s and the drift current density, generated by the depletion region j_{DR}. As per [13], they can be described as:

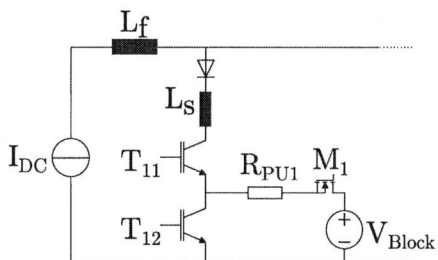

Fig. 5: Circuit Design to load T_{12} with V_{Block}

$$j_r = j_s + j_{DR} = \frac{q \cdot n_i^2}{N_D} \cdot \sqrt{\frac{D_p}{\tau_p}} + \frac{n_i \cdot q \cdot w_{DR}}{\tau_{sc}} , \tag{1}$$

where q represents the electron charge constant, n_i the intrinsic electron concentration, N_D the doping concentration, D_P the hole diffusion coefficient, τ_p the holes lifetime, τ_{sc} the carrier lifetime in the space charge region and w_{DR} is the width of the space charge region under the blocking voltage. The intrinsic carrier concentration of silicon can be stated as [14]:

$$n_i^2(T_j) = C_1 \cdot T_j^3 \cdot e^{-\frac{C_2}{T_j}} . \tag{2}$$

Equation (1) and (2) show that the leakage current is strongly positively related to the junction temperature of the device. The strong dependence makes the leakage current a potent thermo-sensitive parameter (TSP). The $V_{CE}(T_j)$-method is very linear and provides a mean value over all chips. A minor degradation in the solder layers, which leads to a temperature enhancement of a small fraction of the whole chip area results therefore only in a relatively slight increase of the $V_{CE}(T_j)$. The leakage current, however increases over proportional with just a small hot spot.

The indirect measurement is shown in Figure 5. After the LS-Switch T_{12} is switched off, an additional DC-Voltage source (V_{Block}) is connected over the pull-up resistance R_{PU}. After some µs the MOSFET M_1 is switched off. Due to the leakage current the voltage over T_{12} decreases.

This behavior is shown for two temperatures in Figure 6 during power cycling. The figure shows the V_{CE} of T_{12} in the beginning of a temperature ripple at 90 °C (left) and in the end of the ripple at 130 °C (right). First the initial switch-off takes place, marked with the high voltage peak due to the switching inductances. After this the voltage rises to the voltage level of V_{Block}, subsequently the voltage source is disconnected and the voltage over the devices decreases. At the higher temperature the voltage decrease

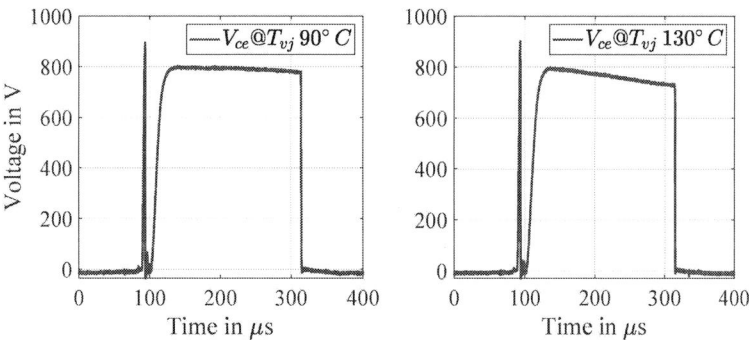

Fig. 6: Measurement: V_{CE} during powercycling, with $V_{Block} = 800 \, \text{V}, f_{sw} = 2 \, \text{kHz}$, left: beginning of temperature ripple with $T_j = 90 \, °\text{C}$, right: end of temperature ripple with $T_j = 130 \, °\text{C}$

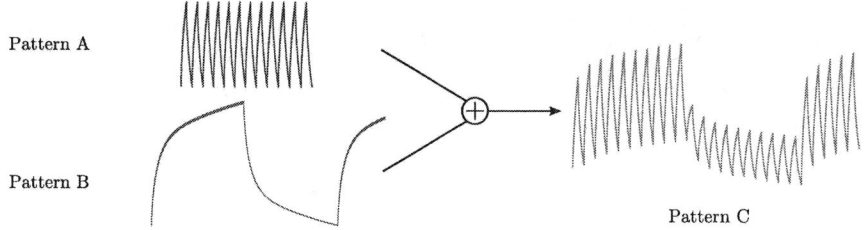

Fig. 7: Patterndesign: the simulataneous superposition of different temperature ripples

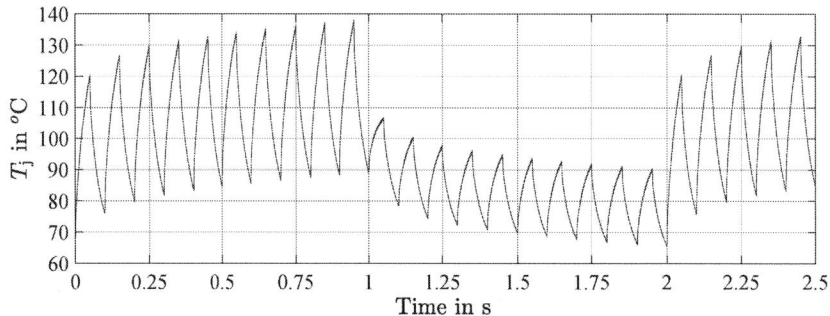

Fig. 8: Simulation: Target pattern **C**, from 0 s to 1 s a 50 K ripple is present, at 1 s the modulation index is changed, this change forms the big superimposed ripple with 72 K with f_{slow}

is much faster due to the increased leakage current.

For the power cycling periodically Z_{th}-Measurements are done with the $V_{\text{CE}}(T_{\text{j}})$-Method. For this fast and potent clamping circuits and measurement-current sources were designed.

Patterndesign

For the first testruns, three different testpatterns have been designed. The first two patterns, pattern **A** and pattern **B**, are classic single-temperature-ripple powercycling patterns. The final pattern is a simultaneous superposition of the prior temperature ripples. In contrast to other tests, the new pattern is not a serial switching between pattern **A** and **B**, but rather a new pattern that includes both ripples simultaneously. Figure 7 illustrates this combination of the patterns. Table I summarizes the three patterns.

Table I: Temperature ripples and patterndesign

	ΔT_1 10 Hz $t_{\text{on}} = 0.05\,\text{s}$	ΔT_2 0.5 Hz $t_{\text{on}} = 1\,\text{s}$
Pattern **A**	50 K	-
Pattern **B**	-	72 K
Pattern **C**	50 K (only half of the time)	72 K

Pattern A only includes a 50 K ripple with a frequency of 10 Hz. Pattern **B** consist of a 72 K ripple with a frequency of 0.5 Hz. The final pattern, pattern **C**, superimposes these two ripples. The final pattern **C** is shown in Figure 8. The definition of the time t_{on} for the superposition of ripples is questionable. A rainflow algorithm, which is state-of-the-art in the lifetime calculation, does not differ if the ripples are superimposed or time serial, so the stated t_{on} time for pattern **C** is how a rainflow algorithm calculates the time.

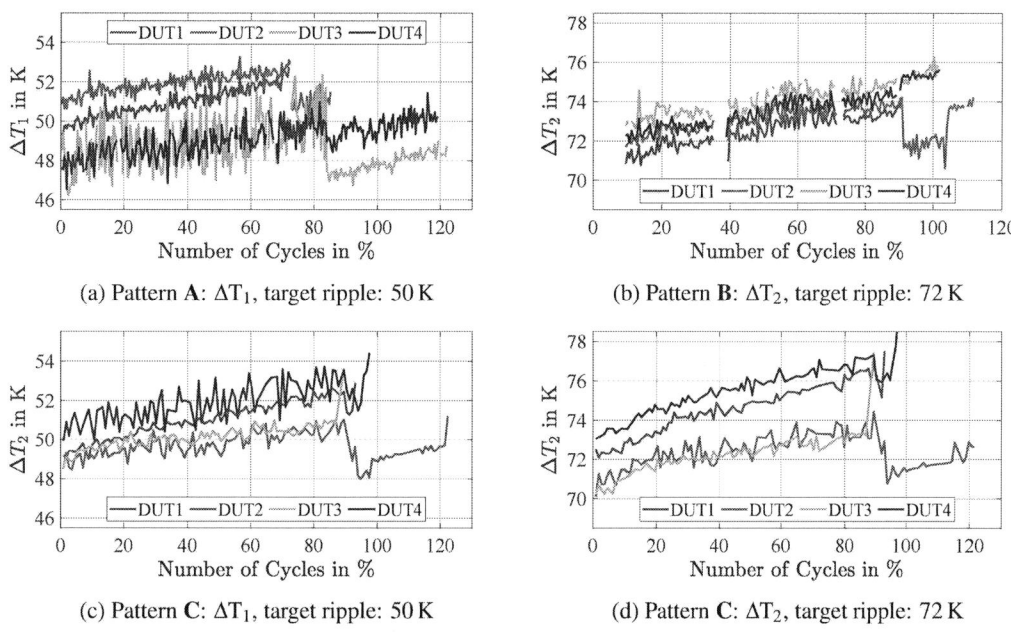

(a) Pattern **A**: ΔT_1, target ripple: 50 K

(b) Pattern **B**: ΔT_2, target ripple: 72 K

(c) Pattern **C**: ΔT_1, target ripple: 50 K

(d) Pattern **C**: ΔT_2, target ripple: 72 K

Fig. 9: Measurement: temperature ripples, ΔT_1 and ΔT_2 for pattern **A**, **B** & **C**

Powercycling Results

Each pattern has been tested with four devices, namely FF650R17 PrimePack modules. The results of the temperature ripple measurements are shown in Figure 9. Two particular findings have to be mentioned. In general, the temperatures tend to rise. However after one device has failed, the values of the remaining devices tend to jump to lower values. After one device has failed the test bench stops, and the particular device is changed manually with a dummy device. This dummy device has a higher power class, though. The exchange seems to influence the losses in the remaining switches. A relative small decrease in the switching inductances L_s can reduce the temperatures in the occurrent values. The lower temperatures result in a prolonged lifetime of the remaining devices.

Another finding is the difference in the absolute values between different devices. This is partly because of the inequality in the switching inductance L_s, small variations in the thermal coupling or simple device variations. The mean value of the measured ripple at the beginning of the test and the target ripples, however show a good agreement.

Similar behavior can also be seen in the maximum junction temperatures of the corresponding ripples. These temperatures are plotted in Figure 10. Again the jumps in the temperature after one device has been changed are visible. In theory, pattern **B** should have the same maximum temperatures as **A** and **C**. For this purpose, the cooling water temperature had to be increased from 15 °C to 45 °C. Unfortunately, the control of the cooling system could not keep this temperature constant, which resulted in a 5 K fluctuation in the cooling water temperature, this fluctuation is also visible in the maximum junction temperature.

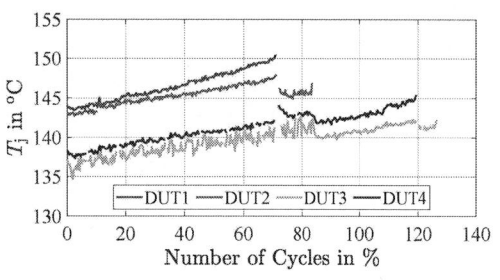
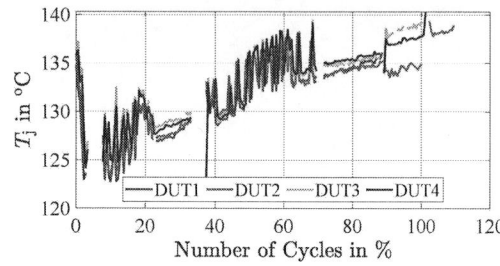

(a) Pattern **A**: maximum junction temperature trend   (b) Pattern **B**: maximum junction temperature trend

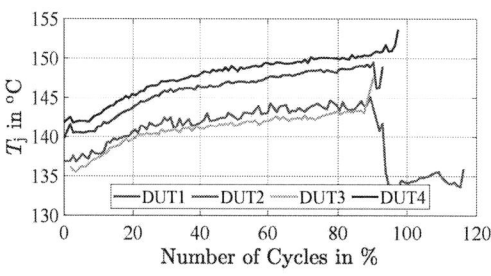

(c) Pattern **C**: maximum junction temperature trend

Fig. 10: Measurement: maximum junction temperature during powercycling for pattern **A**, **B** & **C**

The resulting V_{ce} trends are shown in Figure 11. The EoL (End-of-Life) is denoted as an increase of the forward voltage at load current by 5 %. Another failure criterion is an increase of 20 % of the thermal resistance. During all tests, all devices have reached the defined EoL due to the increase of the forward voltage. This failure mechanism is generally refered to a lift-off of the bondwires, while the degradation of solder layers mainly causes the increase of the thermal resistance. Also here a small voltage change after one device has reached its EoL is noticeable.

Figure 12 puts the results of all tests into perspective. The lifetimes are normalized to the mean lifetime of pattern **C** in realtime, not in cycles. The mean lifetime of pattern **A** was around factor 1.2 the lifetime of pattern **C**. The pattern B was about 0.8 times the lifetime of pattern **C**.
All devices of pattern **C** had lived longer than any device of pattern **B**. And this even with another lifetime relevant ripple (50 K) on top of the ripple of pattern **B** (72 K).
If a linear damage accumulation, in a way the rainflow-algorithm would count the ripples in pattern **C**, is assumed, a lower liftime would be expected. A simplified estimation of this expected lifetime ($L_{PC,expec.}$) of pattern **C** is the inverse of the sum of the lifetimeconsumption of pattern **A** and **B**:

$$L_{PC,expec.} = \left(0.5 \cdot \frac{1}{L_{PA}} + \frac{1}{L_{PB}}\right)^{-1} \approx 0.61 \cdot L_{PC}. \tag{3}$$

The factor 0.5 derives from the fact that $\Delta T_1 = 50\,K$ only occurs half of the time in pattern **C**. The calculated value would be around factor 0.6 of the tested lifetime of pattern **C** L_{PC}, see also black stars in Fig. 12.

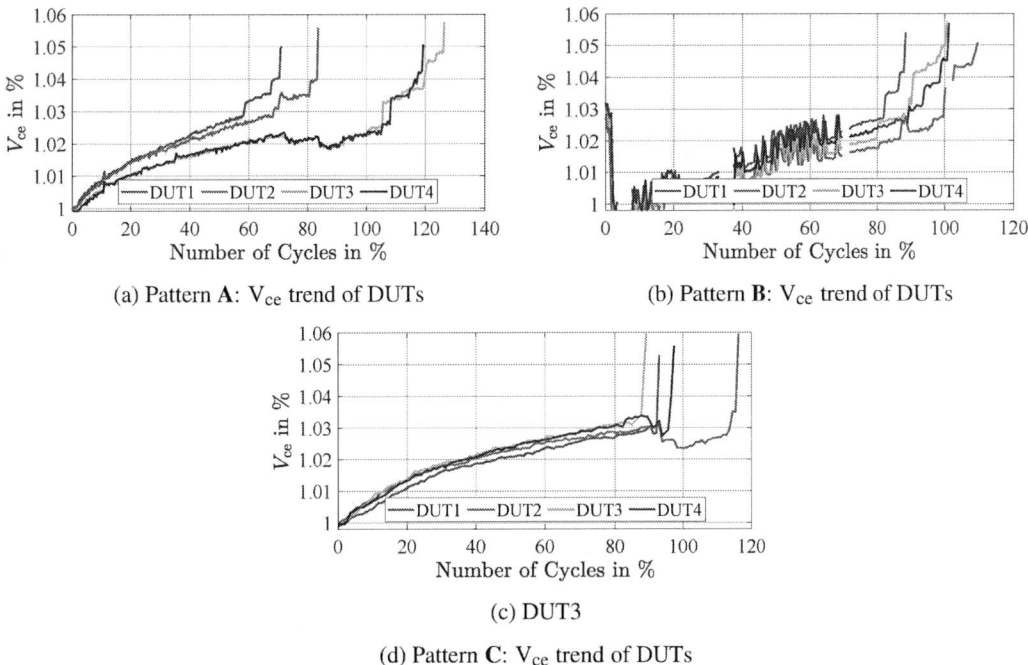

(a) Pattern **A**: V_{ce} trend of DUTs

(b) Pattern **B**: V_{ce} trend of DUTs

(c) DUT3

(d) Pattern **C**: V_{ce} trend of DUTs

Fig. 11: Measurement: V_{ce} trends of DUTs for pattern **A**, **B** & **C**

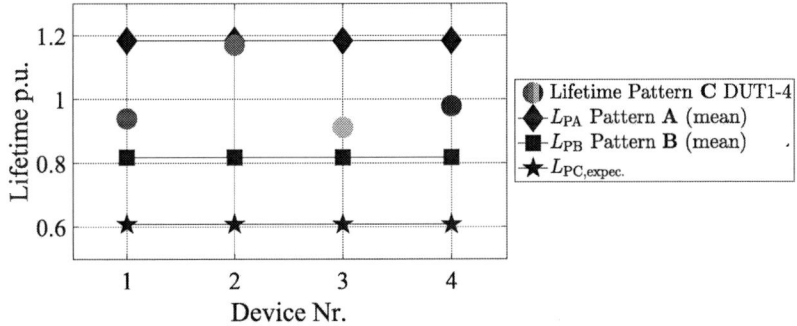

Fig. 12: Lifetimeresults of the different pattern, normalized to the mean Lifetime of pattern **C** in realtime (not cycles)

The results of the different patterns are also plotted in Weibull-Probability plots in Figure 13. Here the spread of the EoL-times for pattern **A** is clearly visible. This spread results in a relatively low β-factor, the slope in the probability plot. Hence also the range of the 90%-Confidence intervals is inflated in contrast to the other results.

For pattern **C** we see that all the EoL-Results are outside of the confidence intervals of pattern **B**, and this even with the additional temperature ripple of pattern **A**.

The final test bench is shown in Fig. 14. The left side is shown with the additional clamping circuits and necessary PCBs for the indirect leakage current measurement, while these PCBs are demounted on the right side to show the DUTs parts below.

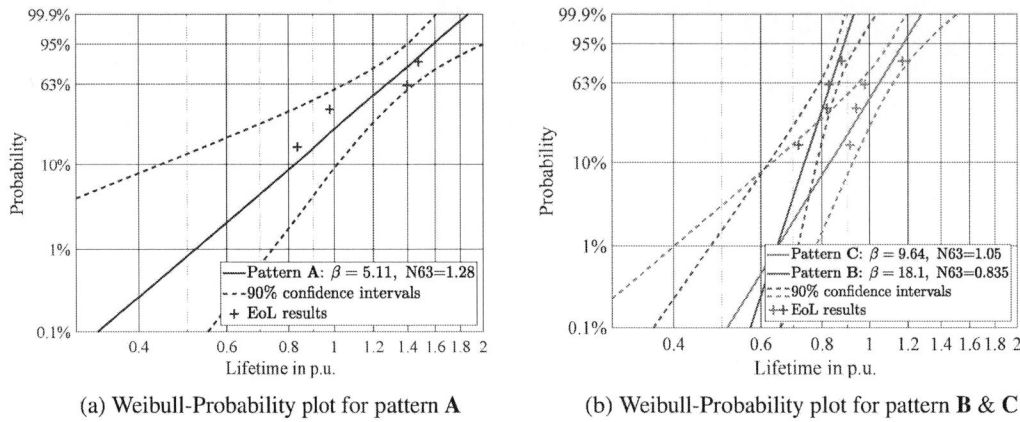

(a) Weibull-Probability plot for pattern **A**   (b) Weibull-Probability plot for pattern **B & C**

Fig. 13: Weibull-Probability plots for pattern **A** and for pattern **B & C**

Conclusion

The presented powercycling test bench has been built up. First powercycling runs have been finished to validate the setup and the measurement systems. The actual measured patterns show a good match with the simulated target patterns.

Since all devices have failed due to bond-wire lift-off and not due to solder delamination, the analysis of the leakage current was unremarkable. To further evaluate the leakage current, different temperature patterns have to be used to address this failure mechanism specifically.

Of course, the results of these first runs must be taken with caution. Slight asymmetries in the switching inductances, especially after the exchange of one device influenced the results. Also the sample size is limited. Nevertheless, given the results the question arises if the linear damage accumulation for simultaneously occurring temperature ripples might be too conservative. Surely, a conservative design is always preferable to a risky variant, hence the results do not imply any risk to existing designs.

More investigations have to be done to further assess the error margin of the lifetime calculation with a state-of-the-art rainflow counting algorithm at mission-profiles with a valid number of superimposed temperature ripples. There might be a relevant lever to further reduce the semiconductor usage for these kind of applications. The presented test bench is capable of producing powercycling results to help answer this question more in detail.

Fig. 14: Powercycling test bench, the image is split to show the uncovered DUTs

Acknowledgments

The project WindVolt (0324256B) is supported by the Federal Ministry for Economic Affairs and Energy on the basis of a decision by the German Bundestag

References

[1] U. Choi, K. Ma and F. Blaabjerg, "Validation of Lifetime Prediction of IGBT Modules Based on Linear Damage Accumulation by Means of Superimposed Power Cycling Tests." in IEEE Transactions on Industrial Electronics, vol. 65, no. 4, pp. 3520-3529, April 2018.

[2] G. Zeng, C. Herold, T. Methfessel, M. Schfer, O. Schilling and J. Lutz, "Experimental Investigation of Linear Cumulative Damage Theory With Power Cycling Test." in IEEE Transactions on Power Electronics, May 2019.

[3] M. Hernes, S. D'Arco, O. C. Spro and D. Peftitsis, "Experimental Validation of Linear Damage Superposition for IGBT Power Modules Under High and Low Temperature Stress Cycles." PCIM Europe digital days 2021; International Exhibition and Conference for Power Electronics, Intelligent Motion, Renewable Energy and Energy Management, 2021.

[4] M. A. Miner, Cumulative damage in fatigue. J. Appl. Mech., vol. 12, pp. 159164, 1945.

[5] Feller, M., Lutz, J., Bayerer, R., "Power cycling of IGBT- modules with superimposed thermal cycles" in Proceedings of PCIM Europe. Nuremberg (2008)

[6] K. Ma, M. Liserre, F. Blaabjerg and T. Kerekes, "Thermal Loading and Lifetime Estimation for Power Device Considering Mission Profiles in Wind Power Converter." in IEEE Transactions on Power Electronics, vol. 30, no. 2, pp. 590-602, Feb. 2015.

[7] D. Weiss and H. Eckel, "Fundamental frequency and mission profile wearout of IGBT in DFIG converters for windpower." 2013 15th European Conference on Power Electronics and Applications (EPE), 2013.

[8] G. Zhang, D. Zhou, F. Blaabjerg and J. Yang, "Mission profile resolution effects on lifetime estimation of doubly-fed induction generator power converter." IEEE Southern Power Electronics Conference (SPEC), 2017.

[9] C. Neumann and H. Eckel, "Comparative Lifetime Estimations for IGBT Modules in Wind Turbine Converters." 2022 24th European Conference on Power Electronics and Applications (EPE'22 ECCE Europe), 2022

[10] C. Herold, P. Seidel, J. Lutz, R. Bayerer, Topologies for inverter like operation of power cycling tests, Microelectronics Reliability, Volume 64, 2016.

[11] P. Seidel, C. Herold, J. Lutz, C. Schwabe and R. Warsitz, "Power cycling test with power generated by an adjustable part of switching losses." 2017 19th European Conference on Power Electronics and Applications (EPE'17 ECCE Europe), 2017.

[12] A. Castellazzi, J. Saiz and M. Mermet-Guyennet, "Experimental characterisation and modelling of high-voltage IGBT modules off-state thermal instability." 2009 13th European Conference on Power Electronics and Applications, 2009, pp. 1-9.

[13] J. Lutz, Halbleiter-Leistungsbauelemente: Physik, Eigenschaften, Zuverlssigkeit , 2nd ed. Berlin, Heidelberg: Springer, 2012.

[14] B. J. Baliga, Material Properties and Transport Physics, in Fundamentals Power Semiconductor Devices. Cham, Switzerland: Springer, 2019.

Design, implementation and characterization of an integrated current sensing in GaN HEMT device by using the current-mirroring technique

Van-Sang NGUYEN, René ESCOFFIER, Stéphane CATELLANI, Murielle FAYOLLE-LECOCQ, Jérémy MARTIN

CEA - French Alternative Energies and Atomic Energy Commission
50 avenue du Lac Léman
Le Bourget du Lac - 73375, France
Tel.: +33 / (0) 4 79 79 27 50
E-Mail: van-sang.nguyen@cea.fr & rene.escoffier@cea.fr
URL: www.cea.fr

Acknowledgements

The Authors acknowledge CEA project "GaN4PV" for funding

Keywords

«Current observer», «Current sensor», «Device characterization», «Double pulse test», «Dynamic R_{on}», «Fast fault detection», «Gallium Nitride (GaN)», «HEMT», «Measurement», «Over-current protection», «Power die», «Power integrated circuit», «Test bench», «Wide bandgap devices»,

Abstract

Based on wide bandgap devices (WBG) characterization constraints, this work presents the design, implementation and characterization of an integrated current sensor in a GaN HEMT (Gallium Nitride High-Electron-Mobility Transistor) by using the current-mirroring technique. Two HEMTs are implemented in this design; the compromised between the size ratio of these two transistors in the current-mirroring circuit and the sensitivity of the sensor are taken into account on the device design phase. In the implementation phase, the auxiliary components are optimized for the operation of the sensor, and then the circuit with the integrated current sensing in GaN power device is characterized with a high temperature double pulse test method, up to 175°C.

Introduction

Nowadays WBG semiconductors facilitate high efficient / compact energy conversion systems design [1-2]. Given their high switching speed, their dynamic characterization requires high bandwidth and low intrusive probes. Current measurement is made difficult by the intrusiveness of the probes, which modify the power switching loop impedance. In general, the measurement of switched currents can be performed by using a coaxial shunt, a current transformer or Rogowski-coil. However, these current sensors implementation introduce significant undesired parasitic inductance in the power loop of the device under test [3-4]. This article presents an integrated current sensor inside a monolithic GaN HEMT device allowing eliminating any intrusive current sensor. Thanks to its dynamic response, this current sensor can be employed with an integrated gate driver as a high-speed protection against short circuits responding within several tens nanoseconds. . The authors present a high temperature DPT (Double-Pulse-Tester) in which the sensitivity and the responses of the integrated current sensor in different temperatures is characterized. Thanks to an infrared beam, the device under test is heated up locally while all the auxiliary circuits are at room temperature.

Current sensing by integrating a current mirror

The current-sense technique principle is based on current mirroring in integrated MOSFET at electronic circuits level [5]. A 100V-10A GaN HEMTs has been designed on a single die prototype as Fig. 1. In this prototype, the drains of the power and sensing devices are separated allowing to verify each device independently. The work ongoing is dedicate to optimize the die's dimensions and implement additional functions by using the current sensor such as short-circuit protection in a new monolithic design.

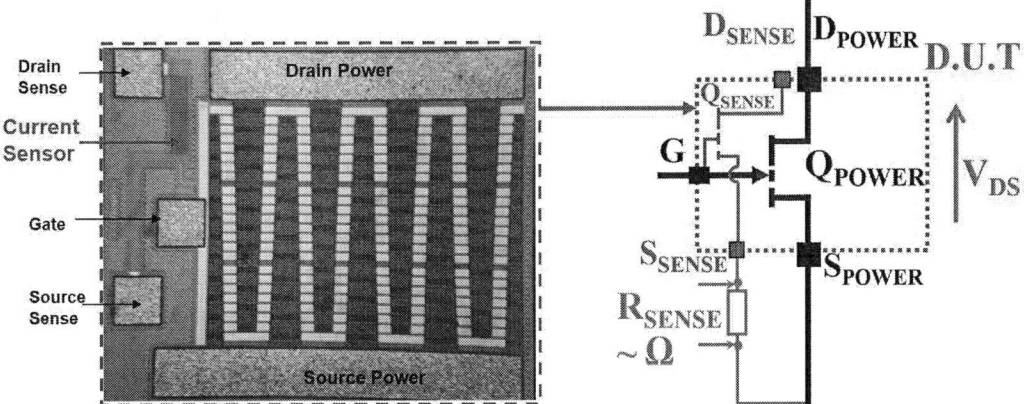

Fig. 1: GaN die (4.6mm x 4.3mm) with integrated current sensor and its schematic. Sense HEMT : 50x smaller Vs power HEMT

The gates of the power HEMT and the sense HEMTs share the same control signal. The current in the sense HEMT copies the current in the power HEMT with the ratio of the ON-state resistances and the value of the sense resistor as in equation (1).

$$\frac{I_{SENSE}}{I_{POWER}} = \frac{R_{DS\_ON\_POWER}}{R_{DS\_ON\_SENSE} + R_{SENSE}} \tag{1}$$

The ON-state resistance of the power HEMT equals several tens of mΩ, the ON-state resistance of the sense HEMT is 50 times higher and the value of R$_{SENSE}$ equals several Ω in order to obtain a high sensitivity of the sensor. In addition, compared to the shunt-based approach [6], for which high temperature calibration can be tedious; the power/sense HEMTs share a similar thermal behavior and the selected R$_{SENSE}$ has a very low tolerance (0.1%) corresponding to a temperature coefficient of resistance of 10 ppm/°C for the high temperature operation.

A DFN (Dual-Flat No-leads) package is used for the implementation of the die; on which can be found three electrodes connected to the power drain and two electrodes to the power source. One electrode is connected to the drain sense and another to the source sense. As mention, a single control electrode is common to the two power/sense transistors. This packaged die is implemented in a DPT circuit Fig. 2.

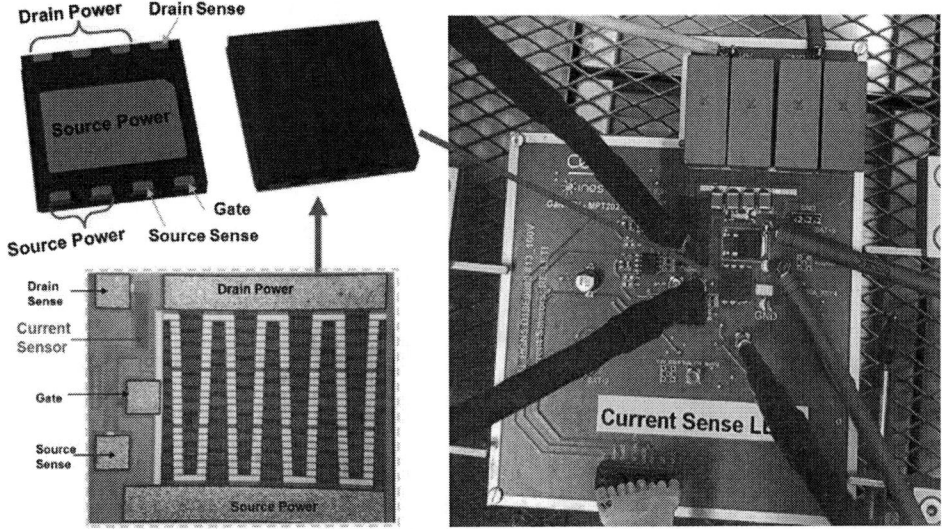

Fig. 2: GaN device with integrated current sensor in the characterization bench

The selected R_{SENSE} is very closely to the source sense pad and source power pad in order to avoid the parasitic inductance. Base on the current ratio in these two devices, the measurement of the voltage drop across R_{SENSE} is calibrated to demonstrate the current on the power device.

The experimental results

As in a recent work on the high temperature dynamic characterization of GaN HEMT [7], the left side of Fig. 3 shows a high temperature test bench in which a specific profile of the infrared beam is applied on the device under test, heated-up to 175°C while avoiding overheating of the auxiliary components. The main difference in this work compared to [7], is that the Fraunhofer-IZM [8-9] sensor allowing to measure the drain-source pulsed current of the DUT is not implemented. The drain source current in this setup is measured by using the current mirror connected to R_{SENSE} (Fig. 3).

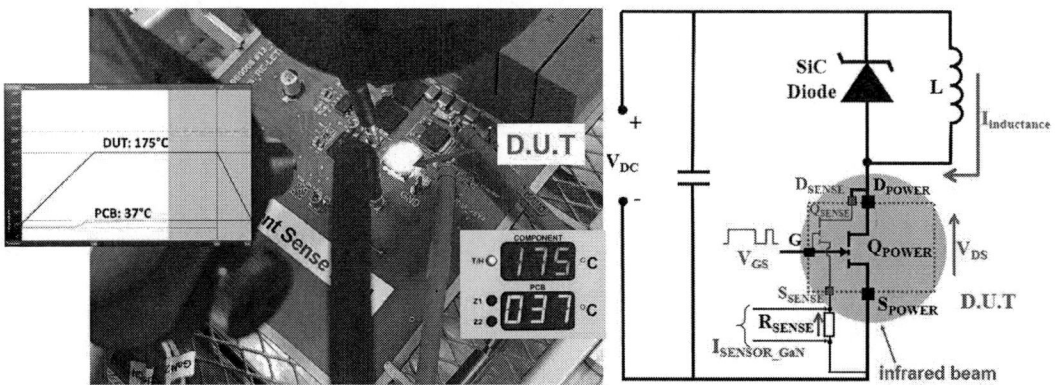

Fig. 3: High temperature set-up with a local heating-up on device under test by using an infrared beam

Fig. 4 shows the experimental results of the sensitivity of the integrated current mirror for three different temperatures (25°C, 125°C and 175°C). The integrated sensor responses with a sensitivity between 60mV/A and 45mV/A for a switched current between 3 A and 8 A. Due to the different temperature coefficients of the GaN HEMT device and R_{SENSE} with a very low temperature variation (10 ppm/°C) the sensitivities depend widely on the temperatures. As the observations under the infrared beam station, the device is auto-heat up when the characterized currents are applied; the higher the applied current is the higher of the ON-state resistance of DUT appears. In the next generations of this integrated current sensor, R_{SENSE} will be integrated inside the device to give a homogeneous behavior to the temperatures.

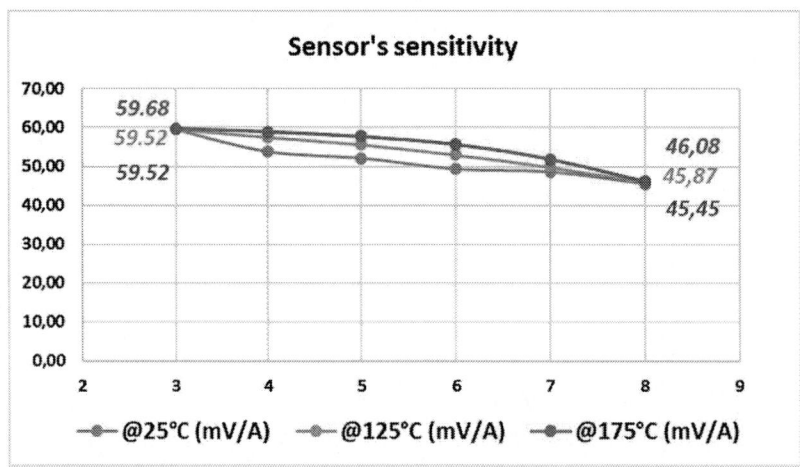

Fig. 4: High temperature characterizations of the sensor's sensitivity

Fig. 5 shows the main waveforms of an experiment in a typical DPT when DUT is heated up to 175°C. As mentioned in Fig. 3, the current on the inductance is on Chanel 1, the drain-source voltage V_{DS} is on Chanel 2, the control signal on the gate V_{GS} is on Chanel 3. With a pre-defined sensitivity, the drain-source current of GaN power device is measured by the integrated sensor on Chanel 4, this current is extrapolated from the measured voltage on R_{SENSE}. In this experiment, the current in the inductance were configured at the value of 8A, the sensor's sensitivity is took at 175°C/8A.

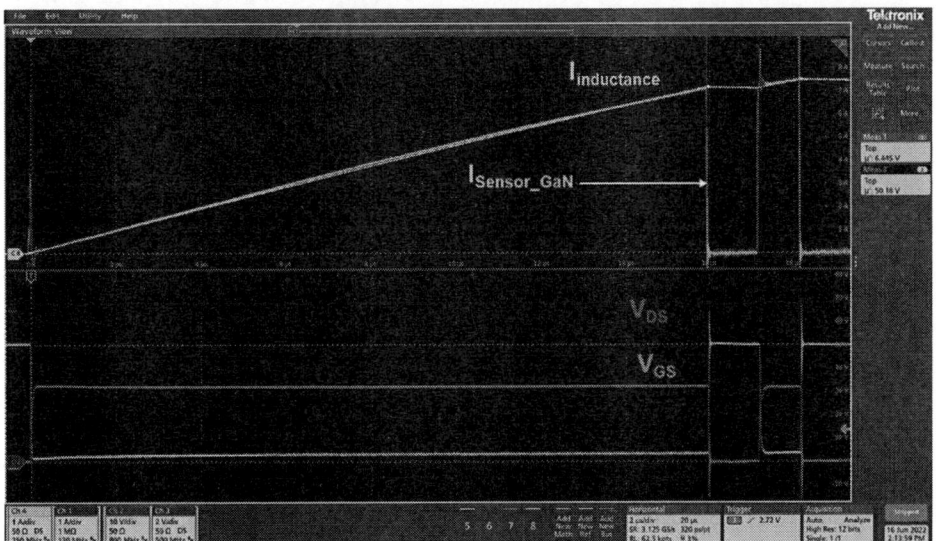

Fig. 5: Experimental waveforms of the DPT at 175°C

Fig. 6 presents the DPT currents in power GaN device by its integrated sensor at six different temperatures from 25°C to 175°C, the current in the inductor (see Fig. 3) is measured separately by a commercial current probe (TCP0030A).

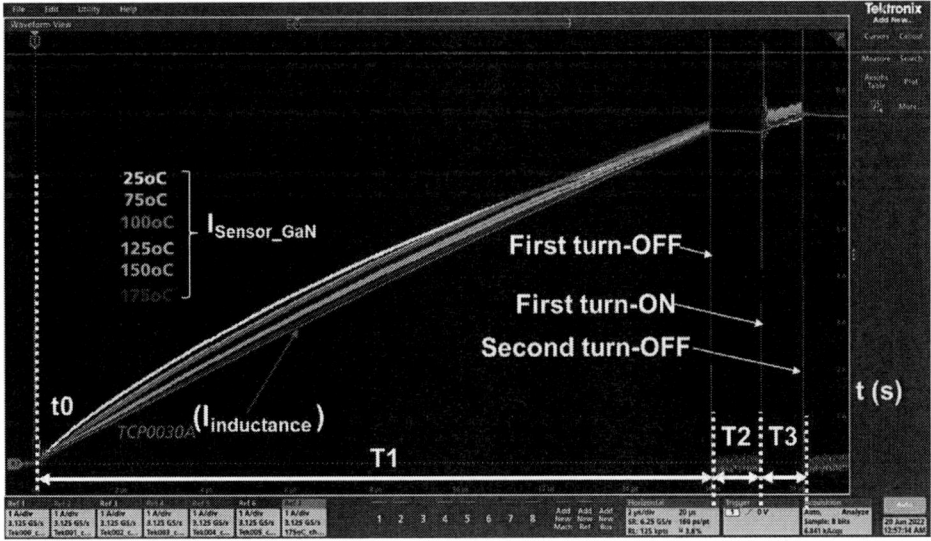

Fig. 6: Superimposed DPT currents by integrated sensor I_{SENSOR_GaN} and inductor's current from 25°C to 175°C

In this typical DPT, there are three interval T1, T2 and T3 in order to turn-ON and turn-OFF the DUT at a pre-configured current value. To reach the configured current for the first turn-OFF during T1, the current in inductance is charged by a constant input DC voltage (V_{DC} in Fig. 3); these first parts of all the measured currents are zoomed in Fig. 7. As we can see, the measure at 175°C is closest to the current by TCP0030A.

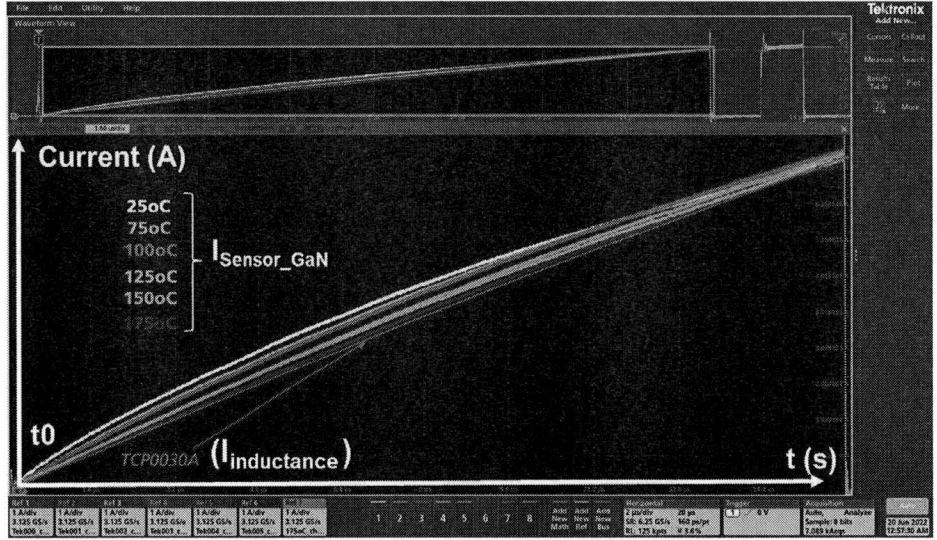

Fig. 7. Zoom in the first parts of measured currents before the first turn-OFF of DPT at different temperatures

The sensor measures this increasing portion of the current as a function of time. Thanks to these increasing curves between 0 A and 8A and at six temperatures, all the useful points characterize the response of the integrated current sensor in GaN device. Then the real transfer function (Isense/Ipower) is presented theoretically in the previous section.

By replacing the R_{DS-ON} value of the current mirror by its proportion with respect to the power part, then the current transfer function is written as:

$$Gain_1 = \frac{I_{SENSE}}{I_{POWER}} \simeq \frac{R_{DS-ON_{POWER}}}{50 . R_{DS-ON_{POWER}} + R_{SENSE}} \simeq 0.005$$

where $R_{DS-ONpower}$ in the order of 80 mΩ at 25°C and R_{SENSE} = 10 Ω

This $Gain_1$ represents the first stage of the integrated sensor in GaN power device without the unit because it is a ratio of currents. The second stage named $Gain_2$ is represented by the R_{sense} that makes a current-voltage conversion (by shunt measurement) which produces a total gain of the sensor as follow

$$G_{SENSOR} = Gain_1 \times Gain_2 = 0.005 \times 10 \, V/A \simeq 50 \, mV/A$$

This equation represents the total Gain of the complete sensor transfer function in Fig. 8.

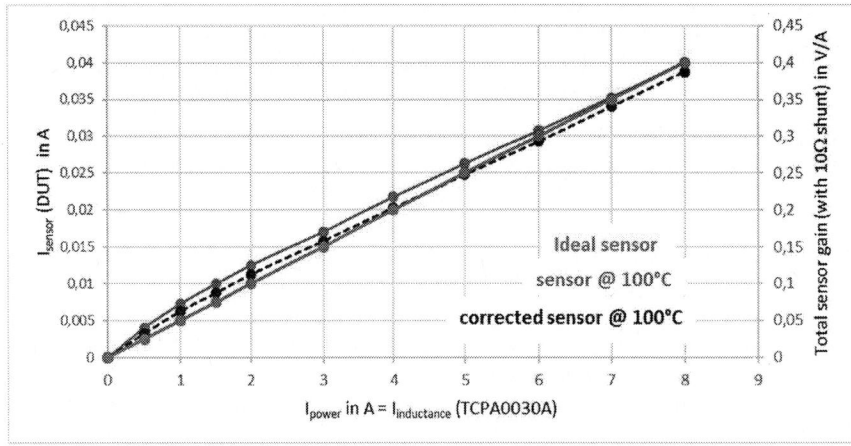

Fig. 8. Linearity of the sensor with $Gain_1$ (left scale) and G_{SENSOR} (right scale) coefficients.

In this figure, a current probe (TCP0030A) with another corrected gain gives the dash line. For these measurements, a smallest error deviation at each point by adjusting precisely the gain of the oscilloscope probe (dash line) can be obtained.

In this way, the final measurement gain could be slightly corrected for the needs of the experiment by a third gain, so called $Gain_3$ with a value close to 1, the final gain of the sensor (dash line in Fig. 8) can be written as:

$$G_{SENSOR-FINAL} = Gain_1 \times Gain_2 \times Gain_3 = 0,005 \times 10\ V/A \times 0.99 = 49.5\ mV/A$$

Follow up the experimental results, the linearity of the current sensor is much better as the temperature increases, the tests at 150°C and 175°C giving the best results.

With this approach, the auto-heatup phenomena is slightly different between the sensor part behavior and the power part behavior on this die that explains the changes of the sensitivity of the sensor in function of the level of measured currents at a fixed temperature. However further studies must be undertaken in order to understand the evolution of self-heating as a function of current levels, and in particular if the improvement in linearity of the sensor at 150°C and 175°C comes from the fact that the contribution of self-heating decreases as the temperature levels increase.

In the next version of this integrated sensor, R_{SENSE} should be integrated and the corrected coefficient must be given at different temperature and current values (by adding a small external electronic correction circuit for example) in order to get the most linear response possible at any configuration. Fig. 9 zooms in the first turn-ON and the second turn-OFF of the measured current at 8 A by the integrated sensor as the second pulse of the DPT, after the calibrations, these observed edges on the sense device share similar behaviors of the power device.

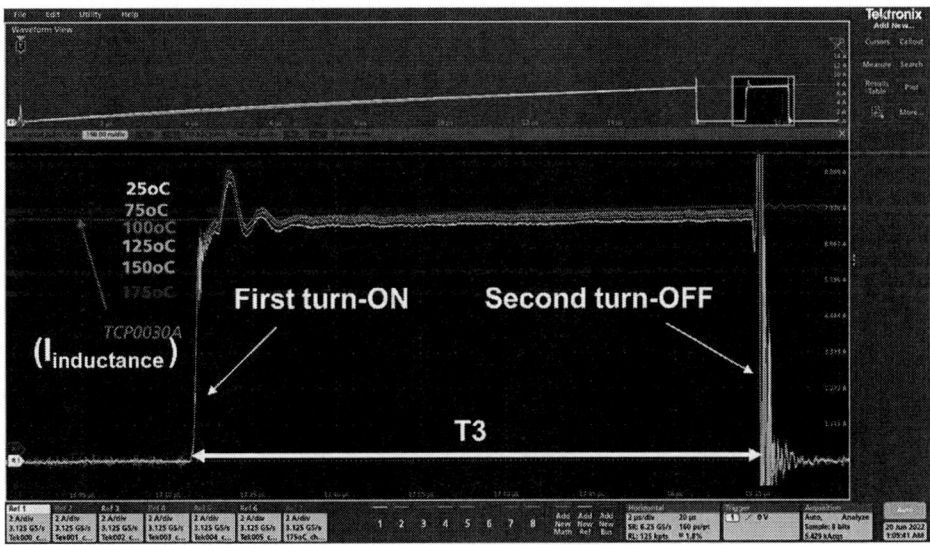

Fig. 9. The second pulse by the integrated sensor during the third interval T3

Fig. 10 and Fig. 11 show the rise times and fall time of the measured current by the integrated sensor at six different temperatures from 25°C to 175°C, the slew-rates of the current are 800 A/μs and 8200 A/μs for the rise time and fall time, respectively.

The rise times t_R of the current from 10% to 90% of the values are 8.045 ns at different temperature. The equivalent bandwidth of the current sensor for the solicitation at a positive step is therefore classically:

$$Bandwidth_{at-3dB} = \frac{0.35}{t_R} = \frac{0.35}{8.045\ ns} = 43.5\ MHz$$

The fall times t_F of the current from 90% to 10% of the values are shorter and the equivalent bandwidth of the sensor for the stress at a negative step varies between 308.7 MHz at 25°C and 274.1 MHz at 175°C.

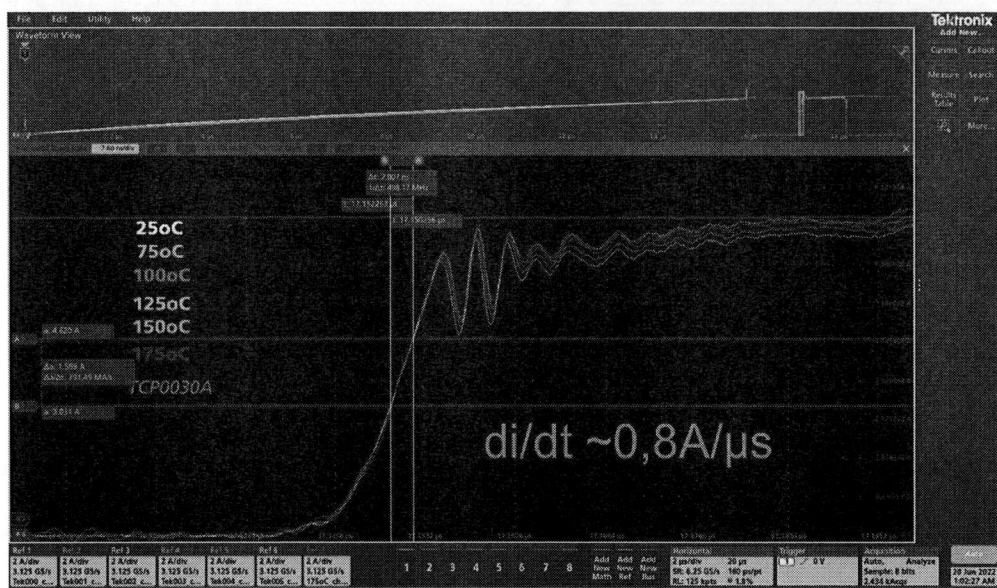

Fig. 10: Rise times of 8ns, the currents at 25°C, 75°C, 100°C, 125°C, 150°C and 175°C

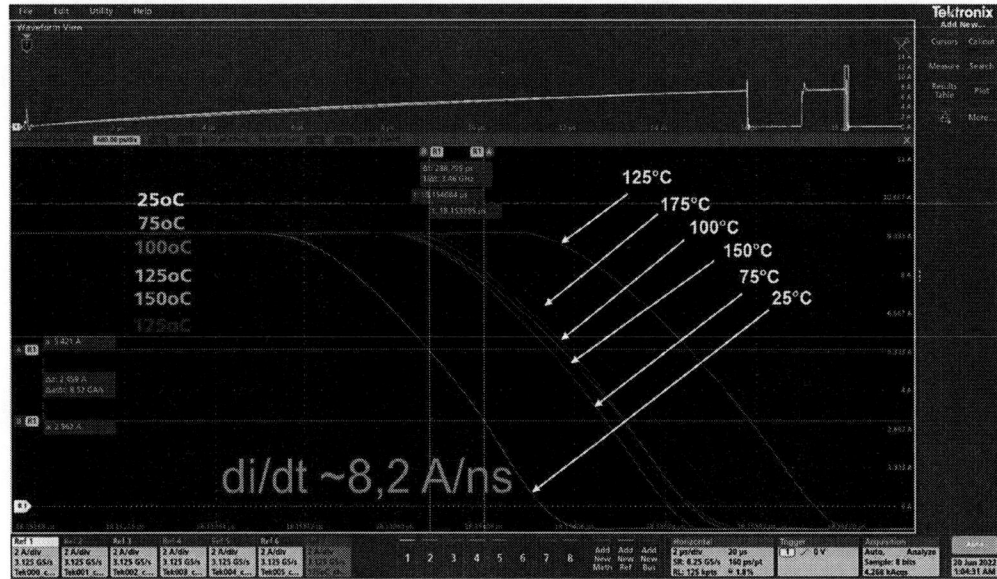

Fig. 11. Fall time of 1.2ns, the current at 25°C, 75°C, 100°C, 125°C, 150°C and 175°C

These fall times increase slightly at the higher temperatures and are represented in Fig. 12.
The current-mirroring technique is implemented in this work, where the measured currents have the similar behavior of the current through GaN power transistor; we could expect the responses at these given values. It is nevertheless necessary to check that the entire measurement chain responds well to the sufficient bandwidth. For this, a shunt and a voltage probe with the bandwidth of 1GHz are used to extract the measured values of the integrated sensor; Fig. 12 shows these measured values on the red line are still in the measureable zone.

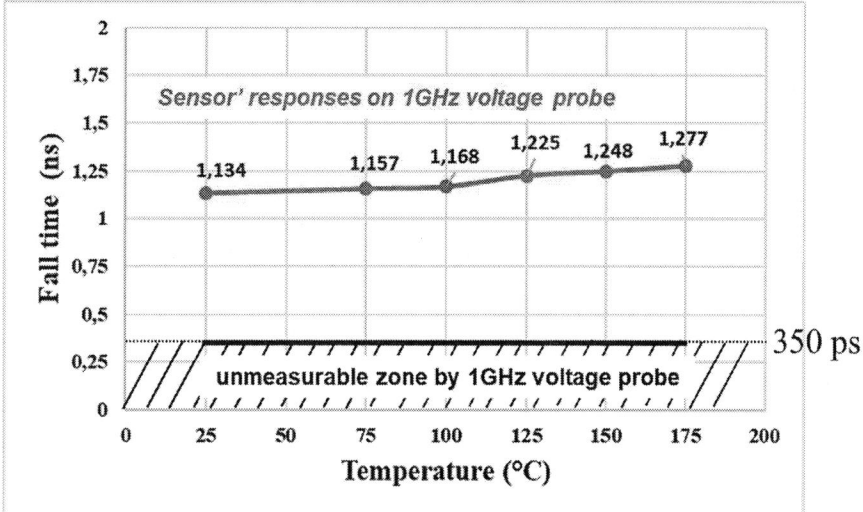

Fig. 12. Fall times at different temperatures from 25°C to 175°C (red line) and the unmeasurable zone

These experimental results show full functions of the integrated current sensor in a monolithic GaN HEMT device. By using this sensor, we can remove the high-speed current sensor that add the parasitic inductance in the power loop. The current-mirroring technique makes the behavior of the measured current similar to the current on the power devices.

In other hand, with a gate driver, this current sensor can use for protect very quickly the GaN device against the short-circuit in a specific application.

However, this integrated device show its limitation on the stability of the sensitivity with the temperature and the characterized value of the current. Authors are going to make a new version of this current sensor with integrated R_{SENSE}, in order to mitigate the dependent of the measures on the temperatures and the currents.

Conclusion

This work presents a full-integrated current sensor inside a GaN power device. The current sensing approach is current mirroring, measuring a copy of the current in a smaller device with similar thermal behavior, making this approach very stable even in high temperature, up to 175°C. The details of the calibration circuit at different temperature and current were demonstrated. And the high temperature characterizations were given. This sensor need to be improved to have a linear response at any value of the temperature. In a dynamic characterization, this sensor could be used to measure the rising and falling edges of the current through the GaN power devices. Moreover, the integrated sensor could be implemented with a gate driver on a monolithic GaN device to protect the device against the short-circuit event within several nanosecond.

References

[1] Yole report on Power SiC 2020: Materials, Devices and Applications, Yole 2020

[2] Yole report GaN power 2021: epitaxy, devices, applications and technology trends, Yole 2021

[3] J. Wang, M. H. Hedayati, D. Liu, S-E. Adami, H. C. P. Dymond, J. O. Dalton, B. H. Stark "Infinity Sensor: Temperature Sensing in GaN Power Devices using Peak di/dt" 2018 IEEE Energy Conversion Congress and Exposition (ECCE), 2018

[4] Z. Zhang, B. Guo, F. Wang, E. A. Jones, L. M. Tolbert, B. J. Blalock "Methodology for Wide Band-Gap Device Dynamic Characterization," IEEE Trans. Power Electron., vol. 32, 2017

[5] M. Biglarbegian, B. Parkhideh "Characterization of SenseGaN current-mirroring for power GaN with the virtual grounding in a boost converter" 2017 IEEE Energy Conversion Congress and Exposition (ECCE), 2017

[6] S. Moench, R. Reiner, P. Waltereit, R. Quay, O. Ambacher, I. Kallfass "Integrated Current Sensing in GaN Power ICs" 2019 31st International Symposium on Power Semiconductor Devices and ICs (ISPSD), 2019

[7] V. S. Nguyen, A. Bier, R. Escoffier, S. Catellani, J. Martin, C. Gillot "A high precision dynamic characterization bench with a current collapse measurement circuit for GaN HEMT operating at 175°C" PCIM Europe digital days 2021, May 2021

[8] K. Klein, D. E. Hoene, and D. K.-D. Lang, "Comprehensive AC Performance Analysis of Ceramic Capacitors for DC link usage," p. 7, 2017.

[9] K. Klein, D. E. Hoene, and D. K.-D. Lang, "Power module design for utilizing of WBG switching performance," p. 8, 2019

GaN-Based Modular Multilevel Converter for Low-Voltage Grid Enables High Efficiency

Philip Kiehnle, Patrick Himmelmann, Marc Hiller
Karlsruhe Institute of Technology
Kaiserstrasse 12
Karlsruhe, Germany
Phone: +49 (0) 721-608 42696
Email: philip.kiehnle@kit.edu
URL: https://www.eti.kit.edu

Acknowledgments

This work was supported by the German Federal Ministry for Economic Affairs and Climate Action (BMWK) as part of the flexQgrid project [grant number 03EI4002F].

Keywords

≪Modular Multilevel Converters (MMC)≫, ≪Gallium Nitride (GaN)≫, ≪Efficiency≫, ≪Grid-connected converter≫

Abstract

Gallium Nitride (GaN) semiconductors with low inductance packages enable low switching losses and high efficiency. In this paper we present a compact arm PCB design with low loop inductance, allowing for fast and efficient switching. The PCB includes four full-bridge cells for a 7 kW Modular Multilevel Converter (MMC) for low-voltage grid applications.

Introduction

Modular Multilevel Converters (MMC) are frequently used in high-voltage DC transmission systems and other applications like medium-voltage motor drives [1]. Even in applications for the low-voltage grid, the advantageous partial load efficiency of unipolar Si-MOSFETs compared to bipolar Si-IGBTs and a reduced filtering effort brings the MMC topology into consideration. But the benefit of reduced conduction losses, is often overcompensated by an increased communication, gate driving and control effort. By using Gallium Nitride (GaN) enhancement-mode high electron mobility transistors (E-HEMTs) instead of Si-MOSFETs, the driving effort can be reduced, due to lower input capacitance and a driving voltage level of only 5 V. GaN also features a lower output capacitance and zero reverse-recovery loss [2].

In order to analyze and validate the expected performance benefits of using GaN-E-HEMTs as power semiconductors in an MMC, a prototype is built. The MMC is being used in a research project to replace conventional photovoltaic (PV) and battery inverters, which are typically based on two- or three-level topologies. The MMC approach offers the possibility to connect batteries to half of the four full-bridge cells of each MMC arm, as it is shown in Fig. 1. Besides the battery connection to the MMC cells, a single high voltage (HV) battery, e.g. from electric cars can be directly connected to the main DC-bus of the MMC with no additional booster stage, as the MMC offers an intrinsic voltage boost capability. This feature can also be used to save the PV booster stage, as present in typical residential PV inverters, when a PV string instead of a HV-battery is connected to the DC side.

Due to its compact size, this MMC can also be used for educational purposes, e.g. to teach the control strategies of an MMC [3].

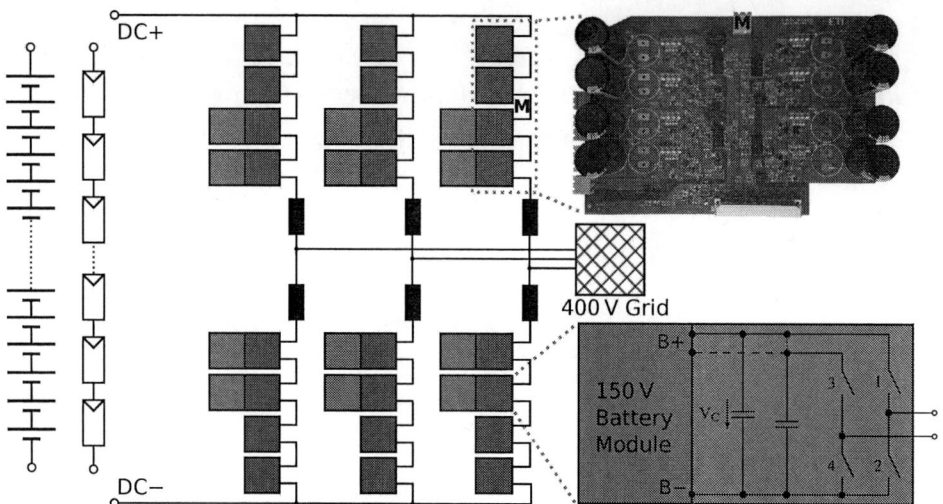

Fig. 1: Hybrid-MMC overview

In this paper, the GaN-based arm PCB with a low internal power consumption is presented. Measurement results of one full-bridge cell and the whole MMC are shown and an overall efficiency estimation is carried out.

Modular Multilevel Converter Arm PCB

The GaN-based arm PCB in Fig. 2 is intended to be used as one arm in an MMC. The four-layer PCB contains four full-bridge cells connected in series on the PCB and has a copper thickness of $70\,\mu m$. In the following, the signal section, the power section and the switching behavior of the developed four-cell arm PCB are described in detail.

Signal Section

The white connector on the bottom of Fig. 2 is used to plug the PCB into the mainboard of the ETI-SoC-System, an in-house-developed signal processing system based on the Zynq 7030 system-on-chip (SoC) from Xilinx [4]. The 16 signal pins of the white connector are directly connected to the Kintex 7 field programmable gate array (FPGA) of the SoC. The white connector is also used to provide power for the voltage and current measurement as well as the gate drivers, thus no dedicated cell supply is necessary. The overall power consumption of the PCB with four cells switching at $100\,kHz$ and active measurements is only $1.0\,W$ from the $5\,V$ mainboard rail.

The FPGA of the Zynq generates the PWM signals with a time resolution of $2\,ns$ and processes the data of the current and voltage measurements. For the cell voltage measurements, four delta-sigma ($\Delta\Sigma$) analog digital converters (ADC) are placed directly on the PCB. The ADC clock of $10\,MHz$ is generated on the PCB, so each measurement channel needs only one FPGA pin. The delta-sigma bitstream can be easily evaluated in the FPGA, since the maximum analog level produces a stream of ones and zeros that are high only $90\,\%$ of the time, so there occur still enough edges for clock recovery. For the arm current measurement, a $\pm50\,A$ AMR current sensor with a bandwidth of $1.5\,MHz$ is connected to a $16\,bit\,/\,5\,MSps$ SAR ADC. Both of these are located inside the blue box (⬤) in Fig. 2.

To provide galvanic isolation, fiber optical transceivers are often used in MMCs. But in favour of a much lower power consumption, isolators based on a SiO_2 isolation barrier are used. They feature $5\,kV_{rms}$ isolation with a common mode transient immunity (CMTI) of $\pm100\,V/ns$. They also shift the FPGA voltage level of $1.8\,V$ to isolated $5\,V$ signals and provide $5\,V$ supply for the cells using an integrated DC-DC converter.

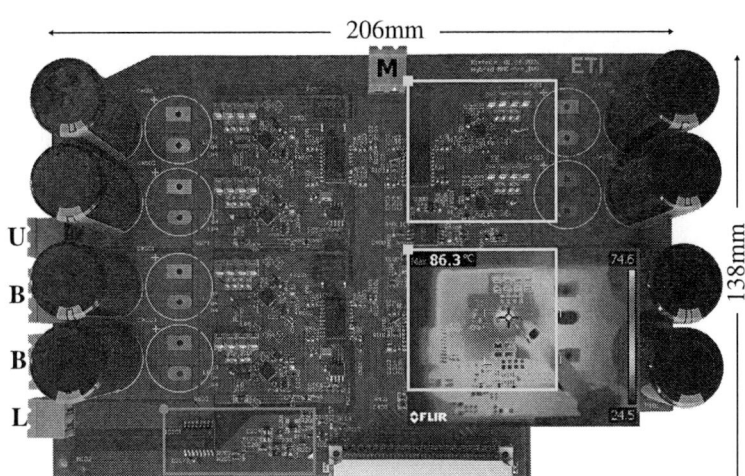

MMC Arm PCB:

- 39 mm height

- current measurement ●

- 4 full-bridge cells:

  · EPC2215 eGaN FETs:
    200 V, 8 mΩ, 32 A

  · with NCP51820 driver ▲

  · with LMG1210 driver ▨

  · V_{cell} measurement

  · 160 V_{nom}, 10 A_{nom}

  · 2.24 mF + 6 µF ceramic

Fig. 2: MMC arm PCB with thermal image at 160 V, 10 A; $C_{cell,installed} = 1.12$ mF

Power Section

Besides the mandatory upper (U) and lower (L) power terminals for an MMC arm, the PCB offers a middle (M) connection, which enables to split the necessary arm inductance into multiple parts. Therefore, the inductors can be smaller and the trace length to the inductors is reduced. This minimizes the polygon areas with switched potential in order to minimize EMI. Additional terminals (B) of the cells on the left in Fig. 2 offer the possibility to connect batteries to half of the cells. Both half-bridges in those cells have their positive terminals separated and the battery terminals therefore have three electrical contacts. This possibility allows the evaluation of circuits, which eliminate AC-phase pulsating battery currents, which would occur with directly connected batteries [5].

All cells use EPC2215 GaN E-HEMTs, but two different gate drivers are used to minimize switching losses on the one hand, while maintaining long ON-state periods on the other hand. Therefore, two cells optimized for minimal switching losses use the LMG1210 half-bridge gate driver, where a very short deadtime between 0.5 ns and 20 ns can be configured. The other two cells with the battery connectors use the NCP51820 half-bridge gate driver, which offers a minimal deadtime of 25 ns. To save power for an extra supply, both use bootstrapping to supply the high side gate, which needs to be charged to 5 V for a low $R_{DS(on)}$. The NCP51820 driver has an integrated LDO-regulator on the high side. Thus, a higher voltage of 9 V is bootstrapped, which allows the battery backed modules to be turned on for a complete half-wave of the 50 Hz grid. During this period, the voltage of the bootstrap capacitor decreases without affecting the gate voltage. This feature requires an additional DC-DC-converter, which provides isolated 9 V from the 5 V mainboard supply.

Switching Behaviour

In the PCB design, it is mandatory to follow the layout rules for fast switching GaN devices [6]. In order to check the design for an acceptable switching behaviour, measurements were made with the NCP51820 driver using its minimal deadtime of 25 ns. A voltage V_c of 160 V was supplied to one half-bridge (see Fig. 3). V_{DS} of the low side E-HEMT was measured with an RL load first connected to the positive DC terminal (DC+) and second to the negative DC terminal (DC-) in order to reverse the current direction through the low side E-HEMT.

With a duty cycle of 50 % and a load current of 10 A, the switching waveform in Fig. 3 was measured with an 1 GHz Oscilloscope and a 500 MHz probe directly soldered onto the PCB. While the load was connected to the negative DC terminal, the rise time (10% to 90%) was much higher compared to the common double pulse test approach [7], where the inductor is connected to DC+. But even with this fast rise time of 3.47 ns, only minimal voltage overshoot of 5.9 V occurs due to the small commutation loop.

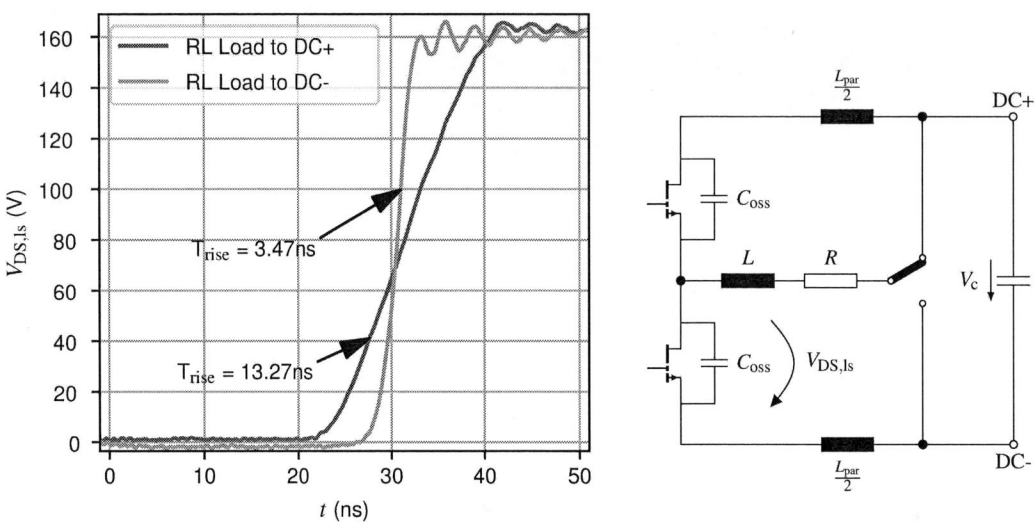

Fig. 3: Turn-off behaviour of low side transistor in GaN half-bridge with $R_{g,on} = 2.7\,\Omega$, $R_{g,off} = 1.0\,\Omega$, $L = 65\,\mu H$, $R = 8\,\Omega$ and $V_c = 160\,V$

In order to measure the parasitic inductance L_{par} of the commutation loop, conventional methods, which require the current through the transistor, cannot be used, as the loop is too small for a current sensor. Therefore the parasitic inductance is estimated by the ringing frequency f_{res} and the equivalent circuit model shown in Fig. 3. In the waveform in Fig. 3, a ringing frequency of 332.7 MHz can be extracted. With a drain-source capacitance C_{oss} of 390 nF according to the data sheet [8], equation (1) can be used to calculate the parasitic inductance of 0.65 nH.

$$L_{par,estim} = \frac{1}{C_{oss}(2\pi f_{res})^2} \tag{1}$$

In order to check these assumptions, an LTspice simulation with the transistor model from EPC was used. The results in Table I show a strong correlation between L_{par} and the estimated inductance. Layout improvements can be easily evaluated with this method.

Table I: Commutation loop parasitic inductance estimation.

	LTspice	LTspice	LTspice	PCB
V_c	100 V	160 V	160 V	160 V
C_{oss}	390 pF	350 pF	350 pF	350 pF
f_{res}	230 MHz	244 MHz	347 MHz	332.7 MHz
L_{par}	1.2 nH	1.2 nH	0.6 nH	unknown
$L_{par,estim}$	1.23 nH	1.22 nH	0.60 nH	0.65 nH

The low inductance of the commutation loop allows the cells to work at even higher DC-link voltages. A cell voltage of $V_c = 180\,V$ was successfully tested, which could be useful in case of grid faults, where the converter has to stay connected to the grid up to voltage levels of $1.25\,V_{nom}$, according to the latest grid codes in Germany [9].

Fig. 4: MMC prototype in a 19-inch rack

MMC Prototype Measurements

In Fig. 4 the assembled MMC is shown. Six Arm PCBs are plugged into the ETI-SoC-Carrier, which is processing all the signals [4]. The arm inductors are visible on top of the PCBs and are connected to the (M) connection point, which was mentioned before. A second ETI-SoC-Carrier board is used for two peripheral cards, which control the AC- and DC-contactors in the system and measure the grid voltage. But those features will be integrated in an enlarged carrier board in the future.

For the integration of batteries into the MMC, it is mandatory to operate the modules at different voltage levels. In the extreme case, which is not recommended for long battery life, the 40-cell, 1.1 kWh battery module operates between 100 V and 168 V.

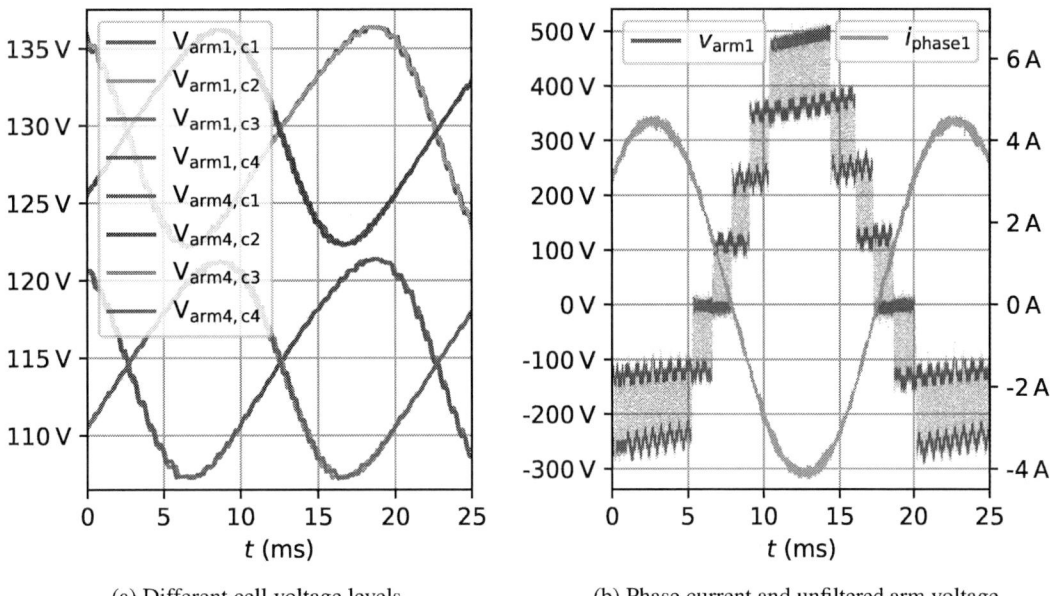

(a) Different cell voltage levels (b) Phase current and unfiltered arm voltage

Fig. 5: MMC measurement data of one phase of the prototype

In Fig. 5(a) the cell capacitor voltages of the upper (arm1) and lower (arm4) of the first phase of the MMC are shown. It can be seen that the control implementation allows different cell voltage levels within the cells of one arm. The cells 1 and 2 are operated at a nominal voltage of 130 V, while the future battery backed cells 3 and 4 are set to 115 V. This level can then be matched to the battery voltage level and the mains frequency related energy ripple of the battery backed cells will mostly disappear. The measurements of Fig. 5(a) were recorded with the TCP/IP interface of the ETI-SoC-System. It is running the MMC energy controller on an ARM processor with an update rate of 10 kHz and allows access to all system variables with the same rate. The current controllers for the internal-, DC-, and grid-currents are implemented in the FPGA of the Zynq SoC, running at a control frequency of 100 kHz.

In Fig. 5(b) the unfiltered output voltage v_{arm1} of the upper MMC arm of phase 1 is shown. It can be seen, that the full-bridge cells allow for positive and negative voltages. In this case an MMC DC voltage of only 200 V was used to drive a resistive load at an RMS voltage level of 400 V and an output power of 2 kW. Those measurement were recorded with an external oscilloscope.

MMC Efficiency

In the upcoming renewable energy based grid, efficient power converters and battery storages are mandatory. The PCB design has been trimmed to an efficient operation. But the arm modules with a total power consumption of 6 W are only one part of an efficient converter. Therefore the signal processing system itself has been taken into account with another 5 W. But the main loss is still caused by the switching and conduction losses of the semiconductors. Those losses have been calculated with the official LTspice model of the EPC2215 GaN transistor at a cell voltage of 160 V.

The temperature dependent increase in $R_{\text{DS(on)}}$ has been taken into account and the temperature was calculated based on thermal measurements (see Fig. 2). The currents have been calculated with a simple MMC model and the formula for the arm current is shown in equation (2).

$$I_{\text{arm,rms}} = \frac{I_{\text{dc}}}{3} + \frac{I_{\text{ac,rms}}}{2} = \frac{P_{\text{in}}}{V_{\text{dc}} \cdot 3} + \frac{I_{\text{ac,rms}}}{2} \tag{2}$$

A MMC DC-bus voltage V_{dc} of 460 V and an effective switching frequency of 200 kHz per arm, which equals to 25 kHz per half-bridge, has been used for the calculation. The arm inductor of Fig. 1 was modelled with a resistance of 13 mΩ. The losses in capacitors and PCB traces have been neglected for now.

In Fig. 6 the converter DC to AC efficiency at different input power levels is shown. Especially at lower power levels, it can be seen that the power consumption P_{control}, which includes the gate signal generation, measurements and the processing system has a major impact on the total efficiency of the converter, as they make up for one third of the total losses. For an MMC in the low power range, this factor would get worse if optical isolation or even dedicated cell microcontrollers would have been used.

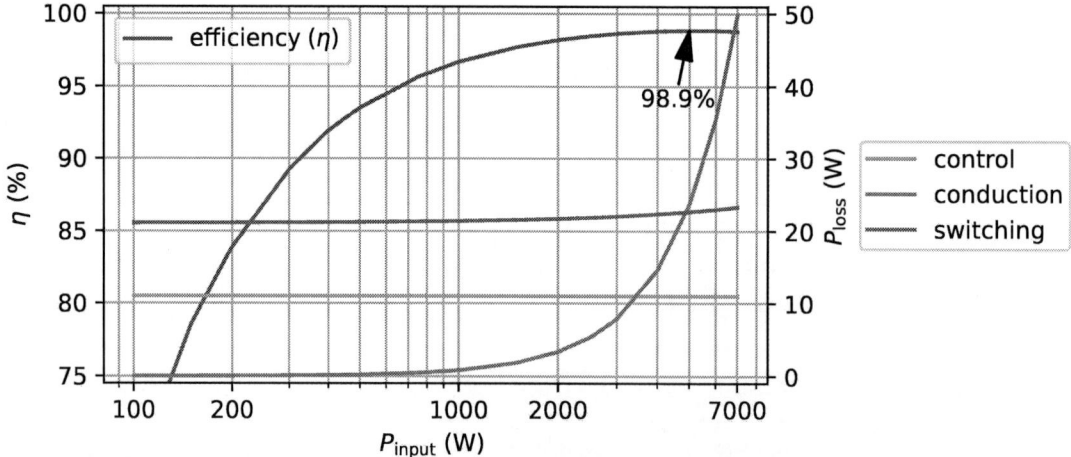

Fig. 6: MMC efficiency and power losses

Conclusion

A GaN-based arm PCB was presented, which enables an MMC with up to 98.9 % theoretical efficiency, although the complex control system for MMCs is taken into account. The switching behaviour of an half-bridge was measured and the parasitic inductance of the commutation loop was calculated to verify a good layout. Further, the assembled MMC prototype with first measurements has been shown. The MMC control implementation supports different cell voltage levels, which allows to connect batteries to half of the full-bridge cells. Finally, the total loss distribution of the MMC was estimated to show the impact of the low internal power consumption of the presented GaN-based arm PCB.

Outlook

There is still room for optimizations, e.g. some of the cells can be configured as half-bridge cells because no negative DC-bus voltage will be necessary for most applications. Thereby the switching and conduction losses can be reduced further. In future research, half of the cells will be connected to batteries.

Practical efficiency will be measured in future work, when methods to minimize cell voltages by capacitor voltage ripple shaping [10] and advanced circulating current controllers, which allow for decrease in capacitor losses [11] have been implemented.

References

[1] P. Himmelmann, M. Hiller, D. Krug, and M. Beuermann, "A new modular multilevel converter for medium voltage high power oil gas motor drive applications," in *2016 18th European Conference on Power Electronics and Applications (EPE'16 ECCE Europe)*, 2016, pp. 1–11.

[2] R. Hou, J. Lu, and D. Chen, "Parasitic capacitance eqoss loss mechanism, calculation, and measurement in hard-switching for gan hemts," in *2018 IEEE Applied Power Electronics Conference and Exposition (APEC)*, 2018, pp. 919–924.

[3] D. Braeckle, P. Himmelmann, L. Gröll, V. Hagenmeyer, and M. Hiller, "Energy pulsation reduction in modular multilevel converters using optimized current trajectories," *IEEE Open Journal of Power Electronics*, vol. 2, pp. 171–186, 2021, 37.12.01; LK 01.

[4] B. Schmitz-Rode, L. Stefanski, R. Schwendemann, S. Decker, S. Mersche, P. Kiehnle, P. Himmelmann, A. Liske, and M. Hiller, "A modular signal processing platform for grid and motor control, hil and phil applications," in *2022 International Power Electronics Conference (IPEC-Himeji 2022 - ECCE ASIA)*, 2022.

[5] M. Gommeringer, F. Kammerer, J. Kolb, and M. Braun, "Novel dc-ac converter topology for multilevel battery energy storage systems," in *PCIM Europe : International Exhibition and Conference for Power Electronics, Intelligent Motion, Renewable Energy and Energy Management, Nuremberg, 14 - 16 May 2013 ; proceedings*, VDE Verlag, 2013, pp. 699–706.

[6] D. Reusch and J. Strydom, "Understanding the effect of pcb layout on circuit performance in a high-frequency gallium-nitride-based point of load converter," *IEEE Transactions on Power Electronics*, vol. 29, no. 4, pp. 2008–2015, 2014.

[7] J.-Z. Fu, G. Kapino, and W.-T. Franke, "Effect investigations of double pulse test on the wide bandgap power devices," in *PCIM Europe digital days 2020; International Exhibition and Conference for Power Electronics, Intelligent Motion, Renewable Energy and Energy Management*, 2020, pp. 1–6.

[8] *EPC2215: 200 V, 162 A enhancement-mode gan power transistor*, EPC, 2020.

[9] VDE. "VDE-AR-N 4105 Anwendungsregel:2018-11 Erzeugungsanlagen am Niederspannungsnetz." (Nov. 2018), [Online]. Available: https://www.vde-verlag.de/normen/0100492/vde-ar-n-4105-anwendungsregel-2018-11.html.

[10] K. Ilves, A. Antonopoulos, L. Harnefors, S. Norrga, L. Ängquist, and H.-P. Nee, "Capacitor voltage ripple shaping in modular multilevel converters allowing for operating region extension," in *IECON 2011 - 37th Annual Conference of the IEEE Industrial Electronics Society*, 2011, pp. 4403–4408.

[11] J. Kolb, F. Kammerer, P. Grabherr, M. Gommeringer, and M. Braun, "Boosting the efficiency of low voltage modular multilevel converters beyond 99%," in *PCIM Europe, Nuremberg, Germany*, 2013.

Energy Management of Smart Homes with Electric Vehicles Using Deep Reinforcement Learning

Xavier Weiss, Qianwen Xu, Lars Nordström
KTH
Teknikringen 33, 10044
Stockholm, Sweden
Telephone: +46(8)790-6830
Email: xavierw@kth.se
URL: https://www.kth.se/

Keywords

≪Energy Management System (EMS)≫, ≪Microgrid≫, ≪Electric Vehicle≫, ≪Energy storage≫, ≪Deep learning≫, ≪Safety≫

Abstract

The proliferation of electric vehicles (EVs) has resulted in new charging infrastructure at all levels, including domestically. These new domestic EVs can potentially provide vehicle to home (V2H) services where EVs are used as energy storage systems (ESSs) for the home when they are not in use. Energy management systems (EMSs) can control these EVs to minimize the electricity cost to the owner but must satisfy constraints. Uncertainty in EV availability and the microgrid environment is also a challenge and can be addressed through real-time operation. Hence this paper formulates the EV charge/discharge scheduling problem as a Markov Decision Process (MDP). A safe implementation of Proximal Policy Optimization (PPO) is proposed for real-time optimization and compared to a day-ahead Mixed Integer Linear Programming (MILP) benchmark. The resulting PPO agent is able to minimize RA and SD costs for a typical EV user 3% better than the MILP solution. It obtains a 39% higher electricity cost than MILP, but unlike MILP does not require accurate forecasting data and operates in real-time.

Introduction

The global supply of EVs is growing and is a crucial factor in decarbonizing the transport sector. The large-scale integration of EVs into the grid will greatly increase electricity consumption, and thus pose significant stress on the grid. To alleviate the stress, a smart home integrated with EV and PhotoVoltaics (PV) is a promising solution.

A smart home is composed of local loads, an ESS and generators and thus forms a natural building block for the smart grid [2]. EVs in a smart home can act as both a load and an ESS. Conventionally, an EV acts as a load with a unidirectional flow of power from home to vehicle (H2V). Increased use of bidirectional converters [3, 4] will allow the EV to provide vehicle to home (V2H) and V2G functionality. This has the potential to save costs for the home owner and provide ancillary services to the grid.

An EMS is key for the optimal operation of smart homes. Conventionally, model-based methods are used. Ref. [7] proposes an optimal control strategy for efficient utilization of available EV for storing PV power and providing grid support. In [8] a smart home with EV and PV is optimized using a dynamic programming successive algorithm to minimize the variance in household load. However, these examples do not consider uncertainty.

EMSs considering uncertainties have thus been proposed. Stochastic optimization is used to forecast uncertainty in [9] for power grids with high penetration of renewables and EVs. Similarly, [6] is able to improve the profit of a charging station and reduce departure delay by using a stochastic Lyapunov drift technique. Robust optimization is proposed in [10] for EV charging in an unregulated electricity market and obtains real-time performance. However, these methods require the distribution of uncertainties to be known - which is difficult to obtain.

Easy access to data and advances in DRL have led to a strong interest in data-driven methods. While these cannot guarantee a global optimum, they can be more computationally efficient and versatile than model-based methods, which is beneficial for real-time applications. In [11], a DQN-based EMS is developed to optimize the cost of a residential microgrid. Ref. [5] develops a two-level actor-critic method to size and schedule components in a smart home. A PPO-based EMS is proposed for demand response in [12]. Finally, in [13] long-short term memory networks are used to build a data-driven forecasting model, which is then combined with a deep deterministic policy gradient agent to reduce charging costs for an EV owner with random arrival times by up to 70.2% - relative to an unmanaged scenario. However, these implementations are not safe. For instance, even when a DRL agent is trained and exhibiting good performance it can still violate constraints - especially in unseen scenarios. Moreover, they do not consider the EV owner's requirement to minimize range anxiety.

To address the above issues, this paper proposes a DRL-based EMS for smart homes which guarantees safety and fulfills customer requirements.

Problem Formulation

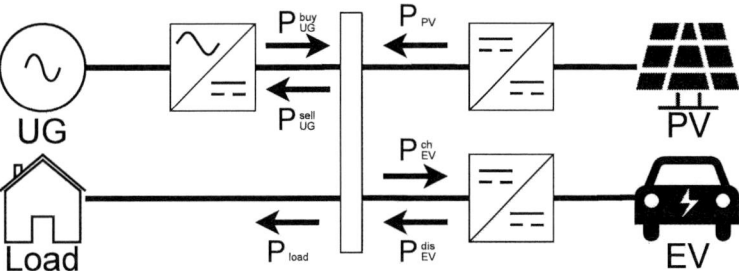

Fig. 1: Layout of residential microgrid

The goal of the EMS considered in this study is to minimize the total operating cost of the smart home shown in Fig.1. The smart home consists of PV, EV, residential load and a Utility Grid (UG) connection. The minimization of the total operating cost of this smart home can therefore be formulated as an optimization problem of the form:

$$\min \sum_{\tau=1}^{T} C_{UG}^{buy}(\tau) \cdot P_{UG}^{buy}(\tau) - C_{UG}^{sell}(\tau) \cdot P_{UG}^{sell}(\tau) + C_{ev}(P_{ev}^{ch}(\tau) + P_{ev}^{dis}(\tau)) \tag{1}$$

subjects to constraints (3)-(13)

where τ is a discrete time slot with a fixed interval (e.g. 15 minutes), C_{UG}^{buy} and C_{UG}^{sell} are the cost of buying and selling power from the Utility Grid (UG), P_{UG}^{buy} and P_{UG}^{sell} are the amounts of power bought and sold to the UG, P_{ev}^{ch} and P_{ev}^{dis} are the amounts of power used to charge/discharge the EV, and C_{ev} is the degradation cost per kWh of using the EV's battery.

The degradation cost of each charge/discharge cycle can be estimated linearly:

$$C_{ev}(P_{ev}(\tau)) = \rho |SoE(\tau) - SoE(\tau - \Delta\tau)| \tag{2}$$

where $SoE = SoC \cdot E_{\max}$ is the State of Energy (SoE) of the EV, $E_{\max}$ is the maximum energy capacity of the battery and P_{ev} is the power transferred from/to the battery. The State of Charge (SoC) is a fraction between 0 and 1 of the EV's maximum energy storage capacity $E_{\max}$.

Power Balance Constraints:

$$P_{load}(\tau) + \frac{1}{\eta_{BIC}} P_{UG}^{sell}(\tau) + P_{ev}^{ch}(\tau) = P_{PV}(\tau) + \eta_{BIC} P_{UG}^{buy}(\tau) + P_{ev}^{dis}(\tau) \,\forall\, \tau \tag{3}$$

where η_{BIC} is the efficiency of the bi-directional converter between the smart home and the utility grid.

Energy Storage Constraints:

$$SoE(\tau) = SoE(\tau-1) + P_{ev}^{ch}(\tau)\eta_{ch} - \frac{P_{ev}^{dis}(\tau)}{\eta_{dis}} \tag{4}$$

$$|SoE(\tau) - SoE(\tau-1)| \leq \Delta SoE_{max} \tag{5}$$

$$E_{\min}(\tau) \leq SoE(\tau) \leq E_{\max}(\tau) \,\forall\, \tau \tag{6}$$

where η_{ch} and η_{dis} are the charging and discharging efficiency of the EV's battery, respectively.

Power capacity constraints:

$$0 \leq P_{ev}^{ch}(\tau) \leq P_{ev}^{ch,max}(\tau) \tag{7}$$

$$0 \leq P_{ev}^{dis}(\tau) \leq P_{ev}^{dis,max}(\tau) \tag{8}$$

$$0 \leq P_{buy}^{UG}(\tau) \leq P_{buy}^{UG,max}(\tau) \tag{9}$$

$$0 \leq P_{sell}^{UG}(\tau) \leq P_{sell}^{UG,max}(\tau) \tag{10}$$

$$0 \leq P_{PV}(\tau) \leq P_{PV}^{\max}(\tau) \tag{11}$$

Bi-directional Flow Constraints:

$$P_{ev}^{ch}(\tau) P_{ev}^{dis}(\tau) = 0 \,\forall\, \tau \tag{12}$$

$$P_{buy}^{UG}(\tau) P_{sell}^{UG}(\tau) = 0 \,\forall\, \tau \tag{13}$$

The optimization problem in Eq.1 with constraints Eqs.3-13 is a MILP problem and can be solved by commercial software like Gurobi. However, uncertainties in PV, EV and the load are not considered which leads to ineffective results. To handle uncertainties a DRL-based EMS is proposed in the next section.

Proposed Method

A safe PPO agent is proposed to manage the EV in the EMS by modelling it as a Markov Decision Process (MDP). First the EMS problem is reformulated as an MDP so it can be solved using DRL methods. A real-time PPO algorithm [14] is then suggested for its simplicity and its robustness in noisy environments. The PPO algorithm is finally equipped with a novel safety layer to guarantee it does not violate any of the constraints in Eq.3-13.

Markov Decision Process

State Space The state of the residential microgrid is given as:

$$s(\tau) = (C_{UG}, P_{load}, P_{PV}, SoE, A_{EV}) \tag{14}$$

where P_{load} and P_{PV} are the energy (in kWh) consumed/generated by the house and the PV installation and A_{EV} is a binary value indicating whether the EV is available.

Action Space The action space is continuous and is generally expressed as a 4-vector:

$$a(\tau) = (P_{buy}^{UG}, P_{sell}^{UG}, P_{ev}^{ch}, P_{ev}^{dis}) \tag{15}$$

In this implementation, $a(\tau)$ is the output of a safety layer. Through the use of a safety layer the agent only directly controls the charge P_{ev}^{ch} and discharge P_{ev}^{dis} of the EV's battery:

$$a'(\tau) = (P_{ev}^{ch}, P_{ev}^{dis}) \tag{16}$$

Reward The reward is simply a reformulation of Eq.1:

$$r(\tau) = C_{UG}^{sell} - C_{UG}^{buy} - C_{ev} - C_{RA} - C_{SD} \tag{17}$$

The agent's objective is to maximize the reward in real-time, hence there is no discount factor to consider. All costs are specified in Table.Ib.

The Range Anxiety (RA) cost reflects the EV owner's anxiety when leaving without enough energy to complete their trip:

$$C_{RA} = \max(K_{RA} \cdot (E_{trip} - SoE_\tau), 0)) \tag{18}$$

where E_{trip} is the energy required to complete the trip, SoE_τ is the SoE at time slot τ and K_{RA} is a constant.

The State Difference (SD) cost rewards the agent for ending each day with the same SoE it started with, or higher. While this is not necessary for a real-time implementation, it encourages the agent to not cause energy shortages when operating across multiple days.

$$C_{SD} = K_{SD}^{\max(T-\tau,1)} \cdot max(SoE_{initial} - SoE_\tau, 0) \tag{19}$$

where K_{SD} is a constant and $SoE_{initial}$ is the SoE at the start of the day. For $\kappa < 1$, c_{SD} increases exponentially as the agent approaches the end of the day unless the condition is met.

DRL Agent

For its performance and simplicity, the standard PPO algorithm is selected. Since the microgrid environment has several sources of noise – including the variation in PV generation due to weather, load and EV usage due to occupant behavior and electricity prices due to market forces – the PPO agent is especially appropriate since it is generally more robust against perturbations to the input s_τ than other DRL algorithms like Deep Deterministic Policy Gradient (DDPG). The full implementation details for PPO are provided in [14, 17]. Parameters for both the actor and the critic components of the PPO agent are given in Table.Ia.

To avoid destroying the learned policy in a single update and maintain computational efficiency, PPO adopts a clipped version of trust regions [15] directly into the surrogate loss function:

$$L^{CLIP}(\theta) = \hat{E}_\tau[min(\delta_\tau(\theta)\hat{A}_\tau, clip(\delta_\tau(\theta), 1 - \varepsilon, 1 + \varepsilon)\hat{A}_l\tau)] \tag{20}$$

where ε puts a trust-region constraint on how much the policy can be updated and $\delta_\tau(\theta) = \frac{\pi_\theta(a(\tau)|s(\tau))}{\pi_{\theta\ old}(a(\tau)|s(\tau))}$ is the ratio of the new policy over the old policy. The advantage $\hat{A}_\tau$ is given by:

$$\hat{A}_\tau = \delta_\tau + (\gamma)\delta_{\tau+1} + \cdots + (\gamma)^{T-\tau+1}\delta_{T-1} \tag{21}$$

where $\delta_\tau = r_\tau + \gamma V(s_{\tau+1}) - V(s_\tau)$, γ is the discount factor and V is the value function. The advantage therefore compares the discounted rewards r_τ to the baseline value estimate from the critic.

The complete loss function for PPO also adds a Mean-Square Error (MSE) and entropy term:

$$L_\tau^{CLIP+MSE+S}(\theta) = \hat{\mathbb{E}}_\tau \left[L_\tau^{CLIP}(\theta) - c_1 L_\tau^{MSE}(\theta) + c_2 S[\pi_\theta](s_\tau) \right] \tag{22}$$

where L_τ^{CLIP} is the clipped surrogate loss function from Eq.20 at time slot τ, $L_\tau^{MSE}(\theta)(= V_\theta(s_\tau) - r_\tau)^2$ is the MSE between the value function $V_\tau(\theta)$ and the reward r_τ and S is the entropy of the agent's output. In this study the constants c_1 and c_2 are set to 0.5 and $c_2 = -0.01$.

The standard deviation of the PPO agent's output layer action $a'(t)$ is also set to gradually decay according to:

$$\sigma(\text{epoch}_{no}T + \tau) = max(\sigma_{min}, \sigma_{initial} - \sigma_{decay\ rate}(\text{epoch}_{no}T + \tau - 1)) \tag{23}$$

Safety Layer

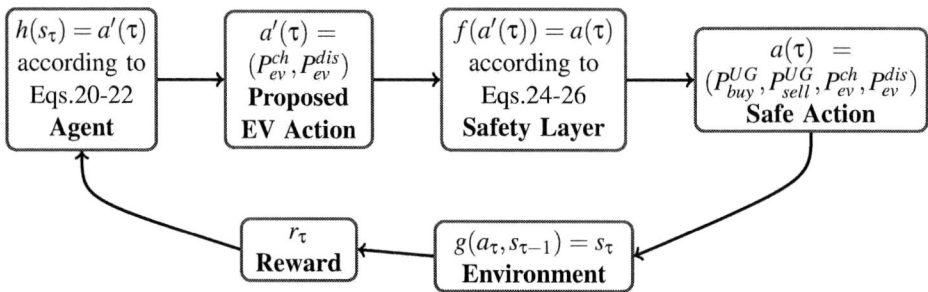

Fig. 2: Architecture of proposed safe PPO agent

In order to guarantee that the PPO agents satisfy the constraints Eqs.3-13 a safety layer is employed. The safety layer takes the proposed action from the agent $a'(t)$ (see Eq.16) and transforms it into a legal action $a(t)$ (see Eq.15) in the environment as shown in Fig.2.

Since the PPO agent's output is passed through a *tanh* activation function, the output is in the $[-1, 1]$ range. This is dynamically re-scaled to match the safe limits of EV charging or discharging:

$$P_{ev}^{ch} = lb' + \frac{(P_{ev}^{ch} - P_e v^{ch,lb})(ub' - lb')}{P_e v^{ch,ub} - P_e v^{ch,lb}}$$
$$P_{ev}^{dis} = lb' + \frac{(P_{ev}^{dis} - P_e v^{dis,lb})(ub' - lb')}{P_e v^{dis,ub} - P_e v^{dis,lb}} \tag{24}$$

where lb' & ub' are the lower and upper bound of the original range ($[-1, 1]$ in this case). For the EV, the SoE, power capacity and availability constraints can be included by taking the minimum of the rate limit, the remaining SoE, and available SoE:

$$P_{ev}^{ch,ub} = min\left(A_{ev}P_{ev}^{ch,max}, E_{max} - SoE\right) \text{ and } P_{ev}^{ch,lb} = -P_{ev}^{ch,ub}$$
$$P_{ev}^{dis,ub} = min\left(A_{ev}P_{ev}^{dis,max}, SoE - E_{min}\right) \text{ and } P_{ev}^{dis,lb} = -P_{ev}^{dis,ub} \tag{25}$$

where the use of a minimum function implies that the safety layer applies a non-linear transformation on the PPO's output.

As suggested by the split in the SoE constraint from Eq.4, the bidirectional-flow constraints from Eq.12-13 are guaranteed in the safety layer by splitting a single input into two. For the range $[-1, 1]$ the split is at 0.

Power balance is maintained at every time step by having the agent only output 1 value corresponding to both P_{ev}^{ch} and P_{ev}^{dis}. The other actions of buying/selling electricity are derived automatically from Eq.3:

$$B = (P_{load} + P_{ev}^{ch} - P_{ev}^{dis} - P_{PV}) \tag{26}$$

$$P_{buy}^{grid} = \frac{1}{\eta_{BIC}}(B) \text{ if } B > 0, 0 \text{ otherwise} \tag{27}$$

$$P_{sell}^{grid} = \eta_{BIC}(B) \text{ if } B < 0, 0 \text{ otherwise} \tag{28}$$

where η_{BIC} is the bidirectional converter's efficiency.

Based on the Eq.24 and Eq.26 the safety layer dynamically transforms the raw and potentially unsafe EV charge/discharge action $a'(t)$ proposed by PPO agent into safe action $a(t)$ that can be executed in the environment.

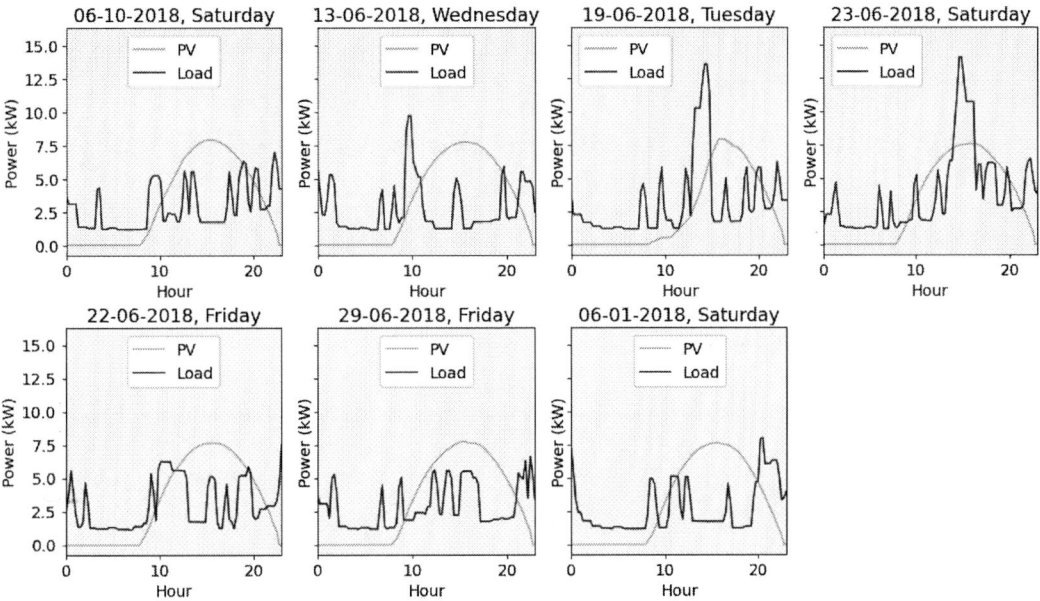

Fig. 3: Sample load, PV and EV profiles in training set

Case Study

Environment		PPO	
Time Resolution	15 minutes	Learning rate$_{actor}$	1e-5
Train Size	15 days	Learning rate$_{critic}$	0.9
Validation Size	6 Days	γ	0.99999
Test Size	8 Days	ε	1.0
Patience	20 epochs	$\sigma_{initial}$	1.0
Grace Period	100 epochs	σ_{min}	0.1
Max Epochs	800 epochs	$\sigma_{decay\ rate}$	0.01
Random Seed	1812	Batch size	5

(a) Hyperparameters

Costs		EV Battery		Microgrid Components	
Parameter	Value	Parameter	Value	Parameter	Value
K_{SD}	0.5	$SoE_{initial}$	4.8 kWh	eff_{bat}	0.95
K_{RA}	0.1 per kWh	E_{max}	24 kWh	eff_{conv}	0.95
c_{ev}	0.01 per kWh	E_{min}	0 kWh	$\Delta Conv_{max}$	200 kWh
c_{buy}	See Fig. 5	ΔSoE_{max}	1.75 kWh	ΔUG_{max}	1000 kWh
c_{sell}	0.5*c_buy	E_{trip}	10 kWh		

(b) Fixed parameters

Table I: Hyperparameters and fixed parameters for the experimental setup

In this case study we consider the DC microgrid setup shown in Fig.1 using data from the Open Energy Data Initiative (OEDI) accessible at [1] and shown in Fig.3. The PPO agent is trained on 30 days of simulated time series data. The data is at a 15 minute time resolution and is from a fictional house in California. The house has a 7.4kW PV installation and a mean load of 2.6 kW. The variation in PV and load profiles will allow an assessment of the PPO agent's ability to plan.

The EV profile was selected to reflect a weekday commuter lifestyle, where the EV departs at 8am and returns at 6pm. The capacity of the battery was set to 24 kWh based on [16], with a maximum charge/discharge rate of 7 kWh every 15 minutes. The battery is assumed to be 95% efficient, and starts with 4.8 kWh of charge, up to a maximum of 24 kWh and a minimum of 0kWh.

The converter is assumed to be 95% efficient and is able to convert up to 200 kWh per time step. Electricity can be bought from/to the utility grid according to the price profile shown in Fig.5. Electricity can also be sold using the same price profile but at 50% of the buying price.

Method	Test Score (price only)	Test Score (w. RA and SD)
PPO	-7.7053	-8.55218
MILP	-4.7332	-8.8287

Table II: Comparison of performances on unseen test data for 3 different methods

Results

The scores of the PPO and MILP algorithms on unseen data show that the DRL agent results in less cost savings for the EV owner than the conventional method. However, the MILP solution was solved by assuming all information for the 24 hour window is perfectly forecast while the PPO solution operates in real-time and hence does not use a forecast. Once the soft constraints of RA and SA are considered, the PPO algorithm even marginally outperforms the MILP algorithm - though these are not explicitly minimized in the MILP. That a similar performance is obtained using data alone and with real-time operation therefore shows the merit of the data-driven technique.

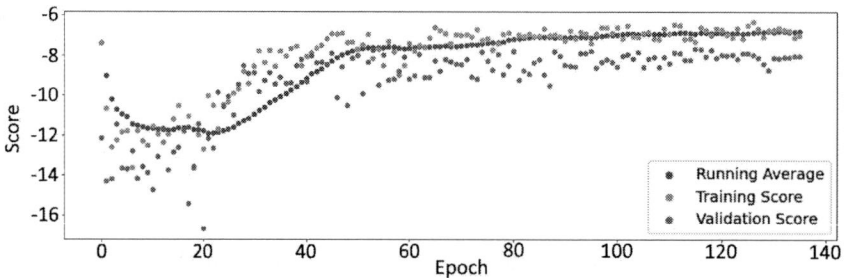

Fig. 4: PPO algorithm training curve

The training performance curve of the PPO agent in Fig.4 shows there is room for further refinement of the agent. The high sample efficiency of the PPO algorithm is clearly visible since it converges in approximately 50 epochs. There is a small gap between the validation set scores and the training set scores, hence the model does not appear to be overfitting. This poor performance may be due to the reward being calculated on the re-scaled safe action rather than the raw output of the agents, hence there is not a clear mapping between the reward and the agent's proposed action. Hence a large penalty could be given for proposed actions that would violate a constraint, but nevertheless are rendered safe through the safety layer.

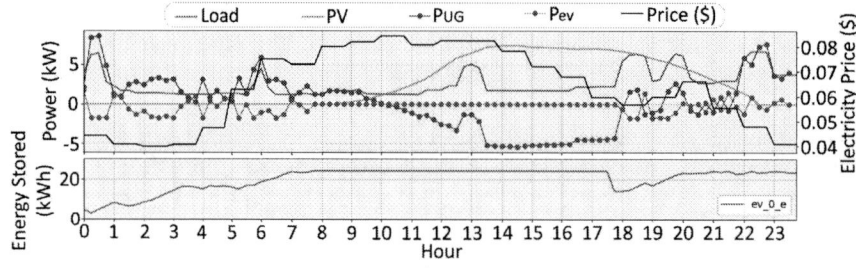

Fig. 5: PPO agent behavior

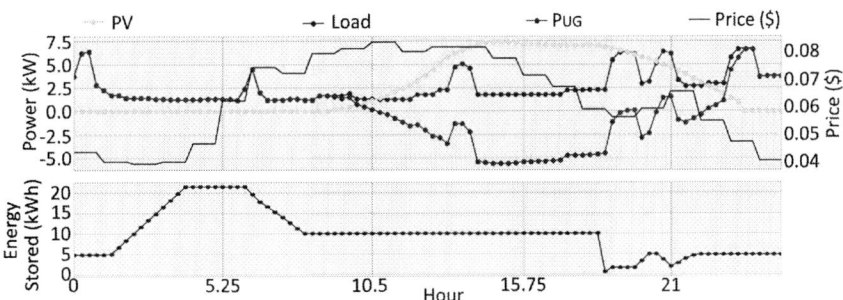

Fig. 6: Behaviour of MILP solution

Based on the optimal behaviour shown by the MILP solution, the data-driven method does make optimal use of the variable electricity price. As shown in Fig.5 the PPO agent appears to charge preferentially in the early hours (midnight to 4am) as electricity is cheaper then. However, more cost savings would be made if the agent discharged the battery when electricity prices were high. Hence the agent appears to have converged to a local minimum and will require further exploration to find the global optimum.

Conclusion

The black-box behaviour and unconstrained output of standard deep reinforcement learning agents can lead to safety violations during operation. To address this concern in the energy management of a smart home with an electric vehicle, a safety layer is appended to the output of a PPO agent to dynamically re-scale any outputs into a safe range, thus guaranteeing the safe operation of the system. The proposed method is able to operate reliably in real-time and on unseen data. It is compared to a day-ahead MILP algorithm, which has perfect information about the next 24 hours of solar production and household demand. The PPO algorithm has a 39% higher electricity cost than the MILP algorithm, but obtains a comparable performance (within 3%) in minimizing the range anxiety of the EV owner and ensuring the battery charge at the end of the day is at least as high as the start of the day. Hence despite the worse savings the shielded PPO is a more practical option than the MILP algorithm, since the latter does not run in real-time.

References

[1] "OEDI: Commercial and Residential Hourly Load Profiles for all TMY3 Locations in the United States", https://data.openei.org/submissions/4520.

[2] Q. Xu, T. Zhao, Y. Xu, Z. Xu, P. Wang and F. Blaabjer, "A Distributed and Robust Energy Management System for Networked Hybrid AC/DC Microgrids", IEEE Transactions on Smart Grid, vol. 11, no. 4, pp.3496-3508, 2020.

[3] Y. Shen, H. Wang, A. Al-Durra, Z. Qin and F. Blaabjerg, "A Bidirectional Resonant DC–DC Converter Suitable for Wide Voltage Gain Range," in IEEE Transactions on Power Electronics, vol. 33, no. 4, pp. 2957-2975, April 2018, doi: 10.1109/TPEL.2017.2710162.

[4] S. Liu, X. Xie and L. Yang, "Analysis, Modeling and Implementation of a Switching Bi-Directional Buck-Boost Converter Based on Electric Vehicle Hybrid Energy Storage for V2G System," in IEEE Access, vol. 8, pp. 65868-65879, 2020, doi: 10.1109/ACCESS.2020.2985772.

[5] S. Lee and D. Choi, "Energy Management of Smart Home with Home Appliances, Energy Storage System and Electric Vehicle: A Hierarchical Deep Reinforcement Learning Approach" in Sensors 2020, Vol. 20, Page 2157.

[6] E. Bagherzadeh, A. Ghiasian and A. Rabiee, "Long-term profit for electric vehicle charging stations: A stochastic optimization approach", Sustainable Energy, Grids and Network, Vol.24, 2020

[7] M. J. E. Alam, K. M. Muttaqi, and D. Sutanto, "Effective utilization of available PEV battery capacity for mitigation of solar PV impact and grid support with integrated V2G functionality," IEEE Trans. Smart Grid, vol. 7, no. 3, pp. 1562–1571, 2016.

[8] F. Hafiz, P. Fajri and I. Husain, "Load regulation of a smart household with PV-storage and electric vehicle by dynamic programming successive algorithm technique," 2016 IEEE Power and Energy Society General Meeting (PESGM), 2016, pp. 1-5

[9] B. Wang, P. Dehghanian and D. Zhao, "Chance-Constrained Energy Management System for Power Grids With High Proliferation of Renewables and Electric Vehicles," in IEEE Transactions on Smart Grid, vol. 11, no. 3, pp. 2324-2336, May 2020

[10] N. Korolko and Z. Sahinoglu, "Robust Optimization of EV Charging Schedules in Unregulated Electricity Markets," in IEEE Transactions on Smart Grid, vol. 8, no. 1, pp. 149-157, Jan. 2017.

[11] Y. Liu, D. Zhang and H. B. Gooi, "Optimization strategy based on deep reinforcement learning for home energy management," in CSEE Journal of Power and Energy Systems, vol. 6, no. 3, pp. 572-582, Sept. 2020

[12] H. Li, Z. Wan and H. He, "A Deep Reinforcement Learning Based Approach for Home Energy Management System," 2020 IEEE Power Energy Society Innovative Smart Grid Technologies Conference (ISGT), 2020, pp. 1-5.

[13] S. Li et al., "Electric Vehicle Charging Management Based on Deep Reinforcement Learning," in Journal of Modern Power Systems and Clean Energy.

[14] Schulman, J., Wolski, F., Dhariwal, P., Radford, A. and Klimov, O., 2017. Proximal policy optimization algorithms. arXiv

[15] J. Schulman, S. Levine, P. Moritz, M. Jordan and P. Abbeel, "Trust Region Policy Optimization", arXiv, 2017

[16] R. Lian, J. Peng, Y. Wu, H. Tan, H. Zhang, "Rule-interposing deep reinforcement learning based energy management strategy for power-split hybrid electric vehicle" Energy, Vol. 197, 2020.

[17] N.Barhate, "Minimal PyTorch Implementation of Proximal Policy Optimization", GitHub, https://github.com/nikhilbarhate99/PPO-PyTorch, 2021.

Simple and Low-Computational Losses Modeling for Efficiency Enhancement of Differential Inverters with High Accuracy at Different Modulation Schemes

Ahmed Shawky
Aswan University, Aswan, Egypt
E-Mail:
Ashawky@apearc.aswu.edu.eg

Mokhtar Aly
Universidad San Sebastián,
Bellavista 7, Santiago, Chile
E-Mail: mokhtar.aly@uss.cl

Emad M. Ahmed
College of Engineering, Jouf
University, Sakaka, Saudi Arabia
E-Mail: emad@eng.aswu.edu.eg

Samir Kouro
Universidad Tecnica Federico Santa Maria
Valparaiso, Chile
E-Mail:samir.kouro@ieee.org

José Rodriguez
Universidad San Sebastián,
Bellavista 7, Santiago, Chile
E-Mail: jose.rodriguezp@uss.cl

Keywords

«Conduction losses», «Core loss», «DC-AC converter», «Device modelling», «Efficiency», «High frequency power converter», «Modulation scheme», «Power losses», «SEPIC converter»,.

Abstract

The modeling of losses at differential inverters is becoming crucial, especially for enhancing efficiency and reliability of their DC-DC modules. Losses modeling based on sinusoidal duty cycle, at line frequency F_L, without considering differential inverters characteristics and switching frequency (F_s) of their DC-DC modules is not accurate, needs high computational demand, and is not applicable for different PWM modulation schemes. In this paper, a simple and accurate losses modeling for differential inverters is proposed based on two-stage calculation process. In first stage, the losses is calculated based on the switching frequencies for DC-DC modules F_s. Then, in the second stage, the losses is averaged according to operating frequency of differential inverters F_L. The decoupling between both frequencies facilitates the easy insertion of differential inverters characteristics such as static linearization approach and low order even harmonics. Also, it easily obtains the RMS currents in terms of module parameters which reduce the required computational calculations. The proposed modeling is applicable for most modulation schemes such as SVMS, CMS and DMS, thanks to the decoupling property of proposed losses modeling. It is generic for single-phase, three-phase, and multi-phase differential inverters and thoroughly supported efficiency improvement even at modular differential inverters. The flow chart of the presented methodology is explained in detail and effectively applied for many DC-DC modules. For verification, a differential inverter based on SEPIC modules is introduced to validate the accuracy of the proposed losses modeling.

Introduction

Differential inverters have shown noteworthy success in power electronics industry due to their modularity, single-stage and bi-directional power with step-up/down ability [1]. Single/three phase differential inverters utilized boost, buck boost, Cuk, SEPIC or flyback converters have been presented for many applications [2-9]. Moreover, improvements in selective low-order harmonics compensation and accurate mathematical modeling for differential SEPIC inverter have been suggested by authors in many works [10-15]. However, the efficiency of all differential inverters requires supplementary efforts to be higher and has not been thoroughly highlighted in advance. In [16], efficiency is enhanced using soft-switched DC-DC modules by adding LC series resonant circuits at conventional single-phase differential inverters for buck, boost and buck-boost converters. It achieves ZVS operation and validates the differential inverter operation at higher switching frequency with small components. In [9], four-switch SEPIC-based differential inverter with lower components counts is suggested. Advanced Wide-Band-Gap (WBG) devices and advanced transformer cores are proposed to enhance efficiency in [12] and [11], respectively. WBG devices such as Gallium Nitride (GaN) and silicon carbide (Sic), have low on resistances and produces low losses and new cores provide low core losses, eddy losses and hysteresis losses. In addition to continuous modulation scheme (CMS), Discontinuous modulation scheme (DMS) and space vector modulation scheme (SVM) were proposed in [17] and [18], respectively, to enhance the efficiency.

Although, previous works effectively enhanced efficiency of differential inverters using the conventional methods, it did not consider the losses of each component and its effect on the cost and complexity. For example, soft-switching eliminates switching losses but requires additional components, as discussed in [16] which may add cost and complexity. In reality, compromising between soft-switching and hard switching requires accurate modeling of switching losses in order to accurately identify the gained efficiency versus added complexity. Moreover, WBG devises instead of SI devices needs prioritizing between added efficiency and extra cost. These issues requires accurate modeling of current differential inverters and identify the percentage of losses at each part. Then, previous suggested proposals of efficiency enhancement are weighted according to efficiency, cost and complexity.

It's worth noticing that most published work at differential inverters discussed the type of internal DC-DC converters, low-order harmonics and different PWM schemes. It did not effectively fill the gap "losses modeling" because they use general losses equations and obtain inaccurate prediction. In [6], buck-boost differential inverters are compared in terms of losses by calculating RMS currents in the passive elements and switching devices. However, assuming only sinusoidal currents have many limitations. Sinusoidal PWM do not match the real PWM of differential inverters which results from merging the differential inverter characteristics and the utilized DC-DC modules which known as static linearization approach [13]. Also, the losses modeling is limited to continuous modulation scheme (CMS) and did not applicable for DMS and SVMS modulation schemes [17-18].

This paper presents an accurate mathematical losses modeling, which considers PWM scheme, static linearization approach and type of DC-DC converters. This model decouples between the line-frequency and switching-frequency states of differential inverter characteristics, which directly solve the issue of static linearization approach. Also, the proposed decoupling is not only accurately predict real PWM, but it also considers the different PWM schemes because it derives the accurate duty cycle and obtained modulation index. Then, the proposed modeling calculates the power losses directly from the utilized DC-DC converters and their parameters and in turn obtains the average power loss differential inverter over line-frequency range. Doing this facilitates application on other DC-DC converters by repeating only the second stage of losses calculations, which validates a straightforward efficiency-improvement tool by knowing the power loss distribution. Finally, the effectiveness of proposed losses modeling has been investigated, compared and verified by using SEPIC DC-DC converter. The results of three-phase differential inverter has only shown due to page limitations.

Working Principle of Differential Inverters

The concept of differential inverter was firstly appeared in 1999 [7]. A new single-stage single-phase boost inverter was proposed using two bi-directional boost converters. Boost converter modules are connected in parallel at input and differentially connected at the output. Output differential connection decouples the DC component of the output voltage, resulted form unipolar operation of DC-DC converters, and let AC voltage component to appear at the output terminal of differential inverter, as illustrated in Figure 1. This configuration is then applied to many DC-DC converter topologies such as isolated and non-isolated converters and have presented thereafter in many works [2-18].

The PWM of differential boost inverter in [7] and other literature work [2-18], is primarily based on the conventional sinusoidal PWM of voltage source inverter (VSI) to drive the switches of their DC-DC modules [13]. The AC variable duty cycle d_x is generated according to the fundamental frequency $F_L = \frac{1}{T_L}$. Then the switching frequency $F_s = \frac{1}{T_s}$ is being high, compared to fundamental frequency to get small passive elements at promoted DC-DC modules and get compact differential inverter. This is accomplished by comparing saw tooth signal at F_s with d_x at F_L, as shown in Figure 2. The shown PWM is currently known as continuous modulation scheme (CMS). This operation is symmetrical for the rest of the modules at other phases because the only difference is the phase-shift. Therefore, authors in [13], suggested a generalized form of differential inverters that based on replacing the traditional leg of VSI with bi-directional converter, as shown in Figure 3.

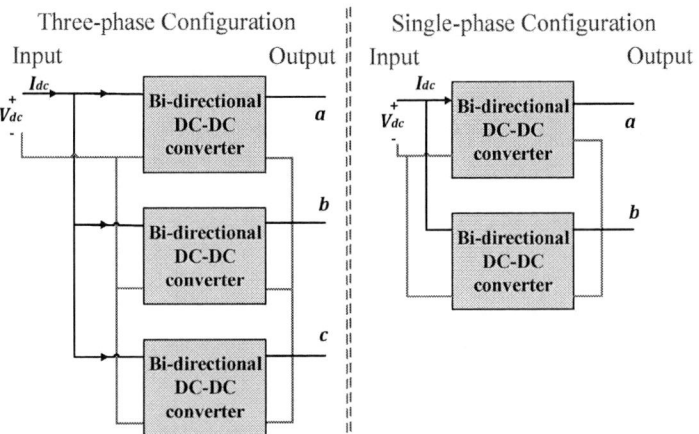

Fig. 1: Three-phase and single-phase differential inverters.

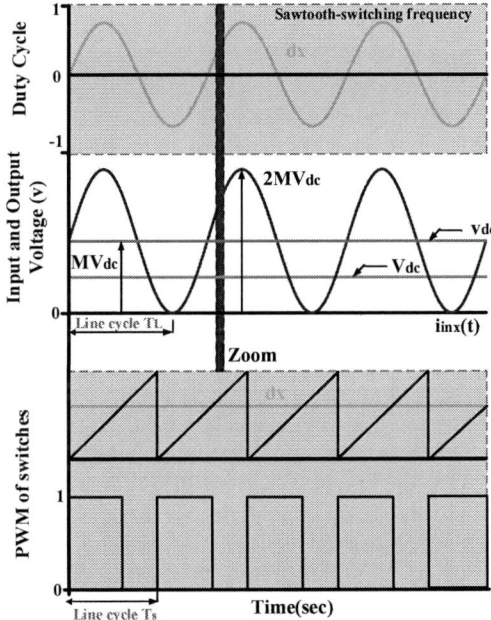

Fig. 2: Sinusoidal PWM of differential inverter that known as CMS.

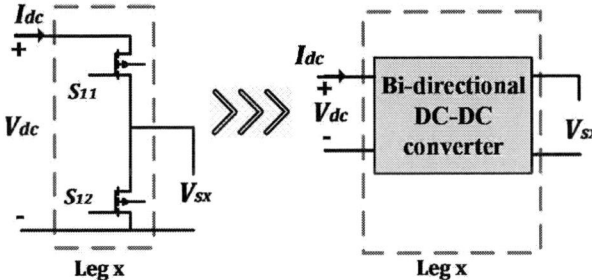

Fig. 3: General configuration of VSI and differential inverters that shows the similarities between them.

Flowchart of Proposed Losses Modeling

The power losses calculation of differential inverter, conducted in [6], is based on the conventional power calculation methods of VSI. Sinusoidal waveform of duty cycle d_x is obtained first by using output voltage of differential inverter (DC and AC component) and input DC voltage and the required voltage gain. Then, the obtained d_x is used with RMS currents of various components of the utilized DC-DC module to calculate average power losses of each component. Although this method considers the characteristics of differential inverter and utilized DC-DC module, it needs high computational burden due to huge math calculation especially using many different integral equations for different waveforms at various components. Also, it have many assumptions that results to considerable mismatch between actual power losses and predictable power losses. Moreover, these calculation should be repeated with different PWM modulation scheme because the obtained power losses calculations is obsessive for CMS. Finally, modeling of power losses at DC-DC modules such as power switches, capacitors and inductors is relatively complicated due to the following deemed differences between differential inverters and VSIs:

- The unipolar operation of DC-DC modules utilized in differential inverters is totally different with VSI legs. Where in VSI, the output voltage on each leg is bipolar.
- The DC component of the output voltage at differential inverter has deemed positive value. However, this value in VSI equal zero.
- The Interact between both frequencies states in differential inverter has different waveform from VSI. Where in VSI, the waveform is generally sinusoidal but in differential inverter it depends on the type of utilized DC-DC module (buck or boost or buck-boost converter)
- The different characteristics between differential inverters and VSIs such as static linearization approach.
- The different PWM schemes such as DMS and SVMS in differential inverters and VSI.

To solve the previous mentioned issues, decoupling between the characteristics of differential inverter and the characteristics of the used DC-DC modules is must. So, in the proposed losses modeling, the power losses calculation has two different stages (Stage-1 & Stage-2). In Stage-1, the utilized DC-DC modules at differential inverter is modeled like conventional DC-DC converters without considering the variable duty cycle d_x. This important notice effectively considers the unipolar operation of conventional DC-DC converters and include the accurate DC component of the output voltage. In Stage-1, the proposed losses modeling for buck converter that presented in [19] is easily used in differential inverters based on buck converter modules. Also, power loss calculation and modeling of other DC-DC converters, such as boost, buck-boost, flyback and SEPIC converters, conducted in [20] is directly used. It is worth noticing that this step (Stage-1) eases power loss calculation by applying direct equations of integrated devices, which were discussed for many DC-DC converters in literature work. Also, the selected DC-DC modules is analyzed for both switching states (on and off) to obtain the connected devices and calculate its power loss based on datasheet parameters and connection intervals.

In Stage-2, the extracted power losses at each interval is averaged on F_L according to the type of PWM scheme and the characteristics of differential inverter. Eq. (1) shows the characteristics of differential inverter at different modulation schemes. Where, h_x is the output voltage component at each module. M_{dc} and M_{ac} are the DC and AC voltage gain of the utilized modules.

$$
\left.
\begin{aligned}
&d_x = \frac{h_x}{1+h_x} = \frac{M_{dc}+M_{ac}sin(\omega t)+M_o}{1+M_{dc}+M_{ac}sin(\omega t)+M_o} \\
&d_x \text{, duty cycle of utilized module at phase } x \\
&x = a, b \text{ for single phase} \\
&x = a, b, c \text{ for three phase}
\end{aligned}
\right\}
\tag{1}
$$

It's worth emphasizing that both component are equal and depend on the conventional voltage gain of the utilized DC-DC module. M_o is the voltage component that results from different modulation scheme, that given in Eq. 2 for CMS, SVMS and DMS, respectively.

$$M_o = 0$$
$$M_o = 0.5 \left[\max(m_a + m_b + m_c) + \min(m_a + m_b + m_c) \right]$$
$$M_o = 0.5 \min(m_a + m_b + m_c)$$

$$(2)$$

Where, m_a, m_b and m_c are the abc component of modulation index at different phases of differential inverter which obtain the static gain signals of the utilized DC-DC modules. Figure 4 illustrates the algorithm of the proposed losses modeling. The link between Stage-1 and Stage-2 is the exact value of duty cycle d_x at each switching interval (T_s), as shown in Figure 4 and Eq. (1). So it's important to define an appropriate time-step T_i for the losses modeling; it is better to use a value small than T_s ten times or more. Increasing the time-step fasten the calculation process but increase the calculation error. On the other hand, decreasing it enhances the accuracy of the model but expands the execution time. Moreover, the actual duty cycle is being higher because the voltage drop in voltage gain (M_{dc} and M_{ac}) which resulted from losses in different components of DC-DC modules. Finally, this process is implemented in Excel sheet and repeated for many variables such as load, as shown in Figure 4.

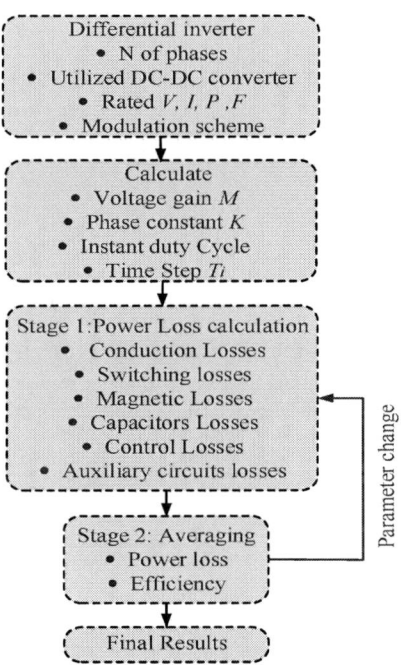

Fig. 4: Flowchart of proposed losses modeling.

Case-Study: Losses Calculation of SEPIC based three-phase Differential Inverter.

A three-phase grid-connected differential inverters based on the three SEPIC modules is presented to validate the proposed losses modeling. The inverter parameters are shown in Table I. The PWM schemes are SVM, CMS and DMS [1]. The gain is $M_{dc} = M_{ac} = 1.5$ and d_x is obtained using Eq. (1). Because $T_s = 20us$, a step time $T_i = 1us$ has chosen in the Excel sheet. The operation modes is then investigated to finalize Stage-1, as illustrated in Figure 5, i.e., for buck-boost SEPIC converter. Doing this at Stage-1, obtains power losses of every component at converters is obtained using traditional equations which used previously at DC-DC converter applications. i.e., the conduction losses of the main switch S_{mx} is obtained from Eq. (3) where, R_{dson} is the on-resistance of the switch, $d_x(t)$ is the duty cycle at this instant. I_D Is the average drain current over one cycle, and, Δi_D is the ripple component of switch current. Also, the switching losses are obtained based on Eq. (4). t_{rise}, t_{fall} and C_{oss}, are the rise time, fall time and drain source capacitance, respectively.

$$P_{cond}(t) = R_{dson}d_x(t)\left[I_D{}^2 + \frac{\Delta i_D{}^2}{12}\right] \tag{3}$$

Table I: System parameters of studied differential inverters

Parameter	Rated power	Input DC voltage	Grid voltage	Line Frequency
	P	V_{dc}	V_{LL}	F_L
Value	$0.2 - 1.6kW$	$100, 120\ V$	$200\ V$	$60Hz$
Parameter	Converter Inductor	Output Capacitor	Grid inductance,	Switching Frequency
	L_x	C_{ox}	L_{gx}	F_s
Value	$200\mu H$	$14\mu F$	$4mH$	$50\ kHz$

$$\left.\begin{aligned}
P_{sw}(t) &= P_{swon} + P_{swof} + P_{coss} \\
P_{swon}(t) &= \frac{1}{6}t_{rise}F_{sw}V_{ds}\left[I_D - \frac{\Delta i_D}{2}\right] \\
P_{swof}(t) &= \frac{1}{6}t_{fall}F_{sw}V_{ds}\left[I_D + \frac{\Delta i_D}{2}\right] \\
P_{coss}(t) &= 0.5\ C_{oss}V_{ds}{}^2
\end{aligned}\right\} \tag{4}$$

Eq. (5) are for body diode losses of S_{mx}. In both, I_F is average forward current over T_s, V_F is the forward voltage, Δi_F is the ripple component, t_{dead} is the dead time between main and synchronous switch, and Q_{rr} is the reverse recovery charge. Losses of other components is covered in [19]. Finally, Stage-2 of proposed modeling averages losses according the different modulation schemes.

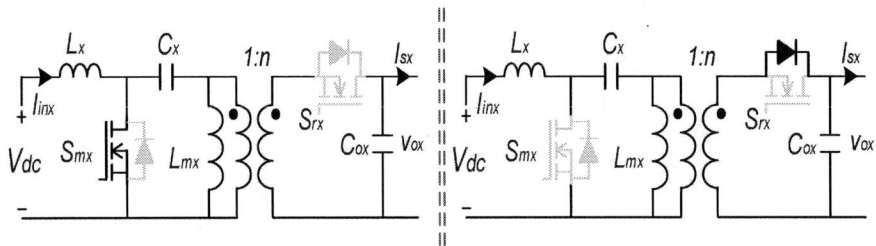

Fig. 5: Switching states of the SEPIC buck-boost converter at differential inverter.

$$\left.\begin{aligned}
P_{cond}(t) &= d_x(t)I_F V_F \\
P_{sw}(t) &= P_{swon} + P_{swof} + P_{Qrr} \\
P_{swon}(t) &= t_{dead}F_{sw}V_F\left[I_F - \frac{\Delta i_F}{2}\right] \\
P_{swof}(t) &= t_{dead}F_{sw}V_F\left[I_F + \frac{\Delta i_F}{2}\right] \\
P_{coss}(t) &= 0.5\ V_{ds}F_{sw}Q_{rr}
\end{aligned}\right\} \tag{5}$$

Results and Discussion

The results of SEPIC based differential inverter is shown here. Figure 6 shows the waveforms of inverter at different modulation schemes. It's worth noticing that d_x waveforms is different and then losses will be different. Therefore, using proposed modeling calculates power losses at Stage-1 for full period then averages the losses based on the type of modulation scheme is easy and direct solution. Figure 7 shows the efficiency of the differential inverter using the proposed losses modeling, simulations and experiments. It's worth noticing that the difference between theoretical, simulated and experimentally measured efficiency is small due to the accuracy of proposed model. Although the error between calculated efficiency between proposed model and experiments results from the unknown parasitic of the circuits and external wires, the proposed model is accurate because it gives the same performance with experimental measured efficiency.

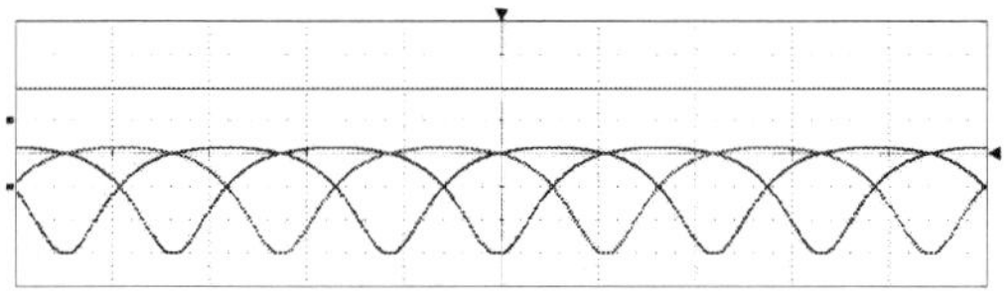

CMS

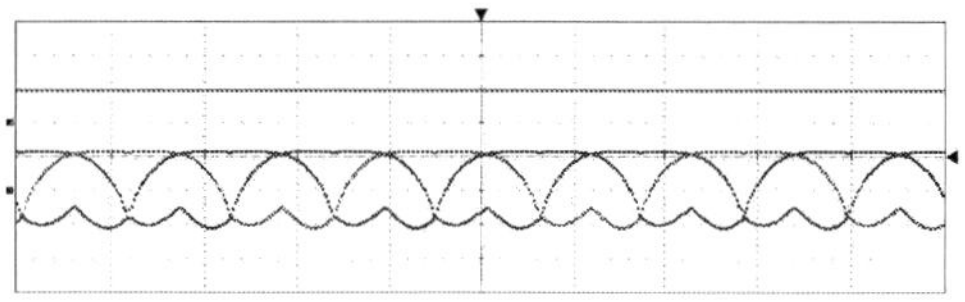

SVMS

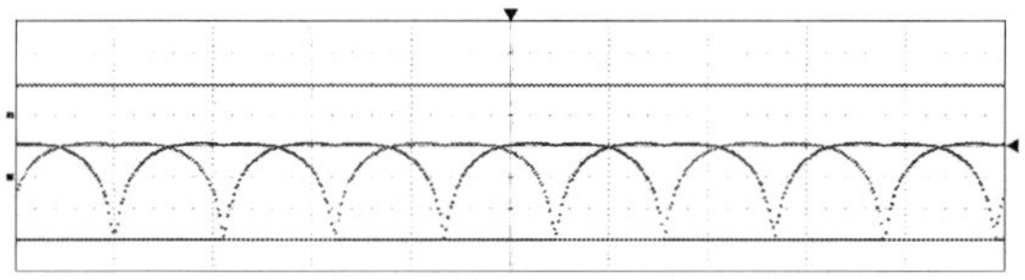

DMS

Fig. 6: Waveforms of DC voltage gain M_{dc} and variable duty cycle d_x (d_a, d_b and d_c) of SEPIC based differential inverter, respectively.

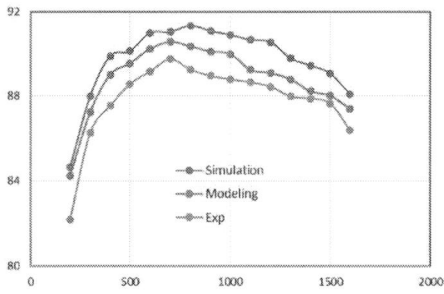

Fig. 7: Efficiency comparison at CMS modulation.

Figure 8 shows the simulation results for CMS, SVMS and DMS using proposed losses modeling. It's worth noticing that the DMS modulation gives the best efficiency which exactly matched with results at [17]. The loss-division is also investigated using proposed modeling for the utilized components, as shown in Figure 9-11, at different input voltage levels and different power rating at CMS modulation. It includes modeling of switches such as MOSFET devices, magnetics such as input inductor and HFT, and passive elements such as capacitors, extra. At CMS, The dominant power losses for $V_{dc} = 100V$ are associated with HFT with 29.19%. Then the power losses of input inductor and switches are 27.9% and 26.95%, respectively. It is worth mentioning that the power losses of main switches (17.05%) are larger than those of synchronous switch (9.9%) due to its real PWM waveform of differential inverter. For higher input DC voltage, the power loss of HFT increased to 29.9% at $V_{dc} = 120V$, respectively. The same finding is for inductor, 30.24%. However, the power losses of switches are reduced to 25.54% and 24.83%. This finding is important to express the effect of input DC voltage on overall efficiency. Figure 10 and 11 present the power losses division at different power ratings, $P_o = 0.8kW$ and $P_o = 0.4kW$, respectively. These findings are significant at the differential inverter because the proposed modeling predicts the sources of power loss at many operating conditions accurately.

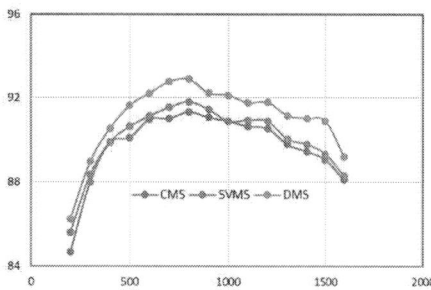

Fig. 8: Efficiency comparison at CMS, SVMS and DMS using proposed modeling.

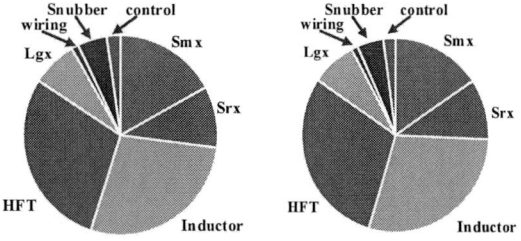

Fig. 9: Loss-distribution at $V_{dc} = 100, 120$ V and $P_o = 16kW$

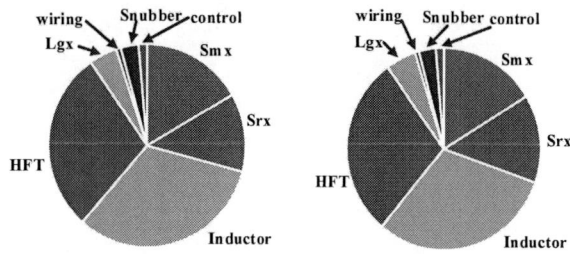

Fig. 10: Loss-distribution at $V_{dc} = 100, 120$ V and $P_o = 0.8kW$

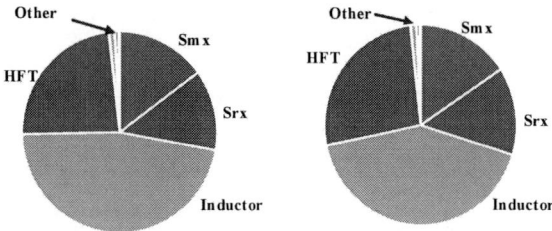

Fig. 11: Loss distribution at $V_{dc} = 100, 120$ V and $P_o = 0.4kW$.

Conclusion

An accurate losses modeling for differential inverters is presented. It's based on decoupling between differential inverter switching states at line-frequency and switching-frequency. It eases the power loss calculation, which does not exist in traditional losses models. The added features are the direct calculations that can be developed quickly, even at different phases and different DC-DC modules for differential inverter topologies. It obtains the current equations through components in terms of module parameters. The proposed model can be applied for single-phase, three-phase, and multi-phase differential inverters because it depends on its integrated DC-DC modules. A case-study using three-phase SEPIC differential inverter is introduced to validate the proposed model's accuracy. Finally, the presented methodology's flow chart is generalized and applied to other DC-DC converters to facilitate the power losses calculation.

Acknowledgement

J. Rodriguez acknowledges the support of ANID through projects FB0008 and 1210208. S. Kouro acknowledges the support of AC3E (ANID/BASAL/FB0008), FONDECYT 1221741 and SERC Chile (ANID/FONDAP/15110019).

References

[1] A. Shawky, M. Ahmed, M. Orabi, and A. E. Aroudi, "Classification of Three-Phase Grid-Tied Microinverters in Photovoltaic Applications," Energies, vol. 13, no. 11, p. 2929, Jun. 2020, doi: 10.3390/en13112929.

[2] C. Cecati, A. Dell'Aquila and M. Liserre, "A novel three-phase single-stage distributed power inverter," in *IEEE Transactions on Power Electronics*, vol. 19, no. 5, pp. 1226-1233, Sept. 2004, doi: 10.1109/TPEL.2004.835112.

[3] A. A. Khan, Y. W. Lu, W. Eberle, L. Wang, U. A. Khan and H. Cha, "Single-Phase Split-Inductor Differential Boost Inverters," in *IEEE Transactions on Power Electronics*, vol. 35, no. 1, pp. 107-120, Jan. 2020, doi: 10.1109/TPEL.2019.2913750.

[4] A. Darwish, D. Holliday, S. Ahmed, A. M. Massoud and B. W. Williams, "A Single-Stage Three-Phase Inverter Based on Cuk Converters for PV Applications," in *IEEE Journal of Emerging and Selected Topics in Power Electronics*, vol. 2, no. 4, pp. 797-807, Dec. 2014, doi: 10.1109/JESTPE.2014.2313185.

[5] A. Shawky, M. E. Ahmed and M. Orabi, "Performance analysis of isolated DC-DC converters utilized in Three-phase differential inverter," *2016 Eighteenth International Middle East Power Systems Conference (MEPCON)*, 2016, pp. 821-826, doi: 10.1109/MEPCON.2016.7836989.

[6] A. Darwish, A. M. Massoud, D. Holliday, S. Ahmed and B. W. Williams, "Single-Stage Three-Phase Differential-Mode Buck-Boost Inverters With Continuous Input Current for PV Applications," in *IEEE Transactions on Power Electronics*, vol. 31, no. 12, pp. 8218-8236, Dec. 2016, doi: 10.1109/TPEL.2016.2516255.

[7] R. O. Caceres and I. Barbi, "A boost DC-AC converter: analysis, design, and experimentation," in *IEEE Transactions on Power Electronics*, vol. 14, no. 1, pp. 134-141, Jan. 1999, doi: 10.1109/63.737601.

[8] H. Soni, S. K. Mazumder, A. Gupta, D. Chatterjee and A. Kulkarni, "Control of Isolated Differential-Mode Single- and Three-Phase Ćuk Inverters at Module Level," in *IEEE Transactions on Power Electronics*, vol. 33, no. 10, pp. 8872-8886, Oct. 2018, doi: 10.1109/TPEL.2017.2779408.

[9] M. S. Diab, A. Elserougi, A. M. Massoud, A. S. Abdel-Khalik and S. Ahmed, "A Four-Switch Three-Phase SEPIC-Based Inverter," in *IEEE Transactions on Power Electronics*, vol. 30, no. 9, pp. 4891-4905, Sept. 2015, doi: 10.1109/TPEL.2014.2363853.

[10] A. Shawky, M. A. Sayed and T. Takeshita, "Selective Harmonic Compensation of Three Phase Grid tied SEPIC based Differential inverter," *2019 IEEE Applied Power Electronics Conference and Exposition (APEC)*, 2019, pp. 396-403, doi: 10.1109/APEC.2019.8721854.

[11] A. I. M. Ali, M. A. Sayed, A. Shawky and T. Takeshita, "Efficient Single-Stage Three-Phase Isolated Differential-Based Flyback Inverter with Selective Harmonic Compensation Strategy for Grid-Tied Applications," *2020 IEEE Applied Power Electronics Conference and Exposition (APEC)*, 2020, pp. 1778-1785, doi: 10.1109/APEC39645.2020.9124255.

[12] A. Shawky, T. Takeshita and M. A. Sayed, "Single-Stage Three-Phase Grid-Tied Isolated SEPIC-Based Differential Inverter With Improved Control and Selective Harmonic Compensation," in *IEEE Access*, vol. 8, pp. 147407-147421, 2020, doi: 10.1109/ACCESS.2020.3014894.

[13] A. Shawky, T. Takeshita, M. A. Sayed, M. Aly and E. M. Ahmed, "Improved Controller and Design Method for Grid-Connected Three-Phase Differential SEPIC Inverter," in *IEEE Access*, vol. 9, pp. 58689-58705, 2021, doi: 10.1109/ACCESS.2021.3072489.

[14] A. Shawky, M. A. Sayed and T. Takeshita, "Single-Stage Three-Phase Step-Up SEPIC Differential Grid-Tied Inverter Features an In-depth Mathematical Analysis for Solving Practical Design Issues," *2021 IEEE Applied*

Power Electronics Conference and Exposition (APEC), 2021, pp. 2650-2656, doi: 10.1109/APEC42165.2021.9487369.

[15] A. Shawky, T. Takeshita and M. A. Sayed, "Analysis and Performance Evaluation of Single-Stage Three-Phase SEPIC Differential Inverter with Continuous Input Current for PV Grid-Connected Applications," *2021 IEEE Applied Power Electronics Conference and Exposition (APEC)*, 2021, pp. 2719-2726, doi: 10.1109/APEC42165.2021.9487141.

[16] B. Koushki, A. Safaee, P. Jain and A. Bakhshai, "Zero voltage switching differential inverters," *2015 IEEE Applied Power Electronics Conference and Exposition (APEC)*, 2015, pp. 1905-1910, doi: 10.1109/APEC.2015.7104606.

[17] S. Mehrnami, S. K. Mazumder and H. Soni, "Modulation Scheme for Three-Phase Differential-Mode Ćuk Inverter," in *IEEE Transactions on Power Electronics*, vol. 31, no. 3, pp. 2654-2668, March 2016, doi: 10.1109/TPEL.2015.2442157.

[18] A. Shawky, M. Aly, E. M. Ahmed, S. Kouro and J. Rodriguez, "Space Vector Modulation Scheme for Three-Phase Single-Stage SEPIC-Based Grid-Connected Differential Inverter," *IECON 2021 – 47th Annual Conference of the IEEE Industrial Electronics Society*, 2021, pp. 1-6, doi: 10.1109/IECON48115.2021.9589574.

[19] M. Orabi and A. Shawky, "Proposed Switching Losses Model for Integrated Point-of-Load Synchronous Buck Converters," in *IEEE Transactions on Power Electronics*, vol. 30, no. 9, pp. 5136-5150, Sept. 2015, doi: 10.1109/TPEL.2014.2363760.

[20] R. W. Erickson and D. Maksimovic, Fundamentals of Power Electronics, 2nd ed. Norwell, MA: Kluwer, 2001.

Estimation of Battery Parameters in Cascaded Half-Bridge Converters with Reduced Voltage Sensors

Nima Tashakor, Bita Arabsalmanabadi, Elham Hosseini, Kamal Al-Haddad, Stefan Goetz
Technische Universität Kaiserslautern
Kaiserslautern, Germany
E-Mail: Tashakor@eit.uni-kl.de

Acknowledgements

The authors acknowledge the financial support by the Federal Ministry of Education and Research of Germany in the project "Open6GHub" (grant number: 16KISK004).

Keywords

Modular battery, parameter estimation, modular multilevel converter

Abstract

Although modular multilevel converters (MMC) and cascaded half-bridge (CHB) converters are an established concept in HVDC, MMCs and CHBs have started to find new applications, including modular converters with integrated energy storage systems. Despite various advantages of the low-voltage modularity, a complex and expensive monitoring/control system can hinder finding a foothold in many emerging applications that are more cost-driven, such as the e-mobility market. Estimators and observers can reduce the monitoring cost and complexity by reducing the number of required sensors and communication bandwidth. However, estimation methods rarely consider MMCs with integrated battery, and most available methods neglect all resistances. This paper fills this gap by developing an online estimation technique for parameters of all battery modules in an MMC. The proposed method exploits the slow dynamics of the battery to use a simpler and less computationally demanding algorithm that can easily be implemented in low-end controllers. Based on the developed model of the system, the iterative algorithm can estimate the voltage and internal resistance of every module through measuring the output voltage and current of the battery pack and avoid direct measurements from the modules. As a result of substantial reduction in the number of monitoring sensors for estimating the battery parameters, the proposed technique is simpler and less costly in comparison with other sensor-based techniques. Furthermore, the proposed technique accelerates convergence using optimal learning rate value. Simulations validate the ability of the proposed estimation technique under different scenarios. The estimation technique can identify both internal resistance and open-circuit voltage of the batteries with approximately 2 % accuracy.

Introduction

As an alternative to conventional fuel, electric transportation has been growing rapidly over the last ten years. One of the main challenges in electric vehicles (EV) arises from the range anxiety [1]. Energy storage systems, usually batteries, are of paramount importance to address this issue by decreasing refilling time and increasing the capacity [2]. In addition, the introduction of fast and ultra-fast electric vehicle chargers increased the required voltage and power levels in an EV [3]. Correspondingly, the EV battery size has increased to hundreds of battery cells including multiple serial/parallel strings in each EV, which cannot be considered as one battery pack anymore, but rather a collection of modules hard-wired together [3]. While manufacturers use larger and larger batteries in cars, hot debates concerning the battery design and monitoring become more significant [4]. Experts believe that fully controlled modular batteries are the next step in the evolution of battery pack design with very high expectations for modular multilevel converters (MMC) and cascaded bridge converters (CBC) [5-10].

MMCs are dominate in high-voltage applications due to simplicity and cost-effectiveness; however, with the falling cost of power electronics, it is expected that their popularity also grows in medium- and low-voltage applications [11, 12]. One major trend in MMCs is to replace the capacitor of each module in a string with a battery, resulting in a fully controlled modular converter/battery pack and similar to normal MMCs. Multiple strings can be connected to form different dc, single-phase, and multi-phase structures [13]. Additionally, on the module level, MMCs can use different topologies, among which half-bridge (HB) is the most popular one at the moment [14, 15]. Requiring only two switches within each module offers the simplest circuit among several alternatives [16, 17]. Although an MMC with battery modules can offer many advantages, it requires constant monitoring of the module's states to ensure a balanced operation. The inherent tolerance of the modules and different thermal or load conditions as well as different aging factors can lead to charge imbalance among the modules [18, 19]. Imbalance between different modules is one of the main issues in any battery system including a modular one, which leads to shorter battery lifetime, lower accessible energy, and lower system efficacy [20, 21]. Therefore, the capacitors or batteries must be monitored constantly to ensure stable operation [22].

Most conventional monitoring systems rely on real-time measurements of each module's parameters [23]. However, in a high-power industrial application with a high number of modules, the collection and processing of all this data is a significant challenge [24, 25]. Additionally, multiple sensors per module increase cost and size of data acquisition system. Minimizing direct measurement of module capacitor voltages has been widely investigated in the recent literature by incorporating observers and estimators of those voltages [26-28]. Abushafa et al. use an exponentially weighted recursive least square technique to estimate the module capacitor voltages with only one voltage sensor per arm [26]. Arco et al. present an algorithm based on the operation of a two-step predictor–corrector scheme, where in the first step the algorithm uses a simple model-based estimation and then corrects the estimation based on the output voltage of the whole string in the second step [27]. The adaptive linear neuron algorithm ([28, 29]) and Kalman filter ([30, 31]) are other types of estimator with varying degree of accuracy that require only the output voltage of the string to estimate the voltage of every module. However, there are also models that focus on reducing the sensors throughout the string rather completely removing them. Rong et al. present a voltage estimation method that compensates the sampling delay effect [32]. The presented method divides the modules into groups and then estimates the capacitor voltages inside each group from the measured output voltage of the group. While most of the above-mentioned methods consider identical capacitor values, Taffese et al. propose an online voltage observer that considers both capacitance variation and time delay [33]. Konstantinou et al. provide a review of the different estimation/observation methods for MMCs [34].

Most of the reported techniques are applied to capacitor-based modules but are also applicable to battery-based modules to estimate the terminal voltage of the batteries. However, whereas the internal resistance of the capacitors is negligible, the internal resistance of the batteries is considerably higher and cannot be neglected [35]. Furthermore, although the terminal voltage of the battery is a particularly good indicator of its charge, this is not the case for batteries [36, 37]. Therefore, even though a wide range of techniques is available for estimating terminal voltage of the modules in MMC, to the best of our knowledge, none of them is able to estimate the internal and the open-circuit voltage of the battery modules, which are crucial in battery management and monitoring functions, such as state of charge and state of health monitoring [37-39]. Another noticeable difference between capacitors and batteries as the energy storage of a module is their dynamics. The voltage of a capacitor changes continuously with fairly fast dynamics, requiring estimators with high convergence speed and fast update rates. However, the battery parameters enjoy a slower dynamic, which can be exploited to reduce the estimator complexity using simpler methods [40].

This paper presents an estimation technique for open-circuit voltage and internal resistance of battery modules in an MMC topology. The proposed method decouples the equations of voltage and resistance through consecutive sampling of the output voltage and current of the pack and then proceeds with updating the voltage and resistance vectors. Furthermore, the learning factor of the system is optimized to minimize the convergence speed. The proposed technique benefits from low computation that can be implemented in most low-end controllers, without any additional sensor for the modules. Moreover, it has the advantage of estimating the internal resistance of the battery, which has previously been neglected.

The proposed MMC Structure

Figure 1 depicts the topology of a dc modular battery structure consisting of N half-bridge battery-based modules, inductor L, capacitor C, and the load/supply. The output voltage and current of the pack are measured using two sensors with the sampling time T_s. Each module includes a battery and two MOSFETs (i.e., S_{iU} and S_{iL}) as well as their body diodes (i.e., D_{iU} and D_{iL}). The modular system is a multilevel dc/dc bidirectional converter, behaving as a buck converter during discharge and a boost converter during charge.

A. Module Operating Modes

As Fig. 2 shows, each module contains a half-bridge and a battery, with four possible operation modes during bidirectional energy transfer as follows: i) S_{1U} is on, S_{1L} is off, and the batteries are charging. The terminal voltage of the battery is $v_{t_i} = V_{oc_i} - (r_{bt_i} + r_{dU_i})i_p$ (see Fig. 2(a), Mode 1); ii) S_{1U} is off, S_{1L} is on, and the batteries are charging. The module is bypassed, and the terminal voltage of the battery is $v_{t_i} = -r_{dsL_i}i_p$ (see Fig. 2(a), Mode 2); iii) S_{1U} is on, S_{1L} is off, and the batteries are discharging. The terminal voltage of the battery is $v_{t_i} = v_{oc_i} - (r_{bt_i} + r_{dsU_i})i_p$ (see Fig. 2(a), Mode 3); iv) S_{1U} is off, S_{1L} is on, and the batteries are discharging. The module is bypassed, and the terminal voltage of the battery is $v_{t_i} = -r_{dL_i}i_p$ (see Fig. 2(a), Mode 4).

For the sake of simplicity, we assume identical resistances for the MOSFETs and their body diodes. Consequently, Fig. 2(b) shows the electrical equivalent circuit of the battery when modules is connected in series to the arm and Fig. 2(c) shows the electrical equivalent circuit when the module is bypassed from the arm.

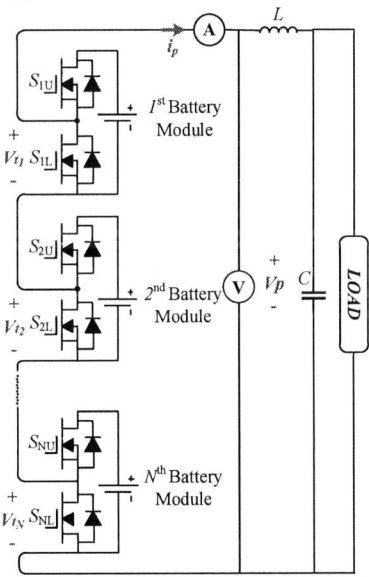

Fig.1 A single arm battery-based dc MMC macro structure

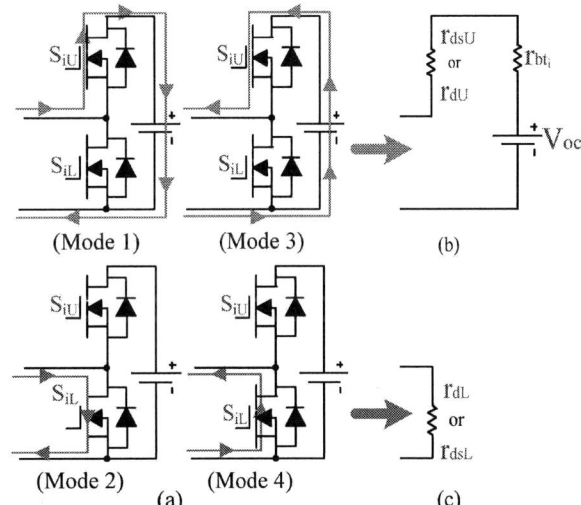

Fig. 2 The module structure (a) different operation modes of each module, (b) electrical equivalent model of a module when inserted in series (c) electrical equivalent of the module when bypassed.

B. Modulation Techniques

While there are many modulation techniques for MMCs, phase-shifted carrier (PSC) modulation is popular because of its stable performance during dynamic changes and ease of implementation [41]. In PSC modulation, the zero point of each carrier is slightly shifted with respect to its neighboring carriers, where the phase-shift of the jth module follows $\varphi_i = 2\pi/N$. The switching signals result from comparing carrier waves with the with the modulation index. Figure 3 provides an intuitive representation of the MMC operation, where at first module two is discharging in Fig. 3(a) and then bypassed in Fig. 3(b). Additionally, Figure 3(c) shows the implementation of PSC modulation.

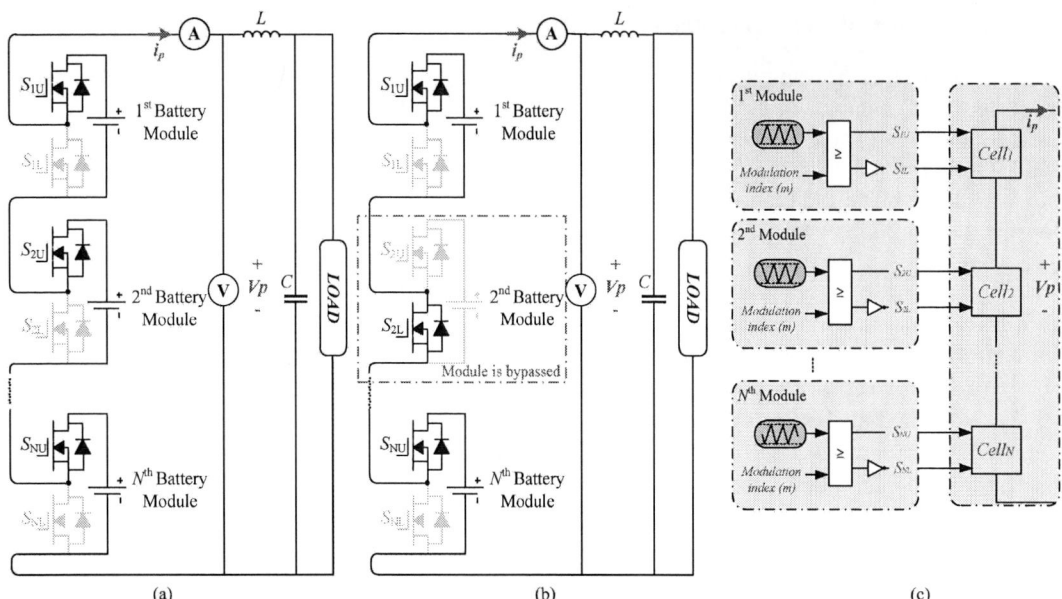

Fig.3 (a) all modules are in series; (b) the second modules is bypassed (c) Overall control block diagram of PSC modulation

Battery Parameter Estimation

In this paper, we propose an iterative algorithm to estimate parameters of each battery module in the battery-based MMC shown in Fig. 1. Using the developed mathematical model in the previous section, the proposed estimator–corrector algorithm can estimate the voltage and internal resistance of each module through measuring only the output voltage and current of the battery pack. Most of the existing reports consider battery as a single DC source with negligible resistance. The technique is the same for both voltage and resistance estimation.

As discussed, there is a nonlinear relation between the system parameters. A BP parameter estimation algorithm is derived to estimate battery parameters in the proposed non-linear dynamic model. As shown in Fig. 6, an error function is defined between the measured and the model outputs.

A. Mathematical model of the MMC

Figure 4 depicts the equivalent model of one arm. We can model the battery in its simplest form with a constant voltage source as the open-circuit voltage in series with a resistance. The open-circuit voltage depends on the battery's state of charge, but for the purpose of modeling, it can be considered constant in brief intervals. In Fig. 4, R_{eq} depicts the effective resistance of the entire arm. The KVL on the arm results in (1.a) and (1.b) for charge and discharge states respectively.

$$v_p(t) = \left(\boldsymbol{S}_U(t)\right)^T \boldsymbol{V}_{OC} - i_p(t) \begin{bmatrix} \left(\boldsymbol{S}_U(t)\right)^T \left(\boldsymbol{R}_{bt} - \boldsymbol{R}_{ds,U}\right) \\ -\left(\boldsymbol{S}_L(t)\right)^T \boldsymbol{R}_{d,L} \end{bmatrix}$$
(1.a)

$$v_p(t) = \left(\boldsymbol{S}_U(t)\right)^T \boldsymbol{V}_{OC} - i_p(t) \begin{bmatrix} \left(\boldsymbol{S}_U(t)\right)^T \left(\boldsymbol{R}_{bt} - \boldsymbol{R}_{d,U}\right) \\ -\left(\boldsymbol{S}_L(t)\right)^T \boldsymbol{R}_{ds,L} \end{bmatrix}$$
(1.b)

where $\boldsymbol{S}_U$ and $\boldsymbol{S}_L$ are the vectors of gate signals for the upper switches and lower switches. $\boldsymbol{R}_{ds}$ and $\boldsymbol{R}_d$ are respectively the switch and diode resistances' vectors, where the subscripts U and L correspond to upper and lower switches in

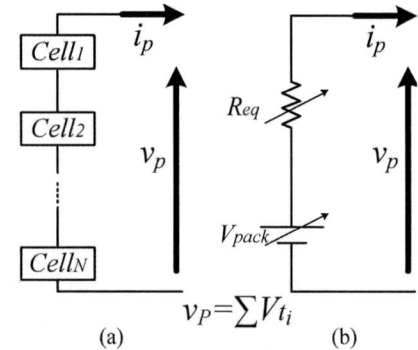

Fig.4. MMC circuit and associated current dynamic loops: (a) Converter schematic. (b) Equivalent circuit of output current and voltage

the half-bridges. Furthermore, V_{oc} and R_{bt} correspond to the open-circuit voltages and internal resistances of the modules. v_p and i_p are the measured voltage and current of the pack terminals. Also, as shown in Fig. 4, the positive direction of the pack current is during discharging. For the sake of simpler representation, we consider the resistances of lower and upper MOSFETs as well as diodes equal, i.e., $R_{ds,U} = R_{ds,L} = R_{d,U} = R_{d,L} = R_{sw}$. Therefore, we can further simplify (1) in discrete time domain as

$$v_p^{(k)} = S_U^{(k)} \times V_{oc} - i_p^{(k)} \cdot \left[\left(S_U^{(k)} \right)^T \times R_{bt} + \left(S_U^{(k)} + S_L^{(k)} \right)^T \times R_{sw} \right], \tag{2}$$

where subscripts k shows the kth sample.

Additionally, $S_U^{(k)}$ and $S_L^{(k)}$ are complementary, and the resultant vector of $S_U^{(k)} + S_L^{(k)}$ is a unity vector. Hence, (2) can be further simplified to

$$v_p^{(k)} = S_U^{(k)} \times V_{oc} - i_p^{(k)} \cdot \left[\left(S_U^{(k)} \right)^T \times R_{bt} + N r_{sw} \right], \tag{3}$$

where r_{sw} is the internal resistance of one switch/diode and the total number of modules is N.

As can be seen, both R_{bt} and V_{oc} can affect the output of the pack. Therefore, it is necessary to somehow decouple the effect of these two vectors in (3). With that goal in mind, we write (3) for two consecutive instances of k and $(k + 1)$ as follows

$$v_p^{(k)} + i_p^{(k)} \cdot N \cdot r_{sw} = \left(S_U^{(k)} \right)^T \times V_{oc} - i_p^{(k)} \cdot \left(S_U^{(k)} \right)^T \times R_{bt}, \tag{4}$$

$$v_p^{(k+1)} + i_p^{(k+1)} \cdot N \cdot r_{sw} = \left(S_U^{(k+1)} \right)^T \times V_{oc} - i_p^{(k+1)} \cdot \left(S_U^{(k+1)} \right)^T \times R_{bt}. \tag{5}$$

Subtracting (4) from (5) when $S_U^{(k)} = S_U^{(k+1)} = S_U$, we can remove V_{oc} from the new equations per

$$\Delta v_p + \Delta i_p \cdot N \cdot r_{sw} = -\Delta i_{pack} \cdot (S_U)^T \times R_{bt}, \tag{6}$$

where $\Delta v_p = v_p^{(k+1)} - v_p^{(k)}$; and $\Delta i_p = i_p^{(k+1)} - i_p^{(k)}$. By doing so, V_{oc} is removed from the equation and R_{bt} is the only variant. We can rewrite (6) as

$$-\frac{\Delta v_p}{\Delta i_p} - N \cdot r_{sw} = (S_U)^T \times R_{bt}. \tag{7}$$

With R_{bt} as the only variable in (7), we can use a back propagation algorithm to solve it [28]. In this step, iterative methods can estimate R_{bt} using

$$W_1^{(k+1)} = W_1^{(k)} + \alpha_1 \, x(y_1 - \hat{y}_1), \tag{8}$$

where α_1 is learning rate, and k and $(k + 1)$ denote the present and next samples, respectively. Also, $y_1 = -\frac{\Delta v_p}{\Delta i_p} - N r_{sw}$, $\hat{y}_1 = (S_U)^T \times R_{bt}$, and $x = S_U \times W_1^{(k+1)}$ is the updated estimate of the internal resistance vector (R_{bt}) and $W_1^{(0)}$ the initial estimate. The vector of internal resistance (R_{bt}) is updated in each-time step based on (8) under the condition that $S_U^{(k)} = S_U^{(k+1)}$. If the $S_U^{(k)} \neq S_U^{(k+1)}$ then the R_{bt} retains its previous value. In the next step, the vector of open-circuit voltages V_{oc} is updated through a similar procedure, either using the new R_{bt} or using its old value. Equation (3) can be rewritten as

$$v_p^{(k+1)} + i_p^{(k+1)} \left(S_U^{(k+1)} \right)^T \times R_{bt} + N r_{sw} = \left(S_U^{(k+1)} \right)^T \times V_{oc}. \tag{9}$$

Similar to the procedure for updating the internal resistance vector, V_{oc} vector is estimated using

$$W_2^{(k+1)} = W_2^{(k+1)} + \alpha_2 x (y_2 - \hat{y}_2), \tag{10}$$

where α_2 is the learning rate, $y_2 = v_p^{(k+1)} + i_p^{(k+1)} \left(S_U^{(k+1)} \right)^T \times R_{bt} + Nr_{sw}$, $\hat{y}_2 = \left(S_U^{(k+1)} \right)^T V_{oc}$, $x = S_U^{(k+1)}$, and $W_2^{(k+1)}$ is the updated vector of open-circuit voltages of the batteries.

The convergence speed of (8) and (10) is determined according to the learning rate. To optimize the convergence speed, a line search algorithm is used to optimize the value of the learning rates α_1 and α_2. The details of optimizing the learning rate using the line search algorithm are presented in Appendix I and II. Based on the optimization results, the optimum learning rates that result into the fastest convergence for (8) and (10) are identical and equal to $\alpha_1 = \alpha_2 = \frac{1}{x^T x}$. Therefore, the optimum estimators are

$$\begin{cases} W_{1,k+1} = W_{1,k} + x^T \frac{(y_2 - \hat{y}_2)}{x\, x^T}, \\ W_{2,k+1} = W_{2,k} + x^T \frac{(y_2 - \hat{y}_2)}{x \times x^T}. \end{cases} \tag{11}$$

After updating both R_{bt} and V_{oc}, the pack voltage is estimated using (3) and compared to the actual measurement, in case the difference is lower than a minimum threshold, the estimator has converged, and the values are usable.

Figure 5 presents the flowchart of the proposed decoupling and estimation method. Based on the algorithm flowchart, the controller receives two consecutive samples (k) and $(k + 1)$ of pack voltage and current as well as the switching states of the modules. If the controller receives identical switching states and detects a variation in current, then the controller will update the previous estimation of the internal resistance vector according to (8), otherwise this step is skipped. Then the controller updates the vector of open-circuit voltages using (10) and goes to the beginning of the algorithm. When the output voltage estimation of the pack is close to the measured value for more than N iterations, then we can conclude that the estimator has converged to its final values.

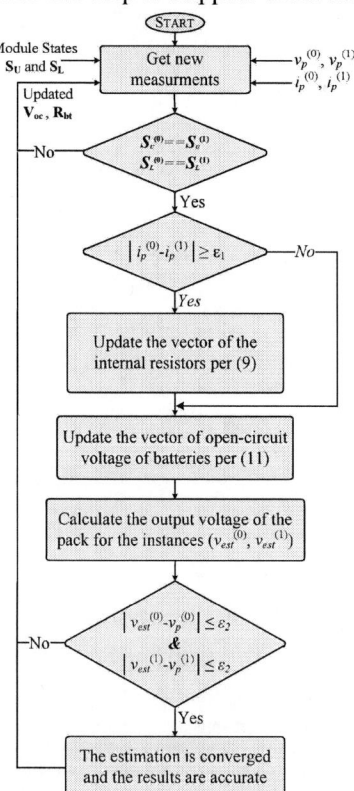

Fig. 5. The proposed back-propagation algorithm

Results and Discussions

Detailed simulations in MATLAB/Simulink study the performance of the proposed estimation technique using a single-arm modular battery system. Table I provides the main parameters of the simulated system, with eight half-bridge modules. In the simulation, batteries are modelled using a simplified electrical equivalent circuit consisting of an internal resistance as well as a constant dc voltage source. The modular battery is feeding a resistive load.

Fig. 6 (a) and (b) present the output voltage and current of the battery pack as well as the load. At $t = 0.25$ s, the modulation index is changed from 0.4 to 0.8. Figure 7 depicts the results of the average estimation error for output voltage and internal resistance of battery cells and the estimation results of output voltage and internal resistance of battery cells for all modules during ideal condition. We consider ideal condition to be the condition where all the modules have identical parameters. While the actual internal resistance of all modules is 5 mΩ, the estimation vector is initialized with every resistance equal to

Table I Parameters of single arm
MMC

PARAMETER	VALUE
V_{dc}	$400 - 800 \, [V]$
C_{dc}	$10 \, [mF]$
L_{dc}	$1 \, [mH]$
R_{ldc}	$10 \, [m\Omega]$
R_{load}	$3.2 \, [\Omega]$
R_{ds}	$1 \, [m\Omega]$
R_d	$1 \, [m\Omega]$
$v_{oc,1-8}$	$120 \, [V]$
$r_{bt,1\sim8}$	$5 \, [m\Omega]$

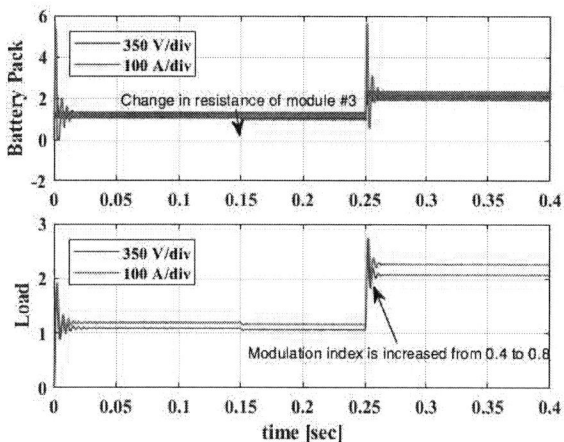

Fig. 6 (a) The output voltage and current of battery pack (b) The output voltage and current of load

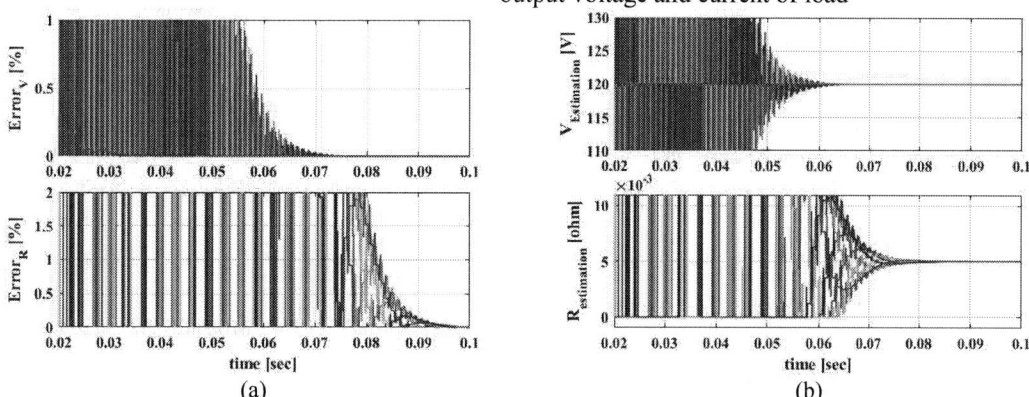

Fig. 7 simulation results for ideal conditions(a) estimation errors; (b) estimation values for open-circuit voltage and internal resistance of battery modules

1 Ω. Consequently, the error is significant at the beginning. However, once the estimator is activated, it converges in less than 90 ms.

Figure 8(a) illustrates the behavior of the estimator when the internal resistance of one module is changed to 0.1 Ω to simulate an abnormal condition. The resistance of module three is changed at $t = 0.2$ s and Fig. 8(b) show that the estimation vector converges to correct values in 5 ms. The results validate the accuracy and high speed of the proposed algorithm which is proper for industrial

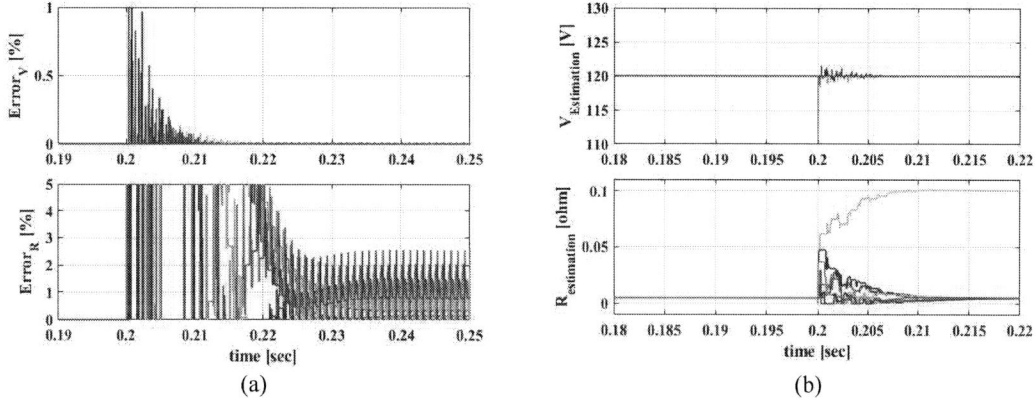

Fig. 8 simulation results for abnormal conditions(a) estimation errors; (b) estimation values for open-circuit voltage and internal resistance of battery modules

applications.

Conclusions

This paper presents a decoupling technique as well as a back-propagation algorithm to estimate open-circuit voltage and internal resistance battery-based modules in a modular battery system. The proposed estimation technique exploits the already available data about the output voltage and current of the pack as well as the switching states of the modules and requires no extra measurement for estimation purposes. The line search optimization maximizes the convergence speed of the algorithm, which is confirmed through simulations. The proposed technique benefits from low-cost implementation as it requires fewer sensors and little computation. Furthermore, the proposed estimation algorithm can be easily extended for modular battery systems with several arms.

APPENDIX I

To have the maximum speed of convergence for estimating R_{bt}, an optimization function $C_1\left(R_{bt}^{(k)}\right)$ is defined based on a least square scheme to calculate the difference between $y_1 = -\frac{\Delta V_p}{\Delta I_p} - Nr_{sw}$, and $\hat{y}_1 = (S_U)^T R_{bt}$ as

$$C_1\left(R_{bt}^{(k)}\right) = \frac{1}{2}\left[(S_U)^T R_{bt}^{(k)} + \frac{\Delta v_p}{\Delta i_p} + Nr_{sw}\right]^2. \tag{12}$$

Gradient of $C_1(R_{bt}^{(k)})$ over $R_{bt}^{(k)}$ is presented by $S^{(k)}$ as

$$S^{(k)} = -\nabla C_1 = -S_U\left((S_U)^T R_{bt}^{(k)} + A_1\right), \tag{13}$$

where $A_1 = \frac{\Delta v_o^{(k)}}{\Delta i_o^{(k)}} + Nr_{sw}$. We can use $S^{(k)}$ to update $R_{bt}^{(k)}$ in the direction that reduces the error (i.e., reduces $C_1(R_{bt})$) as follows

$$R_{bt}^{(k+1)} = R_{bt}^{(k)} + \alpha_1^{(k)} S^{(k)}. \tag{14}$$

The value of α_1 determines the convergence speed and to achieve maximum convergence speed we will use the exact line search algorithm for optimizing the value of α_1. Using the gradient of $C_1(R_{bt}^{(k+1)})$ over α_1 and equating it with zero results into

$$C_1'^{(k+1)}(\alpha^*) = \nabla C_1\left(R_{bt}^{(k)} + \alpha^* S^{(k)}\right)^T S^{(k)} = 0, \tag{15}$$

where α_1^* is the optimum value for α_1.

By substituting (12) and (14), we can rewrite (15) per

$$\left[(S_U)^T\left(R_{bt}^{(k)} + \alpha^* S^{(k)}\right) + A_1\right]^T (S_U)^T\left[S_U\left((S_U)^T R_{bt}^{(k)} + A_1\right)\right] = 0. \tag{16}$$

Substituting (13) in (16) results in

$$\left[(S_U)^T R_{bt}^{(k)} + \alpha^* (S_U)^T\left(-S_U\left((S_U)^T R_{bt}^{(k)} + A_1\right)\right) + A_1\right]^T (S_U)^T\left[S_U\left((S_U)^T R_{bt}^{(k)} + A_1\right)\right] = 0. \tag{17}$$

After some simplifications and rearrangements, (17) turns into

$$\left[(S_U)^T R_{bt}^{(k)} - \alpha^* (S_U)^T S_U\left((S_U)^T R_{bt}^{(k)} + A_1\right) + A_1\right]^T (S_U)^T S_U\left((S_U)^T R_{bt}^{(k)} + A_1\right) = 0, \tag{18}$$

which can be solved for α^* per

$$\alpha^* = \frac{1}{(S_U)^T S_U}. \tag{19}$$

APPENDIX II

To have the maximum speed of convergence to estimate V_{oc}, an optimization function $C_1\left(V_{oc}^{(k)}\right)$ is defined based on least square technique to calculate the difference between $y_2 = v_p^{(k)} + i_p^{(k)}\left[(S_U)^T R_{bt}^{(k)} + N r_{sw}\right]$, and $\hat{y}_2 = S_U V_{oc}^{(k)}$ as

$$C_2\left(V_{oc}^{(k)}\right) = \frac{1}{2}\left[v_p^{(k)} + i_p^{(k)} R_{eq2} - (S_U)^T V_{oc}^{(k)}\right]^2, \tag{20}$$

where $R_{eq2} = (S_U)^T R_{bt} + N r_{sw}$. Similar to Appendix I, gradient of $C_1(V_{oc}^{(k)})$ over $V_{oc}^{(k)}$ is represented by $S^{(k)}$ as follows

$$S^{(k)} = -\nabla C_2 = -S_U^T \left(A_2 - (S_U)^T V_{oc}^{(k)}\right), \tag{21}$$

where $A_2 = v_p^{(k)} + i_p^{(k)} R_{eq2}$. Therefore, the vector of the open-circuit voltages can be updated using

$$V_{bt}^{(k+1)} = V_{oc}^{(k)} + \alpha^{(k)} S^{(k)}. \tag{22}$$

Similarly, the exact line search algorithm can be used to find the optimum value of the learning rate α_2. Gradient of $C_2(V_{oc}^{(k+1)})$ over α_2 is

$$\nabla C_2 \left(V_{oc}^{(k+1)}\right)^T S^{(k)} = 0. \tag{23}$$

Replacing (22) and (23) in (24) results into

$$\left[-S_U A_2 + S_U (S_U)^T \left[V_{oc}^{(k)} - \alpha^* S_U \left(A_2 - (S_U)^T V_{oc}^{(k)}\right)\right]\right]^T \left[S_U \left(A_2 - (S_U)^T V_{oc}^{(k)}\right)\right] = 0. \tag{24}$$

With some simplifications, (24) is rearranged into

$$(S_U)^T \left[-A_2 + (S_U)^T V_{oc}^{(k)} + \alpha^* (S_U)^T S_U \left(A_2 - (S_U)^T V_{oc}^{(k)}\right)\right] \left[A_2 - (S_U)^T V_{oc}^{(k)}\right] S_U = 0, \tag{25}$$

which has two roots per

$$\begin{cases} A_2 = (S_U)^T V_{oc}^{(k)} \\ -A_2 + (S_U)^T V_{oc}^{(k)} + \alpha^* (S_U)^T S_U \left(A_2 - (S_U)^T V_{oc}^{(k)}\right) = 0 \end{cases} \tag{26}$$

While one of the roots is independent of α_2 and only happens if the estimator is converged to the exact open-circuit voltage, the other one can be solved for α_2 per

$$\alpha^* = \frac{1}{(S_U)^T S_U}. \tag{27}$$

References

[1] N. Tashakor, V. Monteiro, T. Ghanbari, and E. Farjah, "An Improved Modular Charge Equalization Structure for Series Cascaded Battery," presented at the 2019 27th Iranian Conference on Electrical Engineering (ICEE), 30 April-2 May 2019, 2019.

[2] A. N. Link, O'Connor, A. C., Scott, T. J., "Battery Technology for Electric Vehicles.," *London: Routledge, ,* 2015, doi: https://doi.org/10.4324/9781315749303.

[3] C. Jung, "Power Up with 800-V Systems: The benefits of upgrading voltage power for battery-electric passenger vehicles," *IEEE Electrification Magazine,* vol. 5, no. 1, pp. 53-58, 2017, doi: 10.1109/MELE.2016.2644560.

[4] J. Fang, Y. Tang, H. Li, and X. Li, "A Battery/Ultracapacitor Hybrid Energy Storage System for Implementing the Power Management of Virtual Synchronous Generators," *IEEE Transactions on Power Electronics,* vol. 33, no. 4, pp. 2820-2824, 2018, doi: 10.1109/TPEL.2017.2759256.

[5] M. Gjelaj, S. Hashemi, C. Traeholt, and P. B. Andersen, "Grid integration of DC fast-charging stations for EVs by using modular li-ion batteries," *IET Generation, Transmission & Distribution,* vol. 12, no. 20, pp. 4368-4376, 2018, doi: 10.1049/iet-gtd.2017.1917.

[6] S. M. Goetz, Z. Li, A. V. Peterchev, X. Liang, C. Zhang, and S. M. Lukic, "Sensorless scheduling of the modular multilevel series-parallel converter: enabling a flexible, efficient, modular battery," presented at the 2016 IEEE Applied Power Electronics Conference and Exposition (APEC), 2016.

[7] M. Quraan, P. Tricoli, S. D'Arco, and L. Piegari, "Efficiency assessment of modular multilevel converters for battery electric vehicles," *IEEE Transactions on Power Electronics,* vol. 32, no. 3, pp. 2041-2051, 2016.

[8] C. Gan, Q. Sun, J. Wu, W. Kong, C. Shi, and Y. Hu, "MMC-Based SRM Drives With Decentralized Battery Energy Storage System for Hybrid Electric Vehicles," *IEEE Transactions on Power Electronics,* vol. 34, no. 3, pp. 2608-2621, 2019, doi: 10.1109/TPEL.2018.2846622.

[9] N. Tashakor, E. Farjah, and T. Ghanbari, "A Bidirectional Battery Charger With Modular Integrated Charge Equalization Circuit," *IEEE Transactions on Power Electronics,* vol. 32, no. 3, pp. 2133-2145, 2017, doi: 10.1109/TPEL.2016.2569541.

[10] J. Fang, S. Yang, H. Wang, N. Tashakor, and S. Goetz, "Reduction of MMC capacitances through parallelization of symmetrical half-bridge submodules," *IEEE Transactions on Power Electronics,* 2021.

[11] M. Priya, P. Ponnambalam, and K. Muralikumar, "Modular-multilevel converter topologies and applications – a review," *IET Power Electronics,* vol. 12, no. 2, pp. 170-183, 2019, doi: 10.1049/iet-pel.2018.5301.

[12] J. Fang, H. Deng, N. Tashakor, F. Blaabjerg, and S. M. Goetz, "State-Space Modeling and Control of Grid-Tied Power Converters with Capacitive/Battery Energy Storage and Grid-Supportive Services," *IEEE Journal of Emerging and Selected Topics in Power Electronics,* pp. 1-1, 2021, doi: 10.1109/JESTPE.2021.3101527.

[13] Y. Li and Y. Han, "A Module-Integrated Distributed Battery Energy Storage and Management System," *IEEE Transactions on Power Electronics,* vol. 31, no. 12, pp. 8260-8270, 2016, doi: 10.1109/TPEL.2016.2517150.

[14] S. Gonzalez, Verne, S., Valla, M. , "Multilevel Converters for Industrial Applications," 2014, doi: 10.1201/b15252.

[15] D. Ronanki and S. S. Williamson, "Modular multilevel converters for transportation electrification: Challenges and opportunities," *IEEE Transactions on Transportation Electrification,* vol. 4, no. 2, pp. 399-407, 2018.

[16] Z. Li, R. Lizana, S. Sha, Z. Yu, A. V. Peterchev, and S. Goetz, "Module Implementation and Modulation Strategy for Sensorless Balancing in Modular Multilevel Converters," *IEEE Transactions on Power Electronics,* 2018.

[17] N. Tashakor, M. Kilictas, E. Bagheri, and S. Goetz, "Modular Multilevel Converter with Sensorless Diode-Clamped Balancing through Level-Adjusted Phase-Shifted Modulation," *IEEE Transactions on Power Electronics,* pp. 1-1, 2020, doi: 10.1109/TPEL.2020.3041599.

[18] A. Ghazanfari and Y. A. I. Mohamed, "A Hierarchical Permutation Cyclic Coding Strategy for Sensorless Capacitor Voltage Balancing in Modular Multilevel Converters," *IEEE Journal of Emerging and Selected Topics in Power Electronics,* vol. 4, no. 2, pp. 576-588, 2016, doi: 10.1109/JESTPE.2015.2460672.

[19] G. Chen, H. Peng, R. Zeng, Y. Hu, and K. Ni, "A Fundamental Frequency Sorting Algorithm for Capacitor Voltage Balance of Modular Multilevel Converter With Low-Frequency Carrier Phase Shift Modulation," *IEEE Journal of Emerging and Selected Topics in Power Electronics,* vol. 6, no. 3, pp. 1595-1604, 2018, doi: 10.1109/JESTPE.2017.2764684.

[20] X. Hui, F. Yatao, and W. Yiying, "Review of equalizing methods for battery pack," presented at the Electrical Machines and Systems (ICEMS), 2014 17th International Conference on, 22-25 Oct. 2014, 2014.

[21] T. Kacetl, J. Kacetl, N. Tashakor, M. Jaensch, and S. Goetz, "Degradation-Reducing Control for Dynamically Reconfigurable Batteries," *arXiv preprint arXiv:2202.11757,* 2022.

[22] N. Tashakor, B. Arabsalmanabadi, F. Naseri, and S. Goetz, "Low-Cost Parameter Estimation Approach for Modular Converters and Reconfigurable Battery Systems Using Dual-Kalman-Filter," *IEEE Transactions on Power Electronics,* pp. 1-1, 2021, doi: 10.1109/TPEL.2021.3137879.

[23] B. Arabsalmanabadi, N. Tashakor, Y. Zhang, K. Al-Haddad, and S. Goetz, "Parameter Estimation of Batteries in MMCs with Parallel Connectivity using PSO," presented at the IECON 2021 – 47th Annual Conference of the IEEE Industrial Electronics Society, 13-16 Oct. 2021, 2021.

[24] H. B. I, S. Rivera, Z. Li, S. Goetz, A. Peterchev, and R. L. F, "Different parallel connections generated by the Modular Multilevel Series/Parallel Converter: an overview," presented at the IECON 2019 - 45th Annual Conference of the IEEE Industrial Electronics Society, 14-17 Oct. 2019, 2019.

[25] N. Tashakor, Z. Li, and S. M. Goetz, "A Generic Scheduling Algorithm for Low-Frequency Switching in Modular Multilevel Converters with Parallel Functionality," *IEEE Transactions on Power Electronics,* pp. 1-1, 2020, doi: 10.1109/TPEL.2020.3018168.

[26] O. S. M. Abushafa, S. M. Gadoue, M. S. A. Dahidah, D. J. Atkinson, and P. Missailidis, "Capacitor Voltage Estimation Scheme With Reduced Number of Sensors for Modular Multilevel Converters," *IEEE Journal of Emerging and Selected Topics in Power Electronics,* vol. 6, no. 4, pp. 2086-2097, 2018, doi: 10.1109/JESTPE.2018.2797245.

[27] S. D. Arco and J. A. Suul, "Estimation of sub-module capacitor voltages in modular multilevel converters," in *2013 15th European Conference on Power Electronics and Applications (EPE)*, 2-6 Sept. 2013 2013, pp. 1-10, doi: 10.1109/EPE.2013.6631931.

[28] M. Abdelsalam, S. Tennakoon, H. Diab, and M. I. Marei, "An ADALINE based capacitor voltage estimation algorithm for modular multilevel converters," in *2016 19th International Symposium on Electrical Apparatus and Technologies (SIELA)*, 29 May-1 June 2016 2016, pp. 1-4, doi: 10.1109/SIELA.2016.7542968.

[29] P. Poblete, G. Pizarro, G. Droguett, F. Nunez, P. Judge, and J. Pereda, "Distributed Neural Network Observer for Submodule Capacitor Voltage Estimation in Modular Multilevel Converters," *IEEE Transactions on Power Electronics,* pp. 1-1, 2022, doi: 10.1109/TPEL.2022.3163395.

[30] M. D. Islam, R. Razzaghi, and B. Bahrani, "Arm-Sensorless Sub-Module Voltage Estimation and Balancing of Modular Multilevel Converters," *IEEE Transactions on Power Delivery,* vol. 35, no. 2, pp. 957-967, 2020, doi: 10.1109/TPWRD.2019.2931287.

[31] G. Pizarro, P. M. Poblete, G. Droguett, J. Pereda, and F. Nunez, "Extended Kalman Filtering for Full State Estimation and Sensor Reduction in Modular Multilevel Converters," *IEEE Transactions on Industrial Electronics,* pp. 1-1, 2022, doi: 10.1109/TIE.2022.3165286.

[32] F. Rong, X. Gong, X. Li, and S. Huang, "A New Voltage Measure Method for MMC Based on Sample Delay Compensation," *IEEE Transactions on Power Electronics,* vol. 33, no. 7, pp. 5712-5723, 2018, doi: 10.1109/TPEL.2017.2748969.

[33] A. A. Taffese, E. d. Jong, S. D'Arco, and E. Tedeschi, "Online Parameter Adjustment Method for Arm Voltage Estimation of the Modular Multilevel Converter," *IEEE Transactions on Power Electronics,* vol. 34, no. 12, pp. 12491-12503, 2019, doi: 10.1109/TPEL.2019.2907178.

[34] G. Konstantinou, H. R. Wickramasinghe, C. D. Townsend, S. Ceballos, and J. Pou, "Estimation Methods and Sensor Reduction in Modular Multilevel Converters: A Review," in *2018 8th International Conference on Power and Energy Systems (ICPES)*, 21-22 Dec. 2018 2018, pp. 23-28, doi: 10.1109/ICPESYS.2018.8626987.

[35] B. G. Carkhuff, P. A. Demirev, and R. Srinivasan, "Impedance-Based Battery Management System for Safety Monitoring of Lithium-Ion Batteries," *IEEE Transactions on Industrial Electronics,* vol. 65, no. 8, pp. 6497-6504, 2018, doi: 10.1109/TIE.2017.2786199.

[36] B. Arabsalmanabadi, N. Tashakor, A. Javadi, and K. Al-Haddad, "Charging Techniques in Lithium-Ion Battery Charger: Review and New Solution," presented at the IECON 2018 - 44th Annual Conference of the IEEE Industrial Electronics Society, 21-23 Oct. 2018, 2018.

[37] D. N. T. How, M. A. Hannan, M. S. H. Lipu, K. S. M. Sahari, P. J. Ker, and K. M. Muttaqi, "State-of-Charge Estimation of Li-ion Battery in Electric Vehicles: A Deep Neural Network Approach," *IEEE Transactions on Industry Applications,* pp. 1-1, 2020, doi: 10.1109/TIA.2020.3004294.

[38] C. R. Gould, C. M. Bingham, D. A. Stone, and P. Bentley, "New Battery Model and State-of-Health Determination Through Subspace Parameter Estimation and State-Observer Techniques," *IEEE Transactions on Vehicular Technology,* vol. 58, no. 8, pp. 3905-3916, 2009, doi: 10.1109/TVT.2009.2028348.

[39] X. Tan *et al.*, "Real-Time State-of-Health Estimation of Lithium-Ion Batteries Based on the Equivalent Internal Resistance," *IEEE Access,* vol. 8, pp. 56811-56822, 2020, doi: 10.1109/ACCESS.2020.2979570.

[40] O. S. H. M. Abushafa, M. S. A. Dahidah, S. M. Gadoue, and D. J. Atkinson, "Submodule Voltage Estimation Scheme in Modular Multilevel Converters with Reduced Voltage Sensors Based on Kalman Filter Approach," *IEEE Transactions on Industrial Electronics,* vol. 65, no. 9, pp. 7025-7035, 2018, doi: 10.1109/TIE.2018.2795519.

[41] N. Tashakor and M. Khooban, "An Interleaved Bi-Directional AC–DC Converter With Reduced Switches and Reactive Power Control," *IEEE Transactions on Circuits and Systems II: Express Briefs,* vol. 67, no. 1, pp. 132-136, 2020, doi: 10.1109/TCSII.2019.2903389.

Method to analyze the influence of switching behavior in hard switching half bridge topologies for traction application

Dominik Nehmer, Michael Gleissner, Lukas Bergmann, Mark-M. Bakran

UNIVERSITY OF BAYREUTH
DEPARTMENT OF MECHATRONICS
CENTER OF ENERGY TECHNOLOGY
Universitaetsstrasse 30
Bayreuth, Germany
Phone: +49 (0) 921 55-7822
Email: dominik.nehmer@uni-bayreuth.de
URL: http://www.mechatronik.uni-bayreuth.de

Acknowledgments

This paper was supported by the Federal Ministry for Economic Affairs and Climate Action based on a decision by the German Bundestag.

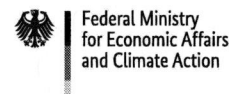

Federal Ministry
for Economic Affairs
and Climate Action

Keywords

≪Harmonics≫, ≪Silicon Carbide (SiC)≫, ≪Traction application≫, ≪Active front-end≫, ≪Voltage Source Converter (VSC)≫.

Abstract

The current harmonics feed back (interference current) of railway traction inverter towards the grid has to be limited. The goal of this paper is to present a method which allows to model harmonics caused by switching behavior and turn on delay. Since simulation is limited by numerical errors due to the step size of the numerical solvers and also takes much time, an analytical method is chosen.

Introduction

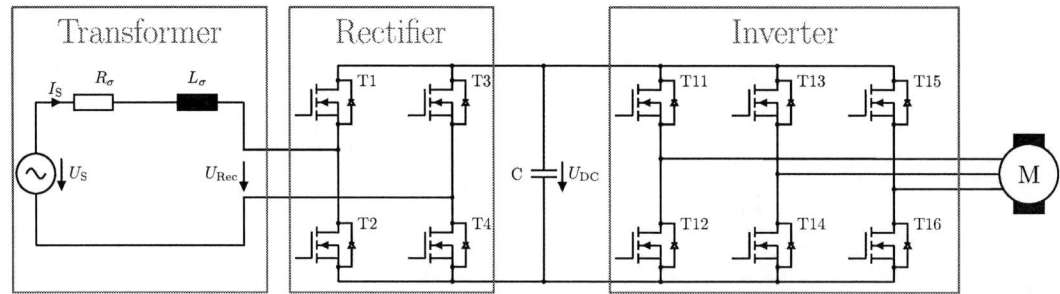

Fig. 1: Simplified circuit diagram of traction inverter

In Figure 1 a simplified circuit diagram of the traction inverter is shown. On the left side the secondary side of a transformer towards the grid is shown. L_σ represents the stray inductance of the transformer related to its secondary side and R_σ its resistance. The transistors T1 to T4 build up an active front end rectifier. Hence the converter is voltage based the intermediate circuit is a capacitor. The transistors

T11-T16 represent an inverter for the electrical machine M. The rectifier and the inverter are based on hard switching half bridges.

In the following the influence of the switching behavior on interference currents towards the grid is analyzed. First the calculation method of harmonics is presented. The next step is the modeling of the switching behavior. After this the method is compared to a simulation model approach. Then effects of the switching behavior on the harmonics based on the voltage spectrum are presented. The influence on the harmonics is compared for a Si IGBT and a SiC MOSFET power module. After this the current spectrum is calculated and the influence of the switching behavior on interference current is evaluated. Finally a conclusion and an outlook is given.

Method

Basic idea

In [1] and [2] an analytical method is presented by using Fourier analysis of carrier signal and superimposition of fundamental wave. For the calculation a Bessel function has to be solved. In [3] even the turn on delay is modeled based on this method. In the following a method is presented to model turn on delay $T_{d,o}$ (to prevent half bridge short circuit) and switching times (equivalent voltage time area of commutation process). Therefore the switching behavior is modeled as a square area and an effective dead time with the same number of volt-seconds as described in [4]. Then the basic PWM signal is calculated using a synchronous sine triangular modulation. After this the switching angles of the PWM are applied by the switching dead times and turn on delay. Finally the Fourier spectrum is calculated.

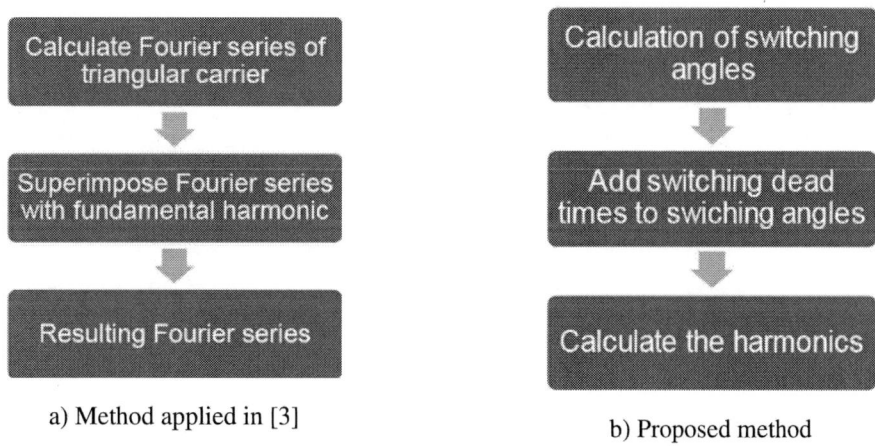

a) Method applied in [3]

b) Proposed method

Fig. 2: Current and proposed method to calculate Fourier coefficients

In Figure 2 b) the method for investigation of the influence on harmonics through switching characteristics is shown. In the first step the switching angles are calculated by using a synchronous sine triangular modulation. In the second step the switching angles are applied by the effective dead time of the switching process. In the last step the harmonics are calculated. Therefore the necessary equations are presented.

Calculation of the switching angles

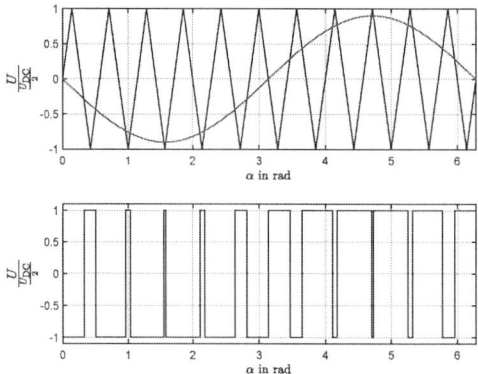

Fig. 3: Synchronous sine triangular modulation with resulting PWM signal

At the top of Figure 3 the fundamental wave (red) and the triangular carrier (blue) are shown. The modulation is a sine triangular modulation with natural sampling. This modulation and sampling scheme is chosen so that there are no harmonics between the carrier bands and the influence of the dead times can be identified clearly. At the bottom of the figure the resulting PWM signal is shown. The edges of the PWM signals are the switching angles without switching dead times.

Adding switching dead times to switching angles

Now the switching angles are added to the effective switching dead time. The dead time includes the turn on delay (to prevent a half bridge short circuit) and the effective voltage-time area of the commutation process. The effective voltage-time area of the commutation process can be expressed as a function of current, temperature and other parameters.

Calculation of the Fourier spectrum

$$
\begin{aligned}
u(\omega t) &= \sum_{k=1}^{\infty} (a_k \cos(k\omega t) + b_k \sin(k\omega t)) = \sum_{k=1}^{\infty} \left(\sqrt{a_k^2 + b_k^2} \cos(k\omega t) - \arctan\left(\frac{b_k}{a_k}\right) \right) \\
&= \sum_{k=1}^{\infty} (A_k \cos(k\omega t - \varphi_k))
\end{aligned}
\tag{1}
$$

Equation 1 expresses the Fourier series of a periodic signal. The series can be described as a combination of sin and cos terms as given in the first row or as a cos term with amplitude and phase as given in the second row. The coefficients a_k and $b_\mathbf{k}$ have to be calculated. Therefore the following integral has to be solved sectionally:

$$
\begin{aligned}
a_k &= \frac{1}{\pi} \int_{-\pi}^{\pi} (u(x)\cos(kx))\, dx \\
&= \frac{1}{\pi} \left[\int_{-\pi}^{\alpha_1} \left(\frac{U_{DC}}{2} \cos(kx) \right) dx - \int_{\alpha_1}^{\alpha_2} \left(\frac{U_{DC}}{2} \cos(kx) \right) dx \pm \ldots - \int_{\alpha_l}^{\pi} \left(\frac{U_{DC}}{2} \cos(kx) \right) dx \right] \\
&= \frac{U_{DC}}{2k\pi} \left[(\sin(k\alpha_1)) - \sin(-k\pi) - (\sin(k\alpha_2)) - \sin(k\alpha_1) \pm \ldots - (\sin(k\pi)) - \sin(k\alpha_l) \right] \\
&= \frac{U_{DC}}{k\pi} \left(\sum_{n=1}^{l} (-1)^{n+1} \sin(k\alpha_n) \right)
\end{aligned}
\tag{2}
$$

The first line shows the definition of the coefficient a_k [5]. In the second line the integral is divided in parts. Each of the parts is separated by the switching angles. In each part the amplitude of the PWM signal is $\pm\dfrac{U_{DC}}{2}$. In the third line the integrals are solved. In the last line the coefficient a_k is expressed as a row of sin terms. The number of terms l equals the number of switching angles.

Now the Fourier spectrum can be calculated analytical. It is not necessary to calculate Bessel functions or to solve differential equations. Additionally, the switching angles can be varied separately, which allows to model switching dead time.

The coefficient b_k can be calculated accordingly and is given as follows:

$$b_{\mathrm{k}} = \frac{1}{\pi} \int_{-\pi}^{\pi} (u(x)\sin(kx))\,dx = \frac{U_{DC}}{k\pi}\left(\sum_{n=1}^{l}(-1)^{n+1}\cos(k\alpha_{\mathrm{n}})\right) \tag{3}$$

Switching dead time

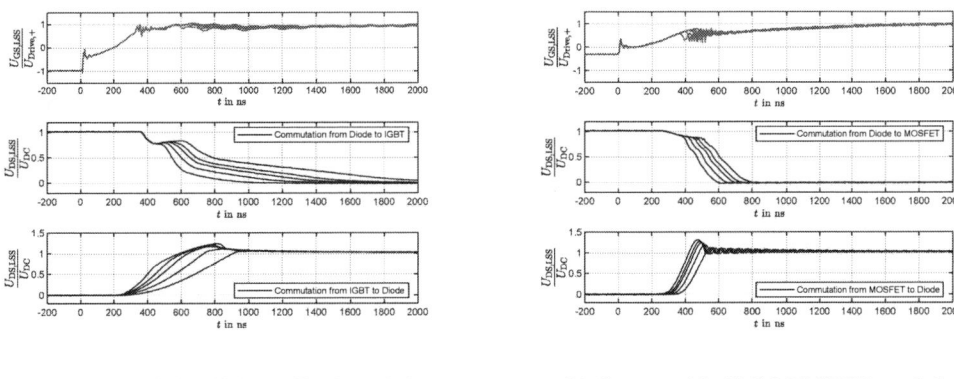

a) Mitsubishi CM600 Si-IGBT module   b) Comparable SiC-MOSFET module

Fig. 4: Voltage curves for different load currents of the traction inverter modules

In Figure 4 the voltage commutations for a Si-IGBT from U_{DC+} to U_{DC-} (middle) and from U_{DC-} to U_{DC+} (bottom) are shown. As already described in [4] the voltage slope can be modeled as a square wave and effective dead time. Both should have the same voltage-time area since it causes the additional harmonics.

Calculation of switching dead times

The effective switching-on time can be expressed as follows:

$$T_{\mathrm{on}} = \frac{1}{U_{DC}} \int_{0}^{t_{\mathrm{switch,end}}} (u(t))\,dt \tag{4}$$

Equivalent to the switching-on time the switching-off time is:

$$T_{\mathrm{off}} = \frac{1}{U_{DC}} \int_{0}^{t_{\mathrm{switch,end}}} (U_{DC} - u(t))\,dt \tag{5}$$

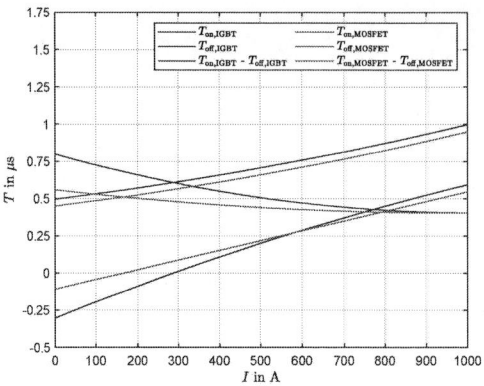

Fig. 5: Equivalent switching times of Si-IGBT and SiC-MOSFET as function of the current

In Figure 5 the switching times of Si IGBT an SiC MOSFET are shown. Both are moduls designed for railway traction application and have the same power rating. Also the difference between the turn-on and turn-off time is shown, since the difference causes an error of the voltage-time areas.

Switching angles with dead times

Table I: Relevant dead times for the different switching edges

	$U_{\mathrm{DC+}} \to U_{\mathrm{DC-}}$	$U_{\mathrm{DC-}} \to U_{\mathrm{DC+}}$
I > 0	T_{off}	$T_{\mathrm{on}} + T_{\mathrm{d,on}}$
I < 0	$T_{\mathrm{on}} + T_{\mathrm{d,on}}$	T_{off}

In Table I the relevant dead times for positive and negative voltage edges as well as positive and negative phase current are shown.

Comparison of methods

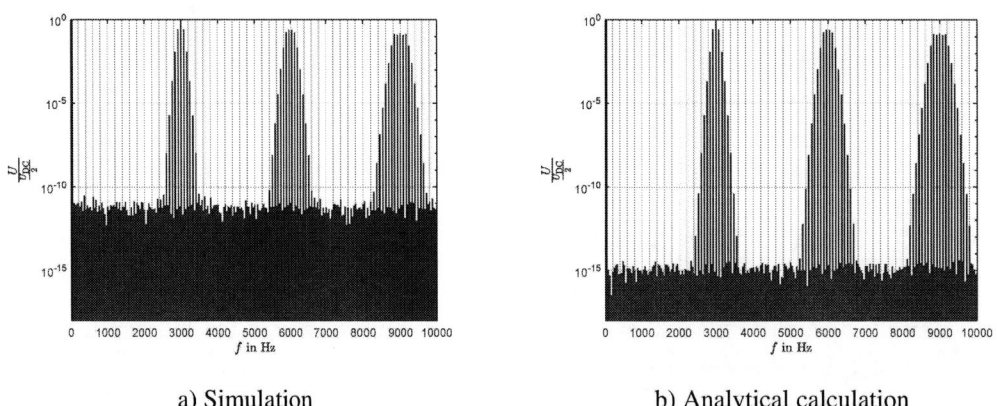

              a) Simulation   b) Analytical calculation

Fig. 6: Comparison of Fourier spectrum error precision of different methods ($f_1 = 200\,Hz$, $f_S = 3\,kHz$, $U = 0.9\,U_{\mathrm{max}}$, $I = I_{\mathrm{N}}$, $\varphi_{\mathrm{I}} = 0°$)

In Figure 6 the Fourier spectrum of a simulation approach and the analytical method without dead times are shown in a logarithmic scale so that the errors are visible. The simulation spectrum was evaluated using PLECS and its FFT function. The simulation approach shows a much higher error harmonics between the carrier bands. This harmonics are caused by the window function [6] and mainly by the step size of the numerical solver. This effect leads to long simulation times and a high numerical error by simulating the inverter. The analytical method shows much lower error. An error is caused by numerical solving of the sine triangular modulation. Since the numerical solver has to calculate the points of intersection and not to solve a differential equation the precision is much more accurate. The calculation of the Fourier coefficients is limited in precision due to sin or cos evaluation and the data type. This error is also very low compared to solving the differential equations numerically.

The analytical method also requires significantly less computing time. While the analytical calculation in a MATLAB script takes about 30 seconds up to 3 minutes, the simulation and FFT in PLECS takes about 5 to 45 minutes. The required computing time for both methods increases with the ratio of f_S to f_1 and the maximum considered frequency of the FFT. In order to increase the accuracy of the simulation, the relative error of the solver must be reduced. This significantly increases the simulation time.

Dead time effects on harmonics

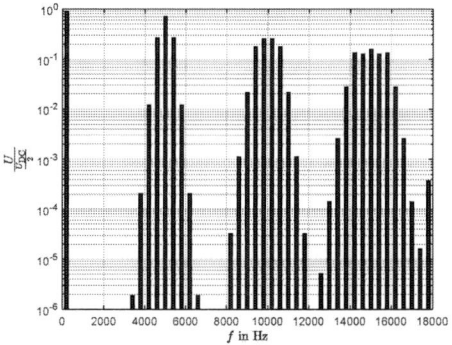
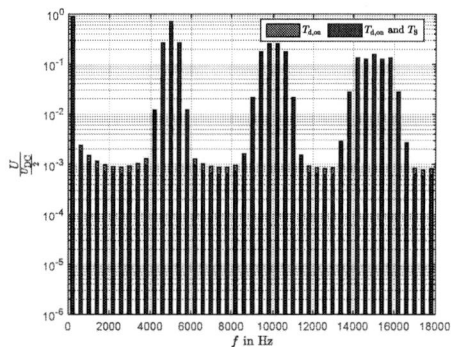

a) Reference without dead times b) With dead times of Si IGBT module

Fig. 7: Fourier spectrum comparison without dead time, turn on delay and turn on delay with switching dead time ($f_1 = 200\,Hz$, $f_S = 5\,kHz$, $U = 0.9\,U_{max}$, $I = I_N$, $\varphi_I = 0°$)

In Figure 7 the Fourier spectrum without dead times, turn on delay as well ($T_{d,on}$) as turn on delay and switching times ($T_{d,on}$ and T_S) are shown. In case of no dead times the fundamental wave, the carrier and sideband harmonics are clearly identifiable. If switching dead times are taken into account, additional harmonics can be seen. Especially new harmonics will arise between the carrier bands. As already mentioned the additional harmonics are caused by voltage time area error.

In Figure 7b) comparison between the harmonics with turn on delay and turn on delay with switching behavior is shown. One can observe the harmonics in case of turn on delay and switching times the harmonics are lower then in case of only taking turn on delay into account.

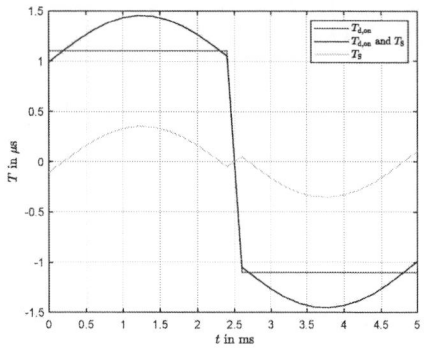 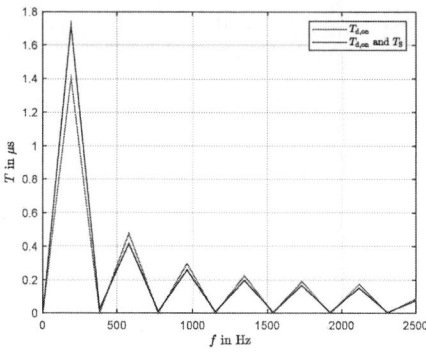

a) Dead time over fundamental wave

b) FFT of dead time

Fig. 8: Explanation for dead times effect ($f_1 = 200\,Hz$, $f_S = 5\,kHz$, $U = 0.9\,U_{\max}$, $I = I_N$, $\varphi_I = 0°$)

In figure 8 the explanation for this effect is shown. On the left side the dead times are plotted over a fundamental wave. The case with turn on delay and switching times is more sinusoidal than the case with turn on delay. Applying a Fast Fourier Transformation (FFT) on the dead times, the effects on harmonics can be explained. The impact in case of turn on delay and an switching times on the fundamental harmonics is great and lower on the other harmonics than in case of only modeling the turn on delay. This could also be observed in the harmonics in 7b).

Comparision of Si IGBT and SiC MOSFET module

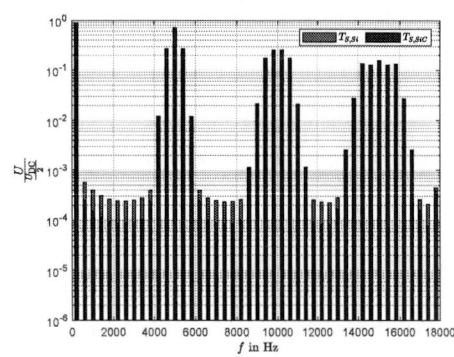 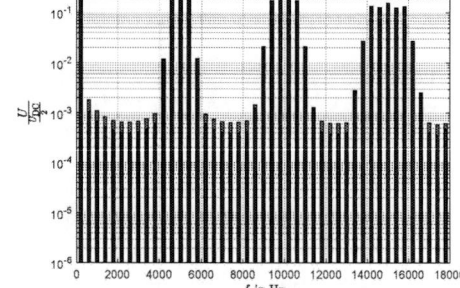

a) Harmonics with switching dead times

b) Harmonics with switching dead times and turn on delay

Fig. 9: Comparison of harmonics with Si IGBT and SiC MOSFET module ($f_1 = 200\,Hz$, $f_S = 5\,kHz$, $U = 0.9\,U_{\max}$, $I = I_N$, $\varphi_I = 0°$)

In Figure 9a) the harmonics for the Si IGBT and the SiC MOSFET modules are compared. The additional harmonics caused by the switching dead times of the Si module are twice as high as for the SiC module. The higher current harmonics are caused by the higher switching dead time of the Si IGBT module as already presented in Figure 5.

Additionally when using the Si IGBT module the switching times are much higher. To prevent a short circuit of the Si module the turn on delay has to be higher as for the fast switching SiC module. This will also cause higher harmonics as shown in Figure 9b). For the Si IGBT module a turn on delay of 1.1 μs and 0.6 μs for the SiC MOSFET module is chosen. The harmonics caused by the turn on delay and switching times of the Si module are also about twice as high as with the SiC module.

Harmonic reduction of parallel rectifiers

For higher power two or more rectifiers have to be switched in parallel. In the following the reduction of harmonics with an appropriate control is presented. Two spectra with a phase offset can be superimposed using the following equation:

$$A_{k,res} = A_k \cos(\omega t - \varphi_{k,1}) + A_k \cos(\omega t - \varphi_{k,2}) = 2A_k \cos\left(\frac{\varphi_{k,1} - \varphi_{k,2}}{2}\right)\cos\left(\omega t - \frac{\varphi_{k,1} + \varphi_{k,2}}{2}\right) \quad (6)$$

The resulting amplitudes of the harmonics normalized to one rectifier can be calculated as follows:

$$|A_{k,res}| = A_k\left(\cos\frac{\varphi_{k,1} - \varphi_{k,2}}{2}\right) \quad (7)$$

Now the phase shift has to be determined. One possibility is the elimination of the carrier harmonics. The condition for the elimination is determined by the equation:

$$\varphi_2 = \varphi_1 + \frac{T_S}{T_1}180° = \varphi_1 + \frac{f_1}{f_S}180° \quad (8)$$

This condition has to be realized with a suitable control.

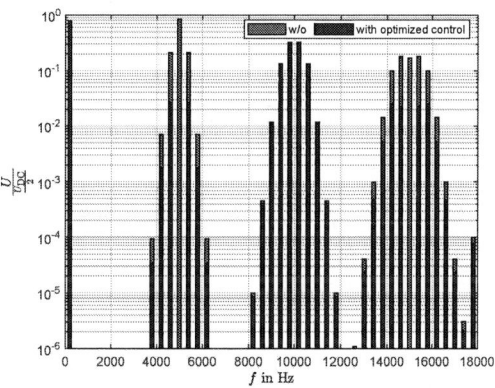

Fig. 10: Harmonic reduction with two parallel rectifiers ($f_1 = 200\,Hz$, $f_S = 5\,kHz$, $U = 0.7\,U_{max}$, $I = I_N$, $\varphi_1 = 0°$)

In figure 10 the harmonics without and with the optimized control are compared. It can be observed that the carrier harmonics in the first and third carrier band are eliminated. Also a reduction of the sideband harmonics can be seen.

Evaluation of interference current

Until now the voltage harmonics was considered. Now an easy approach for calculating the interference current will be presented. Since the harmonics caused by the inverter will be filtered out by the intermediate circuit capacitor, only harmonics caused by the rectifier will be taken into account.

The voltage harmonics can be calculated as presented. With the circuit of the transformer in Figure 1 one can calculate the current harmonics as follows:

$$|\underline{I}_{P,k}| = \frac{|\underline{I}_{S,k}|}{\ddot{u}} = \frac{|\underline{U}_{Rec,k}|}{\ddot{u}\sqrt{(2\pi k f_1 L_\sigma)^2 + (R_\sigma)^2}} \quad (9)$$

To evaluate the influence of the interference current filter bands have to be determined. For this purpose the filter bands 4,75 - 6,25 kHz, 9,5 - 14,5 kHz and 14,51 - 16,5 kHz are chosen (based on the filter bands in [7]).

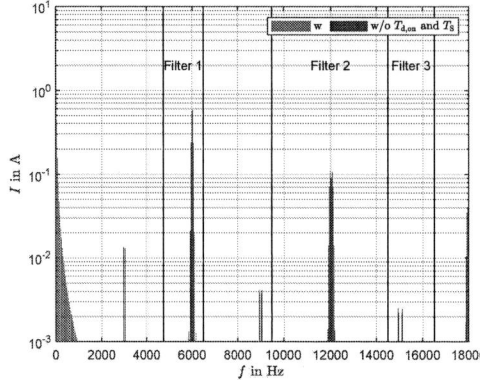

Fig. 11: Interference current ($f_1 = 16.7\,Hz$, $f_S = 3.006\,kHz$, $U = 0.7\,U_{\text{max}}$, $I = I_N$, $\varphi_1 = 0°$)

In Figure 11 the spectrum of the interference current is shown. The harmonics are dominated by the carrier harmonics. The influence of dead times on the harmonics is small. Especially in the filter bands the influence is negligible and thus the filter values are nearly independent of the dead times.

Table II: Filter values for different switching frequencies ($f_1 = 16.7\,Hz$, $U = 0.7\,U_{\text{max}}$, $I = I_N$, $\varphi_1 = 0°$)

f_S	Filter 1	Filter 2	Filter 3
1.002 kHz	260 mA	125 mA	35.8 mA
3.006 kHz	874 mA	219 mA	4.85 mA
5.010 kHz	6.50 mA	525 mA	2.82 mA
7.014 kHz	0.168 mA	376 mA	1.17 mA
9.018 kHz	0.369 mA	0.306 mA	0.624 mA

In table II the filter values for different switching frequencies are shown. For higher switching frequencies several harmonics can be reduced in the different filter bands. Since the limit value for the different filter bands are not equal a appropriate switching frequency has to be chosen. For 9 kHz the carrier harmonics are not inside any filter band and all values are below its limit.

The Si module has about 3.7 times higher switching losses than the SiC module. For the same switching losses a 3.7 higher switching frequency can be applied for the SiC module. This degree of freedom can be used to optimize the filter values and thus not to exceed the permissible limit.

Conclusion and outlook

In this paper an analytic method was presented to evaluate the harmonics caused by turn-on delay and switching dead times. It combines the possibility to model switching behavior as a function of the current and other parameters as well as small numerical error.

Effects caused by the dead times are shown and explained. An evaluation of the dead times and switching behavior on the interference current was presented.

Hence the influence of dead times on interference current is very small, using SiC with faster switching shows no direct benefit. Through lower switching losses with SiC higher frequencies are reachable. This enables a new degree of freedom in which the switching frequency can be chosen to optimize the filter values.

References

[1] D. G. Holmes and T. A. Lipo, "Pulse width modulation for power converters: Principles and practice", vol. 2 of IEEE Press series on power engineering. Piscataway, NJ: IEEE Press, 2003

[2] S. Bernet, "Selbstgeführte Stromrichter am Gleichspannungszwischenkreis". Berlin, Heidelberg: Springer Berlin Heidelberg, 2012

[3] T. W. Rasmussen, A. Vashishtha, and A. Jotwani, "Investigation of harmonics content in pwm natural and regular sampling including dead time and load current phase," in 2020 22nd European Conference on Power Electronics and Applications (EPE'20 ECCE Europe), pp. P.1–P.10, IEEE, 92020

[4] M. Seilmeier, C. Wolz, and B. Piepenbreier, "Modelling and model based compensation of nonideal characteristics of two-level voltage source inverters for drive control application," in 2011 1st International Electric Drives Production Conference, pp. 17–22, IEEE, 092011

[5] L. Papula, "Mathematische Formelsammlung für Ingenieure und Naturwissenschaftler". Wiesbaden: Vieweg+Teubner, 2006

[6] T. Kuttner, "Praxiswissen Schwingungsmesstechnik". Wiesbaden: Springer Fachmedien Wiesbaden, 2015

[7] "Technische Regelung für den Nachweis der elektromagnetischen Verträglichkeit zwischen Schienenfahrzeugen und der Infrastruktur im Geltungsbereich der EBO (TR-EMV) Teil 2 - Nachweis der Einhaltung der Störstromgrenzwerte", https://www.eba.bund.de/SharedDocs/Downloads/DE/Fahrzeuge/Fahrzeugtechnik/EMV/31_Regelung_TR_EMV_Teil_2.pdf?__blob=publicationFile&v=7, 2015

Impact of aluminum casing on high-frequency transformer leakage inductance and AC resistance

Reda BAKRI[1], Xavier MARGUERON[1], Wendell DA CUNHA ALVES[2], Xavier CIMETIERE[1], Frédéric GILLON[1], Antoine BRUYERE[1], Lucian VATAMANU[2]

[1] Univ. Lille, Arts et Metiers Institute of Technology, Centrale Lille, Junia, ULR 2697 - L2EP, F-59000 Lille, France

[2] Valeo Siemens eAutomotive France, 14 avenue des Beguines, 95892, Cergy-Pontoise, France

E-Mail: reda.bakri@centralille.fr

Acknowledgements

This work is supported by the H2020 - KDT JU programme of the European Union under the grant of the TRANSFORM project 'Trusted European SiC Value Chain for a greener Economy' (KDT Grant No. 101007237) and the French investment bank (BPI).

Keywords

«High frequency power converter», «Transformer», «Parasitic inductance », « Eddy current loss », «Finite element analysis»

Abstract

High-Frequency (HF) transformer is a central part of isolated HF power converters. Its parameters, in particular leakage inductance and losses, have a significant impact on the overall converter performances. With the increase of switching frequencies linked to the use of SiC and GaN based active devices, the control of the HF transformer parameters becomes essential. A HF transformer is usually designed without considering its surrounding. However, the latter can have a significant impact on the transformer's performance. In this paper, the effect of an aluminum casing surrounding the transformer is studied and quantified for two parameters: The transformer leakage inductance and the supplementary losses. The goal is to consider the casing effect in the early design stage of power converters.

Introduction

High-power density converters are a global trend in power electronics [1], [2]. Compact and more efficient converters are needed for embedded power electronics in many applications like automotive or avionics for example. In these applications, volume and weight constraints are essential. To reduce the volume and weight of power converters, increasing switching frequencies is necessary. With the emergence of wide band gap semiconductors like SiC and GaN, reaching frequencies up to MHz becomes feasible [3], [4]. These active components present low conduction and switching losses compared to Si active devices. In order to further increase efficiency, soft switching is employed to reduce the switching losses.

In isolated DC-DC converters, high-frequency (HF) transformer is a key component. It insures galvanic isolation and voltage adaptation simultaneously. With this increase of frequency, HF transformers become more and more critical. Indeed, this increase induces a rise of losses in magnetic components due to eddy current and proximity effects. As a result, the component's temperature increases and the cooling system capability needs to be adapted. In general, the converter overall efficiency is penalized [5] and the transformer parasitic elements become more critical. One of the main parameters is the leakage inductance. It has an important effect on the converter performances. In hard switching converters, a high value of leakage inductance could be a source voltage ringing, which results on additional stress and losses on semiconductors. In soft switching converters, a sufficient value of leakage inductance is necessary to discharge the capacitive energy of the switches and guaranty the soft switching condition. If its value is adequate, the use of an additional inductor can be avoided.

Thermal management is also a key element in high density power converters. Good thermal management is mandatory in order to ensure the proper function while not increasing the converter's volume too much. To that end, converter and components can be encased in aluminum container. The latter can have a non-negligible influence on the HF transformer parameters. Nevertheless, the study of the effect of the metallic parts surrounding the component is rarely performed, even if the effect of losses is mentioned in [6] for an automotive application. Therefore, in this paper, the effect of an aluminum casing on HF transformer is investigated, based on Finite Element Analysis (FEA) and measurements. The focus is on leakage inductance variation and supplementary losses induced by the casing. The goal is to highlight phenomena and their impact on the design of HF transformer in DC/DC converters.

The paper is organized as follows: In the first section, the HF transformer and the case test are described. In the second section, FEA is used to highlight and quantify the effect of the aluminum casing on the transformer's losses and leakage inductance. Finally, in the last section, measurements are performed on a prototype to validate the presented results.

High-frequency transformer and case study definition

Fig.1 presents the equivalent circuit of HF transformer. The magnetizing branch is composed from the magnetizing inductance Lm and core loss equivalent resistance R_c. The total leakage inductance L_{lk} and equivalent losses resistance R_{ac} are modeled on the primary side. All the elements of the equivalent circuit depend on frequency f. The effect of casing will be studied on these two specific parameters: L_{lk} and R_{ac}.

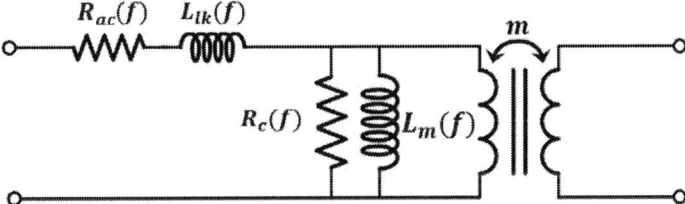

Fig. 1: HF transformer equivalent electrical circuit

For the study, a test case is defined in order to bring out and to quantify the influence of the metallic casing based on FEA and measurements. The studied case is described in Fig. 2a. A 2-winding transformer is designed based on two U-cores (U15/11/16) 3C95 material from Ferroxcube [7]. Primary and secondary windings have each 8 turns of Litz wires with 30 strands of 0.1 mm diameter. A view of the 3D geometry is shown in Fig. 2b. In Fig. 2c, the transformer is put into an aluminum casing. Three casing thickness are considered: 1.5, 2 and 3 mm, with 1 mm margin space between the transformer and the casing's internal faces for each case.

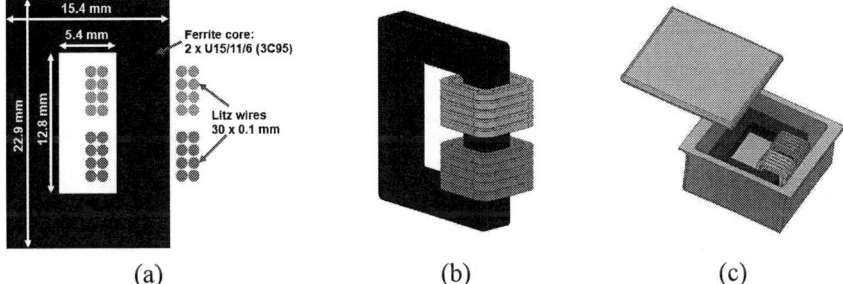

Fig. 2: Case study: (a) 2D view of the transformer, (b) 3D geometry, (c) Transformer inside the aluminum casing

Finite element study of the casing effect

In this section, the FEA based methodology that is performed to analyze the impact of casing is first introduced. Then, the main results are presented and analyzed.

FEA Methodology

As presented in Fig. 2b, the transformer presents a 3D shape that cannot be easily simplified, especially for the leakage flux that follow multiple directions path inside and outside the transformer window. Then, in order to avoid complex 3D FEA with Liz wires, double 2D finite element studies are preferred [8], [9]. A couple of 2D simulations is then needed to reproduce the 3D effects as shown in Fig. 3a. The first simulation is done in the XZ cut plane, containing winding part covered by the core (*i.e.* the transformer window). This is referred as XZ simulation throughout the paper. The second one is done in the YZ cut plane, representing the part of winding outside the window (*i.e.* the end winding). This is referred as YZ simulation in the next parts. The two simulations are combined to compute the global leakage inductance and the lobalwinding resistance. The same analysis is also completed with the aluminum casing (Fig. 3b).

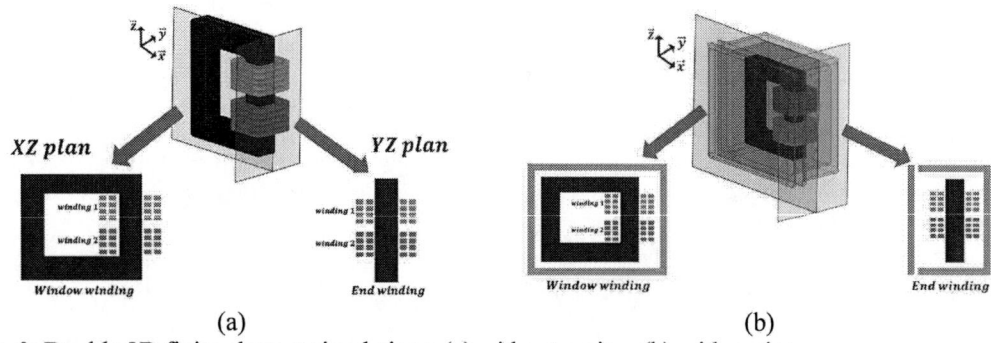

| (a) | (b) |

Fig. 3: Double 2D finite element simulations: (a) without casing, (b) with casing

FEA is performed under Ansys Maxwell Software [10] for several frequencies between 20 Hz and 1 MHz. For these simulations, the geometry of the Litz wires (Fig. 4) is detailed and the 30 strands of every Litz conductor are modeled. Finally, the mesh size is adapted to accurately model HF effects in conductors and casing.

Fig. 4: Litz wire detailed model

The leakage inductance L_{lk} is computed from magnetic energy W_{mag} in short-circuit condition (compensated Amper-turns) and primary current I_{rms} as expressed by equation (1):

$$L_{lk} = \frac{2W_{mag}}{I_{rms}^2} \tag{1}$$

The equivalent R_{ac} resistance is computed from losses P_{losses} in windings and metallic casing (2):

$$R_{ac} = \frac{P_{losses}}{I_{rms}^2} \tag{2}$$

The magnetic energy W_{mag} and losses P_{losses} are deduced by adding the two 2D simulation results (per length unit) weighted by the corresponding mean length as expressed in (3) and (4):

$$L_{lk} = l_{XZ} \cdot L_{lk\_XZ} + l_{YZ} \cdot L_{lk\_YZ} \tag{3}$$

$$R_{ac} = l_{XZ} \cdot R_{ac\_XZ} + l_{YZ} \cdot R_{ac\_YZ} \tag{4}$$

where l_{XZ} and l_{YZ} are the simulations length in the two configurations recalled in Fig.5a. The two simulations length are linked to the core section and the winding thickness as illustrated in Fig.5b. Their values are 10.25mm and 9mm, respectively.

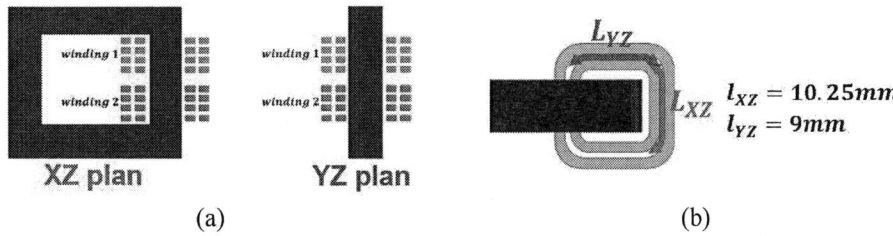

(a)   (b)

Fig. 5: Simulations in XZ and YZ plans: (a) simulation configurations, (b) Simulation lengths

The relative variation of the leakage inductance (5) and winding resistance (6) after adding the casing is computed for every simulated frequency in order to quantify the impact of the metallic casing with the frequency.

$$L_{lk\_variation}(f) = \frac{|L_{lk\_Casing}(f) - L_{lk\_WO\_Casing}(f)|}{L_{lk\_WO\_Casing}(f)} \tag{5}$$

$$R_{ac\_variation}(f) = \frac{|R_{ac\_Casing}(f) - R_{ac\_WO\_Casing}(f)|}{R_{ac\_WO\_Casing}(f)} \tag{6}$$

Double 2D FEA Analysis

Fig. 6a and Fig. 6b presents the leakage inductance for the simulations in XZ and YZ plans without and with casing, respectively. Without casing (Fig. 6a), the XZ plan simulation has higher inductance compared to YZ plan. From Fig. 5b, it could be noticed that the variation due to the casing effect is more significant in YZ plan.

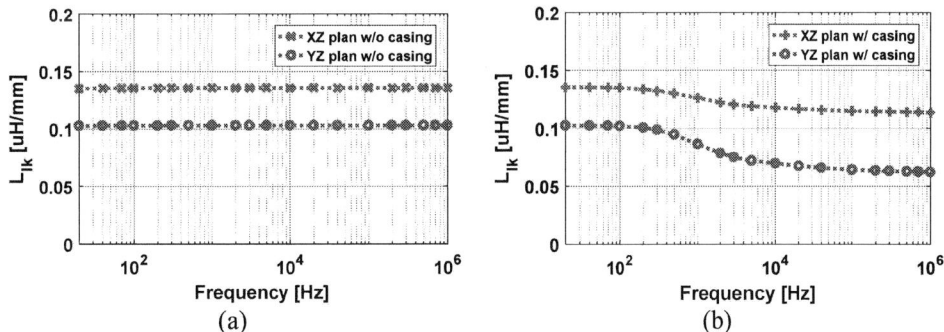

(a)   (b)

Fig. 6: Leakage inductance per unit length for XZ and YZ plans: (a) without casing, (b) with casing

Regarding AC resistance (*i.e.* losses) without casing, the XZ plan simulation has also a higher resistance compared to the YZ plan (Fig. 7a). Adding the casing has more influence on the YZ plan. Then, it becomes close to the XZ plan resistance, especially in HF (Fig.7b).

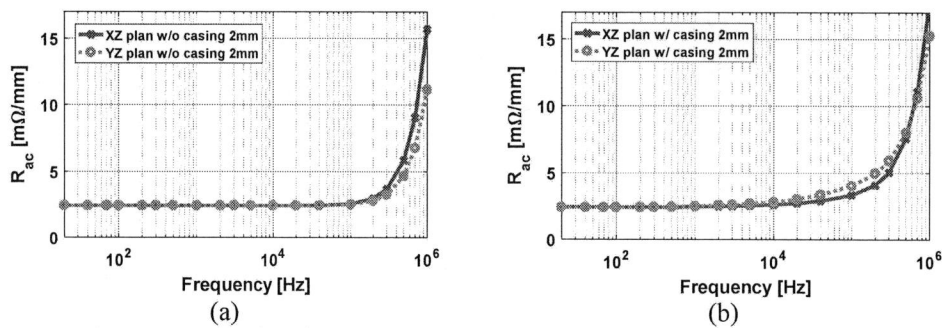

Fig. 7: AC resistance per length unit for XZ and YZ plans: (a) without casing, (b) with casing

L_{lk} and R_{ac} global analysis

From the simulation results and the mean lengths (3) (4), the leakage inductance evolution with frequency is plotted in Fig. 8a, with and without casing. The variation of the leakage inductance obtained for the 3 casing thicknesses is shown in Fig. 8b. It could be noticed that the leakage inductance without casing is very stable with frequency thanks to Litz wire. It significantly drops when the transformer is encased. Indeed, its value is reduced by more than 20% compared to the case without casing. The variation starts from 10 kHz and the leakage value decreases as the frequency increases. The casing impact is then non-negligible on the leakage inductance parameter. From these results one can say that the casing thickness influence is negligible.

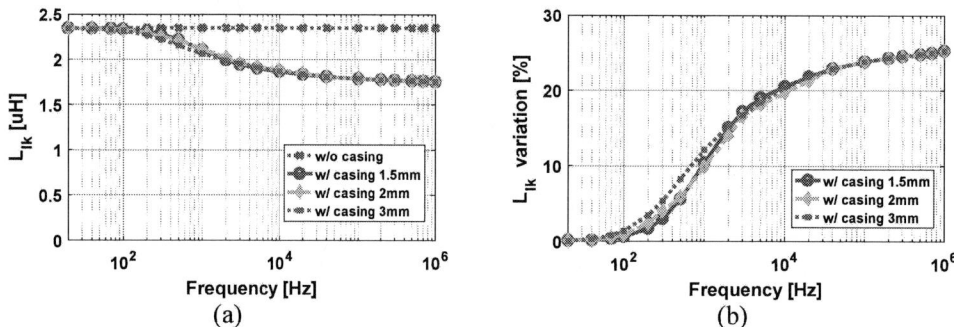

Fig. 8: Leakage inductance vs frequency: (a) values with and without casing, (b) variation (%) after adding the casing

These results on the leakage inductance can be explained by the shielding effect of the aluminum casing in HF. Indeed, in low frequency (LF), the effect of the casing is not significative and the leakage flux pass through it as shown by the flux density distribution in Fig. 9b. Fig. 9a recalls the simulation geometry showing the transformer elements and the casing. In LF, the leakage energy has the same value and is stored in the same space with or without the casing. For higher frequencies, currents are induced in the aluminum casing due to the leakage flux. These currents create an opposite magnetic field. As a consequence, the magnetic energy is only stored in the space inside the casing (Fig. 9c). The magnetic energy is then reduced and so the leakage inductance.

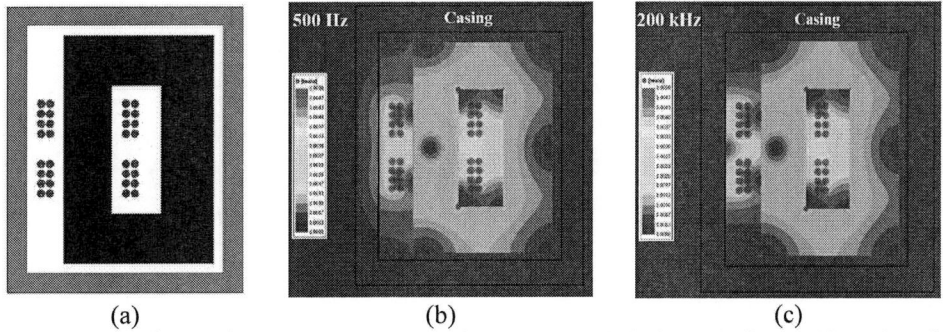

(a)   (b)   (c)

Fig. 9: FEA Simulations: (a) XZ cut plane simulation geometry (window winding) with casing, (b) flux density at 500Hz, (c) flux density at 200kHz

In order to explain the weakness impact of the casing thickness, the skin depth of aluminum is expressed with (7) and plotted for the three-casing thickness (Fig. 10):

$$\delta_{Al} = \frac{1}{\sqrt{\pi \cdot \mu_0 \cdot \sigma_{Al} \cdot f}} \tag{7}$$

where μ_0 is the air permeability $4\pi \cdot 10^{-7}\ H/m$, σ_{Al} is the aluminum electrical conductivity $38 \cdot 10^6\ S$ and f is the frequency.

From 3kHz, the three-casing thickness are greater than aluminum skin depth (Fig. 10). Then, the induced currents have the same distribution for all the casing. As a consequence, from this frequency, the three casing have the same behavior. In LF, the induced currents are negligeable, so the three casing can be assimilated to an air box from electromagnetic point of view.

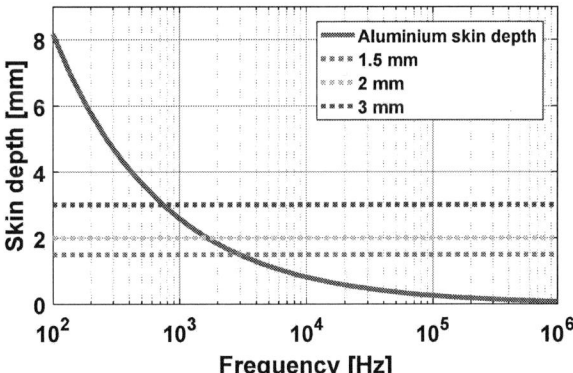

Fig. 10: Aluminum skin depth with frequency

Regarding losses, winding resistance equivalent resistance R_{ac} evolution with frequency is shown in Fig. 11a and its relative variation due to metallic casing in Fig.11b. The presence of the aluminum casing has caused a significant increase on the transformer winding resistance that can exceed 20% for frequencies upper than 100kHz. Above 500 kHz, the effect of the casing is less significant because losses are dominated by the proximity losses of the Litz wires.

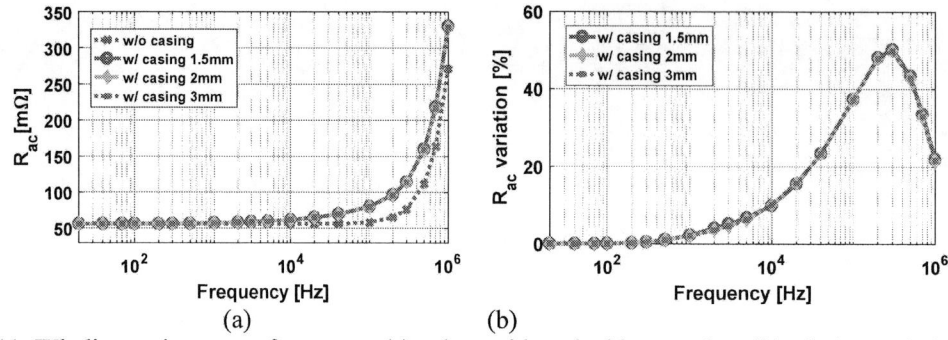

Fig. 11: Winding resistance vs frequency: (a) values with and without casing, (b) relative variation (%) after adding the casing

In order to determine the origin of the AC resistance increase, the latter is split into two parts (8): R_{Cu} represents the copper losses in the transformer due to the Litz windings and R_{Al} represents the induced losses in the aluminum casing. The total resistance R_{ac} is then the sum of the two introduced resistances (8).

$$R_{ac} = R_{Cu} + R_{Al} \qquad (8)$$

Fig. 12a presents the winding copper loss equivalent resistance with and without casing. It can be noticed that the copper losses are not affected by the casing. Hence, the equivalent resistance that increases when the transformer is encased is only due to the induced losses in the casing. To compare the winding copper losses and the induced losses in the casing, the ratio between the resistances linked to each part $\frac{R_{AL}}{R_{Cu}}$ is plotted in Fig.12b. The induced losses in the aluminum casing exceeds 40% of the total copper losses in the frequency range between 100kHz and 500kHz. Those losses are significant and must be evacuated by the thermal management system. Moreover, it impacts negatively the efficiency of the power conversion.

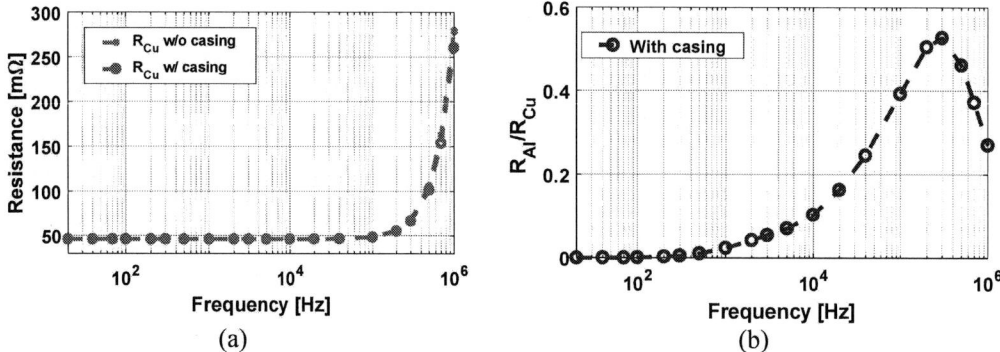

Fig. 12: Resistance vs frequency: (a) winding resistance with and without casing, (b) ratio between casing losses (aluminum) and copper losses

Measurements and experimental validation

A prototype of the defined test case (Fig. 2) has been manufactured in order to validate the FEA results and the L_{lk} and R_{ac} variations. The prototype without and with casing is shown in Fig. 13a and Fig.13b. In addition, three aluminum casings with 1.5mm, 2mm and 3mm thickness respectively, have been manufactured (Fig. 13c).

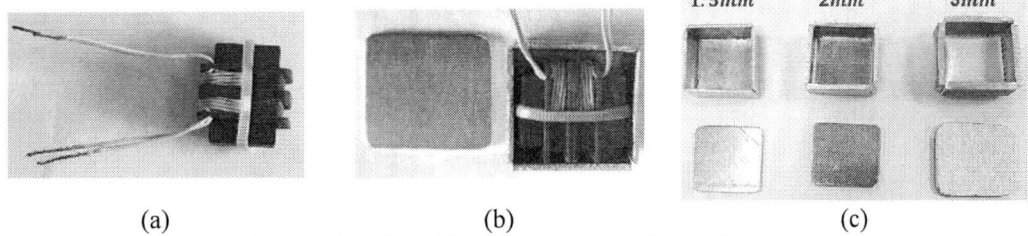

| | | |
| (a) | (b) | (c) |

Fig. 13: Prototype: (a) without casing, (b) with casing, (c) manufactured casings

The transformer characterization is performed using Keysight E4990A high precision impedance analyzer [11] (Fig. 14a). The transformer equivalent leakage inductance and equivalent resistance have been extracted from the impedance Z_{1CC} and phase θ_{1cc} measurements, when the transformer secondary is short-circuited. As an example, Fig. 14b shows the impedance Z_{1CC} and the phase θ_{1cc} of the prototype without casing.

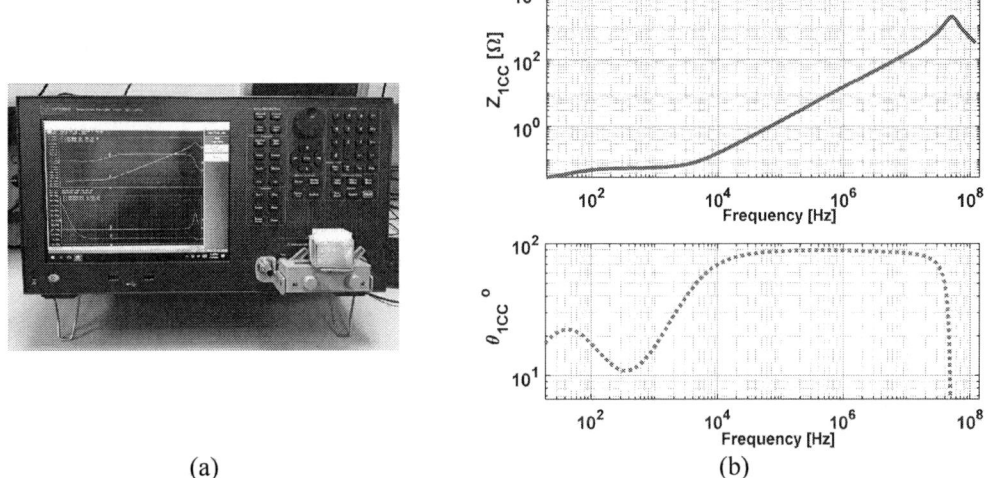

| | |
| (a) | (b) |

Fig. 14: Characterization: (a) Keysight E4990A impedance analyzer, (b) measured impedance and phase with short-circuited secondary

Measurements are performed on the transformer prototype without casing and with the three presented casings (Fig. 13c). The measured leakage inductances are plotted between 10 kHz and 1 MHz in Fig. 15a. Those measurements confirm the results of the simulations. The leakage inductance significantly decreases with frequency when the transformer is put in a casing. Its variation (Fig. 15b) exceeds 20% for the operating frequencies of HF converters.

The winding resistance is strongly increased with the aluminum casing as shown in Fig. 16a. Its variation (Fig.16b) is also consistent with FEA results. As expected, the effect of the casing thickness is not significant. The small difference between the three casings is due to the manufacturing tolerance and the geometrical disparities between the 3 casing prototypes.

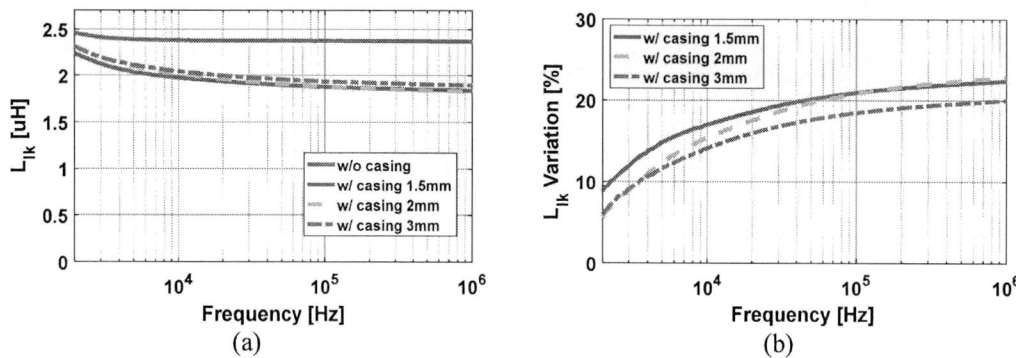

Fig. 15: Measured leakage inductance: (a) value without and with casing, (b) variation after adding the casing

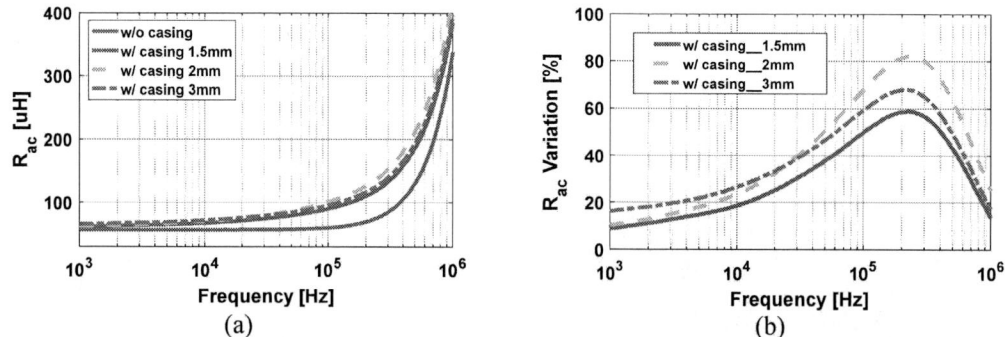

Fig. 16: Measured winding resistance: (a) value without and with casing, (b) variation after adding the casing

Measured and computed leakage inductances are plotted in Fig. 17a for the transformer without and with the 1.5mm casing. The relative error between measured and computed values are plotted in Fig. 17b. The error does not exceed 6% for the leakage inductance without casing and 2% in the presence of casing. As a consequence, the double 2D FEA modeling seems very effective to model such leakage inductance problem and it allows high accuracy results.

Regarding the equivalent resistance, measurements and FEA results are plotted in Fig. 18a without and with the 1.5mm casing. The relative error between measurements and simulations are reported in Fig. 18b. The model resistance is also quite accurate with a relative error which is less than 12% up to 500kHz. For higher frequencies, the relative error is under 20%. For the loss computation, the accuracy of the numerical model is affected by the simplification of Litz wires in 2D FEA.

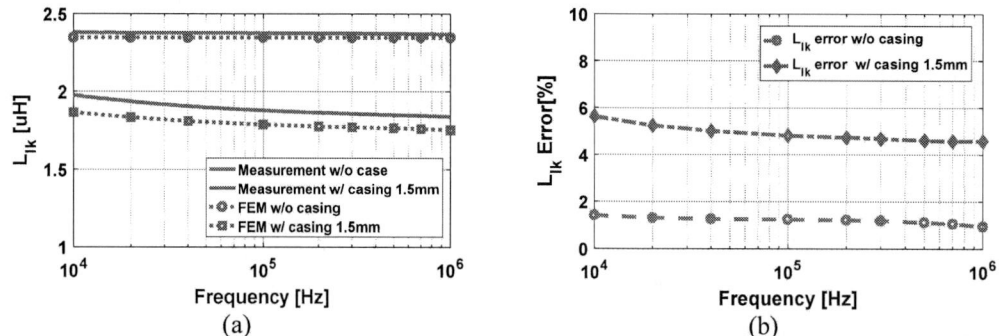

Fig. 17: Comparison of leakage inductance without and with 1.5mm casing: (a) Measurements / FEA results, (b) relative error

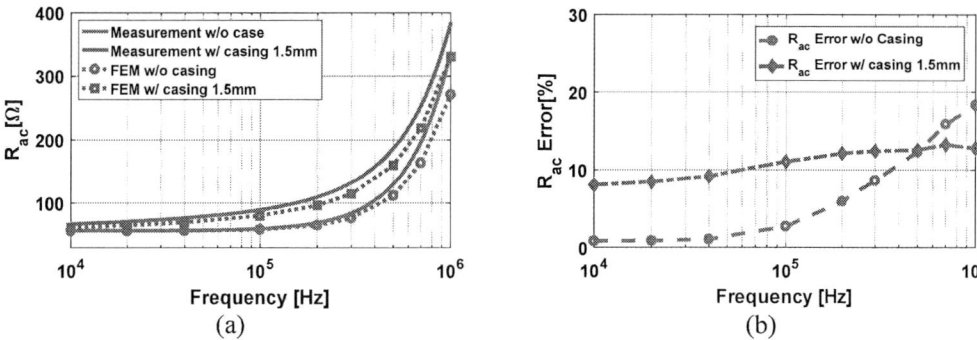

Fig. 18: Comparison of AC resistance without and with 1.5mm casing: (a) Measurements / FEA results, (b) relative error

Conclusion

In this paper the effect of an aluminum casing on the leakage inductance and losses of a HF transformer is investigated. A case study is defined based on a ferrite core and Litz wire. Double 2D FEA and measurements on prototype are performed to highlight and quantify the impact of casing on leakage inductance and losses. According to the results, the casing surrounding the transformer significantly affect the leakage inductance and the losses. Indeed, leakage inductance decreases by 20% and losses increase by more than 40%, particularly in the frequency range commonly used in HF converters. Three casing thickness have also been studied and the thickness influence seems to be negligible because the casing thickness is greater than aluminum skin depth. Casing effects must be considered while designing the converter, especially for high power density where the transformer is in contact with metallic part used for thermal management.

References

[1] J. W. Kolar *et al.*, "PWM Converter Power Density Barriers," in *2007 Power Conversion Conference - Nagoya*, Apr. 2007, p. P-9-P-29. doi: 10.1109/PCCON.2007.372914.

[2] G. Calderon-Lopez *et al.*, "Towards Lightweight Magnetic Components for Converters with Wide-bandgap Devices," in *2020 IEEE 9th International Power Electronics and Motion Control Conference (IPEMC2020-ECCE Asia)*, Nov. 2020, pp. 3149–3155. doi: 10.1109/IPEMC-ECCEAsia48364.2020.9367822.

[3] J. Millán, P. Godignon, X. Perpiñà, A. Pérez-Tomás, and J. Rebollo, "A Survey of Wide Bandgap Power Semiconductor Devices," *IEEE Transactions on Power Electronics*, vol. 29, no. 5, pp. 2155–2163, May 2014, doi: 10.1109/TPEL.2013.2268900.

[4] M. Parvez, A. T. Pereira, N. Ertugrul, N. H. E. Weste, D. Abbott, and S. F. Al-Sarawi, "Wide Bandgap DC–DC Converter Topologies for Power Applications," *Proceedings of the IEEE*, vol. 109, no. 7, pp. 1253–1275, Jul. 2021, doi: 10.1109/JPROC.2021.3072170.

[5] R. Bakri, X. Margueron, J. S. N. T. Magambo, P. le Moigne, and N. Idir, "Power density of planar transformers designed with commercial standard cores," in *2020 22nd European Conference on Power Electronics and Applications (EPE'20 ECCE Europe)*, Sep. 2020, p. P.1-P.10. doi: 10.23919/EPE20ECCEEurope43536.2020.9215747.

[6] M. Gerber, J. A. Ferreira, I. W. Hofsajer, and N. Seliger, "A high-density heat-sink-mounted inductor for automotive applications," *IEEE Transactions on Industry Applications*, vol. 40, no. 4, pp. 1031–1038, Jul. 2004, doi: 10.1109/TIA.2004.830766.

[7] Ferroxcube, "Soft Ferrites and Accessories Data Handbook." Available online on www.ferroxcube.com, 2013.

[8] R. Prieto, J. A. Cobos, O. Garcia, P. Alou, and J. Uceda, "Model of integrated magnetics by means of 'double 2D' finite element analysis techniques," in *30th Annual IEEE Power Electronics Specialists Conference. Record. (Cat. No.99CH36321)*, Jul. 1999, vol. 1, pp. 598–603 vol.1. doi: 10.1109/PESC.1999.789082.

[9] R. Prieto, J. A. Cobos, O. Garcia, P. Alou, and J. Uceda, "Study of 3-D magnetic components by means of 'double 2-D' methodology," *IEEE Transactions on Industrial Electronics*, vol. 50, no. 1, pp. 183–192, Feb. 2003, doi: 10.1109/TIE.2002.807663.

[10] "ANSYS - Simulation Driven Product Development." http://www.ansys.com, accessed 30 March 2022.

[11] "keysight E4990A impedance analyzer, on line avalaible : https://www.keysight.com/fr/en/product/E4990A/impedance-analyzer-20-hz-10-20-30-50-120-mhz.html."

Neural Networks-Generalized Predictive Control for MIMO Grid-Connected Z-Source Inverter Model

Navid Salehi, Herminio Martinez-Garcia, Guillermo Velasco-Quesada
ELECTRONIC ENGINEERING DEPARTMENT, UNIVERSITAT POLITECTICA DE
CATALUNYA – BarcelonaTech (UPC)
Escola d'Enginyeria de Barcelona Est (EEBE), Av. Eduard Maristany, n° 16. E-08019
Barcelona, Spain
Tel.: +34.93.413.72.90
E-Mail: navid.salehi@upc.edu, herminio.martinez@upc.edu, guillermo.velasco@upc.edu
URL: http://www.eel.upc.edu

Acknowledgements

The authors would like to thank the Spanish Ministerio de Ciencia, Innovación y Universidades (MICINN)-Agencia Estatal de Investigación (AEI) and the European Regional Development Funds (ERDF), by grant PGC2018-098946-B-I00 funded by MCIN/AEI/10.13039/501100011033/ and by ERDF A way of making Europe.

Keywords

«Z-source», «MPC», «GPC», «ANN», «Non-minimum phase», «MIMO»

Abstract

This paper presents a neural network-generalized predictive control (NN-GPC) for a single-phase grid-connected z-source inverter. The NN forecasts the predictive horizon, and the conventional GPC algorithm calculates the control horizon. The results verify the proposed NN-GPC effectively enhances the dynamic operation of z-source inverter regarding the non-minimum phase characteristics of these converters.

Introduction

As a sustainable solution to provide the world's electrical energy, distributed energy resources (DERs) integrated with renewable energies (REs) are strongly regarded by researchers. Inverters in microgrids (MGs) and DERs systems provide standard AC electricity for customers and play a special role in the stability and optimal operation of the system. By introducing the concept of impedance source converters, z-source inverters (ZSI) are investigated rapidly in order to improve their performance. ZSI proposes a single-stage inverter with inherent buck-boost ability due to implementing an impedance network into the DC link. Therefore, the ZSI can operate as a voltage source inverter (VSI) and current source inverter (CSI) simultaneously by controlling the duty cycle of the converter. Although the reliability, efficiency, and cost-effectiveness are improved in ZSI, some weak points led to the proposal of different impedance network topologies [1]. The main disadvantages of impedance source converters are the high current and voltage stress of the input rectifier diode and the inability to inject reactive power into the grid [2]. Various topologies are proposed to alleviate these weaknesses. However, inherently impedance source converters suffer from the mentioned problems. In [3], a modified cascaded Z-source high step-up boost converter is proposed in order to obtain a high conversion ratio with low voltage stress of semiconductor devices. In addition, the results verify that the proposed topology operates at continuous current mode (CCM) with low current and voltage stress.

The control techniques in ZSI are normally deployed to control the capacitor voltage and inductor current of the impedance network and the inductor current of the inverter's AC side. However, the control strategies in ZSI are a challenging issue due to the existing right-half-plane (RHP) zero into the control to capacitor voltage transfer function of the impedance network. Accordingly, different control strategies are adopted to ZSI as a non-linear and non-minimum phase system [4]. The proportional-integral-derivative (PID) controller can be applied to the ZSI to control the converter around a specific operating point. Therefore, the converter operation is restricted, and the converter dynamic is affected by operating point variation. In [5], a PID-like fuzzy control strategy is established to control the peak dc-link voltage. The PID parameters are determined according to the fuzzy logic-based rule sets. The rule sets in this paper are modified based on the trajectory performance of the phase plane in order to enhance the transient performance. Sliding mode control, fuzzy logic controller, model predictive control (MPC), and neural networks control are some other controllers that can effectively apply to the ZSI. In [6], a neural networks control technique is exploited to control DC boost and AC output voltage of ZSI. The space vector pulse-width-modulation (SVPWM) is modified in this paper in order to control the shoot-through (ST) duty ratio to boost dc voltage.

Model predictive control (MPC) is also widely applied to the ZSI control scheme. In [7], MPC predicts the capacitor voltage, inductor current, and output load current of a switched-inductor quasi ZSI to compare with the corresponding reference values. Then, the switching states are selected to achieve the minimum cost function. In [8], MPC is used in islanded and grid-connected operation modes to achieve a seamless transition between operation modes, fast dynamic response, and small tracking error under the steady-state condition of controller objectives. In [9], MPC is applied to the quasi ZSI to reduce the inductor current ripple and the output current error. In recent years, different MPC algorithms based on the predictive process model and defined cost function have been introduced. Dynamic matrix control (DMC) and model algorithmic control (MAC) are two MPC algorithms based on system step response and impulse response. Moreover, generalized predictive control (GPC), which is based on the system's discrete transfer function, and predictive functional control (PFC) based on the system's state space are two advanced MPC algorithms. Neural networks MPC (NN-MPC) can also be used specifically in complicated systems that conventional system identification methods such as step response, impulse response, transfer function, or state space of the system cannot be applied straightforwardly. The NN-based MPC algorithms are not fast due to high burden calculations. Therefore, NN-MPC is not developed in recent years compared to other MPC algorithms.

In this paper, neural networks GPC (NN-GPC) is applied to the ZSI in order to predict the capacitor voltage and inductor current of the impedance network and output inductor current. In this algorithm, the forced response of the predictive control is obtained by neural networks. The analysis shows that the proposed MPC algorithm enhances the dynamic response of the ZSI without increasing the calculation burdens. To verify the theoretical analysis, the simulation results are compared with conventional GPC. The rest of the paper is structured as follows. First, the multi-input multi-output (MIMO) model of the ZSI is presented. Then, by obtaining the discrete transfer functions of the ZSI, the proposed NN-GPC is introduced. Eventually, the simulation results are shown in order to verify the theoretical analysis.

MIMO z-source inverter model

As it can be seen from Fig. 1, the single-phase Z-source inverter consists of an impedance network and an H-bridge inverter. The impedance network involves two inductors L_1 and L_2, and two capacitors C_1 and C_2. In addition, a rectifier diode is in series with the input DC voltage source. The impedance network components are symmetrical i.e. the inductors L_1 and L_2, and capacitors C_1 and C_2 are equal ($L_1=L_2=L$ and $C_1=C_2=C$). Therefore, the inductor current and capacitor voltage are identical. Furthermore, the parasitic resistor of the inductors is considered in the inverter model. This inverter has three operating intervals over a complete duty cycle. In the first interval (T_{ON}), the switches Q_1 and Q_4 are conducted in a positive active switching state. During the second interval (T_{OFF}), the switches Q_2 and Q_3 are conducted in a negative active switching state. Finally, the third interval (T_{ST}) is related to the shoot-through (ST) switching state that all switches are turned on simultaneously to boost the input voltage. According to the z-source converter operation and considering the inductor volt-seconds balance, the following relations are established in ZSI [10]:

$$V_{C1} = V_{C2} = V_C = \frac{1 - D_{ST}}{1 - 2D_{ST}} \times V_{DC}$$

(1)

$$V_{INV} = 2V_C - V_{DC} = \frac{1}{1 - 2D_{ST}} \times V_{DC},$$

(2)

where D_{ST} is the shoot-through duty cycle ($D_{ST} = T_{ST}/T$). To obtain the state-space of the ZSI, the average model of the operation modes is considered in this paper. To this end, according to the operation modes of ZSI, the average of three different state-space are evaluated:

$$\dot{X} = A_{AVG}X + B_{AVG}U,$$

(3)

where X is the inductor current and capacitor voltage of impedance network, and inductor current of inverter AC side as the state variables, $X=[I_{L1}, V_{C1}, I_{LS}]$, and U is the input DC voltage and grid voltage, $U=[V_{DC}, V_{Grid}]$. Moreover, the A_{AVG} and B_{AVG} are defined as ($A_{AVG}, A_{ON}, A_{OFF}, A_{ST}, B_{AVG}, B_{ON}, B_{OFF}$, and B_{ST} are matrices):

$$A_{AVG} = \frac{1}{T}\left(T_{ON}A_{ON} + T_{OFF}A_{OFF} + T_{ST}A_{ST}\right)$$

(4)

$$B_{AVG} = \frac{1}{T}\left(T_{ON}B_{ON} + T_{OFF}B_{OFF} + T_{ST}B_{ST}\right),$$

(5)

where $T_{ON}=(\mu_m+1-D_{ST}/2)\times T$, $T_{OFF}=(1-T_{ON}-T_{ST})$, and μ_m is amplitude modulation index. To obtain the state-space of ZSI, presented in Fig 1, the ZSI is simplified into two separated systems. The H-bridge, L_S, and grid are considered a simple switch and constant current source (I_S) in the first system. Therefore, the first system consists of the input voltage source, impedance network, and equivalent circuit of the switch and current source. Moreover, the second system involves the constant DC voltage (V_{INV}), H-bridge, grid impedance, and grid voltage. In [11-12], the state-space ZSI is analyzed with a similar procedure. Accordingly, the state-space of ZSI is obtained as:

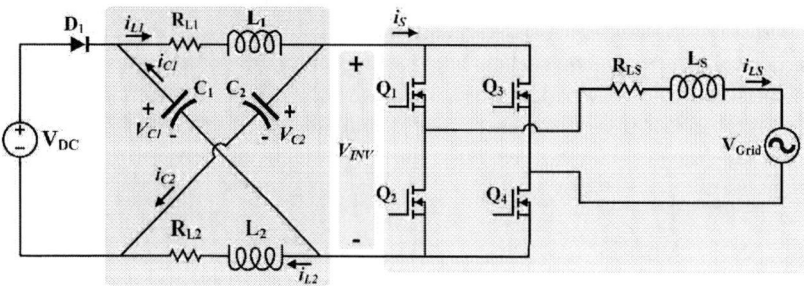

Fig. 1: Single-phase grid-connected z-source inverter

$$\begin{bmatrix} \dot{i}_{L1} \\ \dot{v}_{C1} \end{bmatrix} = \begin{bmatrix} \dfrac{-R_{L1}}{L} & -\dfrac{1}{L} + \dfrac{2}{L}D_{ST} \\ -\dfrac{1}{C} + \dfrac{2}{C}D_{ST} & 0 \end{bmatrix}\begin{bmatrix} i_{L1} \\ v_{C1} \end{bmatrix} + \begin{bmatrix} \dfrac{(2D_{sh}-1)V_{DC} - 2R_{L1}I_S}{L(2D_{ST}-1)^2} \\ \dfrac{2I_S}{C(2D_{ST}-1)} \end{bmatrix}d_{ST}$$

(6)

$$\begin{bmatrix} i_{LS}^{\Box} \end{bmatrix} = \left[-\frac{R_{LS}}{L_S} \right][i_{LS}] + \left[\frac{2V_C - V_{DC}}{L_S} \right]\mu_m \tag{7}$$

Therefore, the small-signal analysis can be performed in order to obtain the transfer functions of ZSI:

$$G_1(s) = \frac{i_L}{D_{ST}} = \frac{C\left(V_{DC} - 2D_{ST}V_{DC} - 2I_S r_L\right)s + \left(2 - 8D_{ST} + 8D_{ST}^2\right)I_S}{\left(2D_{ST} - 1\right)^2\left(CLs^2 + Cr_L s + 4D_{ST}^2 - 4D_{ST} + 1\right)} \tag{8}$$

$$G_2(s) = \frac{v_C}{D_{ST}} = \frac{\left(2D_{ST} - 1\right)\left(2I_S L s + 4I_S r_L - (1 - 2D_{ST})V_{DC}\right)}{C\left(V_{DC} - 2D_{ST}V_{DC} - 2I_S r_L\right)s + 2I_S + 8D_{ST}^2 I_S - 8D_{ST} I_S} \tag{9}$$

$$G_3(s) = \frac{i_{Ls}}{\mu_m} = \frac{2V_C - V_{DC}}{L_S s + r_{Ls}} \tag{10}$$

In $G_1(s)$ to $G_3(s)$, capital letters represents the variables at the operating point. Therefore, I_S and V_{DC} are the constant value at the operating point. As it can be seen, $G_2(s)$ represents a non-minimum phase system due to the existing RHP zero into the transfer function. Fig. 2 represents the pole and zero trajectories of control-to-capacitor with system parameter variations (C is considered constant and L is varied). Consequently, a MIMO ZSI is presented that the input variables (control variables) are shoot-through duty cycle (D_{ST}) and amplitude modulation index (μ_m), and the outputs are capacitor voltage of DC side and inductor current of inverter AC side.

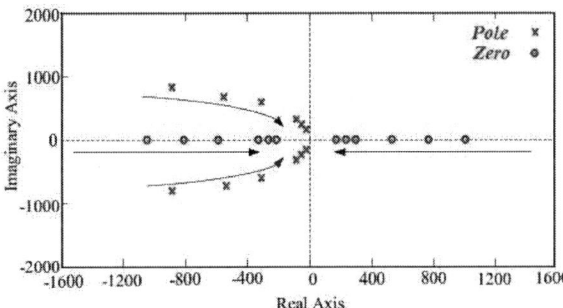

Fig. 2: Pole and zero trajectories with system parameters variations

Proposed neural networks-generalized predictive model

The GPC algorithm is exploited based on the transfer function that can possibly be a non-minimum phase and unstable system by considering the challenges regarding disturbances and noises. In MPC algorithms, the response is consists of free response and forced response. The free response refers to the signals in the past, and the forced response is related to the signals in the future. In this paper, controlled auto-regressive integrated moving average (CARIMA) is used to model the system in GPC:

$$A(z^{-1})y(t) = z^{-d}B(z^{-1})u(t-1) + C(z^{-1})\frac{e(t)}{(1-z^{-1})} \tag{11}$$

where d is the system delay, e(t) is the noise that in this paper noise is not considered into the model, and $A(z^{-1})$ and $B(z^{-1})$ are obtained from the discrete-time model of transfer functions:

$$A(z^{-1}) = 1 + a_1 z^{-1} + a_2 z^{-2} + ... + a_n z^{-n_a} \tag{12}$$

$$B(z^{-1}) = b_0 + b_1 z^{-1} + b_2 z^{-2} + ... + b_n z^{-n_b} \tag{13}$$

According to the Diophantine equation in (13), the $E(z^{-1})$ and $F(z^{-1})$ can be calculated as a recursive relationship in (14) and (15) in order to predict the signals:

$$E_j(z^{-1})(1-z^{-1})A(z^{-1})+z^{-j}F_j(z^{-1})=1 \tag{14}$$

$$E_{j+1}(z^{-1})=E_j(z^{-1})+f_{j,0}z^{-j} \tag{15}$$

$$f_{j+1,i}=f_{j,i+1}-f_{j,0}\tilde{a}_{i+1} \qquad i=0,1,...,n_{\tilde{a}}-1 , \tag{16}$$

where j is the prediction horizon, and i represents the corresponding terms of the $F_j(z^{-1})$. In addition, $\tilde{a}_i$ is i th term of $(1-z^{-1})A(z^{-1})$. Therefore, according to (10) and considering $e(t)=0$, the estimated outputs at the predictive horizon can be obtained:

$$\hat{y}(t+j\,|\,t)=B(z^{-1})E_j(z^{-1})(1-z^{-1})u(t+j-d-1)+F_j(z^{-1})y(t) \tag{17}$$

In (16), $F_j(z^{-1})y(t)$ is dependent on the past control signals. However, $B(z^{-1})E_j(z^{-1})(1-z^{-1})u(t+j-d-1)$ is included both past and future control signals. In conventional GPC, the free response and forced response are separated by the following equation:

$$y=\Phi y_- + \Pi u_- + \Omega u \tag{18}$$

The first two terms in (17) are related to the past control signals, and the third term is related to the future control signals. The matrixes Φ, Π, and Ω are defined as below in the conventional GPC algorithm:

$$\Phi=\begin{bmatrix} f_{d+1,0} & \cdots & f_{d+1,n_a} \\ \cdots & \cdots & \cdots \\ f_{d+N,0} & \cdots & f_{d+N,n_a} \end{bmatrix} \quad \Pi=\begin{bmatrix} g_{d+1,1} & \cdots & g_{d+1,n_{g1}} \\ \cdots & \cdots & \cdots \\ g_{d+N,N} & \cdots & g_{d+N,n_{gN}} \end{bmatrix} \quad \Omega=\begin{bmatrix} g_{d+1,0} & 0 & 0 \\ \cdots & \cdots & \cdots \\ g_{d+N,N-1} & \cdots & g_{d+N,0} \end{bmatrix} \tag{19}$$

In the proposed NN-GPC algorithm, the forced response presented as Ωu in conventional GPC is replaced with feed-forward neural networks in order to predict the control signals. Therefore, the Ω matrix components are obtained by NN. In this algorithm, the GPC calculations reduce, instead the neural networks calculations are put in the algorithm. Fig. 3 represents the NN structure to predict the state variables. More accurate predicted state variables for the GPC enhance the algorithm operation and dynamic response of the ZSI. Fig. 4 shows the proposed NN-GPC. The inductor current and capacitor voltage of the impedance network and inductor current of inverter AC side as the feedback signals are inputs of the feed-forward NN and GPC. After algorithm calculations, the cost function units evaluate the optimum output control signals according to the predicted signals and reference signals. Eventually, the output control signals are applied to the gate drive to switch on or off the switches.

The cost function can be define as:

$$CF=(R-Y)^T(R-Y)+\alpha u^T u , \tag{20}$$

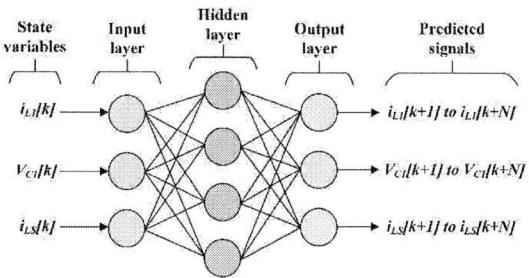

Fig. 3: Neural networks structure to predict the control signals

where R is the reference values $R=[R(t+d+1), R(t+d+2), ..., R(t+d+N)]$, Y is defined in (17), and α is a constant value. Therefore, the optimum control signal can be stated:

$$u = (\Omega^T \Omega + \alpha I)^{-1} \Omega^T (R - \Phi y_- - \Pi u_-) \qquad (21)$$

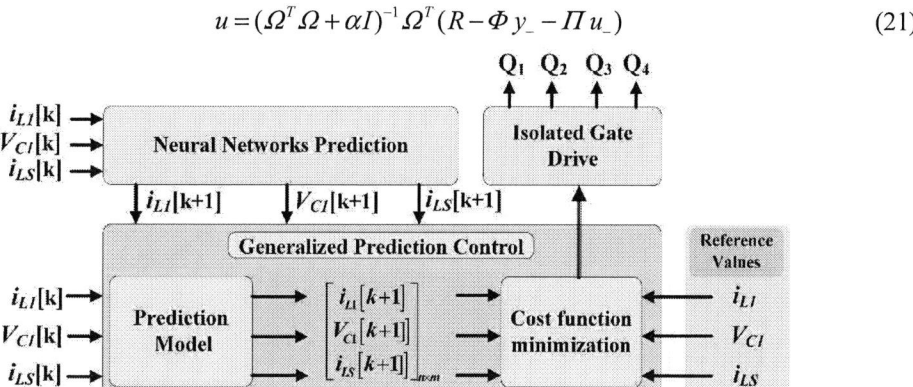

Fig. 4: Proposed NN-GPC

Performance evaluation and simulation results:

To evaluate the performance of the proposed NN-GPC on ZSI control, a ZSI with the specifications in Table I is considered. The NN is trained offline based on the obtained data of the ZSI simulation performance. Consequently, the NN offers the components of matrix Ω that are related to the future signals according to the instant values of state variables. On the other hand, the GPC algorithm evaluates the components of the matrices Φ and Π that are basically referred to the past signals. The NN parameters and GPC algorithm parameters are presented in Table I.

In order to evaluate the ZSI operation with the proposed NN-GPC, the simulation is carried out in MATLAB/ SIMULINK, and the results are compared with the conventional GPC. The neural networks toolbox is utilized to predict the state variables, and the GPC is set up by MATLAB codes. Eventually, the output signals of the cost function minimization are applied to the gate driver. The switching method in this simulation is the unipolar method. Figure 5 and 6 show the capacitor voltage and inductor current of the impedance network, coupling inductor current (L_S), and H-bridge inverter input voltage. It can be observed that the state variables follow their reference values effectively. In order to evaluate the dynamic response of ZSI in transients, the output current reference at 0.1ms is changed from 10A to 20A. The performance of NN-GPC and conventional GPC in transient can be seen in Fig. 5 and 6. The state variables in the conventional GPC are not able to trace the reference values in transition when the current reference is changed. However, the accurate prediction of state variables by feed-forward NN in NN-GPC makes the voltages and current track the reference values effectually. Although the MPC-GPC is also shown a favorable result in general specially comparing with PID controller, the proposed NN-GPC presents an effective performance specifically from the dynamic response perspective. Table II represents a characteristics comparison of GPC and NN-GPC.

Table I: Z-source inverter and NN-GPC specifications

ZSI parameters		NN-GPC parameters		
V_{DC}	100Vdc	NN specifications	Input Data	3×100
V_{Grid}	110Vrms		Output Data	3×100
C_1 & C_2	1mF		Training	70%
L_1 & L_2	5.1mH		Validation	15%
R_{L1} & R_{L2}	0.25Ω		Testing	15%
L_s	5.1mH		No. Hidden Neurons	3
R_{Ls}	0.25Ω	GPC specifications	Sample T	2µs
I_{OUT}	20A		Predictive horizon	4
$VC_{1\text{-ref}}$	230V		Control horizon	4

Table II: Comparison of GPC and NN-GPC characteristics

Specification	GPC	NN-GPC
System control	- Linear system - Nonlinear system - Minimum phase system - Non-minimum phase system	- Linear system - Nonlinear system - Minimum phase system - Non-minimum phase system
Complexity	Lower	Higher
Dynamic response	Lower	Higher

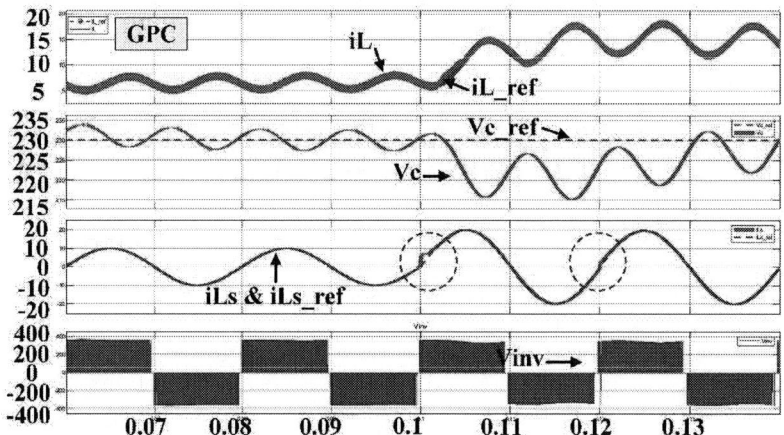

Fig. 5: Simulation results of ZSI control by GPC

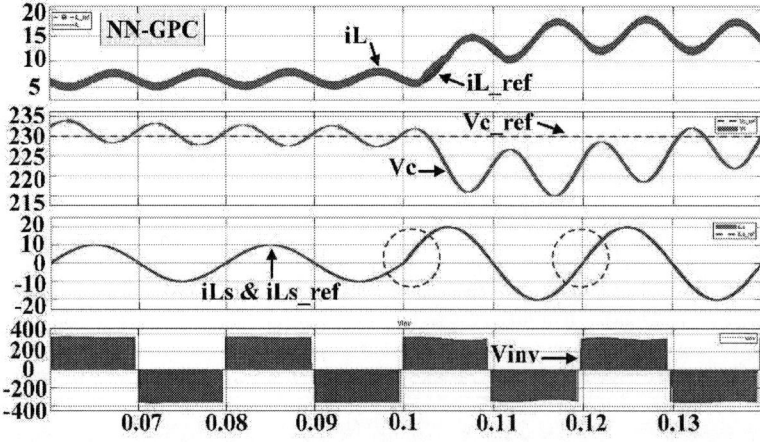

Fig. 6: Simulation results of ZSI control by NN-GPC

Conclusion

In this paper, an NN-GPC algorithm is proposed in order to control a ZSI. As discussed, the ZSI has non-minimum phase characteristics due to existing RHZ zero in control to capacitor voltage transfer function. Therefore, conventional linear control methods such as PID controllers make the ZSI operation unreliable and inefficient. The MPC algorithms as a promising control method for non-linear and non-minimum phase systems is considered recently to optimum control of ZSI. Eventually, the GPC algorithm is applied in this paper according to the MIMO model of ZSI. By deploying a feed-forward NN and improving the system identification to predict the state variables of the converter, the performance of the GPC algorithm is enhanced. Eventually, the simulation results for a specific ZSI are presented to verify the proposed NN-GPC algorithm.

Funding

Grant PGC2018-098946-B-I00 funded by: MCIN/ AEI /10.13039/501100011033/ and by ERDF ERDF A way of making Europe.

References

[1] Hasan Babayi Nozadian, Mohsen, et al. "Switched Z-source networks: a review." IET Power Electronics 12.7 (2019): 1616-1633.

[2] Nguyen, Minh-Khai, Young-Cheol Lim, and Sung-Jun Park. "Improved trans-Z-source inverter with continuous input current and boost inversion capability." IEEE transactions on power electronics 28.10 (2013): 4500-4510.

[3] Salehi, Navid, Herminio Martínez-García, and Guillermo Velasco-Quesada. "Modified Cascaded Z-Source High Step-Up Boost Converter." Electronics 9.11 (2020): 1932.

[4] Ellabban, Omar, Joeri Van Mierlo, and Philippe Lataire. "A comparative study of different control techniques for an induction motor fed by a Z-source inverter for electric vehicles." 2011 International Conference on Power Engineering, Energy and Electrical Drives. IEEE, 2011.

[5] Ding, Xinping, et al. "A direct DC-link boost voltage PID-like fuzzy control strategy in Z-source inverter." 200
8 IEEE Power Electronics Specialists Conference. IEEE, 2008.

[6] Rostami, H., and D. A. Khaburi. "Neural networks controlling for both the DC boost and AC output voltage of Z-source inverter." 2010 1st Power Electronic & Drive Systems & Technologies Conference. IEEE, 2010.

[7] Bakeer, Abualkasim, et al. "Control of switched-inductor quasi Z-Source Inverter (SL-qZSI) based on model predictive control technique (MPC)." 2015 IEEE International Conference on Industrial Technology. IEEE, 2015.

[8] Sajadian, Sally, and Reza Ahmadi. "Model predictive control of dual-mode operations Z-source inverter: Islanded and grid-connected." IEEE Transactions on Power Electronics 33.5 (2017): 4488-4497.

[9] Xu, Yuhao, et al. "Model Predictive Control Using Joint Voltage Vector for Quasi Z-Source Inverter with Ability of Suppressing Current Ripple." IEEE Journal of Emerging and Selected Topics in Power Electronics (2021).

[10] Peng, Fang Zheng. "Z-source inverter." IEEE Transactions on industry applications 39, no. 2 (2003): 504-510.

[11] Zakipour, Adel, Shokrollah Shokri-Kojori, and Mohammad Tavakoli Bina. "Sliding mode control of the nonminimum phase grid-connected Z-source inverter." International Transactions on Electrical Energy Systems 27, no. 11 (2017): e2398.

[12] Loh, Poh Chiang, D. Mahinda Vilathgamuwa, Chandana Jayampathi Gajanayake, Yih Rong Lim, and Chern Wern Teo. "Transient modeling and analysis of pulse-width modulated Z-source inverter." IEEE Transactions on Power Electronics 22, no. 2 (2007): 498-507.

Voltage Estimation for Diode-Clamped MMCs Based on a Simplified Neural Network

Nima Tashakor, Davood Keshavarzi, Shady Banana, and Stefan Goetz
Technische Universität Kaiserslautern
Kaiserslautern, Germany
E-Mail: tashakor@eit.uni-kl.de

Acknowledgements

The authors acknowledge the financial support by the Federal Ministry of Education and Research of Germany in the project "Open6GHub" (grant number: 16KISK004).

Keywords

«Capacitor voltage balancing», « Estimation technique», «Modular converter», «Modular multilevel converter», «Neural network».

Abstract

The modular multilevel converter (MMC) is a popular solution in high-voltage dc application and has significant potential in others. Generally, the MMC's stable operation is at the expense of numerous sensors, communication burdens, and complicated balancing strategies that can suppress its expansion in to cost driven applications. Hence, the introduction of a sensorless voltage balancing strategy with a simple controller is an attractive objective. A diode-clamped MMC offers a simple and yet effective solution by providing a balancing path between two modules through a diode. However, to compensate the lack of bidirectional energy transfer, modifying modulation technique is necessary. The level-adjusted phase-shifted carrier (LA-PSC) modulation introduces a small circulating current that ensures a correct balancing direction. Although the open-loop implementation of LA-PSC might be necessary for cost reduction in some applications, protection and control considerations may still necessitate careful monitoring of the modules' voltages. This paper proposes a voltage estimation strategy based on a simple neural network that does not require any measurement of the modules' voltages. Provided analysis as well as the simulation results confirm that the estimator can track the voltages with above 99% accuracy during balanced and imbalanced conditions.

Introduction

The modular multilevel converter (MMC) is a promising solution for high-voltage applications. Its advantages compared to other multilevel converters are excellent harmonic cancellation through quantized voltage levels, simple scalability, and higher flexibility due to its modularity [1-3]. These advantages make MMCs distinctly appealing for medium- to high-voltage applications [4, 5]. However, there are still challenges that need to be addressed, in which balancing and monitoring functions are among the most critical ones [6-8].

Balancing of MMC modules is pursued through two approaches. The first approach is based on various sorting algorithms, which necessitate constantly measuring the modules' voltages and transmitting the measured values back to the central controller [9-11]. In this approach, direct measurements through multiple sensors at modules terminals as well as a high-bandwidth communication interface increase the cost and complexity of the system [12, 13]. The required cell-sorting procedure further reduces the appeal of these methods, while it complicates safety-critical functions [14, 15]. The second approach modifies the macro- and micro-structure of the MMCs to generate balancing paths between the submodules through parallel connections [16-21]. Such topology modifications have great potential to solve some of the main challenges of MMCs [8, 22-24]. Some well-known examples of such topologies are MMCs with parallel mode [19, 25, 26], Marx MMC [27], and MMC with clamped switches [6, 28] are examples of auto-balancing topologies, which can achieve stable sensorless operation. However, most modifications require extra active switches and driving circuitry that although attractive in some applications, hinder the more widespread acceptance in high-voltage systems due to the cost or extra complexity [29, 30].

The diode-clamped module is the simplest configuration that does not need additional controlled semiconductors and can achieve self-balancing [31, 32]. In such circuits, one or more diodes form parallel connections between adjacent modules [27, 33-35]. Fig. 1 shows the simplest configuration of a diode-clamped topology with a single diode. Since the diode can only conduct in one direction, the formed balancing path will be unidirectional. However, there are other clamping topologies to create a physical bidirectional balancing path with more complex structures [27, 29, 32, 33]. This further complicates the topology and increase the cost [32].

Fortunately, there is a more practical approach to enforce a balancing direction that is compatible with the diode conduction path. This can be easily achieved by manipulating the modulation reference of each module. Liu et al. achieve this by constantly measuring voltage of the module at the highest end of each arm and constantly controlling its modulation index [34]. While this method is inherently interesting, it requires constant measurement of the voltage and will also change the shape of the arm voltage. Zheng et al. control the balancing direction by introducing different delays in switching functions of different modules, i.e., the higher the module is positioned in the arm, the higher its corresponding delay [29]. Since the method is open-loop, it can reduce the balancing efficiency due to overcompensation. In addition, the delays can change the shape of the output voltage. We proposed a simple open-loop balancing method that adjusts the vertical starting point of the carriers, the so called level-adjusted phase-shifted carrier (LA-PSC) modulation [36]. Even though LA-PSC does not change the output voltage shape, it may suffer from overcompensation in certain conditions which can be critical in some applications.

Although there are various sensorless and open-loop balancing approaches, knowing the modules' voltages are vital due to monitoring and protection concerns [37-41]. Therefore, this paper proposes estimation based on a simplified neural network for diode-clamped MMC with LA-PSC modulation. The proposed method benefits from simplicity and does not require any voltage measurement at module levels, which leads to minimizing communications and cost.

Diode-clamped MMC

Fig. 1 illustrates the single-phase diode-clamped MMC topology consisting of two similar arms. The

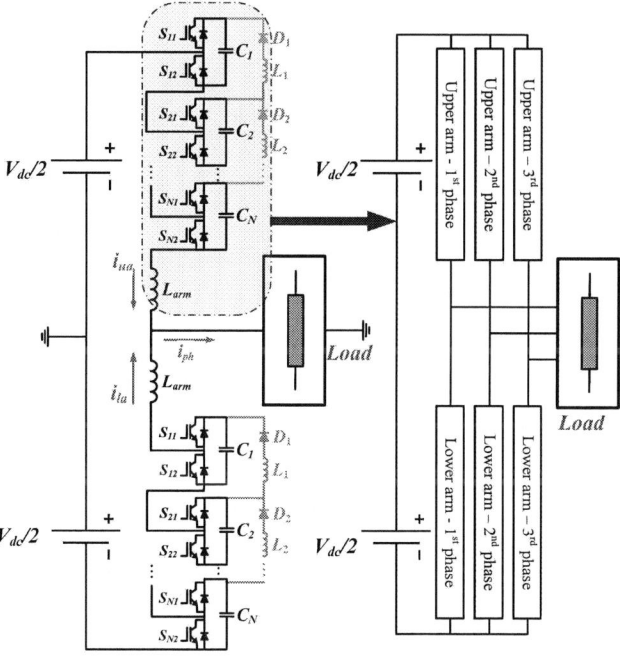

Fig. 1: single and three-phase topology of diode-clamped MMC

three-phase system can be easily developed based on single-phase topology.

Operation of the clamping circuit

In MMC, each arm contains N modules with $(N-1)$ clamping circuits, which consists of a diode and an inductor in series. The inductor restricts the maximum balancing current to protect the diode from large currents. The voltage across the clamping branch (v_{b_i}) includes the diode voltage (V_{fd}) and the inductance voltage (v_{L_i}) as well as the voltage across the total resistance of the path ($v_{R_{sum}}$) per

$$v_{b_i} = V_{\text{fd}} + v_{L_i} + v_{R_{sum}}. \tag{1}$$

Depending on the control signal of the $(i+1)^{\text{th}}$ module, we have three possibilities for the module voltage. When $S_{(i+1)1}$:off and $S_{(i+1)2}$:on, voltage across the clamping branch is $v_{b_i} = v_{c_{i+1}} - v_{c_i}$. If $v_{c_{i+1}}$ is smaller than $v_{c_i} + V_{\text{fd}}$, the diode is open as Fig. 2(a) shows, but if $v_{c_{i+1}}$ is larger than $v_{c_i} + V_{\text{fd}}$, a balancing current flows from C_{i+1} to C_i as Fig. 2(b) illustrates. When $S_{(i+1)1}$:on and $S_{(i+1)2}$:off and the diode is reverse biased, and the balancing current declines to zero as Fig 2(c) depicts and remains zeros as the diode cannot conduct in reverse. With negligible resistive elements ($V_{R_{sum}} = 0$), the derivative of the clamping inductors is

$$\frac{\text{d}i}{\text{d}t} = \frac{-\left(v_{c_i} + V_{fd}\right)}{L_i}. \tag{2}$$

In the case of large imbalances between two modules, multiple switching sequences are necessary,

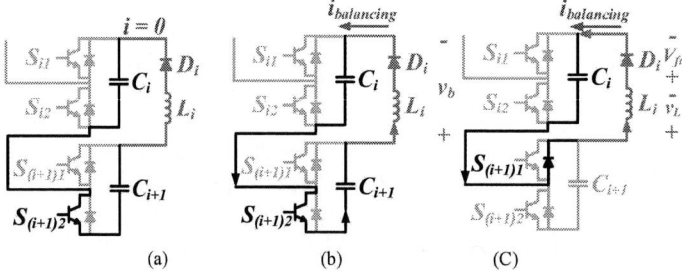

(a)   (b)   (C)

Fig. 2. Different operation modes of the diode-clamped modules

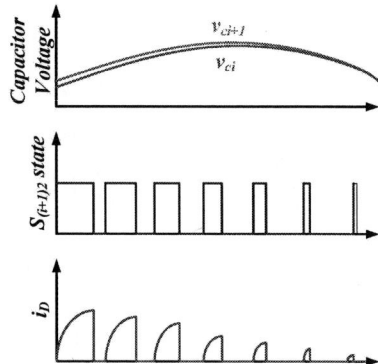

Fig. 3: Balancing process for two adjacent modules

until the voltage difference of the modules becomes negligible ($v_{c_{i+1}} - v_{c_i} \leq V_{\text{fd}}$). Fig. 3 shows an intuitive representation of this procedure. Therefore, ignoring of the initial voltage of each module, the final relation between the $v_{c_{i+1}}$ and v_{c_i} would be $v_{c_i} + V_{\text{fd}} > v_{c_{i+1}}$. The analysis can be extended to all the modules in an arm following

$$v_{c_1} + V_{\text{fd}} \geq \dots \geq v_{c_{(N-1)}} + V_{\text{fd}} \geq v_{c_N}. \tag{3}$$

Circuit analysis

Fig. 4(a) shows the equivalent electrical circuit when $S_{(i+1)2}$ is on (forward-bias state) and $V_{c_{i+1}} > V_{c_i} + V_{fd}$, where R_{L_i} represents the internal resistance of the inductance L_i, and $R_{C_{i+1}}$ as well as R_{C_i} are

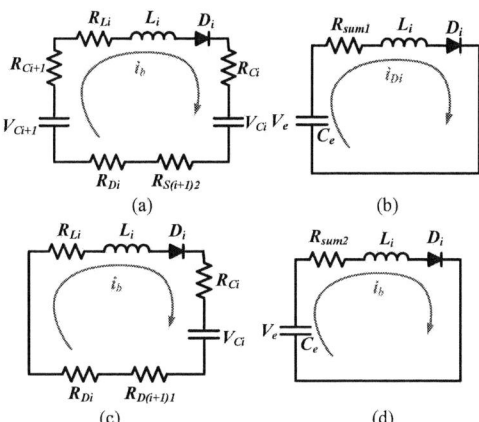

Fig. 4. Electrical circuit during the balancing: (a) $S_{(i+1)2}$ is on and balancing current flows from C_{i+1} to C_i; (b) simplified electrical circuit when $S_{(i+1)2}$ is on; (c) $S_{(i+1)2}$ turns off and balancing current decays to zero; (d) simplified electrical circuit when $S_{(i+1)2}$ is off.

the internal resistances of the capacitors C_{i+1} and C_i. Similarly, $R_{S_{(i+1)2}}$ is the internal resistance of the switch $S_{(i+1)2}$. Additionally, Fig. 4(b) depicts the simplified electrical circuit, V_e is the equivalent voltage, and C_e is the equivalent capacitance. According to the simplified electrical circuit, the total branch resistance $R_{sum1} = R_{C_{i+1}} + R_{L_i} + R_{C_i} + R_{S_{(i+1)2}}$, the equivalent voltage $V_e = V_{C_{i+1}} - V_{C_i}$, and the equivalent capacitance $C_e = 0.5C_i = 0.5C_{i+1}$ and the resulted second order RLC circuit can be readily analyzed. When $S_{(i+1)2}$ turns off, the electrical circuit changes to Fig. 4(c). Similarly, Fig. 4(d) shows the simplified electrical circuit in this mode, where $V_e = V_{C_i}$, $C_e = C_i$, and $R_{sum2} = R_{L_i} + R_{C_i} + R_{S_{(i+1)1}}$. Based on Kirchhoff voltage law in Fig. 4(b) and after some manipulations a second-order differential equation can be derived as follows

$$\frac{\mathrm{d}^2 i_D(t)}{\mathrm{d}t^2} + \frac{R_{sum1}}{L_i}\frac{\mathrm{d}i_D(t)}{\mathrm{d}t} + \frac{1}{L_i C_e} i_D(t) = 0 \tag{4}$$

Applying Laplace transformation and solving it arrives to

$$P_{1,2} = -\frac{R_{sum1}}{2L} \pm \sqrt{\frac{R_{sum1}^2}{4L_i^2} - \frac{1}{L_i C_e}} \tag{5}$$

The equivalent resistance R_{sum1} is relatively small, $R_{sum1} < 2\sqrt{\frac{L_i}{C_e}}$, hence with normal parameters the current will have a damped oscillation given by

$$i_D(t) = \frac{U_{\text{diff}}}{\sqrt{\frac{L_i}{C_e} - \frac{R_{sum1}^2}{4}}} e^{-\alpha t} \sin \omega_d t \tag{6}$$

where the damping factor $\alpha = \frac{R_{sum1}}{2L_i}$, and the frequency of the oscillation is $\omega_d = \sqrt{\frac{1}{L_i C_e} - \frac{R_{sum1}^2}{4L_i^2}}$.

Furthermore, it should be noted that the clamping path only allows for the current to pass in one direction, and the current cannot be negative.

If the maximum permissible voltage difference between modules is $U_{\text{diff\_max}}$, the maximum peak diode current I_{Pmax} follows

$$I_{\text{P\_max}} = \frac{U_{\text{diff\_max}}}{\sqrt{\frac{L_i}{C_e} - \frac{R_{sum1}^2}{4}}}. \tag{7}$$

Therefore, selecting the inductor per

$$L_i \geq \left(\frac{R_{sum1}^2}{4} + \frac{U_{\text{diff\_max}}^2}{I_{\text{P\_max}}^2}\right) C_e, \tag{8}$$

ensures that the current of the clamping diode (D_i) is below its rated value [34]. Further reduction of the current rating of the diode is possible by increasing the size of the clamping inductor. However, larger inductor values reduces the speed of balancing [42]. A more detailed design procedure for the clamping diode and inductor are provided in the literature [36].

Level-Adjusted Phase-Shifted Carrier (LA-PSC)

The conventional phase-shifted carrier (PSC) modulation compares a reference waveform (modulation index) with carriers that are phase-shifted with respect to each other. In PSC modulation, each carrier corresponds to one unique module in the arm and the phase-shift between two successive carriers is fixed to $\frac{2\pi}{N}$. With ideal conditions, PSC should reach a stable operating point, but in practice, the system gradually diverges from the desired operation point unless balancing mechanism exists [24].

A diode-clamped circuit can guarantee that condition (3) is always preserved, and the LA-PSC modulation can ensure the correct balancing direction by a small vertical displacement in the conventionally inline carriers. Fig. 5 illustrates an intuitive representation of the shifted carriers in LA-PSC modulation. A positive level-adjustment (for the ith carrier, $\delta_i > 0$) decreases the average duration that the module is connected in series and a negative level-adjustment ($\delta_i < 0$) increases the average duration of series intervals. Lower duration with series connection reduces the module charge and vice versa due to positive average arm current. If δ_i is the vertical-adjustment of ith carrier, $\delta_1 \geq \delta_2 \geq \cdots \geq \delta_N$ ensures a bottom to top balancing direction at all times. The effective modulation index (m_i) for the ith module in upper arm with a level-adjustment equal to δ_i is

$$m_{u,i} = \frac{1 - m_a \sin(\omega t)}{2} - \delta_{u,i}, \tag{9}$$

where m_a is the normalized amplitude of the phase a.

Since the modulation index of each module is controlled separately, it is possible to directly control the modulation indexes per (9) with similar results regarding the balancing direction. The effective modulation index for the complete arm is the average of all the individual m_i values as

$$m_u = \frac{1 - m_a \sin(\omega t)}{2} - \frac{\sum_i^N \delta_{u,i}}{N}. \tag{10}$$

Defining the level-adjustments for the ith module in the upper arm per

$$\delta_{u,i} = \Delta_a \left(\frac{1}{2} - \frac{j-1}{N-1} \right) \tag{11}$$

results in a zero average for the term corresponding to the level-adjustment in (10), and hence an unchanged effective voltage of the arm. Δ_a in (11) is the total level-adjustment between the first and last carriers of the arm.

Since the balancing direction in the lower arm must also be from bottom to top, $\delta_{l,i}$ should be equal to

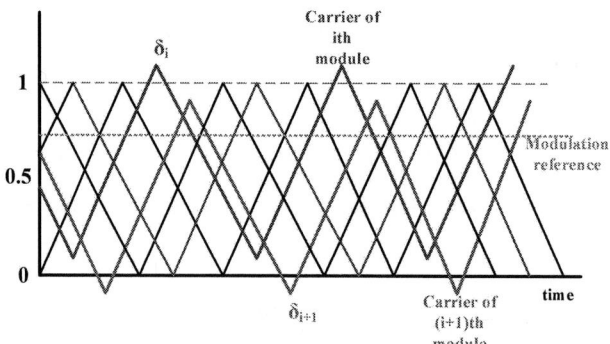

Fig. 5. Intuitive representation of the PSC and the LA-PSC

$\delta_{u,i}$ following (11). However, the phase-shifts' of the carriers in the upper and lower arm should be mirrored, (i.e., the phase-shift of the first module in the upper arm should be equal to the phase-shift of the last module in the lower arm) to cancel out the effects of the phase level-adjustments on the phase voltage, e.g., vector of phase-shifts for the upper and lower arm can be $\phi_u = \left[0, \frac{2\pi}{N}, \dots, \frac{2\pi(N-1)}{N} \right]^T$ and $\phi_l = \left[\frac{2\pi(N-1)}{N}, \frac{2\pi(N-2)}{N}, \dots, 0 \right]^T$. Therefore, the new carriers for the upper and lower arms, respectively, are

$$C_{u,i}^{\text{new}} = \Delta_a \left(\frac{1}{2} + \frac{j-1}{N-1}\right) + C_{u,i}^{\text{old}}, \tag{12}$$

$$C_{l,i}^{\text{new}} = \Delta_a \left(\frac{1}{2} + \frac{j-1}{N-1}\right) + C_{l,N-i}^{\text{old}}. \tag{13}$$

Proposed estimation algorithm

Although, the behavior of the diode-clamped MMC is easy to understand, obtaining accurate analysis considering all the inherent dynamics is challenging. Furthermore, the non-uniform distribution of the mismatches among modules as well as the effect of the proposed level-adjustments would make a detailed analytical solution even more difficult or even unfeasible.

On the other hand, some methods based on machine learning such as neural network (NN) do not suffer from such restrictions. The random nature of parameter tolerances (e.g., self-discharge of the modules) makes deriving a mathematical solution almost impossible while it be easily modelled through proper training of the NN. Furthermore, accurate analytical solutions are challenging, but it is possible to define the output boundaries for each of the module voltages as corrective measures in case a NN starts to deviate from normal behavior.

Proposing a new NN is not the purpose of this paper, and to reduce the computational burden, as an example, the simplest form of a neural network estimates the voltage of the modules. The back propagation (BP) algorithm is selected to train this network. Additionally, a boundary function checks the fidelity of the NN's output.

The main idea of the BP algorithm is to choose the weights of the neural network (NN) in a manner that the error between the desired output and the NN output can be minimized. The method for minimizing the error is based on the gradient descent approach. The NN is responsible for minimizing the error between the estimation of the output voltage and its measurement based on the module states. The output voltage of the arm respecting to the module voltages and their states is

$$V_{arm} = \sum_{j=1}^{N} V_{c_j} S_{j1} = V_c^T \times S_1 \tag{14}$$

where V_c is the vector of module voltages in one arm and S_1 is the vector of the upper switches (S_{j1}) in each module in the same arm.

As Fig. 6 illustrates, the NN is composed of one input vector (module states), one layer with one neuron (representation of single equation for one arm), and one output (arm voltage). Each element of input vector of the NN ($X(k)$) is the switching signal of one module in the arm, $X_j(k) = S_{j1}$, per

$$X(k) = [S_{11} \quad S_{21} \quad \cdots \quad S_{N1}], \tag{15}$$

where $S_{j1} = 1$ and $S_{j1} = 0$, respectively, if the jth module is connected in series or if it is bypassed.

The desired output is the arm voltage $d = y(k)$, which is already measured for higher level controls, and the weighting vector represents the estimated voltage of the modules $W(k) = V_{c,est}$ at the kth step per

$$W(k) = \begin{bmatrix} V_{c_1,est} & V_{c_2,est} & \cdots & V_{c_N,est} \end{bmatrix}. \tag{16}$$

A linear activation function, $f(\alpha) = \alpha$, simplifies implementing the BP algorithm. Therefore, the derivative of the activation function is $f'(\alpha) = 1$. Table I shows the pseudo-code of the BP algorithm. Hence, the BP will determine the best weights (modules' voltages) that reduce the error between the estimated and measured arm voltage. As an improvement to the BP algorithm a momentum term can be used for updating the weights per

$$\Delta w_j(k) = \eta \left((1-\alpha)\delta x_j(k) + \alpha \Delta w_j(k-1)\right), \tag{17}$$

where α is a momentum constant in the range $0 \leq \alpha < 1$ and η is the learning rate. The momentum term reduces the oscillations in the beginning of the estimation. The value of the learning rate (η) influences the learning behavior, i.e., small η values can slow the convergence, and with large η values can cause oscillation and/or even divergence. Although η and α are normally defined heuristically, they depend on the switching frequency of the modules as well as the sampling frequency. We have selected $\eta = 0.001$ and $\alpha = 0.1$.

Table. I: Pseudo-code of the estimtor based on BP

STEP 0	Initialization of the weights	$w_j = V_{c_j,est}$, j = 1, ... N
STEP 1	Compute the output ($\hat{y}$)	$\hat{y} = f(\alpha) = \alpha = X(k) \times W(k)$
STEP 2	Compute the error	$e = y(k) - \hat{y}$
STEP 3	Compute the local gradients	$\delta = f'(\alpha)e$
STEP 4	Compute the weight updates	$\Delta w_j = \eta \, \delta \, x_j(k)$
STEP 5	Compute the new weights	$w_j(k) = w_j(k) + \Delta w_j$
STEP 6	Return to	STEP 1

Simulation results

A single-phase configuration with 16 modules ($N = 8$) and 14 clamping branches verifies the feasibility of the proposed estimation technique. Table. II lists the parameters of the simulation systems. The switching frequency of each module is selected to be 5 kHz, which results in the equivalent switching frequency of 40 kHz for each arm.

Table. II: Parameters of the Simulation and Experimental Setups

Parameters		Simulation
Rated Power		$P_{ac} = 1.14$ MW
Load Inductor		$L_L = 0.1$ mH
Module rated voltage		$V_{sm} = 1.2$ KV
DC link voltage		$V_{dc} = 9.6$ KV
Number of the modules		16
Arm inductance		$L_{arm} = 5$ mH
Arm Resistor		$R_{larm} = 50$ mΩ
Carrier frequency		$f_c = 5$ KHz
Sampling frequency		$f_s = 0.1$ MHz
Output frequency		$f = 50$ Hz
Modules	Capacitance	$C_{cap} = 6$ mF
	Capacitor resistor	$R_{cap} = 2$ mΩ
Modulation index		$m = 0.75{\sim}0.95$
Clamping Circuit	Inductance	$L_{ld} = 10$ μH
	Resistor	$R_{ld} = 0.5$ mΩ

We study the behavior of the estimator under balanced conditions with identical modules and imbalanced conditions with mismatch between module capacitances and self-discharge rates. The estimator is initialized to random values to show that the estimator is stable and can converge even with completely incorrect initial values, however, the steady-state voltages of the modules can be also used for initialization. Fig. 7 studies the behavior of the proposed estimator during balanced condition. As seen, the real voltages of modules have variation around the nominal voltage in and the estimator follows them properly. The estimation accuracy is above 99% and it converges in less than 50 ms. Fig. 8 illustrate the estimator behavior in an imbalanced system caused by inserting a parallel resistor

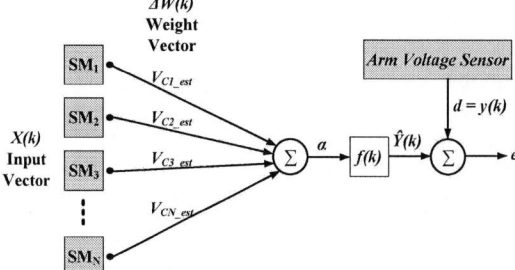

Fig. 6: Different operation modes of the diode-clamped modules

(about a few hundred ohms) with the third module. As shown in Fig. 8, the proposed algorithm can estimate the modules' voltages with above 99% accuracy. In both scenarios, the estimator closely follows the module voltages and can achieve above 99% accuracy, which further verify the capability of the proposed estimator.

Conclusion

This paper proposes an estimator based on a simplified NN that can be integrated into the control system of a diode-clamped MMC with LA-PSC modulation to achieve sensorless operation while fully monitoring the modules' voltages per the protection and safety functions. The proposed estimator is very simple and needs no extra measurements at the modules, which can significantly reduce the number of sensors as well as the communication. The analysis and simulation results verify the feasibility and applicability of the proposed method. The simulation results confirm the estimator can achieve above 99%.

References

[1]   N. Tashakor, F. Iraji, and S. G. Goetz, "Low-Frequency Scheduler for Optimal Conduction Loss in Series/Parallel Modular Multilevel Converters," *IEEE Transactions on Power Electronics*, pp. 1-1, 2021, doi: 10.1109/TPEL.2021.3110213.

[2]   J. Fang, Z. Li, and S. M. Goetz, "Multilevel Converters With Symmetrical Half-Bridge Submodules and Sensorless Voltage Balance," *IEEE Transactions on Power Electronics*, vol. 36, no. 1, pp. 447-458, 2021, doi: 10.1109/TPEL.2020.3000469.

[3]   J. Fang, H. Deng, N. Tashakor, F. Blaabjerg, and S. M. Goetz, "State-Space Modeling and Control of Grid-Tied Power Converters with Capacitive/Battery Energy Storage and Grid-Supportive Services," *IEEE Journal of Emerging and Selected Topics in Power Electronics*, pp. 1-1, 2021, doi: 10.1109/JESTPE.2021.3101527.

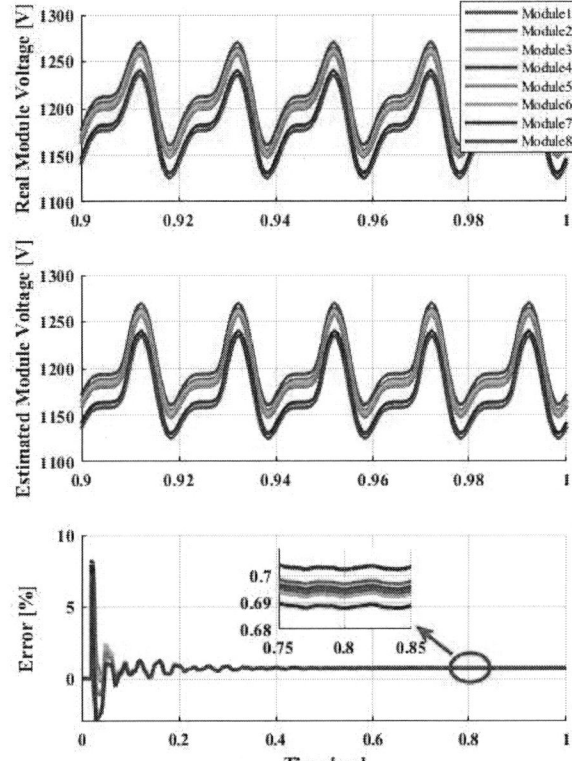

Fig. 7: Real and estimated modules' voltages and average error in balanced condition

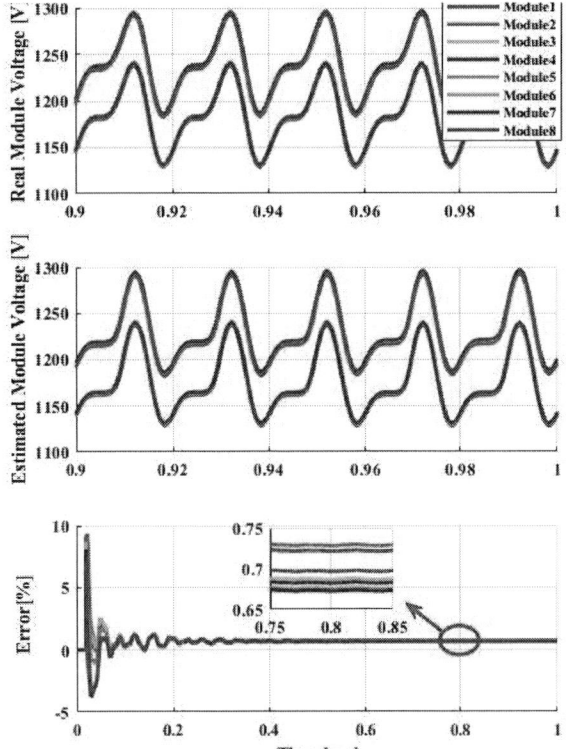

Fig. 8: Real and estimated modules' voltages and average error in unbalanced condition

[4] Q. Xiao, Y. Jin, and J. Wang, "A Novel Fault-Tolerant Control Method for Modular Multilevel Converter with an Improved Phase Disposition Level-Shifted PWM," *IEEE,* 2019.

[5] B. Arabsalmanabadi, N. Tashakor, Y. Zhang, K. Al-Haddad, and S. Goetz, "Parameter Estimation of Batteries in MMCs with Parallel Connectivity using PSO," presented at the IECON 2021 – 47th Annual Conference of the IEEE Industrial Electronics Society, 13-16 Oct. 2021, 2021.

[6] S. Ali, Z. Ling, K. Tian, and Z. Huang, "Recent Advancements in Submodule Topologies and Applications of MMC," *IEEE Journal of Emerging and Selected Topics in Power Electronics,* pp. 1-1, 2020, doi: 10.1109/JESTPE.2020.2990689.

[7] C. Wang, L. Xiao, C. Wang, M. Xin, and H. Jiang, "Analysis of the Unbalance Phenomenon Caused by the PWM Delay and Modulation Frequency Ratio Related to the CPS-PWM Strategy in an MMC System," *IEEE Transactions on Power Electronics,* vol. 34, no. 4, pp. 3067-3080, 2019, doi: 10.1109/TPEL.2018.2849088.

[8] J. Fang, S. Yang, H. Wang, N. Tashakor, and S. Goetz, "Reduction of MMC capacitances through parallelization of symmetrical half-bridge submodules," *IEEE Transactions on Power Electronics,* 2021.

[9] F. Deng, C. Liu, Q. Wang, R. Zhu, X. Cai, and Z. Chen, "A Currentless Submodule Individual Voltage Balancing Control for Modular Multilevel Converters," *IEEE Transactions on Industrial Electronics,* pp. 1-1, 2019, doi: 10.1109/TIE.2019.2952808.

[10] K. Wang, Y. Deng, H. Peng, G. Chen, G. Li, and X. He, "An Improved CPS-PWM Scheme-Based Voltage Balancing Strategy for MMC With Fundamental Frequency Sorting Algorithm," *IEEE Transactions on Industrial Electronics,* vol. 66, no. 3, pp. 2387-2397, 2019, doi: 10.1109/TIE.2018.2813963.

[11] T. Yin, Y. Wang, X. Wang, S. Yin, S. Sun, and G. Li, "Modular Multilevel Converter With Capacitor Voltage Self-balancing Using Reduced Number of Voltage Sensors," presented at the 2018 International Power Electronics Conference (IPEC-Niigata 2018 -ECCE Asia), 20-24 May 2018, 2018.

[12] P. Poblete, G. Pizarro, G. Droguett, F. Nunez, P. Judge, and J. Pereda, "Distributed Neural Network Observer for Submodule Capacitor Voltage Estimation in Modular Multilevel Converters," *IEEE Transactions on Power Electronics,* pp. 1-1, 2022, doi: 10.1109/TPEL.2022.3163395.

[13] G. Pizarro, P. M. Poblete, G. Droguett, J. Pereda, and F. Nunez, "Extended Kalman Filtering for Full State Estimation and Sensor Reduction in Modular Multilevel Converters," *IEEE Transactions on Industrial Electronics,* pp. 1-1, 2022, doi: 10.1109/TIE.2022.3165286.

[14] A. Dekka, B. Wu, and N. R. Zargari, "A Novel Modulation Scheme and Voltage Balancing Algorithm for Modular Multilevel Converter," *IEEE Transactions on Industry Applications,* vol. 52, no. 1, pp. 432-443, 2016, doi: 10.1109/TIA.2015.2477481.

[15] N. Tashakor, B. Arabsalmanabadi, F. Naseri, and S. Goetz, "Low-Cost Parameter Estimation Approach for Modular Converters and Reconfigurable Battery Systems Using Dual-Kalman-Filter," *IEEE Transactions on Power Electronics,* pp. 1-1, 2021, doi: 10.1109/TPEL.2021.3137879.

[16] P. Fang Zheng, "A generalized multilevel inverter topology with self voltage balancing," *IEEE Transactions on Industry Applications,* vol. 37, no. 2, pp. 611-618, 2001, doi: 10.1109/28.913728.

[17] N. Tashakor, Z. Li, and S. M. Goetz, "A Generic Scheduling Algorithm for Low-Frequency Switching in Modular Multilevel Converters with Parallel Functionality," *IEEE Transactions on Power Electronics,* pp. 1-1, 2020, doi: 10.1109/TPEL.2020.3018168.

[18] Z. Li, J. K. Motwani, Z. Zeng, S. Lukic, A. V. Peterchev, and S. Goetz, "A Reduced Series/Parallel Module for Cascade Multilevel Static Compensators Supporting Sensorless Balancing," *IEEE Transactions on Industrial Electronics,* pp. 1-1, 2020, doi: 10.1109/TIE.2020.2965470.

[19] X. Hu, Y. Zhu, J. Zhang, F. Deng, and Z. Chen, "Unipolar Double-Star Submodule for Modular Multilevel Converter With DC Fault Blocking Capability," *IEEE Access,* vol. 7, pp. 136094-136105, 2019, doi: 10.1109/ACCESS.2019.2942137.

[20] H. B. I, S. Rivera, Z. Li, S. Goetz, A. Peterchev, and R. L. F, "Different parallel connections generated by the Modular Multilevel Series/Parallel Converter: an overview," presented at the IECON 2019 - 45th Annual Conference of the IEEE Industrial Electronics Society, 14-17 Oct. 2019, 2019.

[21] M. B. Ghat and A. Shukla, "A New H-Bridge Hybrid Modular Converter (HBHMC) for HVDC Application: Operating Modes, Control, and Voltage Balancing," *IEEE Transactions on Power Electronics,* vol. 33, no. 8, pp. 6537-6554, 2018, doi: 10.1109/TPEL.2017.2751680.

[22] J. Xu, J. Li, J. Zhang, L. Shi, X. Jia, and C. Zhao, "Open-loop voltage balancing algorithm for two-port full-bridge MMC-HVDC system," *International Journal of Electrical Power & Energy Systems,* vol. 109, pp. 259-268, 2019/07/01/ 2019, doi: https://doi.org/10.1016/j.ijepes.2019.01.032.

[23] J. Rodriguez, L. Jih-Sheng, and P. Fang Zheng, "Multilevel inverters: a survey of topologies, controls, and applications," *IEEE Transactions on Industrial Electronics,* vol. 49, no. 4, pp. 724-738, 2002, doi: 10.1109/TIE.2002.801052.

[24] N. Tashakor, V. Monteiro, T. Ghanbari, and E. Farjah, "An Improved Modular Charge Equalization Structure for Series Cascaded Battery," presented at the 2019 27th Iranian Conference on Electrical Engineering (ICEE), 30 April-2 May 2019, 2019.

[25] J. Xu, J. Zhang, J. Li, L. Shi, X. Jia, and C. Zhao, "Series-parallel HBSM and two-port FBSM based hybrid MMC with local capacitor voltage self-balancing capability," *International Journal of Electrical Power & Energy Systems,* vol. 103, pp. 203-211, 2018.

[26] F. Z. Peng, W. Qian, and D. Cao, "Recent advances in multilevel converter/inverter topologies and applications," in *The 2010 International Power Electronics Conference - ECCE ASIA -,* 21-24 June 2010 2010, pp. 492-501, doi: 10.1109/IPEC.2010.5544625.

[27] C. Gao and J. Lv, "A new parallel-connected diode-clamped modular multilevel converter with voltage self-balancing," *IEEE Transactions on Power Delivery,* vol. 32, no. 3, pp. 1616-1625, 2017.

[28] N. Tashakor, M. Kilictas, J. Fang, and S. Goetz, "Switch-Clamped Modular Multilevel Converters with Sensorless Voltage Balancing Control," *IEEE Transactions on Industrial Electronics,* 2020.

[29] T. Zheng et al., "A Novel High-Voltage DC Transformer Based on Diode-Clamped Modular Multilevel Converters With Voltage Self-Balancing Capability," *IEEE Transactions on Industrial Electronics,* vol. 67, no. 12, pp. 10304-10314, 2020, doi: 10.1109/TIE.2019.2962486.

[30] J. Fang, F. Blaabjerg, S. Liu, and S. Goetz, "A Review of Multilevel Converters with Parallel Connectivity," *IEEE Transactions on Power Electronics,* pp. 1-1, 2021, doi: 10.1109/TPEL.2021.3075211.

[31] G. Lu, C. Gao, and X. Li, "Voltage self-balance method for series connected IGBTs by using clamping diodes," presented at the IECON 2017 - 43rd Annual Conference of the IEEE Industrial Electronics Society, 29 Oct.-1 Nov. 2017, 2017.

[32] C. Gao, X. Jiang, Y. Li, Z. Chen, and J. Liu, "A DC-Link Voltage Self-Balance Method for a Diode-Clamped Modular Multilevel Converter With Minimum Number of Voltage Sensors," *IEEE Transactions on Power Electronics,* vol. 28, no. 5, pp. 2125-2139, 2013, doi: 10.1109/TPEL.2012.2212915.

[33] J. Xu, M. Feng, H. Liu, S. Li, X. Xiong, and C. Zhao, "The diode-clamped half-bridge MMC structure with internal spontaneous capacitor voltage parallel-balancing behaviors," *International Journal of Electrical Power & Energy Systems,* vol. 100, pp. 139-151, 2018.

[34] X. Liu et al., "A Novel Diode-Clamped Modular Multilevel Converter With Simplified Capacitor Voltage-Balancing Control," *IEEE Transactions on Industrial Electronics,* vol. 64, no. 11, pp. 8843-8854, 2017, doi: 10.1109/TIE.2017.2682013.

[35] T. Zheng, C. Gao, X. Liao, X. Liu, B. Sun, and J. Lv, "A medium-voltage motor drive based on diode-clamped modular multilevel converters," presented at the 2017 20th International Conference on Electrical Machines and Systems (ICEMS), 11-14 Aug. 2017, 2017.

[36] N. Tashakor, M. Kilictas, E. Bagheri, and S. Goetz, "Modular Multilevel Converter with Sensorless Diode-Clamped Balancing through Level-Adjusted Phase-Shifted Modulation," *IEEE Transactions on Power Electronics,* pp. 1-1, 2020, doi: 10.1109/TPEL.2020.3041599.

[37] H. Givi, E. Hosseini, and E. Farjah, "Estimation of Batteries Voltages and Resistances in Modular Multilevel Converter With Half-Bridge Modules Using Modified PSO Algorithm," presented at the 2021 12th Power Electronics, Drive Systems, and Technologies Conference (PEDSTC), 2-4 Feb. 2021, 2021.

[38] R. Chakraborty, J. Samantaray, A. Dey, and S. Chakrabarty, "Capacitor Voltage Estimation of MMC using a Discrete-Time Sliding Mode Observer Based on Discrete Model Approach," *IEEE Transactions on Industry Applications,* pp. 1-1, 2021, doi: 10.1109/TIA.2021.3124982.

[39] Z. Wang and L. Peng, "Grouping Capacitor Voltage Estimation and Fault Diagnosis With Capacitance Self-Updating in Modular Multilevel Converters," *IEEE Transactions on Power Electronics,* vol. 36, no. 2, pp. 1532-1543, 2021, doi: 10.1109/TPEL.2020.3011131.

[40] Z. Ke et al., "Capacitor Voltage Ripple Estimation and Optimal Sizing of Modular Multi-Level Converters for Variable-Speed Drives," *IEEE Transactions on Power Electronics,* vol. 35, no. 11, pp. 12544-12554, 2020.

[41] B. Arabsalmanabadi, N. Tashakor, S. Goetz, and K. Al-Haddad, "Li-ion Battery Models and A Simplified Online Technique to Identify Parameters of Electric Equivalent Circuit Model for EV Applications," presented at the IECON 2020 The 46th Annual Conference of the IEEE Industrial Electronics Society, 18-21 Oct. 2020, 2020.

[42] Y. Jin et al., "A Novel Submodule Voltage Balancing Scheme for Modular Multilevel Cascade Converter—Double-Star Chopper-Cells (MMCC-DSCC) based STATCOM," *IEEE Access,* 2019.

A Non-cooperative Game-theoretic Distributed Control Approach for Power Quality Compensators

Claudio Burgos-Mellado[1], Victor Bucarey[1], Helmo K. Morales-Paredes[2],
Diego Muñoz-Carpintero[1]

[1]Universidad de O'Higgins, Rancagua, Chile
[2] São Paulo State University UNESP, Sorocaba, Brazil
Email: claudio.burgos@uoh.cl

Acknowledgments

This work was supported in part by the "Agencia Nacional Investigacion y Desarrollo" (ANID) through the Chilean Grant: ANID/concurso de fomento a la vinculación internacional para instituciones de investigación regionales (modalidad corta duración) /FOVI210023, and by the National Council for Scientific and Technological Development (CNPq) under Grant 309297/2021-4 and the São Paulo Research Foundation (FAPESP) under Grants 2016/08645-9.

Keywords

≪Power quality≫, ≪Distributed Generation≫, ≪Active Filter≫, ≪Grid forming≫, ≪Microgrid≫

Abstract

This paper demonstrates that the Game Theory (GT) can be an effective tool for implementing a distributed control scheme for coordinating the compensation efforts of power quality compensators (*PQCs*) feeding a common unbalanced load. A non-cooperative game is formulated where each *PQCs* minimises its own interests, defined as the power losses that incur each *PQC* in carrying unbalanced power from its connection point in the system to the point of common coupling (*PCC*). By doing this, the whole system will move until it reaches a global equilibrium (the so-called Nash equilibrium). The power losses are calculated based on the conservative power theory (CPT), allowing the implementation of the proposal in the natural *abc* reference frame. A comparison (via simulations) between the proposed non-cooperative distributed scheme and a cooperative distributed approach based on the consensus theory shows that the proposed non-cooperative game compensates the *PCC* with fewer overall losses than the consensus-based cooperative approach, improving the efficiency of the whole compensation system.

Introduction

Power quality compensators (*PQCs*) are widely used in electrical systems to improve the power quality at one or more points of electrical systems. This power quality enhancement can be performed in electrical variables such as powers, voltages and currents. Focusing on the compensation of unbalanced and distorted currents, a typical solution is to place power quality compensators (*PQCs*) in shunt connection at the point where the power quality needs to be improved: named the point of common coupling (*PCC*) in this paper. In this case, *PQCs* will inject unwanted currents at the *PCC*; therefore, the compensation of currents at the *PCC* is fulfilled. This configuration is usual in isolated AC microgrids (MGs), as shown in Fig. 1. In this type of electrical system, usually, there is a grid-forming converter in charge of imposing the voltage and frequency in the MG. In contrast, other converters operate in grid-follower mode, meaning their behaviour is approximately like a current source. Note that power converters allow the integration of the distributed generation units into the MG. In this scenario, grid-follower converters are sized to provide the nominal power based on the potential of power harvesting from the distributed energy sources. However, due to the intermittence of natural resources, it is expected that the nominal VA capacity of power converters will not be fully used for long periods; therefore, this available VA capacity can be used for the grid-following converters to compensate for unbalanced and distorted currents at the *PCC*, making them works as *PQCs*. In this case, and considering the system shown in Fig. 1, grid-follower converters can inject unwanted currents at the *PCC* and, therefore, improve the quality of the currents seen by the grid former converter. These grid-follower converters, operating as *PQCs*, need to be coordinated to determine their respective compensation efforts.

The coordination of a compensation system of *PQCs* using a centralised control approach has been widely used and reported in the literature [1, 2, 3]. In this approach, there is a central controller in charge of receiving all the information from the *PQCs* converters, determining the compensation effort of each one of them, and finally sending this information to activate the *PQC* function of the converters. This approach has some disadvantages, such as susceptibility to single-point failures, the computational capability of the central controller increases with the number of compensating devices, and thereby high-cost control platforms are required. For this reason, the distributed control approach has been getting attention from researchers in recent years. This approach does not require a central controller as the control effort is distributed among local controllers (LCs) placed on the converters. In this case, LCs operate autonomously and cooperatively to obtain global objectives. It has advantages over the centralised approach: better reliability, flexibility, scalability, and plug-and-play operation [4]. This type of control scheme can be classified as a cooperative distributed control approach, as all the LCs work collaboratively, achieving global objectives. Recently in [5, 6], cooperative control schemes based on the consensus theory have been proposed to coordinate the compensation effort of converters placed on MGs, showing promising results. However, a collaborative distributed approach may not be the best solution to calculate the compensation efforts for the converters if aspects such as losses in the distribution lines are considered. Based on this point, this paper proposes a non-cooperative distributed control approach based on forcing the *PQCs* to play a Nash Equilibrium [7] of a game with suitable cost functions. In this game, each *PQC* chooses the proportion of the compensation effort while minimising the costs: one cost related to the power losses that face each *PQC* to compensate the PCC and a suitable penalty if the compensation is not achieved. Unlike the cooperative approach, where all the *PQCs* work collaboratively, converters have their own interests in the proposed non-cooperative distributed approach; therefore, they act to minimise their respective objective functions. By doing that, the global system will reach the so-called Nash equilibrium (NE).

Based on the discussion above, this paper shows that the problem of calculating the compensation efforts of the grid-following converters shown in Fig. 1 can be formulated in terms of the Nash Equilibrium of a game, generating a non-cooperative distributed control approach for coordinating the *PQCs*. This approach yields the compensation efforts for each converter, where the losses in the lines can be considered, improving the efficiency in managing the whole *PQC* system. The performance of the proposed non-cooperative approach will be compared with that obtained using the cooperative method discussed in [5, 6]. Finally, the conservative power theory (CPT) [5] is used to calculate the compensation currents that needs to be injected by the *PQCs* into the *PCC* to achieve the compensation.

Non-Cooperative Distributed Approach for Calculating the Compensation Efforts of *PQC* converters

Focusing on the MG shown in Fig. 1 and using the CPT, the load current i_{load}, can be decomposed as the sum of a balanced current and an unbalanced current (see Fig. 1). In this case, the *PQC* system must inject i_u at the *PCC* to achieve that the current injected by the grid forming converter will only be the balanced load current component i_b (see Fig. 1). In this paper, it is assumed that *PQCs* displayed in Fig. 1 have enough VA power capability to inject any unwanted current into the MG. Based on that, and considering that the switch sw_1 is closed in Fig. 1, yield (1).

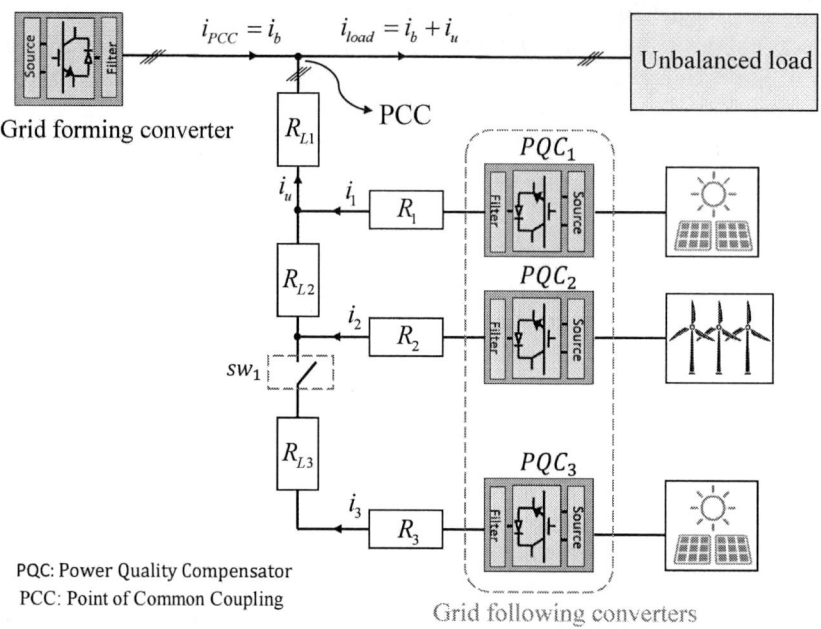

Fig. 1: Isolated three-phase three-wire AC microgrid considered in this work.

$$i_1 + i_2 + i_3 = i_u \cdot (n_1 + n_2 + n_3) = i_u \quad , \quad where \quad \sum_{h=1}^{3} n_h = 1 \tag{1}$$

In this work, the compensation efforts (n_1, n_2 and n_3) of the *PQC* compensation system are calculated in a distributed fashion by the non-cooperative approach proposed in this paper. This is achieved by formulating a non-cooperative game among the *PQCs*. To this end, it is necessary to set the individual cost of each *PQC* of participating or not participating in the game. In this work, the cost associated with a given *PQC* is quantified as the losses incurred by the *PQC* in carrying the compensation power from its place into the MG to the *PCC*. In this sense, the losses are calculated based on the conservative power theory, which provides a framework to calculate unbalanced and distorted currents in the natural *abc* reference frame. For the sake of clarity, this section is subdivided into the following subsections: (i) A brief overview of the Conservative Power Theory (CPT), focusing on its use to calculate losses in the distribution lines, (ii) the introduction of the proposed non-cooperative game considering only two *PQCs*, and finally, (iii) the extension of the proposed non-cooperative distributed control approach considering *N PQCs*. These sections are discussed as follows.

Theoretical Background

According to the CPT [5], the instantaneous vector current (i) is split into, balanced active current (i_b^a), balanced reactive current (i_b^r), unbalanced current (i_u), and void current (i_v). So, the current, i is expressed as:

$$i = i_b^a + i_b^r + i_u + i_v \tag{2}$$

By definition, the collective *RMS* current can be split into:

$$I^2 = I_b^{a2} + I_b^{r2} + I_u^2 + I_v^2 \tag{3}$$

Note that each current term is orthogonal to each others. Thus, multiplying the collective *RMS* current and voltage, the apparent power (*A*) can be decomposed into:

$$A^2 = V^2 I^2 = V^2 I_b^{a^2} + V^2 I_b^{r^2} + V^2 I_u^2 + V^2 I_v^2 \tag{4}$$

where $P = V I_b^a$ is the active power, $Q = V I_b^r$ is the reactive power, $N = V I_u$ is the unbalance power, and $D = V I_v$ is void (distortion) power. More information about the current and power components defined by the CPT can be found in [5]. Apart from P, all power terms characterise a non-ideal aspect of the load behaviour. The global performance index is the power factor:

$$\lambda = \frac{P}{A} = \frac{I_b^a}{I} = \frac{I_b^a}{\sqrt{I_b^{a^2} + I_b^{r^2} + I_u^2 + I_v^2}} \tag{5}$$

Note that, to provide active power (*P*) at a given load, any deviation of power from a balanced purely resistive load increases line losses. Thus, from (4) and active power (*P*) definition, the minimum line losses results:

$$P_{min}^{loss} = R_{line} I_{min}^2 = R_{line} I_b^{a^2} \tag{6}$$

On the other hand, considering the load current (3), the total loss is given by:

$$P_{total}^{loss} = R_{line} I_{load}^2 = R_{line}(I_b^{a^2} + I_b^{r^2} + I_u^2 + I_v^2) = P_{min}^{loss}(\frac{1}{\lambda})^2 \tag{7}$$

Finally, we can define the line utilisation factor as the ratio between the total losses and minimum losses:

$$\rho_{line} = \frac{1}{\lambda^2} = \frac{P_{total}^{loss}}{P_{min}^{loss}} \tag{8}$$

The above equation shows how the harmonics, unbalance and reactive power increase the line utilisation factor, enlarging line losses. Moreover, $\rho_{line} = 1$ only in the case of balanced resistive loads (current waveforms proportional to voltage waveforms).

The MG in Fig. 1 shows the case considered in this paper, i.e., where the grid following converters injected only unbalanced currents to the PCC for compensation purposes. In this scenario, the power loss associated with this compensation is given by: $P^{loss} = R_{line} I_u^2$. In the subsequent sections, the operating principle of the proposed non-cooperative distributed strategy for this situation is explained in detail.

Proposed non-cooperative game considering two *PQC* converters

In this section, we analyse the proposed game with two *PQCs* (the switch sw_1 in Fig. 1 is open). By doing this, the main ideas and concepts behind the proposed game can be easily presented and discussed, and then its extension to more complex networks will be more understandable. It is worth remembering that this game aims to determine n_1 and n_2 associated with PQC_1 and PQC_2 respectively, that satisfy (1) (considering $n_3 = 0$), and at the same time, minimise the losses in carrying the compensation power from the *PQCs* to the *PCC*. The formulation of the proposed non-cooperative game is detailed as follows.

PQC_1: This player feeds the *PCC* with a current equal to $i_u \cdot n_1$ where $n_1 \in [0, 1]$. In this case, the PQC_1 minimises its cost function (9) that quantifies the losses in carrying $i_u \cdot n_1$ to the *PCC* (see Fig. 1) and an additional cost K_1 if the system is not able to supply all the compensating current (i.e. $n_1 + n_2 < 1$). In (9), I_u is the the collective RMS (norm) value of i_u. From here onwards, we use the notation $\mathbb{1}$ as the indicator function, meaning that $\mathbb{1}_{n_1+n_2<1}$ equals to 1 if $n_1 + n_2 < 1$ and 0 otherwise.

$$\underset{n_1 \in [0,1]}{Min} \quad (R_1 + R_{L1}) \cdot I_u^2 \cdot n_1^2 + \mathbb{1}_{\{n_1+n_2<1\}} \cdot K_1 \tag{9}$$

Given that the second term involves the decision of the PQC_2, we compute the reaction curve of player PQC_1 given n_2. This reaction curve corresponds to the optimal response of PQC_1 considering that the PQC_2 plays n_2. This reaction curve is computed as:

$$n_1^*(n_2) = \begin{cases} 0 & \text{if } K_1 < I_u^2 \cdot (R_1 + R_{L1})(1 - n_2)^2 \\ 1 - n_2 & \text{otherwise.} \end{cases} \tag{10}$$

PQC_2: Similarly to player PQC_1, player PQC_2 minimises its cost function, given by (11). This function quantifies the losses in carrying $i_u \cdot n_2$ ($n_2 \in [0,1]$) to the PCC shown in Fig. 1 and the additional cost K_2 if the system is not able to supply all the compensating current (i.e. $n_1 + n_2 < 1$). In (11), I_u corresponds to the collective RMS value of i_u.

$$\underset{n_2 \in [0,1]}{Min} \quad (R_2 + R_{L2} + R_{L1}) \cdot I_u^2 \cdot n_2^2 + \mathbb{1}_{\{n_1+n_2<1\}} \cdot K_2 \tag{11}$$

Analogously, the reaction curve of PQC_2, considering the PQC_1 plays n_1 is given by (12):

$$n_2^*(n_1) = \begin{cases} 0 & \text{if } K_2 < I_u^2 \cdot (R_2 + R_{L2} + R_{L1})(1 - n_1)^2 \\ 1 - n_1 & \text{otherwise.} \end{cases} \tag{12}$$

The Nash Equilibrium (NE) of this game are the points where both reaction curves intersect. Fig. 2 shows the graphical representation of the game between PQC_1 and PQC_2 given by equations (9)-(12). Depending on the parameters of the MG and the cost K_1 and K_2, $PQCs$ may participate or not in this game. Fig. 2 represents the case where there is only one NE: the cost K_1 and K_2 are too high to incentivise the $PQCs$ to inject current to the PCC, so both $PQCs$ do not participate. However, for the proposed game, we manage costs of K to have one or multiple NEs.

In the proposal, we define $\bar{n}_i$ the strategy for player $i = 1, 2$ that makes the other player to be indifferent between participating or not. Note that $\bar{n}_1 = 1 - \frac{1}{I_u}\sqrt{\frac{K_2}{(R_2 + R_{L1} + R_{L2})}}$ is explicitly computed by setting to equality the *if* condition in (12). Following the same procedure, $\bar{n}_2$ is calculated using (10), giving: $\bar{n}_2 = 1 - \frac{1}{I_u}\sqrt{\frac{K_1}{(R_1 + R_{L1})}}$.

Note that the game represented in Fig. 2(a) can be managed via the free parameters K_1 and K_2. In this paper, these parameters are set to ensure that the proposed game shown in Fig. 2(a) has an equilibrium point in which both $PQCs$ participate, producing the game illustrated in Fig. 2(b). In that figure, the Nash equilibrium point (n_1^*, n_2^*) is obtained by imposing the conditions: (i) $n_1(\bar{n}_2) = \bar{n}_1$, and (ii) $n_2(\bar{n}_1) = \bar{n}_2$, as shown in Fig. 2(b). By doing this, the following relationship is got: $\sqrt{\frac{K_1}{R_1 + R_{L1}}} + \sqrt{\frac{K_2}{(R_2 + R_{L2} + R_{L1})}} = I_u$. This equation shows that for a given unbalanced current I_u, a single equilibrium point (for the proposed game) is obtained if K_1 and K_2 are in the surface given by that equation.

It must be highlighted that for the MG studied in this paper (see Fig. 1), it is considered that all the $PQCs$ have the same characteristics and nominal power; therefore, it can be assumed that $K_1 = K_2 = K^*$.

Using this assumption, the relationship discussed in the paragraph above can be rewritten as follows:
$K^* = I_u^2 \cdot \left(\frac{1}{\sqrt{R_1 + R_{L1}}} + \frac{1}{\sqrt{R_2 + R_{L2} + R_{L1}}} \right)^{-2}$. Thus, at using this latter equation on equations (9)-(12), it can be assured that the proposed game has a single-equilibrium point in which both *PQCs* participate, as shown in Fig. 2(b).

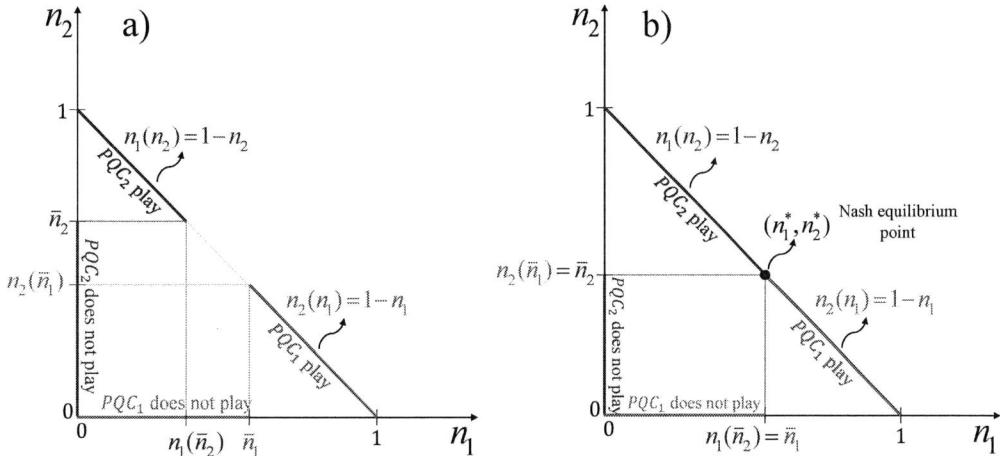

Fig. 2: Reaction curves a) case $K < K^*$, b) case $K = K^*$

Generalisation of the proposed non-cooperative game

Let us consider that the MG displayed in Fig. 1 has "N" *PQCs* converters that need to be coordinated to supply the compensation current I_u at the *PCC*. In this scenario, the proposed non-cooperative game to coordinate these *PQCs* is as follows: (i) the objective function for the ith *PQC* is given by (13), (ii) the reaction curve of PQC_i is shown in (14), (iii) the critical point until the *PQC* play the game $\bar{n}_i$ is given by (15). Note that these equations consider $K_i = K \, (\forall i = 1, ..., N)$, meaning that all the *PQCs* have the same characteristics. In this sense, the parameter K that ensure that the game has only an equilibrium point (Nash equilibrium point) is shown in (16).

$$
\underset{n_i \in [0,1]}{Min} \quad I_u^2 \cdot n_i^2 \cdot \left(R_i + \sum_{j=1, j \leq i}^{N} R_{Lj} \right) + \mathbb{1}_{\left\{ \sum_{j=1}^{N} n_j < 1 \right\}} \cdot K_i \tag{13}
$$

$$
n_i^*(n_{j \neq i}) = \begin{cases} 0 & \text{if } K_i < I_u^2 \cdot \left(R_i + \sum_{j=1, j \leq i}^{N} R_{Lj} \right) \cdot \left(1 - \sum_{j=1, j \neq i}^{N} n_j \right)^2 \\ 1 - \sum_{j=1, j \neq i}^{N} n_j & \text{otherwise.} \end{cases} \tag{14}
$$

$$
\bar{n}_i = \frac{1}{N-1} + \frac{N-2}{N-1} \cdot \frac{\sqrt{K}}{I_u} \cdot \left(R_i + \sum_{j=1, j \leq i}^{N} R_{Lj} \right)^{-0.5} - \frac{1}{N-1} \cdot \frac{\sqrt{K}}{I_u} \cdot \sum_{h=2, h \neq i}^{N} \left(R_h + \sum_{j=1, j \leq h}^{N} R_{Lj} \right)^{-0.5} \tag{15}
$$

$$
K = I_u^2 \cdot \left(\sum_{j=1}^{N} \frac{1}{\sqrt{R_{eq-j}}} \right)^{-2}, \quad \text{where} \quad R_{eq-j} = R_j + \sum_{h=1, h \leq j}^{N} R_{Lh} \tag{16}
$$

Table I: Compensation effort given by the distributed cooperative method [5, 6] (second column), and the proposed distributed non-cooperative approach (third column).

Compensation effort	Cooperative approach [5, 6]	Proposed Non-cooperative approach
n_1	1/3	0.40
n_2	1/3	0.32
n_3	1/3	0.28

Simulation Results

In this section, the proposed non-cooperative game to calculate the compensation efforts for the *PQCs* is validated through simulation work. To this end, the system shown in Fig. 1 is simulated, using PLECS software, considering three *PQCs*, i.e., the switch sw_1 in Fig. 1 is closed during the validation process. The line resistances illustrated in that figure are equal to 1Ω, and the grid forming converter generates a three-phase voltage of 220Vrms and 50Hz. The unbalanced load illustrated in Fig. 1 corresponds to an unbalanced resistive load with the following values: $R_a = 22\Omega$, $R_b = 11\Omega$, and $R_c = 5\Omega$.

Note that, for the case studied in this paper, it is assumed that the three *PQCs* have the same characteristics. Then, if they are managed using the consensus-based distributed control schemes proposed in [5, 6], their compensation efforts will be equal to each other, as shown in the second column of Table I. This result is because all the *PQCs* have the same VA capacity. Therefore, under the cooperative approach [5, 6], all *PQCs* should equally share the compensation effort. In contrast, if the proposed non-cooperative method is used for controlling the *PQCs*, it is found that their compensation efforts are different, as shown in the third column of Table I. In this case, each *PQC* looks to minimise its own interest (losses in carrying compensation power from its connection point to the PCC), reaching the Nash equilibrium, as shown in the third column of Table I.

Fig. 3(a) shows the unbalanced currents that inject each *PQC* to achieve the compensation of i_{PCC} (see Fig. 1), for the case where the compensation effort is calculated by a cooperative approach (second column in Table I). On the other hand, Fig. 3(b) shows the same information but for the case where the compensation efforts are determined by the proposed non-cooperative approach (third column in Table I). From that figure, it can be appreciated that the currents injected by the *PQCs* when they are managed by the cooperative approach [5, 6] are the same (per phase), as they have the same VA power capacity. In contrast, the *PQC* system feeds the PCC with different currents when the proposed non-cooperative approach drives them. In this case, the PQC_1 inject more compensation current to the PCC than the other *PQCs*, as shown in Fig. 3(b). And the PQC_3 inject a lesser compensation current than the rest of the *PQCs*. This latter trend is because the cost, in terms of power losses, is higher for PQC_3 than for others. In addition, the cost for the PQC_1 is smaller than the others; therefore, it feeds the PCC with the highest unbalanced current. The results discussed above show that the proposed distributed non-cooperative approach, differently from the distributed cooperative one, can include criteria of power losses for calculating the compensation efforts of the *PQC* system, which is an advantage of the proposal. This allows achieving the compensation process with an equal loss distribution among the *PQCs*, as shown in Fig. 4(b). In that figure, a comparison of the power losses distribution of the *PQCs* is illustrated for both the cooperative and non-cooperative approaches during the compensation process. While the proposed non-cooperative method compensates the PCC with a similar losses distribution for the *PQCs* (see Fig. 4(b)), the cooperative approach compensates the PCC with an unequal power losses distribution for the *PQCs* (Fig. 4(a)). To sum up, comparing both results shown in Fig. 4, it is concluded that the proposed non-cooperative approach achieves the coordination of the *PQCs* with a better losses distribution than the cooperative approach reported in [5, 6].

Fig. 5 shows the current at the PCC (i_{PCC}) when the *PQCs* are enabled. Note that this behaviour is the same no matter what of the two strategies shown in Fig. 4 is used as in both cases, the compensation of i_{PCC} is fully achieved: The main difference between them is how the *PQCs* are coordinated. In this sense, the non-cooperative allocation (calculation of the compensation efforts n_1, n_2 and n_3) has one main advantage: it is not coordinated by a central controller, so under that allocation, none of the PQC have the incentive of deviating by themselves without any further regulation.

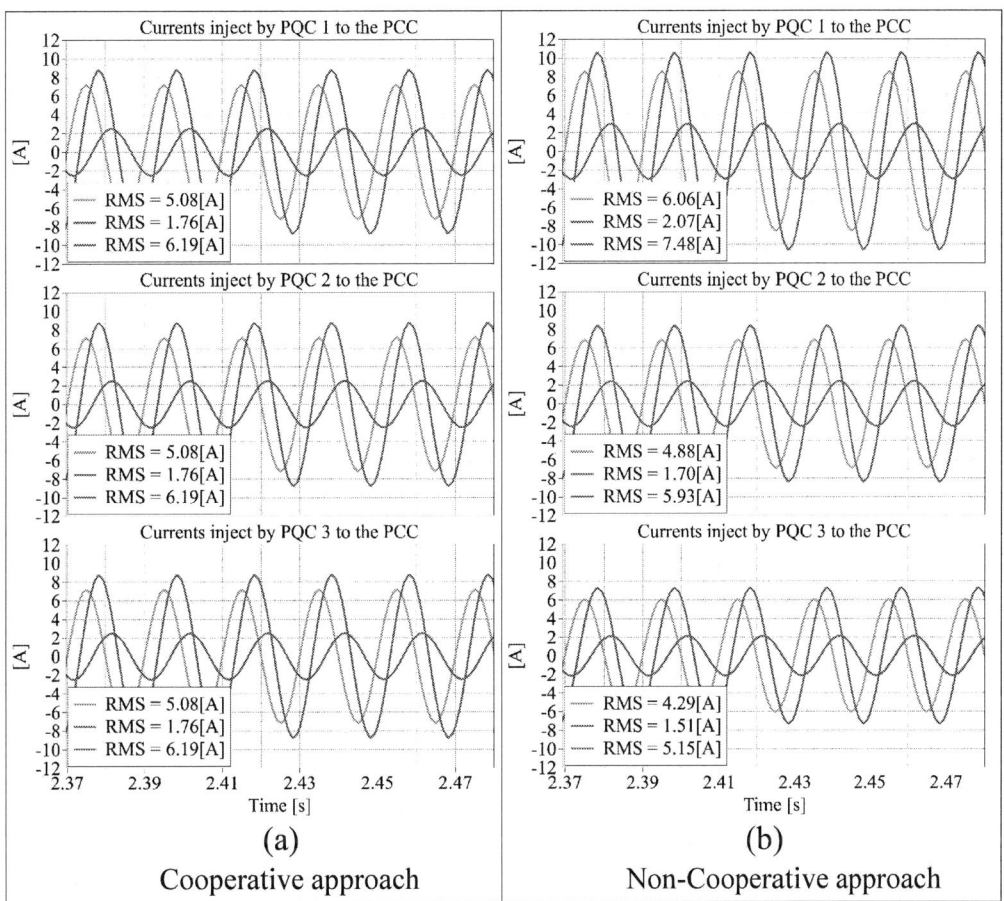

Fig. 3: Currents at the PCC before and after activating the *PQC* system.

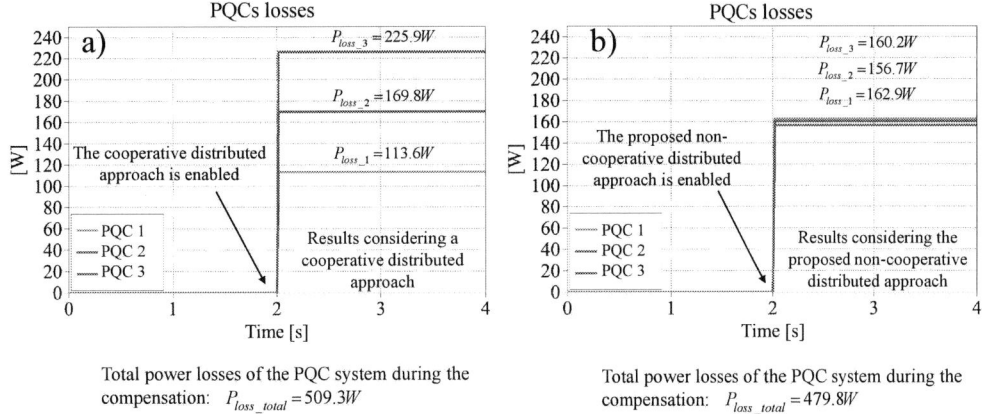

Fig. 4: (a) Results obtained when the cooperative control scheme proposed in [5, 6] is used to coordinate the *PQC* converters, (b) Results obtained when the proposed non-cooperative control is used.

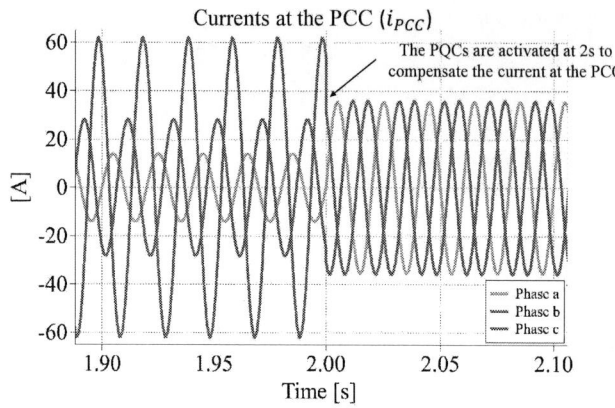

Fig. 5: Currents at the PCC before and after activating the PQC system.

Conclusions

This paper demonstrated that the compensation effort of a set of power quality compensators ($PQCs$) can be calculated using the proposed distributed non-cooperative method based on the game theory. Also, a comparison between the non-cooperative approach and the distributed cooperative approach reported in [5, 6] (based on the consensus theory) was performed. From that comparison, it was found that different from the collaborative method; the non-cooperative method can calculate the compensation effort for the $PQCs$ considering criteria such as the line power losses, which is an advantage of the proposal and could bring economic benefits for the entire PQC system. In future work, we aim to focus on two extensions of this setting. First, to study the case where each PQC has a limited capacity b_i, meaning that each strategy satisfies $n_i \leq b_i$ under the assumption that $\sum_i b_i \geq 1$. Note that, if $b_i \leq \bar{n}_i$ in Equation (15) the analysis above is still valid. Secondly, extending this approach to more complex networks where multiple PCC are considered.

References

[1] Abbas Marini, Mohammad-Sadegh Ghazizadeh, Seyed Saeedallah Mortazavi, Luigi Piegari, "A harmonic power market framework for compensation management of DER based active power filters in microgrids", International Journal of Electrical Power & Energy Systems, vol. 113, pp. 916-931, 2019.

[2] Ali Mortezaei, Marcelo Godoy Simões, Mehdi Savaghebi, Josep M. Guerrero, Ahmed Al-Durra,"Cooperative control of multi-master–slave islanded microgrid with power quality enhancement based on conservative power theory" IEEE transactions on smart grid, vol. 9, pp. 2964-2975, 2016.

[3] Helmo K. Morales Paredes, Alessandro Costabeber, and Paolo Tenti, "Application of conservative power theory to cooperative control of distributed compensators in smart grids", 2010 International School on Nonsinusoidal Currents and Compensation, pp. 190-196, 2010.

[4] Enrique Espina, Jacqueline Llanos, Claudio Burgos-Mellado, Roberto Cardenas-Dobson, Manuel Martinez-Gomez and Doris Saez, "Distributed control strategies for microgrids: An overview", IEEE Access, vol. 8, pp.193412–193448, 2020.

[5] Helmo Morales-Paredes, Claudio Burgos-Mellado, Paulo Jakson Bonaldo, Diego Tardivo Rodriguez and Juan Sebastian Gomez Quintero,"Cooperative Control of Power Quality Compensators in Microgrids", 2021 IEEE Green Technologies Conference (GreenTech), pp. 380-386, 2021.

[6] Juan Sebastian Gomez, Jacqueline Llanos, Enrique Espina, Claudio Burgos-Mellado and Jose Rodriguez, "Cooperative Power Conditioners for Microgrids in Mining." 2021 23rd European Conference on Power Electronics and Applications (EPE'21 ECCE Europe), 2021.

[7] Drew Fudenberg and Jean Tirole, "Game theory", MIT press, 1991.

A Comparative Analysis of Power Converter Topologies for Integration of Modular Batteries in Electric Vehicles

Alberto Cárcamo[1], Aitor Vázquez[1], Alberto Rodríguez[1], Diego G. Lamar[1], Marta M. Hernando[1], Daniel Remón[2]

University of Oviedo, E+ Ecoeficiencia e Ingeniería S.L. [2]

Edificios Departamentales Oeste, Mod. 3, 33204, C. Gregorio Marañón, 1, 33203

Gijón, España

uo278796@uniovi.es

Acknowledgements

This work has been supported by the Principality of Asturias and FICYT under project SV-PA-21-AYUD/2021/51931 and by the Spanish Government under project PID2021-127707OB-C21.

Keywords

«Bi-directional converters», «Dual Active Bridge (DAB) DC-DC converter», «Battery charger», «Power converters for EV».

Abstract

This paper presents a comparative analysis between four proposed DC-DC power converter topologies, for integration of removable batteries into an electric vehicle (EV) that also has a primary energy storage system (ESS). To perform this analysis, first the minimum requirements for the converter are defined, such as volume, power density, and specific density, as well as the system operating condition. The comparison between the proposed topologies is done by evaluating three main aspects: efficiency, volume, and current ripples. The procedure consists of performing a steady-state analysis of each topology to obtain the main operating values and waveforms, validating the results and the proposed control strategies through simulations, calculating the losses for each converter, and estimating their volume. Different components are evaluated in each topology to perform a power loss analysis, considering several options of Si MOSFETs, transformers, and inductors.

Introduction

EVs are becoming a trend in the past years and are foreseen to dominate the future in mobility [1], as new technologies are developed, and prices become competitive in comparison with the Internal Combustion Engine Vehicles (ICE). The commitment of the world with reducing the carbon emissions has contributed to the increase of research and development of EV systems, including the power converter topologies, whose research is oriented at increasing efficiency, reliability, as well as the reduction of their size and weight [2]. All this is supported with the governments' policies which are being updated in order to support this technology, especially in the European Union (EU) [3].

EVs usually has only one ESS, typically a non-removable battery. There have been several proposals regarding EV designs, which consider implementing a primary removable battery, which can be replaced by technicians, rather than the user, using specialized equipment. This is not only due to safety reasons but also due to the size and weight of the battery [4]. The impact on the electrical grid may also be addressed, by charging the batteries at off-peak periods, or even providing grid support [5]–[7], while including the possibility of extending the batteries life by charging them in slow-charging mode [8]. Battery swapping has been offered in 2013 by different companies, but failed due to lack of interest from customers or cooperation from manufacturers [9] [10]. Among the different proposals, this paper studies a system that considers the use of modular removable batteries in addition to a main non-removable battery. These additional batteries can be considered as a secondary ESS for the EV, which allows an increased flexibility of the energy management, and the possibility of increasing the EV range, among other things. Fig. 1 shows the power system that is considered for this study. It comprises a primary non-removable battery connected to the HV-DC bus through a BMS, which is always mandatory as it provides functions such as balancing the charge of the cells [11], monitoring the temperature, providing safety measures, among other functions. The HV-DC bus supplies power to the LV-DC bus, where the electronic control units (ECUs) and the auxiliary electronic systems are. It also supplies power to the electric motors through the inverters. The system has an additional EES connected to the HV-DC bus, the secondary removable batteries, which are integrated by connecting them through a BMS and a power converter, the latter being the converter under analysis in this paper.

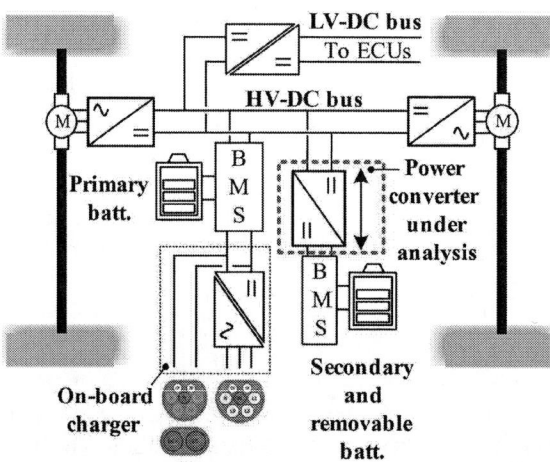

Fig. 1: EV power system, with a main non-removable battery and secondary removable batteries, connected through a converter (red dotted).

The converter should be able to work on a system with certain characteristics and comply with the requirements defined below. One of the characteristics of the system is the range of voltage variations in the secondary batteries (input port), and in the HV-DC bus (output port). This voltage range depends on the State of Charge (SoC) of the removable batteries at the input port, and the HV-DC voltage at the output port, which is set by the primary battery at a nominal voltage that can vary from 200V to 450V [12]. Due to these voltage variations, the converter design should be able to operate in a relatively wide range of voltages in the input and output [13]. The converter should also be bidirectional, to sink energy from the HV-DC bus, either for regenerative breaking, or to allow the charging of the secondary batteries directly from the HV-DC bus, without removing them from the vehicle. The requirements defined for this study are shown in Fig. 2 and are as follows: nominal power of 10 kW, nominal voltage of 60V and 400V for the input and output, respectively, a volume of 4.6 dm³ specified by the maximum dimensions of 575 x 100 x 80 mm and a maximum weight of 10 kg, which gives a resultant minimum power density of 2.174 kW/ dm³ and a specific density of 1 kW/kg.

Four topologies are chosen for this comparative analysis. Two of them are based on a boost converter: the Input-Parallel Output-Parallel boost converter (IPOP), and the Cascaded-Boost converter (CB). The other two topologies are based on a phase-shifted full-bridge converter providing galvanic isolation through a transformer: Current-Fed Phase-Shifted Full-Bridge converter (CF-PS-FB), and the Dual-Active-Bridge converter (DAB).

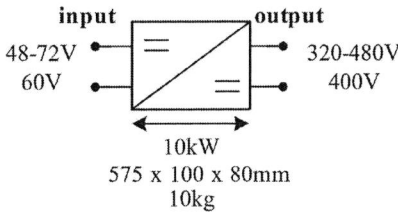

Fig. 2: Converter design requirements and system specifications.

The paper is organized as follows: the proposed topologies are described in Section II, Section III shows the methodology used for the comparison, the results of the analysis are summarized in Section IV, and finally the conclusions are presented in Section V.

Proposed Power Converter Topologies

Input-Parallel Output-Parallel Converter (IPOP)

The IPOP converter is a modular Boost Converter with three modules connected in parallel at the input and at the output, as can be seen in Fig. 3 [14]. Each module of the converter handles one third of the total power, allowing the use of components with lower current rating. The control signals of each module are 120° phase-shifted to reduce the current ripple at the input.

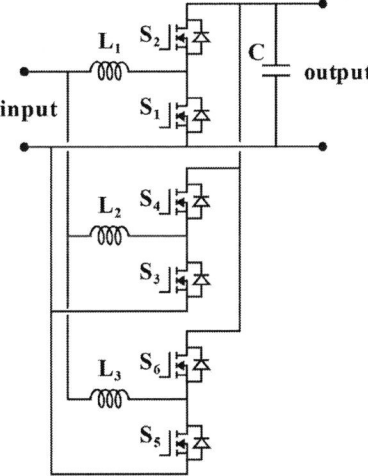

Fig. 3: IPOP converter schematic.

Due to the large voltage ratio between the input and output, it is necessary to work at a high nominal duty cycle in each module, around 85%. This causes the grounded MOSFET power transistors (S1, S3 and S5) to conduct a larger current than the floating ones (S2, S4 and S6). Therefore, the conduction losses in these devices are unbalanced, and entails the selection of their model to be carried out independently for the grounded and for the floating devices. This is further explained in Section III.

This topology has two advantages: the reduction of the current rating of the power transistors and the decrease of the current ripple at the input. These advantages are a consequence of the modular arrangement and the interleaved control. Among the disadvantages, it can be mentioned the high duty cycle and the asymmetry, the increased volume due to the inductors of each module, and the operation with hard switching of the MOSFETs, which lowers its efficiency, specially at high switching frequencies.

Cascaded-Boost Converter (CB)

The CB converter, which is also based on a modular boost converter, is made up of 2 stages that are cascaded through a DC bus [15]. Fig. 4 shows the CB converter schematic. The first stage, which is connected to the input, is a two-module IPOP boost converter that works with a fixed 50% duty cycle and with an interleaved control that is phase-shifted 180° to reduce the current ripple, virtually eliminating it completely. The fixed duty cycle boosts the input voltage to double its value, resulting in a nominal voltage at the DC bus of 120 V. The second stage is a boost converter, which boosts the DC bus voltage to the HV-DC bus voltage (400 V).

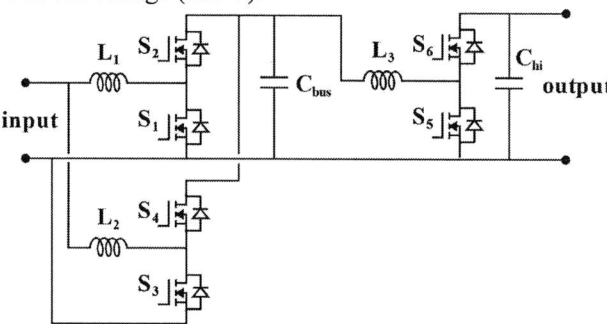

Fig. 4: CB converter schematic.

Besides virtually cancelling all the current ripple at the input, another advantage of working at a 50% duty cycle is that the MOSFETs used in the first stage have the same current flowing through them, making it possible to select the same model for the ground-referenced and the floating-referenced device. The voltage at the DC-bus allows the use of low voltage rated devices in the first stage, while using high voltage rated devices on the second stage. At the second stage, the MOSFET selection may be done independently, although its nominal duty cycle is not as high as in the IPOP, it is around 75%.

An independent control strategy is proposed for each stage. The first stage has a cascaded control with the outer loop controlling the voltage at the DC-bus, and a current controller at the inner loop that regulates and balances the currents in each module. The second stage has a current controller, which receives the reference of the power flow, regulating the dynamic behavior of the converter.

One disadvantage identified on this topology is the volume, as it needs an extra capacitor for the DC-bus, and besides the two inductors of the first stage, it needs an inductor at the second stage that is rated for the full 10 kW of power. The control is also more complex than the one for the IPOP.

Current-Fed Phase-Shifted Full-Bridge Converter (CF-PS-FB)

The CF-PS-FB converter is the first proposed topology that offers galvanic isolation, and it is based on two Full-Bridges (FBs), a current-fed FB at the input and a voltage-fed FB at the output. The input FB is connected to the secondary batteries through an inductor, and has an active clamp made from a transistor and a capacitor for achieving soft-switching [16]. Both FBs are connected through a transformer, and optionally an inductor can be added in series. Fig. 5 shows its schematic.

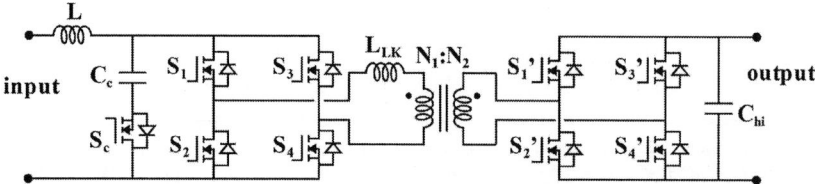

Fig. 5: CF-PS-FB schematic.

This converter has two operating modes depending on the direction of the power flow: boost mode (input to output) and buck mode (output to input). In boost mode, the input FB operates as a current-fed FB converter, while the output FB works as a synchronous rectifier. In buck mode, the output FB operates as a voltage-fed converter while the input FB operating as a synchronous rectifier. The active clamp allows the possibility of the converter to operate under Zero-Voltage Switching (ZVS) and Zero-Current Switching (ZCS) [17].

One disadvantage of this converter is the need of changing the operating mode when the direction of the power flow changes, which besides increasing the complexity of the control, it also takes some time, which can limit the dynamic response of the converter. Also, this converter has more components, including an additional circuit for demagnetizing the input inductor at the turning on of the converter, as well as in case of failure. This extra circuitry is not shown in Fig. 5.

Dual Active Bridge Converter (DAB)

The DAB is also based on two FBs, connected through a transformer, using an inductance to transfer power, naturally behaving as a current source. Usually, this inductance is added as an external inductor, however, the transformer leakage inductance may also be used. The DAB converter schematic is shown in Fig. 6. Although the power transfer in this topology depends on several factors, it is important to note that given a previously defined and fixed switching frequency, the value of the inductance, L_{LK}, is the one that defines the maximum power the converter is capable of transferring, being this inductance inversely proportional to the power.

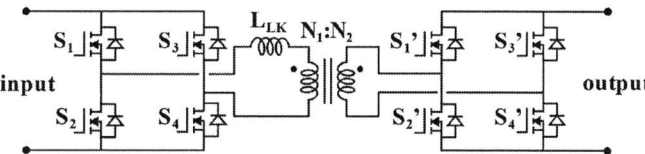

Fig. 6: DAB converter schematic.

There are several control techniques [18] [19], but the simplest and the one considered for this study is the Single Phase-Shift (SPS) control. This technique consists in using a fixed duty cycle of 50% in both FBs, and introducing a phase-shift between them, which is the control variable for the power flow.

One of the advantages of this topology is the ZVS operation, which allows to transfer relatively high power with high efficiency. Note that ZVS is not always guaranteed, as it is lost when operating at low power, or at wide range of voltages [20]. However, there are several solutions to extend the ZVS operation point, even throughout all range of power transfer, and a wide voltage range. Some solutions include the use of more complex control strategies, such as Triple Phase-Shift (TPS), or using the transformer magnetizing inductance to increase the reactive current [21]. Considering SPS control, the design approach followed is oriented to maximize its efficiency at a certain power, although this implies reducing the efficiency at other operating points. The disadvantage of this converter is mainly the high current ripple.

Methodology

Analysis Parameters

To perform the analysis between the proposed topologies, the following parameters are evaluated:

- Current Ripple: The peak-to-peak value of the current ripple, at both the input and the output, is evaluated. Topologies with high current ripples may require the addition of external filters, which will add volume and weight to the final design. It is important to mention that these filters are not included in this analysis.

- Efficiency: An estimation of the transistor power losses is carried out. The magnetic elements efficiency performance is also evaluated through simulation. Losses due to gate drivers, control circuits, or capacitors series resistance are not considered in this analysis.
- Volume: For the volume estimation, only the magnetic elements (inductors and transformers) are considered. Heat dissipation is done by employing the chassis as heatsink, hence, no additional heat sink is considered. The volume of the capacitors is also neglected.

Analysis Procedure

The procedure for the comparative analysis is summarized as follows:

Steady-State Analysis and Simulation

A steady-state analysis is performed on each of the proposed topologies to obtain the theoretical operating values, which includes the current ripples, the average and rms currents, peak values, among other. This analysis is performed at the nominal power of 10 kW and nominal input and output voltages of 60 and 400 V, respectively. It is important to note that for this analysis, four different frequencies are evaluated on each topology: 25, 50, 75, and 100 kHz.

The result from the steady-state analysis is then validated by comparing them to the simulation results, using the software PSIM® and SIMULINK®. The simulations are also used to validate the control strategies exposed in Section II.

Component Selection

The component selection is a crucial part of this analysis as the converter performance is directly related to their characteristics. This analysis is carried out for the switching devices (MOSFETs), and the magnetic elements (inductors and transformers). Regarding the selection of the switching devices, several part numbers are evaluated for each topology, selecting the ones that provide the best efficiency for the converter. The MOSFETs are divided in two groups: high voltage and low voltage devices. The former are super-junction MOSFETs with a maximum rated break-down voltage of 600-650 V, while the latter are MOSFETs with a maximum break-down voltage of 100-200 V. As mentioned earlier in the paper, there are cases in which the devices are evaluated independently, e.g., the same half-bridge might have 2 different part numbers. For the IPOP, the ground referenced MOSFETs are evaluated independently from the floating-referenced ones, with all of them being high voltage devices. For the CB, the first stage devices are all the same, while in the second stage, the devices are evaluated independently. Also, the first stage considers only low voltage devices, while high voltage devices are considered for the second stage. For the CF-PS-FB and the DAB, the high voltage devices are evaluated on the input FB and the high voltage devices at the output FB, where in both topologies there are no independent evaluations performed. All the devices are rated for automotive applications, and only Silicon (Si) MOSFETs are considered, excluding technologies such as Silicon Carbide (SiC) and Gallium Nitride (GaN), due to their higher cost. Another aspect taken into consideration is the parallelization of the MOSFETs, as it allows the inclusion of devices with good characteristics, but that do not comply with the current rating requirements.

Regarding the magnetic elements, design tools Magnetics Inductor Designer® and Ansys PExprt® are used for the design, selection, and losses estimation for the inductors, while the transformers are selected from commercial options.

Power Losses Analysis

This analysis is divided in two parts: the power losses due to the switching devices, and the power losses in the magnetic components. Regarding the losses on the MOSFETs, only the conduction and switching losses are considered, neglecting the losses due to the gate drivers and the body diode reverse recovery, among others. Although there are several models for calculating the MOSFET switching losses, some of them very precise and complex, a simplified model is used, defined by equation (1), where, P_{sw} are the switching losses, f_{sw} is the switching frequency, and E_{on} and E_{off} are the turn-on and turn-off losses, respectively. Equation (2) is an expansion of equation (1), where V_{DS} and i_{DS} are the drain to source voltage and the drain current, respectively. $t_{on/off}$ is the time it takes the MOSFET to turn-on/turn-off, and is usually given by the manufacturer [22].

$$P_{sw} = f_{sw}\left(E_{on} + E_{off}\right) \tag{1}$$

$$E_{on/off} = \frac{1}{2}\cdot V_{DS}\cdot i_{DS}\cdot t_{on/off} \tag{2}$$

Note that for the CF-PS-FB and the DAB topologies, because they work with soft-switching, the turn-on losses are neglected and only the turn-off losses are considered. Regarding the magnetic elements, the inductor losses are estimated through the simulations, while the transformer losses are estimated using their datasheet.

Volume Estimation

To estimate the volume of the magnetic components, design tools are used for estimating the volume of the inductors while the transformer dimensions are obtained from the datasheet, as it is a commercial device. In the next section, a more detailed analysis is presented regarding the inductors, as well as the transformers.

Results

Current Ripple

The current ripple value at the input and output for each converter is presented in Table I. The IPOP and CB presents the lowest input current ripple, due to the interleaving, with the CB presenting the best results. However, both topologies present a high output current ripple, with the CF-PS-FB converter presenting the most balanced results at the input and output, with 34.48 and 37 A, respectively. The DAB presents the highest current ripple, both at the input and at the output, due to the reactive current, a characteristic of this topology that is necessary to achieve ZVS.

Table I: Peak-to-Peak Current Ripple

	IPOP	CB	CF-PS-FB	DAB
Input Current Ripple [A]	5.8	0.4	34.48	523.77
Output Current Ripple [A]	60.12	80.84	37.36	244.72

Efficiency

The efficiency results mainly depend on the component selection and the switching frequency at which the converter operates. Table II shows the selected components (part number) for each converter, as they exhibited the best efficiency results for their respective topology. This table also shows the number of parallelized MOSFETs and distinguishes the ones that are evaluated independently. Table III shows the selected magnetic components for each topology in function of the switching frequency. It also shows the core material and part number for the inductors (L1, L2, and L3), and the type and part number for the transformer (T1), which is a 15 kW planar transformer from HIMAG PLANAR® for all frequencies.

Fig. 7 shows a comparison of the power losses and efficiency between the four topologies, where the power losses on the switching devices are shown in Fig. 7(a), the power losses on the magnetic elements are shown in Fig. 7(b), and the total efficiency is shown in Fig. 7(c). As expected, in Fig. 7(a), the MOSFET power loss increase as frequency increases, as the switching losses are directly proportional to the frequency. However, they increase with a steeper slope on the IPOP and CB topologies, due to hard switching. For the CB, the magnetic losses are reduced at higher frequencies, and for frequencies above 75 kHz, it has the best results after the IPOP. Note that in Table III, the inductor's core material changes according to the switching frequency, however, this does not necessarily imply the reduction of the losses as frequency increases, as it is seen for the DAB in Fig. 7(b), where losses increase from 50 kHz to 75 kHz. A deeper analysis should be performed to understand this behavior, as it is out of the scope of this paper. Regarding the total efficiency, which is shown in Fig. 7(c), the DAB presents the best results, followed by the CF-PS-FB. This is due to their ability to operate at ZVS condition, noting that at 25 kHz, the difference between the four topologies is very small, presenting an efficiency of 93.92%, 94.74%, 94.89% and 96.51 %, for the IPOP, CB, CF-PS-FB, and DAB, respectively.

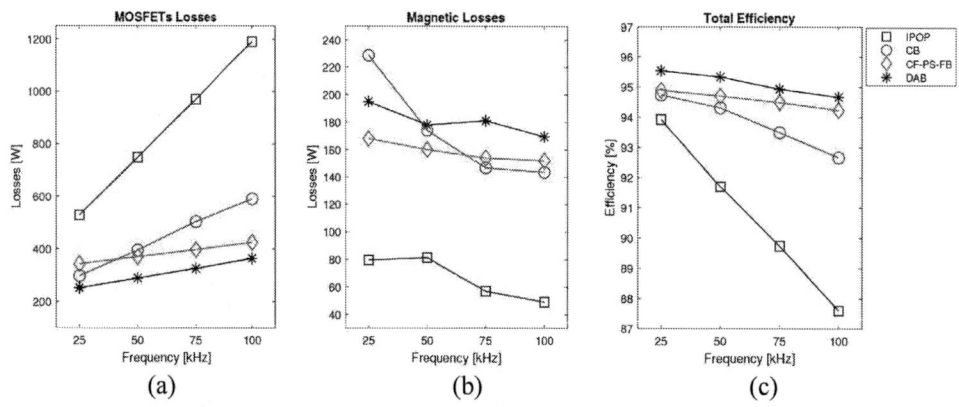

Fig. 7: Comparative analysis between the proposed topologies: IPOP (blue), CB (red), CF-PH-FB (magenta), and DAB (black), evaluated in function of the switching frequency. (a) MOSFET Losses, (b) Magnetic Losses, (c) Total Efficiency.

Table II: Component Selection (MOSFETs)

		MOSFET	Parallel MOSFETs
IPOP	**Ground-referenced**	STW68N65DM6-4AG	2
	Floating-referenced	IPB65R099CFD7A	1
CB	**1st Stage**	IRF7779L2TRPBF	1
	2nd Stage	IPB65R099CFD7A	5
CF-PS-FB	**Input**	NVHL082N65S3F	1
	Output	IPB044N15N5ATMA1	1
DAB	**Input**	IRF7779L2TRPBF	3
	Output	IPW65R035CFD7A	1

Table III: Component Selection (Magnetic Components)

	Magnetic Components		25 kHz	50 kHz	75 kHz	100 kHz
IPOP	L1, L2, L3	Material	MPP	High Flux	High Flux	Kool Mu HF
		Core	TVH61134A	0058091A2	C058076A2	0076071A7
CB	L1	Material	High Flux	High Flux	High Flux	High Flux
		Core	C058076A2	C058076A2	C058076A2	C058076A2
	L2	Material	High Flux	MPP	Kool Mu Max	Kool Mu Max
		Core	0058072A2	0055777A2	0079074A7	0079716A7
CF-PS-FB	L_{Lk}	Material	MPP	High Flux	High Flux	Kool Mu Max
		Core	C055438A2	C058071A2	C058584A2	0079894A7
	T1		Planar Transformer: HI-MAG 540			
DAB	L_{Lk}	Material	3C81	3C81	3C92	3C95
		Core	T140	T87	E65	ETD49
	T1		Planar transformer: HI-MAG 540			

Volume

The results from the volume analysis are shown in Table IV, where the volume is estimated in function of the switching frequency. The volume is reduced as the switching frequency increases, due to the reduction in size of the magnetic elements. The DAB presents the lowest volume throughout the evaluated switching frequency range. It is important to note that, although all four topologies comply with the maximum volume requirement of 4.6 dm³, it is also important to evaluate the longitudinal measurement requirements. Fig. 8 shows a graphical representation of the maximum size of the converter and the magnetic elements, to evaluate the compliance with the dimension requirements. Only two topologies at two different frequencies are presented in Fig. 8, showing the front, top, and side views, where the converter's maximum dimensions (575x100x80 mm) are represented by a black rectangle, while the inductors and the transformers are shown in colors blue and cyan, respectively. For the inductor, the toroids are piled horizontally. The IPOP converter at 25 and at 75 kHz frequencies is shown in Fig. 8(a) and Fig. 8(c), respectively, where the 3 inductors overlap with the maximum length at 25 kHz but complies with requirements at 75 kHz. It can be appreciated that the DAB converter complies with the dimension requirements at both 25 and at 75 kHz, as it is appreciated in Fig. 8(b) and Fig. 8(c), respectively. The results of the volume analysis, which are not shown completely in this paper, conclude that the CF-PS-FB and the DAB comply throughout the whole frequency range, while the IPOP and the CB only fails to comply at 25 kHz, where they present the best efficiency results.

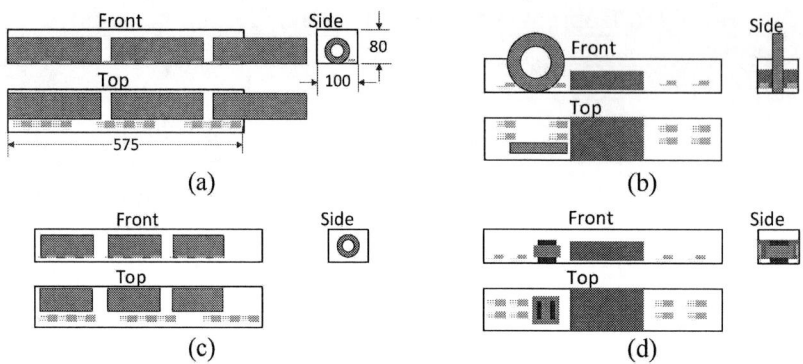

Fig. 8: Graphical representation of the converter maximum dimensions and the magnetic components. All the dimensions are in mm. (a) IPOP at 25 kHz. (b) DAB at 25 kHz. (c) IPOP at 75 kHz. (d) DAB at 75 kHz.

Table IV: Volume Estimation

Volumen [dm3]	25 kHz	50 kHz	75 kHz	100 kHz
IPOP	2.43	1.6	1.2	1.08
CB	3	2.96	2.17	1.23
CF-PS-FB	1.41	1	0.99	0.95
DAB	0.95	0.95	0.88	0.85

Conclusions

The IPOP and CB topologies present good efficiency results at low switching frequency, and also present the lowest input current ripple, due to their modular arrangement and interleaved control. However, at low frequencies, their length does not comply with the requirements, due to the number of inductors and their dimensions.

The CF-PS-FB and DAB, on the other hand, are able to operate under soft switching conditions, which allows them to increase the switching frequency without penalizing their efficiency. It is important to note that their dimensions comply with the requirements throughout all the frequency range.

The results show that the most adequate topology for this application is the DAB converter, as it presents the highest efficiency and power density throughout all the frequency range. Its main disadvantage is the high current ripple on both, the input and output. This implies the need of including filters, which will negatively affect its weight and volume. The CF-PS-FB is also an option, as its efficiency is around 1% less than the DAB and has a balanced current ripple at both ports. Both topologies have galvanic isolation, and even though is not a requirement, it benefits this application for safety reasons and because it allows to boost the voltage in a natural way.

As a result of this study, a prototype of the DAB converter, designed to operate at 75 kHz, is being built for testing the integration of removable batteries, where the future work will be focused on possible variations of the topology, control techniques, using wide-bandgap materials for 800 V HV-DC bus, among others.

References

[1]   F. Blaabjerg, H. Wang, I. Vernica, B. Liu, and P. Davari, "Reliability of Power Electronic Systems for EV/HEV Applications," *Proc. IEEE*, pp. 1–17, 2020.

[2] D. Dell'Isola, M. Urbain, M. Weber, S. Pierfederici, and F. Meibody-Tabar, "Optimal Design of a DC-DC Boost Converter in Load Transient Conditions, Including Control Strategy and Stability Constraint," *IEEE Trans. Transp. Electrif.*, vol. 5, no. 4, pp. 1214–1224, 2019.

[3] European Commission, "Paris Agreement | Climate Action." [Online]. Available: https://ec.europa.eu/clima/policies/international/negotiations/paris_en. [Accessed: 17-Jan-2022].

[4] H. Ko, S. Pack, and V. C. M. Leung, "An Optimal Battery Charging Algorithm in Electric Vehicle-Assisted Battery Swapping Environments," *IEEE Trans. Intell. Transp. Syst.*, pp. 1–10, 2020.

[5] Y. Song, J. Li, G. Ji, and Z. Xue, "Study on the typical mode of EV charging and battery swap infrastructure interconnecting to power grid," *China Int. Conf. Electr. Distrib. CICED*, vol. 2016-Septe, no. Ciced, pp. 10–13, 2016.

[6] R. P. Twiname, D. J. Thrimawithana, U. K. Madawala, and C. A. Baguley, "A Dual-Active Bridge Topology with a Tuned CLC Network," *IEEE Trans. Power Electron.*, vol. 30, no. 12, pp. 6543–6550, 2015.

[7] Z. Xian and G. Wang, "Optimal dispatch of electric vehicle batteries between battery swapping stations and charging stations," *IEEE Power Energy Soc. Gen. Meet.*, vol. 2016-Novem, 2016.

[8] Y. Zheng, Z. Y. Dong, Y. Xu, K. Meng, J. H. Zhao, and J. Qiu, "Electric vehicle battery charging/swap stations in distribution systems: Comparison study and optimal planning," *IEEE Trans. Power Syst.*, vol. 29, no. 1, pp. 221–229, Jan. 2014.

[9] M. A. H. Rafi, R. Rennie, J. Larsen, and J. Bauman, "Investigation of fast charging and battery swapping options for electric haul trucks in underground mines," in *2020 IEEE Transportation Electrification Conference and Expo, ITEC 2020*, 2020, pp. 1081–1087.

[10] Z. Chen, "The combination of battery swapping system and connected vehicles technology in intelligent transportation," *Proc. - 2020 Int. Conf. Intell. Transp. Big Data Smart City, ICITBS 2020*, pp. 72–75, 2020.

[11] S. Chowdhury, M. N. Bin Shaheed, and Y. Sozer, "State-of-Charge Balancing Control for Modular Battery System with Output DC Bus Regulation," *IEEE Trans. Transp. Electrif.*, vol. 7, no. 4, pp. 2181–2193, Dec. 2021.

[12] J. Reimers, L. Dorn-Gomba, C. Mak, and A. Emadi, "Automotive Traction Inverters: Current Status and Future Trends," *IEEE Trans. Veh. Technol.*, vol. 68, no. 4, pp. 3337–3350, Apr. 2019.

[13] O. C. Onar, J. Kobayashi, and A. Khaligh, "A Fully Directional Universal Power Electronic Interface for EV, HEV, and PHEV Applications," *Ieee Trans. Power Electron.*, vol. 28, no. 12, pp. 5489–5498, 2013.

[14] A. Vazquez, A. Rodriguez, D. G. Lamar, and M. M. Hernando, "Advanced Control Techniques to Improve the Efficiency of IPOP Modular QSW-ZVS Converters," *IEEE Trans. Power Electron.*, vol. 33, no. 1, pp. 73–86, 2018.

[15] F. H. Makarim, B. Antares, A. Rizqiawan, and P. A. Dahono, "Optimization of Multiphase Cascaded DC-DC Boost Converters," *ICEVT 2019 - Proceeding 6th Int. Conf. Electr. Veh. Technol. 2019*, pp. 285–289, 2019.

[16] R. S and S. Chellappan, "2-kW , 48- to 400-V , > 93 % Efficiency , Isolated Bidirectional DC-DC Converter Reference Design for UPS," no. June, pp. 1–30, 2017.

[17] K. Wang, C. Y. Lin, L. Zhu, D. Qu, F. C. Lee, and J. S. Lai, "Bi-directional dc to dc converters for fuel cell systems," *IEEE Work. Power Electron. Transp.*, vol. 11, pp. 47–51, 1998.

[18] S. Chi, P. Liu, X. Li, M. Xu, and S. Li, "A Novel Dual Phase Shift Modulation for Dual-Active- Bridge Converter," *2019 IEEE Energy Convers. Congr. Expo. ECCE 2019*, pp. 1556–1561, 2019.

[19] B. Zhao, Q. Song, and W. Liu, "Power characterization of isolated bidirectional dual-active-bridge dc-dc converter with dual-phase-shift control," *IEEE Trans. Power Electron.*, vol. 27, no. 9, pp. 4172–4176, 2012.

[20] C. D. A. Bridge, S. Bal, S. Member, D. B. Yelaverthi, and S. Member, "Improved Modulation Strategy Using Dual Phase Shift Modulation for Active Commutated," *IEEE Trans. Power Electron.*, vol. 33, no. 9, pp. 7359–7375, 2018.

[21] K. Martín Diaz, "Análisis , diseño y construcción de un proveedor de bus para sistemas de distribución en corriente continua domésticos," Universidad de Oviedo, 2018.

[22] Y. H. Abraham, H. Wen, W. Xiao, and V. Khadkikar, "Estimating power losses in Dual Active Bridge DC-DC converter," *2011 2nd Int. Conf. Electr. Power Energy Convers. Syst. EPECS 2011*, pp. 4–8, 2011.

Design of a High-Dynamic Test Bench for Accelerated Dielectric Lifetime Testing with Adjustable Voltage Slopes and Temperatures

Hendrik Schefer[1,2], Lucas Hanisch[1], Tim-Hendrik Dietrich[1], Regine Mallwitz[1,2], Markus Henke[1,2]

INSTITUTE FOR ELECTRICAL MACHINES, TRACTION AND DRIVES (IMAB)[1]

CLUSTER OF EXCELLENCE SE²A – SUSTAINABLE AND ENERGY-EFFICIENT AVIATION[2]

TU Braunschweig

Braunschweig, Germany

Email: h.schefer@tu-braunschweig.de

URL: https://www.tu-braunschweig.de/imab

URL: https://www.tu-braunschweig.de/se2a

Acknowledgments

This work was supported by the Deutsche Forschungsgemeinschaft (German Research Foundation, DFG) through the Germany's Excellence Strategy-EXC 2163/1-Sustainable and Energy Efficient Aviation under Grant 390881007.

Keywords

≪Test bench≫, ≪Insulation≫, ≪Wide bandgap≫, ≪Reliability≫, ≪Electrical Machine≫

Abstract

Upcoming future mobility will require high power densities; therefore, wide bandgap semiconductors (WBGs) and HVDC supply voltage could be one solution. One crucial design criterion is the insulation coordination in all drive train components. This paper presents and discusses a test bench design to emulate dielectric stress due to fast switching and hard switching WBGs at high voltages and various environmental conditions.

I Introduction

Future applications in electromobility will require power-dense drives that operate reliably even under changing environmental conditions. Electric cars are increasingly using fast-switching wide bandgap semiconductors and a voltage level of 800 V [1]. Drives in electrified aircraft are also exposed to specific environmental conditions like voltage levels up to 7500 V [2]. The resulting increased stress focuses attention on the electrical insulation system as the percentage of insulation faults in electric motors increases [3].

The increased voltages and specific environmental conditions also place additional stress on the power electronics. In [4] a well-used industrial-based dataset about failure mechanisms in power electronic converters is presented. Out of this dataset, printed circuit boards (PCBs) are responsible for 26 % of faults, but the dataset doesn't distinguish between sub-fault mechanisms. [5] illustrates finite element method-based (FEM) insulation coordination of a high-blocking silicon carbide (SiC) stage. Additionaly, air gaps or silicone are different design elements, but creepage distances must be understood.

More recent research has tended to focus not on the reliability of a particular part or component, but on developing a deeper understanding of the various damage mechanisms [6]. The different damage mechanisms belong to different physical disciplines and can represent both constant and transient loads. In addition, the significance of each damage mechanism varies depending on the individual load spectrum, so reliability studies must be assigned to a specific mission profile. The combination of multiple damage mechanisms necessitates the research question about the interactions between the individual damage mechanisms. So a lot of tests has to be done. An efficient and frequently used methodology to investigate the significance of individual loads in a load spectrum is the Design of Experiments (DoE) approach [7]. Parameter studies following a mission profile require a test bench that allows dynamic and independent adjustment of the multi-physical parameters. In this paper, the development of such a test bench is presented. For this purpose, first the considered loading factors and then the concept and the setup of the first test bench version are presented. Subsequently, first test results are presented and evaluated. The resulting optimization approaches are then implemented in an improved test bench version and the measurement results are presented.

The paper is structured in four sections. An introduction contains the motivation and state of the art. Ch. II present an analysis of stress factors for PCBs, inductor windings and windings in electrical machines. Additionaly, in Ch. III and Ch. IV the paper discusses two test bench designs in detail and compares different objectives. In the first test bench concept in Ch. III, the power electronics are placed outside the warming cabinet. Since the correspondingly long cables favor transient overvoltages, the power electronics are placed inside the warming cabinet in the second test bench concept in Ch. IV. The last section concludes the presented experiences and gives a future outlook.

II   Stress Factors and Methodical Implementation of the Test Bench

Various devices under test (DUT) representing both power electronics and electrical machine components will be examined in the test benches presented in Ch. III and Ch. IV. For example, twisted pair DUTs according to the IEC 60172 standard are used to represent the winding of electrical machines. For creepage distance tests on PCBs, own DUTs have been developed. Investigations at component level offer the possibility to analyze individual weak points or damage mechanisms. In addition, DUTs that represent the entire system will provide more realistic measurement results. In this chapter, the load requirements for the power electronics and the electric machine are derived for this purpose, followed by the resulting requirements for the test bench.

The test methodology is based on the destruction (breakdown) of the DUTs by adaptable stress factors and recording the lifetime. Based on a DoE, the influence of the parameters can be identified and the lifetime extrapolated. An accelerated lifetime test with increased stress factors is practicable to minimise the testing time. Finally, validation measurements prove the extrapolation of the lifetime.

A   Failure Modes of PCBs and Inductor Windings

Accelerated altering of printed circuit boards can distinguish in proofing the dielectric strength of FR4 or the creepage distances between different traces. Future studies will face investigations on creepage distances under several conditions. [8] compares diverse electrode models of PCBs and illustrates the influence of several parameters, e.g. temperature, pressure, duty cycle and distances, on the breakdown voltage. Like the IEC 60 664, standards give regulation about the degree of pollution and sine wave voltage changes. The dielectric breakdown time is an important parameter. The mode of action of various aging parameters is shown in Fig 1. This illustration applies to the damage mechanisms in PCBs as well as in electrical machines and will be described in more detail in the next section.

B   Failure Modes of Electrical Machines

As mentioned before, the stress on the machine side increases. By failure investigations of electrical machines in [3] it becomes clear that with increasing electrical stress the number of electrical failures in the stator winding of electrical machines with 66 % prevails over other failure phenomena such as bearing or rotor damage (13 % each). The failure mechanisms of the stator winding are shown in Fig. 1. Although materials science has developed high performance thermoplastics, temperature continues to be a dominant aging factor, first studied over 70 years ago [9]. Other dominant damage parameters added by increased electrical stress, and whose influence is being investigated in recent studies, are the voltage signal, consisting of the voltage amplitude and the voltage slope, and switching frequency [10]. It should be noted that in addition to the primary aging factors implemented in the test bench presented, there are also mechanical aging factors. As already noted, these have a less significant effect on the failure of electrical machines in percentage terms, especially with increasing electrical loads in the future. These are shown in dashed lines in Fig. 1. The aging factors cause erosion processes that damage the insulation. Aging factors can favor several erosion processes. For example, a higher temperature indirectly favors electrical erosion processes, since the permittivities of insulating materials are temperature-dependent and therefore field strength increases occur at higher temperatures. Therefore, coupling mechanisms between aging factors and erosion processes are also of great interest. With greater damage to the insulation materials, erosion trees form and finally an electrical breakdown leads to failure of the machine winding. A detailed analysis of the above-mentioned dominant damage mechanisms from Fig. 1 was carried out in [11] on DUTs with hairpin windings.

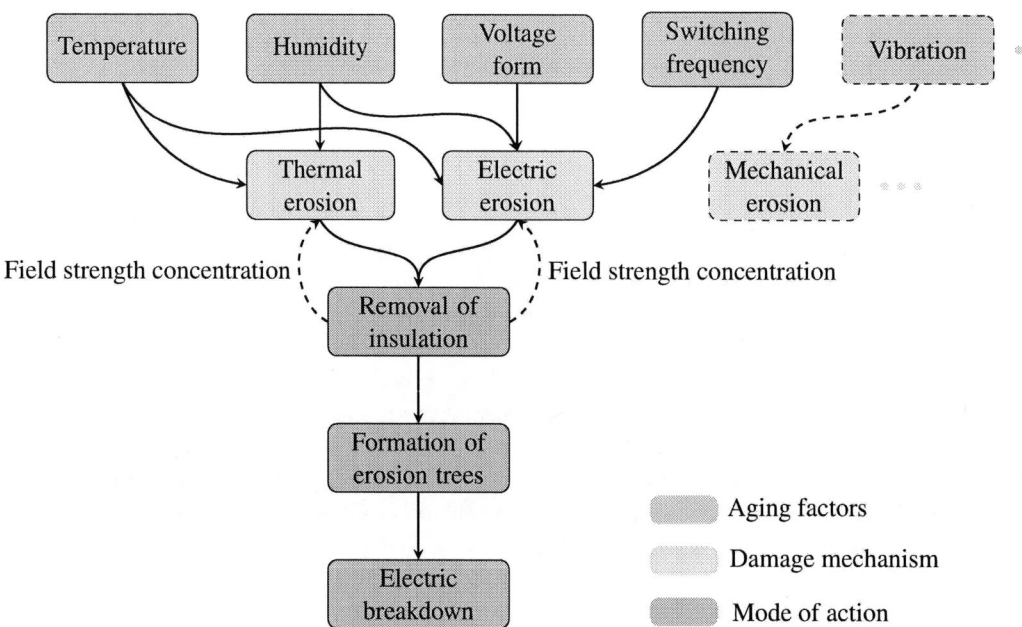

Fig. 1: Damage mechanisms and their modes of action

C Test Bench Requirements

To enable larger parameter studies, new test bench designs must provide independent adjustment of parameters such as rectangular voltage changes, voltage levels, rise- and fall times, frequencies, duty cycle, temperature, humidity and pressures. The independently controllable aging parameters of the test bench are shown in Table I. The number of parameters leads to a DoE to save time, and the test bench has to mature many DUTs at a time [7].

Table I: Independently controllable aging parameters of the test bench and their limits [* climate control]

Parameter	DC Voltage V	Voltage Slope $\frac{kV}{\mu s}$	Frequency kHz	Duty cycle %	Temperature °C	Humidity * %
Value	< 800	$< 100 \frac{kV}{\mu s}$	< 100	$5 - 95$	$-70 - 300$	$10 - 95$

III Test Bench Version 1: Warming Cabinet

In order to achieve the high voltage slopes required for the test bench, modern semiconductors in silicon carbide technology are available, with whose short switching times the necessary voltage slopes can be impressed into the DUTs.

For the first set-up of the test bench (Fig. 2), a proven six-strand inverter [12] is used, which can be modified to generate independent voltage slopes on six channels (outside the warming chamber: left side).

All DUTs are inside the warming chamber ($T = 0..300\,°C$) and connected to each half-bridge of the six-strang inverter via an approx. $40\,cm$ long cable. The amplitude of the voltage pulses to be applied to the DUTs is influenced by the selection of the DC link voltage (Voltage Source: above the generator), and the slope of the voltage edges can be set individually for each channel by selecting different gate resistors. Especially when testing several similar DUTs, the influence of different voltage slopes can be investigated in parallel under otherwise identical test conditions.

As long as the insulation of the DUT is intact, only the parasitic capacitance of the insulation is charged when the voltage slope is applied. Ideally, no permanent current occurs for the remaining duration of the pulse. If the pulse is subsequently switched off, this capacitance is discharged again. However, if insulation damage occurs during the test, a short-circuit current may flow through the DUT, endangering the semiconductors of the inverter and possibly causing a voltage dip in the feeding high-voltage source, so that the other channels of the test setup are also affected. In order to be able to switch off the affected channel immediately in the event

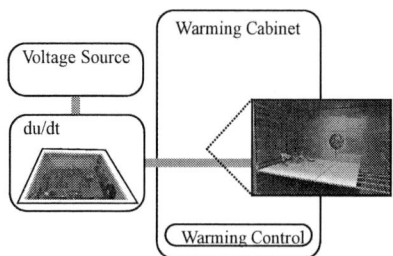

Fig. 2: du/dt test bench V1

of a short circuit, the output current of the channel must be monitored. Since in this case the knowledge of the absolute amplitude of the DUT current is not relevant, but only the occurrence of an overcurrent event must be detected with the shortest possible delay time, the desaturation (DESAT) detection of the gate driver for the high side switches in the pulse generator is used for this purpose. If a short circuit occurs in the DUT, it is detected by the DESAT detection and the corresponding semiconductor is immediately blocked. The overcurrent detection is thus independent of the driving controller and has a very short reaction time.

IV Test Bench Version 2: High du/dt combined with Temperature / Humidity

For the next test bench generation, specially adapted power electronics were developed to enable higher voltage slopes. First, a prototype with a half-bridge is built to test the electrical properties (Fig. 3).

The focus of the development is on the reduction of the commutation mesh and the use of semiconductor switches, which are primarily selected for this application under the aspect of high switching speed. Since no continuous currents have to be carried by the switches in the experimental setup, semiconductors can also be used that have a higher on-resistance, but in return have a high switching frequency. For the test carrier shown, SiC MOSFETs from Infinion's IMZ120R series in the TO-247 package are used, within which different current-carrying capacities and channel resistances are available. To optimise the PCB layout, the high and low side switches are arranged face-to-face, which increases the design effort for cooling, but allows a reduction of the conductor path meshes on the PCB. After initial tests with this setup have been successful, the circuit is expanded for use in the test stand and equipped with the necessary peripherals.

The mentioned requirements demand climate adjust-

Fig. 3: Pre-study du/dt generator V2

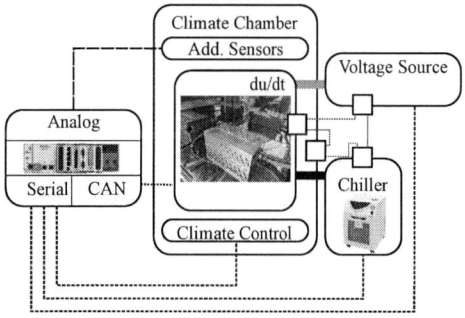

Fig. 4: du/dt test bench V2

ing, voltage and high voltage changings. Additionally, the DoE requires a decoupling of different parameters, such as du/dt and the voltage. Due to the understanding of test bench version 1, small impedances between DUT and generator are needed to ensure the decoupling between parameters. The cable impedance leads to high voltage changes due to mismatch, so decoupling of parameters and gate resistance adjusting are difficult. Therefore, small impedances between the generator and DUT can solve these problems and pre-study archives high voltage changing by good blocking characteristics.

Fig. 4 illustrates the test bench design. The new design demands a water-cooled generator inside the climate chamber with low cable impedances. A real-time embedded industrial controller (Ni cRIO, left side) enables the overall test bench control and capture metadata (time, temperature, humidity, voltage, semiconductor temperature, etc.). A voltage source supplies the six-strand inverter in the climate cabinet.

A   Electrical Design

Version 2 (Fig. 5 a)) is fitted with six clamping opportunities based on six SiC half-bridges to conduct a DoE, and it is included in a cooler. The DUTs can be clamped to the test bench on the shortest possible path, and the DUTs has to face environmental conditions (humidity and temperature). Inside the housing, SiC half-bridge fed by drivers enables the high required switching speeds. The driver design and the semiconductors are designed in the pre-study (one half-bridge configuration, Fig. 3). Version 2 has slight differences in the electrical design to the pre-study; the sigma-delta modulator captures the DC bus voltage, the used isolated driver measures the semiconductor temperature, and other DC capacitors enable various capacitive loads. Additional functionalities are a Controller Area Network (CAN) interface, multiple isolated in- and outputs, an input for the emergency loop and a potential-free switch.

A field-programmable gate array (FPGA, type: Intel Max 10) enables the highest flexibility and guarantees the calculation of the pulse-width modulation (PWM), duty cycle, bus voltage evaluation, semiconductor temperature verification, high precision timer, self-designed CAN interface, and safety functionalities. All functionalities are implemented, such as the PWM calculation, duty cycle comparison, the DESAT-triggered timers, semiconductor PWM-based thermal measurements, sigma-delta filter for DC Voltage, the safe handling and the CAN interface.

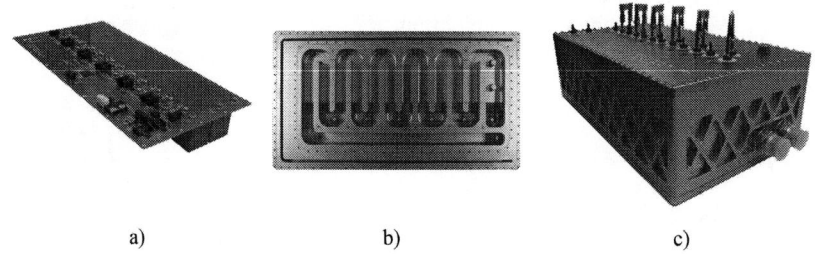

a)   b)   c)

Fig. 5: du/dt generator V2 [a) commissioned electronic b) heat sink design c) assembled generator]

B   Thermal Design

The thermal and mechanical design is crucial because the construction has to protect the electronic devices from harsh environmental conditions and cool SiC semiconductors down. The targeted environmental condition is 150 °C, and the design of the cooler is very challenging. Also, the housing design limits the switching frequency (a stress factor). The critical electronic elements, such as the power electronic semiconductors and the capacitors, are thermally connected to the coolant. A meandering cooling channel directs the fluid to the appropriate places, illustrated in the Fig. 5 b). The liquid is accelerated and decelerated at the corresponding points using various tightening and loosening points. Furthermore, so-called feed-through channels supply cool liquid to the last half-bridges. The chiller offers a cooling power of around 1200 W. A verification of the cooler design is done by a Computer Fluid Design (CFD) study. All electrical power connections are led out through a polytetrafluoroethylene (PTFE) insulator so that the temperature is limited to $200\,°C$. At $200\,°C$, dangerous vapours occur due to fluorinated compounds. A small D-SUB 15 aerospace connector enables the electronic power supply and communication.

Tab. II shows the thermal design verification with a switching frequency of $50\,kHz$, $600\,V$ DC voltage link and twisted pair wires. Two additional $1\,k\Omega$ platinum resistor (PT1000) sensors are implemented in the housing to capture air temperature inside the housing $T_{a,HS}$. The double-walled heat sink possesses a good thermal decoupling to the climate chamber temperature T_a; this is observable by the two PT1000 sensors. The average temperature of all 12 negative temperature coefficient (NTC) sensors $\overline{T_{NTC1..12}}$ (NTCs connected to source potential) illustrates a linear increase over the climate chamber temperature. Two sensors are a little bit warmer (approx. $29\,°C$ at $T_a = 150\,°C$); therefore, the thermal interface could be checked or improved. The highest measured temperature of the 12 NTCs ($\mathrm{Max}(T_{NTC1..12})$) is $107\,°C$ under the maximal junction temperature of $175\,°C$, and the air temperature inside the housing reaches only $< 50\,°C$. Also, the chiller can cool the system with its cooling power of $1200\,W$. For environmental-friendly long-term studies, thermal insulation with stone wool panels is feasible and can increase the thermal design point to higher temperatures to $< 200\,°C$. Moreover, the thermal design enables also an increase in the switching frequency (a stress factor).

Table II: Temperature verification [T_a : ambient climate chamber temperature, T_c: coolant inlet temperature, $T_{a,HS}$: ambient temperature in the heat sink, $T_{NTC1..12}$: NTC measuring points of the 12 semiconductors]

| T_a | T_c | $T_{a,HS,1}$ | $T_{a,HS,1}$ | $\mathrm{Max}(T_{NTC1..12})$ | $\overline{T_{NTC1..12}}$ |
°C	°C	°C	°C	°C	°C
25	20	26.8	26.8	55	45.5
50	20	29.9	31.2	60	49.25
75	20	33.7	35.3	70	56.3
100	20	37.4	39.5	75	61.3
125	20	41	43.5	92	69.6
150	20	45.5	48.4	107	78.3

C   Verification of the Switching Behaviour

The Fig. 6 shows the measured voltage change by a $400\,MHz$ single-ended passive voltage probe. The left graph illustrates voltages changes defined by an inductive load. Due to the help of a double pulse test, different voltage changing velocities are captured under various gate and voltage settings. The in-

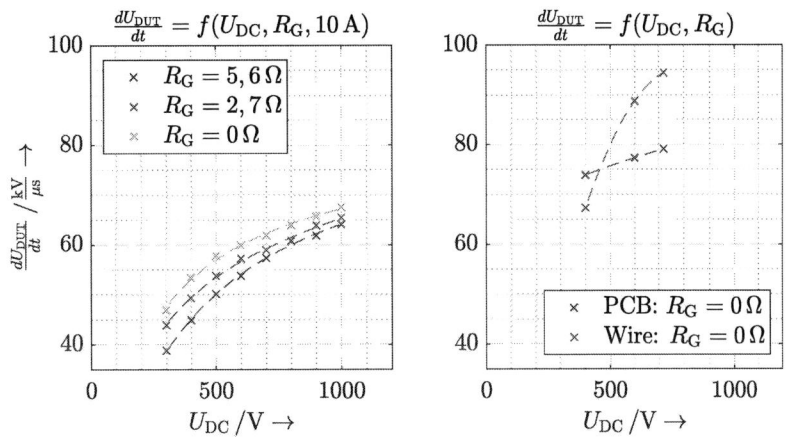

Fig. 6: Verification of the voltage slope of test bench V2 [PCB: electrode distance $500\,\mu m$]

ductive load and the various first puls length enable a comparable current level. Moreover, the double pulse test ensures the verification of the DESAT functionality. A Zener diode can easily adjust the drain threshold current to various capacitive loads. In addition, the right graph in Fig. 6 illustrates the changing speeds from two different DUTs. It can be recognized that the wire-based DUT has much higher voltage changes $\frac{dU_{DUT,max}}{dt} = 95\,\frac{kV}{\mu s}$ compared to the PCB-based DUT $\frac{dU_{DUT,max}}{dt} = 79\,\frac{kV}{\mu s}$. Nevertheless, it also has a higher overshoot, which is due to the changed impedance.

V Comparison of Test Bench Versions

In order to compare the two test benches, the voltage load was investigated on a twisted pair wire, which is often used to evaluate the conductor insulation of electrical machines. Fig. 7 shows the slopes of the voltage signals on the DUT at an DC link voltage of 700 V. It can be seen that the transient overvoltage U_σ at the test specimen could be massively reduced by the second test bench version. In the first test bench version, a transient overvoltage of 609 V was measured and the DUT was loaded with a maximum total voltage of $1,309$ V. This shows that under unfavorable boundary conditions and long cables between the power electronics and the electrical machine, the maximum voltage load at the machine terminals can be 1.87 times the DC link voltage. Through further development of the test bench, the transient overshoot at 700 V DC link voltage could be reduced to only 139 V in the second test bench version, thus reducing the maximum total load by almost 36 %. at 700 V DC link voltage of the two test bench versions V1 and V2 on a twisted pair DUT.

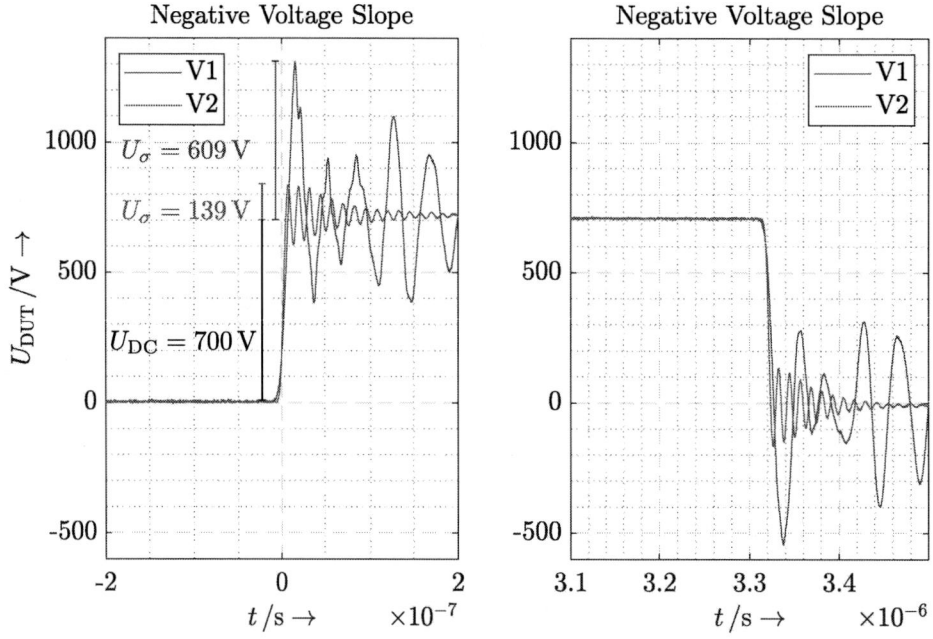

Fig. 7: Comparison of the voltage signals $[R_G = 0\,\Omega]$

Fig. 8 shows the values for transient overvoltages U_σ and voltage slopes $\frac{dU_{DUT}}{dt}$ at different DC link voltages. The results from Fig. 7 are confirmed, namely that the voltage load could be significantly reduced regardless of the DC link voltage. At the same time, it can be seen in the figure on the right that the voltage slope could be increased. The voltage slope up to the DC link voltage and up to the transient overvoltage are indicated in each case. It is noticeable that the voltage slope seems to saturate at higher DC link voltages. Reducing the voltage load by lowering transient overvoltages has a significant influence on the lifetime of the winding of electrical machines. To confirm this, lifetime measurements were performed with the twisted pair DUTs that were also used for the voltage measurements in Fig. 7

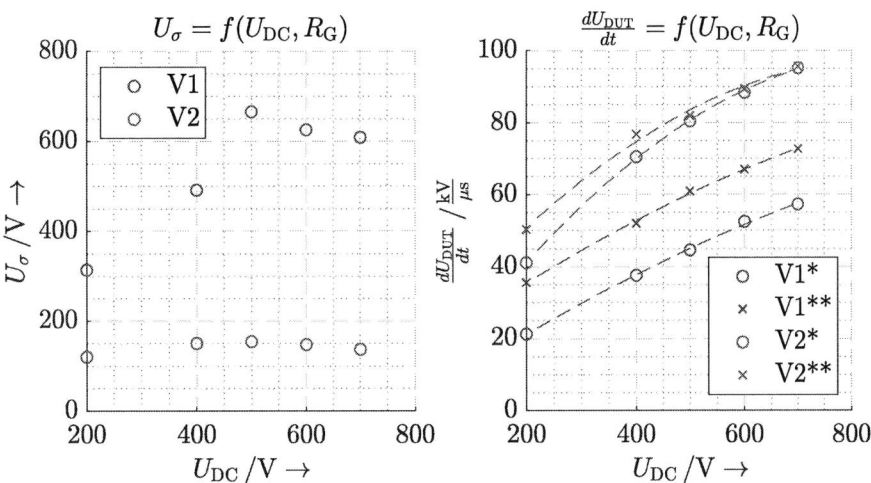

Fig. 8: Comparison of transient overvoltages U_σ and voltage slopes $\frac{dU_{DUT}}{dt}$ for the two test bench versions $[R_G = 0\,\Omega, * 10/90\,\%U_{DC}, **10/90\,\%(U_{DC} + U_\sigma)]$

and Fig. 8. The DUTs were loaded with a DC link voltage of 600 V at a frequency of 30 kHz. In order to investigate the electrical stress individually, environmental conditions such as temperature and humidity were not taken into account. In order to be able to make a statistically significant statement, nine DUTs were subjected to the lifetime measurements. The end of life of the DUTs is reached when the DESAT detection registers a breakdown between the wires and switches off the phase.

Fig. 9 shows the results of the lifetime measurements and how the failure probability of the sample can be described by means of a Weibull distribution function. The Weibull distribution function gives the

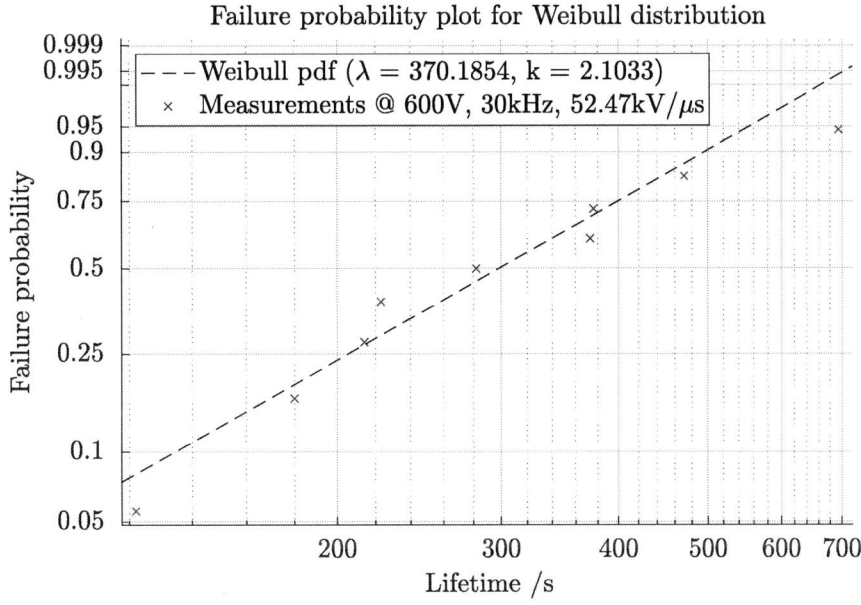

Fig. 9: Lifetime measurements and statistical fit by Weibull distribution function for test bench version V1 (Mean= 327.87s; Standard deviation = 168.8s)

failure probability F as a function of the lifetime L with the scale λ and shape k parameter according to equation (1):

$$F(L) = 1 - e^{-\left(\frac{L}{\lambda}\right)^k} \tag{1}$$

On average, the nine test specimens failed after 327 s. The variation in the measurement results was quite high with a standard deviation of 168.8 s. In contrast, no failure was detected in the second test bench version after 15 h of ageing at 600 V DC link voltage. Since the voltage slope of the second test bench version is even higher than that of the first version, it can be concluded that it is not the voltage slope but the transient overvoltage that significantly damages the insulation and leads to an electrical breakdown.

VI   Conclusion

This paper discusses the design of a high dynamic rectangular du/dt generator to stress dielectrics in electrical machines and power electronic devices. A requirement-oriented design study is represented. The first version (Fig. 2) and a pre-study (Fig. 3) have gained a lot of experience. These experiences have influenced the design of V2. A water-cooled du/dt generator within the climate chamber is proposed, and all preparations for metadata analysis are implemented. The measurement results show that the transfer to the climatic chamber and the short cables were able to massively reduce the transient overvoltages. At the same time, higher voltage slopes were achieved with the new design (V2). Furthermore, lifetime measurements could show that the transient overvoltage strongly stresses the insulation and has a more significant influence on the lifetime of the winding of electrical machines than the voltage slope.

References

[1] D. Esmaeil Moghadam and J. Lange, "The future of e-cars – will highvoltage systems become a new standard?," JEC Compos. Mag., Special Issue Mobility, pp. 30–31, May 2020.

[2] H. Schefer, L. Fauth, T. H. Kopp, R. Mallwitz, J. Friebe and M. Kurrat, "Discussion on Electric Power Supply Systems for All Electric Aircraft," in IEEE Access, vol. 8, pp. 84188-84216, 2020, https://doi.org//10.1109/ACCESS.2020.2991804

[3] J. He, C. Somogyi, A. Strandt and N. A. O. Demerdash, "Diagnosis of stator winding short-circuit faults in an interior permanent magnet synchronous machine," 2014 IEEE Energy Conversion Congress and Exposition (ECCE), 2014, pp. 3125-3130, https://doi.org//10.1109/ECCE.2014.6953825

[4] E. Wolfgang, "Examples for failures in power electronics systems,"presented at ECPE Tutorial 'Rel. Power Electron. Syst.', Nuremberg, Germany, Apr. 2007

[5] Y. Xu, M. Ghessemi, J. Wang, R. Burgos and D. Boroyevich, "Electrical Field Analysis and Insulation Evaluation of a 6 kV H-bridge Power Electronics Building Block (PEBB) using 10 kV SiC MOSFET Devices," 2018 IEEE Energy Conversion Congress and Exposition (ECCE), 2018, pp. 2428-2435, https://doi.org//ECCE.2018.8557489

[6] H. Wang et al., "Transitioning to Physics-of-Failure as a Reliability Driver in Power Electronics," in IEEE Journal of Emerging and Selected Topics in Power Electronics, vol. 2, no. 1, pp. 97-114, March 2014, https://doi.org//10.1109/JESTPE.2013.2290282

[7] L. V. Hanisch and M. Henke: "Lifetime Modelling of Electrical Machines using the Methodology of Design of Experiments", Simulation Notes Europe SNE-Journal 31(2), 2021, 95-100, https://doi.org/10.11128/sne.31.tn.10568.

[8] Q. Zhou, M. Wen, T. Xiong ,T. Jiang, M. Zhou,X. Ouyang and L. Xing,"Study on Insulation Breakdown Characteristics of Printed Circuit Board under Continuous Square Impulse Voltage", Energies 2018, 11(11), 2908, https://doi.org/10.3390/en11112908

[9] T. W. Dakin, "Electrical Insulation Deterioration Treated as a Chemical Rate Phenomenon," in Transactions of the American Institute of Electrical Engineers, vol. 67, no. 1, pp. 113-122, Jan. 1948, doi: 10.1109/T-AIEE.1948.5059649.

[10] F. Pauli, L. Yang, M. Schröder and K. Hameyer, Lebensdauerabschätzung von Wicklungsisolierstoffsystemen in SiC-betriebenen elektrischen Niederspannungsmaschinen. Elektrotech. Inftech. 136, 175–183 (2019), https://doi.org/10.1007/s00502-019-0711-2

[11] L. V. Hanisch, T. -H. Dietrich and M. Henke, "Analysis of Partial Discharges and Failure Mechanism in Electrical Machines with Hairpin Winding," 2021 IEEE 13th International Symposium on Diagnostics for Electrical Machines, Power Electronics and Drives (SDEMPED), 2021, pp. 1-7, doi: 10.1109/SDEMPED51010.2021.9605500.

[12] T. Fricke, C. Uzlu, R. Mallwitz, J. Ries, J. Wussow, J. Brockschmidt, M. Kurrat, B. Engel, P. Jungklass and F. Grieger, "NetProsum2030: A Contribution to the Solution for Distributed Energy Supply in 2030," PCIM Europe digital days 2020; International Exhibition and Conference for Power Electronics, Intelligent Motion, Renewable Energy and Energy Management, 2020, pp. 1-8.

Novel modulation method for common-mode noise reduction in Solid-State Transformer based on ISOP configuration

Naoto Kikuchi, Hiroki Watanabe, Keisuke Kusaka, Jun-ichi Itoh
Nagaoka University of Technology
1603-1, Kamitomioka-machi, Nagaoka, Niigata, JAPAN
Tel.: +81 / (258) –47.9533.
E-Mail: s195028@stn.nagaokaut.ac.jp
URL :http:// itohserver01.nagaokaut.ac.jp/itohlab/index.html

Keywords

«Solid-State Transformer», «Converter control», «Power factor correction», «Resonant converter», «EMC/EMI»

Abstract

This paper proposes a modulation method for the reduction of the common-mode noise in Solid-State Transformer (SST) with the ISOP connection. Modular multilevel configurations based on the ISOP connection increase the common-mode noise path because of increasing the switching components. Moreover, the cancellation of the common-mode noise using active common-mode canceler is not able to be applied because of complex common-mode noise path in SST with the ISOP connection. In the proposed method, one of the cells is driven by PWM in order to compensate for the harmonic component. The other cells are driven by square-wave operation in order to share the load. Therefore, the common-mode voltage is suppressed by the switching state of PFC. The advantage of the proposed method is that the additional EMC filter is not necessary for the reduction of the common-mode noise. The operation waveform with the proposed method is shown in the experimental results. The conducted emission is reduced by 11 dB and 7 dB in 200 kHz-band and 1 MHz-band, respectively.

Introduction

Recently, the DC microgrid has been attracted in terms of widespread use of the renewable energies. Solid-State Transformer (SST) is the one of the key components of the distributed system proposed by the Future Renewable Electric Energy Delivery and Management Systems Center [1-3]. SSTs are a power electronic interface between a medium voltage system and a low voltage system, which provides galvanic isolation with medium-frequency transformer [4]. The role of the power converter is the power flow control and the compensation for the reactive power. The medium frequency transformer enables to increase in the power density of the converter. Focusing on improvement of efficiency and power density, various circuit topologies of SST have been proposed in this decade [1-10]. Modular multilevel configurations in based on input series and output parallel (ISOP) have been widely used. The advantage of the ISOP connection is that low on-resister and low-switching loss devices are available because the applied voltage on each cell is divided by the number of cells. For these topologies, SiC-MOSFETs have been widely used because of their fast switching and low on-resistance. However, these wide-bandgap devices increase the common-mode noise due to the rapid switching behavior [11]. The emission limit for information technology equipment is defined in the International Special Committee on Radio Interference (CISPR) 11. Therefore, the EMC filter is designed for compliance with CISPR 11/EN55011 standard [12]. The active common-mode canceler (ACC) based on the voltage cancellation method or the current cancellation has been proposed in [13-20]. These filters have the advantage of suppressing sufficiently the common-mode noise with a push-pull emitter follower amplifier. However, the additional ACC is limited by the rated voltage of the amplifier. Therefore, It is not able to apply the additional ACC to the medium-voltage system such as SST.

In this paper, a novel modulation method is proposed in order to reduce the common-mode noise in SST with the ISOP connection. The advantage of the proposed method is that the additional EMC filter is not necessary for the reduction of the common-mode noise. The originality of this paper is that the modulation method uses the deference driving frequency in order to suppress the common-mode voltage. Owing to the propose modulation, the leakage current is suppressed in comparison with the conventional modulation. Moreover, the new contribution of this paper is that the reduction of the common-mode noise by replacing high-frequency SiC devices to low frequency Si devices.

This paper is organized as follows: firstly, the system configuration of SST with the conventional method and the proposed method is described. Secondly, the principle of the proposed method for the reduction of the common mode noise is explained with the common-mode equivalent circuit of SST. Finally, the conducted emission is compared between the conventional method and the proposed method in the experiment.

System Configuration

Circuit configuration

Fig. 1 shows the circuit configuration of the single-phase SST when the number of cell is three. Note that the number of cell is decided by the grid voltage and voltage rating of the power devices. Each cell has a PFC stage and an isolated resonant DC/DC converter. The input of the PFC stage in the cells is connected in series. The outputs of the isolated resonant DC/DC converters, which ensure the galvanic isolation, are connected in parallel. The input diode rectifier is common for all cells in order to reduce the number of components. The high-frequency operation contributes to minimizing the isolation transformer. The transformer of the isolated resonant DC/DC converter is smaller than the commercial frequency transformer because the switching frequency is higher than the grid frequency. Besides, the resonant capacitor C_r is connected to the primary side of the isolated transformer in series. The DC/DC converter is controlled with open-loop control due to the constant voltage transfer ratio. The switching frequency is a little higher than the resonant frequency for zero-voltage switching (ZVS) operation.

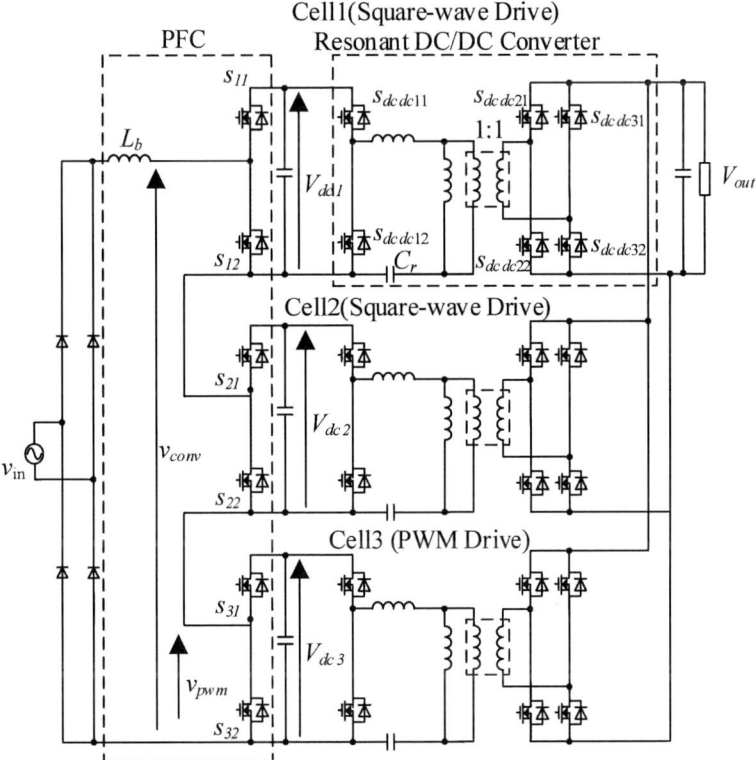

Fig. 1. Circuit configuration of the single-phase SST.
The proposed modulation method is applied into the PFC stage.

Conventional method

Fig. 2 shows the block diagram of the conventional control method of the PFC stages. The conventional method controls the inductor current with PWM in all cells. The overall PFC circuits controls phase and amplitude of the input current to correct the input power factor. Then, reference of the input current is given by

$$i_L^* = I_{amp} \left| \sin\left(\omega t\right) \right| \tag{1}$$

where I_{amp} is amplitude of the input current command. The phase of the input current is generated by PLL from the phase of the grid voltage. Note that the ripple current is reduced by operating phase shifted carrier in the PFC converter. Thus, the input voltage is equally divided because the switching timing is equally shifted.

Fig. 3 shows the switching pulse generation of the secondary side rectifier. The full-bridge converter on the secondary side operates as a synchronous rectifier. The switching pulse is the same as the pulse of the primary side. In the primary side, the resonant current i_{re} is positive when S_{dcdc11}, S_{dcdc21}, S_{dcdc32} are turn-on. Note that the phase of the primly side resonant current is a little different in order to design resonant DC/DC converter. Similarly, S_{dcdc12}, S_{dcdc22}, S_{dcdc31} are turn-on when the resonant current is negative.

Proposed modulation method

Fig. 4 shows the relationship between the input voltage and the output voltage of the cell. The switching state of the cell3 is followed by the boost mode. The cell3 is driven by the PWM operation in order to compensate for the harmonic component of the other cells. The cell1 and cell2 are driven by square-wave operation. The output voltage of the cell operated with square-wave is determined by the comparison with the voltage command of the PFC stage and the DC-link voltage of each cell. The switching device on the upper arm of the cell1 turns on when the output voltage command v_{conv} is higher than DC-link voltage of the cell1. Similarly, the switching device on the upper arm of the cell2 turns on

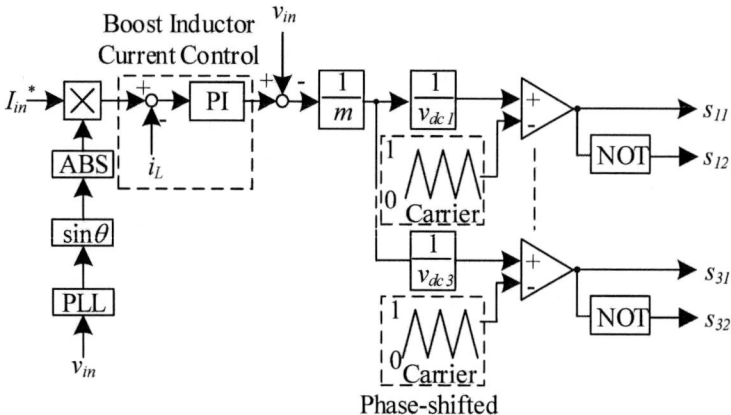

Fig. 2. Control block diagram of the PFC with the conventional method. The conventional modulation of the PFC is driven by PWM.

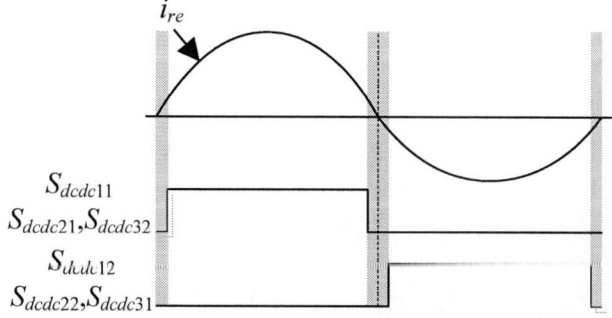

Fig. 3. Pulse generation for secondly side rectifier in DC/DC converter.

when the output voltage command is higher than the sum of the DC link voltages of the cell1 and the cell2. Thus, the output voltage of the cell1 and the cell2 is double of the grid frequency.

Fig. 5 shows the block diagram for the control of the input current with the proposed method. The output voltage of the square-wave cells behave to the current controller as the disturbance because the output value of the PI controller is the total output voltage of all the cell converters. Thus, the input voltage of the cell3 v_{in3} is given by

$$v_{in3} = |v_{in}| - v_{dc1}s_{sqr11} - v_{dc2}s_{sqr21} \tag{2}$$

where v_{dc1} and v_{dc2} are the DC-link voltage of cell1 and cell2, respectively. Moreover, s_{sqrn1} is the switching function of the square-wave cell in the PFC stage. When the upper switch s_{n1} is on-state, s_{sqrn1} equals 1. When the upper switch s_{n1} is off-state, s_{sqrn1} equals 0. As shown in (2), the current controller compensates the output voltage of the square-wave cells with the feed-forward control. Then, the output voltage compensated is the output voltage of the cell3, the gate signal of the PFC circuit is determined with standardizing and comparing triangle carrier. Note that the output power of each square-wave cell is balanced by the sorting the switching state in SST [21].

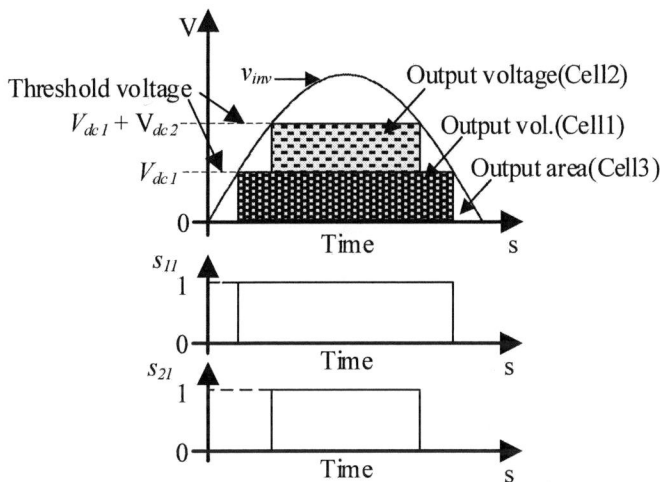

Fig. 4. Operation principle of the proposed method.
The relationship between the input voltage and the output voltage of cell #1 and #2, #3. (m = 3).

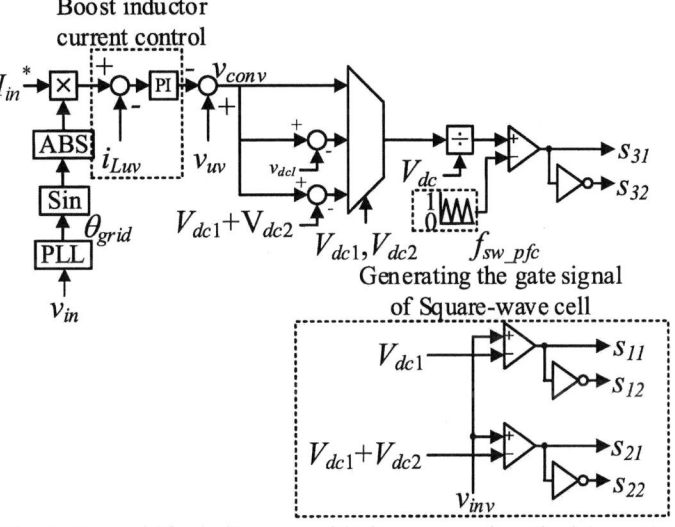

Fig. 5. Control block diagram with the proposed method (m = 3).

Common-mode equivalent Circuit

Equivalent circuit model of PFC

Fig. 6 shows the PFC circuit based on the input series connection with the parasitic components. This model consists two cell converters for the simplification of the following analysis. Note that this model does not consider the different values for the parasitic capacitance, the parasitic inductance and the parasitic resistance. The circuit model considers the parasitic capacitance between the switching component and the heatsink in order to simulate the leakage current generated by the switching operation of the PFC stage. Moreover, the circuit model is symmetric in order to design the equivalent model of the PFC stage.

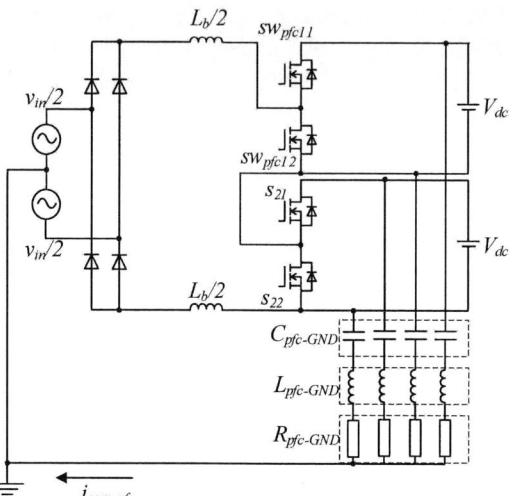

Fig. 6. Circuit configuration of the PFC stage with the parasitic components.

Fig. 7 shows the equivalent circuit model of PFC. C_{pfc_gnd} is the parasitic capacitance. L_{pfc_gnd} is the parasitic inductance. R_{pfc_gnd} is the parasitic resistance. L_b is boost inductance. The common-mode voltage-source v_{com_pfc1}, v_{com_pfc2} are given by

$$v_{com\_pfc1} = sw_{pfc11}V_{dc1}/2 + sw_{pfc21}V_{dc2}/2 \qquad (3)$$

$$v_{com\_pfc2} = -sw_{pfc11}V_{dc1}/2 + sw_{pfc21}V_{dc2}/2 \qquad (4)$$

where V_{dc1} and V_{dc2} are DC-link voltage of each cell, sw_{pfcn1} is the switching state of the upper arm device, sw_{pfcn2} is the switching state of the lower arm device. Hence, the common-mode voltage is changed by the switching state of each cell. As the shown in Fig. 7, the leakage current of the PFC is given by

$$i_{com\_pfc} = (2v_{com\_pfc1} + v_{com\_pfc2})/(2Z_1 + Z_2) \qquad (5)$$

where Z_1 is the impedance of the boost inductor L_b. Z_2 is the combined impedance of the parasitic capacitor C_{pfc_gnd}, the parasitic inductor L_{pfc_gnd} and the parasitic resistance R_{pfc_gnd}. Here, the common-mode voltage v_{com_pfc1} is twice impact of the common-mode voltage v_{com_pfc2} as shown in (5). This means that suppressing the common-mode voltage v_{com_pfc1} is effective in order to attenuate the leakage current in PFC. In the conventional method, the switching operation of the cell1 is caused by PWM. On the other hand, in the proposed method, the switching operation of the cell1 is double of the grid frequency in order to be driven by square-wave. Thus, the common-mode voltage v_{com_pfc1} is suppressed by the proposed method.

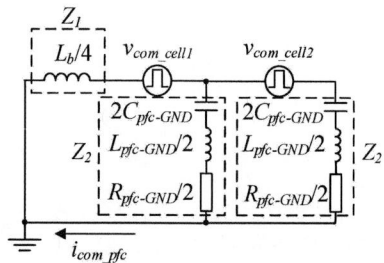

Fig. 7. Common-mode equivalent model of the PFC stage.

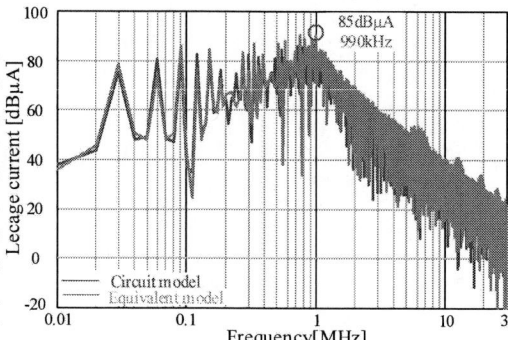

Fig. 8. Comparison of the leakage current between the circuit model and the equivalent model in the PFC stage.

Fig. 8 shows the comparison of the leakage current with the circuit model and the equivalent model in the PFC stage with the switching frequency of 30 kHz. The envelope of the leakage current in the circuit model almost agrees with the equivalent model. Each model has the spectrum of maximum value 85 dBµA at the 990 kHz. This means that the frequency is the resonant frequency between the parasitic capacitor and the boost inductor.

Equivalent circuit model of resonant DC/DC converter

Fig. 9 shows the circuit model of the resonant DC/DC converter with the parasitic components. This model consists one cell converter because the gate signal of each cell is same with the open-loop control. The circuit model considers the parasitic components between the switching component and the heatsink in order to simulate the leakage current generated by the switching operation of the DC/DC converter. Moreover, the circuit model is symmetric in order to design the equivalent model of DC/DC converter.

Fig. 10 shows the common-mode equivalent model of DC/DC converter. Here, C_{dcdc_gnd} is the parasitic capacitance. L_{dcdc_gnd} is the parasitic inductance. R_{dcdc_gnd} is the parasitic resistance. Note that the magnetizing inductance L_m does not consider because the indicator is shorted in the equivalent model. The common-mode voltage-source v_{com_dcdc}, is given by

$$v_{com\_dcdc} = sw_{dc1}V_{dc1}/2 - sw_{dc2}V_{dc1}/2 \qquad (6)$$

where V_{dc1} and V_{dc2} are the DC-link voltage of each cell, sw_{dc1} is the switching state of the upper arm device and sw_{dc2} is the switching state of the lower arm device. Hence, the common-mode voltage is changed by the switching state of each cell.

Fig. 11 shows the comparison of the leakage current with the circuit model and the equivalent model in the resonant DC/DC converter with the switching frequency of 50 kHz. The envelope of the leakage current in the circuit model almost agrees with the equivalent model. Each model has the spectrum of the maximum value 101 dBμA at the 2.3 MHz. This means that the frequency is the resonant frequency between the parasitic capacitor and the parasitic inductance.

Equivalent circuit model of SST

Fig. 12 shows the SST circuit model based on the ISOP connection with the parasitic components. This model uses three cell converters for the comparison with experimental model. Note that this model does not consider the different values for the parasitic capacitance, the parasitic inductance and the parasitic resistance.

Fig. 13 shows the common-mode equivalent model of SST. This model considers the parasitic component between the switching devices and the heatsink in PFC and the DC/DC converter. Note that the parasitic component of the secondly side rectifier does not consider because the conventional method and the proposed method are driven by the same operation with open-loop control.

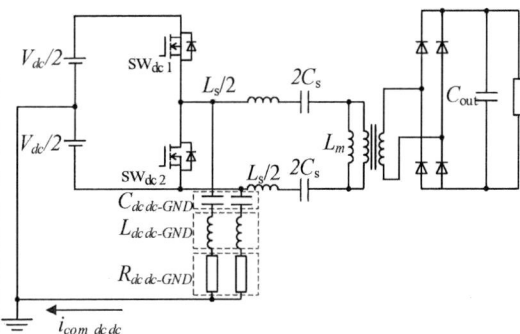

Fig. 9. Circuit configuration of resonant DC/DC converter with the parasitic components.

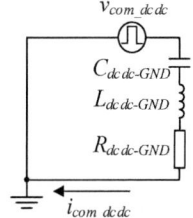

Fig. 10. Common-mode equivalent model of resonant DC/DC converter.

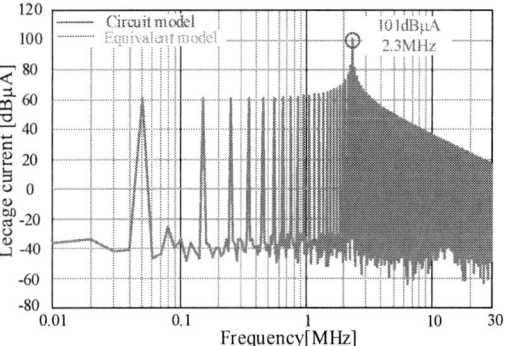

Fig. 11. Comparison of the leakage current with the circuit model and the equivalent model in resonant DC/DC converter.

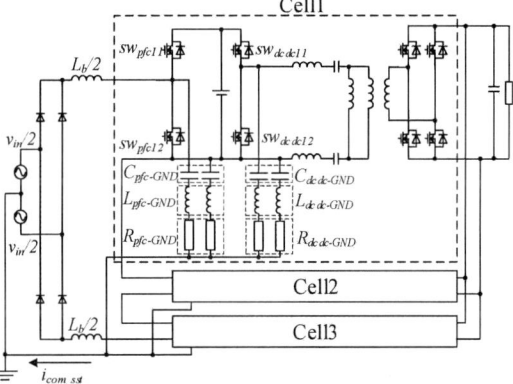

Fig. 12. Circuit configuration of SST with the parasitic components.

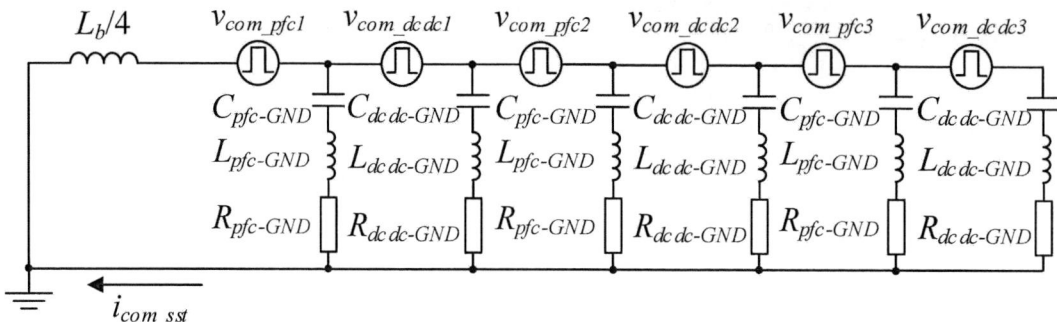

Fig. 13. Common-mode equivalent model of SST.

Fig. 14 shows the comparison of the leakage current between the circuit model and the equivalent model in SST with the switching frequency of 30 kHz in the PFC part using the conventional method. The switching frequency of DC/DC converter is 50 kHz with open-loop control. The envelope of the leakage current in the circuit model almost agrees with the equivalent model. Each model has the spectrum of the maximum value 103 dBμA at the 2.3 MHz.

Fig. 15 shows the comparison of the leakage current between the simulation model and the experimental model. The leakage current has 74.4 dBμA in 120 kHz-band and 87.6 dBμA in 180 kHz-band. Thus, the switching frequency components of PFC is large impact in SST. Moreover, the simulation model is the appropriate model because of almost the agreement with the experimental result.

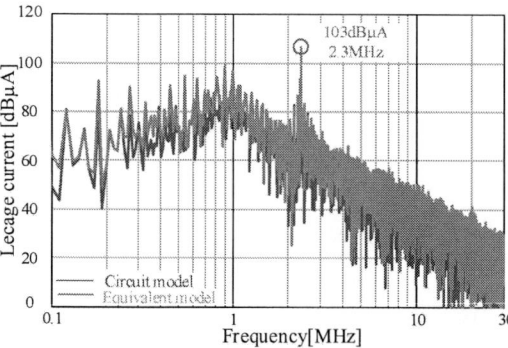

Fig. 14. Comparison of the leakage current with the circuit model and the equivalent model in SST.

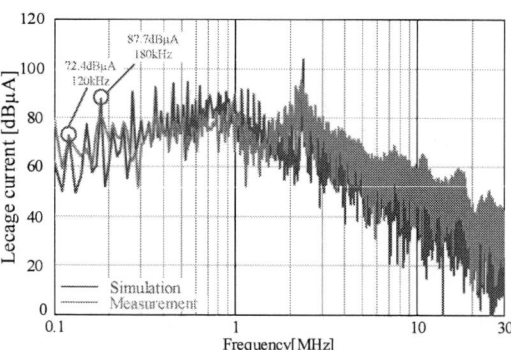

Fig. 15. Comparison of the leakage current with simulation and experiment in SST.

Experimental results

Experimental result with the conventional method

Table I shows the experimental parameters. The input voltage is 200 V, the rated power is 1.0 kW, and the number of cells is three. Moreover, the prototype is operated with the conventional control block diagram, as shown in Fig. 3, and the proposed control, as shown in Fig. 5.

Fig. 16 shows the operation waveforms of the conventional method. Fig. 16(a) shows that the input current THD is 3.18%, and the input power factor is 0.99. Fig. 16(b) shows that each cell is driven by PWM against grid voltage.

Experimental result with the proposed modulation

Fig. 17 shows the operation waveforms of the proposed method. In the proposed method, the switching frequency of the PFC circuits is 30 kHz. Fig. 17(a) shows that the input current THD is 2.91%, and the input power factor is 0.99. In

Table I. Experimental parameters.

Input voltage	V_{in}	200 V_{rms}
Rated output power	P_{out}	1.0 kW
Rated output voltage	V_{out}	50 V
Switching Device		SCT2080KE
Grid Freqency	f_s	50 Hz
Primary side capacitor	C_1	1500 μF
Resonant capacitor	C_r	204 nF
Leakage inductor	L_r	50 μH
Secondary side capacitor	C_2	3600 μF
Trans turns ration	$N_1:N_2$	1 : 1
Number of cells	m	3
Switching frequency of PFC	f_{sw_pfc}	30 kHz
Switching frequency of LLC	f_{sw_llc}	50 kHz

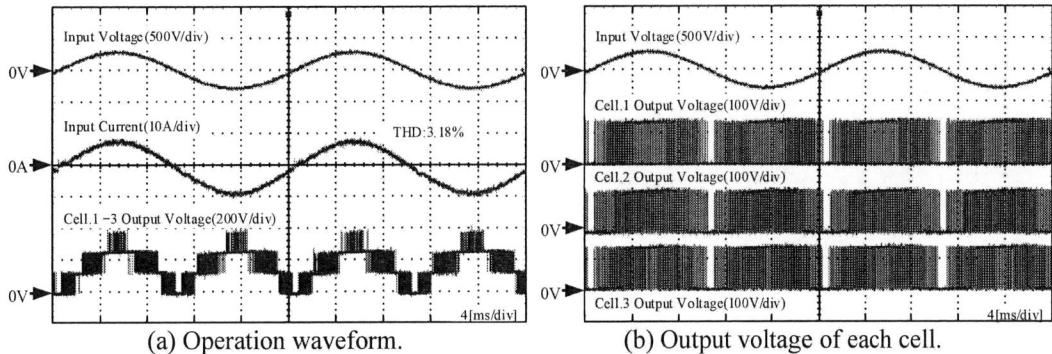

(a) Operation waveform. (b) Output voltage of each cell.

Fig. 16. Operation waveform with the conventional method.

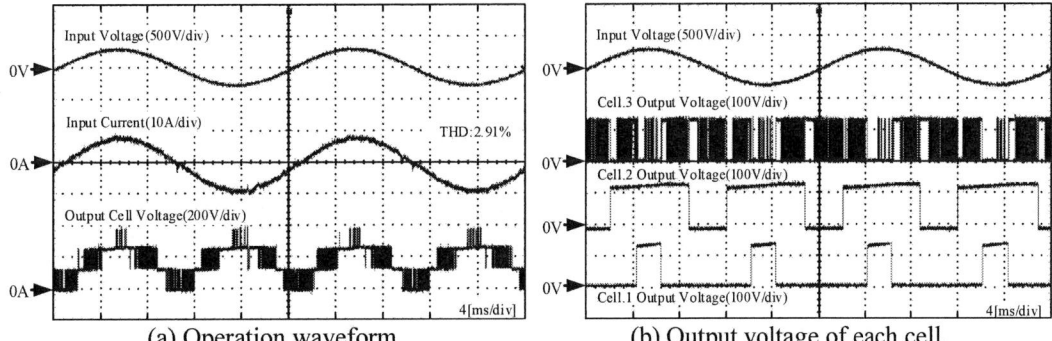

(a) Operation waveform. (b) Output voltage of each cell.

Fig. 17. Operation waveform with the proposed method.

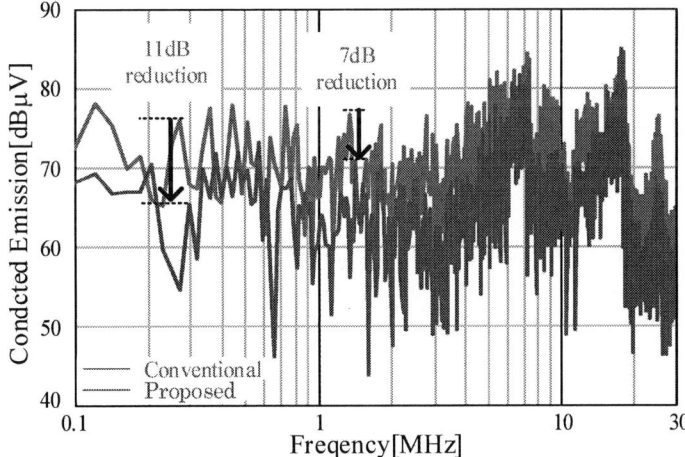

Fig. 18. Frequency spectrum of the Conducted emission
with the conventional or the proposed method.

addition, Fig. 17(b) shows the output voltage of the square-wave cells in double of the grid frequency. The PWM cell compensates the harmonic components of the square-wave cell.

Comparison of the conducted emission with the conventional and the proposed method.

Fig. 18 shows the analysis of the conducted emission with the conventional and the proposed method. The proposed method reduces the carrier frequency component produced by PWM. Moreover, the attenuation of almost 11dB and 7dB in 200 kHz-band and 1.2 MHz-band are achieved, respectively.

Fig. 19 shows the comparison of the frequency spectrum with applying PWM to PFC of each cell using the proposed method. Focusing on the frequency spectrum of Cell1 and Cell3, the attenuation of almost 13 dB and 7 dB in 200 kHz-band and 2 MHz-band are achieved, respectively. Thus, the proposed method is effective in order to reduce the conducted emission by applying PWM to the lower cell.

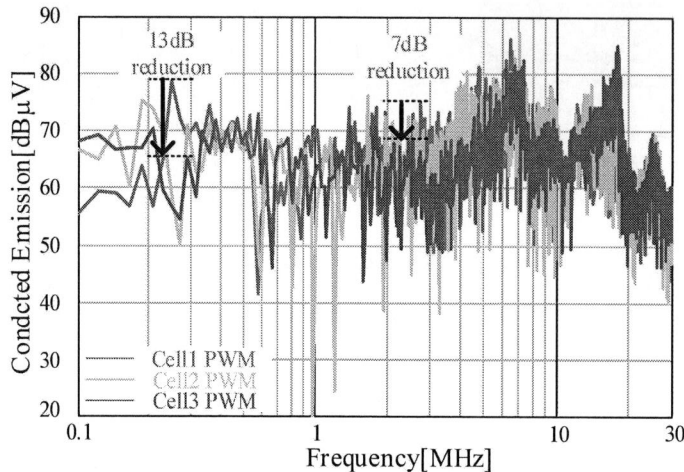

Fig. 19. The comparison of the frequency spectrum with applying PWM to each cell with the proposed method.

Conclusion

This paper had proposed a novel modulation method in order to reduce the common-mode noise in SST based on the ISOP connection. In the proposed method, one of the cells is driven by PWM in order to compensate for the harmonic component. This paper showed the reduction of the common-mode voltage with the proposed method at the switching frequency. The principle of the proposed method was shown by the equivalent circuit of SST. As the experimental result, the conducted emission was reduced by 11 dB at 200 kHz-band, 7 dB at 1 MHz-band with the proposed method. Moreover, the proposed method is effective in order to reduce the conducted emission by applying PWM to the lower cell.

References

[1] J. E. Huber and J. W. Kolar: "Applicability of Solid-State Transformers in Today's and Future Distribution Grids", IEEE Transactions on Smart Grid, Vol. 10, No. 1, pp. 317 -326, (2019).

[2] L. Ferreira Costa, G. De Carne, G. Buticchi and M. Liserre: "The Smart Transformer: A solid -state transformer tailored to provide ancillary services to the distribution grid", IEEE Power Electronics Magazine, Vol. 4, No. 2, pp. 56-67,(2017).

[3] J. E. Huber and J. W. Kolar, "Solid-State Transformers: On the Origins and Evolution of Key Concepts," in IEEE Industrial Electronics Magazine, vol. 10, no. 3, pp. 19-28, Sept. 2016.

[4] J. E. Huber and J. W. Kolar, " Volume/weight/cost comparison of a 1MVA 10 kV/400 V solid-state against a conventional low-frequency distribution transformer," in Proc. IEEE Energy Convers. Congr. Expo. (ECCE), Pittsburgh, PA, USA, Sep. 2014, pp. 4545–4552.

[5] J. E. Huber, J. Böhler, D. Rothmund and J. W. Kolar, "Analysis and cell-level experimental verification of a 25 kW all-SiC isolated front end 6.6 kV/400 V AC-DC solid-state transformer," in CPSS Transactions on Power Electronics and Applications, vol. 2, no. 2, pp. 140-148, 2017.

[6] X. Cai et al., "Fluctuation Power Control Strategy for MMC-based SST to Reduce the Submodule Capacitor Voltage Oscillation," 2019 10th International Conference on Power Electronics and ECCE Asia (ICPE 2019 - ECCE Asia), Busan, Korea (South), 2019, pp. 2430-2435.

[7] T. M. Parreiras, A. P. Machado, F. V. Amaral, G. C. Lobato, J. A. S. Brito and B. C. Filho, "Forward Dual-Active-BridIEEE Transactions on Industry Applications, vol. 54, no. 6, pp. 6353-6363, Nov.-Dec. 2018.

[8] L. Zhang, J. Qin, Q. Duan and W. Sheng, "Component Sizing and Voltage Balancing of MMC-based Solid-State Transformers Under Various ACLink Excitation Voltage Waveforms," 2019 IEEE Applied Power Electronics Conference and Exposition (APEC), Anaheim, CA, USA, 2019, pp. 371-375.

[9] T. Liu et al., "Design and Implementation of High Efficiency Control Scheme of Dual Active Bridge Based 10 kV/1 MW Solid State Transformer for PV Application," in IEEE Transactions on Power Electronics, vol. 34, no. 5, pp. 4223-4238, May 2019.

[10] A. Rodriguez et al., "Auxiliary power supply based on a modular ISOP flyback configuration with very high input voltage," 2016 IEEE Energy Conversion Congress and Exposition (ECCE), Milwaukee, WI, 2016, pp. 1-7.

[11] N. Oswald, B. H. Stark, D. Holliday, C. Hargis and B. Drury: "Analysis of Shaped Pulse Transitions in Power Electronic Switching Waveforms for Reduced EMI Generation", IEEE Transactions on Industry Applications, Vol. 47, No. 5, pp. 2154-2165, (2011).

[12] IEC CISPR 11 Edition.5.0:Industrial, Scientific And Medical Equipment -Radio-Frequency Disturbance Characteristics - Limits And Methods Of Measurement, IEC Standard,May,2009.

[13] S. Ogasawara, H. Ayano and H. Akagi: "An active circuit for cancellation of common-mode voltage generated by a PWM inverter", IEEE Transactions on Power Electronics, Vol. 13, No. 5, pp. 835-841, (1998).

[14] M. C. D. Piazza, A. Ragusa, and G. Vitale, "Effects of common-mode active filtering in induction motor drives for electric vehicles," IEEE Trans. Veh. Technol., vol. 59, no. 6, pp. 2664–2673, Jul. 2010.

[15] W. Chen, X. Yang, J. Xue, and F. Wang, "A novel filter topology with active motor CM impedance regulator in PWM ASD system," IEEE Trans. Ind. Electron., vol. 61, no. 12, pp. 6938–6946, Dec. 2014.

[16] Yingliang Huang, Yongxiang Xu, Yong Li, Guijie Yang, Jibin Zou, "PWM Frequency Voltage Noise Cancelation in Three-Phase VSI Using the Novel SVPWM Strategy", IEEE Transactions on Power Electronics, vol.33, no.10, pp.8596-8606, 2018.

[17] Ziyou Lim, Ali I. Maswood, Gabriel Heo Peng Ooi, "Common-Mode Reduction for ANPC With Enhanced Harmonic Profile Using Interleaved Sawtooth Carrier Phase-Disposition PWM", IEEE Transactions on Industrial Electronics, vol.63, no.12, pp.7887-7897, 2016.

[18] D. Shin et al., "Analysis and design guide of active EMI filter in a compact package for reduction of common-mode conducted emissions," IEEE Trans. Electromagn. Compat., vol. 57, no. 4, pp. 660–671, Aug. 2015.

[19] H. Peng, B. Narayanasamy, A. I. Emon, Z. Yuan, M. Ul Hassan and F. Luo, "Design and Implementation of Selective Active EMI Filter with Digital Resonant Controller," 2020 IEEE Energy Conversion Congress and Exposition (ECCE), Detroit, MI, USA, 2020, pp. 5855-5861.

[20] Yingliang Huang, Yongxiang Xu, Wentao Zhang, Jibin Zou, "PWM Frequency Noise Cancellation in Two-Segment Three-Phase Motor Using Parallel Interleaved Inverters", IEEE Transactions on Power Electronics, vol.34, no.3, pp.2515-2525, 2019.

[21] N. Kikuchi, J. -I. Itoh, K. Kusaka and H. N. Le, "Hybrid Multiple Chopper Cells of PWM and Square-wave Operation for Solid-state Transformer," 2020 22nd (EPE'20 ECCE Europe), 2020, pp. P.1-P.10.

Modular STATCOM for compensation of reactive power and voltage asymmetry in medium-voltage distribution power grids

Josef Štengl, Tomáš Kormska, Jakub Talla, Zdeněk Peroutka
UNIVERSITY OF WEST BOHEMIA
Univerzitní 2732/8, 301 00
Pilsen, Czech Republic
jstengl@fel.zcu.cz

Acknowledgements

This research has been supported by the Ministry of Education, Youth and Sports of the Czech Republic under the project No. SGS-2021-021 and by the Technology Agency of the Czech Republic under the project No. TN01000007.

Keywords

«Static Synchronous Compensator (STATCOM)», «Medium voltage», «Grid-connected converter», «Zero sequence voltage», «Reactive power»

Abstract

A cost-effective STATCOM is proposed and discussed in this paper. Its modular topology enables to scale the power to target systems. Besides reactive power, the proposed STATCOM enables to generate a zero-sequence component and compensate the voltage asymmetry, e.g., in widely spread medium-voltage distribution power grids with resonant grounding.

Introduction

Power consumption is constantly growing which brings several problems; many power grids are operated close to the limit of their transmission or distribution capacity. To ensure stable and reliable power supply, it is necessary to increase the robustness of power grids. In the case of medium-voltage distribution power grids, there is increasing pressure on the power quality. Reactive power (inductive or capacitive) is to be compensated within the distribution power grid. Its transmission to a superior power system might expose the operator to financial penalty. Also, a voltage symmetry is an issue. A lot of distribution power grids are operated as resonant-grounded systems [1]. In these systems, arc suppression coils are tuned in resonance with stray capacitance to ground to effectively compensate earth faults. However, tuning the coil in the resonance point or nearby causes significant voltage asymmetry [2].

Modern systems based on power semiconductor converters enable to solve problems mentioned above. Reactive power can be effectively compensated by STATCOM systems. However, these systems operate with positive sequence component only [3-7]. This is enough for the reactive power compensation, but not for the compensation of the voltage asymmetry. Moreover, their design is often based on multilevel converters, which, on one hand, generate a high-quality voltage waveform, on the other hand, their design is rather complicated and may suffer by high costs and unreliability [3-5].

This paper deals with a modular STATCOM designed for medium-voltage distribution power grids. Its design pursues two main goals: i. compensation of both, reactive power, and voltage asymmetry, ii. robustness, cost effective design and modularity for power scaling.

System topology

The topology of the proposed STATCOM is depicted in Fig. 1. The whole system is connected to a three-phase distribution power grid of 22 kV using a three-phase transformer 22/0.4 kV with star-connected primary winding. The star point (the neutral) is earthed which enables to operate with the zero-sequence component. Secondary windings are connected to the power converter. The modular design of the converter is based on unified basic power units of 150 kVA. Each unit is a air-cooled single-phase voltage-source inverter (full bridge, see Fig. 2), based on inexpensive IGBT modules with blocking voltage of 1200 V. These basic power units are connected in parallel via filtering inductors (see Fig. 1) and all are interconnected via a common dc link. Its voltage is controlled to 730 V.

In our installation, there are altogether 9 power units of 150 kVA and the total power of the whole installation is 1,35 MVA. However, the total power can be easily scaled by adding or removing power units and the installation can be adapted to the final power system.

To achieve a high-quality voltage and current, the converter is connected to the transformer via an LCL filter, where the installed components L_f and C_f are completed by the transformer leakage inductance L_σ (see Fig. 1 and 2).
The start procedure of the STATCOM includes precharging from the additional precharge circuit consisting of a conventional diode rectifier, while the power converter is disconnected from the grid (the contactor K_1 is switched off, see Fig. 1). Once the precharging is completed, the contactor K_2 is switched off, and the power converter starts to generate three-phase voltage on the filter capacitor C_f. Once this voltage is synchronized to the transformer secondary, the contactor K_1 is switched on and the power converter is connected to the grid voltage.

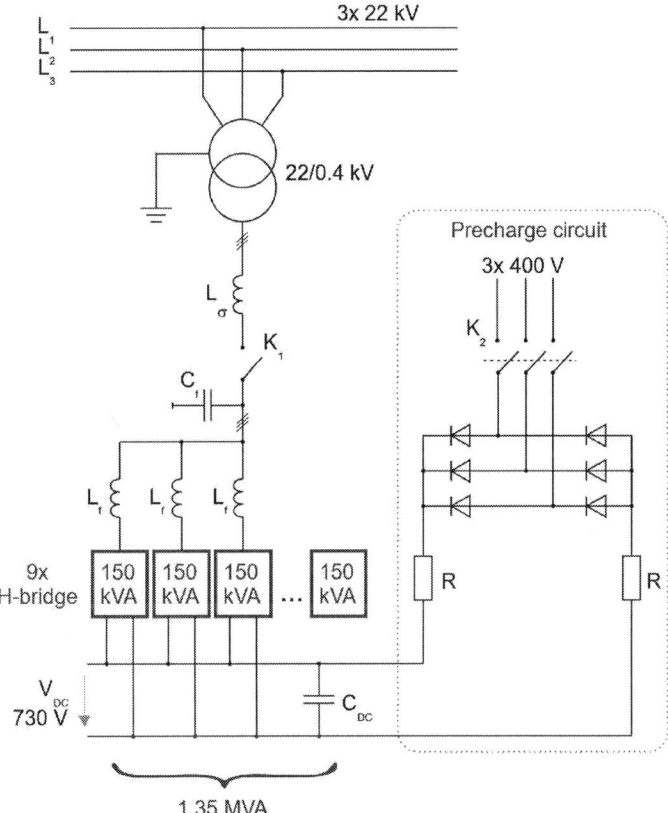

Fig. 1: Topology of the proposed air-cooled modular STATCOM controlling both, positive sequence and zero sequence component

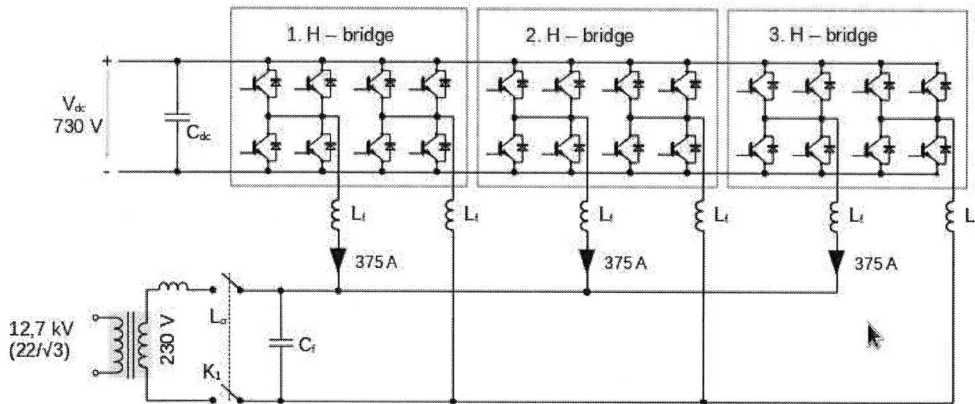

Fig. 2: Detail of the secondary: Three or more power units (single-phase bridge inverters) supplying one phase of transformer

Control algorithm

The employed control algorithm of the whole STATCOM is depicted in Fig. 3. It controls all power units in the same time; the carriers of the PWMs are shifted to minimize the current ripple.

The control algorithm consists of two main loops, one for positive and one for zero sequence component. Two PI controllers control the positive sequence component using the currents I_d, I_q in the rotating reference frame d-q. The I_d control loop is employed to control active power which is necessary to keep the dc-link voltage at the required constant value of 730 V. To linearize the system, the error related to energy $(\frac{1}{2}C_{dc}V_{dc}^{w\,2} - \frac{1}{2}C_{dc}V_{dc}^2)$ is employed instead the direct voltage error (see Fig. 3). If necessary, the positive sequence component can be utilized to supply some load connected to the dc-link voltage.

The compensation of the reactive power of the grid is controlled using the I_q current component. Thus, the converter can generate either inductive or capacitive reactive power. Since the installation includes the LCL filter, the derivative feedback $(k_d(I_m - I_s) = k_d I_{Cf})$ is employed to stabilize the control loop and to avoid oscillations.

The second part of the algorithm is designed for the zero sequence component (see Fig. 3, bottom part), which in fact is a single-phase current control problem. Therefore, it is advantageous to use a resonant controller operating directly in the stationary reference frame, and thus no coordinate transformation is required. This control loop is employed to compensate the voltage asymmetry in the power grid.

In the final step, all required voltage components (V_d^w, V_q^w and V_0^w) are transformed back to the abc coordinates and the carrier-based PWM is used to generate the required phase voltage by the converter (see Fig. 3).

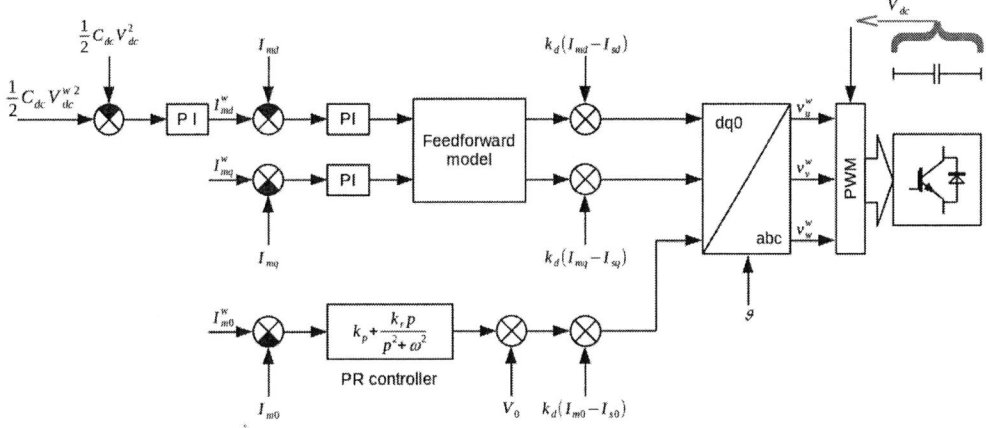

Fig. 3: Control algorithm of the STATCOM consisting of two control loops, one for positive sequence and one for zero sequence current component

Synchronization with power grid voltage

Grid connected converters must be usually synchronized to the grid voltage. Although the phase locked loop (PLL) is probably the most common solution, a computationally simple "sliding" discrete Fourier transform (DFT) algorithm is adopted for this system. This method extracts a single spectral component, namely the fundamental harmonic of 50 Hz, from the voltage waveform, sampled in a sliding window (single period of the fundamental). In principle, the voltage phasor of the fundamental $\bar{S}^1 = |\bar{S}^1|e^{j\omega t}$ rotating in the complex plane is shifted by an angle of $\frac{2\pi}{N}$ in every sampling period:

$$\bar{S}(k) = \bar{S}(k-1)e^{\frac{2\pi}{N}}, \tag{1}$$

where k is sampling time and N number of samples of the fundamental. For the real and imaginary part, we can write

$$S_{re}(k) = S_{re}(k-1)\cos\left(\frac{2\pi}{N}\right) - S_{im}(k-1)\sin\left(\frac{2\pi}{N}\right) - x(k-N) + x(k), \tag{2}$$

$$S_{im}(k) = S_{re}(k-1)\sin\left(\frac{2\pi}{N}\right) + S_{im}(k-1)\cos\left(\frac{2\pi}{N}\right), \tag{3}$$

where $x(k)$ is the actual voltage sample and $x(k-N)$ is the oldest sample in the sliding window. From (2) and (3), the actual angle ϑ can be calculated, which is crucial for the synchronization with the grid (see Fig. 3).

$$\vartheta(k) = \arctan\left(\frac{S_{im}(k)}{S_{re}(k)}\right)$$

Fig. 4 shows the performance of the DFT-based synchronization in a test bed using the sine-wave (Fig. 4a) and square-wave inputs (Fig. 4b) with amplitude step change.

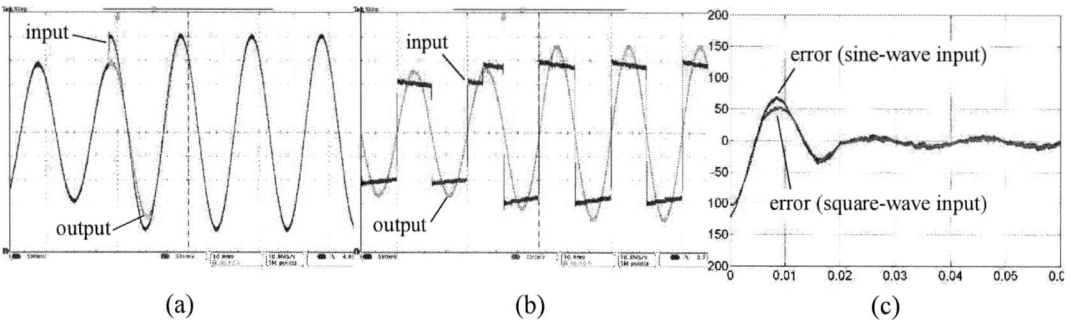

| (a) | (b) | (c) |

Fig. 4: Synchronization to the input signal using the DFT method: (a) Sine-wave input, (b) Square-wave input, (c) Error related to the fundamental component.

As can be seen from Fig. 4, the DFT method provides a pure sine wave on its output. Since the length of the sampling window corresponds to the single period of the fundamental (50 Hz), the transient takes 20 ms when a change of the input signal occurs. As the main advantages of this synchronization method, we can mention low computational complexity, robustness to the input distortion level resulting from the principle of spectral analysis and the absence of parameter tuning which is necessary for PLL-based and other methods.

In resonant-grounded distribution power grids [1], the phase to ground voltage may differ from the phase voltage of the grid transformer (related to the neutral) and the zero-sequence component v_0 is usually present. This leads to asymmetry of the phase to ground voltage. Fig. 6 shows the voltage phasor diagram based on measurements taken in the medium-voltage distribution power grid of 22 kV with resonant grounding. Due to this reason, it is preferable to synchronize the converter to the phase-phase voltage, which is stable, rather than to phase to ground voltage, which is directly measured. However, for the synchronization, the V_α component can be used (see Fig. 5):

$$V_\alpha = \frac{2}{3}\left(v_u - \frac{1}{2}v_v - \frac{1}{2}v_w\right). \tag{4}$$

As shown in (4), its computation is based on the phase voltage difference. Thus, the V_α is related to phase-phase voltage and stable.

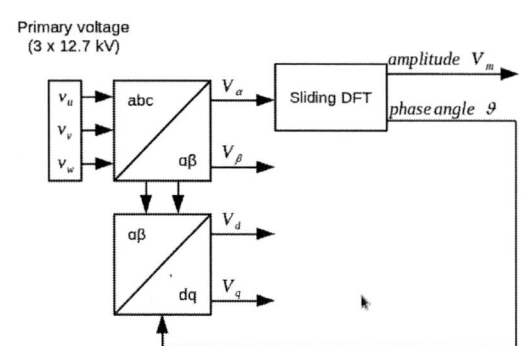

Fig. 5: Synchronization to the V_α component of the grid voltage using the sliding DFT method

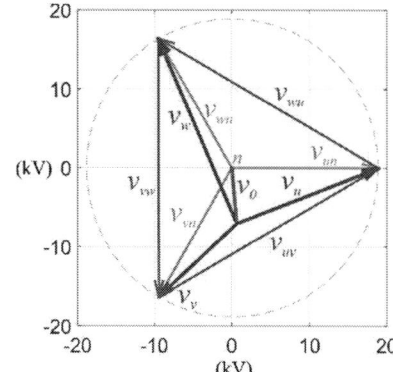

Fig. 6: Experimental results: Voltage asymmetry in resonant-grounded distribution power grid of 22 kV

For safety reasons, the sliding DFT discussed above is applied to the V_β component as well. It is used for faster and more reliable detection of the power outage of the grid.

$$V_\beta = \frac{2}{3}\left(\frac{\sqrt{3}}{2}v_v - \frac{\sqrt{3}}{2}v_w\right) \qquad (4)$$

Experimental results

Fig. 7 and 8 show the start-up sequence of the STATCOM. Once the precharging is finished and the contactor K_2 is disconnected (see Fig. 1), the converter starts to generate ac voltage (see Fig. 8, bottom, time sequence c) synchronized to the voltage on the transformer secondary (see Fig. 8, top). The dc-link voltage is slightly decreasing (see Fig. 7, time sequence c). Next, the contactor K_1 is switched on and the power converter is connected to the transformer secondary. The connection time corresponds to the border between time sequence c and d in Fig. 7 and 8. Finally, control algorithm is activated; the active current component I_d is increased, and the dc-link voltage converges to the required level of 730 V (see Fig. 7, time sequence d).

Fig. 9 illustrates the STATCOM operation in the resonant-grounded medium-voltage distribution power grid. For better readability, the waveforms of phase v and w are not displayed. It can be seen that the STATCOM generates inductive or capacitive reactive power.

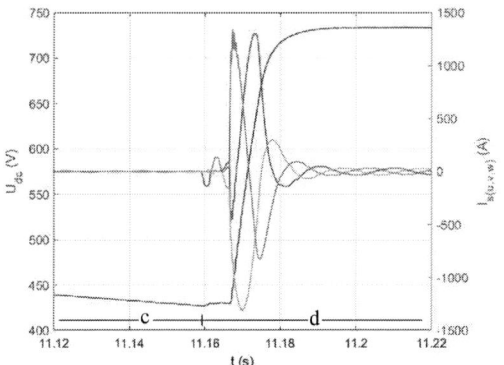

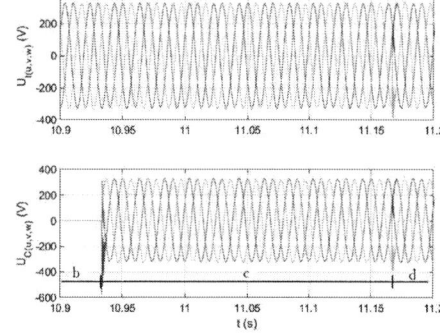

Fig. 7: Experimental results, start-up sequence: Activation of the control algorithm and convergence of the dc-link voltage to its required level (blue: dc-link voltage, green, orange and yellow: three-phase currents generated by converter on secondary side of transformer)

Fig. 8: Experimental results, start-up sequence: Generation of three-phase voltage (v_{cu}, v_{cv}, v_{cw}, bottom figure) synchronized with voltage on secondary side of transformer (top figure) and converter connection to the transformer (K_1 switched on in t =11.16 s)

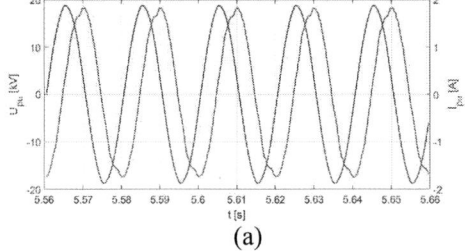

(a)

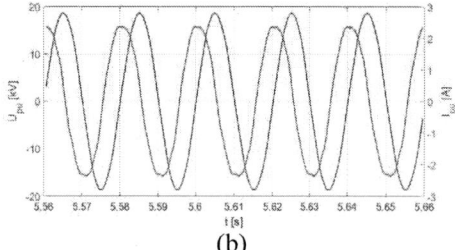

(b)

Fig. 9: Experimental results, operation in power grid of 22 kV: Compensation of reactive power in (a) inductive and (b) capacitive mode (voltage (blue) and current (orange) of phase u).

The final installation in the distribution substation of 110/22 kV is shown in Fig. 10.

Fig. 10: Final installation of proposed STATCOM of 1.35 MVA in distribution substation of 110/22 kV

Conclusion

A STATCOM of 1.35 MVA designed for widely spread resonant-grounded medium-voltage distribution power grids is discussed in the paper. The proposed cost-effective air-cooled modular topology of the power converter enables to simply scale the total power to a target power system. Besides reactive power, the STATCOM enables to generate zero sequence component and compensate the grid voltage asymmetry as well. For the operation with both components, the proposed control algorithm includes two current control loops based on two PI controllers operating in rotating the reference frame and single PR controller operating in the stationary coordinates. To achieve robustness and high reliability, the sliding DFT is employed to synchronize the control algorithm to the grid voltage without undesired parameter tuning. Since the proposed STATCOM design avoids complicated structures of multilevel converters, it saves costs, it is robust and suitable for operation in medium-voltage power grids, demanding on reliability.

References

[1] "IEEE Guide for Protective Relay Applications to Distribution Lines," *IEEE Std C37.230-2020* pp. 1-106, 2021.

[2] G. Kaufmann and R. Vaitkevičius, "Sensitive ground fault detection in compensated systems (arc suppression coil). What is influencing the sensitivity?," *The Journal of Engineering,* vol. 2018, no. 15, pp. 971-977, 2018.

[3] V. Spudić and T. Geyer, "Model Predictive Control Based on Optimized Pulse Patterns for Modular Multilevel Converter STATCOM," *IEEE Transactions on Industry Applications,* vol. 55, no. 6, pp. 6137-6149, 2019.

[4] H. A. Pereira, M. R. Haddioui, L. O. M. d. Oliveira, L. Mathe, M. Bongiorno, and R. Teodorescu, "Circulating current suppression strategies for D-STATCOM based on modular multilevel converters," *2015 IEEE 13th Brazilian Power Electronics Conference and 1st Southern Power Electronics Conference (COBEP/SPEC),* pp. 1-6, 2015.

[5] Y. Jin *et al.*, "A Dual-Layer Back-Stepping Control Method for Lyapunov Stability in Modular Multilevel Converter based STATCOM," *IEEE Transactions on Industrial Electronics,* pp. 1-1, 2021.

[6] R. K. Varma and R. Salehi, "SSR Mitigation With a New Control of PV Solar Farm as STATCOM (PV-STATCOM)," *IEEE Transactions on Sustainable Energy,* vol. 8, no. 4, pp. 1473-1483, 2017.

[7] R. K. Varma and H. Maleki, "PV Solar System Control as STATCOM (PV-STATCOM) for Power Oscillation Damping," *IEEE Transactions on Sustainable Energy,* vol. 10, no. 4, pp. 1793-1803, 2019.

Novel Method for Active Short Circuit (ASC) Tests of Power Module in Automotive Traction Application

Tobias Appel, Arne Bieler
Danfoss Silicon Power
Husumer Str. 251
24941 Flensburg, Deutschland
Phone: +49 (0) 461 430140
Email: tobias.appel@danfoss.com, arne.bieler@danfoss.com
URL: www.siliconpower.danfoss.com

Keywords

≪Permanent magnet motor≫, ≪Short circuit≫, ≪Machine emulation≫, ≪Control methods for electrical systems≫, ≪Traction application≫ .

Abstract

Active short circuit is a special operation for power modules in automotive traction applications. For safety reasons, it must be qualified in detail during the design phase. Therefore, this article presents a novel method allowing power module manufacturers to test the ASC robustness of the device before an inverter unit is built.

Introduction

E-mobility, with its demand for efficiency, cost-efficiency, and safety, requires the development of new products, processes, and test methods because there are different drive systems in electrically powered vehicles. In this article, the focus is set on those inverters with batteries, inverters with DC link capacitors, and permanent magnet synchronous motor (PMSM), which are also operated in the field weakening range. This is one of the most popular systems in vehicles, as the PMSM is very efficient and requires little maintenance.

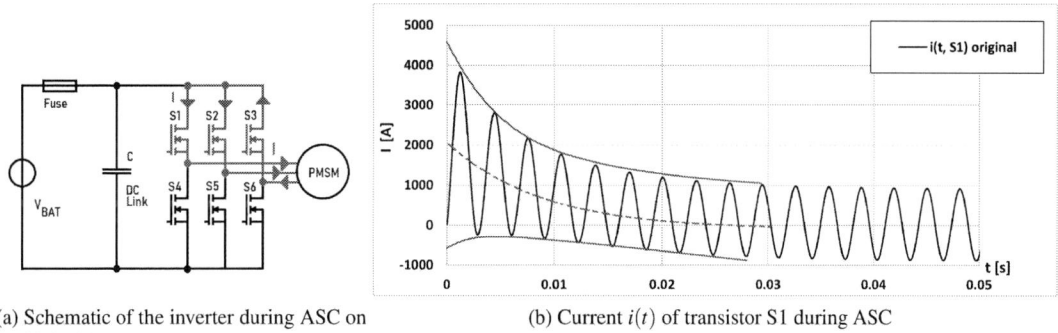

(a) Schematic of the inverter during ASC on the high side

(b) Current $i(t)$ of transistor S1 during ASC

Fig. 1: ASC currents in the inverter

Due to the permanent excitation of the machine, this drive requires an inverter with a sophisticated safety concept [1] in case the main controller of the inverter fails, for example. There are two modes in which the system can be set in that case. One is the state with all switches open, called free-wheeling (FW), where only the diodes or body diodes are conducting current. The other one is the case where all switches of the high (Fig.: 1a) or low side are conducting current, called active short circuit (ASC).

As long as the connection between the inverter and the battery is intact and the battery is not fully charged, the FW is one potential safety mode. Due to the voltage (EMF) of the machine in full speed, it needs to be considered that FW means that a negative torque is applied and so the car brakes. The ASC can be used in various cases and is, therefore, one of the standard solutions in case the controller fails. If the ASC is applied during normal operation, a typical current known from short-circuits of 3-phase systems near a generator occurs. A detailed calculation of the current, according to IEC 60909, is given in [2]. The worst-case current for the power module is an asymmetric current, as shown in Figure 1b. Initially, sub-transient currents with high peak values occur; these sub-transient currents subside and change into a continuous short-circuit current [2]. With its high amplitude and high dissipated energy, the sub-transient current is a challenge for the semiconductors and the power module, even though no switching losses occur. The current could lead to overheating of the semiconductor switches, as well as critical hot spots.

Potential hot spots can be enlightened with a 3D finite element analysis (FEA). The first step is to evaluate the average temperature of the semiconductors. With the use of bipolar devices, such as IGBT, the so-called virtual junction temperature (T_{vj}) is simulated or calculated. To perform this in the most efficient way is to use a calculation as described in [3]. The approach is to apply the instantaneous dissipation power stepwise to the transient thermal resistances Z_{th}. The resulting T_{vj} is the base for every next time step and used for the calculation. This method is validated by measurement and FEA. The Z_{th} for the tested modules is known from measurements. With this input data, the calculation can provide the $T_{vj}(t)$ with the same accuracy as a transient FEA. Therefore, this method is used for all shown temperature simulations in this article.

Despite all scientific efforts, the validation of the ASC mode must be done in the final application. This is possible for OEMs, but not for the component manufacturers due to confidentiality and availability of the drive train. In order to conduct adequate tests during the design phase of the power modules, basic test methods are needed.

One of these is a 50 Hz surge current test as described in [4] and [5] where a half-sine wave of a 50 Hz current is applied to the device. The focus of interest in these tests is the maximum temperature of the junction of an IGBT. A test series with increasing current shows the limits of the module. The pulse before the device is destroyed is the last good pulse. Its peak temperature must not be exceeded at ASC, according to [4] and [5]. For a more general description of the last good pulse, the I^2t value is introduced. It is a metric to characterize the energy a semiconductor can withstand, as described in [5] for an RC-IGBT inverter.

An improvement of this approach by introducing a laboratory setup emulating the ASC current is shown in the following sections.

Test Setup and Parameters for Testing with Oscillating Current

In this section, the test setup and a method for dimensioning the components is described. The experiment is designed to emulate the ASC current of the application, as closely as possible, by using a resonant circuit. A picture and a schematic of the circuit is given in Figure 2.

The resonant circuit as shown in Figure 2 consists of the choke L (Fig. 2b) (2 (Fig. 2a)), a capacitor bank C (3), and their parasitic resistances R. The capacitors are charged by a voltage source (V_{dc}). The used voltage source needs a sufficient high internal resistance (R_i) not to influence the signal. To switch the transistor, a gate driver (5) and interface are needed. To perform the tests with elevated temperature, a heating plate (4a) with a controller (4b) is used. To measure the temperature, the option of the controller to connect the module internal PT1000 is used. The electrical signals are measured with passive voltage probes, a rogowski coil (6), and an oscilloscope.

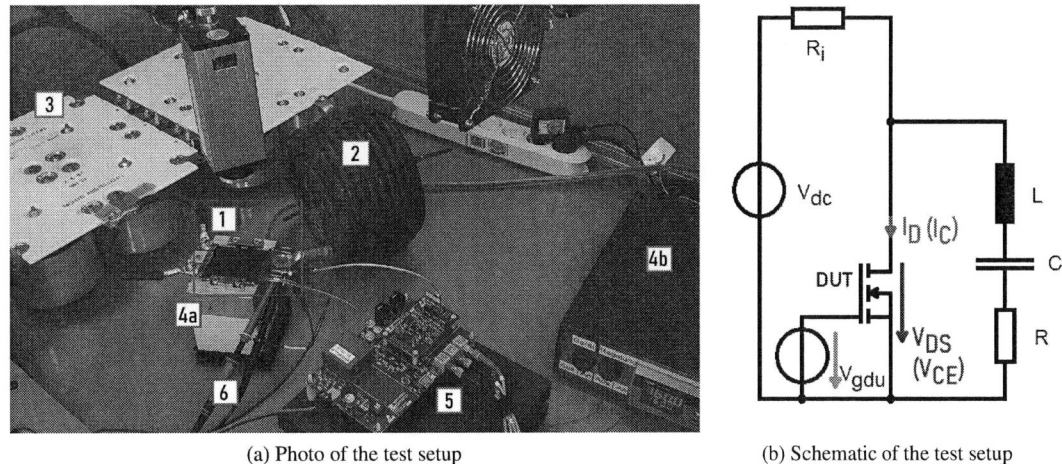

(a) Photo of the test setup   (b) Schematic of the test setup

Fig. 2: Test setup and schematic

The resonant circuit can be dimensioned with the set of equations 1 - 3 and the elements of the setup are dimensioned targeting the specified parameters i_{max}, I^2t and f. A detailed discussion of the parameters is given in the section below.

The blocking capability of the power module can be used as V_{c0}. With the equations 1 and 2, L and C can be determined. Equation 1 is derived from the stored energy in L and C and provides the starting current $\hat{i}$ of the damped oscillation. However, the maximum current i_{max} during ASC has to be calculated by equation 3 at $sin(\omega \cdot t) = 1$.

$$\hat{i} = V_{c0}\sqrt{\frac{C}{L}} \tag{1}$$

$$\omega = \sqrt{\frac{1}{L \cdot C}}, \qquad f_0 = \frac{1}{2\pi}\sqrt{\frac{1}{L \cdot C}} \tag{2}$$

$$i(t) = \hat{i} \cdot e^{-\delta \cdot t} \cdot \sin(\omega \cdot t), \qquad \delta = \frac{R}{2L} \tag{3}$$

$$I^2t = \int_0^\infty (i^2(t))dt = \hat{i}^2 \frac{1}{4}(\frac{1}{\delta} - \frac{\delta}{\delta^2 + \omega^2}) \tag{4}$$

Once L and C are determined, the parameter I^2t is only influenced by the damping ratio (δ) of the resonant circuit. If high I^2t is needed, it is advisable to dimension the inductance of the coil as high as possible because the parasitic resistance decreases disproportionately.

The I^2t value of the total current can be calculated by equation 4. The envelope of the damped oscillation is almost symmetric with respect to the x-axis compared to the current in Figure 1b. A slightly asymmetric envelope is caused by the damping. Since the current signal starts with a positive peak and due to the damping, the amplitude of positive peaks is always higher than the one of negative peaks. When this effect is neglected the I^2t of the positive current through the IGBT is half of the value calculated by equation 4. The equation 4 is derived by the integration of the damped oscillation (eq. 3).

To perform an ASC test with this method, the following steps after calculation of parameters and assembling the circuit are needed: The switch under test (DUT) is blocked by a negative voltage of the gate driver (5). The capacitor C (3) 2 is charged by the voltage source (Vdc). With a V_{gdu} pulse of a sufficient length, the test is performed. A long "ON"-time is important so that the switch does not turn off the oscillating current.

Laboratory Testing Results

In a laboratory test using the novel method to evaluate the ASC robustness, one 1200V SiC and two different 750V Si power module types have been investigated.

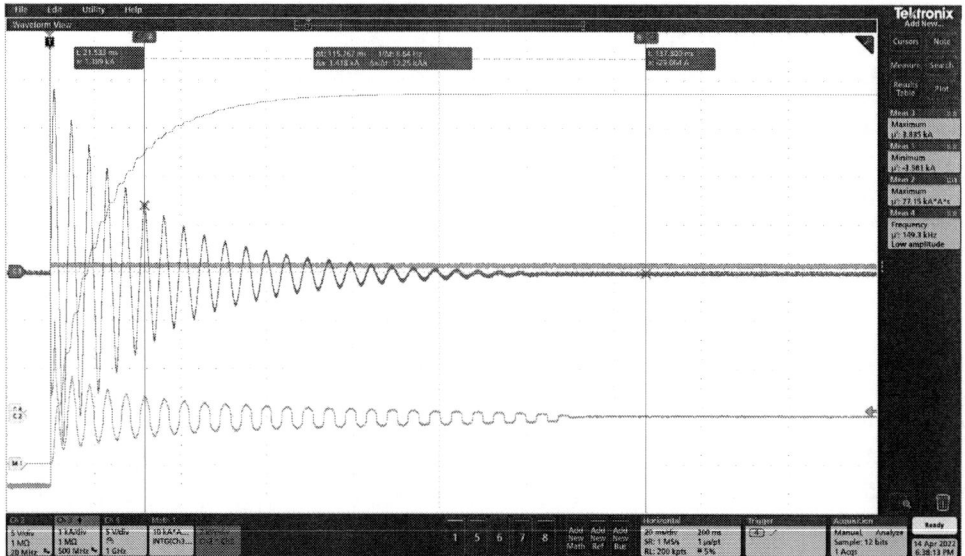

Fig. 3: Oscilloscope screen shot during the $77kA^2s$ test; (blue (Ch 2): V_{CE}, red (Ch 3): I_C, green (Ch 4): V_{gdu}, gold (Math 1): I^2t)

An exemplary scope picture is shown in Figure 3. It is the test with highest I^2t of the Table I. Besides waveforms of current and voltage, it includes the evolution of the $I^2t(t)$.

The test evaluates the ASC case shown in Fig 1 by the scope snapshot. All parameters are chosen to hit the evolution of $I^2t(t)$ best, as discussed in the next section. Tests were performed with elevated temperature. Specifically, 750V-900A-Module test is done at 165°C. This is the temperature of the last working point before the ASC occurs in the application.

Table I: Parameters of Test for different Power Modules

	i_{max} [A]	I^2t [kA^2s]	f [Hz]	L [μH]	C [mF]	Vdc [V]	$R[m\Omega]$
750V-900A-Module	3835	77	240	110	4	660	9
750V-650A-Module	1980	30.7	175	204	4	490	13
1200V-660A-Module	1675	6.642	415	72	2	342	17.5

In Table I, the values and parameters of the three tests are shown. The I^2t values of Table I are calculated from the application ASC currents. The calculated values and the actual values in the experiment slightly differ due to the damping of the forward voltages of the transistor and the diode.

With the components used, it is possible to test up to an I^2t of $77kA^2s$ at a peak current of 4 kA. To test beyond this value, the damping of the oscillating circuit must be decreased. For this purpose, the coil must be designed by an expert. Increasing the copper cross-section beyond the skin depth is usually counterproductive. The AC-resistance of an inductor also depends on number of layers, due to the proximity effect. For testing with lower I^2t values, the damping has to be increased by introducing longer cables or washers.

The power modules that were tested using the novel method show results which can help to investigate the robustness during ASC, as well as determine limits of the semiconductors and power modules. For this purpose, all tested modules were checked for deviations in an end-of-line test. The limits of the deviations correspond to those of a reliability test. In addition, a reliability test (e.g. H3TRB) can be performed with these modules.

Discussion of the Novel Test Method and Surge Current Tests

It is shown that with the novel test method, a current close to the ASC current in the application can be applied to the module. With the set of equations, the needed parameters can be calculated accurately. The following section will show why the new method is an improved test approach. It is also discussed on which parameters the main focus should be set.

To analyze the difference between the surge-current [4] test and the novel test method, a comparison of the current, the calculated $I^2t(t)$, and the resulting simulated junction temperature is given in Figure 4.

Framework of displayed curves

The current curves are measured, except for the "initial ASC current (240 Hz)" (Fig. 4b). This is a fictive ASC current, due to the lack of measurement of the "emulated current (320 Hz)". The $I^2t(t)$ is calculated on the basis of the currents, it is the integration of the square of the current function. The temperature is simulated by the method described in the introduction and in [3].

The currents as shown in Figure 4a and 4b are only positive due to usage of an IGBT module test. The ASC current requirement as shown in Fig 1b is the basis for the investigation and is given in Fig 4, indicated as "i(t, IGBT) initial". Comparing the initial and the emulated current by the previously described test setup, some differences can be observed. The main difference is that the emulated current shows pure sine half-waves, whereas the initial current has a DC-offset.

To verify whether the emulated ASC testing is valid, thermal simulations with ASC, emulated, and the initial currents were made. The results (Fig.: 4e, 4f) show the junction temperature of the six different currents. The curves, "T(t,IGBT) initial," are the result of ASC during a full working inverter. If the junctions of the IGBT are heated up to 165°C by this working inverter, through which the 65°C coolant flows, the modules experience different thermal fields.

The resulting temperature development of the currents "T(t,IGBT) emulation" is simulated in the same way with a 165°C pre-heated module as starting condition. This is needed to perform a lab test without a full functional inverter including cooling and sufficient power-supplies. The surge current curves, "T(t,IGBT) surge," are simulated under the same starting conditions. Due to the pre-heating, a homogeneous temperature field is present in the module.

Testing ASC with focus on the I^2t value

The emulated current of Figure 4a is tuned to get the best matching $I^2t(t)$ curve. This is the approach of the left side. Therefore, the frequency and the damping of the emulated current is lowered to 240 Hz. A longer first sine half-wave and higher following ones cause best tracking of the $I^2t(t)$. With the standard surge current test [4], a 50 Hz half-sine wave is applied to the power module.

Looking at the pure I^2t values in Figure 4c at the end of the tests, there is no difference between all method and the one of the ASC in the application. The time progression of the $I^2t(t)$ curve is also significant, due to different peak temperatures. The initial and emulated $I^2t(t)$ curves are looking similar despite the smaller slope of the emulated one. In contrast, the shape of the $I^2t(t)$ curve of the 45 Hz surge current test clearly deviates from the others.

As can be seen in the figures 4e, the temperature curves show deviations from those of the ASC application. It is noticeable that the surge current causes an enormous overshoot. Unrealistic high temperatures can destruct the semiconductors or the module.

Testing ASC with focus on the temperature development

To determine the parameters of the laboratory test, thermal simulations (Fig. 4f) are carried out upfront with various emulated currents (surge current and pre-heated modules. The goal is to hit the temperature curve of the ASC in the application.

The emulated current in Figure 4b is the same as on the left and the "initial ASC current" is changed. The simplest way is to cover the current envelope and the parameters i_{max}, and f of the initial current. This approach is shown here because of its reasonably good temperature matching and its simplicity. With the surge current test, a 45 Hz half-sine wave with lower i_{max} is applied to the power module to hit the envelope of the temperature.

The $I^2t(t)$ and temperature curves on the right are calculate in the same way as on the left. The matching of $I^2t(t)$ curve plays a minor role. The $I^2t(t)$ curves shown in Fig. 4d have bigger differences at the end.

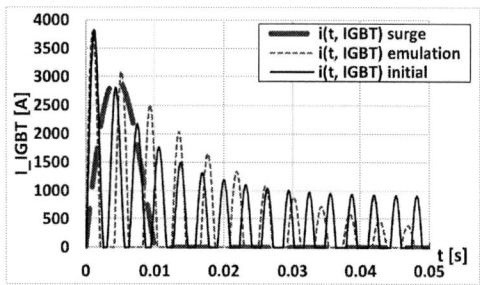

(a) Initial ASC current (320 Hz), Surge-current (50 Hz) and the emulated current (240 Hz)

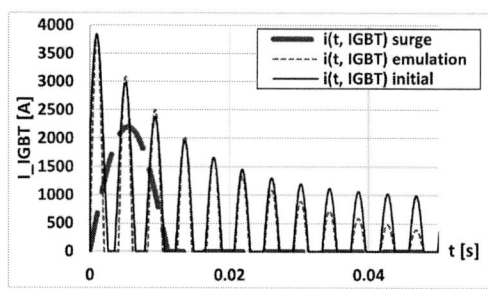

(b) Initial ASC current (240 Hz), Surge-current (45 Hz) and the emulated current (240 Hz)

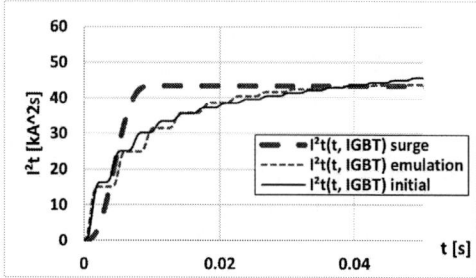

(c) $I^2t(t)$ values of ASC current (320 Hz), Surge-current (50 Hz) and the emulated current (240 Hz)

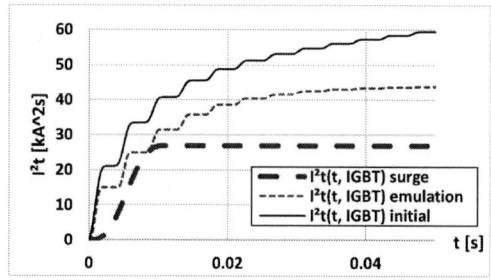

(d) $I^2t(t)$ values of initial ASC current (240 Hz), Surge-current (45 Hz) and the emulated current (240 Hz)

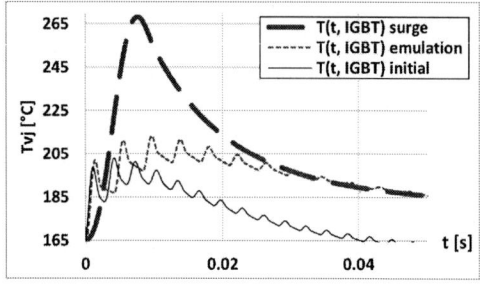

(e) Module heating with Initial ASC current (320 Hz), Surge-current (50 Hz) and the emulated current (240 Hz)

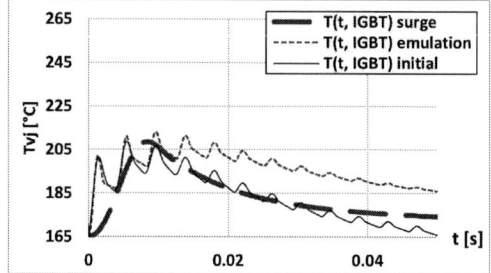

(f) Module heating with initial ASC current (240 Hz), Surge-current (45 Hz) and the emulated current (240 Hz)

Fig. 4: Comparison of $I^2t(t)$, time-dependent currents, and the virtual junction temperatures with original ASC , LC resonator and surge current tests.

Comparing the different focal points

As seen in Figures 4e and 4f, the temperatures on the right side (Fig.4f) are matching better to each other despite different I^2t values than those on the left side (Fig. 4e). Because of different thermal fields due to different cooling and heat generation conditions, the amount of stored heat can also be different without significant temperature rises.

The emulated current of Figure 4b causes the same junction temperature as the initial ASC current in the first sine half-waves. However, after reaching similar peak values, the temperature in the test with emulated current, "T(t,IGBT) emulation," decays much slower. The temperature development of the surge current shows in the beginning a bigger tracking error because temperature rise is delayed.

The analysis of the temperature shows that the parameters have to be set to hit the best possible match of the temperatures. Therefore, the emulated current could cover at least the envelope of the initial current. A finer tuning of the damping of the emulated current or a turn off the DUT in the right moment would lead to the best result.

The $I^2t(t)$ and I^2t are only sufficient values for tests if the cooling and heating is identical.

The correlation of temperature and current (Fig.:4) has to be considered, as well. As shown in Fig 4f, the corresponding thermal responses and the emulated current are almost identical to those of the initial current for the first sine half-waves. This is the range of interest, due to occurring simultaneously very high current amplitudes and temperatures.

It is also possible to test the peak current and the correlated heating of the semiconductor for a SiC MOSFET or a diode-IGBT pair in only one test.

Summary

A novel test method for testing the ASC fault condition, without the original inverter and drive train, in the laboratory is described here. It is shown how the original ASC current can be emulated with a resonant circuit and how the elements of the resonant circuit and the pre-charge voltage are calculated. In the laboratory test described in this article, it is shown that the novel method can test a wide range of different ASC and different module types such as Si and SiC (Tab. I).

Conclusion

It is possible to test all critical values of the ASC ($i_{max\ D,IGBT}$, T_{max}). Therefore, the current shape must be emulated accurately and the I^2t value can be neglected.

It is shown that the method allows a more accurate reproduction of the ASC temperature profile in the range of interest as compared to surge current tests.

Because of similar currents and temperatures in the range of interest, the novel method can test the ASC in the lab best. The new method benefits from the similarity with the ASC in application.

Outlook

In order to strengthen the evidence shown in this method, further investigations will be conducted. Some of these investigations include FEM simulations with dynamic ASC currents, further reliability studies, testing of the limits, and analysis of the destruction patterns and measurement of T_{vj}.

References

[1] Aravind Ramesh Chandran, Martin D. Hennen, Antero Arkkio and Anouar Belahcen, Safe Turn-off Strategy for Electric Drives in Automotive Applications, Published in: IEEE Transactions on Transportation Electrification (Early Access), DOI: 10.1109/TTE.2021.3104461
[2] Poulain, Christophe & Metz-Noblat, B. & Dumas, F.. (2005). Calculation of short-circuit currents.
[3] Arne Bieler, Ole Mühlfeld, Analytical Modelling of Dynamic Power Losses Inside Power Modules for 2-Level Inverters, PCIM Europe 2018, 5 – 7 June 2018, Nuremberg, Germany
[4] Antoni Ruiz et al.: Active Short Circuit Capability of Half¬-Bridge Power Modules Towards E¬-Mobility Applications, PCIM Europe digital days 2021, 3 – 7 May 2021
[5] Hayato Nakano et al.: Impact of I2t capability of RC-IGBT and Leadframe combined structure in xEV active short circuit survival, PCIM Europe 2018, 5 – 7 June 2018, Nuremberg, Germany

Short Circuit Performance and Current Limiting Mode of a Monolithically Integrated SiC Circuit Breaker for DC Applications up to 800 V

Norman Boettcher[*], Taro Takamori[‡], Keiji Wada[‡], Wataru Saito[§], Shin-ichi Nishizawa[§] and Tobias Erlbacher[*,¶]

[*] Fraunhofer Institute for Integrated Systems and Device Technology (IISB)
Schottkystraße 10, 91058, Erlangen, Germany
Phone: +49 9131-761 605
Email: norman.boettcher@iisb.fraunhofer.de
URL: https://www.iisb.fraunhofer.de

[‡] Tokyo Metropolitan University, Dep. of Electrical Engineering and Computer Science
1-1 Minami-Osawa, Hachioji-shi, 192-0397, Tokyo, Japan

[§] Kyushu University, Research Institute for Applied Mechanics, Renewable Energy Center
6-1 Kasuga-koen, Kasuga-shi, Fukuoka 816-8580, Fukuoka, Japan

[¶] Friedrich-Alexander University, Chair of Electron Devices (LEB)
Cauerstraße 6, 91058, Erlangen, Germany

Acknowledgments

The authors would like to thank H. Mitlehner, M. Rommel, A. Hürner, N. Kaminski, A. Würfel, J. Erlekampf and the π-Fab personnel for fruitful discussions and their efforts towards fabrication of these novel SSCB devices. Moreover, the authors would like to gratefully acknowledge that this work was made possible by sponsorships of Japan Society for the Promotion of Science (JSPS) and German Ministry of Education and Research (BMBF) under grant 03INT501BC "SiC-DCBreaker".

Keywords

≪Solid-State Circuit Breaker (SSCB)≫, ≪Current limiter≫, ≪Short circuit≫, ≪Self-sensing control≫, ≪Power semiconductor device≫, ≪Silicon Carbide (SiC)≫, ≪JFET≫.

Abstract

This paper presents the short circuit performance of a novel SiC circuit breaker device, which is based on the "thyristor dual" functionality. The developed device structure is motivated with regard to manufacturing aspects and electrical requirements. Furthermore, the basic "thyristor dual" operation is elaborated on the basis of quasi-static electrical measurements of a fabricated prototype. The proposed self-sensing and self-triggering devices make auxiliary circuitry like sensors and micro-controllers expendable and practically have no propagation delay. As a result, short circuit clearance within 122 ns at 800 V is demonstrated in experiments. Moreover, by utilisation of a third device terminal, a temporary current limiter functionality can be obtained. The scalability of the current limit value is discussed on the basis of measurements in time domain. The maximum current limit value achieved is 7.4 times higher than the trigger current level of the circuit breaker device. Additionally, the same circuit configuration which is used for the current limiting mode, allows to remotely reset the circuit breaker after it turned to blocking-state. This opens up a wide range of possibilities to enhance the circuit breaker with intelligent functionalities.

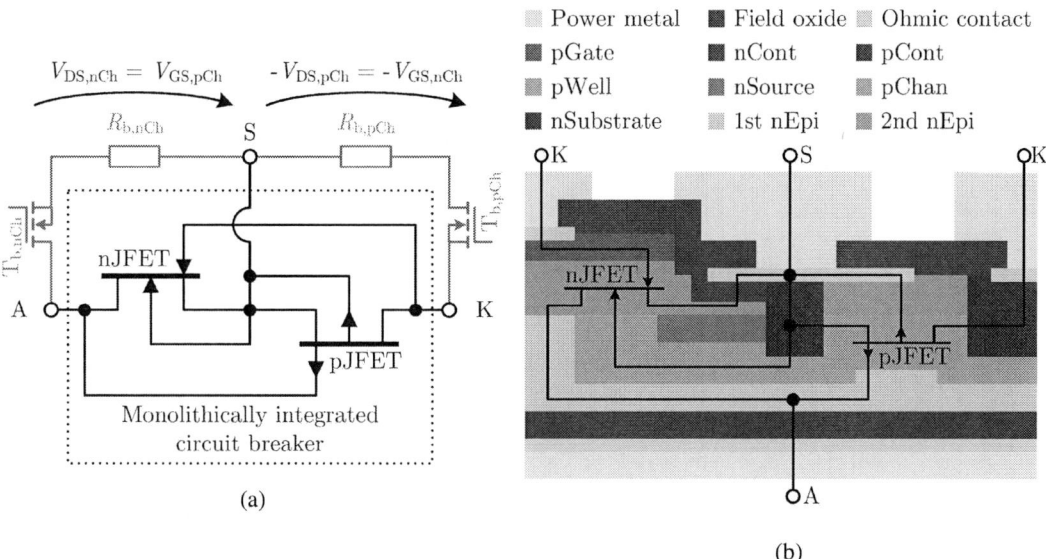

Fig. 1: Equivalent circuit diagram (a) and schematic cross section (b) of the proposed monolithically integrated circuit breaker. The bypasses formed by the transistors $T_{b,nCh}$ and $T_{b,pCh}$ and the resistors $R_{b,nCh}$ and $R_{b,pCh}$ in (a), are used for the proposed current limiter functionality and are not present in default circuit breaker configuration. The indices "nCh" and "pCh" represent nJFET and pJFET, respectively. In (b), the prefixes "n" and "p" indicate n-type and p-type regions, respectively.

Motivation

In recent years, a trend towards power electronic DC-applications exhibiting several hundred volts DC-link voltage is observed. The development of adequate solid state circuit breakers (SSCB) pursued various approaches to overcome challenges regarding arching, complexity and response time [1, 2]. However, even highly integrated state-of-the-art SSCB concepts require sense, supply or drive circuits in addition to the semiconductor switch [3, 4]. Moreover, especially in DC micro-grid applications, inrush currents much higher than the nominal current level occur during the charging period of power converter input capacitances. These currents must be tolerated by the circuit breaker without triggering, which poses further difficulties in terms of system level design [5]. With the discovery of the "thyristor dual" functionality, a potential two-pole device level solution with an intrinsic trigger mechanism has been introduced [6]. Promising characteristics have been observed in experiments at up to 400 V using discrete SiC JFET devices in cascode configuration [7–9]. In an effort towards monolithic integration, TCAD modelling was utilised on a topology suitable for 900 V DC-applications to discuss design and fabrication implications [10, 11]. Finally, the first physical demonstration of a simple two-pole "thyristor dual" device capable of several hundred volts blocking voltage has been realised in a 4H-SiC JFET technology [12]. In this work, we investigate the transient performance of those devices in a short circuit experiment at 800 V DC-link voltage. Furthermore, a temporarily activatable and scalable current limiting function by utilisation of a third device terminal is introduced, which also allows for remote reset of the SSCB.

Monolithically Integrated SiC Circuit Breaker Technology

As shown in Fig. 1a, the "thyristor dual" concept is realised by employing an n-channel JFET (nJFET) and a p-channel JFET (pJFET). Note, that the bypasses formed by the transistors $T_{b,nCh}$ and $T_{b,pCh}$ and the resistors $R_{b,nCh}$ and $R_{b,pCh}$, respectively, are not present in default configuration. The JFETs are arranged in a series configuration, where each gate is interconnected to drain of the other JFET. In this arrangement, intrinsically over-current triggered and self-sustained blocking operation is achieved, which is caused by the positive feedback of forward and control voltage (V_{DS} and V_{GS}) of both JFETs [11].

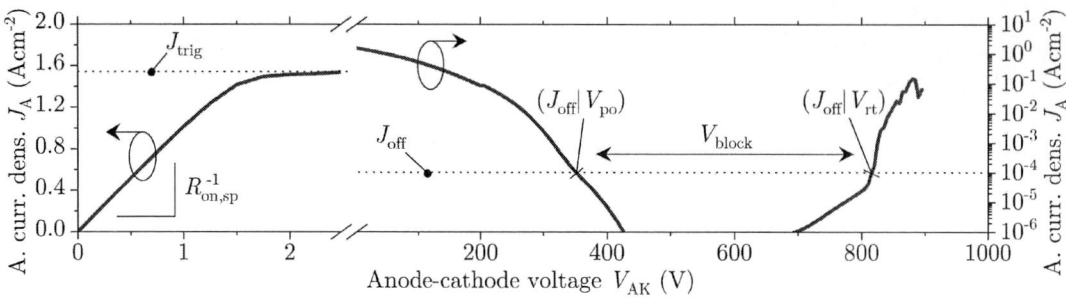

Fig. 2: Measured quasi-static output characteristics $I_A(V_{AK})$ of a proposed monolithically integrated SSCB device. In the low voltage region ($V_{AK} \leq 1\,V$), a static on-state resistance R_{on} is observed. For $I_A \leq I_{off}$, the SSCB is considered in blocking-state. Note, that $I_{off} = 1\,mA$ is chosen rather arbitrarily.

With respect to 800 V DC-applications, several hundred volts must be expected at the JFET gates after the SSCB turns to blocking-state. The schematic cross section of the device topology depicted in Fig. 1b is specifically developed to provide sufficient voltage sustainability at the pJFET gate structure. Regarding fabrication, the employed 4H-SiC technology is based on ion implantation, re-epitaxy and SiC dry etching. Ion implantation of nitrogen and aluminium is used to create n-doped (nSource and nCont) and p-doped regions (pWell, pChan, pGate and pCont), respectively. After finishing the implantations in the first n-type epitaxial system (1st nEpi), a second epitaxial layer is grown (2nd nEpi). The nJFET MESA gate structure is created by ion implantation of pGate and subsequent SiC dry etching. The ohmic contact region in the center of the unit cell allows carriers to overcome the natural pn-junction at the common source terminal (S) of both JFETs. Ultimately, the high potential anode terminal (A) is located at the device bottom side, whereas the low potential cathode (K) and the source terminal (S) are accessible from the top side. As can be obtained from the inserted equivalent circuit diagram, the pJFET gate is facing towards the 1st nEpi drift region. Since the majority of the blocking voltage can be expected to drop across the drift region, high pJFET gate voltage sustainability is obtained.

The measured curves $J_A(V_{AK})$ shown in Fig. 2 represent the quasi-static output characteristics of the SSCB devices investigated in this study. For anode current density values J_A below a certain trigger current density value J_{trig}, the devices operate in a current controlled linear on-state region. The emphasised linear region reveals a static specific on-state resistance $R_{on,sp}$. By exceeding J_{trig}, the intrinsic pinch-off mechanism is triggered and the device turns to self-sustained blocking-state, blocking the DC-link voltage V_{bat}. After triggering, the $J_A(V_{AK})$ trajectory strongly depends on the circuit configuration [13]. However, as long as V_{bat} remains within the blocking voltage window V_{block} between the pinch-off voltage V_{po} and the reach-through voltage V_{rt}, the blocking-state is maintained. Note, that $J_{off} = 100\,\mu A\,cm^{-2}$ is chosen rather arbitrarily.

Experimental

Measurement Environment

For the experiments in this study, the circuit configuration depicted in Fig. 3a is used. By turning on transistor T_P, the load current I_L is ramped up over time according to

$$I_L(t) = \frac{V_{bat}}{R_{load}} \cdot \left(1 - e^{-\frac{t \cdot R_{load}}{L_{load}}} \right), \tag{1}$$

where $R_{load} = 183\,\Omega$ and $L_{load} = 9.6\,mH$ represent load resistance and load inductance, respectively. The on-state resistances of T_P and the SSCB can be assumed much smaller than R_{load} and are neglected with respect to eq. 1. During current ramping, the SSCB carries the full load current. Therefore, the blocking mechanism is triggered as soon as I_L reaches the absolute trigger current I_{trig}. After triggering, the current

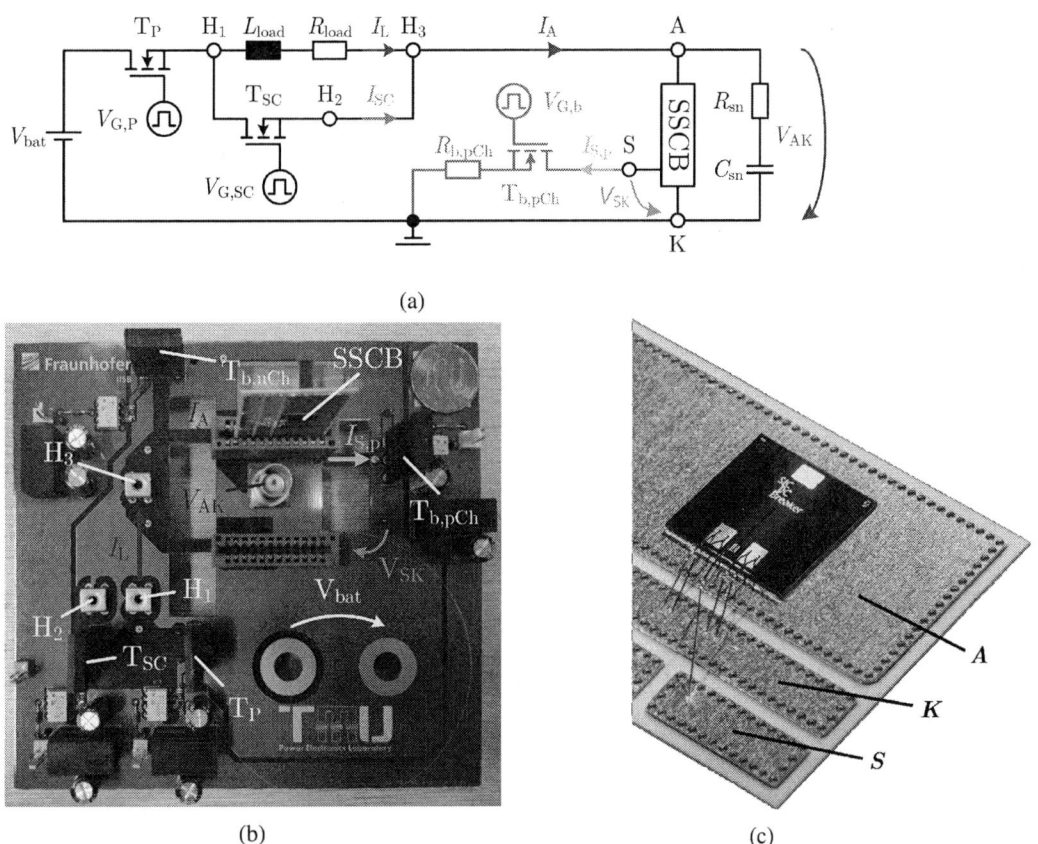

(a)

(b)   (c)

Fig. 3: Simplified circuit diagram (a) and corresponding circuit board (b) used for the experiments. The load is determined to $L_{load} = 9.6\,mH$ and $R_{load} = 183\,\Omega$. $1.2\,kV$ CoolSiC MOSFETs (IMW120R220M1H) are used for T_P, T_{SC}, $T_{b,nCh}$ and $T_{b,pCh}$. In (c) a photo image of a proposed SSCB chip mounted on DBC substrate is shown.

through the SSCB cannot increase any further and the device turns to blocking-state.

However, in case of the short circuit experiment, transistor T_{SC} is turned on before I_{trig} is reached, creating a short circuit in parallel to the load. For the experiments, the SSCB devices are mounted on DBC substrates as presented in Fig. 3c. The corresponding circuit board shown in Fig. 3b, is designed for investigation of up to two circuit breaker devices in parallel by utilisation of card edge connectors. The two-device option is not used in this study. Note, that an empty card edge connector socket represents an open circuit. The load is connected to the screw terminal hubs H_1 and H_3. The short circuit is established between the screw terminal hubs H_2 and H_3, with the aid of a copper wire of several centimeters length. The bypass transistors $T_{b,nCh}$ and $T_{b,pCh}$ are arranged as depicted in Fig. 1a but are only used to activate the proposed current limiting mode or for remote reset. The bypass resistors $R_{b,nCh}$ and $R_{b,pCh}$ are located at the circuit board back side. Furthermore, an RC-snubber can be found on the circuit board backside, close to the card edge connectors. The snubber parameters $R_{sn} = 4.7\,\Omega$ and $C_{sn} = 220\,pF$ are solely chosen to suppress oscillations in the very high frequency band. Due to the rather small value of C_{sn}, no snubber impact on switching speed and surge voltage is expected.

Short Circuit Experiment

At the beginning of the short circuit experiment, I_L is ramped according to eq.1. The short circuit is applied at $t = 0$, which corresponds to $V_{bat} = 800\,V$ and $I_L = 1\,A$. Fig. 4 shows the transient response of the SSCB device to the short circuit event. After T_{SC} is activated, the absolute Anode current $I_A = I_L + I_{SC}$ increases at a rate of $\frac{dI_A}{dt} = 0.66\,A\,ns^{-1}$ and reaches a maximum value of $I_{react} = 18.1\,A$ at approximately

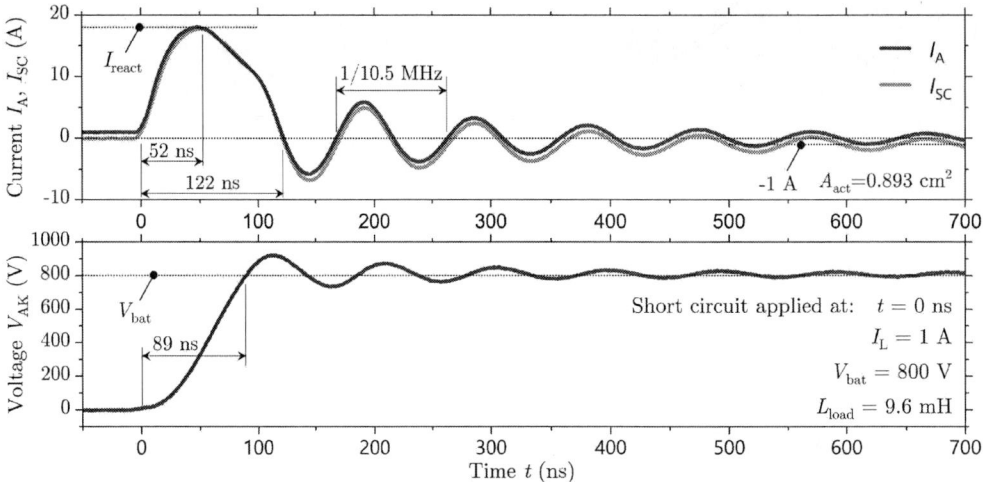

Fig. 4: Measured short circuit response of the proposed SSCB at a battery voltage of $V_{bat} = 800\,\text{V}$. The top graph shows the anode current $I_A(t)$ and the short circuit current $I_{SC}(t)$. The anode-cathode voltage $V_{AK}(t)$ is depicted in the bottom graph.

$t = 52\,\text{ns}$. Notably, I_{react} is more than 10 times higher than the trigger current level $I_{trig} = J_{trig} \cdot A_{act}$, where $A_{act} = 0.893\,\text{cm}^2$ represents the active area of the specific SSCB under test (see also Fig. 2). Considering that I_{react} is reached at the inflection point of $V_{AK}(t)$, the origin of these high current values is concluded to a displacement current. As soon as the SSCB device starts blocking the DC-link voltage, every capacitance between nodes A and K is charged to 800 V, including the snubber capacitance C_{sn}. The remaining parasitic capacitances C_{par} related to assembly, packaging and the SSCB chip are estimated to

$$C_{par} = I_{react} \cdot \frac{\partial t}{\partial V_{AK}} - C_{sn} = 1.24\,\text{nF}, \tag{2}$$

which seems reasonable, taking into account $C_{sn} = 220\,\text{pF}$ and the rather simple packaging approach.

Triggered by the high short circuit current, the SSCB blocks V_{bat} within 89 ns. The first zero-crossing of I_A is obtained 122 ns after the short circuit is applied. Subsequently, oscillations of 10.5 MHz occur for all investigated signals, where V_{AK} and I_A converge to V_{bat} and 0 A, respectively, verifying that the SSCB is in steady and self-sustained blocking mode. Since the SSCB is completely pinched off after triggering, I_L finds the free wheeling path through the body diode of T_{SC}. Therefore, the oscillation of I_{SC} converges to $-1\,\text{A}$ and, subsequently, declines over time until the energy stored in L_{load} is depleted.

Current Limiting Mode

Since the common source terminal of both JFETs (S) is not connected in default configuration, the "thyristor dual" circuit breaker resembles a two-pole device. As a consequence, the trigger current level cannot be externally controlled but is solely determined by the device design, which can be scaled by adequate variation of design parameters [11, 12]. Nevertheless, the source terminal (S) can be utilised to tolerate higher current values for a dedicated period of time (e.g. during current inrush). By employing a simple series connection of a transistor and a resistor as proposed in Fig. 1a between the terminals (A) and (S) or (S) and (K), a bypass in parallel to either of the JFETs is established. The bypass is activated by turning on either $T_{b,nCh}$ or $T_{b,pCh}$, whenever a tolerable over current event is estimated. Notably, only one bypass is supposed to be used at a time, while the other one represents an open circuit. In this configuration, the intrinsic trigger mechanism is suppressed, because the positive feedback of the JFET forward voltage V_{DS} and control voltage V_{GS} is prevented. Consequently, instead of turning to blocking-state, the SSCB is limiting the current I_A to the JFET saturation current level, which is controlled by the resistance values of either $R_{b,nCh}$ or $R_{b,pCh}$.

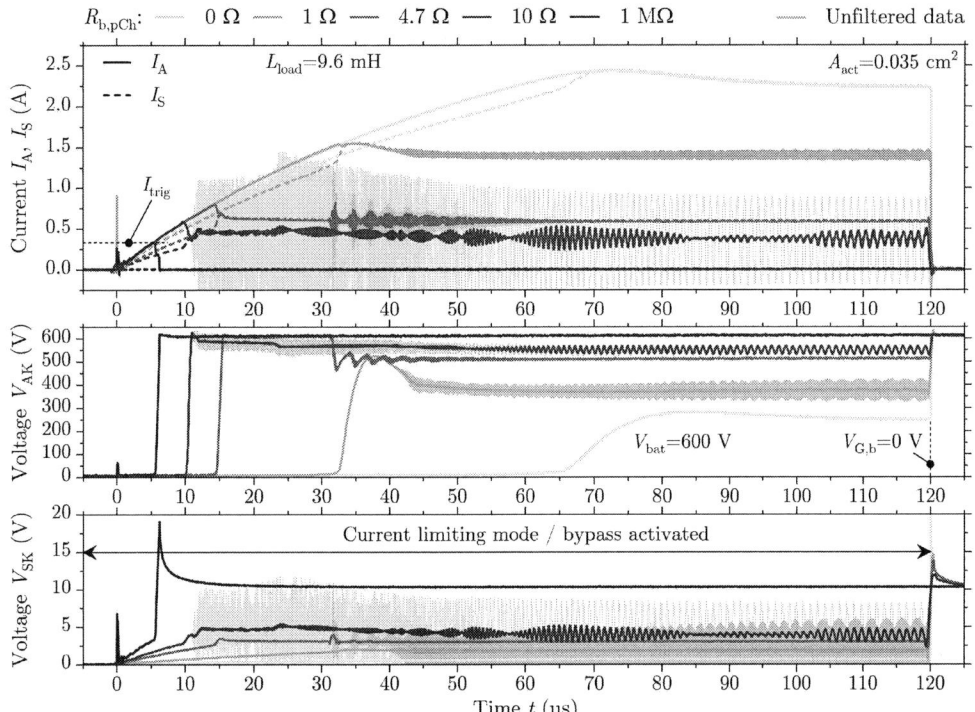

Fig. 5: Illustration of the current limit mode of the proposed SSCB in dependence of the bypass resistance value $R_{b,pCh}$. The graph in the top shows the anode current $I_A(t)$ and the bypass current measured at the source terminal (S) $I_S(t)$. In the middle the anode-cathode voltage $V_{AK}(t)$ is depicted. The course of the floating source-cathode $V_{SK}(t)$ can be obtained from the bottom graph.

The influence of $R_{b,pCh}$ on the current limit level is depicted in Fig. 5. Here, the course of I_A, I_S, V_{AK} and V_{SK} over time is shown in case of an activated bypass over $T_{b,pCh}$. In this experiment, the measurement set-up depicted in Fig. 3a is used as well. However, T_{SC} remains turned off. The default SSCB configuration is represented by $R_{b,pCh} = 1\,M\Omega$, showing the intrinsic trigger at $I_A \approx 330\,mA$, which corresponds to the design of this particular SSCB device. Therefore, the SSCB blocks the supply voltage of 600 V as soon as $I_A = I_{trig}$. For this case, the SSCB immediately stops conducting the current after triggering and I_L commutates to T_{SC} as discussed above.

By using an active bypass with a lower $R_{b,pCh}$ value, the SSCB keeps I_A at a constant value, instead of stopping the current flow. At $t = 120\,\mu s$, the bypass is deactivated and the SSCB device is turning to blocking-state since I_A is higher than the trigger current level. Please note, that heavy oscillation occurred during the current limiting experiment. Therefore, a moving average is applied to particular sets of the measurement data. The unfiltered data is depicted in grayscale. Nevertheless, the results sufficiently serve the purpose of illustrating the general behaviour in current limiting mode. As can be observed, the value to which I_A is limited increases as $R_{b,pCh}$ decreases, up to 2.44 A for $R_{b,pCh} = 0\,\Omega$, which is 7.4 times higher than I_{trig}. During the current ramping, the majority of the current, which is measured at the anode terminal (A) is leaving the SSCB at the source terminal (S). The difference $I_A - I_S$ is assumed to flow through the pJFET according to the current divider given by the pJFET channel resistance and $R_{b,pCh}$. As soon as I_A reaches the maximum value, the SSCB forward voltage V_{AK} significantly increases and both JFETs can be considered saturated. In saturation, the pJFET channel resistance is much higher than $R_{b,pCh}$. Consequently, the entire current flows through the bypass and I_S becomes equal to I_S.

In general, the current limit value I_{lim} for an ohmic current limiter can be calculated to

$$I_{lim} = \frac{V_{bat}}{R_{lim}}, \tag{3}$$

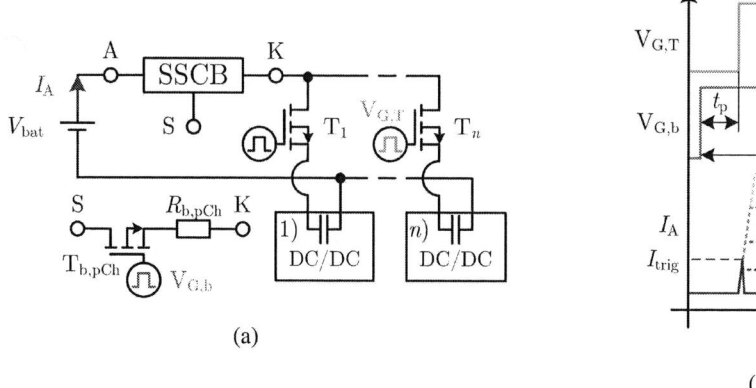

(a)

(b)

Fig. 6: Schematic circuit diagram (a) and corresponding schematic gate signals (b) to illustrate the utilisation of the proposed current limiting mode.

where R_{lim} is the current limiting resistance. In the proposed case, R_{lim} results from the series connection of R_{load}, $R_{\text{b,pCh}}$ and the channel resistance of the nJFET. The on-state resistance of T_P and $T_{\text{b,pCh}}$ may be neglected with respect to eq. 3. Note, that in current limiting mode, the JFETs are operating in saturation region. Therefore, the nJFET channel resistance is determined by V_{SK} and, hence, strongly depends on $R_{\text{b,pCh}}$. As a result, I_{lim} is not linearly scaling with $R_{\text{b,pCh}}$.

Such functionality is particular useful during inrush current events in DC-networks. In Fig. 6a, a schematic circuit diagram of a DC-network consisting of the proposed SSCB with applied (S)-(K) bypass and a number of n DC/DC converters. By turning on the power switch T_n, the input capacitance of the corresponding DC/DC converter is charged to V_{bat}. During this charging period, current values much higher than I_{trig} can occur, which may cause the SSCB to turn to blocking-state. This case is illustrated in Fig. 6b by the purple I_A curve. Applying the bypass for the current limit mode as discussed above, allows the SSCB to tolerate these high currents without triggering. Moreover, the duration of the charging period can be manipulated by variation of $R_{\text{b,pCh}}$ as shown by the dotted blue curves. As illustrated by the curves of $V_{\text{G,T}}$ and $V_{\text{G,b}}$, the current limiting mode must be activated shortly before the inrush event is expected to occur and last at least for the time period t_{CL}, which corresponds to the estimated charging duration. For the pre biasing period t_p, several hundred nanoseconds are sufficient as will be demonstrated in the next chapter.

Remote Circuit Breaker Reset

In addition to the current limiting functionality, the proposed bypass configuration is capable to remotely reset the circuit breaker from blocking-state to on-state by applying a $V_{\text{G,b}}$ pulse of several micro seconds. In Fig. 7, the gate signals $V_{\text{G,P}}$, $V_{\text{G,SC}}$ and $V_{\text{G,b}}$, the current signals I_L, I_A and I_{SC} and the voltage signal V_{AK} are shown for a sequence of an intrinsic trigger, followed by a remote reset and a subsequent short circuit event at $V_{\text{bat}} = 800\,\text{V}$. Turning on T_P in phase I leads to a significant current spike due to current inrush, which potentially triggers the SSCB as discussed above. For this reason, the current limiting mode is activated by turning on $T_{\text{b,pCh}}$ for a period of $t_{\text{CL}} = 600\,\text{ns}$, $t_p = 100\,\text{ns}$ before T_P is turned on, preventing the trigger event. In phase II, I_L is ramped according to eq. 1. At the beginning of phase III, I_L reaches I_{trig} and the SSCB starts blocking V_{bat}. As can be obtained, I_A declines rather slowly during the charging period of the parasitic capacitances to $V_{\text{bat}} = 800\,\text{V}$ as discussed above. However, as soon as $V_{\text{AK}} = V_{\text{bat}}$, the body diode of T_{SC} is polarised in forward direction. Consequently, I_L commutates to T_{SC} and I_A drops abruptly. The corresponding maximum transients are determined to $\frac{dV_{\text{AK}}}{dt} = 1.72\,\text{kV}\,\mu\text{s}^{-1}$ and $\frac{dI_A}{dt} = -63.5\,\text{A}\,\mu\text{s}^{-1}$, respectively. The current turn-off in phase III corresponds to a clamped inductive switching event, which is completed within $0.86\,\mu\text{s}$ after the trigger current level is reached.

The demagnetisation of L_{load} over a period of approximately $75\,\mu\text{s}$ is emphasised in phase IV. Here, the SSCB remains in steady and self-sustained blocking mode. In phase V at $t = 100\,\mu\text{s}$, I_L declined to

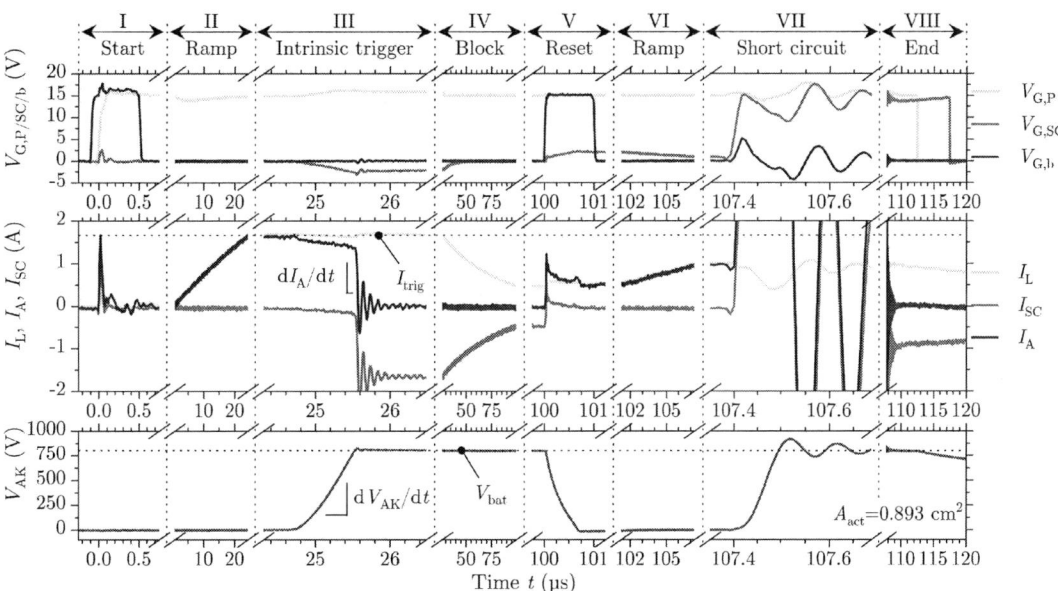

Fig. 7: Measured sequence of events including the intrinsic trigger, remote reset of the proposed SSCB and a short circuit event at $V_{bat} = 800\,\text{V}$. The graph in the top shows the gate signals $V_{G,P}(t)$, $V_{G,SC}(t)$ and $V_{G,b}(t)$. In the middle the current signals $I_L(t)$, $I_A(t)$ and $I_{SC}(t)$ are depicted. The course of $V_{AK}(t)$ can be obtained from the bottom graph.

approximately 500 mA. At this point a $V_{G,b}$ gate voltage pulse of 1 µs is applied to the bypass transistor $T_{b,pCh}$. Subsequently, I_L commutates back to the SSCB and V_{AK} drops to a corresponding on-state voltage level. The SSCB is reset and conducts I_L in the linear operating region (see Fig. 2). In phase VI, I_L ramps up over time again. The subsequent short circuit depicted in phase VII is identical to the short circuit event discussed above. In phase VIII, T_P is turned off to end the experimental sequence.

Conclusion

The short circuit performance of a novel self-sensing and self-triggering solid state circuit breaker device is presented in this work. A reaction time of 52 ns to a short circuit event at 800 V DC-link voltage and 1 A short circuit current is demonstrated with the aid of experiments. Short circuit clearance is achieved within 122 ns. These results represent the fastest response among comparable studies published so far. Notably, the two-pole devices ensure safe circuit breaker operation without the necessity of an additional power supply or other auxiliary circuitry. However, in order to obtain additional functions (e.g. scalable current limiter, remote reset), a simple series connection of a transistor and a resistor can be added between the floating source terminal and the anode or cathode terminal. With regard to the oscillation behaviour during current limit mode and considering the rather low trigger current values of the prototypes investigated in this study, the potential of this technology is yet to be exploited in order to be competitive on the market. However, by scaling the trigger current value to several 10 A, from our findings we believe, the proposed device topology is a promising candidate for DC applications in e-mobility and DC micro-grid environments.

References

[1] S. M. Sanzad Lumen, R. Kannan and N. Z. Yahaya, "DC Circuit Breaker: A Comprehensive Review of Solid State Topologies," *IEEE Intern. Conf. on Power and Energy (PECon)*, Penang, Malaysia, Dec. 2020, pp. 1-6, doi: 10.1109/PECon48942.2020.9314300.

[2] R. Rodrigues, Y. Du, A. Antoniazzi and P. Cairoli, "A Review of Solid-State Circuit Breakers," *IEEE Trans. on Power Electronics*, vol. 36, no. 1, pp. 364-377, Jan. 2021, doi: 10.1109/TPEL.2020.3003358.

[3] D. He, Z. Shuai, Z. Lei, W. Wang, X. Yang and Z. J. Shen, "A SiC JFET-Based Solid State Circuit Breaker With Digitally Controlled Current-Time Profiles," *IEEE Journal of Emerging and Selected Topics in Power Electronics*, vol. 7, no. 3, pp. 1556-1565, Sept. 2019, doi: 10.1109/JESTPE.2019.2906661.

[4] Z. Miao, G. Sabui, A. Moradkhani Roshandeh and Z. J. Shen, "Design and Analysis of DC Solid-State Circuit Breakers Using SiC JFETs," *IEEE Journal of Emerging and Selected Topics in Power Electronics*, vol. 4, no. 3, pp. 863-873, Sept. 2016, doi: 10.1109/JESTPE.2016.2558448.

[5] D. Marroquí, J. M. Blanes, A. Garrigós and R. Gutiérrez, "Self-Powered 380 V DC SiC Solid-State Circuit Breaker and Fault Current Limiter," in IEEE Transactions on Power Electronics, vol. 34, no. 10, pp. 9600-9608, Oct. 2019, doi: 10.1109/TPEL.2019.2893104.

[6] J. L. Sanchez et al., "A new high-voltage integrated switch: the "thyristor dual" function," 11th International Symposium on Power Semiconductor Devices and ICs. ISPSD'99 Proceedings (Cat. No.99CH36312), 1999, pp. 157-160, doi: 10.1109/ISPSD.1999.764086.

[7] A. Würfel, J. Adler, A. Mauder and N. Kaminski, "Over Current Breaker Based on the Dual Thyristor Principle," *Proc. 28th Intern. Symp. on Power Semiconductor Devices and ICs (ISPSD)*, Prague, Czech Republic, June 2016, pp. 143-146, doi: 10.1109/ISPSD.2016.7520798.

[8] Wuerfel A., J. Adler, A. Mauder and N. Kaminski , "High Speed Electronic Over Current Breaker for DC-Grids without Additional Sensing," *Proc. Intern. Exhib. and Conf. for Power Electronics, Intelligent Motion, Renewable Energy and Energy Management (PCIM)*, May Nuremberg, Germany, 2016, pp. 1-8.

[9] M. Albrecht, A. Hümer, T. Erlbacher, A. J. Bauer and L. Frey, "Experimental verification of a self-triggered solid-state circuit breaker based on a SiC BIFET," 2016 European Conference on Silicon Carbide & Related Materials (ECSCRM), 2016, pp. 1-1, doi: 10.4028/www.scientific.net/MSF.897.665.

[10] A. Huerner, T. Erlbacher, A. J. Bauer and L. Frey, "Monolithically Integrated Solid-State-Circuit-Breaker for High Power Applications," *Mat. Sci. Forum*, vol. 897, pp. 661-664, May 2017, doi: 10.4028/www.scientific.net/msf.897.661.

[11] N. Boettcher and T. Erlbacher, "Design Considerations on a Monolithically Integrated, Self Controlled and Regenerative 900 V SiC Circuit Breaker," 2020 IEEE Workshop on Wide Bandgap Power Devices and Applications in Asia (WiPDA Asia), 2020, pp. 1-6, doi: 10.1109/WiPDAAsia49671.2020.9360279.

[12] N. Boettcher and T. Erlbacher, "A Monolithically Integrated SiC Circuit Breaker," in IEEE Electron Device Letters, vol. 42, no. 10, pp. 1516-1519, Oct. 2021, doi: 10.1109/LED.2021.3102935.

[13] N. Boettcher, T. Takamori, K. Wada, W. Saito, S. -i. Nishizawa and T. Erlbacher, "Fabrication Aspects and Switching Performance of a Self-Sensing 800 V SiC Circuit Breaker Device," 2022 IEEE 34th International Symposium on Power Semiconductor Devices and ICs (ISPSD), 2022, pp. 261-264, doi: 10.1109/ISPSD49238.2022.9813628.

Application of a HV bipolar square-wave voltage generator for qualification and assessment of energy equipment

Rico Fischer-Baeumer[1], Kai Göhrmann[1], Konrad Domes[2], Benjamin Sahan[1], Christian Staubach[1]

[1]Hochschule Hannover
University of Applied Sciences and Arts
Ricklinger Stadtweg 120
30459 Hannover, Germany

[2]SAXOGY POWER ELECTRONICS
Dittesstr. 15
09126 Chemnitz, Germany

E-Mail: kai.göhrmann@hs-hannover.de, rico.fischer-baeumer@hs-hannover.de,
benjamin.sahan@hs-hannover.de, christian.staubach@hs-hannover.de, info@saxogy.de

Keywords

Partial discharge, Modular Multilevel Converters (MMC), Cascaded H-Bridge, Insulation, Condition monitoring

Abstract

The increasing use of wideband gap devices poses a major challenge for the insulation system of motor windings due to its steep dv/dt voltage slopes. This paper focuses on the application of a modular bipolar square-wave voltage generator, which was built at the Hochschule Hannover and is now used in the high-voltage lab. Exemplary measurement results related to qualification and condition monitoring are presented. It is shown that the stress of a bipolar square-wave with high frequency and steep voltage slopes on the insulation system is much higher compared to a standard sinusoidal 50 Hz wave.

1 Introduction

Due to ongoing developments in power electronics [1] and new power conversion technologies [2, 3], the insulation system of energy equipment, such as rotating machines, cables, transformers, bushings, etc, is increasingly stressed by inverter voltage pulses. Especially the advancing of fast switching wide bandgap devices must be considered by the providers of isolation materials such as lacquer ectara. It is known, that the repetition frequency, peak-to-peak-voltages and voltage gradient during switching have a major influence on specific parts of the insulation system [4]. In comparison to LV-insulation systems, only limited knowledge is present regarding performance and aging of MV- and HV-insulation systems. However, the steep dv/dt voltage slopes may lead to an accelerated aging of the insulation system [5, 6]. Partial discharges may occur, which not only damage the machine or lead to spontaneous failure, but also may represent a health hazard for people [7]. Therefore, the need to investigate ageing behaviour of insulation materials in MV- and HV-Inverter applications is increasing. Consequently, a test device is needed that simulates the pulsed voltages in inverter operation realistically. This paper focuses on isolation materials of twisted pairs and insulation systems for rotating machines. In a first step a short introduction to a built bipolar square-wave voltage generator is given and the challenges with capacitive loads and their short circuit behaviour are investigated. Further the effects of a square-wave voltage on twisted pairs and insulation systems of generator bars are discussed. The differences between sinusoidal and square-wave stress on these insulations systems are shown.

2 Modular bipolar square-wave voltage generator

The topology used for the following tests is based on the well-known concept of cascaded H-Bridges as shown in Fig. 1 [8]. Each of these power stages uses a DC-Link voltage of $V_d = 1200$ V and a 1700 V IGBT H-Bridge module (see [9]) to generate a bipolar square-wave with a peak-to-peak voltage of $V_{pp} = 2400$ V. This is achieved by switching the individual IGBTs in the H-Bridge module complementary. Multiple of these power stages can be connected in series to increase the output voltage to the desired level. This way the DC- potential of one power stage matches the DC+ potential of the stage below it. The output voltage of the inverter consequently results in the sum of all the DC-link voltages: $\pm V_d \cdot n$, with n being the total number of power stages. The stages need to be isolated to each other and to earth. To do so the AC power supply which powers the DC-Link capacity of all the stages is isolated through transformers followed by rectifiers. Both in combination provide the desired DC voltage to the DC-Link Capacitors of each power stage. Handling the coupling capacities of the transformers can be challenging as they lead to high stray currents, but this topic is out of scope for this paper.

All power stages are controlled by a single microcontroller, which generates the necessary control signals to switch the IGBTs of each power stage complementary and at the same time. These control signals are transmitted via an optical fiber to each power stage to maintain the former mentioned electrical isolation. The square-wave generator used in this paper consist of four power stages. This results in a maximum output voltage of $V_{out,max} = \pm 4800$ V respectively $V_{pp,max} = 9600$ V. The risetime is adjustable by varying the gate resistors of each driver stage.

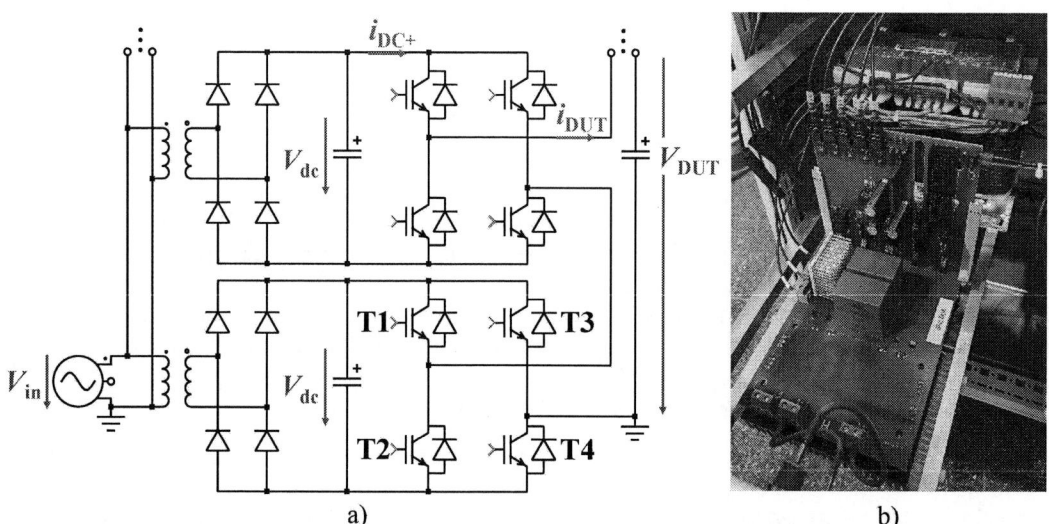

a)   b)

Fig. 1: a) Schematic of Modular square-wave generator and b) picture of one power stage

2.1 Switching characteristics under capacitive load

Typical datasheets of power semiconductors only list switching losses for currents as low as 10 % of the nominal current and inductive load. When testing insulations systems, the DUT typically behaves like a capacitive load with high ohmic resistance. Therefore, the currents experienced by the power semiconductor in this use case are not represented by the datasheet values. However precise values are needed to conduct simulations in development and for later thermal management of the power semiconductors.

To obtain better understanding of the switching losses under capacitive loads the IGBT module of one power stage is characterized with discrete capacitors with 55 pF, 100 pF, 1 nF and 3,3 nF as a load. The pF capacitors demonstrate the load of multiple twisted pair DUTs. The nF capacitors are in range of the capacitance of a generator bar. The rise time is set to approximately 150 ns and V_d to 1200 V. Furthermore, the IGBT module is tested at 100 °C to simulate operating conditions.

To analyze the dynamic losses the DC-link current (i_{DC+}, see fig. 1a) is measured with a Rogowski coil [10]. Additionally, the DUTs current are measured with a Pearson probe and the voltage above the IGBT and the DUT by differential probes [11–13]. To determine the needed Energy of the H-Bridge the product of the DC-link current and DC-link voltage of 1200 V is integrated. The switching losses are then estimated by evaluating the energy balance. Therefor the energy stored on the Capacitor $\int |V_{DUT}| \cdot i_{DUT}$ is subtracted from the previously determined DC-link energy. The difference is the switching losses of the IGBTs. The results of this procedure are shown in figure 2. The DC-link currents of the different capacitances are shown in figure 2a) and the corresponding energies in 2b). It is visible that the currents during the switching process are not negligible and therefore neither are the resulting losses. While aiming for a high frequent switching frequency these losses add up rapidly, even if they are in order of magnitude 10 times smaller than the minimum losses in the datasheet. This shows that an appropriate cooling system is needed at high switching frequencies and capacitive load.

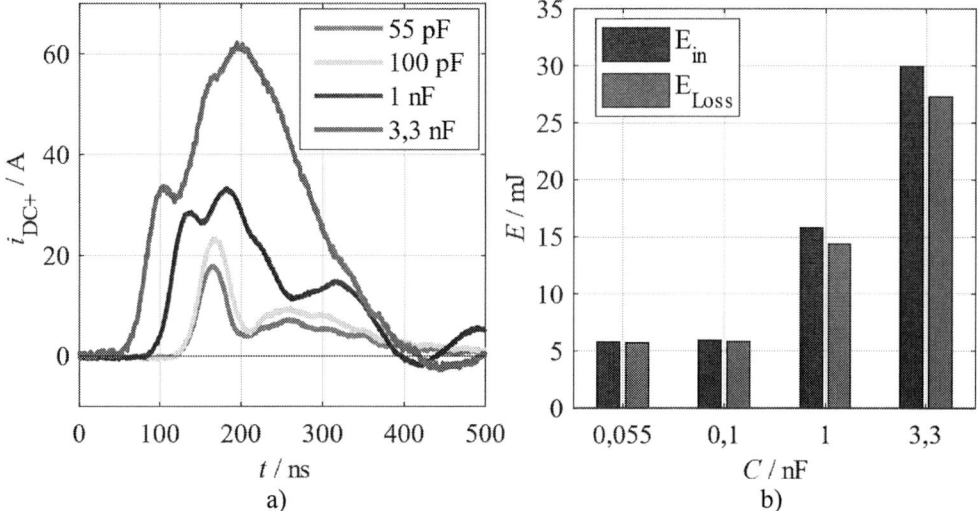

Fig. 2: a) DC-link current i_{DC+}, and b) corresponding input energy and loss energy per switching cycle of one H-Bridge

2.2   Breakdown behavior of the capacitive load

To analyze the lifetime of the test specimens, their breakdown behavior must be examined. In the worst case, they fail in a low ohmic short circuit. While the first laboratory prototype still employs standard IGBTs the usage of new SiC-MOSFET is planned. However, these devices show much lower short-circuit robustness. Typical allowable short circuit pulse duration is in the range of 2-3 μs.

As short circuit protection a shunt to measure the current in the DC- path in combination with a fixed voltage comparator is used as suggested in [10]. In contrast to typical gate drivers with DESAT-detection this approach allows a faster and more flexible protection.

An alternative over current protection is the use of melt-down fuses. This approach comes with ease of use but has several disadvantages: Each DUT needs its own separate fuse, and each fuse must be capable of withstanding the high frequency and high du/dt testing voltage in its entirety. This voltage can be several thousand Volt with rise times of 100 ns or less and as well as more than 20 kHz. Another disadvantage is that each fuse added to the testing setup also adds more parasitic inductance and capacity.

To investigate the breakdown behavior the following test procedure is used: Initial measurement with a double pulse to show the behavior under square-wave voltage stress when the DUT is known to be good. Afterwards the DUT is tested until end of life using accelerated aging with a high voltage AC source. Finally, the destroyed DUT is again tested with a double pulse square-wave voltage. For this test a single SiC-MOSFET H-Bridge is build using two 1200 V half-bridge modules [14]. This procedure is repeated for five test specimens.

Figure 3a) displays the initial measurement with a DC-link voltage of 900 V. All five DUTs show very similar characteristics here. The current wave form experiences a spike when the H-Bridge switches its polarity from -900 V to 900 V. The output current shows a short oscillation formed by the resonant tank of the capacitive twisted pair DUT and the cable inductance (length ~ 20 cm). This behavior represents the expectations.

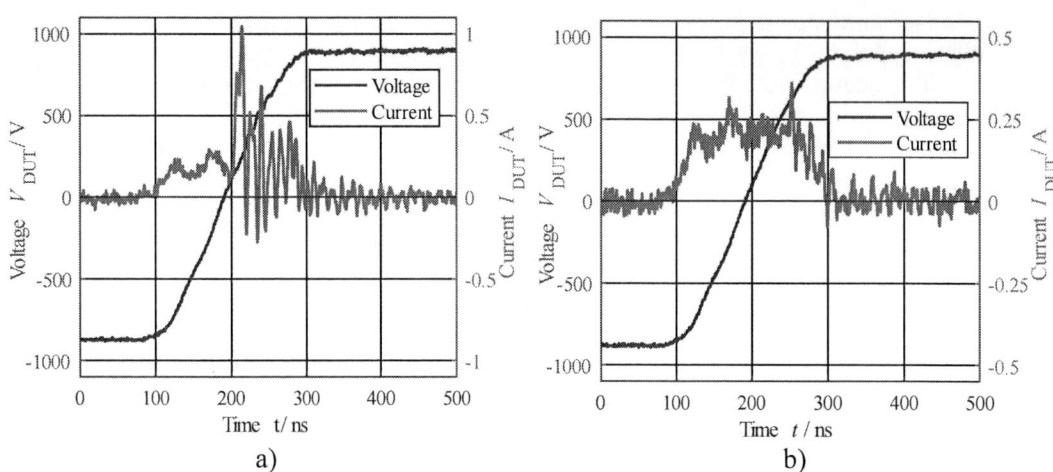

Fig 3: Voltage and current wave forms a one DUT a) before and b) after destruction but below breakdown voltage

After destroying the five DUTs a strong variation of the breakdown voltage has been observed. This can be explained by the length of the gap between the two strains of wire of the DUT where the defect accrued. Depending on the position of the defect the two strains of wire can be closer together or wider apart and the corresponding length of this gap and the air humidity while testing results in different breakdown voltage of the defect site of each DUT. Figure 3b) shows the waveforms of the DUT after the end-of-life test. No major abnormalities in the voltage and current wave forms compared to its intact state in Fig 3a) can be observed when setting the DC-link voltage to 900 V.

In contrast to this Fig. 2 shows the waveforms of a different DUT after destruction at a DC-link voltage $V_d = 645$ V. The insulation of the DUT breaks down and a low ohmic short circuit occurs. The current rises fast and reaches a peak value of 870 A before being shut down by the short circuit protection. The di/dt is only limited by parasitic inductance of the setup. It can be observed that the short circuit pulse duration is below 1µs which is well within the specification of the datasheet. However, no specific allowable number of short circuit events is specified in the datasheet and a case specific assessment between supplier and end user is necessary as stated in [15]. All tests are carried out at an ambient temperature of 22.5 °C and with a relative humidity of 31.3 %.

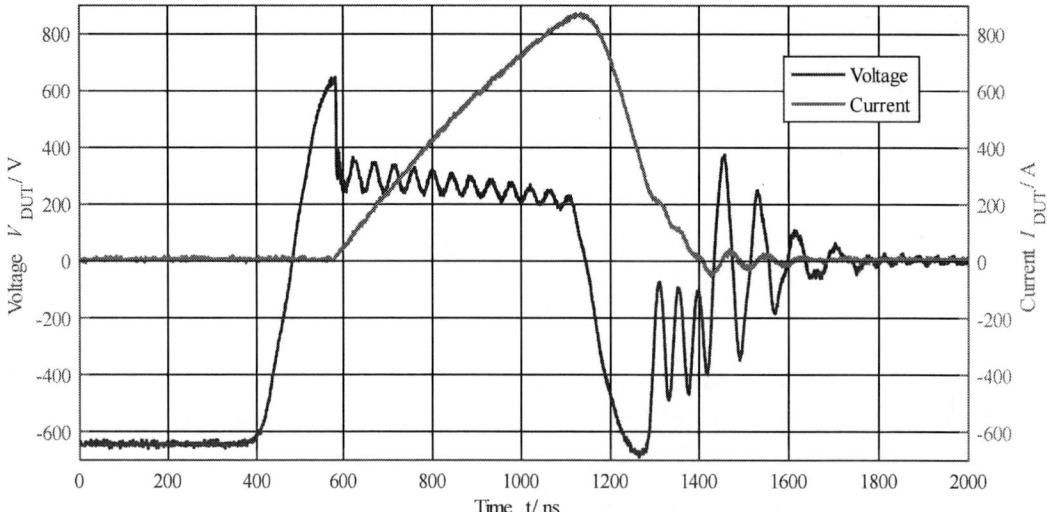

Fig 4: Current and voltage wave forms of short circuit operation of a twisted pair DUT

3 Influence of square-wave voltage with high dv/dt on the insulation of rotating machine windings

The characteristic inverter parameters used to drive rotating machines, i.e. repetition frequency, peak-to-peak-voltages and voltage gradient during switching can have a severe influence on specific parts of the insulation system [16, 17]. In comparison to LV-insulation systems, only limited knowledge regarding performance and aging of MV- and HV-insulation systems is available. Following some examples for application of the developed generator are presented.

A main challenge is to adjust the resulting stress for the individual DUT´s load. The resulting capacity and the parasitic inductivities can lead to oscillation voltages and currents and therefore imposes extra stress on the DUTs. It must be pointed out that the voltage of the unloaded square-wave generator is not comparable to the voltage with DUTs connected to it regarding voltage gradient and amplitude. In addition, the destruction of one or more DUT during the test may change the circuits behaviour as well and thereby changing the test conditions for the remaining DUTs.

3.1 PD-resistant HV-insulation system qualification – IEC 60034-18-42

In this case study a qualification of the stress grading system according IEC 60034-18-42 is presented [17]. It is known from other works, that especially the stress grading system of HV-insulation systems is exposed to high thermal and electrical stresses compared to AC operation [18–20]. This is mainly due to the conductive and semi-conductive materials used for electric field control in the slot and end-winding region [21]. Figure 5 shows the measurement setup for the stress grading system qualification according to IEC 60034-18-42. In this example 4 stages of the square-wave generator are used, which enables a maximum peak-to-peak voltage of about 9.6 kV.

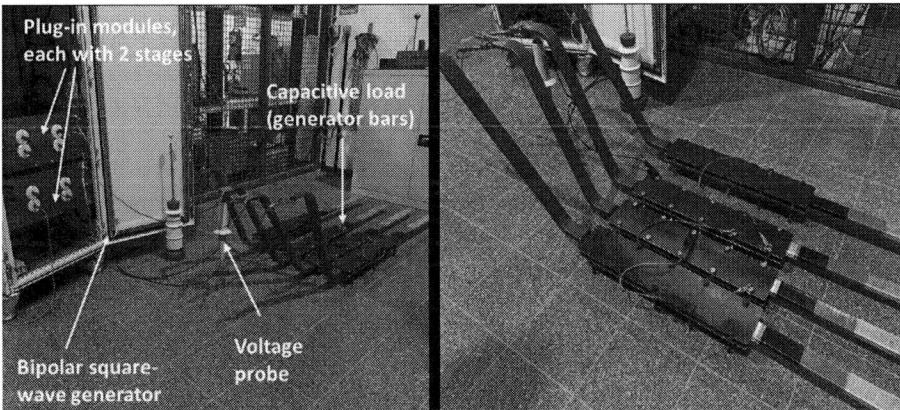

Fig. 5: Measurement setup for qualification according to IEC 60034-18-42

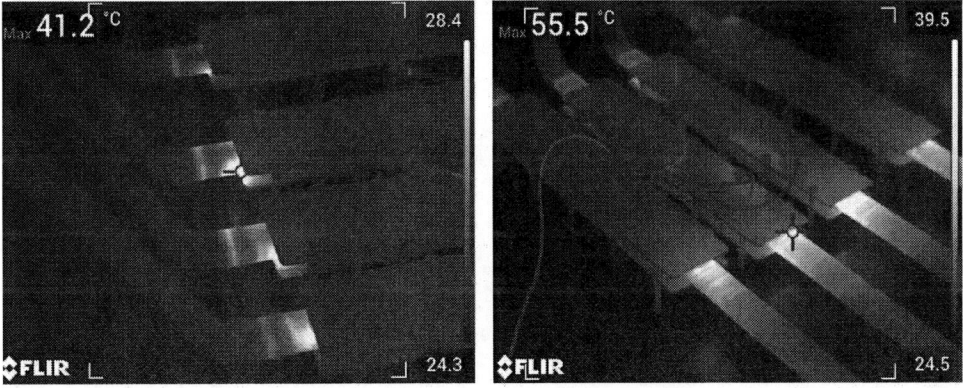

Fig. 6: Thermal images of both bars end indicating the areas exposed to the highest electric and thermal stresses caused by the inverter voltages at 1 kHz

Figure 6 presents thermal images of the test samples with clear hot-spots generated at the slot-exit area. This is mainly due to the high displacement currents generating ohmic loses in the conductive and semi-conductive materials even at quite low switching frequencies around 1-2 kHz.

As a result of these thermal stresses the insulation system is highly stressed and can age quite fast. By means of an UV-camera partial discharge activity is observable at the locations with the highest stress. This will result in visible deterioration due to chemical reactions and, see Figure 7.

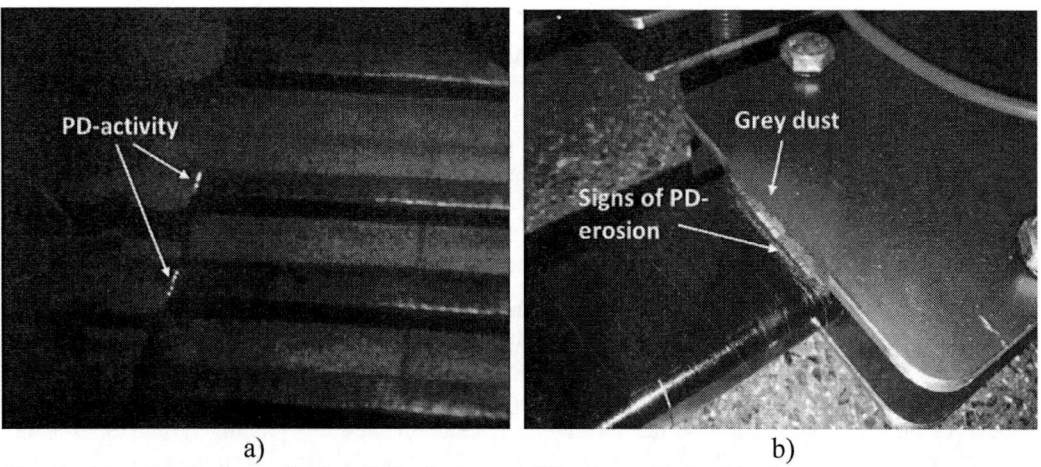

a)   b)

Fig. 7: a) Partial discharge (PD) activity during qualification with the bipolar square-wave generator detected via an UV-camera and b) visible deterioration of the insulation system

Additionally, due the characteristic of the developed generator, the influence of the most important parameters, such as peak-to-peak-voltage and switching frequency, on the resulting electric and thermal stress can be investigated. The graph in Figure 8 gives the results of the temperature increase at the bar hot-spot area depending on the peak-to-peak-voltage for different repetition frequencies. At higher voltages it is only possible to apply smaller frequencies due to the risk of burning caused by excessive loses in the HV-insulation system. The graph in figure 8 points out, that there is a quite significant influence of repetition frequency on the resulting temperature increase i.e., thermal stress.

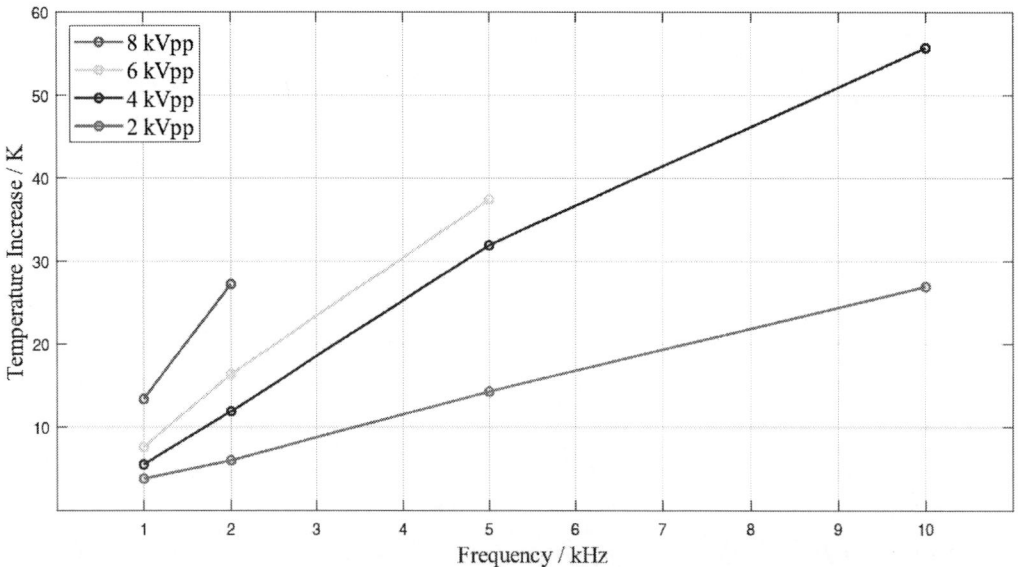

Fig. 8: Influence of repetition frequency on the temperature increase in the stress grading area

3.2 LV-wire insulation lifetime testing and PDIV of twisted pairs under square-wave and sinusoidal voltage

Another application of the developed bipolar square-wave generator is the wire insulation lifetime testing according to IEC 62068 [22]. twisted pairs are stressed with repetitive voltage impulses. The objective is to quantify the electric lifetime of the wire insulation for a given electric stress, the repetition frequency goes up to 20 kHz. Recommendations for appropriate voltage parameters are currently in discussion [23]. The twisted pair with the connections and the applied voltage is shown in Figure 9. Due to the field concentration between the wires quite significant PD-activity is generated at the triple-junction locations. Figure 10 presents visible and figure 11b) UV-pictures of the twisted pair during testing.

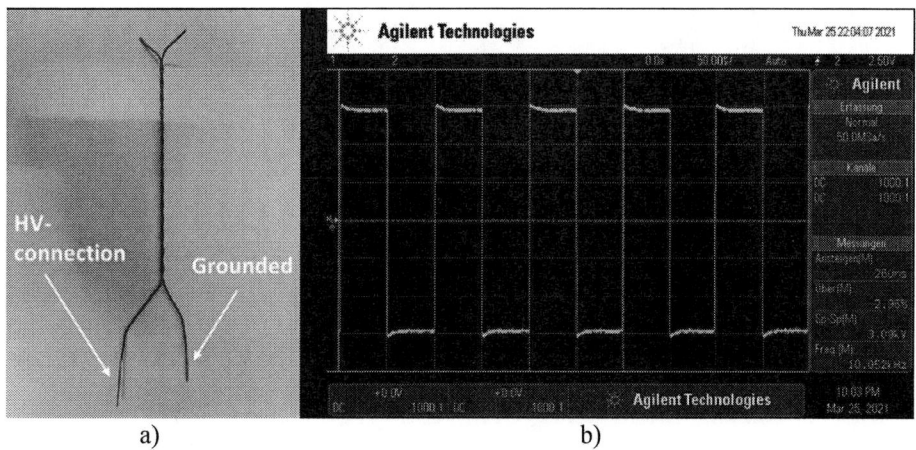

a)   b)

Fig. 9: a) Twisted pair and b) applied repetitive voltage impulses

To investigate the influence of square-wave voltage on PDIV five DUTs were placed on a retainer and stressed with square-wave voltage with 100 ns risetime and 20 kHz switching frequency. The PDIV for each of the five DUTs is noted separately. To have a comparison the same test is carried out again but this time using 50 Hz sinusoidal voltage.

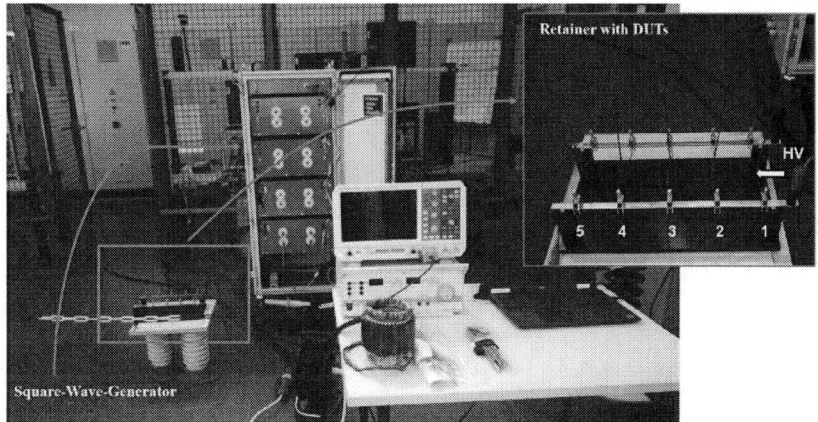

Fig. 10: Testing setup used for application of square-wave voltage

Under square-wave voltage stress the DUTs show vastly different PDIV. The peak-to-peak value of the PDIV is ranging from 1.46 kV to 1.67 kV. In comparison the PDIV while sinusoidal voltage is applied is uniform at the same voltage of 1.78 kV. It is noteworthy to point out that the PDIV is from 0.11 kV to 0,32 kV lower when the DUTs are stressed with square-wave voltage. Figure 11a) shows the PDIV for each of the five DUTs. Another noteworthy discovery is that in addition to the fact that each DUT has a different PDIV the corona discharge of each twist of one individual DUT is also not uniform when a square-wave voltage is applied. This can be seen in figure 11b). A voltage of 1.6 kV is applied to the retainer and therefore to each DUT. In accordance to figure 11a) DUT number 1 – 3 show no corona discharge as they are below the PDIV. DUT 4 and 6 show nonuniform corona discharge and even spots in the middle where no corona can be observed.

This concludes in the realization that the inherent structure of a twisted pair may not be ideal for lifetime testing of insulation systems and another testing method hast to be developed and adopted to achieve conclusive results.

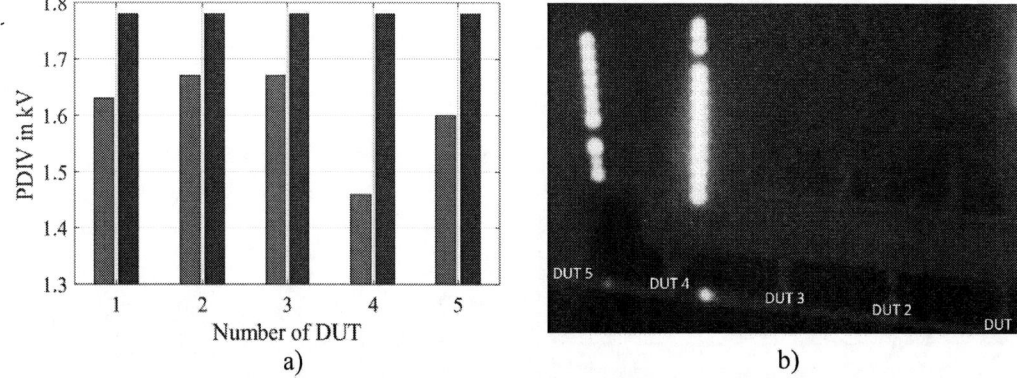

a)   b)

Fig. 11: a) PDIV of all five twisted pair DUTs under square-wave (red) and sinusoidal stress (blue), b) Nonuniform corona discharge of individual twists of a twisted pair DUT (same Voltage applied)

3.3 Offline PD-measurement with repetitive, impulse voltage

Comparable to the IEC standard for offline partial discharge measurements of rotating machines with sinusoidal voltages for condition assessment [24] currently a new standard related to repetitive impulse voltage excitation is discussed [25]. First investigations in our lab were conducted, see Figure 12. However, the challenge is not to generate but to measure the partial discharge activity accurately. This is due to the fact, that the expected frequency content of the partial discharge events is in similar range to the repetitive impulse voltage. Also, oscillation is likely to occur distorting the measurement results may occur. Therefore, further work is needed regarding this topic. Especially, precise boundary conditions and general guidelines in the relevant standard might be necessary.

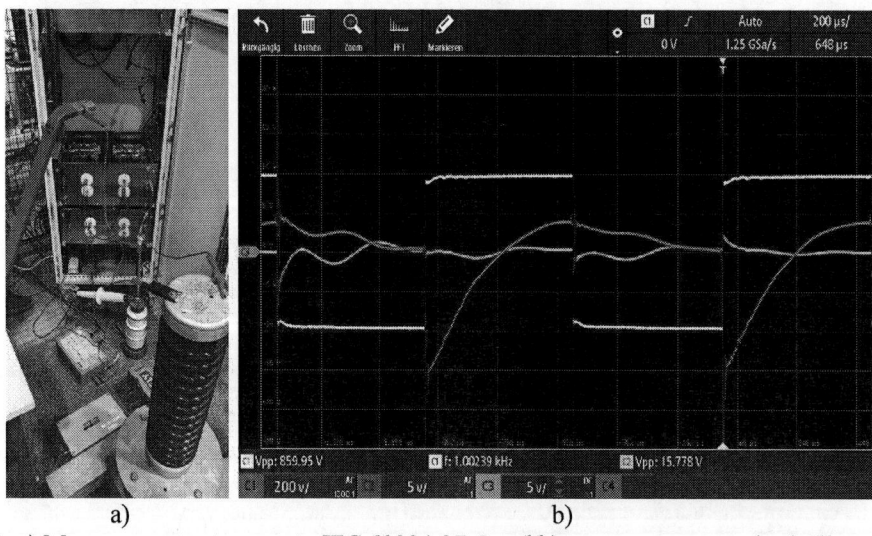

a)   b)

Fig. 12: a) Measurement setup acc. to IEC 60034-27-5 and b) measurement results (yellow: square-wave voltage, green: measured voltage of the measurement quadrupole, orange: output of PD activity channel of the measurement quadrupole

4 Conclusion

The approach of testing different insulation materials using a modular, cascaded H-Bridge voltage generator shows clear results. The stress caused by using square wave voltage differs from the stress caused by sinusoidal voltage. Especially generator bars experience a high-level thermal stress, which gets visible through the thermal imaging camera as well as the resulting grey dust. This increased stress is caused by high voltage slopes and the high switching frequency of the square-wave voltage. Standard partial discharge measurement is not possible without problems. This results in the need for further work regarding this topic.

Additionally, requirements for building a square-wave voltage generator are shown. At first, a superior short circuit detection is needed. Otherwise save operating conditions cannot be guaranteed and the low ohmic short circuit can easily destroy the power semiconductors. In addition, an adequate cooling solution for the power semiconductors is highly recemented as the switching losses under capacitive load cannot be ignored. Especially while testing generator bars with high capacity and high frequency. All these findings show clear trends and challenges, which will require further investigations.

5 Acknowledgements

This research project was carried out in the framework of the industrial collective research program ZIM. It was supported by the Federal Ministry for Economic Affairs and Climate Action (BMWK) through the AiF (German Federation of Industrial Research Associations eV) based on a decision taken by the German Bundestag. The authors would like to thank Stefan Reddig for his support.

6 References

[1] K. Bae et. al., *Current State and Development Trends of Insulation Systems in BEV Traction Motors Steered by Electric Powertrain Innovation: International Exhibition and Conference for Power Electronics, Intelligent Motion, Renewable Energy and Energy Management Proceedings, 3 - 7 May 2021.* Berlin: VDE Verlag, 2021.

[2] T. Hildinger, "Frades II- Europe's Largest and Most Powerful Doubly Fed Induction Machine, HydroVision International," Charlotte, NC, USA, 2018.

[3] C. Staubach and T. Hildinger, "Innovative Technologie eines drehzahlvariablen Pumpspeicherkraftwerks unter Berücksichtigung des hochspannungstechnischen Isoliersystems, Generatoren in konventionellen Kraftwerken, Windparks und Wasserkraftwerken," 14. Essener Tagung, 2020.

[4] *Rotating electrical machines – Part 18-42: Partial discharge resistant electrical insulation systems (Type II) used in rotating electrical machines fed from voltage converters - Qualification tests,* IEC 60034-18-22, 2017.

[5] Institute of Electrical and Electronics Engineers, *2020 IEEE Electrical Insulation Conference (EIC).* Piscataway, NJ: IEEE, 2020.

[6] Marco Denk and Mark-M. Bakran, "Partial Discharge Measurement in a Motor Winding fed by a SiC Inverter – How critical is high dV/dt really?: PCIM Europe 2018; International Exhibition and Conference for Power Electronics, Intelligent Motion, Renewable Energy and Energy Management," pp. 1–6, Jun. 2018,

[7] R. Bartnikas, "Partial discharges: Their mechanism, detection and measurement," vol. 9, no. 5, pp. 763–808, 2002.

[8] Peter W. Hammond, "MEDUMVOLTAGE PWM DRIVE AND METHOD," 5,625,545, Apr 29, 1997.

[9] Infineon, Ed., "F4-100R17N3E4 Datasheet: EconoPACK™3modulewithTrench/FieldstopIGBT4andEmitterControlleddiodeandNTC," May. 2015.

[10] PEM Datasheet, "CWT Mini," 2020.

[11] Sapphire Instruments Datasheet, "SI-9010A Specifications,"

[12] PMK Datasheet, "BumbleBee®," 2020.

[13] Pearson Electronics, INC. Datasheet, "Pierson_Sonde_2877,"

[14] Infineon, Ed., "FF11MR12W1M1_B11 Datasheet: EasyDUALmodulewithCoolSiC™TrenchMOSFETandPressFIT/NTC," Jul. 2018.

[15] Z. Yuan, I. Voss, P. Salmen, T. Aichinger, R. Elpelt, and P. Friedrichs, "How Infineon controls and assures the reliability of SiC based power semiconductors (Whitepaper): Whitepaper,"

[16] *Rotating electrical machines: Part 18-42: Partial discharge resistant electrical insulation systems (Type II) used in rotating electrical machines fed from voltage converters - Qualification tests*, IEC 60034-18-41, 2017.

[17] *Rotating electrical machines: Part 18-42: Partial discharge resistant electrical insulation systems (Type II) used in rotating electrical machines fed from voltage converters - Qualification tests*, IEC 60034-18-42, 2017.

[18] C. Staubach and T. Hildinger, *Stress grading system evaluation for a converter feed hydro generator winding: IEEE Electrical Insulation Conference (EIC)*. Knoxville, TN, USA: IEEE, 2020.

[19] E. Sharifi-Ghazvini, *Analysis of Electrical and Thermal Stresses in the Stress Relief System of Inverter Fed Medium Voltage Induction Motors*. PhD Thesis, University of Waterloo. Ontario, Canada, 2010.

[20] J. Wheeler, "Effects of converter pulses on the electrical insulation in low and medium voltage motors," *IEEE Electrical Insulation Magazine*, vol. 21, no. 2, pp. 22–29, 2005, doi: 10.1109/MEI.2005.1412216.

[21] B. Marusic, "Efficiency Evaluation of Semiconducting Stress Control System: IEEE Electrical Insulation Conference (EIC)," Pittsburgh, PA, USA, 1994.

[22] *Electrical insulating materials and systems - General method of evaluation of electrical endurance under repetitive voltage impulses*, IEC 62068:2013.

[23] *Winding wires - Test methods - Part 7: Electrical endurance under high frequency voltage impulses*, IEC 60851-7.

[24] *Rotating electrical machines – Part 27-1: Off-line partial discharge measurements on the winding insulation*, IEC 60034-27-1, 2017.

[25] *Rotating electrical machines – Part 27-5: Off-line partial discharge measurements on winding insulation of rotating electrical machines during repetitive impulse voltage excitation*, IEC 60034-27-5, Draft.

A Decentralized and Communication-free Control Algorithm of DC Microgrids for the Electrification of Rural Africa

Lucas Richard[1,2], David Frey[1], Marie-Cécile Alvarez-Herault[1], Bertrand Raison[1]

[1]Univ. Grenoble Alpes, CNRS, Grenoble INP,* G2Elab, France

[2]Nanoé, France

Email: lucas.richard@g2elab.grenoble-inp.fr

Keywords

≪Smart microgrids≫, ≪Nanogrid≫, ≪Renewable energy systems≫, ≪Test bench≫, ≪Decentralized control structure≫, ≪DC-DC converter≫, ≪Control methods for electrical systems≫

Abstract

Following the United Nation Sustainable Development Goals of ensuring universal access to basic and modern electric services by 2030, a strong research interest has emerged in the power grid community to design microgrids adapted for rural electrification of Sub-Saharan countries. Their optimal topology, their stability as well as their control architecture are still open to debate. However, DC microgrids with decentralized storage and production are increasingly gaining attention as they enable the progressive building of electric infrastructure in a bottom-up manner, which increases the economic viability, the modularity and the scalability of such rural electrification scheme. In addition, there is a growing consensus in terms of control for decentralized and communication-free algorithms. The absence of a centralized controller is crucial to avoid a single point of failure and to enable plug & play feature within the microgrid. Moreover, DC microgrids must be robust and affordably deployable even in areas with limited or no telecommunication signals. Therefore, this paper proposes the design through software simulations of a decentralized and communication-free control algorithm of DC microgrids adapted for the rural electrification of Africa. The proposed control is then thoroughly validated on a lab test bench and extensive results are presented.

Introduction

Rural electrification is one of the biggest challenges that Sub-Saharan Africa and South-East Asia are facing nowadays to improve living conditions of millions and foster socio-economic development [1, 2, 3]. Indeed, despite the United Nation (UN) goals of ensuring universal access to basic and modern clean energy services by 2030 [4], almost one billion people are still lacking access to electricity, trapping them in energy poverty [5]. The vast majority of unelectrified people reside in rural places of Sub-Saharan Africa or South-East Asia, where abundant resources of solar energy are available [5, 6].

In the past decade, DC microgrids have gained attention as a promising solution to tackle rural electrification problematics [1, 2, 3, 6, 7]. Unlike Solar Home Systems (SHS), microgrids offer the possibilities of building medium-scale power grids, allowing productive use of energy, without necessarily the huge investment costs needed for national grid extension. However, their optimal topology, sizing and control architecture remain unanswered questions open to debates [8, 9]. Centralized power architectures have recently been widely installed whereas decentralized topologies have shown lower upfront costs and better efficiency than centralized ones [2, 10]. In addition, decentralized architectures enable the progressive building of electric infrastructures in a bottom-up manner, growing with the needs of the communities [6]. This also breaks down large initial investment costs in successive small parts, increasing the economic sustainability of rural electrification projects. This is therefore the belief of the authors that DC microgrids with decentralized production and storage is preferable, as advocated by the swarm

---

*Institute of Engineering Univ. Grenoble Alpes

electrification concept [3]. For instance, such DC microgrids can be build on already installed SHS or nanogrids (NG), i.e. a solar panel, a lead-acid battery for 4 to 6 households as installed by Nanoé, a French-Malagasy social venture, in Madagascar [6].

However, such DC microgrids are entirely based on power electronic converters. The topology of DC microgrids as well as their operation and cost-effectiveness are unavoidably intertwined with their control algorithm. To avoid a single point of failure where any problem would impact the whole microgrid and to enable plug & play feature on the microgrid, centralized controllers are less and less favored, even if they facilitate the proper operation of the DC microgrids. Therefore, there is a strong research interest on distributed [11, 12] and fully decentralized control algorithms [10, 13, 14, 15, 16, 17, 18, 19]. However, distributed control schemes always require communication between adjacent controllers, which dramatically reduces the economic viability of the proposed solution while increasing its technical complexity. For those reasons, distributed schemes are often put aside for rural electrification projects.

Most research works propose State-of-Charge (SoC) based droop control for DC microgrids with decentralized production and storage, but they either implement voltage droop control (V-I droop) [16, 17, 18] or current droop control (I-V droop) [10, 13, 14, 15]. However, I-V droop control offers faster dynamics in comparison to V-I droop and only necessitate the setting of one PI regulator [10]. The objectives of the proposed control algorithms vary, from SoC equalizing or balancing between the batteries distributed over the microgrids to communal load supporting. All research papers design and validate their proposed control algorithm through software simulation but only [10, 13, 14, 15, 19] (at the Center for Research on Microgrids (CROM) facilities and at the University of Manitoba) carry out experimental validation.

Based on those observations, this paper proposes a decentralized and communication-free control algorithm to interconnect NGs with the objective to form a village-wide DC microgrid. This control algorithm is adapted for the progressive building of electric infrastructures designed for rural Sub-Saharan Africa and is validated through software simulations and a lab test bench.

Solar DC Microgrid for Swarm Electrification

Microgrid Topology

The microgrid under study is shown in Fig. 1. This microgrid is designed to interconnect 12 or 24 V NGs installed by Nanoé in Madagascar and composed of one solar panel (between 150 and 300 W), one lead acid battery (between 90 Ah and 260 Ah) for 4 to 6 houses [6]. The NGs are usually 20 to 80 meters away from each other within a village and would be connected through 16 or 25 mm^2 electric lines. The NGs are interconnected to a 60 V DC bus through bidirectional DC-DC buck-boost converters.

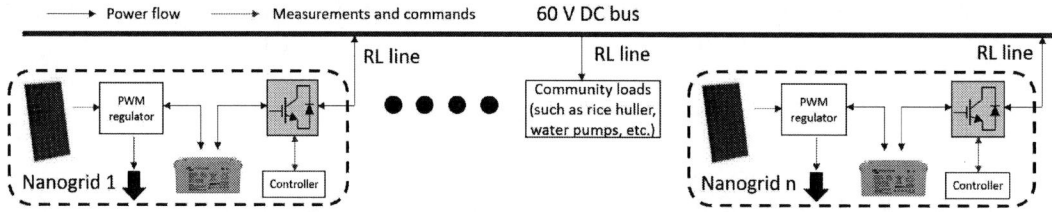

Fig. 1: Topology of the proposed microgrid.

The objectives of forming a village-wide microgrid are twofold. Firstly, by mutualizing installed production and storage capacities, the hardware resources will be used more efficiently, improving the economic and environmental sustainability of those rural electrification projects. Secondly, such DC microgrids will increase the electrical services brought to the communities through enhanced reliability and high-power communal loads (e.g. agro-processing machines, water pumps, etc.), enabling the end-users to progressively climb the energy ladder. Such communal loads can be connected directly to the 60 V DC bus or through a DC-DC converter depending on their nature.

Decentralized and Communication-free Control Algorithm

The proposed topology implies that the control algorithm of the microgrid is decentralized to avoid a single point of failure and enable plug & play feature and communication-free to be affordably deployable even in areas where telecommunication signals are inexistent or unreliable. The plug & play feature is of particular importance for the swarm electrification approach as the basic power units (e.g. NGs, SHS) are supposed to be able to operate in full autonomy and must then be able to connect and disconnect from the microgrid without any impact on the rest of the microgrid. In addition, even if a microgrid is already deployed, new power units could be installed and later connected to the microgrid.

Due to its decentralized nature, the control algorithm can only rely on local variables. Therefore, the control algorithm must pilot energy sharing (i.e. the magnitude of current injection or absorption) based on the DC bus voltage, representing the global level of available energy on the microgrid, and on the SoC of the NG battery, indicating the level of local available energy. The higher the DC voltage, the more globally charged is the microgrid and vice versa. High and low limits (arbitrarily set at $\pm$ 10%) are imposed on the DC bus voltage so that the control algorithm enables relevant power flows while guarantying stability and maintaining the DC bus voltage within a pre-defined zone. This relationship between the DC bus voltage level and the global level of energy on the microgrid is crucial for further expansion of the electric infrastructure (e.g. microgrid interconnection or connection to an AC grid). In addition, three levels of battery SoC are defined, weak from 0 to 60% SoC, medium for 60 to 80% SoC and strong for above 80% SoC. The control algorithm must ensure that any NG with a higher SoC range supports the other NGs with a lower SoC range, with respect to their respective energy reserve. The control algorithm, inspired by [10] with additional modifications on the mode equations, defines different modes of current injection or absorption depending on the DC bus voltage and the local SoC, as shown in Fig. 2 and described below. Table I summarises the different parameters of the control algorithm, whose values have been tuned through software simulations and could easily be modified if necessary.

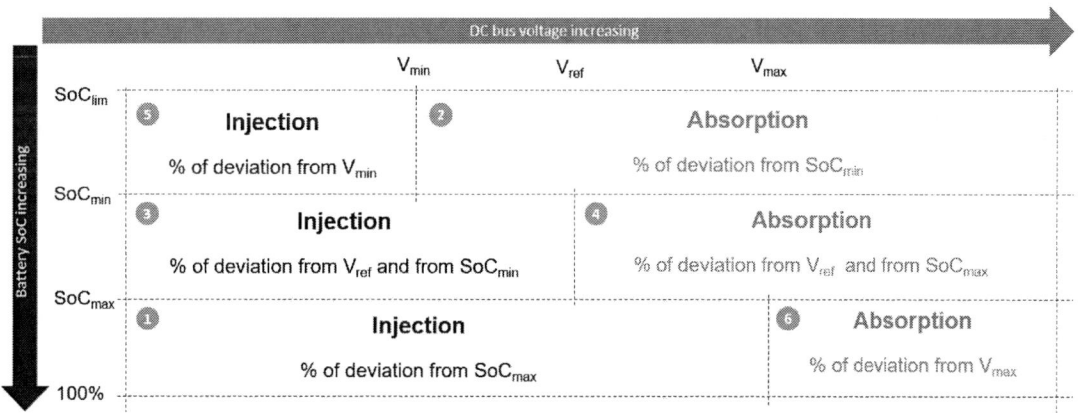

Fig. 2: Decentralized and communication-free control algorithm.

1. Pure Injection

The NG is strong with a SoC above 80% whereas, within the microgrid, some NGs can absorb current, as the DC bus voltage is below V_{max}. The NG therefore injects current with respect to its own SoC, and the higher the SoC, the greater the current injected. A limit is set on the maximum injected current with I_{rated} equal to $C_{bat}/10$. The α exponent enables faster power sharing by increasing the current reference value at a given SoC in comparison with a linear evolution. Lastly, to avoid any edge or pumping effect when the control algorithm suddenly changes mode or is blocked around a DC bus voltage at the boundary of two modes (or around a SoC value for modes 3 and 4), the current reference function is made continuous with respect to the DC bus voltage through the introduction of hyperbolic tangents.

$$I_{ref} = I_{rated} \cdot \left(\frac{SoC - SoC_{max}}{1 - SoC_{max}} \right)^{\alpha} \cdot \tanh\left(\gamma_v \cdot (V_{max} - V_{bus}) \right) \tag{1}$$

2. Pure Absorption

On the opposite to Pure injection, the NG is weak with a SoC below 60% whereas, within the microgrid, some NGs can inject current, as the DC bus voltage is above V_{min}. The NG therefore absorbs current with respect to its SoC, and the lower the SoC, the greater the current absorbed. A limit is set on the maximum absorbed current at I_{rated} at a SoC equal or below SoC_{lim}. The roles of α and of the hyperbolic tangent are similar than in the Pure Injection mode.

$$I_{ref} = max\left(-I_{rated}, -I_{rated} \cdot \left(\frac{SoC - SoC_{min}}{SoC_{lim} - SoC_{min}} \right)^{\alpha} \cdot \tanh\left(\gamma_v \cdot (V_{bus} - V_{min}) \right) \right) \tag{2}$$

3. Voltage-controlled Injection

The NG is in the medium zone with its SoC between SoC_{min} and SoC_{max} and the microgrid is globally discharged as indicated by a DC bus voltage below V_{ref}. Therefore, the NG injects current with a SoC-based droop control with an injected current proportional to the DC bus voltage deviation to V_{ref}. The droop coefficient varies between $1/R_d$ at SoC_{min} and $2/R_d$ at SoC_{max} so that the higher the SoC, the greater the current injected to the microgrid. The maximal current injected is also limited to I_{rated}. An additional coefficient $\frac{C_{bat}}{C_{max}}$ is included to take into account the battery capacity so that the higher the battery capacity, the greater the current injected. For similar reasons to modes 1 and 2, hyperbolic tangents are introduced to make the current reference function continuous with respect to the SoC.

$$\begin{cases} k_d = \frac{1}{R_d} \cdot \left(1 + \frac{SoC - SoC_{min}}{SoC_{max} - SoC_{min}} \right) \cdot \tanh\left(\gamma_s \cdot (SoC_{max} - SoC) \right) \cdot \tanh\left(\gamma_s \cdot (SoC - SoC_{min}) \right) \\ I_{ref} = min\left(I_{rated}, k_d \cdot \frac{C_{bat}}{C_{max}} \cdot (V_{ref} - V_{bus}) \right) \end{cases} \tag{3}$$

4. Voltage-controlled Absorption

The NG is in the medium zone with its SoC between SoC_{min} and SoC_{max} and the microgrid is globally charged as indicated by a DC bus voltage above V_{ref}. Therefore, the NG absorbs current with a SoC-based droop control with an absorbed current proportional to the DC bus voltage deviation to V_{ref}. The droop coefficient varies between $2/R_d$ at SoC_{min} and $1/R_d$ at SoC_{max} so that the lower the SoC, the greater the current absorbed from the microgrid. In a similar fashion to the Voltage-controlled Injection, a current limit, a coefficient $\frac{C_{bat}}{C_{max}}$ and hyperbolic tangents are introduced.

$$\begin{cases} k_c = \frac{1}{R_d} \cdot \left(2 - \frac{SoC - SoC_{min}}{SoC_{max} - SoC_{min}} \right) \cdot \tanh\left(\gamma_s \cdot (SoC_{max} - SoC) \right) \cdot \tanh\left(\gamma_s \cdot (SoC - SoC_{min}) \right) \\ I_{ref} = max\left(-I_{rated}, k_c \cdot \frac{C_{bat}}{C_{max}} \cdot (V_{ref} - V_{bus}) \right) \end{cases} \tag{4}$$

5. Voltage-regulated Injection

The NG is in the weak zone with a SoC below SoC_{min} as well as the rest of the microgrid, as indicated by a voltage below V_{min}. Therefore, the NG must support the DC bus voltage to bring it back to V_{min} through a voltage-droop control with a V_{min} setpoint. A limit at I_{rated} is set. This Voltage-regulated Injection mode is counter-intuitive as a weak NG must inject to the microgrid but is necessary to maintain the microgrid on and to guarantee that the DC bus voltage stays within a pre-defined zone (between V_{min} and V_{max}). However, the NGs should rarely be in this mode or at least they should be between Mode 2 and 5 with a DC bus voltage settled at V_{min} (with I_{ref} then at 0 A), indicating an overall weak microgrid.

$$I_{ref} = min\left(I_{rated}, \frac{V_{min} - V_{bus}}{R_d} \right) \tag{5}$$

6. Voltage-regulated Absorption

The NG is in the strong zone with a SoC above SoC_{max} as well as the rest of the microgrid, as indicated by a voltage above V_{max}. Therefore, the NG must support the DC bus voltage to bring it back to V_{max} through a voltage-droop control with a V_{max} setpoint. A limit at I_{rated} is set. This Voltage-regulated Absorption mode is counter-intuitive as a strong NG must absorb current from the NG but is necessary to guarantee the DC bus voltage stays within a pre-defined zone (between V_{min} and V_{max}). However, the NGs should rarely be in this mode or at least they should be between Mode 1 and 6 with a DC bus voltage settled at V_{max} (with I_{ref} then at 0 A), indicating an overall strong microgrid.

$$I_{ref} = max\left(-I_{rated}, \frac{V_{max} - V_{bus}}{R_d} \right)$$

$$(6)$$

Table I: Parameters of the control algorithm.

Parameters	V_{ref}	V_{min}	V_{max}	SoC_{max}	SoC_{min}	SoC_{lim}	R_d	α	γ_v	γ_s	C_{bat}	C_{max}	I_{rated}
Value	60 V	54 V	66 V	80%	60%	30%	0.5	1/3	5	50	Capacity of the NG	180 Ah	$\frac{C_{bat}}{10}$

A selfish behavior can also be added to the control algorithm to favor self-recharging before supporting the other NGs in the medium zone, by staying a longer time in mode 3 or 4. To this end, SoC_{max} can be first set at 95%. Then, once the SoC of the NG has reached 95% (i.e. self-recharging can be considered almost complete), SoC_{max} is changed to 80% to fully support the other NGs by shifting to mode 1. However, this selfish behavior still enables the support of the weak NGs through modes 3 and 4.

Software Validation

High-level Simulation Model

A simulation model of the microgrid is needed to validate and tune the proposed control algorithm. A microgrid interconnecting 5 NGs is modelled in Matlab-Simulink, with averaged models for the converters [20] to enable long-term simulation of a few days of operation in a few minutes. Each NG is fully modelled with PV production, household consumption and lead-acid battery storage. Then, each NG is connected to the 60 V DC bus through a current-controlled bidirectional buck-boost converter whose current reference is given by the control algorithm presented in the previous section. Field data from Nanoé [6] are used for the geographical layout of the microgrid, shown in Fig. 3, and for the NGs consumption and production. The SoC of each NG battery is evolving according to its local production/consumption balance and its exchange with the microgrid. The resistances of the power lines are taken into account with a resistance per km of 1.465 Ω/km (i.e. for 25 mm aluminium cable). A communal load can be included within the simulation with an adjustable power consumption. In addition, this model can easily be extended to a higher number of NGs or communal loads.

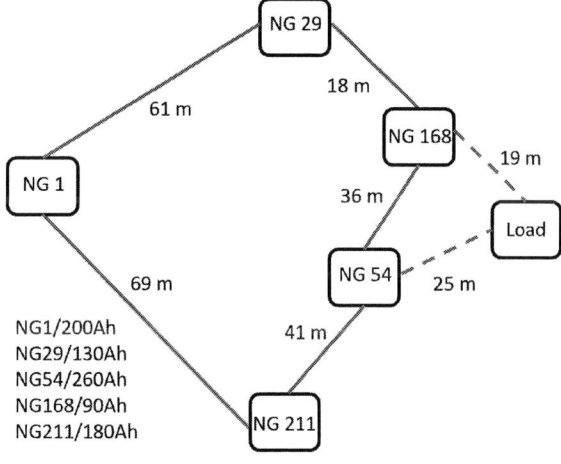

Fig. 3: Topology of the simulated microgrid.

Software Results

Fig. 4 and Fig. 5 show the evolution of the current exchanged between the NGs, of the DC bus voltage and of the SoC of each NG with and without a microgrid for more than 3 days of operation. For illustration purposes, the consumption of NGs 29 and 168 (from Nanoé field data [6]) have been doubled, which makes them undersized, and the simulation is started with unusually low SoC at 3:40 pm. The selfish feature of the control algorithm is enabled. Note that a positive current is injected on the microgrid and vice versa. It can be seen that NGs 1, 54 and 211 support most of the time NGs 29 and 168, either in Voltage-controlled mode 3 or 4 or in Pure Injection mode 1, depending on their SoC level. In addition, the DC bus voltage does traduce well the overall energy availability on the microgrid. Most importantly, Fig. 5 proves that interconnecting NGs enables to supply an overall higher load demand than with isolated NGs. Without the microgrid, to support a twice bigger demand, NGs 29 and 168 would need a bigger installation. The microgrid enables to optimise the use of hardware resources of the NGs already installed, both to allow for higher electrical services and to enhance the economic sustainability of the proposed rural electrification scheme.

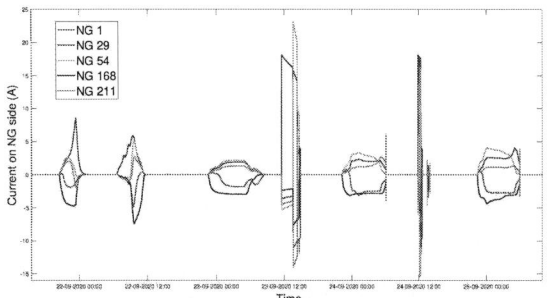 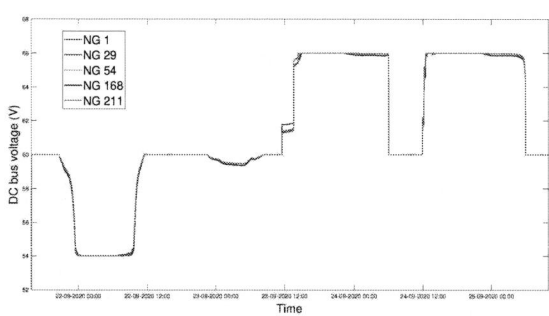

Fig. 4: Long-term current exchange and DC bus voltage evolution.

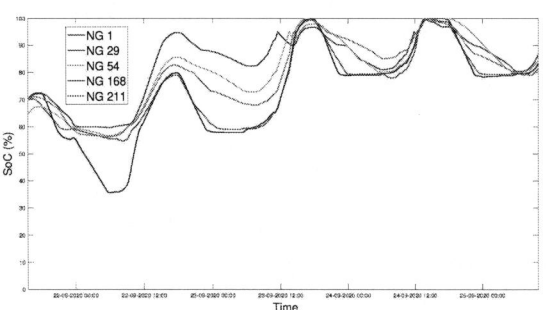 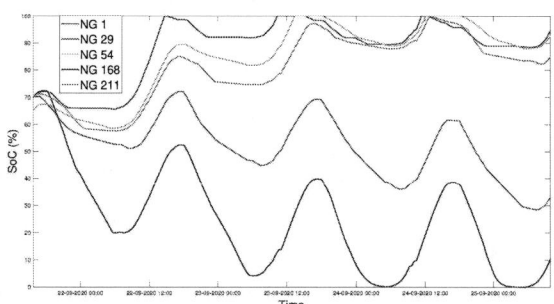

Fig. 5: Long-term State-of-Charge evolution with (left figure) and without a microgrid (right figure).

The different modes of the control algorithm are illustrated in Fig. 6 and Fig. 7, where the SoC of each NG is artificially changed within the simulation, to illustrate different operating points. At t=0 s, NG 1 is strong with a SoC at 100%, NGs 29, 54 and 168 are in the medium zone with SoCs respectively at 70%, 75% and 65%, and NG 211 is weak with a SoC at 40%. Therefore, NG 1 is in Pure Injection Mode, NG 29, 54 and 168 are in Voltage-controlled Absorption Mode whereas NG 211 is in Pure Absorption Mode. However, as the DC bus voltage is close to 60 V, NG 29, 54 and 168 do not absorb much current. From t=1000 s, as the SoC of NG 211 starts to increase, NG 211 switches to Voltage-controlled Absorption from t=1400 s to t=2200 s, then to Pure Injection Mode. Note than NG 1 is then injecting more current on the microgrid than NG 211 due to its greater battery capacity. Similarly, NG 54 is absorbing more current from t=1400 to t=2800 s than NG 29 and 168. As the SoC of NG 54, 168 and 29 start to increase respectively at t=2500 s, t=3500 s and t=4500 s, the NGs switch modes from Voltage-controlled Absorption modes to Pure Injection. Once all the NG have reached the strong zone, the DC bus voltage stabilizes at 66 V, as the NGs are at the equilibrium between the Pure Injection Mode and the Voltage-regulated absorption to keep the DC bus voltage within its pre-defined zone. It can be noted than due to the voltage drops on the lines, the closer a strong NG is from a weak NG, the more current it will inject on the DC bus as it sees a more accurate image of the DC bus voltage. The farther a NG is from a consumption point, the higher the voltage drop, and the less accurate is the DC bus voltage with respect to the consumption point.

In addition, the proposed DC microgrid enables to power communal loads by all the NGs, with respect to their respective SoC and to their proximity to the communal load. Fig. 8 illustrates the operation of the microgrid with a communal load. All the NGs are strong with their SoC at 100%. At t=1000 s, a 300 W load is connected to the DC bus and it is respectively increased to 600 and 900 W at t=2500 s and t=4000 s, then disconnected at t=5500 s. The communal load is installed close to NG 54 and 168, as shown in Fig. 3. Therefore, NG 54 and 168 are injecting the most current on the microgrid with respect to their battery capacity, respectively 260 Ah and 90 Ah. On the opposite, NG 1, the second strongest NG on the microgrid is injecting very little current to power the communal load as it is located far away from it. This can also be seen with the DC bus voltage witnessed by NG 1 and NG 211, which are significantly greater than the DC bus voltage witnessed by the other NGs. Thus, due to the voltage drops on the lines, NG 1 and 211 do not know there is a high power demand on the microgrid. This shows that the geographical dissemination of strong and weak NGs as well as of communal loads must be carefully planned to obtain a balanced microgrid.

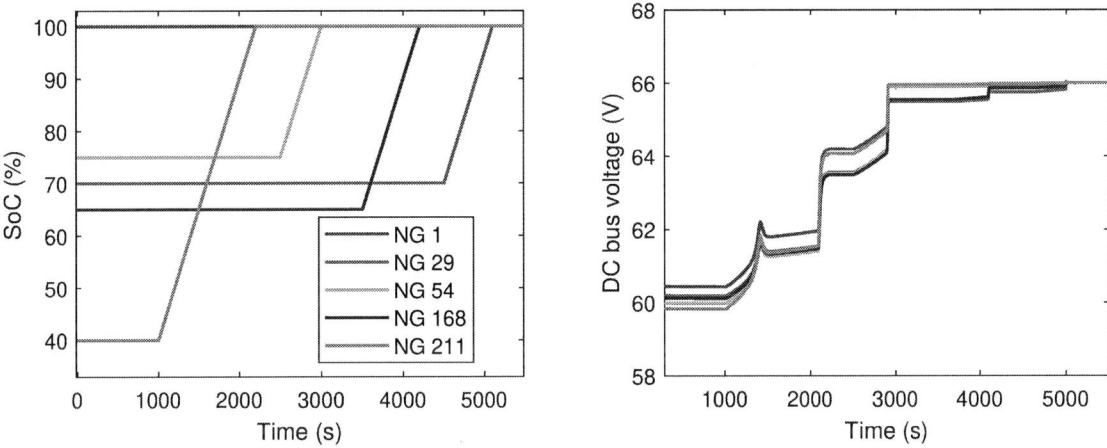

Fig. 6: Different SoC and DC bus voltage operating points.

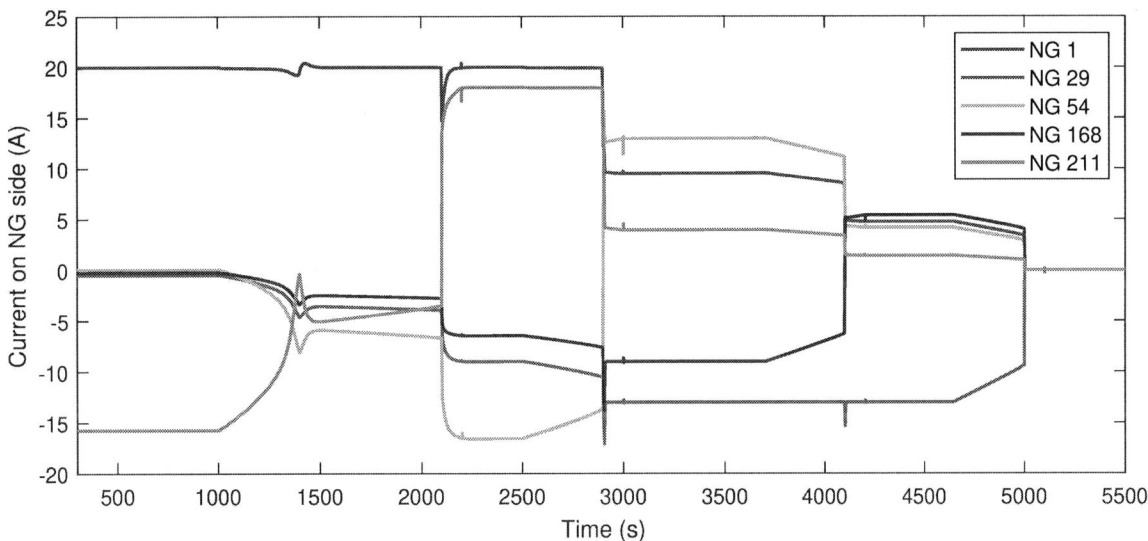

Fig. 7: Current exchanged between the nanogrids at different operating points.

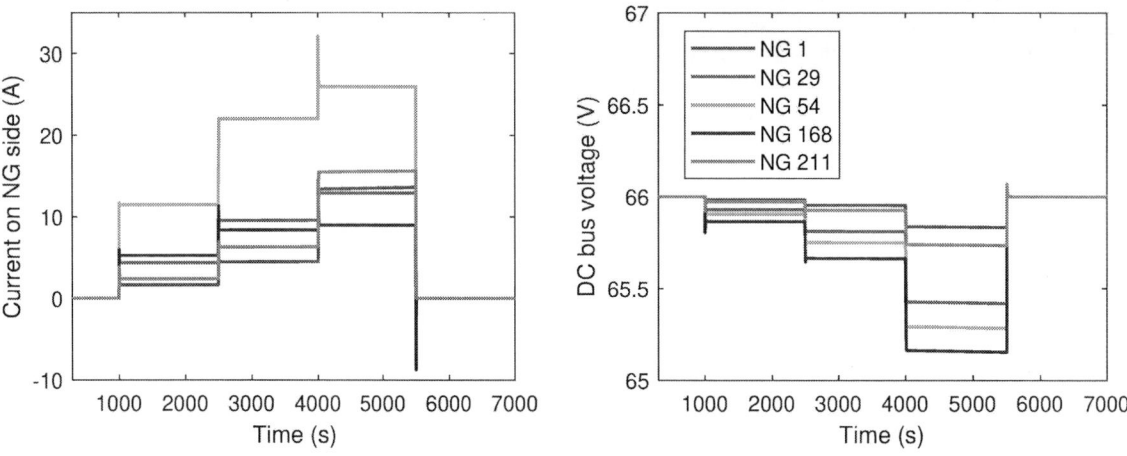

Fig. 8: Communal load operation.

Experimental Validation

Lab Test Bench

A test bench has been developed in the lab to experimentally validate and further tune the proposed control algorithm. The test bench contains 3 bidirectional buck-boost converters developed in-house, interconnected through RL lines emulating the impedances of the electric lines on the field. At the input of each converter, a power supply in parallel to an electronic load or a battery is connected. Each converter is controlled by a MyRio Embedded Controller, from National Instruments, programmed in LabVIEW [21]. The MyRio controller associated with LabVIEW enables to implement the proposed control algorithm, to emulate a SoC estimator (and associated scenario of SoC evolution), to record data every 25 ms and to ease monitoring through a Graphic User Interface (GUI). A communal load, emulated by a power resistor, can be connected to the DC bus. In addition, grid reconfiguration is easily doable in order to test different grid topologies (i.e. radial or meshed). Fig. 9 shows the schematic of the lab test bench and its actual set-up.

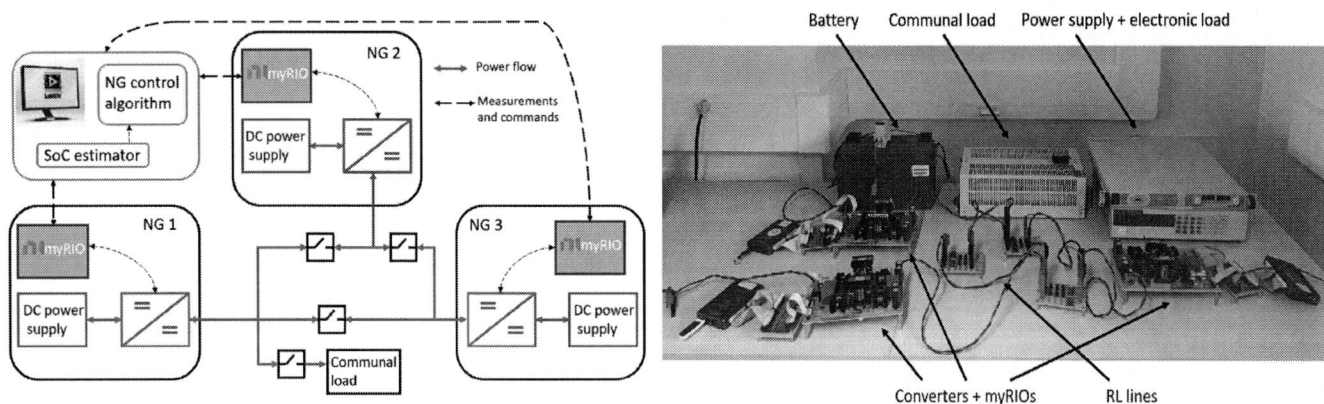

Fig. 9: Schematic and actual set-up of the DC microgrid test bench.

Experimental Results

The test bench permits complete monitoring of the power flows on the DC microgrid, tuning of the control algorithm parameters and extensive testing of its capabilities. Fig. 10 a) shows the evolution of the DC bus voltage and the current exchanged between the 3 converters (named NG 1, NG 2 and NG 3) at different operating points. The SoCs are artificially changed within the LabVIEW GUI to replicate

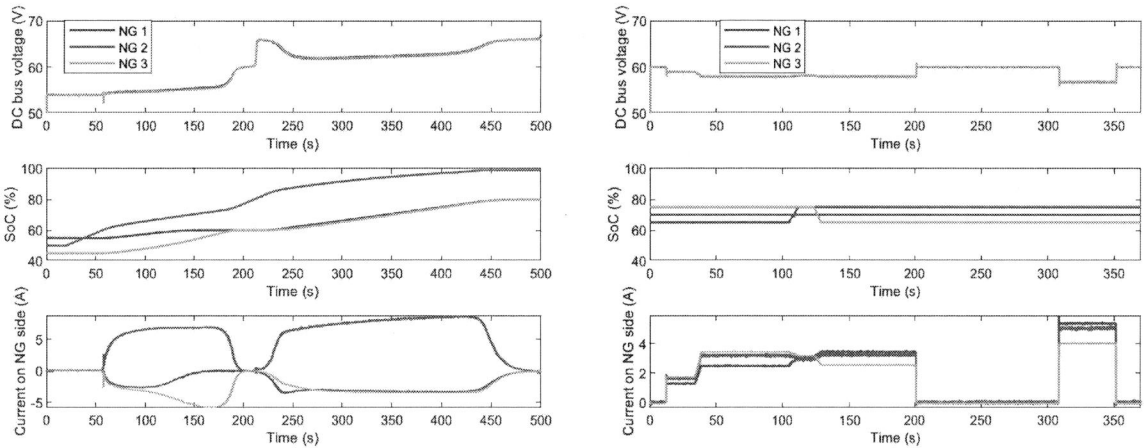

Fig. 10: (a) Experimental results at different operating points, (b) with communal load operation.

a potential real SoC evolution within a short time frame. The 3 NGs are initialized in the weak zone (i.e. SoC<60%), therefore they are operating in Voltage-regulated Mode, stabilizing the DC bus voltage at 54 V. As the SoC of NG 2 reaches 60%, NG 2 starts to operate in Voltage-controlled Mode and, the DC bus voltage being low, it injects current on the DC bus that NG 1 and NG 3 absorb, being now in Pure Absorption Mode. At t=180 s, the 3 NGs are in the medium zone, therefore the DC bus voltage is maintained at 60 V. Once NG 2 reaches a SoC above 80%, it switches to Pure Injection Mode and injects current on the DC bus. As a response to the increase of the DC bus voltage, NG 1 and 3, in Voltage-controlled Mode, absorb currents until their SoC reaches 80%. Note that here the selfish behavior of the control is not implemented and that I_{rated} is equal to 9 A.

Furthermore, Fig. 10 b) illustrates the operation of a communal load on the DC microgrid. The SoC of each NG is initialized in the medium zone. At t=10 s, the power resistor is connected to the DC bus with low power consumption, which is increased at t=30 s. The NGs are in Voltage-controlled Mode and therefore respond to the DC bus voltage deviation to V_{ref}, i.e. 60 V, by injecting current to the DC bus, with respect to their own SoC. Within the medium zone, the higher the SoC, the higher the current injected (with some limitations when approaching 80% SoC due to the hyperbolic tangents). The load is then disconnected at t=200 s and reconnected at a greater power at t=310 s until t=350 s, to observe different operating points. At t=120 s, the SoCs are artificially changed to illustrate the current sharing scheme. Note here that the current injected by NG 2 is higher than the others with respect to its SoC due to is close proximity to the communal load. A zoom on the DC bus voltage indicates that NG 2 sees a DC bus voltage 0.1 to 0.15 V lower than NG 1 and 3, and thus injects more on the DC bus. Those results validate the operation of the microgrid with a communal load and confirm the importance of carefully selecting the location of a high-power communal load, as the neighbouring NGs will be the ones contributing the most.

Those experimental results have enabled to further tune the control algorithm. In particular, the slope of the voltage hyperbolic tangent (i.e. γ_v) has been decreased to avoid oscillations on the current reference, which should not evolve too fast. This illustrates the crucial need to confront theoretical control algorithms to the reality of test benches. Furthermore, additional studies can be easily performed on this test bench (e.g. different grid topologies and DC bus voltage levels, start-up and protection schemes [22]).

Conclusion

This paper presents a decentralized and communication-free control algorithm for DC microgrids adapted to the progressive building of electric infrastructures in rural Sub-Saharan Africa. The control algorithm is first designed through simulations and then experimentally validated and tuned on a test bench developed in-house.

Future works will focus on the enhancement of the control algorithm to include the control of the solar panel production to optimise the power flows within the NG and the charging of the NG battery. In addition, designing a reliable and precise SoC estimator is not straightforward, therefore, it would be of interest to study control algorithms based on the battery voltage instead of the SoC. Thus, accurate battery modelling is needed. Lastly, the impact of each local control algorithm on the overall stability of the DC microgrid should be thoroughly analysed.

References

[1] Moner-Girona, M., Bódis, K., Morrissey, J., Kougias, I., Hankins, M., Huld, T., Szabó, S. (2019). Decentralized rural electrification in Kenya: Speeding up universal energy access. Energy for Sustainable Development, Vol. 52, pp. 128–146.

[2] Nasir, M., Khan, H. A., Zaffar, N. A., Vasquez, J. C., Guerrero, J. M. (2018). Scalable Solar DC microgrids: On the Path to Revolutionizing the Electrification Architecture of Developing Communities. IEEE Electrification Magazine, Vol. 6, No. 4, pp. 63–72.

[3] Groh, S., Philipp, D., Lasch, B. E., Kirchhoff, H. (2015). Swarm Electrification: Investigating a Paradigm Shift Through the Building of Microgrids Bottom-up. Chapter 1 of Decentralized Solutions for Developing Economies, pp. 25-44.

[4] "United Nation 17 Sustainable Development Goals." (2022), [Online]. Available: https://sdgs.un.org/goals.

[5] "World Energy Outlook (WEO)." (2021), [Online]. Available: https://www.iea.org/reports/world-energy-outlook-2021.

[6] "Nanoé presentation website" (2022), [Online]. Available: https://www.nanoe.net/en/

[7] Jhunjhunwala, A., Lolla, A., Kaur, P. (2016). Solar-DC Microgrid for Indian Homes: A Transforming Power Scenario. IEEE Electrification Magazine, Vol. 4, No. 2, pp. 10–19.

[8] Dragicevic, T., Lu, X. Vasquez, J.C., Guerrero, J. M. (2016). DC Microgrids — Part I : A Review of Control Strategies and Stabilization Techniques. IEEE Transactions on Power Electronics, Vol. 31, No. 7, pp. 4876–4891.

[9] Meng, L., Shafiee, Q., Trecate, G. F., Karimi, H., Fulwani, D., Lu, X., Guerrero, J. M. (2017). Review on Control of DC Microgrids and Multiple Microgrid Clusters. IEEE Journal of Emerging and Selected Topics in Power Electronics, Vol. 5, No. 3, pp. 928–948.

[10] Nasir, M., Jin, Z., Khan, H. A., Zaffar, N. A., Vasquez, J. C., Guerrero, J. M. (2019). A Decentralized Control Architecture Applied to DC Nanogrid Clusters for Rural Electrification in Developing Regions. IEEE Transactions on Power Electronics, Vol. 34, No. 2, pp. 1773–1785.

[11] Shafiee, Q., Dragicevic, T., Andrade, F., Vasquez, J. C., Guerrero, J. M. (2014). Distributed consensus-based control of multiple DC-microgrids clusters. IECON Proceedings, pp. 2056–2062.

[12] Shafiee, Q., Dragičević, T., Vasquez, J. C., Guerrero, J. M. (2014). Hierarchical Control for Multiple DC-Microgrids Clusters. IEEE Transactions on Energy Conversion, Vol. 29, No. 4, pp. 922–933.

[13] Nasir, M., Anees, M., Khan, H. A., Guerrero, J. M. (2019). Dual-loop control strategy applied to the cluster of multiple nanogrids for rural electrification applications. IET Smart Grid, Vol. 2, No. 3, pp. 327–335.

[14] Li, D., Ho, C. N. M. (2021). A Module-Based Plug-n-Play DC Microgrid with Fully Decentralized Control for IEEE Empower a Billion Lives Competition. IEEE Transactions on Power Electronics, Vol. 36, No. 2, pp. 1764–1776.

[15] Nasir, M., Anees, M., Khan, H. A., Khan, I., Xu, Y., Guerrero, J. M. (2019). Integration and Decentralized Control of Standalone Solar Home Systems for Off-Grid Community Applications. IEEE Transactions on Industry Applications, Vol. 55, No. 6, pp. 7240–7250.

[16] Samende, C., Bhagavathy, S. M., McCulloch, M. (2019). State of Charge Based Droop Control for Coordinated Power Exchange in Low Voltage DC Nanogrids. Proceedings of the International Conference on Power Electronics and Drive Systems, July 2019.

[17] Samende, C., Bhagavathy, S. mothilal, Gao, F., McCulloch, M. (2021). Decentralized Voltage Control for Efficient Power Exchange in Interconnected DC Clusters. IEEE Transactions on Sustainable Energy, Vol 12, No. 1, pp 103-115.

[18] Lu, X., Sun, K., Guerrero, J. M., Vasquez, J. C., Huang, L. (2015). Double-Quadrant State-of-Charge-Based Droop Control Method for Distributed Energy Storage Systems in Autonomous DC Microgrids. IEEE Transactions on Smart Grid, Vol. 6, No. 1, pp. 147–157.

[19] Nasir, M., Khan, H. A., Hussain, A., Mateen, L., Zaffar, N. A. (2018). Solar PV-Based Scalable DC Microgrid for Rural Electrification in Developing Regions. IEEE Transactions on Sustainable Energy, Vol. 9, No. 1, pp. 390–399.

[20] "Average-Value chopper (Mathworks help)." (2022), [Online]. Available: https://fr.mathworks./help/physmod/sps/ref/averagevaluechopper.html.

[21] "LabVIEW presentation website." (2022), [Online].Available: https://www.ni.com/en-za/shop/labview.html.

[22] Richard, L., Derbey, A., Frey, D., Alvarez-Hérault, M.C., Raison, B. (2022) Experimental Design of Solar DC Microgrid for the Rural Electrification of Africa. In Proceedings of the PCIM Europe 2022, pp. 1–10

Universal Real-Time Model for Active Rectifiers in Versatile Totem-Pole PFC Configurations

Axel Kiffe, Thorben Hoffstadt
dSPACE GmbH
Rathenaustraße 26, 33102 Paderborn, Germany

Tel.: +49 / (0) – 5251 1638-0
Fax: +49 / (0) – 5251 161980
E-Mail: {akiffe, thoffstadt}@dspace.de
URL: http://www.dspace.com

Keywords

Real-time simulation, Modelling, Power factor correction, Field Programmable Gate Array, Battery charger

Abstract

On-board chargers in electric vehicles often include a totem-pole PFC, as this compact active rectifier provides bidirectional energy flow and can be easily adapted to single-phase or three-phase grid operation. To test its control unit in terms of HIL simulations, a real-time capable model of the topology is required. This paper deals with the modeling and realization of a universal real-time model for versatile totem-pole configurations in single-phase or three-phase grid operation.

The proposed model uses the ideal switch representation. An extended approach to detect switch-state changes caused by Dirac impulses is introduced. Additional focus is placed on the implementation of the final model on an FPGA. Simulation results prove the validity of the proposed model.

I. Introduction

Aside from the increasing use of renewable energies, electrifying vehicles is an important task to reach the targets of reducing emissions and preventing significant global warming. Due to the high number of power electronics and electric drives in electric and hybrid-electric vehicles, the interest in hardware-in-the-loop (HIL) simulation for testing is increasing. HIL simulation is a well-known approach for testing control units in the automotive indus-try: The real plant is replaced by a real-time system that captures the output of the control unit, computes the reaction of the plant, and simulates the input of the control unit. This enables testing in the laboratory under reproducible con-ditions and reduces the risk of accidents when using the real plant.

HIL simulation of power electronic cir-cuits is an ambitious task due to the typ-ically high eigenvalues compared to mechanical systems, the high switching frequencies, and the fast structural changes of the system based on the semiconductor switches.

Furthermore, the variety of the topolo-gies is quite high, which inspires the wish for a generic and topology-based

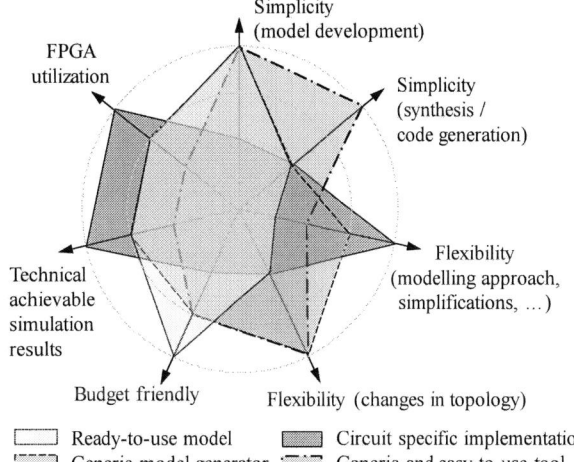

Fig. 1: Relative comparison of approaches for realizing a real-time simulation of power electronic circuits, [1].

tool to develop real-time capable models. However, a trade-off between an easy-to-use generic tool, less synthesis time, resource-efficient computation, and a cutting-edge simulation exists, especially in respect to high switching frequencies. Therefore, the approaches for realizing a FPGA-based HIL-simulation of a power electronic circuit can be distinguished between:

- Generic and easy-to-use tool (topology-based modeling)
- Generic model generator (topology-based modeling)
- Circuit-specific implementation by engineering services
- Ready-to-use model available on the market

The different properties and a relative comparison of the approaches for realizing a FPGA-based HIL-simulation can be found in Fig. 1, [1].

A generic and easy-to-use tool as well as a generic model generator use the topology of the circuit directly as modeling basis. Changes in the topology can be considered with low effort. But the performance of the real-time model will be comparable low. In strict contrast, high performance will be achieved with a circuit-specific implementation. However, the development of such a model is usually difficult and it might be impossible to adapt it to a different topology because simplifications might become untenable. A compromise is given by a ready-to-use model for a common topology. In advance, they should provide mechanisms to satisfy multiple applications, instead of one specific application. An additional advantage of ready-to-use models is the reduced cost caused by a broader user group in different applications.

The focus of this paper is placed on a flexible ready-to-use model. It is referred to [1] for further comparisons and a detailed introduction of all approaches.

II. Application

A typical structure of an on-board charger (OBC) is depicted in Fig. 2a). It consists of an active rectifier (AC/DC) providing power factor correction capability and a DC/DC converter to charge the high voltage (HV), [2]. If the vehicle includes a low voltage (LV) battery as well, the OBC comprises a second DC/DC. For the DC/DC converters, resonant converters are often used. A real-time capable model for an LLC converter was presented in [1].

For the active rectifiers, totem-pole PFC topologies are one of the most popular and promising bidirectional topologies in two-stage bidirectional OBCs for electric vehicles, [2]. A schematic of a totem-pole

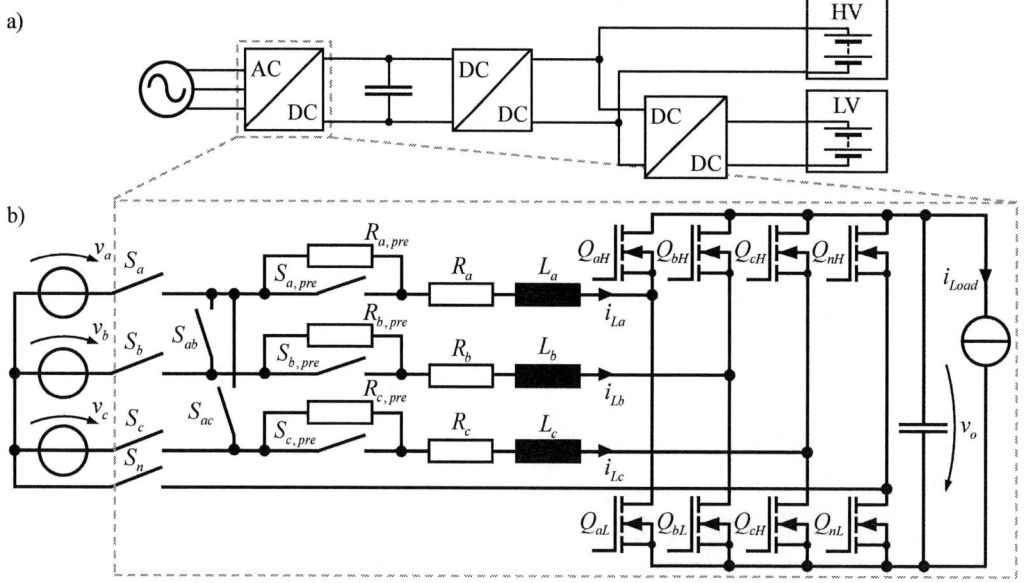

Fig. 2: General structure of an on-board charger (a) and schematic of a totem-pole rectifier (b), which can be used for versatile configurations and operation modes.

rectifier is shown in Fig. 2b). While the semiconductor switches are used for the basic functionality of the converter, the relays are used for pre-charging and configuration of versatile operation modes for single-phase and three-phase grid. A list of configurations can be found in Table I.

Therein, character a stands for active operation, which means the semiconductor is controlled by the electric control unit and its digital outputs, e.g., PWM signals. The character p stands for passive operation of the semiconductor devices, i.e., the diodes of the MOSFETs are considered but the MOSFETs are not controlled by an external gate signal. The symbol $-$ stands for a forced off state of the corresponding switch.

Table I: Configurations of switching devices in Fig. 2 for realizing different topologies.

Topology		Grid relays				Phase relays		Pre-charge relays			Half-bridges				Sources
		S_a	S_b	S_c	S_n	S_{ab}	S_{ac}	$S_{a,pre}$	$S_{b,pre}$	$S_{c,pre}$	Q_{ai}	Q_{bi}	Q_{ci}	Q_{ni}	
Totem-pole	single-phase	a	$-$	$-$	a	$-$	$-$	a	$-$	$-$	a	$-$	$-$	a/p	v_a
	2-phase interleaved	a	a	$-$	a	a	$-$	a	a	$-$	a	a	$-$	a/p	$v_a = v_b$
	3-phase interleaved	a	a	a	a	a	a	a	a	a	a	a	a	a/p	$v_a = v_b = v_c$
Full-bridge	3-phase	a	a	a	$-$	$-$	$-$	a	a	a	a	a	a	$-$	v_a, v_b, v_c

p passive operation, a active operation, $-$ forced to an off state

III. Modeling Concept and Simplification

Because the transient behavior of the currents and voltages are of interest, an oversampling model is investigated in this paper. Oversampling models keep switch-states fixed for one simulation step. Thus, their sample time T must be several times smaller than the switching period T_S. According to [3] and from practical experience, a minimum oversampling factor $\kappa = T_S/T \approx 20$ is required. Under consideration of typical switching frequencies of totem-pole rectifiers in the range of $50\,\text{kHz}$ up to $150\,\text{kHz}$, a simulation step-size T smaller than $333\,\text{ns}$ has to be obtained.

Due to the switching behavior of the n_{sw} semiconductor devices, all $2^{n_{sw}}$ switch-states and their mathematical representations must be considered. In addition to the simulation of the nominal operation, this also enables failure simulation. In respect to the topology shown in Fig. 2, 2^{17} mathematical representations need to be considered. This requires too many resources to directly apply a state-space approach. Thus, simplifications need to be exploited, to overcome this resource issue.

Figure 3 illustrates the realized concept. Usually, the pre-charging relays $S_{i,pre} \forall i \in \{a, b, c\}$ are switching slowly and not frequently. Due to this, there is enough time to update the mathematical representations on the FPGA by a corresponding processor interface depending on the switch-state of the relays. Hence, the three relays can be replaced by variable resistors $R_i \forall i \in \{a, b, c\}$ that change their values between $R_{i,off} = R_{i,pre} + R_i$ and $R_{i,on} = R_i$.

Furthermore, the switches S_{ab} and S_{bc} are used to select between a three-phase operation and an interleaved single-phase operation. In the model, their influence is considered

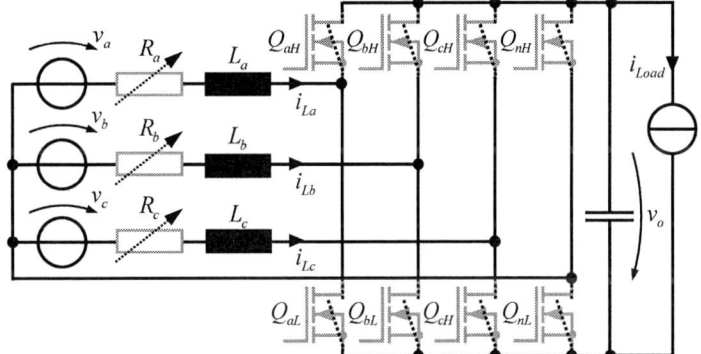

Fig. 3: Topology of the totem-pole PFC after applying the conceptional simplification to the overall circuit in Fig. 2.

by defining the voltage sources appropriately. For example, in three-phase full-bridge mode according to Table I, S_{ab} and S_{bc} are off and v_a, v_b, v_c represent the phase voltages of the grid. In case of a two-phase interleaved operation in totem-pole configuration with S_{ab} and S_{bc} both on, the voltages are chosen identical, $v_a = v_b = v_c$.

The switches of the grid relays (S_a, S_b, S_c and S_n) are realized by exploiting failure simulation. This means that the switch-logic for the MOSFETs is extended to force a switch-state independently of the active switching by gate-signals or passive switching caused by internal currents or voltages. Based on that, if, for example, a single-phase totem-pole configuration with S_b and S_c in off-state is investigated, the MOSFETs Q_{bH}, Q_{bL}, Q_{cH} and Q_{cL} must be forced to off-state, consistently, to avoid any inductor currents i_{Lb} and i_{Lc}.

Based on these simplifications, a model for the topology in Fig. 3 can be developed. This universal model can simulate all configurations listed in Table I, by forcing the switch-state of the switches indicated with − to off-state with the previously mentioned failure simulation feature.

IV. Basics of Power Electronics Simulation with Ideal Switch Model

Some switch models, e.g., resistive switch model [4] or inductive-capacitive switch model [5], for representing the semiconductor devices cause high eigenvalues or impair the simulation accuracy. Thus, the ideal switch model is applied here, e.g. [1], [6], [7]. Herein, the approach from [1] is utilized, adjusted, and extended to satisfy the requirements from the application.

By using an ideal switch model, every semiconductor is replaced by a short-circuit in on-state or by an open-circuit in off-state. After the semiconductors are replaced according to their switch-state, only linear devices remain. Based on well-known approaches from the literature, e.g. [8] or [9] with the conversion of [10], the corresponding state-space representations can be determined systematically. This yields a switch-state dependent state-space representation:

$$\dot{\mathbf{x}} = \mathbf{A}_i \mathbf{x} + \mathbf{B}_i \mathbf{u} , \quad \mathbf{y} = \mathbf{C}_i \mathbf{x} + \mathbf{D}_i \mathbf{u} . \tag{1}$$

While $\mathbf{x}$, $\mathbf{u}$ and $\mathbf{y}$ are the state-, input- and output-vector, $\mathbf{A}_i$, $\mathbf{B}_i$, $\mathbf{C}_i$ and $\mathbf{D}_i$ are the system, input, output, and feedthrough matrix of the i-th switch-state. The selection of the correct switch-state for the next simulation step is done by a switch-state logic. For that, event-based and switching event-oriented conditions were formulated in [1]. These conditions become true only if a switching event currently occurs. Furthermore, the conditions provide information about the type of switching event, by having independent conditions for natural switching events (also called passive or internal switching events) and forced switching events (also called active or external switching events). While forced switching events are triggered by external signals, like a changed gate signal of a relay, natural switching events are triggered on changes of internal currents and voltages of the circuit only, e.g., a diode. An extract of the switch conditions is given in Fig. 4 for a MOSFET, which is required for the investigated totem-pole topologies. Depending on the additional subscript N or F, the condition is related to a natural or forced switching event, respectively.

In addition to the advantages of avoiding high eigenvalues and undesirable inaccuracies in the simulation, the ideal switch model requires a more complex switch-state detection compared with other switch models, e.g., a resistive switch model. The reason lies in the ideal connection or disconnection according to the switch-state, which influences the dependency between the storage devices, i.e., currents of inductors and voltages of capacitors. As an example, the small circuit in Fig. 5a is investigated, where the voltage v as well as the inductor current i_L are assumed to be positive. The diode was replaced by its ideal switch representation. At the beginning, the switch S is closed so that the current i_L flows through it. Then, the

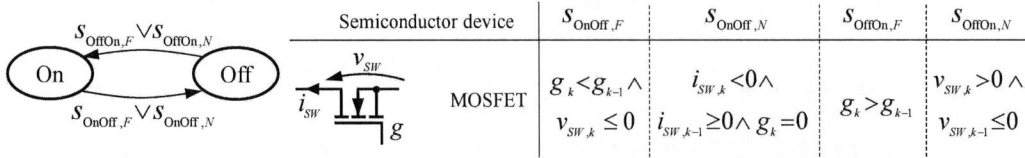

Fig. 4: Extract of the event-based and switching event-oriented switching conditions from [1].

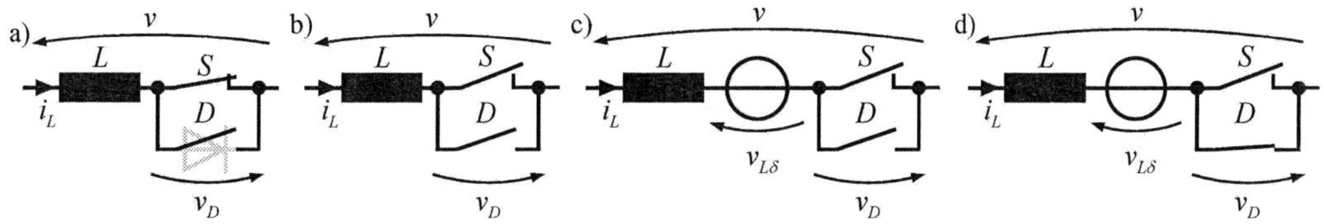

Fig. 5: Illustration of switching transitions caused by a Dirac impulse.

switch S is forced to off-state – see Fig. 5b. The switch for the diode D would change into on-state, if the voltage v_D became positive. However, due to a positive voltage v and inductor current i_L, voltage $v_D = -v$ is negative. In a real circuit, the positive inductor current continuity is ensured by the diode (see Fig. 5d). In the simulation, this is a conditionally caused switching event, which was triggered by trying to force the inductor current from a value i_L^- to another i_L^+. In the example, i_L^+ is zero (Fig. 5c). From mathematical perspective, a step in a state-variable can be explained by the occurrence of a Dirac impulse:

$$\left[x_{sw}^+ - x_{sw}^-\right] \cdot \sigma\left(t - t_{sw}\right) = \int_{-\infty}^{t} \left[x_{sw}^+ - x_{sw}^-\right] \cdot \delta\left(\tau - t_{sw}\right) d\tau, \quad x_{sw}^- = x\left(t_{sw}^-\right), \quad x_{sw}^+ = x\left(t_{sw}^+\right) \tag{2}$$

Consequently, the theoretical occurrence of a Dirac impulse would lead to infinite voltages or currents for inductances or capacitances, respectively:

$$v_{L\delta}(t) = L \cdot \left(i_L^+ - i_L^-\right) \cdot \delta\left(t - t_{sw}\right), \quad i_{C\delta}(t) = C \cdot \left(v_C^+ - v_C^-\right) \cdot \delta\left(t - t_{sw}\right). \tag{3}$$

These voltages and currents can be considered by adding voltage sources in series with inductances and current sources in parallel to capacitances. For the example, this is illustrated in Fig. 5c and leads to

$$z_{SWE,D} = v_D = -v_{L\delta} - v = \underbrace{-L \cdot \left(0 - i_L\right) \cdot \delta\left(t - t_{sw}\right)}_{z_{SW\delta,D}} \underbrace{-v}_{z_{SW,D}} \tag{4}$$

for the diode voltage. This type of equation can be separated into an impulsive part $z_{SW\delta} \cdot \delta\left(t - t_{sw}\right)$, which includes the Dirac delta function, and a non-impulsive part z_{SW}. Due to the infinite value of the Dirac delta function, the impulsive part $z_{SW\delta} \cdot \delta\left(t - t_{sw}\right)$ overrules the non-impulsive part z_{SW}, [7]. Thus, for the example in Fig. 5, the term $z_{SW,D} = -v$ does not matter, and the impulsive part $z_{SW\delta,D} \cdot \delta\left(t - t_{sw}\right)$ becomes greater than zero. To avoid approximations of the Dirac delta function, it is suitable to separate this equation type into the factor of the Dirac delta function $z_{SW\delta}$ and the non-impulsive part z_{SW}. Based on this, a simple logic can decide if z_{SW} would be overruled by $z_{SW\delta}$ and how this influences the switch-state, [7]. Thus, the initial state-space representations according to (1) with additional sources for considering Dirac impulses in input-vector $\mathbf{u}$ must be prepared for the simulation by a preprocessing. Therein, the

- sources for Dirac impulses were substituted appropriately,
- the output vector $\mathbf{y}$ provides the non-impulsive z_{SW} as well as the impulsive part $z_{SW\delta} \cdot \delta\left(t - t_{sw}\right)$ of all switches,
- the state-space representation is discretized.

Without further explanations, the following discrete state-space representation results from the preprocessing:

$$\mathbf{x}_{k+1} = \boldsymbol{\Phi}_i \mathbf{x}_k + \mathbf{H}_i \mathbf{u}_{Sk} = \begin{bmatrix} \boldsymbol{\Phi}_i & \mathbf{H}_i \end{bmatrix} \begin{bmatrix} \mathbf{x}_k \\ \mathbf{u}_{Sk} \end{bmatrix}, \quad \begin{bmatrix} \mathbf{z}_{SWi,k} \\ \dot{\mathbf{z}}_{SWi,k} \\ \mathbf{z}_{SW\delta i,k} \\ \mathbf{y}_{Li,k} \end{bmatrix} = \begin{bmatrix} \mathbf{C}_{dSWi} \\ \mathbf{C}_{ddSWi} \\ \mathbf{C}_{dSW\delta i} \\ \mathbf{C}_{dLi} \end{bmatrix} \mathbf{x}_k + \begin{bmatrix} \mathbf{D}_{dSWi} \\ \mathbf{D}_{ddSWi} \\ \mathbf{D}_{dSW\delta i} \\ \mathbf{D}_{dLi} \end{bmatrix} \mathbf{u}_{Sk} = \begin{bmatrix} \mathbf{C}_{dSWi} & \mathbf{D}_{dSWi} \\ \mathbf{C}_{ddSWi} & \mathbf{D}_{ddSWi} \\ \mathbf{C}_{dSW\delta i} & \mathbf{D}_{dSW\delta i} \\ \mathbf{C}_{dLi} & \mathbf{D}_{dLi} \end{bmatrix} \begin{bmatrix} \mathbf{x}_k \\ \mathbf{u}_{Sk} \end{bmatrix} \tag{5}$$

In addition to the measurement values $\mathbf{y}_{Li,k}$, the non-impulsive part ($\mathbf{z}_{SWi,k}$), and the factor of the impulsive part ($\mathbf{z}_{SW\delta i,k}$) of the switch values, the output vector contains the derivative of the non-impulsive switch values ($\dot{\mathbf{z}}_{SWi,k}$) as well. It is used to detect toggling between switch-states caused by natural switching events. However, the preprocessing and the toggling detection are not essential for understanding the content of this paper. Thus, it is referred to [1] for further details of these topics.

V. Dirac Impulse Detection

In offline simulations, the occurrence of the theoretical Dirac impulses can be detected by variable step-size solvers to detect if the left-hand limit x_0^- (in the example i_L^-) is different from the right-hand limit x_0^+ (in the example i_L^+). In contrast, real-time simulation requires a fixed step-size or at least a fixed calculation time. This avoids the classical usage of variable-step zero-crossing detection mechanisms, so that a different approach is required. A Dirac detection logic was developed in [1], that evaluates if a state variable disappears because of linear dependencies caused by a forced switching event. This seems suitable for the example in Fig. 5. However, for the to-

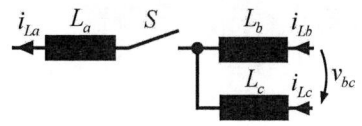

Switch-state	Number of state variables	Dependencies
on	2	$i_{La} = 0 \qquad 0 = i_{Lb} + i_{Lc}$
off	2	$i_{La} = i_{Lb} + i_{Lc}$

Fig. 6: Example of a topology where a Dirac impulse occurs, but the number of independent state variables remains constant.

tem-pole topology shown in Fig. 3, several situations can occur, where the number of state variables is constant, but a Dirac delta impulse must be considered. The necessity can be understood by investigating the example in Fig. 6. Two state variables exist independently from the switch-state of S. However, when a forced switching event occurs and S changes its switch-state, the equations for the linear dependencies change, which represents a step in the inductor current i_{La}. Thus, a Dirac impulse occurs in the simulation, i.e.

$$v_{La\delta} = L_a \cdot \left[\underbrace{0}_{i_{La}^-} - \underbrace{\left(i_{Lb} + i_{Lc} \right)}_{i_{La}^+} \right] \cdot \delta\left(t - t_{sw} \right). \tag{6}$$

when the switch S changes from off- to on-state. Therefore, a Dirac detection logic requires to consider the equation of the linear dependencies instead of evaluating the number of linear independent state variables, only. To take this into account, an extended Dirac detection logic can be derived. The basis is the Boolean function (7), which becomes true when the evaluated switch-state is included in the set $M_{p,q}$.

$$L_{p,q} = \sum_{m \in M_{p,q}} \prod_{j=1}^{n_{sw}} \tilde{s}_j \quad \text{with } \tilde{s}_j = \begin{cases} \overline{s}_j & \text{for } s_{m,j} = 0 \\ s_j & \text{for } s_{m,j} = 1 \end{cases} \tag{7}$$

The set $M_{p,q}$ comprises all switch-states, where the q-th dependency equation is used to force the p-th state-variable. An example for illustrating the set definition and the coherence to the different dependency equations can be found in Fig. 7. In the switch-states $(000)_2$, $(010)_2$ and $(100)_2$, the p-th state-variable x_p is forced by equation f_0. Apply (7) to the example with $q = 0$ yields:

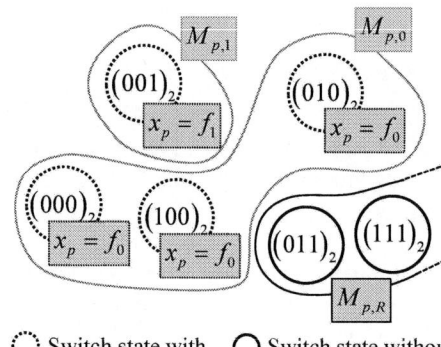

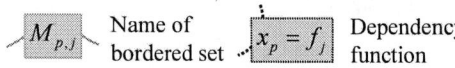

Fig. 7: Example for illustrating the set definition and the coherence to the different dependency equations.

$$L_{p,0} = \sum_{m \in M_{p,0}} \prod_{j=1}^{3} \tilde{s}_j = \left(\overline{s}_1 \wedge \overline{s}_2 \wedge \overline{s}_3 \right) \vee \left(\overline{s}_1 \wedge s_2 \wedge \overline{s}_3 \right) \vee \left(s_1 \wedge \overline{s}_2 \wedge \overline{s}_3 \right), \ M_{p,0} = \left\{ (000)_2, (010)_2, (100)_2 \right\}. \tag{8}$$

To determine the switches, which can cause Dirac impulses when they change their switch-state due to a forced switching event, $L_{p,q}$ is negated and simplified to the minimal disjunctive normal form (MDNF), e.g., by applying the Quine-McCluskey algorithm, [11].

$$\overline{L}_{p,q} \overset{MDNF}{=} \sum_{j=1}^{n_{seg}} h_{p,q,j}\left(s_1, \ldots, s_{n_{sw}} \right) = h_{p,q,1}\left(s_1, \ldots, s_{n_{sw}} \right) \vee \ldots \vee h_{p,q,n_{seg}}\left(s_1, \ldots, s_{n_{sw}} \right) \overset{e.g.}{=} \underbrace{s_3}_{h_{p,0,1}(s_3)} \vee \underbrace{\left(s_1 \wedge s_2 \right)}_{h_{p,0,2}(s_1,s_3)} \tag{9}$$

The terms $h_{p,q,j}\left(s_1,\ldots,s_{n_{sw}}\right)$ describe the topologies between the switches, which are crucial for the enforcement of the p-th state-variable by the dependency equation f_q. Based on the knowledge that a switch-state belongs to a certain set and the feedback about the type of switching event from the logic in Fig. 4, a Boolean function can be formulated to detect a Dirac impulse in respect to f_q:

$$D_{p,q} = L_{p,q,k} \cdot \sum_{j=1}^{n_{swg}} h_{p,q,j}\left(s_{1,k-1},\ldots,s_{n_{sw},k-1}\right) \cdot \overline{h}_{p,q,j}\left(s_{1,k},\ldots,s_{n_{sw},k}\right) \cdot g_{p,q,j}\left(s_{1F,k},\ldots,s_{n_{sw}F,k}\right)$$

$$\overset{e.g.}{=} L_{p,0,k} \wedge \left[\left(h_{p,0,1}\left(s_{3,k-1}\right) \wedge \overline{h}_{p,0,1}\left(s_{3,k}\right) \wedge \left(s_{3F,k}\right) \right) \vee \left(h_{p,0,2}\left(s_{1,k-1},s_{2,k-1}\right) \wedge \overline{h}_{p,0,1}\left(s_{1,k},s_{2,k}\right) \wedge \left(s_{1F,k} \vee s_{2F,k}\right) \right) \right] \quad (10)$$

$$= L_{p,0,k} \wedge \left[\left(s_{3,k-1} \wedge \overline{s}_{3,k} \wedge s_{3F,k} \right) \vee \left(\left(s_{1,k-1} \wedge s_{2,k-1}\right) \wedge \overline{\left(s_{1,k} \wedge s_{2,k}\right)} \wedge \left(s_{1F,k} \vee s_{2F,k}\right) \right) \right]$$

Here, $g_{p,q,j}\left(s_1,\ldots,s_{n_{sw}}\right)$ represents the OR-operated switch-states s_i which occur in $h_{p,q,j}\left(s_1,\ldots,s_{n_{sw}}\right)$ and s_{iF} represents the feedback from the event- and switch-type oriented switching condition for forced switching events, see Fig. 4. To comprehend (10), $L_{p,q,k}$ evaluates if for the switch-state in the current simulation step (t_k) the p-th state-variable is forced by the dependency function f_q. Furthermore, the sum evaluates for each essential switch topology $h_{p,q,j}\left(s_1,\ldots,s_{n_{sw}}\right)$ if it was true in the last simulation step (t_{k-1}) so that $\overline{L}_{p,q}$ was true, $h_{p,q,j}\left(s_1,\ldots,s_{n_{sw}}\right)$ is false for the current simulation step (t_k) and if a forced switching event occurred for switches, which are involved in $h_{p,q,j}\left(s_1,\ldots,s_{n_{sw}}\right)$. Using (10), the comprehensive Dirac impulse detection for the p-th state-variable is given by

$$D_p = \sum_{\forall q} D_{p,q} . \tag{11}$$

In conclusion, if D_p becomes true, the Dirac impulse source for the p-th state-variable needs to be considered. Thus, the impulsive conditions of the switch-values, which correspond to the involved switches (see $g_{p,q,j} \forall i,j$), must be evaluated and can overrule the results of the non-impulsive conditions.

VI. Failure simulation

One intention of failure simulation is to test the reaction of the circuit and the control unit, when one or more semiconductors do not work properly or when the connections for the gate-signal are interrupted. This requires that all $2^{n_{sw}}$ switch-states for the discrete state-space representation in (5) are considered, although not all of them might be relevant for the nominal operation of the topology.

Here, the failure simulation is also used to set constraints for the switches to simulate different totem-pole configurations with one and the same model, see Table I.

An interruption of a gate signal can be easily reproduced, by setting the corresponding model input for the gate-signal to continuously low or high. To force a switch into a certain switch-state, e.g., to consider a faulty semiconductor, requires the extension of the event-based and switching event-oriented switching conditions from [1] (extraction shown in Fig. 4). For the failure simulation feature, the results of the corresponding switching conditions $S_{OnOff,F}$, $S_{OffOn,F}$, $S_{OnOff,N}$ and $S_{OffOn,N}$ are discarded when failure simulation is activated. They are replaced by $S_{OnOff,F} = c_k < c_{k-1}$, $S_{OffOn,F} = c_k > c_{k-1}$ and $S_{OffOn,N} = S_{OnOff,N} = 0$ where c is the control signal for the failure simulation. This signal defines if the switch should be forced in off-state ($c = 0$) or in on-state ($c = 1$).

VII. Integration on FPGA

Beside the general modeling approach introduced in the previous sections, its implementation on an FPGA is important as well. Amongst others, FPGAs are well known for their high flexibility and the capability of massive parallel processing. However, resources like logic cells or DSPs to execute these operations are limited. This requires the consideration of pipelining and multiplexing for the implementation of huge and complex models like the one proposed within this paper for the totem-pole PFC.

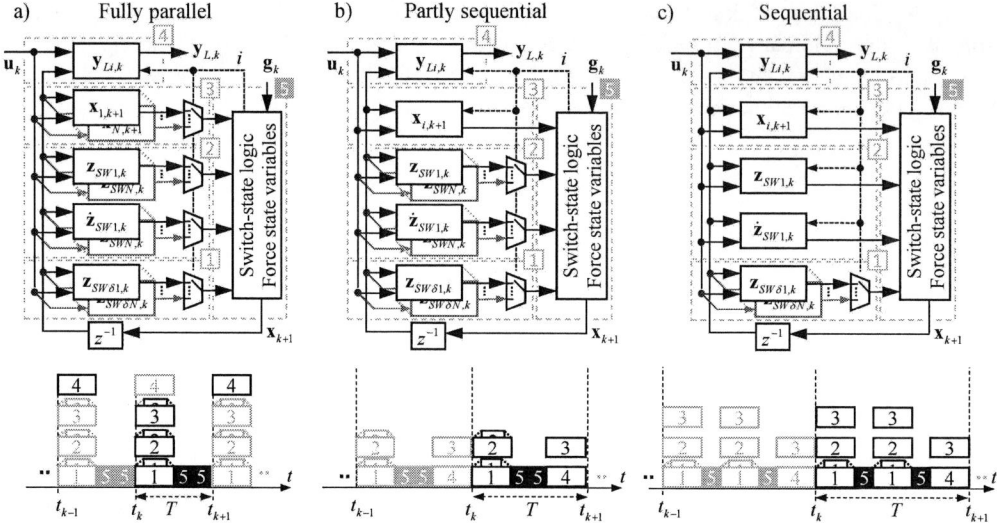

Fig. 8: Different model structures and their timing sequences for the subcomponents considering two iterations of the switch-state logic.

In [1] a systematic approach for the implementation of a matrix times vector multiplication was introduced. It allows adjustment of the trade-off between utilized hardware resources, routing effort, and latency required for the multiplication. In addition to this, Fig. 8 depicts three different model structures and their calculation sequences that can be applied to implement models according to the proposed modeling approach.

Table II compares the structures in terms of required number of matrix multiplications and expected model step-size. For the fully parallel structure in Fig. 8a), the state space representation from (5) must be calculated for all $2^{n_{sw}}$ switch-states in parallel. All required quantities for the switch-state logic are available at the same time so that they just need to be evaluated in an iterative manner. Unless some switch-states can be described with the same state space representation as other switch-states, this requires that $2^{n_{sw}}$ matrix times vector multiplications, e.g., to update $\mathbf{x}_{k+1}$, have to be implemented and executed in parallel. For the outputs $\mathbf{y}_{Li,k}$, model pipelining is applied by calculating the output equation in parallel to the next simulation step – see calculation sequence under the block diagram of Fig. 8a). As only one matrix multiplication is required in this case, the resources can be saved without affecting the step-size. Compared to the other model structures, the fully parallel structure requires a lot of hardware resources. However, it will result in the smallest model step-size T, i.e., the highest PWM resolution or oversampling factor κ can be achieved with this structure. It was applied for the model of the LLC converter with a step-size of $T = 144\,\text{ns}$ in [1].

As mentioned above, it is possible to adjust the required hardware resources for the matrix and vector multiplications to a certain extent with the multiplexing approach from [1]. However, if not enough resources are available or if the latency for the multiplications becomes too high due to a high multiplexing grade, the partly sequential model structure from Fig. 8b) can be used. Here, the switch

Table II: Comparison of the computational effort of the model structures in Fig. 8.

model structure	required number of matrix times vector multiplications for					minimum model step-size T/T_{FPGA}
	$\mathbf{x}_{k+1}$	$\mathbf{y}_{Li,k}$	$\mathbf{z}_{SW,k}$	$\dot{\mathbf{z}}_{SW,k}$	$\mathbf{z}_{SW\delta,k}$	
fully parallel	$\leq 2^{n_{sw}}$	1	$\leq 2^{n_{sw}}$	$\leq 2^{n_{sw}}$	$\leq 2^{n_{sw}}$	$\max\left(l_{x_{k+1}}, l_{z_{SW}}, l_{\dot{z}_{SW}}, l_{z_{SW\delta}}\right) + n_{Iter} \cdot l_{Logic}$
partly sequential	1	1	$\leq 2^{n_{sw}}$	$\leq 2^{n_{sw}}$	$\leq 2^{n_{sw}}$	$\max\left(l_{z_{SW}}, l_{\dot{z}_{SW}}, l_{z_{SW\delta}}\right) + n_{Iter} \cdot l_{Logic} + l_{x_{k+1}}$
sequential	1	1	1	1	$\leq 2^{n_{sw}}$	$\left(\max\left(l_{x_{k+1}}, l_{z_{SW}}, l_{\dot{z}_{SW}}, l_{z_{SW\delta}}\right) + l_{Logic}\right) \cdot n_{Iter} + l_{x_{k+1}}$

n_{sw} number of switches, l_j latency for operation j, n_{Iter} number of iterations for switch-state logic quantities $\mathbf{z}_{SWi}$, $\dot{\mathbf{z}}_{SWi}$ and $\mathbf{z}_{SW\delta i}$ are also calculated in parallel so that they can be directly evaluated by the switch-state logic. However, the calculation of $\mathbf{x}_{k+1}$ is postponed compared to the parallel structure. This increases the step-size by the latency $l_{x_{k+1}}$ for this operation but reduces the number of matrix multiplications from $2^{n_{SW}}$ to just one for updating $\mathbf{x}_{k+1}$.

A further significant reduction of required hardware resources can be achieved by applying the sequential model structure as shown in Fig. 8c). In contrast to the afore mentioned structures, the switch quantities $\mathbf{z}_{SWi}$ and $\dot{\mathbf{z}}_{SWi}$ are calculated for the actual valid switch-state i, only. Usually, the impulsive switch quantities in $\mathbf{z}_{SW\delta i}$ are equal to zero for the most switch-states. Furthermore, often state-variables like an inductor current represent these impulsive parts (see Fig. 6) so that $\mathbf{C}_{\mathbf{d}SW\delta i}$ and $\mathbf{D}_{\mathbf{d}SW\delta i}$ from (5) are typically sparse matrices. Thus, the number of required multiplications is typically much smaller than for the other quantities. Due to this, all $\mathbf{z}_{SW\delta i}$ are calculated in parallel so that even with the sequential model structure fast model step-sizes are achieved.

Based on $\mathbf{z}_{SWi}$ and $\dot{\mathbf{z}}_{SWi}$ for the actual valid switch-state i the switch-state logic determines a new temporary switch-state. For this new state the corresponding impulsive quantities $\mathbf{z}_{SW\delta i}$ are selected and evaluated, followed by an update of the switch quantities for the resulting temporary switch-state within the next iteration step. Compared to the parallel model structure, this sequential approach yields an increase of the step-size by a factor almost equal to the number of iterations n_{Iter}. However, significantly less matrix and vector multiplications need to be implemented. This enables the implementation of models for huge and complex topologies like the totem pole PFC from Fig. 2 and 3.

VIII. Simulation results

For the topology in Fig. 3 with $n_{sw}=8$ switches, a model was realized based on the approaches presented in the previous sections. For all $2^{n_{sw}}=256$ switch-states, a state-space representation according to (5) was determined with four state-variables $\mathbf{x}=\begin{bmatrix} i_{La} & i_{Lb} & i_{Lc} & v_o \end{bmatrix}^T$, inputs $\mathbf{u}_S=\begin{bmatrix} v_a & v_b & v_c & i_{Load} \end{bmatrix}^T$ and outputs $\mathbf{y}_L=\begin{bmatrix} i_{La} & i_{Lb} & i_{Lc} & v_o \end{bmatrix}^T$. To achieve an oversampling factor κ of at least 20 for switching frequencies up to 150 kHz and to reduce the required hardware resources under consideration of the mentioned dimensions, the sequential model structure is applied.

The model is implemented on a dSPACE HIL simulator with a high-end DS6602 FPGA board that includes a Xilinx Kintex Ultrascale+ FPGA. The clock period for the DS6602 is $T_{FPGA}=8\,\text{ns}$. The model uses $n_{Iter}=2$ iterations and is calculated within 40 sample clocks. This results in a simulation step-size of $T=320\,\text{ns}$. It fulfills the specifications for the oversampling factor κ for switching frequencies up to 150 kHz.

Simulation results are shown in Fig. 9 for a single-phase grid operation. For comparison, a simulation of the topology in Fig. 2 with Simscape Electrical™ Specialized Power Systems is used as reference.

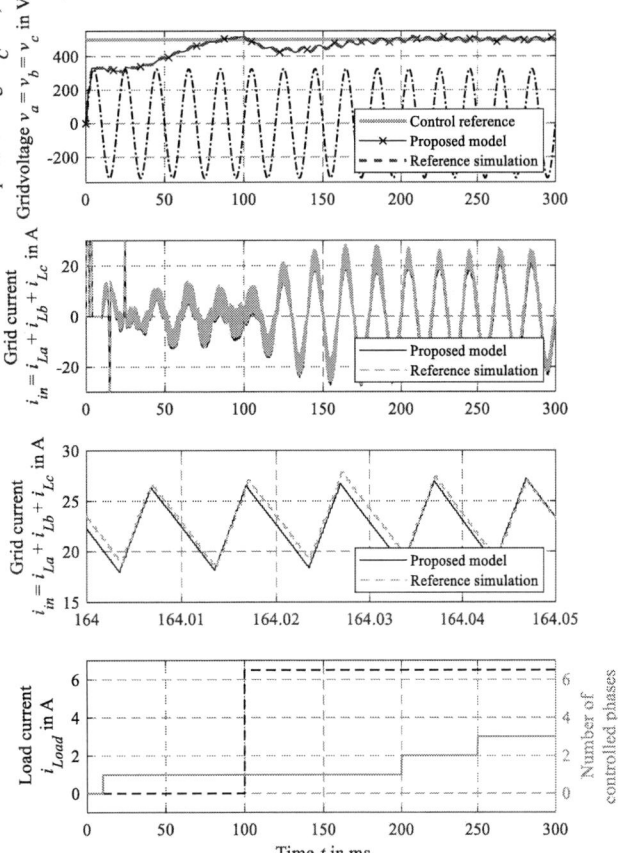

Fig. 9: Simulation results of the totem-pole PFC rectifier model.

The simulation starts with passive operation. At $t = 10\,\text{ms}$, the current control for i_{La} of the first phase as well as the voltage controller are activated. After the voltage reaches the reference value, a load step occurs at $t = 100\,\text{ms}$. Later, the current controllers of the second (i_{Lb}) and third (i_{Lc}) interleaved phases are activated sequentially at $t = 200\,\text{ms}$ and $t = 250\,\text{ms}$. This shows the appropriate consideration of the grid relays S_a, S_b, and S_c with the simplified topology in Fig. 3 combined with failure simulation. In general, the simulation results of the reference and the model show good agreement.

IX. Conclusion

Within this paper, the development of a ready-to-use model for an FPGA-based HIL simulation of a totem-pole topology was presented. The model is able to simulate several totem-pole configurations, like two-phase or three-phase interleaved operation, without loading different specific models to the FPGA.

The considered totem-pole circuit as well as the possible operations modes were introduced in section II. Afterwards, the concept and simplifications required to realize the FPGA-based HIL simulation for such a complex topology were described in section III. As a result, a simplified topology with equivalent behavior but only eight switches was obtained.

Section IV recapped basics of the applied modeling approaches based on ideal switches, while in section V an extended and systematic approach to detect switch-state changes caused by Dirac impulses was presented. An approach to force certain switches into a specific switch-state was introduced in section VI. It can be used for failure simulation, but here it was also applied to define constraints for the switches depending on the investigated totem-pole configuration.

For the implementation on an FPGA, the available resources must be considered as well. To adjust the trade-off between required resources and obtained step-size, three different model structures were compared in section VII. Finally, the proposed real-time model was validated by a simulation in section VIII. The results were in good agreement with the reference simulation.

References

[1] A. Kiffe, T. Hoffstadt, F. Puschmann: Modeling approach for real-time capable oversampling models applied to a fast switching LLC resonant converter, 2021 23rd European Conference on Power Electronics and Applications (EPE'21 ECCE Europe), 2021, pp. P.1-P.11.

[2] J. Yuan, L. Dorn-Gomba, A. D. Callegaro, J. Reimers, A. Emadi: A Review of Bidirectional On-Board Chargers for Electric Vehicles, IEEE Access, vol. 9, pp. 51501-51518, 2021.

[3] H. F. Blanchette, T. Ould-Bachir and J. P. David, "A State-Space Modeling Approach for the FPGA-Based Real-Time Simulation of High Switching Frequency Power Converters," in IEEE Transactions on Industrial Electronics, vol. 59, no. 12, pp. 4555-4567, Dec. 2012.

[4] R. Champagne, L.-A. Dessaint, H. Fortin-Blanchette, G. Sybille: Analysis and validation of a real-time AC drive simulator, in IEEE Transactions on Power Electronics, vol. 19, no. 2, pp. 336-345, March 2004.

[5] P. Pejovic and D. Maksimovic, "A method for fast time-domain simulation of networks with switches," in IEEE Transactions on Power Electronics, vol. 9, no. 4, pp. 449-456, July 1994.

[6] K. De Cuyper: "Automated modeling and implementation of power converters on a real-time FPGA based emulator.", PhD Thesis, 2015.

[7] Allmeling, J. H., Hammer, W. P.: PLECS-piece-wise linear electrical circuit simulation for Simulink, Proceedings of the IEEE 1999 International Conference on Power Electronics and Drive Systems. PEDS'99, pp. 355-360 vol.1, 1999.

[8] E. S. Kuh, R. A. Rohrer: The state-variable approach to network analysis, Proceedings of the IEEE, vol. 53, no. 7, pp. 672-686, July 1965.

[9] Unbehauen, R.: Elektrische Netzwerke. Springer Verlag, 1981.

[10] S. Natarajan, "A systematic method for obtaining state equations using MNA," in IEE Proceedings G - Circuits, Devices and Systems, vol. 138, no. 3, pp. 341-346, June 1991.

[11] W. V. Quine (1955) A Way to Simplify Truth Functions, The American Mathematical Monthly, 62:9, 627-631, DOI: 10.1080/00029890.1955.11988710.

Investigation of core-loss mechanisms in large-scale ferrite cores for high-frequency applications

Michael Baumann, Christoph Drexler, Jonas Pfeiffer, Jens Schueltzke, Erwin Lorenz and Michael Schmidhuber

SUMIDA COMPONENTS & MODULES GMBH
Dr. Hans-Vogt-Platz 1
94130 Obernzell, Germany
Tel.: +49 8591 937 – 441
Fax: +49 8591 937 – 9403
E-Mail: cdrexler@eu.sumida.com
URL: https://www.sumida.com

Acknowledgements

The authors would like to thank the Federal Ministry for Economic Affairs and Climate Action. Parts of the work presented here were done within the project with the funding code 03EI4014C.

Keywords

«Ferrite», «Finite-element analysis», «Magnetic device», «Core loss», «Core loss modelling», «Automotive component», «Dielectric loss»

Abstract

In this paper, core-loss mechanisms in ferrite cores are investigated. It is demonstrated that ferrite materials showing almost identical core-losses in the datasheets exhibit a completely different loss-behavior when cores with large cross-sections are manufactured. Therefore, there is a high risk that the wrong ferrite material is selected in the design process of a magnetic component for power electronic applications when only the losses from the datasheets are considered. Consequently, a model is developed which leads to a better understanding of the origin losses in ferrite cores and can be used as powerful tool to design and optimize magnetic components for high-frequency applications.

With this model, formulas are derived which allow the calculation of two additional loss contributions which are the electrical polarization losses and the volume eddy current losses. In a detailed experimental study of various ferrite core shapes and sizes the model is verified and clearly shows that for example at 100kHz with a magnetic cross-section of 500mm², the additional losses cannot be neglected. The dependencies and input parameters are discussed in detail. Finally, a FEM-based workflow is presented and verified by calorimetric measurements. In the future, this workflow can be used to precisely predict the losses of virtual prototypes in a development process i.e., for automotive applications.

Introduction

With the availability of SiC- and GaN-switches, a lot of effort has been spent to increase the switching frequencies of power electronic systems. While on the one hand the converted power can be increased in such novel systems, on the other hand it was often said to be obvious that the size of the magnetic components can be decreased in the same manner. However, it was discovered and reported recently [1] that the core losses drastically increase in the range of several hundreds of kHz making ferrite cores thermally unstable and the design process of such magnetic components much more difficult.

The development process of inductive components has been improved over decades and is summarized e.g., in [2-4]. A lot of effort has been made to develop methods to calculate the losses of ferrite cores in such components. The most popular approach is known as the Steinmetz-formalism, where sets of measurements are evaluated by phenomenological equations. Methods to estimate additional contributions due to eddy-currents have been addressed recently as well [5,6]. By contrast, models with a physical origin like the Play-model [7], which is an improved version of the Preisach-model, or the Jiles-Atherton-Model [8], among many others not mentioned here, have been developed over decades. However, all these models are typically benchmarked with ring-cores and often show weaknesses when it comes to a comparison with a large-scale ferrite-core. Thus, in a conventional design process of a magnetic component, the lack of accuracy of the models lead to iterative loops of assembly and testing of hardware prototypes which cost time and money.

In this paper, the mechanisms of core-losses in ferrite cores are presented and quantified. Based on 3D-finite-element-method (FEM) simulations, the core-loss density of PQ50-cores and other frequently used core shapes manufactured of SUMIDA high-frequency power ferrites is determined and experimentally verified. The findings give a better understanding about the loss-mechanisms in ferrite cores and close the gap between measurement and calculation. The presented method might serve as a powerful tool to investigate virtual prototypes in a modern development process for magnetic components.

Motivation

The focus of this paper is on the quantitative comparison of core-losses derived from finite-element-method simulations and calorimetric measurements. With the tendency addressed above to shrink the size of magnetic components for power electronic systems, the understanding of the loss-mechanisms in the core i.e., in ferrite-cores, are becoming more and more important. It is a known secret that there are huge discrepancies between losses in small size toroidal cores used for datasheet value measurements and large-scale cores used for applications. This manifests itself in the comparison of two ferrite materials presented in Fig. 1.

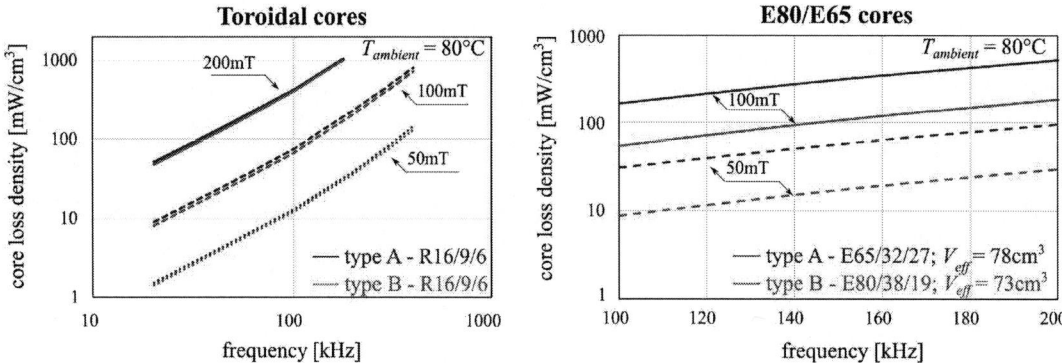

Fig. 1: Power loss density measurement of toroidal cores (left) and E-cores (right) for two SUMIDA ferrites at different peak flux densities: type A - high saturation flux density ferrite (blue) and type B - low loss high frequency ferrite (red). Both sets of measurements have been measured following standard DIN EN62044-3 for core temperatures of 80 °C.

Therein, the comparison of core-loss density measurements for two different SUMIDA power ferrite materials ($\mu_{eff} \approx 2000$) are presented. The first one is a ferrite with a high saturation flux density and is in the following called type A. The second one is a material which is optimized towards low losses at frequencies up to 400 kHz and is called type B. From those measurements it follows, that for small toroidal cores (left-hand side) the losses are almost the same with less than 5 % deviation for the considered temperature and frequency range. Hence, the materials look almost identical regarding the expected core losses. However, measurements of an E80-core of type B and an E65-core of type A (right-hand side) yield a difference in the core loss density of a factor of 3 to 4. It has to be mentioned that the E-cores have not a perfectly identical shape but have almost the same effective volume. Consequently, similar core-losses would be expected by using the Steinmetz formalism. Therefore, there is a significant risk that the less-effective material could be selected in a development process when relying just on data derived from small-scale toroidal measurements. The findings give rise to having a closer look on the microscopic mechanisms behind the core losses and to deriving methods to predict the losses of magnetic components upfront in virtual prototypes.

Core-loss mechanism in ferrite cores

A common approach for the losses of ferrite cores is the separation in hysteresis, eddy-current and residual losses [4]:

$$P_{core} = P_{hyst} + P_{eddy} + P_{res} \tag{1}$$

Therein, the hysteresis losses are ascribed to losses due to the change of the magnetic moments with time. The eddy-currents are losses due to the non-zero conductivity of the ferrite and residual losses arise due to a phase shift between the external magnetic field and the oscillating magnetic moments at high frequencies being often neglected. In [1], a correction factor to the Steinmetz-approach was derived which accounts to the eddy-current losses and closes the gap between simulated and measured core-losses. However, this approach is non-physical and can be seen rather as a rule-of-thumb.

In this work, an additional core-loss contribution is investigated which will be in the following entitled as dielectric volume losses. Therefore, Eq. (1) can be re-written:

$$P_{core} = P_{hyst} + P_{(eddy,grain)} + P_{res} + P_{dielectric} \tag{2}$$

Herein, $P_{eddy,grain}$ accounts to the eddy current losses on grain-level which depend on the material compound. These losses do not depend on the geometry of the core and are already considered in the approach of the Modified Steinmetz Equation (MSE) for datasheet values. The additional term $P_{dielectric}$ is related to dielectric volume losses and will be derived and explained in the following chapter.

Deriving of dielectric volume loss

In [2] a derivation of eddy current losses for a rectangular cross-section based on the field equations is shown. However, the part of losses due to electric polarization is not considered. In the following, a simple method is shown, which also includes the electrical polarization loss component. The following derivation is only valid for Power-MnZn-Ferrite such as type A and type B where the penetration depth is still fully given due to the high resistivity and limited magnetic cross-section [9].

For a sinusoidal magnetic flux density B, an angular frequency ω and with the magnetic cross-section A_e (Fig. 2) a voltage U_{eff} is induced:

$$U_{eff} = \frac{\omega \hat{B} A_e(x)}{\sqrt{2}} = \frac{\omega \hat{B} \cdot 2x(l-d+2x)}{\sqrt{2}} \tag{3}$$

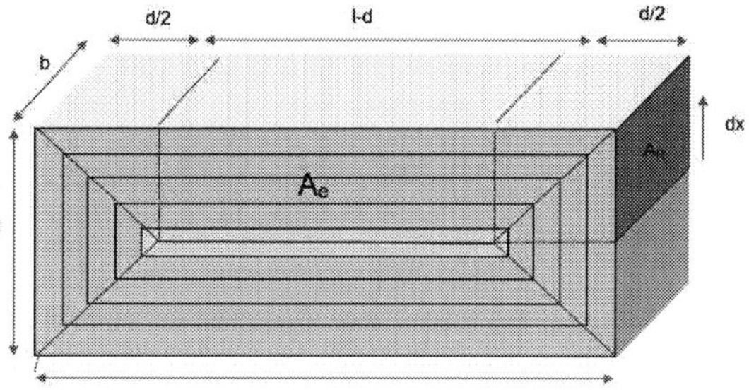

The resistant R of concentric loops of the geometric shape shown in Fig. 2 can be described with the specific resistant ρ as:

$$R = \frac{2\rho l}{dA_R(x)} = 2\rho \frac{(l-d)+8x}{bdx} \tag{4}$$

Fig. 2: Magnetic cross-section with integration loops

The movement of electrons due to the conductivity results in power losses:

$$dP = \frac{U_{eff}^2}{R} = \frac{4\pi^2 f^2 \hat{B}^2 b(l-d+2x)^2 x^2}{\rho(l-d+4x)} dx \tag{5}$$

Dividing by volume $V_e = lbd$ and integrating over dx, the specific power losses P_v can be calculated:

$$P_v = \frac{\pi^2 f^2 \hat{B}^2}{16\rho} \cdot \frac{64}{ld} \int_0^{d/2} \frac{2lx^2 - 2dx^2 + 4x^3}{l - d + 4x} dx = \frac{\pi^2 f^2 \hat{B}^2}{16\rho} \left[\frac{(l-d)^4}{4ld} \ln\frac{l+d}{l-d} - \frac{(2l-2d)^2}{4} \right] \tag{6}$$

By introducing the effective magnetic cross section $A_e = ld$ and a geometric factor of the shape F_G

$$F_G = \frac{(l-d)^4}{4(ld)^2} \ln\frac{l+d}{l-d} - \frac{l^2 - 4dl + d^2}{2ld} \tag{7}$$

Eq. (6) can be written:

$$P_v = \frac{\pi^2 f^2 \hat{B}^2 A_e F_G}{16\rho} \tag{8}$$

A good simplification for F_G gives following equation with the help of the aspect ratio $F = l/d$:

$$F_G \approx \frac{8}{3F} - \frac{5}{3F^{1,7}} \tag{9}$$

In first approximation F_G can be further simplified with:

$$F_G \approx \frac{3}{F+2} \tag{10}$$

If, in addition to the electrical conductivity losses, dielectric polarization losses also occur, both loss mechanisms can be combined into an effective dielectric loss ε_r'' [10]:

$$\varepsilon_r'' = \varepsilon_{dipol}'' + \frac{\sigma}{\varepsilon_0 \omega} \tag{11}$$

The effective dielectric loss can be regarded as a high-frequency conductivity σ_{hf} [10]. The frequency-dependent part of conductivity can be neglected in case of MnZn-ferrite and used frequencies.

$$\sigma_{hf} = \varepsilon_0 \omega \varepsilon_{dipol}'' + \sigma \tag{12}$$

Since the electric field not only acts on the ionic conductivity, but also on the dipolar conductivity, high-frequency conductivity can be used in the loss formula (8):

$$P_{dielectric} = \frac{\pi^2 f^2 \hat{B}^2 A_e F_G (\varepsilon_0 \omega \varepsilon_{dipol}'' + \sigma)}{16} = \frac{\varepsilon_0 \pi^3}{8} \left(\varepsilon_{dipol}'' + \frac{\sigma}{2\pi f \varepsilon_0} \right) f^3 \hat{B}^2 A_e F_G \tag{13}$$

The simplest form with $\varepsilon'' = \varepsilon_0 \varepsilon_r''$ and the help of (10) is:

$$P_{dielectric} = \frac{3\pi^3 \varepsilon''}{8} f^3 \hat{B}^2 \frac{A_e}{F+2} \tag{14}$$

If the magnetic cross-section or the flux density along the magnetic path is not constant, then the individual volume elements must be described in sections where the most general form is:

$$P_{dielectric} = \frac{\pi^3 \varepsilon'' f^3}{8V_e} \sum_{i=1}^n \frac{\hat{\emptyset}_i^2 l_i}{F_{G_i}} \tag{15}$$

In (12), one contribution is related to the material's conductivity σ and is in the following referred to as $P_{eddy,volume}$ and describes eddy currents which spread in the whole core-volume due to the conductive connection of single grain boundaries. Another contribution depends on the imaginary part of the dielectric function ε''_{dipol} of the core material and will be called P_{dipol} in the following. Both parameters are dependent from temperature and frequency. A_e is the effective cross section of the core and F describes the length-ratio of the core's cross section and will be called aspect-factor in the following. F is equal to one for circular and square-shaped cross-sections and increases for rectangular shapes with the increasing difference of the side-lengths.

Experimental investigation of the dielectric volume loss

In order to investigate the additional losses experimentally, several tens of toroidal cores and U-cores of material type A and B were produced. The effective cross section of these samples ranges from 25 to 890 mm^2 and the geometric form-factor F is between 1 (circular or quadratic) and 4 (rectangular). The core-losses for all these designs were measured for different frequencies, flux-densities and temperatures following standard DIN EN62044-3 (see Fig. 3). Several hundreds of measurement points were acquired with that method. In parallel, the core-losses were calculated, in a first step, without the additional losses and are plotted against the measured losses in Fig. 4 (red points).

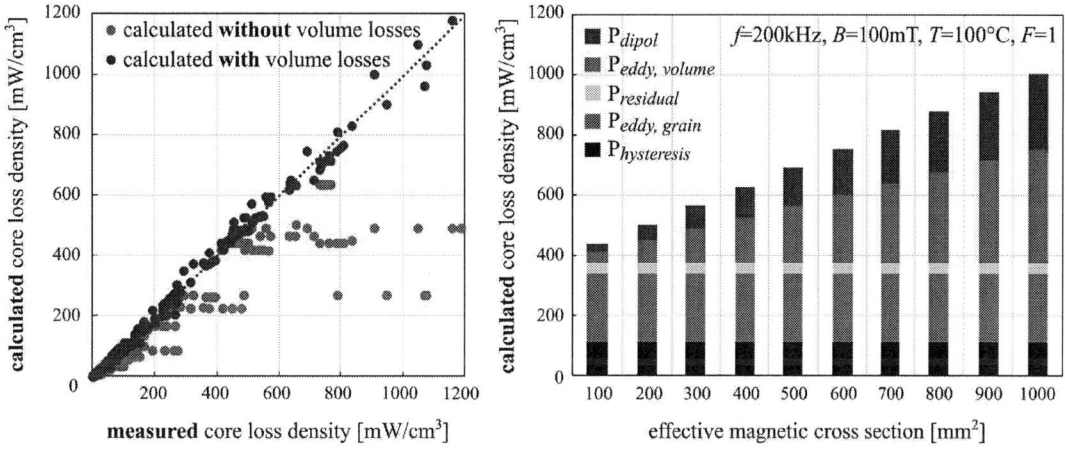

Fig. 3: Toroidal cores with different effective cross section in the experimental setup of the core-loss measurement

Fig. 4: left: Comparison of calculated and measured core-losses with (blue) and without (red) dielectric losses; each point represents either a toroidal core or a U-core with different cross-sections and produced of different ferrite materials; each core was measured/calculated at different points of operation under variation of frequency, flux-density and temperature; the dashed line reflects sufficient match of calculation and measurement; right: magnitudes of each contribution for ferrite-type B at a single point of operation and different cross-sections.

In this presentation method each point appearing on a line with a slope equal to one (dashed blue line) has a perfect agreement between measurement and calculation. It follows that a huge number of designs can be found far of that line when the dielectric volume losses are not considered (red points). It is not shown in the framework of this paper that especially at high-frequencies all designs with large cross-sections and/or low geometric form-factors show deviations up to a factor of three.

It is obvious from Eq. (14) and (15) that for the calculation of the additional losses the numbers for the effective dielectric loss $\varepsilon_r^{''}$ needs to be known. Hence, measurements need to be performed in order to extract these values which are presented in the next chapter.

Determination of the parameters for the dielectric volume loss

The effective dielectric loss $\varepsilon_r^{''}$ is a material parameter depending on temperature, frequency and conductivity and can be determined for several ferrite materials via measurements of small ferrite samples according to DIN 53483-2 reported by [11]. In the diagram Fig. 5 (left) an example for a Low-loss-MnZn-Ferrite is given. For simplification the Eq. (11) can be used to separate $\varepsilon_r^{''}$ in a part of conductivity loss $\sigma/(\omega\varepsilon_0)$ and in a part of polarization loss $\varepsilon_{dipol}^{''}$ shown in Fig. 5 (right). In this case the advantage is that $\varepsilon_{dipol}^{''}$ is in a first approximation independent of frequency f, temperature T and magnetic flux density B in the range between 50 kHz and 400 kHz and between 80 °C and 125 °C.

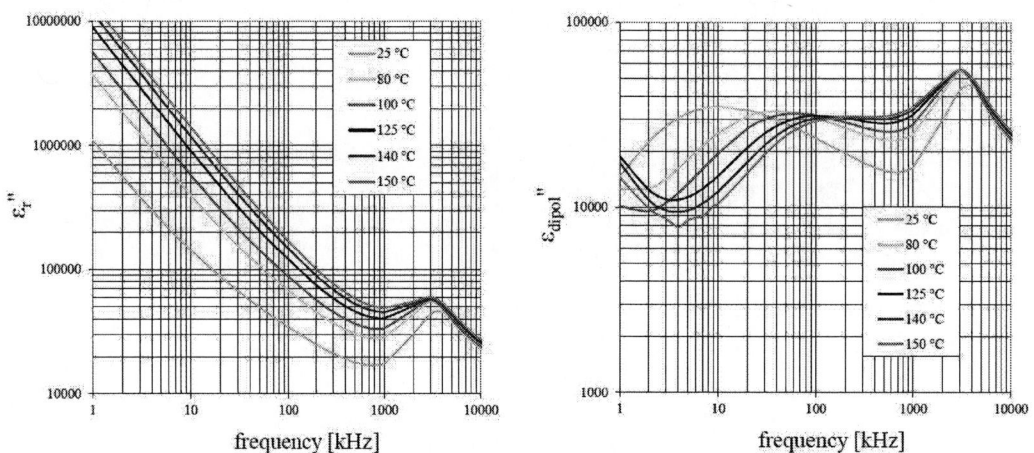

Fig. 5: Example of effective dielectric losses and electric polarization; left: Effective dielectric loss $\varepsilon_r^{''}$ in dependence on frequency and temperature according to DIN 53483-2; right: Electric polarization loss $\varepsilon_{dipol}^{''}$ in dependence of frequency and temperature according to Eq. (11)

The temperature dependence of the conductivity $\sigma(T)$ respectively the specific resistance $\rho(T)$ can be expressed by using the well-known Arrhenius-equation:

$$\rho(T) = \rho_\infty e^{E/k_B T} \tag{16}$$

with the Boltzmann constant k_B and the activation energy $E = 0,2\ eV$ which is in first approximation constant for Power-MnZn-ferrites [12]. The limit resistance ρ_∞ is given by using (16) with the usually available resistant $\rho(25°C)$. In Table I a comparison between the determination of ρ_{25} and $\varepsilon_{dipol}^{''}$ via small signal measurement and high-power measurement is done.

Table I: Dielectric polarization loss $\varepsilon_{dipol}{}''$ in dependence of frequency and temperature according to Eq. 10

	Small signal DIN53483-2		High power Regression (Fig. 5Error! Reference source not found.)	
	$\rho_{25°C}$	$\varepsilon_{dipol}{}''$	$\rho_{25°C}$	$\varepsilon_{dipol}{}''$
Type A ferrite	4 Ωm	60.000	5 Ωm	70.000
Type B ferrite	13 Ωm	30.000	10 Ωm	35.000

FEM-analysis and calorimetric measurements of core-losses

In the final chapter of this paper, the quantitative comparison of core-losses derived from finite-element-method simulations and calorimetric measurements is presented. Therefore, PQ50-cores of SUMIDA made of Fi395 medium-frequency power-ferrite are investigated. The DUTs are assembled with an airgap in order to improve the signal quality of the frequency generator. A litz-wire with 630 strands with a diameter of 0.1 mm is used for the winding to keep the winding losses low. In the experiment, a rectangular voltage is applied to the DUT and the corresponding peak flux-density is calculated from

$$B_{max} = \frac{U}{4A_{eff}Nf} \tag{17}$$

The temperature of the core is measured with type-K sensors at the center and the outer leg, respectively. More details about the calorimetric measurements can be found in [1].

In the simulations, a current source is used which directly injects a triangular current to the winding model which coincides with the amplitudes from the experiments to enable a one-by-one comparison of simulation and measurement. Both the spatial and the temporal distribution of the magnetic flux density is determined in a transient magnetoquasistatic analysis using the 3D-FEM-simulation tool JMAG. The hysteresis losses are evaluated with the formalism of the Modified Steinmetz Equation (MSE) following [2]. The volume eddy current and the electric polarization called dielectric losses are calculated element by element with the formulas derived above. The temperature rise is determined in an iterative thermal and magnetic simulation loop with a modified loss density distribution. For the thermal simulations, a simplified thermal equivalent circuit is used which assumes heat dissipation by free convection only (Fig. 6). The heat transfer coefficients are calculated following [12] or by empirically determined values. The simulation workflow is presented in Fig. 6. For more precision of the thermal simulation, a conjugate heat transfer solver can be used to accurately describe the convection of the sample. However, this step is too time-consuming for the presented workflow and therefore, out of scope of this paper.

Two plots with the results of simulated and measured losses of a PQ50-core are shown in Fig. 7. The first set of data demonstrates the increase of the losses with the magnetizing current and resulting flux density at a constant frequency of 200 kHz. It should be noted that for higher flux densities than 110 mT the core becomes thermally unstable due to the excessive self-heating. In another plot of Fig. 7 the frequency of the magnetizing current was changed but the amplitude was adjusted to keep the losses and thus, the temperature almost constant. The simulation results clearly show that the hysteresis losses calculated by the MSE (red points) underestimate the measured losses by almost 40 %. It must be noted that the component's winding is made of litz-wire in order to keep the winding losses (yellow circles) low which are calculated as described in [1]. From the authors point of view, the huge discrepancy cannot be explained with the often-mentioned process deviations of ferrite cores or any statistical variations of the component's electrical parameters or any calculation errors.

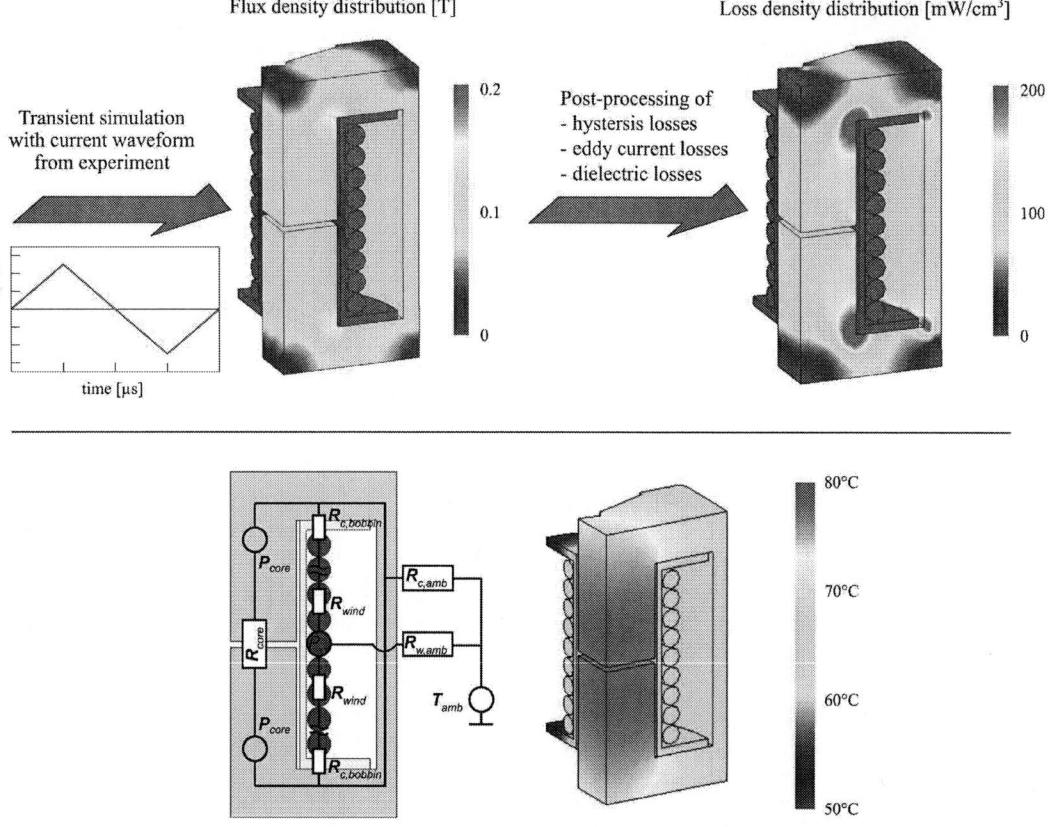

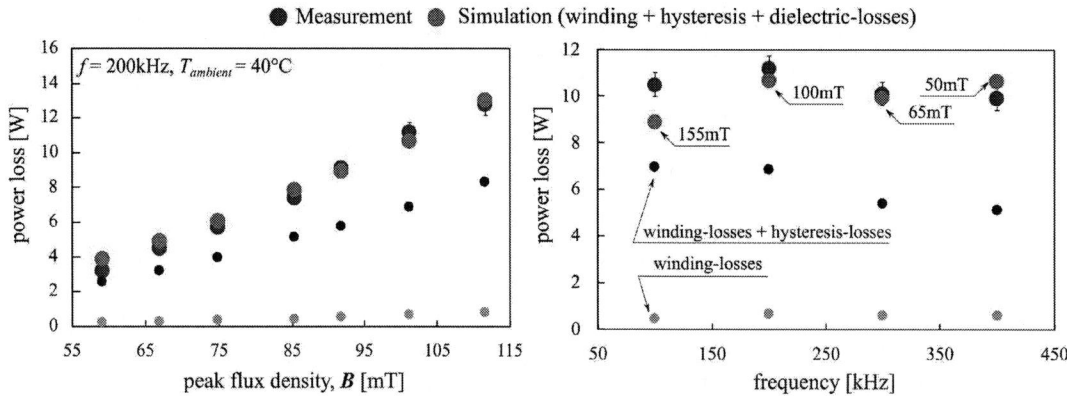

Fig. 6: 3D-FEM-Simulation results of a PQ50-core; top: Workflow of the simulation based on calculation of the core-losses; bottom: Simplified equivalent thermal circuit and temperature distribution of the component.

Fig. 7: Comparison of simulated (dark red) and measured (blue) losses for different flux-densities and frequencies. The error bars indicated a 5 % deviation of the measured values.

The accuracy can be increased when the dielectric volume losses introduced above are added to the core losses indicated by the dark red circles in Fig. 7. Therefore, the polarization loss ($\varepsilon''_{\text{dipol}} = 35000$) is assumed to be independent of the frequency and temperature which can be seen from

Fig. 5. The resistivity/conductivity loss (ρ = 1.66 Ωm) of the material is calculated for 100 °C with the constant taken from Table I. The precision of the simulation results is now discussed for the point of operation with a maximum flux-density of 100 mT and a frequency of 200 kHz. Here the difference between the measured and calculated losses is about 4.3 W when considering classical low volume losses only. The resulting additional losses due to conductivity and polarization losses are calculated to 2.3 W and 1.6 W, respectively by using the flux-density distribution of the FEM-simulation. The form-factor of the PQ-core is taken from the round-shaped magnetic cross-section valid in the center-leg. Thus, the total calculated losses in the component are approximately 10.8 W and, hence, about 0.5 W lower than the measured value of 11.3 W. This procedure is repeated for all operating points and the overall achievable accuracy in the investigated range is approximately 85 % to 90 %.

Finally, the uncertainties of the approach are discussed. The specific resistivity of the material is taken for a temperature of 25 °C. It is known that it decreases with the temperature and consequently. The temperature and frequency dependence of ε_{dipol}'' needs to be investigated further and implemented in the simulation workflow. However, the results presented here show that the used numbers are reasonable in the investigated range. Consequently, the work presented here shows that the losses of cores with non-toroidal shapes can be predicted with significant accuracy. This gives rise to using FEM-simulations as a powerful tool in an industrial development process for magnetic components for automotive applications. In the future, accurate virtual prototypes can be generated and can substitute cost- and time-consuming hardware-loops.

Conclusion

To conclude, core-loss mechanisms for large-scale ferrite cores are investigated in this paper. It is demonstrated that for cores with cross-sections much higher than that used for datasheet measurements and at frequencies above hundred kHz additional loss mechanisms become significant. In order to accurately predict the losses of such cores, a model is presented which allows the calculation of the dielectric losses which vanish for small cores.

The model is analyzed in the framework of an extensive experimental study of test-structures of different geometries in wide range of frequency, temperature, and flux-density. By splitting the dielectric loss ε_r'' in a part of conductivity loss $\sigma/(\omega\varepsilon_0)$ and in a part of polarization loss ε_{dipol}'' the calculated data can be fitted to the measurement results to determine these parameters. Surprisingly, a good agreement between the measurement and the model can be achieved just with single numbers for each material.

In another measurement of test structures, the polarization loss is determined experimentally. The results give rise that the polarization losses can be calculated in a wide frequency and temperature range with a material specific but constant parameter. In addition, it is shown that the conductivity losses are temperature dependent and can be modeled using the Arrhenius-equation. Consequently, the findings presented in this paper give rise to calculate the additional dielectric volume losses over a wide frequency and temperature range with just two additional parameters.

The model is finally tested with the calculation of the core losses based on 3D-FEM-simulation results and compared to calorimetric measurements. The essential parameters for the FEM-simulation can be used from the experiments presented above. The simulations are in good agreement with the calorimetric measurements and show that core-losses can be predicted with an accuracy of several per cent by considering all loss-mechanisms.

References

[1] C. Drexler, M. Wohlstreicher, P. Wrensch, H. Jungwirth and M. Schmidhuber. "Calculation and verification of high-frequency losses in power inductors for automotive application", PCIM Europe digital days, Nuremberg, Germany, 2021, Part of ISBN: 978-3-8007-5515-8.

[2] M. Albach. "Induktivitäten in der Leistungselektronik", Springer Fachmedien Wiesbaden GmbH, Wiesbaden, Germany, 2017, DOI: 10.1007/978-3-658-15081-5.

[3] P. Zacharias. "Magnetic Components", Springer Nature, Wiesbaden, Germany, 2022, ISBN: 978-3-658-37205-7.

[4] E.C. Snelling. "Soft Ferrites: Properties and Applications", Iliffe Books Ldt, London, UK, 1969, ISBN: 978-0-592-02790-6

[5] M. Kacki, M. S. Ryłko, J. G. Hayes and Ch. Sullivan. "A Study of Flux Distribution and Impedance in Solid and Laminar Ferrite Cores", IEEE Applied Power Electronics Conference and Exposition (APEC), Anaheim, CA, USA, 2019, DOI: 10.1109/APEC.2019.8722252.

[6] T. Dimier and J. Biela. "Eddy Current Loss Model for Ferrite Ring Cores based on a Meta-Material Model for the Core Properties", IEEE Transactions on Magnetics, Vol. 58, No. 2, 2022, DOI: 10.1109/TMAG.2021.3084812.

[7] J. Kita, K. Hashimoto, Y. Takahashi, K. Fujiwara, Y. Ishihara, A. Ahagon, T. Matsuo. "Magnetic Field Analysis of Ring Core Taking Account of Hysteretic Property Using Play Mode", IEEE Transactions on Magnetics, Vol. 48, No. 11, pp. 3375-3378, 2012, DOI: 10.1109/TMAG.2012.2204045.

[8] D. C. Jiles and D. L. Atherton. "Theory of ferromagnetic hysteresis", Journal of Magnetism and Magnetic Materials, Vol. 61, No. 1-2, pp. 48-60, 1986, DOI: 10.1016/0304-8853(86)90066-1

[9] A. Stadler. "Messtechnische Bestimmung und Simulation der Kernverluste in weichmagnetischen Materialien", Ph.D. thesis, Friedrich-Alexander-Universität Erlangen-Nürnberg, Erlangen, Germany, 2009.

[10] J. R. Macdonald. "Impedance Spectroscopy - Emphasizing Solid Materials and Systems", John Wiley & Sons Inc, New York, NY, USA, 1987, ISBN: 978-0-471-83122-8

[11] M. Kacki, M. S. Ryłko, J. G. Hayes and Ch. Sullivan. "A Practical Method to Define High Frequency Electrical Properties of MnZn Ferrites", IEEE Applied Power Electronics Conference and Exposition (APEC), New Orleans, LA, USA, 2020, DOI: 10.1109/APEC39645.2020.9124101

[12] W. Kampczyk and E. Röß. "Ferritkerne : Grundlagen, Dimensionierung, Anwendungen in der Nachrichtentechnik", Siemens AG Berlin München, Berlin, Germany, 1978, ISBN: 978-3-800-91254-4

[13] https://e-magnetica.pl/thermal_resistance_of_ferrite_cores, 31.05.2022, 10:16.

Generation of methodology for making benchmark microgrids and application in ESUSCON microgrid

Oscar Dörner, Patricio Mendoza-Araya
UNIVERSITY OF CHILE
Energy Center, Electrical Engineering Department
Av. Tupper 2007, Santiago, Chile
Tel.: +56 (2) 2978 4768
E-Mail: oscardornersb@gmail.com, pmendoza@ing.uchile.cl

Acknowledgements

This work has been partially supported by the Energy Center, FCFM, University of Chile, and ANID/FONDAP/15110019. The results of this work are shared in the following repository: *https://github.com/OscarDorner-UCH/ESUSCON-Isolated-Microgrid-Benchmark.*

Keywords

Microgrid, Test bench, Smart grids, Renewable energy systems, Energy management system

Abstract

A novel methodology for the generation of benchmark microgrids is proposed. For validation purposes, the methodology is applied to the case of the Chilean microgrid ESUSCON. More than 220,000 operating data points are generated, and subsequently validated according to the proposed methodology. Finally, all the estimates and assumptions are published.

Introduction

Over the years, microgrids have increasingly been implemented in the world. An example is the case of the European project "MORE MICROGRIDS", where the number of microgrids projects on the continent was significantly increased. From the information generated by this type of projects, particular implementations, and other investigations, various studies on microgrids have been published by research and development centers, studies that can be distributed in the following categories [1]: 1) Operation and system control; 2) Planning and design; 3) Quality of supply; 4) Protections and 5) Stability. From such studies, several have been carried out in various benchmark microgrids (or test networks), which have different sizes, architectures, and volumes of information. This renders the comparison of aspects such as performance challenging [2].

Existing benchmark microgrids can be divided into two categories: a) Benchmarks made from real data, such as those of an existing microgrid, obtaining data with a deterministic approach [3,4] and b) Benchmark made from fictitious or reconstructed data [1,5], such as those that can be obtained through assumptions and stochastic models already known. In the first category, these data can be formed from historical demand profiles, composition, and topology of the microgrid, failure history, among others. In the second category, the real data of a microgrid are not considered, so the benchmark is generated from topologies available in the literature and variations of these. In addition, the operation data can be modeled from data from other studies, such as line parameters, Distributed Generation (DG) parameters, among others. Demand profiles, failure rates, and meteorological conditions can also be modeled using a stochastic approach.

Due to the difficulties raised when comparing benchmark microgrids that are not easily comparable to each other, having benchmarks that allow comparative studies to be carried out, where optimal approaches can be identified and evaluated for specific studies, is necessary. For this purpose, this work proposes the use of the data sets that the ESUSCON microgrid project has generated [6], to create a microgrid benchmark. In the ESUSCON microgrid, the isolated characteristic stands out, thus imposing that the focus of this work is this type of microgrid. By the use of these data sets, a robust benchmark is generated that meets the conditions already indicated for isolated microgrids, making full use of such data. The data to be used for the benchmark are provided by the University of Chile due to its participation in the ESUSCON project. This benchmark enables multiple studies, including power flows, voltage/frequency control, reliability, and energy management. It is intended to be used for steady-state and dynamic simulations of the operation of an isolated microgrid. In this work, the benchmark is validated with an application of an energy management system.

Benchmark generation methodology

The proposed methodology uses as a reference the workflow proposed in [3], which is complemented with the lessons learned from the operation of the ESUSCON microgrid [6]. The methodology guides the designer of the benchmark in initial instances, such as the compilation of general characteristics of a microgrid, until the final steps conclude with the generation of the microgrid benchmark. The proposed methodology is presented in Fig. 1.

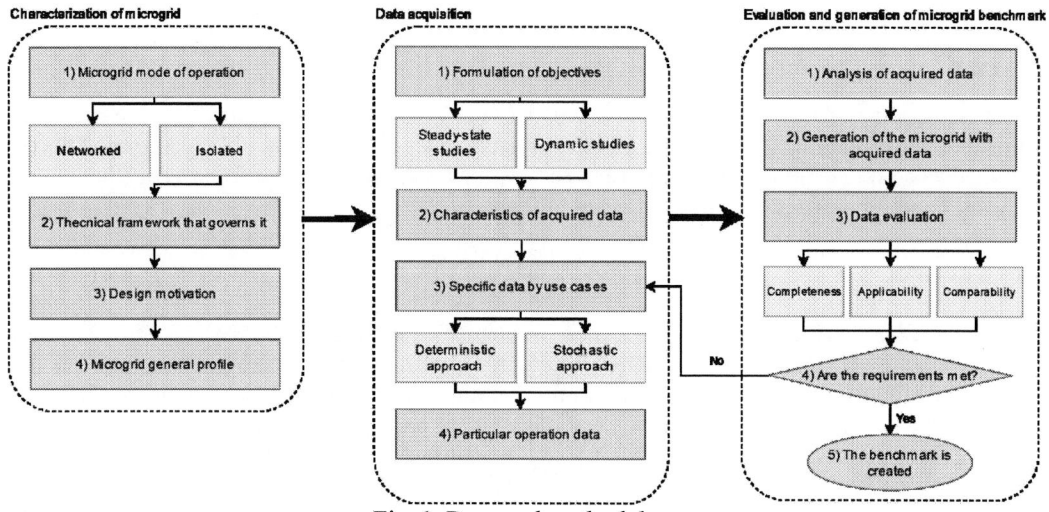

Fig. 1: Proposed methodology.

Within Fig. 1, three main stages stand out:
1. Characterization of the microgrid.
2. Data acquisition.
3. Evaluation and generation of the microgrid benchmark.

The proposed methodology, although it is intended to be applicable for the specific case of isolated microgrids, can be adapted to future iterations where it is possible to identify all the necessary measures to characterize a benchmark microgrid connected to the grid. This is possible because the stages of "Data acquisition" and "Evaluation and generation of microgrid benchmark" consider a data collection according to the specific objectives that are raised, so it is only necessary to change the initial structure of the first stage.

The first stage of the methodology, although it proposes a characterization of the microgrid to be generated, also serves to publish important information that corresponds to the technical operation data required in the second and third stages. This information serves to give context to what the microgrid benchmark proposes and what it is designed for.

The second stage of the methodology corresponds to an iteration where the data to be published are obtained. In this stage, some steps allow for the generation of useful and related data for the benchmark. In this stage the work that exists before the step "Specific data by use cases" is noteworthy. In the first and second step of this stage, the objectives of the benchmark are formulated, and the characteristics of the data collected are made explicit. Therefore, the data obtained are subject to what is decided in these first two steps, which highlights the relevance of the proposed objectives.

In the third and last stage of the methodology, the benchmark is evaluated, to be subsequently generated. In this stage of the methodology, attention is paid to the consideration that is taken to return to the second stage in case of having a negative evaluation in some aspect of the microgrid to be generated. This part of the stage allows for new or updated data to be obtained from the previous stage, so that the subsequent evaluations are positive, allowing for the correct generation of a benchmark microgrid.

In summary, the methodology is presented with the necessary steps to make a microgrid benchmark of the isolated type. This methodology has room for various criteria to be used for obtaining its data, as well as evaluations on them. The order of the methodology steps aims to be coherent with the idea of going from general to specifics: it begins with general characteristics of the microgrid, goes through specific data, and culminates with the evaluation and generation of the proposed benchmark.

Application of the methodology in the ESUSCON microgrid

Characterization of the microgrid

ESUSCON's microgrid is of the isolated type. From the technical point of view, this network corresponds to a three-phase system, feeding the demand through single-phase connections to households. The rated voltage of the network is 220/380 [V]. This network is classified as a low voltage isolated microgrid with overhead lines. Finally, the operating frequency of the microgrid is 50 [Hz].

The isolated microgrid of the ESUSCON project is in the north of Chile, specifically in Huatacondo, where an existing distribution network was upgraded to a renewable energy based microgrid that allows for uninterrupted supply of electricity. This project fostered the development of basic needs as well as entrepreneurial activities that required 24-hour electricity use, significantly improving the quality of life of the inhabitants [7]. The parameters of the microgrid [8] of this project are presented in Table I. Fig. 2 shows a diagram of the local distribution network.

Table I: ESUSCON microgrid design parameters.

Parameter	Value	Unit
Photovoltaic panel power	22.68	kW
Wind turbine power	3.00	kW
Maximum storage power	40	kW
Storage capacity	150	kWh
Diesel generator maximum power	120	kVA
Diesel generator minimum power	10	kVA

The main motivation in the generation of this benchmark microgrid is the use of the large amount of data generated in the almost 9 years in which the interdisciplinary team of the University of Chile carried out the design and operation of the microgrid in Huatacondo [7], called by the community ESUSCON (*Energía Sustentable Cóndor*).

Another motivation is to make better use of all the lessons learned and documentation made during the ESUSCON project. Due to the extensive work that was carried out by the team, there is a large amount of information on the operation, failures events, and growth of the demand in the electrical system of the Huatacondo community. The present work is intended to give added value to the work carried out by students, engineers, and professors in the design and operation of the microgrid present in Huatacondo.

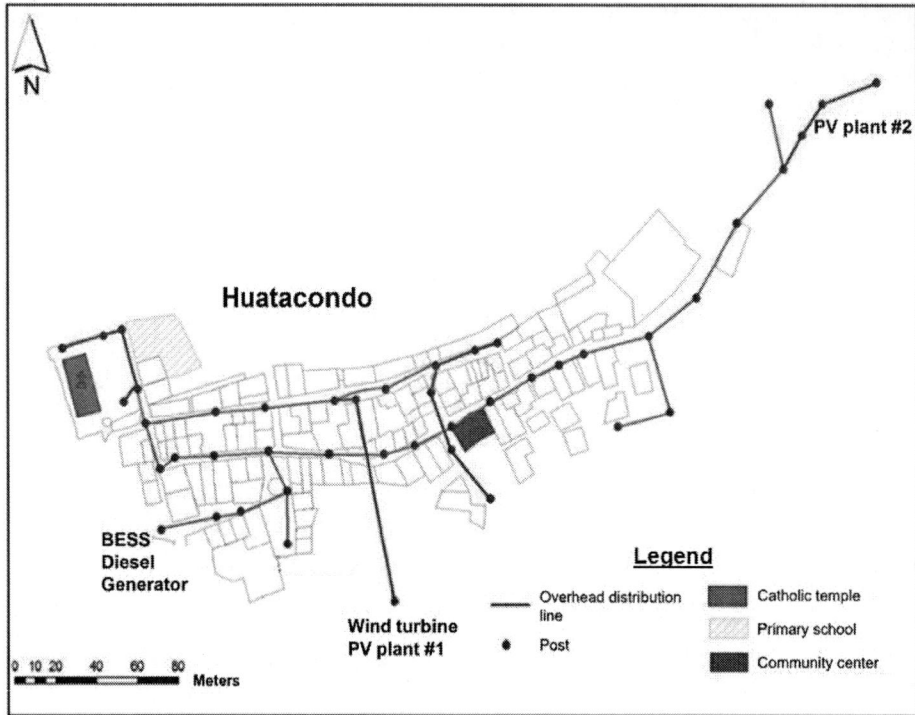

Fig. 2: Huatacondo distribution network.

The University of Chile supported the operation of the microgrid from 2011 to 2019. There is a wide set of operational data corresponding to this period (close to 2,220,000 data points) covering generation and demand, smart metering, and battery state of charge, among others.

Data acquisition

For the utilization of the generated benchmark microgrid, the following study objectives are formulated: network architecture planning, energy management, power flow, supply reliability, and network resilience. Furthermore, for the set of proposed studies, various parameters are required to generate an adequate benchmark microgrid. The necessary data sets are microgrid topology, demand profiles, loads within the network, generation and storage units, distribution line parameters, generation profiles, and fault/disturbance events.

We proceed to present the most relevant data sets for the work carried out, which correspond to microgrid topology, demand profiles and generation attributes.

Microgrid topology and electricity demand profiles

The microgrid topology corresponds to that of a radial distribution network. This is presented in Fig. 3.a). The operational data of the ESUSCON benchmark microgrid are in a temporary resolution of 15 minutes. As an example of the evolution of demand through the years, a histogram is shown in Fig. 3.b), where the "estimated data" is created for the purposes of the benchmark.

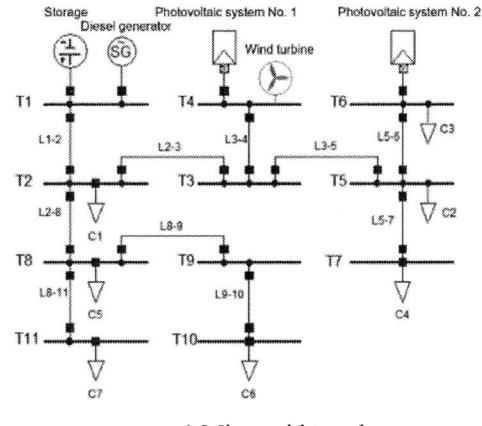

a) Microgrid topology

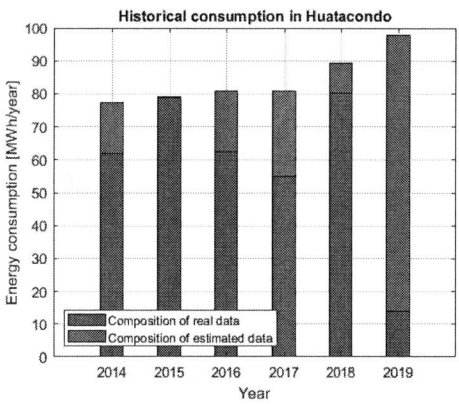

b) Historical electricity consumption

Fig. 3: ESUSCON benchmark microgrid schematic and historical electricity consumption.

The data used for this set are those of solar/wind generation, diesel generation, and power from the battery energy storage system (BESS). The data comprise both active and reactive power. The demand is obtained from computation using the aforementioned data, hence, already includes power losses in the overhead lines.

Evaluation and generation of the benchmark

As a summary, the benchmark completeness metric is presented, where all the annual operational data are counted and contrasted with those expected and estimated. These results are presented in

Table II. The publication of these data is carried out in a repository of public access.

Table II: Comparison of the number of expected samples with samples obtained (15-minute sampling).

Year	No. hours	Expected samples	Existing samples	Estimated samples	Completeness (existent+estimated) [%]
2014	8,760	35,040	28,129	6,911	100
2015	8,760	35,040	34,464	576	100
2016	8,784	35,136	25,151	9,985	100
2017	8,760	35,040	323,809	11,231	100
2018	8,760	35,040	31,488	3,552	100
2019	8,760	35,040	5,087	29,953	100

Benchmark validation proposal

To perform a validation on the data generated for the benchmark, it is proposed to simulate the behavior of an Energy Management System (EMS) with the generated data, to later compare it with the real operation data in Huatacondo, where the configuration of the microgrid contemplates its operation under the same EMS [9]. A simplified EMS is implemented for the benchmark, as presented in Fig. 4.

The configurations that were used for the simulation of the EMS are presented in the next section. These simulation options take into consideration a variety of performance indicators, to be used for comparing the benchmark with the real operation, where results such as diesel consumption, solar generation and renewable penetration can be obtained in the proposed evaluation.

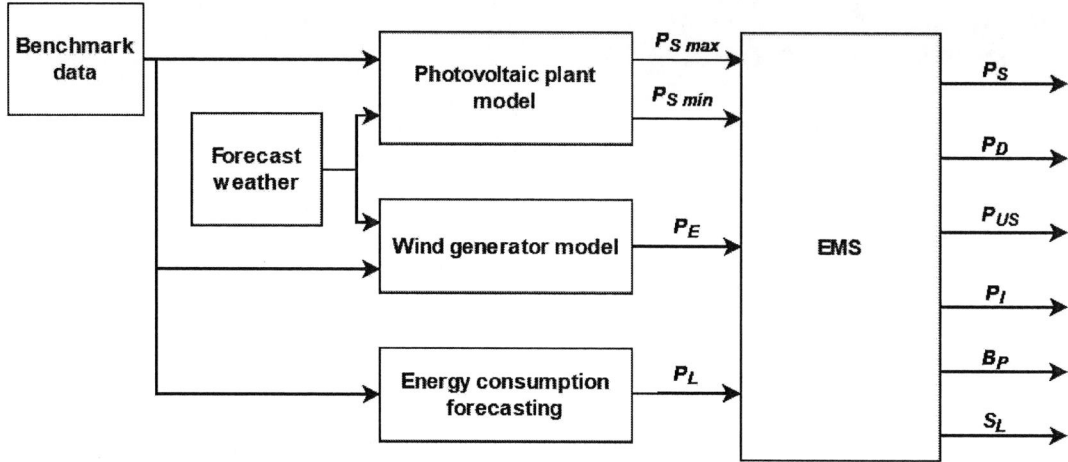

Fig. 4: EMS used in operation simulation according to ESUSCON design.

It is expected that the results of the EMS simulation show similarities with the behavior of the historical operation of the microgrid. The EMS implemented in ESUSCON contemplates the use of data sampled every 15 minutes, and the benchmark can easily adapt to that dynamic (i.e., it will also use a 15-minute sample). The results to be presented consider the forecasts for diesel generation, power setpoints for the BESS, and expected renewable generation.

The input and output variables with which the EMS operates are the following:

- Input variables:
  - $P_{S\,max}$: prediction of maximum possible solar power.
  - $P_{S\,max}$: prediction of minimum possible solar power.
  - P_E: wind power.
  - P_L: expected demand.
- Output variables:
  - P_S: expected solar power:
  - P_D: reference power for diesel generator.
  - P_{US}: fault power.
  - P_I: potencia del inversor del banco de baterías.
  - B_P: battery bank inverter power.
  - S_L: signs of demand slippage.

The objective (cost) function to be minimized in the proposed EMS is the following:

$$J = \delta_t \sum_{t=1}^{T} C_{com}(t) + \sum_{t=1}^{T} C_{par}(t) + \delta_t C_{man} T_{man} + \delta_t C_f \sum_{t=1}^{T} P_f(t) + \delta_t C_{ver} \sum_{t=1}^{T} P_{ver}(t) + C_{inv} SoH_{per} \quad (1)$$

where direct costs are:

- $\delta_t \sum_{t=1}^{T} C_{com}(t)$: represents the fuel cost per use of the thermal unit.
- $\sum_{t=1}^{T} C_{par}(t)$: represents the starting costs of the thermal unit, and estimation of fuel used for a cold start.
- $\delta_t C_{man} T_{man}$: represents the costs associated with the maintenance of the thermal unit based on the hours of use. Here T_{man} is the number of intervals in which the thermal unit was operating.

In addition, there are indirect costs:

- $\delta_t C_f \sum_{t=1}^{T} P_f(t)$: represents the cost of loss load.
- $\delta_t C_{ver} \sum_{t=1}^{T} P_{ver}(t)$: represents the cost associated with renewable energy surplus due to generation in excess.
- $C_{inv} SoH_{per}$: represents the cost of lost battery life (degradation).

The simulations will be carried out for 2 scenarios:

- Short-term simulation: considers a time window of 2 days (December 27 and 28, 2017).
- Long-term simulation: considers a time window of 6 months (second half of 2017).

The variables to study will be the energy supplied by the diesel generator, the photovoltaic generator, the battery bank, and the total energy injected into the system.

Validation of results

Short-term results

The results obtained in the short-term for the costs associated with the operation are shown in Table III, where C_{die} is the diesel generator fuel cost, C_{par} is the starting cost, C_f is the cost of loss load, C_{ver} is the cost of renewable energy surplus, C_{bat} is the cost of battery life, and C_{tot} is the total operating cost. In addition, the results obtained for the penetration of solar energy in the short term are presented in Fig. 5. It can be seen in this figure that the EMS simulation closely matches the real operation in terms of total generated energy, although the demand ("Total Energy Supplied") is lower on the real operation, due to excess solar energy not used, hence the higher C_{ver}. In terms of cost, diesel generation shares basically the same costs. As the real operation has lower demand, it also has lower battery usage, hence lower battery degradation cost.

Table III: Summary of short-term operating costs.

Summary of short-term costs [USD]						
Scenario	C_{die}	C_{par}	C_f	C_{ver}	C_{bat}	C_{tot}
Real Operation	47,08	4,87	0	0,95	14,26	67,17
Benchmark EMS	47,11	4,87	0	0,17	26,77	78,92

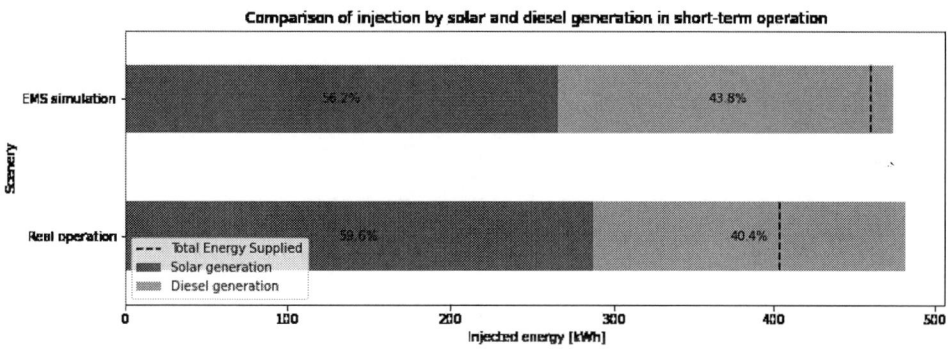

Fig. 5: Renewable penetration for EMS simulation scenario and real operation of the microgrid in the short-term.

A comparison between the real operation and the simulated operation by EMS is presented using the magnitudes and proportions of the amount of energy supplied through diesel generation and by the battery bank in Fig. 6. It can be readily seen in this figure that the real operation and benchmark simulation, although slightly differing in magnitudes, match in their proportions.

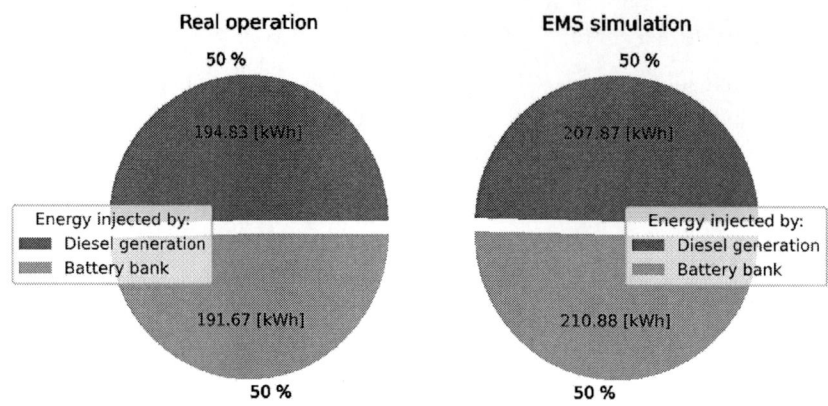

Fig. 6: Comparison of injected energy for both scenarios in the short-term.

Long-term results

The results obtained in the long-term for the costs associated with the operation are shown in Table IV. In addition, the results obtained for the penetration of solar energy in the long term are presented in Fig. 7. In the long term, although individual costs differ, overall total cost is similar. Also, as seen in Fig. 7, the demand ("Total Energy Supplied") is basically identical in both cases.

Table IV: Summary of long-term operating costs.

Summary of long-term costs [USD]						
Scenery	C_{die}	C_{par}	C_f	C_{ver}	C_{bat}	C_{tot}
Real Operation	6538,41	717,98	0	613,81	1083,07	8953,28
Benchmark EMS	5301,59	489,2	0	564,59	2482,92	8838,29

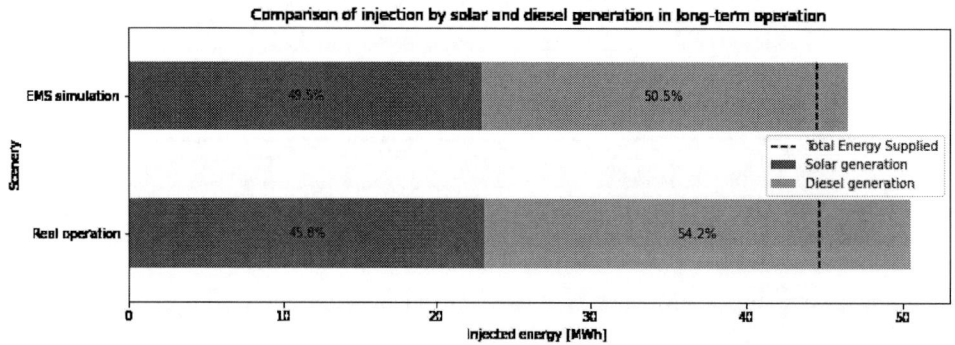

Fig. 7: Renewable penetration for EMS simulation scenario and real operation of the microgrid in the long-term.

Subsequently, a comparison between the real operation and the operation simulated by EMS is presented in Fig. 8, with the magnitudes and proportions of the amount of energy supplied by diesel generation and by the battery bank. The proportion of battery use is larger on the EMS simulation, which coincides with a bigger C_{bat} in Table IV. Conversely, the real operation does not use the battery as intensively, hence a larger use of the diesel generator.

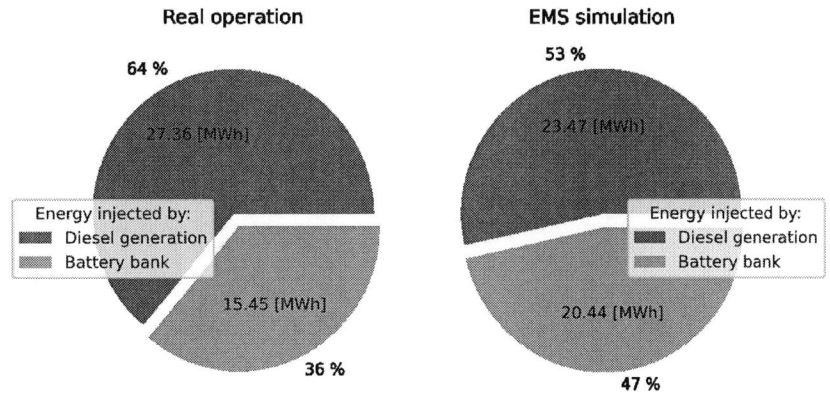

Fig. 8: Comparison of injected energy for both scenarios in the long-term.

Overall comparison

As a general analysis, the percentage differences for the variables studied in the two proposed scenarios are presented in Fig. 9. In general, the long-term results show less difference for the renewable generation and demand, although the simulated EMS performs a more efficient operation (with 15% lower diesel and battery usage). In the long-term, the variability is not perceived as in the short-term, where differences are more noticeable. Nevertheless, the benchmark demonstrates its usefulness for the evaluation of the EMS strategy.

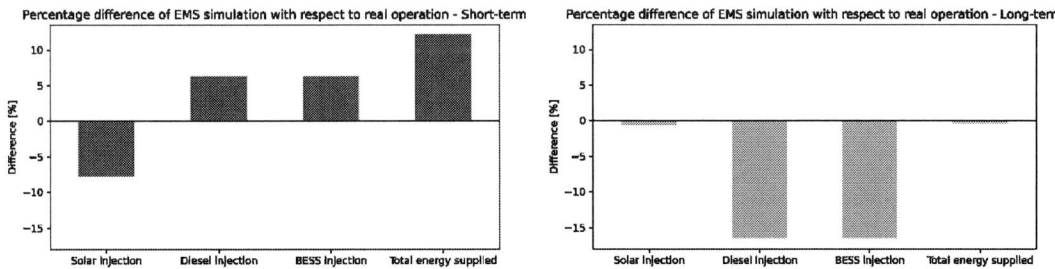

Fig. 9: Percentage differences for the two scenarios studied.

Conclusion

The application of the methodology is carried out in the ESUSCON microgrid, culminating in the generation of a benchmark microgrid with satisfactory performance. In this benchmark, more than 210,000 operational data corresponding to 6 years (2014 to 2019) were generated, together with complementary information that allows the proposed studies to be carried out. The generalized nature of the iterative steps of the methodology stands out, where the maximum use of the available data is sought for each generated benchmark.

Regarding the validation on the benchmark, the simulations carried out present satisfactory results, having as metrics the different energy injections that are in the system. In these simulations, it happens that for cases where there is a certainty of real operation under EMS, the BESS and diesel injections have similar behavior, while for solar injection and total energy supplied are not as similar. It is estimated that these variables, in the long term, approach each other for the two scenarios. Although it should be pointed out that the EMS always operates with a 48-hour rolling horizon.

In summary, the preparation of a methodology that allows the generation of benchmark isolated microgrids is achieved, thus fulfilling the main objective of this work. It is possible to fully carry out the steps proposed in the methodology. As future work, the generation of gird-connected benchmark microgrids should be addressed, where data not seen in the isolated case (such as energy prices and outages) as well as different operation philosophy (e.g. maximizing revenue rather than minimizing costs) should be considered.

References

[1] Alam, M. N., Chakrabarti, S., & Liang, X. (2020). A benchmark test system for networked microgrids. IEEE Transactions on Industrial Informatics, 16(10), 6217-6230.

[2] CIGRE Task Force C6. 04.02. (2014). Benchmark Systems for Network Integration of Renewable and Distributed Energy Resources.

[3] Meinecke, S., Bornhorst, N., & Braun, M. (2018, September). Power system benchmark generation methodology. In NEIS 2018; Conference on Sustainable Energy Supply and Energy Storage Systems (pp. 1-6). *VDE.*

[4] Ross, M., Abbey, C., Brissette, Y., & JOÓS, G. (2014). Real-time microgrid control validation on the Hydro-Québec distribution test line. *CIGRE.*

[5] Papathanassiou, S., Hatziargyriou, N., & Strunz, K. (2005, April). A benchmark low voltage microgrid network. In Proceedings of the CIGRE symposium: power systems with dispersed generation (pp. 1-8). *CIGRE.*

[6] Palma-Behnke, R., Ortiz, D., Reyes, L., Jimenez-Estevez, G., & Garrido, N. (2011, July). A social SCADA approach for a renewable based microgrid—The Huatacondo project. In 2011 IEEE Power and Energy Society General Meeting (pp. 1-7).

[7] Álvarez, M. & Garrido, N. (2011). Informe Área Social - Proyecto ESUSCON. Centro de Energía, Facultad de Ciencias Físicas y Matemáticas. 85p

[8] Lanas, F. (2011). Desarrollo y validación de un modelo de optimización energética para una microrred. Memoria Ingeniero Civil Electricista. Santiago, Chile: Facultad de Ciencias Físicas y Matemáticas, Universidad de Chile. 98p.

[9] Palma-Behnke, R., Benavides, C., Lanas, F., Severino, B., Reyes, L., Llanos, J., & Sáez, D. (2013). A microgrid energy management system based on the rolling horizon strategy. IEEE Transactions on smart grid, 4(2), 996-1006.

An Overview of Grid-Connection Requirements for Converters and Their Impact on Grid-Forming Control

Paul Imgart[1], Mebtu Beza[1], Massimo Bongiorno[1], Jan R. Svensson[2]

(1) Chalmers University of Technology, Dept. of Electrical Engineering, Göteborg, Sweden
(2) Hitachi Energy, Hitachi Energy Research, Västerås, Sweden
paul.imgart@chalmers.se

Keywords

≪Grid forming≫, ≪Virtual synchronous machine≫, ≪Grid-connected converter≫,
≪Converter control≫, ≪Renewable energy systems≫.

Abstract

The increasing share of converters in the power system results in the need to revise grid-connection requirements and a shift in converter control strategies towards grid-forming control. This paper analyzes and compares existing standards and future trends in specifications for grid-connected converters and highlights commonalities and differences. The key consequences for converter control performance requirements are presented to facilitate the selection of the most suitable grid-forming control structures.

Introduction

The transformation towards a sustainable power system will result in fewer synchronous machines in the power system, and grid-forming converters are suggested as one solution for the associated challenges [1]. As the grid-forming concept is still relatively new, discussions between grid operators, legislators and manufacturers on the design and implementation of grid-forming control are still ongoing today. Technical requirements and standards define the functionality required from grid-forming converters and are accordingly an important part of these discussions.

Since the first proposal of the virtual synchronous machine (VSM) concept, a large variety of grid-forming converter controls has been proposed [2], with some relevant selections shown in Fig. 1 and compared in [3]. Analyzing the capability of different control strategies to fulfill the grid-connection requirements is an important part in selecting the most promising designs for further research and development.

This paper compares and analyzes technical requirements for grid-connected converter systems from a selection of different legal regulations, grid codes and standards with focus on grid-forming control. In

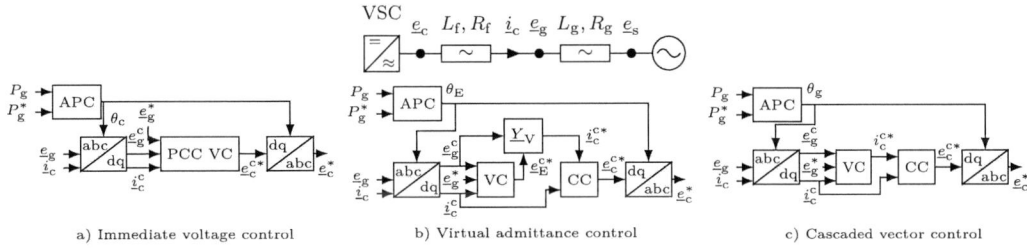

a) Immediate voltage control b) Virtual admittance control c) Cascaded vector control

Fig. 1: Single-line diagram of grid-connected converter system and block-schemes for three different types of grid-forming converter control [3]. Quantities with index E refer to the virtual back EMF. APC: Active power controller; VC: Voltage controller; CC: Current controller; $\underline{Y}_V$: Virtual admittance.

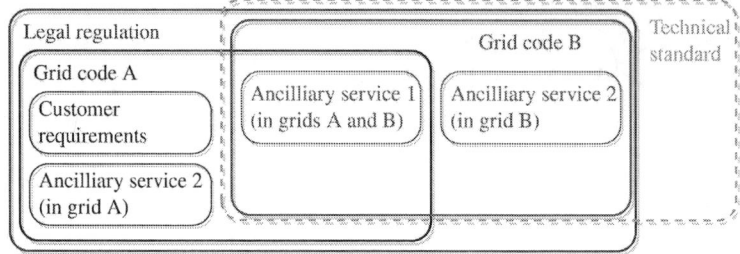

Fig. 2: Illustration of different technical requirement document types and their relation. Grid code A and B serve as examples for two independent grid code documents from different system operators, each also specifying the exemplary ancilliary services 1 and 2.

addition, possible differences and contradictions among the various documents are investigated. The final goal is to highlight implications of the different requirements on converter control strategy and to facilitate the choice of suitable control structures.

Categorization of requirement documents and selection of examples

Documents defining technical requirements for grid-connected converters can be classified according to their author, scope, bindingness and detailedness in the following categories: legal regulations, grid codes, specifications of particular ancilliary services, technical standards and customer requirements. Figure 2 displays how requirements of the different categories are inter-related. Legal regulation is obligatory, but typically less detailed. For connection to a particular grid, additionally the corresponding grid codes have to be followed. The specifications for ancilliary services can be part of the general grid code or provided through additional documents; customers can also provide further requirements. Following technical standards is often voluntary, but can be a requisite for a certain control implementation to be successful.

As illustrated in the figure, the requirements given in different documents are not necessarily the same, but can differ between countries and in some cases even conflict [4]. Moreover standards affect not only the jurisdiction they are valid for, but have an effect on others as well, in particular when concerning development- and testing-intensive equipment such as wind turbines, FACTS and HVDC systems. If one standard explicitly bans a particular implementation, manufacturers are likely to not develop a special solution, but adapt their product to comply with this specification, even if this might result in higher prices and in some cases suboptimal performance for all customers. This is due to the amount of development, testing and tuning needed for these systems, making it uneconomic to maintain a large variety of implementations. While on first glance it might make sense for system operators to tailor requirements to their particular system, a harmonization of requirements could reduce costs, enable share of experience and stimulate innovation by freeing up resources otherwise needed for local adaptations [4]. An exception to this can be certain limits, thresholds, settings or ratings. In conclusion, a harmonization of requirements with option for small local adaptions (as in [5]) is not only in the interest of manufacturers, but even academia and system operators. Consequentially, the regulations shown in Fig. 2 would ideally all coincide or at least be subsets of others. The ENTSO-E network codes for generators [5] and HVDC [6] are recent examples of successful harmonization efforts. However, specifications and requirements for grid-forming converters are still under discussion and development, starting with the definition of grid-forming properties itself. This presents a challenge for all involved parties to find a generally accepted set of required functionalities for grid-forming converters.

The following paragraph describes the grid codes and standards selected for the overview in this paper. The European Commission's network code with requirements for generators (NC RfG) [5] and HVDC systems and DC connected generation (NC HVDC) [6] are drafted by ENTSO-E and are the underlying legal regulation for grid codes in the European Union. They are concretized in national laws and regulations, e.g. in the case of Sweden [7] for RfG (SE RfG) and [8] for HVDC (SE HVDC), or national standards such as the VDE standard VDE-AR-N 4131 for HVDC in Germany [9]. A relevant technical standard

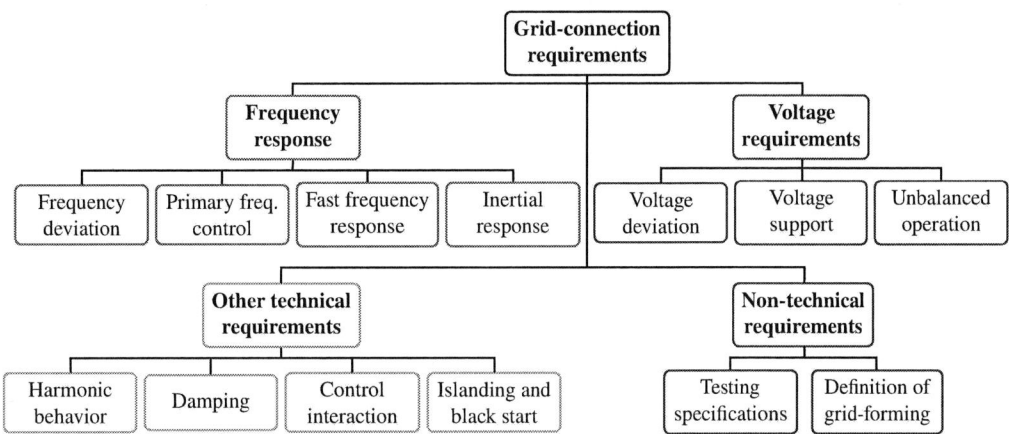

Fig. 3: Categories of typical requirements in regulation, grid codes and standards.

for DC interfaced generation is the draft IEEE P2800 [10]. The UK grid code modification GC0137 (NG GC0137) [11] containing requirements for grid forming capability is an example for an ancilliary service specification. The final investigated document is the ENTSO-E report on the contributions of grid-forming converters [12]. It is not a technical standard but reflects the debate about which features are discussed as a part of grid-forming capabilities. These documents have been chosen due to their relevance to the current and future state of requirements for grid-forming converters and to represent different countries and the categories mentioned above. It should be noted that the used standards in some cases apply to HVDC systems, in some cases to power-electronic interfaced generation and in some cases to either. None of the selected documents is formulated exclusively for FACTS, but often they partly follow HVDC regulations. Different systems such as e.g. wind power parks, grid-connected battery energy storage and HVDC systems will have different purposes and capabilities and comparing the requirements they have to fulfill is consequently meaningful only up to a certain degree. However, as all of these are grid-connected converter systems, they can all be equipped with grid-forming control, which motivates a comparison of the required capabilities.

Comparison of requirements

To facilitate the comparison and analysis, the categorization of a selection of the requirements included in the investigated documents is suggested as shown in Fig. 3. This section describes how the requirements belonging to these categories are treated in the selected documents, with Table I presenting a summary.

Frequency response

Frequency deviation. Nearly all of the compared standards contain a range of steady-state frequency deviations from nominal frequency with corresponding minimum operating times, which are summarized in Table II. The frequency range and times given for NC RfG & HVDC are minimum requirements, which can be further increased by the relevant system operator. Observe that NG GC0137 and the ENTSO-E report do not explicitly indicate low-frequency ride-through, but provide requirements for grid-forming behavior in case of frequency deviations, which implicitly requires operation at lower frequencies. The compared standards provide the requirements in a similar format, but there are significant differences in the required time spans, with the low-inertia systems Great Britain and Ireland placing highest requirements. Apart from steady-state variations, some standards also define the transient conditions which the converter system has to be able to ride through in the form of maximum phase-angle jump and rate of change of frequency. As detailed in Table II, NC HVDC, VDE-AR-N 4131 and IEEE P2800 give requirements for maximum rate of change of frequency, where the IEEE norm is considerably stricter than the others. IEEE P2800 even contains the requirement to withstand sub-cycle phase-jumps of 30°. The ability to withstand high frequency gradients can be affected by requirements for high virtual inertia, as the inertia emulation will slow the ability of the converter to track changes in the system frequency.

Table I: Selection of requirements from the documents. Green: explicitly specified, required by standard or on demand of the TSO; Yellow: unspecific/implicitly specified; Red: not mentioned.

Type of requirement	NC RfG & HVDC	SE RfG & HVDC *	VDE-AR-N 4131	IEEE P2800	NG GC0137	ENTSO-E report
Frequency response						
Frequency deviation	✓	✓	✓	✓	○	○
Primary frequency control	✓	✓	✓	✓	✕	✕
Fast frequency response	✕	✕	✓	✓	✕	✕
Inertial response	✓	✕	✓	✕	✓	✓
Voltage requirements						
Voltage deviation	✓	✓	✓	✓	○	○
Voltage support	✓	✕	✓	✓	○	○
Unbalanced operation	✕	✕	✓	○	✕	✓
Other technical requirements						
Harmonic behavior	✕	✕	✓	✓	○	✓
Damping	✓	✓	✓	○	✓	✓
Control interaction	○	○	✓	○	✓	✓
Islanding and black start capability	✓	✕	✓	✕	✕	○
Non-technical requirements						
Testing specifications	✓	✕	✓	✓	✓	○
Definition of grid-forming capability	✕	✕	✕	○	✓	✓

*: As SE RfG & SE HVDC are national implementations of NC RfG & NC HVDC, respectively, they include the requirements mentioned there. Here is indicated where these documents contain substantial concretizations or additions.

Primary frequency control. All investigated documents require the capability to control active power in dependency of the deviation of the system frequency from nominal with the exception of NG GC0137 and the ENTSO-E report. In case of NC RfG, the requirements vary with converter rating, with smaller generators only required to act in case of overfrequency. All specifications are based on a deadband around nominal frequency and a proportional gain and take the fluctuating character of renewable resources into consideration, allowing for adjustments based on the availability of primary energy.

Fast frequency reserve. Converters are capable of controlling their active power output significantly faster than required for primary frequency control. This can be used to counter the effects of the decline in synchronous inertia caused by the replacement of synchronous machines with converter interfaced generation [13]. In consequence, requirements for fast frequency reserve (FFR) are contained in VDE-AR-N 4131 in general terms and in IEEE P2800 with details about performance and possible FFR variants. Even though NC RfG & HVDC do not contain FFR, it has in the meanwhile been defined by ENTSO-E as an auxiliary service in a separate document and is procured as a service e.g. in Sweden [14].

Inertial response. To counteract the effects of the reduction of mechanical inertia synchronously connected to the grid, converters can emulate inertia by modulating the active power output to counteract either changes in the frequency or phase angle of the voltage at the point of common coupling (PCC). NC RfG & HVDC as well as VDE-AR-N 4131 specify that the TSO can require inertial response capability from connected converter system with the relevant control principle and parameters to be agreed upon by TSO and owner of the converter system in that case. In the case of NG GC0137, both frequency-derivative proportional inertial support as well as a response to grid voltage phase jumps are obligatory requirements. Finally, in the ENTSO-E report inertial response is discussed as one of the key capabilities constituting grid-forming control.

Provision of primary frequency control, fast frequency reserve and inertial response requires active power, but none of the investigated documents stipulates the availability of energy storage or prime energy. This means that the control system needs to be able to provide the response, but additional requirements not currently covered by the standards need to be fulfilled for the active participation in this services.

Table II: Minimum connection time for specific frequency ranges according to selected example requirements. "Gen." denotes requirements for generation, while "HVDC" is for HVDC converter stations.

Frequency range	NC RfG Con. Europe, Nordic & Baltic [5]	NC RfG Great Britain & Ireland [5]	NC HVDC [6]	SE HVDC [8]	VDE AR-N 4131 [9]	IEEE P2800 [10]
0.94 p.u. − 0.95 p.u.	−^{a}	20 s^b	> 20 s^c	60 s	60 s^c	−^{a}
0.95 p.u. − 0.97 p.u.	30 min^c	90 min	90 min^c	100 min	90 min^c	5 min^c
0.97 p.u. − 0.98 p.u.	30 min^c	90 min^c	90 min^c	100 min	90 min^c	5 min^c
0.98 p.u. − 1.02 p.u.	unlimited	unlimited	unlimited	unlimited	unlimited	unlimited
1.02 p.u. − 1.03 p.u.	30 min^d	90 min	90 min^c	100 min	90 min^c	5 min^c
1.03 p.u. − 1.04 p.u.	−^{a}	15 min^b	15 min^c	30 min	15 min^c	−^{b}
Minimum RoCoF	−^{e}	−^{e}	2.5 Hz/s	2.5 Hz/s^f	2.5 Hz/s	5 Hz/s

^{a}: No connection requirement is given; ^{b}: Great Britain only, no connection requirement given for Ireland.; ^{c}: To be specified by the relevant TSO, but not less than the shown time; ^{d}: In Baltic region, ^{c} applies; ^{e}: To be specified by the relevant TSO; ^{f}: Not explicitly stated, but as national implementation of NC HVDC those requirements apply.

Voltage requirements

Voltage deviation. This category includes the behavior required from the converter system by some standards during transient disturbances such as faults and voltage dips. These fault ride-through (FRT) requirements are commonly provided as a voltage-time-profile, which represents a minimum requirement, i.e. the converter system must be able to stay connected when subject to a voltage than does not go below the defined curve. This way of defining low-voltage ride-through requirements is common to the European standards, but is not present in IEEE P2800, where instead minimum connection times for different voltage ranges are defined. The FRT requirements from a selection of standards are listed in Table III, with the shape of the corresponding voltage profile given in Fig. 4. For the summary in Table III, the IEEE requirements have been rearranged as shown in the dashed profile in Fig. 4.

In some standards FRT is not only a connection requirements, but contains even the obligation of fast fault current injection. NC RfG & HVDC provide the possibility for the TSO to define such requirements. The national implementations SE RfG & HVDC do not concretize this further, while VDE-AR-N 4131 states priority for reactive current and contains specifications for voltage support and unbalanced operation, which can even have an impact on fault current injection. IEEE P2800 contains a very detailed description of different fault modes and expected fault current contributions. None of the mentioned standards provide fault current magnitudes, but place this responsibility on the relevant system operator. In NG GC0137, no minimum connection times are given as they are part of the main grid code, but requirements for grid-forming behavior are mentioned in terms of fast fault current and reactive power injection in case of grid voltage magnitude transients. Emphasis is put on the immediate response, reflecting the voltage source behavior of the grid-forming converter. The same is true, even though less specific, for the analysis in the ENTSO-E report, where fault current contribution is defined as one of the characteristics of grid-forming converters. All fault current requirements take the limited over-current capability of converters into account and allow for appropriate over-current limitation and protection.

Apart from the fault ride-through in response to transient disturbance, the standards also specify connection requirements regarding steady-state voltage deviations. These are summarized in Table IV.

Voltage support. Converter systems are commonly required to contribute to the control of the voltage magnitude at their PCC by exchanging reactive power. For NC RfG & HVDC and their national implementations, this requirement is given in the form illustrated by Fig. 5. The standard defines an outer envelope. The TSO then determines an inner envelope within it, and with a predetermined maximum voltage and reactive power range. Within that envelope, the TSO specifies a profile for the exchange of reactive power in dependence of the voltage. The converter system is required to have the capability of providing the desired reactive power exchange non-withstanding the active power operating point. The reactive power exchange can either be controlled in voltage, reactive power or power factor control

Table III: FRT requirements.

Quantity	NC RfG < 110kV[a]	NC RfG ≥ 110kV[a]	NC HVDC[a]	SE HVDC	VDE AR-N 4131	IEEE P2800[b]
V_{ret}	0.05 p.u. – 0.15 p.u.	0 p.u.	0 p.u. – 0.3 p.u.	0 p.u.	0 p.u.	0 p.u.
V_{clear}	V_{ret}	V_{ret}	V_{ret}	V_{ret}	V_{ret}	0.25 p.u.
V_{rec1}	V_{clear}	V_{clear}	0.25 p.u. – 0.85 p.u.	0.85 p.u.	0.85 p.u.[c]	0.5 p.u.
V_{rec2}	0.85 p.u.	0.85 p.u.	0.85 – 0.9 p.u.	0.85 p.u.	V_{rec1}	0.7 p.u.
t_{clear}	0.14 s – 0.15 s[d]	0.14 s – 0.15 s[d]	0.14 s – 0.25 s	0.25 s	0.15 s	0.32 s
t_{rec1}	t_{clear}	t_{clear}	1.5 s – 2.5 s	2 s	3 s	1.2 s
t_{rec2}	t_{rec1}	t_{rec1}	t_{rec1}	t_{rec1}	t_{rec1}	3 s
t_{rec3}	1.5 s – 3.0 s	1.5 s – 3.0 s	t_{rec1} – 10 s	10 s	–	6 s

[a]: TSO determines parameters within specified range; [b]: Only requirements for voltages < 0.9 p.u. given here. Requirements are shown for plants without auxiliary equipment limiting FRT and rearranged to comply with the form in Fig. 4; [c]: 0.894 p.u. (340 kV) if rated voltage is 380 kV; [d]: Can be 0.14 s – 0.25 s if required by system protection and secure operation.

Table IV: Minimum connection time for specific voltage ranges according to selected requirements.

Voltage range	NC RfG Con. Europe[a]	NC RfG Nordic[a]	NC RfG Great Britain[a]	NC HVDC Con. Europe[a]	NC HVDC Nordic[a]	NC HVDC Great Britain[a]	VDE AR-N 4131	IEEE P2800[b]
0.85 p.u. – 0.90 p.u.	60 min	–[c]	–[c]	unlimited	–[c]	–[c]	unlimited[d]	–[c]
0.90 p.u. – 1.05 p.u.	unlimited	unlimited	unlimited	unlimited	unlimited	unlimited	unlimited	unlimited
1.05 p.u. – 1.10 p.u.	20 min – 60 min[e]	< 60 min[e]	15 min	60 min[f]	< 60 min[e]	15 min	unlimited	30 min
1.10 p.u. – 1.20 p.u.	–[c]	–[c]	–[c]	–[c]	–[c]	–[c]	60 min[g]	1 s

[a]: Applies for a rated voltage of 300 kV to 400 kV, and in case of NC RfG type D power park modules only; [b]: Connection requirements for voltage of 0.9 p.u. and above are shown here, lower voltages in Table III; [c]: No connection requirement is given, see Table III for FRT; [d]: For rated voltage of 380 kV, this applies for $V > 0.894$ p.u. (340 kV); [e]: To be specified by the TSO withing the stated range; [f]: To be specified by the TSO as not less than 60 min for 1.05 p.u. < V < 1.0875 p.u., 60 min for 1.0875 p.u. < V < 1.1 p.u.; [g]: Requirement of 60 min applies for 1.1 p.u.<V<1.158 p.u.

mode, which are further specified in NC RfG & HVDC. While SE RfG & HVDC do not add any further requirements regarding voltage support, VDE-AR-N 4131 contains detailed specifications of the control mode and performance requirements as well as inner envelopes for the voltage control profiles. In IEEE P2800, a reactive power requirement is established, which lies within the outer envelope from NC RfG. It also includes detailed control mode and performance requirements for voltage, reactive power and power factor control. NG GC0137 does not add any further specifications regarding voltage support beyond general grid code apply. However, both here and in the ENTSO-E report the capability to create a system voltage is required, i.e. the ability to behave as a voltage source independent of a grid voltage.

Unbalanced operation. Of the studied standards, NC RfG, NC HVDC, SE RfG & SE HVDC do not make any specifications regarding the expected behavior under unbalanced operating conditions apart from FRT during asymmetric faults to be detailed by the TSO. In IEEE P2800, no requirements for unbalanced but otherwise normal operating conditions are given, but negative sequence current injection is required during asymmetric faults. NG GC0137 does not contain requirements of unbalanced operation. In VDE-AR-N 4131 on the other hand, voltage control is required not only for positive, but also for negative and zero sequence, and the ENTSO-E report demands the converter system to act as sink for unbalances, i.e. provide a low-impedance path for negative sequence currents. This indicates a paradigm shift towards more system responsibility, similar to the one described below for harmonic behavior.

Other technical requirements

Harmonic behavior. Neither NC RfG & HVDC nor SE RfG & HVDC include requirements for the harmonic behavior of the converter system. In VDE-AR-N 4131, maximum values for the total harmonic distortion as well as the magnitude of each of the relevant harmonics in the PCC voltage are specified

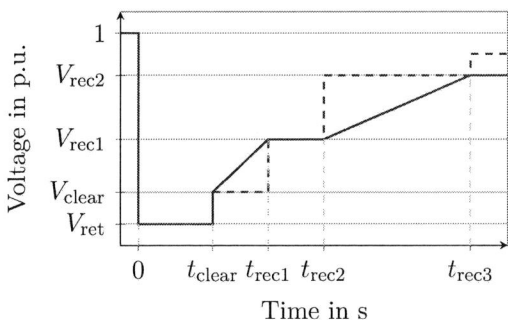
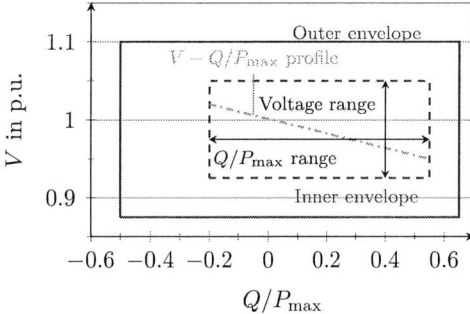

Fig. 4: Structure of FRT minimum voltage profile. Solid blue: European standards ([5]–[9]); dashed red: IEEE P2800 minimum connection times.

Fig. 5: Reactive power capability as defined by NC RfG. The outer envelope is as shown in the standard, the inner and $V-Q$ profile are indicative.

for DC connected generation, but not for HVDC. In contrast, IEEE P2800 gives requirements for the harmonic distortion of the current, but explicitly does not limit the voltage harmonics. The standard contains a detailed discussion of converter voltage harmonics and liability in the annexes. NG GC0137 does not specify harmonic behavior explicitly, but the required behavior as a voltage source behind a reactance will affect this. In the ENTSO-E report, the ability to act as a sink for harmonics is named as a key capability of a grid-forming converter. This means that the converter should provide an inductive or resistive-inductive path for harmonic current components, improving the quality of the voltage at its PCC. Here, a paradigm shift becomes visible, which already has started to come into effect e.g. in VDE-AR-N 4131: Instead of only being responsible for their own harmonic pollution as previously, converters and grid-forming converters in particular can be expected to be required in the future to increasingly act as sinks for harmonics, actively improving the grid voltage quality.

Damping. This category includes requirements regarding the damping of power oscillations and sub-synchronous resonances. Specifications of control actions involving active power (frequency and inertial response) or reactive power (voltage support) can also contain requirements regarding the damped behavior of the control action, but are not considered here. Contribution to the damping of power oscillations is required in NC RfG & HVDC but not specified further, which is complimented with requirements for damping capability for subsynchronous torsional interaction in NC HVDC. The effective range for the required power system stabilizer is further specified for generation in SE RfG. In VDE-AR-N 4131, HVDC systems are required to contribute to the damping of subsynchronous and power oscillations. While not a formal requirement in IEEE P2800, power oscillation damping and subsynchronous instability are discussed in the annex. The supply of damping power analogous to synchronous machines is a requirement for grid-forming converters in NG GC0137 and the ENTSO-E report.

Control interaction. Adverse control interaction is not a new concern, but before the focus has been on interactions in the sub-synchronous frequency range. However, the increasing amount of converter systems in the grid and the decreasing electrical distance between them has resulted in a much wider frequency range for possible interactions, up to several kHz. This is reflected by the increasing amount of detail in requirements of this category, when compared to earlier versions of the investigated documents. IEEE P2800 does not name control interactions as part of the requirements, but contains an in-detail discussion of control instability in its informative annexes. VDE-AR-N 4131 requires a study about possible interactions of the HVDC system with other grid components. The study's method is not prescribed explicitly, but measurement or simulation of grid and HVDC system input admittance as well as frequency domain and EMT studies are suggested as examples. The ENTSO-E report includes a in-depth discussion of control interactions and the same study methods as in VDE-AR-N 4131 are suggested. Two contradictory viewpoints are discussed; one deeming general specifications able to rule out harmonic interaction in any situation as too conservative and unrealistic, calling for specific case studies instead; the other prescribing a bandwidth limitation with a frequency range in which the converter has to behave as a

Thévenin source behind an impedance. The latter perspective is incorporated in NG GC0137, specifying that the grid forming system must appear passive above a frequency of 5 Hz. Even though none of the studied documents accept adverse interaction between the converter system to be connected and any other component of the grid, NG GC0137 is the only one specifying concrete requirements for how the absence of control interactions should be demonstrated. The discussions of this topic contained in VDE-AR-N 4131, IEEE P2800, NG GC0137 and the ENTSO-E report demonstrate that methods and requirements for ensuring absence of control interactions during high penetration of power electronics in the grid are still under investigation and development.

Islanding and black start capability. According to NC RfG, HVDC and VDE-AR-N 4131, the TSO can request islanding capability and ask for a quotation for black start capability. This capability is not a requirement in IEEE P2800 or NG GC0137, but in the latter grid-forming capability is mentioned as a prerequisite for converters wanting to provide black start capability. The ENTSO-E report does not mention islanding or black start capability explicitly, but places a great emphasis on the ability to create a system voltage as a core capability of grid-forming converters.

Non-technical requirements

Testing specifications. The provision of testing specification is not only crucial for ensuring compliance of constructed converter systems with the requirements, but also assists researchers and industry in developing and testing prospective design ideas beforehand in simulation and under lab conditions. NC RfG & HVDC describe compliance test in detail regarding the functions to be tested and the responsibilities of the different actors in these tests. However, no concrete testing scenarios are provided. The same applies for VDE-AR-N 4131, but in this case a supplementary document is available that provides a large number of simulation scenarios for the required capabilities [15]. IEEE P2800 contains a verification matrix listing the type of test and the responsible party for each of the requirements. However, no details are given how simulations and tests are to be conducted. In NG GC0137, simulation, testing and online monitoring are specified as compliance requirements, with three simulation scenarios provided to test compliance with specific parts of the grid-forming capability. Testing and benchmarking, as well as simulation studies for some requirements, are discussed in detail in the ENTSO-E report, but due to the general character of the document no concrete scenarios are suggested.

Definition of grid-forming capability. As discussed in the ENTSO-E report, not all converter systems need to fulfill the same requirements, and grid-forming capability might only be required by a small share of must-run units. For this reason it is instrumental to divide converters in different classes, each coming with a specific set of requirements. The report defines classes 1, 2A, 2B, 2C and 3, where the first four correspond to the type A, B, C and D in NC RfG. Each class or type should fulfill all requirements of the ones below, with further control requirements added subsequently. Common criteria for assigning the class are power and voltage rating of the converter system. In the definition given in the ENTSO-E report, class 3 corresponds to full grid-forming capability on top of the requirements applying for class 2C, which are in essence those specified in NC RfG for type D power park modules or in NC HVDC for HVDC. In this definition, grid-forming means the ability to create system voltage, contribute to fault level, act as a sink for harmonics and unbalances, contribute to inertia, support system survival to enable under-frequency load shedding and prevent adverse control interaction. NG GC0137 likewise defines grid-forming capability as fulfillment of a number of technical requirements, but puts an even stronger emphasis on the voltage-source-behind-a-reactance character of the converter. From the other investigated documents, only IEEE P2800 contains a discussion of the concept grid-forming capability in the annex, while the others do not mention the concept. This is even the case for VDE-AR-N 4131, which contains most of the capabilities typically associated with grid-forming control (e.g. inertial response) as optional requirements.

Qualitative comparison of the requirements

As visualized in Table I, the extent and detail of the studied documents vary widely. While the high-level, legally binding documents such as NC RfG & HVDC remain vague in a lot of aspects and rely on underlying standards and agreements for concretion, low-level norms such as VDE-AR-N 4131 and

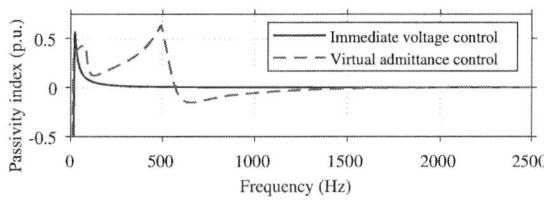

Fig. 6: Passivity index over dq-frequency for immediate voltage control (blue) and virtual admittance control (dashed red) [18].

IEEE P2800 contain an extensive description of requirements with great detail. Additionally there are specifications for specific ancilliary services such as NG GC0137, which contain detailed requirements for the service they concern, but otherwise rely on the underlying grid code. Finally, the ENTSO-E report is a report reviewing and discussing challenges, practices and research and has therefore a different character than the normative documents. It serves as an indication for trends emerging in the standards and shows very clearly in which areas development in the requirements is discussed the most.

Demand for grid-forming capabilities in existing and future standards

Table I shows that none of the obligatory requirements currently in effect mention grid-forming converter control or contain a definition of it. IEEE P2800 discusses grid-forming control in its informative appendices, and only NG GC0137 and the ENTSO-E report define the concept.

When discussing grid-forming converter control as an emerging concept, it is important to be aware of how the meaning of this term has shifted. In 2012, grid-forming control defined a control strategy solely determining the grid voltage in magnitude and frequency, like it is the case in micro- and island grids [16]. Converters with these properties have been in operation for more than 20 years [17]. In the current discussion however, the term grid-forming is commonly not longer used for converters serving as a slack bus, but instead for converters participating in voltage and frequency control by acting as a voltage source behind an impedance [1], [11], [12].

Even though none of specifications currently in effect requires grid-forming converter control, VDE-AR-N 4131 contains a number of optional requirements that are part of the definition of grid-forming capabilities in the ENTSO-E report and typically associated with grid-forming converters, such as inertial response, acting as a sink for harmonics and islanding capability, albeit without utilizing the label grid-forming. To a lesser extent, this also applies to IEEE P2800. With the introduction of NG GC0137, grid-forming capability has become an ancilliary service in Great Britain. Even though standards do not yet require full grid-forming capability, step by step more of the capabilities associated with grid-forming control are becoming part of them as optional or mandatory requirements.

Impact of requirements on grid-forming control topologies

It is important to underline that the compared standards in general are technology agnostic, i.e. not prescribing how the requirements have to be implemented in hard- and software, but instead posing performance requirements to be fulfilled. This allows for the development of different solutions and drives a competition of ideas leading towards the most efficient implementation. However, there can still be consequences for which control strategies are suitable, as not all implementations might be able to comply with the given performance requirements. One notable exception from this is NG GC0137, which explicitly excludes the usage of a virtual impedance with reference to the high bandwidths required.

Even though not explicitly ruled out, some control topologies might not be able to fulfill specific performance requirements. One example for this is non-passive behavior caused by a current controller, as demonstrated by [18] and shown in Fig. 6. In consequence this means that the grid-forming control schemes shown in Fig. 1 (b) and (c) cannot fulfill the passivity requirements formulated in NG GC0137, and that the topology from (a) cannot rely on a current controller for FRT if this requirement is to be met. This illustrates how the grid code requirements currently in place or under discussion can be very challenging to fulfill with existing grid-forming converter control topologies.

Conclusions

This paper has compared and analyzed current technical requirements for grid-connected converters and future trends. It has shown qualitative differences between different standards and discussed the need for harmonization. Furthermore, the impacts of specific requirements on the control topology of converters were demonstrated. Grid-forming control is not a requirement in any of the existing standards, but has been recently included as an ancilliary service in the UK grid code and specific capabilities associated with grid-forming control are appearing as optional requirements in other standards currently in effect as well. This demonstrates the importance of further development of grid-forming converter control as well as the need for harmonized specifications in line with both capabilities of existing control structures and the grid operator's needs.

References

[1] J. Matevosyan *et al.*, "Grid-Forming Inverters: Are They the Key for High Renewable Penetration?" *IEEE Power and Energy Magazine*, vol. 17, no. 6, Nov. 2019.

[2] R. Rosso *et al.*, "Grid-Forming Converters: Control Approaches, Grid-Synchronization, and Future Trends — A Review," *IEEE Open Journal of Industry Applications*, vol. 2, 2021.

[3] A. Narula, M. Bongiorno, and M. Beza, "Comparison of Grid-Forming Converter Control Strategies," in *2021 IEEE Energy Conversion Congress and Exposition (ECCE)*, Oct. 2021.

[4] M. Alt *et al.*, "Overview of Recent Grid Codes for Wind Power Integration," presented at the 12th International Conference on Optimization of Electrical and Electronic Equipment, 2010.

[5] European Commission, *COMMISSION REGULATION (EU) 2016/ 631 - establishing a network code on requirements for grid connection of generators*, Apr. 16, 2016.

[6] ——, *COMMISSION REGULATION (EU) 2016/ 1447 - establishing a network code on requirements for grid connection of high voltage direct current systems and direct current-connected power park modules*, Aug. 26, 2016.

[7] Energimarknadsinspektionen, *Energimarknadsinspektionens föreskrifter om fastställande av generellt tillämpliga krav för nätanslutning av generatorer*, Dec. 5, 2018.

[8] ——, *Energimarknadsinspektionens föreskrifter om fastställande av generellt tillämpliga krav för nätanslutning av system för högspänd likström och likströmsanslutna kraftparksmoduler.* Mar. 19, 2019.

[9] Verband der Elektrotechnik Elektronik Informationstechnik e. V., *Technical requirements for grid connection of high voltage direct current systems and direct current-connected power park modules (VDE-AR-N 4131 TAR HVDC)*, Mar. 2019.

[10] IEEE, *P2800/D6.0 (March 2021) - Draft Standard for Interconnection and Interoperability of Inverter-Based Resources Interconnecting with Associated Transmission Systems*, Mar. 2021.

[11] National Grid ESO, *Draft Final Modification Report GC0137: Minimum Specification Required for Provision of GB Grid Forming (GBGF) Capability*, Nov. 11, 2021.

[12] ENTSO-E, "High Penetration of Power Electronic Interfaced Power Sources and the Potential Contribution of Grid Forming Converters," 2020.

[13] ENTSO-E *et al.*, "Future system inertia 2," 2018.

[14] ENTSO-E *et al.*, "FFR Design of Requirements – External document," Feb. 12, 2020.

[15] Verband der Elektrotechnik Elektronik Informationstechnik e. V., *VDE FNN Guideline FNN Guideline: Grid forming behaviour of HVDC systems and DC-connected PPMs*, 2020.

[16] J. Rocabert *et al.*, "Control of Power Converters in AC Microgrids," *IEEE Transactions on Power Electronics*, vol. 27, no. 11, Nov. 2012.

[17] Hitachy Energy, "It's time to connect," 2021.

[18] M. Beza and M. Bongiorno, "Impact of converter control strategy on low- and high-frequency resonance interactions in power-electronic dominated systems," *International Journal of Electrical Power & Energy Systems*, vol. 120, Sep. 2020.

Modular Battery-Integrated Power Electronics—Modelling, Advantages, and Challenges

Nima Tashakor[1], Jan Kacetl[1], Tomas Kacetl[1], Stefan Goetz
Technische Universität Kaiserslautern
Kaiserslautern, Germany
E-Mail: tashakor@eit.uni-kl.de, jan.kacetl@porsche-engineering.de, tomas.kacetl@porsche-engineering.de

Acknowledgements

The authors acknowledge the financial support by the Federal Ministry of Education and Research of Germany in the project "Open6GHub" (grant number: 16KISK004).

Keywords

«Batteries», «Estimation technique », «Modular Multilevel Converter», «Reconfigurable Batteries», «Fault Tolerance», «Average State-Space Model», «Parameter Estimation», «Charge Balancing», «Smart Batteries»

Abstract

Modular battery-integrated converters (dynamically reconfigurable modular batteries) are expanding into emerging applications. Although widely popular, we are yet to fully exploit their potential. This paper provides a critical discussion of the more neglected aspects with particular focus on electro mobility applications. It also provides insight on the challenges and/or concerns.

Introduction

Despite renewable energy generation expansion, growth of the electromobility sector, environmental incentives, and intensified research on the battery technology, many of the traditional challenges still persist. Today, an electric vehicle (EV) is powered by the series and parallel connection of literally hundreds of cells [1]. In addition to the increased capacity, a trend toward higher voltage levels is observed that results in increasing the share of series connections in batteries with the same energy capacity [2-4]. The advantages behind a higher voltage battery pack (i.e., 800 V) include lower weight, better efficiency, and faster charging [5]. However, higher pack voltages and more serial connections can also lead to stricter protection and safety requirements, more complex monitoring and balancing sub-systems, as well as lower efficiency of the inverters at partial load [6-8].

Recent developments in power electronics have enhanced the performance of the low-voltage transistors while continuously reducing their prices, paving the way for emergence of new concepts such as battery-integrated modular multilevel converters, also known as dynamically reconfigurable batteries [9-13]. These systems break the conventionally hard-wired battery pack into multiple modules and integrate them with power electronics to achieve dynamic reconfiguration. There are quite well-known advantages to a modular battery system, such as better balancing, fault tolerance, and controllability as well as faster output regulation [14-16]. Furthermore, it is proven that the overall efficiency can be improved [17, 18]. However, considering the available degree of freedom in such systems, they have yet to fully achieve their potential. Additionally, as popular as these systems are, there are still many aspects that are not fully investigated such as their effect on battery aging, the current profiles and its effect, other topologies, and optimal module sizing.

---

[1] The first three authors have contributed equally to the paper.

Discussion of Possible Macro and Micro Topologies

Reconfigurable batteries and battery-integrated cascaded converters are not a new concept and there are many different macro and micro topologies available in the literature [19]. Different string connections can provide DC, single-phase, and multiphase structures with specific features [10]. It is possible to place only one connection between every two adjacent modules (e.g., in case of half-bridge or bi-directional full-bridge) allowing for merely bypass and series connection in the string or there can be more than one point of connection (e.g., diode-clamped [20, 21], FET-clamped [22],

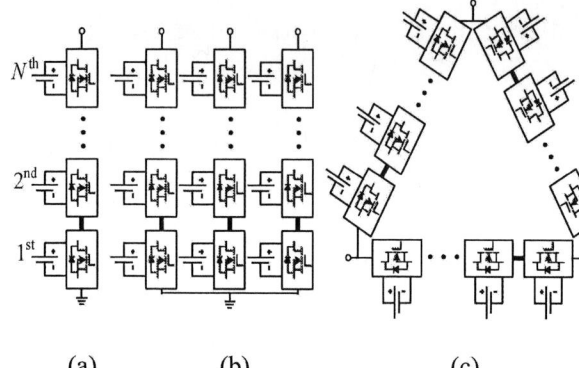

Fig. 1. (a) Single-phase single-arm structure, (b) three-phase single-arm structure with star connection, (c) three-phase single-arm structure with delta connection

or topologies with higher number of switches [23]). Since a review of different topologies is not the purpose of this work, we only discuss the critical points of the more feasible approaches to battery-integrated systems with particular focus on electromobility, but more detailed reviews of different topologies can be found in the literature [24-26].

AC Structures

In general, all the available AC topologies for the modular multilevel converters are also usable with reconfigurable AC batteries, and indeed a large body of research focusses on developing or modifying available control and algorithms for battery-integrated inverters [9, 27, 28]. However, while the double-arm MMC topology is very useful in capacitor-based modules to create a buffer between the main source of energy (the DC links), it offers almost no practical advantage in case of battery-integrated modules, needs higher numbers of battery modules, and increases the isolation voltage requirement as well as cost of the monitoring/balancing/protection subsystems.

The main idea behind the dual-arm topology is to provide a floating AC output, while the voltage sum of both strings remains constant, and minimize the size of the module capacitances as the intermediary energy storage. However, with batteries inside the modules, the main sources of energy are the batteries, and there is no need for a double-arm structure. Comparingly, the cascaded full-bridge topology in a single-arm structure can achieve the same output voltage levels, with the same number of semiconductors, while reducing the isolation voltage, the number of battery modules, and monitoring as well as protection requirements to half. A duality of this condition also holds true for the dual-arm cascaded full-bridge topologies with parallel function [29] and the single-arm double full-bridge structure [12]. It is possible to connect multiple strings in different structures, e.g., star and delta structures as Fig. 1 illustrates. Furthermore, Fig. 2 depicts the two interesting module topologies for AC application. The double-full-bridge topology can offer parallel connectivity, whereas the single-full-bridge can only provide bypass and series connections [30].

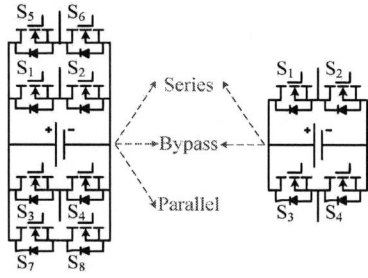

Fig. 2. Recommended module topologies for single-arm structures

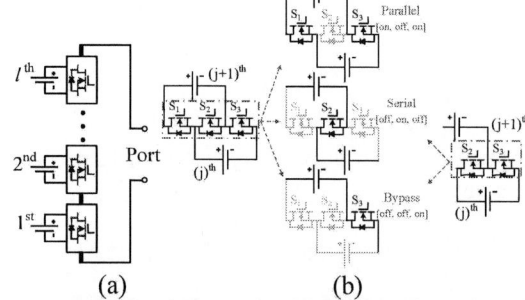

Fig. 3. (a) Generic structure of a string; (b) simplest module topologies with possible operation modes

DC Structures

Many applications, also in electromobility, do not aim for a reconfigurable AC battery as the overall architecture, including one or more separate DC outputs for auxiliaries, poses different constraints. A more cost-effective approach can be a reconfigurable DC battery modulating the DC-link voltage, while a conventional inverter generates necessary AC voltage [31]. For a DC structure with integrated batteries, normally only a single string suffices, but there are still many debates about the sizing (which will be discussed in Section III) as well as the module topology.

For DC structures, the minimum available connection modes are bypass and series, which can be achieved with the minimum of two switches per module in half-bridge topology. However, parallel connectivity in addition to the previous connection modes is preferred in many applications, since it can offer better efficiency, reduce ripple issues, increase the rated power, average the load better across modules, and contribute to balancing of the modules. Although it is possible to achieve parallel function in some cases with an additional diode in diode-clamped topologies, the resulting efficiency is not ideal. For more efficiency, parallel connection as well as previous bypass and series modes and thus at least three switches are necessary. Fig. 3(a) illustrates a generic form of a DC string, and Fig. 3(b) depicts the two simplest module topologies [32]. More complex circuits can offer additional capabilities, e.g., fault protection [33, 34].

Average State-Space Model of a String

Modelling each battery with a series resistance as well as multiple RC networks, the string voltage at each instance can be calculated per

$$V_{\text{string}}(t) = S^T(t) \times \left(V_{oc}(t) - R_0(t) i_{\text{string}} - V_{RC,1} - V_{RC,2} - \cdots \right), \tag{1}$$

where S^T is the modules state vector and can be $+1$ for positive series, -1 for negative series, and 0 for bypass/parallel [35]. Additionally, $V_{oc}^T = [V_{oc,1}, \dots, V_{oc,N}]$ and $R_0^T = [R_{0,1}, \dots, R_{0,N}]$ are respectively the vectors of open-circuit voltage as well as the internal resistance, and $V_{RC,i}$ is the vector of voltages across the ith RC network. Lastly, $\times$ denotes the cross product. The derivation of $V_{RC,i}$ is

$$\frac{dV_{RC,i}}{dt} = S.\frac{1}{C_i}.i_{\text{string}} - \frac{V_{RC,i}}{R_i C_i}, \tag{2}$$

where $R_i^T = [R_{i,1}, \dots, R_{i,N}]$, and $C_i^T = [C_{i,1}, \dots, C_{i,N}]$ are respectively the vector of the RC network resistances and capacitances. Additional RC networks can similarly be included into the model for better accuracy.

As (1) and (2) suggest, the equations have a discrete form depending on the state vector S. However, a continuous derivation is possible through averaging the states as

$$V_{\text{string}}(t) = M^T(t) \times \left(\underbrace{V_{oc}(t) - R_0(t).i_{\text{string}}}_{V_t} - V_{RC,1} - V_{RC,2} - \cdots \right), \tag{4}$$

$$\frac{dV_{RC,i}}{dt} = M.\frac{1}{C_i}.i_{\text{string}} - \frac{V_{RC,i}}{R_i C_i}, \tag{5}$$

where $M^T = [m_1, m_2, \dots, m_N]$ is the vector of the modulation indices of the modules, $\left(M.\frac{1}{C_i} \right)^T = [\frac{m_1}{C_{1,1}}, \frac{m_2}{C_{1,2}}, \dots, \frac{m_N}{C_{1,N}}]$ represents the dot product of the vectors M and $\frac{1}{C_i}$, and $V_t^T = [V_{t,1}, \dots, V_{t,N}]$ is the vector of the terminal voltage of the battery modules.

With a LC low-pass filter (L_{LPF} and C_{LPF}) across the string as Fig. 4 illustrates, it is possible to write the average state space equations per

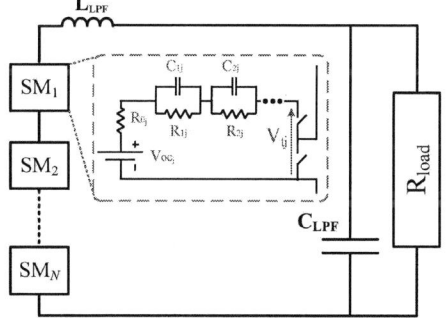

Fig. 4. Circuit of the modeled string

$$\dot{X} = A \times X + B \times V_t, \tag{6}$$

where $X^T = [i_{\text{string}}, V_o]$ is the vector of the state variables, and A and B are defined per

$$A = \begin{bmatrix} \left(-\dfrac{1}{L_{dc}}\right)(Nr_{\text{sw}} + R_{\text{LPF}}) & \dfrac{-1}{L_{\text{LPF}}} \\ \dfrac{1}{C_{\text{LPF}}} & -\dfrac{1}{(R_{\text{Load}}C_{\text{LPF}})} \end{bmatrix}, \tag{7}$$

$$B = \begin{bmatrix} M^T \dfrac{1}{L_{LPF}} \\ 0 \dots 0 \end{bmatrix}. \tag{8}$$

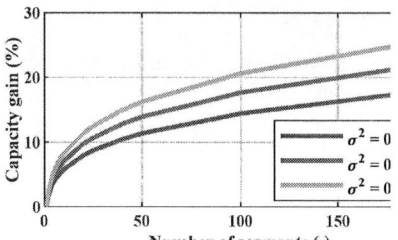

Fig. 5. Average capacity gain of cell string segmenting ($\mu = 4$ Ah, $\sigma^2 = 0.05$, $\sigma^2 = 0.07$, $\sigma^2 = 0.09$)

Therefore, at each instance, first the $V_{RC,i}$ values are updated using (5), then the new terminal voltages of each module are calculated using (4), and lastly, the new operating point of the string is calculated per (6).

Unknown Caveats, Challenges, Trade-Offs, and Potentials

Sizing Statistics and Necessary Trade-offs

The capacity of hard-wired battery packs is limited by the weakest cell in the string. Thus, the capacity follows the law of minimum [36, 37]. This limitation to the weakest element leads to a large loss of capacity in large batteries with many cells in series. It grows with the number of cells and therefore aggravates for high-voltage packs such as in grid storage and electric vehicles. The battery pack mean usable capacity is calculated per

$$E(Q_n) = \int_{-\infty}^{\infty} Q \cdot f(Q, \mu, \sigma) \cdot \left(1 - F(Q, \mu, \sigma)\right)^{n-1} dQ, \tag{9}$$

where Q is the ampere-hour capacity, f is the cell capacity distribution function with Gaussian distribution, F is the cell capacity cumulated distribution function, μ is the mean value, σ is the standard deviation, and n is the number of cells in series. To minimize the effect, long strings require high-quality cells with minimal parameter variation. However, low manufacturing tolerances and cell pre-selection increases the overall cost. Still, even initially perfectly matching cells age differently, leading to even further increase of the tolerances in a battery pack [38].

Although multiple active balancing techniques focused on increasing the usable battery pack capacity exist, they can only balance charge but not power, heating, or ageing. Furthermore, simple passive balancing still dominates in the industry. Reconfigurable battery packs can fully control the current flow in the battery, thus actively balance the pack on the module level. Such a battery pack splits the long hard-wired cell string into shorter segments which leads to gain in the usable capacity, as Fig. 5 depicts. With additional balancing capability, the usable life of the overall battery pack can be increased, as further extreme tolerances between the modules' capacities can be neutralized.

Active regulation of each module can provide the opportunity to integrate the battery management functions into the control loops of each module and offset the cost of additional power electronics circuitry. However, increasing the segments can also increase complexity along with cost, and the additional electronics can reduce the efficiency of the energy storage system. Therefore, there is still a trade-off between the number of segments and the gain in usable capacity.

Parallel Connection, Better Load share

In contrast to traditional module topologies with series and bypass connectivity, topologies with parallel connectivity offer substitution of bypassing by clustering in parallel groups [39]. Elimination of inactive states and distribution of load by paralleling effectively reduces root-mean square (RMS) value of the module load, which consequently decreases string impedance and increases efficiency [40]. Furthermore, parallel distribution prevents temporal overloading of the battery at phase currents exceeding their maximum discharge rates [41].

With simplification, the average equivalent resistance of the reconfigurable battery, respectively, with and without parallel mode can be approximated per

$$R_{\text{eq,without}} \approx M^T \times R_{bt} + N_1 r_{\text{sw}}, \tag{10}$$

$$R_{\text{eq,with}} \approx \frac{M^T}{(1-M^T).N} \times R_{bt} + N_2 r_{\text{sw}}, \tag{11}$$

which confirm that with comparable die area for the transistors, the parallel mode can help reducing the conduction loss. In (10) and (11), $R_{bt} = R_0 + R_1 + \cdots$ is the vector of equivalent resistance of the battery model shown in Fig. 4. Additionally, N_1 and N_2 are the number of semi-conductors derated by their current at each moment and can depend on the topology as well as the modulation. However, in many cases assuming $N_1 \approx N_2$ and equal to total number of modules (N) is a reasonable assumption [29].

The distribution of the load among paralleled modules is governed by voltage differences between modules, which are typically kept minimal through controller's effort, and proportions between battery impedance and their interconnection paths, typically low enough for sufficiently low differences [42]. Dependent on the resistance ratio of interconnection and batteries in the modules paralleling a high number of modules pushes the majority of the load towards outer modules of the parallel group, whereas the inner modules close to middle of the group remain lightly loaded [43].

Fig. 6 displays distribution of the load share in a reconfigurable battery with ten modules, where the outer modules (Modules 1, 2, 9, and 10) can take significantly more load than modules in the center of the parallel cluster (Modules 5 and 6). The parabolic character of the current distribution is discussed in greater detail in [43]. Higher load share will discharge the outer modules faster, which consequently increase the balancing current (i.e., a circulating current between two adjacent modules), until an equilibrium is achieved.

Higher Output Quality/Better Output Discretization/Lower Filter

Unlike conventional two-level converters, modular reconfigurable systems provide the output voltage in discrete steps. The voltage discretization itself brings about lower voltage derivations resulting in output filter reduction. In addition, modularization allows the implementation of low-voltage semiconductor switches with high switching dynamics to increase the module switching rate. Furthermore, proper modulation techniques, such as phase-shifted carrier (PSC) modulation further increase the effective output frequency N times [30, 44]. The lower voltage deviation, high module switching dynamic, and PSC modulation can reduce the requirement on the output filter, as described by

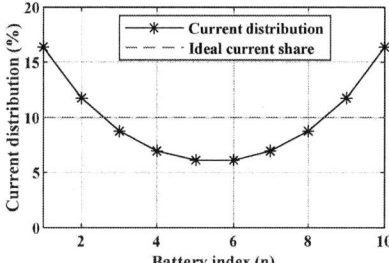

Fig. 6. Distribution of load in parallel group

$$L \approx \frac{V_{BP}}{N^2 \cdot f_{sw}}, \tag{12}$$

where V_{BP} is the battery pack maximal voltage and f_{sw} is the module switching rate.

Such a feature is easily evaluated by comparing the reconfigurable system (as shown in Fig. 4) to a conventional buck converter. Typically, the filtering inductor is designed with respect to the switching rate, the duty cycle (D), input (V_{in}) and output (V_{out}) voltages to restrict the certain current ripple (ΔI_L) below a certain threshold per

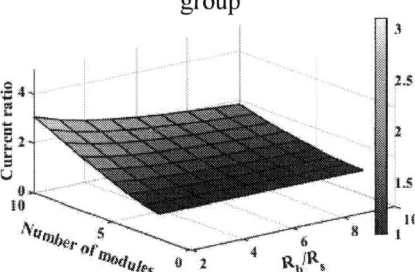

Fig. 7. Influence of impedances and number of moduels

$$\Delta I_L \leq \frac{D(V_{in} - V_{out})}{f_{sw} L}. \tag{13}$$

The highest current ripple in a conventional buck converter supplied from a battery pack ($V_{in} = V_{BP}$) occurs when $D = 0.5$ at $V_{out} = V_{in}/2$. In a reconfigurable battery, the highest ripple occurs when the difference between the input and the output voltage is equal to the half of the module voltage, e.g., $V_{out} - V_{in} = V_m/2 = V_{BP}/2N$. At such an operating point, the reconfigurable

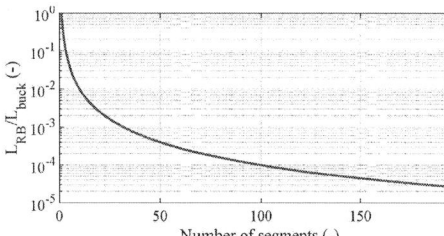

Fig. 8. Output filter reduction with battery pack segmenting

system switches between two voltage levels nearest to the output voltage with equal conduction times t_{on} and t_{off}, thus exhibits an equivalent duty cycle of 50% ($D = 0.5$). Comparing the current ripples for both topologies, and considering the same maximum threshold and the switching rate, we obtain

$$\frac{0.5\left(V_{\text{BP}}-\frac{V_{\text{BP}}}{2}\right)}{f_{\text{sw}}L_{\text{buck}}} = \frac{0.5\frac{V_{\text{BP}}}{2N}}{Nf_{\text{sw}}L_{\text{RB}}}, \tag{14}$$

which can be simplified to

$$\frac{L_{\text{RB}}}{L_{\text{buck}}} = \frac{1}{N^2}. \tag{15}$$

Equation (15) confirms that the size of the filtering inductor has an inversed squared relation with the number of modules. Consequent to the improved quality and increased effective switching frequency at the output, the size of the output filter is reduced. Fig. 8 illustrates the ratio of such a reduction compared to a normal hard-wired battery pack supplying a buck converter.

Estimation Techniques to Simplify Monitoring

Reconfigurable batteries offer many advantages over conventional battery packs. However, some functions such as the protection and monitoring are still necessary, which added to the more complex circuitry necessitate more sensors as well as higher-bandwidth communication channels [30, 41]. On the other hand, similar to MMCs with capacitors, knowing the exact state of each module in addition to the output voltage and current can provide sufficient information to estimate the state of each module without any direct measurement at the module level. However, on the contrary to modules with capacitors, with batteries the terminal voltage is not the only necessary parameter that needs monitoring [45]. Two important parameters that deserve monitoring are the equivalent resistance for age and under-load state-of-charge (SOC) estimation as well as the open-circuit voltage, which is a direct indicator of SOC [46].

Theoretically, it is possible to use the available methods for MMCs with capacitors to estimate the terminal voltage of the batteries and then use the terminal voltage, the module state, and the arm current to estimate other parameters of the battery such as open-circuit voltage as well as its electrical equivalent

Table I: Generic algorithms for parameter estimation of reconfigurable batteries

Case	Updatable parameters	Process / Comment
Battery has been idle for a long time (few hours)	V_{oc}	- After a long period of time V_{oc} can be measured through terminal voltage $V_{oc} = V_t$
Fast current variations (few milliseconds)	R_0, τ_1, R_1	- There is an immediate jump due to R_0: $R_0 = \frac{\Delta v_1}{\Delta i}$ - The time-constant τ_1 can be calculated from V_t shape - After few ms, the RC network reaches steady state, and R_1 can be calculated per $R_1 = \frac{\Delta v_2}{\Delta i} - R_0$
Slow current variations (few seconds)	R_2, τ_2	- With slow variations of current, the first RC network response is negligible, and τ_2 can be approximated from V_t - Neglecting SOC variations, the resistance can be calculated per $R_2 = \frac{\Delta v_3}{\Delta i_3} - R_0 - R_1$
$\vdots$		
Constant current, in steady-state condition	V_{oc}	- The only variation is due to the SOC changes, and open-circuit voltage is $V_{oc} = V_t - iR_0 - iR_1 - \cdots$

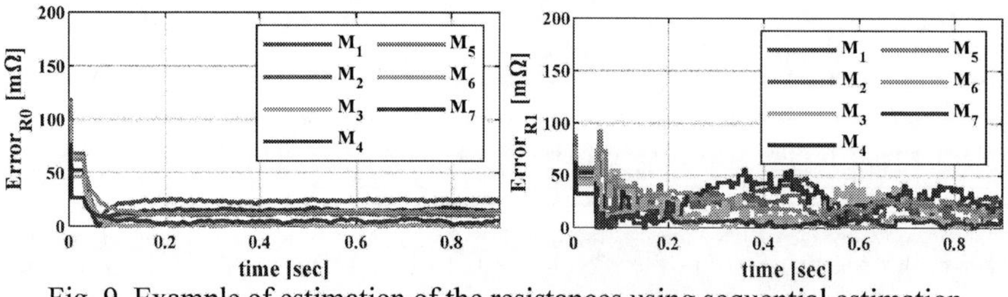

Fig. 9. Example of estimation of the resistances using sequential estimation

circuit [47]. As long as the switching frequency of the modules is considerably higher than the battery time constants, it is possible to decouple the terminal voltage of each module by just measuring the output voltage. However, the effect of the voltage sensor must be compensated [48] per

$$v_{\text{comp}} = v_{\text{meas}} + \frac{\Delta v_{\text{meas}}}{1 - \exp\left(-\frac{\Delta t}{\tau}\right)}. \tag{16}$$

Table I presents a generic algorithm for a sequential parameter estimation approach, while Fig. 9 offers an example result of applying the provided procedure in an experimental data to estimate R_0 vector depicted in Fig. 4.

Current Ripple Profile and Its Effect on Aging

The spectrum of the module load is strongly affected by the character of the terminal load. Batteries with AC output feature dominant components of the fundamental phase frequency, especially the second harmonic, which is further accompanied by an increased spectrum around the switching rate. The current ripple is typically subject to filtering in supplemented DC link capacitors on each module [49]. Alternatively, the current ripple suppression might be incorporated in the control objectives of the reconfigurable battery [40].

Highly rippled load of the module, as shown in Fig. 10, eventually increases losses in the battery cells and accelerates their ageing [50]. Recent studies show, however, that high-frequency ripple can mostly flow through the electrode capacitance without extensive activation of faradaic processes on the electrode-electrolyte interface, which reduces cycling of the battery and consequently ageing [51]. The behavior of the electrode capacitance is described in Randles' model in Fig. 11, where faradaic processes are modeled as a constant-phase element (Warburg impedance Z_W) and a charge transfer resistance R_{ct}. The faradaic path is bypassed by the electrode double layer capacitance C_{dl}, which forms the high pass characteristic of the cell impedance. For higher frequencies, the charge cumulated in the capacity C_{dl} is sufficient to feed the load ripple without extensive fluctuation of the voltage across the charge transfer resistance R_{ct} (electrode-electrolyte interface respectively). Voltage across the interface drives kinetics of the primary electrochemical reaction, but also increases kinetics of side reactions. Therefore, reduced fluctuation of the interface potential, decreases the additional ageing of the battery call connected to the rippled load [52]. In summary, shifting the ripple profile of the load current to higher frequencies can lead to a reduction in aging.

Conclusion

This paper investigates key aspects of the reconfigurable batteries through statistical analysis, discussion, simulations, and experiments. The paper investigates the trade-offs between module size as well as the number of serial cells and analyzes the output voltage quality as well as filter size of reconfigurable batteries. Furthermore, it discusses the effect of parallel connection in reducing the effective impedance of the system, and the uneven distribution of the load current among paralleled modules. The possibility of state estimation to reduce the cost of monitoring system is proposed, with an example case of the experimental results. Lastly, we debate the concept of exploiting the double-layer capacitance of the battery to reduce the aging.

References

[1]   A. N. Link, O'Connor, A. C., Scott, T. J., "Battery Technology for Electric Vehicles.," *London:*

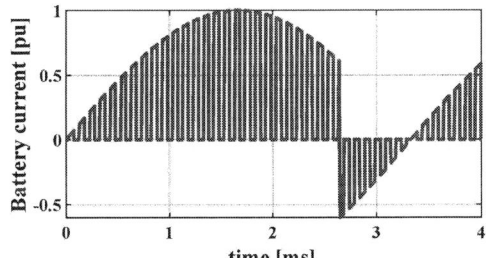

Fig. 10. Current waveform of a battery cell in a cascaded full-bridge topology

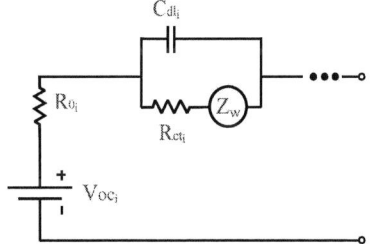

Fig. 11. Randles' equivalent circuit of the battery model

Routledge, , 2015, doi: https://doi.org/10.4324/9781315749303.

[2] S. Absar, W. Taha, and A. Emadi, "Efficiency Evaluation of Six-Phase VSI and NSI for 400V and 800V Electric Vehicle Powertrains," presented at the IECON 2021 – 47th Annual Conference of the IEEE Industrial Electronics Society, 13-16 Oct. 2021, 2021.

[3] I. Aghabali, J. Bauman, P. J. Kollmeyer, Y. Wang, B. Bilgin, and A. Emadi, "800-V Electric Vehicle Powertrains: Review and Analysis of Benefits, Challenges, and Future Trends," *IEEE Transactions on Transportation Electrification,* vol. 7, no. 3, pp. 927-948, 2021, doi: 10.1109/TTE.2020.3044938.

[4] A. Allca-Pekarovic, P. J. Kollmeyer, P. Mahvelatishamsabadi, T. Mirfakhrai, P. Naghshtabrizi, and A. Emadi, "Comparison of IGBT and SiC Inverter Loss for 400V and 800V DC Bus Electric Vehicle Drivetrains," in *2020 IEEE Energy Conversion Congress and Exposition (ECCE)*, 11-15 Oct. 2020 2020, pp. 6338-6344, doi: 10.1109/ECCE44975.2020.9236202.

[5] C. Jung, "Power Up with 800-V Systems: The benefits of upgrading voltage power for battery-electric passenger vehicles," *IEEE Electrification Magazine,* vol. 5, no. 1, pp. 53-58, 2017, doi: 10.1109/MELE.2016.2644560.

[6] J. Zhang, Z. Wang, P. Liu, and Z. Zhang, "Energy consumption analysis and prediction of electric vehicles based on real-world driving data," *Applied Energy,* vol. 275, p. 115408, 2020.

[7] M. Jaensch, J. Kacetl, T. Kacetl, and S. Götz, "Modulation index improvement by intelligent battery," Patent US10784698B2, 2020.

[8] Y. Yang, Q. Ye, L. J. Tung, M. Greenleaf, and H. Li, "Integrated Size and Energy Management Design of Battery Storage to Enhance Grid Integration of Large-Scale PV Power Plants," *IEEE Transactions on Industrial Electronics,* vol. 65, no. 1, pp. 394-402, 2018.

[9] C. Gan, Q. Sun, J. Wu, W. Kong, C. Shi, and Y. Hu, "MMC-Based SRM Drives With Decentralized Battery Energy Storage System for Hybrid Electric Vehicles," *IEEE Transactions on Power Electronics,* vol. 34, no. 3, pp. 2608-2621, 2019, doi: 10.1109/TPEL.2018.2846622.

[10] N. Tashakor, E. Farjah, and T. Ghanbari, "A Bidirectional Battery Charger With Modular Integrated Charge Equalization Circuit," *IEEE Transactions on Power Electronics,* vol. 32, no. 3, pp. 2133-2145, 2017, doi: 10.1109/TPEL.2016.2569541.

[11] Y. Li and Y. Han, "A Module-Integrated Distributed Battery Energy Storage and Management System," *IEEE Transactions on Power Electronics,* vol. 31, no. 12, pp. 8260-8270, 2016, doi: 10.1109/TPEL.2016.2517150.

[12] S. M. Goetz, Z. Li, X. Liang, C. Zhang, S. M. Lukic, and A. V. Peterchev, "Control of modular multilevel converter with parallel connectivity—Application to battery systems," *IEEE Transactions on Power Electronics,* vol. 32, no. 11, pp. 8381-8392, 2016.

[13] Z. Zedong, W. Kui, X. Lie, and L. Yongdong, "A Hybrid Cascaded Multilevel Converter for Battery Energy Management Applied in Electric Vehicles," *Power Electronics, IEEE Transactions on,* vol. 29, no. 7, pp. 3537-3546, 2014, doi: 10.1109/TPEL.2013.2279185.

[14] S. Ali, Z. Ling, K. Tian, and Z. Huang, "Recent Advancements in Submodule Topologies and Applications of MMC," *IEEE Journal of Emerging and Selected Topics in Power Electronics,* pp. 1-1, 2020, doi: 10.1109/JESTPE.2020.2990689.

[15] T. Zheng *et al.*, "A Novel Z-Type Modular Multilevel Converter with Capacitor Voltage Self-Balancing for Grid-Tied Applications," *IEEE Transactions on Power Electronics,* pp. 1-1, 2020, doi: 10.1109/TPEL.2020.2997991.

[16] S. Xu, "A New Multilevel AC/DC Topology Based H-Bridge Alternate Arm Converter," *IEEE Access,* vol. 8, pp. 57997-58005, 2020, doi: 10.1109/ACCESS.2020.2982202.

[17] M. Quraan, P. Tricoli, S. D'Arco, and L. Piegari, "Efficiency assessment of modular multilevel converters for battery electric vehicles," *IEEE Transactions on Power Electronics,* vol. 32, no. 3, pp. 2041-2051, 2016.

[18] N. Tashakor, B. Arabsalmanabadi, L. O. Cervera, E. Hosseini, K. Al-Haddad, and S. Goetz, "A Simplified Analysis of Equivalent Resistance in Modular Multilevel Converters with Parallel Functionality," presented at the IECON 2020 The 46th Annual Conference of the IEEE Industrial Electronics Society, 18-21 Oct. 2020, 2020.

[19] M. Quraan, T. Yeo, and P. Tricoli, "Design and Control of Modular Multilevel Converters for Battery Electric Vehicles," *IEEE Transactions on Power Electronics,* vol. 31, no. 1, pp. 507-517, 2016, doi: 10.1109/TPEL.2015.2408435.

[20] T. Zheng *et al.*, "A Novel High-Voltage DC Transformer Based on Diode-Clamped Modular Multilevel Converters With Voltage Self-Balancing Capability," *IEEE Transactions on Industrial Electronics,* vol. 67, no. 12, pp. 10304-10314, 2020, doi: 10.1109/TIE.2019.2962486.

[21] N. Tashakor, M. Kilictas, E. Bagheri, and S. Goetz, "Modular Multilevel Converter with Sensorless Diode-Clamped Balancing through Level-Adjusted Phase-Shifted Modulation," *IEEE Transactions on Power Electronics,* pp. 1-1, 2020, doi: 10.1109/TPEL.2020.3041599.

[22] Y. Jin *et al.*, "A Novel Submodule Voltage Balancing Scheme for Modular Multilevel Cascade Converter—Double-Star Chopper-Cells (MMCC-DSCC) based STATCOM," *IEEE Access,* 2019.

[23] J. Xu, J. Li, J. Zhang, L. Shi, X. Jia, and C. Zhao, "Open-loop voltage balancing algorithm for two-port full-bridge MMC-HVDC system," *International Journal of Electrical Power & Energy Systems,* vol. 109, pp. 259-268, 2019/07/01/ 2019, doi: https://doi.org/10.1016/j.ijepes.2019.01.032.

[24] J. Fang, F. Blaabjerg, S. Liu, and S. M. Goetz, "A Review of Multilevel Converters With Parallel Connectivity," *IEEE Transactions on Power Electronics,* vol. 36, no. 11, pp. 12468-12489, 2021, doi: 10.1109/TPEL.2021.3075211.

[25] A. K. Bhattacharjee, N. Kutkut, and I. Batarseh, "Review of Multiport Converters for Solar and Energy Storage Integration," *IEEE Transactions on Power Electronics,* vol. 34, no. 2, pp. 1431-1445, 2019, doi: 10.1109/TPEL.2018.2830788.

[26] S. Debnath, J. Qin, B. Bahrani, M. Saeedifard, and P. Barbosa, "Operation control and applications of the modular multilevel converter: A review," *IEEE Trans. Power Electron,* vol. 30, no. 1, pp. 37-53, 2015.

[27] N. Li, F. Gao, T. Hao, Z. Ma, and C. Zhang, "SOH Balancing Control Method for the MMC Battery Energy Storage System," *IEEE Transactions on Industrial Electronics,* vol. 65, no. 8, pp. 6581-6591, 2018, doi: 10.1109/TIE.2017.2733462.

[28] L. Baruschka and A. Mertens, "Comparison of Cascaded H-Bridge and Modular Multilevel Converters for BESS application," presented at the 2011 IEEE Energy Conversion Congress and Exposition, 17-22 Sept. 2011, 2011.

[29] N. Tashakor, F. Iraji, and S. G. Goetz, "Low-Frequency Scheduler for Optimal Conduction Loss in Series/Parallel Modular Multilevel Converters," *IEEE Transactions on Power Electronics,* pp. 1-1, 2021, doi: 10.1109/TPEL.2021.3110213.

[30] N. Tashakor, Z. Li, and S. M. Goetz, "A Generic Scheduling Algorithm for Low-Frequency Switching in Modular Multilevel Converters With Parallel Functionality," *IEEE Transactions on Power Electronics,* vol. 36, no. 3, pp. 2852-2863, 2021, doi: 10.1109/TPEL.2020.3018168.

[31] J. Kacetl, J. Fang, T. Kacetl, N. Tashakor, and S. Goetz, "Design and Analysis of Modular Multilevel Reconfigurable Battery Converters for Variable Bus Voltage Powertrains," *IEEE Transactions on Power Electronics,* pp. 1-1, 2022, doi: 10.1109/TPEL.2022.3179285.

[32] N. Tashakor, Z. Li, and S. M. Goetz, "A Generic Scheduling Algorithm for Low-Frequency Switching in Modular Multilevel Converters with Parallel Functionality," *IEEE Transactions on Power Electronics,* pp. 1-1, 2020, doi: 10.1109/TPEL.2020.3018168.

[33] D. Ronanki and S. S. Williamson, "Modular multilevel converters for transportation electrification: Challenges and opportunities," *IEEE Transactions on Transportation Electrification,* vol. 4, no. 2, pp. 399-407, 2018.

[34] Z. Li, R. Lizana, S. Sha, Z. Yu, A. V. Peterchev, and S. Goetz, "Module Implementation and Modulation Strategy for Sensorless Balancing in Modular Multilevel Converters," *IEEE Transactions on Power Electronics,* 2018.

[35] S. M. Goetz, Z. Li, A. V. Peterchev, X. Liang, C. Zhang, and S. M. Lukic, "Sensorless scheduling of the modular multilevel series-parallel converter: enabling a flexible, efficient, modular battery," presented at the 2016 IEEE Applied Power Electronics Conference and Exposition (APEC), 2016.

[36] M. A. Hannan, M. M. Hoque, A. Hussain, Y. Yusof, and P. J. Ker, "State-of-the-Art and Energy Management System of Lithium-Ion Batteries in Electric Vehicle Applications: Issues and Recommendations," *IEEE Access,* vol. 6, pp. 19362-19378, 2018, doi: 10.1109/ACCESS.2018.2817655.

[37] W. Waag, C. Fleischer, and D. U. Sauer, "Critical review of the methods for monitoring of lithium-ion batteries in electric and hybrid vehicles," *Journal of Power Sources,* vol. 258, pp. 321-339, 2014/07/15/ 2014, doi: https://doi.org/10.1016/j.jpowsour.2014.02.064.

[38] M. Baumann, L. Wildfeuer, S. Rohr, and M. Lienkamp, "Parameter variations within Li-Ion battery packs – Theoretical investigations and experimental quantification," *Journal of Energy Storage,* vol. 18, pp. 295-307, 2018/08/01/ 2018, doi: https://doi.org/10.1016/j.est.2018.04.031.

[39] S. M. Goetz, A. V. Peterchev, and T. Weyh, "Modular Multilevel Converter With Series and Parallel Module Connectivity: Topology and Control," *IEEE Transactions on Power Electronics,* vol. 30, no. 1, pp. 203-215, 2015, doi: 10.1109/TPEL.2014.2310225.

[40] Z. Li, R. Lizana, S. M. Lukic, A. V. Peterchev, and S. M. Goetz, "Current Injection Methods for Ripple-Current Suppression in Delta-Configured Split-Battery Energy Storage," *IEEE Transactions on Power Electronics,* vol. 34, no. 8, pp. 7411-7421, 2019, doi: 10.1109/TPEL.2018.2879613.

[41] B. Arabsalmanabadi, N. Tashakor, Y. Zhang, K. Al-Haddad, and S. Goetz, "Parameter Estimation of Batteries in MMCs with Parallel Connectivity using PSO," presented at the IECON 2021 – 47th Annual Conference of the IEEE Industrial Electronics Society, 13-16 Oct. 2021, 2021.

[42] G. Gunlu, "Dynamically Reconfigurable Independent Cellular Switching Circuits for Managing Battery Modules," *IEEE Transactions on Energy Conversion,* vol. 32, no. 1, pp. 194-201, 2017, doi: 10.1109/TEC.2016.2616190.

[43] Y. Zhu, W. Zhang, J. Cheng, and Y. Li, "A novel design of reconfigurable multicell for large-scale battery packs," presented at the 2018 International Conference on Power System Technology (POWERCON), 6-8 Nov. 2018, 2018.

[44] N. Tashakor and M. Khooban, "An Interleaved Bi-Directional AC–DC Converter With Reduced Switches and Reactive Power Control," *IEEE Transactions on Circuits and Systems II: Express Briefs,* vol. 67, no. 1, pp. 132-136, 2020, doi: 10.1109/TCSII.2019.2903389.

[45] D. N. T. How, M. A. Hannan, M. S. H. Lipu, K. S. M. Sahari, P. J. Ker, and K. M. Muttaqi, "State-of-Charge Estimation of Li-Ion Battery in Electric Vehicles: A Deep Neural Network Approach," *IEEE Transactions on Industry Applications,* vol. 56, no. 5, pp. 5565-5574, 2020, doi: 10.1109/TIA.2020.3004294.

[46] J. Meng *et al.*, "An Overview and Comparison of Online Implementable SOC Estimation Methods for Lithium-ion Battery," *IEEE Transactions on Industry Applications,* vol. 54, no. 2, pp. 1583-1591, 2018.

[47] B. Arabsalmanabadi, N. Tashakor, S. Goetz, and K. Al-Haddad, "Li-ion Battery Models and A Simplified Online Technique to Identify Parameters of Electric Equivalent Circuit Model for EV Applications," presented at the IECON 2020 The 46th Annual Conference of the IEEE Industrial Electronics Society, 18-21 Oct. 2020, 2020.

[48] F. Rong, X. Gong, X. Li, and S. Huang, "A New Voltage Measure Method for MMC Based on Sample Delay Compensation," *IEEE Transactions on Power Electronics,* vol. 33, no. 7, pp. 5712-5723, 2018, doi: 10.1109/TPEL.2017.2748969.

[49] T. Sarkar and S. K. Mazumder, "Optimum input-filter-capacitor sizing for fuel-cell based single-phase inverter for current-ripple mitigation," presented at the Proceedings of the 2011 14th European Conference on Power Electronics and Applications, 30 Aug.-1 Sept. 2011, 2011.

[50] S. De Breucker, K. Engelen, R. D'hulst, and J. Driesen, "Impact of Current Ripple on Li-ion Battery Ageing," *World Electric Vehicle Journal,* vol. 6, no. 3, 2013, doi: 10.3390/wevj6030532.

[51] F. Chang, F. Roemer, and M. Lienkamp, "Influence of Current Ripples in Cascaded Multilevel Topologies on the Aging of Lithium Batteries," *IEEE Transactions on Power Electronics,* vol. 35, no. 11, pp. 11879-11890, 2020, doi: 10.1109/TPEL.2020.2989145.

[52] T. Kacetl, J. Kacetl, N. Tashakor, M. Jaensch, and S. Goetz, "Degradation-Reducing Control for Dynamically Reconfigurable Batteries," *arXiv preprint arXiv:2202.11757,* 2022.

Design of Triple-Active Bridge Converter with Inherently Decoupled Power Flows

Dong-Uk Kim, ByengJoo Byen, ByungHwang Jeong, and Sungmin Kim
Hanyang University, Hyosung Corporation
Ansan, Anyang, Korea
Tel.: +82/31-400-5172.
E-Mail: ksminmoon@hanyang.ac.kr

Keywords

«DC-DC converter», «Tri-port isolated converter», «Bi-directional converters», «Efficiency», «Design».

Abstract

For the realization of DC grid, the high-power multi-port isolated DC/DC converter is necessary to integrate various DC sources and loads. Triple-Active Bridge (TAB) converter is one of the most popular topologies which has three-port high frequency transformer and three active bridges. It can realize the bi-directional power flow among different sources and loads. However, the intrinsic power coupling in each port of three-port transformer leads to the unwanted voltage fluctuation of DC side. The power coupling among three ports can be minimized by a low leakage inductance at one port. However, it can decrease the power efficiency in some power flow conditions of TAB converter. This paper has analyzed the reason for the degradation of power efficiency in various power flow condition. To estimate the power loss components in TAB converter, the power loss breakdown method is used. With the two experimental configuration of TAB converter: the conventional leakage inductance and a low leakage inductance, the power loss is estimated. From the experimental results, it can be concluded that the designing method of leakage inductance with a low leakage inductance at one port is appropriate for the TAB converter which has fixed power flow condition.

Introduction

As the increasing demand on DC power, the high-power DC-DC converter which can combine various DC sources/loads has been developed. For the integration of various DC sources/loads, a Triple-Active Bridge (TAB) converter is an appropriate candidate for its 3-way and bi-directional power control [1]. The TAB converter controls the power via three-port High Frequency Transformer (HFT). At each port, various type of DC sources/loads can be connected with half or full-bridge converters. The power flow of TAB converter can be controlled by adjusting the phase of three port voltages. Generally, the phase of one port is fixed and those of the others are shifted. Therefore, TAB converter has two controllable phase shift angles to control three port power. However, TAB converter has a coupled power among three ports since the three ports of HFT have magnetically coupled structure. It makes an intrinsic power coupling of three ports. This power coupling always exists in multi-port high frequency transformer. Therefore, it is hard to control each port power independently. In previous studies, various decoupling control algorithms are suggested to minimize the effect of power coupling [2],[3]. However, these algorithms are complicated to be applied, and the decoupling performance is definitely dependent on the inductance values of the TAB converter. Moreover, the system parameters are not easy to recognize exactly. As another approach, the decoupling method that uses a low leakage inductance at one port of HFT is suggested in [4]. With a low leakage inductance at one port, the power flows between the other ports are minimized. This method requires only two external series inductors for customizing the phase shift angle guaranteeing the confident decoupling performance with no complex control algorithm. In addition, the size and cost of magnetic components can be saved. However, the total loss

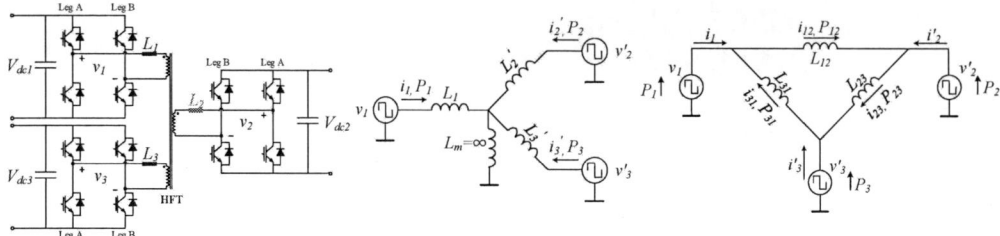

Fig. 1: Triple active bridge converter (a) schematic (b) Wye-model (b) Delta-model

of TAB converter can be increased due to the high peak current at the low leakage inductance port. Therefore, the careful design of leakage inductance and configuration is required.

In this paper, the brief review of TAB converter according to the leakage inductance is described. Then, the two methods designing leakage inductance are suggested: the conventional three leakage inductances configuration and the two leakage inductances with a low leakage inductance. The efficiencies of two cases are compared in various power flow condition and the reason for efficiency degradation is analyzed. Using the power loss breakdown method, each power loss component is estimated in various power condition. In addition, the guideline for selecting the position of low leakage port guaranteeing the decoupled power control and high efficiency is suggested. To verify the analysis and design methods, the experiments with 2kW TAB converter are conducted.

TAB converter with a low leakage inductance at one port

The schematic of TAB converter can be seen in Fig. 1(a). Each port of HFT consists of leakage inductance and full-bridge converter. The leakage inductance, L_1, L_2, L_3 includes the intrinsic leakage inductance of HFT, $L_{\sigma 1}$, $L_{\sigma 2}$, $L_{\sigma 3}$, and the external series inductance, L_{s1}, L_{s2}, L_{s3}. To control the power flow of each port, two controllable phase shift angles of port voltage, ϕ_{12}, ϕ_{13} are used. The phase shift angle of port-1 voltage is fixed to zero, and ϕ_{12}, ϕ_{13} are shifted. Fig. 1(b) shows the Wye-model of TAB converter. The voltage, current, inductances are referred to the port-1 side. To analyze the power transfer between two ports, Delta-model is used as shown in Fig. 1(c). The inductors, L_{12}, L_{23}, L_{31} are equivalent leakage inductance between two ports in Delta-model. The power delivered between two ports via equivalent leakage inductance of Delta-model, P_{12}, P_{23}, P_{31} can be derived as follows [5]:

$$
\begin{cases}
P_{12} = \dfrac{V_1 V_2'}{2\pi^2 f_{sw} L_{12}} \phi_{12}(\pi - \phi_{12}) \\[2mm]
P_{23} = \dfrac{V_2 V_3'}{2\pi^2 f_{sw} L_{23}} \phi_{23}(\pi + \phi_{23}) \\[2mm]
P_{31} = -\dfrac{V_1 V_3'}{2\pi^2 f_{sw} L_{31}} (\phi_{23} + \phi_{12})(\pi - (\phi_{23} + \phi_{12}))
\end{cases}
\tag{1}
$$

Through (1), the port power from each port can be derived as follows:

$$
P_1 = P_{12} - P_{31}, \ P_2 = -P_{12} + P_{23}, \ P_3 = P_{31} - P_{23}
\tag{2}
$$

In (2), two phase shift angles determine port power. It means that the control loops that controls each port power cannot be configured independently. The control loops have inherent coupling term, and it is configured in Two Input Two Output (TITO) form. The input parameters are two phase shift angle, and output parameters are two port power. The decoupling control methods have been studied in many papers to control in Single Input Single Output (SISO) form. Recently, a method has been proposed to minimize the power decoupling between two output power through the design of component values, not algorithm: One inductance is designed to be very small and other two inductance to be proper values to transfer the power among three ports. If the leakage inductance at port-2 is very low ($L_2 \approx 0$), the equivalent leakage inductance can be simplified as follows:

$$
L_{12} = L_1 + L_2' + \frac{L_1 L_2'}{L_3'} \approx L_1, \ L_{31} = L_1 + L_3' + \frac{L_1 L_3'}{L_2'} \approx \infty, \ L_{23} = L_2' + L_3' + \frac{L_2' L_3'}{L_1} \approx L_3'
\tag{3}
$$

By the simplified equivalent inductance, the power transfer between the port-1 and port-3, P_{31} is almost zero even though phase shift angle ϕ_{12}, ϕ_{13} are varying. Therefore, the power from each port can be simplified as follows:

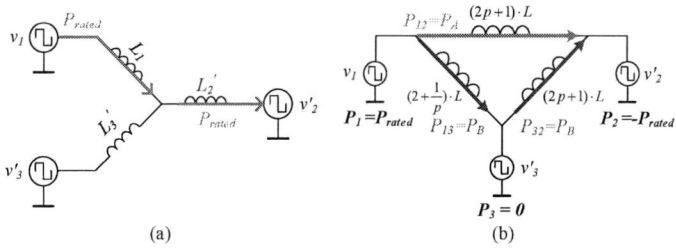

Fig. 2: The power flow in case of $P_1 = -P_2$ and $L_1=L'_2=L'_3$ (a) in Y-model (b) in Delta-model

$$P_1 = P_{12}, \; P_2 = -P_{12} + P_{23}, \; P_3 = -P_{23} \tag{4}$$

From (1) and (4), the port-1 power, P_1 is controlled by only ϕ_{12} and the port-3 power, P_3 is controlled by ϕ_{23}, independently. Therefore, two control loops that controls the power of each port can be configured in SISO form. This method guarantees the confident decoupling performance with no complex control algorithm. In addition, the external series inductor is not necessary to customize the phase shift angle. However, the power efficiency can be decreased in some power flow conditions due to the high peak current at the low leakage inductance port.

Design of Leakage inductance with a low leakage inductance at one port

In practical, the leakage inductance cannot be zero. In this paper, p is the ratio of a low leakage inductance to inductance of other ports. And the low leakage inductance is positioned at port-2. The leakage inductances can be defined as follows:

$$L_1 = L, L'_2 = p \cdot L, L'_3 = L \tag{5}$$

The equivalent leakage inductance can be calculated as (6).

$$L_{12} = \frac{L_1 L'_2 + L'_2 L'_3 + L'_3 L_1}{L'_3} = (2p+1)L, L_{23} = \frac{L_1 L'_2 + L'_2 L'_3 + L'_3 L_1}{L_1} = (2p+1)L, L_{13} = \frac{L_1 L'_2 + L'_2 L'_3 + L'_3 L_1}{L'_2} = \left(2 + \frac{1}{p}\right)L \tag{6}$$

In case of the bi-directional TAB converter, it can have various power flow conditions. Therefore, the leakage inductance should be designed considering these various power flow conditions. In general, the inductance should be designed enough to transfer the rated power between the interfacing active bridge ports.

In following design procedure, it is assumed that the rated power is delivered from port-1 to port-2 as shown in Fig. 2. In Wye-model, the rated power, P_{rated} is delivered via L_1 and L_2. In Delta-model, the power from port-1 is divided into two power terms (P_A and P_B) according to the impedance of equivalent inductance. In this condition, the active power delivered by port-3 is zero, and P_{13} is same with P_{32}. Therefore, power terms in Delta-model can be determined as follows:

$$P_A + P_B = P_{rated} \tag{7}$$

$$P_A = \frac{2p+\frac{1}{p}+3}{4p+\frac{1}{p}+4} \cdot P_{rated}, P_B = \frac{2p+1}{4p+\frac{1}{p}+4} \cdot P_{rated} \cdot \tag{8}$$

Assuming the maximum phase shift angle, $\phi_{12,max}$, the P_A is delivered and the power equation is as follows:

$$P_A = \frac{V_1 V'_2}{2\pi^2 f_{sw} L_{12}} \phi_{12,max}(\pi - \phi_{12,max}) \tag{9}$$

From (6), (8) and (9), the maximum value of L in (5), L_{max} can be derived.

$$L_{max} = \frac{K}{(2p+1) \cdot P_{rated}} \cdot \frac{4p+\frac{1}{p}+4}{2p+\frac{1}{p}+3} \qquad \left(K = \frac{V_1 V'_2}{2\pi^2 f_{sw}} \phi_{12,max}\left(\pi - \phi_{12,max}\right)\right) \tag{10}$$

In the conventional three inductance design, the ratio, p is designed to be 1. Three active bridge ports in TAB converter have same leakage inductance. The maximum leakage inductance and equivalent leakage inductance are (11) and (12), respectively.

$$L_{\max,p=1} = \frac{1}{2} \cdot \frac{K}{P_{rated}} \qquad (11)$$

$$L_{12} = L_{23} = L_{31} = 3 \cdot L_{max,p=1} \qquad (12)$$

In the two-leakage inductance design with a low inductance at one port, the ratio, p is designed to be very small value such as 0.05, which is the only 5% of other two leakage inductance values. The maximum leakage inductance and equivalent leakage inductance are (13) and (14), respectively.

$$L_{max,p=0.05} \approx 0.95 \cdot \frac{K}{P_{rated}} \qquad (13)$$

$$L_{12} = L_{23} = 1.1 \cdot L_{max,p=0.05}, \quad L_{13} = 22 \cdot L_{max,p=0.05} \qquad (14)$$

Power loss analysis of TAB converter in various power flow path

A. Variation of peak current according to the configuration of leakage inductance

The current waveforms of each port can be determined by the leakage inductance and the phase shift angle. Especially, the peak current of active port having a low leakage inductance can be significantly increased in some power flow conditions. The increased peak current leads to the increase of power loss components which are related to the peak current. To predict the power loss variation according to the leakage inductance, the peak current value is expected by deriving the current equation. Among various power flow conditions, the two cases are analyzed, and the peak current value is compared.

Fig. 3 shows power flow diagrams in Delta-model and steady-state waveforms in case of $P_1=-P_3$. In this case, the relationship of phase shift angle is $\phi_{21}<0$, $\phi_{23}>0$. Therefore, the power is delivered from port-1 to port-3. When the leakage inductance of all ports exists as shown in Fig.3(a), which is $p=1$, the power transferred via L_{12} is as follows:

$$P_{12,p=1} = \frac{1}{3}P_{rated} = \frac{V_1 V_2'}{2\pi^2 f_{sw} L_{12}} \phi_{12,p=1}. \qquad (16)$$

When the leakage inductance at port-2 is very low as shown in Fig. 3(b), which means p is nearly 0, the power transferred via L_{12} is as follows:

$$P_{12,p=0} = P_{rated} = \frac{V_1 V_2'}{2\pi^2 f_{sw} L_{12}} \phi_{12,p=0}. \qquad (17)$$

From (16) and (17), the relationship of phase shift angle in two cases can be obtained.

$$\phi_{12,p=0} = 2 \cdot \phi_{12,p=1} \qquad (18)$$

Using (18), the peak current value at the port-2 where the active power is not delivered can be calculated as follows:

$$i_{2,p=1}(t_1) = \frac{2V_1}{2\pi f_{sw} L_{23}} \phi_{12,p=1} \qquad (19)$$

$$i_{2,p=0}(t_1) = \frac{6V_1}{2\pi f_{sw} L_{23}} \phi_{12,p=1} \qquad (20)$$

In case of using low leakage inductance at one port, the peak current at low leakage inductance port increase three times as obtained in (20). As shown in Fig. 3(c), the phase shift angle and peak current relationship are verified in simulation results. In this case, the peak current significantly increases, which is the reason for high turn-off switching loss in full-bridge converter at port-2. On the other hand, due to the absence of external series inductance, $L_{2,series}$, the core and copper loss of inductor does not exists. However, the increase of switching loss at low leakage inductance side is large enough to degrade the power efficiency of TAB converter.

In Fig.4, the power is delivered from port-1 and 3 to port-2, equally. In this case, the relationship of phase shift angle can be calculated as same procedure with (16)-(18).

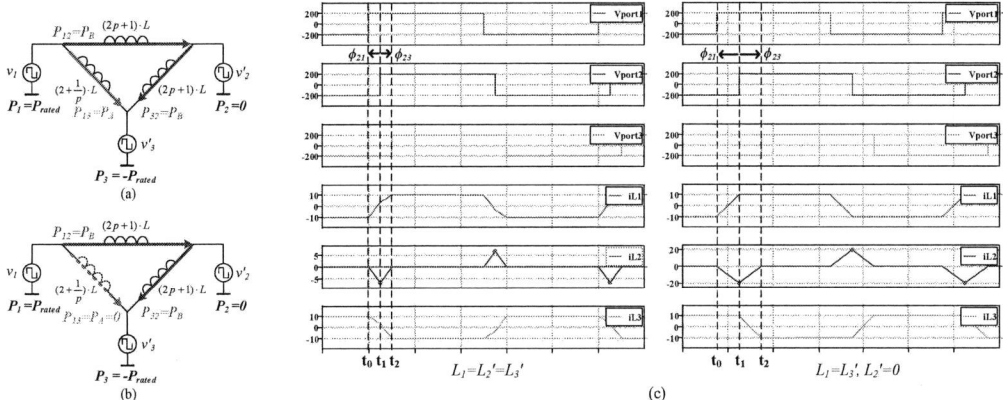

Fig. 3: Power flow of TAB converter in case of $P_1 = -P_3$ (a) $L_1 = L'_2 = L'_3$ (b) $L_1 = L'_3$, $L'_2 \approx 0$ (c) current and voltage waveforms

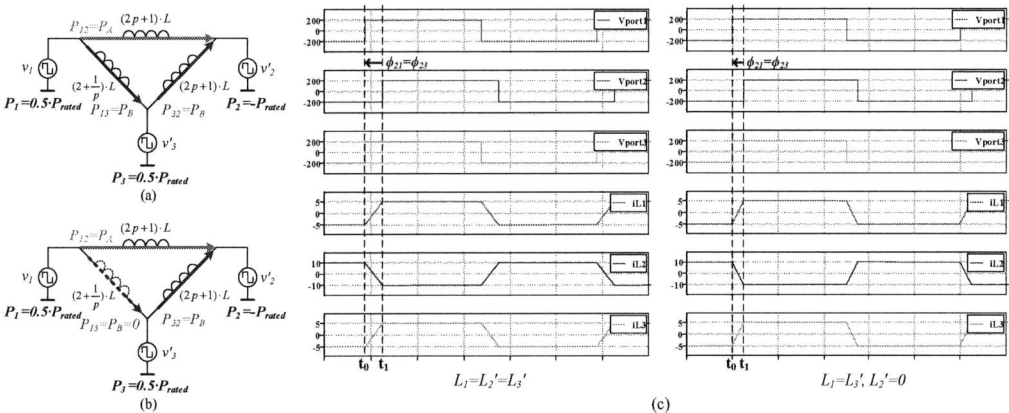

Fig. 4: Power flow of TAB converter in case of $P_1 = P_3 = 0.5 \cdot P_{rated}$, $P_2 = -P_{rated}$ (a) $L_1 = L'_2 = L'_3$ (b) $L_1 = L'_3$, $L'_2 \approx 0$ (c) current and voltage waveforms

$$\phi_{12,p=0} = \frac{2}{3}\phi_{12,p=1} \tag{21}$$

Using (21), the peak current value at the port-2 where the low leakage inductance side can be calculated as follows:

$$i_{1,p=0}(t_1) = i_{1,p=1}(t_1) = \frac{V_1}{L_{12}\omega}\phi_{21}. \tag{22}$$

In this power flow condition, the peak current is same as calculated in (22). It can be concluded that the turn-off switching loss of full-bridge converter at port-2 is also same. In addition, due to the absence of external series inductance of port-2, the inductor core and copper losses does not exist. As a result, the total power loss can be reduced. The results of two power flow in Fig. 3 and 4 show that the peak current increases in some power flow condition and the switching loss can be also increases, which can affect to the total power efficiency. Therefore, the position of a low leakage inductance should be carefully selected. To analyze the effect of variation of peak current to the power loss of TAB converter, the power loss breakdown is conducted in next section.

B. Power loss breakdown procedure of TAB converter

In this section, the power loss breakdown method is used for TAB converter. There are many previous studies about the method of power loss breakdown for the DC/DC converter [6]-[9]. In the DC/DC converter such as TAB converter, there are three main power loss components: 1) power devices, 2) high-frequency transformer 3) series inductors. In this paper, it is assumed that the leakage inductances of transformer of three ports are minimized. The leakage inductance which is necessary for delivering power is realized by connecting additional series inductors. The power losses in controller and gate drivers are excluded in this case.

The power loss of power devices is composed of conduction loss, P_{cond} and switching loss, P_{sw}. The total conduction loss of the three full-bridge converter of TAB converter is calculated as follows:

$$P_{cond} = 4 \cdot R_{on} I_{1,rms}^2 + 4 \cdot R_{on} I_{2,rms}^2 + 4 \cdot R_{on} I_{3,rms}^2 \qquad (23)$$

where R_{on} is on-state resistance of power device. The switching loss of power device occurs during turn-on and off state. The switching loss is proportional to the current flowing through the switch at the time of switching. In case of TAB converter, Zero-Voltage Switching (ZVS) enables the nearly zero switching loss during turn-on of switch. However, the DC/DC converter that uses phase-shift modulation suffer from the high-peak current during turn-off of switch. Therefore, the switching loss can be calculated as follows:

$$P_{sw} = P_{sw,Turn-on} + P_{sw,Turn-off} \approx P_{sw,Turn-off}. \qquad (24)$$

In many manufactures of power device, the switching loss information according to the switch current and power loss models are supported. The switching loss can be easily obtained even with various power flow condition.

The power loss of transformer is determined by the winding loss and iron loss. The copper loss of winding is winding loss, $P_{wire,TR}$ and the winding loss of TAB converter can be calculated as follows:

$$P_{wire,TR} = R_{wire,1} I_{RMS,1}^2 + R_{wire,2} I_{RMS,2}^2 + R_{wire,3} I_{RMS,3}^2 \qquad (25)$$

where $R_{wire,1}$, $R_{wire,2}$, and $R_{wire,3}$ are winding DC resistances of each port at TAB converter. In case of iron loss, it is determined by the electro-magnetic force and magnetization current of high-frequency transformer. Since these values are constant even different power flow condition, the iron loss of high-frequency transformer also has constant value. By injecting magnetization current at one port of high-frequency transformer, the iron loss can be measured experimentally.

The power loss of series inductor is similar with the that of high-frequency transformer: winding and iron loss. The winding loss of series inductor is calculated as follows:

$$P_{wire,L} = R_{wire,L} I_{RMS,L}^2 \qquad (26)$$

where $R_{wire,L}$ is winding DC resistance of series inductor. The iron loss of inductor can be estimated by the Steinmetz equation, or the datasheet of core provided by the manufacturer.

$$P_L = K \cdot f^a \cdot B_{peak}^b \qquad (27)$$

where K, a, and b are Steinmetz coefficient of the core, f is switching frequency, and B_{peak} is peak value of magnetic flux density. The peak magnetic flux density is calculated based on the peak inductor current value which is obtained experimentally and the B-H curve of the core. Using the described power loss breakdown method, the comparison of power losses with two leakage inductances configuration and three leakage inductances configuration are analyzed with experimental results in next section.

Experimental results

To verify the proposed method, the TAB converter setup is configured as shown in Fig. 5. To compare the power efficiency according to the leakage inductance, two different cases are set: the ratio, $p=1$ and $p=0.01$. First, the leakage inductance of all ports is equal ($L_1=L'_2=L'_3=52$uH), which is $p=1$. Second, the leakage inductance at port-2 is minimized ($L_1=L'_3=104$uH, $L'_2=1$uH), which is $p=0.01$. The output power of each port is controlled to the power reference of controller. The TAB converter is operated in six modes as shown in Fig. 5 to consider the various application of TAB converter. The power efficiency is experimentally measured using Newtons4th PPA4530 power analyzer. Each component of power loss is calculated based on the described power loss breakdown method.

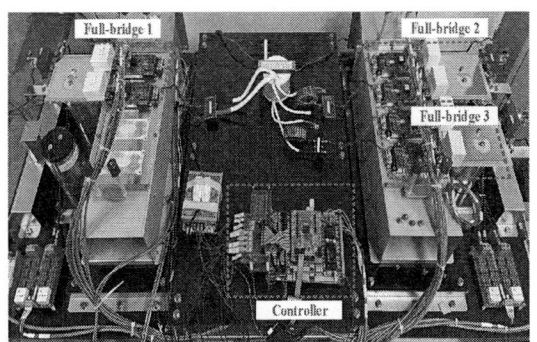

Fig. 5: TAB converter setup for experiment

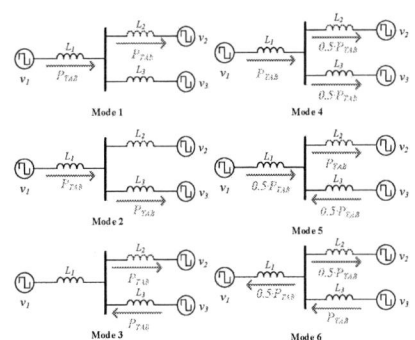

Fig. 6: six-operation mode of TAB converter

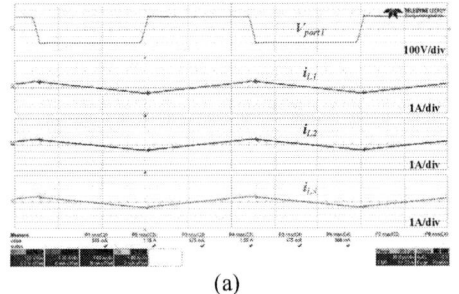

(a)

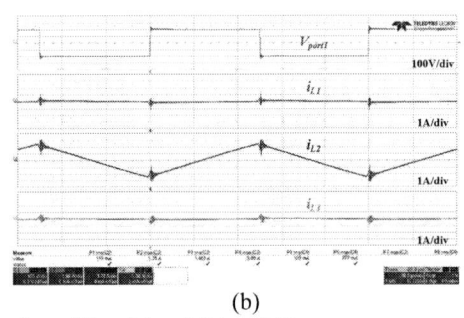

(b)

Fig. 7: Steady-state voltage and current waveforms in no-load condition (a) p=1 (b) p=0.01

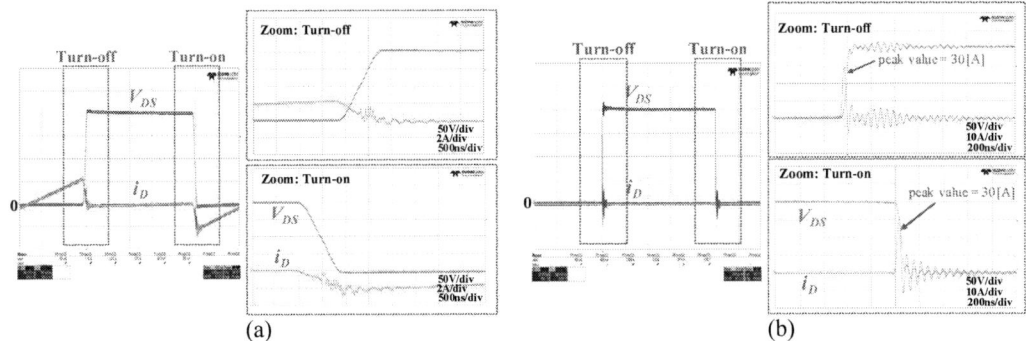

(a)   (b)

Fig. 8: MOSFET drain-source voltage (V_{DS}) and drain current (i_D) waveforms in case of p=0.01 (a) upper switch of leg A at port-2 (b) upper switch of leg A at port-1

A. No-load condition

Fig. 7 shows the output voltage of port-1 and three port currents. In no load condition, the port currents are magnetization current. In case of p=1, the same magnitude of current flows to the three-port due to the equal leakage inductance distribution and the power loss is 10.4W. On the other hand, in case of p=0.01, most of the magnetization current is provided from the port-2 in which a low leakage inductance is located, and the power loss is 31.8W. The increase of power loss in case of p=0.01 is due to the current imbalance of three-port. The current imbalance of three-port during No-load condition can lead to the increase of switching loss. As shown in Fig. 7(b), the magnetization current is provided from the port-2 where the low leakage inductance is located. Fig. 8(a) and (b) show the drain-source voltage and drain current during turn-on and off of MOSFET in case of p=0.01. The drain current of switch is measured using Rogowski coil. The switch at port-2 satisfies the ZVS condition as shown in Fig. 8(a). However, in case of switch at port-1, the hard-switching occurs during turn-on because the current flowing through the switch is not enough to satisfy ZVS condition as shown in Fig. 7(b). In addition, as shown in Fig. 8(b), the current oscillation with high peak current during turn-on and off leads to the high turn-on and

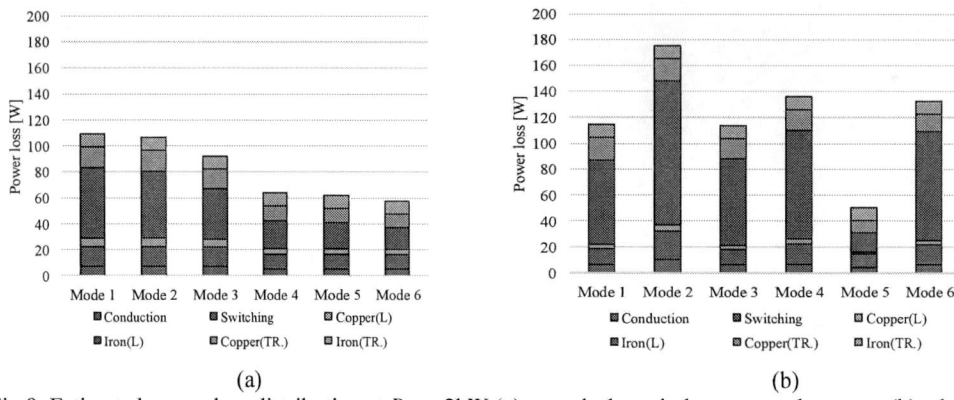

Fig.9: Estimated power loss distribution at P_{TAB}=2kW (a) same leakage inductance at three-port (b) a low leakage inductance at one-port

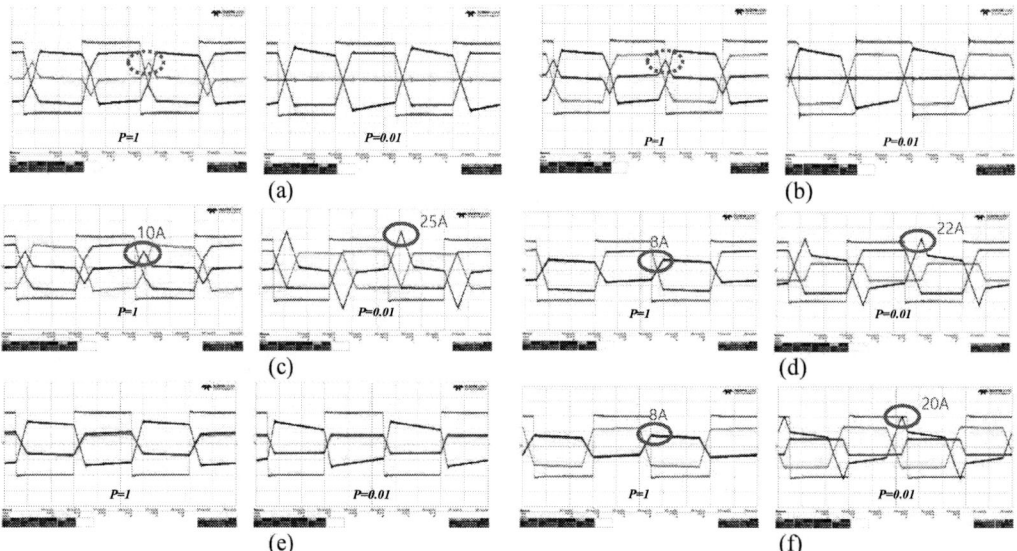

Fig. 10: Steady-state waveform of TAB converter at *p=1* and *p=0.01*
(a) Mode 1 (b) Mode 2 (c) Mode 3 (d) Mode 4 (e) Mode 5 (f) Mode 6

off switching losses. Even though the power losses in series inductors and winding losses of port-1 and port-3 is reduced, the increase of switching losses is dominant. Eventually, it degrades the power efficiency of TAB converter with low leakage inductance at one port during No-load condition.

B. Six operation mode condition

To analyze the difference of power loss in various power flow condition, the six-operation modes are set, and the power losses are measured. The power loss breakdown method suggested in previous section is used to analyze the power loss components. Fig. 9 (a) and (b) represent the estimated power loss distribution at P_{TAB}=2kW in different leakage inductance configurations. Except the operation Mode 5, the power losses of two leakage inductances configuration are larger than those of three leakage inductances configuration. The major reason is that the peak current of inductor is increased due to the uneven leakage inductance distribution. As expected in (21) and (22), the peak current is equal in both case ($p=1$ and $p=0.01$) and can be checked in Fig. 10(e). Therefore, the switching loss is same. Instead, the phase-shift angle is decreased in the case of $p=0.01$ as expected in (21) so that the conduction loss is slightly decreased. In addition, the elimination of series inductor at port-2, where the rated current flows, results in the significant power loss reduction in series inductors. With the low leakage inductance at port-2 ($p=0.01$), the power efficiency at Mode 5 is increased to 97.6% at P_{TAB}=2kW, which is a 1% increase from the case of evenly distributed leakage inductance ($p=1$). In the Mode 1, the equivalent leakage inductance, L_{31} is infinite if the leakage inductance at port-2 is zero. Therefore, the current at L_{31}, i_{31} is also zero, which results in the ideally zero switching loss at port-3. However, the core loss and winding loss at port-2 are increased as shown in Fig. 9 because the leakage inductance is doubled ($L_{1,p=0.01}=L'_{3,p=0.01}=2 \cdot L_{1,p=1}=2 \cdot L'_{3,p=1}$). In case of Mode 3, the power flow is similar with the Mode 1 since the same leakage inductance at port-1 and port-3. Therefore, the power efficiency of Mode 1 and Mode 3 is estimated as same value. The operation mode 2, 4, and 6 are cases where the power delivered through the port with a low leakage inductance is below the half of the rated power. Due to the low leakage inductance, the di/dt of inductor current increases so that the peak current at the port also increases at the time of turn-off as shown in Fig. 10(c).

The power loss analysis of TAB converter in six operation modes shows that the effect of power loss reduction is valid only in mode 5. Therefore, it can be concluded that the proposed two leakage inductances configuration is appropriate for the TAB converter which has fixed power flow condition. In addition, to minimize the power losses and utilize the power decoupling, the low leakage inductance should be located at the port that the most active power is transferred. If the power flow of the three ports of the TAB converter is not determined or is inconstant, the leakage inductance should be designed as evenly distributed form in the respect of power efficiency. For the TAB converter which has fixed or symmetric power flow, the proposed design method has many advantages in terms of power efficiency, intrinsic decoupling control, cost, and etc.

Conclusion

The TAB converter can control each port power independently with two leakage inductances configuration. However, it decreases the power transfer efficiency in some power flow condition. In this paper, leakage inductance design procedure with a low leakage inductance at one port is suggested and compared with conventional configuration. In addition, the equations of peak inductor current are derived. In some power flow conditions, the peak current increases significantly. It leads to the increase of switching losses. Therefore, the position of low leakage inductance should be selected in consideration of power flow. To minimize the increase in turn-off switching loss, a low leakage inductance should be placed on the port that the delivers the most power of TAB converter.

References

[1] H. Tao, A. Kotsopoulos, J. L. Duarte and M. A. M. Hendrix.: Transformer-Coupled Multiport ZVS Bidirectional DC–DC Converter With Wide Input Range, IEEE Transactions on Power Electronics Vol. 23 no 2, pp 771-781

[2] J. Zeng, W. Qiao and L. Qu.: An Isolated Three-Port Bidirectional DC–DC Converter for Photovoltaic Systems With Energy Storage, IEEE Transactions on Industry Applications Vol. 51 no 4, pp 3493-3503

[3] Y. Chen, P. Wang, H. Li and M. Chen.: Power Flow Control in Multi-Active-Bridge Converters: Theories and Applications, 2019 IEEE Applied Power Electronics Conference and Exposition (APEC), pp. 1500-1507

[4] S. Bandyopadhyay, P. Purgat, Z. Qin and P. Bauer.: A Multi-Active Bridge Converter With Inherently Decoupled Power Flows, IEEE Transactions on Power Electronics Vol. 36 no. 2, pp 2231-2245

[5] L. F. Costa, G. Buticchi and M. Liserre.: Optimum Design of a Multiple-Active-Bridge DC–DC Converter for Smart Transformer, in IEEE Transactions on Power Electronics Vol. 33 no. 12, pp. 10112-10121

[6] S. Inoue and H. Akagi: A Bidirectional Isolated DC–DC Converter as a Core Circuit of the Next-Generation Medium-Voltage Power Conversion System, IEEE Transactions on Power Electronics, Vol. 22 no. 2, pp. 535-542

[7] F. Krismer and J. W. Kolar.: Accurate Power Loss Model Derivation of a High-Current Dual Active Bridge Converter for an Automotive Application, IEEE Transactions on Industrial Electronics, Vol. 57 no. 3, pp. 881-891

[8] H. Akagi, T. Yamagishi, N. M. L. Tan, S. -i. Kinouchi, Y. Miyazaki and M. Koyama.: Power-Loss Breakdown of a 750-V 100-kW 20-kHz Bidirectional Isolated DC–DC Converter Using SiC-MOSFET/SBD Dual Modules, IEEE Transactions on Industry Applications, Vol. 51 no. 1, pp. 420-428

[9] R. Haneda and H. Akagi.: Power-Loss Characterization and Reduction of the 750-V 100-KW 16-KHz Dual-Active-Bridge Converter With Buck and Boost Mode, IEEE Transactions on Industry Applications, vol. 58 no. 1, pp. 541-553

Application of a multi-winding magnetic component characterization method to optimize cross-regulation performances in DCM flyback converters

Denis MOTTE-MICHELLON[1,2], Brahim RAMDANE[2], Yves LEMBEYE[2],
Bruno COGITORE[1]
EXXELIA/G2ELAB
[1]Exxelia R&D center: 137 rue Mayoussard, ZA Centr'Alp, 38 430 Moirans, France
[2]Univ. Grenoble Alpes, CNRS, Grenoble-INP*, G2Elab, 38 000, Grenoble, France
Tel.: +33 / (0) – 476.35.05.92
E-Mail: denis.motte-michellon@g2elab.grenoble-inp.fr
URL: https://exxelia.com

Keywords

Multiple secondary windings, Device characterization, Flyback converter, Transformer, DC voltage control.

Abstract

In this paper, output voltage deviations issues in multi-output DCM cross-regulated flybacks are studied, with a particular focus on the transformer. Using the extended Cantilever model, a general explanation of the influence of the transformer on voltage deviations is developed. Then, an application of an analytical model based on a formulation of the magnetic vector potential allowing to characterize the transformer around 1,000 faster than its finite-element method (FEM) equivalent is presented. This characterization method, fast and accurate, offers the possibility to create a design aid tool for the transformer manufacturer, allowing to define the winding layout with the best cross-regulation performances for a transformer taking place in a given converter.

Introduction

Multi-output Discontinuous Conduction Mode (DCM) flybacks have several advantages when it comes to powering several loads with a total power not exceeding ~100W: its low number of components makes it cheap and reliable, and the embedded transformer offers voltage level adaptation thanks to the number of turns, as well as galvanic insulation. Thanks to all these advantages, multi-output DCM flybacks are appreciated by converter manufacturers for low-power space-bound applications.

However these advantages do not come without drawbacks: The simplest and most widely spread method to regulate output voltages in this type of converters is cross-regulation. It means that only one output voltage is closed-loop regulated with action on the duty cycle, while the other output voltages are let free. The idea behind this method is that the regulated output voltage has approximately the same behavior (i.e. the same transfer function V/α, α being the duty cycle) than all the other outputs. Thus, regulating the voltage of this single output should be sufficient to regulate the voltage of all the others, at least approximately.

This hypothesis is sometimes wrong, and cross-regulated outputs (i.e. outputs that are not directly regulated) experience important steady-state voltage deviations, with known examples of voltage excursions up to almost twice the nominal values. The current solution to this problem is to implement linear regulators on outputs for which voltage deviations are not acceptable, such as sensitive high-end sensors. Nevertheless, this solution is especially unsatisfying for space industry because it leads to adding extra subsystems to the converter, consequently increasing cost, complexity, weight and possible failure sources.

*Institute of Engineering Grenoble-Alpes

Transformers were known to be part of the causes of cross-regulation failures. Thus, our work aimed at understanding how the magnetic component influences output voltages deviations in order to propose design recommendations, and ultimately a design optimization tool for the transformer designer.

In this paper, an analysis of the influence of the transformer on cross-regulation is first presented. Then an analytical modeling method for characterizing the transformer is developed for the purpose of creating a reliable and fast design optimization tool for the transformer designer.

I – Global model of cross-regulation

I.1 – Context, method and objectives

In this study, the objective was to find out how to design multi-output flyback DCM in order to reduce the magnitude of voltage deviations due to cross-regulation imperfections. Focus was put on the magnetic component, since it is known to have a significant influence on cross-regulation performances, especially through the set of all its magnetic couplings – which can also be interpreted as leakage inductances. From a behavioral model of the transformer, an analysis of the influence of the magnetic phenomena was conducted through analytical calculations and a consistent correlation was found.

In order to confirm the conclusions that were drawn, characterizations of the behavioral model of several versions of one transformer were conducted on a finite elements method (FEM) software. All the versions of the transformer were featuring the same winding set. Number of turns and copper section are identical for each winding from one version to another. Only their positions in the frame were different. This method allowed the modification of only the magnetic couplings between the windings while keeping the other parameters unchanged between the several versions. Then, simulations of the operation of the converter with the models of the several versions were conducted in a circuit simulation software (PSIM), which allowed to compute output voltages on several operating points, highlighting the influence of the transformer on output voltage deviations. This approach allowed to find a link between the winding layout and the transformer influence on the cross-regulation performances. Thus, transformer design recommendations were formulated in accordance with the analysis of the phenomena.

However, the FEM characterization method, albeit precise, is much too slow to explore all the possible layouts or to run an optimization method in the general case. Indeed, the number of possible versions of the transformer is equal to N!, with N being the number of elements whose positions can be permutated. Let us take the example of a transformer featuring 10 different layers of windings. The number of versions that can be obtained for this transformer is 10! = 3 628 800, which is way too much to be comprehensively explored in a reasonable time. Thus, another goal was to find an analytical model of the magnetic energy in the transformer window that could be used to characterize the equivalent circuit model, in order to reach faster computation time than the FEM formulation. Two analytical magnetic energy models were compared to the FEM formulation, taken as the reference. One of these two models revealed to be accurate enough and around one thousand times faster than the FEM model, unlocking the possibility to develop a winding layout optimization tool to achieve transformer design with better cross-regulation performances.

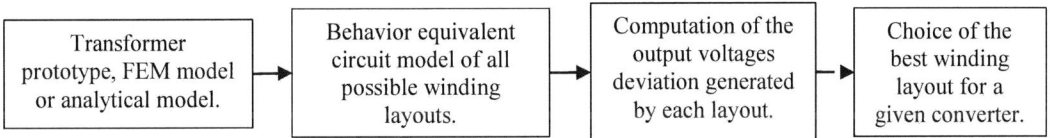

Fig. 1: Method used to analyze cross-regulation performances of different winding layout on a transformer

I.2 – Model of the transformer

In order to represent the behavior of the transformer (i.e. the relationships between the voltages across the windings and the current flowing through them), we chose to use a magnetic quasi-static model excluding energy losses in the transformer and capacitive effects.

For stray capacitances, this simplification was made under the assumption that magnetic phenomena have a much greater impact on transformer behavior than capacitive effects. Electric energy is of course stored along with magnetic energy, since electric field develops between windings. However flybacks typically operate under 500kHz, which is under the typical no-load resonant frequencies of windings (around some MHz), meaning the modules of the impedances of the inductances are lower than the modules of the impedances of the stray capacitances. On the other hand, simulations showed that the winding resistances had little influence on the cross-regulation phenomenon. Thus, a model including only inductances seemed adapted to model properly the transformer behavior.

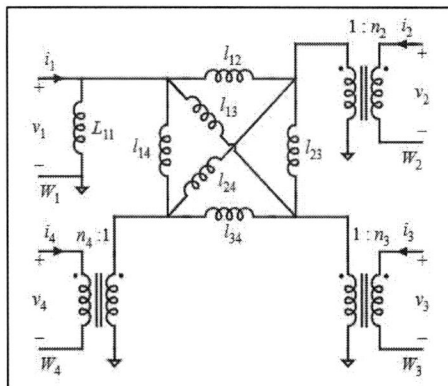

Fig. 2: Extended Cantilever model, four-winding example

The extended Cantilever model [1] was chosen. This model is equivalent to the inductance matrix, meaning it contains the same information about the transformer behavior – i.e. it has the same number of independent parameters. Fig. 2 shows an example of the structure of the model for a four-winding magnetic component. The particularity of this model is that it models all the couplings between all windings thanks to leakage inductances placed between every couple of windings. One of its advantages is that it allows an easy comparison of the values of the couplings between windings, since all the leakage inductance are placed on the side of a single winding. What is more, the formulation of the couplings by leakage inductances allows to write easily formulas describing energy flows between windings, leading to the simplified analytical model presented in §II.1

II – Simplified model of the voltage deviations on the lowest power output

II.1 – Observations and hypotheses

In DCM cross-regulated multi-output flybacks, voltage deviations are most often observed on voltage rises on low-power secondaries. Thus, focus was put on modeling this particular phenomenon with an analytical formula that could give a physical interpretation of the phenomenon.

A four-winding flyback transformer was characterized to obtain the values of the parameters of its Cantilever model. This allowed us to run simulations of a simple three-output flyback converter.

In order to work with expressions of reasonable complexity, a number of classical simplifying hypotheses were made: output voltages are constant with no voltage ripple – steady state is reached and the output capacitors have very high values compared to the power consumption of the loads they feed. Converter components are ideal, and there is no energy dissipation in any of the converter's component neither in the transformer. In addition, there is no snubber circuit on the primary of the converter, hence the current is transferred instantaneously from the primary to the secondaries when the MOS opens at αT – with α being the duty cycle and T the duration of the period of the converter. A two-output example is displayed on Fig. 3:

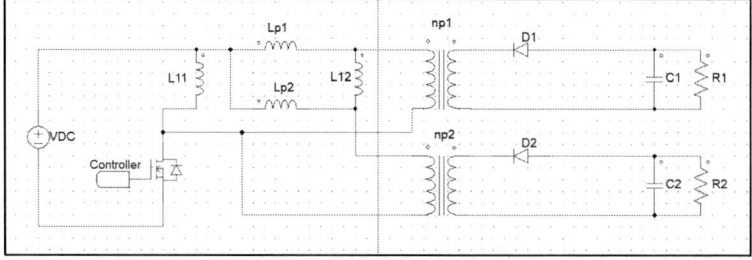

Fig. 3: Structure of a simplified two-output flyback

From the simulations, we obtain the secondary current waveforms shown on fig. 4. These broken lines current waveforms are also shown in [3] and [4]. In [3] the author uses the same hypotheses – except for the primary snubber – to obtain analytical formulas of the voltages outputs, but they are extremely cumbersome for a flyback with only two outputs, and we could not deduce a physical explanation from them.

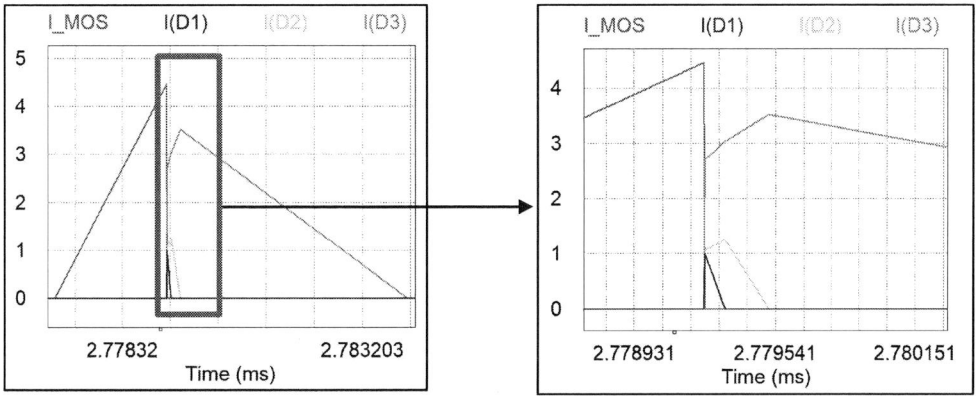

Fig. 4: (a) Current waveforms in the primary MOS and in the diodes of secondary windings obtained in a PSIM simulation. (b) Zoom on the beginning of the demagnetization phase.

An important thing to notice is that with these hypotheses, the current in a secondary k at αT^+ – just after the MOS switches to its blocked state – does not depend of the power of the load of the output k in any way. Instead, it depends on i_p, the current in the primary at αT^- and on the leakage inductance between the primary and the secondary k compared to the other leakage inductances. It is expressed in (1):

$$i_k(\alpha T^+) = i_p(\alpha T^-) \cdot \frac{\dfrac{1}{L_{pk}}}{\sum_{j=1}^{N} \dfrac{1}{L_{pj}}} \tag{1}$$

II.2 – Equivalent circuit of the magnetic phenomena

From the PSIM simulations and the waveforms of the fig. 4 (b), one can notice that the current of the lower-power secondary #1 is the first to fall to zero. Since the secondary #1 drives current for a very short duration, it has little time to exchange energy with the two other secondaries. We make the hypothesis that there is no direct exchange of energy between this secondary and the others, since the conduction time of the secondary #1 is very short. In the Cantilever model, energy exchanges between secondary windings would be modelled by currents flowing through the leakage inductance connecting the secondary windings. Since they are assumed negligible, the equivalent circuit describing the situation of the low-power output during its conduction time is presented on Fig. 5:

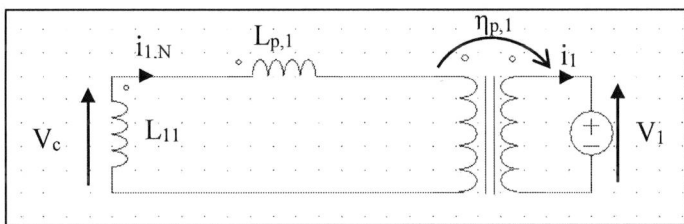

Fig. 5: Circuit describing the situation of the lowest power output during its conduction time.

V_c is the voltage at the terminals of the magnetizing inductance of the transformer. Physically speaking, it expresses the variation of the magnetic flux common to all windings. With the Cantilever model, its expression is found by expressing the equality of the current in the magnetizing inductance L_{11} and the sum of the currents in the leakage inductances:

$$\forall\, t, i_{L11}(t) = \sum_{k=1}^{M} i_{k,N}(t) \tag{2}$$

Then, deriving (2) with respect to time (in (3)) and expressing the derivatives of the currents as a function of the voltages across the inductances (in (4)) yields an expression linking V_c and the voltages of all the outputs:

$$\forall\, t, \frac{d\, i_{L11}(t)}{dt} = \sum_{k,\ i_k(t)\neq 0} \frac{d\, i_{k,N}(t)}{dt} \tag{3}$$

$$\forall\, t, \frac{V_c(t)}{L_{11}} = \sum_{k,\ i_k(t)\neq 0} \frac{\dfrac{V_k(t)}{\eta_{p,k}} - V_c(t)}{L_{p,k}} \tag{4}$$

It is important to note that only the secondary windings in which the current is non-zero are taken into account in this expression. If the current in a secondary winding falls to zero, the diode of its outputs becomes blocked, thus the current remains zero and its derivative is zero. Consequently the secondary has no more influence on the voltage across the magnetizing inductance. Finally, transforming (4) gives us the expression of V_c as a function of the output voltages, the magnetizing inductance and the leakage inductances between the primary and all the secondary windings:

$$\forall\, t, V_c(t) = \frac{\sum \dfrac{V_k(t)}{\eta_{p,k} \cdot L_{p,k}}}{L_{11} + \sum L_{p,k}} \tag{5}$$

II.3 – Analytical expression of the output voltage

Now that V_c is expressed, let us proceed to the expression of V_1 as a function of the parameters of the converter. With the assumption we have made, at the time αT^+ the current i_1 has a non-zero value, and so does the current $i_{1,N}$. Since the current $i_{1,N}$ is flowing through the inductance $L_{p,1}$, its derivative with respect to time can be expressed in (6):

$$\forall\, t, \frac{di_{1,N}(t)}{dt} = \frac{V_c(T) - \dfrac{V_1(t)}{\eta_{p,1}}}{L_{p,1}} \tag{6}$$

Since V_c and V_1 are constants over the period of the converter that is considered, the current $i_{1,N}$ (and thus the current i_1) decreases at a constant pace ($V_1/\eta_{p,1}$ is greater than V_c) until it falls to zero. Thus, the current i_1 has the shape of a triangle, of base $(\beta_1 - \alpha)T$ and a maximal value at αT, given by (1). $\beta_1 *T$ is the time when the current i_1 falls to zero, so β_1 is the duty cycle of the diode of the output #1, D1. It is given by (7):

$$\beta_1 \cdot T = \alpha \cdot T + \frac{i_1(\alpha T^+)}{\dfrac{V_c - V_{1,N}}{L_{p,1}}} \tag{7}$$

Expressing the duty cycle of D1 allows us to express the average value of the current through the secondary winding. Assuming that the load of the output is a resistor of value R_{load1} under the voltage V_1, and that the steady state of the output #1 is reached, the average current in the load and the average of the current in the secondary #1 over a period of the converter are the same. The equality between the average current in the winding and the average current in the load is written in (8). Solving the obtained 2nd-order polynomial with V_1 as the unknown yields the result exposed in (9):

$$\frac{1}{T} \cdot \frac{1}{2} \cdot i_1(\alpha T^+) \cdot (\beta_1 T - \alpha T) = \frac{V_1}{R_{Load1}} \tag{8}$$

$$V_1 = \frac{\eta_{p1}}{2} \cdot \left(V_c + \sqrt{V_c^2 + \frac{2 \cdot R_{load1} \cdot i_p(\alpha T^-)^2 \cdot \frac{1}{L_{p,1}}}{T \cdot \left(\sum_{k=1}^{N} \frac{1}{L_{p,k}} \right)^2}} \right) \tag{9}$$

II.4 – Physical interpretation of the formula

Even though this formula is obtained from strong simplifying hypotheses, we deduce the following physical interpretation: The power distributed to the output #1 by the flyback strongly depends on the magnetic couplings in the transformer – expressed in the previous formulas under the form of leakage inductances. Thus, if the transformer has a set of couplings that makes the output #1 receive more power than it is supposed to consume in nominal conditions, the voltage of the output #1 will experience a steady-state overvoltage. The bigger R_{load1} is, meaning the smaller the power consumed by the load of the output #1 is, the higher the output voltage is likely to rise over its nominal value.

This formula also illustrates the influence of the outputs between them: if the total power supplied by the flyback does increase, meaning the duty cycle of the MOS increases, the value of $i_p(\alpha T^-)$ will also increase. If the power consumption of the load of the output #1 remains the same, its voltage will rise due to the increased power received.

For this particular case the solution is to increase the leakage inductance L_{p1} – meaning that the magnetic coupling between the primary and the secondary 1 is weakened. This effect is due to the fact that increasing the inductance L_{p1} leads to a lower value of $i_1(\alpha T^+)$, meaning that the secondary #1 receives a smaller amount of energy after the MOS opens. Hence, cross-regulation performances can be improved by designing the transformer in order to obtain appropriate magnetic couplings between all windings, so the energy is distributed to outputs according to their respective needs.

III – Extended explanation of the influence of the transformer

III.1 – Energy exchanges between secondary windings

According to the models of the converter and the transformer presented above, works in the converter as an energy dispatcher, and its effects depend on the magnetic couplings between all the windings, since they all can exchange energy. Consequently, the energy dispatch can only be correctly modeled if the transformer model contains as much information as the inductance matrix of the transformer – i.e. the model of a transformer with N windings must feature at least $N(N+1)/2$ independent parameters.

§II exposes the influence of the couplings between primary and secondary windings, since they control the energy dispatch when the current is transferred from the primary to the secondary windings. Another effect that influences output voltage deviations is the exchange of energy between the secondary windings during the demagnetization phase. Indeed, if two secondary windings both drive current during a time $T_{j,k}$, they may exchange energy. Since the secondary windings stop driving current one after another, $T_{j,k}$ is equal to the duration of the conduction time of the secondary that stops driving current first among j and k, as it is expressed in (11).

In order to simplify the notations, let us use the secondary voltage of the output k referred to the primary $V_{k,prim}$, defined by (10):

$$V_{k,prim}(T) = \frac{V_k(t)}{\eta_{p,k}} \tag{10}$$

With V_k being the voltage of the output k and $\eta_{p,k}$ the turn ratio between the primary and the secondary k. If there is a difference between the voltages of two outputs j and k referred to the primary, the secondary j will give to the secondary k an energy $W_{j,k}$, given by (11):

$$W_{j,k} = \frac{(V_{j,prim} - V_{k,prim})^2 \cdot (\min(\beta_j, \beta_k) \cdot T_c)^2}{L_{jk}} \tag{11}$$

With $L_{jk} = L_{kj}$ being the effective leakage inductance between the winding j and the winding k in the Cantilever model. The energy goes from the output having the highest normalized voltage to the output having the lowest. Thus, this effect tends to bring normalized voltages towards a common value, meaning it reduces voltage deviations. What comes from the formula (11) is also that decreasing the value of the inductance L_{jk} – i.e. improving the magnetic coupling between the windings j and k – increases this effect. As a result, another guideline to reduce voltage deviations is to improve as much as possible the coupling between all the secondary windings. This requirement can be contradictory with the need to make magnetic couplings between primary and a secondary proportional to the power of the output fed by this secondary. Thus, our goal is to create a design aid software to help the transformer designer to identify the winding layout with the best cross-regulation performances for a transformer to be put in a given converter. §IV presents the application of an analytical formulation allowing characterizing the inductance matrix in a much faster way than a FEM software, allowing to explore quickly a great number of winding layouts.

III.2 – Validation of the extended explanation

In order to validate the explanation exposed in the paragraph II.3, the method presented in the paragraph I was applied on a four-winding transformer meant to be implemented in a three-output flyback. The transformer is built with an EQ25 magnetic circuit. The windings are wound on a frame on five concentric layers that occupy its whole volume. The primary and the higher power secondary are both wound on two layers, while the two low power secondaries are wound on the same layer as shown in Fig. 6. This distribution of the windings is such that by swapping layers, $5!/(2!*2!) = 30$ different versions can be obtained. The Cantilever models of the 30 versions were characterized through FEM simulations.

Then, PSIM simulations of the operation of the converter with the models of the 30 different versions of the transformer were successively made. The three outputs of the simulated converter have very different nominal power consumption, respectively 0.07W for the output #1, 0.35W for the output #2 and 34W for the output #3. During the tests, the voltage of the output #3 was closed-loop regulated to be kept at 28V, its nominal value. Two operating points were tested, one with the output #3 consuming 30% of its nominal power, and the other with the output #3 consuming its nominal power. The load resistors of the outputs #1 and #2 were kept constant and their steady-state voltages monitored, to be compared to their nominal values, 14V for both.

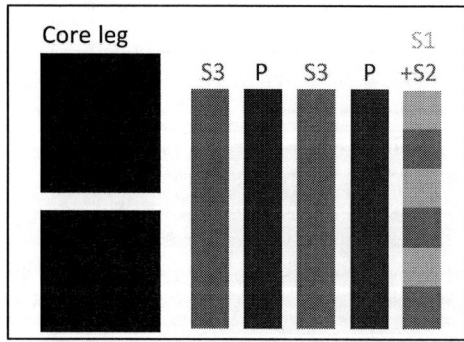

Fig. 6: Example winding layout with average cross-regulation performances.

This approach allowed to characterize the cross-regulation performances of the 30 possible winding layouts for this application. It allowed us to confirm the explanation of the energy dispatch taking place in the multi-output flyback: winding layouts that tended to dispatch power accordingly to each output's need during MOS turn-off and also during demagnetization phase generate the lowest voltage deviations on cross-regulated outputs. Oppositely, those tending to dispatch too much power to cross-regulated outputs resulted in steady-state voltage rises on these outputs. For confidentiality issues, the best and the worst winding layouts for this application cannot be disclosed. Nonetheless, Fig. 7 shows a comparison of the cross-regulation performances of the remarkable winding layouts. These results show that the transformer plays a crucial role in the multi-output flyback, and has a major impact on cross-regulation performance: the best winding layouts generate close to zero voltage deviations (about +1% on

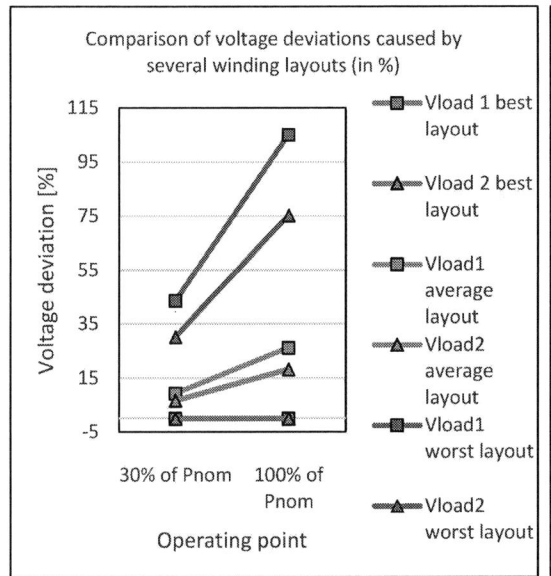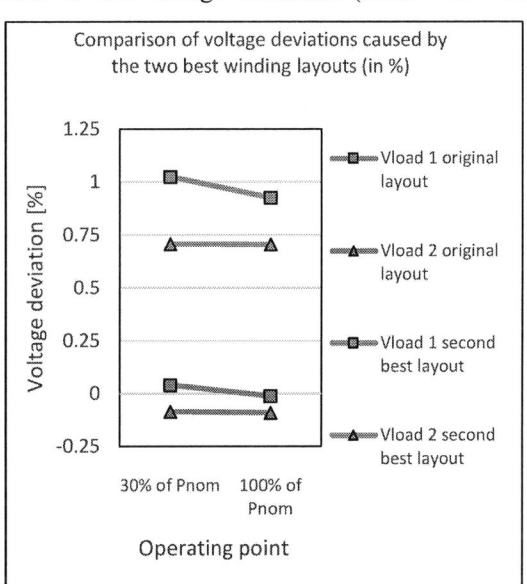

Fig. 7: Comparison of the performances of several winding layouts.

Vload1 for the second best layout), while the worst one practically twice the nominal voltage on the least power output (+105% on Vload1 at full power). Consequently, the flyback transformer design should be made with the greatest care, with a great attention given to all the magnetic couplings and not only the number of turns.

IV – An analytical magnetic vector potential model to bypass FEM characterizations

IV.1 – Principle of the model and application

This model is presented in [6]. It is aiming to compute the magnetic energy stored in the transformer during a short-circuit test by computing the interaction between all the conductors thanks to formulas expressing the vector potential generated by the current in the conductors. The values of all short-circuit inductances can be obtained – for tests with a single winding being fed and a single other one being shorted. It is based on the formula of the magnetic energy expressed by magnetic vector potential:

$$W_{mag} = \frac{1}{2} \iiint_{\Omega} \vec{A} \vec{J} \, dV \tag{12}$$

With $\vec{A}$ the magnetic vector potential and $\vec{J}$ the current density. The integral is limited to conductors, since the integrated term is worth zero where the current density is zero.

The hypotheses are the following: the problem is considered two-dimensional, the current densities are homogeneous on the section of the conductors, and the sum of ampere-turns in the transformer window equals strictly zero. This situation is equivalent to a short-circuit test with a negligible leakage energy.

Since the problem is considered two-dimensional, the formula (12) is simplified in (13), which expresses the energy per length unit:

$$Wl_{mag} = \frac{1}{2} \iint_{\Omega} \vec{A}\vec{J} \, dS \tag{13}$$

For implementing this computation, a formula expressing the magnetic vector potential field generated by a surface current distribution is provided. Since this model was basically developed to compute the short-circuit inductances in planar transformers, conductors are supposed to be of square shape. It is possible to use the same principle than in [6] to compute magnetic energy with round conductors, it would only take to compute new integrals adapted to this new shape. Nonetheless, we chose to try with formulas developed for square conductors, as it does little difference on the current distribution.

IV.2 – Implementation and results

Based on the formulas provided in [6], a program was created to compute all the short-circuit inductance of one version of the transformer which all possible versions were characterized with a FEM method. Since the goal was to find a characterization method that was faster than a FEM software but still accurate, a comparison of the two methods was realized. The voltage deviations found on PSIM simulations were used as point of comparison, as it is the interesting values for optimizing the cross-regulation performances. The version generating the biggest voltage deviations was used, in order to maximize the differences and obtain a proper comparison between the FEM and the analytical method.

For the analytical method, two different versions were tested: in the first one, all the separate conductors were considered one by one and included in the computations. In the second, winding layers containing the conductors of a single winding – such as P and S3 on Fig. 6 – were modelled as large rectangular conductors. Thus, the first version featured 66 different conductors while the second featured only 10.

For the calculation, all the possible short-circuit tests with one winding fed and one winding shorted are computed in order to obtain the mutual inductances between all the windings. For one short-circuit test, the MATLAB script that was made successively computes the energy generated by the interaction between one conductor and all the others, including itself. What is computed is the interaction between the magnetic vector potential field generated by the conductor k and the current flowing in all the conductors. Every interaction is a piece of the integral expressed in (13) and represents a part of the total magnetic energy stored in the system. Then, when the total magnetic for a short-circuit test is known, the short-circuit inductance seen from the fed winding $L_{shorted}$ is obtained thanks to (14):

$$W_{mag} = \frac{1}{2} L_{shorted} \cdot I_{fed}^2 \tag{14}$$

With I_{fed} the current in the fed winding. Then, from the formula (15) given in [2], the formula (16) is obtained and allows to compute the mutual inductance between the winding j and k – with j being the fed winding and k the shorted winding:

$$L_{shorted} = L_j - \frac{M_{j,k}^2}{L_k} \tag{15}$$

$$M_{j,k} = \sqrt{L_j \cdot L_k - L_k \cdot L_{shorted}} \tag{16}$$

With L_j and L_k being respectively the self-inductances of the winding j and k. Once the inductance matrix is filled, the extended Cantilever model parameters are computed from formulas given in [1]. Then, a PSIM simulation allows to obtain the output voltages generated by the model of the transformer characterized through the analytical potential vector model. The voltage deviations found in PSIM simulations with the two versions of the transformer model generated by the potential vector model – one with all conductors modelled and the other with complete layers modelled by single conductors – are compared to the voltage deviations generated by the model obtained with FEM characterization, taken as reference on Fig. 8.

These results suggest that even a simplified version of the potential vector model offers a decent precision to characterize the inductance matrix of a transformer. The version where all the conductors are modelled offers an excellent precision, with very little error on the output voltages (maximum 1.13% for the model with all wires and 8.69% for the model with full layers modelled as plates) compared to the results obtained with the model from a FEM characterization with a mesh of about 70k degrees of freedom. What is more, the analytical model proves to be very fast compared to the FEM characterization, about 1,000 faster for the version with all wires modelled, and about 10,000 times faster for the simplified geometry. Thus, this analytical method unlocks the possibility to create a design aid tool exploring a great number of winding version in a reasonable amount of time.

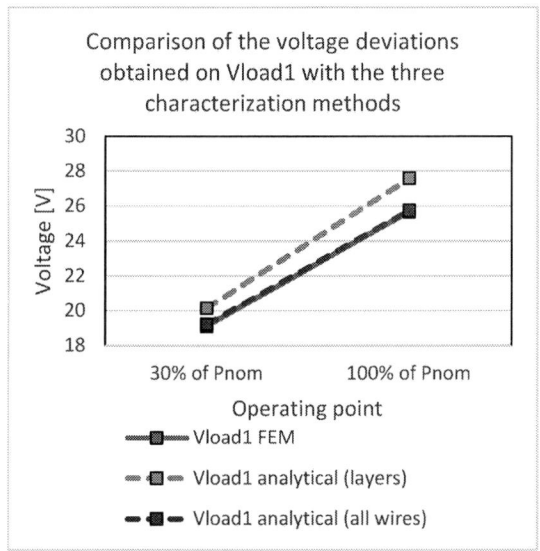 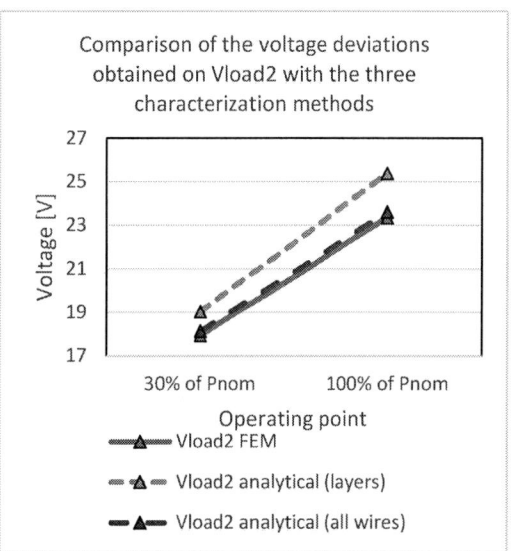

Fig. 8: Comparison of the precision of two versions of the analytical model.

Conclusion

A physical analysis of the influence of the transformer on cross-regulation is presented, as well as guidelines to design transformers that generate less voltage deviations. The analysis is confirmed by the results on a four windings transformer implemented in a three outputs flyback with a great power heterogeneity among the outputs. In order to use it to create an optimization tool, a model allowing to compute short-circuit inductances in transformers thanks to analytical formulas is tested. This model, based on the computation of the magnetic energy thanks to a formulation of the magnetic vector potential in the transformer window has proved to be both fast and accurate. This model allows the creation of a design aid tool for improving cross-regulation performance of transformer in DCM flyback converters in the future.

References

[1] Erickson R. W.: A multiple-winding magnetics model having directly measurable parameters, PESC 98 Record. 29th Annual IEEE Power Electronics Specialists Conference Vol. 2 pp. 1472-1478

[2] Keradec J.P. : Transformateurs HF à enroulements : schémas à constants localisées, Techniques de l'ingénieur, TIP301WEB, D3058 V1

[3] Ji C.: Cross Regulation in Flyback Converters: analytic model and solution, IEEE Transactions on Power Electronics Vol. 16 no 2 pp. 231-239

[4] Hu Y.Q.: Mathematical Modeling of Cross-Regulation Problem in Flyback Converters, 2001 IEEE 32nd Annual Power Electronics Specialists Conference, paper 01CH37230

[5] Chen M.: A Systematic Approach to Modeling Impedances and Current Distribution in Planar Magnetics, IEEE transaction on power electronics Vol. 31 no 1 pp. 560-580

[6] Margueron X.: Complete Analytical Calculation of Static Leakage Parameters: A Step Toward HF Transformer Optimization. Industry Applications, IEEE Transactions on Industry Applications Vol. 46 no 3 pp 1055-1063

Application of an electrostatic machine in a low-voltage micro-grid

Gabriel Ramos Huerta, Patricio Mendoza-Araya
UNIVERSITY OF CHILE
Energy Center, Electrical Engineering Department
Av. Tupper 2007, Santiago, Chile
Phone: +56 (2) 2978 4768
Email: gabriel.ramos.h@ug.uchile.cl, pmendoza@ing.uchile.cl

Keywords

≪Electrical machines≫, ≪Microgrid≫, ≪Synchronous motor≫, ≪Stability≫, ≪Fault tolerance≫.

Acknowledgments

The authors would like to thank the the Electrical Engineering Department and the Energy Center of the University of Chile for the financial aid.

Abstract

In this work, dq axis models and equations for an electrostatic machine (ESM) are implemented. Speed and excitation controllers are adapted for this machine, and tests are run to check their performance. Finally, this machine is inserted into a micro-grid environment (CIGRE low voltage benchmark micro-grid) to find its possible contributions to the system regulation and short circuit currents.

Introduction

Back in the 1830's decade, Michael Faraday discovered what would become one of the most important laws in electromagnetism: the *Electromagnetic Induction* law. Later, this discovery would lead the scientific communities to study and model one of the most recognized kind of machines: the electromagnetic machine (EMM). This machine, thanks to its sturdiness, efficiency and stability, would then become the protagonist of electrical power systems.

At the same time, Jean Peltier was making another big discovery: the *Electrostatic Induction* law. Though, his discovery would pass rather unnoticed until the 1920's, when Van de Graaff created his famous electrostatic machine (ESM) named after him. This machine was rather inefficient and hard to build; and since electromagnetic machines had taken over the electrical systems, they were left aside.

No theoretical basis for the ESM would be developed until 1960's, when Dominique Gignoux [1], along with F. J. McCoy and W. R. Bell [2] revisited it. They concluded that this machine's principle of work was, as a generation method, inefficient; but it could eventually work if there were any kind of fluid development that would tolerate the existence of strong electric fields with no electric arcs, i.e. strong dielectrics.

Recently, dielectric fluid development has reached new goals [3]. This encouraged the study of the ESM as an energy generation method. Mainly, two relevant works have been made [4] [5]; in these, electrostatic motor prototypes are built, considering a strong mathematical modelling. These prototypes feature a high amount of poles, lack of iron core, small power capacity and the use of a dielectric fluid in order to avoid electrical discharges.

Most of the recent research has been focused on the application of this machine as a motor. However, its application as a generator is appealing, particularly in a microgrid, considering the size of current prototypes [4]. Given this machine's characteristics, which behaves in steady state as a current source, its incorporation as a generator might not be as usual. In this line, the main goal of this work is to explore and show what are the extents of this machine's benefits in a micro-grid environment, working as a generator.

The rest of the document is structured as follows. First, the main characteristics and equations of this machine will be provided; these will be then implemented using PLECS. Classical controller models will be adapted and implemented, in order to give this machine some sort of regulation. Finally, this machine will be inserted into CIGRE low voltage benchmark micro-grid, in order to evaluate its dynamic behavior. Finally, this work concludes giving some insight about this machine's properties, benefits and challenges.

Electrostatic Machine Modelling

As opposed to the conventional electromagnetic machine, this machine works with the electrostatic field induction principle. This means, instead of using an electromagnetic field to induce current, an electrostatic field is created to induce voltage. In order to reach a meaningful electrostatic field magnitude, high voltages are usually required.

A great work in ESM modeling and designing is provided in [4], where a small prototype was built. For the purposes of this paper, the most relevant equations and models are extracted and analyzed from [4], while the complete mathematical derivation of such models can be found in the cited reference.

dq axis models

The dq axis models for an ESM are presented in Figs. 1a and 1b (extracted from [4]), and the dq axis models for an electromagnetic machine are shown in Figs. 2a and 2b (extracted from [6]). These models do not include dampers:

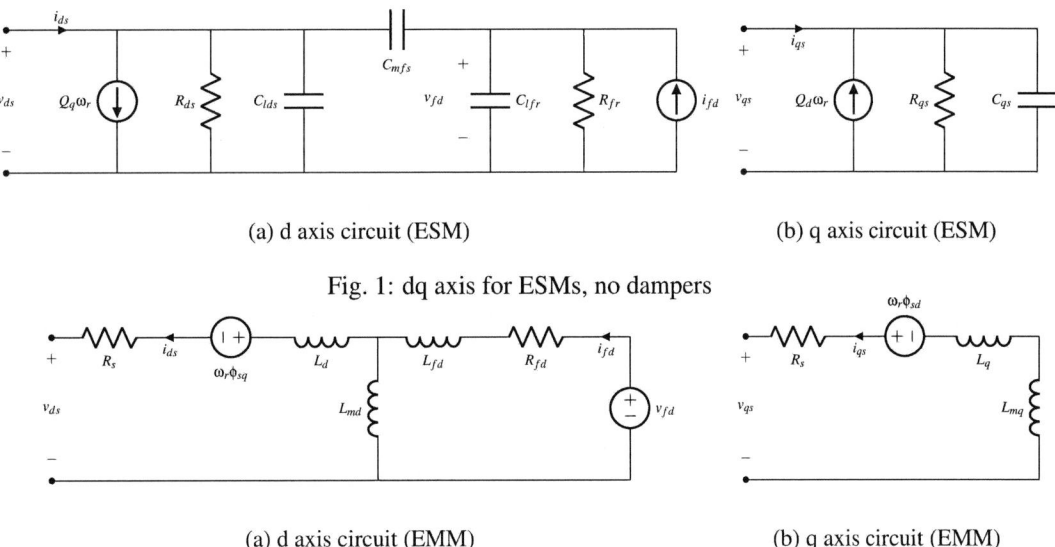

(a) d axis circuit (ESM)　　　　　　　　　　　　(b) q axis circuit (ESM)

Fig. 1: dq axis for ESMs, no dampers

(a) d axis circuit (EMM)　　　　　　　　　　　　(b) q axis circuit (EMM)

Fig. 2: dq axis for electromagnetic machines, no dampers

Table I: Subscript list

Subscript	Meaning
f	Field quantity
r	Rotor quantity
s	Stator quantity
d,q	d or q axis quantity
m	Stator - Rotor (mutual) quantity
l	Leakage quantity

The meaning of the subscript in Figs. 1 and 2 are listed in Table I. These show a duality between the electromagnetic and electrostatic machines:

- The existence of a *speed current* in the ESM, as a dual of the well known *speed voltage* present in electromagnetic machines, due to Park's transform.
- The presence of a key element that couples the rotor and stator circuits, represented as a mutual inductance in the electromagnetic machine. In the ESM, this element is depicted as a mutual capacitance C_{mfs}.
- In general, the ESM's dq circuits are shaped as the Norton's equivalent circuit of the electromagnetic's, although series RL circuits are replaced by parallel RC circuits.

Although the full model for the ESM can be readily obtained from the equivalent circuit in Fig. 1, only the most relevant equations are presented below. The output current of the machine I_{out} is

$$I_{out} = V_{fd} \cdot \omega_r \cdot C_m \tag{1}$$

where V_{fd} is the excitation voltage, ω_r is the angular frequency of the machine and C_m is the stator-rotor mutual capacitance. This equation shows the lineal dependence of the output current to the machine frequency.

The electrical power and torque provided by this machine are:

$$P_e = \frac{3}{2} \cdot (v_{ds}^r i_{ds}^r + v_{qs}^r i_{qs}^r) + 3 v_{fr} i_{fr}' \tag{2}$$

$$T_e = \frac{3P}{2} \cdot (C_m v_{qs}^r v_{fr}) \tag{3}$$

where P stands for the pole number. Also, the following capacitance values must be defined:

$$C_{lds} = C_{ds} - C_{mfs} \tag{4}$$

$$C_{lfr} = C_{fr} - C_{mfs} \tag{5}$$

where C_{ds} can be approximated by C_s for a cylindrical shape, hence $C_{mfs} = C_m$; these values will be later extracted from the short and open circuit tests performed in [4]. Finally, this machine is, like any other synchronous machine, subject to the Swing equation:

$$J \cdot \frac{d^2 \theta_m}{d\theta^2} = T_m - T_e \tag{6}$$

where θ_m is the mechanical angle of the rotor, T_m, T_e are the mechanical and electrical torque and J is the machine inertia.

Metodology

In order to evaluate the ESM's dynamic behavior, the following methodology is proposed.

First, the machine is modeled in a simulation environment. Both dq axis circuits and relevant equations are implemented using PLECS software. In order to validate the machine implementation, short-circuit and open-circuit experiment results reported in [4] will be reproduced. It must be noted that dampers won't be modelled, as there is a lack of information about their electrical parameters.

After the validation, the CIGRE low-voltage benchmark micro-grid is modelled (Fig. 3). Since the original machine prototype built in [4] does not meet the power level required to participate in this micro-grid (tens of [kW]), a per-unit scaled model of the prototype is used. Then, this machine is introduced at the end of the residential feeder of the micro-grid, replacing the fuel cells (modelled as current source inverters).

In order to participate in the primary control of the system, classical controllers will be adapted for this machine; these consider an excitation controller and a speed controller. After their implementation, three tests are performed: load step, line disconnection and three-phase to ground short-circuit. Finally, the results are analyzed in an attempt to discover this machine's possible contributions to a micro-grid's transient response.

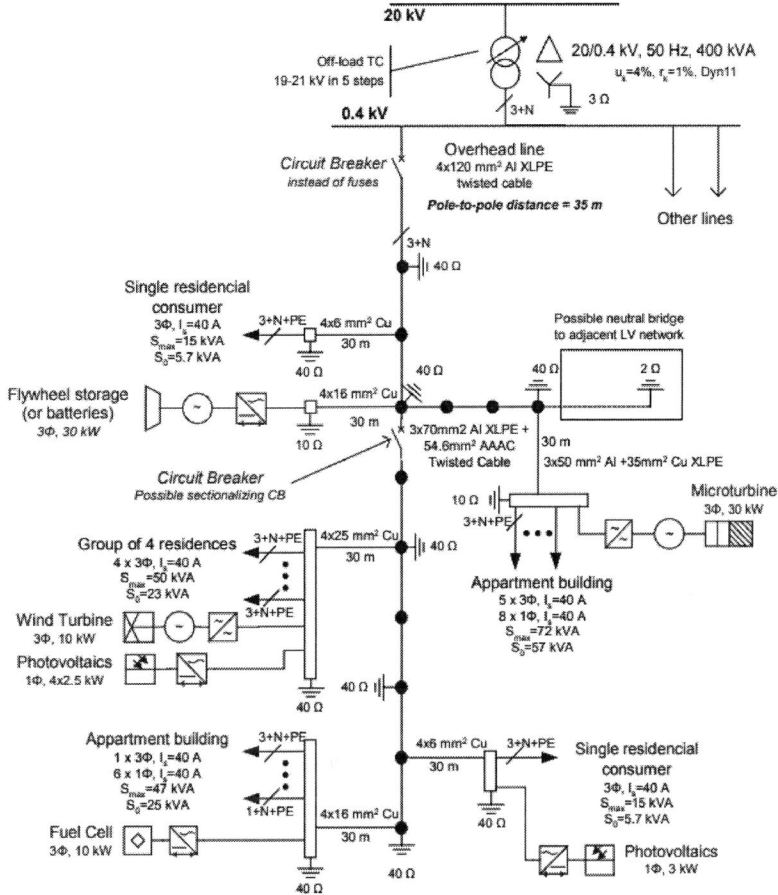

Fig. 3: CIGRE low voltage microgrid ([7])

Study Cases

The original CIGRE micro-grid (with the fuel cell) will be addressed as Case A. The case where this fuel cell is replaced with an ESM will be called Case B. On both setups, three experiments will be performed:

- Load step: a load increase of 40% is applied to the fuel cell / machine's busbar. This includes an increase in both active and reactive power consumption. The main goal of this experiment is to check the machine's primary control response, and the dynamic behavior during the system stabilization.
- Line disconnection: in this experiment, the line that connects both the residential consumer and the photovoltaics is disconnected via a breaker in the $4x6mm^2Cu$, 30 $[m]$ line. The goal of this experiment is to provide insight on possible coordination's modifications on the protection scheme of the micro-grid.
- Short-circuit: the last experiment applies a three-phase short circuit with a small ground impendance. The objective is to detect the machine's contribution to short circuit currents.

Results

Machine model validation

The prototype machine parameters are listed in table II.

Table II: Parameters of the implemented electrostatic machine model

Parameter	Description	Value	Unit
C_f	Rotor capacitance	13.8	nF
C_m	Rotor-Stator capacitance	2.0	nF
C_s	Stator capacitance	13.8	nF
R_f	Rotor resistance	50	$M\Omega$
R_s	Stator resistance	1.7	$M\Omega$
p	Pole number	96	-

The results of performing the short and open circuit tests are shown in figures 4a and 4b.

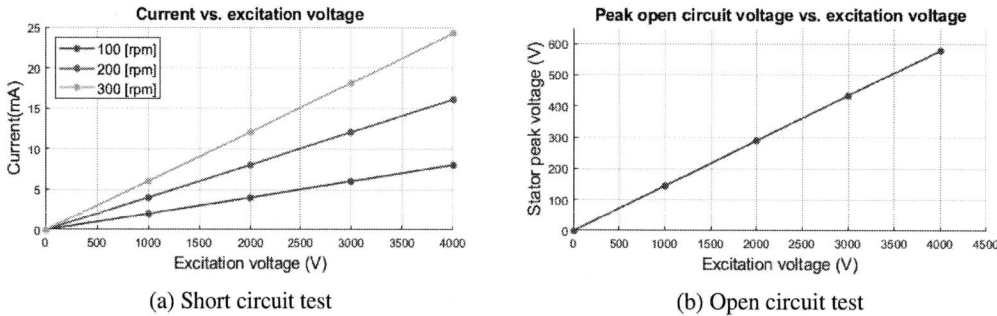

(a) Short circuit test

(b) Open circuit test

Fig. 4: Results for short and open circuit tests

These results closely match the ones obtained in [4], during the practical tests of the machine. Hence, the electrostatic machine model in the simulation environment is successfully verified.

Re-sizing of the machine

Since this machine's nominal values are not an adequate fit for the power levels required by the CIGRE microgrid, a resizing is performed using the electrical parameters provided by [4] and CIGRE's nominal voltage and a chosen power capacity. New impedances are rescaled by the following factor:

$$Z_{re-scaled} = Z_{original} \cdot \frac{Z_{base}^{new}}{Z_{base}^{old}} \tag{7}$$

Where Z_{base}^{new} and Z_{base}^{old} are calculated with the nominal parameters:

$$Z_{base}^{new} = \frac{(230 \cdot \sqrt{2})^2}{15000} \tag{8}$$

$$Z_{base}^{old} = \frac{3500^2}{300} \tag{9}$$

where $Z = V^2/S$ was used. The power value for the prototype (300 [W]) was never explicitly listed, but described as *fraction of a horsepower* in [5]. Also, the pole number of the machine is reduced to 30, in order to keep the machine working at a reasonable speed (100 rpm, as mentioned in [4]).

To estimate this machine's inertia, a small diesel generator is considered as reference [8]. Since the electromagnetic's machine inertia considers an iron core and this machine is of a smaller capacity (12.5 [kW]), an estimated value of 2 [kgm^2] is used.

Controllers

A DC2A AVR is adapted for current control (ACR from now on, as in *Automatic Current Controller*), along with a DEGOV1 governor for speed control. DEGOV1 features no modifications; DC2A presents a slight modification in its reference input. Since this machine must be able to adapt to the requirements of the system, its output current cannot be a fixed value. A dynamic reference current is calculated using a power factor as a reference:

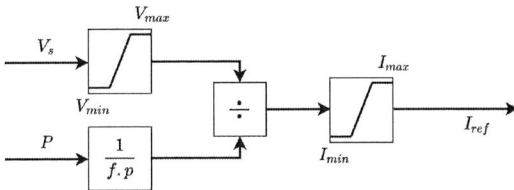

Fig. 5: Reference generation diagram for the ACR

In Fig. 5, the output active power of the machine is used to calculate a value for the apparent power S, using a power factor $f.p$. This, along with the measured voltage on the machine's point of connection, allows for a dynamic reference current calculation. Saturators are in place in order to ensure well defined divisions and reasonable reference current limits. The rest of the original AVR model remains the same.

The controllers' parameters are listed in table III; these were obtained from the range values listed in [9] and [10], respectively.

Table III: Values for adapted DC2A and DEGOV1

ACR			Governor		
Parameter	Value	Unit	Parameter	Value	Unit
TR	0.02	s	K	12	p.u
KA	150	p.u	T1	0.8	s
KE	0.3	p.u	T2	0.1	s
KF	0.3	p.u	T3	0.5	s
TA	0.02	s	T4	0.16	s
TB	0.02	s	T5	0.25	s
TC	0.1	s	T6	0.01	s
TE	1	s	TD	0.01	s
TF	0.05	s	TMax	-0.5	p.u
IMax	1.2	p.u	TMin	1.5	p.u
IMin	0.7	p.u			
f.p	0.8	-			

CIGRE microgrid experiments

As described in the methodology, three experiments on the CIGRE microgrid are performed. The base values used for this microgrid are listed in table IV. Nominal values for the converters and its controllers are shown in table V and VI.

Table IV: Base values for CIGRE low voltage micro-grid

Parameter	Nominal value	Unit
Base rms voltage (single phase)	230	V
Base rms current (single phase)	21.739	A
Base power (three phase)	15000	W

Table V: Nominal parameters: CIGRE microgrid

Unit	Operating Mode	Active power [kW]	Reactive power [kVAr]
Microturbines	Grid Forming (VSI)	45	20
PV	Grid Following	3	1.5
Wind Turbines	Grid Following	10	5
Batteries	Grid Following	30	15
Fuel cell	Grid Forming (CSI)	15	8

Table VI: Inner and outer loop control values for CIGRE microgrid unit

Unit	Voltage loop		Current loop		Droop control	
	Kp	Ki	Kp	Ki	Mp (%)	Mq (%)
Microturbines	3	10	0.8	3	5	3
PV	-	-	0.9	0.8	-	-
Wind turbines	-	-	0.7	0.6	-	-
Batteries	-	-	0.5	0.6	-	-
Fuel cell	-	-	-	-	5	3

Load step

The results of applying a load increase of 40% in the apartment building load that shares busbar with the fuel cell are shown in Fig. 6. In that figure, the transient responses of both the fuel cell and the electrostatic machine (named "study unit") during the application of the load step is shown. Since the fuel cell droop control and the implemented ACR are fundamentally different, the steady state values are different; this is noticeable in Fig. 6b and 6d.

Fig. 6a shows the different frequency responses of the study units. It is clear that case B has a slower response, mainly due to the machine's inertia. This inertia is also noticeable in Figs. 6b and 6d; since this machine's output current varies linearly with the machine's frequency, the oscillations are even more evident in the output current response (6e).

Fig. 6c shows the amount of active power delivered by the study units. Just like the frequency response, it is rather slow, taking up to 4 [s] to reach a steady state value. During this transient, there must be another unit dispatching the required active power; in this case, it is the converter of the microturbines, operating in grid forming mode (Fig. 6f).

Since this machine operates as a current source, its transient response does not directly affect the system's voltage. The oscillations seen in Fig. 7 suggest that the machine's inertial response only affects the local busbar's voltage and the machine's output current frequency, and does not have a direct impact on the microturbines' output voltage frequency. All of these frequencies do eventually reach the same point, with damped oscillations that converge to the fundamental frequency component, determined by the microturbine's droop control. This may suggest that, the closer an electrostatic machine is to a voltage controlled inverter, the more impact it will have on the grid's voltage frequency.

This experiment proves that, even with classical controllers, this machine is able to participate in the grid's primary frequency control, although this controller does not allow for primary voltage control. The adapted controller, nevertheless, allows for a constant output power factor.

Line disconnection

The results for a line disconnection are shown in Fig. 8.

The main goal of this experiment is to detect similarities and differences between waveform responses during a fast-paced perturbation. It can be seen in Figs. 8c and 8d that the responses are very similar, and none of the currents feature high peaks. This suggests that, compared to a current controlled inverter, protection systems in the microgrid would not need to be tuned again, as the response of the machine is highly damped and of low amplitude. The peaks seen in Figs. 8a and 8b belong to that small window of time where the rms value, calculated with a very high sampling time, does vary due to the current peaks seen in the waveform figures.

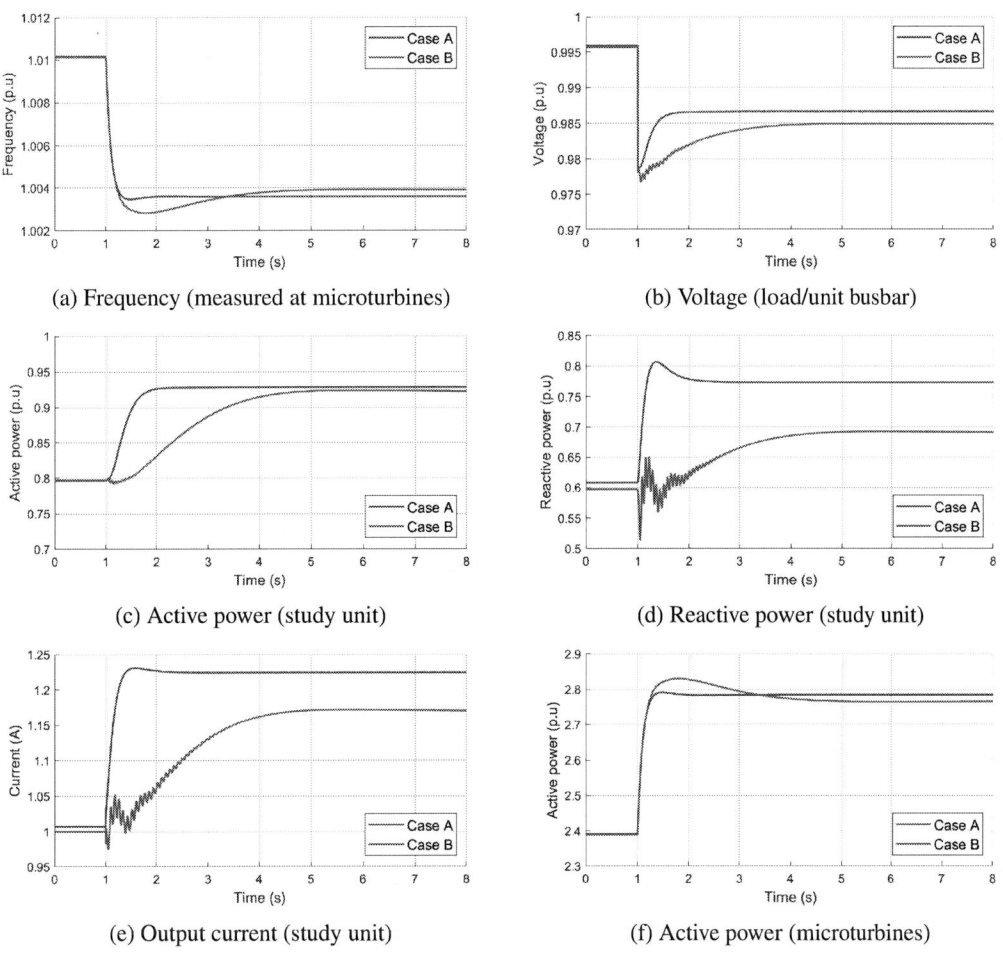

(a) Frequency (measured at microturbines)

(b) Voltage (load/unit busbar)

(c) Active power (study unit)

(d) Reactive power (study unit)

(e) Output current (study unit)

(f) Active power (microturbines)

Fig. 6: Results: load step

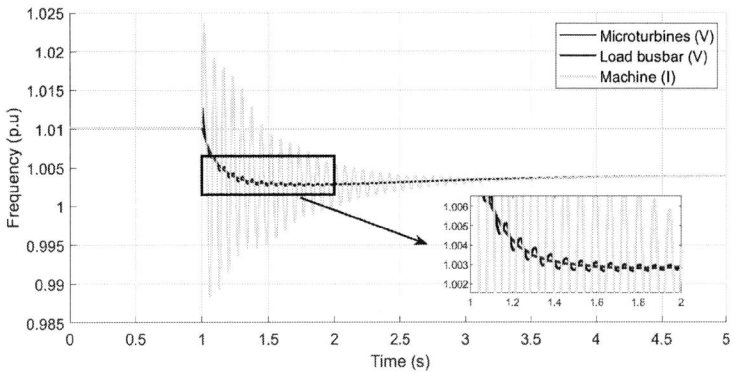

Fig. 7: Frequencies in the grid during the load step (V: frequency of voltage measurement, I: frequency of current measurement)

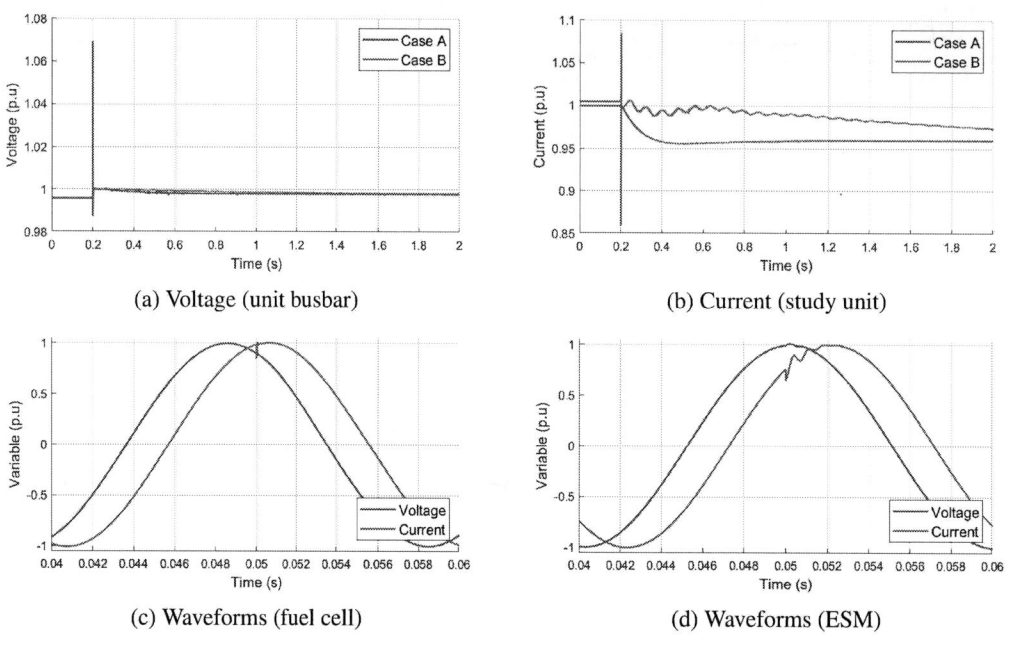

(a) Voltage (unit busbar)

(b) Current (study unit)

(c) Waveforms (fuel cell)

(d) Waveforms (ESM)

Fig. 8: Results: line disconnection

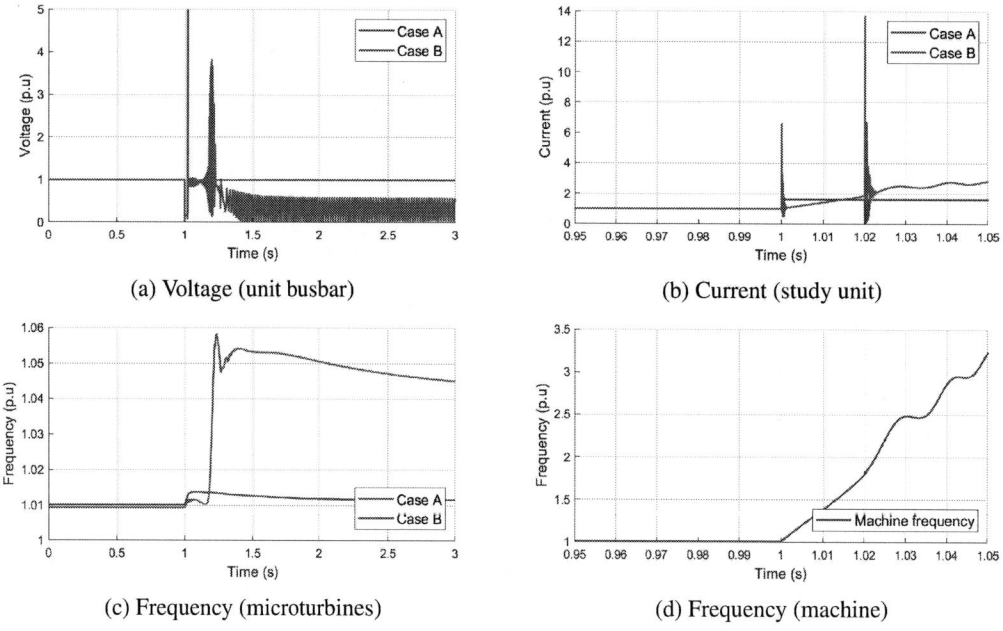

(a) Voltage (unit busbar)

(b) Current (study unit)

(c) Frequency (microturbines)

(d) Frequency (machine)

Fig. 9: Results: short circuit

Short circuit

The results of applying a short circuit at time 1 [s], and clearing it after 20 [ms], are shown in Fig. 9. These results show that the system is unstable for the case B. The electrostatic machine is unable to evacuate its input mechanical torque (shown separately in Fig. 10); this happens because, due to the sudden voltage drop, the amount of electrical torque perceived by the machine drops as well, in accordance to (3). The governor is unable to react as fast, causing the machine to drastically increase its

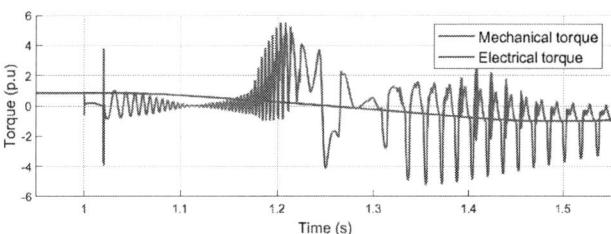

Fig. 10: Short circuit: machine torques

speed, as seen in Fig. 9d. After the short circuit clearance, the governor -that was trying to lower the machine's speed- is not providing the machine with enough mechanical torque to take on the upcoming electrical torque, that goes up after the short circuit's clearance. On top of that, the unstable current coming from the machine heavily distorts the local busbar voltage, making it harder for the governor to stabilize the rotor.

Conclusions

In this work, an implementation of the electrostatic machine model in a simulation software was achieved. Then, this machine was incorporated into a low voltage microgrid, along with a classical governor model and a modified AVR controller, in order to evaluate its operation as a generator. The results prove that this machine is able to participate in primary frequency control with a classical governor model, although with a slow response. In addition, the adapted ACR was able to keep the output power of the machine at a constant power factor. While this feature does not allow for direct participation in primary voltage control, it proved useful as a means of keeping the machine working in a certain operating condition.

The line disconnection experiment shows that, compared to a current source inverter, there is no need of re-tuning for the microgrid's protection scheme. This was expected, as the machine is modelled as a controlled current source; and given its inertia, fast variations are damped.

During the short circuit experiment, it was shown that the adapted controllers have a hard time dealing with such an event. This happens mainly due to the inability of the machine to provide torque, given the quick drop in the grid's voltage. This suggests that the machine might need a faster governor and a slower current controller, as a means to maintain synchronism.

Future work

This work's main goal was to explore the incipient development of electrostatic machines, and detect its possible contributions and challenges. As such, the experiments prompt for a lot of future research. First, the implemented controllers were not specifically designed for an electrostatic machine; they were slightly adapted. Therefore, there is a lot of room for improvement. As mentioned previously, and suggested by the experimental results, faster governors might have a significant impact on the machine's transient behavior, and a well designed ACR might let this machine participate in primary voltage control.

Machine dampers were not implemented in this machine's models. As of today, no physical implementation of these circuits has been published. Depending on their characteristics, these might have a great impact on the machine's behavior.

Finally, there is a lot of analytic work to be done in regards to this machine's operating conditions. Following the work in [4], power transfer equations can be extracted, and a capability curve can be obtained. These might give great insight on the machine's stable operating points and controller design.

References

[1] D. Gignoux, "Electrostatic generators in space power systems," AIAA 1964-450. 1st Annual Meeting, June 1964.

[2] F. J. McCoy and W. R. Bell, "Electrostatic generators for power production," Conference on Electrical Insulation & Dielectric Phenomena — Annual Report 1967, 1967, pp. 138-146.

[3] B. Ge and D. C. Ludois, "Dielectric liquids for enhanced field force in macro scale direct drive electrostatic actuators and rotating machinery," in IEEE Transactions on Dielectrics and Electrical Insulation, vol. 23, no. 4, pp. 1924-1934, August 2016.

[4] B. Ge, A. N. Ghule and D. C. Ludois, "High Torque Density Macro-scale Electrostatic Rotating Machines: Electrical Design, Generalized $d-q$ Framework, and Demonstration," in IEEE Transactions on Industry Applications, vol. 55, no. 2, pp. 1225-1238, March-April 2019.

[5] A. N. Ghule, P. Killeen and D. C. Ludois, "Synchronous Electrostatic Machine Torque Modulation via Complex Vector Voltage Control With a Current Source Inverter," in IEEE Journal of Emerging and Selected Topics in Power Electronics, vol. 8, no. 2, pp. 1850-1857, June 2020.

[6] A. Cuculić, D. Vučetić, R. Prenc, and J. Ćelić, "Analysis of Energy Storage Implementation on Dynamically Positioned Vessels," Energies, vol. 12, no. 3, p. 444, Jan. 2019.

[7] Papathanassiou, S., Hatziargyriou, N., & Strunz, K. "A benchmark low voltage microgrid network." In Proceedings of the CIGRE symposium: power systems with dispersed generation (pp. 1-8), CIGRE, April 2005.

[8] Diesel turbine datasheet, model n. P13.5-, 12.5 [kVA], by FGWilson™. (Online) Available: `https://www.fgwilsonpowerworks.com/brochures/uXQgbPAuCw.pdf/P13%205-6.pdf`, Visited on June 26, 2022.

[9] Exciter Models: Standard Dynamic Excitation Systems in NEPLAN Power System Analysis Tool. NEPLAN AG, Oberwachtstrasse 2, 8700 Küsnacht ZH, Switzerland. (Online) Available: `https://www.neplan.ch/wp-content/uploads/2015/08/Nep_EXCITERS1.pdf`. Visited on June 26, 2022.

[10] Turbine Governor Models: Standard Dynamic Turbine-Governor Systems in NEPLAN Power System Analysis Tool. NEPLAN AG, Oberwachtstrasse 2, 8700 Küsnacht ZH, Switzerland. Available (Online): `https://www.neplan.ch/wp-content/uploads/2015/08/Nep_TURBINES_GOV.pdf` Visited on June 26, 2022.

Influences of Parasitic Capacitances in Wide Bandwidth Rogowski Coils for Commutation Current Measurement

Philipp Ziegler, Tobias Festerling, Jörg Haarer, Philipp Marx, David Hirning, Jörg Roth-Stielow
INSTITUTE FOR POWER ELECTRONICS AND ELECTRICAL DRIVES
University of Stuttgart
Pfaffenwaldring 47
Stuttgart, Germany
Phone: +49 / (711) -685 67383
Email: philipp.ziegler@ilea.uni-stuttgart.de
URL: http://www.ilea.uni-stuttgart.de

Keywords

≪Integrated Rogowski coils≫, ≪Current sensor≫, ≪Component for measurements≫, ≪Parasitic elements≫, ≪Noise≫

Abstract

This paper focuses on the capacitive coupling of PCB integrated Rogowski coils for commutation current measurement. It investigates the cause and the influence of parasitic capacitances on the measurement signal. It discusses the advantages of a differential winding arrangement and the reasons why shielding is not an option in this application. The parasitic capacitance of three different winding arrangements are measured and compared to a Rogowski coil approach with minimized parasitic capacitance. On the base of sensor prototypes, the influence on the measurement is investigated in a pulsed current source.

Introduction

The use of wide bandgap (WBG) semiconductor devices made from silicon carbide (SiC) and gallium nitride (GaN) as semiconductor material is strongly increasing. Their ability to operate at higher switching frequencies, due to lower switching and conduction losses compared to silicon devices, is the key factor for high efficient and compact power electronic converter systems [1]. The switching speed of these devices is progressively increasing and an accurate measurement of the switching current becomes more challenging due to higher requirements. The bandwidth of current sensors for the characterization of WBG semiconductor devices needs to keep up with the steeper current transients. Due to progress in the packaging technology of power semiconductor devices, the parasitic inductance and capacitance of switching cells is decreasing [2]. For a commutation current sensor the smallest possible insertion impedance is desired for a non-invasive measurement. With higher possible integration levels and higher changing rates of the drain-source voltage during the switching transient (dV/dt) also the measurement environment becomes more challenging.

Measuring the voltage drop across a resistive shunt is the simplest way to measure a current. The sensing resistance is implemented in the current path and can provide a wide bandwidth measurement signal [3, 4]. For a precise measurement the resistance value needs to be known exactly and the parasitic inductance should be as low as possible. Otherwise the influence of the parasitic inductance needs to be compensated in a post processing step [4]. Besides these advantages, shunt based current sensing leads to additional losses and can not provide a galvanically isolated measurement signal, which is often required. For an isolated current measurement magnetic field sensors or a high frequency current transformer can be used. Magnetic fields sensors like Hall, fluxgate or tunneling magnetoresistance sensors suffer from a limited bandwidth. High frequency current transformers suffer from their size and insertion inductance. An alternative option is a Rogowski coil. The concept of the Rogowski coil is more than 100 years old and has been the subject of research ever since [5]. The sensing technique offers many degrees of freedom, which can be utilized for the different kind of applications, for instance in current probes [6,7], compact current sensors [8] or fully integrated Rogowski coils in the PCB for commutation current measurement [9, 10]. The concept of a hybrid current sensor combines a Rogowski coil (high frequency current measurement) with magnetic field sensors (low frequency current measurement) to extend the bandwidth of the sensing techniques [11]. In this way galvanic isolated current sensors with a bandwidth from DC up to several hundred MHz can be realized. Such current sensors can be used for the control of power electronic circuits [12,13], fully integrated in a chip [14] or as a commutation current sensor, which is integrated with a low inductive coaxial housing into the switching cell [15–18]. This paper focuses on the design of PCB integrated wide bandwidth Rogowski coils for commutation current measurement. It investigates the influence of parasitic coupling capacitances on the measurement system and why shielding is not an option in this application. Three state-of-the-art winding arrangements are compared in terms of their design parameters, bandwidth and the parasitic coupling capacitance to a coaxial housing. A new developed Rogwoski coil approach is presented, which aims for a minimized coupling capacitance in this kind of application. This minimized approach is compared to the state-of-the-art designs. The influence of the coupling capacitance is investigated in sensor prototypes in a pulsed current source.

Basic Principle of a Rogowski Coil

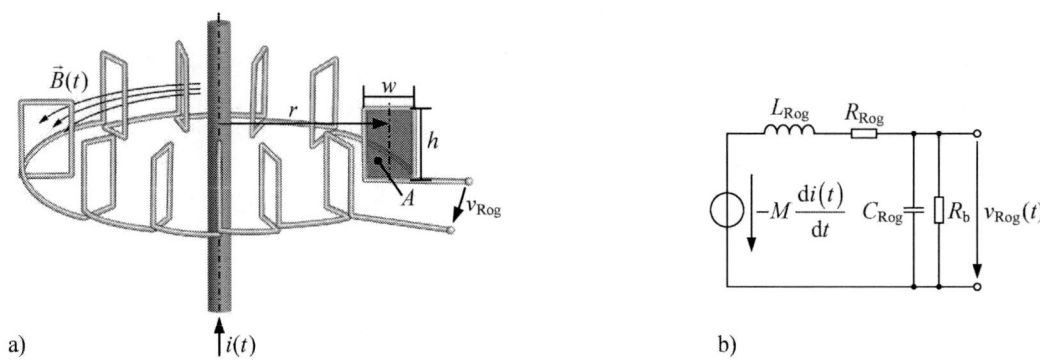

Fig. 1: Basic realization (a) and the equivalent circuit diagram (b) of a Rogowski coil

The principle of a Rogowski coil is based on Faraday's law of induction [3]. A change of the magnetic flux density $\vec{B}$ in a conductor loop induces a voltage v. Hence, a Rogowski coil uses a number of N conductor windings, which are arranged around the current carrying conductor, see Fig. 1 a). With the use of Ampère's law the relation between the derivative in the current $i(t)$ and the voltage at the Rogwoski coil terminals v_{Rog} is described by:

$$v_{\mathrm{Rog}}(t) = -N \cdot A \cdot \frac{\mathrm{d}|\vec{B}(t)|}{\mathrm{d}t} = -\frac{\mu_0 \cdot N \cdot A}{2\pi \cdot r} \frac{\mathrm{d}i(t)}{\mathrm{d}t} = -M \frac{\mathrm{d}i(t)}{\mathrm{d}t} \tag{1}$$

Besides the number of turns, the induced voltage is depending on the winding area A and the radius r between the center of the conductor and the winding. These parameters form the mutual inductance M of the Rogowski coil, which describes the sensitivity. In an equivalent circuit diagram, see Fig. 1 b), this behavior is modeled by means of a voltage source. In addition, the parasitic elements of a Rogowski coil, the self-inductance L_{Rog}, the parasitic interwinding capacitance C_{Rog} and the winding resistance R_{Rog} complete the model. The bandwidth of the Rogowski coil is defined by the resonance frequency f_{res}

resulting from L_{Rog} and C_{Rog}. This resonance pole can be damped with an additional burden resistance R_{b}. In the Laplace domain the damped Rogowski coil is described with the following transfer function [19, 20]:

$$\frac{v_{\text{Rog}}(s)}{i(s)} = \frac{-sMR_{\text{b}}}{s^2 L_{\text{Rog}} C_{\text{Rog}} R_{\text{b}} + s(L_{\text{Rog}} + C_{\text{Rog}} R_{\text{Rog}} R_{\text{b}}) + R_{\text{Rog}} + R_{\text{b}}} \tag{2}$$

The resonance frequency f_{res} is determined by (3) and defines the upper frequency limit of the bandwidth of the Rogowski coil.

$$f_{\text{res}} = \frac{1}{2\pi} \cdot \sqrt{\frac{R_{\text{Rog}} + R_{\text{b}}}{L_{\text{Rog}} C_{\text{Rog}} R_{\text{b}}}} \tag{3}$$

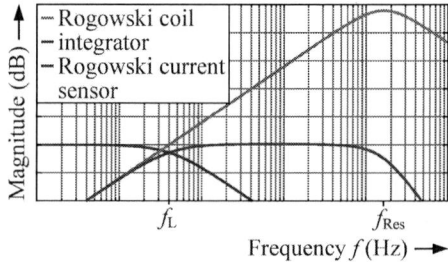

Fig. 2: Frequency response of a Rogowski coil current sensor

To reconstruct the temporal course of the current from the induced voltage, an integrator at the Rogowski coil terminals is required. The integrator can be either realized in a passive or an active way and acts as a low pass filter. This low pass filter defines the lower frequency limit f_{L}. The combination of the frequency response of the Rogowski coil (green) and the integrator (blue) results in the bandpass characteristic of the current sensor (purple), see Fig. 2.

Influence of the parasitic capacitances

There are two ways to integrate a Rogowski coil into the commutation path for current measurement. The direct way integrates the winding arrangement in the PCB of the switching cell [10]. For an indirect way, a measurement point for the Rogowski coil is added for example with a coaxial housing, replacing a through hole connection in the switching cell [21], see Fig. 3. In any case there will be a capacitive coupling of the current sensor and the switching cell. Especially the windings of the Rogowski coil represent a large area, which causes a parasitic coupling capacitance. Due to the high dV/dt during the switching transient, these parasitic capacitances have a strong impact on the measurement signal of a commutation current sensor. In Fig. 4 a) this effect is taken into account in the equivalent circuit diagram of a Rogowski coil. The switching potential is represented by an additional voltage source $v_{\text{sp}}(t)$ resulting in a current through the parasitic capacitances $C_{\text{P,n}}$. This current leads to an additional voltage $v_{\text{CC}}(t)$ at the terminals of the Rogowski coil caused by the capacitive coupling. This effect causes additional oscillations on the measurement signal, which result in

Fig. 3: Current sensor integration with a coaxial housing into the commutation path

a distortion. In that case the measurement signal of a current transient during the characterization of a device can be flatter or even steeper then the actual current. The parasitic capacitance is therefore a major source for measurement errors. Unfortunately, shielding the Rogowski coil is not an option for a commutation current sensor, because of the required bandwidth [22]. Additional shielding layers significantly increase the interwinding capacitance and decrease the bandwidth of the Rogowski coil. Other ways must be found to reduce the influence of the parasitic capacitance of the Rogowski coil in this application. One common optimization approach is to realize a fully differential Rogowski coil by splitting

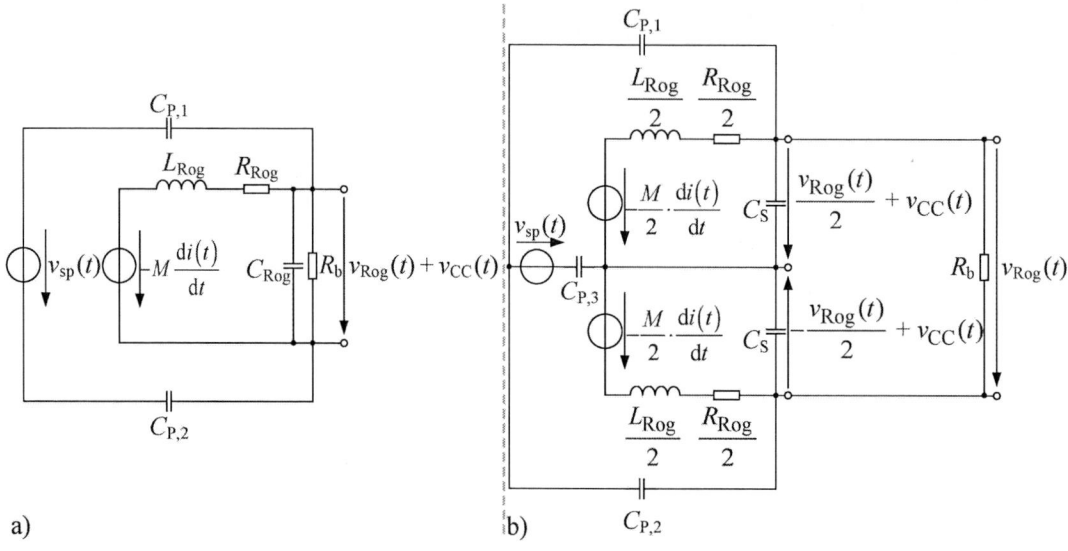

Fig. 4: Influence of capacitive coupling on (a) a single winding and (b) a differential winding

up the winding in two coils with different winding directions and a common ground. This also splits up the coupling path and the resulting additional voltage $v_{CC}(t)$, see Fig. 4 b). In an ideal symmetrical realization this leads to a cancellation of the additional voltage at the output terminals due to the opposite winding directions [22]. Due to technical and manufacturing reasons an ideal differential coil can not be realized and there will still be an influence of the capacitive coupling. The coupling capacitances form a resonance circuit with the self-inductance of the Rogowski coil. This resonance circuit leads to an oscillation and additional noise on the measurement signal in the switching transition. For this reason different winding arrangements are investigated and compared.

Comparison of different winding arrangements

In this work three state-of-the-art winding arrangements (sawtooth, zickzack and double winding [9]) are investigated and compared to a new approach, which aims to minimize the parasitic capacitances. To investigate the influence of shielding, the sawtooth structure is also modified with a shielding layer on the top and bottom layer, which also acts as return conductor. To make the shielding possible, the winding structure has to be realized in the two inner layers of the PCB. The realization of the winding arrangements of the state-of-the-art Rogowski coils are depicted in Fig. 5. The minimized concept is

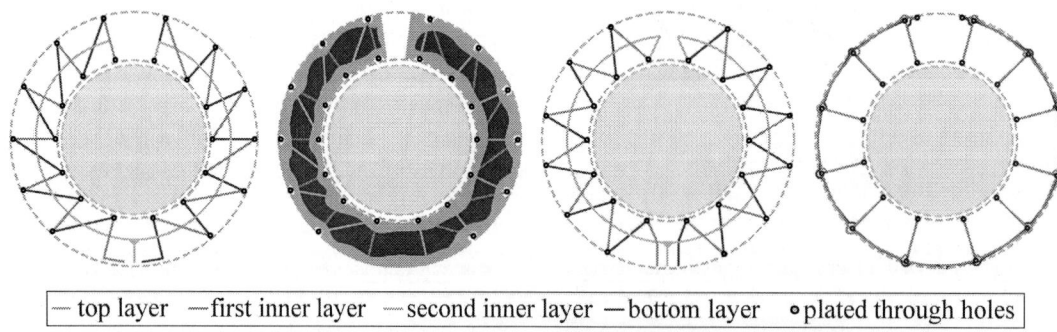

— top layer — first inner layer — second inner layer — bottom layer ● plated through holes

Fig. 5: Winding arrangements: sawtooth, shielded sawtooth, zickzack, double winding

designed for a fixed setup in a coaxial housing, whereby the number of turns can be reduced due to the fixed placement. To achieve the same mutual inductance and sensitivity, as the other approaches, the height of the windings is increased and adjusted to the usable height of the coaxial housing. This approach allows a reduction of the parasitic elements C_P, C_{Rog} and L_{Rog} by reducing the number of turns and the overall trace length to a minimum. This reduces the effect of the parasitic coupling and increases the resonance frequency and therefore the bandwidth of the Rogowski coil. The realization of the minimized approach is depicted in Fig. 6. Each winding is manufactured on a separate PCB with

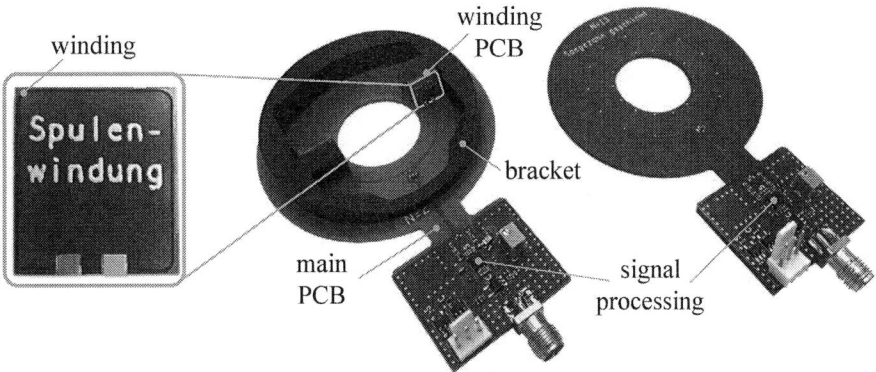

Fig. 6: Realization of the minimized approach (left) and one of the state-of-the-art sensors (right)

minimal trace widths. The two winding PCBs are soldered to a main PCB, which carries the signal processing. In order to guarantee a 90 degree angle between the winding PCB and the main PCB and to give the structure more stability, the windings are additionally fixed by a bracket. This approach is compared with state-of-the-art winding arrangements, all manufactured in a four layer PCB. In Tab. I the parameters of the different Rogowski coils are presented. In the following sections the design aspects of the approaches are discussed as well as the measurement method.

Table I: Parameters of the Rogowski coils

	Sawtooth	Sawtooth (shielded)	Zickzack	Double winding	Minimized approach
Number of turns N	13	13	13	12	2
Inner radius r_i (mm)	11	11	11	11	11
Outer radius r_o (mm)	16.5	17.5	16.5	17.1	17.5
Height h (mm)	1.4	1.2	1.4	1.4	8.1
Trace length l (mm)	261	199	248	343	172
Disturbance area A_d (mm²)	54.7	-	33.4	1.2	3.2
Self-inductance L_{Rog} (nH)	108.8	62.2	99.7	131.1	57.2
Interwinding capacitance C_{Rog} (pF)	1.61	5.31	1.66	1.72	1.17
Coupling capacitance C_P (pF)	40.5	307.6	40.4	55.1	3.5
Resonance frequency f_{res} (MHz)	380.4	276.1	391.8	335.6	610

Dimensions

An attempt is made to set the parameters of the Rogowski coils as equal as possible for the comparison, but due to the winding structure of the double winding a turns number of $N = 13$ is not possible. Therefore, the turns number is reduced to $N = 12$ and the outer radius r_o is slightly increased to achieve the same mutual inductance. To realize the shielding of the sawtooth winding the top and bottom layer is used. Therefore, the Rogwoski coil is routed in the two inner layers. To compensate the decreased height, the outer radius r_o is slightly increased. All Rogowski coils are designed in a fully differential way with the focus on the design to be as symmetrical as possible and a mutual inductance of $M = 1.45\,\mathrm{nH}$. The double winding concept focuses on the reduction of the disturbance area A_d. Therefore, the conductors are placed above each other resulting in the longest trace length l. The minimized approach also significantly reduces the disturbance area compared to the sawtooth and zickzack winding. This approach also offers the shortest trace length, which also leads to the smallest parasitic elements and the highest resonance frequency f_{res}.

Mutual inductance

The mutual inductance M of the different designs is also determined to evaluate the design parameters as well as manufacturing tolerances. For this purpose, the induced voltage v_{Rog} at the terminals of the Rogwoski coil is measured during a rising edge of a current pulse. With the value of the derivative of the current over time the mutual inductance is calculated with (4) .

$$M = \left| \frac{v_{Rog}}{\frac{di}{dt}} \right| \tag{4}$$

The results are listed in Tab. II. The resulting values are in the range of the desired mutual inductance of $M = 1.45\,\mathrm{nH}$. In addition to manufacturing tolerances, the deviation can be caused by measurement inaccuracies in the voltage measurement as well as the determination of the current derivation.

Table II: Measured mutual inductance M

	Sawtooth	Sawtooth (shielded)	Zickzack	Double winding	Minimized approach
Mutual inductance M (nH)	1.6	1.39	1.6	1.65	1.54
Deviation (%)	9.6	4.1	9.6	13.8	6.3

Frequency response

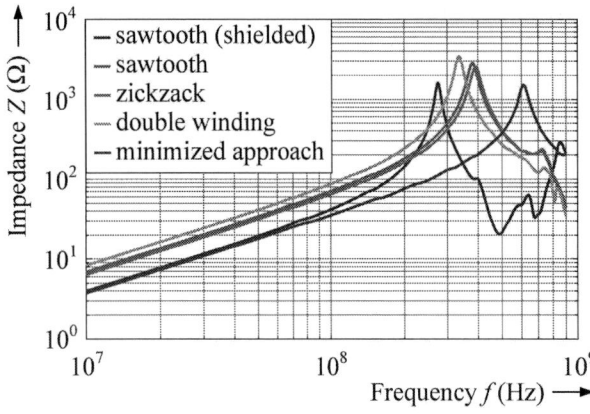

Fig. 7: Measured frequency response of the Rogowski coil designs

With a network analyzer (Keysight Technologies E5061B) the frequency response of the different winding arrangements is measured and depicted in Fig. 7. It is seen that the minimized approach leads to the highest resonance frequency f_{res}. The sawtooth and zickzack structures lead to almost the same course due to only slight differences in the winding arrangement and trace length. In the double winding the trace length is increased compared to the other state-of-the-art structures, causing the lowest resonance frequency. Only the shielded version of the sawtooth structure leads to an even lower resonance frequency. The self-inductance L_{Rog} is determined with the use of the gradient of the impedance course during the linear section below 100 MHz.

With the knowledge of the self-inductance and the resonance frequency the interwinding capacitance C_{Rog} is calculated. Two main aspects of the winding arrangements can be identified. First, with increasing trace length also the self-inductance increases. The second aspect is that the interwinding capacitance is in the same range in all structures except in the shielded structure. The shield leads to a significant increase of the interwinding capacitance and therefore, to a decrease of the resonance frequency. Depending on the desired bandwidth of the Rogowski coil a shielding layer on the top and bottom layer is therefore acceptable or not. For commutation current measurement a bandwidth as high as possible is desired and prevents the use of a shielding.

Disturbance area

The principle of the Rogowski coil is based on the induced voltage caused by a change of the magnetic flux density $\vec{B}$ going through the winding structure. Not only the current in the center conductor, which is intended to be measured, causes a changing flux density. Also external sources, like nearby conductors, can induce a voltage with a changing flux density $\vec{B}_d$ through the disturbance area A_d of the winding structure. For the sawthooth winding and double winding this area is depicted in Fig. 8. There are

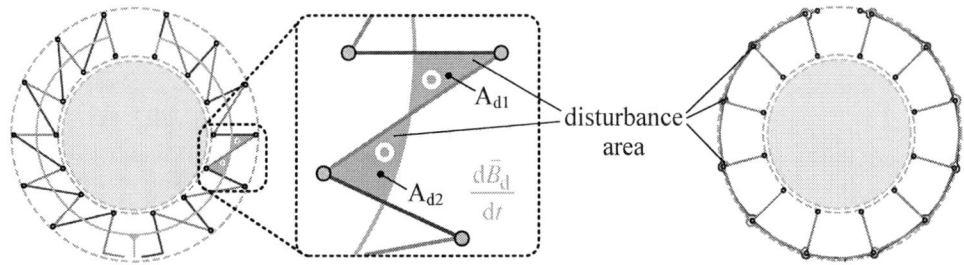

Fig. 8: Measured frequency response of the Rogowski coil designs

two options to reduce the influence of the disturbance area. The double winding concept focuses on a minimization of the area [9]. Due to the overlapping traces only a very small area around through hole connections is left. A second option is to design the disturbance area in a way, which leads to an cancellation due to a superposition. In the example in Fig. 8, the areas A_{d1} and A_{d2} lead to opposite signs of the induced voltage. A design with the focus on the same area can reduce the effect. In this work same areas are not realized in the state-of-the-art winding. But the effect of the superposition is considered in the values of the disturbance area stated in Tab. I. An optimization to realize the same areas could be the change of the return conductor in the second inner layer from a circular form to a polygon with its corners under the conductors on the top and bottom side. The radius of this polygon needs to be adjusted in a way that the resulting areas are identical. In the minimized approach the disturbance area is also small due to connection lines routed close to each other. In the shielded version of the sawtooth structure, the shield prevents a disturbance area.

Coupling capacitance

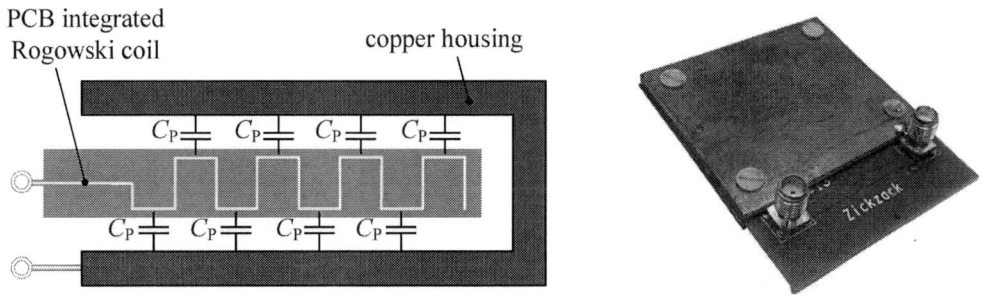

Fig. 9: Measured frequency response of the Rogowski coil designs

For the measurement of the coupling capacitance C_P each Rogowski coil is placed in a copper housing, see Fig. 9. With a network analyzer (Omicron Lab Bode 100) the capacitance is measured between the winding and the copper housing using a high impedance measurement setup. The values in Tab. I show that the coupling capacitance of the minimized approach is significantly reduced compared to the state-of-the-art winding arrangements.

Measurement results

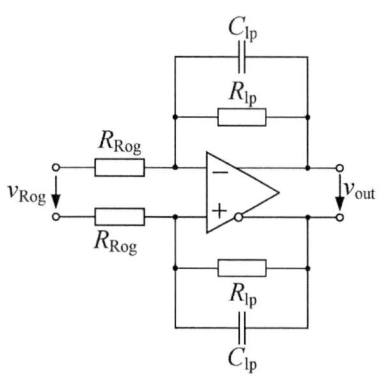

Fig. 10: Realization of the signal processing

For every winding arrangement a prototype of a current sensor is built and characterized. Due to the highest bandwidth and the similar parasitic capacitance of all state-of-the-art prototypes, the zickzack winding is used as reference to the minimized approach. The signal processing in all prototypes is designed in the same way. It consist of an active low pass filter realized with one fully differential operational amplifier (Texas Instruments THS4513) in the configuration depicted in Fig. 10 [18]. The signal level of v_{out} then is increased with a second amplifier (Analog Devices AD8009) resulting in the sensor signal v_{Sen}. The measurement range of all sensors is set to $I_{\text{max}} = \pm 110\text{A}$ with a sensitivity of $18\,\frac{\text{mV}}{\text{A}}$. The sensors are tested in a pulsed current source with current transients similar to the occurring commutation current of fast switching semiconductor devices [23]. The current sensors are placed in a coaxial chamber, which carries the current. The advantage of this chamber is the low inductance, which allows steep rise and fall times. For the prototypes of the zickzack winding and the minimized approach, the measurement results of a current edge are depicted in Fig. 11 a). A high frequency current transformer (Bergoz Instrumentation, CT-B1.0-S) with a bandwidth from 200 Hz up to 500 MHz is used as a reference. The measured current courses of both sensors are compared to the course of the Bergoz current transformer. The occurring deviation is related to the measurement range of $I_{\text{max}} = \pm 110\text{A}$. The deviation of both sensors is less then $\pm 1\%$.

To investigate the influence of the coupling capacitances, the sensor signal is measured with the voltage supply of the signal processing turned off. In this case the sensor signal represents the voltage across the burden resistance, which is damped by the resistors of the signal processing, forming a voltage divider with the 50Ω input impedance of the oscilloscope. The measured sensor signals for the zickzack winding arrangement $v_{\text{Sen,zz}}$ and the minimized approach $v_{\text{Sen,ma}}$ are depicted in Fig. 11 b). To put the sensor courses into context, the current course i of the pulsed current source is depicted as well. The course of the minimized approach almost follows an ideal trapezoidal course during the current rising and falling edge. Only a small oscillation is superimposed. In contrast, the course of the zickzack winding arrangement shows almost double the amplitude in the current transients, which occurs due to an oscillation of the resonance circuit formed by the coupling capacitance and the self-inductance. Due to the higher values of the parasitic elements in the zickzack winding, the oscillation is larger compared to the minimized approach and decays more slowly. This oscillation also leads to more noise in the sensor signal and measurement deviation.

Conclusion

The ongoing development of wide bandgap semiconductor devices leads to shorter switching times. This increases the requirements for the commutation current measurement in terms of bandwidth and immunity against high dV/dt. In this paper the influence of a parasitic coupling capacitance between the winding of the Rogowski coil and a coaxial housing is investigated. Three state-of-the-art winding arrangements and a shielded version are compared to a new minimized approach in terms of their bandwidth, the parasitic elements and an occurring disturbance area. The new minimized approach is designed for a fixed setup in a commutation current sensor and focuses on a reduction of the coupling capacitance. In addition this approach leads to an increased bandwidth. With the investigated winding arrangements, current sensors are realized and tested in a pulsed current source. It is seen that the coupling capacitance is forming a resonance circuit with the self-inductance leading to measurement deviation.

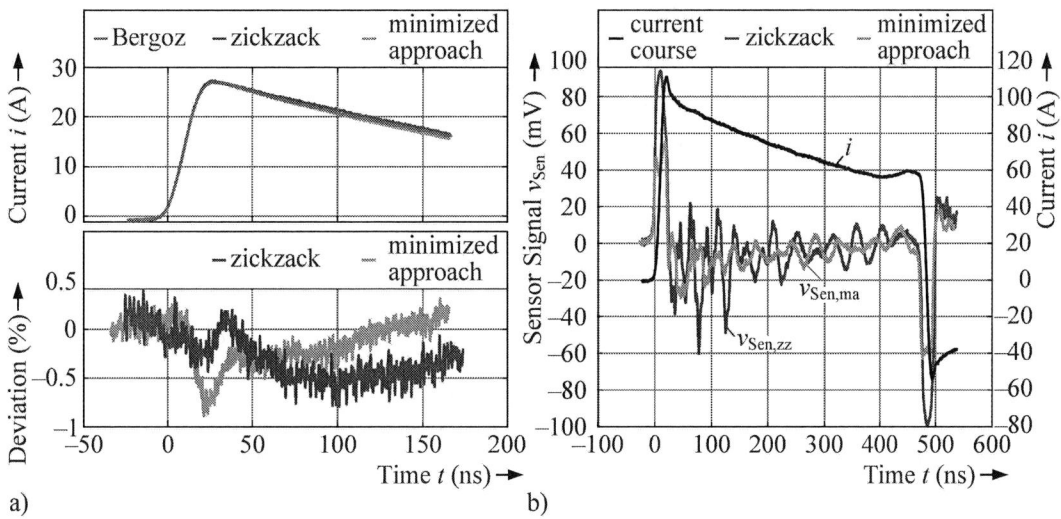

Fig. 11: Measured current rise and the occuring deviation (a); Sensor signal with turned off voltage supply (b)

References

[1] J. Millan, P. Godignon, X. Perpina, A. Perez-Tomas, and J. Rebollo, "A survey of wide bandgap power semiconductor devices," *IEEE Transactions on Power Electronics*, vol. 29, no. 5, 2014.

[2] E. Hoene, A. Ostmann, and C. Marczok, "Packaging very fast switching semiconductors: Cips 2014: 8th international conference on integrated power electronics systems," in *CIPS 2014: 8th International Conference on Integrated Power Electronics Systems*, 2014.

[3] S. Ziegler, R. C. Woodward, H. H.-C. Iu, and L. J. Borle, "Current sensing techniques: a review," *IEEE Sensors Journal*, vol. 9, no. 4, 2009.

[4] W. Zhang, Z. Zhang, F. Wang, E. V. Brush, and N. Forcier, "High-bandwidth low-inductance current shunt for wide-bandgap devices dynamic characterization," *IEEE Transactions on Power Electronics*, vol. 36, no. 4, pp. 4522–4531, 2021.

[5] W. Rogowski and W. Steinhaus, "Die messung der magnetischen spannung," *Archiv für Elektrotechnik*, vol. 1, no. 4, pp. 141–150, 1912. [Online]. Available: http://dx.doi.org/10.1007/bf01656479

[6] C. Hewson and J. Aberdeen, "An improved rogowski coil configuration for a high speed, compact current sensor with high immunity to voltage transients," in *APEC 2018*, APEC, Ed. Piscataway, NJ: IEEE, 2018.

[7] L. Ming, Z. Xin, W. Liu, and P. Chiang Loh, "Structure and modelling of four-layer screen-returned pcb rogowski coil with very few turns for high-bandwidth sic current measurement," *IET Power Electronics*, vol. 13, no. 4, pp. 765–775, 2020.

[8] D. Koch, H. Bantle, J. Acuna, P. Ziegler, and I. Kallfass, "Design methodology for ultra-compact rogowski coils for current sensing in low-voltage high-current gan based dc/dc-converters," in *2021 23rd European Conference on Power Electronics and Applications (EPE'21 ECCE Europe)*, 2021.

[9] J. N. Fritz, C. Neeb, and R. W. de Doncker, "A pcb integrated differential rogowski coil for non-intrusive current measurement featuring high bandwidth and dv/dt immunity," in *Power and Energy Student Summit*, 2015.

[10] A. Rafiq, S. K. Pramanick, and R. Maheshwari, "Design of pcb coil based high bandwidth current sensor with power-loop stray inductance characterization," *IEEE Transactions on Industrial Electronics*, vol. 68, no. 12, pp. 12 791–12 801, 2021.

[11] N. Karrer and P. Hofer-Noser, "A new current measuring principle for power electronic applications," in *ISPSD '99: Proceedings of the 11th International Symposium on Power Semiconductor Devices and ICs*, 1999, pp. 279–282.

[12] S. J. Nibir, M. Biglarbegian, and B. Parkhideh, "A non-invasive dc-10-mhz wideband current sensor for ultra-fast current sensing in high-frequency power electronic converters," *IEEE Transactions on Power Electronics*, vol. 34, no. 9, pp. 9095–9104, 2019.

[13] P. S. Niklaus, D. Bortis, and J. W. Kolar, "High-bandwidth high-cmrr current measurement for a 4.8 mhz multi-level gan inverter ac power source," in *APEC 2021: IEEE Applied Power Electronics Conference and Exposition*. IEEE.

[14] T. Funk and B. Wicht, "A fully integrated dc to 75 mhz current sensing circuit with on-chip rogowski coil," in *2018 IEEE Custom Integrated Circuits Conference (CICC)*. Piscataway, NJ: IEEE, 2018.

[15] N. Troester, J. Ruthardt, M. Nitzsche, and J. Roth-Stielow, "Wide bandwidth current sensor combining a coreless current transformer and tmr sensors," in *PCIM Europe 2018: International Exhibition and Conference for Power Electronics, Intelligent Motion, Renewable Energy and Energy Management*, 2018.

[16] P. Ziegler, N. Tröster, D. Schmidt, J. Ruthardt, M. Fischer, and J. Roth-Stielow, "Wide bandwidth current sensor for commutation current measurement in fast switching power electronics," in *2020 22nd European Conference on Power Electronics and Applications (EPE'20 ECCE Europe)*, 2020, pp. P.1–P.9.

[17] P. Ziegler, F. Stjepandic, J. Ruthardt, P. Marx, M. Fischer, and J. Roth-Stielow, "Wide bandwidth current sensor for characterization of high current power semiconductor modules," in *2021 23rd European Conference on Power Electronics and Applications (EPE'21 ECCE Europe)*, 2021, pp. 1–9.

[18] P. Ziegler, Y. Zhao, J. Haarer, J. Ruthardt, M. Fischer, and J. Roth-Stielow, "Compact design of a wide bandwidth high current sensor using tilted magnetic field sensors," in *2021 IEEE Energy Conversion Congress and Exposition (ECCE)*. IEEE, 10/10/2021 - 10/14/2021, pp. 5528–5534.

[19] M. H. Samimi, A. Mahari, M. A. Farahnakian, and H. Mohseni, "The rogowski coil principles and applications: A review," *IEEE Sensors Journal*, vol. 15, no. 2, pp. 651–658, 2015.

[20] H. Li, Z. Xin, X. Li, J. Chen, P. C. Loh, and F. Blaabjerg, "Extended wide-bandwidth rogowski current sensor with pcb coil and electronic characteristic shaper," *IEEE Transactions on Power Electronics*, vol. 36, no. 1, 2021.

[21] N. Troster, J. Wolfle, J. Ruthardt, and J. Roth-Stielow, "High bandwidth current sensor with a low insertion inductance based on the hoka principle," in *EPE'17 ECCE Europe: 19th European Conference on Power Electronics and Applications*. IEEE, 2017.

[22] S. Hain and M.-M. Bakran, "New rogowski coil design with a high dv/dt immunity and high bandwidth," in *2013 15th European Conference on Power Electronics and Applications (EPE)*. IEEE, 9/2/2013 - 9/6/2013, pp. 1–10.

[23] N. Troester, D. Bura, J. Woelfle, M. Stempfle, and J. Roth-Stielow, "Design of a 300 amps pulsed current source with slopes up to 27 amps per nanosecond for current probe analysis," in *PCIM Europe 2017: International Exhibition and Conference for Power Electronics, Intelligent Motion, Renewable Energy and Energy Management*, 2017.

Systematic analysis of oscillations in DC-links of fast switching power electronics

Tobias Fricke, Regine Mallwitz
INSTITUTE FOR ELECTRICAL MACHINES, TRACTION AND DRIVES (IMAB)
TU Braunschweig
Braunschweig, Germany
Email: tobi.fricke@tu-braunschweig.de
URL: https://www.tu-braunschweig.de/imab

Acknowledgments

This work was supported by the german Federal Ministry of Education and Research (BMBF) through the funded project CODAPE under Grant 16ME0357.

Keywords

≪Capacitors≫, ≪DC-link≫, ≪Design optimization≫, ≪Hardware≫, ≪Harmonics≫, ≪High frequency power converter≫, ≪Impedance analysis≫, ≪Impedance measurement≫, ≪MOSFET≫, ≪Parasitic elements≫, ≪Passive component≫, ≪Test bench≫

Abstract

Increasingly faster switching and higher switching frequencies amplify the impact of parasitics of the DC-link causing higher voltage overshoots, oscillations, losses and also EMI. Higher-frequency resonances due to polarization effects of dielectrica and oscillations between either different capacitors or capacitors and switches (or even other components) due to a mismatch of pcb parasitics and component impedances are just some of the possible causes. This paper marks the beginning of a series of studies with the aim of a set of design rules for DC-links of high switching frequency power electronics. Therefore circuit simulations, impedance measurements (of capacitors and pcb's) and experiments using an evaluation platform are carried out. Some of the results are presented in this paper.

I Introduction

The increase of switching speeds and frequencies caused by the evolution of wide-bandgap-semiconductors like Siliconcarbide (SiC) and Galliumnitride (GaN) also leads to an increase of the influence of parasitics. First of all the assembly technologies and pcb layouts represented a first hurdle. Decreasing the stray inductances of both in the power loop was a necessary step towards the utilization of the new semiconductors properties, since it limits the $\frac{di}{dt}$ and causes voltage overshoots and oscillations [1]. But due to the higher switching frequencies of SiC and GaN the requirements for magnetic elements like chokes and transformers also increased, which brought them more into focus again [2]. New core materials, winding configurations, braids and even design methods were and are still investigated [3]. In addition the influence of capacitors and DC-links on voltage overshoots and oscillations emerges more and more often [4]. One reason for this is as well the stray inductance, but in this case the ones of capacitors and the DC-link-assembly including the pcb. For this reason the parasitic elements of the DC-link are currently in the focus. Optimizing single capacitor types seems to bee a promising option [5][6]. But also the investigation of pcb layouts and interconnections are being discussed more and more often [7][8]. Finally, the stray inductance is not the only important parameter that can cause the problems mentioned above. Oscillations between capacitors and switches or even between different capacitors are possible.

Many of this investigations regard a specific application and not the DC-link in general (related to high switching frequency applications). The presented paper therefore forms the beginning of a series of studies which will help to understand the influence of capacitors, their combinations, circuit topologies and their interaction on voltage overshoots, oscillations, losses and EMI. Based on this findings design rules are to be created, which will support during the design process of DC-links for fast switching power electronics. Therefore this paper will present three parts of this studies. Impedance measurements of different electrolytic and ceramic capacitors regarding the DC-bias, the realization of a GaN-based test platform for evaluation purposes and simulations of fundamental correlations, which will be described first.

II Simulations

The dc-link's behaviour (more precisely: its impedance) mainly depends on the characteristics of three parts of the system. The capacitors, the switches and the PCB layout. The combination of these parts defines the DC-link's behaviour, with the parasitics playing an important role since they are responsible for the most unintended oscillations. They can cause oscillations between capacitors and switches or even between different capacitors. This interactions can become complex which is why they will be subject of more detailed studies in the future. Nevertheless there are a few general questions that often come up when discussing about capacitors and DC-links:

1. Which effect has the DC-links main resonance frequency f_{link} (e.g. the resonance of C_{link} and L_{link} defined in Fig. 1)?
2. Where should the switching frequency be in relation to this resonance frequency?
3. Is it better to have some ESR (R_{link} in Fig. 1) e.g. for dampening in terms of EMI?

This questions are suitable for gaining some fundamental knowledge and testing the simulation models described next. The simulation model implemented a single-phase synchronous buck-converter using LTspice. All parameters were adopted from the DC-link evaluation platform described in Section IV. To start with, the DC-link was modeled by using a simple capacitor-model which consists of R_{link}, L_{link} and C_{link} (see Fig. 1).

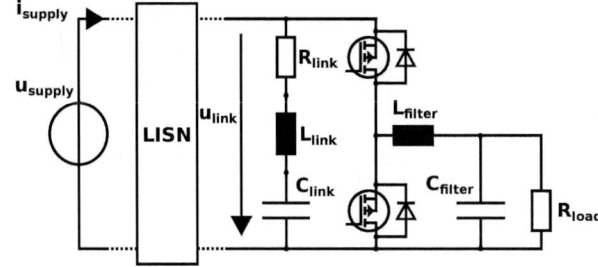

Fig. 1: Schematic representation of the implemented simulation model

The simulations neglected the influences of the pcb layout and the switches (ideal switches were used) to keep the number of parameters low. The key parameters are:

- Switching frequency $f_{sw} = 1\,\mathrm{MHz}$
- DC-link supply voltage $u_{supply} = 150\,\mathrm{V}$
- Duty cycle $50\,\%$
- Load resistance $R_{link} = 5\,\Omega$
- DC-link capacity $C_{link} = 10\,\mu\mathrm{F}$

To replicate real conditions the line was implemented as HV-LISN (Line Impedance Stabilization Network, also known as Artificial Mains Network or AMN) with values according to CISPR 25 [9]. This ensures comparable conditions. To receive significant data a specifically implemented Python-based simulation-controller automatically varied the main resonance frequency f_{link} of the DC-link (by modifying L_{link}) and the ESR (here: R_{link}). This leads to a result matrix that allows for a separate analysis of the impacts of f_{link} and R_{link} on characteristic values like the DC-link voltage and the supply current, which will be described next.

A DC link voltage

Since the DC-link itself was the subject of the investigations the influences of the switches and their parasitics were consciously neglected. This was realized by the usage of ideal switches (except for resistances).

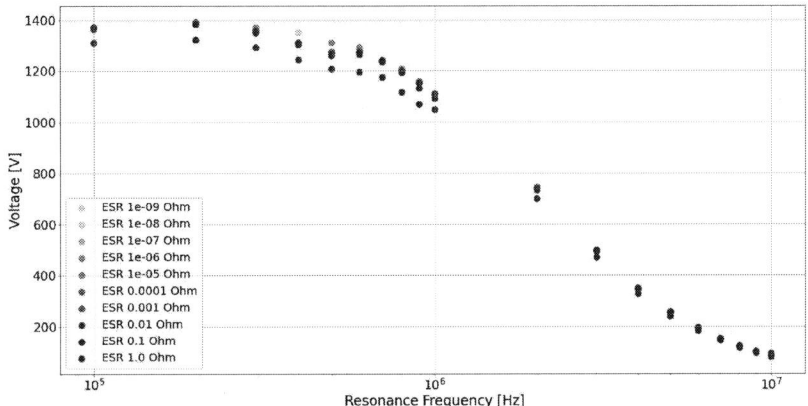

Fig. 2: Max. positive voltage overshoot related to the mean value of u_{link} during variation of f_{link} and R_{link}

Fig. 2 shows the maximum positive voltage overshoot of the DC-link voltage u_{link}. Several dots at the same frequency mark the values of different R_{link} settings while the dot's course along the x-axis shows the behaviour of the voltage overshoots during variation of f_{link}. The voltage overshoot starts to drop the closer f_{link} gets to the switching frequency of 1 MHz. The inflection point of the curve lies slightly higher than f_{sw}. Finally at $f_{\text{link}} \approx 10 \cdot f_{\text{sw}}$ the gradient of the curve becomes small and the region of the lowest voltage overshoot is reached.

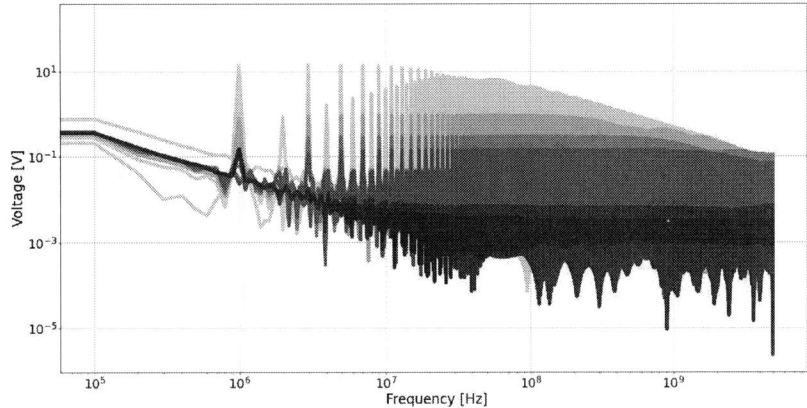

Fig. 3: Spectra of u_{link} during variation of f_{link} (darkest color marks highest f_{link}) and a fixed R_{link} at 1 nΩ

In Fig. 3 the resistance R_{link} stays fixed at 1 nΩ. By doing so, the impact of f_{link} on the harmonics of u_{link} can be evaluated. This comparison shows significantly lower amplitudes the higher the resonance frequency of the DC-link gets (Note that the resonance frequency f_{link} is marked by the colors with the darkest blue marking the highest frequency while the x-axis shows the frequency of the spectra). This behaviour correlates with the course of the voltage overshoot analyzed before. In sum this results showed that a main resonance frequency of the DC-link above the switching frequency decreases the harmonics and the voltage overshoot significantly. A good indicative value seems to be $f_{link} \geq 10 \cdot f_{sw}$. Since the switching frequencies of SiC and GaN can be up to several MHz this might be a tough requirement.

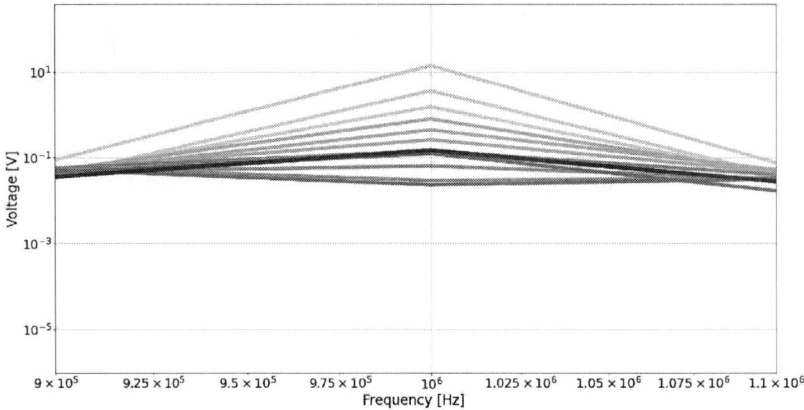

Fig. 4: Spectra of u_{link} during variation of f_{link} (darkest blue marks highest f_{link}) and a fixed R_{link} at $1\,n\Omega$, orange marks resonance frequencies close to f_{sw}

Another striking aspect of the spectra is the behaviour around the switching frequency (see Fig. 4). If the main resonance of the DC-link lies close to the switching frequency the two impedances $\underline{X}_{C,link}$ and $\underline{X}_{L,link}$ become almost equal and cancel each other out. In consequence only R_{link} is effective. If this resistance is low enough, the resulting harmonic of u_{link} will be lower than one of a DC-link with an even higher main resonance frequency. To highlight this, frequencies close to f_{sw} are marked in orange (see Fig. 4). The significance of this effect (e.g. regarding the capacitors losses) will be subject of further studies.

B Supply Current

The current i_{supply} sourced by the supplying voltage source shows a different characteristic than the DC-link voltage analyzed above. Fig. 5 shows the RMS-value of the AC-part of the current i_{supply} for different combinations of f_{link} and R_{link} (the plot configuration is identical to the one of Fig. 2).

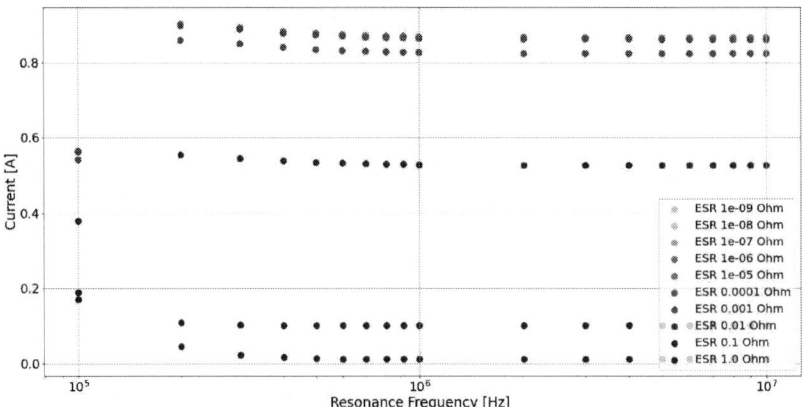

Fig. 5: RMS-value of the AC-part of the supply current i_{supply} during variation of f_{link} and R_{link}

In contrast to the DC-link voltage the main resonance frequency f_{link} has only an effect at very low frequencies, approximately below $0.1 \cdot f_{sw}$. Above this frequency the impact of f_{link} can be neglected. The situation is different for the impact of the resistance R_{link} as it shows a greater impact on the supply current than it did on the DC-link voltage. A higher resistance leads to a lower AC-RMS-value of i_{supply}.

C Summary

In sum the simulations lead to the following observations:

- f_{link} has great impact on the overshoot and the harmonics of u_{link}. Higher frequencies lead to lower overshoot and amplitudes of the harmonics
- R_{link} has small impact on the overshoot and the harmonics of u_{link}. The higher f_{link} the lower the impact of the resistance
- f_{link} has small impact on the AC-RMS of the supply current i_{supply}. The impact above $0.1 \cdot f_{sw}$ is neglectable
- R_{link} has significant impact on the AC-RMS of the supply current i_{supply}. The higher the resistance the lower the AC-RMS

Whether this observations are generally valid or not, what might be possible restrictions to them and how they have to be customized for specific applications will be subject of further studies. Nevertheless this results showed that the characteristics of the DC-links capacitor can have major impact on voltage overshoot and harmonics. This motivates the investigation of different capacitors described next.

III Capacitors

The simulations described in II used a fairly simple capacitor model consisting of only one resonance circuit. Nevertheless they showed a major impact of the capacitors impedance (resonance frequency and ESR) on important characteristics like voltage overshoots and EMI. To evaluate the influence of real capacitors more complex models are needed, some are already available [11]. The complexity of these models is needed because of the highly non-ideal characteristics of most capacitor types. The following are only the most significant reasons for those non-idealities [10]:

- different behaviour of the dielectric materials (like frequency- and temperature-dependencies of the permittivity ε_r)
- different polarization mechanisms (e.g. electron polarization, ionic polarization and oriental polarization)
- differences between polar and non-polar materials
- hysteresis effects like they appear in ferroelectric materials (for example Barium Titanate $BaTiO_3$)
- piezoelectric effects (ceramic capacitors)

To evaluate these models and to be able to simulate the behavior of real capacitors impedance measurements are needed. In terms of comparability among each other and the usage in the evaluation platform described in IV the voltage rating of most capacitors was set to 250 V.

A Electrolytic capacitors

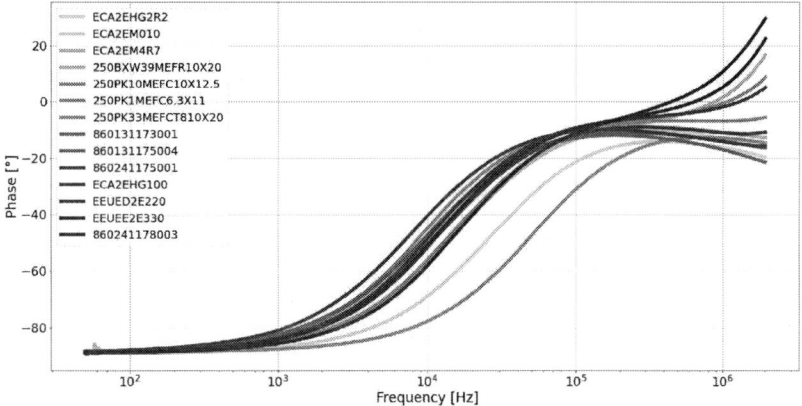

Fig. 6: Impedances of different electrolytic capacitors, all 250 V rated, applied DC-bias 150 V, measured with Keysight E4990A and 16065A DC bias fixture

Due to their high capacitance density electrolytic capacitors (especially Aluminium-electrolytic capacitors, Al_2O_3) are often the first choice for many applications in power electronics. However, since these capacitors typically show a comparatively high ESR and low resonance frequencies their usage is at least questionable in applications with high switching frequencies and fast switching times. Measurements were taken from several different capacitors (all 250 V rated) within the range of $0.68\,\mu F$ to $33\,\mu F$. The results of the impedance measurements are shown in Fig. 6. Characteristic is the early (at comparatively low frequencies) beginning of the capacitors main resonances at frequencies between 1 kHz and 10 kHz. Compared to the known resonance of a simple capacitor model like the one used in II the impedances courses are very wide (referred to the frequency axis) and show an extensive flat region around 0°. Some of them even turn around to a negativ gradient before they get up towards 90°. In addition the impedances show no relevant voltage dependency, which was the expected behaviour. It can be noted that the wide course and the flat region of the impedance seems to be characteristic for electrolytic capacitors.

B Multi-Layer ceramic capacitors

Multi-layer ceramic capacitors (MLCC's) are often used in modern power electronics with higher switching frequencies since they show comparatively low ESR and ESL. The partially very high capacitance densities are another reason for their usage especially in volume critical applications.

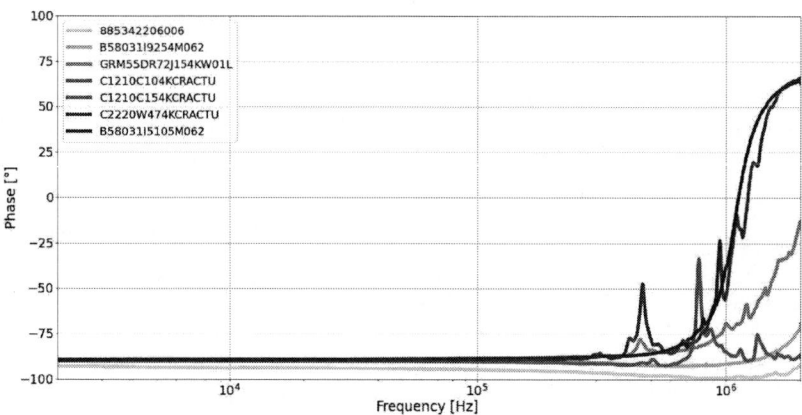

Fig. 7: Impedances of different ceramic capacitors, applied DC-bias 150 V, measured with Keysight E4990A and 16065A DC bias fixture

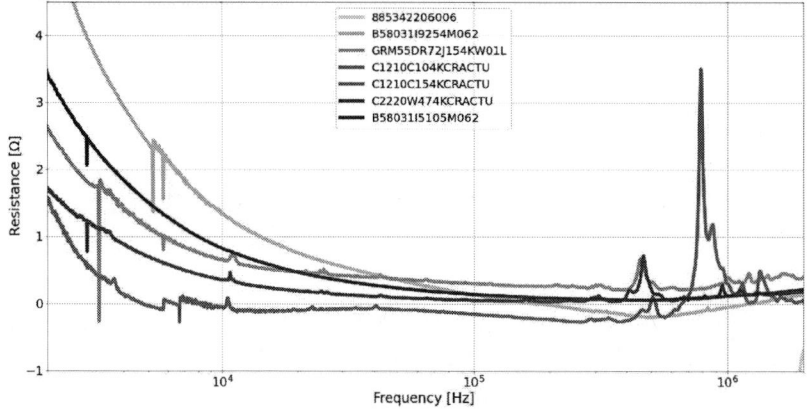

Fig. 8: ESR of different ceramic capacitors, applied DC-bias 150 V, measured with Keysight E4990A and 16065A DC bias fixture

Fig. 7 shows the impedances of different ceramic capacitors. Not all of them are, like the electrolytics, 250 V rated (e.g. the B58031I9254M062 and B58031I5105M062 which are due to their PZLT-ceramic not available with 250 V rating). The highest capacity of all MLC-capacitors was $2.2\,\mu F$, which is lower than those of the electrolytic ones. Nevertheless there are some $1.0\,\mu F$ capacitors in both measurements (compare ECA2EM010 in Fig. 6 and B58031I5105M062 in Fig. 7) which shows, that the MLC's main resonance frequency lies much higher than those of the electrolytic ones. Thinking of the simple capacitor model used in II the ESL of the ceramic capacitors has to be much smaller (as long as the capacities are equal).

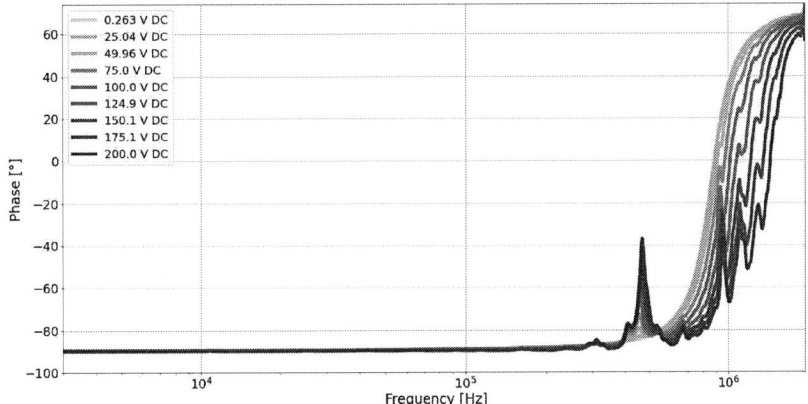

Fig. 9: Impedance of KEMET C2220W474KCRACTU 500 V $0.47\,\mu F$ X7R, measured with different DC-bias voltages using the Keysight E4990A and 16065A DC bias fixture

Unfortunately these type of capacitors also suffers from disadvantages like a high dependency of the capacity from the electric field strength and therefore the voltage. Fig. 9 shows this behaviour using the KEMET C2220W474KCRACTU X7R-capacitor as an example. The higher the DC-bias voltage the higher the main resonance frequency. In addition the course of the impedance gets rougher the higher the DC-bias gets, which is a sign of additional resonances that get stronger with the increase of the DC-bias. What strikes in this plot is a phase peak before the main frequency that also increases with the DC-bias.

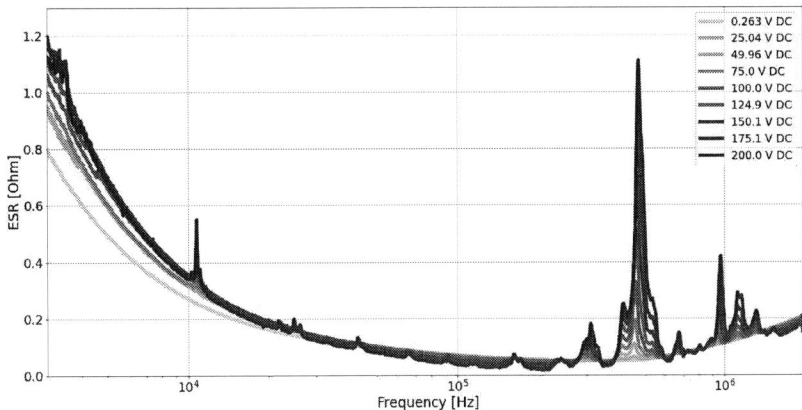

Fig. 10: ESR of KEMET C2220W474KCRACTU 500 V $0.47\,\mu F$ X7R, measured with different DC-bias voltages using the Keysight E4990A and 16065A DC bias fixture

This peak causes a temporary increase of the ESR, which can be seen in Fig. 10. This peak can cause a significant increase in losses when current harmonics match it's area. Even oscillations between capacitors due to a mismatch of their impedances in combination with those of the pcb are possible. Since all other ceramic capacitors showed this spikes it seems to be characteristic (see Fig. 8). Whether these behaviour of the capacitors is significant or not will be subject of future studies. One opportunity to evaluate all described effects is the DC-link evaluation platform described in the next section.

IV DC-link evaluation platform

The evaluation platform is based on a synchronous buck converter using EPC2034C® eGaN-FET's. The system covers a frequency range from 50 kHz to 1 MHz at a maximum $i_{load} = 15\,A$ and a maximum DC-link voltage of 200 V (see Fig. 11). The special feature is the easy interchangeability of the DC-link.

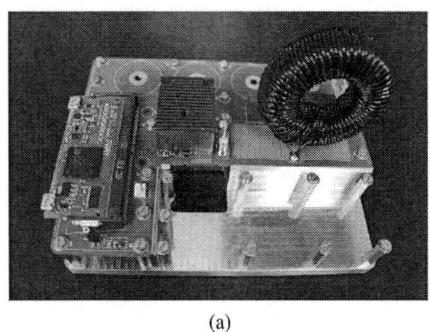
(a)

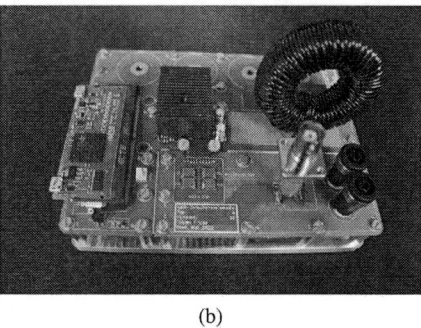
(b)

Fig. 11: (a) GaN-converter-module mounted on the milled baseplate without a DC-link-pcb connected, (b) module with a connected DC-link-pcb and a special copper clamp between the two pcb's

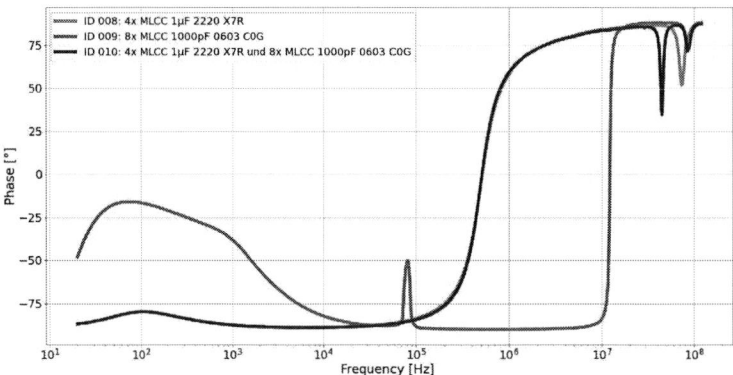

Fig. 12: DC-link impedance of evaluation platform using different ceramic capacitor combinations

Impedances of different capacitor combinations are shown in Fig. 12. The results are a combination of the capacitors impedances and those of the setup. While the bigger capacitors ($4 \times 1\,\mu F$) define the main resonance frequency the smaller ones add higher resonances, further correlations are being investigated.

V Conclusion

Section I described the increasing influence of parasitics in DC-links when using high switching frequencies and fast switching times. The emergence of oscillations and overshoots of the DC-link voltage was mentioned as well as the impact on harmonics of the supply currents and therefore the increase of EMI. To investigate this systematically, a series of sudies was announced whose inception is marked by the sections II, III and IV. This began with the description of simulations which investigated fundamental correlations in DC-links. They lead to observations regarding the impact of the DC-links main frequency and the ESR on voltage overshoots, harmonics and RMS-values (see section II C for details). This section clarified the impact of the capacitors impedance, which is why they were thematized in section III. It pointed out the reasons for the capacitors non-ideal behaviour before impedance measurements were analyzed. This showed some specific characteristics of different capacitor types as well as the impact of the DC-bias voltage. In addition, the emergence of ESR-spikes of ceramic capacitors was discussed. Finally section IV described a DC-link evaluation platform and some first results regarding the combination of different sized MLC-capacitors, switching tests to study oscillations will be carried out.

References

[1] Miller G.: New semiconductor technologies challenge package and system setups, 2010 6th International Conference on Integrated Power Electronics Systems, 2010

[2] Gerfer A.: Magnetics in the GaN/SiC Power Electronics World, 2019 31st International Symposium on Power Semiconductor Devices and ICs (ISPSD), 2019

[3] T. Schobre, R. G. Aríztegui and R. Mallwitz: Genetic Algorithm Based Multi Objective Optimization for Inductor Design, 2020 22nd European Conference on Power Electronics and Applications (EPE'20 ECCE Europe), 2020, pp. 1-9, DOI 10.23919/EPE20ECCEEurope43536.2020.9215902

[4] Fischer D., Rohn R., Mallwitz R.: Comparative Implementation of a two-stage DC-Link, PCIM Europe 2022, DOI 30420/565822268

[5] Schnack J., Brückner S., Süncksen H., Schümann U., Mallwitz R.: Analysis and Optimization of Electrolytic Capacitor Technology for High-Frequency Integrated Inverter, IEEE Transactions on Components, Packaging and Manufacturing Technology, Vol. 11, No. 6, June 2021

[6] Schnack J., Golev V., Schümann U., Mallwitz R.: An Investigation in DC-link Film Capacitors for reduced Parasitic Inductances, PCIM Europe 2019, ISBN 978-3-8007-4938-6

[7] Schnack J., Golev V., Gördes J.P., Schümann U., Mallwitz R., Stahl S., Süncksen H.: Low-Inductance DC-link Design dedicated to SiC-based Highly Integrated Inverters, CIPS 2020

[8] Neemann H., Schobre T., Mallwitz R., Obernolte U.: Impact of Wide-Bandgap Semiconductors on DC-Link Considerations in Servo-Drive Applications, PCIM Europe digital days 2020, ISBN 978-3-8007-5245-4

[9] Schwarzbeck Messelektronik OHG: Single path vehicle AMN (LISN) NNBM 8124 (Datasheet), https://www.schwarzbeck.de

[10] Stiny L.: Passive Elektronische Bauelemente: Aufbau, Funktion, Eigenschaften, Dimensionierung und Anwendung, ISBN 978-3-658-24732-4

[11] Myoung-Gyun K., Byoung L., Tae-Yeoul Y.: Equivalent-Circuit Model for High-Capacitance MLCC Based on Transmission-Line Theory, DOI 10.1109/TCPMT.2011.2170990

[12] Meisser M., Haehre K., Kling R.: Impedance Characterization of High Frequency Power Electronic Circuits, DOI 10.1049/cp.2012.0240

[13] Tiggelman M. P. J., Reimann K., Schmitz J.: Reducing AC impedance measurement errors caused by the DC voltage dependence of broadband high-voltage bias-tees, DOI 10.1109/ICMTS.2007.374483

EMI Mitigation Induced by An IGBT Driver Based on A Controlled Gate Current Profile

Daniel S. Martinez-Padron[1], Nicolas Patin[1], Eric Monmasson[2]

[1] Université de technologie de Compiègne,
Roberval (Mechanics, energy and electricity)
Centre de recherche Royallieu CS 60319-60203
Compiègne Cedex - France

[2] SATIE laboratory, CY Cergy Paris Université
Cergy-Pontoise, France
Phone: +33 (0)-3 234727
Email: dmartine@utc.fr; nicolas.patin@utc.fr; eric.monmasson@cyu.fr

Acknowledgments

The authors would like to thank The National Council of Science and Technology of Mexico (CONA-CyT) for the academic scholarship under the grant number 705759.

Keywords

≪Gate driver≫, ≪IGBT≫, ≪Electromagnetic interference (EMI)≫, ≪Switching losses≫, ≪EMI/EMC≫.

Abstract

The transistors used in power electronics applications are source of electromagnetic interference(EMI) during switching process. In this work a gate current profile to reduce conducted EMI is proposed. It is based on the gate charge curve and it allows to control the conducted EMI generation with two degrees of freedom. In order to evaluate the performance of proposed method, it is compared with a validated method in the literature, namely, CATS method. It is shown that, for the same level of power switching losses, the conducted EMI generation is less with the proposed method.

1 Introduction

In last decades, power transistors such as field-effect transistor (MOSFET) and insulated gate bipolar transistor (IGBT) have been widely used in industrial and consumer electronics applications. They are suitable devices due to their capability to handle high currents at high switching frequencies. However, they are source of electromagnetic interference(EMI) during switching since the high levels of dI/dt and dV/dt are sources of radiated and conducted EMI, which affects normal operation of surrounding equipment [1]. The EMI generation and power losses are strongly related, since the switching losses occur during the overlap of non-zero voltage and current. Low switching losses implies short switching time but at the same time it increases the EMI levels. For this reason, a suitable transistor driver has to guarantee an adequate trade-off between power losses and EMI. In order to address this problem, several driving methods have been proposed in the literature. For instance, control the voltage fall injecting gate current during turn-on transistor and during turn-off, changing the gate resistance value to satisfy the EMI requirements is proposed in [3] and [2] to obtain an acceptable trade-off between switching speed and EMI. This driving method is improved in [4] in turn-off process to avoid spurious turn-on on IGBT by applying a negative bias. A driving method based on a closed-loop approach is presented in [6][7]. In [7], a closed loop gate driver is reported that feedbacks the voltage, and as consequence its slope measure, and controls the voltage and current transients by the gate current. In the same way, in [6] a closed loop IGBT driver is proposed. It feedbacks the voltage and current transient to shape them into smoothest waveforms with a PI controller. Driving method called CATS is proposed in [10] to reduce the EMI generation smoothing dV/dt and dI/dt waveform applying an intermediate gate voltage level. (In french: *Commande Autour de la Tension de Seuil*, that could be translated by "Control around the Threshold Voltage"). In this sense, an active voltage control is reported in [12] to shape the dV/dt of an IGBT. The EMI reduction is achieved with an "S"-shaped waveform that has smooth corners. It is reported that the high order time derivatives of dV/dt and dI/dt waveform can reduce the EMI generation [13] and it is theoretically proved in [14] that a Gaussian shape switching pattern ensure the optimal trade-off between power losses and EMI generation. Most of the proposed methods are based on gate charge approach reported in[19]. However they don't consider the non-linear behaviour during the Miller's Plateau 'knee', which occurs during the interval between the end of the dI/dt and the beginning of dV/dt. Recently, in [15, 16, 17, 18] drivers based on gate current with a fine resolution haven been proposed. They allow to shape the transient waveform injecting steps of current during gate charge. However, there is not a systematic method to choose the level and the duration of the steps. For these reasons, a driving method based on a gate current profile is proposed. It consists on applying a gate current profile formed by steps that is based on the gate charge curve [19] and it allows to shape the voltage and its derivative to reduce the conducted EMI. In order to evaluate its performance, the proposed method is compared with a previously evoked methodology, namely, CATS because it is an effective strategy validated in practice that can serve as reference. Additionally, a benchmark which considers the switching power losses and the frequency response as two performance criteria is also proposed. The article is organized as follows: Section 2 describes the proposed methodology and the proposed gate profile. In section 3, the benchmark and the comparison between CATS and the proposed method are described. Finally in Section 4 conclusions are drawn.

2   Proposed methodology

In power electronics, the high levels of dV/dt and dI/dt during switching are sources of conducted and radiated EMI. Slowing down the switching duration can reduce the EMI generation but it tends to increase the power losses. As mentioned previously, smoothing the transients waveform can improve the trade-off between EMI and switching losses. For insulated gate transistors, such as MOSFET and IGBT, switching transient waveforms are strongly related to the charge supplied to gate input capacitance and the switching behaviour can obtained from gate charge curve. In order to shape the voltage/current transient waveform and reduce the EMI generation, a driving method based on a gate current profile is proposed. The method consists of injecting different current levels during gate charging. In order to describe the proposed method, the transistor gate charge is detailed in the following paragraph.

Transistor gate charge

The gate charging behaviour of insulated gate power transistors, as IGBT and MOSFET, determines the shape and duration of dV/dt and dI/dt and consequently the EMI generation. This behaviour is caused by transistor internal structure, the internal capacitances, the wiring resistance and inductance, the gate resistor, the collector voltage and current, load current, among others. The internal capacitances are the main elements involved in this process and the complexity of this mechanism is due to their non linear behaviour. The gate charging dynamics can be simplified considering that it is conformed by different values during gate charging shown in Fig. 1. The V_{ge} vs charge curve is obtained by injecting a constant current into the gate and it can be split in several stages. In Fig. 2 (a) a typical test circuit with an IGBT and their internal capacitances is presented. Injecting a constant current in the gate it is possible to obtain the dynamic curve shown in Fig. 2 (b) which presents the turn-on behaviour by gate charge. This dynamic curve can be divided in 5 stages, influenced by the different gate charge values, that are described below

- **Stage 1.** Before the gate voltage V_g reaches the threshold voltage V_{th} the collector current I_c is strongly limited (only few μA). The slope of V_g is fixed by the gate-emitter capacitance C_{ge}.

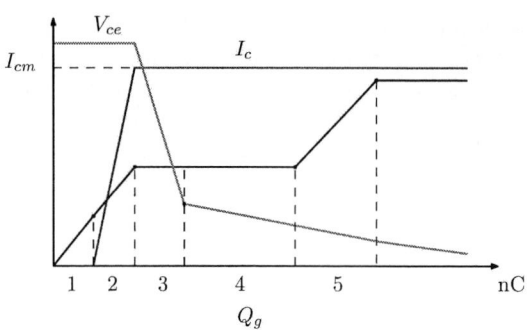

Fig. 1: Gate charge curve.

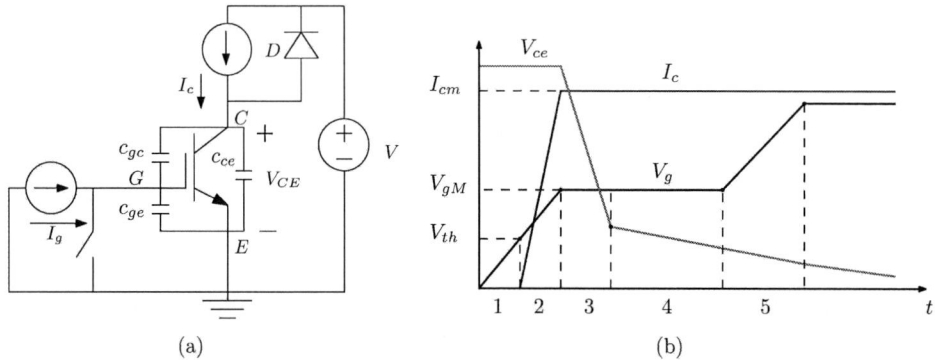

(a)   (b)

Fig. 2: (a) Test circuit and (b) dynamic charge during a turn-on transient.

- **Stage 2.** Once V_{gth} is reached, the device is turned-on and the collector current flows reaching its maximum value I_{cm}, while the collector voltage V_{ce} remains unchanged at its maximum value. V_g reaches the Miller's plateau voltage V_{gM}.
- **Stage 3 and 4.** During this stage the Miller effects starts, the input appears to be infinite since the V_g remains equal to V_{gM} even though the gate circuit is supplying current to the gate. V_{ce} starts dropping with different slopes determined by the charges of Miller's capacitance [19]. During the Miller's plateau, remains due the non linear gate-collector capacitance C_{gc} (Miller's capacitance).
- **Stage 5.** As soon as the Miller's effect ends, V_g starts to increase again towards its final value.

Proposed gate current profile

As presented before, the switching transient behaviour of insulated gate transistors is determined by the gate charge supplied. The total charge is the same during switching and it can be controlled by the gate current amplitude and duration. These parameters allow to control the duration and waveform of the transient, and at the same time the EMI generation and power switching losses. In this sense, a driving method to reduce the EMI generation based on the gate charge curve is proposed in this article. The proposed method consists of injecting a current step profile (CSP) to control the voltage transient switching and, consequently, to reduce the conducted EMI. The proposed CSP is shown in Fig. 3 and it is formed by 4 steps I_j with $1 \leq j \leq 4$. Notice that this kind of profile can be implemented using the drivers presented in [15, 16, 17, 18]. Each step of CSP allows to control the behaviour of switching transient by their amplitude and duration. In this article the duration is fixed and the amplitude step is calculated with the correspondent equations of each stages from the dynamic curve presented in Fig. 2. The procedure to calculate the amplitude of each step of CSP is described as follows.

- **Step 1**. This step is a mixture of stage 1 and 2 in gate charge curve. During stage 1, the V_{th} is reached. The gate current is determined by the charge for C_{ge} then

$$I_{g1} = C_{ge} \frac{dV_{ge}}{dt} \approx C_{ge} \frac{V_{th}}{\Delta t_{s1}}, \tag{1}$$

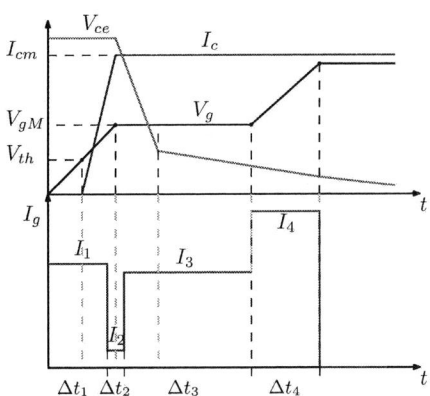

Fig. 3: Current step profile proposed.

where I_{g1} is the gate current for Stage 1, Δt_{s1} is the time that V_{ge} takes to reach V_{gth}. During stage 2, I_c increases until I_{cm} during Δt_{s2} and C_{gc} is charged by the current

$$I = C_{gc} \frac{dV_{gc}}{dt}. \tag{2}$$

Considering that $V_{ge} = V_{ge} - V_{ce}$, V_{ce} remains constant and the transconductance of the IGBT is $g_m = \Delta I_c / \Delta V_{ge}$, the gate current can be calculated by

$$I_{g2} = \frac{C_{ies}}{g_m} \frac{dI_c}{dt} \approx \frac{C_{ies}}{g_m} \frac{\Delta I_c}{\Delta t_{s2}}, \tag{3}$$

where I_{g2} is the gate current for Stage 2. Taking $I_1 = I_{g1} = I_{g2}$, then step 1 duration can be calculated from equations (1) and (6) and $\Delta t_1 = \Delta t_{s1} + \Delta t_{s2}$.
- **Step 2**. This step considers the end of stage 2 and the beginning of stage 3, during V_{ce} and I_c intersection and the 'knee' of Miller's Plateau [19]. Due to the non linear behaviour of the input capacitance [19] and the effect of the diode recovery, a low step I_2 amplitude is proposed.
- **Step 3**. During this step, the dV/dt is controlled. From stage 3, the gate current can be estimated by

$$I_3 = C_{gc} \frac{dV_{ce}}{dt} \approx C_{gc} \frac{\Delta V_{ce}}{\Delta t_3}. \tag{4}$$

Then, fixing Δt_3, which corresponds to the step I_3 duration, the dV/dt can be controlled by step I_3 amplitude. The duration of step current I_3 is selected by trial and error test.
- **Step 4**. In order to finish the gate charge, the amplitude of the step I_4 is calculated using (1) considering the equivalent capacitance during this time.

Proposed method evaluation

In order to verify the effectiveness of CSP, it is evaluated in the test circuit shown in Fig. 4. The evaluation is performed by simulation using Ltspice software and the IGBT IKW40N65ET7 SPICE model [21]. The evaluation consists on injecting the CSP and varying the amplitude of I_3, which shape the voltage waveform, to evaluate the impact of amplitude value on conducted EMI generation and power losses. The I_3 amplitude values chosen in this evaluation are: $I_3 = 10mA$, $I_3 = 20mA$, $I_3 = 30mA$, $I_3 = 40mA$ and $I_3 = 50mA$. The simulations are performed with a minimal step simulation of 100 ps and the parameters in Tab. I. The selection of the steps of the proposed profile is described as follows.
- **Step 1**. First, considering the value of V_{th} and C_{ies} shown in Tab.Iand proposing $t_{s1} = 1\mu s$, the

Table I: Simulation Parameters

Parameter	Value
Input capacitance (C_{ies})	2.475 nF
Reverse transfer capacitance (C_{res})	25pF
Switching period (T)	50ns
Transconductance (g_m)	10.25 S
Threshold voltage (V_{th})	5.8 V
Collector voltage (V)	130 V
Collector current (I_c)	40 A

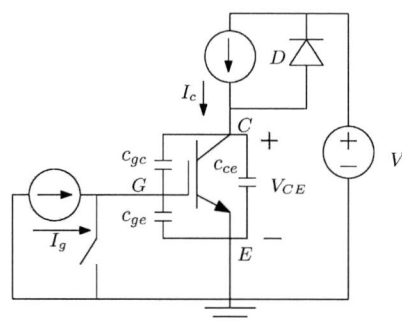

Fig. 4: Test circuit for the CSP evaluation.

current gate to reach the V_{th} can be calculated by

$$I_1 = C_{ies}\frac{V_{th}}{\Delta t_{s1}} = 14.32mA. \tag{5}$$

For the second part, the maximum collector current I_{cm}=40 A. In order to slowdown the current slope during Miller plateau, it is proposed reach 80% of I_{cm} during this step. Since $I_1 = 14.32mA$, the duration of this part is given by

$$\Delta t_{s2} = \frac{C_{ies}}{g_m}\frac{\Delta I_c}{I_1} = 538.49ns. \tag{6}$$

In this sense, the step 1 amplitude is $I_1 = 14.32mA$ with a time duration of $\Delta t_1 = 1.538\mu s$.

- **Step 2**. For this step, in order to finish the current slope and start the 'knee' of Miller's Plateau, a low step I_2 amplitude is proposed. In this case $I_2 = 1.5mA$ is selected and the value of C_{ies} estimated is $250pF$, then the time duration is calculated by

$$\Delta t_2 = \frac{C_{ies}}{g_m}\frac{\Delta I_c}{I_2} = 200ns. \tag{7}$$

The step 2 amplitude is $I_2 = 1.5mA$ with a time duration of $\Delta t_2 = 200ns$.

- **Step 3**. For this step, five different values are selected: $I_3 = 10mA$, $I_3 = 20mA$, $I_3 = 30mA$, $I_3 = 40mA$ and $I_3 = 50mA$. The time duration is set to $\Delta t_3 = 500ns$ for all cases.
- **Step 4**. In order to finish the gate charge, the time duration $\Delta t_4 = 500ns$ is proposed, the capacitance estimated from datasheet gate charge curve is $C = 18nF$ and the amplitude of the step 4 is calculated by

$$I_4 = C\frac{\Delta V}{\Delta t_4} = 0.204A. \tag{8}$$

Then, the step 4 amplitude is $I_4 = 0.204mA$ with a time duration of $\Delta t_4 = 500ns$.

The obtained results, shown in Fig. 5 and Fig. 6, present the V_{ce} waveform during falling and rising edge at different amplitude of I_3 respectively. Additionally, the frequency spectra of each test are presented in Fig. 7 that shows their EMI generation can be reduced by reducing the amplitude of I_3 that can be used as degree of freedom to improve the conducted EMI requirements. However, as shown in Tab.II the power losses increase when the EMI generation is reduced.

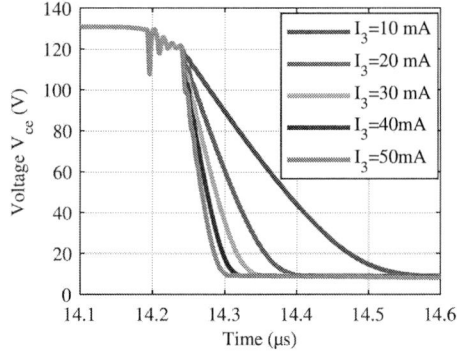

Fig. 5: Simulation results for falling edge.　　Fig. 6: Simulation results for rising edge.

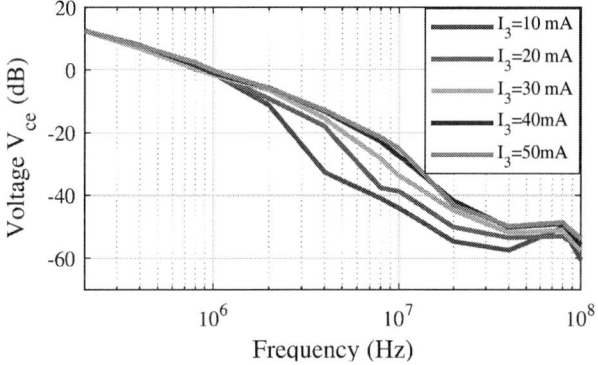

Fig. 7: Frequency spectra at different levels of I_3.

Table II: Power losses for different level of I_3

I_3 amplitude	Power losses P_{sw}
10 mA	114.37 W
20 mA	103.14 W
30 mA	99.10 W
40 mA	96.87 W
50 mA	95.37 W

3 Comparison

The performance of the CSP is evaluated by a comparison with an effective method reported in the literature. For this purpose, CATS methodology is chosen as a reference and it is described as follows.

CATS methodology

CATS methodology is proposed in [10] and it consists in reducing the dI/dt and dV/dt by introducing an intermediate gate voltage level close to the threshold voltage V_{th}. During turn-on, an intermediate voltage $V_{th} < V_{int} < V_g$ is injected during T_{int} unlike classical driving as shown in Fig. 8(a). During turn-

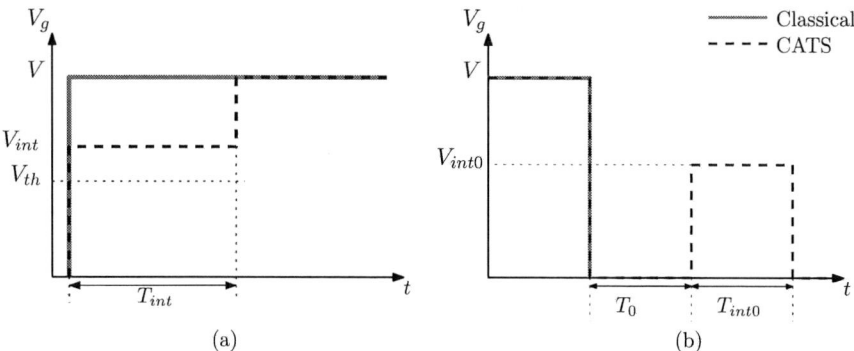

Fig. 8: Operating principle of CATS: (a) Turn-on and (b)Turn-off.

off, a zero or negative voltage is forced to evacuate the excess of charges, which maintain the transistor in conduction during a time interval T_o. Then a second level $V_{int0} < V_{int}$ is injected during T_{int0} [11] as shown in Fig. 8(b).

Benchmark

In order to perform an adequate comparison between the CSP and CATS methods, an appropriate benchmark has to be established. Since, the relationship between the power losses and EMI is an important issue, a benchmark test is proposed with the following considerations:

- The analysis frequency bandwidth considered is 200 kHz to 20 MHz because it is the typical range of propagation of conducted EMI [20].
- The switching power losses are fixed to the same value for both strategies and they are calculated as reported in the transistor datasheet [21] as:

   1. For turn-on, the switching power losses are measured during the time interval between 10 % of V_{ge} and 2% of V_{ce}.

   2. For turn-off, the switching power losses are measured during the time interval between 90 % of V_{ge} and 2% of I_c.

- The switching duration is proposed to achieve V_{ge}=15V and ensure the IGBT saturation.

Results

The comparison is performed using the proposed benchmark and the test circuit shown in Fig. 9 with the IGBT IKW40N65ET7 [21]. The transistor parameters are

- C_{ies}=2.475 nF,
- C_{res} = 25pF,
- g_m= 10.25 S,

and the following parameters are taken into account

- V=130 V,
- f_{sw}=20 kHz,
- I_c=40 A.

The comparison consists of determine the EMI generation performance of CSP and CATS methodology at the same level of switching power losses. Three different levels of power losses are chosen: $P_{sw} = 90W$, $P_{sw} = 93W$ and $P_{sw} = 99W$. The comparison is performed with Ltspice software considering a maximum simulation step of 100 ps. For CATS methodology the parameters T_{int}, V_{int0}, T_0 are fixed and V_{int} is chosen as degree of freedom with the following values

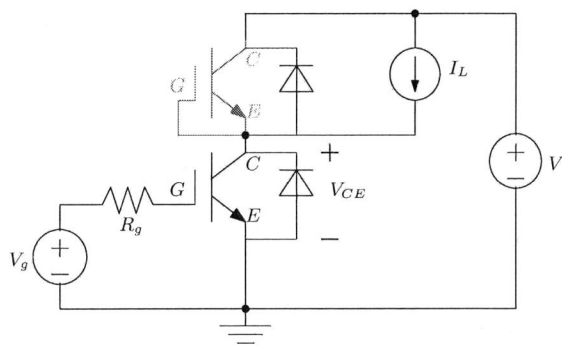

Fig. 9: Test circuit for the comparison.

- $T_{int} = 2\mu$ ns,
- $V_{int0} = 625$ns,
- $T_0 = 1.875\mu s$,
- $V_{int} = 7.5V$ for $P_{sw} = 90W$,
- $V_{int} = 7.53V$ for $P_{sw} = 93W$,
- $V_{int} = 7.605V$ for $P_{sw} = 99W$.

For the CSP, the source V_g in Fig. 9 is replaced by a current source $I_g(t)$ and the step I_3 becomes the degree of freedom to vary the switching power losses. The CSP is formed by

- $I_1 = 14.32$ mA and $\Delta t_1 = 1.538\mu s$,
- $I_2 = 1.5$ mA and $\Delta t_2 = 200ns$,
- $I_3 = 60$ mA and $\Delta t_3 = 150ns$ for $P_{sw} = 90W$,
- $I_3 = 40$ mA and $\Delta t_3 = 76ns$ for $P_{sw} = 93W$,
- $I_3 = 20$ mA and $\Delta t_3 = 50.85ns$ for $P_{sw} = 99W$,
- $I_4 = 0.204$ mA and $\Delta t_4 = 650ns$.

For both methods the gate voltage is V_{ge}=15V which implies a gate charge of $Q_G = 228$ nC. The comparison results are presented in Fig. 10, Fig. 11 and Fig. 12. They show the frequency spectra envelope of CSP and CATS for the different level of switching power losses. The CSP has the lowest response after 10 MHz for the three cases. The average difference between the two methodologies in the frequency band of 10 MHz to 20 MHz is 7.51 dBV for $P_{sw} = 90W$, 12.02 dBV for $P_{sw} = 93W$ and 23.32 dBV for $P_{sw} = 99W$. It implies an interesting reduction of conducted EMI at the same level of power losses.

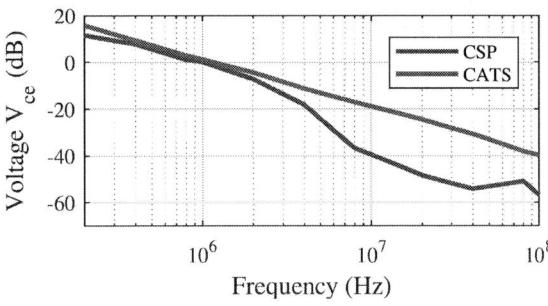

Fig. 10: Result of the comparison at $P_{sw} = 99W$.

4 Conclusion

In this paper, an IGBT driving technique based on a gate current profile is presented. Its aim is the mitigation of conducted EMI under switching losses constraints. the proposed CSP allows us to control the dV_{ce}/dt by a step current amplitude and its time duration. The CSP performance is evaluated by

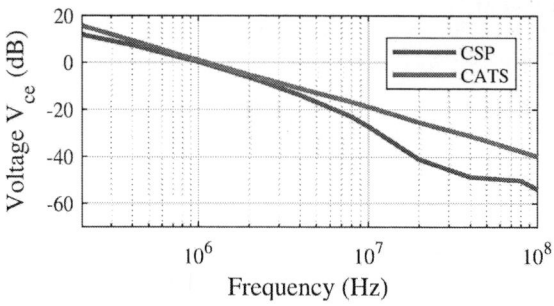

Fig. 11: Result of the comparison at $P_{sw} = 93W$.

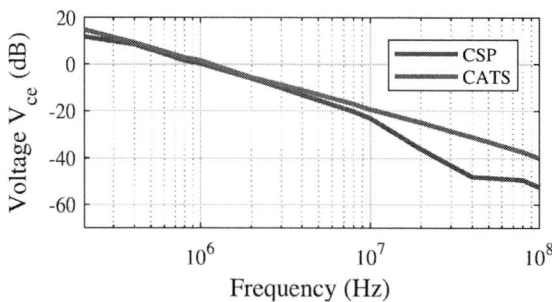

Fig. 12: Result of the comparison at $P_{sw} = 90W$.

simulation with different levels of I_3 that shows the feasibility of the method of reducing the generation of EMI. This method is compared with a reference control scheme (CATS) and simulation results show that for the same switching losses, in three different cases, the CSP exhibits the lowest level in the frequency domain which implies that is generates less conducted EMI. A disadvantage of proposed method with respect to methods such as CATS, is that it is necessary to know some transistor parameters which made more complicated its implementation. An advantage is that the CSP is a systematic method that allows control the conducted EMI generation with two different degrees of freedom. This leads to the possibility of split the step 3 into several levels to shape voltage transient to get a more sophisticated waveform, for instance, Gaussian waveform to ensure the optimal trade-off between EMI and power losses.

References

[1] F. Costa and D. Magnon.: Graphical analysis of the spectra of EMI sources in power electronics, in IEEE Transactions on Power Electronics, vol. 20, no. 6, pp. 1491-1498.

[2] A. Galluzzo, M. Melito, G. Belverde, S. Musumeci, A. Raciti and A. Testa.: Switching characteristic improvement of modern gate controlled devices, 1993 Fifth European Conference on Power Electronics and Applications, vol.2, pp. 374-379, 1993.

[3] A. Consoli, S. Musumeci, G. Oriti and A. Testa.: An innovative EMI reduction design technique in power converters, in IEEE Transactions on Electromagnetic Compatibility, vol. 38, no. 4, pp. 567-575, Nov. 1996.

[4] S. Musumeci, A. Raciti, A. Testa, A. Galluzzo and M. Melito.: Switching-behavior improvement of insulated gate-controlled devices, in IEEE Transactions on Power Electronics, vol. 12, no. 4, pp. 645-653, July 1997.

[5] Shihong Park and T. M. Jahns.: Flexible dv/dt and di/dt control method for insulated gate power switches, in IEEE Transactions on Industry Applications, vol. 39, no. 3, pp. 657-664, May-June 2003.

[6] Y. Lobsiger and J. W. Kolar.: Closed-Loop $di/d\,t$ and dv/dt IGBT Gate Driver, in IEEE Transactions on Power Electronics, vol. 30, no. 6, pp. 3402-3417, June 2015.

[7] H. Riazmontazer, A. Rahnamaee, A. Mojab, S. Mehrnami, S. K. Mazumder and M. Zefran.: Closed-loop control of switching transition of SiC MOSFETs, 2015 IEEE Applied Power Electronics Conference and Exposition (APEC), 2015, pp. 782-788.

[8] P. R. Palmer and H. S. Rajamani.: Active Voltage control of IGBTs for high power applications, in IEEE Transactions on Power Electronics, vol. 19, no. 4, pp. 894-901, July 2004.

[9] I. Baraia, J. A. Barrena, G. Abad, J. M. Canales Segade and U. Iraola.: An Experimentally Verified Active Gate Control Method for the Series Connection of IGBT/Diodes, in IEEE Transactions on Power Electronics, vol. 27, no. 2, pp. 1025-1038, Feb. 2012.

[10] H. Sawezyn.: Etude de la commande autour de la tension de seuil (CATS) des transistors de puissance à grille isolée et de ses applications, Université des Sciences et technologies de Lille, PhD Thesis, 2003.

[11] N. Idir, R. Bausière, J.J. Franchaud, H. Sawezyn.: Contrôle des commutations des transistors à grille isolée : commande CATS, Revue internationale de génie électrique 2004,49-74.

[12] X. Yang, Y. Yuan, X. Zhang and P. R. Palmer.: Shaping High-Power IGBT Switching Transitions by Active Voltage Control for Reduced EMI Generation, in IEEE Transactions on Industry Applications, vol. 51, no. 2, pp. 1669-1677.

[13] N. Oswald, B. H. Stark, D. Holliday, C. Hargis and B. Drury.: Analysis of Shaped Pulse Transitions in Power Electronic Switching Waveforms for Reduced EMI Generation, in IEEE Transactions on Industry Applications, vol. 47, no. 5, pp. 2154-2165, Sept.-Oct. 2011.

[14] N. Patin and M. L. Viñals.: Toward an optimal Heisenberg's closed-loop gate drive for Power MOSFETs, IECON 2012 - 38th Annual Conference on IEEE Industrial Electronics Society, 2012, pp. 828-833.

[15] S. Fukunaga, H. Takayama and T. Hikihara.: A Study on Switching Surge Voltage Suppression of SiC MOSFET by Digital Active Gate Drive, 2021 IEEE 12th Energy Conversion Congress and Exposition - Asia (ECCE-Asia), 2021, pp. 1325-1330.

[16] H. Takayama, T. Okuda and T. Hikihara.: A Study on Suppressing Surge Voltage of SiC MOSFET Using Digital Active Gate Driver, 2020 IEEE Workshop on Wide Bandgap Power Devices and Applications in Asia (WiPDA Asia), 2020, pp. 1-5.

[17] R. Morikawa, T. Sai, K. Hata and M. Takamiya.: New Gate Driving Technique Using Digital Gate Driver IC to Reduce Both EMI in Specific Frequency Band and Switching Loss in IGBTs, 2020 IEEE 9th International Power Electronics and Motion Control Conference (IPEMC2020-ECCE Asia), 2020, pp. 644-651.

[18] K. Miyazaki et al.: General-purpose clocked gate driver (CGD) IC with programmable 63-level drivability to reduce Ic overshoot and switching loss of various power transistors, 2016 IEEE Applied Power Electronics Conference and Exposition (APEC), 2016.

[19] F. Portuese and M. Melito.: Gate charge leads to easy drive design for Power Mosfet circuits, PCIM '90 Proceedings of the 18th International Intelligent Motion Conference, 1990, pp. 237-243.

[20] C. R. Paul.: Introduction to Electromagnetic Compatibility, 2nd ed. Hoboken, NJ: Wiley, 2006.

[21] Infineon.: Low Loss Duopack: IGBT with Trench and Fieldstop technology IKW40N65ET7, TRENCH-STOP™, 2020,Datasheet version 2.2.

An Accurate and Fast Model of Three-Level Three-Phase Dual-Active Bridge Converters in Real-Time Simulation

Ming Jia, Philipp Joebges, Rik W. De Doncker
Institute for Power Generation and Storage Systems
E.ON Energy Research Center
RWTH Aachen University
Mathieustrasse 10, 52074
Aachen, Germany
Email: post_pgs@eonerc.rwth-aachen.de

Acknowledgments

This work is supported by the European Regional Development Fund (EFRE-0500029).

Keywords

≪Dual Active Bridge (DAB)≫, ≪Modelling≫, ≪Real-time simulation≫, ≪Stability analysis≫

Abstract

This paper proposes a C-code average model for an efficient real-time simulation of a three-level three-phase dual-active bridge (DAB) dc-dc converter. To evaluate the model and the control algorithms, the simulations and control hardware-in-the-loop emulators are used, for example, testing a 5000 V dc-link DAB converter with up to 30-degree phase shift between the primary and secondary sides of a transformer. From the results of the hardware-in-the-loop test, it can be concluded that the model proposed in this work is numerically stable and is executed considerably faster than the state-of-the-art proposed auto-generated models.

Introduction

Reliable power electronic systems with dynamic behavior combined with shorter development cycles are increasingly demanded. This necessitates the development of a flexible evaluation tool for the control algorithm. Hence, a platform is needed for its execution, particularly to access its behavior under fault situations. In control-hardware-in-the-loop (CHIL) scenarios, a physical controller is connected to a virtual plant in a real-time (RT) simulator instead of a physical plant [1]. Performing offline simulations requires results as accurate as possible. A variable-step solver is often implemented, so that the step size is decreased when the states of the model change rapidly, whereas the step size increases when the model changes slowly. Ultimately, the accuracy of the simulation results is determined solely by the computational power and the mathematical model of the system. In a real-time simulation environment, however, discrete execution typically has a fixed step size. The accuracy of the results and the time required to simulate the system decreases as the step size increases. Consequently, the accuracy of the computations is also determined by the step time used to produce the results.

Average models are already implemented as basic circuit units in the software PLECS. However, the generated code is extensive and complex, and the execution time of this model in the RT simulator is not optimized. Furthermore, these average models are not investigated for dc-dc converters. [2] proposed a self-developed C-code model for a two-level converter. It considered a half-bridge consisting of two

Insulated Gate Bipolar Transistor (IGBT) modules as the modeling unit. The results demonstrated that the C-code model is more efficient and has a similar level of accuracy as the auto-generated models. In this work, a self-developed C-code-based average model for a three-level three-phase dual-active bridge dc-dc converter (3L-DAB3) is further investigated. The model uses neutral-point clamped half-bridge as the modeling unit to simulate the converter. The following sections focus on the calculation of these switching signals.

Another improvement in this work is that both the simplified transformer model and the three single-phase models are implemented. In [2], only the simplified transformer model is inverstigated. Incorporating a three single-phase transformer into the model significantly enhances the precision of the proposed model. Both models are compared with the corresponding PLECS model to validate the accuracy and operating speed in the RT simulator. The implementation of self-defined C-code model reduces the size of the model description and expedites the calculation and execution of the RT simulation. The efficiency of the RT simulation is improved, and the numerical stability is investigated.

C-Code Average Model of a 3L-DAB3

Modeling Methodology

The PLECS average model of a half-bridge consists of controlled current and voltage sources and a pair of diodes. It is able to process switching gate signals in decimal format. Fig. 1 exemplifies a commutation from the bottom switch to the top switch of one phase-leg. The red lines S_{top} and S_{bot} indicate the actual upper and lower gate signals of the switch model. In a fixed time-step simulation, it is impossible to capture the precise moment at which the current changes direction during a sampling period. The current values are updated only at the beginning of each time step T. The blue lines with shaded areas $\overline{S}_{top}$ and $\overline{S}_{bot}$ show the averaged gate signals over a time step with dead time.

The RT simulator AixControl XRS7070 consists of a Field Programmable Gate Arrays (FPGA) and a Digital Signal Processor (DSP), thus combining the speed of sampling the switching signals in an FPGA with the flexibility and computational capability of a DSP. Fig. 2 presents the whole PWM processing. The FPGA receives the PWM signals from the external control module simultaneously, which is either 0 or 1. The FPGA samples the continuous PWM signals every 20 ns and then passes the average value over a 10 μs time step to the DSP for calculation. After that the FPGA samples the signals, the PWM signals are real numbers within the range [0,1] instead of 0 and 1.

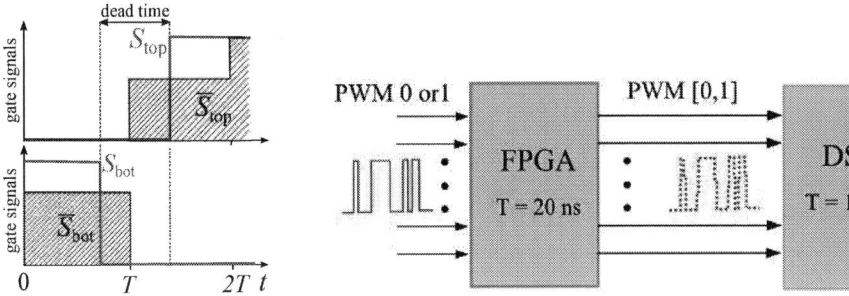

Fig. 1: A switching state of one phase-leg

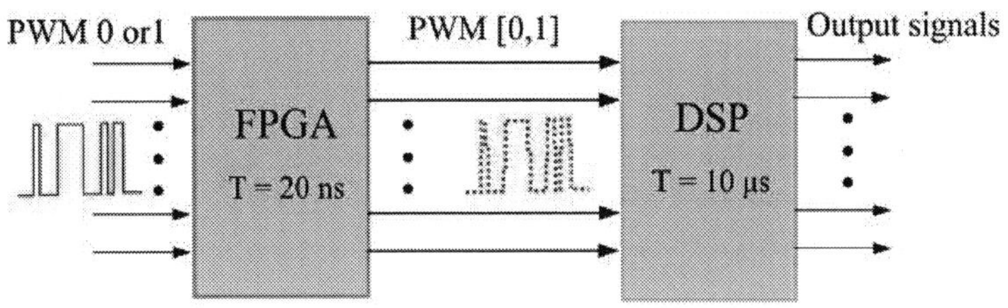

Fig. 2: PWM signals in an RT simulator

Modeling for 3L-DAB3

A 3L-DAB3 consists of two three-level neutral-point-clamped (NPC) converters with a transformer as the ac link [3][4]. A mathematical model for the 3L-DAB3 is derived from the half-bridge model introduced in [5]. The parameter k_{xp} is introduced to represent the direction of the current in the average model. It is either one for a positive ac currents or zero for a negative ac current. Fig. 4 depicts the calculation procedure of the C-code average model. The parameter $x \in \{a,b,c\}$ represents the phase-leg a, b and *c.* The calculation begins with the capacitor voltages and the phase currents. With the knowledge of gate signals and phase current directions, the phase voltages and arm currents can be obtained. The capacitor voltages and phase currents in the subsequent time step can be determined by integrating the phase voltages and arm currents.

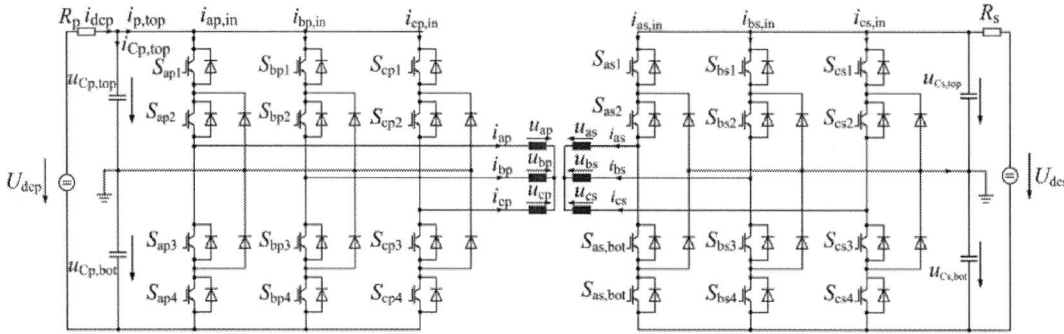

Fig. 3: Circuit diagram of a 3L-DAB3 converter

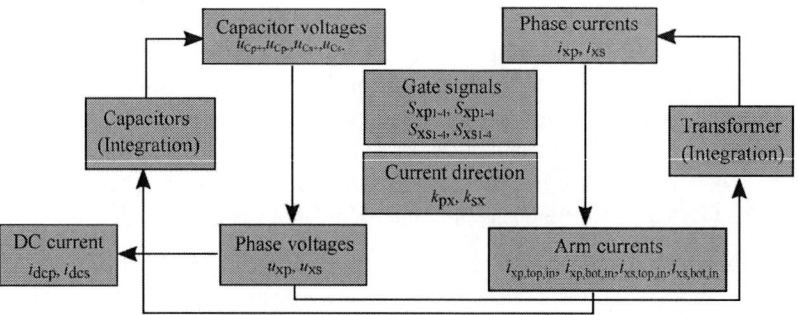

Fig. 4: Flowchart of the C-code realization for a 3L-DAB3 converter

The ac voltages at the primary side can be expressed as

$$
\begin{aligned}
u_{xp} =& [\min(\overline{S}_{xp1}, \overline{S}_{xp2}) \cdot u_{Cp,top} - (1 - \overline{S}_{xp2}) \cdot u_{Cp,bot}] \cdot k_{xp} \\
& - [(1 - \overline{S}_{xp3}) \cdot u_{Cp,top} - \min(\overline{S}_{xp3}, \overline{S}_{xp4}) \cdot u_{Cp,bot}] \cdot (1 - k_{xp}) \\
=& \, \overline{S}_{uxp,top} \cdot u_{Cp,top} + \overline{S}_{uxp,bot} \cdot u_{Cp,bot}.
\end{aligned}
\tag{1}
$$

In the voltage calculation, the parameters $\overline{S}_{uxp,top}$ and $\overline{S}_{uxp,bot}$ represent the averaged switching signals for the top and bottom switches, respectively, in one phase-leg of the primary side. They can be calculated as follows:

$$
\overline{S}_{uxp,top} = \min(\overline{S}_{xp1}, \overline{S}_{xp2}) \cdot k_{xp} - (1 - \overline{S}_{xp3})(1 - k_{xp})
\tag{2}
$$

$$
\overline{S}_{uxp,bot} = (1 - \overline{S}_{xp2}) \cdot k_{xp} - \min(\overline{S}_{xp3}, \overline{S}_{xp4}) \cdot (1 - k_{xp}).
\tag{3}
$$

The switching signals $\overline{S}_{uxp,top}$ and $\overline{S}_{uxp,bot}$, which are expressed in an *abc* reference frame, are implemented in the model. However, for the evaluation of the stability of the model, a transformation to the $\alpha\beta$ reference frame is required to decouple the variables. The corresponding transformed switching signals can be derived as:

$$
\overline{S}_{u\alpha p,top} = \overline{S}_{ap,top} - 0.5 \cdot \overline{S}_{bp,top} - 0.5 \cdot \overline{S}_{cp,top}
\tag{4}
$$

$$
\overline{S}_{u\beta p,top} = \overline{S}_{bp,top} - 0.5 \cdot \overline{S}_{cp,top}.
\tag{5}
$$

The expression is likewise applicable to the bottom switches. The current flowing through the phase-leg at the primary side of the converter can be expressed as

$$i_{xp,in,top} = [\min(\overline{S}_{xp1}, \overline{S}_{xp2}) \cdot k_{xp} + (1 - k_{xp}) \cdot (1 - \overline{S}_{xp3})] \cdot i_{xp} = \overline{S}_{ixp,top} \cdot i_{xp} \tag{6}$$

$$i_{xp,in,bot} = [(1 - \overline{S}_{xp3}) \cdot k_{xp} + (1 - k_{xp}) \cdot \min(\overline{S}_{xp1}, \overline{S}_{xp2})] \cdot i_{xp} = \overline{S}_{ixp,bot} \cdot i_{xp}. \tag{7}$$

The parameters $\overline{S}_{ixp,top}$ and $\overline{S}_{ixp,bot}$ represent the average switching signals for the top and bottom switches in one phase-leg of the primary side in the current calculation.

$$\overline{S}_{ixp,top} = \min(\overline{S}_{xp1}, \overline{S}_{xp2}) \cdot k_{ixp} - (1 - \overline{S}_{xp3})(1 - k_{ixp}) \tag{8}$$

$$\overline{S}_{ixp,bot} = \min(\overline{S}_{xp1}, \overline{S}_{xp2})(1 - k_{ixp}) - (1 - \overline{S}_{xp3}) \cdot k_{ixp} \tag{9}$$

They can be transformed in the $\alpha\beta$ frame as follows:

$$\overline{S}_{i\alpha p,top} = \overline{S}_{iap,top} - 0.5 \cdot \overline{S}_{ibp,top} - 0.5 \cdot \overline{S}_{icp,top} \tag{10}$$

$$\overline{S}_{i\beta p,top} = \overline{S}_{ibp,top} - 0.5 \cdot \overline{S}_{icp,top}. \tag{11}$$

Simplified Transformer with Infinite Mutual Inductance

The power transfer in the 3L-DAB3 converter is determined by the voltages across the leakage inductance of the transformer. Thus, the transformer is simplified as series-connected inductors between the primary and secondary sides. The main inductances between the primary and secondary sides are omitted, as shown in Fig. 5. The transformer is converted to the primary side with the total leakage inductance $L_{x\sigma} = L_{xh} + L'_{xh}$, where L_{xh} is the leakage inductance of the primary side and L'_{xh} is the normalized leakage inductance of the secondary side. This is analogous for the copper resistance $R_{xw} = R_{xs} + R'_{xs}$. To simplify, the total leakage inductances and copper resistances of three phases are identical to be L_σ and R_w. The following equation can be derived for the α component of the phase current.

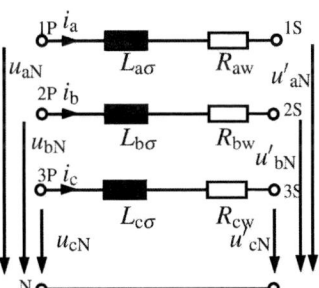

Fig. 5: Equivalent circuit of a simplified three-phase transformer

$$
\begin{aligned}
L_\sigma \cdot \frac{di_\alpha}{dt} &= u_{\alpha p} - N \cdot u_{\alpha s} - R_w \cdot i_\alpha \\
\leftrightarrow \frac{di_\alpha}{dt} &= \frac{\overline{S}_{u\alpha p,top}}{L_\sigma} \cdot u_{Cp,top} + \frac{\overline{S}_{u\alpha p,bot}}{L_\sigma} \cdot u_{Cp,bot} \\
&\quad - N \cdot \frac{\overline{S}_{u\alpha s,top}}{L_\sigma} \cdot u_{Cs,top} - N \cdot \frac{\overline{S}_{u\alpha s,bot}}{L_\sigma} \cdot u_{Cs,bot} - \frac{R_w}{L_\sigma} \cdot i_\alpha
\end{aligned} \tag{12}
$$

It can be derived analogously for the β component of the phase current.

$$\frac{di_\beta}{dt} = \frac{\overline{S}_{u\beta p,top}}{L_\sigma} \cdot u_{Cp,top} + \frac{\overline{S}_{u\beta p,bot}}{L_\sigma} \cdot u_{Cp,bot} - N \cdot \frac{\overline{S}_{u\beta s,top}}{L_\sigma} \cdot u_{Cs,top} - N \cdot \frac{\overline{S}_{u\beta s,bot}}{L_\sigma} \cdot u_{Cs,bot} - \frac{R_w}{L_\sigma} \cdot i_\beta \tag{13}$$

Three Single-Phase Transformers in Y-Y Connection

To include the influence of both the main inductances L_{xh} and the leakage inductances L_{xs}, three single-phase transformers in Y-Y connection is built, as shown in Fig. 6.

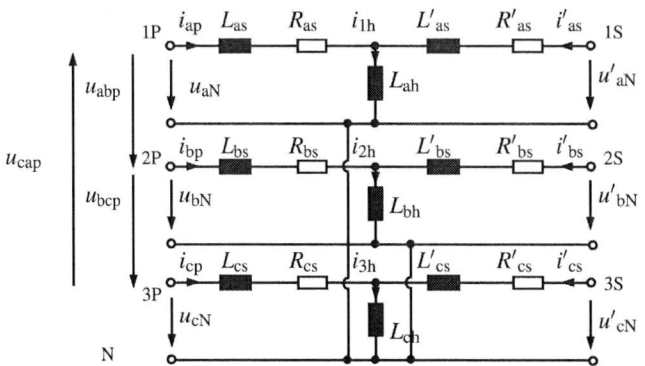

Fig. 6: Equivalent circuit of three single-phase transformers in Y-Y connection

By applying Kirchhoff's voltage law (KVL) to the primary side of the equivalent circuit, the following equations can be derived in the *abc* reference frame.

$$\overline{S}_{uap,top} \cdot u_{Cp,top} + \overline{S}_{uap,bot} \cdot u_{Cp,bot} - N \cdot \overline{S}_{ubs,top} \cdot u_{Cs,top} + N \cdot \overline{S}_{ubs,bot} \cdot u_{Cs,bot}$$
$$= i_{ap} \cdot R_{as} + (L_{as} + L_{ah}) \cdot \frac{di_{ap}}{dt} + L_{ah} \cdot \frac{di'_{as}}{dt} - i_{bp} \cdot R_{bs} - (L_{bs} + L_{bh}) \cdot \frac{di_{bp}}{dt} - L_{bh} \cdot \frac{di'_{bs}}{dt} \tag{14}$$

$$\overline{S}_{ubp,top} \cdot u_{Cp,top} + \overline{S}_{ubp,bot} \cdot u_{Cp,bot} - N \cdot \overline{S}_{ucs,top} \cdot u_{Cs,top} + N \cdot \overline{S}_{ucs,bot} \cdot u_{Cs,bot}$$
$$= i_{bp} \cdot R_{bs} + (L_{bs} + L_{ah}) \cdot \frac{di_{ap}}{dt} + L_{ah} \cdot \frac{di'_{as}}{dt} - i_{bp} \cdot R_{as} - (L_{bs} + L_{bh}) \cdot \frac{di_{bp}}{dt} - L_{bh} \cdot \frac{di'_{bs}}{dt} \tag{15}$$

The parameter N represents the turn ratio of the transformer. The parameters R_{xp} and i_{xp} represent the parasitic resistances and the currents at the primary side of the transformer, respectively. The parameters R_{xs} and i_{xs} are correspondingly the variables at the secondary side. By applying Kirchhoff's circuit law (KCL) to the primary side of the equivalent circuit, the following equation can be derived.

$$0 = \frac{di_{ap}}{dt} + \frac{di_{bp}}{dt} + \frac{di_{cp}}{dt} \tag{16}$$

DC-Link Capacitor

According to Kirchhoff's voltage law (KVL), the current on the top side $i_{Cp,top}$ can be expressed by the dc current i_{dcp} and the sum of the half-bridge currents $i_{p,top}$ as,

$$i_{Cp,top} = C_p \cdot \frac{du_{Cp,top}}{dt} = i_{dcp} - i_{p,top}. \tag{17}$$

The expressions for i_{dcp} and $i_{p,top}$ in (17) are as follows:

$$i_{dcp} = \frac{U_{dcp} - u_{Cp,top} - u_{Cp,bot}}{R_p} \tag{18}$$

$$i_{p,top} = i_{\alpha p} \cdot \overline{S}_{i\alpha p,top} + i_{\beta p} \cdot \overline{S}_{i\beta p,top} \tag{19}$$

with $i_{\alpha p}$ and $i_{\beta p}$ equal to i_α and i_β. Substituting (18) and (19) into (17) results in

$$
\begin{aligned}
\frac{\mathrm{d}u_{\mathrm{Cp,top}}}{\mathrm{d}t} &= -\frac{1}{C_{\mathrm{p1}} \cdot R_{\mathrm{p}}} \cdot (U_{\mathrm{dcp}} - u_{\mathrm{Cp,top}} - u_{\mathrm{Cp,bot}}) - \frac{1}{C_{\mathrm{p1}}} \cdot (i_{\mathrm{ap}} \cdot \overline{S}_{\mathrm{iap,top}} + i_{\mathrm{bp}} \cdot \overline{S}_{\mathrm{ibp,top}} + i_{\mathrm{cp}} \cdot \overline{S}_{\mathrm{icp,top}}) \\
&= -\frac{1}{C_{\mathrm{p1}} \cdot R_{\mathrm{p}}} \cdot u_{\mathrm{Cp,top}} - \frac{1}{C_{\mathrm{p1}} \cdot R_{\mathrm{p}}} \cdot u_{\mathrm{Cp,bot}} \\
&\quad - \frac{1}{C_{\mathrm{p1}}} \cdot \overline{S}_{\mathrm{i\alpha p,top}} \cdot i_\alpha - \frac{1}{C_{\mathrm{p1}}} \cdot \overline{S}_{\mathrm{i\beta p,top}} \cdot i_\beta + \frac{1}{C_{\mathrm{p1}} \cdot R_{\mathrm{p}}} \cdot U_{\mathrm{dcp}}
\end{aligned}
\tag{20}
$$

$$
\begin{aligned}
\frac{\mathrm{d}u_{\mathrm{Cp,bot}}}{\mathrm{d}t} &= -\frac{1}{C_{\mathrm{p2}} \cdot R_{\mathrm{p}}} \cdot (U_{\mathrm{dcp}} - u_{\mathrm{Cp,top}} - u_{\mathrm{Cp,bot}}) - \frac{1}{C_{\mathrm{p2}}} \cdot (i_{\mathrm{ap}} \cdot \overline{S}_{\mathrm{iap,bot}} + i_{\mathrm{bp}} \cdot \overline{S}_{\mathrm{ibp,bot}} + i_{\mathrm{cp}} \cdot \overline{S}_{\mathrm{icp,bot}}) \\
&= -\frac{1}{C_{\mathrm{p2}} \cdot R_{\mathrm{p}}} \cdot u_{\mathrm{Cp,top}} - \frac{1}{C_{\mathrm{p2}} \cdot R_{\mathrm{p}}} \cdot u_{\mathrm{Cp,bot}} \\
&\quad - \frac{1}{C_{\mathrm{p2}}} \cdot \overline{S}_{\mathrm{i\alpha p,bot}} \cdot i_\alpha - \frac{1}{C_{\mathrm{p2}}} \cdot \overline{S}_{\mathrm{i\beta p,bot}} \cdot i_\beta + \frac{1}{C_{\mathrm{p2}} \cdot R_{\mathrm{p}}} \cdot U_{\mathrm{dcp}}
\end{aligned}
\tag{21}
$$

The equations above can be applied analogously to the secondary side of the converter.

Simulation Results

The optimal integration approach for the modulation method provided in [8] has been validated by simulations using the software PLECS. The simulation parameters are listed in Table I.

Table I: Parameters of the simulation

Parameters	Values	Parameters	Values
dc-link voltage at the primary side	5 kV	dc-link voltage at the secondary side	5.5 kV
Main inductance	10 mH	Leakage inductance	67.5 μH
dc-link capacitor	5 mF	Switching frequency	1 kHz
Phase shift	30°	Resistance	0.1 mΩ
Sampling time	1 μs	Duty cycle	0.5

The PLECS average model using the simplified transformer model is compared with the self-defined code average model, in respect of phase voltages and ac currents, as shown in Fig. 7. Both errors of the phase voltages are small enough to be neglected. The errors of the ac currents are smallest with the trapezoidal integration method.

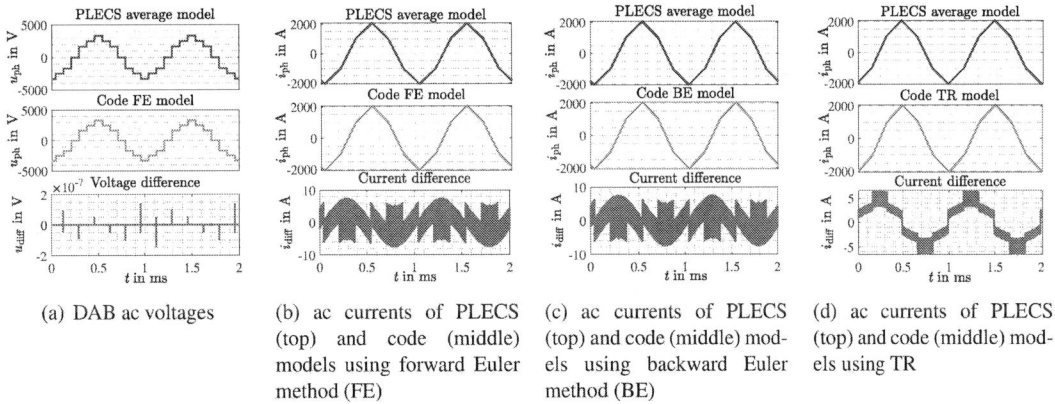

(a) DAB ac voltages

(b) ac currents of PLECS (top) and code (middle) models using forward Euler method (FE)

(c) ac currents of PLECS (top) and code (middle) models using backward Euler method (BE)

(d) ac currents of PLECS (top) and code (middle) models using TR

Fig. 7: Simulation results of the implemented 3L-DAB3 models with the simplified tranformer model

The PLECS average model compares to the self-defined code average model using the three single-phase transformer in Y-Y connection, as shown in Fig. 8. The phase voltages using this model differ from those generated using the simplified model. During the switching transition for 1 µs, there are maximum 800 V and 1800 V differences for TR and FE, respectively. The voltages of the simplified model are estimated based on the state of the switches without integrating the currents. In contrast, the three single-phase transformer is determined by integrating the phase currents. The application of the integration methods only has an impact during switching transitions. Throughout all other time periods, the errors are less than 0.01%.

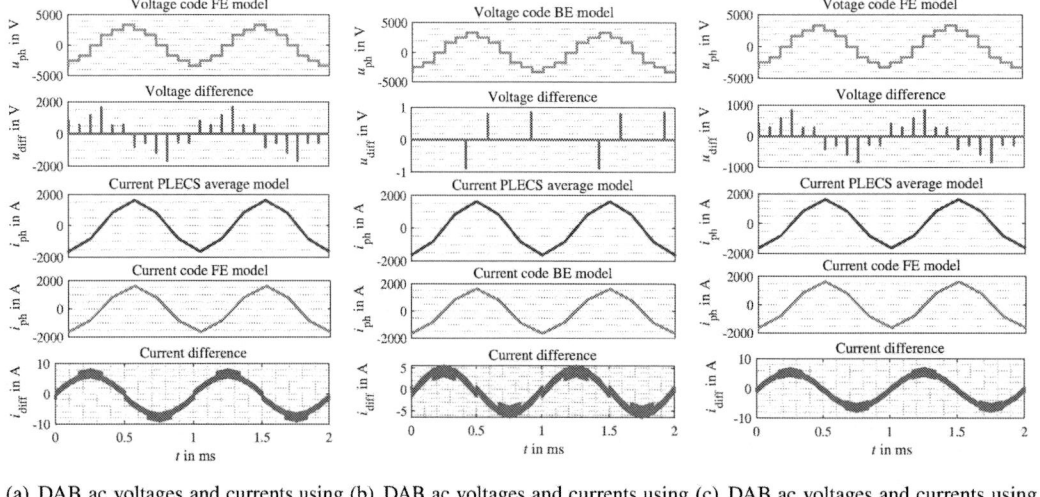

(a) DAB ac voltages and currents using FE

(b) DAB ac voltages and currents using BE

(c) DAB ac voltages and currents using TR

Fig. 8: Simulation results of the implemented 3L-DAB3 models with three single-phase transformers in Y-Y connection

HIL Verification

An HIL test setup for the models using an RT simulator XRS7070 of AixControl is shown in Fig. 9. The RT controller sends PWM signals to the RT simulator. The simulator calculates the voltages and currents and sends the results back to the RT controller. The 3L-DAB3 average model can be implemented either directly by the PLECS average model or by the self-defined C-code average model using the aforementioned modeling methodology. The C-code is automatically generated from PLECS models and can be subsequently downloaded into the RT simulator [6]. The test parameters are the same as that used in the simulation, as summarized in Table I.

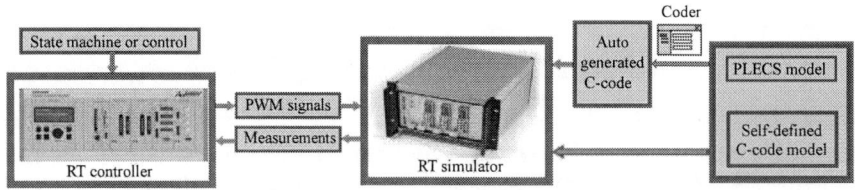

Fig. 9: Setup of the HIL test for a 3L-DAB3 converter

The HIL results with both the PLECS models and the self-defined C-code average models with an RT

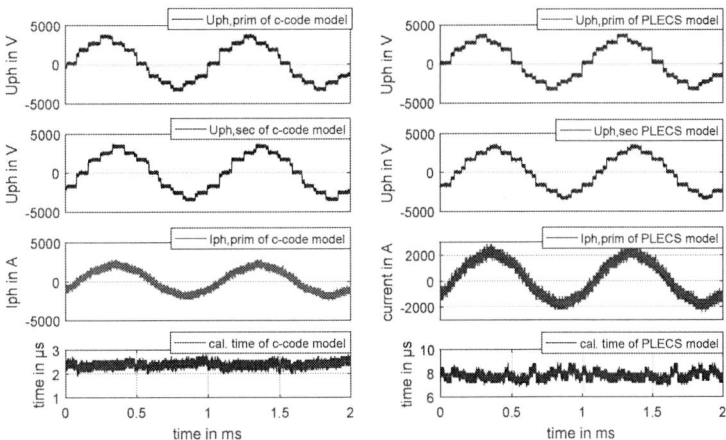

Fig. 10: Results of simulator XRS7070 using the simplified transformer model and TR method

simulator are shown in Fig. 10. The phase voltages and ac currents of both models coincide with each other. The calculation time for the C-code average model is approximately 2.5 s, which is 75% faster than that one of the PLECS model. An overall comparision of the applied models is provided in Table II. The auto-generated C-code from the PLECS model comprises 70,000 lines of code, whereas the self-defined C-code model has only less than 600 lines of code. The total run time includes the time spent communicating input and output as well as for the computation of the model. To ensure that the entire procedure is completed within one time step, the step size T should be greater than the total time. This prevents the RT simulation from overrunning. As a result, the minimum step time of the PLECS model is 25 µs, which is 47% longer than for the C-code model. In the RT simulator, the calculation time of the C-code model is 68% faster than that of the PLECS model. The time for input and output communication does not differ significantly, since they are determined almost entirely by the number of ports and slots used. Therefore, the difference in total time is not significantly higher than the difference in the calculation time.

Table II: Comparison between the C-code model and the PLECS model of the 3L-DAB3 in the RT simulator (in µs)

	C-code model	PLECS model
Code lines	<600	>70000
Time-step	17 µs	25 µs
Solver	fixed-step RADAU	fixed-step RADAU
Integration method	TR	unknown
Calculation time	<3 µs	<8.5 µs
Total time	<15 µs	20 µs

Numerical Stability Analysis of C-Code Average Model

In this section, the numerical stability of the C-code average model for 3L-DAB3 using different integration is analyzed. By using the preceding equations, a state-space model representing the 3L-DAB3 topology can be obtained. The top and bottom dc-link voltages of the primary side $u_{Cp,top}$ and $u_{Cp,bot}$, the top and bottom dc-link voltages of the secondary side $u_{Cs,top}$ and $u_{Cs,bot}$ and the ac currents i_α and i_β are taken as the state variables. Input variables are the dc-link voltages of the primary and secondary sides

U_{dcp} and U_{dcs}, respectively. A state-space model of the 3L-DAB3 can be expressed as

$$\dot{x} = A \cdot x + B \cdot u \tag{22}$$

$$y = C \cdot x + D \cdot u \tag{23}$$

with the state variables $x = \begin{bmatrix} u_{Cp,\text{top}} \\ u_{Cp,\text{bot}} \\ u_{Cs,\text{top}} \\ u_{Cs,\text{bot}} \\ i_\alpha \\ i_\beta \end{bmatrix}$, the inputs $u = \begin{bmatrix} U_{\text{dcp}} \\ U_{\text{dcs}} \end{bmatrix}$ and the outputs $y = \begin{bmatrix} u_{\alpha p} \\ u_{\beta p} \\ u_{\alpha s} \\ u_{\beta s} \end{bmatrix}$.

$$A = \begin{bmatrix}
-\frac{1}{C_{p1}R_p} & -\frac{1}{C_{p1}R_p} & 0 & 0 & -\frac{\overline{S}_{i\alpha p,\text{top}}}{C_{p1}} & -\frac{\overline{S}_{i\beta p,\text{top}}}{C_{p1}} \\
-\frac{1}{C_{p2}R_p} & -\frac{1}{C_{p2}R_p} & 0 & 0 & -\frac{\overline{S}_{i\alpha p,\text{bot}}}{C_{p2}} & -\frac{\overline{S}_{i\beta p,\text{bot}}}{C_{p2}} \\
0 & 0 & -\frac{1}{C_{s1}R_s} & -\frac{1}{C_{s1}R_s} & \frac{\overline{S}_{i\alpha s,\text{top}}}{C_{s1}} & \frac{\overline{S}_{i\beta s,\text{top}}}{C_{s1}} \\
0 & 0 & -\frac{1}{C_{s2}R_s} & -\frac{1}{C_{s2}R_s} & \frac{\overline{S}_{i\alpha s,\text{bot}}}{C_{s2}} & \frac{\overline{S}_{i\beta s,\text{bot}}}{C_{s2}} \\
\frac{\overline{S}_{u\alpha p,\text{top}}}{L_\sigma} & \frac{\overline{S}_{u\alpha p,\text{bot}}}{L_\sigma} & -N \cdot \frac{\overline{S}_{u\alpha s,\text{top}}}{L_\sigma} & -N \cdot \frac{\overline{S}_{u\alpha s,\text{bot}}}{L_\sigma} & -\frac{R_w}{L_\sigma} & 0 \\
\frac{\overline{S}_{u\beta p,\text{top}}}{L_\sigma} & \frac{\overline{S}_{u\beta p,\text{bot}}}{L_\sigma} & -N \cdot \frac{\overline{S}_{u\beta s,\text{top}}}{L_\sigma} & -N \cdot \frac{\overline{S}_{u\beta s,\text{bot}}}{L_\sigma} & 0 & -\frac{R_w}{L_\sigma}
\end{bmatrix}, \quad B = \begin{bmatrix}
\frac{1}{C_{p1} \cdot R_p} & 0 \\
\frac{1}{C_{p2} \cdot R_p} & 0 \\
0 & \frac{1}{C_{s1} \cdot R_s} \\
0 & \frac{1}{C_{s2} \cdot R_s} \\
0 & 0 \\
0 & 0
\end{bmatrix},$$

$$C = \begin{bmatrix}
\overline{S}_{u\alpha p,\text{top}} & \overline{S}_{u\alpha p,\text{bot}} & 0 & 0 & 0 & 0 \\
\overline{S}_{u\alpha p,\text{top}} & \overline{S}_{u\alpha p,\text{bot}} & 0 & 0 & 0 & 0 \\
0 & 0 & \overline{S}_{u\beta s,\text{top}} & \overline{S}_{u\beta s,\text{bot}} & 0 & 0 \\
0 & 0 & \overline{S}_{u\beta s,\text{top}} & \overline{S}_{u\beta s,\text{bot}} & 0 & 0
\end{bmatrix}, \quad D = \begin{bmatrix}
0 & 0 \\
0 & 0 \\
0 & 0 \\
0 & 0
\end{bmatrix}$$

The model above must be discretized to be processed on a digital processor. In this work, FE, BE, and TR are implemented to discretize the state-space model in the format:

$$x[n+1] = F \cdot x[n] + G \cdot u[n] \tag{24}$$

Based on the stability criterion, a discrete-time system is analytical stable if and only if all of its eigenvalues are located inside a unit circle. The three integration methods have the same matrix A, but they have different stability regions according to A. To normalize the stability region, the continuous state-space matrix A is transferred to the discrete state-space matrix F. The stable region for each integration method is transferred to be a unit circle about the origin point with the matrix F. The relationship between the current and voltage variables in the next time step using the FE, BE, and TR integration are respectively derived in (25), (26) and (27). The matrix F for each integration method can be concluded as:

$$\begin{aligned}
x(t+h) &= x(t) + \dot{x}(t) \cdot T \\
\leftrightarrow x(t+T) &= (1 + h \cdot A) \cdot x(t) \Rightarrow F_{\text{FE}} = 1 + T \cdot A
\end{aligned} \tag{25}$$

$$\begin{aligned}
x(t+T) &= x(t) + \dot{x}(t+T) \cdot T \\
\leftrightarrow x(t+T) &= (1 - T \cdot A)^{-1} \cdot x(t) \Rightarrow F_{\text{BE}} = (1 - T \cdot A)^{-1}
\end{aligned} \tag{26}$$

$$\begin{aligned}
x(t+T) - x(t) &= \frac{1}{2} \cdot [\dot{x}(t+T) + \dot{x}(t)] \cdot T \\
\leftrightarrow x(t+T) &= \frac{1 + \frac{1}{2} \cdot T \cdot A}{1 - \frac{1}{2} \cdot T \cdot A} \cdot x(t) \Rightarrow F_{\text{TR}} = \frac{1 + \frac{1}{2} \cdot T \cdot A}{1 - \frac{1}{2} \cdot T \cdot A}.
\end{aligned} \tag{27}$$

Apart from an optimized simulation, the derived mathematical description further allows for an analysis of the numerical stability for different discretization methods and time steps [7]. The eigenvalues can be obtained from the matrix F of the state-space model and used to analyze the stability of the system. Fig. 11 depicts the eigenvalues of discrete matrices of the 3L-DAB3 converter for different discretization approaches with the extended parameter range. The location of the poles is indicative of the dynamics and stability of the system. Fig. 11 illustrates, that both BE and TR are stable with an extended range of parameters.

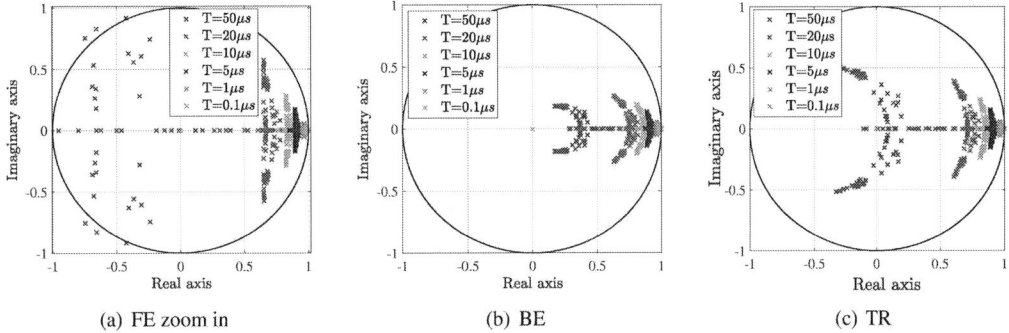

(a) FE zoom in   (b) BE   (c) TR

Fig. 11: Stability test of the 3L-DAB3 C-code average model with an extended range of specifications

The natural frequency (around the boundary) and the damping ratio (along the radius) are depicted in Fig. 11. There are more unstable poles with a larger time-step (light colour in Fig. 11) than with a smaller time-step (dark blue in Fig. 11). There are 1,824 out of 28,416 poles unstable with step size 1 µs and FE integration. The unstable poles are far away from the unit circle and are therefore not completely included in the Fig. 11. No poles with BE or with TR are unstable. Although the model with a smaller step time has poles intuitively closer to the unit circle boundary, it has a better dynamic performance than the system of larger time steps. Both BE and TR are suitable for the code average model of the 3L-DAB3 converter with the specified parameters.

Conclusion

The C-code average model of the 3L-DAB3 converter is more efficient than the PLECS average model. In the RT simulation, the model is valid with a reduced time step and requires less calculation steps and offers shorter execution time. Of the three integration algorithms, the trapezoidal integration offers the best precision. The T-equivalent transformer model is more accurate than the simplified transformer model but requires more computation time on the simulator. By introducing a self-defined C-code average model, the real-time model of the 3L-DAB3 converter can be used for CHIL testing of DAB converters with higher switching frequencies.

References

[1] Seung Tae, Cha, Qiuwei Wu, Arne Hejde Nielsen, Jacob: Real-Time Hardware-in-the-Loop (HIL) Testing for Power Electronics Controllers

[2] Ming Jia and Philipp Joebges and R. W. De Doncker: A Fast and Robust Model of Dual-Active Bridge Converters in Real-Time Simulation, 2020 22nd European Conference on Power Electronics and Applications, 2020

[3] Hafiz Abu Bakar Siddique: The Three-Phase Dual-Active Bridge Converter Family Modeling, Analysis, Optimization and Comparison of Two-Level and Three-Level Converter Variants, Dissertation, RWTH Aachen, DOI: 10.18154/RWTH-2020-01205, 2019

[4] Stephan Thomas: A Medium-Voltage Multi-level DC/DC Converter with High Voltage Transformation Ratio, Dissertation, RWTH Aachen, 2014

[5] Jost Allmeling, Niklaus Felderer and Min Luo: High Fidelity Real-Time Simulation of Multi-Level Converters, The 2018 International Power Electronics Conference, 2018

[6] Philipp Joebges: Distributed Real-Time Simulation of Modular Bidirectional DC-DC Converters for Control-Hardware-in-the-Loop, Dissertation RWTH Aachen University, DOI: 10.18154/RWTH-2021-10060, 2021

[7] Robert Uhl and Amir Arasteh and Antonello Monti and Arne Hinz and Rik W. De Doncker: Nodal-Reduced Modeling of Three-Phase Dual-Active Converters for EMTP-type Simulations, 2017 IEEE 26th International Symposium on Industrial Electronics (ISIE), 2017

[8] Philipp Joebges and Anton Gorodnichev and R. W. De Doncker: Modulation and Active Midpoint Control of a Three-Level Three-Phase Dual-Active Bridge DC-DC Converter under Non-Symmetrical Load, 2018 International Power Electronics Conference, 2018

A Calorimetric and Electrical Method for Measuring Loss Energies of Half-Bridges

Jörg Haarer, Mattea Eckstein, Philipp Ziegler, Philipp Marx, David Hirning, Jörg Roth-Stielow
INSTITUTE FOR POWER ELECTRONICS AND ELECTRICAL DRIVES
University of Stuttgart
Pfaffenwaldring 47
Stuttgart, Germany
Phone: +49 / (711) - 685-67372
Email: joerg.haarer@ilea.uni-stuttgart.de
URL: http://www.ilea.uni-stuttgart.de

Keywords

≪Device characterisation≫, ≪Switching losses≫, ≪Conduction losses≫, ≪Silicon Carbide (SiC)≫

Abstract

For the optimal design of power electronic systems the exact knowledge about the different loss mechanisms of the used semiconductor devices is essential. However, conventional measurement methods based on electrical parameters are facing their limits, as switching speed of modern power devices is steadily increasing enabled by the use of wide bandgap semiconductor materials. For this reason, calorimetric measurement techniques are becoming more and more popular. However, due to the nature of their principle, calorimetric measurement methodologies can usually only determine the total power losses of the device under test. To overcome this disadvantage a methology which combines different calorimetric and electrical measurements to separately determine the switching and conduction energies in a half-bridge with an ohmic-inductive load while maintaining the accuracy of calorimetric measurement methods is developed. In addition, the presented methodology identifies the switching energies of the two different switching transitions within one switching period, depending on the load current as well as the dead time, considering thermal influences. A hardware setup for the presented methology, is realized. Using this test setup, the loss energies of a half-bridge based on silicon carbide MOSFETs are investigated. The resulting measurements are presented and verified by measurements with a power analyzer.

Introduction

Half-bridges are fundamental building blocks of numerous topologies in power electronics. Their design is crucial in terms of efficiency, volume and cost for a power electronic system. The most decisive factor is the choice of appropriate power semiconductor devices, depending on the specific application. Whereas conduction losses can be estimated using data sheet values or measured by means of DC measurements, the estimation of switching energies based on data sheet values is only possible to a very limited extent. Also the characterization of switching energies by measuring electrical variables in a double pulse test can no longer be regarded as a reliable and accurate method due to the increased use of power semiconductor devices made from semiconductor materials such as silicon carbide (SiC) and gallium nitride, because of the limited bandwidth, linearity and time synchronization of available measurement equipment [1] [2] [3]. For this reason, calorimetric measurement methods for measuring switching energies are becoming more and more popular, as they achieve a significantly better accuracy in power dissipation measurement [4] [5]. However, a major drawback of calorimetric measurement methods is, that only the overall power losses of the device under test (DUT) can be measured [3]. In order to separate the loss energies of different loss mechanisms while maintaining the measurement accuracy of calorimetric measurement methods, a methology, which combines a series of electrical ad calorimetric measurements, is developed. This method allows the determination of the energy loss generated by different loss mechanisms during the operation of a half-bridge with ohmic inductive load. By operating a half-bridge in various operating modes, resulting in different loss mechanisms for each of those operating modes, it is subsequently possible to determine different loss mechanisms by a suitable combination of the measured power losses. In addition to the separation between switching and conduction losses, the presented method allows a further subdivision of the switching losses for the two different switching operations within a switching period. Moreover, the conduction losses can be separated into power MOSFET and body diode losses. In order to understand how the different operating modes are chosen and how the resulting losses have to be combined, it is necessary to analyze the relevant loss mechanisms in half-bridge operation of MOSFETs.

Loss mechanisms

Loss mechanisms in SiC power MOSFETs can fundamentally be categorized into conduction and switching losses. The origin of these different loss mechanisms can be attributed to the different circuit elements in the equivalent circuit in Fig. 1. However, the temporal occurrence during a switching period differs between the different mechanisms depending on the operating conditions of the circuit [6].

The origin of conduction losses can be found in the substrate, the drift layer and, in case of the power MOSFET conducting, in the channel and in the junction field-effect transistor (green) in case of the body diode conducting, in its pn junction (blue). If the transistor is turned on, the current is conducted by the power MOSFET. If the transistor is turned off, the body diode can conduct current in reverse direction. In both cases the resulting conduction losses are calculated by (1).

$$p_{\text{cond}}(t) = i_{\text{D}}(t) \cdot v_{\text{DS,int}}(t) \tag{1}$$

Besides conduction losses, especially switching losses occur in the power MOSFET, caused by the temporal overlap of current and voltage. In the anti-parallel body diode, reverse recovery losses occur when the excess charge is removed. Further origins of switching losses are the charging/discharging process of the output capacitance C_{OSS} (orange) and oscillations (purple), induced by the resonance circuit of parasitic inductances and capacitances between drain and source of the MOSFET. In SiC-MOSFETs the reverse-recovery charge Q_{rr}, is significantly reduced in comparison to Si-MOSFETs. Oscillations take place in any switching process and are highly dependent on the parasitic inductance of the switching cell. Therefore, the impact of losses, induced by Q_{rr} and oscillations, on the overall switching losses are not considered separately. In addition to the losses of the actual power loop of the semiconductor device, gate driving losses (yellow) and ohmic losses (grey), which result from the various packing technologies, occur. As these loss mechanisms are just a mere fraction of the overall transistor losses, the independent loss mechanisms, gate drive losses and conduction losses of the package, are neglected.

The commutation in a half-bridge always involves two transistors. During this commutation the load current i_{L} either commutates from high side to low side or from low side to high side. Each of those commutations can be subdivided into the switching transition of the prior conducting transistor (PCT) which is turned off and the switching transition of the subsequently conducting transistor (SCT) that is turned on. Also, the dependence on the direction of the load current i_{L} has to be taken into account. This results in the four cases with their corresponding switching energies listed in Tab. I.

Tab. I shows, that it is sufficient to consider only two of these cases to determine the total switching energy for a single commutation of a half-bridge as only the sign of the current through the PCT is decisive for the total switching energy of the half-bridge. As shown in Fig. 2, depending on the boundary conditions, for each of the possible cases, a tuple of switching energies results. These different switching

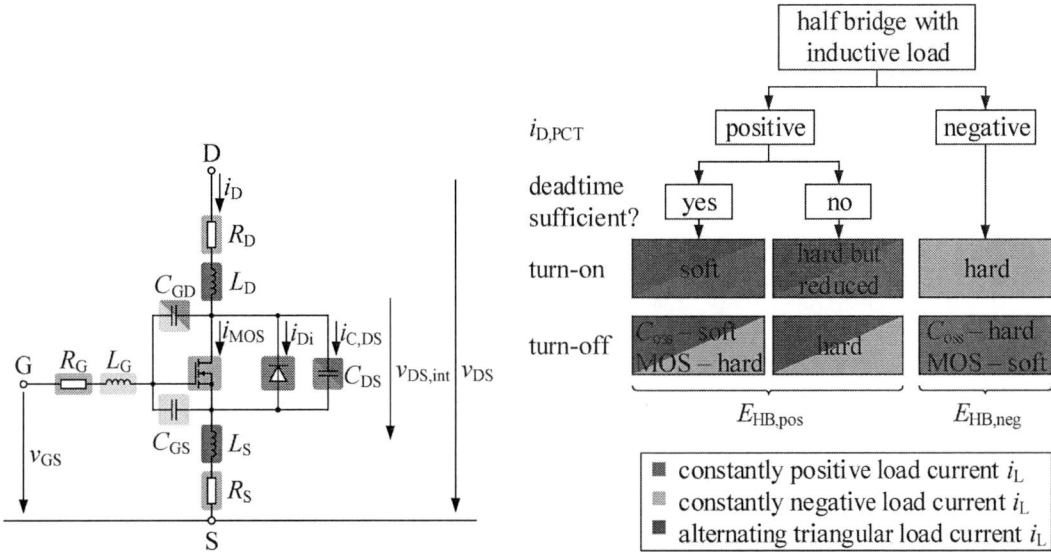

Fig. 1: Equivalent circuit of the investigated SiC-MOSFET

Fig. 2: Dependency of the switching losses on the direction of the PCT's drain current

Table I: Possible cases of commutations in a half-bridge with an ohmic-inductive load

commutation from	sign of i_L	sign of $i_{D,PCT}$	high side losses	low side losses	total losses
high side to low side	$i_L > 0$	$i_{D,PCT} > 0$	$E_{off,pos}$	$E_{on,pos}$	$E_{HB,pos}$
low side to high side	$i_L < 0$	$i_{D,PCT} > 0$	$E_{on,pos}$	$E_{off,pos}$	$E_{HB,pos}$
low side to high side	$i_L > 0$	$i_{D,PCT} < 0$	$E_{on,neg}$	$E_{off,neg}$	$E_{HB,neg}$
high side to low side	$i_L < 0$	$i_{D,PCT} < 0$	$E_{off,neg}$	$E_{on,neg}$	$E_{HB,neg}$

energies cover the entire range of switching energies that are relevant for half-bridge operation with ohmic inductive load.

In case of a negative drain current $i_{D,PCT}$ through the PCT, switching energies are independent from the dead time, assuming a constant load current i_L for the duration of the dead time [7]. The commutation is initiated by turning off the PCT. Because of the negative drain current $i_{D,PCT}$ in the PCT and the continuity condition of the load current i_L, the diode anti-parallel to the PCT starts to conduct the drain current. The output capacities C_{OSS} keep their voltages and are not yet charged or discharged. As the voltage across the power MOSFET is clamped by the forward biased diode during the entire turn-off transition of the PCT, the turn off transition is a zero voltage transition regarding the power MOSFET. By turning on the SCT, the actual commutation process of the load current i_L from the PCT to the SCT is initiated. First, the SCT takes the load current from the PCT, resulting in power losses due to the temporal overlap of the voltage $v_{DS,int}$ and the current i_{MOS}. Once the commutation of the load current i_L is completed, the anti-parallel diode of the PCT is reverse biased. Hence, the output capacitance C_{OSS} of the SCT is discharged abrupt as it is shorted by the power MOSFET. This results in a high discharge current. The whole energy E_{OSS}, stored in the output capacitance C_{OSS}, dissipates in the form of thermal energy. As the gradient of the drain-source voltage $|dv_{DS}/dt|$ is similar in both transistors, charging the output capacitance of the PCT comes along with a large current as well. On top off the load current and the current resulting from discharging the SCT's output capacitance, an additional current that is required to charge the output capacitance of the PCT, has to be conducted by the SCT. Due to the increased current i_{MOS} in the power MOSFET, additional conduction losses occur in the power loop and leads of the transistor.

In the case of a commutation with positive drain current $i_{D,PCT}$ through the PCT, the commutation is initiated by turning off the PCT. In contrast to the commutation with a negative current $i_{D,PCT}$, in this case, the charging/discharging of the output capacitances of the two transistors starts immediately when the PCT is turned off. This is due to the inductance of the load, which forces the load current into/out of the switching node of the half-bridge. Although the turn off transition of the PCT is passively relieved by its output capacitance, there are switching losses in the power MOSFET of the PCT which result from the temporal overlap of the voltage $v_{ds,int}$ and the current i_{MOS}. The magnitude of these losses depends on the switching speed of the PCT and the load current. If the dead time and the energy stored in the inductance of the load are sufficient, the output capacitances of the transistors are fully charged/discharged by the time the SCT is turned on. The SCT can be turned on without loss at a drain-source voltage of zero volts. In this case, the energy E_{OSS} stored in the output capacitances is not dissipated into heat but recovered [8]. Furthermore, the conduction losses in the leads and interconnects also decrease due to the reduced charging currents through the PCT compared to the hard switched transition with negative drain current in the PCT.

If the SCT is not turned on immediately after the charging/discharging process, the anti-parallel diode of the SCT conducts the load current until the SCT is turned on. However, for the time period the anti-parallel diode conducts the current, conduction losses in the diode occur, which are typically higher than the conduction losses of the power MOSFET. If the dead time or the energy in the inductance of the load is too low to completely charge/discharge the output capacitances, the result is a partial zero voltage switching of the SCT.

Measurement Methology

In a half-bridge three possible combinations of drain currents through the PCT are possible for the individual commutations within one switching period. To investigate them, a suitable configuration of a half-bridge must be found for each combination of drain currents in the PCT. Further, the DUT must be the low side transistor S_{A2} in all of these configurations to keep the boundary conditions constant for all measurements. A constantly positive load current can be realized by the circuit in Fig. 3 . The realization of a constantly negative load current is achieved by the circuit in Fig. 4. A triangular alternating load current can be obtained from the circuit in Fig. 5.

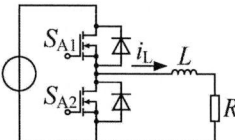 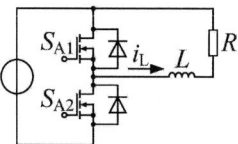 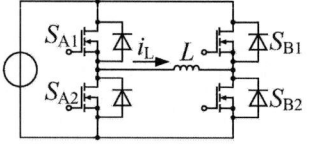

Fig. 3: Configuration for a constantly positive load current

Fig. 4: Configuration for a constantly negative load current

Fig. 5: Configuration for a triangular alternating load current

The overall losses E_{DUT} of the DUT are measured calorimetrically. To separate the switching energies from the conduction energies, the temporal course of the drain-source voltage v_{DS} of the DUT as well as the load current i_L are measured using an oscilloscope. The duration of power MOSFET and diode conduction is determined from the drain-source voltage. The conduction losses are calculated by (1). The voltage drop $v_{DS,int}$ at the power MOSFET and the anti-parallel diode are predetermined by stationary DC measurements imposing a DC current into the power device [9]. To take the influence of the temperature into account, the DUT is preheated to the temperature that applies at the corresponding operating point by imposing the same overall losses by a DC current. The dependency of the power MOSFET's on state resistance $R_{DS,on}$ on the drain current is modeled according to the datasheet values. Diode conduction losses are estimated under the assumption of a continuous drain current during dead time. The junction temperature is regarded as constant within a switching period as the duration of the switching period is in the microsecond range. In this way, the losses in a transistor of the half-bridge can be determined for the different operating modes. The switching energies for the different cases are composed by the sum of the components marked in the same color in Fig. 2. Furthermore, the switching energies $E_{\mathrm{HB,pos}}$ and $E_{\mathrm{HB,neg}}$ that occur during commutation in the half-bridge can be determined from these measurements. As shown in Fig. 2, the switching energy of a commutation in the half-bridge with positive current through the PCT is equivalent to the switching energies of the DUT for a triangular alternating current into the load. Although there is no operation mode where the sum of the turn-on and turn-off energies $E_{\mathrm{DUT,on+off}}$ equals the switching energies for a negative drain current in the PCT, the switching energies for those commutations can be estimated by (2), where $E_{\mathrm{DUT,iL+}}$, $E_{\mathrm{DUT,iL-}}$ and $E_{\mathrm{DUT,iL\sim}}$ are the DUT's switching energies for constantly positive, negative or alternating triangular load current.

$$E_{\mathrm{HB,neg}} = E_{\mathrm{DUT,iL+}} + E_{\mathrm{DUT,iL-}} - E_{\mathrm{DUT,iL\sim}} \tag{2}$$

Measurement Setup

The measurement setup is depicted in Fig. 6. The power electronic system consists of a full bridge with exchangeable ohmic-inductive load and can realize all three of the configurations from Fig. 3 to 5 [10]. The system is powered by a remotely controllable power supply (Delta Elektronika SM 500-CP-90) with an output regulation of better than 5 mV and 10 mA, which provides the voltages and currents required for the various measurements. The signal electronics consists of a control computer which implements a sequence control for the measurement procedure. A FPGA is used to generate the PWM signals. A multichannel thermometer (Pico Technology TC-08) is used to measure the DUT and the ambient temperature. An oscilloscope (Tektronix MSO 4034), a high voltage differential probe (PMK Bumblebee) and a current probe (Tektronix TCP0030A) measure the temporal courses of the drain-source voltage of the DUT and the load current. The trigger signal of the oscilloscope is generated by the control computer. For the DC voltage measurements a precision multimeter (Fluke 8845A) that provides an accuracy of 0.4 mV is used.

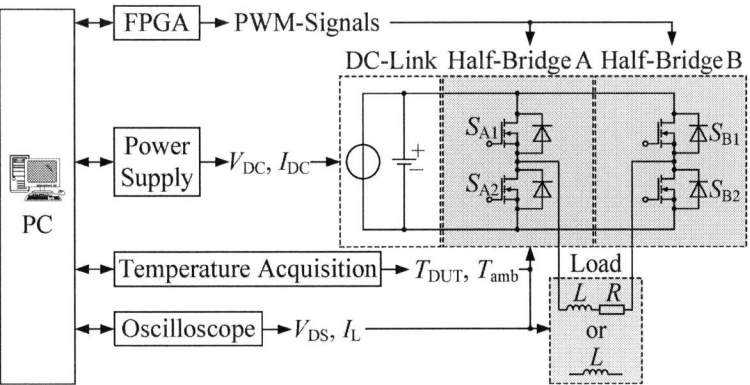

Fig. 6: Overview diagram of the measurement setup

For the realization of the calorimetric measurement system, all transistors of the full bridge are placed on identical heat sinks. In addition, the entire setup is placed in a wind tunnel to achieve the same forced convection for all transistors. By means of the multi-channel thermometer, the DUT temperature as well as the ambient temperature in the wind tunnel are measured. From the temperature difference, the total power dissipation of the DUT is determined by means of a lookup table.

In order to determine the accuracy of the calorimetric measurement method, an arbitrary sequence of measurements is performed in which well-known power losses are imposed by means of a constant DC current. The losses generated in this way are subsequently determined by the calorimetric measurement method. The setup equals the pulsed calorimetric setup, to ensure the same thermal conditions in both, the reference, and the actual loss measurement. Fig. 7 shows the results of a calorimetric validation measurement. A sequence of operating points with different constant power losses are set one after the other in the automated process. When the thermal steady state is reached the temperature difference as well as the actual power loss are measured. The green bars show the actual DUT power loss. The red bars show the estimated DUT power loss. Most operation points show a relative deviation less than 1 percent. Only operation point three, with overall power loss less than 5 W, exceeds this limit.

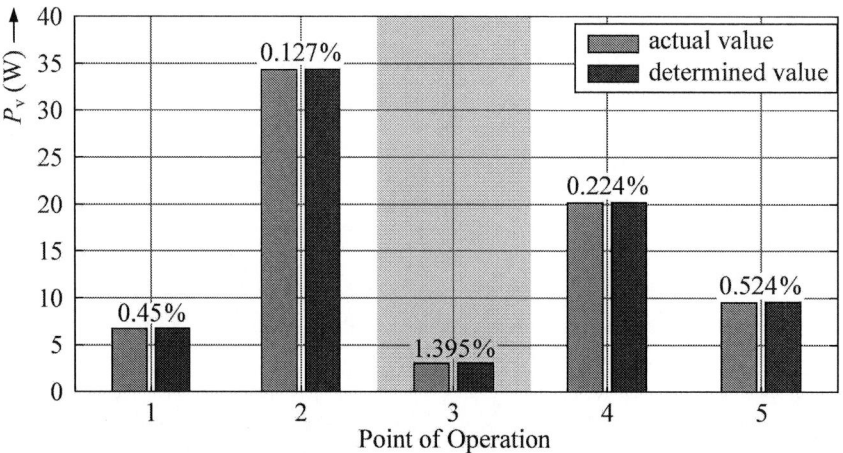

Fig. 7: Comparison between actual imposed and calorimetrically determined losses

Results

Fig. 8 shows the power MOSFET's conduction energies $E_{c,MOS}$, the diode-conduction energies $E_{c,diode}$ and the switching energies $E_{DUT,on+off}$ of the DUT for different dead times in the range of 5 ns to 100 ns for a constant load current of 5 A.

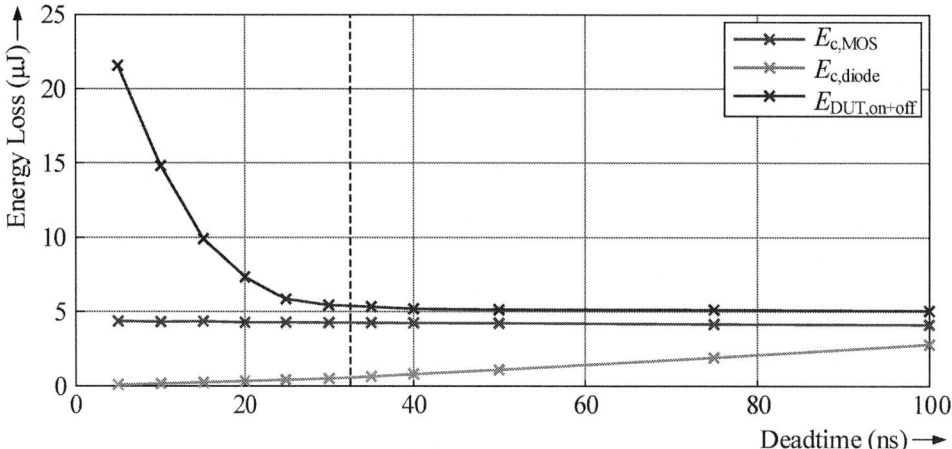

Fig. 8: Switching and conduction energies over dead time for a constant load current of 5 A through the PCT.

Power MOSFET conduction losses stay almost constant for different dead times. In contrast, diode-conduction losses increase steady over longer dead times. If the dead time is insufficient, which results in no diode conduction after turn-on, the diode conducts before turn-off. At the transition from zero voltage switching to partially zero voltage switching, the gradient of $E_{c,diode}$ changes. For sufficient dead times, switching losses remain roughly constant as well. As soon as dead time is insufficient, $E_{\text{DUT,on+off}}$ starts to increase significantly. The shorter the dead time, the sooner the load current induced charging/discharging process is interrupted, the higher the power loss.

The switching energy of the half-bridge for a single commutation of the half-bridge at different load currents, dead times and operation modes is depicted in Fig. 9.

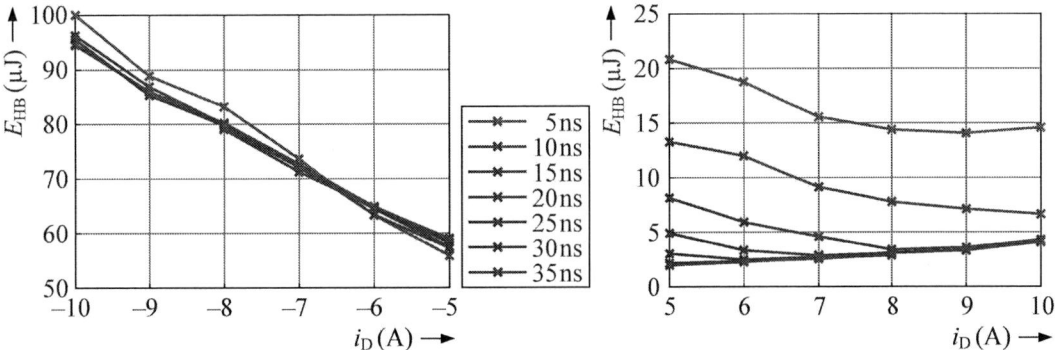

Fig. 9: Loss of a commutation over the drain current through the PCT for different dead times.

The higher the absolute value of the drain current i_{D}, the higher the switching energy E_{HB}, holds true for both directions of drain current in PCT and sufficient dead time. As the left figure indicates, there is no dependency of the switching energy on the dead time for negative drain currents in the PCT. In contrast, a strong dependency exists in case of a positive drain current of the PCT. A comparison of the switching energies for different positive drain currents in the PCT shows, that a higher drain current, which equals the load current right before the switching process, results in a smaller minimum sufficient dead time. In case of 5 A load current even a dead time of 25 ns is not sufficient to ensure zero voltage switching. In contrast, even 15 ns are sufficient to achieve zero voltage switching at 9 A load current.

To validate the presented methology, a comparison between the presented methology and measurements with a power analyzer is conducted. For this purpose, the total losses of a half-bridge for different load currents and dead times are calculated from the loss energies determined by means of the presented methology. Afterwards, measurements are performed with a power analyzer (Zimmer LMG500). Figure 10 shows the comparison of the estimated and measured losses of a half-bridge for an constantly positive and constantly negative load current plotted over the absolute value of the inductor current for different dead times.

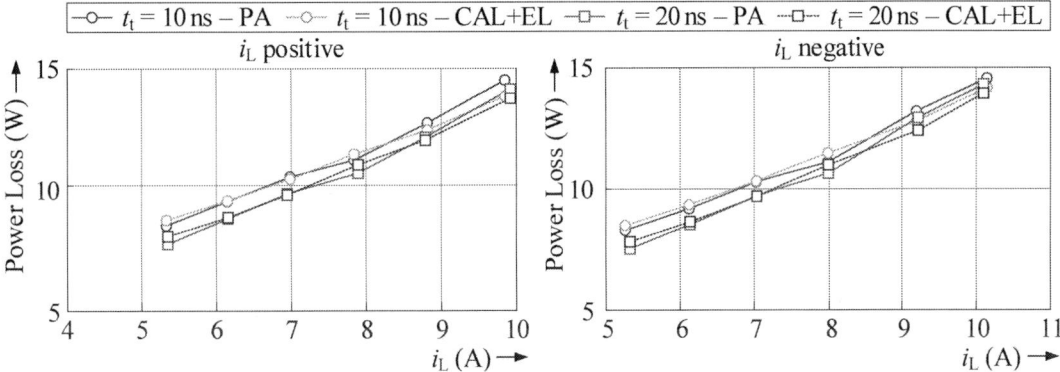

Fig. 10: Comparison between losses determined by means of the presented methology (CAL+EL) and losses measured with a power analyzer (PA)

The measured power losses show good correlation with the estimated losses. Only minor measurement errors in the range of the measurement accuracy of the calorimetric measurement method and the power analyzer are found.

Conclusion

The presented measurement method for the loss energies of half-bridges is simple but robust compared to exclusively electrical measurement methods such as the double pulse test. By combining calorimetric measurement methods and static electrical measurements, the complex measurement of the commutation current can be omitted. The presented method is successfully validated on the example of a SiC half-bridge. A comparison of the losses estimated by the presented method and measurements with a power analyzer are shown to validate the presented method. The measurement results show that the presented method allows to determine loss energies from which the losses for a switching operation with a given current through the PCT and the given dead time can be estimated. If operating points with a power dissipation of less than 5 W are to be measured, the thermal resistance of the DUT can be increased to improve the accuracy of the calorimetric measurement. Further the method can be improved by measuring the hysteresis losses of the transistor by means of a Sawyer Tower for more detailed analysis of the loss energies [11].

References

[1] M. Nitzsche, M. Zehelein, N. Troester, and J. Roth-Stielow, "Precise voltage measurement for power electronics with high switching frequencies," in *PCIM Europe 2018; International Exhibition and Conference for Power Electronics, Intelligent Motion, Renewable Energy and Energy Management*, 2018, pp. 1–6.

[2] P. Ziegler, N. Tröster, D. Schmidt, J. Ruthardt, M. Fischer, and J. Roth-Stielow, "Wide bandwidth current sensor for commutation current measurement in fast switching power electronics," in *2020 22nd European Conference on Power Electronics and Applications (EPE'20 ECCE Europe)*, Sep. 2020, pp. P.1–P.9. [Online]. Available: https://ieeexplore.ieee.org/document/9215686

[3] J. Weimer, D. Koch, R. Schnitzler, and I. Kallfass, "Determination of hard- and soft-switching losses for wide bandgap power transistors with noninvasive and fast calorimetric measurements," in *2021 33rd International Symposium on Power Semiconductor Devices and ICs (ISPSD)*, 2021, pp. 327–330.

[4] J. Weimer and I. Kallfass, "Soft-switching losses in gan and sic power transistors based on new calorimetric measurements," in *2019 31st International Symposium on Power Semiconductor Devices and ICs (ISPSD)*, 2019, pp. 455–458.

[5] M. Amyotte, E. S. Glitz, C. G. Perez, and M. Ordonez, "Gan power switches: A comprehensive approach to power loss estimation," in *2018 IEEE Energy Conversion Congress and Exposition (ECCE)*, 2018, pp. 1926–1931.

[6] A. Wintrich, U. Nicolai, W. Tursky, and T. Reimann, *Application manual power semiconductors*, 2nd ed. Ilmenau: ISLE Verlag, 2015.

[7] J. Gareau, R. Hou, and A. Emadi, "Review of loss distribution, analysis, and measurement techniques for gan hemts," *IEEE Transactions on Power Electronics*, vol. 35, no. 7, pp. 7405–7418, 2020.

[8] S. K. Roy and K. Basu, "Analytical model to study turn-off soft switching dynamics of sic mosfet in a half-bridge configuration," *IEEE Transactions on Power Electronics*, vol. 36, no. 11, pp. 13 039–13 056, 2021.

[9] B. Kohlhepp, D. Kuebrich, R. Schwanninger, and T. Duerbaum, "Switching loss estimation of gan-hemts by thermal measurement procedure," in *PCIM Europe digital days 2021; International Exhibition and Conference for Power Electronics, Intelligent Motion, Renewable Energy and Energy Management*, 2021, pp. 1–9.

[10] J. Brandelero, B. Cougo, T. Meynard, and N. Videau, "A non-intrusive method for measuring switching losses of gan power transistors," in *IECON 2013 - 39th Annual Conference of the IEEE Industrial Electronics Society*, 2013, pp. 246–251.

[11] D. Bura, T. Plum, J. Baringhaus, and R. W. De Doncker, "Hysteresis losses in the output capacitance of wide bandgap and superjunction transistors," in *2018 20th European Conference on Power Electronics and Applications (EPE'18 ECCE Europe)*, 2018, pp. P.1–P.9.

Condition Monitoring Approach of a SiC Power Semiconductor using Turn-Off Delay with an Integration in a SiC Driver

Victor Golev, Ulf Schümann, Rando Raßmann, Jan Bockholt
THE AUTHOR'S COMPANY / INSTITUTION
University of Applied Sciences Kiel
Kiel, Germany
Tel.: +49 / (431) – 210.4226.
Fax: +49 / (431) – 210.64212.
E-Mail: victor.golev@fh-kiel.de
URL: https://www.fh-kiel.de/

Keywords

«Condition Monitoring», «Silicon Carbide (SiC)», «Smart Gate Drivers», «Intelligent Gate Driver», «Driver Concepts»

Abstract

This paper deals with a condition monitoring approach based on the turn-off delay measurement and its integration in a SiC driver to measure the temperature-sensitive parameters in a continuous switching operation of a SiC power semiconductor. The presented approach requires a small area for integration and can be implemented with little effort.

Introduction

With the increasing demand for highly efficient power conversion systems, especially in the field of e-mobility and renewable energy, modern power semiconductors such as SiC and GaN are gaining in importance. Due to their small size, the modern semiconductors enable the development of high-speed switching and compact systems. This creates an additional demand for small state of health monitoring systems for wide bandgap semiconductors, which can significantly increase the operational reliability and reduce maintenance costs. The detection of a fault or the predicted service life of a power module can protect both higher-level systems from damage and the user from injury [2] [6]. In addition, it enables the user to perform predictive maintenance.

Often a higher working temperature is used as an indicator for a not optimal functioning semiconductor. Temperature changes are well known as an indicator of degradation processes in a power semiconductor. They are mainly caused by the thermomechanical effects in the layer assembly under the chips and the connections. The power cycling in a power module leads initiation and propagation of fractures in the solders and generate metallurgical damage to wire bonds and emitter metallization [7]. A common approach for condition monitoring systems is the measurement of the temperature of the semiconductor itself or dependent electrical parameters [3]. For common state of health monitoring systems of a semiconductor, various variables such as currents, voltages, resistances and temperatures are measured in real time. Other condition monitoring systems are using dynamic parameters such as switching gradients and time delays of a semiconductor which is a particular challenge [8][9]. The approach to determine a parameter is individual and requires technical effort which is depending on the method and the area of application. For example, high-resolution analog-digital-converter (ADC) and high-speed digital signal processor (DSP) are often used for condition monitoring based on the switching characteristics of the semiconductors. The cost for solutions of this kind can make such a system uneconomical.

This paper presents a possibility to detect the temperature changes of a semiconductor by integrating a turn-off-delay-based circuit into a SiC driver using a Time-to-Digital-Converter (TDC). The turn-off delay time is measured as a function of the chip temperature respectively of a switch and the drain current of a SiC semiconductor using a self-built driver. The main aspect of this measurement is the use of the TDC device, which measures the time between the input and the output of the turn-off edge of a semiconductor with a high resolution.

Basic Measurements

One possible dynamic method for monitoring the temperature change behavior of a semiconductor is the turn-off delay measurement [1]. It has been proven that the change in turn-off delay is directly related to the chip temperature and the drain current through the chip. This relationship arises from the temperature dependence of the parasitic capacitances of the SiC semiconductor. The main parasitic capacitances of a SiC semiconductor are C_{GD}, C_{DS} and C_{GS} (see fig. 1). The first two of the three parasitic capacitances are mainly responsible for the temperature dependence of the turn-off delay [4]. The following measurements in fig. 1 clearly show a correlation between the turn-off delay of a SiC semiconductor, the chip temperature and the drain current. The setup for the measurement of this experiment is based on a double-pulse experiment. The SiC semiconductors were heated from the outside using a heating plate. All signal traces were triggered to the same gate signal. A change in temperature of 150 °C causes a turn-off delay of 30 ns [1].

The effect described above can be used as a condition observer for online monitoring of a SiC-MOSFET. As already mentioned above, a significant change in the chip temperature is often an indicator for a degradation of the semiconductor.

The technical prerequisite for the use of the measuring system based on the effect described above is the temperature insensitivity of electrical components. Furthermore, the results achieved have no absolute reference point and can only be evaluated in reference-free systems. This is required because of the variation of values of the parasitic capacitances of such SiC MOSFETs depending on the manufacturer and the manufacturing processes.

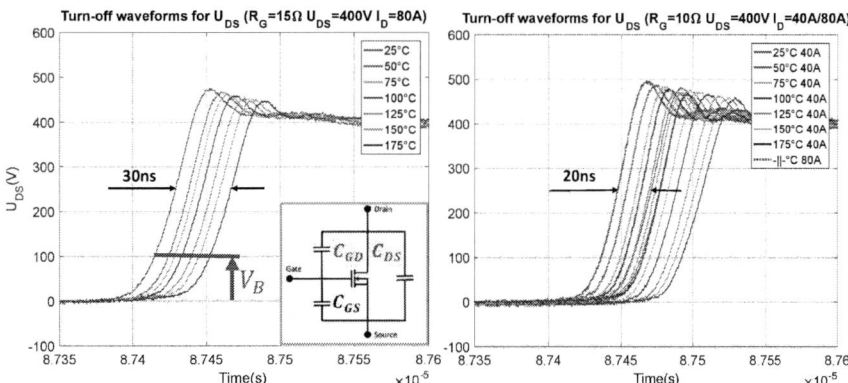

Fig. 1: Turn-off delay dependence on the temperature and drain current in the SiC semiconductor (C2M0025120D) [1]

It should be mentioned that the turn-off delay depends additionally on other parameters such as the gate resistance of the drivers, which are not always related to the chip temperature. Therefore, it is recommended to use gate resistors with a low temperature coefficient for such applications. In this way, the temperature drift can be reduced.

Another challenge to describe the temperature change of the semiconductor using the effect described above is the voltage and current dependence. The separation of these dependencies from each other in a practical application of the described effect is described in further chapters of this paper.

Measurement Circuit and Setup

This chapter describes the circuit for measuring the turn-off delay based on a TDC module. The TDC module measures the time between the turn-off edge of the trigger signal (driver) and the drain-source voltage. Most TDCs work with a Transistor-Transistor-Logic (TTL) and for this reason the level must be adjusted to the TTL level [5]. The main requirement for such a circuit is to transmit the measured drain-source voltage to the input of the TDC device without any loss of information. Therefore, the signal has to be reduced in the amplitude and protected against distortion.

The drafted interface circuit shown in the center block of fig. 2 consists of the following components:
- high voltage blocking diode circuit,
- filter
- impedance converter
- voltage divider

The high voltage blocking diode circuit with an integrated voltage source ensures that the high drain-source voltage of the switch is cut of at the level of the blocking voltage V_B. In fig. 1 we can see that not all positions of the switching edge signal are suitable for measuring the time delay because of the non- linear signal in the lower and upper areas. The blocking voltage should not be selected too small in order to remain in the linear range of the turn-off edge.

When the power semiconductor is switched on the diode D_1 conducts and the drain-source voltage is measured via the input impedance converter. It should be mentioned that the input impedance converter, used in the circuit, is made up of J-FETs. This is a cheaper approach for an alternative impedance converter based on an operational amplifier with a low input capacitance. The fig. 2 shows the schematic measuring principle of the system [1].

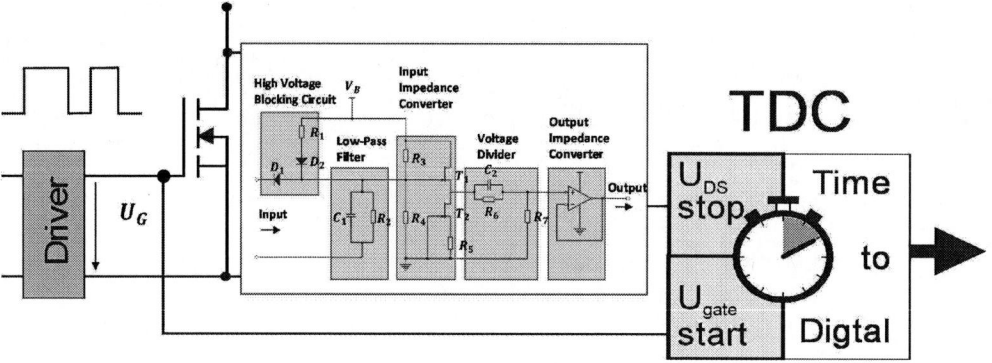

Fig. 2: Block diagram of the turn-off delay measuring system

The turn-off delay measuring system described above has been integrated into a SiC driver. A driver circuit is located on the top side of the board and the TDC measuring circuit is placed on the bottom side (see fig. 3c/b). The integration area of the TDC measuring circuit is approx. 2 cm^2 and is located directly at the high-voltage side. As shown in fig. 3, the circuit is decoupled from the low voltage side via a digital isolator. The communication with the TDC module is implemented via a serial interface. The power supply is also implemented with an isolated 2.5W DC/DC converter in a small package. The layout of the individual component groups can be found in fig. 3a.

The time to digital converter module TDC7200 from TI is used for the time measurement. This device has an internal self-calibrated time base which compensates the drift over time and temperature. Self-calibration enables the time-to-digital conversion accuracy in the order of picoseconds [5].

The driver itself is based on the UCC217100 chip and is a fully integrated SiC driver for a half bridge topology.

Fig. 3: Driver with TDC measurement circuit

The measurement tolerance with the presented TDC driver becomes smaller if the stop signal is interference-free and properly prepared for the measurement with TDC. In addition, the measurement tolerance increases proportionally with the increase of the drain-source voltage. For this reason, it is important to design the input of the interface circuit with as little capacitance as possible. The circuit shown in this paper was developed based on an active probe.

A SiC full bridge with power semiconductors has been set up to test the developed driver with the TDC-based measurement system. The complete measurement setup is shown in fig. 4. The central part of it consists of a heating plate which enables the heat up and cooling of the semiconductors independently of the drain current and the DC-link board which carries the SiC power modules. The setup has operated with an inductive load and the current has been measured by using a current probe (I-Prober 520). The turn-off delay time has been measured across of all four semiconductors by using two half-bridge drivers each with two TDC measuring circuits. The two SiC power modules are attached below the DC-link board and are thermally coupled to the heating surface via a heat-conducting pad. A phase shift control is implemented on a Aurix TC277 microcontroller board to control the semiconductors. The evaluation of the measurement results and reading out the TDCs is also done with the Aurix. In addition, a fan was used to cool the driver.

The Temperature measurement in this measurement setup is realized by several NTC resistors, which are placed close to the SiC semiconductors. The values of the read-out temperature are processed directly into the test rig software via the microcontroller.

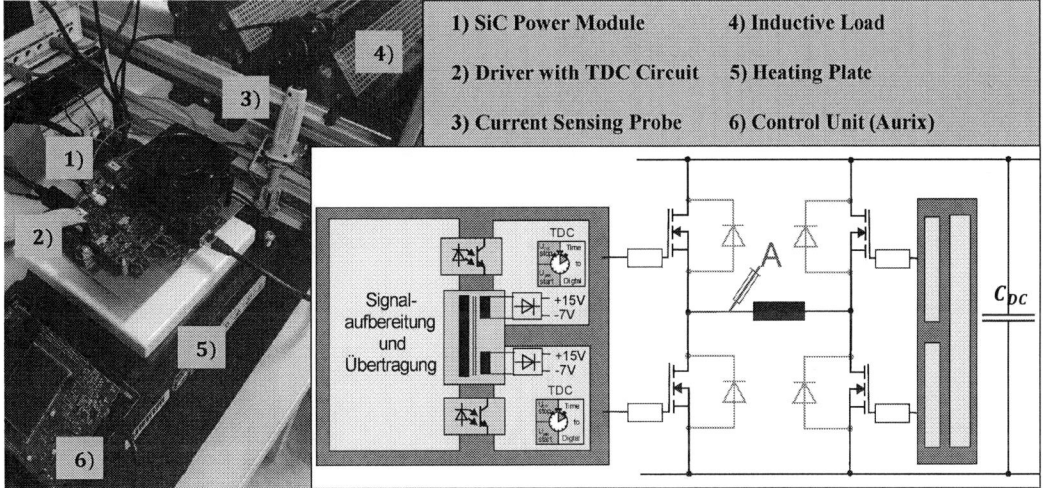

Fig. 4: Turn-off-delay measurement setup realized using the SiC full bridge with the delf-developed driver

The measuring process on the measurement setup is as follows.

The value of the current value which flows through the measured object (semiconductor) is set via the adjustable load or phase shift between two half-bridges. If the phase shift and the load remain the same for several periods, the same current remains the same for every turn-off edge. In this way, the temperature can be adjusted with the heating plate independently of the current. This TDC measurement setup can record the turn-off time delay either continuously on every turn-off edge or on demand.

Measurements

This chapter analyses the results of the measurement. The TDC driver is tested on two different systems. The measurements on the full bridge are made over several periods with a constant current and a small amplitude of up to 5 A_{PK}. In this way, the self-heating of the power module was minimized by the losses. The driver was then put into operation on a frequency converter system.

Validation of the system on a full bridge

The fig. 5 shows the turn-off delay recorded with the developed driver in the measurement setup on the full bridge as described above. The temperature measured by the NTC resistor with uniform heating over a longer time period correspond to a mean chip temperature and is not the junction temperature. Temperature measured in this way is nevertheless directly related to junction temperature. The diagrams in fig. 5 show the dependency of the turn-off delay to the chip temperature with variation of the current in the range from 1 A_{PK} to 5 A_{PK}. Here, the variant a) shows the nearly linear temperature dependency and variant b) shows the non-linear dependency on the current. This means that the current and the temperature have different influences on the turn-off delay time in terms of linearity.

This measurement has been recorded with a measurement accuracy of $\pm 250\ ps$. Changing the temperature by 10 °C causes a turn-off delay of approx. 4 ns. This results in a measurement tolerance of approx.1 °C.

In contrast to the temperature dependency, the current dependency is more significant, especially with small currents. As the current increases, the turn-off delay saturates. For the practical application of the effect, this means that the temperature-dependent measurement should be carried out at a high current. In this case the temperature dependence will dominate over the current dependence. In this way, the measurement accuracy of the turn-off delay can be increased as a function of the chip temperature.

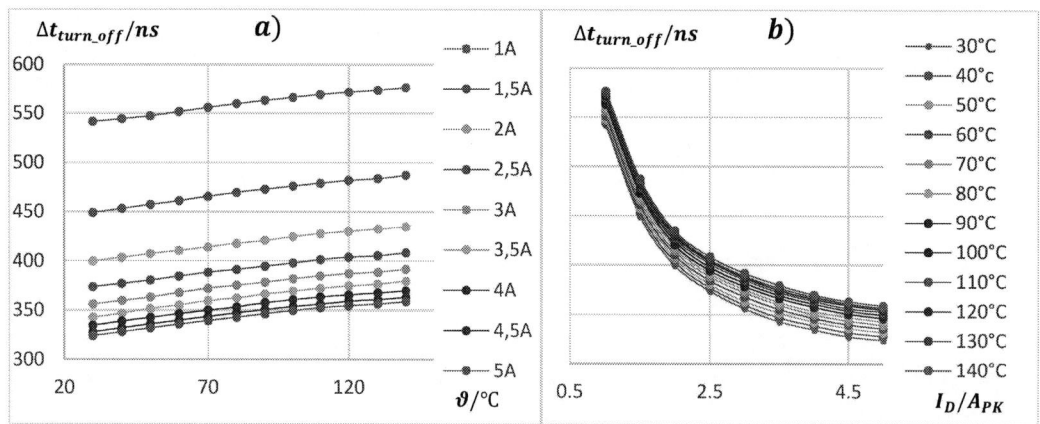

Fig. 5: Turn-off delay measurement on a H-bridge made of SiC power semiconductors with an inductive load ($V_{ds} = 100$ V)

Revalidation of the system in a 3-phase frequency converter

A 3-phase SiC frequency converter was used to test the measurement electronics on the frequency converter during operation. This measurement setup is designed for the use of the designed drivers with the TDC measuring system. To process the measurement data, a software has been developed and implemented on the microcontroller (Aurix) together with a software for the control of the converter.

The converter system has been operated with an inductive load and a ventilated heat sink. The drivers were also cooled with a fan. In the ideal case, the control electronics should be thermally decoupled from the semiconductors to minimize the thermal feedback into the driver circuit. Such a decoupling cannot be realized in a practical application. To minimize this effect, heat-insulating materials have been used in combination with a thermal management.

The following measurement in fig. 6 has carried out on converter system under load. The measurement was recorded at a current switching frequency of 1 kHz over 2.5 periods. The measured values cannot be recorded with every switching process. This is because the current does not flow through the same switch with every switching process in a half bridge operation. The current flows through different switches every second period or, due to the dead time, the current flows through the body-diode. In these cases, the turn-off delay of the measured switch is constant or not defined.

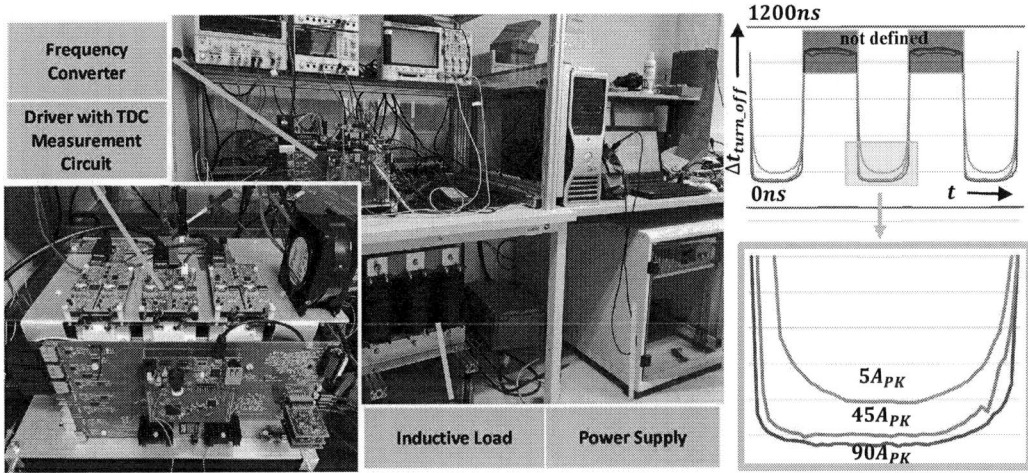

Fig. 6: Turn-off-delay measurement and measurements setup in frequency converter operation

For technical reasons, it was not possible to record the temperature during this measurement. However, the results do correlate with the measurements in a full bridge. It can be seen, that the turn-off delay saturates with large currents. It must be considered that large flowing currents cause additional heating of the SiC semiconductor due to losses at the on-resistance. Therefore, it is assumed that the current has a quadratic dependency on losses [10].

$$P_{loss} = P_{sw} + I_{D(on)}^{2} \cdot R_{DS(on)} \cdot D_{Duty\ Cycle} \tag{1}$$

Another aspect that influences the turn-off-delay measurement in a converter system is the cooling system. This means that such a curve as in fig. 5a for the temperature dependency is not possible with a conventional converter due to the cooling capacity of the cooling system and the AC current. The heating and cooling process of a cooled semiconductor are different in their slope. Based on the difference in the slope of the warm-up and cool-down phases the influence of the thermal capacity of the cooler becomes visible in the measurement (see fig. 6) [11].

Evaluation of the measurement Results

The main idea of this paper is to realize a state of health monitoring of a power module. The practical application of the measurement system presented to determine the absolute chip temperature is difficult by various factors such as current dependencies. The application of the turn-off delay measurement system to a power module or a half-bridge due to the same manufacturing and operating conditions as the same cooling system or the use of the same chips minimizes the effects that cause the scattering of the measurement results. The temperature drift of electrical components such as the gate resistor also has an impact on the charging time of the parasitic capacitances in a power module. In a power module both switches are usually controlled by the same driver board with the same gate resistors. This allows a direct comparison between the high- and the low-side switches with regard to the temperature-sensitive switching characteristics.

The turn-off delay time must be measured and evaluated for high-und low side switches periodically at the same current level. A significant difference between two values indicates the malfunction of a switch respectively a power module. The method can also be applied to a full bridge, half bridge, B6 module or multi-level converter systems. These values must be recorded for all semiconductors within a system and compared with each other. The accuracy of the measurement increase with the number of observed switches and the failure of a single switch can be predicted more accurate.

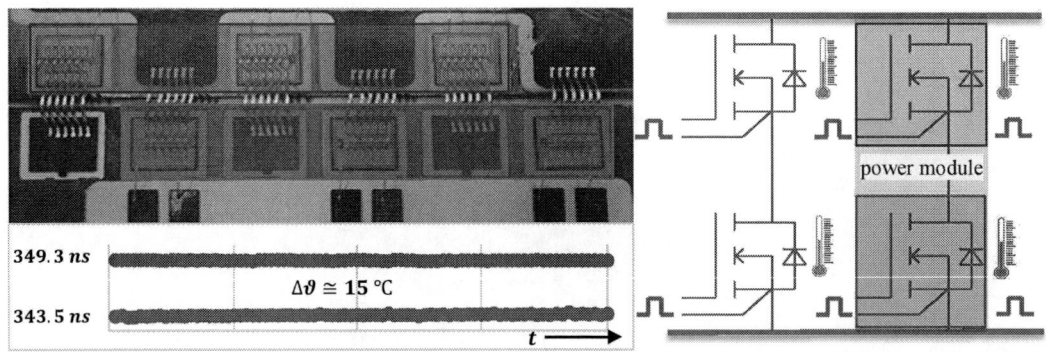

Fig. 7: Temperature comparison between the upper and the lower semiconductor inside a SiC power module

To prove the function of the method, the switches in a half-bridge of a power module were heated to different temperatures. This represents the same effect which would occur during an error state. In this case, both switches had a temperature difference of 15 °C. The tests were carried out in the measurement setup shown in fig. 4. The tested power module consists of a half bridge with 3 chips in parallel per switch. The results for such a measurement can be found in fig. 7. The set the temperature difference of 15 °C between the high- and low-side switch has been measured and validated in continuous mode of operation in the full bridge setup.

Conclusion

As part of this work, a measurement system based on the TDC module was presented. In addition, the measuring circuit was implemented in a SiC driver and tested on two different systems. The measurements have shown that the measuring principle as well as the measuring system itself represent a functioning condition observer for a temperature-sensitive parameter while the converter is in operation. It was metrologically demonstrated that the system can measure temperature differences between two switches within a power module. Using this measurement method can be developed within a frequency converter or energy converter. Malfunctions of a switch can be detected based on turn of delay time differences for example between switches in a half bridge configuration.
The measuring circuit itself is compact and inexpensive and can be implemented in small areas without great effort.

References

[1] Golev et al., "Measurement of Temperature-Sensitive Parameters of SiC Power Semiconductors during Turn-Off using a Time-to-Digital Converter", Kiel, University of Applied Sciences Kiel, Germany, 2020

[2] Yang et al., "Condition Monitoring for Device Reliability in Power Electronic Converters: A Review," IEEE Trans. Power Electron., vol.25, no.11, pp.2734-2752, 2010

[3] N. Baker et al., "Junction Temperature Measurements via Thermo-Sensitive Electrical Parameters and their Application to Condition Monitoring and Active Thermal Control of Power Converters", IECON Conference, Vienna, 2013

[4] Li et al., "Analysis of Voltage Variation in Silicon Carbide MOSFETs during Turn-On and Turn-Off", Energies, Volume 10, Issue 10, 2017

[5] Texas Instruments, "Datasheet TDC7200 Time-to-Digital Converter for Time-of-Flight Appplications in LIDAR, Magnetostrictive and Flow Meters", Dallas, Texas, 2019

[6] Han et al., "Condition monitoring techniques for electrical equipment - a literature survey," IEEE Trans. Power Del., vol. 18, no. 1, pp. 4–13, Jan. 2003

[7] Smet V. et Al., "Ageing and failure modes of IGBT modules in high-temperature power cycling", IEEE Trans. on Ind. Electronics, vol. 58, no. 10, pp. 4931 – 4941, 2011

[8] Gonzalez et Al., "Temperature Sensitive Electrical Parameters for Condition Monitoring in SiC Power MOSFETs", 8th IET International Conference on Power Electronics, Machines and Drivers (PEMD 2016), 19-21 April 2016

[9] Zhang et Al., "Online Junction Temperature Monitoring Using Intelligent Gate Drive for SiC Power Devices ", IEEE Transactions on Power Electronics, vol.34, no.8, pp.7922-7932, 2018

[10] Wrzecionko et Al., "A 120°C Ambient Temperature Forced Air-Cooled Normally-off SiC JFET Automotive Inverter Systems", IEEE Transactions on Power Electronics, vol.29, no.5, pp.2345-2358, 2014

[11] Ghaisas G. et Al., "A Critical Review and Perspective on Thermal Management of Power Electronics Moduls for Inverters and Converters", Indian National Academy of Engineering, 2021

Measurement results of Multilevel Hysteresis Control for paralleled Two-Level Converters

Magdalena Gierschner[1], Yves Hein[1] and Hans-Günter Eckel[1], Christian Heien[2]
[1]University of Rostock,[2]Wobben Research an Development GmbH
Albert-Einstein-Straße 2
Rostock, Germany
Email: magdalena.gierschner@uni-rostock.de
URL: https://www.uni-rostock.de/

Keywords

≪Converter control≫, ≪DC-AC converter≫, ≪Interleaved converters≫, ≪Parallel operation≫, ≪Wind energy≫

Abstract

The multilevel hysteresis control strategy was adjusted to four paralleled two-level converters. Simulations were done to compare it with uncoordinated hysteresis control and current control with pulse width modulation. Although this novel control strategy has its greatest advantages for several parallel converters and was originally designed for 14 parallel converters, the results for four converters are still significantly better than state of the art hysteresis cotrol. Furthermore, a test bench with four paralleled converter was build up to prove the simulation results.

Introduction

Paralleled two-level converters can be used to increase the power rating of grid-side converters. The redundancy is advantageous in case of converter failures, too. Moreover, different control strategies can reduce the switching losses and the Total Harmonic Distortion (THD) of the output current. Uncoordinated hysteresis control and PI current control with flattop pulse width modulation (PWM) and interleaved carrier signals are state of the art and widely used in industry [1]. The most significant advantage of hysteresis control is the fast dynamic response, but depending on the hysteresis bandwidth, either switching losses are high, or power quality is poor. In contrast, PI current control is slow but characterized by a high quality of the grid current [1, 2]. In [3] a novel control concept, called multilevel hysteresis control, is introduced for 14 paralleled converters. It is based on a multilevel hysteresis modulator and combines the benefits of a fast dynamic response and a low THD. The coordinated hysteresis control is adjusted to four paralleled inverters, and a test bench was designed to prove the simulation results. The paper is structured as follows: The first part explains the theoretical control concept. In the second part, the coordinated and uncoordinated hysteresis control simulation results are compared with PI current control. The test bench and the measurement results are shown in the third part.

General Concept of Multilevel Hysteresis Control with Current Balancing

The system consists of four parallel converters that are sharing one DC-link (Fig. 1). They are connected via chokes L at the AC-side, and at the Point of Common Coupling (PCC), the sum of the individual converter currents $i_{a1}, i_{b1}, i_{c1}, ..., i_{a4}, i_{b4}, i_{c4}$ gives the grid currents i_a, i_b, i_c. The currents i_a, i_b, i_c are fed via a transformer into the grid. The transformer is modeled with the inductances, L_T, and a constant voltage source provides the DC-link voltage. These four converters can be operated as a multilevel converter with $4 + 1$ voltage levels [4].

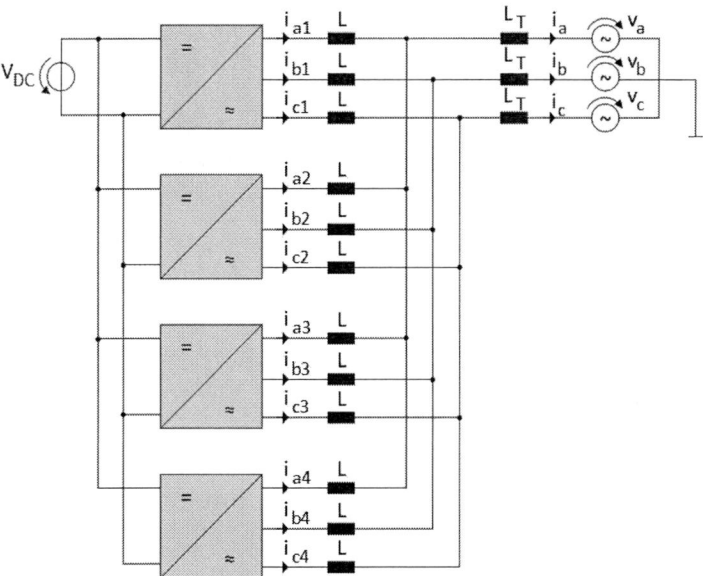

Fig. 1: Structure of the simulated system

The control concept is shown in Fig. 2. The grid current is controlled in the rotating dq-reference frame with Proportional (P) controllers to avoid windup because of the discrete output values of the following hysteresis controller. The angle θ of the measured voltage v_m is used to transform the DC-values into the three-phase abc system and vice versa. The sum v_i of the measured voltages v_m and the required voltage across the chokes v_F are the multilevel hysteresis modulator's input. The output of the multilevel hysteresis modulator is the desired voltage level k. This voltage level can be generated with multiple switching states of the parallel converters. Therefore, it is necessary to choose the optimal switching state. This happens in dependence of the individual current. To reach a one step higher voltage level, the converter with the lowest individual current is switched on and to reach a lower voltage level, the converter with the highest individual current is switched off.

Simulations have shown that this kind of balancing is insufficient to keep the individual currents in a specified band. Therefore, an additional balancing is implemented in the state machine without affecting the actual voltage level k. This concept switches the on/off state of the converter with the highest phase current and the converter with the lowest phase current (explained in detail in [3]). If it is impossible to switch converters because all four are on respectively off, the balancing must change the voltage level. For protection reasons, the individual current must be limited, which can disturb the voltage level, too. This is a real drawback, if only four parallel converters are used, because a state where all converters are switched on respectively off is reached more often than with 14 paralleled converters.

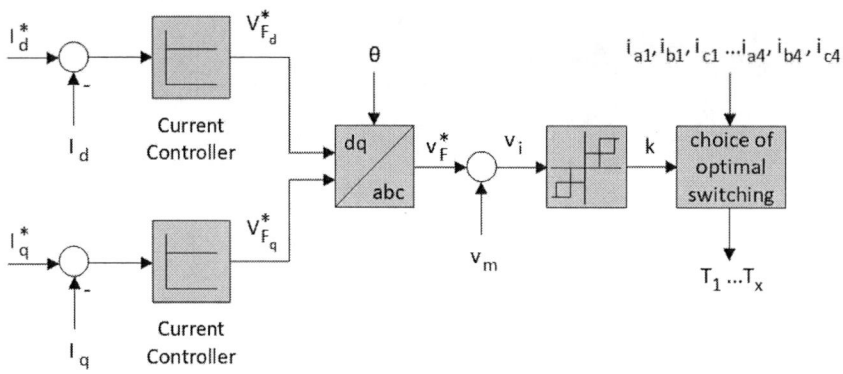

Fig. 2: Multilevel Hysteresis Control with Current Balancing

Simulation Results of Multilevel Hysteresis Control with Current Balancing

The grid-side converter of a wind-energy plant with a nominal power of $P_n = 4.26\,\text{MW}$ was modelled. For nominal power (pure active power) the reference output current is $i_d^* = 4638\,\text{A}$ and $i_q^* = 0\,\text{A}$. To reduce the converter losses, the DC-link voltage is set to the minimal needed voltage with a reserve of 5 % [5]. To calculate the converter losses, the electrical parameters of the traction converter FF1800XTR17T2P5 are used.

The gain of the P controller for d- and q-current and the hysteresis bandwidth i_{diff} influence the performance of the current control, with a higher controller gain, the dynamic increases. However, the switching frequency and the switching losses increase, too. The smaller the hysteresis bandwidth, the smaller is the current error, and the THD decreases, but the switching frequency respectively the switching losses will increase. To find the pareto optimal outcome the P gain was varied between 1 and 4 and the current hysteresis bandwidth between 1000 A and 1100 A. The results are plotted in the pareto chart of Fig. 3. The pareto optimum was defined as the smallest distance to coordinate origin.

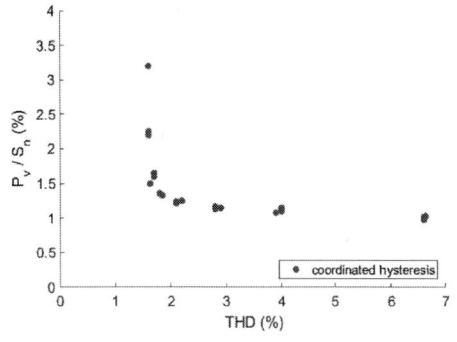
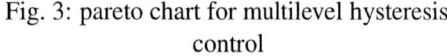

Fig. 3: pareto chart for multilevel hysteresis control

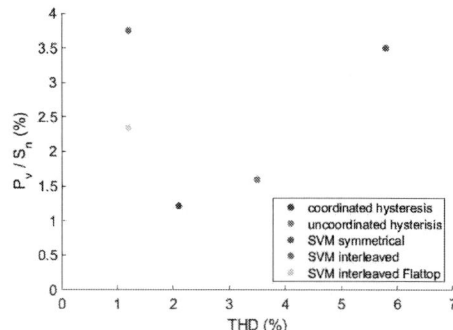

Fig. 4: comparison of different current control and modulation strategies

In Fig. 4 the pareto optimum of coordinated multilevel hysteresis control (blue) is compared with the pareto optimum of uncoordinated hysteresis control (green) and the converter losses and THD of the output current for PI current control with different PWM strategies. PWM strategies (red) can be optimezed by use of interleaved switching, which leads to a better THD (orange) and Flattop Space Vector Modulation, which saves additionally round about one third less losses (yellow). However, multilevel control can reduce the THD of the output current as well as the converter losses, too. Of course, this effect is even higher for more paralleled converters but is still significant for four paralleled converters.

Measurement Results of Multilevel Hysteresis Control with Current Balancing

The test bench is shown in 5 and 6. It consists of ten paralleled back-to-back converters with identical parameters. The shared DC-link has a nominal voltage of $V_{DC} = 50\,$V and a maximum voltage of $V_{DCmax} = 80\,$V. The nominal current of each converter is $i_n = 5\,$A and the maximum current is $i_{max} = 15\,$A. The thermal design is for a switching frequency of $f_s = 5000\,$Hz.

Fig. 7 shows the measurement results of a setup with four paralleled converters. The reference output current is $i_d^* = 28\,$A. At $t = 36\,$ms a step in the set point i_d^* from $100\,\%$ to $50\,\%$ of the current occurs. The maximum allowed current difference is chosen to $i_{diff} = 5\,$A. The measurement results prove the excellent dynamical behaviour of the multilevel hysteresis control. The measured current follows the reference value nearly instantaneously without any overshoot and the current limits are not exceeded.

Fig. 5: Test bench:
DC-link, filter and grid

Fig. 6: Test bench:
converter

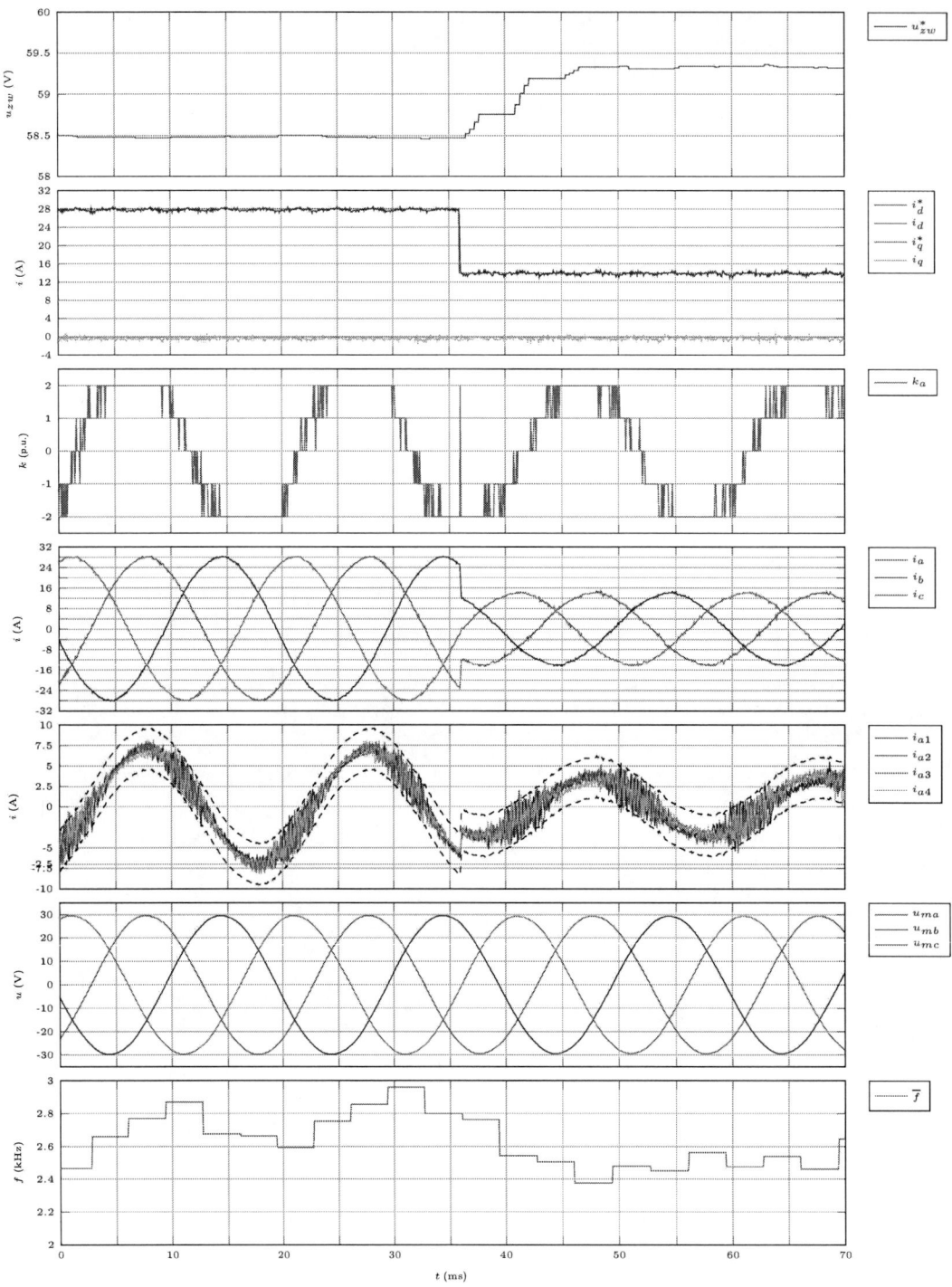

Fig. 7: step in the reference current

Conclusion

The coordinated multilevel hysteresis control strategy was adjusted to four paralleled two-level converters. Simulation results are promising. The THD of the output current, as well as the overall losses of the converter, are minimized. Furthermore, the fast dynamic response of the coordinated multilevel hysteresis control is comparable with the excellent dynamic behaviour of conventional hysteresis control. A test bench was built, which confirms the expected advantages. This novel current control and modulation strategy can be used for every converter control strategy with an inner current control loop, and it is not limited to grid- side converters with state of the art grid-feeding control concepts.

References

[1] A. Gensior, "Approximated sliding-mode control of parallel-connected grid inverters," 2020 22nd European Conference on Power Electronics and Applications, Lyon, France, 2020

[2] M. Meyer and A. Sonnenmoser, "A hysteresis current control for parallel connected line-side converters of an inverter locomotive," 1993 Fifth European Conference on Power Electronics and Applications, Brighton, UK, 1993

[3] Y. Hein, H. Eckel and C. Heyen, "Multilevel Hysteresis Control with Current Balancing of paralleled Two-level Converters," 2021 23rd European Conference on Power Electronics and Applications (EPE'21 ECCE Europe), 2021

[4] Vijay Kannan, "Operation of Parallel Connected Converters as a Multilevel Converter," PHD Dissertation, Technical University of Dresden, 2018

[5] M. Schröder, S. Gierschner and H. Eckel, " Influence of variable dc link voltage for grid side VSC control techniques in terms of current harmonics", PCIM 2013

Design and Development of a Short-Circuit Test Bench for Low-Voltage Direct Current Protection Devices

Simon Ravyts*,+, Thomas Vandenbussche*, Koen Stul*,+ and Jan Cappelle*,+
*KU Leuven - Department of Electrical Engineering (ESAT) - ELECTA GENT
*Gebroeders De Smetstraat 1, 9000 Gent, Belgium
+EnergyVille, Thor Park 8310, Genk, Belgium
Email: simon.ravyts@kuleuven.be
URL: https://www.energyville.be

Keywords

≪AC-DC microgrid≫, ≪DC circuit breaker≫, ≪DC power supply≫, ≪LVDC≫

Abstract

In this paper, the design and development of a short-circuit test bench for LVDC protection devices will be presented. The popularity of LVDC microgrids is increasing due to the better compatibility with DC loads, DC storage and DC generation. However, the protection of these grids remains one of the key hurdles to be overcome before a further industrialization is possible. An important aspect to be considered is the fact that the testing procedures that are described in today's standards do not align with the short-circuit currents that are encountered in converter dominated LVDC grids. The standard IEC test procedures are based on a strong, inductive LVDC grid. However, a converter dominated LVDC grid is more capacitive than inductive and the short-circuit currents that flow contain a fast transient peak that might go unnoticed. Therefore, a test bench was developed to test LVDC circuit breakers under realistic fault conditions.

Introduction

The quest for higher efficiencies in electrical networks has motivated researchers to investigate Low-Voltage Direct Current (LVDC) networks. LVDC networks offer several advantages compared to an AC distribution. First, there is a higher compatibility with DC loads (LED, IT equipment, VSDs, ...), DC generators (PV) and DC storage (batteries). If these native DC loads are interconnected on a DC network, the total amount of conversion steps can be reduced, which leads to an efficiency increase [1]. Further more, the components that are required to convert AC to DC and back can be omitted from the system, which leads to a reliability increase [2] and a reduction in the required amount of components and thus raw materials. Second, LVDC grids offer a higher power transfer capability for a given conductor cross section, compared to an AC system [3]. Third, DC grids are inherently controlled by power electronics converters. This allows to actively control the power flow within the DC system, a distinct requirement in a smart grid.

One of the main arguments against LVDC networks is the difficulty to protect the system when overcurrents, such as short-circuits occur. The arc that is formed in the protection mechanism is more difficult to quench due to the lack of zero-crossings [4]. Three short-circuit protection systems can be distinguished, namely fuses, mechanical Circuit Breakers (CBs) and Solid State Circuit Breakers (SSCBs) [5]. The use of fuses in LVDC grids has been treated in detail in [6], where a protection methodology has been proposed. Also SSCBs are intensively investigated and many topologies have been proposed in literature [7]. However, research towards the use of mechanical circuit breakers in LVDC systems is missing in literature. Their drawbacks, such as a slow response and the need for a sufficiently large steady state

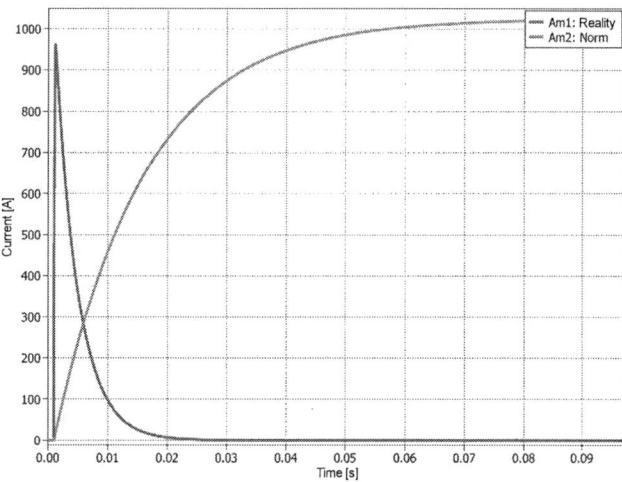

Fig. 1: Comparison of the DC short-circuit behavior according to IEC norms (blue) and the behavior in a converter dominated DC grid (red).

short-circuit current [4], are typically described but, to the author's best knowledge, no guidelines exist for their correct use in LVDC grids. Furthermore, it will be shown in this paper that there is a discrepancy between the standard testing methods for DC CBs (as prescribed by the IEC standards) and the actual current that needs to be interrupted in a converter dominated LVDC grid. The structure of this paper is as follows: First, the standard testing procedures will be described and they will be compared to the expected fault currents in a LVDC grid. Then, the design and development of a short-circuit test bench will be discussed to test the CBs under more realistic conditions for LVDC grids. Finally, the results of the short-circuit tests for CBs will be discussed.

Standard testing procedures of DC MCBs

There are two IEC standards that cover the testing methods for both AC and DC Circuit Breakers (CBs), namely IEC 60947-2 and IEC 60898. The first defines methods for CBs that will be used in industrial applications, whereas the latter is intended for CBs that will be used in residential applications. Although the end use is different, the described testing procedure is similar: A short-circuit needs to be made in a circuit consisting of a strong DC voltage source V_{DC}, an inductor L and a resistor R. The values are not specified separately but the time constant $\tau = L/R$ of this first-order circuit is given. For the IEC 60898, single pole circuit breakers up to 220 V DC and two pole circuit breakers up 440 V DC can be used. The ripple current needs to be below 5% and a time constant of 4 ms (+0%/-10%) for $I_{sc} < 1500$ A is used. Furthermore, there is a separate classification 'T15', where a time constant of 15 ms is used for testing. For the IEC 60947-2, circuit breakers up to 1500 V DC are defined and the time constants for the tests are 5, 10 or 15 ms, depending on the magnitude of the short-circuit current.

Short-circuits in converter dominated LVDC grids

The short-circuit behavior of converter dominated LVDC grids has been discussed in many publications. In case of a grid-connected VSC, the short-circuit behavior is typically described in three stages [8]: First, the VSC output capacitor discharges while the voltage in the DC system collapses. Then, diode freewheeling might occur, but the occurrence of this stage mainly depends on the amount of circuit inductance [9]. The third stage is the steady state fault current. This fault current runs from the AC grid, through the anti-parallel diodes of the VSC, towards the fault. To avoid a complete voltage collapse in the DC system, the fault needs to be interrupted before the capacitor is completely discharged, i.e. during the first stage. From the discussions above, it becomes apparent that the standard testing procedures do not provide a realistic scenario for converter dominated LVDC applications. This is also shown in Fig.

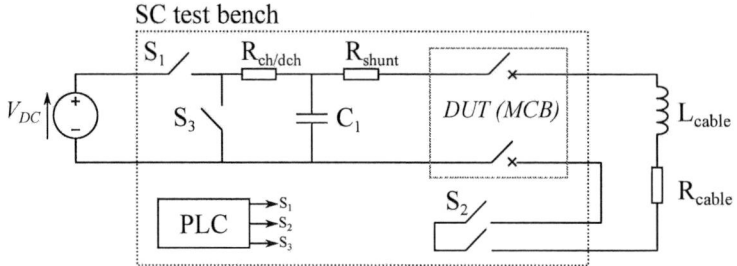

Fig. 2: Schematic overview of the short-circuit test bench.

1, where the short-circuit current of a realistic converter controlled LVDC scenario is compared to the standard testing procedure by means of a PLECS simulation. The capacitive discharge is a very rapid second-order phenomenon that dies out after 20 ms, whereas SC current of the standard is a first-order system that reaches its final value after approximately $5\tau = 5* 15$ ms $= 75$ ms.

In the remainder of this paper, it will be tested if DC MCBs are capable of tripping during the capacitive discharge, a test which is currently not included in standard testing procedures. However, this is important because tripping during the first fault stage avoids overheating and potential failure of the anti-parallel diodes, which can occur during fault stages two and three. As a consequence, complete converter breakdown can occur, leading to a complete shutdown of the DC grid.

Design of the short-circuit test bench

As the IEC testing methods are very different from the expected fault currents, the question arises: Will DC MCBs also trip under these conditions? To answer this question, a DC short-circuit test bench was developed to experimentally investigate this research question. A simplified schematic overview of the short-circuit test bench is shown in Fig. 2 and the Device Under Test (DUT) is highlighted in red. The different parts will be discussed below. A picture of the actual test bench is shown in Fig. 3. The short-circuit test bench was designed with following criteria in mind:

Voltage and current range

Depending on the application, different DC voltage levels are used nowadays but 350 and 380 V are very common for applications such as offices and datacenters [11, 10]. Therefore, a maximum DC voltage level of 400 V was chosen as an upper limit. Typical LVDC MCBs have a rated service breaking capacity (Ics) of 10 kA. The capacitors of the test bench were selected to deliver a maximum short circuit current up to 12 kA.

Measurement system

The measurements are done using a Tektronix DPO2014 oscilloscope, in combination with two TESTEC TT-SI 9101 100 MHz differential probes. One probe is used to measure the voltage across the capacitor bank, the second measures the voltage across a 2.4 mΩ shunt resistor R_{shunt} in the main current path. The shunt resistor current measurement was chosen for its simplicity, high bandwidth and low cost.

Safety

All components are placed inside a metal electrical cabinet to achieve a high level of electrical safety during the tests. The only components that are placed externally are the DC voltage source, used for charging the capacitors, and the oscilloscope for the measurements. Furthermore, an emergency stop button and a door safety switch are included to immediately discharge the capacitors in case of emergency.

Easy variation of test parameters

The short-circuit current in LVDC grids was discussed in the previous section. It is determined by the circuit parameters (R, L, C) and the initial voltage of the capacitor. The voltage level is set by using an

Fig. 3: Picture of the developed short-circuit test bench.

external DC voltage source V_{DC} and can vary between 0 and 400 V. The total installed capacitance C_1 is 16050 μF, divided between 9 film capacitors of 650 μF and 15 electrolytic capacitors of 680 μF. The capacitance C_1 can thus vary stepwise. The mix of film and electrolytic capacitors was explicitly chosen to investigate the influence of the capacitor ESR on the fault behavior. Film capacitors are known to have a low ESR, whereas the ESR of electrolytic capacitors is typically higher [12]. For the chosen capacitors, the ESR of the film capacitor is about 100 times lower. An external inductance is included in the design as well. To increase the resistance, an external cable can be connected in series with the fault. This is indicated in Fig. 2 with L_{cable} and R_{cable}. By varying the length or cross section, both the resistance and inductance will vary, which is useful to represent faults near the end of a feeder. An overview of the used capacitors is given in Table I.

Contact bounce

The short-circuit needs to be initiated by a contactor. Both solid-state as mechanical contactors are commercially available. The main drawback of solid state contactors in this application, is the forward voltage drop V_f of the internal bipolar semiconductor device, typically a thyristor. The large DC short-circuit current leads to unacceptable losses in the solid state contactor and therefore, a mechanical contactor was chosen. However the losses are ohmic, mechanical contactors have another issue, namely contact bounce. Contact bounce is a phenomenon that can be encountered when mechanical contacts

Table I: Overview of the used capacitors.

Type	Capacitance (μF)	ESR (mΩ)	V (V)
KEMET ALS30A681KF500 Aluminum Electrolytic	680	133	500
KEMET C44UHGT6650M81K Metallized Polypropylene	650	1.2	600

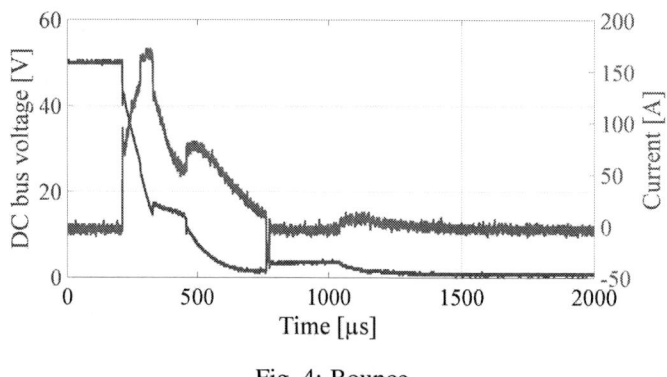

Fig. 4: Bounce.

Fig. 5: No bounce.

close. Due to the elasticity of the system, the contact will rapidly change between open and closed and consequently arcing will occur. This is shown in Fig. 4: due to the contact bounce, arcs form and the measurements are far from the ideal scenario. To reduce the occurrence of contact bounce, multiple contactors can be placed in parallel. By doing so, it becomes less likely that the contactors are in an open position at the same instant of time. This is clearly visible in Fig. 5 where two contactors are combined and the impact of contact bounce is reduced to a minimum, leading to a strongly improved performance.

Reproducibility

An automated test procedure was developed and implemented on a Siemens LOGO! PLC. By doing so, the chance on human errors during the tests is reduced to a minimum, which benefits the general safety of the system, the robustness and reproducibility of the results. The test procedure is explained with respect to Fig. 2. The test sequence is initiated by pressing the START button on the door of the electrical cabinet. First, contactor S1 is closed. The voltage source now charges the internal capacitor bank through the resistance $R_{ch/dch}$. When the capacitors are fully charged, S1 opens and S2 closes. The capacitors discharge, similar to the first phase of the short-circuit behavior, which was discussed above. The current and voltage profile are measured and one can inspect if the MCB has tripped or not. Depending on the instant of tripping, the capacitor bank can still be partially charged. To finalize the test, the capacitor bank is discharged in a controlled way by closing S3. Three measurements are shown in Fig. 6, indicating a high reproducibility of the results.

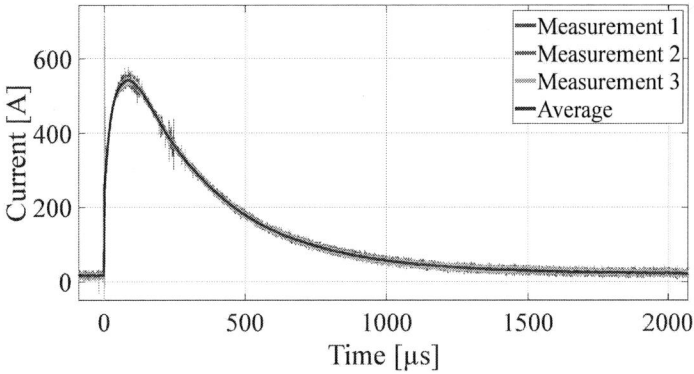

Fig. 6: Three example measurements of a capacitive short-circuit current. The results are nearly identical, indicating a large degree of reproducibility.

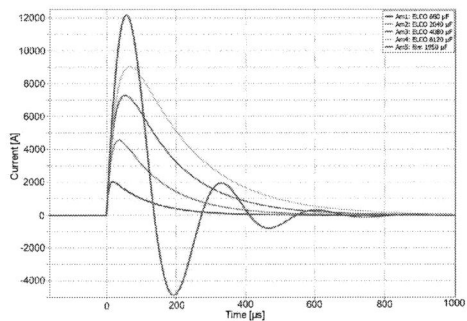

Fig. 7: Simulated capacitor discharges for different capacitors.

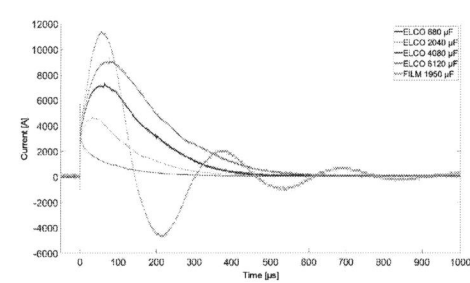

Fig. 8: Measured capacitor discharges for different capacitors.

Parasitics

An important aspect in the design of the test bench is to achieve a system with a very low parasitic inductance and resistance of the cabling inside the electrical cabinet. For this reason, insulated copper bus bars were used to connect the capacitors from one another towards the DUT. This allows to have a minimal distance between both conductors, without increasing the chance on short-circuits or flash-over. The parasitic inductance was measured to be 0.9 μH and the parasitic resistance was measured to be 20 mΩ. Figs. 7 and 8 show the comparison between simulated capacitor discharges and the actual measurement (without protection device). They indicate a strong correspondance between both, highlighting the neglible effect of the parasitics.

Short-circuit measurements with MCB

A 16A MCB with a C curve from Schneider Electric (C60H-DC series), tested according to IEC 60947-2 standard, was tested. The tripping curve is shown in Fig. 9 and the let-through energy (I^2t value) as a function of the peak current is shown in Fig. 10.

This MCB was subjected to a number of tests with varying capacitance, resistance, inductance and voltage level. By varying these parameters, it was found that two parameters seem important for correct tripping, namely ΔI, which is the current amplitude above the magnetic trip limit I_m and Δt, which is the time that the current is larger than the magnetic trip limit. The results of this test are shown in Fig. 11, including how ΔI and Δt are measured. Three cases are distinguished: First, if both ΔI and Δt are sufficiently large, tripping occurs and there is a remaining voltage on the capacitors. These measurements are indicated in blue. Secondly, there are cases where I_m is exceeded but the MCB did not trip. These measurements are indicated in yellow. Note that the remaining voltage on the capacitors is zero. Such

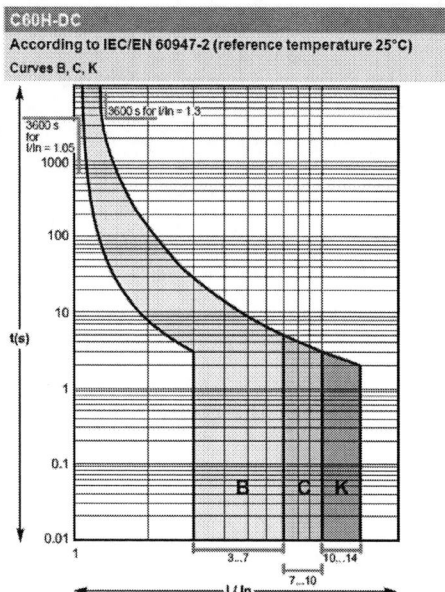

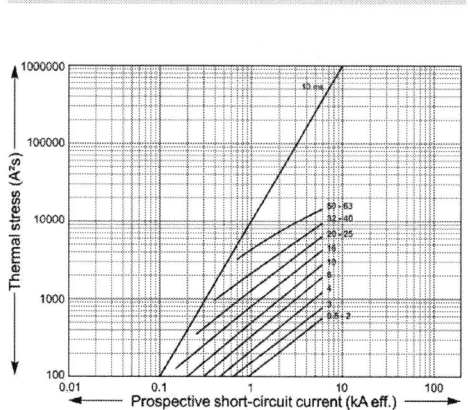

Fig. 10: Let-through energy as a function of peak current of the DUT, retrieved from the datasheet [13].

Fig. 9: Time-current characteristic of the DUT, retrieved from the datasheet [13].

an example measurement is shown in Fig. 12. However the magnetic trip limit was exceeded by more than 100 A, no tripping did occur. The two 'bumps' in the measurement are probably a consequence of internal arcing, indicating that the contacts started opening but the internal magnetic force was not sufficient to completely open the MCB. Thridly, measurements where the MCB did trip but the remaing voltage on the capacitor was approximately zero, are shown in orange and indicated as 'Limit tripping'. Note that this latter defines a limit between tripping and non-tripping. It can be seen from Fig. 9 that this limit tends to follow a hyperbolic trendline (not shown). For large ΔI, the required time Δt to assure tripping can be relatively small. In contrast, when ΔI is low, the required time Δt to guarantee tripping becomes much larger.

In contrast to fuses in LVDC grids, where tripping can be accomplished based on the system I^2t value [6], the correct use of MCBs cannot be based on the same metric. This is also visible in Fig. 13 where the let-through energy of the tests were tripping occured is calculated and displayed as a function of the measured peak current I_{peak}. When comparing this this plot to Fig. 10, a strong discrepancy between the datasheet and the measured values is visible. According to the datasheet, which is measured according to the specifications in IEC 60947-2, i.e. with an inductive short-circuit, the I^2t value is around 700 A^{2}s for a 1 kA short-circuit. The measured values are much lower and do not only rely on the peak current but also on the used capacitance.

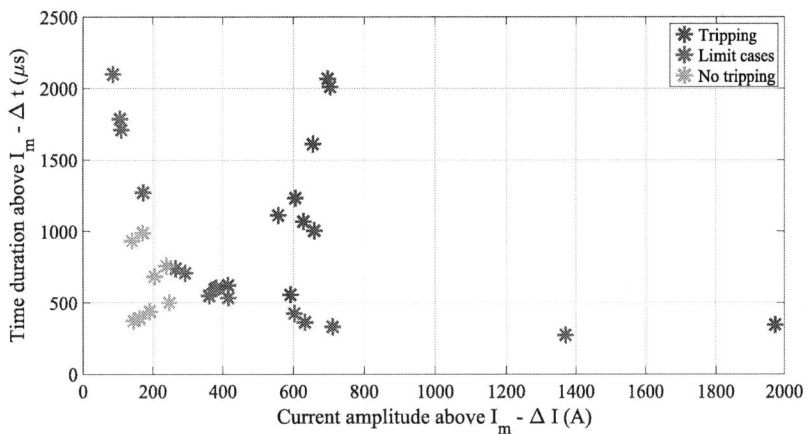

Fig. 11: Measured ΔI and Δt for a 16 A MCB with C curve.

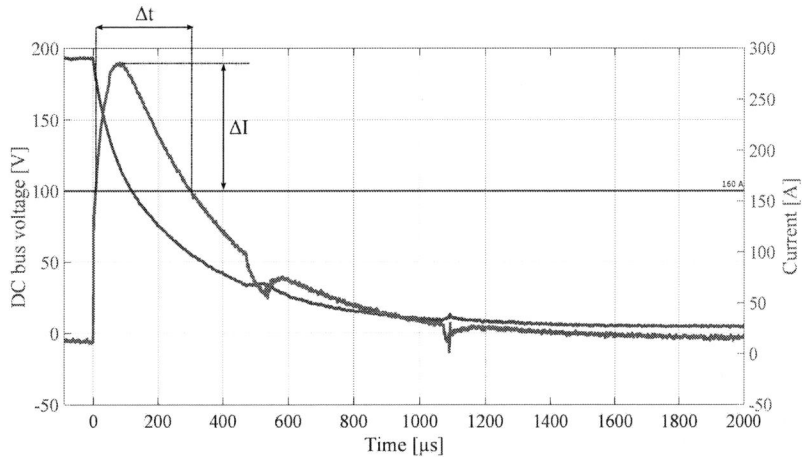

Fig. 12: Voltage and current measurement of a SC current. The peak SC current is approximately 280 A. The SC current exceeds the magnetic trip limit of the MCB (160 A for a 16 A MCB, class C, shown as a black line) but the MCB does not trip.

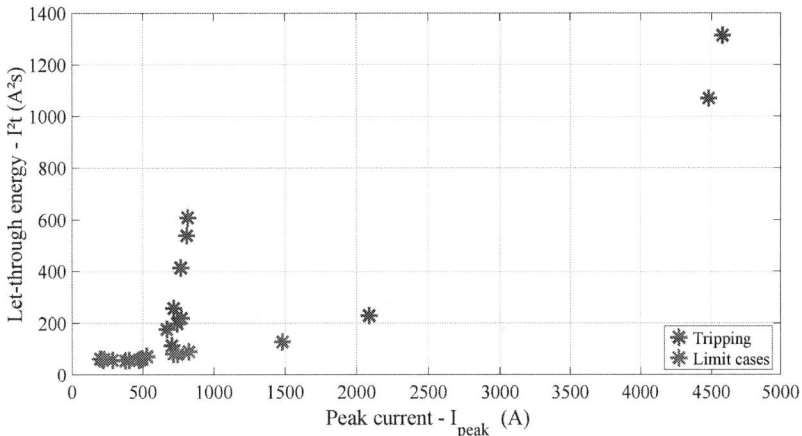

Fig. 13: Measured let-through energy as a function of the peak current.

Conclusions

The test procedures described in IEC 60947-2 and IEC 60898 are based on DC short-circuit behavior in strong, inductive DC networks. However, converter-dominated LVDC grids behave as weak, capacitive grids and their short-circuit behavior is very different. More specifically, the peak short-circuit current is a transient phenomenon that lasts for a much shorter time period than described by the standard testing procedures. The duration might not be sufficiently long to make DC MCBs trip. Therefore, a LVDC short-circuit test bench was designed and developed to test DC MCBs under realistic conditions for converter dominated LVDC grids. It was shown that the test bench provides repeatable results. Furthermore, the tests show that the hypothesis is indeed correct since DC SC breakers do not necessarily trip, although the magnetic current limit was strongly exceeded. Example measurements are done on a 16 A MCB highlight that two conditions need to be met to guarantee MCB tripping, the current needs to be sufficiently high above the magnetic trip limit and it needs to stay above this limit for a sufficiently long time. As a consequence, the design procedure for using MCBs in LVDC grids is different than when fuses are implemented. Furthermore, the reported MCB datasheet values, such as the let-through energy, are not representative for use in converter dominated LVDC networks.

References

[1] R. Weiss, L. Ott and U. Boeke, "Energy efficient low-voltage DC-grids for commercial buildings," 2015 IEEE First International Conference on DC Microgrids (ICDCM), 2015, pp. 154-158, doi: 10.1109/ICDCM.2015.7152030.

[2] A. Kwasinski, "Quantitative Evaluation of DC Microgrids Availability: Effects of System Architecture and Converter Topology Design Choices," in IEEE Transactions on Power Electronics, vol. 26, no. 3, pp. 835-851, March 2011, doi: 10.1109/TPEL.2010.2102774.

[3] K. A. Smith, S. J. Galloway, A. Emhemed and G. M. Burt, "Feasibility of Direct Current street lighting & integrated electric vehicle charging points," 6th Hybrid and Electric Vehicles Conference (HEVC 2016), 2016, pp. 1-6, doi: 10.1049/cp.2016.0987.

[4] R. M. Cuzner and G. Venkataramanan, "The status of DC microgrid protection," in Proc. IEEE Ind. Appl. Soc. Annu. Meet., 2008, pp. 1–8.

[5] L. Hallemans et al., "Fault Identification and Interruption Methods in Low Voltage DC Grids — A Review," 2019 IEEE Third International Conference on DC Microgrids (ICDCM), 2019, pp. 1-8, doi: 10.1109/ICDCM45535.2019.9232856.

[6] S. Ravyts, G. V. d. Broeck, L. Hallemans, M. D. Vecchia and J. Driesen, "Fuse-Based Short-Circuit Protection of Converter Controlled Low-Voltage DC Grids," in IEEE Transactions on Power Electronics, vol. 35, no. 11, pp. 11694-11706, Nov. 2020, doi: 10.1109/TPEL.2020.2988087.

[7] W. Javed, D. Chen, M. E. Farrag, and Y. Xu, "System configuration, fault detection, location, isolation and restoration: A review on LVDC microgrid protections," Energies, vol. 12, no. 6, 2019, doi: 10.3390/en12061001.

[8] J. Yang, J. E. Fletcher, and J. O'Reilly, "Short-circuit and ground fault analyses and location in VSC-based DC network cables," IEEE Trans. Ind. Electron., vol. 59, no. 10, pp. 3827–3837, Oct. 2012.

[9] S. -Y. Lee, Y. -K. Son, H. -J. Cho and S. -K. Sul, "Normalization of Capacitor-Discharge I2t by Short-Circuit Fault in VSC-Based DC System," in IEEE Transactions on Power Electronics, vol. 37, no. 1, pp. 843-854, Jan. 2022, doi: 10.1109/TPEL.2021.3096075.

[10] D. L. Gerber, V. Vossos, W. Feng, C. Marnay, B. Nordman and R. Brown, "A simulation-based efficiency comparison of AC and DC power distribution networks in commercial buildings", Applied Energy, Volume 210, 2018, Pages 1167-1187, ISSN 0306-2619

[11] G. Alee and W. Tschudi, "Edison Redux: 380 Vdc Brings Reliability and Efficiency to Sustainable Data Centers," in IEEE Power and Energy Magazine, vol. 10, no. 6, pp. 50-59, Nov.-Dec. 2012, doi: 10.1109/MPE.2012.2212607.

[12] H. Wang and F. Blaabjerg, "Reliability of Capacitors for DC-Link Applications in Power Electronic Converters—An Overview," in IEEE Transactions on Industry Applications, vol. 50, no. 5, pp. 3569-3578, Sept.-Oct. 2014, doi: 10.1109/TIA.2014.2308357.

[13] Schneider Electric, Multi9 catalogue 2020, Multistandard protection for OEM

A Novel Modified-TOGI based PLL for the Three-Phase Unbalanced and Distorted Grid Conditions

Khanh-Hung Nguyen, Ahmad Ali Nazeri, Xiao Yu, Peter Zacharias
UNIVERSITY OF KASSEL
Wilhelmshöher Allee 71-73
Kassel, Germany
Phone: +49 (0) 561-804-6305
Email: khanh-hung.nguyen@uni-kassel.de
URL: https://www.uni-kassel.de/eecs/evs/startseite

July 12, 2022

Acknowledgments

Khanh-Hung Nguyen is very thankful for the scholarship from the Vietnamese Government Scholarship and the support of the KDEE-EVS department, the University of Kassel to accomplish this research.

Keywords

≪Unbalanced Grid≫, ≪Asymmetric Grid≫,≪Decoupled-Double Synchronous Reference Frame (DDSRF)≫, ≪Modified Second-Order Generalized Integrator (MSOGI)≫, ≪Multi Second-Order Generalized Integrator (MSOGI)≫, ≪Second-Order Generalized Integrator≫

Abstract

This paper proposes a novel modified third-order generalized integrator phase-locked loop (TOGI-PLL) for balanced, unbalanced, and distorted grid conditions. This technique also eliminates the error factors that cause in experimental such as DC-offset and removing the second-order oscillation noise by adding the low-pass filter when generating the estimated filtered frequency. A step-by-step controller design procedure, system stability analysis, and comparison with the traditional second-order generalized integrator (SOGI), dual second-order generalized integrator (DSOGI), and decoupled double synchronous reference frame PLL (DDSRF-PLL) are carried out. The proposed PLL approach is simulated and experimentally implemented on a three-phase 7.5kW grid-connected inverter system.

Introduction

In grid-connected inverters, the phase and the amplitude of the grid voltage are extracted employing phase-locked-loop (PLL) techniques to synchronize the injected current into the grid voltage and must meet modern grid codes. Significantly, in an extensive control system, many control capabilities can burden the working of the microprocessor. It is necessary for the programmer to curtail and optimize the execution time as well as the code capacity in RAM or Flash of the DSP by simplifying the instructions to catch the results agile with the highest accuracy.

The unbalanced and distorted grid makes it difficult to extract the phase of the utility voltage for the generation of the references. Hence, the implementation of the exact phase-locking under unbalanced and distorted grid voltage conditions has recently been investigated [1]. A synchronous reference frame

PLL (SRF-PLL) has been extensively implemented in grid-connected systems due to its simplicity in structure, fast dynamic response, and can easily been implemented [2]. However, the SRF-PLL will result in inaccuracy if the harmonics and imbalance in the grid voltage presents, which results in an inexact tracking of the fundamental positive sequence component (FPSC) of the grid voltage and results in instability of the whole system. To overcome the drawbacks of the SRF-PLL, a decoupled double SRF-PLL (DDSRF-PLL) has been proposed in [3]. The DDSRF-PLL can decouple the positive and negative sequence components of the grid voltage by using a decoupling network to diminish the oscillations but the presence of the low-pass filter (LPF) introduces the time delay, which results in a slower dynamic response. A second-order generalized integrator (SOGI-PLL) has been widely implemented for the single and three-phase system due to its simplicity and has the capability of exact locking even under non-ideal grid conditions [4]. The SOGI-PLL produces errors in the extraction of the FPSC when the grid voltage contains the dc offset. A dual SOGI PLL (DSOGI-PLL) with frequency adaptive positive sequence detection technique based on the stationary reference frame ($\alpha\beta$) for the extraction of the positive sequence is proposed in [5, 6, 7, 8] for the unbalanced, and distorted grid conditions. The DSOGI-Frequency Locked Loop (DSOGI-FLL) was integrated in the DSOGI-PLL system to determine without any error the instantaneous positive- and negative-sequence components of the abnormal conditions grid voltage [9]. For A mixed second-and third-order generalized integrator PLL (MSTOGI-PLL) has been proposed in [10] to accurately lock the phase in the presence of the harmonics, dc offset elimination, and voltage imbalance. A frequency adaptive tracking block has also been implemented for the compensation of the grid frequency variations. In this paper, an extensive comparison, performance analysis of SOGI, DSOGI, and DDSRF with the proposed PLL technique for the three-phase balanced, asymmetric and distorted grid conditions are presented. The proposed modified-TOGI-PLL (MTOGI) structure is simple to implement and works better in performance than other PLLs for the unbalanced grid conditions. A step-by-step design procedure of the controller with the closed-loop stability analysis is presented. The traditional PLLs with the new PLL structure known as MSOGI is simulated in MATLAB/Simulink® and experimentally implemented on a 7.5 kW Three-phase grid-connected inverter system with the DSP from TI C2000 F28335 to verify the analytical results.

Controller design of the PLL

The controller design parameters play a vital role in achieving zero-steady state error. Proper selections of the PLL gains have always been a challenge for the exact tracking of the grid voltage. The traditional design of the controller parameters are given in [2], [5], and [11]. The problem with the design procedure explained in [2], [5], and [11] is the optimum tuning of the controller parameters and has to be redesigned depending on the type of the PLL and the operating conditions. This paper proposes a step-by-step design procedure of the controller parameters, which works well for balanced, unbalanced, and distorted grid conditions and satisfies the stability margins of the system. Taking the loop filter (LF) with the voltage-controlled oscillator (VCO) and considering the sampling delay, the open-loop transfer function $G_{OL}(s)$ is given as follows:

$$G_{OL}(s) = K_{pi}(\frac{1+sT_{ni}}{sT_{ni}})(\frac{1}{1+sT_s})(\frac{V_s}{s}) \tag{1}$$

where T_s is the sampling time, V_s is the peak grid voltage, K_{pi} is the proportional gain and T_{ni} is the time constant of the PI controller. The symmetrical optimum (SO) method in [12] was selected to calculate the controller gains. According to the SO method, the controller gains should be calculated in such manners that the amplitude and the phase plot of the $G_{OL}(s)$ are symmetrical about the crossover frequency (ω_c), which is the geometric mean of the two frequencies of the $G_{OL}(s)$ [13]. Considering a normalizing factor, alpha (α), the closed-loop transfer function $G_{CL}(s)$ is compared with the third-order standard equation, which yields to

$$G_{CL}(s) = \frac{\alpha\omega_c^2 s + \omega_c^3}{s^3 + \alpha\omega_c s^2 + \alpha\omega_c^2 s + \omega_c^3} \tag{2}$$

where $\omega_c = \omega_n$ and α is used to compensate the damping factor and the bandwidth (ω_c) of the system. The PLL controller gains can be then determined as [13],

$$\omega_c = \frac{1}{\alpha T_s}, \qquad\qquad T_n i = \alpha^2 T_s, \qquad\qquad K_p i = \frac{1}{\alpha V_s T_s} \qquad\qquad (3)$$

Substituting (3) into (2) results in the factor of α which is related to the damping factor (ξ) as:

$$\xi = \frac{\alpha - 1}{2} \qquad\qquad (4)$$

By changing α the bandwidth of the system and damping can be controlled.

Second-Order Generalized Integrator (SOGI) PLL

Fig.1a shows the block diagram of the SOGI. The SOGI is responsible to generate the in-phase and in-quadrature signals from the grid voltage. The SOGI block uses the angular frequency (ω_g) before computing the integrators. The value of the SOGI gain (k) is important, which affects the bandwidth for the closed-loop system. The SOGI deteriorates the double frequency components in the measured fundamental frequency. The LPF is used to drop the noise, which may appear in the error signal [14]. The SOGI forward integrator and backward integrator need to be carefully discretized to avoid an algebraic loop as shown in Fig.1a. The discretization methods of forward and backward integrators can be found in [5]. The SOGI integrators have been discretized by with trapezoidal method and implemented as a third-order IIR filter. The simplest structure of the SOGI is shown in Fig.1b.

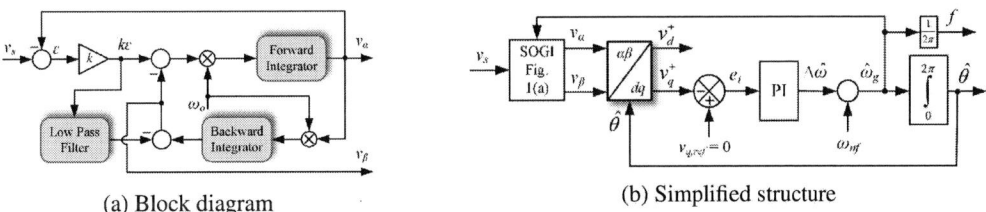

(a) Block diagram

(b) Simplified structure

Fig. 1: Second-Order Generalized Integrator (SOGI) PLL

The presented SOGI is based on the second-order integrator, which is defined as:

$$GI = \frac{\omega s}{s^2 + \omega^2} \qquad\qquad (5)$$

where ω is the resonant frequency of the SOGI. The closed-loop transfer function of the α and β voltages are given as

$$H_\alpha = \frac{v_\alpha(s)}{v_\beta(s)} = \frac{k\omega s}{s^2 + k\omega s + \omega^2} \qquad\qquad\qquad H_\beta = \frac{v_\beta(s)}{v_\alpha(s)} = \frac{k\omega^2}{s^2 + k\omega s + \omega^2} \qquad\qquad (6)$$

Dual Second Order Generalized Integrator (DSOGI) PLL

The DSOGI works in the SRF frame based on the instantaneous symmetrical component (ISC) method. Based on the SOGI given in Fig.2, the filtered outputs of two signals in SRF are obtained with 90° of phase shift are used to calculate the symmetrical components proposed given in [15]. The frequency adaptive positive sequence detection method is implemented on the SRF frame to extract the positive sequence for the unbalanced, and distorted grid conditions in DSOGI-PLL as given in Fig.2 [16]. The decomposition of the voltage vector into three symmetrical components as and the subscripts 0, 1, and 2

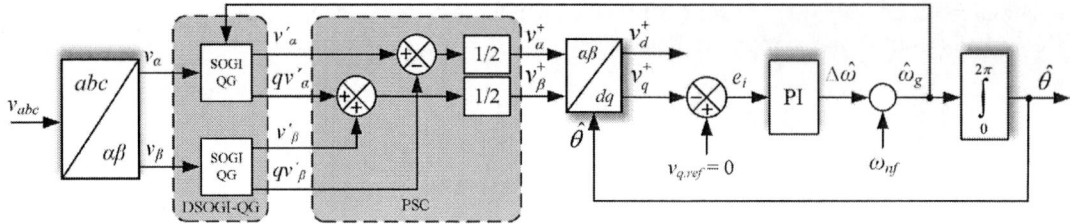

Fig. 2: Schematic diagram of the dual second-order generalized integrator (DSOGI) PLL

refers to the zero, positive and negative voltage sequence components respectively.

$$
\begin{bmatrix} V_a \\ V_b \\ V_c \end{bmatrix} = \begin{bmatrix} V_{a,0} \\ V_{b,0} \\ V_{c,0} \end{bmatrix} + \begin{bmatrix} V_{a,1} \\ V_{b,1} \\ V_{c,1} \end{bmatrix} + \begin{bmatrix} V_{a,2} \\ V_{b,2} \\ V_{c,2} \end{bmatrix}
$$
(7)

$$
V_{abc,1} = \begin{bmatrix} V_{a,1} \\ V_{b,1} \\ V_{c,1} \end{bmatrix} = \frac{1}{3} \begin{bmatrix} 1 & a^2 & a \\ a & 1 & a^2 \\ a^2 & a & 1 \end{bmatrix}
$$
(8)

where $a = e^{j(-120°)}$. The positive sequence in the orthogonal SRF can be determined as:

$$
\begin{bmatrix} V_{\alpha,1} \\ V_{\beta,1} \end{bmatrix} = \frac{2}{3} \begin{bmatrix} 1 & -\frac{1}{2} & -\frac{1}{2} \\ 0 & \frac{\sqrt{3}}{2} & \frac{\sqrt{3}}{2} \end{bmatrix} \frac{1}{3} \begin{bmatrix} 1 & a^2 & a \\ a & 1 & a^2 \\ a^2 & a & 1 \end{bmatrix} \begin{bmatrix} V_a \\ V_b \\ V_c \end{bmatrix}, \qquad \begin{bmatrix} V_{\alpha,1} \\ V_{\beta,1} \end{bmatrix} = \frac{1}{2} \begin{bmatrix} 1 & -q \\ q & 1 \end{bmatrix} \begin{bmatrix} V_\alpha \\ V_\beta \end{bmatrix}
$$
(9)

where $q = e^{-j\frac{\pi}{2}}$ is the phase shift operator and provides a 90° lagging phase voltage from the original phase.

Decoupled Double Synchronous Reference Frame (DDSRF) PLL

The DDSRF-PLL contributes a critical role in detecting the positive component of the fundamental frequency of the grid voltage under unbalanced and distorted grid conditions. The DDSRF-PLL is superior to the SRF-PLL because of the positive sequence detector, which cancels the detection error in the latter one. This is achieved by transforming both positive and negative sequence components of the grid voltage into the double synchronous reference frame (SRF). Then, a decoupling network is developed to extract clean and separate positive and negative sequence components [3]. The decoupling network cancels out the double frequency oscillations at 2ω in dq^{+1} and dq^{-1} reference frame signals. Therefore, there is no need to decline the bandwidth of the PLL to attenuate such oscillations and the real amplitude of the unbalanced input voltage sequence components are indeed exactly detected [5].

Modified TOGI (MTOGI) PLL

The third-order generalized integrator (TOGI) is added to the traditional SOGI structure given in Fig. 3. The TOGI eliminates the DC offset present in the utility grid. The symmetrical positive sequence in SRF is extracted by implementing two TOGI blocks for each sequence. The authors proposed a novel-based PLL technique known a modified-third-order generalized integrator (MTOGI), which is much easier to implement, does not need for transformations, less computational burden, and better transient response as compared to the traditional SOGI, TOGI, DDSRF-PLLs. The three-phase grid voltages in abc frame can be given to the MTOGI and directly extracts the symmetrical components as given in (8). The matrix A and B implemented in the MTOGI are given in (13). The V_{abc} are transformed to the Synchronous reference frame (SRF) and is given to the PI controller to generate the estimated theta ($\hat{\theta}$). A second-order LPF is used to attenuate the second-order oscillations to generate the estimated filtered frequency and is given back to the MTOGI as given in Fig.3. The 120° phase shift operator will be separated into

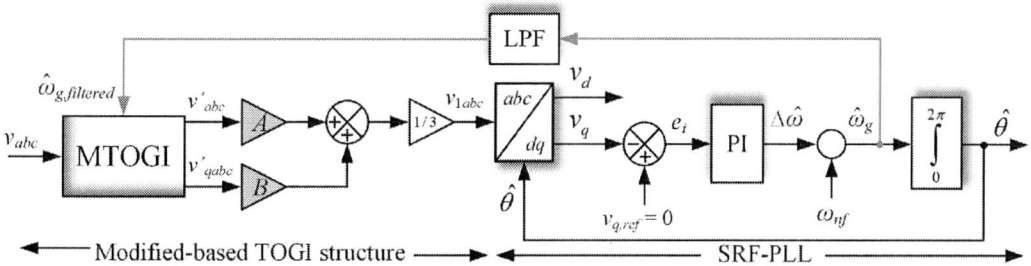

Fig. 3: The simplified structure of the modified-third-order generalized integrator (MTOGI) PLL

two parts as

$$a = e^{-j\frac{2\pi}{3}} = e^{-j(\frac{\pi}{2}+\frac{\pi}{6})} = -\sin(\frac{\pi}{6}) + \cos(\frac{\pi}{6})e^{-j\frac{\pi}{2}} = s_1 + c_1 q \tag{10}$$

$$a^2 = e^{-j\frac{4\pi}{3}} = e^{j(\frac{\pi}{2}+\frac{\pi}{6})} = -\sin(\frac{\pi}{6}) - \cos(\frac{\pi}{6})e^{-j\frac{\pi}{2}} = s_1 - c_1 q \tag{11}$$

where $s_1 = \sin(\frac{\pi}{6})$; $c_1 = \cos(\frac{\pi}{6})$

$$\begin{bmatrix} V_{a,1} \\ V_{b,1} \\ V_{c,1} \end{bmatrix} = (\frac{1}{3})(A.V'_{TOGI} + B.V'_{TOGI}) \tag{12}$$

$$A = \begin{bmatrix} 1 & s_1 & s_1 \\ s_1 & 1 & s_1 \\ s_1 & s_1 & 1 \end{bmatrix}, \qquad B = \begin{bmatrix} 1 & -c_1 & c_1 \\ c_1 & 1 & -c_1 \\ -c_1 & c_1 & 1 \end{bmatrix}, \qquad LPF = \frac{w_c^2}{s^2 + 2\xi\omega_c s + \omega_c^2} \tag{13}$$

Simulation analysis

A simulation model was instituted to classify the performance and achievement of the new Modified TOGI (MTOGI) PLL. The simulation was established under these abnormal conditions namely: (1) in Fig.4, PLL is connected with distorted grid with 10% 17th and 10% 23rd harmonics at $t = 0.2s$; (2) the grid voltage has the frequency variations, changing from 50Hz to 55Hz at $t = 0.15s$ and coming back to 50Hz at $t = 1.5s$ in Fig.6; (3) the three-phase voltage becomes to asymmetric at $t = 0.5s$ with $V_A = 1.2V_m$, $V_B = V_m$, and $V_C = 0.7V_m$ is shown in Fig.7. The parameters of PI controller was determined from the symmetrical optimum (SO) that are presented in the above subsection "Controller design of PLL" are $k_p = 0.6505$, $k_i = 29.9248$, and $k_{sogi} = \sqrt{2}$.

Fig.4 demonstrates the comparative simulation results of our Modified-TOGI PLL against the others: SOGI, DSOGI, and DDSRF-PLL in the transient time. With the reasonable and satisfactory parameters controller design, the output theta (θ) is very accurate at zero-crossing in all cases as shown in Fig.4a. In regard to the q-sequence value in transient phase, it can be seen clearly that MTOGI PLL has the better response compared with DSOGI and DDSRF, and after approximately 0.3s, q-component becomes almost zero, the single SOGI has the best q value but the overshoot is very high. It could be not exceptional when implemented in an experimental setup. Concerning the output frequency, MTOGI and SOGI still have a very positive assessment, but SOGI has quite a high overshoot.

Fig.5 exhibits the operations of the PLL controller when the grid becomes too distorted with harmonics. Although MTOGI-PLL does not have the best output achievement, it, along with other PLLs, is also swift to stabilize the system and track to the set-point fundamental frequency of 50Hz precisely. This confirm that MTOGI-PLL can work very well in case of a distorted grid. Fig.6 expresses completion of PLLs during the frequency jump of three-phase voltage. At $t = 0.15s$ it has a frequency change (from 50Hz to 55Hz), and $t = 1.5s$ diminishes to 50Hz. After less than 1s, all of four PLL types can bring the system back to the desired frequency, but MTOGI-PLL and DDSRF-PLL have excellent results, with smaller overshoot in the transient phase and faster to the steady-state. We can see obviously here in this

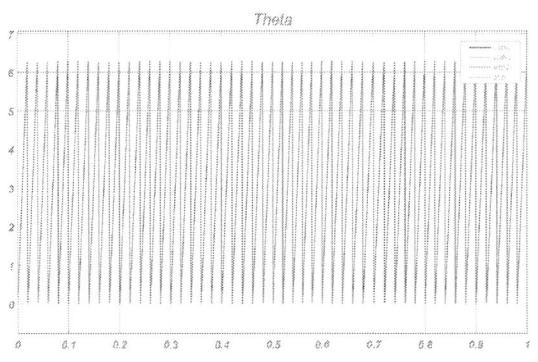

(a) Output theta

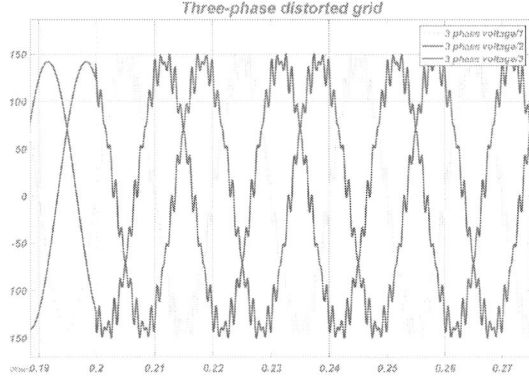

(a) Distorted grid voltages

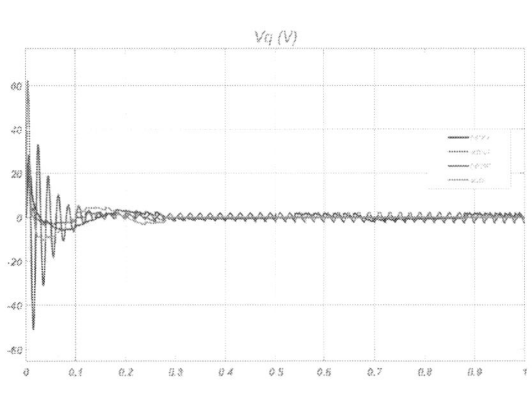

(b) *q*-component of instantaneous voltage

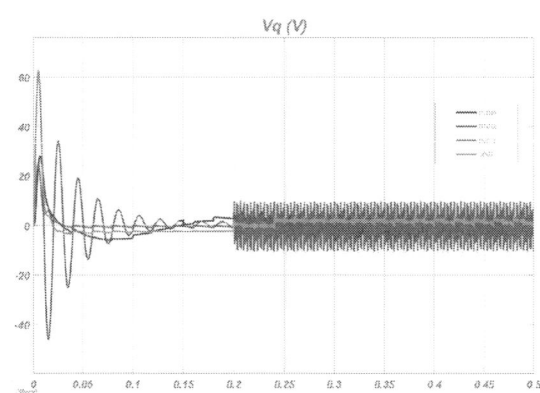

(b) *q*-component of instantaneous voltage

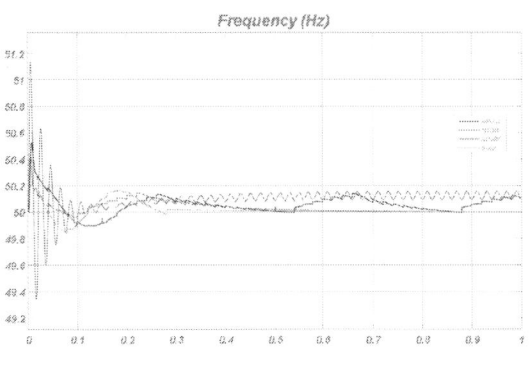

(c) Output frequency

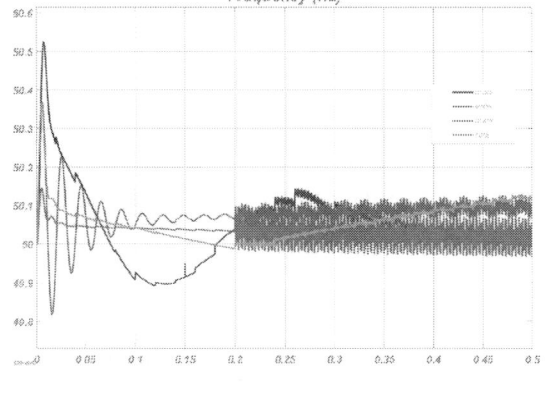

(c) Output frequency

Fig. 4: Transient response comparison of every type of PLL

Fig. 5: Response when connecting with distorted grid

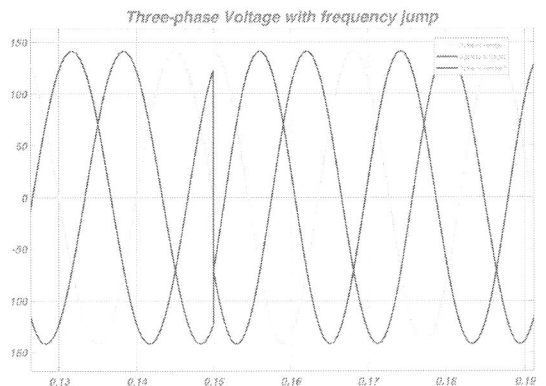

(a) Three-phase voltage with frequency jump

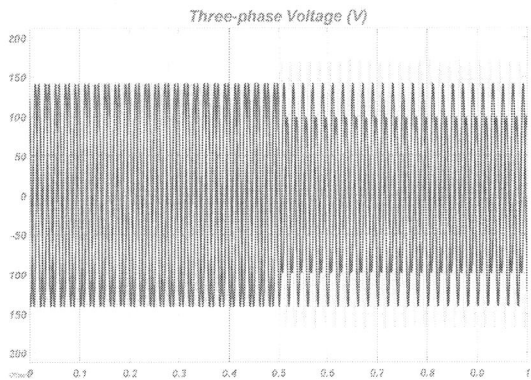

(a) Three-phase grid voltage with amplitude jump

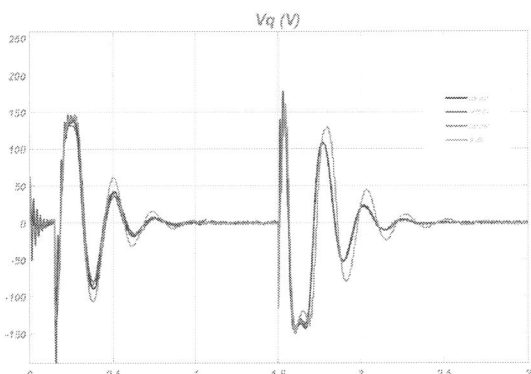

(b) q component of instantaneous voltage

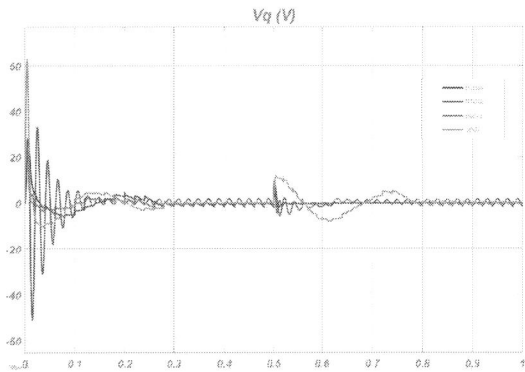

(b) q component of instantaneous voltage

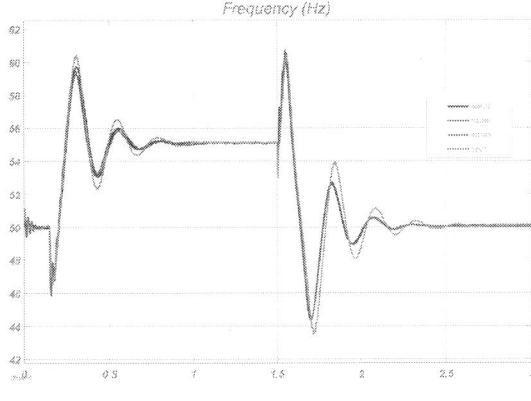

(c) Output Frequency

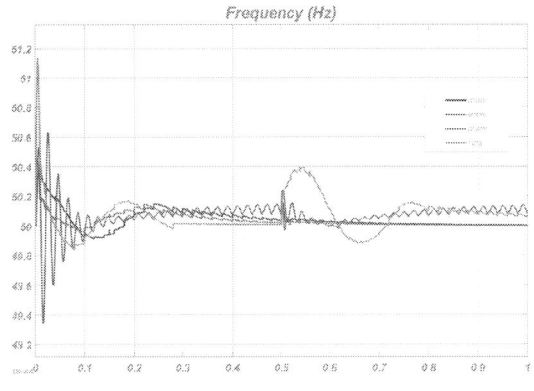

(c) Output Frequency

Fig. 6: PLL performance under frequency jump

Fig. 7: PLL efficiency under amplitude and phase jump

circumstance that the overshoots of all PLL controllers are huge. It is necessary to identify and design the total system more precisely to have a very valuable development.

When the three-phase voltage has an amplitude jump at $t = 0.5s$ in Fig.7, the voltage of phase A boosts to $1.2V_m$ (V_m - peak value) and phase C declines to $0.7V_m$, MTOGI-PLL controller accomplishes extremely superb. MTOGI controller has the best performance (after a very short time it goes to the steady-state and lower overshoot). Therefore, MTOGI-PLL is absolutely suitable for the unbalanced and asymmetric condition grid.

Experimental Results

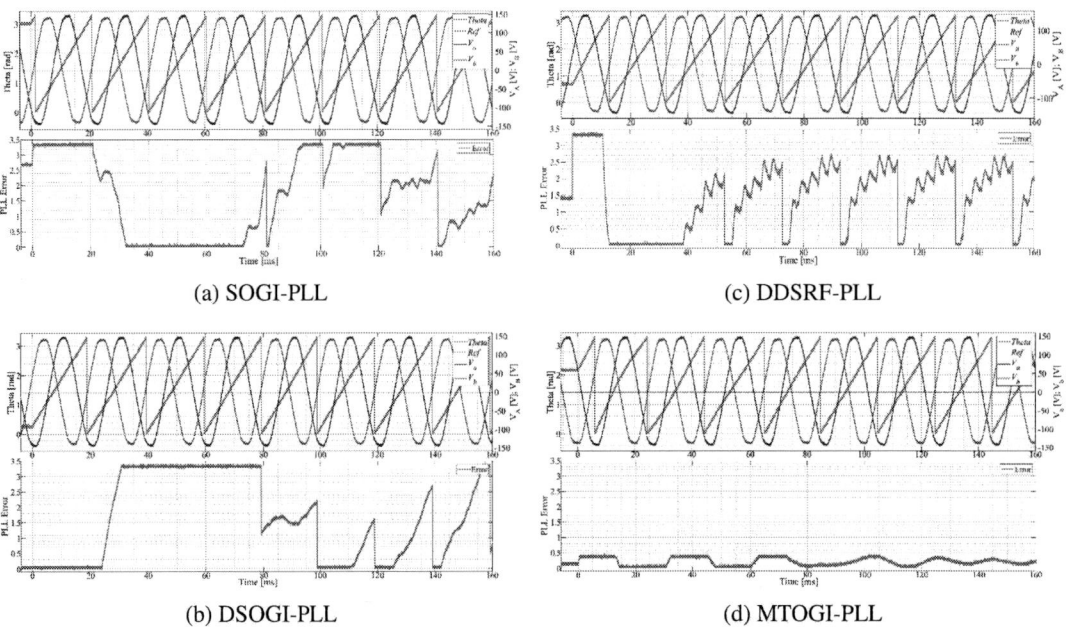

(a) SOGI-PLL   (c) DDSRF-PLL

(b) DSOGI-PLL   (d) MTOGI-PLL

Fig. 8: The transient response of the SOGI

The proposed MTOGI PLL has been implemented on the experimental bench TMS320F28335 from TI that controlling a 7.5 kW grid-connected inverter and compared with the traditional PLLs for the balanced, distorted, and artificial unbalanced grid condition. Fig.8a and Fig.8b illustrate the transient response of the controller based on the SOGI and DSOGI-PLL respectively. The grid voltages V_a and V_b are depicted above with the controller error (red-line) below in Fig.8. It can be seen that the steady-state error in DSOGI is achieved in 100ms compared to the SOGI, which is around 140ms respectively. The PLL signal is synchronized in 3∼4 cycles. Fig.8c and Fig.8d indicate the transient response of the controller based on the DDSRF and MTOGI-PLL respectively. The steady-state error of the controller in MTOGI is attained in 80ms where the DDSRF PLL takes 100ms. respectively The PLL signal is synchronized in around 1∼2 cycle(s) in MTOGI where the DDSRF PLL takes 2∼3 cycle(s). It can be noticed that the steady-state performance and dynamic response of the MTOGI are better than SOGI, DSOGI, and the DDSRF-PLL. Fig.9 presents the dynamic response of the MTOGI-based PLL for the (a) grid voltage amplitude and phase jump, where (b) outlines the frequency jump from 50Hz to 52Hz and (c) illustrates the distorted grid conditions. It can be observed that the dynamic response of the MTOGI shows superior performance under different grid conditions. The proposed MTOGI PLL has been implemented on a 7.5 kW grid-connected inverter and compared with the traditional PLLs for the balanced, distorted, and artificial unbalanced grid condition as shown in Fig.9d.

Conclusion

This paper proposes a novel MTOGI-PLL (Modified-Third Order Generalized Integrator PLL) for the balanced, unbalanced, and distorted grid conditions. The MTOGI-PLL has been compared with SOGI,

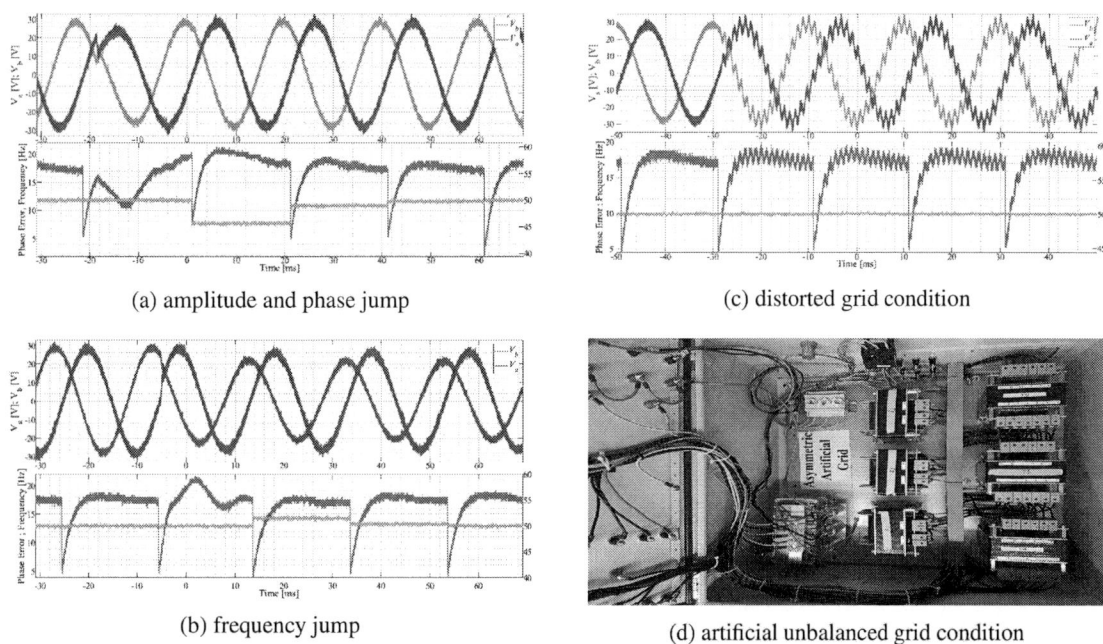

(a) amplitude and phase jump

(c) distorted grid condition

(b) frequency jump

(d) artificial unbalanced grid condition

Fig. 9: The dynamic response of the MTOGI

DSOGI, and DDSRF-PLL. The MTOGI based PLL is better in dynamic performance with faster locking of the fundamental frequency in $1 \sim 2$ cycle(s). Specially, the MTOGI-PLL is easy to implement, has no need for transformations, and has less computational burden. The simulation and experimental results verify the steady-state and dynamic performance of the proposed PLL for the unbalanced and distorted grid conditions.

References

[1] F. a. D. L. a. L. L. a. L. X. Xiao: A frequency-fixed sogi-basedpll for single-phase grid-connected converters, IEEE Transactions on Power Electronics, pp. 1713-1719, 2016.

[2] S. a. M. M. a. F. F. D. a. G. J.M. Golestan: Dynamics assessment of advanced single-phase PLL structures, IEEE transactions on industrial electronics, pp. 2167-2177, 2012.

[3] P. a. P. J. a. B. J. a. C. J. I. a. B. Rodriguez: Decoupled double synchronous reference frame PLL for power converters control, IEEE Transactions on Power Electronics, pp. 584-592, 2007.

[4] Remus Teodorescu, Marco Liserre and Pedro Rodríguez: GRID CONVERTERS FOR PHOTOVOLTAIC AND WIND POWER SYSTEMS, 2011 John Wiley & Sons, Ltd. ISBN: 978-0-470-05751-3

[5] R. a. L. M. a. R. P. Teodorescu: Grid converters for photovoltaic and wind power systems, John Wiley & Sons, 2011.

[6] Rofiatul Izah, Subiyanto, Dhidik Prastiyanto: Improvement of DSOGI PLL Synchronization Algorithm with Filter on Three-Phase Grid-Connected Photovoltaic System, Jurnal Elektronika dan Telekomunikasi (JET), Vol. 18, No. 1, August 2018, pp. 35-45

[7] Mehmet Emin MERAL, Dogan CELIK: DSOGI-PLL Based Power Control Method to Mitigate Control Errors Under Disturbances of Grid Connected Hybrid Renewable Power Systems, POWER ENGINEERING AND ELECTRICAL ENGINEERING VOLUME: 16 — NUMBER: 1 — 2018 — MARCH

[8] Elie Habboub, Jean Sawma, Nizar Daou, Flavia Khatounian: Experimental Validation of a Combined Multi-Variable Filter – Dual Second Order Generalized Integrator Phase-Locked Loop Technique, 2018 IEEE International Multidisciplinary Conference on Engineering Technology (IMCET)

[9] Pedro Rodríguez, Alvaro Luna, Raul Santiago Muñoz-Aguilar, Ion Etxeberria-Otadui, Remus Teodorescu, and Frede Blaabjerg: A Stationary Reference Frame Grid Synchronization System for Three-Phase Grid-Connected Power Converters Under Adverse Grid Conditions, IEEE TRANSACTIONS ON POWER ELECTRONICS, VOL. 27, NO. 1, JANUARY 2012

[10] C. a. Z. X. a. W. X. a. C. X. a. Z. Z. a. G. X. Zhang: A grid synchronization PLL method based on mixed second-and third-order generalized integrator for DC offset elimination and frequency adaptability, IEEE Journal of Emerging and Selected Topics in Power Electronics, pp. 1517-1526, 2018.

[11] S. a. M. M. a. F. F. D. a. G. J. M. Golestan: Design and tuning of a modified power-based PLL for single-phase grid-connected power conditioning systems, IEEE Transactions on Power Electronics, pp. 3639-3650, 2012.

[12] W. Leonhard: Control of electrical drives, Springer Science & Business Media, 2001.

[13] V. a. B. V. Kaura: Operation of a phase locked loop system under distorted utility conditions, IEEE Transactions on Industry applications, pp. 58–63, 1997.

[14] T. D. C. a. Z. K. a. P. A. a. O. V. V. a. S. M. G. Busarello: Designing a second order generalized integrator digital phase locked loop based on a frequency response approach, 2019 IEEE PES Innovative Smart Grid Technologies Conference-Latin America (ISGT Latin America), 2019.

[15] C. L. Fortescue: Method of symmetrical co-ordinates applied to the solution of polyphase networks, JTransactions of the American Institute of Electrical Engineers, pp. 1027-1140, 1918.

[16] A. a. F. S. a. W. B. Salamah: Three-phase phase-lock loop for distorted utilities, IET Electric Power Applications, pp. 937-945, 2007.

Comparison of Two and Three-Level AC-DC Rectifier Semiconductor Losses with SiC MOSFETs Considering Reverse Conduction

Guangyao Yu, Thiago Batista Soeiro, Jianning Dong and Pavol Bauer
Dept. Electrical Sustainable Energy, DCE&S group, Delft University of Technology
Delft, The Netherlands
Email: G.Yu-1@tudelft.nl

Keywords

≪AC-DC converter≫, ≪Analytical losses computation≫, ≪Conduction losses≫, ≪Silicon Carbide (SiC)≫, ≪Shunt current≫.

Abstract

This paper presents the semiconductor losses analytical equations in closed form for two-level voltage source converter, three-level neutral point clamped (NPC) and three-level T-Type PFC topologies in high power applications. The reverse parallel current conduction between the SiC MOSFETs channel and body diode is considered. A circuit simulation model is built in PLECS to estimate the semiconductor losses and to verify the accuracy of the developed analytical model. A calculation example of the semiconductor losses of a 200 kW three-phase rectifier is shown.

Introduction

With the popularity of electric vehicles, there is a growing need to have fast charging infrastructures. The power rating for DC fast charger is typically rated at 50 kW, and chargers with power rating up to 350 kW is also available from ABB [1]. Power factor correction circuits are used as a front-end converter in these applications.

Semiconductor losses of two and three-level converters with IGBTs are well studied in literature such as in [2], therefore, it is not covered here. In order to increase the power density, SiC MOSFETs (modules) are preferred due to its superior switching performance. The semiconductor losses study of these SiC MOSFETs based three-level converters when the reverse conduction current is distributed between the channel and body diode is not commonly seen. Reference [3] shows the analysis of SiC technology used in two and three-level converters without considering the reverse parallel current conduction. References [4, 5, 6, 7] consider the reverse conduction and blanking time influence in a two-level converter. Reverse parallel current conduction of different modulation schemes in SiC-based inverters is discussed in [8] while this phenomenon is also described in high-power bidirectional converters [9].

The main contribution of this paper is to extend this analysis to three-level rectifiers to provide an accurate losses calculation model, especially for the conduction losses calculation. Besides, a simulation circuit will also be given which can be easily modified to simulate semiconductor losses under more complicated situations such as deadtime influence, third harmonic injection, etc.

AC-DC PFC Topologies

Fig. 1 shows one arm of the AC-DC converters studied in this paper with MOSFETs as active switches. A complete three-phase converter will have three arms. Sinusoidal Pulse-Width-Modulation (SPWM) scheme is assumed in this paper. With SPWM applied to two-level, three-level T-Type and NPC rectifier, Fig. 2 could be used to show the circuit operation [10].

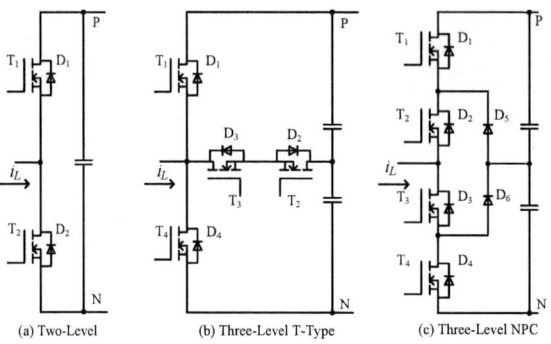

Fig. 1: PFC topologies studied in this paper.

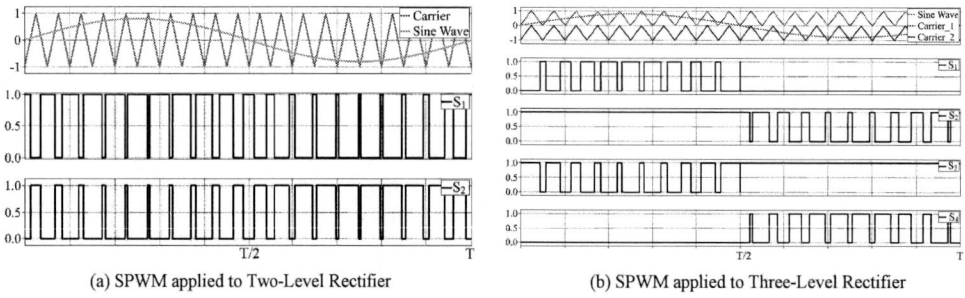

Fig. 2: SPWM applied to two-level and three-level rectifiers.

In Fig. 2, S_1-S_4 are PWM signals applied to the switches T_1-T_4.

Semiconductor Losses Analysis

The semiconductor losses can be divided into conduction losses and switching losses. The MOSFET channel's conduction can also be turned on by applying a gate-source voltage above the threshold voltage to reduce the reverse conduction losses which is also called active or synchronous rectification technique. When the current is high enough to a certain extent, the current will be distributed between the body diode and channel. The circuit model of a MOSFET is shown in Fig. 3. Fig. 4 shows the current through the MOSFET channel and body diode, the deadtime is not considered. T_1 and T_2, D_1 and D_2 are given in Fig. 1.

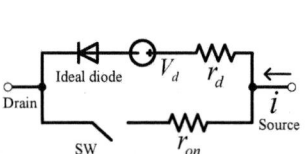

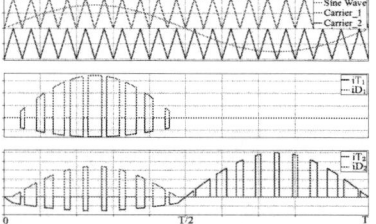

Fig. 3: MOSFET model, during reverse conduction Fig. 4: Parallel current conduction between MOS-
SW should be closed.   FET channel and diode, T-Type rectifier.

For MOSFETs operating in reverse conduction state, when $r_{on}i > V_d$, the current through r_{on} and r_d are:

$$i_{r_{on}} = \frac{r_d i + V_d}{r_{on} + r_d}, \ i_{r_d} = \frac{r_{on}i - V_d}{r_{on} + r_d}.$$ (1)

r_{on} and r_d are the channel and body diode on-resistances, V_d is the body diode forward voltage, i is the summed reverse current shown in Fig. 3. Note, that when $r_{on}i < V_d$, the anti-parallel diode does not conduct current.

Semiconductor Conduction Losses

To calculate the conduction losses of MOSFETs, the rms and average current value of the body diode and rms current value of the MOSFET channel need to be known. To simplify the formulae expression, the three functions below will be adopted:

$$\int_\alpha^{\pi-\alpha} \sin^2\theta d\theta = \frac{\pi - 2\alpha + \sin 2\alpha}{2} = f_1(\alpha), \ \int_\alpha^{\pi-\alpha} \sin^3\theta d\theta = \frac{3}{2}\cos\alpha - \frac{1}{6}\cos 3\alpha = f_2(\alpha)$$
$$\int_0^\alpha \sin^3\theta d\theta = \frac{1}{12}\cos 3\alpha - \frac{3}{4}\cos\alpha + \frac{2}{3} = f_3(\alpha).$$ (2)

α is a variable with a value between 0 and $\frac{\pi}{2}$. Several conditions were assumed for the calculation: the switching frequency is much higher than the grid frequency, the line current is sinusoidal with small ripple and the deadtime is neglected.

Two-Level Rectifier

Below, I_{avg_X} is average current while I_{rms_X} is rms current, $\hat{i}_L$ is the peak input current shown in Fig. 1 and M is the modulation index.

$$\begin{aligned}
I_{avg\_D_1,D_2} &= \frac{1}{2\pi}\int_\varphi^{\pi-\varphi} \frac{r_{on}\hat{i}_L\sin\theta - V_d}{r_{on}+r_d}\frac{1+M\sin\theta}{2}d\theta \\
&= \frac{1}{4\pi(r_{on}+r_d)}[2\cos\varphi(r_{on}\hat{i}_L - MV_d) + r_{on}\hat{i}_L M f_1(\varphi) + V_d(2\varphi - \pi)].
\end{aligned}$$ (3)

In (3), $\varphi = \arcsin\frac{V_d}{r_{on}\hat{i}_L}$. If $V_d > r_{on}\hat{i}_L$, then $\varphi = \frac{\pi}{2}$. The definition of φ applies to the rest of this two-level rectifier section.

$$I_{rms\_D_1,D_2} = \sqrt{\frac{1}{2\pi}\int_\varphi^{\pi-\varphi}\left(\frac{r_{on}\hat{i}_L\sin\theta - V_d}{r_{on}+r_d}\right)^2\frac{1+M\sin\theta}{2}d\theta} =$$
$$\sqrt{\frac{1}{4\pi(r_{on}+r_d)^2}[k^2 M f_2(\varphi) + (k^2 - 2kV_d M)f_1(\varphi) + 2\cos\varphi(V_d^2 M - 2kV_d) + V_d^2(\pi - 2\varphi)]}.$$ (4)

In (4), $k = r_{on}\hat{i}_L$.

Through piecewise integral,

$$I_{rms\_T_1,T_2} = \sqrt{\frac{1}{2\pi}[\hat{i}_L^2(\frac{\varphi}{2} + \frac{\pi}{4} - \frac{\sin 2\varphi}{4} - \frac{2M}{3}) + \hat{i}_L^2 M f_3(\varphi)] + \frac{1}{4\pi(r_{on}+r_d)^2}[q^2 M f_2(\varphi) +}$$
$$\sqrt{(q^2 + 2qV_d M)f_1(\varphi) + 2\cos\varphi(2qV_d + V_d^2 M) + V_d^2(\pi - 2\varphi)]}.$$ (5)

In (5), $q = r_d\hat{i}_L$.

Three-Level T-Type Rectifier

Suppose the MOSFET parameters of $T_1(D_1)$ and $T_4(D_4)$ are r_{on1}, r_{d1} and V_{d1}. The parameters of $T_2(D_2)$ and $T_3(D_3)$ are r_{on2}, r_{d2} and V_{d2}.

$$I_{avg\_D_1,D_4} = \frac{1}{2\pi}\int_\varphi^{\pi-\varphi}\frac{r_{on1}\hat{i}_L\sin\theta - V_{d1}}{r_{on1}+r_{d1}}M\sin\theta d\theta = \frac{1}{2\pi(r_{on1}+r_{d1})}[r_{on1}\hat{i}_L M f_1(\varphi) - 2\cos\varphi V_{d1}M].$$ (6)

In (6), $\varphi = \arcsin \frac{V_{d1}}{r_{on1}\hat{i}_L}$.

$$
\begin{aligned}
I_{avg\_D_2,D_3} &= \frac{1}{2\pi}\int_{\beta}^{\pi-\beta}\frac{r_{on2}\hat{i}_L\sin\theta - V_{d2}}{r_{on2}+r_{d2}}(1-M\sin\theta)d\theta \\
&= \frac{1}{2\pi(r_{on2}+r_{d2})}[2\cos\beta(r_{on2}\hat{i}_L+MV_{d2})-r_{on2}\hat{i}_LMf_1(\beta)+V_{d2}(2\beta-\pi)].
\end{aligned} \tag{7}
$$

In (7), $\beta = \arcsin \frac{V_{d2}}{r_{on2}\hat{i}_L}$. φ and β have the same definition below for T-Type rectifier section.

$$
\begin{aligned}
I_{rms\_D_1,D_4} &= \sqrt{\frac{1}{2\pi}\int_{\varphi}^{\pi-\varphi}\left(\frac{r_{on1}\hat{i}_L\sin\theta - V_{d1}}{r_{on1}+r_{d1}}\right)^2 M\sin\theta d\theta} \\
&= \sqrt{\frac{1}{2\pi(r_{on1}+r_{d1})^2}[k_1^2Mf_2(\varphi)-2k_1V_{d1}Mf_1(\varphi)+2\cos\varphi V_{d1}^2M]}.
\end{aligned} \tag{8}
$$

$$
\begin{aligned}
I_{rms\_T_1,T_4} &= \sqrt{\frac{1}{2\pi}[2\int_0^{\varphi}(\hat{i}_L\sin\theta)^2M\sin\theta d\theta + \int_{\varphi}^{\pi-\varphi}\left(\frac{r_{d1}\hat{i}_L\sin\theta + V_{d1}}{r_{on1}+r_{d1}}\right)^2 M\sin\theta d\theta]} = \\
&\sqrt{\frac{1}{\pi}\hat{i}_L^2Mf_3(\varphi)+\frac{1}{2\pi(r_{on1}+r_{d1})^2}[k_2^2Mf_2(\varphi)+2k_2V_{d1}Mf_1(\varphi)+2\cos\varphi V_{d1}^2M]}.
\end{aligned} \tag{9}
$$

$$
\begin{aligned}
I_{rms\_D_2,D_3} &= \sqrt{\frac{1}{2\pi}\int_{\beta}^{\pi-\beta}\left(\frac{r_{on2}\hat{i}_L\sin\theta - V_{d2}}{r_{on2}+r_{d2}}\right)^2(1-M\sin\theta)d\theta} = \\
&\sqrt{\frac{1}{2\pi(r_{on2}+r_{d2})^2}[-k_3^2Mf_2(\beta)+(k_3^2+2k_3V_{d2}M)f_1(\beta)-2\cos\beta(2k_3V_{d2}+MV_{d2}^2)+V_{d2}^2(\pi-2\beta)]}.
\end{aligned} \tag{10}
$$

$$
\begin{aligned}
I_{rms\_T_2,T_3} &= \sqrt{\frac{1}{2\pi}\hat{i}_L^2(\frac{\pi}{2}+\beta-\frac{\sin2\beta}{2})-\frac{1}{2\pi}M\hat{i}_L^2[\frac{4}{3}+2f_3(\beta)]+\frac{1}{2\pi(r_{on2}+r_{d2})^2}[-k_4^2Mf_2(\beta)+} \\
&\sqrt{(k_4^2-2k_4V_{d2}M)f_1(\beta)+2\cos\beta(2k_4V_{d2}-MV_{d2}^2)+V_{d2}^2(\pi-2\beta)]}.
\end{aligned} \tag{11}
$$

In (8), (9), (10), (11), $k_1 = r_{on1}\hat{i}_L$, $k_2 = r_{d1}\hat{i}_L$, $k_3 = r_{on2}\hat{i}_L$, $k_4 = r_{d2}\hat{i}_L$.

Three-Level NPC Rectifier

Suppose the four MOSFETs in one NPC rectifier arm are of the same type with parameters r_{on}, r_d and V_d.

$$
I_{avg\_D_5,D_6} = \frac{1}{2\pi}\int_0^{\pi}\hat{i}_L\sin\theta(1-M\sin\theta)d\theta = \hat{i}_L(\frac{1}{\pi}-\frac{M}{4}). \tag{12}
$$

$$
I_{avg\_D_1,D_2,D_3,D_4} = \frac{1}{2\pi}\int_{\varphi}^{\pi-\varphi}\frac{r_{on}\hat{i}_L\sin\theta - V_d}{r_{on}+r_d}M\sin\theta d\theta = \frac{1}{2\pi(r_{on}+r_d)}[r_{on}\hat{i}_LMf_1(\varphi)-2\cos\varphi V_dM]. \tag{13}
$$

In (13), $\varphi = \arcsin \frac{V_d}{r_{on}\hat{i}_L}$. φ has the same definition below for NPC rectifier section.

$$
I_{rms\_D_5,D_6} = \sqrt{\frac{1}{2\pi}\int_0^{\pi}(\hat{i}_L\sin\theta)^2(1-M\sin\theta)d\theta} = \hat{i}_L\sqrt{\frac{1}{4}-\frac{2M}{3\pi}}. \tag{14}
$$

$$
\begin{aligned}
I_{rms\_D_1,D_2,D_3,D_4} &= \sqrt{\frac{1}{2\pi}\int_{\varphi}^{\pi-\varphi}\left(\frac{r_{on}\hat{i}_L\sin\theta - V_d}{r_{on}+r_d}\right)^2 M\sin\theta d\theta} \\
&= \sqrt{\frac{1}{2\pi(r_{on}+r_d)^2}[k_1^2Mf_2(\varphi)-2k_1V_dMf_1(\varphi)+2\cos\varphi V_d^2M]}.
\end{aligned} \tag{15}
$$

In (15), $k_1 = r_{on}\hat{i}_L$.

$$I_{rms\_T_1,T_4} = \sqrt{\frac{1}{2\pi}[2\int_0^\varphi (\hat{i}_L\sin\theta)^2 M\sin\theta d\theta + \int_\varphi^{\pi-\varphi} \left(\frac{r_d\hat{i}_L\sin\theta+V_d}{r_{on}+r_d}\right)^2 M\sin\theta d\theta]} =$$
$$\sqrt{\frac{1}{\pi}\hat{i}_L^2 M f_3(\varphi) + \frac{1}{2\pi(r_{on}+r_d)^2}[k_2^2 M f_2(\varphi) + 2k_2 V_d M f_1(\varphi) + 2\cos\varphi V_d^2 M]}. \tag{16}$$

$$I_{rms\_T_2,T_3} = \sqrt{\frac{\hat{i}_L^2}{4} + \frac{1}{\pi}\hat{i}_L^2 M[f_3(\varphi) - \frac{2}{3}] + \frac{1}{2\pi(r_{on}+r_d)^2}[k_2^2 M f_2(\varphi) + 2k_2 V_d M f_1(\varphi) + 2\cos\varphi V_d^2 M]}. \tag{17}$$

In (16) and (17), $k_2 = r_d\hat{i}_L$.

Note, when φ and β are $\frac{\pi}{2}$ when there is no parallel current distribution between the MOSFETs channel and body diode, then, the above equations degrade into the situation that all the reverse conduction current only goes through the MOSFETs channel.

Semiconductor Switching Losses

The reverse recovery losses of the SiC MOSFETs body diode will be neglected in this paper. The turn-on (E_{on}) and turn-off (E_{off}) losses are curve fitted by a second order polynomial at a reference drain source voltage:

$$E_{on} = a_1 i^2 + b_1 i + c_1, E_{off} = a_2 i^2 + b_2 i + c_2. \tag{18}$$

i is the current through the MOSFET channel at the transient moment.

Two-Level Rectifier

The switching losses of T_1 and T_2 are:

$$\begin{aligned} P_{s\_T_1,T_2} &= \frac{f_s V_{dc}}{2\pi V_{ref}} \int_0^\pi [(a_1+a_2)\hat{i}_L^2\sin^2\theta + (b_1+b_2)\hat{i}_L\sin\theta]d\theta + \frac{f_s V_{dc}}{2V_{ref}}(c_1+c_2) \\ &= \frac{f_s V_{dc}}{2V_{ref}}[\frac{(a_1+a_2)\hat{i}_L^2}{2} + \frac{2(b_1+b_2)\hat{i}_L}{\pi} + (c_1+c_2)]. \end{aligned} \tag{19}$$

In (19), f_s is the switching frequency, V_{ref} is the reference voltage [11] [12], V_{dc} is the dc link voltage between P and N shown in Fig. 1.

Three-Level T-Type Rectifier

There are no switching losses of T_1 and T_4. Similar to (19), the switching losses of T_2 and T_3 are:

$$P_{s\_T_2,T_3} = \frac{f_s V_{dc}}{4V_{ref}}[\frac{(a_1+a_2)\hat{i}_L^2}{2} + \frac{2(b_1+b_2)\hat{i}_L}{\pi} + (c_1+c_2)]. \tag{20}$$

In (20), a_1, a_2, b_1, b_2, c_1 and c_2 are the parameters of T_2 (T_3).

Three-Level NPC Rectifier

Diodes D_5 and D_6 have no reverse recovery losses if SiC diodes are used. The switching losses of T_2 and T_3 are:

$$P_{s\_T_2,T_3} = \frac{f_s V_{dc}}{4V_{ref}}[\frac{(a_1+a_2)\hat{i}_L^2}{2} + \frac{2(b_1+b_2)\hat{i}_L}{\pi} + (c_1+c_2)]. \tag{21}$$

Simulation Model

The simulation model for semiconductor losses in PLECS is given in Fig. 5 which is a modification and simplification of [13].

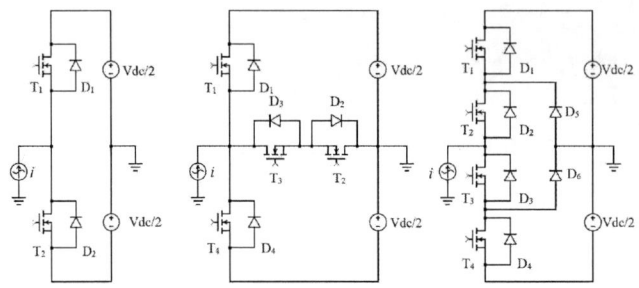

Fig. 5: Semiconductor losses simulation model in PLECS.

In order to simulate the losses in a simpler way, the input source is set to be a current source. The PWM strategy is given in Fig. 2. Due to the thermal analysis principle adopted by PLECS, the operation status of the MOSFETs channel and diode needs to be known to calculate their losses, therefore, the electrical parameters of r_{on}, r_d and V_d should be applied first and will keep the same during simulation. Device parameters from datasheets will be applied to the software thermal domain as a form of a lookup table.

Results

As an analysis example, the DC link voltage is set to be 1400 V which is adopted at applications when a high DC link voltage is needed such as high-power wireless power transfer systems, three-phase power is 200 kW and line-to-line rms voltage is set to be 650 V which means the modulation index M is 0.758. The devices selected for each topology are given in Table I. The semiconductor losses may be different with different devices, the intention of this example analysis is to show the losses based on today's SiC technology and to benchmark the losses between different topologies.

Table I: Selected devices for analysis in this paper with voltage and current rating at T_{case}=25 °C

Topology	Device Selection	Device Name	Manufacturer	V_{rating}(V)	$I_{rating-T}$	$I_{rating-D}$
Two-Level	$T_{1,2}$,$D_{1,2}$	CAS300M17BM2	CREE	1700	325	556
T-Type	$T_{1,4}$,$D_{1,4}$	CAS300M17BM2	CREE	1700	325	556
	$T_{2,3}$,$D_{2,3}$	C3M0016120K	CREE	1200	115	112
NPC	T_{1-4},D_{1-4}	CAS300M12BM2	CREE	1200	423	-
	$D_{5,6}$	SKM125KD12SC	Semikron	1200	-	264

For T-Type rectifier, both T_2 (D_2) and T_3 (D_3) have three C3M0016120K MOSFETs in parallel to increase current capacity.

Conduction losses difference between different models

This subsection will show the conduction losses difference with the derived model (method 1) and also the one assuming that all the reverse conduction current passes through the channel (method 2). To verify the correctness of the derived equations for current stress calculations, the simulation results from PLECS are given in the tables with switching frequency at 10 kHz, maximum step size is 1 μs, relative tolerance is 1e-4, pure sinusoidal current is used as input source. The device parameters at junction temperature of 150°C through curve fitting are (In Table II, $a = a_1 + a_2$, $b = b_1 + b_2$, $c = c_1 + c_2$):

Table II: Device parameters to estimate the semiconductor losses through curve fitting

	r_{on} (mΩ)	r_d (mΩ)	V_d (V)	a	b	c	V_{ref} (V)
CAS300M17BM2	19.59	5.13	0.78	5.628e-8	9.077e-5	2.791e-3	1200
CAS300M12BM2	8.43	4.59	0.77	3.560e-8	2.440e-5	1.411e-3	600
C3M0016120K	39.8	16.85	3.15	1.104e-7	7.532e-6	1.910e-4	600
SKM125KD12SC	-	5.65	0.79	-	-	-	-

Table III-V show the calculated rms and average current of the device and the three phase conduction losses.

Table III: Calculated current value and three phase conduction losses of two-level rectifier

	$I_{rms\_T_1,T_2}$ (A)	$I_{rms\_D_1,D_2}$ (A)	$I_{avg\_D_1,D_2}$ (A)	Losses (W)
Method 1	67.49	72.89	39.27	883
Method 2	125.62	0	0	1855
Simulation	67.49	72.89	39.27	-

Table IV: Calculated current value and three phase conduction losses of NPC rectifier

	$I_{rms\_T_1,T_4}$ (A)	$I_{rms\_T_2,T_3}$ (A)	$I_{avg\_D_{1-4}}$ (A)	$I_{rms\_D_{1-4}}$ (A)	$I_{avg\_D_5,D_6}$ (A)	$I_{rms\_D_5,D_6}$ (A)	Losses (W)
Method 1	63.66	98.37	16.88	38.01	32.35	75.00	1270
Method 2	100.77	125.62	0	0	32.35	75.00	1656
Simulation	63.66	98.37	16.88	38.01	32.35	75.00	-

Table V: Calculated current value and three phase conduction losses of T-Type rectifier

	$I_{rms\_T_1,T_4}$ (A)	$I_{rms\_D_1,D_4}$ (A)	$I_{avg\_D_1,D_4}$ (A)	$I_{rms\_T_2,T_3}$ (A)	$I_{avg\_D_2,D_3}$ (A)	$I_{rms\_D_2,D_3}$ (A)	Losses (W)
Method 1	36.05	65.08	30.15	35.22	0.057	0.382	1316
Method 2	100.77	0	0	35.35	0	0	2089
Simulation	36.05	65.08	30.15	35.22	0.057	0.382	-

Note in Table V, the current for T_2 (T_3) and D_2 (D_3) are the current through each device, since three are put in parallel to increase the current rating.

From the above results, it can be seen that firstly, the simulation results match the theoretical equations as expected which proves the correctness of the derived formulae. Secondly, the conduction losses between these two methods differ much. If defining error percentage as: $error = \frac{\text{Losses(Method 2)} - \text{Losses(Method 1)}}{\text{Losses(Method 1)}}$, then the error is 110%, 30.4% and 58.74%. Therefore, the more accurate models should be used to estimate the losses if there is reverse current distribution between the MOSFET channel and its paralleling diode.

Simulation and theoretical semiconductor losses

The switching losses will be included. Based on the previous equations derived for the switching losses, the results are given in Fig. 6 below at 10 kHz switching frequency:

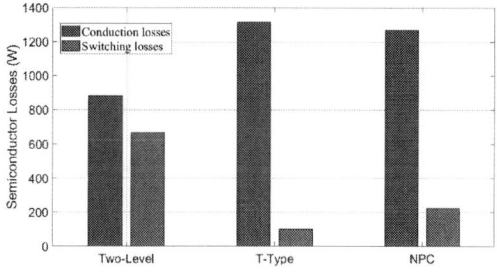

Fig. 6: Semiconductor losses of different rectifier types.

As a verification example, the simulation result of the two level rectifier is shown in Fig. 7 below:

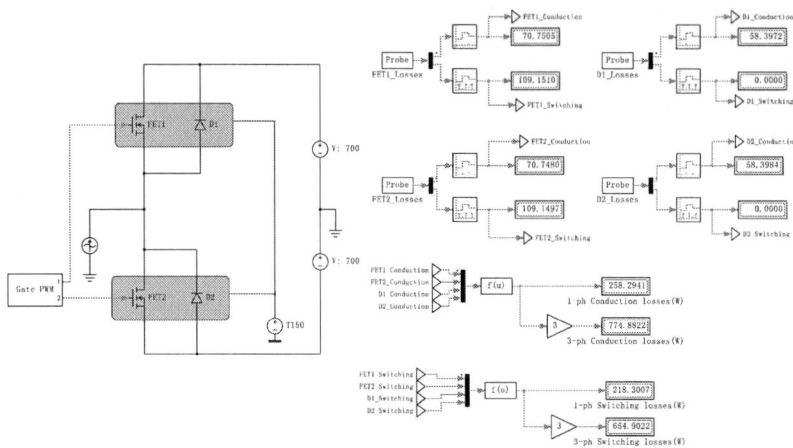

Fig. 7: Semiconductor losses simulation results of a two-level rectifier.

From the simulation results, the switching losses match well with the theoretical values while the conduction losses have some differences. As mentioned previously, in order to calculate the losses of each component, the electrical operation status of each component needs to be known first due to the losses calculation mechanism of PLECS itself, therefore, the on-resistance of the channel is applied while the losses simulation is based on lookup table, so, this difference is expected and acceptable.

Conclusion

Analytical equations considering the reverse current parallel conduction between MOSFETs channel and body diodes are given to facilitate a quick calculation of semiconductor losses for high power three-phase rectifiers. The situation that the MOSFET channel conducts all the current will then become a special case. A losses simulation model is also provided which can be adapted easily to study more complicated situations as discussed in introduction. The studied example shows the necessity to use a more accurate model for conduction losses estimation.

References

[1] H. Tu, H. Feng, S. Srdic and S. Lukic, "Extreme Fast Charging of Electric Vehicles: A Technology Overview," in IEEE Transactions on Transportation Electrification, vol. 5, no. 4, pp. 861-878, Dec. 2019, doi: 10.1109/TTE.2019.2958709.

[2] M. Stecca, T. B. Soeiro, L. R. Elizondo, P. Bauer and P. Palensky, "Comparison of Two and Three-Level DC-AC Converters for a 100 kW Battery Energy Storage System," 2020 IEEE 29th International Symposium on Industrial Electronics (ISIE), 2020, pp. 677-682, doi: 10.1109/ISIE45063.2020.9152545.

[3] R. Yapa, A. J. Forsyth and R. Todd, "Analysis of SiC technology in two-level and three-level converters for aerospace applications," 7th IET International Conference on Power Electronics, Machines and Drives (PEMD 2014), 2014, pp. 1-6, doi: 10.1049/cp.2014.0498.

[4] Rąbkowski, Jacek and Płatek, Tadeusz. "Comparison of the power losses in 1700V Si IGBT and SiC MOSFET modules including reverse conduction," 2015 17th European Conference on Power Electronics and Applications (EPE'15 ECCE-Europe), 2015, pp. 1-10, doi: 10.1109/EPE.2015.7309444.

[5] A. Acquaviva, A. Rodionov, A. Kersten, T. Thiringer and Y. Liu, "Analytical Conduction Loss Calculation of a MOSFET Three-Phase Inverter Accounting for the Reverse Conduction and the Blanking Time," in IEEE Transactions on Industrial Electronics, vol. 68, no. 8, pp. 6682-6691, Aug. 2021, doi: 10.1109/TIE.2020.3003586.

[6] A. Acquaviva and T. Thiringer, "Energy efficiency of a SiC MOSFET propulsion inverter accounting for the MOSFET's reverse conduction and the blanking time," 2017 19th European Conference on Power Electronics and Applications (EPE'17 ECCE Europe), 2017, pp. P.1-P.9, doi: 10.23919/EPE17ECCEEurope.2017.8099052.

[7] Amirpour, Sepideh, Torbjörn Thiringer, and Dan Hagstedt. "Energy Loss Analysis in a SiC/IGBT Propulsion Inverter over Drive Cycles Considering Blanking time, MOSFET's Reverse Conduction and the Effect of Thermal Feedback." 2020 IEEE Energy Conversion Congress and Exposition (ECCE). IEEE, 2020.

[8] G. Su, "Loss Modeling for SiC MOSFET Inverters," 2018 IEEE Vehicle Power and Propulsion Conference (VPPC), 2018, pp. 1-6, doi: 10.1109/VPPC.2018.8604972.

[9] R. A. Wood, D. P. Urciuoli, T. E. Salem and R. Green, "Reverse conduction of a 100 A SiC DMOSFET module in high-power applications," 2010 Twenty-Fifth Annual IEEE Applied Power Electronics Conference and Exposition (APEC), 2010, pp. 1568-1571, doi: 10.1109/APEC.2010.5433440.

[10] Mohan, D., and Sreejith B. Kurub. "A comparative analysis of multi carrier SPWM control strategies using fifteen level cascaded H-bridge multilevel inverter." International Journal of Computer Applications 41.21 (2012).

[11] Ulrich Nicolai, Arendt Wintrich. Determining switching losses of SEMIKRON IGBT modules, AN1403, Aug. 2014.

[12] G. Yu, T. B. Soeiro, J. Dong and P. Bauer, "Study of Back-end DC/DC Converter for 3.7 kW Wireless Charging System according to SAE J2954," 2021 IEEE 15th International Conference on Compatibility, Power Electronics and Power Engineering (CPE-POWERENG), 2021, pp. 1-8, doi: 10.1109/CPE-POWERENG50821.2021.9501207.

[13] B. J. D. Vermulst and J. L. Duarte, "Losses evaluation of two-level and three-level PFC topologies based on semiconductor measurements," 2013 IEEE ECCE Asia Downunder, 2013, pp. 1263-1267, doi: 10.1109/ECCE-Asia.2013.6579271.

Measurement Method for Simple Determination of Sinusoidal Large Signal Losses in Inductive Components

Peter Zacharias[1], Alejandro Aganza-Torres[2]

[1]University of Kassel, Wilhelmshoeher Allee 71, D-34131 Kassel, Germany, Tel.: +49 / 561 – 804 6344 Fax: +49 / 561 – 804 6351. E-Mail: peter.zacharias@uni-kassel.de.com

[2]Autonomous University of San Luis Potosi, Multidisciplinary Academic Unit Middle Zone, Rioverde-San Ciro Road Km. 4, 79617, Rioverde, Mexico, e-mail: alejandro.aganza@uaslp.mx

Keywords

«Magnetic device», «Magnet loss», «Passive component», «Permeability», «Standardization

Abstract

In this work, a resonance-based losses measurement method is shown suitable for the power dissipation measurement of small and large inductive components under sinusoidal test voltage. In a resonance-based measurement method, the magnetic losses of inductive devices are transformed to the measurement of electrical RMS magnitudes through the device under test. The sinusoidal shape of the test voltage is therefore inherently guaranteed in the measurement procedure.

Introduction

Magnetic components are largely responsible for the weight and volume of power electronic converters. The decisive factor in their use is that certain limit temperatures are not exceeded. This enables long service lives to be achieved. In addition to the saturation of the core, losses during operation thus represent important limits that must not be exceeded. A distinction must be made between winding and core losses, because of the nonlinear properties and the sometimes very high quality factors that can be achieved with ferromagnetic core materials, reliable measurement of losses in cores is difficult. Data on losses in core materials are given mainly for sinusoidal excitation in the small-signal range on small toroidal cores [1]. These values cannot be applied to arbitrary large cores and non-sinusoidal quantities for several reasons. One reason is the non-constant magnetic flux density for many core shapes. Because of the nonlinearity of the relationships, the superposition principle cannot be applied to the individual sections. Then other ways have to be searched [2, 3, 4]. If it is possible to provide sinusoidal voltages with sufficient quality or low THD, electrical measurement methods can be used which are characterized by very short measurement times [1, 6]. For non-sinusoidal voltages at the inductor, only calorimetric measurement methods [2, 7, 8] are reasonably applicable. To a large extent, one can dispense with precision measurement equipment if, instead of phase angle measurement, bridge balancing is used [8], as described in the following article. Electrical measurement methods have the great advantage that they provide results several orders of magnitude faster than calorimetric measurement methods. The paper presents a resonance-based measurement method to map losses of an inductive device to the relatively simple measurement of RMS voltages. Alternatives to approximately separate winding losses from total losses are also shown.

Parallel resonance as the basic idea of the measurement method

If you connect a capacitance in parallel to an inductance, you get a parallel resonant circuit. At its resonant frequency, the impedance becomes maximum and the imaginary part of the impedance becomes zero. Only the power to cover the losses of the resistive component is supplied to the terminals of the resonant circuit. The reactive power, which is usually much higher, is exchanged between the inductive component and the capacitance. Even with non-sinusoidal current, harmonics are strongly suppressed by the integrating behavior of the capacitance in the voltage. The sinusoidal shape of the voltage, which is an important secondary condition for the comparability of the measurement conditions, is thus maintained over a wide voltage and frequency range. Due to the clearly dominating fundamental oscillation at the resonance frequency, the classical methods of alternating current calculation can be used for loss determination with good justification. Fig. 1a shows the principle of the measurement. E_1 represents a voltage source with variable frequency and sinusoidal voltage. This source also has to supply the total losses in the measurement circuit. The resonant capacitor C_1 is in parallel with the DUT (Device Under Test) and supplies the required reactive power for it. The resistor R_1, like the capacitor C_1, must be selected as precision components, as the measured losses and inductance values of the DUT are determined in reference to these components. At resonant frequency f_{res}, the phase shift between voltages V_1 and V_2 is zero. This is easy to detect. In the resonance case (Fig. 1b), the complexity of the circuit is reduced if all losses are lumped to a resistor R_{loss}.

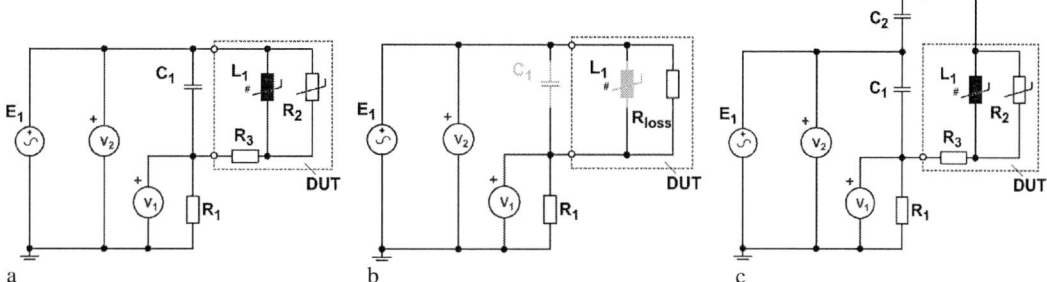

a   b   c

Fig. 1: Principle structure for determining the losses of an inductive component in the large signal range: a basic circuit, b remaining structure in the resonance case, c - extension of the circuit to achieve higher voltages at the DUT.

Thus, the resistance R_{loss} can be determined from the known quantities.

$$R_{loss} = \left(\frac{U_2}{U_1} - 1 \right) R_1 \quad \text{therefore:} \quad P_{loss} = \frac{\left(U_2 - U_1 \right)^2}{R_{loss}} = \frac{\left(U_2 - U_1 \right) U_1}{R_1}$$

In the resonance case, equal amounts of magnetic energy are exchanged with amounts of electrical energy. This results in the effective inductance L_{eff} via the formula

$$L_{eff} = \frac{1}{\left(2\pi f_{res} \right)^2 \cdot C_1}$$

The modification in Fig. 1c allows the test voltage at the DUT to be increased. The calculations of P_{loss} and L_{eff} must then be adjusted accordingly.

Realization of a measurement setup according to the described principle

Fig. 2 shows how such a measurement setup can be realized with simple means. Fig. 2a shows an overview of the interaction of the individual components. Fig. 2b shows the laboratory setup. In order to be able to record series of measurements at the same frequency, the capacitance C_1 was designed to be variable via switchable capacitance decades. The capacitances must have low tolerances and low losses, since they serve as reference elements. The sinusoidal voltage is generated with an adjustable frequency generator and amplified from a level of 1Vrms to 100Vrms with a high-fidelity power amplifier. The nominal power of the amplifier is 100W. It should be noted that this power only covers the power dissipation requirements of the DUT and R_1. The reactive power in the resonant circuit can be many times higher. The capacitors and switches must therefore be designed for the possible currents and voltages at resonance. In the case shown, this is 6A and 100V as a continuous load.

The oscilloscope is used in the measurement setup to determine the resonance case or resonance frequency and to measure the voltages V_1 and V_2. The measurements are then further used for evaluation according to the presented algorithms. The frequency range of the power amplifier is 10kHz - 12MHz. Currently, based on this basic design, a measurement setup is being implemented with air rotary capacitors in the range of 25pF - 1.5nF, mica capacitors up to 100nF and film capacitors up to 10x10uF. This is to cover the widest possible range of measurement tasks of interest. With the resistor R_1 it is possible to switch between different measuring ranges for the power dissipation. Since current flows continuously during adjustment and measurement to cover the losses in the parallel resonant circuit, R_1 must be able to dissipate this power loss. This means that low inductance and easy to cool footprints are considered for R_1. Both the leads and the junctions must be suitable for the relatively large currents flowing through the resistor.

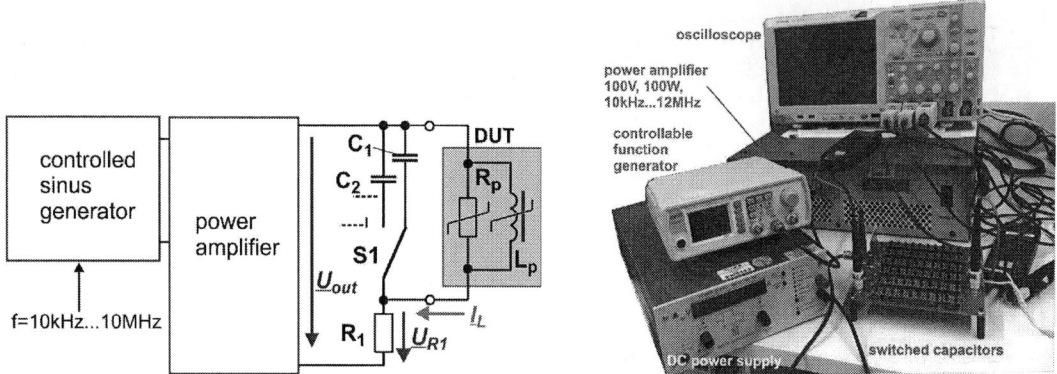

Fig. 2: Set-up of the laboratory test bench for determination of losses in inductive components: a - principle set-up, b - laboratory realization with devices and manually switched capacitors.

Separation of winding and core losses

Even simple inductive components have a considerable complexity. Fig. 3 shows an electrical equivalent circuit for a winding with a ferromagnetic core. Due to the induced electric field, the magnetic field of the winding leads to losses via dielectric and galvanic currents in the core. These are represented in the model by the parallel resistance Rcore(f,U). For both, it can be assumed that they are dependent on both the induced voltage and the frequency.

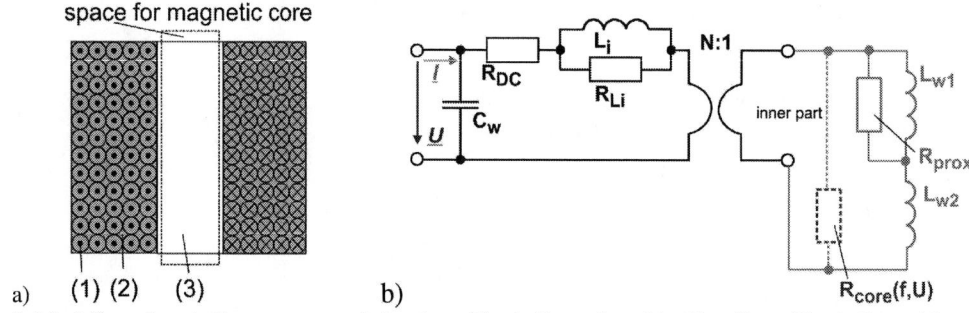

Fig. 3: Modeling of a winding on a core: a) drawing: (1) winding wire with skin effect, (2) winding with proximity effect and (3) space filled with magnetic field, which may be filled with core material; b) equivalent circuit derived from a) with consideration of the individual loss components.

Through the transformer with the number of turns N, the voltage U applied to the terminals of the component becomes the voltage induced on a turn in the core. The winding without a core has a DC resistance R_{DC} and an intrinsic inductance, which is formed by the internal inductance L_i, the winding inductance L_{w1}, which is formed by the linked magnetic field of the winding penetrated by the stray field, and the inductance L_{w2}, which is formed by the linked magnetic flux in the space not filled by magnetic materials. The transformation ratio of the intrinsic ideal transformer transforms these quantities to the terminals of the device (Fig. 4). The internal inductance L_i leads to losses due to eddy currents. At the same time, field displacement takes place, resulting in a reduction of the internal inductance and an increase of the effective resistance. If core material is added, the equivalent circuit for the entire device is obtained. This is in accordance with the rules of structural analysis for magnetic circuits. Since the magnetic field of the component enters the winding material, it causes eddy current losses there. Because of the transformer coupling of these losses, a parallel resistor R_{prox} is physically appropriate as their representative.

The series connection of L_{w1} and L_{w2} also corresponds to the physical conditions. If a core material is inserted, the resulting losses can be represented by a resistance R_{core} parallel to L_{w2}. This is not only dependent on the frequency but also on the voltage. This equivalent circuit reflects the basic physical relationships and approximately the quantitative relationships. With a determination of the properties of the winding without core, the winding losses can thus be determined approximately for a balancing calculation.

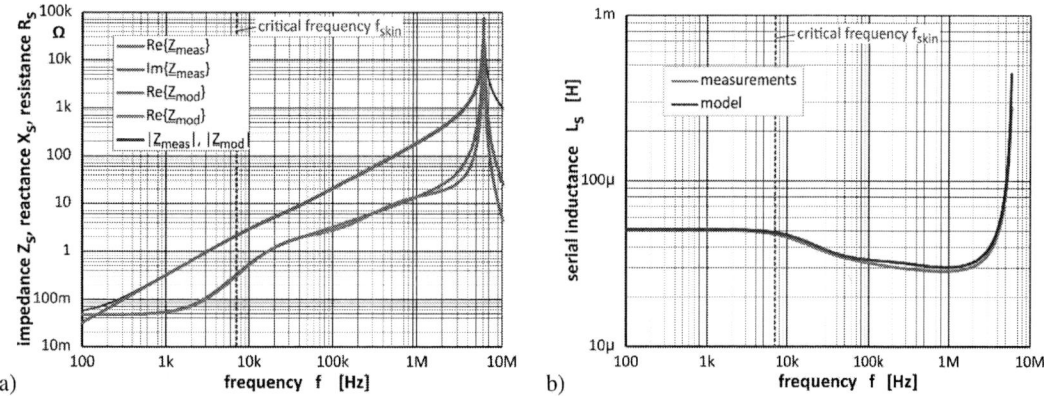

a)   b)

Fig. 4: Measured and calculated impedance and real components of the impedance Z_s (a); Measured and calculated course of the inductive part of the impedance (DUT: solenoid coil, d_{Al} = 2mm, 38x12x18mm, N = 60; model parameters related to the terminals: R_{DC} = 477mOhm, L_i =4.5µH, R_{Li} = 13Ohm; L_{w1}=18µH, R_{prox} = 2.2Ohm, L_{w2} = 28.6µH, R_{core} = ∞, C_w = 24pF).

Fig. 4 compares the measurement results for a short coreless cylindrical coil (solenoid coil; D x d x l = 38mm x 10mm x 18mm) made of 2mm thick aluminium wire with the calculations for a model according to Fig. 3b. It is clear that the increase in losses (represented by the measured real component of the impedance) already start well below the critical frequency f_{skin}. Here f_{skin} is the frequency at which the formal skin thickness of the current is equal to the radius of the wire. Above this frequency, one observes a decrease in the effective inductance L_s=Im$\{Z_s\}$/(2πf). Physically, this is due to the onset of displacement of the magnetic field not only from the winding wire (skin effect) but also from the interlinked region of the winding. The apparent increase of the inductance at approx. 6MHz is caused by the parallel resonance of the effective inductance with the winding capacitance C_w. This leads to an apparent increase of the inductance by the calculation rule.

Actually, the field displacement leads to a loss increase by a factor of ~$f^{0.5}$. A component with such properties does not exist as a basic element in electrical engineering. Although it can be simulated by an LR cascade arrangement. Consequently, the example shown is only an approximation. The found agreement of the model data with the measured data is however already quite appropriate in the frequency range <f_{res}, which is interesting for the application of the component. The parameter found for L_i represents both the skin effect in the wire and the shielding effect between the layers in the proximity effect. It is therefore significantly larger than the actual internal inductance of the wire used (112nH).

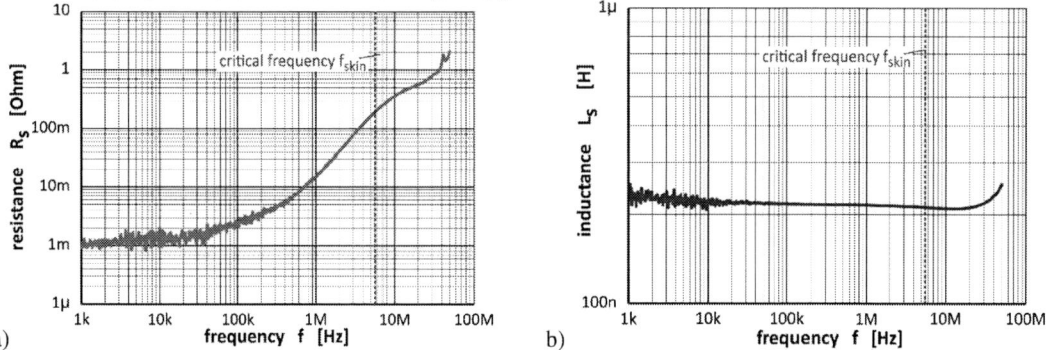

a)   b)

Fig. 5: Measurement results for the impedance components of a coil without core according to Fig. 6 with plotted critical frequency for the skin effect: a) serial resistance Rs; b) serial inductance Ls (fitted model values: R_{DC} = 1mOhm, L_i = 18nH; R_{Li} = 25Ohm; L_{w1}=10nH, R_{prox} = 0.22Ohm, L_{w2} = 190nH, R_{core} = ∞, C_w = 9pF).

The extent of the proximity effect depends on the structure of the winding and the associated stray field in the winding. In Fig. 5, the measured frequency-dependent curves of the real and imaginary parts of the impedance of a single-layer cylindrical coil are shown in Fig. 6 for comparison with Fig. 4. In this case, the magnetic field lines are drawn comparatively long. For this reason, the magnetic field strength tends to be lower than for a solenoid coil. This is associated with lower eddy current losses in the conductor material. In addition, the structure shown uses stranded copper wire 4x (1575 x 0.071mm) as the conductor material. This leads to a much higher critical skin frequency than in Fig. 5. It can also be seen here that the proximity losses become dominant at frequencies much lower than the critical skin frequency.

The approach of taking the losses in the winding material into account is that, in addition to the described measurement of the losses with nearly sinusoidal voltage waveforms, the resulting winding losses are calculated with the help of the model. The parameterization of a model for a winding without a core then allows the expected losses in the entire inductive component to be determined by combining it with a (partially linearized) loss model of the core $R_{core}(U, f)$ in a simulation.

Measurement example

Fig. 6 shows some first measurement results with the previously described principle. Fig. 6a shows the mechanical design of the DUT. On a UR64 core there are 8 x 3 windings of HF stranded wire 1575 x 0.071mm. This keeps the loss components of the winding very low by design. All partial windings are connected in parallel. This "forces" the flux into the core. For the measurement results shown, no air gap was inserted between the core halves. Measurement results for this no air gap condition and for parallel premagnetization will be reported in the main paper. As the UR64 core has a hole for mounting, this is used for a flexible line for a premagnetization current Id to be supplied. This premagnetization current is oriented parallel to the alternating magnetic field of the coils. This means that the main magnetic field and the field for premagnetization are perpendicular to each other. This is called orthogonal premagnetization, and in this case the fields add geometrically. When the result of this superposition enters the saturation region, small and large signal inductance of the main coil change. One of the questions to be investigated was to what extent the losses or the quality factor Q also change. Figs. 6b and 6c show some results of the evaluation of the measurements.

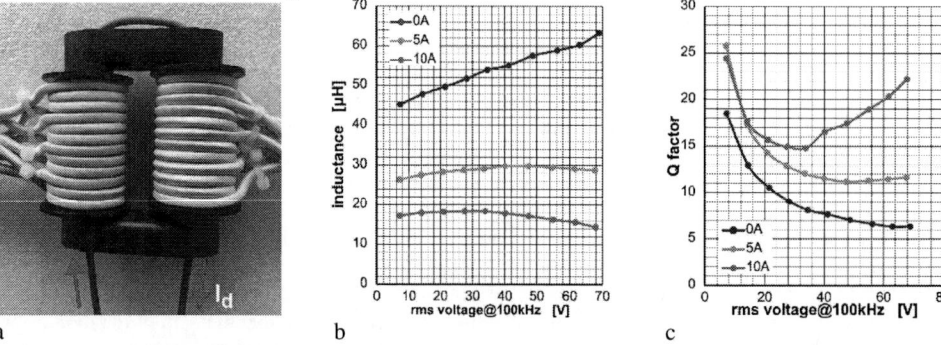

a   b   c

Fig. 6: Measurement results for a Ferroxcube UR64 ferrite core orthogonally premagnetized with a current Id. a-Mechanical structure; b - Effective inductance as a function of RMS voltage and premagnetization current Id and c - Quality factor Q as a function of RMS voltage and premagnetization current Id.

The premagnetization obviously has a significant influence on the inductance of the winding and on the quality factor Q. As expected, the inductance decreases with increasing premagnetization current. An ever increasing current Id means that an ever increasing proportion of the core cross-section around the premagnetization wire becomes saturated and is no longer available for the "concentration" of the magnetic field. This tendency becomes even more pronounced at higher voltages.

If only the losses are considered, after a mathematical subtraction of the winding losses, the results can be plotted in a double logarithmic representation. Then Fig. 7a is obtained. Each of the measurement series is approximated by a power function according to the Steinmetz formula. It can be seen that each of the measurement series yields almost straight lines at this scale. This impression is also confirmed when the measured values are plotted relative to the power function for $I_d = 0A$ (Fig. 7b). Nevertheless, it can be observed that in the lower voltage range the relative deviations are larger than in the upper voltage range. These observations can be used in conjunction with the other measurements for further conclusions. One advantage of the method is that the oscillating circuit capacitor with its high Q factor supports the voltage at the inductive component intrinsically. This makes measurement results easily comparable with each other. Moreover, the change at each measurement point is made instantly, which in comparison with calorimetric measurements, are much faster and less dependable of environmental conditions of the operation of the DUT.

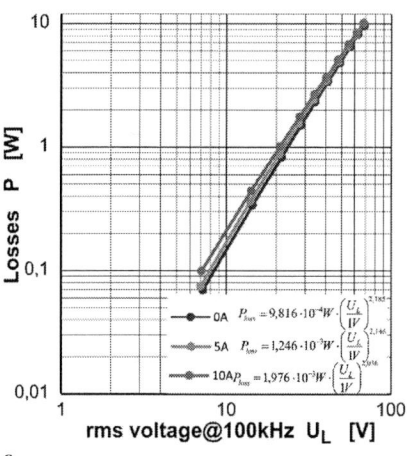 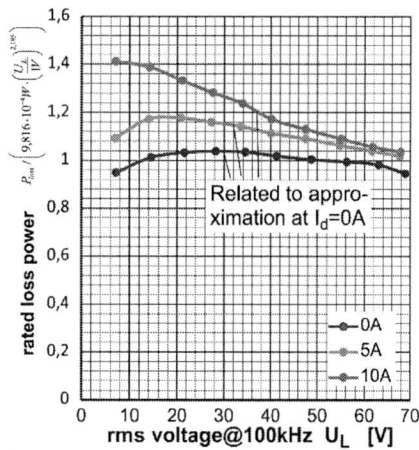

a   b

Fig. 7: Measured power losses as a function of RMS voltage and bias current I_d: a - determined core losses; b - relative deviation of measured values compared to Steinmetz approximation at $I_d=0$

Design of a measurement setup for a wide range of values in general laboratory applications

In large-signal measurements, one usually has relatively high RMS values for voltages and currents. This must be taken into account when selecting and constructing a measuring arrangement. Fig. 8 shows a revised measurement arrangement according to the presented principle. Here, too, the 100W broadband amplifier has to cover only the losses in the DUT and the measurement resistors. In addition to these losses, there are very small contributions from the capacitive elements. These must therefore be selected with the highest possible Q values. In the range up to 1100pF, variable capacitors with air insulation are used. These show very low losses compared to ceramic or film capacitors. To maximize the stability of the capacitance of these capacitors, they are operated in a hermetically sealed room with dried air. Fig. 8b shows the structure of the two switchable variable capacitors in the open state. The capacitance value of each of these capacitors is set manually by means of a fine gear connected to a scale for displaying the current capacitance value.

Further capacitance values can be set by means of changeover switches with mica capacitors up to the range of 100nF. The set capacitance values are required for the evaluation of the measurement results according to the procedure described above. Above 100nF, film capacitors with selected low series loss resistances are used. The frequency-dependent permissible values of current and voltage for the capacitors set limits for measurement ranges. One must not forget that at frequencies >100kHz high reactive powers and thus high currents are already reached even with small capacitance values. For the time of the measurement these should not cause any noticeable heating. Then the loss resistances of the capacitors measured in the small signal range can also be considered in the power loss balance.

The entire setup shown in Fig. 8 is intended for manual operation. By using capacitance values in decades, it is not possible to set an "exact" value for the capacitance due to the discretization. Even considering the possibility of setting fine capacitance values with the variable capacitors, such a goal would be limited by the tolerances of the capacitance values. The individual measurement of each capacitor before installation in the measuring device, mitigates this problem. The sometimes high currents also flow through the switches, which must also be suitable for this purpose. It is therefore advisable to start with small voltages for large-signal measurements and then increase them step by step. Then it is also possible to switch over without problems during the measurement. Typically, the effective inductance then changes somewhat and the resonance ($\Delta\varphi = 0$) must always be readjusted at the new test voltage by adjusting the resonance capacitance value.

The measuring range switching is done with low inductive precision resistors. Since these resistors generate up to 50W of heat, they are mounted on a heat sink which is forced cooled by fans Fig. 8a.

Fig. 8: Structure of the described measuring arrangement before covered: a) Circuit board with the switched and cooled resistors and switched capacitors for the measurement, b) Illustration of 2 variable capacitors (10pF...210pF and 50pF...1100pF) with high Q which is extended to 6100pF with 3 mica caps (1000pF and 2x 2000pF) for the higher frequency range.

Finally, Fig. 9 shows the setting and measuring ranges resulting from the selected setup. To measure a specific core, a suitable number of turns is applied to it. Suitable here means that the measurements of interest can be carried out in the frequency and capacitance ranges available in the setup in Fig. 8.

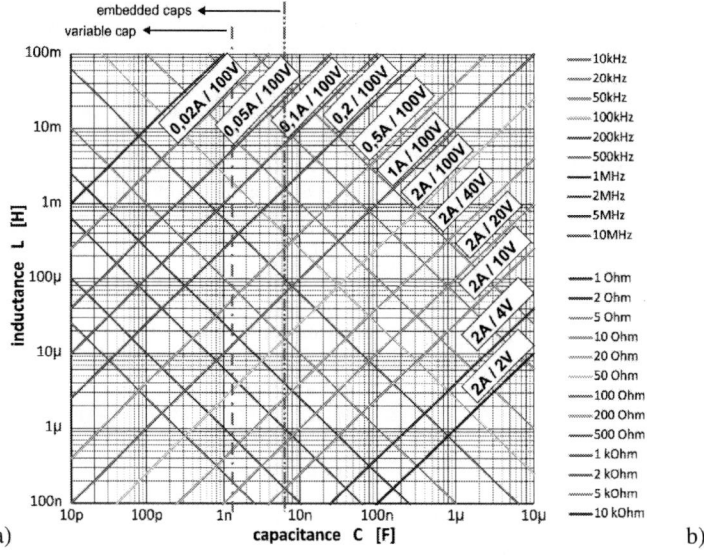

Fig. 9: Relationship between inductance, capacitance, characteristic impedance and measuring frequency for the measuring device (current and voltage values are only orientations here)

Conclusions

In this paper, a resonance-based measurement method is suitable for determining the losses of practically arbitrary cores shapes under sinusoidal voltage loading. The original problem of balancing active and reactive power components, which is usually subject to considerable measurement uncertainties, especially for high-quality inductive components, is transformed into a problem where inherent averaging is performed. Thus, effects of individual measurement errors are strongly suppressed. It is possible to measure both the energetically effective large-signal inductance and the large-signal losses up to relatively high magnetic field strengths. Due to the exclusive use of electrical quantities, the measurement process takes only a very short time compared to the self-heating of the inductive components using the calorimetric measurement method. Application possibilities and statements of the measuring method flexibility are shown on the example for different types of pre-magnetization. A setup for measurement in manual mode was presented. Next steps are to automate this setup.

References

[1] International Electrotechnical Comission, "IEC 62044-1:2002. Cores made of soft magnetic materials - Measuring methods - Part 1: Generic specification ", May 2002.

[2] C. Xiao, G. Chen and W. G. H. Odendaal, "Overview of Power Loss Measurement Techniques in Power Electronics Systems," in IEEE Transactions on Industry Applications, vol. 43, no. 3, pp. 657-664, May-june 2007, doi: 10.1109/TIA.2007.895730.

[3] Lopez-Lopez, J.; Fernandez, C.; Barrado, A; Zumel, P. Comparison of Different Large Signal Measurement Setups for High Frequency Inductors. Electronics 2021, 10, 691. https://doi.org/10.3390/ electronics10060691

[4] Zhang, Yu & Alatawneh, Natheer & Cheng, Ming-Cheng & Pillay, P.. (2009). Magnetic core losses measurement instrumentations and a dynamic hysteresis loss model. 10.1109/EPEC.2009.5420918.

[5] F. Dong Tan, Jeff L. Vollin, and Slobodan M. Cuk : "A Practical Approach for Magnetic Core-Loss Characterization", IEEE TRANSACTIONS ON POWER ELECTRONICS, VOL. 10, NO. 2. MARCH 1995.

[6] ZES Zimmer, "Measurement of magnetic characteristics of transformer-cores and coil materials. Application Note 103 (Rev. 2.0)", https://www.zes.com/en/Service/Downloads/Documents/Application-Notes/103-Measurement-of-the-magnetic-properties-of-transformers-and-coil-cores, Accessed: May 2022.

[7] F. Zámborszky, D. Tóth, Z. Palánki and E. Csizmadia, "Electrical and Calorimetric Power Loss Measurements of Practically Ideal Soft Magnetic Cores," in *IEEE Transactions on Magnetics*, vol. 50, no. 4, pp. 1-4, April 2014, Art no. 6300604, doi: 10.1109/TMAG.2013.2286700.

[8] P. Zacharias: Magnetische Bauelemente. Springer-Nature 2020. ISBN 978-3-658-24741-6, https://doi.org/10.1007/978-3-658-24742-3

A Novel Technique for the Suppression of the Displacement Current through Power Module Base-plate Capacitance

Mahmoud Saeidi, Ahmad Ali Nazeri, Rufad Zilic, and Peter Zacharias
Centre of Competence for Distributed Electric Power Technology
Faculty of Electrical Engineering / Computer Science
University of Kassel, Kassel, Germany
Email: {mahmoud.saeidi, peter.zacharias}@uni-kassel.de
{ahmad.nazeri, uk038019}@student.uni-kassel.de

July 11, 2022

Keywords

≪Baseplate capacitance (BPC)≫, ≪Common mode choke (CMC)≫, ≪Displacement current≫, ≪Multichip power module≫, ≪Wide-bandgap≫

Abstract

The voltage gradient is increased to reduce the switching losses in the wide-bandgap (WBG) semiconductors, which causes the higher power density of the system. Fast rise and fall times during switching of WBG semiconductors result in capacitive non-linearities and displacement currents through power module parasitic capacitances, which needs an appropriate filter design During the switching in the power module, the capacitive current flows through the parasitic capacitance of the power module, which causes the disturbances in the system. This leads to the need for a more oversized filter design, which then increases the overall cost and volume of the system and reduces the efficiency of the system. This paper proposes a novel technique to suppress the capacitive displacement currents without switching speed reduction. The proposed method reduces the volume of the common-mode choke (CMC) and addresses the electromagnetic interference (EMI) and electromagnetic compatibility (EMC) issues. The system was experimentally tested using the commercial 1700 V silicon carbide (SiC) half-bridge power module BSM300D12P2E001 to validate the proposed scheme.

Introduction

Achieving a higher power density with lower switching losses has always been a challenge in WBG semiconductor devices. A higher switching speed results in an increase in the displacement current through multi-chip power module (MCPM) parasitic capacitances in the WBG semiconductors. The significant increase in the displacement current causes more electromagnetic interference (EMI) and electromagnetic capability (EMC) problems in high-frequency applications. Therefore, the EMI/EMC problem is a great challenge in WBG-based applications. To overcome this problem, many researchers have already proposed different methodologies to design the common-mode choke (CMC) filters. The switching behavior of semiconductors has to be considered, which affects the overall filter design procedure. Minimizing the size, volume, and cost of the EMI filter is known as the dominant target in the EMI/EMC considerations. Moreover, the EMI filter components must qualify for the EMC standards and should be avoided to increase the cost and complexity of the overall system. Otherwise, it degrades the convertor efficiency and may need active cooling for high-power applications.

The parasitic capacitance between power module chips and the ground is one of the major contributors to the common-mode (CM) EMI noise in power electronic systems with high dv/dt during switching transients. The capacitive CM or displacement current is generated from the higher dv/dt through the parasitic capacitance and the ground. The parasitic capacitance is intrinsic to the power module and cannot be modified. It is desired to minimize the module parasitic capacitances inside of power modules and smaller direct bonded copper (DBC) dimensions and reduced track widths reduce the total output coupling capacitance [1]. The multi-chip power module packages carry more current and are used for high-power applications. The capacitive coupling between chips on DBC and mounting base-plate is inevitable in the packages and leads to parasitic capacitances, which become charged or discharged during switching. The parasitic capacitances in SiC power modules are due to the chip area being smaller than silicon-based semiconductors. Despite small parasitic capacitances, SiC power modules have higher dv/dt, which causes more capacitive current through capacitances. The displacement current flows through the parasitic capacitance between DBC and the mounting base plate of the power module. The higher the dv/dt will be increased the value of capacitive displacement current significantly. It requires a large EMI filter due to such increasing [2]-[6]. Therefore, proper suppression of the capacitive displacement current is needed to avoid the bulky design of the EMI filters. In the literature review, different methods have been suggested to mitigate the displacement current, which is either complicated to implement or increases the overall cost of the system.

In [7], the authors proposed to place CM filtering capacitors inside the power module close to the noise source, which offers a desirable decrease in the CM noise. The integrated capacitors inside the power module for reducing the CM noise have a major drawback of not withstanding higher junction temperatures of the power module. A separate physical heat sink was used to cancel the displacement current flowing into the grounding system [8], which then increases the overall cost of the system and is not desirable for the power module. This solution is a better choice for the discrete switch where independent heat sinks are needed. The optimized filter design was performed by [9], which has investigated the parasitic parameters in the module and switching cell. The parasitic parameters in the SiC power module have been identified by balancing coupling between the base-plate and the terminals of the module [10]. The CM and differential mode (DM) EMI noises can be effectively attenuated in low and high-frequency ranges by adding decoupling capacitors in the voltage source converter (VSC) [11]. A CM model for a multi-chip power module has been proposed for the EMI issues to identify a method to mitigate the displacement current [12]. The proposed method in [12] placed capacitances between the neutral point and the grounding system. The study refers to the analysis of the CM behavior of an ungrounded WBG-based converter by varying the impedance between the multi-chip module and the grounding structure of the system [13]. A method to design the EMI filter is proposed in [13], once the noise source impedance is determined. Moreover, the model constructs the relationship between the noise source impedance and the passive elements where the equivalent values of the noise source impedance are calculated based on the experimental results [13].

An equivalent modeling approach of the CM is presented in [14], which can mitigate the conducted emissions in WBG-based systems in the frequency range between 10 kHz and 30 MHz. This suggested scheme in [14] is based on the circuit analysis methods and can be implemented with the data obtained from the impedance analyzer. The goal of the approach is to reduce the CM emissions of the power electronic systems [14]. Five cancellation methods for parasitic capacitances have been reviewed and analyzed in [15] where the cancellation frequency ranges are identified and the effective frequency range for each cancellation method is derived based on the constraints. Moreover, using the mutual capacitance the method is suitable for the design of the DM and CM filters [15].

This paper proposes a novel and a simple technique to mitigate the displacement current through the parasitic capacitance between power module chips and the mounting base-plate. The suggested methodology includes the various types of inductors connected in parallel to the parasitic capacitance DC- terminal of the power module and the mounting base-plate. This results in the effective reduction of the displacement current by one-third. Moreover, the implementation of the CMC is compared with the combination of the CMC and inductor connected in parallel to the parasitic capacitances of the power module. In addition, the experimental results have been compared with the reference in [11]. In [11], where the authors have used an array of ≈ 39 capacitors to attenuate the displacement current, which is bulky and costly. The proposed method is implemented on the double-pulse test (DPT) for the measurement of the displacement current to 800 V drain-source voltage.

Proposed Methodology

To achieve effective suppression of the displacement current, one must consider an equivalent circuit, which shows the origin of the displacement current and its path to flow. The high slew rate of voltage during SiC power module switching is generated more the capacitive displacement current. If the switching speed is kept fixed, one must be concentrated on other possible solutions for more generated displacement current. Multi-chip power module got a large chip area, which results in higher parasitic capacitances, which play a key role in the effect of the displacement current. The proposed method is ongoing about the improvement path of displacement current. In [11], was showed an analytical modeling of a common-mode equivalent circuit for power module and approved, how displacement current changes with total parasitic capacitance adjustment.

As shown in Fig. 1a, is proposed to placement an inductor L_{lb} between the DC- terminal and the base-plate as shown, which affects on displacement current path. The parallel connection of the inductor and base-plate parasitic capacitance expresses a new combination of L and C parameters, which provides a different impedance equivalent circuit and resaults in a new displacement current path. The displacement current exhibits high-frequency oscillations which could affect the performance of inductors. Since some inductors are not suitable for high frequencies and will saturate depending on the core material. Because of that, the inductors are selected from different types, inductances, and core materials. According to their frequency-dependent equivalent circuit, the inductor selection in table I is chosen to use in the setup test. Suppressor chokes $L1$ -$L3$ have different inductances values from 60 to 220 μH and different constructions. $L4$ and $L6$ are called air core inductors, which have a difference in the spacing between the windings. $L4$ and $L6$ have different values of parasitic capacitance between the individual windings due to different distance, resulting in an equivalent impedance circuit for each case. Refer to Fig. 1b, which is the placement of a CMC between DC link and power supply. Furthermore, the CMC windings are rebuilt to use as differential mode Choke $L7$ in Table I. The new proposed path can be used in a real application, whereas the proposed method in [11] how far away is from real hardware application.

Fig. 2 shows the value of the parasitic capacitance between the outputs of the MCPM and arbitrary point P is [11]

$$C_{bp} \approx C_{ub} + c_{ab} + c_{lb} \tag{1}$$

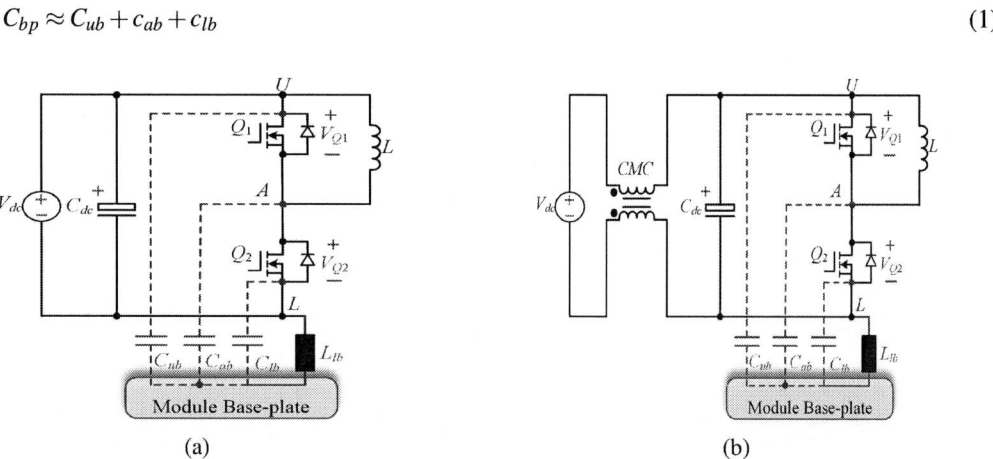

(a)   (b)

Fig. 1: Equivalent circuit of the SiC multi-chip power module (MCPM) with base-plate parasitic capacitance; (a) without CMC, (b) with CMC

Table I: Inductor Values

Inductors	Values
Suppressor chokes $L1$-$L3$	$(60, 150, 220)\,\mu H$
Big Air core $L4$	$4.769\,\mu H$
Ferrite core inductor $L5$	$2.662\,\mu H$
Air core $L6$	$2.275\,\mu H$
CMC as DM inductor $L7$	$3.142\,mH$

where C_{bp}: total parasitic capacitance between MCPM and mounting base plate.
C_{ub}: parasitic capacitance between DC+ terminal and mounting base plate.
C_{ab}: parasitic capacitance between DC- terminal and mounting base plate.
C_{lb}: parasitic capacitance between AC terminal and mounting base plate.

The total impedance of the parasitic capacitance without using the inductor in parallel to the DC- and the mounting base-plate is

$$Z \approx \frac{Z_{Cbp} \times R_{SHUNT}}{Z_{Cbp} + R_{SHUNT}}. \tag{2}$$

where R_{SHUNT} is used to measure the current. The proposed method suggested is to use the inductor in series to the R_{SHUNT} illustrated in Fig. 3, which changes the total impedance of the parasitic capacitance between the mounting base plate and ground.

$$Z_{new} \approx \frac{Z_{Cbp} \times Z_1}{Z_{Cbp} + Z_1} \tag{3}$$

where $Z_1 = Z_L + R_{SHUNT}$ and Z_L is the impedance of the selected inductors.

Experimental results

Fig. 4 shows the experimental setup of the DPT. The SiC N-canal BSM300D12P2E001 multi-chip power module was chosen for the test. The air core inductor of 245 μH as clamped inductive load was connected for the high side switch. The pulse was given to the low side switch. The displacement current is measured first without any selected inductors using a shunt resistor between the DC- terminal and the base plate as shown in Fig. 1a. The displacement current was measured with and without CMC as shown in Fig. 1 to see the effectiveness of CMC in relation to the use of selected chokes. The DC link voltage

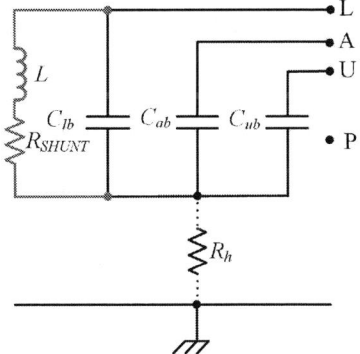

Fig. 2: Parasitic capacitance of MCPM with selected inductor

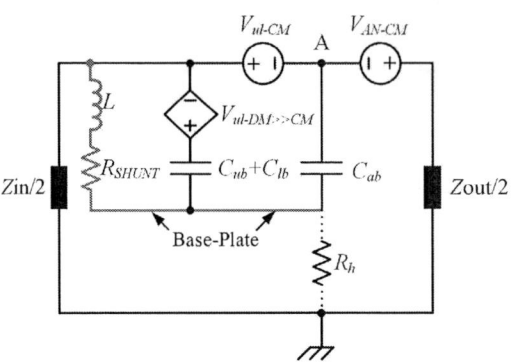

Fig. 3: Common-mode equivalent of half-bridge

was increased in steps of 100 V up to 800 V, which changed dv/dt during power module switching. Fig. 5 shows the experimental results of the drain-source voltage versus the displacement current without the CMC as shown in Fig. 1a.

It shows the implementation of different inductors connected in parallel to the DC- terminal and the power module mounting base plate. The system parameters and the selected inductors are shown in Table I. Fig. 5 shows that the dv/dt is increased and that results in the increase of the displacement current with the higher drain-source voltage (V_{ds}). The displacement current is well suppressed with the inductors of 150 μH and 220 μH compared with other inductors as depicted in Fig. 5. It can be seen that the inductors connected in parallel have mitigated the displacement current from 618 mA to 243 mA at 800 V.

Fig. 6 illustrates the experimental results of the drain-source voltage versus the displacement current where CMC has also been implemented as shown in Fig. 1b. It can be seen that the inductor of 150 μH and 220 μH have mitigated the displacement current better than the other inductors. The combination of the CMC with the inductors in parallel has attenuated the displacement current from 628 mA to 155 mA at 800 V. The displacement current is well attenuated in Fig. 6 as compared to Fig. 5 because of the suppression capability of the CMC filter. As the displacement current is well suppressed maintaining the higher dv/dt with lower switching losses, which results in lower EMI/EMC emissions. Lowering the EMI/EMC emissions assists in the smaller design of the CMC filter, which then helps in reducing the total volume and cost of the system. The results of the proposed technique can be well compared to the results given in [11] wherein [11], they have used an array of around 39 capacitors to suppress the displacement current where we proposed to use a single selected inductor in parallel to the DC- terminal and the intrinsic parasitic capacitance of the base-plate module.

(a)

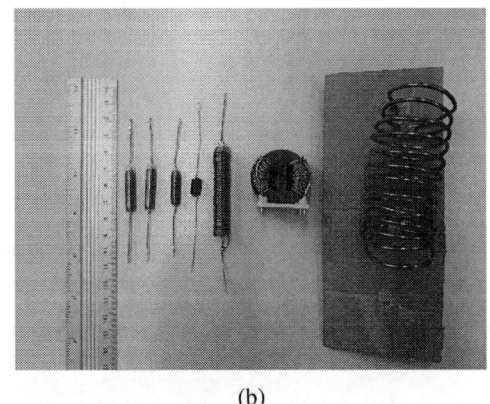
(b)

Fig. 4: Experimental setup; (a) implementation of different inductors under test, (b) selected inductors

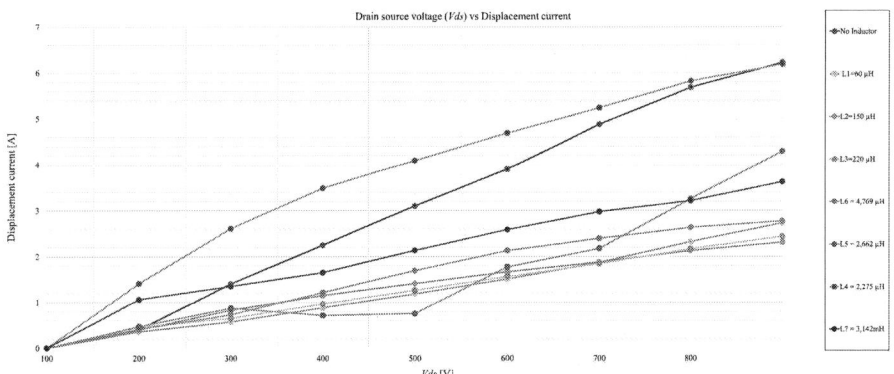

Fig. 5: Drain-source voltage versus displacement current without CMC with the selection of different inductors

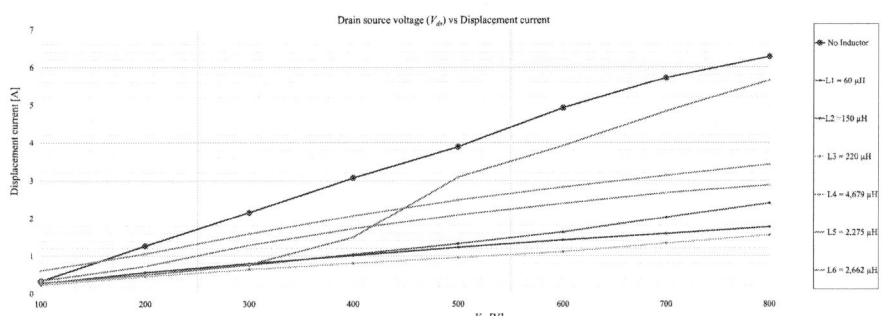

Fig. 6: Drain-source voltage versus displacement current with CMC with the selection of different inductors

Conclusion

The WBG semiconductors such as SiC and GaN are fast switching devices with lower switching losses. The faster switching of the WBG makes higher dv/dt. On the other hand, we have a larger parasitic capacitance base-plate due to the larger area of the power module chip. This results in a higher displacement current, which produces higher EMI/EMC emissions. The suppression of the displacement current needs a bigger filtering system. The proposed technique of connecting a single selected inductor in parallel to the DC-terminal of the module and the base-plate. The experimental results show that the displacement current is well attenuated up to 62.5% without CMC and 75% with CMC. The implementation of the proposed methodology is simple and helps in reducing the overall size, volume, and cost of the system.

References

[1] A. B. Jørgensen et al., "Reduction of parasitic capacitance in 10 kV SiC MOSFET power modules using 3D FEM," 19th European Conference on Power Electronics and Applications (EPE'17 ECCE Europe), 2017, pp. P.1-P.8, 2017.

[2] D. Boroyevich et al., "High-density system integration for medium power applications," 6th International Conference on Integrated Power Electronics Systems, 2010, pp. 1-10, 2010.

[3] J. -J. Yun, H. -J. Choe, Y. -H. Hwang, Y. -K. Park and B. Kang, "Improvement of Power-Conversion Efficiency of a DC–DC Boost Converter Using a Passive Snubber Circuit," in IEEE Transactions on Industrial Electronics, vol. 59, no. 4, pp. 1808-1814, 2012.

[4] X. Gong and J. A. Ferreira, "Investigation of Conducted EMI in SiC JFET Inverters Using Separated Heat Sinks," in IEEE Transactions on Industrial Electronics, vol. 61, no. 1, pp. 115-125, 2014.

[5] P. Godignon et al., "SiC Schottky Diodes for Harsh Environment Space Applications," in IEEE Transactions on Industrial Electronics, vol. 58, no. 7, pp. 2582-2590, 2011.

[6] J. Biela, M. Schweizer, S. Waffler and J. W. Kolar, "SiC versus Si—Evaluation of Potentials for Performance Improvement of Inverter and DC–DC Converter Systems by SiC Power Semiconductors," in IEEE Transactions on Industrial Electronics, vol. 58, no. 7, pp. 2872-2882, 2011.

[7] R. Robutel et al., "Design and Implementation of Integrated Common Mode Capacitors for SiC-JFET Inverters," in IEEE Transactions on Power Electronics, vol. 29, no. 7, pp. 3625-3636, 2014.

[8] X. Gong and J. A. Ferreira, "Investigation of Conducted EMI in SiC JFET Inverters Using Separated Heat Sinks," in IEEE Transactions on Industrial Electronics, vol. 61, no. 1, pp. 115-125, 2014.

[9] A. D. Brovont and A. N. Lemmon, "Utilization of Power Module Baseplate Capacitance for Common-Mode EMI Filter Reduction," IEEE Electric Ship Technologies Symposium (ESTS), pp. 403-408, 2019.

[10] G. Feix, E. Hoene, O. Zeiter and K. Pedersen, "Embedded Very Fast Switching Module for SiC Power MOSFETs," Proceedings of PCIM Europe 2015; International Exhibition and Conference for Power Electronics, Intelligent Motion, Renewable Energy and Energy Management, pp. 1-7, 2015.

[11] A. D. Brovont, A. N. Lemmon, C. New, B. W. Nelson and B. T. DeBoi, "Analysis and Cancellation of Leakage Current Through Power Module Baseplate Capacitance," in IEEE Transactions on Power Electronics, vol. 35, no. 5, pp. 4678-4688, 2020.

[12] Q. Liu, S. Wang, A. C. Baisden, F. Wang and D. Boroyevich, "EMI Suppression in Voltage Source Converters by Utilizing dc-link Decoupling Capacitors," in IEEE Transactions on Power Electronics, vol. 22, no. 4, pp. 1417-1428, 2007.

[13] Y. Liu, S. Jiang, H. Wang, G. Wang, J. Yin and J. Peng, "EMI filter design of single-phase SiC MOSFET inverter with extracted noise source impedance," in IEEE Electromagnetic Compatibility Magazine, vol. 8, no. 1, pp. 45-53, 2019.

[14] A. N. Lemmon, A. D. Brovont, C. D. New, B. W. Nelson and B. T. DeBoi, "Modeling and Validation of Common-Mode Emissions in Wide Bandgap-Based Converter Structures," in IEEE Transactions on Power Electronics, vol. 35, no. 8, pp. 8034-8049, 2020.

[15] S. Wang and F. C. Lee, "Analysis and Applications of Parasitic Capacitance Cancellation Techniques for EMI Suppression," in IEEE Transactions on Industrial Electronics, vol. 57, no. 9, pp. 3109-3117, 2010.

Analysis and Implementation of Effective Placement of EMC Capacitors for WBG Modules

Mahmoud Saeidi, Ahmad Ali Nazeri, Firas Jenhani, and Peter Zacharias
Centre of Competence for Distributed Electric Power Technology
Faculty of Electrical Engineering / Computer Science
University of Kassel, Kassel, Germany
Email: {mahmoud.saeidi, peter.zacharias}@uni-kassel.de
ahmad.nazeri@student.uni-kassel.de, Jenhani.firas1@gmail.com

July 11, 2022

Keywords

≪Common mode choke (CMC)≫, ≪Displacement current≫, ≪Electromagnetic capability (EMC)≫, ≪Wide band-gap (WBG)≫

Abstract

This paper proposes an optimum placement of the electromagnetic capability (EMC) capacitors for the wide band-gap (WBG) devices (SiC). To minimize electromagnetic interference (EMI), the solution of using EMC capacitors near the power module is proposed. Two different EMC capacitor placement scenarios are examined to show how the displacement current is affected. The suggested method reduces the capacitive displacement current during the high slew rate of the drain-source voltage (dv/dt), which then helps to minimize the size and overall cost of the common mode choke (CMC). The proposed methods are simple to construct and easy to implement with a faster switching response. The proposed method is experimentally implemented for a half-bridge SiC power module with the heat sink.

Introduction

The common-mode (CM) EMI noise is introduced due to the parasitic capacitance between high dv/dt nodes and the ground in the power electronic system. The CM current is generated because of the charging and discharging of the parasitic capacitance. The heat generated from the conduction power loss is dissipated to the air via heat sinks. The power module terminals and the heat sinks have introduced the large parasitic capacitance, which plays a key role in EMI filter design. The larger parasitic capacitance leads to the higher displacement current between the high dv/dt nodes and the heat sink. For safety reasons, the heat sinks are connected directly to the ground and cannot be reached from the outside and most of the noise current can flow directly back to the circuits. The goal of minimizing the capacitive displacement current is to reduce the overall size, volume, and cost of the EMI filters and increase the power density according to the standards and keep a higher switching speed. In [1], the authors propose two methods of generating the negative capacitance to cancel the parasitic capacitance, which is based on the mutual capacitance theory and the second one is the mutual inductance. The separated heat sinks combined with the damping snubbers are used in [2] to improve both CM and differential mode (DM) conducted for the EMI/EMC performance for a SiC JFET-based inverter. A CM quantitative model is proposed in [3], to identify the simpler and alternative method to attenuate the baseplate current in the system, which involves placing a single capacitor between the neutral point and the system ground near the module baseplate. This method significantly reduces the CM current through the module baseplate as expected by the module [3].

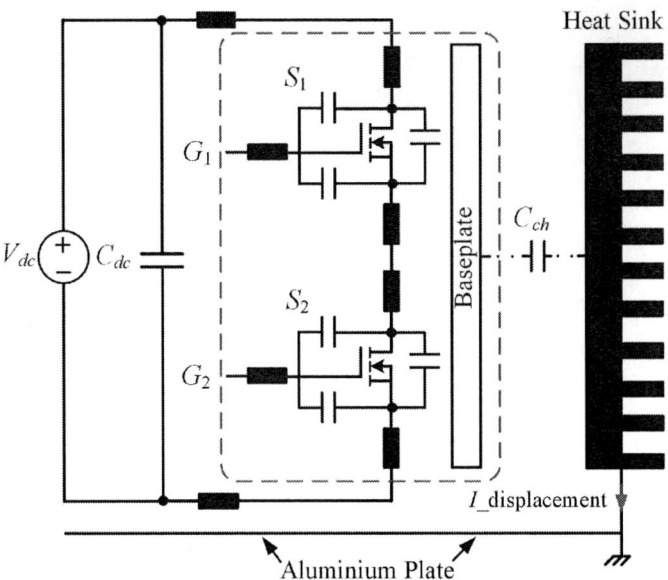

Fig. 1: The equivalent configuration of the H-bridge SiC Module with the parasitic components and the heat sink

A model-based approach is proposed in [4] to identify the effect of the anti-resonance formed by the AC-side of the baseplate capacitance of the power modules and the inductive load. This technique results in low inductance and low capacitance filters where the low capacitance can be beneficial for the ungrounded systems to minimize the line-to-ground capacitance [4]. An extra output filter stage is implemented to reduce the leakage current proposed in [5, 6]. Two complex modulations are proposed to mitigate the common-mode leakage current for a three-phase inverter as reported by [7, 8]. The conducted CM EMI emission of a pulse width modulation inverter-based motor drive investigates and quantifies the increase in [9]. The numerical predictions are used in [10] to generate a set of design guidelines for heat sinks of various sizes and emphasized that changing heat-sink gasket materials as an EMI mitigation strategy is limited to cases in which the heat-sink patch resonance constitutes a significant part of the overall coupling mechanism. The impacts of the printed circuit board (PCB) [11] and system layouts, such as shielded cables and unshielded cables on CM EMI noise of SiC electric vehicle powertrains are introduced in [12, 13].

This paper proposes the effective placement, implementation, and analysis of the EMC capacitors inside the power module. Moreover, a pair of two capacitors in parallel are implemented outside the power module between the DC+/DC- power module terminals with the heat sink. A comparative analysis is presented to reduce the displacement current between the parasitic capacitance of the high dv/dt nodes and the heat sink for the above-mentioned approach.

Minimization of the capacitive displacement current

Fig. 1 illustrates the equivalent circuit configuration of the parasitic capacitance and stray inductance of the half-bridge power module with the heat sink. Fig. 1 shows the typical configuration of the H-bridge where parasitic capacitance C_{dh} is between the heat sink and the power module baseplate and C_{gh} is between the heat sink and the ground. Moreover, the parasitic capacitance C1 is between DC+ terminal and ground, C2 is AC terminal, and ground where C3 is between DC- and ground. The combination of these parasitic capacitance results in a path for the flow of the displacement current. This coupling path has a strong influence on generating the EMI emission during the switching of the power electronic devices. The displacement current needs to be minimized, keeping the high dv/dt, which then helps in smaller switching loss and results in smaller filter sizes with higher power density. Two methods have been proposed to reduce the displacement current. The first method is the placement of the two

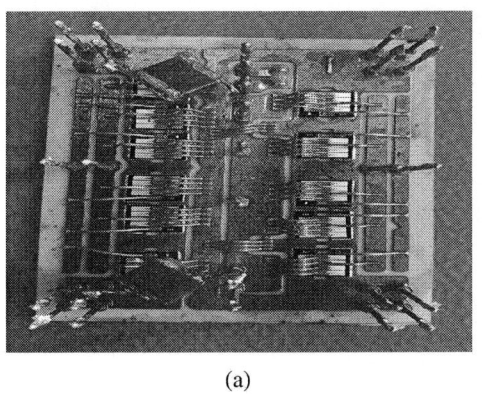
(a)

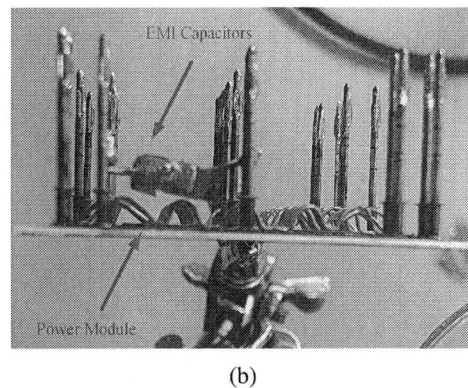
(b)

Fig. 2: EMI capacitors; (a) inside the power module, (b) inside the power module with the smallest space to the chips

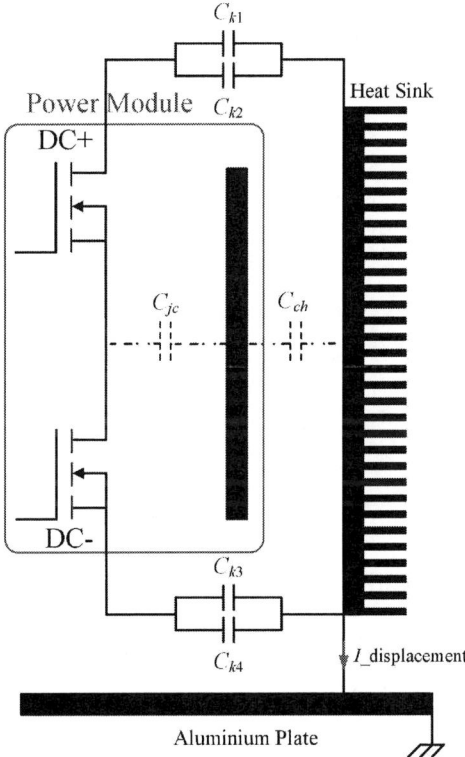

Fig. 3: Placement of two capacitors of 100nF in parallel between the DC+/DC- and heat sink

capacitors of 100nF between DC+ and DC-.

Fig. 2 depicts the experimental implementation of the placement of the capacitors inside the power module. It can be seen that the capacitors are in full contact with the power module with negligible space between them. The goal is to reduce the stray inductance as small as possible and the effect is analyzed on the displacement current. The second method depicted in Fig. 3 is the placement of two capacitors of 100nF in parallel between the DC+/DC- and heat sink. Fig. 4 shows the experimental implementation of the proposed approach. The objective is to create two loops for the displacement current between the DC+/DC- and the heat sink. This approach results in the reduced flow of the displacement current through the heat sink and the ground and forces the displacement current to flow in two loops. This method helps in attenuating the displacement current.

Fig. 4: The placement of the EMI capacitors outside the power module

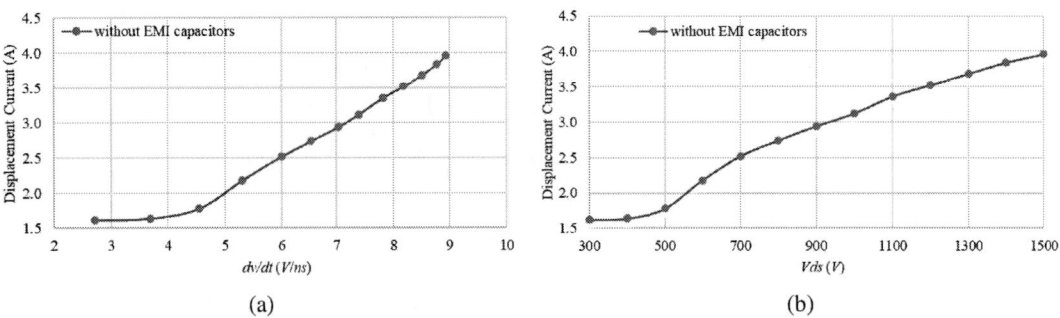

Fig. 5: Displacement current; (a) by voltage gradient, (b) by drain-source voltage

Experimental implementation

Without any EMI capacitors

Fig. 5 shows the experimental results of the voltage gradient and the displacement current. In Fig. 5a, it can be seen that with the higher dv/dt (V/ns), the displacement current also increases. In Fig. 5b, the displacement current increases with the higher drain-source voltage (V_{ds}). The operating V_{ds} voltage is set from 300 V to 1500 V. In this V_{ds} operating range, the measured voltage gradient dv/dt is between 2.71 and 8.49 V/ns at turn-off and turn-on respectively. In Fig. 5b, at the 1500 V_{ds}, the displacement current is 3.96 A (peak to peak). The displacement current flows from the heat sink to the system ground (aluminum plate). It can be seen that the higher displacement current at higher V_{ds} flows, which then results in higher EMI emissions with a larger EMI filter, higher cost, and size of the power electronic devices. This paper proposes two methods to mitigate the displacement current by keeping higher dv/dt.

Ceramic capacitors inside power module

Fig. 6 depicts the experimental results of the voltage gradient and the displacement current, which has been implemented as in Fig. 2. Two ceramic capacitors of 0.1 μF are placed inside the power module between the DC+ and DC- terminals. The ceramic capacitors are selected because of their smaller size, which results in faster energy transfer, and are available at the higher voltage of 1.5 kV. As already explained, this approach is executed to reduce the stray inductance and the distance between the

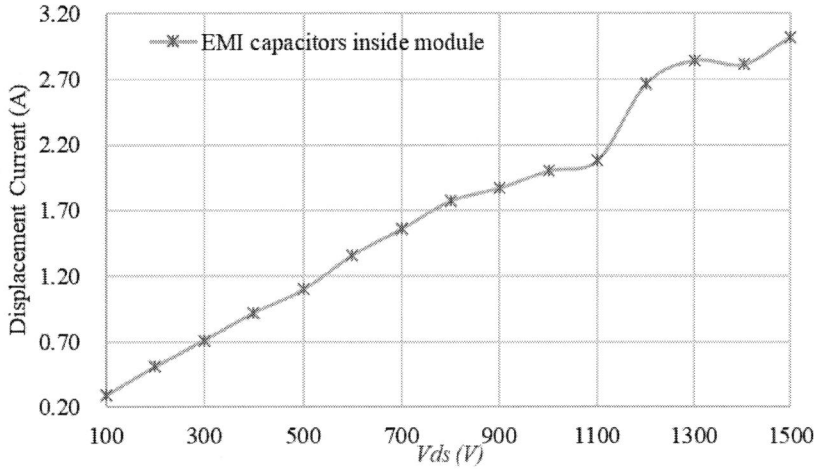

Fig. 6: Displacement current with EMI capacitors inside the power module

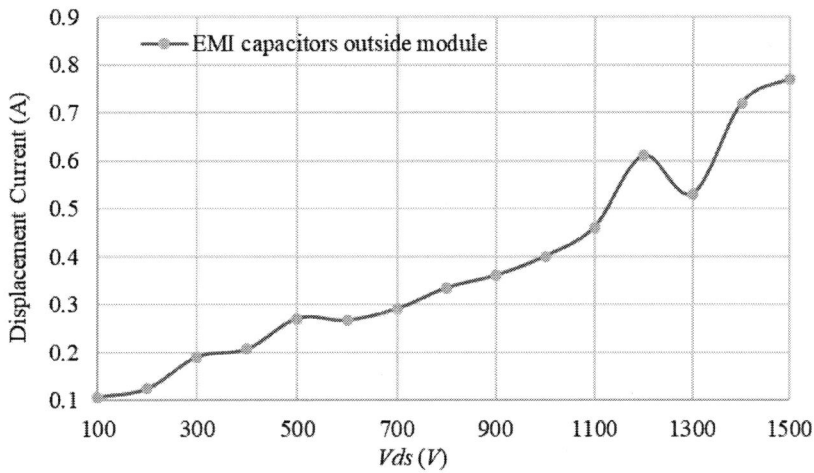

Fig. 7: Displacement current with EMI capacitors outside the power module

connected capacitors and the SiC chips module is less than 1.5mm. The displacement current is reduced to 3.1 A (peak to peak) at 1.5 kV in comparison to Fig. 5a. The displacement current is attenuated up to 22% in comparison to the typical experimental results without any EMI capacitors. The drawback of this approach is that the SiC module has a higher operating temperature ($150\,°C$-$175\,°C$), where the operating temperature of the ceramic capacitors is in the range of ($55\,°C$ to $+125\,°C$). The ceramic capacitors can not be used for a longer time due to their higher difference in operating temperature ranges.

Film capacitors near power module

Fig. 7 presents the experimental results of the voltage gradient and the displacement current, which has been implemented as in Fig. 4. In this configuration, two Polypropylene (PP) film capacitors of $100nF$ are connected in parallel to the DC+/DC- and the heat sink. The most important advantage of the PP film capacitors is the reduced resistive losses (ESR) and the parasitic inductance of the capacitors (ESL) due to their internal structure. The reason for connecting two capacitors in parallel is to create a path of low impedance for the flow of the displacement current. The displacement current is reduced to 0.72 A at 1.5 kV at the proposed method as shown in Fig. 7. This approach is more effective than the other two methods mentioned in sections 1 and 2. This proposed approach attenuates the displacement current up to 80% compared to the no connection of the EMI capacitors.

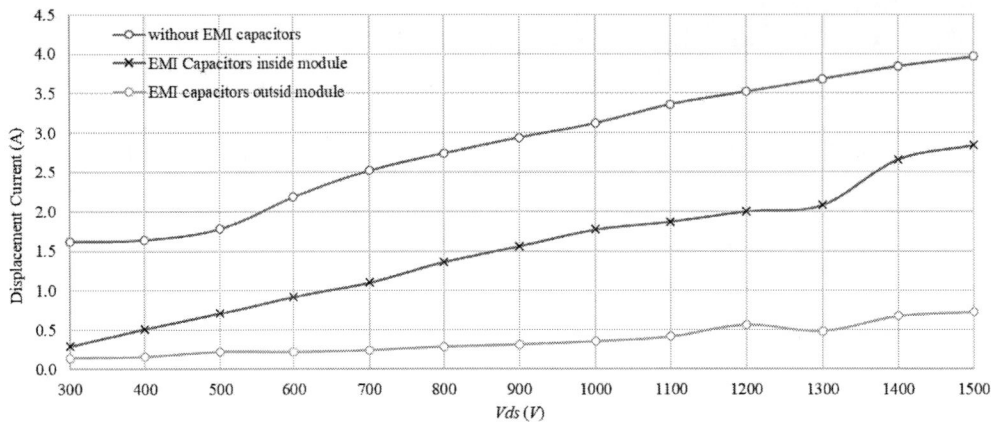

Fig. 8: Comparison of the displacement current for the proposed approach

This method is easier to implement, cost-effective, and results in a smaller EMI filter, which then helps in lower switching losses at higher dv/dt and higher power density. Fig. 8 compares the displacement current to the V_{ds}, where the proposed method of implementing the EMI capacitors in section 3 (green) has higher damping power than the capacitors connected inside the power module (blue) and no EMI capacitors (red) respectively.

Conclusion

This paper proposed a novel approach for the optimum placement of the EMI capacitors. The EMI capacitors were placed inside the power module and are compared with the no placement of the EMI capacitors, which shows the reduction of the displacement current. The novel approach is to place two EMI capacitors in parallel between the DC+/- and the heat sink. The experimental results of the suggested methodology are compared with the results of placing no capacitors and the placement of the EMI capacitors inside the power module. The indicated results of the parallel configuration of the EMI capacitors show sufficient mitigation of the displacement current by 80% up to 0.72 A at 1.5 kV compared to the capacitors placed inside the power module, which is 3.1 A at 1.5 kV respectively. The attenuation of the displacement current helps in the overall reduction of the EMI filters keeping higher dv/dt with lower switching losses and realizing higher power density for the wide-bandgap modules.

References

[1] S. Wang and F. C. Lee, "Analysis and Applications of Parasitic Capacitance Cancellation Techniques for EMI Suppression," in IEEE Transactions on Industrial Electronics, vol. 57, no. 9, pp. 3109-3117, 2010.

[2] X. Gong and J. A. Ferreira, "Investigation of Conducted EMI in SiC JFET Inverters Using Separated Heat Sinks," in IEEE Transactions on Industrial Electronics, vol. 61, no. 1, pp. 115-125, 2014.

[3] A. D. Brovont, A. N. Lemmon, C. New, B. W. Nelson and B. T. DeBoi, "Analysis and Cancellation of Leakage Current Through Power Module Baseplate Capacitance," in IEEE Transactions on Power Electronics, vol. 35, no. 5, pp. 4678-4688, 2020.

[4] A. D. Brovont and A. N. Lemmon, "Utilization of Power Module Baseplate Capacitance for Common-Mode EMI Filter Reduction," 2019 IEEE Electric Ship Technologies Symposium (ESTS), pp. 403-408, 2019.

[5] H. Akagi and T. Doumoto, "A passive EMI filter for preventing high-frequency leakage current from flowing through the grounded inverter heat sink of an adjustable-speed motor drive system," in IEEE Transactions on Industry Applications, vol. 41, no. 5, pp. 1215-1223, 2005.

[6] Y. Zhou and H. Li, "Analysis and Suppression of Leakage Current in Cascaded-Multilevel-Inverter-Based PV Systems," in IEEE Transactions on Power Electronics, vol. 29, no. 10, pp. 5265-5277, 2014.

[7] C. Bharatiraja, J. L. Munda, R. Bayindir and M. Tariq, "A common-mode leakage current mitigation for PV-grid connected three-phase three-level transformerless T-type-NPC-MLI," 2016 IEEE International Conference on Renewable Energy Research and Applications (ICRERA), 2016.

[8] T. K. S. Freddy, N. A. Rahim, W. Hew and H. S. Che, "Modulation Techniques to Reduce Leakage Current in Three-Phase Transformerless H7 Photovoltaic Inverter," in IEEE Transactions on Industrial Electronics, vol. 62, no. 1, pp. 322-331, 2015.

[9] D. Han, S. Li, Y. Wu, W. Choi and B. Sarlioglu, "Comparative Analysis on Conducted CM EMI Emission of Motor Drives: WBG Versus Si Devices," in IEEE Transactions on Industrial Electronics, vol. 64, no. 10, pp. 8353-8363, 2017.

[10] Yu Huang et al., "EMI considerations in selecting heat-sink-thermal-gasket materials," in IEEE Transactions on Electromagnetic Compatibility, vol. 43, no. 3, pp. 254-260, 2001.

[11] B. Archambeault, C. Brench and S. Connor, "Review of Printed-Circuit-Board Level EMI/EMC Issues and Tools," in IEEE Transactions on Electromagnetic Compatibility, vol. 52, no. 2, pp. 455-461, 2010.

[12] R. Zhu, T. Liang, V. Dinavahi and G. Liang, "Wideband Modeling of Power SiC mosfet Module and Conducted EMI Prediction of MVDC Railway Electrification System," in IEEE Transactions on Electromagnetic Compatibility, vol. 62, no. 6, pp. 2621-2633, 2020.

[13] X. Jia, C. Hu, B. Dong, F. He, H. Wang and D. Xu, "Influence of system layout on CM EMI noise of SiC electric vehicle powertrains," in CPSS Transactions on Power Electronics and Applications, vol. 6, no. 4, pp. 298-309, 2021.

AUTHOR INDEX

Abdalrahman, Adil 2241, 3282, 3757
Abdullah, Ahmed.. 554
Abedini, Hossein... 865
Aceña, Javier Cañas.. 484
Adabi, Jafar.. 2537
Addin, Ali Sharaf.. 1824
Afonso, Luciana C. 4018
Aganza-Torres, Alejandro.................................... 1328
Agarwal, Ritika... 3615
Agirrezabala, Eneko 3327
Aguglia, D. 1955
Ahmed, Emad M... 1015
Aiello, Giuseppe 2628
Aillerie, Michel.. 315
Aizpuru, I... 2903
Aizpuru, Iosu 325, 3327, 3574, 3750
Akuru, Udochukwu B.. 2958
Al-Haddad, Kamal 1025
Alaluss, Mohamed 1424
Alatise, Olayiwola 1497, 2477
Albrecht, Fabian 2726
Aldarmon, Mohamed... 2574
Ali, Mohammad 2392, 3022
Ali, Ramy 390
Ali, Rana Asad... 698
Allard, Bruno... 169, 3862
Allioua, Abdelmoumin 2835
Alvarez, Asier... 279
Alvarez-Herault, Marie-Cecile 1147
Alves, Wendell Da Cunha...................................... 1046
Alvi, Muhammad H 1692
Aly, Mokhtar 1015
Andersen, Michael A. E... 1561
Ando, Y. 1785
Andresen, Jan... 1684
Ansari, Sajad A. 3440
Antonopoulos, Antonios 297, 432
Anzola, J. 2903, 2967
Anzola, Jon 3574
Apostolidou, Nena 1796
Appel, Tobias.. 1121
Apte, Pramod 2773
Arabsalmanabadi, Bita... 1025
Arias, Manuel 152
Arrizabalaga, Antxon.. 325
Arrozy, Juris 681
Arruti, Asier 3574, 3750
Artal-Sevil, J. S..................................... 2903, 2967

Arza, Joseba.. 484, 2011
Asllani, Besar... 2515
Asoodar, Mohsen 2843
Atzler, Frank 3391
Aunsborg, Thore Stig 825
Aviñó, Oriol 2715
Ayarzaguena, Ibán.................................... 1765, 3336
Aztiria, Jon 325
Baars, Nico 2788
Babin, Anthony 3696
Baburske, Roman 1424
Bacha, S. 2422, 3179
Bacha, Seddik.. 3140, 3928
Bacheti, Gabriel Gaburro 421
Bachmann, Matthias... 3501
Badenhop, Niklas....................................... 145, 1939
Baek, Seung-Hyuk 2877
Bagaber, Bakr... 3037, 3711
Baimel, D. 3254
Baimel, N. 3254
Bak, Claus Leth 2504
Bakhos, Gianni 3928
Bakran, Mark-M.......................... 805, 1036, 2744
Bakri, Reda 1046
Balachandran, Arvind 1456
Balasubramanian, Sridhar 2030
Ballestín-Bernad, V.................................. 2903, 2967
Banana, Shady... 1064
Banavath, Satish Naik 730
Banda, Joseph... 187, 289
Barba, V. 1975
Barbi, Eli 3254
Barg, Sobhi 361
Barman, Subhranil... 2462
Barón, Kevin Muñoz... 2698
Bashar, Erfan... 2477
Basic, Duro 125
Basler, Michael 242
Basler, Thomas................... 1424, 1713, 1733, 3373
Bauer, Luca 971
Bauer, Pavol.............................. 1319, 3607, 3729
Baumann, Michael 1167
Baumann, Timm Felix 2355
Bäumler, Christian 1733
Bayer, Markus 115
Bayhan, Sertac 3518
Bayram, Islam Safak 3518
Beck, Simon.. 1434, 2038

Beckemeier, Christian..2327
Beczkowski, Szymon Michal..............................2661
Beineke, Stephan 3501
Beiranvand, Hamzeh...................833, 3092, 3846, 3966
Belhaouane, Mohamed Moez 582
Benchaib, Abdelkrim..3928
Bendfeld, Christian 1620
Benech, Philippe... 169
Bensetti, Mohamed 3883
Bergmann, Lukas..1036
Bergveld, Henk Jan...3796
Bermejo, Jose Manuel..............................1765, 3336
Bernal, Carlos..3327
Bernal-Agustín, J. L...2967
Bernal-Ruiz, Carlos 3750
Bernichon, Thomas...3920
Bertilsson, Kent 361
Bertin, Matthieu 534
Beukes, Johan..3112
Beye, Mamadou Lamine.......................................2736
Beza, Mebtu 1187
Bezerra, Vinicius Freire.......................................2689
Bhatnagar, Pallavee 3804
Bhattacharya, Arghyadip 178
Bhoi, Sachin Kumar..3031
Biadene, Davide... 865
Biela, Juergen 1402
Biela, Jürgen 651, 662, 933, 1391, 1434, 2038, 2544
Bieler, Arne 1121
Bier, Anthony 922, 2736
Billa, Laxma R..2301
Bimmel, Luc 2736
Binder, Andreas..2316
Bitsi, Konstantina 3246
Blaabjerg, Frede.................2110, 2182, 2496, 2504, 2939
Blanes, J. M. 3382, 3401
Blank, Thomas.. 232
Blanquez, Francisco R. 2189, 2451
Blasco-Gimenez, Ramon 2189, 2451
Blasuttigh, Nicola 3846
Blatsi, Zoe...2824, 3813
Blömeke, Alexander 4025
Böcker, Joachim 2276, 2432, 2754, 3625, 3686
Bockholt, Jan 1286
Boettcher, Norman 1128
Bohllaender, Marco 4016
Bohne, David... 514
Boige, Francois.. 944
Boisson, Guillaume Piquet.................................... 960
Bolzoni, A. 1371
Bongiorno, Massimo..1187
Bonten, Remco 634

Böorngen, Hannes... 1754
Borcherding, Holger.. 2852
Börngen, Hannes.. 3362
Boroyevich, Dushan 2806
Bosch, Swen.. 2219
Bosga, Sjoerd G. 3246
Bouscayrol, Alain.. 2175
Boutleux, Emmanuel 251
Boutry, Arthur.. 2515
Brabetz, Ludwig.. 2383
Branco, Cesar Augusto Santana Castelo 2948
Braun, Gerrit 2205
Braz, Cesar 1445
Briff, Pablo.. 451
Bringezu, Thilo 662
Brinker, Tobias... 2977
Brogioli, Doriano Constantino 833
Brommer, Volker 2726
Bronstein, S. 3254
Brooks, Michael 279
Brückner, Thomas... 1824
Brulin, Pierre-Yves 3831
Brunner, Andreas 593
Brunner, Frank 3775
Brüns, Michael 474
Bruyere, Antoine 1046
Bruyere, Paul... 960
Bucarey, Victor 1074
Budo, Kohei...213, 351
Bueno, Emilio José.. 421
Bueno-Mariani, Guilherme 3272
Bugarski, Stevan 2334
Bünte, Andreas... 380
Burgos, Rolando..1692, 2806
Burgos-Mellado, Claudio..................................1074, 3429
Burkart, Ralph M. 203
Burke, Richard 3696
Bushra, Rehnuma...2392, 3022
Busquets-Monge, Sergio 2715
Buticchi, Giampaolo 3014
Buttay, Cyril...2049, 2515
Byen, Byengjoo.. 1207
Caarls, Esin Ilhan 681
Cabrera, Michel.. 169
Cacciato, Mario.. 2628
Caillierez, Antoine 3883
Cajander, D. 1955
Cakal, Gokhan.. 3947
Caldognetto, Tommaso 865
Camargo, Renner Sartório 421
Camurca, Luis 3101
Can, Görkem 3092

Cano, Tania C. 335
Cao, Jingming 3215, 3225
Cao, Yongtao 2003
Cappelle, Jan 1300
Cárcamo, Alberto 1083
Carcouet, S. 843
Carpita, Mauro 1543
Carrasco, Miguel 370
Casado, P. 3382, 3401
Castellazzi, Alberto 689, 2156, 2285, 2402, 2893, 3084
Castelli-Dezza, Francesco 1476
Castro, Ignacio 335
Catalán, Pedro 2011
Catellani, Stéphane 922, 990
Ceccarelli, Lorenzo 681
Chakraborty, Sajib 2101, 3031
Chang, Che-Wei 1692
Charkaoui, Abdelmouneim 442
Chatterjee, Kishore 178, 2462
Chen, Zhe 2011
Chen, Zhu 3235
Chevalier, Florian 3582
Chida, Makoto 1580
Chinthavali, Madhu Sudhan 344
Chiumeo, Riccardo 3206
Choksi, Kushan 344
Choudhury, Soham 1966
Chub, Andrii 730
Cimetiere, Xavier 1046
Clerc, Guy 251
Clerici, Alessio 3206
Cobaleda, Diego Bernal 2581
Cogitore, Bruno 1216
Colmenero, Manuel 2189, 2451
Cosso, Simone 2919
Coumont, Martin 1966
Crovetti, Paolo 554
Cui, Yi 3986
Czerwenka, Philipp 593
Dahmen, Christopher 1824, 1855
Damian, Ioan Catalin 2266
Damm, Gilney 3590
Danielsson, Christer 2843
Dargahi, Vahid 2073
Davari, Pooya 2496
Davidson, Jonathan N. 3440
De Bernardinis, Alexandre 315
De Carne, Giovanni 3014
De Cesaris, Ivan 223
De Donato, Giulio 1569
De Doncker, Rik W. 709, 1266, 2119, 3599, 3676,
.... 3740, 3766, 3893

De Lillo, Liliana 3450
De Matos, Jose Gomes 2948
De Oliveira, Eduardo Facanha 2441
De, Dipankar 689
Deb, Arkadeep 1497
Deblecker, Olivier 504
Deboy, Gerald 3984
Deck, Patrick 514
Deckers, Martijn 2795
Degaa, Laid 3696
Delette, Gérard 922
Deng, Kai 3235
Dennetiere, Sébastien 582
Derammelaere, Stijn 3344
Despouys, Olivier 2486
Dick, Christian P. 514
Dickmann, Stefan 758
Dieckerhoff, Sibylle 1466, 2596, 2607, 2644, 3775
Dieng, A. 2092, 2930
Dierks, Rebecca 1533
Dietrich, Tim-Hendrik 1094
Disselkamp, Simon 2912
Domae, Shinichi 3084
Domes, Daniel 2744
Domes, Konrad 1137
Dong, Chaoyu 3215, 3225
Dong, Dong 1692, 2515
Dong, Jianning 1319
Dong, Tenghui 3084
Dorner, Oscar 1177
Dos Santos, Pedro Leal 604
Dragicevic, Tomislav 2496, 2939, 3429
Drexler, Christoph 411, 1167
Driesen, J. 3655
Driesen, Johan 2795
Drimizi, Youssef 2869
Drissi, Khalil El Khamlichi 3786
Duarte, Jorge L. 681, 798
Duarte, Jorge 2788
Duchamp, Jean-Marc 169
Dujic, Drazen 2049
Dumtzlaff, Jacob 1865
Duquesne, Thierry 3582
Dürbaum, Thomas 88, 307
Duun, Sune Bro 825
Dworakowski, P. 2422
Dworakowski, Piotr 2049
Ebel, Thomas 3130
Ebner, Kathrin 4015
Eckart, Martin 3646
Eckel, Hans-Guenter 3460

Eckel, Hans-Günter 11, 59, 70, 980, 1294, 1703,
.... 1744, 1885, 1895, 2308, 4003
Eckstein, Mattea 1277
Effenberger, Thomas 1754
Eggers, Malte 1466
Ehlich, Martin 2852
El Baghdadi, Mohamed 2101, 2293, 3031
El Sherif, Alaa 3796
El-Refaie, Ayman 719, 1692
Ellinger, Thomas 2885
Emmers, G. 3655
Emmers, Glenn 2795
Empringham, Lee 3450
Encarnação, Lucas Frizera 421
Endo, Yusuke 2285
Epping, Daniel 749
Erckrath, Tobias 1350, 1620
Eremia, Mircea 2266
Eriksson, Lars 1456
Erlbacher, Tobias 1128
Ernst, Alexander 3149, 3159
Es-Seghier, Hajar 922
Escoffier, René 990
Etoz, Burhan 1497
Faber, Samuel 307
Falchi, Daniele 2486
Faramehr, Soroush 3822
Farhangi, Shahrokh 787
Fauth, Leon 2003, 2638, 3838
Fayolle-Lecocq, Murielle 990
Fazli, Nastaran 11
Fehr, Hendrik 49, 3391
Felgemacher, Christian 442, 4004
Fernández, Arturo 152
Ferreyra, Fabio 554
Festerling, Tobias 1237
Finney, Stephen 80, 3470, 3813
Fischer, Katharina 1674, 1804
Fischer, Manuel 749
Fischer-Baeumer, Rico 1137
Fölkel, Lorandt 279
Formentini, Andrea 2919, 3975
Forouzesh, Mojtaba 1590, 1601
Forsstrom, Ville 3301
Förster, Nikolas 2432
Foster, Martin P. 3353, 3440
Foteinopoulos, Georgios 1985
Fräger, Lukas 145, 641, 1939, 2588, 2773
Frank, S. R. 3411
Franzki, Jonas 261
Frey, David 1147
Fricke, Tobias 1247

Fricke, Torben 1381
Friebe, Jens 1914, 2003, 2327, 2392, 2588, 2638,
.... 2655, 2689, 2773, 2977, 3022, 3059, 3545, 3838
Fritze, Eric 758
Fröhling, Sören 1674
Fuchs, Simon 1434, 2038
Fuhrmann, Jan 980
Fukunaga, Shuhei 108
Ganeshpure, Dhanashree Ashok 3729
Gao, Xiang 3014
Gaona, Daniel 2441
Garces, Santiago Ramos 3344
Garcia, Raul Murillo 2355
Garrigós, A. 3382, 3401
Gaubert, Jean-Paul 1525
Gauthier, Jean-Yves 3862
Gavelle, Mathieu 2618
Gehl, Adrian 2912
Geiss, Michael 2554
Gemma, Filippo 3975
Geng, Weiwei 3722
Geng, Xiaomeng 2596, 2644, 3775
Gennaro, Francesco 2628
Gensior, Albrecht 49, 370, 3391
Gerges, Tony 169
German, Ronan 2175
Germishuizen, J. J. 3318
Geury, Thomas 2101
Gholami, M. 3179
Gholami, Mehrdad 3140
Ghumman, Sukhjit S 2763
Gieraths, Antje 767
Gierschner, Magdalena 1294
Gierschner, Sidney 11
Gillon, Frédéric 1046
Girona-Badia, Jaume 3704
Glaser, Martin 4020
Gleissner, Michael 805, 1036
Gnärig, Lasse 370
Goetz, Stefan 1025, 1064, 1197, 3636, 3665
Gohler, Katherina 1804
Gohrmann, Kai 1137
Golev, Victor 1286
Goller, Maximilian 1733
Gomes, Lucas Vinícius De Araújo 3059
Gomes, Zariff Meira 3590
Gómez, Alexis A. 1765, 3336
Gomis-Bellmunt, Oriol 2486, 3704
Gonzalez, Jose Ortiz 1497
Gonzalez-Hernando, Fernando 3938
Gonzalez-Torres, Juan-Carlos 3928
Götz, Georg Tobias 709

Gräber, Hendrik 2977
Grabs, Volker 97
Gradinger, Thomas B. 203
Grant, Thomas 2301
Grass, Norbert 2366
Grau, Vivien 854
Gremme, Florian 4021
Griepentrog, Gerd 160, 2780, 2835
Grodnichev, Anton 624
Groke, Holger 3169
Groon, Fabian 3092
Groten, Jonas 279
Gruson, François 582
Guerrero, Bruno 944
Gui, Qiuye 49
Guillaud, Xavier 582
Günes, Ece Olcay 1361
Gupta, Kirti 2110
Gupta, Krishna Kumar 3615, 3804
Gutierrez, Alonso 2618
Haag, Felix 2726
Haake, Daniel 624
Haarer, Jörg 971, 1237, 1277
Habersetzer, Antoine 4015
Hably, A. 3179
Hably, Ahmad 3140
Hackl, Philipp 39
Haederli, Christoph 3282
Häfner, Ying-Jiang 2241, 3282, 3757
Hagedorn, Maximilian 1875
Hajar, K. 3179
Hajar, Khaled 3140
Hajian, Masood 468
Hakkila, Akseli 297
Hald, Alex 380
Hameyer, Kay 3005, 3235
Hammes, David 11
Handt, Karsten 2607
Hanf, Michael 3169
Hanisch, Lucas Vincent 261
Hanisch, Lucas 1094
Hänsel, Stefan 572
Hansen, Sandra 3966
Hanson, Alex J. 1722
Hanson, Jutta 1966
Hao, Chuantong 80, 3470
Hardan, Faysal 468
Harmand, Souad 2996
Hasan, Md. Mahamudul 3031
Hasler, J. P. 1371
Hassan, Tayssir 1466
Hatori, K. 1785

Hatori, Kenji 777
Hattori, Takato 739
Hauenschild, Philipp 1506
Haug, Martin 279, 698
Hayes, John G. 2470
Hegazy, Omar 2101, 2293, 3031
Heide, Daniel 3711
Heien, Christian 1294
Heimler, Patrick 1713
Hein, Yves 1294
Helmholdt-Zhu, Ting 97, 854
Hembel, Ahmed 3947
Henke, Markus 261, 1094, 2030
Henkenjohann, Jonas 1684
Henn, Jochen 3599
Henneberg, Dustin 2885, 3491
Herbold, Johannes 749
Hernando, Marta M. 1083, 1765, 3336
Herzog, Hans-Georg 952
Heydari, Rasool 2682
Hikihara, Takashi 108
Hiller, M. 3411
Hiller, Marc 115, 999
Hillmer, Hartmut 2383
Hilt, Oliver 2596, 2644, 3775
Himker, Niklas 1631
Himmelmann, Patrick 999
Hiraki, Eiji 2164
Hirning, David 971, 1237, 1277, 3536
Hissel, Daniel 315
Hjerrild, Jesper 2504
Hoerner, Michael 1754
Hofer, Heimo 1445
Hofer, Matthias 2251
Hoff, Bjarte 3198
Hoffmann, Klaus F. 758, 2726, 3188
Hoffmann, Madlen 3262
Hoffstadt, Thorben 1157
Hofmann, Viktor 195, 400
Hofmann, Wilfried 3957
Hofstetter, Patrick 195, 400
Hölscher, Jonas 2432
Holtje, Pauline 1665
Holzke, Wilfried 3149, 3159, 3169
Horn, Markus 2383
Hortans, Magnus 3309
Hoshi, Nobukazu 1776, 1844
Hosseinabadi, Farzad 3031
Hosseini, Elham 1025
Hou, Jingning 3722
Houwen, Simon 3344
Hridya, I 187

Hu, Anliang	651
Hu, Bin	2182
Hu, Xiaowei	3722
Huang, Jiasheng	1561
Huerta, Gabriel Ramos	1226
Huesgen, Till	2230
Huisman, Henk	634, 673, 681
Hutzler, Michael	1445
Idir, Nadir	2996, 3582, 3822
Igic, Petar	3822
Iida, Masaki	2164
Iman-Eini, Hossein	787
Imgart, Paul	1187
Incurvati, Maurizio	223, 268
Inoue, Michiko	3420
Iraola, Unai	3327
Ishihara, Mastaka	2164
Itoh, Jun-Ichi	902, 1104, 2127
Ittamveettil, Hridya	289
Izurza, Pedro	484
Jaber, Hamzeh J.	2156, 2285, 3084
Jacques, Dries	3344
Jagannath, Sriram	3362
Jahdi, Saeed	1497, 2477
Jain, Anekant	3615
Jain, Sanjay K.	3615, 3804
Jamal, Adeel	2780
Jaman, Shahid	3031
Jankovic, Marija	442
Jayathurathnage, Prasad	1947
Jena, Kasinath	3804
Jenhani, Firas	1343
Jeong, Byunghwang	1207
Jeschke, Sebina	3235
Jha, Kapil	187, 289
Jia, Hongjie	3215, 3225
Jia, Ming	1266
Joebges, Philipp	1266
Johansson, N.	1371
Johnson, C. Mark	3450
Jonsson, Tomas	1456
Jordà, Xavier	2715
Jørgensen, Asger Bjørn	825, 1641, 2661
Jöst, Dominik	4025
Jovanovic, Raka	3518
Juchem, Ralf	4023
Judge, Paul	80
Junemann, Lennart	1665
Jung, Marco	624, 1515, 1611, 1620
Junghans, Christoph	3460
Junyent-Ferre, Adria	2574
Kabbara, Wassim	3883

Kacetl, Jan	1197, 3636, 3665
Kacetl, Tomáš	1197, 3636, 3665
Kacki, Marcin	2470
Kadem, Karim	3590
Kaerst, Jens Peter	544
Kaiser, Jeremias	307
Kallfass, Ingmar	2698, 3565
Kamel, Tamer	468
Kaminski, Nando	2230, 3149, 3169
Kamm, Simon	2698
Kampen, Dennis	145, 1939, 2588
Kamper, Maarten J.	2958
Karakasli, Vefa	2835
Karamanakos, Petros	297, 1476, 1754
Karau, Fabian	3292
Karnehm, Dominic	767
Karwatzki, Dennis	195
Kasten, Henning	3501
Kayser, Felix	59, 4003
Keilmann, Robert	891
Kempchen, Malte	2912
Kemper, Philipp	749
Kennel, Ralph	1754, 2366, 3362
Kerekes, Tamas	1933
Keshavarzi, Davood	1064
Khader, Meriem	2655
Khan, Basit Ali	2537
Khan, Mohammed Ali	135
Khan, Nameer	3796
Khan, Siam Hasan	484
Khanzadeh, Babak	2344
Khenfri, Fouad	3831
Kiehnle, Philip	999
Kiffe, Axel	1157
Kikuchi, Naoto	1104
Kim, Dong-Uk	1207
Kim, Sungmin	1207, 2877
Kinzer, Dan	3987
Kirsch, Andreas	380
Kitagawa, Wataru	739
Kjærsgaard, Benjamin Futtrup	825
Klee, Matthias	1515
Klever, Severin	3676
Klötzer, Sebastian	4011
Knebusch, Benjamin	1665, 3048
Ko, Youngjong	3014
Kobayashi, Hiroyasu	1580
Kocewiak, Lukasz	2504
Koch, Jan-Niklas	2852
Koczy, Dawid	3149
Kohlhepp, Benedikt	88, 307
Kojima, Tetsuya	3740

Kondo, Keiichiro 1580
Kondratenko, Dmytro 1906
Kopp, Tobias 912
Kormska, Tomáš 1114
Körner, Patrick 2021
Korthauer, Bastian 3625
Kosesoy, Yusuf 634
Kostka, Benedikt 1649
Kostynski, Daniel 3855
Koteich, Mohamad 534
Kouro, Samir 1015
Koutroulis, Eftychios 1985
Kowal, Julia 4014
Kragl, Robert 2554
Krick, Alexander 3989
Krigar, Tim 2375
Krishnamoorthy, Harish Sarma 730
Krüger, Helge 3966
Krümpelmann, Marcel 1631
Kubulus, Pawel Piotr 2661
Kuder, Manuel 767
Kumar, Amit 451
Kumar, Kaushik Naresh 1486
Kumar, Manish 3511
Kuperman, A. 3254
Kuprat, Johannes 3067
Kuring, Carsten 2596, 2644, 3775
Kurrat, Michael 912
Kurukuru, V S Bharath 135
Kusaka, Keisuke 1104, 2127
Kusche, Stephan 3704
Kusebauch, Manuel 3491
Küster, Pierre 411
Kwak, Jaedon 2893
Kyyrä, Jorma 1947
La Mantia, Fabio 833
Labonne, A. 3179
Labonne, Antoine 3140
Labrousse, D. 843
Lacerda, Vinícius Albernaz 3704
Laclaverie, Julien 944
Laforet, David 1445
Lamar, Diego G. 335, 1083, 1765, 3336
Lange, Jarren 2276
Lange, Yannic 2644
Langfermann, Sascha 1939, 2588
Lanzarotto, D. 2564
Larrañaga, Uxue 3938
Larrazabal, Igor 1765, 3336
Larsson, Anders 1456
Lataire, Philippe 2293
Laumen, Michael 3766

Lauri, Andrea 865
Laza, Saioa Burutxaga 370
Lazkano, Markel Zubiaga 484
Le Leslé, Johan 2526
Le Métayer, Pierre 2049
Lee, Jaehong 2877
Lee, Seung-Hwan 2877
Lee, Yonghwa 2402
Lefebvre, Bruno 2515
Lefevre, Guillaume 2526
Legay, Florian 3529
Lehn, Peter W. 1995, 2084, 2145, 2763
Leifert, Torsten 4013
Lemaire-Semail, Betty 2175, 2996
Lembeye, Yves 1216
Lenz, Kevin 442
Lenzen, Patrick 2413
Leuer, Michael 3292
Leuzzi, Riccardo 3975
Lévy, PE 843
Lewicki, Arkadiusz 1906
Lexow, Daniel 1744
Li, Feifei 3235
Li, Ke 3822
Li, Marui 3215, 3225
Li, Qiang 3722
Li, Weihan 4025
Li, Xiang 2301
Li, Xupeng 3373
Li, Zheming 2744
Liang, Mincui 3786
Lichtenstein, Timo 1674
Liebfried, Oliver 2726
Liegmann, Eyke 1754, 3362
Lievre, Aurelien 2175
Lin, Siqi 1914, 2638
Lin-Shi, Xuefang 3862
Lindemann, Georg 3555
Linder, Stefan 3992
Lippold, Florian 1506
Liserre, Marco 421, 833, 3014, 3067, 3092, 3101,
.... 3846, 3966
Liu, Chao 1561
Liu, Steven 604
Liu, Xing 1733, 3373
Liu, Yan-Fei 1590, 1601
Liu, Yining 1947
Llanos, Jacqueline 3429
Löfgren, Jonas 3920
Lombard, Philippe 169
López, Abraham 152
Lorenz, Andreas 814

Lorenz, Erwin 1167
Lorenz, Malte 1875
Lorenz, Oscar 873
Loudot, Serge 3883
Lu, Xuyang 3822
Lu, Yizhou 883
Luan, Shaokang 3309
Luckert, Franz 2706
Luecke, Stefan 3075
Luh, Matthias 232
Luo, Fang 344, 2860
Lusardi, Federico 3975
Lutsch, Michael 88
Lutz, Josef 1713
Lutzen, Hauke 2230
Ma, Wenhao 80
Maamri, Nezha 1525
Maibach, Philippe 3282
Maier, Robert W. 2744
Maitra, Abhishek 1424
Mallwitz, Regine 891, 912, 1094, 1247, 1506
Mambetow, Arthur 145
Manthey, Tobias 2655, 2689, 3059
Marca, Ygor Pereira 798
Marcaide, Inko 3920
Marcault, Emmanuel 2618
Marchesoni, Mario 2919
Margreiter, Thomas 223
Margueron, Xavier 1046
Marks, Hendrik 2030
Marquardt, Rainer 1855
Marroquí, D. 3382, 3401
Martin, Jérémy 990, 2736
Martinez, Wilmar 1914, 2197, 2581
Martinez-Garcia, Herminio 1056
Martinez-Padron, Daniel S. 1256
Martnez, Wilmar 2638
Marx, Philipp 1237, 1277, 3536
März, Martin 493, 3262
Mashaly, Aly 442
Mashayekh, Ali 767
Mathúna, Cian Ó 4006
Mattavelli, Paolo 865
Matthies, David 3159
Maussion, Pascal 2869
Maynard, X. 843
Mazuela, Mikel 325, 3327, 3574
Meddour, Aissam Riad 3696
Mehran, Kamyar 614, 3353
Mehrasa, M. 3179
Mehrasa, Majid 3140
Meier, Hans 2021

Meinert, Janus Dybdahl 825
Meissner, Michael 758, 3188
Mellor, Phil 2477
Mendoza-Araya, Patricio 1177, 1226
Meng, Qingchao 933
Menzel, Steffen 3169
Merlin, Michael M. C. 2824, 3813
Merlin, Michael 80, 3470
Mersche, Stefan 115
Mertens, Axel 641, 1350, 1533, 1631, 1649, 1665,
..1684, 1865, 1875, 2003, 2066, 2392, 2706, 3022, 3037, 3048,
3075, 3555, 3711
Miaja, Pablo F. 152
Mijatovic, Nenad 2496, 2939
Miller, T. J. E. 3318
Minami, Masataka 2285
Mir, Tabish Nazir 468
Mirza, Abdul Basit 344
Mirzadeh, Mina 1350
Mirzaeva, Galina 3903
Miskiewicz, Rafal 1486
Mistretta, C. 1975
Mita, Salvatore 2628
Mo, Wai Keung 3130
Möckel, Andreas 3391
Moench, Stefan 242
Mogorovic, Marko 203
Mohanta, MK Kharabela 689
Möhlenkamp, Georg 3993
Mohsenzade, Sadegh 614, 3353
Moldenhauer, Deniz-Heinz 2205
Mondal, Gopal 572
Mondzik, Andrzej 3804
Monmasson, Eric 1256
Mönninghoff, Sebastian 3005
Montero, E. Rodriguez 1834
Morales-Paredes, Helmo K. 1074
Morand, Julien 2526
Morel, F. 2422, 2564
Morey, Philippe 1543
Morshed, Muhammad 2301
Motte-Michellon, Denis 1216
Mouselinos, Theodoros P. 1551
Moussa, Hassan 3590
Movagharnejad, Hedieh 3048
Mu, Yunfei 3215
Müller, Jonas 2230
Müller, Tankred 474
Munk-Nielsen, Stig 825, 1641, 2661, 3309
Muñoz-Carpintero, Diego 1074, 3429
Muruaga, Endika Bilbao 3529
Musolino, Francesco 554

Mustafeez-Ul-Hassan 2860
Musumeci, S. 1975
Muyllaert, Koenraad 2383
Mysore, Madhu Lakshman 1424
Naeve, Tomasz 1445
Nagayasu, Kiwa 2164
Naghibi, Javad 614, 3353
Nahalparvari, Mehrdad 2843
Najjar, Mohammad 2682
Nakamura, Keiichi 777
Nakamura, Taketsune 3084
Nami, Ashkan 2241, 3757
Nannen, Hauke 160
Nassurdine, B. Mohamed 843
Nayak, Khirod Kumar 2241, 3757
Nayampalli, Vishwas Acharya 1703
Nazeri, Ahmad Ali 1309, 1336, 1343, 2670, 3871
Neal, Harley 2301
Nee, Hans-Peter 2843
Nehmer, Dominik 1036
Neira, Sebastian 2824, 3813
Neuland, Tanja 3991
Neumann, Christian 1895
Neumann, Ingmar 1445
Neumeister, Matthias 572
Nguyen, Allen 1722
Nguyen, Khanh-Hung 562, 1309
Nguyen, Van-Sang 922, 990
Nguyen, Xuan Viet Linh 169
Nian, Heng 2182
Niasar, Mohamad Ghaffarian 3729
Nie, Shuang 2145
Niedernostheide, Franz-J. 2744
Niedernostheide, Franz-Josef 1424
Nielebock, Sebastian 493, 2607
Niemetz, Michael 2021
Niggemann, Oliver 3545
Nikowitz, Mario 2251
Nishio, Atsushi 351
Nishitani, Yota 3420
Nishizawa, Shin-Ichi 1128
Noboru, Wakana 777
Noisette, Philippe 3910
Nooshabadi, Morteza Tadbiri 787
Nordström, Lars 883, 1006
Nymand, Morten 2682
O'Donnell, Terence 390
O'Driscoll, Seamus 4006
Obernolte, Urs 854
Odeh, Charles 1906
Okada, Ryohei 1776, 1844
Olbrich, Markus 2912

Oliveira, Hercules Araujo 2948
Orbay, Raik 3920
Orchard, Marcos 3429
Orfanoudakis, Georgios I. 1985
Örgüt, Osman 1361
Orlik, Bernd 3149, 3159, 3169
Ortega, David 1765, 3336
Ortiz-Gonzalez, Jose 2477
Orts, C. 3382, 3401
Oshnoei, Arman 2939
Ota, Ryosuke 1776, 1844
Ouyang, Ziwei 1413, 1561
Owzareck, Michael 1939, 2588
Oyarbide, Estanis 3327
Paasch, Kasper M. 3130
Pace, Loris 3582
Páez, J. D. 2422
Pagnani, Daniela 2504
Panigrahi, Bijaya Ketan 2110, 3511
Papadopoulos, Georgios 1391
Papadopoulos, Theofilos 432
Papafotiou, George 2788
Papanikolaou, Nick 1796, 2257
Papastergiou, Konstantinos 2355
Pascal, Yoann 3067
Pasquier, Christophe 3786
Passalacqua, Massimiliano 2919
Passmore, Brandon 4005
Pathmanathan, Mehanathan 1995, 2084, 2145, 2763
Patin, Nicolas 1256
Patti, Dario 2628
Patzelt, Nikolaus 1923
Paul, Arup Ratan 178
Pauls, Denis 2441
Pavone, Mario 554
Pedroso, Douglas 335
Peftitsis, Dimosthenis 1486, 2355
Pelletier, Sebastien 223
Penczek, Adam 3804
Peng, Hujun 3235
Péra, Marie-Cécile 315
Pereda, Javier 2824
Pereira, Thiago 3014, 3092, 3101, 3846
Perez, Gaëtan 960
Perez-Cebolla, Francisco Jose 3574, 3750
Peroutka, Zdenek 1114
Perpiñá, Xavier 2715
Perrin, Rémi 2526
Perrin, Remi 3272
Petritz, Andreas 279
Petzoldt, Jürgen 2885, 3491
Peyghami, Saeed 2939

Pfeiffer, Jonas 411, 1167
Pfost, Martin 2375, 2413
Phanse, Ajinkya 1722
Phulpin, Tanguy 3883
Pichon, Pierre-Yves 2526
Pickert, Phil Leon 1381
Piepenbrock, Till 2432
Pietrzak-David, Maria 2869
Pigott, John 3796
Pinheiro, José Renes 3590
Piqué, Gerard Villar 3796
Piróg, Stanislaw 3804
Placzek, Julius M. 833
Plat, Arnaud 3862
Plötz, Till-Mathis 980
Pogulaguntla, Aditya 730
Pohlmann, Sebastian 767
Polezhaev, Vladimir 2230
Ponick, Bernd 1381, 1665, 3048, 3711
Poormohammadi, Fereshteh 2795
Pöschke, Florian 3704
Pouresmaeil, Edris 2537
Pouresmaeil, Kaveh 2788
Pouresmaeil, Mobina 2537
Pramanick, Sumit 1658, 3511
Pree, Elias 1445
Prenleloup, Pierre 3529
Prieto-Araujo, Eduardo 2486, 3704
Puls, Simon 2852
Puschmann, Frank 749
Qin, Zian 3607
Quabeck, Stefan 3893
Quade, Katharina Lilith 4025
Quay, Rüdiger 242
Rabkowski, Jacek 1486, 3938
Rädel, Uwe 2885, 3491
Radha, Krishna Moorthy 344
Rafiq, Aamir 1658
Raggini, Diego 3206
Raghavendra, I Venkata 730
Rahmani, Mehdi 2496
Raison, Bertrand 1147
Rajabian, Amir Azam 614
Ramdane, Brahim 1216
Ramirez, Fernando 289
Rasekh, Navid 3120
Rasool, Haaris 2101, 2293
Raßmann, Rando 1286
Rathjen, Kai-Uwe 758
Rault, Pierre 582
Ravyts, Simon 1300
Raya, Mariana 2715

Razi, R. 3179
Razi, Reza 3140
Regnat, Guillaume 2526
Rehlaender, Philipp 2432, 2754, 3625
Reimann, René 3159
Reincke-Collon, Carsten 370, 3391
Reindl, Andrea 2021
Reiner, Richard 242
Reißenweber, Lukas 525
Reitmeier, Dominik 2211
Remón, Daniel 1083
Rettner, Cornelius 4019
Reyes-Chamorro, Lorenzo 3429
Reynaud, Jean-François 3529
Ribeiro, Luiz Antonio De Souza 2948
Richard, Lucas 1147
Rickert, Kai 115
Rigbers, Klaus 4023
Rigogiannis, Nick 2257
Ringbeck, Florian 4025
Risch, Raffael 651
Rizoug, Nassim 3696, 3831
Robinson, Jonathan 572
Rocha, Gabriel Silva 2948
Roche, Jan-Philipp 3545
Rodríguez, Alberto 335, 1083, 1765, 3336
Rodriguez, Daniel C. 3893
Rodriguez, Joan Marc 2574
Rodriguez, José 1015
Roes, Maurice G. L. 798
Roes, Maurice 2788
Roß, Tilo 3391
Rossi, Mattia 1476
Rothenburger, Max 2383
Roth-Stielow, Jörg 971, 1237, 1277, 3536
Rouphael, Rosalie 1525
Rudolph, Christian 474
Rueß, Manuel 3565
Rufer, Alfred 30
Ruppert, Lukas A. 3766
Ruthardt, Johannes 971
Rylko, Marek S. 2470
Sadarnac, Daniel 3883
Saeidi, Mahmoud 1336, 1343, 3871
Safdarzadeh, Omid 2316
Sah, Gyanendra Kumar 1885
Sahan, Benjamin 1137
Sahin, Ilker 1361
Sahoo, Subham 2110, 2182
Sahu, Malaya Kumar 2241, 3757
Sahu, Silpashree 689
Said, Nasri 2618

Saito, Wataru 1128
Sakai, J. 1785
Salehi, Navid 1056
Samples, Ben 4005
Sanchez, Juan 873
Sanchez-Ruiz, Alain 484
Santos, Francisco 3101
Sanusi, Bima Nugraha 1413
Sanz-Alcaine, José Miguel 3750
Sarlioglu, Bulent 3947
Sato, Kota 1580
Sato, Takashi 3420
Sauer, Dirk Uwe 4012, 4025
Sauerland, Henning 3159
Sawicki, Jean–paul 315
Scarcella, Giuseppe 1569
Scelba, Giacomo 1569, 2628
Schäffner, Philipp 279
Schafmeister, Frank 2432, 2754, 3625, 3686
Schanen, Jean-Luc 787
Schanen, JL 843
Schefer, Hendrik 891, 912, 1094
Schellekens, Jan 634
Schierle, Guido 3188
Schiestl, Martin 223, 268
Schillinger, Tobias 3646
Schillingmann, Henning 2030
Schlegel, Christian 1923
Schlegel, Ludwig 3957
Schmid, Markus 268
Schmidhuber, Michael 411, 1167
Schmies, Dominik 2276
Schmitz, Laurids 3599
Schnabel, Fabian 624, 1515
Scholjegerdes, Moritz 3005
Schön, André 814
Schrödl, Manfred 2251
Schueltzke, Jens 1167
Schuerhuber, Robert 39
Schuhmann, Thomas 3646
Schullerus, Gernot 593, 2334
Schulte, Horst 3704
Schulz, D. 3411
Schulze, Gerold 2383
Schulze, Hans-Joachim 1424
Schumann, Christian 2058
Schumann, Sven 4022
Schümann, Ulf 1286
Schupp, Jan 3309
Schütt, Michael 1885, 2308
Schwarz, Babette 1381
Schwendemann, R. 3411

Scohier, Martin 504
Scrimizzi, F. 1975
Sebastián, Javier 1765, 3336
Seibel, Axel 1515, 1620
Seitz, Arne 4015
Seliger, Norbert 22
Semail, Eric 2996
Sen, Paresh C. 1590, 1601
Sepehr, Amir 2537
Serdyuk, Yuriy 2344
Sergentanis, Grigorios 3450
Serra, Amiron Wolff Dos Santos 2948
Seybold, Felix 3536
Shahparasti, Mahdi 2682
Sharma, Kanuj 2698
Shawky, Ahmed 1015
Shen, Chengjun 2477
Shen, Xiaobing 1914, 2197
Shinoda, Kosei 3928
Shintani, Michihiro 3420
Shousha, Mahmoud 279, 698
Shuqin, Wang 1815
Siala, Sami 125
Siemaszko, Daniel 3910
Siemieniec, Ralf 1445
Sievers, Markus 3855
Singh, Rupam 135
Singh, Shashank Shekhawat 279
Singh, Sukhjit 2084
Skala, Aleksander 3804
Skibin, Stanislav 3301
Soeiro, Thiago Batista 1319, 3729
Solomentsev, Michael 1722
Solovyov, Vyacheslav 2860
Soltau, N. 1785
Soltau, Nils 777
Sönmez, Ertugrul 593, 2334
Soundararajan, Ajeeth Phrassanna 3729
Soupremanien, Ulrich 922
Spieler, Matthias 1692
Sprunck, Sebastian 1611
Sreekanth, T 730
Stadler, Alexander 525
Stadlober, Barbara 279
Staiger, Jochen 2219
Stala, Robert 3804
Stalleicken, Frederik 2607
Stallmann, Frederik 641
Stärz, Ronald 223, 268
Stathis, Spyridon 1402
Staubach, Christian 1137
Steckler, P. B. 2564

Stefanski, L.	3411
Steffen, Jonas	1515
Steinhart, Heinrich	2219
Štengl, Josef	1114
Stevic, Marija	2985
Stewart, Joshua	2806
Steyn, Kyle	3112
Stille, Karl Stephan	2276
Stock, Alexander	1
Stöckl, Thomas	952
Stone, David A.	3440
Strunk, Robin	1350
Stul, Koen	1300
Stutz, Christian	493
Suberski, Martin	2885, 3491
Sujeeth, Arjun	2628
Sullivan, Charles R.	2470
Svensson, Jan R.	1187
Tabrizi, Gholamreza	1611
Takamori, Taro	1128
Takayama, Hajime	108
Takeshita, Takaharu	213, 351, 739
Talla, Jakub	1114
Tang, Chengjun	2813
Tang, Zhongting	1933
Tashakor, Nima	1025, 1064, 1197, 3636, 3665
Tatakis, Emmanuel C.	1551
Tegtmeier, Bernd	1674
Teske, Peter	1466
Thiringer, Torbjörn	2344, 2813, 3920
Thoma, Jürgen	2554
Thönelt, Nick	1713
Thönnessen, André	3676
Tian, Fanghao	2581
Tillmann, Philipp	3740
Tiwari, Arvind Kumar	289
Tiwari, Arvind	187
To, Pham Ha Trieu	59, 70, 4003
Tornello, Luigi Danilo	1569
Torres, C.	3382, 3401
Torrico, Grover	361, 1815
Tournez, Florian	2175
Tran, Dai Duong	2293
Tran, Manh Tuan	2101, 2293
Tresca, Giulia	3975
Trescases, Olivier	3796
Tricoli, Pietro	468
Trochimiuk, Przemyslaw	1486
Tschepp, Andreas	279
Turrisi, Gaetano	1569
Tzanakis, Athanasios	3920
Uicich, Simon	3862

Ulbing, Alexander	3855
Ulmer, Sabrina	593, 2334
Ulrich, Burkhard	459
Umetani, Kazuhiro	2164
Unruh, Peter	1620
Unruh, Roland	3686
Urkizu, June	325
Vaccaro, Luis	2919
Vaessen, Peter	3729
Vagg, Christopher	3696
Vagnon, Eric	2515
Vahid, Sina	719
Vala, Sama Salehi	344
Valderrama, Carlos	504
Valenzuela, Rodrigo Alonso Alvarez	814
Van Cappellen, Leander	2795
Van Mierlo, Joeri	2101
Van Oosterwyck, Nick	3344
Van Tuan, Mai	351
Vandenbussche, Thomas	1300
Vanfretti, Luigi	3928
Vanwalleghem, Bart	3344
Vasiladiotis, Michail	1923
Vatamanu, Lucian	1046
Vázquez, Aitor	1083
Vázquez, Francisco	1765, 3336
Velasco-Quesada, Guillermo	1056
Velazco, Diego	251
Vellvehi, Miquel	2715
Venkataramanan, Giri	3480
Venugopal, Ravinder	2985
Verdier, Jacques	169
Vermeerch, Pierre	582
Veroni, Alessandro	3206
Vershinin, K.	2564
Viana, Caniggia	1995, 2084
Viarouge, I.	1955
Viarouge, P.	1955
Vidal-Albalate, Ricardo	2189
Videau, Nicolas	944
Videt, Arnaud	3822
Villar, Irma	3529, 3938
Vitorino, Montiê Alves	2689, 3059
Vogelsberger, M.	1834
Volzer, Benjamin	2554
Von Hoegen, Anne	3740
Wada, Keiji	1128
Wagner, Valentin	514
Wakelin, Bruce	3309
Wallart, Francois	251
Wallscheid, Oliver	2276, 2432
Waltereit, Patrick	242

Wang, Chu	3722	Yadav, Sachin	3607
Wang, Jun	2136, 3120	Yamaguchi, Masamichi	2127
Wang, Kangan	3014	Yamashita, Shota	213
Wang, Rui	673, 1641	Yamauchi, Kohei	2119
Wang, Xiaoya	3722	Yang, Huoming	1466
Wang, Xin	315	Yang, Jiajun	3014
Wang, Yanbo	2011	Yang, Juefei	2477
Wang, Yangang	2301	Yang, Yinghui	3993
Waradzyn, Zbigniew	3804	Yang, Yongheng	2257
Watanabe, Hiroki	1104	Yaqoob, M.	1815
Wattenberg, Martin	873	Yasuda, Takumi	902
Weicker, Martin	2316	Yeganeh, Mohammad Sadegh Orfi	2496, 2939
Weires, Jonas	604	Yu, Guangyao	1319
Weiser, Mathias C. J.	3565	Yu, Xiao	562, 1309, 2383
Weiss, Xavier	1006	Yu, Xiaodan	3225
Wenzel, Johannes C.	2066	Yuan, Xibo	2136, 3120
Werlig, Christian	3966	Zacharias, Peter	411, 562, 1309, 1328, 1336, 1343,
Weyh, Thomas	767		2383, 2670, 3871
Wicht, Bernhard	2912	Zacher, Benjamin H.	2058
Wieczorek, Nick	3775	Zampardi, Giorgia	833
Wiemer, Adrian	2544	Zanchetta, Pericle	3975
Wiesemann, Julius	1865	Zatocil, Heiko	160
Wiesner, E.	1785	Zdanowski, Mariusz	3938
Wiesner, Eugen	777	Zhang, Bo	1733
Wijnands, Korneel	673, 798, 2788	Zhang, Shimin	709
Wilkowski, Matt	4008	Zhang, Yaqian	2182
Willer, Felix	3838	Zhang, Zhe	1561
Willich, Viktor	3555	Zhang, Zhuoqi	1776
Wohlrath, Fritz	525	Zhang, Ziqian	39
Wolbank, T.	1834	Zhao, Hongbo	1641, 3309
Wolf, Mihaela	2596, 2644, 3775	Zheng, Zhixue	315
Wolfstädter, Simon	4017	Zhetessov, Aidar	3480
Wölk, Alexander	279	Zhu, Zi-Qiang	2958
Wouters, Hans	2197	Ziani, Adel	944
Woywode, Oliver	758	Ziegler, Philipp	971, 1237, 1277, 3536
Wu, Weimin	1985	Zilic, Rufad	1336
Wu, Xiangqiang	1933	Zocher, Markus	2366
Wu, Yuxuan	2860	Zolfi, Pouya	719
Wunsch, Bernhard	3301	Zou, Zhixiang	3014
Würfl, Joachim	2596, 2644, 3775	Zsurzsan, Tiberiu Gabriel	1561
Würsig, Andreas	3966		
Xia, Peizhou	3470		
Xiao, Qian	3215, 3225		
Xiao, Xiong	1966		
Xie, Jun	2885, 3491		
Xie, Lihong	2136		
Xu, Huihui	709		
Xu, James	3796		
Xu, Qianwen	883, 1006		
Xu, Wei	2136		
Xu, Zhongqing	912		
Xu, Zixiao	2182		

IEEE
445 Hoes Lane
Piscataway, NJ 08854-4141

ISBN 978-1-6654-8700-9